Wastewater Hydraulics

Second Edition

Willi H. Hager

Wastewater Hydraulics

Theory and Practice

Second Edition

Springer

Dr. Willi H. Hager
VAW, ETH Zurich
CH-8092 Zürich
Switzerland
hager@vaw.baug.ethz.ch

ISBN 978-3-642-11382-6 e-ISBN 978-3-642-11383-3
DOI 10.1007/978-3-642-11383-3
Springer Heidelberg Dordrecht London New York

Library of Congress Control Number: 2010923828

Cover design: WMXDesign GmbH, Heidelberg

Printed on acid-free paper

Springer is part of Springer Science+Business Media (www.springer.com)

To my father

Preface to Second Edition

Eleven years have passed since the first edition of Wastewater hydraulics was published. Given that the first edition is currently no more available, I was asked by Springer to prepare a second edition, thereby improving the original version in terms of recent research, corrections and outline. The general contents of the book was kept unchanged, however, because no substantial addition was made.

Additions to the first edition were mainly made in the chapters dealing with the hydraulic jump, in which now also jumps in sewers are included, based on research conducted at VAW. Further, important additions to the design of special manholes subjected by supercritical approach flow are included, allowing for a straightforward design of these important elements of a sewer system. Sideweirs were also reconsidered based on experimentation. The second edition was increased by almost 100 pages, therefore.

A number of mistakes, mainly in the Examples, was removed, based on detailed reading and also on the Japanese edition of this book, whose authors have recomputed all examples, and asked for advice if things still remained obscure. I would particularly like to thank my colleague Prof. Dr. Youichi Yasuda, Nihon University, Tokyo, Japan, for having initiated this large work. An Italian version of this book is by the way also in preparation, and should become available in 2011. Altogether, with the original German version, Wastewater hydraulics will then be written in four languages, demonstrating the practical relevance of a text in this field.

Up to now, the figures were prepared with an old system. Thanks to Walter Thürig, VAW, all figures were remade, particularly including now the correct lettering as in the main text. I would like to thank you Walter for this large work. All equations written with formula editor in the original version had to be replaced by a modern tool. This task was masterly made by Mrs. Cornelia Auer, Secretary VAW, including a detailed reading of the entire text. I also owe thanks to Prof. Dr. Corrado Gisonni, Seconda Università di Napoli, Italy, who made a careful re-reading of the entire text.

Two persons have passed away within the 10 past years. We lost our father Willi Hager in 2000, following a serious health problem. He was the true initiator of wastewater hydraulics for me. Whereas practice was his main interest, I attempted to include also a research background. Further, Prof. Dr. N. Roy (1928–2000) has also passed away due to health reasons. He accepted to translate parts of the original German version into English, and thus contributed to the original book significantly. My final thanks go again to Springer-Verlag, presenting an outstanding work.

Zürich, Switzerland
November 2010

Willi H. Hager

Preface to First Edition

Because sewage is similar to water from the hydraulic point of view, wastewater hydraulics does not really vary from common engineering hydraulics. However, wastewater hydraulics differs significantly from neighbor water sciences in terms of undissolved matter. Combined sewer discharge consists of polluted water and solid matter that may either clog a sewer or settle and lead to undesirable consequences such as odour nuisance, increased maintenance or performance problems. Sewer hydraulics is thus not only related to wastewater transport but also to system security, economy and maintenance.

Sewers are normally closed channels with a free surface flow. If a sewer chokes, i.e. if pressurized flow occurs due to limitations in capacity, complex flow conditions may arise involving often multiphase flow with solid and gaseous phases. More dramatically, steep sewers may choke and sewage may overtop a manhole and thus lead to unacceptable flow conditions. This extreme case of sewer failure has to be strictly avoided because raw sewage would be discharged out of the system and pollute urban areas. This aspect of sewage technology is mainly dealt with in sewer hydraulics, and is also presented in this book.

Compared to mechanical, biological and chemical sewage treatment, wastewater hydraulics has received little attention. Currently, no text book seems to be available that covers wastewater hydraulics, although hydraulic design is required by most wastewater authorities. This book thus aims at filling this gap. Particular attention is given to sewer hydraulics, and typical hydraulic structures of treatment plants are also discussed. Basin hydrodynamics with respect to currents in storage chambers, settling or aeration tanks, are not treated, however.

Sewer hydraulics is still a national concern, and practitioners are often unaware of contributions from other languages. This is mainly due to the importance of regional and national water authorities. International collaboration has started to develop in the last two or three decades, and Austria, Germany and Switzerland for instance have organized some collaboration within the German language. This books aims also to bring together knowledge from English, German, French, Italian and Swedish literature. By doing so, regional approaches to a certain problem can be highlighted, and the advantages of various designs combined eventually into an optimum solution.

Numerical modelling has not yet had a real impact on wastewater hydraulics, except for stormwater computations involving rainfall-runoff processes. This book is based on both computation and experiments, and the solutions are presented either analytically or graphically. Many of these solutions could be incorporated into standard software, for example to predict backwater curves.

The original book was published in German (1994) because of the noteworthy contributions of the German speaking countries to this branch of hydraulics. A second edition appeared already in 1995. The present English version was updated and includes a number of new results such as two-phase flows, supercritical flows in bends and sewer sideweirs. Important expressions have been translated from English to German and French to facilitate international use.

This book is not only a reference for researchers but also a basis for practising engineers. It can be used as a text book for graduate students, although it has the characteristics of a reference book. It addresses mainly the sewer hydraulician who has to tackle many a problem in daily life, and who will not always find an appropriate solution. Each chapter is introduced with a summery to outline the contents. To illustrate application of the theory, examples are presented to explain the computational procedures. Further, to relate present knowledge to the history of hydraulics, some key dates on noteworthy hydraulicians are quoted. A historical note on the development of wastewater hydraulics is also added. References are given at the end of each chapter, and they often are helpful starting points for further reading. Each notation is defined when introduced, and listed alphabetically at the end of each chapter.

Wastewater hydraulics was once my profession, and it has gradually turned into one of my hobbies. My father introduced me to this field, and I would like to acknowledge his steady interest in my work. My brother Kurt has accompanied me in terms of common interest, discussion and hobby. I would also like to thank Dr. N. Roy, formerly professor at the Banaras Hindu University, India, for translating parts of the book, to Robert Boes for useful and critical comments, to Dr. Corrado Gisonni for detailed reading and other pleasures, and to Dr. Brian McArdell for additional comments. Prof. H.-E. Minor, Director of VAW, has supported my work, and Mrs. E. Weber has greatly contributed to the details of manuscript preparation. I would also like to thank to Springer-Verlag, Heidelberg and Berlin, for the excellent collaboration.

March 1999 Willi H. Hager

Contents

Chapter 1
Basic Equations

Abstract The basis of hydraulic calculations are three conservation principles. In this chapter these are discussed with reference to their application in hydraulic practice.

The continuity equation ensures the conservation of mass. Notations with respect to discharge and mean velocity are defined.

The specific force principle is based on the momentum principle and calls for force balance. This important principle is discussed and applications are explained.

Lastly, the energy principle is presented and a few simple conclusions with regard to the energy flux are derived. The discussion of results refers, in particular, to the similarity between the hydraulic and the hydrodynamic approaches.

1.1 Introduction

It is possible to describe the flow of fluids through a mathematical representation. The formalisms, built upon physically conforming laws, were essentially developed by Isaac Newton (1642–1727) in the 17th century. Even at present, classical mechanics as also hydromechanics are founded on the four *Newtonian axioms*.

Hydromechanics refer to flow phenomena in which three local flow components may be present. In the general case of unsteady discharge, the flow structure together with its underlying temporal variations must be considered in the four mutually independent coordinates in respective dimensions. Such flows can today be handled in principle, but the solutions of the equations describing the flow system are normally complex and obtained only with highly-developed numerical computations. Such solutions are attractive for scientific investigations of complex problems, but are unsuited for practical problems involving large domains of time and space.

In contrast to the multidimensional computational models of hydrodynamics, *hydraulics* places the flow events in an one-dimensional framework. The simplified calculation procedures apply to cases where the flow takes place along a distinct direction. The latter occurs, for example, in flows in a channel in which the local

W.H. Hager, *Wastewater Hydraulics*, 2nd ed., DOI 10.1007/978-3-642-11383-3_1,

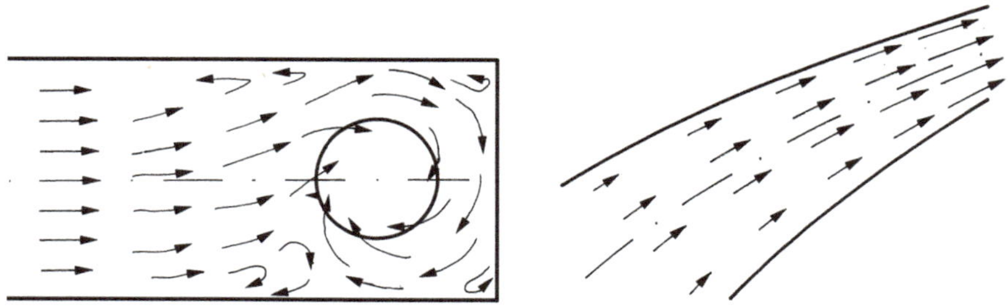

(a) (b)

Fig. 1.1 Two-dimensional flow (**a**) with large rotational component, (**b**) almost one-dimensional flow structure

variation in the channel geometry is small and all streamlines run almost parallel to the channel axis. Figure 1.1 compares a two-dimensional sink flow with an almost one-dimensional channel flow and illustrates the differences between the two. It is clearly evident, that a knowledge of the flow conditions along the axis allows conclusions to be drawn on the structure of the neighbouring streams; however, such information is not indicated in the case of Fig. 1.1a.

Although all fluid flows are invariably three-dimensional in nature, conduit and channel flows in particular can be approximated with an *one-dimensional approach*. Through this simplification, the solution of the model equations is decidedly facilitated; at the same time, however, the information regarding the multi-dimensional nature of the flow is lost. These characteristics, for example, the vortex shedding, or the growth of the boundary layer as the thin zone near the flow boundaries, are not always relevant in practice. The hydraulic solution represents a *reasonable compromise* between engineering calculation and accuracy of information. In the following, therefore, preference is given to the hydraulic concepts.

Fluid flows can be investigated both as a hydraulic and a hydrodynamic phenomenon through three fundamental laws of physics. These are the three *conservation principles*, namely, those involving mass, momentum and energy. The mechanical quantities such as pressure and velocity are functions of position and time and are obtained as solutions of the local and temporal fluid motion.

1.2 Continuity Equation

In flows of homogeneous fluids no difference in the particle property exists. The principle of conservation of mass is therefore expressed with the continuity equation. Before this relationship is introduced the concept of *control volume* is explained. A control volume encloses an arbitrarily considered but fixed volume and can be, simply expressed, represented through a continuously drawn closed curve. Further, by *steady flow,* one understands a time-invariant

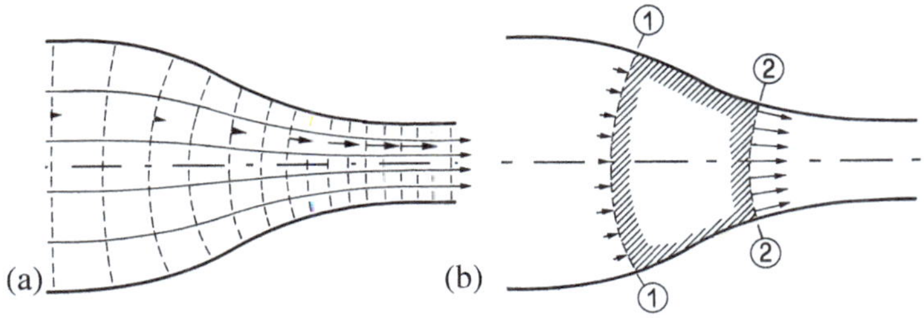

Fig. 1.2 Streamlines and control volumes (**a**) streamlines and equipotential lines in a gradual constriction with corresponding velocity vectors, (**b**) equipotential lines at sections ① and ② with appropriate velocity distributions

motion in contrast to unsteady flow. In the latter case, a different flow pattern appears at each instant of time. In the following, the steady flow is considered exclusively.

Steady flows can be described with *streamlines* (German: Stromlinie; French: Ligne de courant) each one of which, for example, is obtained by drawing a continuous line so that it has the direction of the velocity vectors of Fig. 1.2 at every point. As shown in Fig. 1.2, the velocity vectors are always tangential to the streamline. Therefore, the streamlines can not intersect each other. However, if the streamlines come closer, the flow is accelerated and if they diverge, the flow is retarded. All streamlines including the boundary lines constitute the streamtube (German: Stromröhre; French: Tube de courant). The streamtubes are particularly suitable for describing one-dimensional flow processes. The band of curves perpendicular to all the streamlines is called *equipotential lines*. By drawing streamlines it is often possible to detail the two-dimensional flow pattern.

In Fig. 1.2b a *control volume* is drawn for the gradual flow constriction as shown in Fig. 1.2a. The velocity vectors, which for plane flows define the velocities in magnitude and direction relative to the axis, represent in this figure a nearly uniform velocity distribution. Since the vectors are tangential to the streamlines, their direction relative to the equipotential lines is also perpendicular. Usually one selects the control volume in the manner of Fig. 1.2b, i.e. delineates the boundary of the control volume out of streamlines and equipotential lines.

The *discharge* Q (German: Durchfluss; French: Débit) is defined as the scalar product of the velocity and the area vectors. Simply expressed, Q equals the product of velocity and the surface area normal to this vector. Accordingly, the discharge Q_1 across the equipotential surface ① equals the sum of all the velocity vectors times the length elements of the equipotential line times the depth. This statement holds also for equipotential line ②.

The *continuity equation* for steady flow states that the discharges Q_1 and Q_2 are equal in case that there are neither lateral inflow into, nor outflow from the control volume. In general, it is expressed, through this principle, that mass is neither created nor destroyed (as, for example, by atomic explosion) and therefore all inflows into a control volume equal the sum of all outflows from it. It is observed that the elementary discharge equals the product of velocity and the flow area perpendicular to it.

In applications, discharge Q and cross-sectional area F, of a conduit for example, are often known. The *average velocity* V is then calculated as

$$V = Q/F. \tag{1.1}$$

The average velocity V calculated from Eq. (1.1) agrees with the local velocity the better that the observed flow is simplified as an one-dimensional flow. For a so-called *uniform velocity distribution* the calculated mean velocity agrees exactly with the velocity distribution.

Usually neither inflow nor outflow occur, as is typical for a junction manhole or a stormwater outlet. The continuity equation then simplifies to

$$Q_1 = Q_2\,. \tag{1.2}$$

The inflow Q_1 into the control volume then equals the outflow Q_2 from it. According to Eq. (1.1) the discharge is $Q = VF$, or

$$V_1F_1 = V_2F_2. \tag{1.3}$$

It follows that the larger the cross-section of flow, the smaller is the average velocity, and vice-versa.

As against the preceding approach, the mass balance for the cases sketched in Fig. 1.3 must be modified. For the combining flow of two channel branches, one gets with Q_z as the *lateral inflow*

$$Q_1 + Q_z = Q_2. \tag{1.4}$$

On the other hand, for the intake of water, for example, from a distributed inflow or from a pumping station, it follows with Q_a as the *lateral outflow*

$$Q_1 = Q_2 + Q_a. \tag{1.5}$$

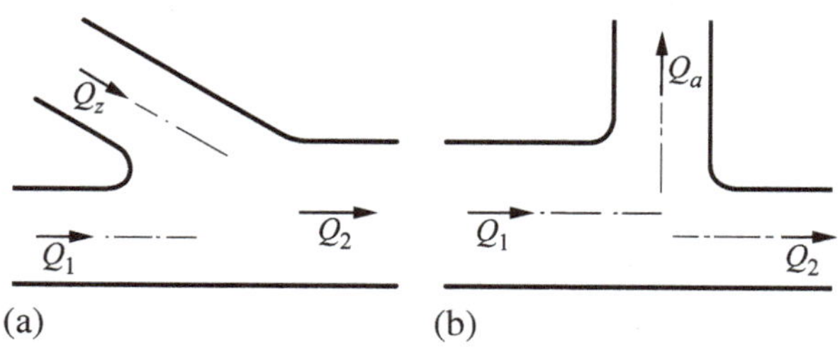

Fig. 1.3 Discharge for (**a**) lateral inflow Q_z, and (**b**) lateral outflow Q_a

The generalisation of Eq. (1.2) is, therefore,

$$Q_1 + Q_z = Q_2 + Q_a. \tag{1.6}$$

The mass balance is the simplest of the three balance equations. It admits a simple explanation and offers no problem in applications.

1.3 Specific Force Principle

By the term *momentum* (German: Impuls; French: Quantité de mouvement) one understands the vector $\boldsymbol{I}$ obtained as the product of mass m and its velocity vector $\boldsymbol{V}$. According to Newton, the sum of all external forces acting on the body of an element must equal the time rate of change of momentum of the element. This relationship gives rise to the so-called *equation of motion*. Although the principle has often been successfully employed in solving practical problems, it experiences, even today, certain rejections in practice. The main reason for this is the non-elementary application of functions producing somewhat formal expressions. For flows with a substantial energy loss, the momentum principle does, in general, provide advantage. Simplified considerations involve the introduction of the so-called *specific force* (German: Stützkraft; French: Impulsion totale).

The streamtube of Fig. 1.4 has an inlet cross-section ① and an outlet cross-section ②. The various forces $\boldsymbol{K}$ pertaining to the momentum principle act on the surface described by this control volume. There are, on the one hand, the *pressure forces* $\boldsymbol{K}_D$ which, simply expressed, always apply where the fluid either enters or leaves the control volume. On the other hand, there are the *tangential forces* $\boldsymbol{K}_T$ resulting from viscous friction. They disturb the flow pattern mathematically as well as physically by distorting the flow with extraneous forces. The last fact operates on hydraulic models as well.

Other forces that are not considered in the following and are enumerated here only for the sake of completeness, are the capillary force and the Coriolis force due to the earth's rotation. More important is the recognition that pressure forces are normally to be considered, which must eventually be supplemented by friction forces. External forces include the body forces, as the weight of the fluid enclosed

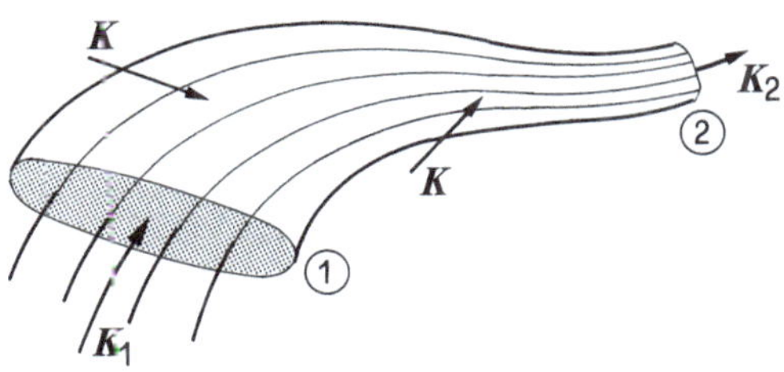

Fig. 1.4 Stream tube with external forces acting on its surface

within the control volume under consideration. Not to be considered are the internal dissipative forces. Therefore, all *external* (subscript a) forces acting on the control volume are known, although this is not always a simple matter. This knowledge enables the momentum principle to be employed without knowing the flow pattern inside the control volume. The simplicity and the functional clarity have substantially contributed to the success of the Newtonian axioms.

By *momentum* (German: Impuls; French: Impulsion) one usually understands the expression $\boldsymbol{I} = m\boldsymbol{V}$ whose variation with time t is, according to the rules of differential calculus,

$$\frac{\mathrm{d}\boldsymbol{I}}{\mathrm{d}t} = \frac{\mathrm{d}}{\mathrm{d}t}(m\boldsymbol{V}) = \boldsymbol{V}\frac{\mathrm{d}m}{\mathrm{d}t} + m\frac{\mathrm{d}\boldsymbol{V}}{\mathrm{d}t}. \tag{1.7}$$

If mass is expressed as the product of density and volume, the time rate of change of mass of a fluid of constant density ρ is given by the expression $\mathrm{d}m/\mathrm{d}t = \rho Q$. Eq. (1.7) then takes the form

$$\frac{\mathrm{d}\boldsymbol{I}}{\mathrm{d}t} = \boldsymbol{V}(\rho Q) + m\frac{\mathrm{d}\boldsymbol{V}}{\mathrm{d}t}. \tag{1.8}$$

In rigid body mechanics a momentum change can occur only through a change in the velocity of the body. For fluids with a mass flow ρQ, a momentum change can also occur in steady flow situation. In that case, the *time* rate of change of velocity $\mathrm{d}\boldsymbol{V}/\mathrm{d}t$ is zero and only the expression $\boldsymbol{V}(\rho Q)$ remains on the right hand side of Eq. (1.8). Because Newton's law must also hold for steady flows

$$\rho Q\boldsymbol{V} = \Sigma \boldsymbol{K}_a. \tag{1.9}$$

This vector equation has in general three component equations of which only one is relevant for one-dimensional consideration in hydraulics. The *direction* to which this component equation refers can be arbitrarily chosen, only, it must be the same for all forces appearing in Eq. (1.9). Figure 1.5 shows two applications of these principles. In Fig. 1.5a the boundaries of the control volume follow those of a manhole and are, at the branch faces in particular, perpendicular to the streamlines. The discharge faces are so far away from the center of the manhole that the corresponding streamlines are almost parallel. The direction to be considered is the direction of

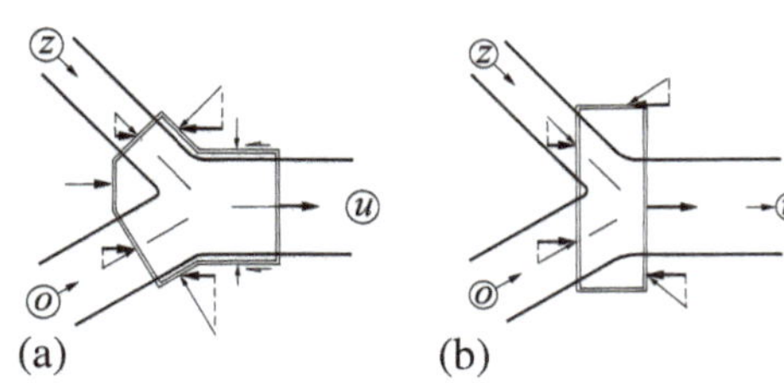

Fig. 1.5 Application of momentum principle to combining flow with commensurate forces for a (**a**) reasonable and (**b**) slightly less sensible demarcation of the control volume

the main stream so that the accompanying wall pressure forces remain unimportant. It may be noted, however, that the tangential forces involve one component in the principal direction. To apply the momentum principle, the sum of all external forces, their values in the one-dimensional case therefore provided with appropriate positive or negative signs, must be set equal to the time rate of change of momentum. For the steady state case, this yields

$$\rho Q_o \boldsymbol{V}_o + \rho Q_z \boldsymbol{V}_z - \rho Q_u \boldsymbol{V}_u = \Sigma \boldsymbol{K}_a. \tag{1.10}$$

The right side for Eq. (1.10) consists essentially of *pressure forces* which act, on the one hand, on the discharge faces and, on the other hand, on the curved surfaces of the control volume. The forces on the inflow and the outflow cross-sections are always obtained as the product of the cross-sectional pressure p_s acting at the centroid and the cross-sectional vector $\boldsymbol{F}$ corresponding to a static force $\boldsymbol{P}_s = p_s \boldsymbol{F}$. The quantity designated as flux of force or *specific force* $\boldsymbol{S}$ is represented by the sum of static and dynamic momentum, divided by specific weight, thus

$$\boldsymbol{S} = \frac{p_s}{\rho g}\boldsymbol{F} + \frac{Q\boldsymbol{V}}{g}. \tag{1.11}$$

The dimension of this normalised specific force is that of a volume unit. This representation is advantageous in hydraulic practice, because neither the density ρ nor the gravitational constant g undergo any change for homogeneous fluids. It remains to be observed at this point that the specific force $\boldsymbol{S}$ represents a vector quantity.

Equation (1.11) permits a modified representation of the momentum principle: If the static pressure force is added to the corresponding dynamic force on the left hand side of Eq. (1.10) the resulting equation divided by (ρg) gives

$$\Sigma \boldsymbol{S} = \Sigma \boldsymbol{K}_B. \tag{1.12}$$

This relationship states that the sum of all the specific forces equals the sum of all external forces acting on the boundary surfaces (subscript B) of the control volume. The latter forces of interest in open channel hydraulics are:

- *Reaction force* on walls, its direction differing from the main stream; non-prismatic and curved boundary surfaces fall under this category of wall surfaces,
- *Body force* due to fluid weight acting in the main flow direction,
- *Friction forces* due to viscosity and surface roughness, and
- Forces arising from *junctions* with lateral inflow or outflow.

Applications of the momentum principle have already been described in the foregoing section. Here the essential points are summarized:

- The momentum principle or its simplified version, the specific force principle, is amenable to exact application provided all *external* forces are known; the fact that the forces performing internal work remain unconsidered in the application of this principle allow conclusions about energy losses (Sect. 1.4) to be drawn.
- The main difficulty in applying the momentum principle lies in the determination of the external forces acting on the boundary surfaces. For a 'Black Box' type calculation where merely the known values of the inflow and outflow cross-sections are of concern, plausible assumptions regarding the *pressure distributions* over the boundary surfaces in the governing flow directions must be available.
- *Simplifications* with regard to body and frictional forces are usually needed. Either they are approximated by some average values or, they are altogether neglected vis-à-vis the specific forces.
- For special manholes, it may even be presumed that the body force is compensated by the friction force. This assumption allows decisive *simplifications*, resulting in simple and therefore often relevant solutions.
- The momentum principle can be applied in the *integral form* as represented in Eq. (1.12) as well as in the *differential form* such that a continuous variation of one or several parameters is possible. Both types of representations are considered in what follows.

1.4 Energy Principle

The principle of energy conservation states that the change in *energy* (German: Energie; French: Energie) of a physical system equals the addition of heat reduced by the work performed by the system. This statement complicates the application in hydraulic practice. Let the *hydraulic energy* E be first defined as

$$E = \rho g Q[z + p/(\rho g) + V^2/(2\,g)], \tag{1.13}$$

with z as elevation above a fixed datum, p as pressure and V as the velocity magnitude. For steady flow, the temporal change in heat supply minus the performed and the accompanying dissipated works in the reach between two cross-sections ① and ② equal the hydraulic energy difference $E_1 - E_2$. The decrease in the mechanical energy content of the flow in the direction of motion becomes evident. Because, in the context of mechanical energy form, the dissipated energy is lost, one speaks of *energy loss*. Between cross-sections ① and ②, one can therefore write

$$E_1 - E_2 = E_d > 0. \tag{1.14}$$

Fundamentally, *fluid viscosity* (German: Flüssigkeitsviskosität; French: Viscosité du fluide) is the source of energy loss. In the momentum principle the wall friction losses are regarded as wall reactions whereas the *form losses* must not be introduced in the estimation of the external forces. Also, for form losses, the viscosity gives rise to the velocity gradient which partially influences a flow.

Often, discharge Q remains constant between two cross-sections. Also, for water with densities $\rho_1 = \rho_2$, and after division by the product of specific weight (ρg) and discharge Q, the energy definition (1.13) can be expressed as

$$H = E/(\rho g Q) = z + p/(\rho g) + V^2/(2g). \tag{1.15}$$

The quantity H is designated *energy head* (German: Energiehöhe, French: Charge) since the expression on the right hand side indicates a length in dimension and physically expresses a height. One distinguishes among (Fig. 1.6):

- z vertical elevation called the *datum head* relative to an arbitrary but fixed level called the datum,
- $p/(\rho g)$ *pressure head* in meters of water column, and
- $V^2/(2g)$ *velocity head* with V as the velocity magnitude.

The energy head H at an arbitrary location can be evaluated if the datum head, the pressure head and the velocity head at that point are known (Fig. 1.6a). As one moves along a streamline, the energy head continuously decreases according to Eq. (1.14). At station ②, the head loss is ΔH with respect to station ①.

One may consider instead of an arbitrary streamline the streamtube, so that the energy principle can approximately also be referred to it. One-dimensional hydraulics applies to the case sketched in Fig. 1.6b for which the energy head H of the axial streamline is almost identical with the energy head of any arbitrary streamline in the same cross-section. Therefore, the velocity distribution is uniform and the pressure is hydrostatically distributed over the cross-section. Though the velocity in the vicinity of the wall is not uniform due to the effect of viscosity, technical flow problems are characterized with a *practically* uniform velocity distribution.

If h_p denotes the *pressure head at the centroid* of the area of the pressure profile, for a constant velocity distribution near the cross section, it follows that

$$h_p = H - V^2/(2g) = z + p/(\rho g). \tag{1.16}$$

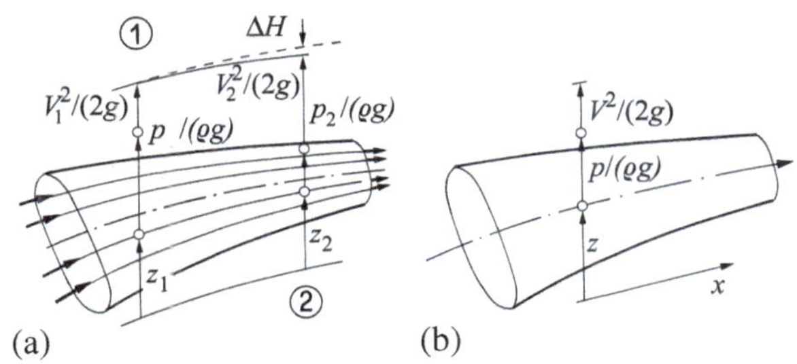

Fig. 1.6 (**a**) Energy head at stations ① and ② of a streamline, with energy head loss ΔH, (**b**) energy head of a streamtube

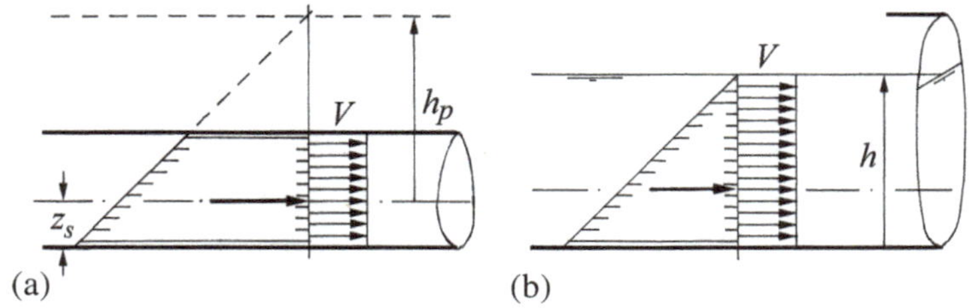

Fig. 1.7 Pressure and velocity distributions in (**a**) pressure conduit, (**b**) open channel

If this relationship is solved for the pressure head normalized to atmospheric pressure $p(z = h_p) = 0$ the *hydrostatic pressure distribution* over the cross-section is

$$p/(\rho g) = h_p - z. \tag{1.17}$$

This representation allows decoupling between pressure and velocity, and permits simplified analysis of one-dimensional flow. Figure 1.7 shows the velocity and pressure distributions for flows in pressure conduits and in open channels. In principle, both can be treated similarly except that:

- For *pressurized conduit flow* the pressure line does not coincide with the conduit soffit, and
- In *open channel flow* the free surface is not known in advance.

If z_s denotes the vertical distance of the centroid of the cross sectional area from the invert of either the channel or the conduit and the pressure head at the centroid is designated h_p, the energy head H for *pressurized conduit flow* (German: Druckabfluss; French: Ecoulement en charge) is

$$H = z_s + h_p + V^2/(2g), \tag{1.18}$$

where z_s as well as h_p can vary with the length coordinate. With the water depth $(z_s + h_p = h)$ of *open channel flow*, the energy head H_* relative to the channel invert is

$$H_* = h + V^2/(2g). \tag{1.19}$$

With these developments, relations analogous to Eq. (1.11) are now available for energy considerations. If, in addition, the mean velocity $V = Q/F$ defined in Eq. (1.1) and the bed elevation $z = z(x)$ above a reference level are introduced, the energy head $H = H_* + z$ in free-surface channels obtains

$$H = z + h + Q^2/(2gF^2). \tag{1.20}$$

In Eq. (1.20) the cross-sectional area F, the water depth h and the bed height geometry $z(x)$ are all dependent on the length coordinate x.

From the preceding formulations the so called *generalized energy principle* between two cross-sections ① and ② of the streamtube can therefore be written as

$$H_1 - H_2 = \Delta H, \tag{1.21}$$

where $\Delta H > 0$ is the mechanically lost energy head. If these losses are negligible, i.e., $\Delta H \to 0$, the Bernoulli relation (Daniel Bernoulli, 1700–1782) results in

$$H_1 = H_2. \tag{1.22}$$

The *Bernoulli relation* states, in effect, that for steady flow with a spatially constant discharge the sum of datum head z, pressure head $p/(\rho g)$ and velocity head $V^2/(2g)$ is invariant. If the energy head is known at a point, conclusions on the rest of the flow domain can immediately be drawn.

In practice the energy principle is preferred to the momentum principle because the "energy" in the hydraulics sense is conceived as a head. One then speaks of *energy line* (German: Energielinie; French: Ligne de charge) as a line $H(x)$ drawn above a fixed reference level which at every point reflects the energy content of the flow. For steady flow with a constant discharge, the energy line may not rise in the direction of the flow.

1.5 Discussion of Results

1.5.1 *Correction Factors*

The foregoing sections have introduced three conservation principles of mechanics, namely, statements on the mass, the momentum and the energy fluxes. In the present day practice the *streamtube theory* is almost exclusively used; that is, the entire flow field is based on:

- *Hydrostatic* pressure distribution, and
- *Uniform* velocity distribution

This reduces the continuity equation to the statement that the sum of the inflows into a control volume equals the total outflow from it.

As regards the momentum principle it is defined as an auxiliary measure of the specific force; the specific force principle derived from it states that the sum of all the specific forces equals the sum of all forces acting on the enveloping surface.

The energy principle finally states that the energy content in the inflow cross-section equals the sum of the energy content in the outflow cross-section and the energy dissipated between the two cross-sections. For constant discharge through a cross-section the energy head is the significant flow parameter.

The assumption of *uniform velocity distribution* is only approximately realized in practice. If streamtube flow is admitted to compute momentum and energy, so-called *correction factors* are introduced. Based on the equations for a single streamline, the correction factor β for the momentum equation is obtained by integration over the cross-section as

$$\beta = \frac{F}{Q^2} \int u_i^2 \mathrm{d}F_i, \tag{1.23}$$

with u_i as the streamwise component of the velocity vectors. Analogously, for energy considerations, the kinetic energy correction factor α is obtained as

$$\alpha = \frac{F^2}{Q^3} \int V_i^3 \mathrm{d}F_i, \tag{1.24}$$

where V_i are the magnitudes of the velocity vectors. Consequently, the generalization of Eq. (1.20), for example, has the form

$$H = z + h + \alpha Q^2/(2gF^2). \tag{1.25}$$

It can be shown that in general $1 < \beta < \alpha$. The values of these correction factors are normally close to unity, such that in comparison to other assumptions of the streamtube theory, they can be neglected, and $\alpha = \beta = 1$.

1.5.2 *Streamline Curvature Effects*

For flows in which the pressure distribution is not hydrostatic, a pressure correction factor α_p must be introduced in the term h of Eq. (1.25). This depends largely on the *streamline curvature* (German: Stromlinienkrümmung; French: Courbure des lignes de courant) and has been discussed by Press and Schröder (1966). If the center of curvature lies above the flow then $\alpha_p > 1$. A typical flow with $\alpha_p > 1$ pertains below a chute, whereas for overfalls with a center of curvature below the flow, $\alpha_p < 1$.

A further influence occurs in *steep* channels with practically straight streamlines. Figure 1.8a shows both the bottom (subscript b) and the surface (subscript o) streamlines as also the vertical water depth t_w, the normal length N_w and the pressure head h. The form of the normal is such that it can be approximated by a circular arc of radius R. Then, $t_w = R \cos\theta_b(\tan\theta_o - \tan\theta_b)$, $N_w = R(\theta_o - \theta_b)$ and $h = R\cos\theta_b(\sin\theta_o - \sin\theta_b)$ from geometrical considerations. Because $\theta_b = \mathrm{d}z/\mathrm{d}x = z'$ and $\theta_o = (\mathrm{d}z + \mathrm{d}h)/\mathrm{d}x = z' + h'$, it follows

$$\frac{h}{N_w} = \frac{\sin\theta_o - \sin\theta_b}{\theta_o - \theta_b} \cong 1 - \frac{3z'^2 + 3z'h' + h'^2}{6} \text{ and} \tag{1.26}$$

$$\frac{t_w}{N_w} = \cos\theta_b \frac{\tan\theta_o - \tan\theta_b}{\theta_o - \theta_b} \cong 1 + \frac{3z'^2 + 6z'h' + 2h'^2}{6}. \tag{1.27}$$

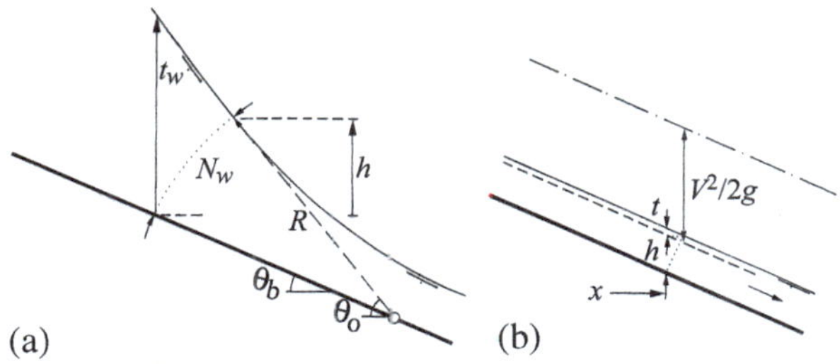

Fig. 1.8 Flow in sloping channel for practically straight streamlines and (**a**) steady non-uniform flow, (**b**) uniform flow

The approximations on the right sides are valid for small values of h' and z' and small streamline slopes.

Using these expressions, Eq. (1.25) can be developed for channels of rectangular cross-section of cross-sectional area $F = bN_w$. Substituting the approximation for h/N_w in Eq. (1.25) yields a first approximation for H as

$$H = z + h + \frac{Q^2}{2gb^2h^2}\left[1 - z'^2 - z'h' - \frac{1}{3}h'^2\right]. \tag{1.28}$$

For uniform flow, the correction expression reduces to $1–z'^2$ and for large bottom inclination to $\cos^2\theta_b$. For known channel bed level geometry $z(x)$ as well as a given energy head and known discharge, $h(x)$ can be determined from a first order differential equation.

If *curvature effects* (German: Krümmungseinfluss; French: Effet de courbure) are taken into account, the terms hh'' and zz'' are to be incorporated in the derivations. This leads to the so-called *Boussinesq equations* (Joseph V. Boussinesq, 1842–1929), the solutions of which require an elaborate procedure. Hager and Schleiss (2009) assumed a linear variation of the streamline curvature along an equipotential line from channel bed to the free surface streamline to obtain

$$H = z + h + \frac{Q^2}{2gb^2h^2}\left[1 + \frac{2hh'' - h'^2}{3} + hz'' - z'^2 - z'h'\right]. \tag{1.29}$$

This equation permits analytical solution of selected flow cases. Usually, a numerical solution is required. The well-known example is the *solitary wave* shown in Fig. 1.9 and discussed in Sect. 7.7. This wave consists of only a single hump and is propagated on both sides of a still water of depth h_o. It appears typically as an impulse wave phenomenon and can only be derived with the assumption of non-hydrostatic pressure distribution. Representative examples follow in Chaps. 6 and 12. Also presented are generalizations for non-rectangular cross-sections and procedures to determine velocity and pressure distributions for a known pressure line profile $h(x)$.

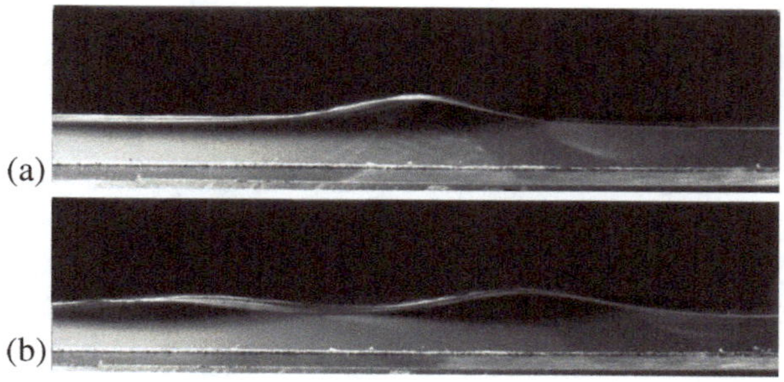

Fig. 1.9 (**a**) Solitary wave and (**b**) cnoidal wave in rectangular laboratory channels

The comprehensive works cited under the literature references below cover the general aspects under which the fundamental equations are derived and their applications to complex flows are presented in the usual manner of classical hydrodynamics. For additional information these may be selectively consulted.

Notation

b	[m]	channel width
E	[Nm]	energy
F	[m^2]	cross-sectional area
g	[ms^{-2}]	acceleration due to gravity
h	[m]	water depth
h_p	[m]	pressure head
H	[m]	energy head
H_*	[m]	energy head relative to channel bed
$\boldsymbol{I}$	[N]	momentum
$\boldsymbol{K}$	[N]	force
$\boldsymbol{K}_a$	[N]	external force
L	[m]	length
m	[kg]	mass
N_w	[m]	normal length
p	[Nm^{-2}]	pressure
P	[N]	pressure force
Q	[m^{3}s^{-1}]	discharge
$\boldsymbol{S}$	[m^3]	specific force
t	[s]	time
t_w	[m]	vertical water depth
u	[ms^{-1}]	velocity component
V	[ms^{-1}]	velocity magnitude
x	[m]	length coordinate

z	[m]	vertical height coordinate
z_s	[m]	vertical height of centroid
α	[–]	energy correction factor
α_p	[–]	pressure head correction factor
β	[–]	momentum correction factor
ΔH	[m]	loss of head
ρ	[kgm^{-3}]	density
θ	[–]	streamline slope

Subscripts

a	lateral outflow
b	bottom, bed
B	boundary surface
d	dissipated
D	pressure
o	inlet, upstream
s	static
T	tangential
u	outlet, downstream
z	inlet, lateral
1	inflow cross-section
2	outflow cross-section

Further Readings

Blevins, R.D. (1984). *Applied fluid dynamics handbook*. Van Nostrand Reinhold: New York.

Chanson, H. (2004). *The hydraulics of open channel flow – An introduction*. Elsevier Butterworth-Heinemann: Oxford, UK.

Daily, J.W., Harlemann, D.R.F. (1966). *Fluid dynamics*. Addison-Wesley: Reading, Mass.

Eck, B. (1966). *Technische Strömungslehre* (Technical hydraulics). Wien [in German].

Graf, W.H., Altinakar, M.S. (1995). *Hydrodynamique – une introduction*. Presses Polytechniques et Universitaires Romandes: Lausanne [in French].

Hager, W.H., Schleiss, A.J. (2009). *Constructions hydrauliques – Ecoulements stationnaires (Hydraulic structures – steady flows)*, ed. 2. Presses Polytechniques et Universitaires Romandes: Lausanne [in French].

Ivicsics, L. (1975). *Hydraulic models*. Research Institute for Water Resources Development. VITUKI: Budapest.

Liggett, J.A. (1994). *Fluid mechanics*. McGraw-Hill: New York.

Lugt, H.J. (1983). *Vortex flow in nature and technology*. Wiley: New York.

Martin, H., Pohl, R. (2000). *Technische Hydromechanik*. 4. Verlag Bauwesen: Berlin [in German].

Montes, S. (1998). *Hydraulics of open channel flow*. ASCE Press: Reston, VA.

Naudascher, E. (1992). *Hydraulik der Gerinne und Gerinnebauwerke* (Hydraulics of channels and channel structures). Springer: Wien, New York [in German].

Novak, P., Guinot, V., Jeffrey, A., Reeve, D.E. (2010). Hydraulic modelling: An introduction. Spon: London.

Prandtl, L. (1965). *Führer durch die Strömungslehre* (Guide across fluid flow). 6. Aufl. Vieweg: Braunschweig [in German].

Press, H., Schröder, R. (1966). *Hydrodynamik im Wasserbau* (Hydrodynamics in hydraulic structures). Ernst & Sohn: Berlin [in German].

Schlichting, H. (1965). *Grenzschicht-Theorie* (Boundary layer theory). 5th edition. G. Braun: Karlsruhe [in German].

Singh, V.P., Hager, W.H., eds. (1996). *Environmental hydraulics*. Kluwer Academic Publishers: Dordrecht, NL.

Townson, J.M. (1991). *Free surface hydraulics*. Unwin Hyman: London.

Truckenbrodt, E. (1968). *Strömungsmechanik*. (Fluid mechanics). Springer-Verlag: Berlin [in German].

White, F.M. (1991). *Viscous fluid flow*. McGraw-Hill: New York.

Chapter 2
Losses in Flow

Abstract Flow losses take place either as friction loss due to wall friction and viscosity or as local flow loss depending on conduit or channel geometry. Both types of losses are described in detail for conduit flow.

The *friction losses* are given by the transition law developed by Colebrook and White and by the solution of the equation of Gauckler-Manning-Strickler. The application criteria of the two formulae are given and the roughness values are presented in tables.

The *local losses* are ascertained for a number of conduit geometries and channel configurations and optimal dimensional ratios are indicated. Finally, flows in pressure conduits and free-surface channels are compared.

2.1 Introduction

It is a well-known fact that mechanical processes are always accompanied by the dissipation phenomenon. It is not possible, therefore, to maintain the initial energy over a certain temporal duration or a local distance without spending additional energy, because these motions are accompanied by physical transformation into thermal energy. One speaks therefore of *energy loss* (German: Energieverlust; French: Perte de charge), whereby only the loss relative to the original mechanical energy is considered.

In hydraulics two kinds of energy losses are distinguished in principle. On the one hand there are losses owing to wall boundaries, referred to as the *boundary layer* and having close causal connection with the viscosity of the fluid and the condition of the wall. On the other hand losses also appear due to changes in the cross-section of the flow and in the direction of the motion. Whenever the flow accelerates or, in particular, decelerates, *separation* occurs and mechanical energy from the flow is withdrawn by large scale vortex motion.

The first kind of loss is closely associated with the boundary layer development and is designated, in accordance with the nomenclature of classical works, as *wall friction loss*. The second kind of loss is intimately linked with the flow geometry and is therefore designated as *form loss* or additional loss (additional to wall friction

W.H. Hager, *Wastewater Hydraulics*, 2nd ed., DOI 10.1007/978-3-642-11383-3_2,

loss). The friction loss is produced by shear stresses along the boundary walls and accounted for in the momentum equation, whereas the additional loss arises from internal shear stresses and may be determined only from indirect considerations.

In the following sections the two kinds of losses are treated individually. The *total loss* ΔH_g is defined as the sum of frictional (subscript R) and additional losses (subscript L for local) as

$$\Delta H_g = \Delta H_R + \Delta H_L. \tag{2.1}$$

Thus the hydraulic losses are estimated using the *principle of superposition*. This principle may naturally be applied with confidence only if the sources of the additional losses lie sufficiently apart. It is on this premise that the combining principle is accounted for.

2.2 Friction Losses

2.2.1 *Equation of Colebrook and White*

As already mentioned, the term "friction loss" is the current customary designation for a loss resulting from the boundary layer development. Although not quite correct, this expression has become popular and shall be so used here.

Consider first a very long cylindrical *pipe of circular cross-section* containing a liquid of constant temperature. Let, for this flow, the discharge be Q, the diameter of the pipe be D, the mean velocity $V = Q/(\pi D^2/4)$ and the kinematic viscosity be ν. If the friction loss is ΔH_R per elementary length Δx, then the *friction gradient* is designated as $S_f = \Delta H_R/\Delta x$. It follows, therefore, that S_f participates in the flow process as a dependent variable.

The English scientist Osborne Reynolds (1842–1912) established that the dimensionless number, which today bears his name,

$$\mathsf{R} = VD/\nu \tag{2.2}$$

essentially describes the *flow regime*. Guided by his experiments with various fluids, Reynolds showed that the *laminar*, i.e. the layered motion, passes into *turbulent* motion if a Reynolds number of about 2300 is exceeded. For water of usual quality, corresponding to about 10°C, the transition velocity (subscript t) is about $V_t = 2\cdot 10^{-3}$ ms^{-1} for $D = 1$ m. Thus, in hydraulic practice, *turbulent flow* appears almost exclusively.

The friction gradient S_f increases almost quadratically with the velocity head ($V^2/2g$) and decreases somewhat linearly with the pipe diameter D. These characteristics led both Henry Darcy (1803–1858) and Julius Weisbach (1806–1871) to propose

$$S_f = \frac{V^2}{2g} \cdot \frac{f}{D}, \tag{2.3}$$

where the so-called *friction factor* f (written as λ in German texts) must nearly be constant. Detailed measurements in the beginning of the twentieth century showed that f depends, essentially, on the *Reynolds number* R, as defined in Eq. (2.2), and the so called *relative wall roughness* $\varepsilon = k_s/D$. The parameter k_s here denotes the wall roughness height of a conduit which produces the same friction loss as is produced by a Nikuradse's sand-roughened pipe (Johann Nikuradse 1894–1979). The roughness measure is also described as the *equivalent sand roughness*.

The idea that a very complex surface irregularity of a conduit can be replaced by a well defined quantity is ingenious and was suggested by Ludwig Prandtl (1875–1953). The sand-roughened pipes were artificially pasted with uniform diameter sand grains and the losses resulting from experiments were measured. Each pair of values (R; ε) therefore yielded a value for the friction factor f. The preceding procedure was repeated for a pipe with arbitrary roughness matrix so that practically the same combination of parameters occurs although another pipe indicates an altogether different wall surface condition. Nevertheless, it would then possess an identical *relative sand roughness* for this discharge. The value of k_s varies in general with the discharge Q and consequently with R. The reason for this lies in the fact that sand-roughened and commercial pipes behave differently with regard to boundary layer development.

In the year 1937 the Englishmen Colebrook and White analyzed the results of experiments on turbulent flow in so-called *smooth pipes* (German: Glatte Rohre; French: Conduites lisses) for which the effect of the Reynolds number predominates, and *rough pipes* (German: Rauhe Rohre; French: Conduites rugueux) in which the parameter ε is of primary importance. For arbitrary turbulent conduit flows they developed a universal law for the friction factor f as a function of relative wall roughness $\varepsilon = k_s/D$ and the Reynolds number R. This *universal friction law* is given by the relation

$$\frac{1}{\sqrt{f}} = -2\log\left[\frac{\varepsilon}{3.7} + \frac{2.51}{\mathsf{R}\sqrt{f}}\right], \quad \mathsf{R} > 2300. \tag{2.4}$$

Figure 2.1 shows graphically the solution of Eq. (2.4) proposed by Moody (1880–1953) in his Moody diagram. One recognizes the various curves $f(\mathsf{R})$ which, depending on the relative roughness ε, approach asymptotically a particular value of f. The larger the value of ε, the higher is this level of f. In the *rough flow regime* there exists, for every relative roughness ε for $\mathsf{R}\to\infty$, a minimum value of the friction factor f. According to Eq. (2.4), the asymptotically approached minimum value of f is

$$f_{\min}^{-1/2} = -2\log(\varepsilon/3.7). \tag{2.5}$$

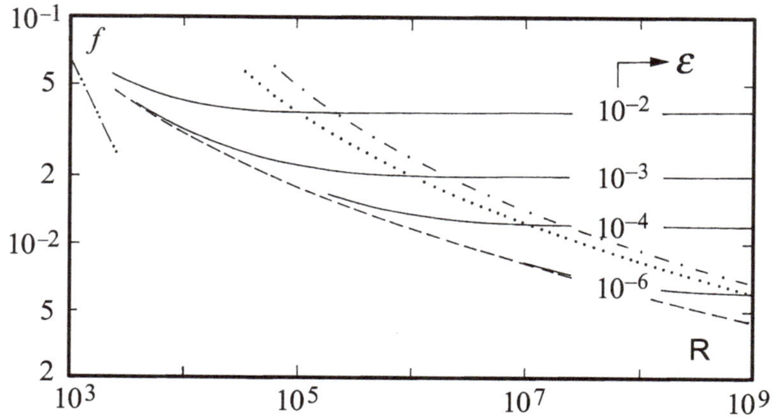

Fig. 2.1 Moody-Diagram, friction factor f as function of Reynolds number $\mathsf{R} = VD/\nu$ for various values of relative wall roughness $\varepsilon = k_s/D$ after Eq. (2.4). (-··· -) laminar flow, (- · · -) 0.75% and (...) 1.5% deviation from fully developed turbulent flow

Hager (1987) observed a deviation in f from $f_{\min}$ of about 1.5% as the limit of the so-called *practical rough flow regime*. This yields, for the limiting relative roughness

$$\varepsilon_r = \frac{1050}{\mathsf{R}}. \tag{2.6}$$

If $\varepsilon > \varepsilon_r$, as is usually found in practice for flows in the rough regime, Eq. (2.5) is to be used. If, however, $\varepsilon < \varepsilon_r$ one has to fall back upon the generalized Eq. (2.4) of Colebrook and White.

In contrast to the rough regime, the flow in the *smooth flow regime* differs only insignificantly from the case $\varepsilon \to 0$. In this case the effect of the wall roughness becomes small in comparison to that of viscosity and one speaks of the *practical smooth flow regime* (subscript s). This is the case if $\varepsilon < \varepsilon_s$ with (Hager 1987)

$$\varepsilon_s = (3.475\mathsf{R})^{-0.9}. \tag{2.7}$$

The range $\varepsilon_s < \varepsilon < \varepsilon_r$ defines the turbulent *transition regime* (German: Übergangs-regime; French: Régime de transition) for which the effects of both viscosity *and* wall roughness are to be considered.

Channel flows, in practice, usually take place under Reynolds numbers R in the range of values between 3×10^4 and 3×10^7 and relative roughnesses ε between 10^{-5} and 10^{-1}. Figure 2.1 shows that the 'smooth' flow regime is rather an unlikely phenomenon in practice. In the following only the transition regime and, particularly, the rough flow regime are considered.

2.2.2 *Transition Regime*

The flow processes in conduits of *circular cross-section* are described by five independent parameters:

- Friction gradient S_f,
- Discharge Q,
- Diameter D,
- Equivalent sand-roughness height k_s, and
- Kinematic viscosity ν.

In problems arising in practice, the determination of the first three parameters is relevant. Values of k_s and ν (Table 3.1) are known in advance.

For the *first type of problem*, the Reynolds number $\mathsf{R} = 4Q/(\pi \nu D)$ as well as the roughness characteristic $\varepsilon = k_s/D$ are known and f can be found either graphically from Fig. 2.1 or iteratively by Eq. (2.4). This implicit relation in f has also been approximated explicitly. For example, Zigrang and Sylvester (1985) proposed

$$f = \frac{1}{4}\left[\log\left(\frac{\varepsilon}{3.7} + \frac{13}{\mathsf{R}}\right)\right]^{-2}. \tag{2.8}$$

In the practical ranges of R and ε, the values of f calculated from Eq. (2.8) and those obtained from the Colebrook and White relationship differ only by about 2%. Other proposals (of explicit representation) have been explained by Chen (1985).

The *second type of problem*, in which the discharge Q is determined, can be solved explicitly by carrying out a transformation of Eq. (2.4) as

$$\hat{q} = Q/Q_o, \quad N = Q_o/(D\nu) \tag{2.9}$$

with $Q_o = (gS_f D^5)^{1/2}$ as the normalizing discharge. Substitution of these in Eq. (2.4) yields the discharge equation (Sinniger and Hager 1989)

$$\hat{q} = -\frac{\pi}{\sqrt{2}} \log\left[\frac{\varepsilon}{3.7} + \frac{2.51}{\sqrt{2}N}\right]. \tag{2.10}$$

Use of Eq. (2.10) is facilitated with Fig. 2.2.

The *third type of problem* is most frequently encountered in practice. It involves the determination of the pipe diameter D. For this, (*) denoted parameters are defined as

$$D^* = D/D_o, \quad k_s^* = k_s/D_o, \quad \nu^* = D_o\nu/Q \tag{2.11}$$

where $D_o = [Q^2/(gS_f)]^{1/5}$ and the unknown diameter D is separated. Introducing these dimensionless variables into Eq. (2.4) yields the implicit expression

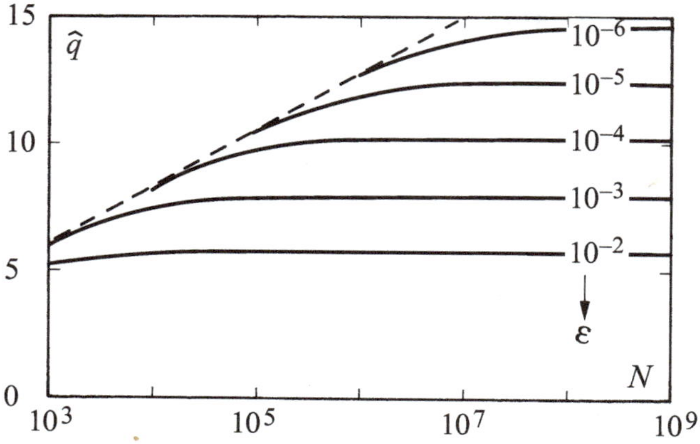

Fig. 2.2 Relative discharge $\hat{q} = Q/(gS_fD^5)^{1/2}$ as function of $N = (gS_fD^3)^{1/2}\nu^{-1}$ for various relative roughnesses $\varepsilon = k_s/D$. (- - -) Smooth flow regime

$$\nu^* = \left[10^{-\sqrt{2}/(\pi D^{*5/2})} - \frac{k_s^*}{3.7D^*}\right]\frac{D^{*3/2}}{1.776}. \tag{2.12}$$

Figure 2.3 shows the smooth flow regime below the envelope curve marked (- - -), while the rough regime is covered by the horizontal lines of the relationship D^* against k_s^*. The plot also shows that D^* varies only between 0.35 and 0.55. An estimated value of $D^* = 0.40$, corresponding to $D = 0.4[Q^2/(gS_f)]^{1/5}$ gives $Q = [gS_f(2.5D)^5]^{1/2}$. The discharge therefore varies with $D^{5/2}$.

The preceding solution procedures involve for all three types of problems an iterative solution with a simple parameter combination. Appropriate values of the *equivalent sand-roughness height* k_s (German: Äquivalente Sandrauhigkeit; French:

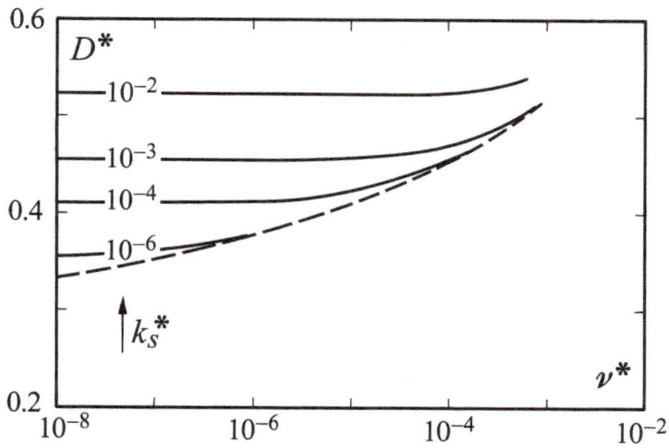

Fig. 2.3 Relative pipe diameter $D^* = D/D_o$ as function of relative kinematic viscosity $\nu^* = \nu D_o/Q$ for various values of $k_s^* = k_s/D_o$ with $D_o = [Q^2/(gS_f)]^{1/5}$. (- - -) Smooth flow regime

Hauteur de rugosité equivalente) in [mm] are presented in Tables 2.1, 2.2 and 2.3. Numerous extraordinary examples of roughness height problems have been reported by Schröder (1990). He also presented methods for roughness determination as well as of the equivalent sand roughness.

Table 2.1 Equivalent sand roughness k_s (excerpted from Richter 1971)

Material and conduit types	Condition	k_s [mm]
drawn and pressed conduit of copper and brass, glass tube	technically smooth, also conduits with metal plating (copper, nickel, chromium)	0.00135–0.00152
plastic pipes	new	0.0015–0.0070
seamless steel pipes rolled and drawn (commercial) new	typical rolled skin	0.02–0.06
	corroded	0.03–0.04
	uncorroded	0.03–0.06
	stainless steel with injection coating of metals	0.08–0.09
	clean zinc coated	0.07–0.10
	commercial zinc coated	0.10–0.16
welded sheet steel, new	typical rolled skin	0.04–0.10
	bitumen coated	0.01–0.05
	cement coated	about 0.18
	galvanized, for pressure charging pipe	about 0.008
used steel pipe	symmetrical rust scars	about 0.15
	moderately rusted, light crusting	0.15–0.40
	moderate crusting	about 0.15
	heavily crusted	2–4
	cleaned after long use	0.15–0.20
	bitumen coated, partly damaged, rusted	about 0.1
	after many years of service	about 0.5
	deposition in sheet form	about 1.1
	25 years in service, irregular tar and naphthalene deposits	about 2.5
cast iron pipes	new, typical cast skin	0.2–0.3
	new, bitumen coated	0.1–0.13
	used, rusted	1–1.5
	crusted	1.5–4
	cleaned after many years of service	0.3–1.5
	city sewers	about 1.2
	heavily rusted	4.5
concrete conduits	new, commercial, smooth tracts	0.3–0.8
	new, commercial, medium rough	1–2

Table 2.1 (continued)

Material and conduit types	Condition	k_s [mm]
	new, commercial, rough	2–3
	new, reinforced concrete, smooth	0.1–0.15
	new, centrifugally cast concrete, smooth	0.1–0.15
	new, centrifugally cast concrete, without plaster	0.2–0.8
	smooth conduit after many years of service	0.2–0.3
	mean value for pipe extension without joints	0.2
	mean value for pipe extension with joints	2.0
asbestos cement pipe	new, smooth	0.03–0.10
earthenware pipe	new, drainage pipe	about 0.7
	new, made from crude clay brick	about 9

Table 2.2 Equivalent sand-roughness height k_s for conduits of various surface characteristics (excerpted from Idel'cik 1979)

Group	Conduit and material type	Condition of surface and type of utilisation	k_s [mm]
I	drawn pipes of		
	brass and copper	technically smooth	0.0015–0.01
	aluminium	technically smooth	0.015–0.06
II	drawn steel pipe	new, unused	0.02–0.10
	without solder joints	cleaned after some years of use	up to 0.04
		with bitumen lining	up to 0.04
		hot water pipe	0.20
		oil pipe line, normal	0.20
		medium corrosion with small tar deposits	≈0.40
		water pipe line after long use	1.2–1.5
		with heavy tar deposits	≈3.0
		conduit with poor surface condition	≥5.0
III	welded steel pipe	new or old, in good condition, transition pieces welded or riveted	0.04–0.10
		new, bitumen lined	≈0.05
		long in service, corroded	≈0.10
		long in use, uniform corrosion	≈0.15
		good transition pieces but poor surface condition	0.3–0.4

Table 2.3 Equivalent sand-roughness height k_s (excerpted from ASCE 1969)

Conduit material	Condition	k_s [mm]	$1/n$ [$m^{1/3}s^{-1}$]
CONDUITS			
Asbestos-cement		0.3–3	67–91
brick		1.5–6	58–77
cast iron pipe	new, unlined	0.25	–
	new, bitumen coated	0.12	–
	new, cement coated	0.3–3	67–91
concrete, monolithic	smooth	0.3–1.5	70–83
	rough	1.5–6	58–67
concrete conduit		0.3–3	67–91
corrugated pipe	rough	30–60	38–45
	invert coated	10–30	45–55
	tar lined	0.3–3	67–90
plastic pipe	smooth	3	70–90
burnt clay pipe		0.3–3	70–90
CHANNELS			
lined with			
asphalt		–	60–77
brick		–	55–83
concrete		0.3–0.10	50–90
rip rap		6	30–50
vegetation		–	25–33
excavated			
earth, straight		3	33–50
winding		–	25–40
rock		–	22–33
unmaintained		–	7–20
natural			
small streams		30–1000	–
fairly symmetrical		–	14–30
unsymmetrical, with puddles		–	10–25

2.2.3 *Rough Turbulent Regime*

In the fully developed turbulent flow regime (German: Vollständiges Rauhregime; French: Régime rugueux developpé), the viscous effect is negligible compared to the effect of wall roughness. In this case the friction factor given in Eq. (2.4) simplifies to

$$f = \frac{1}{4}\left[\log(\varepsilon/3.7)\right]^{-2}, \tag{2.13}$$

and the normalized discharge from Eq. (2.10) is

$$\hat{q} = -\frac{\pi}{\sqrt{2}}\log(\varepsilon/3.7). \tag{2.14}$$

The corresponding diameter from Eq. (2.12) is

$$k_s^* = 3.7D^* \cdot 10^{-\sqrt{2}}/(\pi D^{*5/2}). \tag{2.15}$$

From these, the following explicit *approximations* were found by Sinniger and Hager (1989)

$$D^* = k_s^{*0.03}/1.853, \quad \text{for} \quad 10^{-8} < \varepsilon < 7.10^4; \tag{2.16}$$

$$D^* = k_s^{*1/16}/1.422, \quad \text{for} \quad 7 \cdot 10^{-4} < \varepsilon < 7 \cdot 10^{-2}. \tag{2.17}$$

Equation (2.17) is of special interest because the range of ε for which it holds coincides with the typical *application range* of ε in hydraulic practice. Its solution for discharge and velocity yields

$$Q = 2.56(gS_f)^{1/2}k_s^{-1/6}D^{8/3}, \quad \text{or} \quad V = 3.256(gS_f)^{1/2}k_s^{-1/6}D^{2/3}. \tag{2.18}$$

The discharge, in this special case, depends strongly on the diameter, rather weakly on the hydraulic gradient S_f and quite insignificantly on the equivalent sand-roughness height k_s.

Historical *flow formulae* (German: Fliessformel; French: Formule d'écoulement) have a form which differ only insignificantly from Eq. (2.18). A comparison of the customary expression $V = (2gS_f)^{1/2}(D/f)^{1/2}$ for the velocity obtained from Eq. (2.3) with Eq. (2.18) yields $f = (1/21.2)(k_s/D)^{1/3}$. Relating the diameter D to the hydraulic radius $R_h = D/4$, and combining with the expression

$$1/n = K = 6.51g^{1/2}k_s^{-1/6} \tag{2.19}$$

where $1/n = K$ [$\mathrm{m^{1/3}s^{-1}}$] is a *dimensional roughness factor*, one gets, after Philippe Gauckler (1826–1905), Robert Manning (1826–1897) and Albert Strickler (1887–1963) the well-known GMS relation

$$V = (1/n)S_f^{1/2}R_h^{2/3}. \tag{2.20}$$

In Switzerland one frequently speaks only of the *Strickler formula* (2.20) although it was first proposed by the French scientist Gauckler in 1867. The same Eq. (2.20) was also proposed in 1889 by the Irish Manning on the basis of data measured by Darcy and Bazin and some new experiments carried out by him. In 1895, Manning got this equation modified by another relationship. However, his first formula, Eq. (2.20), although set rather hastily, is still today recognized as the Manning's formula. Albert Strickler, a Swiss engineer, analyzed in 1923 a large number of actual measurements in pressurized pipe and natural stream flows and recommended Eq. (2.20). The merit of his development lies undoubtedly in introducing a formula of the type of Eq. (2.19). Strickler based this on the mean grain diameter of the boundary material of a natural watercourse and found the numerical value to be 6.72 instead of 6.51 obtained from the preceding consideration. In the following, Eq. (2.20) is designated as the *formula of Manning and Strickler*.

Equation (2.19) allows direct correspondence between K- (or $1/n$) and k_s-values for flows in the fully developed turbulent regime. According to Sinniger and Hager (1989) the following two *application criteria* have to be satisfied:

- Flow in fully developed turbulent regime synonymous with a negligible viscosity effect as

$$k_s > 30\nu(g^2 S_f^2 Q)^{-1/5}; \tag{2.21}$$

- Relative roughness ε in the range

$$7 \times 10^{-4} < \varepsilon < 7 \times 10^{-2}, \tag{2.22}$$

ensuring that neither very smooth nor very rough surfaces are considered.

Even today, 70 years after the proposal of Colebrook and White and nearly 90 years after the basic research of Blasius and Prandtl in Gottingen, the formula of Manning and Strickler still is relevant. This is particularly established by two facts:

- First, as shown below, the difficulty in the determination of a lumped $1/n$-value for the boundary surface does not allow an "exact calculation" of discharge, and
- Second, a simple exponential formula offers advantages for the practical calculation of typical parameters such as the discharge Q, the velocity V or the diameter D, particularly for free surface flow.

Basically, a *minimum value* of $1/n = 20\ \text{m}^{1/3}\text{s}^{-1}$ satisfies the application criteria of Eqs. (2.21) and (2.22). The *maximum value* is $1/n = 90\ \text{m}^{1/3}\text{s}^{-1}$, and the minimum friction slope is $S_f = 0.1\%$. Further, it is repeated that Eq. (2.4) or also Eq. (2.10) are strictly valid only for prismatic conduits with uniform distribution of surface roughness. Table 2.4 gives the numerical values for the roughness factor. From this, a distinction in the status of the wall condition as good, normal and bad may be observed. Table 2.4 does not indicate the temporal development of the $1/n$-values, that is, their change with age.

Tables 2.4 and 2.5 specify the accuracy of the flow formula. According to Eq. (2.20) the velocity and consequently the discharge are directly proportional to the $1/n$-value. Usually, a $1/n$-value is given within $\pm 5\%$ accuracy for surfaces of known manufacture. For old channels, deviations in the estimation of $1/n$ of at least $\pm 10\%$ occur provided no detailed measurements are available. The *accuracy of the flow formulae* is therefore related to the difficulty in the estimation of the $1/n$-value. Despite these practical shortcomings, the determination of an exact $1/n$-value shall follow the computation for conduits with the most exact disposal of the relations.

Table 2.4 Friction coefficient $1/n$ [$m^{1/3}s^{-1}$] for formula of Manning and Strickler in relation to boundary condition (excerpted from Chow 1959)

	Wall condition		
Type of wall	good	medium	poor
cast iron, uncoated	85	70	65
cast iron, coated	90	85	75
steel	75	65	60
earthware	90	75	65
clay (drainage pipe)	85	70	60
concrete	85	65	60
polished brick	90	75	65
brick masonry	85	65	60
finished concrete	90	85	75
with concrete coating	85	65	55
broken stone in mortar	60	50	35
dry stone rubble	40	30	25
masonry work	75	65	60

Table 2.5 Friction factor $1/n$ [$m^{1/3}s^{-1}$] for formula of Manning and Strickler as function of wall condition (excerpted from Naudascher 1987)

Channel description	$1/n$ [$m^{1/3}s^{-1}$]
SHEET METAL DUCTS	
smooth conduits with countersunk rivet heads	90–95
new cast iron conduits	90
riveted conduits, rivets not countersunk	65–70
CONCRETE CHANNELS	
with smooth cement finish	100
with steel shuttering	90–100
with smooth plaster	90–95
with smoothened concrete	90
with smooth undamaged cement plaster	80–90
using timber shuttering, without plaster	65–70
rammed concrete with smooth surface	60–65
old concrete, clean faces	60
coarse concrete lining	55
unsymmetrical concrete faces	50
MASONRY CHANNELS	
made of brick masonry work, also clinker, well jointed	80
made of dressed stone	70–80
made of carefully laid broken stone masonry work	70
made of masonry work (normal)	60
crudely made broken stone masonry work	50
made of broken stone walls, slopes plastered	45–50

Table 2.5 (continued)

Channel description	$1/n$ [$m^{1/3}s^{-1}$]
EARTHEN CHANNELS	
In rigid material, smooth	60
In non-mobile sand with some soil and gravel	50
with bed of sand and gravel, plastered sides	45–50
In fine gravel, about 10/20/30 mm	45
In medium gravel, about 20/40/60 mm	40
In coarse sized gravel, about 50/100/150 mm	35
In loamy soil	30
laid with coarse stone aggregate	25–30
In sand, clay or gravel, heavy growth	20–25
NATURAL WATERCOURSES	
with rigid bed, regular	40
with moderate rubble	33–35
with vegetal cover	30–35
with rubble, irregular	30
with strong sediment transport	28
with coarse rubble, about 0.2–0.3 m	25–28
with coarse rubble, with sediment transport	19–22

The Manning and Strickler formula is normally sufficient for fully-developed turbulent flow, whereas Eqs. (2.21) and (2.22) have to be checked for larger structures, and the generalized formula of Colebrook and White has to be applied otherwise.

As is yet to be shown, the formulae of Colebrook and White and of Manning and Strickler can be generalized simply for flows in *free-surface channels*. Therefore, the surface characteristics of such conveyance systems are also represented in Tables 2.1 and 2.2. A particular generalization of the Manning and Strickler formula is presented in Chap. 3.

2.3 Local Losses

2.3.1 *Description*

Local losses occur wherever the streamlines are directed away from the axial direction of flow due to either a change in the wall geometry, or lateral discharge addition or reduction. The main flow then either accelerates or is retarded. Specially for decelerated flow, particles close to the boundary cannot follow the average velocity of the one-dimensional streamtube and separate laterally from the main flow. In the regions of large direction change – as the example in Fig. 2.4 shows – both primary and secondary flow regions exist. For a *diffusor*, essentially an enlargement of the cross-section (Fig. 2.4a), the main current does not follow the walls and two marked zones of *separated flow* appear on the two sides of the main stream in the widened downstream part of the diffusor, where the main stream has not yet fully expanded.

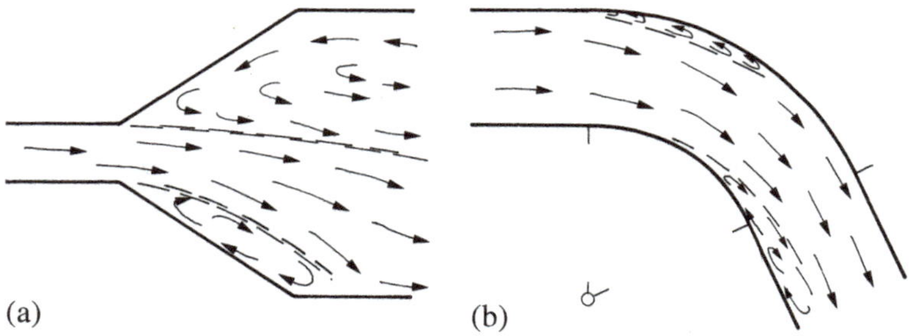

Fig. 2.4 Local losses (**a**) diffuser flow, (**b**) bend flow. Schematic subdivision into primary and secondary flow zones, with (- - -) shear boundary

The separated flows extract considerable energy from the main stream, contributing thereby substantially to the local losses.

In *bend flow* shown in Fig. 2.4b, complicated phenomena can appear as for diffusor flow. Owing to the high pressure at the outer bend side, the particles at the interface do not follow the primary stream but are drawn laterally into the secondary flow zone. Further downstream, there exists an analogue situation along the inner bend wall contributing essentially to the bend losses. These are caused mostly by flow separation and not due to the reattachment of the separated flow (Eck 1991).

Hydraulic separation zones are characterized by high turbulence phenomena, i.e. all flow parameters have a marked dynamic flow component. According to Reynolds, any one of these parameters, e.g., the local pressure intensity p, can be represented as the sum of a *temporal mean value* $\bar{p}$ and a *turbulent fluctuation component* p'. Consequently, for steady discharge, the integrated fluctuating component $\int p' dt$ over a sufficient time period must equal zero. The turbulence number T represents the ratio of a convenient value of the fluctuating component p' (such as the standard deviation) and the temporal mean value $\bar{p}$ corresponding, for example, to $(\overline{p'^2})^{1/2}/\bar{p}$ ensuring a positive number. The turbulence number remains small in the main flow region; in the zone of secondary flow, however, it increases to large orders of magnitude. The maximum value, relative to either pressure or velocity, appears normally at the shear layer as also along the fluctuating boundary line or boundary surface between the primary and the secondary flows (Fig. 2.4). The complex phenomena inside a separation zone are frequently of little interest; more important from the practical standpoint is the answer to the question: How can the flow losses be reduced? The answer requires fundamental knowledge of the flow mechanisms so that different kinds of elements can be compared with each other. This requires defining the so-called dimensionless loss coefficients or resistance values (Eck 1991).

For turbulent flows occurring in practice, the viscous losses are often insignificant. The local losses are then strongly dependent on the velocity field only. Based on Bernoulli's equation (1.15) the energy head H is represented as the sum of the static and the dynamic pressure heads $H = (p_s + p_d)/(\rho g)$, where p_s and p_d are measured, respectively, perpendicular and tangential to the streamline direction.

As *dynamic pressure* one therefore designates the expression $p_d/(\rho g) = V^2/(2g)$. The total (subscript g) pressure is equal to the sum $p_g = p_s + p_d$.

The local energy loss ΔH_L is closely linked with the dynamic pressure p_d. If the fluid is not in motion ($p_d = 0$), no losses are expected. As the velocity of the fluid increases, larger losses occur. The preceding discussion indeed shows that the local losses are proportional to the dynamic pressure, resulting in an almost constant number for the ratio of energy loss and dynamic pressure head. This number is referred to as the *loss coefficient* (German: Verlustbeiwert; French: Coefficient de perte de charge)

$$\xi = \Delta H_L/(V_i^2/2g), \tag{2.23}$$

where V_i is a well defined reference velocity. In practice, it is often stated that:

- Either the quantity V_i is poorly or not at all defined, or
- Calculations are carried out using incorrect reference velocities.

It is therefore important to correctly define ξ-values as well as the *reference velocity* V_i. In particular, all basic quantities are to be so selected that no problems arise for their determination. Frequently, V_i is set equal to the nominal velocity, for example, the mean value of the incoming or the outgoing velocities in the flow elements being investigated. Usually reference to the larger of the two values is made, e.g. the approach flow velocity for the case of Fig. 2.4a. For combining flow, V_i is usually taken as the downstream velocity whereas, for flow division, one refers to the cross-section of the approach flow. Consequently, the reference velocity remains always positive and larger than zero.

Further characteristics of the loss coefficients are explained with examples by Naudascher (1987). A close connection between the loss coefficient ξ and the resistance coefficient c_w is indicated. Flows around *bridge piers* and *trash rack* structures are explained by:

- Reynolds number and pier or bar shapes,
- Roughness and turbulence effects,
- Bottom and ends of piers,
- Wave build up and resulting structural vibration, and
- Constrictions and arrangement of adjacent structures.

In these cases, extensive experimental material exists which is not always available for other structures. For practical purposes this abundance of data is not always available, and reference to *guidance values* of ξ is made. For further information, one should consult the standard works of Richter (1971), Miller (1971, 1978), Idel'cik (1979, 1986), Ward-Smith (1980), and Blevins (1984). The abundance of data material contained in these sources cannot be discussed here.

In the following, a selection of values for the loss coefficient ξ is presented, refering to standard cases and valid, basically, only for *pressurized flow*. The transformation of these results to free surface flow is discussed in 2.4. The flow elements considered are bends, contractions, expansions, inlets and outlets, combining and dividing junctions and special sewer elements. For the cross-sectional form of these elements, mainly the *circular profile* is considered.

2.3.2 *Conduit Bend*

The wall pressure along a *circular bend* (German: Kreiskrümmer; French: Coude circulaire) changes as shown in Fig. 2.5. On both sides of the element the pressure line has a gradient S_f, which steepens along the bend and merges with the gradient S_f further downstream.

The computation of the *energy loss* ΔH_L along the bend is simplified by assuming that the distributed losses are locally concentrated and measured as the vertical distance between the undisturbed energy lines. For an experimental investigation, first the bend is not considered and the equivalent sand roughness height of the conduit material is ascertained. Then, the bend is accounted for, and the value of ΔH_L is separately determined. For the computational determination of ΔH_g, the procedure is reversed. Regarding the friction loss, the presence of the bend is first neglected, the term ΔH_R calculated along the conduit and at the end, the local loss is added with Eq. (2.1). This procedure therefore considers that the loss producing element involves no length. Also, the substantial loss occurs not along the bend but manifests itself in the *downstream reach*, as is clear from Figs. 2.4b and 2.5.

The loss coefficient ξ_k for the circular bend (subscript k) depends essentially on the ratio of the bend mid-radius R and the pipe diameter D, the angle of deviation δ and the Reynolds number $\mathsf{R} = VD/\nu$. Figure 2.6a shows experimental curves after Ito

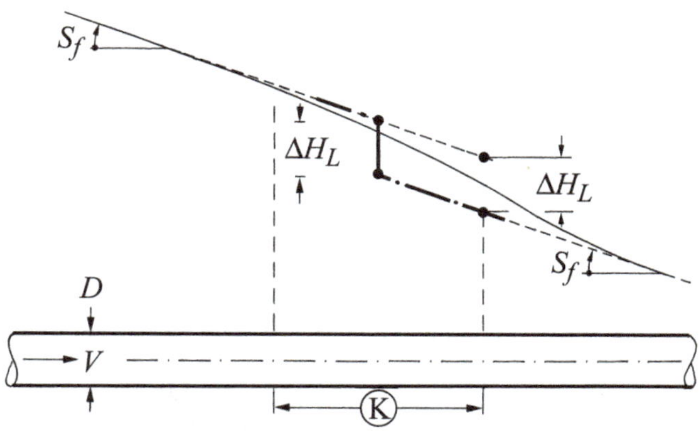

Fig. 2.5 (—) *Pressure line* along a *circular bend* K, (- - -) gradient S_f, (-·-) computational curve with locally concentrated bend loss ΔH_L

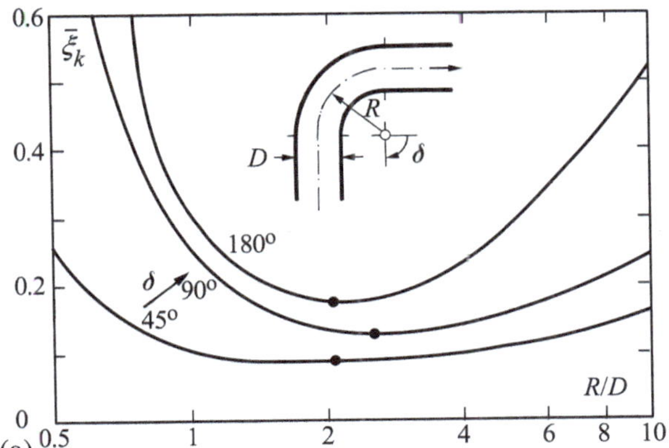

(a)

(b)

Fig. 2.6 (**a**) Total loss coefficient $\bar{\xi}_k = \Delta H_k/(V^2/2g)$ as a function of relative bend radius R/D and angles of curvature δ for $\mathsf{R} \geq 10^6$ after Ito (1960) and Blevins (1984), (•) minimum value, (**b**) Typical open channel bend structure on treatment station

(1960), for a Reynolds number $\mathsf{R} = 10^6$ corresponding to a velocity around 1 ms^{-1} for a typical pipe diameter of 1 m. This study includes the influence of the friction loss along the curve. It is evident from Fig. 2.6a that for every angle δ there exists a *minimum value* (subscript m) of the total loss coefficient ξ_{km} for R/D of about 2. This behaviour reflects the fact that:

- For small bend radius R/D, the separation zone is large, and
- For large bend radius R/D, the influence of friction is dominant.

Between these two extremes lies the optimum value of R/D between 2 and 3. The curves represented in Fig. 2.6a are minimum values. If the Reynolds number R is smaller than 10^6, the value ξ_k obtained from Fig. 2.6 are multiplied by $(10^6/\mathsf{R})^{1/6} > 1$.

Figure 2.7 represents the appropriate *loss coefficient* ξ_k after Miller (1971). There appears likewise an optimum curvature R/D between 1 and 3 in this figure. For $R/D = 2$ the loss coefficient ξ_k does not exceed 0.2 even for 180°. For practical purposes the mid-radius R of the bend should be chosen to be about *double* the pipe diameter. On the other hand, the representation also points to the result that ξ_k values increase steeply for deviations greater than 60°. Therefore, an angle of 90° should be considered the upper limit of a hydraulically well designed conduit.

Figure 2.7 was derived for a Reynolds number $\mathsf{R} = VD/\nu \cong 10^6 (\log \mathsf{R} = 6)$. For a different Reynolds number R, the correction to be applied is (Sinniger and Hager 1989)

$$\xi_k/\xi_k(\mathsf{R} = 10^6) = \frac{3.7}{\log \mathsf{R} - 2.3}. \tag{2.24}$$

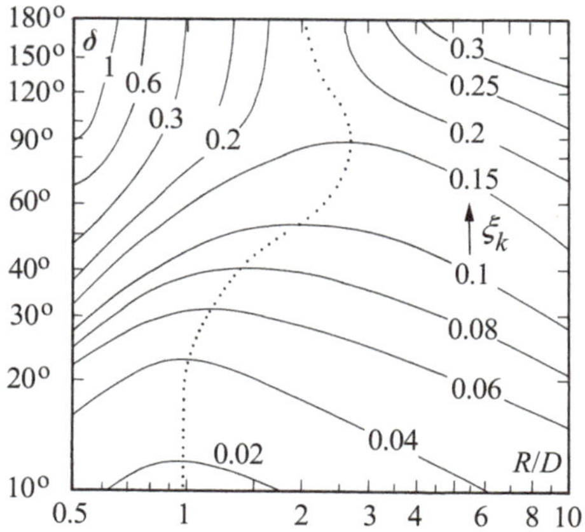

Fig. 2.7 Loss coefficient ξ_k for conduit bend after Miller (1971) as function of relative radius of curvative R/D and angle of deviation δ for circular cross-section and $\mathsf{R} = 10^6$, (...) minimum value

Double bends consisting of two 90°-degree bend elements lying a distant L_k apart were considered by Blevins (1984). The ratio Ξ of the total ξ_k-value to the sum (subscript tot) of the individual values $\xi_{k\text{tot}}$ varies with the relative curvature R/D and with the relative distance L_k/D, where L_k is measured from the end of the first to the beginning of the second bend. For $L_k/D > 20$, $\Xi = 1$, for smaller relative distance $\Xi < 1$, however, such as $\Xi = 0.85$ for $L_k/D = 0$. With the usual assumption $\Xi = 1$, the resulting loss is overestimated. The values of the coefficient for two 90°-degree bends located at two different levels are given similarly. There appears to be practically no difference with the levels of installation of the double bend. Table 2.6 gives the detailed results.

Mitre-bends (German: Kniekrümmer; French: Coude abrupt) in which the change of direction is abrupt, are not generally recommended for sewers. For direction changes as small as $\delta = 40°$, the loss coefficient is already as high as $\xi_k = 0.25$, and for $\delta = 90°$, ξ_k is about 1.2, much larger than the minimum loss coefficient for

Table 2.6 Loss coefficient $\Xi = \xi_{k\text{tot}}/\Sigma\xi_{ki}$ for two 90°-bends in tandem arrangement (**a**) in one plane, (**b**) in two planes

		L_k/D				
	R/D	0	4	10	20	30
(**a**)	1.85	0.86	0.72	0.82	0.95	0.96
	3.3	0.84	0.82	0.86	0.96	1.0
	7.5	0.93	0.96	0.97	1.0	1.0
(**b**)	1.85	0.88	0.73	0.86	0.96	0.97
	3.3	0.86	0.81	0.88	0.97	1.0

circular bends. A series of detailed information on mitre bends have been presented by Sinniger and Hager (1989) and Hager (1992).

2.3.3 *Expansion*

The degree of an expansion (German: Erweiterung; French: Expansion) is usually described with the angle of expansion δ and the cross-sectional area ratio F_1/F_2 of the expansion (subscript e). The *loss coefficient* $\xi_e = \Delta H_{12}/(V_1^2/2g)$ involves the approach flow velocity V_1 (Fig. 2.8).

Conduit expansions are characterized by asymmetrical flow behaviour. For the diameter ratio $D_2/D_1 > 1.4$, the jet entering the expansion attaches to one of its two downstream walls (Fig. 2.8a) and, depending on the expansion angle, the Reynolds number and the expansion ratio, the jet becomes either oscillating or stably asymmetric. The velocity ratios, the wake structure and the resulting consequences for *diffusors* are summarized by Blevins (1984) and Hager (1990). These are not pursued further here.

With regard to expansion losses, the *abrupt* 90°-*diffusor* has special significance because the loss coefficient may be determined from elementary hydraulic considerations. Assuming that the pressure acting on the expansion face is the same

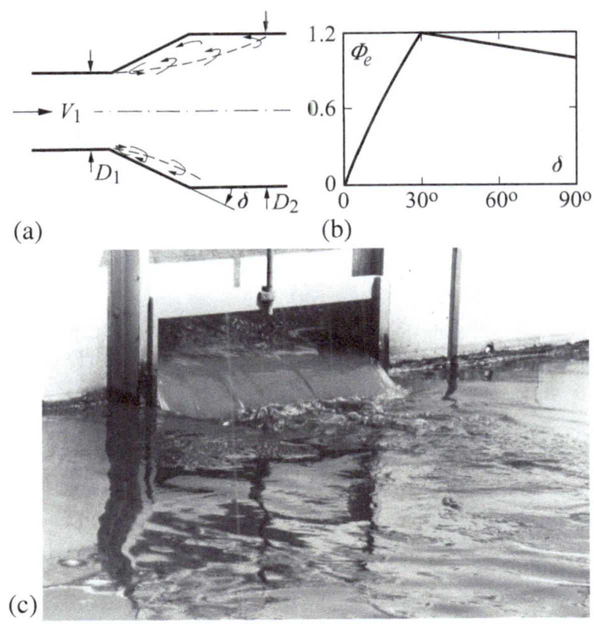

Fig. 2.8 Conduit expansion (**a**) definition plot and flow structure, (**b**) loss coefficient $\Phi_e = \Delta H_{12}/(V_1^2/2g)$ as function of the expansion angle δ. (**c**) Expanding jet from an outlet into an aeration basin

as the inlet pressure p_1 and neglecting the wall resistance, the application of the momentum principle yields the Borda-Carnot expression

$$\xi_{e90^\circ} = \Delta H_{12}/(V_1^2/2g) = [1 - (F_1/F_2)]^2. \quad (2.25)$$

ξ_{e90° is defined with respect to the approach flow velocity V_1. For circular conduit diffusors, the loss coefficient depends on the ratio of diameters (D_1/D_2) only.

In analogy to this special case, the effect of the *expansion angle* δ on ξ_e is accounted for with the relation

$$\xi_e = \Phi_e(\delta) \cdot \xi_{e90^\circ}. \quad (2.26)$$

Equation (2.26) splits the loss coefficient ξ_e in two parts, one depending only on the angle δ and the other depending only on the area ratio F_1/F_2. The experimentally measured values of $\Phi_e(\delta)$ can be represented as (Sinniger and Hager 1989)

$$\Phi_e(\delta) = \frac{\delta}{90^\circ} + \sin(2\delta), \qquad 0 \le \delta \le 30^\circ; \quad (2.27)$$

$$\Phi_e(\delta) = \frac{5}{4} - \frac{\delta}{360^\circ}, \qquad 30^\circ \le \delta \le 90^\circ. \quad (2.28)$$

These two relations satisfy the limit conditions $\Phi_e(\delta = 0) = 0$ and $\Phi_e(\delta = 90^\circ) = 1$ (Fig. 2.8b). The maximum value occurs around $\delta = 30^\circ$. Tests indicate that expansions behave hydraulically similar for δ greater than about 30°. Low values for the losses occur only for small expansion angles ($\delta < 10^\circ$). The maximum ξ_e value is somewhat larger than 1, and the entire approach velocity head is dissipated.

A *conduit outlet* (German: Rohrauslauf; French: Exutoire de conduite) can be considered a special case of an expansion. For example, if an *outfall conduit* discharges into a basin or into the sea, the expansion ratio tends to $F_1/F_2 \rightarrow 0$. Eq. (2.25) then gives $\xi_{e90^\circ} = 1$ and the loss coefficient follows $\xi_e = \Phi_e(\delta)$. For $\delta > 30^\circ$ and expansions with a large downstream section, all kinetic energy is dissipated.

The relationships derived previously apply only for an outfall conduit discharging under water. For an outflow conduit discharging into *air* as a compact jet, there is practically no loss. Further information on conduit expansions together with a comprehensive literature review is presented by Sinniger and Hager (1989).

2.3.4 *Contraction*

Although the contraction geometry equals a reversal of flow direction in a conduit expansion, the flows in the two elements are expected to be fundamentally different, owing to the differences in the separation structures. As schematically represented in (Fig. 2.9a), the flow in a sharp-edged contraction (German: Verengung; French: Contraction) experiences an additional contraction K and expands downstream to the full cross-sectional area F_2. As is well known, a contracted flow is nearly free of energy loss and the loss in the contraction element must be attributed to the *expansion of flow* downstream of the contracted flow. Contraction (subscript v) flow is essentially influenced by the angle of contraction δ and the area ratio $\phi = F_2/F_1 < 1$.

Few experimental studies exist for conduit contractions. Of particular relevance is a paper by Gardel (1962). The relationship derived for the loss coefficient ξ_v is

$$\xi_v = \Delta H_{12}/(V_2^2/2g) = \frac{1}{2}(1-\phi)(\delta/90°)^{1.83(1-\phi)^{0.4}}. \tag{2.29}$$

It is noted that the contraction loss coefficient is always smaller than the corresponding expansion loss coefficient. Also, ξ_v increases substantially with the contraction angle δ; and Eq. (2.29) yields moderate losses for contraction angles below $\delta < 30°$. For $\delta = 90°$, $\xi_v = (1-\phi)/2$, i.e., the loss increases linearly with the decrease in the contraction ratio. Additional information on this topic has been reported by Benedict et al. (1966).

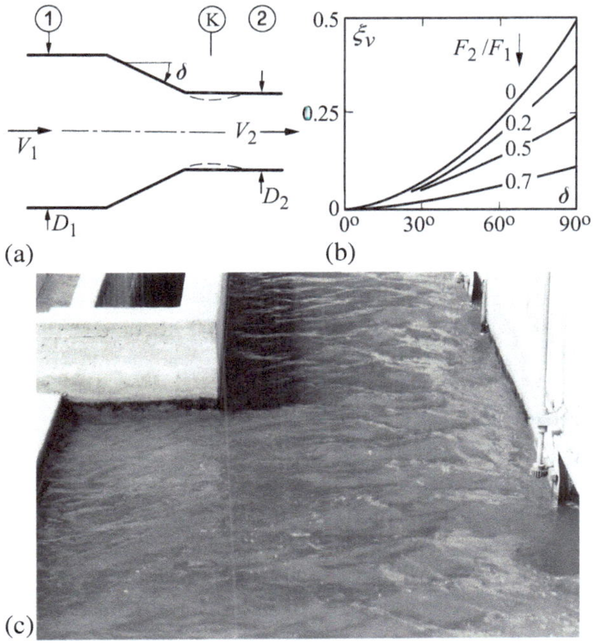

Fig. 2.9 Conduit contraction (**a**) definition and flow structure, (**b**) loss coefficient $\xi_v = \Delta H_{12}/(V_2^2/2g)$ as function of contraction angle δ for various values of area ratio $\phi = F_2/F_1$ after Gardel (1962). (**c**) Contracting main channel due to laterals in a distribution channel

A *conduit inlet* (German: Rohreinlauf; French: Pertuis d'entrée) can be treated as a special case of a contraction element with the area ratio tending to zero. It follows from Eq. (2.29) that $\xi_v(\phi = 0) = (1/2)(\delta/90°)^{1.83}$; for the usual case $\delta = 90°$, the loss coefficient is thus $\delta_v = 0.5$. For a sharp-edged conduit inlet, therefore, the head loss is half the pipe velocity head $(1/2)\,[V_2^2/(2g)]$. This high value of the conduit velocity head may be reduced by *rounding* (German: Ausrundung; French: Courbure) the contraction inlet with the radius r_v. The data by Idel'cik (1979) can be approximated as

$$\xi_v = \frac{1}{2}\exp(-15r_v/D). \tag{2.30}$$

Figure 2.10b shows that the loss coefficient decreases sharply with the relative rounding radius. Comparing Eq. (2.30) with the work of Knapp (1960) one observes for $r_v/D > 1/6$ practically no inlet loss. Similarly, not much pressure difference exists in this case between the conduit axis and the boundary wall streamline. The rounding of sharp edges has a strong hydraulic effect but the *point of separation* is no longer clearly defined. Rounded transition elements may further be quite expensive.

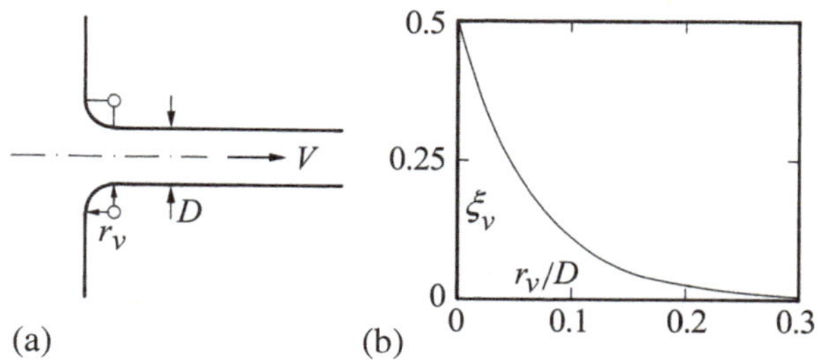

Fig. 2.10 Rounded conduit inlet (**a**) definition sketch, (**b**) loss coefficient $\xi_v = \Delta H_v/(V^2/2g)$ as function of the relative rounding radius r_v/D

2.3.5 *Combining Conduit Junction*

In conduit systems, combining and dividing conduits are frequently encountered (see also 2.3.6). The corresponding loss coefficients may be quite high and it is important to consider their numerical magnitudes, therefore.

Figure 2.11a shows a definition sketch for the combining conduit (German: Rohrvereinigung; French: Combinaison de conduites). In principle, the flow continuing from the upstream branch (subscript o) is distinguished from the lateral branch (subscript z) that enters the flow system. The combining angles are referred to as δ_o and δ_z, and the cross-sectional areas are F_o and F_z, respectively. The discharge

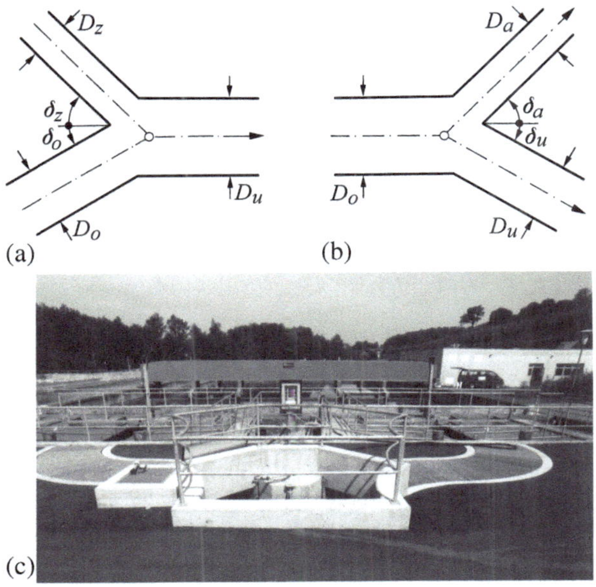

Fig. 2.11 Schematic representation and notation for (**a**) combining conduit, (**b**) dividing conduit. (**c**) Junction structure on treatment station

ratio remains always between zero and unity and the outflow is designated Q_u. The relative inflow is, consequently, $q = Q_z/Q_u$ such that $Q_o = Q_u - Q_z = Q_u(1-q)$. For the area ratios, one considers likewise the downstream cross-sectional area F_u as the reference quantity which gives

$$m = F_z/F_u, \quad n = F_o/F_u, \quad q = Q_z/Q_u. \tag{2.31}$$

The *loss coefficients* for the lateral and the upstream branches, respectively, are defined as

$$\xi_z = \frac{H_z - H_u}{V_u^2/2g}, \quad \xi_o = \frac{H_o - H_u}{V_u^2/2g}. \tag{2.32}$$

Both values are non-dimensionalized with respect to the downstream velocity $V_u > 0$.

Combining flows are analyzed using the basic principles of pressure distribution together with the momentum equation. Vischer (1958) obtained the loss coefficients for the conduit junction involving *sharp edges* as

$$\xi_z = 1 - 2m^{-1}\, q^2 \cos\delta_z - 2n^{-1}(1-q)^2 \cos\delta_o + (m^{-1}q)^2, \tag{2.33}$$

$$\xi_o = 1 - 2m^{-1}\, q^2\, \cos\delta_z - 2n^{-1}(1-q)^2\, \cos\delta_o + [n^{-1}(1-q)]^2. \tag{2.34}$$

These two relations differ only in their last terms. The loss coefficients depend on five independent parameters δ_o, δ_z, m, n and q. *Special cases* result for $n = 1$ ($F_o = F_u$) and $\delta_o = 0$ (upstream and downstream branches connected with a straight line). Then, the results of Favre (1937) apply

$$\xi_z = -1 + 4q + (m^{-2} - 2m^{-1}\, \cos\delta_z - 2)q^2, \tag{2.35}$$

$$\xi_o = q[2 - (1 + 2m^{-1}\, \cos\delta_z)q]. \tag{2.36}$$

These relationships are also developed by Idel'cik (1986). The evaluation of Eqs. (2.33) and (2.34) shows that the influence of the junction angle δ_z on both ξ_z and ξ_o is small. Figure 2.12 shows the representation for $\delta_z = 45°$, which changes only slightly for other angles δ_z. The figure also shows the large influence of both the remaining parameters m and n on ξ_o and ξ_z. With the exception of small values of q, ξ_z is always positive and the energy head of the lateral inflow is always larger than the downstream energy head H_u for large values of the lateral inflow. In contrast, for all large values of q, ξ_o is negative, i.e. the water from the upstream branch is sucked in by the incoming lateral flow, similar to a water jet pump. For the upstream branch, therefore, the mechanical energy increases, resulting in a *negative* value for the loss coefficient ξ_o. The total energy budget corresponding to the sum of energies in the upstream and lateral branches is naturally always larger than the energy in the downstream branch.

Of practical interest are the conduit junctions of minimum energy loss. Apart from the case with $\delta_o \to 0$ and $\delta_z \to 0$ in which both the incoming flows enter the main stream with minimum possible lateral component, the expressions for the *optimum area ratios* (subscript opt) are (Vischer 1958)

$$m_{\text{opt}} = q/\cos\delta_z, \quad n_{\text{opt}} = (1-q)/\cos\delta_o. \tag{2.37}$$

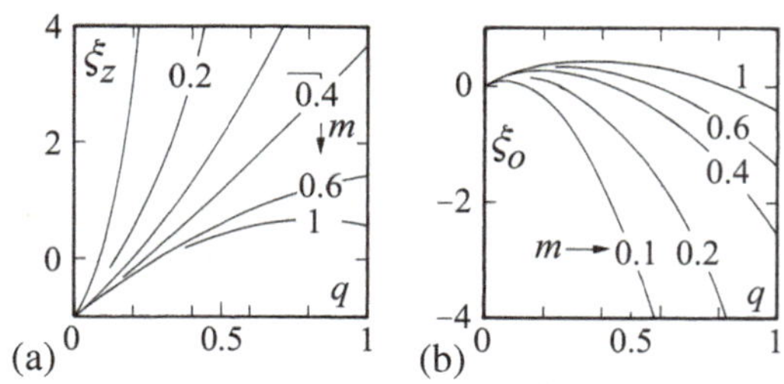

Fig. 2.12 Loss coefficient in combining conduit junction for $\delta_z = 45°$ and $\delta_o = 0$ for (**a**) lateral branch ξ_z and (**b**) upstream branch ξ_o in relation to discharge ratio $q = Q_z/Q_u$ for various area ratios $m = F_z/F_u$ and $n = 1$

These relations express that the projections of the cross-sectional areas F_o and F_z on the downstream conduit correspond to the respective quantity distribution in the two branches of inflow. For variable discharge ratio q, the optimum values are retained for the design discharges.

Small values for the loss coefficients result for comparable velocities in the three branches and sharp edges in the junction element are eliminated by appropriate *rounding*. Further results can be derived from the extensive measurements reported by Gardel (1957) as well as by Gardel and Rechsteiner (1970). Ito and Imai (1973) summarized these studies and presented empirical equations. The rounded conduit junctions are treated in Sect. 16.2.2.

2.3.6 *Dividing Conduit Junction*

Whereas the combining flow in a conduit junction has a flow structure similar to that in a conduit contraction (and thereby remaining amenable to elementary calculations), a conduit junction for flow division (German: Abflusstrennung; French: Séparation d'écoulement) is analogous to a conduit expansion. Losses in both elements are governed by *flow separation* from the walls, a phenomenon that becomes more pronounced with decreasing velocity.

Designating the flow from the upstream branch with subscript o, the lateral flow branch with subscript a and the downstream flow branch with subscript d (through-flow), the loss coefficients for flow division (Fig. 2.11b) are defined analogous to those for the combining flow as

$$\xi_a = \frac{H_o - H_a}{V_o^2/2g}, \quad \xi_d = \frac{H_o - H_d}{V_o^2/2g}. \tag{2.38}$$

The loss coefficients are defined with reference to the non-zero inflow velocity V_o. Assuming a plausible distribution of the lateral division angle δ_a, Hager (1984) calculated for $F_o = F_a = F_d$ and $\delta_d = 0$ with $\bar{q}=Q_a/Q_o$

$$\xi_a = 1 - 2\bar{q}\cos\left(\frac{3}{4}\delta_a\right) + \bar{q}^2, \tag{2.39}$$

$$\xi_d = \frac{4}{5}\bar{q}\left(\bar{q} - \frac{1}{2}\right). \tag{2.40}$$

Idel'cik (1979) studied configurations with $\delta_d = 0$ and $F_a + F_d = F_o$, i.e. equal cross-sectional areas of the approach and downstream branches. Denoting the velocity ratios

$$\mu_a = V_a/V_o, \quad \mu_d = V_d/V_o, \tag{2.41}$$

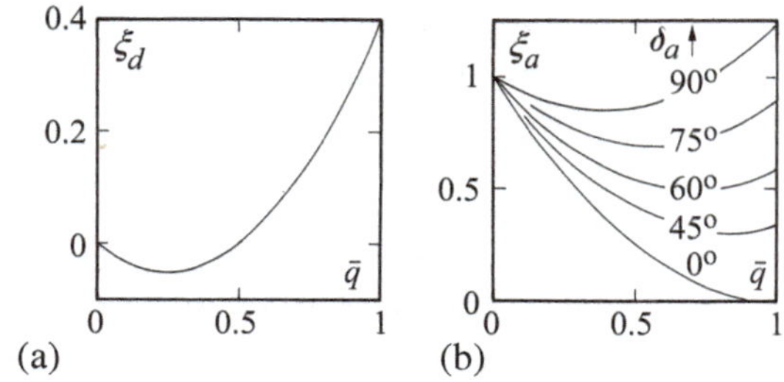

Fig. 2.13 Loss coefficients for conduit division with $F_o = F_a = F_d$, (**a**) through branch and (**b**) lateral branch as functions of the through discharge ratio $\bar{q} = Q_a/Q_o$ and lateral division angle δ_a

one obtains approximately

$$\xi_a = 1 - 2\,\mu_a \cos\delta_a + (1 - \sin^3\delta_a)\mu_a^2, \tag{2.42}$$

$$\xi_d = (1 + \mu_d)(1 - \mu_d)^2. \tag{2.43}$$

In general, the loss coefficient ξ_d for the through branch is *independent* of the off-take angle δ_a. Further, for small relative lateral discharge $\bar{q} << 1$, the value of ξ_a is nearly unity. Therefore, if almost the entire fluid flows downstream, the velocity head of the upstream branch is dissipated. The effect of rounding the sharp edges is substantial and has been investigated by Idel'cik (1986).

2.3.7 *Y-Junction*

By the term Y-junction one understands a symmetric conduit junction for either combining or dividing flow, as shown in Fig. 2.14. In a *T-junction*, the two approach branches are perpendicular to the downstream branch, or the two downstream branches are perpendicular to the approach flow branch. The following considers the cases $\delta_z = \delta_o$ for combining flow, and $\delta_a = \delta_d$ for flow division. First, three branches of equal cross-sectional area are considered, and second the total inflow and total outflow cross-sectional areas are assumed equal. The relevant information originates from Miller (1978).

For *combining flow* in T-junctions (subscript T) the results of Idel'cik (1979) are considered. The loss coefficient $\xi_{Tv} = \Delta H_{Tv}/(V_u^2/2g)$ is expressed in terms of the velocity head of the downstream branch, $q_z = Q_z/Q_u$ is the discharge ratio and $\phi = F_u/F_z$ the ratio of the cross-sectional areas of the downstream to the lateral branches. Then, for $\delta_z = 90°$,

$$\xi_{Tv} = 1 + \phi^2 + 3\phi^2(q_z^2 - q_z). \tag{2.44}$$

The ξ_{Tv} curves are symmetrical with respect to the optimal ratio $q_z = 0.5$ and the loss coefficient is always larger than unity (Fig. 2.15a). The loss coefficient is considerably reduced by inserting a *division wall* (Fig. 2.15b) and then amounts to

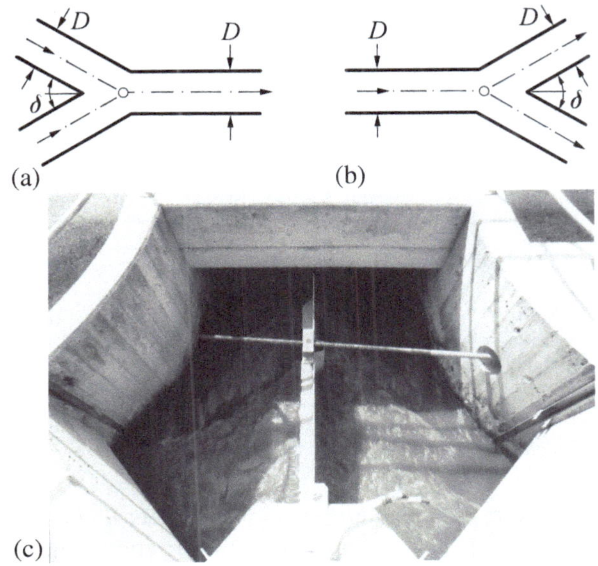

Fig. 2.14 Y-junction for (**a**) combining flow and (**b**) flow division. (**c**) Typical junction structure on wastewater treatment station

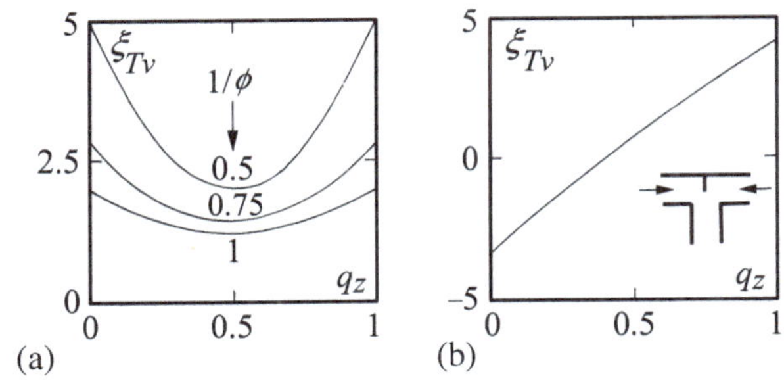

Fig. 2.15 Loss coefficient of combining flow in T-junction, $\xi_{Tv} = \Delta H/(V_u^2/(2g))$ as a function of discharge ratio $q_z = Q_z/Q_u$ and the area ratio $1/\phi = F_z/F_u$ for $\delta_z = 90°$, (**a**) without division wall, (**b**) *with* division wall

$$\xi_{Tv} = 7(q_z - 0.4). \tag{2.45}$$

The loss coefficients for the other branches are determined with the corresponding discharge ratios.

The *flow division* in a Y-junction was studied by Miller (1978). Figure 2.16 shows the experimental results for all three branches of equal cross-sectional area as also with equal cross-sectional areas for the upstream and downstream branches. In order to keep the losses within reasonable limits, one should limit the dividing angle to less than 60° and use symmetrical flow conditions.

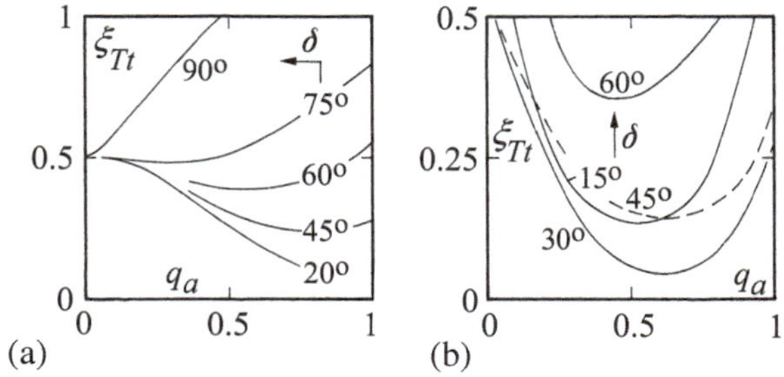

Fig. 2.16 Flow division in Y-junction, $\xi_{Tt} = \Delta H_{Tt}/(V^2/2g)$ as function of discharge ratio $q_a = Q_a/Q_o$ for various division angles δ (Miller 1978). (**a**) identical cross-sectional area, (**b**) equal cross-sectional area of up- and downstream branches

2.3.8 *Trash Racks*

Figure 2.17a shows a trash rack (German: Einlaufrechen; French: Grille d'entrée) set in an open channel at an angle δ_{Re} relative to the direction of flow. The approach flow velocity is V_o and the rack (subscript *Re*) loss is ΔH_{Re}. The loss coefficient expressed in terms of the approach flow velocity V_o is (Sinniger and Hager 1989)

$$\xi_{Re} = \beta_{Re}\ \zeta_{Re} c_{Re} \sin\delta_{Re} \tag{2.46}$$

with β_{Re} as the rack coefficient (Table 2.7) shown in Fig. 2.18. This figure also shows the rack dimensions, namely, $\bar{a}$ the clear spacing of the rack bars, $\bar{b}$ the center to center bar spacing and $\bar{L}$ and $\bar{d}$ the width and the thickness of the bars, respectively.

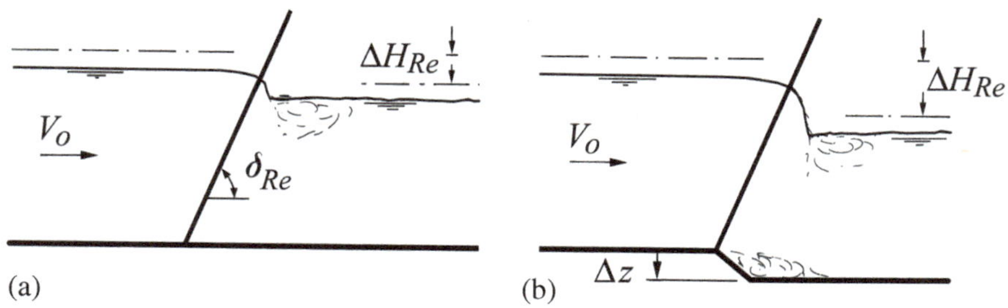

Fig. 2.17 Rack (schematic) in open channel (**a**) without and (**b**) *with* depressed downstream floor

Table 2.7 Rack factor β_{Re} in relation to the rack geometry shown in Fig. 2.18

Type	1	2	3	4	5	6	7
β_{Re}	1	0.76	0.76	0.43	0.37	0.30	0.74

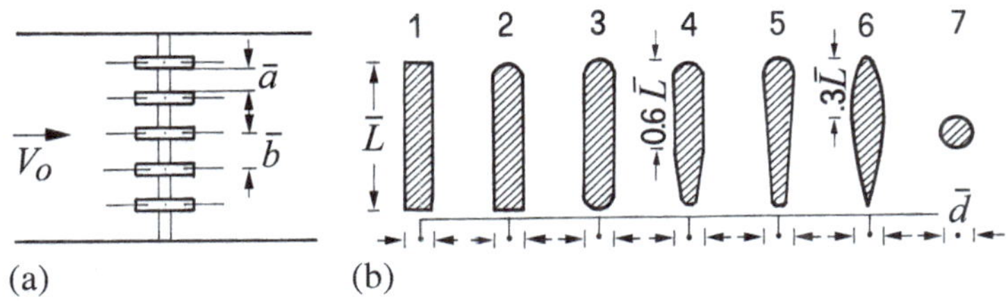

Fig. 2.18 (**a**) Plan of the rack bars, (**b**) Types of rack bars

For *clean racks* $c_{Re} = 1$, for mechanically cleaned racks c_{Re} varies between 1.1 and 1.3 and for the manually cleaned racks c_{Re} varies between 1.5 and 2. The rack geometry enters with the factor ζ_{Re} (Idel'cik 1979). Customarily, ζ_{Re} amounts to around 1. In the following, a simplified procedure is presented.

Equation (2.46) simplifies if the ratio $\bar{L}/\bar{d}$ equals to about 5 and $\bar{a}/\bar{b} > 0.5$, i.e. the rack bars are slender. Then, according to Idel'cik (1979)

$$\xi_{Re} = \frac{7}{3}\beta_{Re}c_{Re}\left[\frac{\bar{b}}{\bar{a}} - 1\right]^{4/3} \sin\delta_{Re}. \tag{2.47}$$

For rough estimates of the rackloss, *ad hoc* values of 5 cm for mechanically cleaned racks and a maximum of 10 cm for manually cleaned racks are adopted for a velocity of around 1 ms^{-1}. To compensate for the head loss $\Delta H_{Re} = \xi_{Re} \cdot V_o^2/(2g)$, the channel floor downstream of the rack is often depressed by the amount $\Delta z = \Delta H_{Re}$ (Fig. 2.17b).

2.3.9 *Slide Gate*

Slide gates (German: Plattenschieber; French: Vanne à tiroir) are installed in pressure conduits either fixed or adjustable in position. Figure 2.19 shows three main types with a) straight bottomed closing edge, b) semi-circular bottom geometry of radius R_s as well as c) with a guide ring. Let the conduit diameter be D and the axial opening height s. Further, let r_v be the rounding radius of the bottom edge and t_p the gate thickness (Fig. 2.19d).

In the following the customary circular sharp-edged slide gate of cross-sectional area $F_o = \pi D^2/4$ and with a semi-circular bottom edge of relative diameter $2R_s/D \cong 1.2$ are considered. The relations between the relative cross-sectional area F/F_o and the gate opening factor $S = s/D$ are then approximately

$$\text{Straight bottom edge} \quad F/F_o = 1.70S^{3/2}\left[1 - \frac{1}{4}S - \frac{4}{25}S^2\right], \tag{2.48}$$

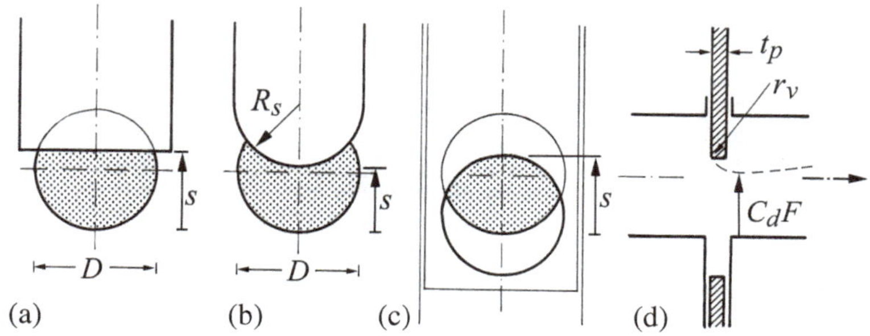

Fig. 2.19 Slide gate, definition sketches for (**a**) straight bottom gate, (**b**) semi-circular bottom geometry, (**c**) gate with guide ring, (**d**) streamwise flow pattern

$$\text{Semi-circular bottom edge} \quad F/F_o = 1.20S\left[1 - \frac{1}{6}S^3\right], \tag{2.49}$$

$$\text{With ring guide} \quad F/F_o = S^{4/3}. \tag{2.50}$$

Experimental studies by Schedelberger (1975) show that the contraction coefficient C_d (Fig. 2.19d) and the corresponding loss coefficient ξ_p depend only on the edge rounding radius r_v and the area ratio $\phi = F/F_o$. In contrast, no effects neither of the relative gate thickness t_p/D, nor of the gate shape or even of the Reynolds number $\mathsf{R} = V_oD/\nu$ in the downstream conduit are detected. The test results are characterized by the minimum (subscript o) value

$$C_{do} = 0.61 + \frac{2}{3}\rho_p^{1/2}, \tag{2.51}$$

where $\rho_p = r_v/D \leq 0.2$ is the degree of edge rounding. For $\rho_p = 0$, Eq. (2.51) yields the basic contraction ratio 0.61. The dependence of the *coefficient of discharge* C_d on the area ratio ϕ follows the relation

$$C_d = C_{do} + 0.73\left[\phi - \frac{2}{3}\rho_p^{1/7}\right]^2. \tag{2.52}$$

For $\phi = 1$ all values of C_d tend eventually to unity.

The *loss coefficient* $\xi_p = \Delta H/\left[V_o^2/(2g)\right]$ expressed in terms of the approach flow velocity $V_o = Q/(\pi D^2/4)$ follows the modified Borda-Carnot relation

$$\xi_p = \left[\frac{1}{C_d\phi} - 1\right]^2. \tag{2.53}$$

This is true only for pressurized flow. For *free surface flow* $\xi_p = 0$, potential flow prevails, and the *discharge* Q is with Eq. (2.52)

$$Q = C_d F[2g(h_o - h_u)]^{1/2}, \tag{2.54}$$

where h_o is the upstream pressure head and h_u the mean downstream flow depth. Often h_u is negligible as compared to h_o. An alternative derivation is presented in Chap. 4.

2.4 Discussion of Results

2.4.1 *Free Surface and Pressurized Flows*

The preceding discussion concerning the additional losses involves primarily *pressurized conduit flow* (German: Druckrohrabfluss; French: Ecoulement en charge). The influence of the free surface is discussed here because pressurized conduit flow can be regarded as a special case with the Froude number (Chap. 6) tending to zero.

The results on loss coefficients pertain only to important cases of sewer flow. There exists, however, an enormous amount of data on loss coefficients which have been lucidly summarized, for example, by Idel'cik (1979), Blevins (1984) or Miller (1994). The flow elements mentioned here are explained with reference to pressurized conduits but can be applied also for *free surface flows* (German: Freispiegelabfluss; French: Ecoulement à surface libre). For example, bend or junction elements are frequently met in sewer systems. The dominant emphasis in the available data relates to pressurized conduits whereas relatively few studies with the corresponding open channel flow have been undertaken. The question therefore arises as to whether it is possible to transform the loss values obtained in pressurized conduits to comparable flow situations in free surface channels and, if so, what limits restrict their applicability?

To answer these questions, first the equations of closed conduit and free surface flows are established. For this purpose, the energy line slope is designated S_E and the slope of the conduit axis on the channel bed is designated S_o. With $h_p = p/(\rho g)$ as the pressure head, the system of equations for the *pressurized conduit* at location $x = x^*$ is (Fig. 2.20)

$$H = h_p + \frac{V^2}{2g}, \quad \frac{dH}{dx} = S_o - S_E, \tag{2.55}$$

and for the *free-surface channel* with the water depth h

$$H = h + \frac{V^2}{2g}, \quad \frac{dH}{dx} = S_o - S_E. \tag{2.56}$$

The *cross-sectional area* F of a pressure conduit is given by $F = F(x)$ thereby accounting for the cross-sectional variation along the direction of flow (Fig. 2.20a).

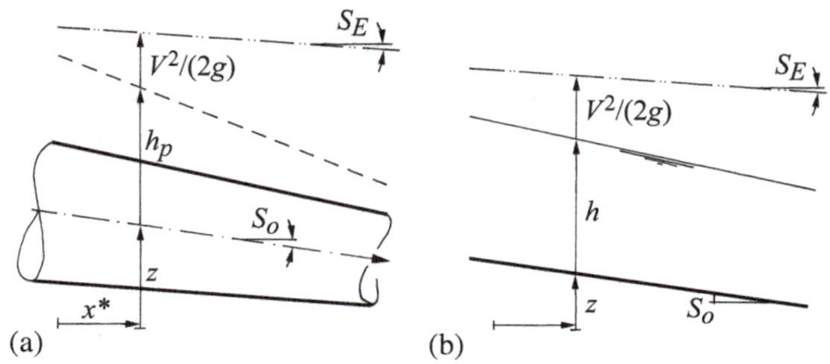

Fig. 2.20 Similarity between flow in (**a**) pressurized conduit and (**b**) free-surface channel

For a free-surface channel, the cross-sectional area F at a section $x = x^*$ can vary with both the local water depth h *and* the distance along the channel, expressed as $F = F(x,h)$. If V is set equal to Q/F in Eqs. (2.55) and (2.56), and the equations are then differentiated with respect to x assuming constant discharge Q, it follows for *pressurized conduit flow*

$$\frac{\mathrm{d}h_p}{\mathrm{d}x} - \frac{Q^2}{gF^3}\frac{\mathrm{d}F}{\mathrm{d}x} = S_o - S_E, \tag{2.57}$$

and for *free-surface flow*

$$\frac{\mathrm{d}h}{\mathrm{d}x} - \frac{Q^2}{gF^3}\frac{\mathrm{d}F}{\mathrm{d}x} = S_o - S_E. \tag{2.58}$$

For given conduit geometry, including area function $F(x)$, inclination of the conduit axis $S_o(x)$ as well as known discharge Q and friction slope S_E, Eq. (2.57) can be solved for the local pressure head function $h_p(x)$. In contrast, the solution of Eq. (2.58) needs further development. Given that the *total derivative* of a function $F = F(x,h)$ equals

$$\frac{\mathrm{d}F}{\mathrm{d}x} = \frac{\partial F}{\partial x} + \frac{\partial F}{\partial h}\cdot\frac{\mathrm{d}h}{\mathrm{d}x} \tag{2.59}$$

the analogy to Eq. (2.58) is

$$\frac{\mathrm{d}h}{\mathrm{d}x}[1 - \mathsf{F}^2] - \frac{Q^2}{gF^3}\frac{\partial F}{\partial x} = S_o - S_E. \tag{2.60}$$

The similarity of Eqs. (2.57) and (2.60) shows that pressurized and free-surface flows are identical provided $\mathsf{F}^2 = Q^2/(gF^3)(\partial F/\partial h) = 0$, i.e. if the *Froude number* (German: Froudezahl; French: Nombre de Froude) equals zero. As discussed in Chap. 6, the dimensionless number F characterizes the dynamics of free-surface flows. From the preceding derivations it is now clear that pressurized flow represents a special case of free-surface flow and that the loss coefficients (obtained for pressurized flows) can be carried over to the corresponding open channel flow situation provided the Froude number remains small. The absolute upper limit is $\mathsf{F} = 1$, i.e. so-called critical flow, for which the pressure term in Eq. (2.60) completely vanishes. In practice, the transformation concept of the loss coefficients to free-surface channels can be applied up to about $\mathsf{F} = 0.7$. Beyond this value, this simplified concept ceases to be valid owing to streamline curvature effects in the form of *standing surface waves* (Chap. 1).

2.4.2 *Transformation Principle*

The preceding observation that the loss coefficients determined for pressure conduits may also be applied to open channel flow is of practical importance. Were this not so, many extensive series of investigations would have to be repeated with the additional parameter F. Various studies point to the transferability of the loss coefficients. In practice, this calculation procedure has been in use for a long time primarily because of the absence of any other data basis. A proof was however still lacking. If the *transformation principle* from pressurized to free-surface flows is accepted, solutions of many engineering problems are sufficiently accurate. This is verified in Chap. 16 for the selected example of junction manholes.

At this point a further proposition may be introduced for which the derivations that follow are essentially simple. Starting from the notion of energy head

$$H = z + h + \frac{Q^2}{2gF^2} \tag{2.61}$$

pertaining to a cross-section at location $x = x^*$ (Fig. 2.21) and denoting the average slope of the energy line S_E the following holds for two adjacent cross-sections ① and ②

$$z_1 + h_1 + \frac{Q_1^2}{2gF_1^2} = z_2 + h_2 + \frac{Q_2^2}{2gF_2^2} + S_E L_{12}. \tag{2.62}$$

The *energy loss* $\Delta H_{12} = S_E L_{12}$ is set equal to the sum of the friction loss $\Delta H_R = S_f L_{12}$ and the additional loss $\Delta H_L = \xi_{12}(V_1^2/2g)$ If the bottom slope is S_o, the bottom level difference $(z_1 - z_2) = S_o L_{12}$, then (Fig. 2.21)

$$(S_o - S_f)L_{12} = h_2 - h_1 + \frac{Q_2^2}{2gF_2^2} - \frac{Q_1^2}{2gF_1^2}(1 - \xi_{12}). \tag{2.63}$$

Equation (2.63) is the basic equation for calculating *water surface profiles* over an element of length L_{12} with additional losses included. For subcritical flow, the

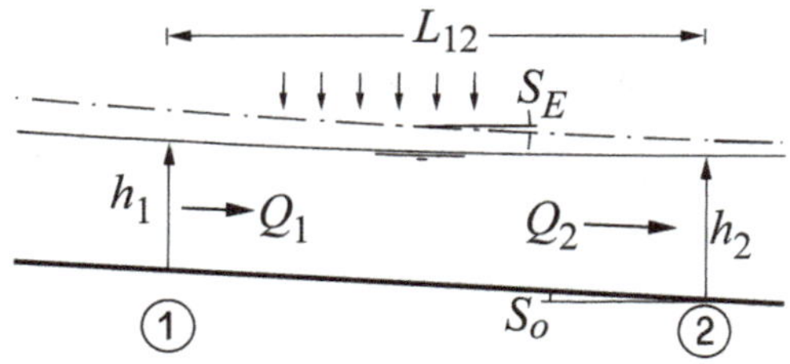

Fig. 2.21 Flow in open channel with hydraulic losses

parameters at the downstream cross-section ② are known, and those in the upstream cross-section ① are sought (Chap. 8). The upstream flow depth h_1 can therefore be determined from Eq. (2.63) provided the channel geometry is defined.

Subcritical flows ($\mathsf{F} < 1$) are characterized by a large static pressure portion h and a relatively small dynamic portion $V^2/2g$. The energy head with respect to the channel bottom is

$$H_* = H(z = 0) = h + \frac{Q^2}{2gF^2}. \tag{2.64}$$

For so-called *weak* subcritical flows ($\mathsf{F} < 0.5$) that are of primary concern here, one gets $Q^2/(2gF^2h) << 1$. Accordingly, by an approximate method the dynamic portion is slightly modified. The calculation is simplified significantly if assuming a constant velocity head over the length L_{12}. This transforms Eq. (2.63) to

$$h_1 - h_2 = (S_o - S_f)L_{12} - \frac{Q_2^2}{2gF_2^2}\xi_{12}. \tag{2.65}$$

This relation gives explicitly the unknown upstream flow depth h_1 if the average friction slope $S_{fa} = (S_{f1}+S_{f2})/2$ is replaced by the *constant* energy line slope $S_f = S_{f2}$, and

$$h_1 - h_2 = (S_o - S_{f2})L_{12} - \frac{Q_2^2}{2gF_2^2}\xi_{12}. \tag{2.66}$$

This last simplification is again valid since the energy gradient strongly depends on the Froude number and tends to zero for small values of F. Besides, the length L_{12} is often small in comparison with a typical length and consequently the friction effect is not significant. Equation (2.66) possesses a form convenient for practical calculations and allows *explicit* determination of flow depth h_1 in the upstream cross-section. For a channel in which the friction gradient S_f is nearly compensated for by the bottom slope S_o, Eq. (2.66) reduces to

$$h_1 = h_2 + \frac{Q_2^2}{2gF_2^2}\xi_{12}. \tag{2.67}$$

Channels which satisfy the condition $S_o = S_f$ are often found in practice, because the friction gradient S_f normally lies between 0.1% and 1%. Eq. (2.67) emphasizes that the upstream flow depth h_1 then equals the downstream flow depth h_2 plus the product of the loss coefficient ξ_{12} and velocity head $V_2^2/(2g)$. Applications of these relations are discussed subsequently.

Notation

$\bar{a}$	[m]	clear spacing of rack bars
$\bar{b}$	[m]	centre to centre spacing of rack bars
c_{Re}	[–]	coefficient for rack logging
c_w	[–]	resistance coefficient
C_d	[–]	coefficient of discharge
$\bar{d}$	[m]	thickness of rack bars
D	[m]	diameter
D_o	[m]	diameter relative to S_f and D
D^*	[–]	non-dimensional diameter relative to D_o
f	[–]	friction factor
F	[m^2]	cross-sectional area
F	[–]	Froude number
g	[ms^{-2}]	acceleration due to gravity
h	[m]	flow depth
h_p	[m]	pressure head
H	[m]	energy head
k_s	[m]	equivalent sand roughness height
k_s^*	[–]	sand roughness relative to D_o
$\bar{L}$	[m]	length of rack bars
L_k	[m]	length of bend
L_{12}	[m]	distance between two cross-sections
m	[–]	area ratio for conduit junction (combining flow)
n	[–]	area ratio for conduit junction (combining flow)
$1/n$	[m$^{1/3}$s^{-1}]	roughness coefficient
N	[–]	viscosity parameter
p	[Nm^{-2}]	pressure
$\bar{p}$	[Nm^{-2}]	mean pressure
p'	[Nm^{-2}]	fluctuating pressure component
p_d	[Nm^{-2}]	dynamic pressure
p_g	[Nm^{-2}]	total pressure
p_s	[Nm^{-2}]	static pressure
q	[–]	discharge ratio for conduit junction (combining flow)
$\hat{q}$	[–]	relative discharge
Q	[m^{3}s^{-1}]	discharge

Q_o	[m^{3}s^{-1}]	discharge normalized with S_f and D
r_v	[m]	radius of rounding
R	[m]	centre line radius of conduit bend
R_s	[m]	radius of semi-circular bottom gate
R	[–]	Reynolds number
s	[m]	height of gate opening
S	[–]	gate opening ratio
S_E	[–]	energy gradient
S_f	[–]	friction gradient
S_o	[–]	bottom slope
t	[s]	time
t_p	[m]	gate thickness
T	[–]	turbulence number
V	[ms^{-1}]	average velocity
x	[m]	length coordinate
z	[m]	elevation coordinate
β_{Re}	[–]	rack coefficient
δ	[–]	angle
δ_{Re}	[–]	rack angle
ε	[–]	relative wall roughness height
ϕ	[–]	area ratio
Φ_e	[–]	influence of expansion angle
μ	[–]	velocity ratio
ν	[m^{2}s^{-1}]	kinematic viscosity
ν^*	[–]	viscosity relative to D_o and Q
ρ	[kgm^{-3}]	density
ρ_p	[–]	degree of rounding
ξ	[–]	loss coefficient
ξ_{12}	[–]	loss coefficient between two cross-sections
ζ_{Re}	[–]	rack coefficient
Ξ	[–]	ratio of loss coefficients

Subscripts

a	lateral outflow branch
d	through branch
e	expansion
g	total, sum
k	bend
L	additional, local
m	minimum
o	lateral branch, approach flow
opt	optimum

p	slide plate (gate)
R	friction
r	rough
Re	rack
s	smooth
t	transition
T_t	T-piece, flow division
T_v	T-piece, combining flow
tot	total
u	downstream branch
v	contraction
z	lateral inflow
1	inflow
2	outflow

References

American Society of Civil Engineers ASCE (1969). Design and construction of sanitary and storm sewers. *Manuals and Reports of Civil Engineering Practise* 37. ASCE: New York.

Benedict, R.P., Carlucci, N.A., Swetz, S.D. (1966). Flow loss in abrupt enlargements and contractions. *Journal of Engineering for Power* 88(1): 73–81.

Blevins, R.D. (1984). *Applied fluid dynamics handbook*. Van Nostrand Reinhold: New York.

Chen, J.J.J. (1985). Systematic explicit solutions of the Prandtl and Colebrook-White equations for pipe flow. *Proc. Institution Civil Engineers* 79: 383–389; 81: 159–165.

Chow, V.T. (1959). *Open channel hydraulics*. McGraw-Hill: New York.

Eck, B. (1991). *Technische Strömungslehre* (Technical fluid flow). 9th edition. Springer: Berlin [in German].

Favre, H. (1937). Sur les lois régissant le mouvement des fluides dans les conduites en charge avec adduction latéral (On the laws governing the fluid motion in pressurized conduits with lateral discharge). *Revue Universelle des Mines* 80: 502–512 [in French].

Gardel, A. (1957). Les pertes de charge dans les écoulements au travers de branchements en té (The head losses for flows in T-junctions). *Bulletin Technique de la Suisse Romande* 83(9): 123–130; 83(10): 143–148 [in French].

Gardel, A. (1962). Perte de charge dans un étranglement conique. (Head loss in a conical contraction). *Bulletin Technique de la Suisse Romande* 88(21): 313–320; 88(22): 325–337 [in French].

Gardel, A., Rechsteiner, G.F. (1970). Les pertes de charge dans les branchements en té des conduites de section circulaire (Head losses in T-junctions of circular conduits). *Bulletin Technique de la Suisse Romande* 96(25): 363–391 [in French].

Hager, W.H. (1984). An approximate treatment of flow in branches and bends. *Proc. Institution of Mechanical Engineers* 198C(4): 63–69.

Hager, W.H. (1987). Die Berechnung turbulenter Rohrströmungen (The computation of turbulent pipe flows). *3R-International* 26(2): 116–121 [in German].

Hager, W.H. (1990). Strömungsverhältnisse in Rohr- und Kanal-Erweiterungen (Flow patterns in pipe and channel expansions) *Österreichische Wasserwirtschaft* 42(11/12): 305–312 [in German].

Hager, W.H. (1992). Kniekrümmer (Mitre bends). *3R-International* 32(2/3): 94–100 [in German].

Idel'cik, I.E. (1979). *Memento des pertes de charge* (Handbook of energy losses) 2nd edition. Eyrolles: Paris [in French].

Idel'cik, I.E. (1986). *Handbook of hydraulic resistance*. Hemisphere Publishing Corporation: Washington.

Ito, H. (1960). Pressure losses in smooth pipe bends. *Journal of Basic Engineering* 82: 131–143.

Ito, H., Imai, K. (1973). Energy losses at 90° pipe junctions. *Journal of the Hydraulics Division* ASCE 99(HY9): 1353–1368; 100(HY8): 1183–1185; 100(HY9): 1281–1283; 100(HY100): 1491–1493; 101(HY6): 772–774.

Knapp, F.H. (1960). *Ausfluss, Überfall und Durchfluss im Wasserbau* (Outflow, overflow and throughflow in hydraulic structures). G. Braun: Karlsruhe [in German].

Miller, D.S. (1971). *Internal flow*. BHRA: Cranfield-Bedford.

Miller, D.S. (1978). *Internal flow systems*. BHRA Fluid Engineering: Cranfield-Bedford.

Miller, D.S., ed. (1994). *Discharge characteristics*. IAHR Hydraulic Structures Design Manual 8. Balkema: Rotterdam.

Naudascher, E. (1987). *Hydraulik der Gerinne und Gerinnebauwerke* (Hydraulics of channels and channel structures). Springer: Wien, New York [in German].

Richter, H. (1971). *Rohrhydraulik* (Pipe hydraulics). Springer: Berlin [in German].

Schedelberger, J. (1975). Schliesscharakteristiken von Einplattenschiebern (Closing characteristics of singular plate gates). *3R-International* 14(3): 174–177 [in German].

Schröder, R.C.M. (1990). Hydraulische Methoden zur Erfassung von Rauheiten (Hydraulic methods for the definition of roughnesses). *DVWK Schrift* 92. Parey: Hamburg, Berlin [in German].

Sinniger, R.O., Hager, W.H. (1989). *Constructions hydrauliques – Ecoulements stationnaires* (Hydraulic structures – steady flows). Presses Polytechniques Romandes: Lausanne [in French].

Vischer, D. (1958). Die zusätzlichen Verluste bei Stromvereinigung in Druckleitungen (The additional losses due to junction flow in pipes). *Dissertation* TH Karlsruhe, appeared also as *Arbeit* 147, Th. Rehbock-Laboratorium: Karlsruhe [in German].

Ward-Smith, A.I. (1980). *Internal fluid flow*. Clarendon Press: Oxford.

Zigrang, D.J., Sylvester, N.D. (1985). Discussion to "A simple explicit formula for the estimation of pipe friction factor", by J.J.J. Chen. *Proc. Institution Civil Engineers* 79: 218–219.

Chapter 3
Design of Sewers

Abstract Sewer design in practice is of fundamental significance because the dimensions of sewer systems are specified. Two particular discharges are considered, namely, the minimum and the maximum discharges. The *minimum discharge* influences the channel slope and therefore the depositional behaviour, whereas the *maximum discharge* establishes the sewer dimensions. A design is flexible if the minimum and the maximum discharges are close. The circumstances however become more difficult for large differences between the two extreme discharges. At the end of this chapter, the optimum cross-sectional sewer shape is also discussed.

3.1 Introduction

The design of sewers is of paramount importance. Usually the design procedure to be formulated must be simple and explicit. Calculations must include at least the extreme discharges.

For the *maximum discharge* (Sect. 3.2) the conveyance capacity of a particular sewer reach for uniform flow must be ascertained. Then, the numerical calculation for the flowing full condition becomes simple. Normally, the circular cross-section is considered. Because the design does not account for the flow types, *uniform flow* (Chap. 5) is taken into consideration for the selection of the sewer, and abrupt changes of bottom slope, sewer diameter and discharge are avoided. Computations of backwater and drawdown curves (Chap. 8) involving a significant flow acceleration or flow deceleration indicate that the capacity of a sewer reach is guaranteed.

For the *minimum discharge*, problems arise for solid transport. It must consequently be pointed out that particularly for large ratios of maximum to minimum discharges, the tractive force necessary for the transport of solids in the sewer is mobilised. In Sect. 3.3 various procedures are outlined first and two proposals are recommended. Common to both approaches is the concept of minimum wall shear stress which increases the minimum velocity as the sewer diameter increases.

W.H. Hager, *Wastewater Hydraulics*, 2nd ed., DOI 10.1007/978-3-642-11383-3_3,

Even today the effects of sewer abrasion and sewer aging are evaluated only in exceptional cases. Both aspects cannot be treated in a generalized manner but must be analyzed from local data. Although maximum allowable velocities up to about 5 ms^{-1} may be adopted according to Pfeiff (1960), steep sewers with velocities up to 15 ms^{-1} are in service without any lasting effect due to abrasion.

3.2 Maximum Discharge

3.2.1 *Flowing Full Condition*

The basis of sewer design is the 'flowing full' condition, the state of transition between the free-surface and the pressurized flow. As explained in Chap. 5, this surcharge free condition is not realized in experiments. Because about 85% part-full flow corresponds to the full flow discharge, the 'flowing full' condition is physically set at around 85% part-full stage.

The flowing full condition (subscript v) is distinguished from the part-full flow by its simple geometry. For the common circular cross-section, the cross-sectional area is $F_v = (\pi/4)D^2$, where D is the pipe diameter, $P_v = \pi D$ the wetted perimeter and $R_{hv} = F_v/P_v = D/4$ the hydraulic radius (Chap. 5). Using the resistance law of Colebrook and White (Chap. 2), the relative discharge $q_r = Q_v/(S_E g D^5)^{1/2}$ obtains

$$q_r = -\frac{\pi}{\sqrt{2}} \log\left[\frac{2.51\nu}{(2gS_E D^3)^{1/2}} + \frac{k_s}{3.71D}\right]. \tag{3.1}$$

Here ν is the kinematic viscosity, g acceleration due to gravity, S_E the energy line slope and k_s the equivalent sand-roughness height. Eq. (3.1) can be rewritten as

$$q_r = -\frac{\pi}{\sqrt{2}} \log\left[1.77 q_r \mathsf{R}_r^{-1} + 0.27\kappa_s\right]. \tag{3.2}$$

Consequently, the relative discharge q_r depends on the diameter Reynolds number $\mathsf{R}_r = Q/(\nu D)$ and on the relative roughness height $\kappa_s = k_s/D$. The Reynolds number reflects the influence of viscosity and the relative sand-roughness height accounts for the sewer surface boundary material.

The viscosity ν for pure water varies essentially with the sewage temperature T_s. The values adopted by ATV (1988) are reproduced in Table 3.1. Normally the estimated value $\nu = 1.31\times10^{-6}$ m^2s^{-1} is taken for sewers. This includes the usually higher sewage temperature and the presence of sewage instead of pure water.

Table 3.1 Kinematic viscosity ν for pure water as a function of sewage temperature T_s

T_s [°C]	5	10	15	20	25	30
$\nu\cdot10^6$ [m^2s^{-1}]	1.52	1.31	1.15	1.01	0.90	0.80

Equation (3.1) involves six parameters, namely, Q_V, S_E, g, D, ν and k_s. If five of these are given, the sixth can be determined. In the design procedure for the diameter, D appears in all three relative parameters q_r, R_r and κ_s of Eq. (3.2). If the energy slope S_E equals the bed slope S_o, the corresponding flow is uniform (Chap. 5). As described in Sect. 2.2, the diameter can then be determined closely with an explicit formulation. Further details on computational procedures are also presented in Chap. 5.

3.2.2 *Operative Roughness*

Whether the equivalent roughness is known exactly or not depends on whether the influence of viscosity or of the relative roughness is dominant (Chap. 2). If the effect of viscosity dominates, as in the *smooth turbulent flow regime*, one does not need to know the exact value of the equivalent sand roughness height k_s. If, however, the flow is in the *rough turbulent regime* for which the effect of viscosity becomes negligible, exact information on the equivalent sand roughness has to be available in advance. For the transition regime which often is determining in sewer design, the knowledge of the equivalent sand roughness k_s is naturally important.

The wall friction due to fluid viscosity and roughness of boundary material makes up an important part of the energy transfer in sewer flow. Other influences have also to be taken into account. There is an additional loss, as discussed in Chap. 2, resulting from non-prismatic sewer reaches. According to the energy principle, the energy losses are, in calculation, additive to a comprehensive loss coefficient in which the friction and all other individual losses are taken into account. Conversely, one can combine all the individual losses with the friction loss and obtain a mean higher roughness, the so-called *operative roughness* k_b (German: Betriebliche Rauheit; French: Rugosité opérationelle) to give the total energy loss. By this model representation the individual losses are differently expressed and distributed over the entire calculation reach and retained by a representative mean value. The concept of operative roughness (ATV 1988; Howe 1989) has the advantage to lump the loss effects, and the necessity to specify detailed information on individual effects is avoided. It has, however, the disadvantage that for long nearly prismatic sewers, higher losses are accounted for not always corresponding to hydraulically economic solutions, therefore. On the whole, the effort required for working out the individual information may be offset by the increased cost of conduit material.

The operative roughness k_b contains the following influences (ATV 1988):

- Wall roughness,
- Inexact and altered layout,
- Conduit joints,
- Shape of inlet transition, and
- Manhole structures.

For standard sewers the operative roughness height is $k_s = 0.1$ mm for the effective wall roughness, including the operational conditions of the sewer system against the original new conduit conditions. This lumped value does not contain the influences of:

- Reduction of nominal width,
- Combining flow in junction manholes, and
- Inlet and outlet structures of throttle conduits, pressurized conduit flows and inverted siphons.

Reductions of the nominal diameter are to be estimated, according to ATV (1988), from the effective mean discharge measurement. Small amounts of reduction can be ignored. The evaluation of performance, however, has to be carried out with basically 95% of the nominal diameter provided the actual effective diameter cannot be ascertained individually.

With regard to junction structures (Chap. 16), the losses are to be individually estimated. According to ATV (1988) the proof can be renounced in case either a bottom drop of height *D*/20 is built on the channel floor or the discharge in the downstream sewer is limited to 85% instead of 90% of the flowing full discharge, as mentioned in Sect. 3.2.1.

Deviations from these prescribed values for the lumping concept is permitted by the k_b values presented in Table 3.2 in the form of an *individual design concept.* The respective losses due to the equivalent wall roughness k_s and the individually arising losses are assigned in proportion of their contributions with a lower limit value of $k_s \geq 0.1$ mm. Changes from the original conditions are to be taken into account. ATV (1988) recommends the lumping concept both for the size determination and the discharge capacity estimation using the k_b values listed for standard sewers. This procedure is to be considered as the normal case and further reference to individual items is not required. For non-standard sewers described in Chap. 6 and *in-situ* concrete channels without specific information, the effective wall roughness corresponds to the operative wall roughness $k_b = 1.5$ mm.

Table 3.2 Lumped values for the operative roughness k_b [mm] after ATV (1988)

Application	k_b [mm]
Conduit throttle,[a] Pressure conduit,[a,b] Inverted siphons[a] and relined reaches (re-inspected) without manholes	0.25
Conveyance sewers with manholes	0.50
Intercepting sewers and pipelines with manholes, likewise with unformed manholes as also conveyance sewers with special manholes	0.75
Intercepting channels and pipelines with special manholes, masonry channels, *in-situ* concrete channels, sewers of non-standard conduits without specific information about roughness	1.50

[a]Without outlet and bend losses, [b]without pressurized network

3.3 Minimum Discharge

3.3.1 *Design Considerations*

Vicari (1916) suggested that two conditions must be satisfied for the minimum (subscript m) discharge Q_m in sewers:

- Minimum flow depth $h_m = 3$ cm and
- Minimum tractive force $S_m = 2.5$ N/m^2.

It is from these considerations that the required maximum diameter is calculated.

Ackers et al. (1964) conducted an extensive study of aging of sewers. Their most essential conclusions were:

- The sewer coating development varies strongly with the condition and the material content of the sewage. The coating develops rapidly to an end thickness,
- For higher velocities but otherwise similar conditions, the sewer coating is thinner than for lower velocities,
- For sewer coating thicknesses smaller than 3 mm, the same resistance as for new conduits may be adopted. Beyond this limit thickness, the resistance effect increases sharply,
- Sewers whose inverts are covered with gravel and which carry discharges with Froude numbers of about 0.5 develop standing waves leading to considerably higher resistance,
- The recommended equivalent roughness height amounts to $k_s = 1.5$ mm for sewer coatings smaller than 5 mm thick in normal condition, for good condition $k_s = 0.5$ mm, and for bad condition $k_s = 3$ mm. These values become as high as about 25 mm for crusted conditions and even about ten times higher for sewers with gravel deposition.

According to Smith (1965) a sewer should pass the maximum discharge safely as well as maintain self-cleansing for the minimum discharge. His design procedure is based on a minimum velocity.

3.3.2 *Yao's Procedure*

According to Yao (1974) the *minimum sewer velocity* depends on the characteristics of the flow boundary surface, the deposited material and the depth of flow. The concept of minimum velocity is replaced by the concept of minimum wall shear stress.

In order that a particle in the lower zone of the cross-section settles, the angle of inclination is to be smaller than the angle of repose of the particle material. According to Lysne (1969) the average natural angle of repose may be taken as 35° which would cause deposition below the 10% part-full depth in a circular cross-section.

The *mean shear stress* τ_o for uniform flow in open channels is

$$\tau_o = \rho g R_h S_o \tag{3.3}$$

with ρ as fluid density, g as gravitational acceleration, R_h as hydraulic radius and S_o as bottom slope. The local boundary shear stress τ varies along the perimeter, from the maximum at the invert to the minimum at the free surface. For sewers with a small filling ratio, the effect of the variation in the value of τ is negligible.

The *minimum shear stress* τ_{om} relates to the initiation of motion and is estimated for sand from *Shields' diagram* (Graf 1971, Raudkivi 1993). Yao recommends in separate sewer systems for particles of diameters from 0.2 mm to 1 mm a minimum bed shear stress $\tau_{om} = 1$ to 2 N/m^2. For combined sewer systems, values between 3 and 4 N/m^2 are mandatory. Applying the formula of Manning and Strickler with $K = 1/n$ as the roughness coefficient then gives, for the full flowing sewer (subscript v),

$$Q_v = 0.62(1/n)(\tau_o/\rho)^{1/2} D^{13/6}. \tag{3.4}$$

A comparison of the concepts of constant minimum velocity V_m and the constant bottom shear stress results, for a particular value of V_m, in smaller self-cleansing for larger sewer diameters.

Equation (3.3) can also be expressed as a relationship between the bottom slope S_o, the sewer diameter D and the bottom shear stress τ_o as

$$S_o = \tau_o/(\rho D). \tag{3.5}$$

For constant value of τ_o/ρ, the larger the diameter the smaller is the bottom slope required. A relationship between the flowing full velocity V_v, the sewer diameter D and the bottom shear stress (τ_o/ρ) obtains as

$$\frac{V_v}{(1/n)D^{1/6}(\tau_o/\rho)^{1/2}} = 0.79. \tag{3.6}$$

Example 3.1 What is the minimum velocity in a 500 mm sewer, for $1/n = 85$ m$^{1/3}$s^{-1}, if a minimum bed shear stress of 2 N/m^2 must be guaranteed at the flowing full stage?
From Eq. (3.6) follows $V_v = 0.79 \cdot 85 \cdot 0.5^{1/6}(0.2/1000)^{1/2} = 0.85$ ms^{-1}. The corresponding discharge is $Q_v = 0.85\,(\pi/4)0.5^2 = 0.167$ m^{3}s^{-1}.

For *part-full discharge* (subscript t) the hydraulic radius must be used and according to Eq. (3.3), the part-full velocity is

$$V_t = 1/n(\tau_o/\rho)^{1/2} R_h^{1/6}. \tag{3.7}$$

For small part-full conditions defined by $y = h/D < 1/2$, the relation

$$R_h/D = \frac{2}{3}y\left(1 - \frac{1}{2}y\right) \tag{3.8}$$

approximates the hydraulic radius to within 1% of the exact value. From Eqs. (3.7) and (3.8) one obtains for the part-full velocity V_t

$$\frac{V_t}{(1/n)D^{1/6}(\tau_o/\rho)^{1/2}} = 0.935y^{1/6}(1 - 0.08y) \tag{3.9}$$

Example 3.2 How large is the minimum velocity for a 15% part-full flow in Example 3.1?
With $y = 0.15$, the right hand side of Eq. (3.9) yields $0.935 \cdot 0.15^{1/6}$ $(1\text{–}0.08 \cdot 0.15) = 0.67$ and for $V_t = 0.67 \cdot 85 \cdot 0.5^{1/6}(0.2/1000)^{1/6} = 0.72\ \text{ms}^{-1}$. This is only about 15% less than the flowing full velocity.

Table 3.3 gives the ratios of the velocities for part-full and flowing full conditions for a representative number of part-full stages. It can be observed that beyond 40% part-full filling there is practically no further influence on the minimum velocity.

Table 3.3 Ratio of minimum velocities $\mu_t = V_t/V_v$ for part-full and full flow conditions in circular sewer

Part-full stage y	0.05	0.1	0.2	0.4	0.6	0.8	1
Ratio μ_t	0.72	0.80	0.89	0.98	[1.09]	[1.13]	1

For a minimum part-full flow of 5%, it follows from Eq. (3.9)

$$\frac{V_t}{(1/n)D^{1/6}(\tau_o/\rho)^{1/2}} = 0.57. \tag{3.10}$$

For a minimum bottom shear stress of $\tau_o = 2\ \text{N/m}^2$ and $(1/n) = 85\ \text{m}^{1/3}\text{s}^{-1}$, the values of V_t calculated from Eq. (3.10) for a range of diameters are compiled in Table 3.4. The velocity of $0.50\ \text{ms}^{-1}$ for the smallest possible diameters increases

Table 3.4 Minimum velocity V_t for 5% part-full flow with $1/n = 85\ \text{m}^{1/3}\text{s}^{-1}$ according to Eq. (3.10)

D [mm]	150	200	250	300	400	500	600	700	800
V_t [ms^{-1}]	0.50	0.52	0.54	0.56	0.59	0.61	0.63	0.65	0.66
	900	1000	1200	1400	1500	1600	1800	2000	2500
	0.67	0.69	0.71	0.72	0.73	0.74	0.76	0.77	0.80

by only a small amount to 0.80 ms^{-1} for the largest size recommended in practice. For usual sewer diameters one may adopt roughly a minimum velocity of about 0.60 to 0.70 ms^{-1}, therefore.

Further research on the minimum slope required to avoid deposition of sand and gravel in sewers has been conducted by Novak and Nalluri (1978), Mayerle et al. (1991), Butler et al. (1996), and Nalluri and Ab Ghani (1996). A review of recent works is also available (Hager 1998).

3.3.3 *ATV Procedure*

The ATV (1988) refers mainly to the works of Macke (1980, 1983) and Sander (1994). A table for various sewer diameters and 50% part-full flow containing the minimum velocities V_m and the corresponding minimum bed slopes S_{om} to avoid deposition of material is presented. The relation between the sewer diameter D, V_m and S_{om} is reproduced in Table 3.5. The relation between D and V_m can also be expressed by

$$V_m[\mathrm{ms}^{-1}] = 0.5 + 0.55D\,[\mathrm{m}]. \tag{3.11}$$

Values of V_m obtained from Eq. (3.11) agree with those of Table 3.5 for nominal diameters smaller than 300 mm. Beyond that, the ATV recommends higher velocities. For 10% part-full flow, the minimum velocities obtained from Eq. (3.11) should be increased by an additional amount of about 10%.

Schütz (1985) underlined the effect of backwater. All relations derived previously are valid strictly for uniform flow; for control manholes with lateral inflow in particular, attention should be paid to backwater effects into the lateral branch. Schütz further recommended the empirical formula

$$S_{om}[-] = 1/D\,[\mathrm{mm}], \tag{3.12}$$

according to which the minimum sewer slope for a diameter of 1000 mm amounts to 0.1%. For a sewer diameter of 250 mm a minimum sewer slope $S_{om} = 0.4\%$ is sufficient, however.

Table 3.5 Minimum velocity V_m and corresponding minimum slope S_{om} as functions of conduit diameter D for 50% part-full flow. For 10–20% part-full flow, V_m increases by about 10% (ATV 1988)

D [mm]	150	200	250	300	400	500	600	800	1000
V_m [ms^{-1}]	0.48	0.50	0.52	0.56	0.67	0.76	0.84	0.98	1.12
S_{om} [%]	0.27	0.20	0.16	0.15	0.14	0.14	0.14	0.13	0.13
	1200	1400	1500	1600	1800	2000	2200	2400	3000
	1.24	1.34	1.39	1.44	1.54	1.62	1.72	1.79	2.03
	0.12	0.12	0.12	0.12	0.12	0.11	0.11	0.11	0.11

Sander (1994) observed an average particle size of $d = 0.35$ mm. Further, the determining minimum filling ratio should be 10%, and a minimum bottom shear stress is 0.8 Nm^{-2}. For sewers smaller than $D = 1$ m, the minimum bottom slope required against long term sedimentation is

$$S_{om} = 1.2‰\, D^{-1}\ [\mathrm{m}]. \tag{3.13}$$

For $D > 1$ m, the *absolute minimum bottom slope* $S_o = 1.2‰$ should be used. Equation (3.12) is thus modified for both small and large sewer diameters. Equation (3.13) has been tested with selected sewer systems and found appropriate for common sewage.

3.4 Sewer Cross-Sections

Sewers built in older times had been developed in a variety of shapes. Carson et al. (1894) described, as an example, the pipe handle cross-section (similar to a horseshoe cross-section with vertical intermediate walls and semi-circular soffit), the gothic section with a pointed arch and the egg-shaped cross section besides the circular cross-section.

French (1915) compared the above mentioned four sections regarding the velocities by application of Kutter's formula for equal discharge. The egg-shaped section was identified as the *best capacity* cross-section up to 35% part-full flow and for larger flow depths all sections are hydraulically similar within 5%. Overall, the circular section performs best. The egg-shaped section shows in service more deposition than the corresponding circular section. The study defined thirty of the sections then in use in the USA and specified their cross-sectional characteristics for full flow conditions.

Donkin (1937) compared the circular with the egg-shaped and the U-shaped sections both hydraulically and economically. The U-shaped section constructed in brickstone was found to be narrowly optimum compared to the other two sections. The techniques used in these investigations may not be considered sufficient in current standards.

Thormann (1941) introduced the *standardisation* of cross-sectional types of sewers. Fifteen cross-sections were proposed all of which are axis-symmetric and either egg-shaped or horseshoe shaped. Denoting the width of the section as B and the height from invert to soffit T, six different axis ratios $B{:}T = 2{:}\alpha_p$ were proposed, with $\alpha_p = 3.5, 3, 2.5, 2, 1.5$ and 1. The cross-sectional forms include:

- Extra-high, normal, transposed, depressed and transposed-depressed egg-shaped sections,
- Extra-high and normal circular sections,
- Cap- or hood-shaped cross-section with $\alpha_p = 2.5$ and 2,
- Parabolic cross-section 2:2,
- Kite-shaped cross-section 2:2, and
- Horseshoe-shaped cross-sections for $\alpha_p = 1.5$ and 1.

These cross-sections form the basis of ATV 110 (1988). The standard construction technique for the cross-sections was established by Schoenefeldt et al. (1943). Thormann (1944) defined the cross-sectional geometry of fifteen standard sections. Roske (1958) referred to the dimensionless representation of the cross-sectional sizes only to the circular, the egg-shaped and the horseshoe-shaped sections.

Kuhn (1976) concluded that neither the circular nor the egg-shaped nor the horseshoe sections possess definite advantage over the other sections recommended, so that no general recommendation can be given. Because of the industrial finish technique the *circular cross-section* is employed in a wide variety of situations for which the section is often referred to as the standard sewer cross-section.

Schmidt (1976) compared the standard egg-shaped section with the circular section. For the same cross-sectional area the relation between the diameter D_E of the normal 2:3 egg-shaped section and the diameter D_k of the circular section is given by $D_k = 1.2D_E$. As long as the discharge is $Q/Q_v \leq 0.22$, the velocity in the egg-shaped section is higher than that in the corresponding equal area circular section. It is stated that for a night minimum discharge of around 1% of the storm water flow, around 7% part-full stage establishes in the egg-shaped section while it is only about 4% in the circular section. To produce the same velocity, the circular section would require about 30% more bottom slope than the egg-shaped section. According to Schmidt (1976), the *standard egg-shaped* section is suited for slopes which are unfavourably placed with regard to the avoidance of deposition during dry weather flow.

Sartor and Weber (1990) followed the opinion of Schmidt and recommended specially the egg-shaped section because of its advantages in maintenance and water

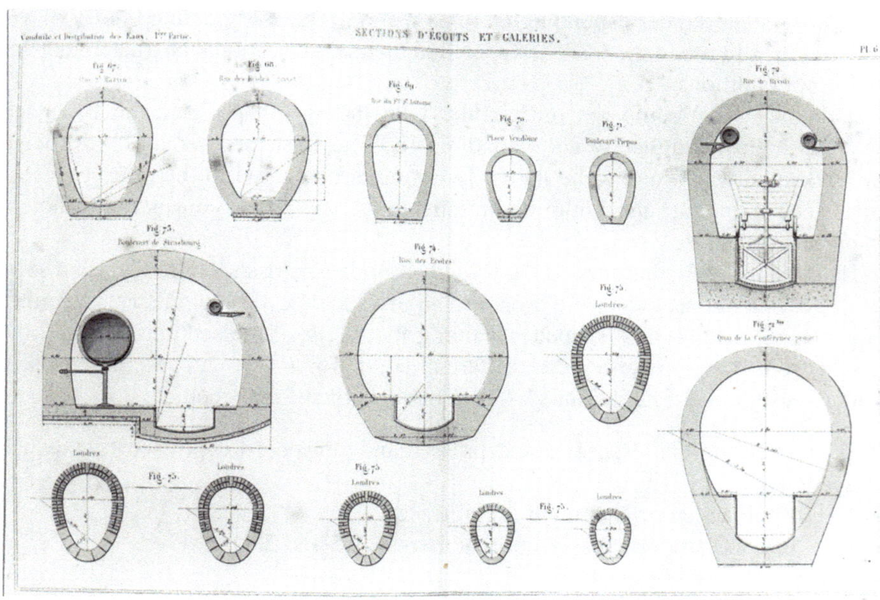

Fig. 3.1 Cross-sections of sewers used in the city of Paris, after Dupuit (1854)

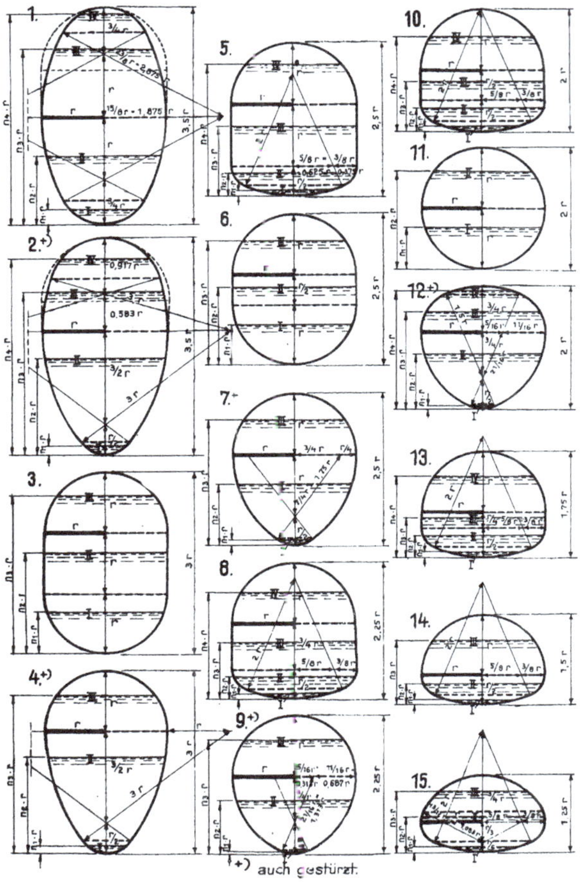

Fig. 3.2 Sewer cross-sections of Thormann (1944)

quality. A quantification of these advantages requires comparative accounting of the pollutions carried by the sewers. Egg-shaped sections having cross-sectional dimensions smaller than 500/750 are of special interest.

According to ATV (1988) the *standard cross-sections* are (Chap. 5):

- Circular section,
- Egg-shaped section 2:3, and
- Horseshoe section 2:1.5.

The remaining twelve cross-sectional forms, standardized by Thormann referred to previously and described in Chap. 5, can be alternatively described with the part-full flow characteristic curves of their normalized sections. For the determination of the full flow quantities, the so-called form factor must be known. The form factor describes the influence of cross-sectional geometry on the discharge. What the relevant reports in the literature do not state is whether in the future only the three standard sections have to be considered. Having this in mind, only the circular section is usually considered herein. On questions of sewer flow, the standardized egg-shaped and the horseshoe sections are also accounted for. Pecher et al. (1991) follow this treatment also, whereas Unger (1988) considers solely the circular and the standard egg-shaped sections.

Notation

B	[m]	width of cross-sectional profile
d	[m]	particle size diameter
D	[m]	pipe diameter
F	[m^2]	cross-sectional area
g	[ms^{-2}]	acceleration due to gravity
h	[m]	flow depth
k_b	[m]	operative roughness height
k_s	[m]	equivalent sand roughness height
K	[$m^{1/3}s^{-1}$]	coefficient of roughness
P	[m]	wetted perimeter
q_r	[−]	relative discharge
Q	[m^3s^{-1}]	discharge
R_h	[m]	hydraulic radius
R_r	[−]	Reynolds number with respect to D
S_E	[−]	energy line slope
S_o	[−]	bottom slope
T	[m]	height from invert to soffit of cross-section
T_s	[°C]	sewage temperature
y	[−]	filling ratio
α_p	[−]	ratio of the axes
κ_s	[−]	relative roughness
μ_t	[−]	velocity ratio
ν	[m^2s^{-1}]	kinematic viscosity
ρ	[kgm^{-3}]	density
τ	[Nm^{-2}]	shear stress

Subscripts

m	minimum
o	bed, bottom part-full
v	full flow condition

References

Ackers, P., Crickmore, M.J., Holmes, D.W. (1964). Effects of use on the hydraulic resistance of drainage conduits. *Proc. Institution Civil Engineers* 28: 339–360; 34: 219–230.

ATV (1988). Richtlinien für die hydraulische Dimensionierung und den Leistungsnachweis von Abwasserkanälen und -leitungen (Guidelines for the hydraulic design of sewers). Regelwerk Abwasser – Abfall, *Arbeitsblatt* A110. Abwassertechnische Vereinigung: St. Augustin [in German].

Butler, D., May, R.W.P., Ackers, J.C. (1996). Sediment transport in sewers. Proc. Institution Civil Engineers *Water, Maritime & Energy* 118(6): 103–120.

Carson, H., Kingman, H., Haynes, T., Collison, H.N. (1894). Cross-sections of sewers and diagrams showing hydraulic elements of four general types, Metropolitan Sewerage Systems. *Engineering News* 30(5): 121–123 (incl. supplement).

Donkin, T. (1937). The effect of the form of cross-section on the capacity and cost of trunk sewers. *Journal Institution of Civil Engineers* 7: 261–279.

Dupuit, J. (1854). *Traité théorique et pratique de la conduite et de la distribution des eaux* (Theoretical and practical treatment of the pipe and water distribution). Carilian-Goeury et Dalmont: Paris [in French].

French, R. de L. (1915). Circular sewers versus egg-shaped, catenary and horseshoe cross-sections. *Engineering Record* 72(8): 222–223; 72(20): 608–610.

Graf, W.H. (1971). *Hydraulics of sediment transport.* McGraw-Hill: New York.

Hager, W.H. (1998). Minimalgeschwindigkeit und Sedimenttransport in Kanalisationen (Minimum velocity and sediment transport in sewers). *Gas Wasser Abwasser* 78(5): 346–350 [in German].

Howe, H. (1989). Grundzüge des neuen ATV-Arbeitsblattes A110 (Fundamentals of the new ATV A110 guideline). *Korrespondenz Abwasser* 36(1): 28–29 [in German].

Kuhn, W. (1976). Der manipulierte Kreis – Gedanken zur Profilform bei Abwasserkanälen (The manipulated circle – Thoughts on the sewer profile shape). *Korrespondenz Abwasser* 23(2): 30–37 [in German].

Lysne, D.K. (1969). Hydraulic design of self-cleaning sewage tunnels. *Journal Sanitary Engineering Division* ASCE 95(SA1): 17–36.

Macke, E. (1980). Über Feststofftransport bei niedrigen Konzentrationen in teilgefüllten Rohrleitungen (On sediment transport for low concentration in partially filled pipes). *Mitteilung* 69. Leichtweiss-Institut für Wasserbau, TU Braunschweig: Braunschweig [in German].

Macke, E. (1983). Bemessung ablagerungsfreier Strömungszustände in Kanalisationsleitungen (Design of depositionless flows in sewers). *Korrespondenz Abwasser* 30(7): 462–469 [in German].

Mayerle, R., Nalluri, C., Novak, P. (1991). Sediment transport in rigid bed conveyances. *Journal Hydraulic Research* 29(4): 475–495.

Nalluri, C., Ab Ghani, A. (1996). Design options for self-cleansing storm sewers. *Water Science and Technology* 33(9): 215–220.

Novak, P., Nalluri, C. (1978). Sewer design for no-sediment deposition. *Proc. Institution Civil Engineers* 65(2): 669–674; 67(2): 251–252.

Pecher, R., Schmidt, H., Pecher, D. (1991). *Hydraulik der Abwasserkanäle in der Praxis* (Hydraulics of sewers in practise). Parey: Hamburg, Berlin [in German].

Pfeiff, S. (1960). Mindest- und Höchstfliessgeschwindigkeiten in Entwässerungsnetzen (Minimum and maximum velocities in sewer networks). *gwf Wasser/Abwasser* 101(4): 83–85 [in German].

Raudkivi, A.J. (1993). *Sedimentation – Exclusion and removal of sediment from diverted water.* IAHR Hydraulic Structures Design Manual 6. Balkema: Rotterdam.

Roske, K. (1958). Dimensionslose Grössen in der Hydrodynamik der offenen Gerinne (Dimensionless quantities in the hydrodynamics of open channels). *Stuttgarter Bericht* 5. Inst. Industrie und Siedlungswasserwirtschaft, TU Stuttgart. Oldenbourg: München [in German].

Sander, T. (1994). Zur Dimensionierung von ablagerungsfreien Abwasserkanälen unter besonderer Berücksichtigung von neuen Erkenntnissen zum Sedimentationsverhalten (The design of depositionless sewers under particular attention of new results relative to sedimentation). *Korrespondenz Abwasser* 32(5): 415–419 [in German].

Sartor, J., Weber, J. (1990). Die Wiederentdeckung des Eiprofils aufgrund von Schmutz-Frachtbetrachtungen (The re-discovery of the egg-shaped profile based on contaminant discharge considerations). *Korrespondenz Abwasser* 37(6): 689–693 [in German].

Schmidt, H. (1976). Die Verwendung von Eiprofilen aus hydraulischer Sicht (The use of egg-shaped sewers based on hydraulic considerations). *Korrespondenz Abwasser* 23(7): 209–212 [in German].

Schoenefeldt, O., Thormann, E., Conrads, A. (1943). Einheitliche Leitungsquerschnitte für die Stadtentwässerung (Uniformized sewers for city drainage). *Gesundheits-Ingenieur* 66(16): 192–200 [in German].

Schütz, M. (1985). Zur Bemessung weitgehend ablagerungsfreier Strömungszustände in Kanalisationsleitungen nach Macke (On the design of practically depositionfree flows in sewers based on Macke). *Korrespondenz Abwasser* 32(5): 415–419 [in German].

Smith, A.A. (1965). Optimum design of sewers. *Civil Engineering and Public Works Review* 60(2): 206–208; 60(3): 350–353; 60(9): 1279–1283.

Thormann, E. (1941). Einheitliche Leitungsquerschnitte für Entwässerungsleitungen (Filling curves of drainage pipes). *Gesundheits-Ingenieur* 64(8): 103–110 [in German].

Thormann, E. (1944). Füllhöhenkurven von Entwässerungsleitungen (Filling curves for drainage pipes). *Gesundheits-Ingenieur* 67(2): 35–47 [in German].

Unger, P. (1988). *Tabellen zur hydraulischen Dimensionierung von Abwasserkanälen und -leitungen, DN*100-4000 *und DN*300/450 –1400/2100 (Tables for hydraulic design of sewers). Ingwis-Verlag: Lich [in German].

Vicari, M. (1916). Kleinste Sohlgefälle für Schmutzwasserkanäle (Minimum bottom slopes for sewers). *Gesundheits-Ingenieur* 39(51): 537–540 [in German].

Yao, K.M. (1974). Sewer line design based on critical shear stress. *Journal Environmental Engineering Division* ASCE 100(EE2): 507–520; 101(EE1): 179–181; 101(EE4): 668–669.

(a)

(b)

Screw pumps have an optimum performance for sewage pumping. Large pumping station (**a**) side view, (**b**) front view

Chapter 4
Sewage Pumping – Throttling Devices

Abstract Although pumps and pumping stations are a specialized domain of technology, the most salient features of sewage pumping are introduced. From these the pre-eminent position of the screw pumps is established.

The hydraulic characteristics of the more important types of throttling devices, such as the vortex throttle, the gate valve or the throttle valve are also described and, in particular, the requirements of ATV with respect to such devices are discussed. Reference is also made to automatic discharge and level control.

4.1 Introduction

A pumping station (German: Pumpwerk; French: Station de pompage) can often improve the economy of a drainage system because it lifts the inflows from various directions to a central *sewage treatment plant.* As lifting equipment, both centrifugal and screw pumps are considered.

In order that an automatic, disturbance-free and safe pumping is guaranteed:

- The water must flow directly to the pumps so that pump suction problems are reduced, and
- A trash rack be installed so that expensive and troublesome pump cleaning is avoided.

For smaller pumping stations the determining factor is not the discharge but the *security against clogging* (German: Verstopfung; French: Engorgement). Usually a minimum passage of a sphere of 100 mm diameter has to be guaranteed. As pump types, single impeller wheel, free stream wheel and especially screw pumps are installed. Use of vacuum pumps is treated by Fass (1970). With a shaft diameter of 200 mm, the outer screw diameter amounts to at least 500 mm.

For sewage lifting by pressure, a *minimum velocity* of 0.5–1.0 ms^{-1} has to be ensured as otherwise clogging may take place. The upper economic limit of the velocity is about 2.2–2.5 ms^{-1}. Within these ranges both the pressure loss and the required pressure remain within reasonable limits. The hourly number of operations

W.H. Hager, *Wastewater Hydraulics*, 2nd ed., DOI 10.1007/978-3-642-11383-3_4,

of the pumps should not exceed 10–12. General information regarding sewage pumps and pumping stations may be obtained from the publications of Tuttahs (1987a and 1987b).

Similarly, caution against clogging must be considered for all the control members because only a sewage treatment plant with maximum possible uniform inflow is reliable. Currently, a selection can be made from a variety of control devices. They are based on various principles, quite distinguishable in cost and their control action, varying from deficient to quite remarkable. In view of this multiplicity both in the physical sense as also from the point of view of the reliability of operations, the ATV endeavours to make the market position clearer by some comparative assessment. The supplier is required to submit a *claim list* giving detailed specification of their products. In this list the supplier has to furnish information regarding the functions as well as the required operational criteria. Recently, *inspection* by a member of the *Gruppe W des Güteschutzes KANALBAU* has been allowed. Remedial measures may be taken for unauthorized manipulation on throttling devices installed.

4.2 Types of Pumps

As a general rule, at least *two* pumps are to be installed so that an alternative pump remains if one fails or is under inspection. A general description of sewage pumping stations is provided by Lessmeier (1983).

4.2.1 *Centrifugal Pumps*

Centrifugal pumps can be installed both above (dry) and below (wet) sump water level. One also differentiates between horizontal and vertical settings.

For above-ground installations no difference exists between dry and wet settings. In installations below ground level, however, the suction and the machine portions are in separate chambers in the *dry setting*. According to ATV A134 (1980) a basic decision in favour of either of the two settings is not possible. Operational, hygienic and safety considerations suggest dry setting, whereas the installation cost is then higher.

Horizontally set pumps have few advantages over vertically set pumps such as simpler repairs and lower sensitivity to vibrations. Figure 4.1 shows centrifugal pumps. Details on the design of pump-sumps of centrifugal pumps are given by Dasek (1989).

The *lift height* of a pump is set as the sum of:

- Difference H in height between the highest point on the delivery pipe and the sump water level, and
- Total of all head losses between the above two points.

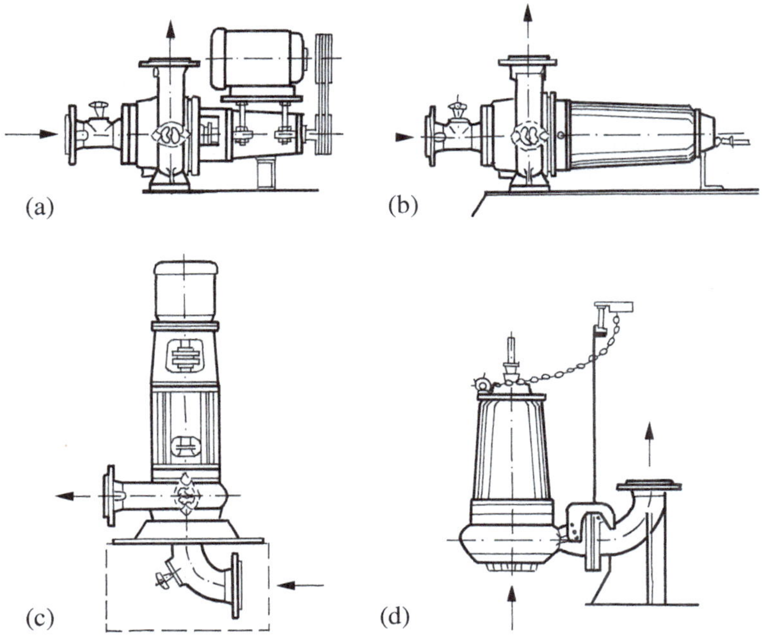

Fig. 4.1 Centrifugal pumps after ATV A134 (1980) (**a**) Dry and horizontal setting with separately mounted motor, (**b**) dry and horizontal setting with coupled motor, (**c**) dry and vertical setting, (**d**) wet and vertical setting

Figure 4.2 shows the *Q-H* diagram of a centrifugal pump for one steep and another flat throttle curve for two different conduit characteristic curves. The *operation range* lies between the throttle and the conduit characteristic curves. It may be seen that the operation zone is longer for the flat throttle curve. Here, *Q* is discharge.

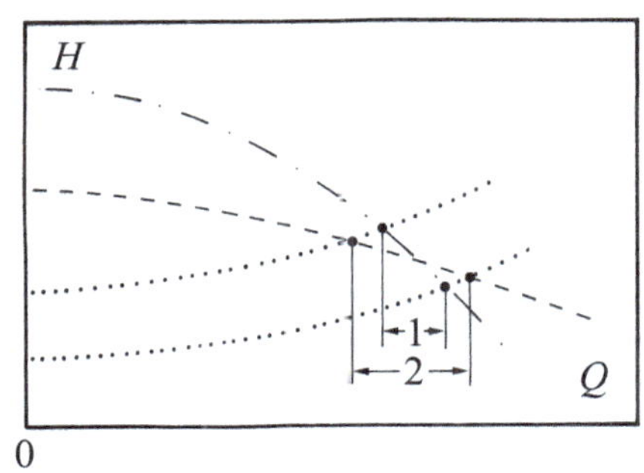

Fig. 4.2 Lift diagram for the centrifugal pump. (– –) Steep and (- - -) flat throttle curve, and (...) conduit characteristic curves for minimum and maximum geodetic heights with corresponding operational reaches

4.2.2 *Screw Pumps*

The lift screw is suitable for the overhead pressure free-lifting of sewage into open channels. It does not have destructive effects on the solid matter in the sewage and the *operational reliability* is guaranteed because of the slow rotational speed. The lifting begins at the so-called touch point and attains the maximum at the filling point. The screw pump (German: Schneckenpumpe; French: Vis d'Archimède) distinguishes itself by a flexible working range making continuous pumping possible.

The *installation angle* of screws lies between 30° and 40° relative to the horizontal. With regard to the static loading, the lengths of the screws are restricted to 8–12 m so that *lift heights* of 6–8 m are obtained. The number of gears is from 1 to 3 and the number of rotations 50–75 per minute. The assessment of correct upstream and downstream water levels is essential for the optimum layout of screw pumps. These considerations are more important than for centrifugal pumps, as subsequent changes are often difficult.

As shown in Fig. 4.3, the filling point is located at about the level of the water surface in the approach flow channel. The *touching point* is then obtained from the size of the screws. For the construction of the inlet structure, recirculation spaces are to be avoided. The *fall point* has to be above the maximum water surface level in the downstream channel so that water does not flow back into the inlet chamber.

It should be possible to shut off each screw pump by a bulkhead or *stop log*. It is advantageous to have stairs in the aisles between the adjacent guide troughs housing the screw pumps for convenient cleaning. Screw pumps installed outdoors are to be protected against sunshine.

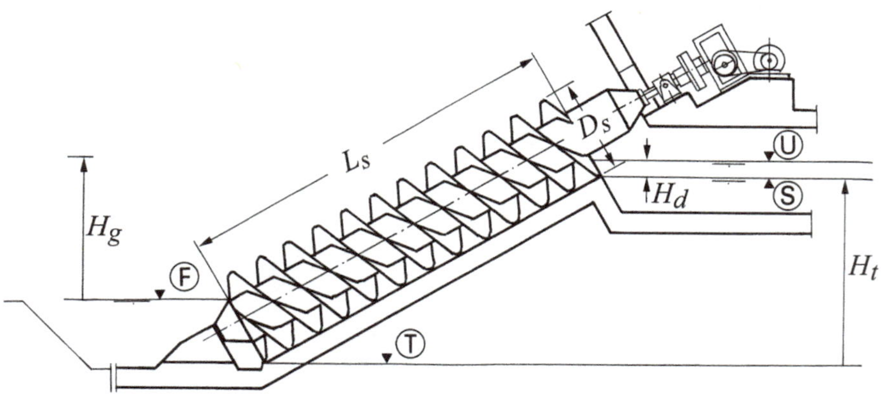

Fig. 4.3 Screw pump with D_s screw diameter, L_s screw length, H_g difference in geodetic heights, H_t total head difference and H_d variable level difference, for complete stop of flow $H_d = 0$. T touching point, F filling point, S fall point and U downstream water level

According to Witschi (1970) screw pumps are employed:

- For lifting sewage to treatment station,
- To overcome level differences in sewers,
- As reversible mud pumps,
- To drain flooded areas and
- For irrigation of fields.

The *advantages* of screw pumps are:

- A good working efficiency of around 75% for variable discharge,
- A high operational reliability,
- An independent adaptation to the inlet conditions,
- Reliable transport of solid matter,
- No thrust loading to favour settling process in clarifiers,
- Without large peaks in energy consumption because the screw gets filled first,
- Low wear and tear of the bearings due to low rational speed, and
- Simple maintenance.

As *disadvantages*, Witschi alleges:

- Possible odour nuisance due to open installation,
- Higher cost than for centrifugal pumps, and
- Expensive structural installation.

If the inflow falls below about 30% of the design inflow discharge, the pump should work intermittently. To reduce the *number of turnovers* below 10 per hour, the *pump sump* must be sufficiently large. To hold the leakage losses within the guide troughs, the gap between the screw diameter D_s and the side of the pump trough should be about $0.005D_s^{1/2}$ [m].

4.3 Throttling Devices

4.3.1 *General*

In combined sewer systems only a certain multiple of the dry weather flow is discharged to the treatment station so that during rain-weather a large portion of the inflow must be bypassed to the receiving water. In other words, the inflow generated in a catchment basin is so diverted that only the *critical treatment discharge* is led to the treatment station. By the term critical (subscript k) we understand here the limiting discharge Q_k and not the critical discharge Q_c discussed in Chap. 6.

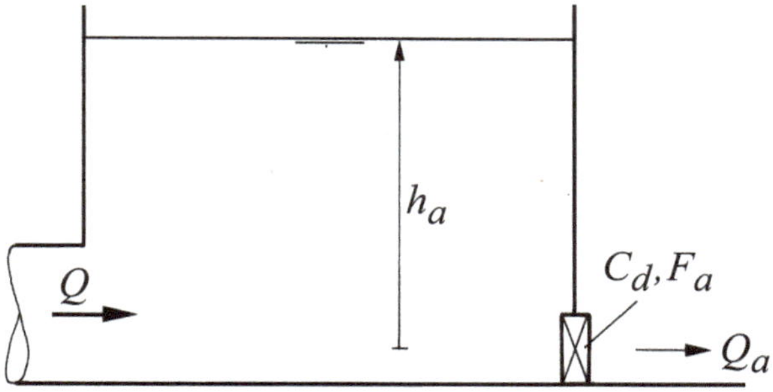

Fig. 4.4 Schematic outflow from a reservoir

Hydraulically, the problem may be posed thus: How is the outflow from a basin controlled so that for $Q < Q_k$ all discharge reaches the treatment station, whereas it receives for $Q \geq Q_k$ only the critical discharge?

Figure 4.4 shows a basin which is fed with the inlet conduit and whose outflow is regulated by a control device. The *outflow discharge* Q_a follows the relation

$$Q_a = C_d F_a (2gh_a)^{1/2}, \tag{4.1}$$

where C_d is the discharge coefficient, F_a the cross-sectional area of the appropriate outflow section and h_a is outflow head, i.e. $(2gh_a)^{1/2} = V_a$ is the appropriate outflow velocity. The coefficient C_d includes the effects of all additional factors, such as contraction losses and curvature of the outflow jet, that reduce the outflow discharge. For variable discharge $Q > Q_k$, the outflow Q_a can be kept constant either by:

- Keeping the *pressure head* h_a constant for constant values of C_d and F_a (type A float),
- Reducing the *outflow area* F_a as h_a increases (type B hose throttle), or
- Reducing C_d for increasing h_a (type C vortex throttle).

A combination of the three effects is also possible. The throttling pipe discussed in Chap. 9 belongs to type B and is not mentioned here since it cannot be considered as a device.

Most of the throttling devices have strongly constricted cross-sections. Because they discharge sewage containing solid matter, attention has to be paid to clogging and thus to the *operation reliability* of the entire system (Kaul 1986).

The ATV (1993) distinguishes between throttling devices and outlets, regulators as also control devices. The hose throttle and the vortex throttle belong to the category of throttling devices.

4.3.2 *Vortex Throttle*

The vortex throttle (German: Wirbeldrossel; French: Soupape de vortex) is a flow reducing device which was invented in the 1920s, developed in Stuttgart, Germany, in the 1970s and is currently produced by UFT, Bad Mergentheim, Germany (Brombach 1982b). The device contains no moving parts and the throttling effect rests only on the deceleration losses of the *tangential flow* in the vortex chamber. In the center of its housing an air pocket is formed that is directly connected with the atmosphere by ventilation. Figure 4.5 shows the usual horizontally placed base ① of the vortex chamber. The tangential inlet ② joins at the bottom. The cover ③ can be opened and carries the ventilation nipple ④. The replaceable outflow diaphragm ⑥ at the bottom ⑤ of the device is particularly flow resistant.

The water leaves the vortex chamber, practically without any loss of pressure, as a rotating *hollow jet* of high velocity. The vortex flow is stable and low in turbulence for the minimum discharge, and the vortex throttle performs well under full load. The kinetic energy downstream of the vortex throttle is dissipated in a stilling basin.

The smallest inlet diameter of a vortex throttle is 200 mm, and the outlet ring has at least this size. For the same spherical passage, the vortex throttle allows about half the minimum flow passed by the corresponding conduit throttle. The vortex throttle uses a height of fall equal to the nominal clearance and another height is necessary for the drop at the inlet sump. Vortex throttles are *optimum* for relatively small basins having sufficient bottom slope.

The vortex throttle possesses a high degree of *operational reliability* since foreign bodies in the sewage are washed down without causing *clogging*. This self-cleaning action fails only if bodies clog the passage such as roof laths or pieces of reinforced concrete which enter the vortex throttle. Fibre material, however, does not pose any problem and does not lead to any tress formation. The replaceable diaphragm permits a simple subsequent adjustment of the flow resistance. Since a

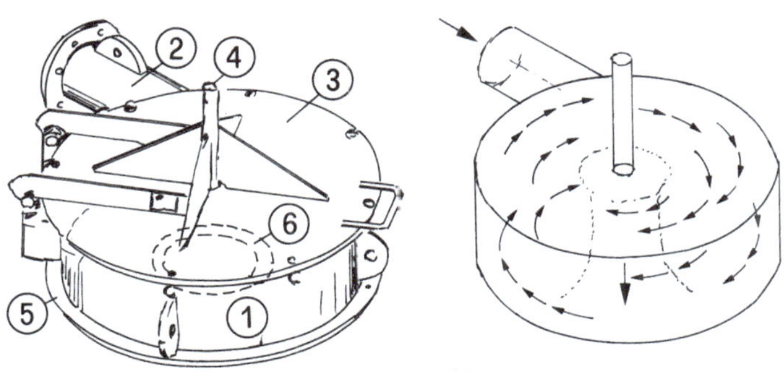

(a) (b)

Fig. 4.5 Vortex throttle (**a**) assembly and (**b**) flow pattern (Brombach 1982a). ① Vortex chamber, ② inlet conduit, ③ detachable cover, ④ ventilation, ⑤ base, ⑥ diaphragm ring complete *self-cleansing* cannot be achieved, regular flushing of the mud and slime combined with inspection is recommended.

4.3.3 *Regulating Devices*

Gate Valve

The gate valve (German: Drosselschieber; French: Robinet-vanne), shown in Fig. 4.6a, is rather rarely used for storm water drainage basins but finds application in the control of *storm water overflows*. According to Brombach (1982a), the reason for this may be the lack of characteristic discharge curves. The equipment consists of a circular passage of diameter D which is uncovered by lifting an U-shaped or rectangular diaphragm.

The action of the gate valve is different for upstream water levels below or above the upper *diaphragm* edge. As long as the water does not head up behind the diaphragm, the high velocity jet shoots below it and a peak appears in the $Q(h_p)$-curve. As the device gets pressurized, the discharge Q first reduces and then increases only slowly with increasing pressure head h_p (Fig. 4.6b). The design discharge depends primarily on the nominal size of the slide and the maximum pressure head. The application domains of gate valves are storm water overflows and storm water drainage basins. The minimum gate opening should be 10–15 cm for the smallest 20 cm diameter gate valve. Improper installation of the diaphragm slide must be prevented. Sliding plate gates described in Sect. 2.3.9 can be considered a special case of gate valves.

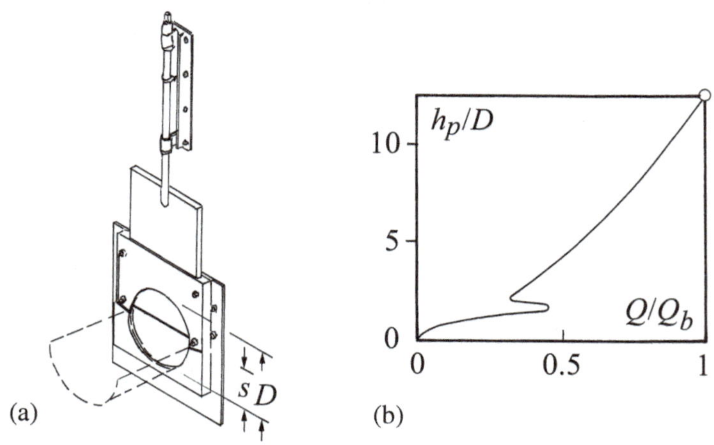

Fig. 4.6 Gate valve (**a**) valve assembly and (**b**) typical discharge-head curve (schematic)

The gate valve *Fluid Gate®* developed by UFT appears to have the following advantages (Borcherding and Brombach 1995):

- Fixed directly on to a vertical wall by dowels,
- Exact and continuous adjustment of discharge,
- Indication of opening height on a graduated scale,
- Compact construction and low head losses,
- Favourable cross-section for water passage,
- Corrosion free PVC and stainless steel construction, and
- Hoist above water level.

UFT offers nominal sizes from 250 to 1000 mm with capacities of 0.105–1.9 m^3s^{-1}.

The hydraulics of gate valves was investigated by Hager (1987a). Figure 4.7 shows the idealized geometry of a *circular diaphragm* of radius R_2 uncovering partly a circular or an U-shaped conduit of radius R_1. For a circular diaphragm in a circular conduit, let $r = R_2/R_1 > 1$ be the diameter ratio and $S = s/R_1$ the relative gate opening. The *cross-sectional area* F_D of the valve opening is then

$$F_D/R_1^2 = 1.9S \qquad \text{for} \quad r \cong 1.05, \tag{4.2}$$

$$F_D/R_1^2 = \arccos(1-S) - (1-S)(2S-S^2)^{1/2} \qquad \text{for } r \to \infty. \tag{4.3}$$

The value $r = 1.05$ is typical, whereas $r \to \infty$ corresponds to the straight-edged diaphragm (Fig. 2.19). Similar expressions can also be derived for the U-shaped cross-section (Hager 1987a).

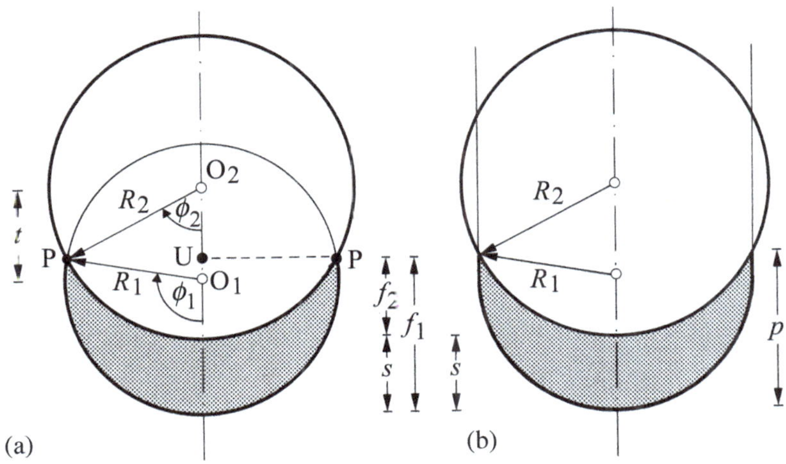

Fig. 4.7 Valve geometry for circular diaphragm in (**a**) circular pipe and (**b**) U-shaped section

The *discharge* of a gate valve is

$$Q_a = C_d F_D[2g(H-s)]^{1/2} \tag{4.4}$$

where $H = h_o + V_o^2/(2g)$ is the approach (subscript o) flow energy head measured relative to the invert level of the outlet, and s is the axial diaphragm opening height. Because the upstream inflow is usually from a reservoir, h_o is often set for H.

In model tests no influence of the parameter s/H on C_d was found. Except for small diaphragm openings (which are not even accurately measurable), the *discharge coefficient* is approximately $C_d = 0.69 \pm 0.02$. Other characteristics of the outflow jet are described by Hager (1987b).

Throttle Valve

O. Schulze GmbH, Gladbeck (Germany) offers a *float-throttle valve* (German: Schwimmer-Drosselklappe; French: Vanne papillon flottante) which can be used to regulate the outflow from a storage reservoir. The installation consists of a float-controlled throttle valve and a sliding outlet diaphragm both of which are housed in a common manhole (Fig. 4.8). The float-controlled *outlet valve* is said to guarantee a minimum discharge of about 30 ls^{-1}. The outlet conduit is provided with two floats which, for the rising water level, throttle the outflow by a valve. The water level in a common manhole is maintained with the sliding diaphragm.

The $Q(h_o)$-line involves the gap width s as parameter. Depending on the width of the sliding diaphragm, outflow discharges of 10 to 80 ls^{-1} are obtained for pressure heads ranging form 1 to 13 m. The hydraulic characteristics of such valves were analyzed by Burrows (1984). Outlet valves have the following *advantages*:

- Unhindered discharge characteristics up to the allowable discharge limit,
- Discharge independent of basin level,
- Simple regulation and maintenance,
- Little danger of clogging and depositions, and
- Electricity supply not necessary.

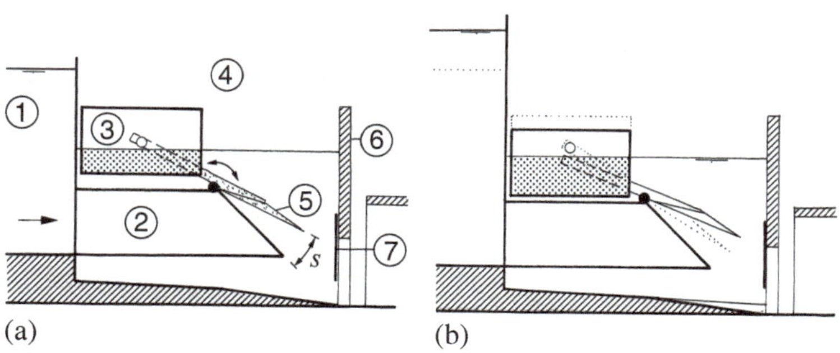

Fig. 4.8 Float-controlled outlet valve (**a**) notation, ① reservoir, ② outlet valve, ③ float, ④ valve chamber, ⑤ valve flap, ⑥ bulkhead wall, ⑦ sliding diaphragm, (**b**) effect of rise in float water level

In comparison to other throttling devices discussed here, fewer detailed reports on throttle valves exist.

Discharge Limiter ALPHEUS

This float-controlled apparatus of the Firma Gfw, Taunusstein, Germany, produces *constant outflow*. The flap inside the flow limiter causes a greater water depth than in a storage chamber. The float rises with the water surface and thereby draws down the shutter. The device is fixed with dowels on the waterside upstream of the outlet. The discharge limiter can pass a minimum discharge of 7 ls^{-1}, the upper limit discharge is about 100 ls^{-1}.

Hose Throttle

The hose throttle (German: Schlauchdrossel, French: Vanne de tuyau souple) allows for a nearly *constant outflow* from a reservoir of variable reservoir level (Vischer 1979). The device is specially suitable for controlling the small and the smallest of the outflows for pressure heads of only a few meters in storm water drainage basins of small to medium volume. Basically, the hose throttle can be installed either outside (dry, type U), or inside the basin (wet, type I).

For a *type U* installation, the throttle is enclosed by a piece of coaxially mounted transparent envelope pipe (Fig. 4.9). Besides the two lateral recesses for the throttle membrane, the throttle pipe has two further openings which are covered by a *filter cloth*. As the reservoir gets filled, the water enters through the pressure openings, fills the annular space and contracts the hose. The throttle starts because the cross-section of the recesses is somewhat smaller than the pipe cross-section. The air trapped in the upper part of the envelope pipe forms an *air cushion*. Accordingly, a small volume of water enters the annular space with every start of the throttle and flows out again. Although this dirties the annular space with time, there is no loss of function. Figure 4.10a shows a typical arrangement in a special manhole.

Hose throttle *type I* (Fig. 4.10) applies for the wet-installation concept. The cantilevered pipe piece with the stretched hose membrane protrudes into the reservoir chamber. The required depressed floor is permanently under water and serves also

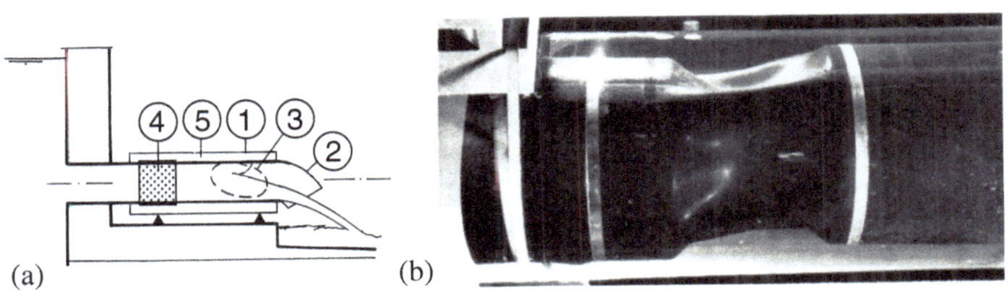

Fig. 4.9 *U-type hose throttle* (**a**) Working principle. ① Envelope pipe, ② throttle pipe, ③ recesses, ④ filter cloth, ⑤ air cushion. (**b**) Hose throttle in operation

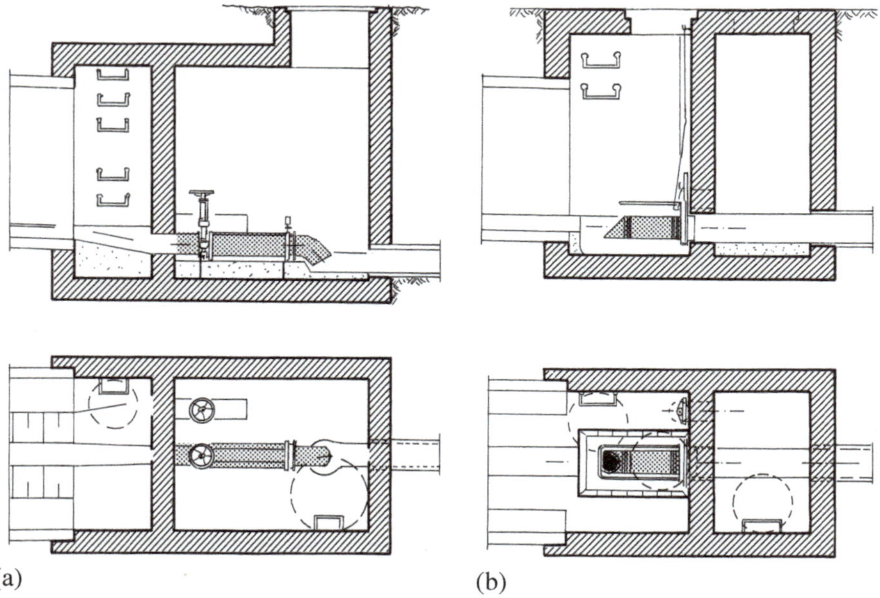

Fig. 4.10 Hose throttle mounted in special manhole. (**a**) Type U and (**b**) Type I. Longitudinal section (*top*) and plan (*bottom*).

as a rubble catcher. For dry weather flow the throttle pipe lies free and the sloping cut at its free end hinders clogging. The outflowing jet gets ventilated in the adjacent manhole chamber. For higher discharges the reservoir pressure acts directly on the membrane and constricts the cross-section. The hydraulic head required for a hose throttle is one pipe diameter.

The *discharge characteristic* of the hose throttle (Brombach 1987) is determined by the geometry and the size of the two lateral recesses in the throttle pipe as also the elasticity of the hose membrane. Subsequent modification is possible. The outflow discharge characteristic above the part-full flow is practically vertical. For a particular valve of nominal size, the hose throttle can be freely dissembled and refitted, within the extreme discharge ratio $Q_{\min}/Q_{\max} = 1{:}3$. Outflow discharges between 10 ls^{-1} and 60 ls^{-1} are possible for nominal sizes between $D = 200$ and 250 mm.

Regarding *maintenance*, the hose throttle has to be checked particularly for *clogging*. In case of clogging, the piled up water in the storm water drainage basin can be evacuated by a parallel arrangement without outside help. The throttle membrane is sensitive and should not be subjected to any external shock loading. Volkart and de Vries (1985) describe the static and dynamic deformations as also the hysteresis effect.

Other makes are offered by *BgU* in Bretzfeld, Germany (Balance-throttle), *Metallbau Nill* in Winterthur, Switzerland, *Steinhardt* in Taunusstein (HydrOslide), *BAP*, Germany (Slider valve), *ASA Technik GmbH* in Krefeld-Forstwald, Germany (Control Equipment Division), among many others.

4.4 Requirements of ATV

A number of throttling devices suitable for practical purposes and environmental safety are available. In order to assess their appropriateness and comparative advantages in a particular situation, ATV A111 (1993) proposed a regulation. *Regulating* and *control devices* can be installed in storm water overflows and storm water drainage basins. Their design follows from the specifications of suppliers. The absolute minimum discharge is 10 ls^{-1}. The instructions prescribed in the *ATV Criteria List* (Table 4.1) correspond to special requirements. The information has to be checked by one of the ATV or any other member units of the European Water Pollution Control Association (EWPCA) recognized Institute.

For fabrication and installation, only suitable construction material should be used. For movable parts of the devices, information about their behaviour under corrosive environment is necessary. All equipment has to be maintained at least once a year. For lift weirs, movable shutters, vortex throttles and regulators, a *maintenance contract* has to be settled with the manufacturer or a member of the *Gütezeichen W des Güteschutzes* KANALBAU, unless the user agency does assign suitable personnel of their own. The subsequent *adjustment* is permissible exclusively under the supervision of the respective manufacturer or of otherwise authorized personnel.

4.5 Hinged Flap Gate

4.5.1 *Description*

The Hinged Flap Gate is a novel design for a simple *hydraulic level control*. It can be positioned into a rectangular *storage sewer* and allows to keep a certain *storage level*. The features of this laboratory tested device are (Raemy and Hager 1998):

- Plane flap gate suspended at top crest,
- Movable sealing strip along both walls,
- Free gate outflow configuration,
- Working range from $0 \leq \delta \leq 45°$,
- Gate regulation with a counter weight,
- Excellent dynamic performance (no vibrations), and
- Easily adjustable.

Table 4.1 Criteria for regulating devices (ATV 1993)

A	Statement of function by supplier	
	A.1	Type of device
	A.2	Information regarding size
	A.3	Metering element
	A.4	Operating mechanism
	A.5	Final control element
B	*Information regarding function*	
	For all devices characteristic curves are to be supplied from a recognized test center; for changes in the design, new characteristic curves have to be supplied. A test centre can group together devices of neighbouring nominal sizes.	
C	*Selectivity*	
	C.1	Maximum ±5% selectivity of devices for drainage basins
	C.2	Maximum ±10% selectivity of devices for storm overflows
D	*Selection criteria by supplier*	
	D.1	External energy supply
	D.2	Additional bottom drop downstream of the device
	D.3	Rated capacity to be observed
	D.4	Rated capacity change
	D.5	Limiting downstream water depth
	D.6	Lowest discharge and minimum width of passage
	D.7	Materials used
E	*Operational criteria*	
	E.1	Inspection The required inspection interval is to be specified by the supplier and to be guaranteed by the user-operator (Reference to *Gütezeichen Gruppe W des Güteschutz* KANALBAU).
	E.2	Maintenance measures Instructions for the necessary maintenance measures are to be prepared by the supplier. Changes at established installations shall only follow agreement between the operators and the supplier.
	E.3	Removal of clogging Instructions for clogging removal are to be prepared by the supplier, including a warning against the clogging danger and indicating a bypass of the devices displayed at the installation sites.

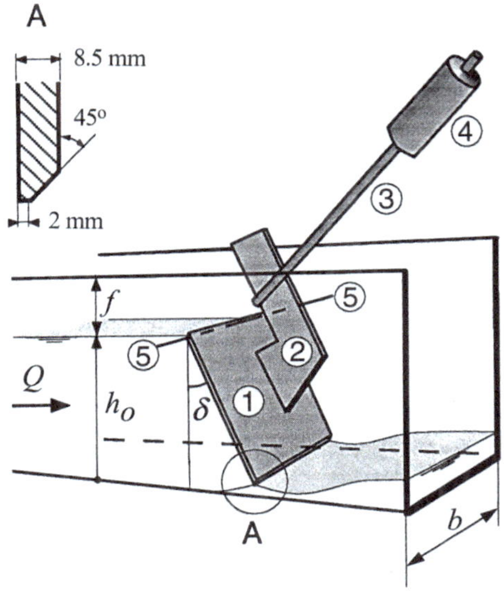

Fig. 4.11 Schematic of Hinged Flap Gate, for details see text

Figure 4.11 shows a schematic view of the device. The Hinged Flap Gate ① is connected with a ② suspension and a ③ rod, on which the counter weight ④ is attached. Its position can be varied along the suspension rod. The gate is hinged on the top crest such that the channel bottom is touched at its resting position ($\delta = 0$). The gate angle δ is measured from the resting position positively in the downstream direction. The approach flow depth is h_o and the freeboard required f. The flow is normally discharged by a gate having a standard crest, and overflows the structures if the flap gate would be blocked.

4.5.2 *Hydraulic Characteristics*

The *discharge Q* for a gate structure can be expressed as (Bos 1976)

$$Q = C_d ab(2gh_o)^{1/2} \tag{4.5}$$

where a is the outflow depth, b the channel width, and C_d the discharge coefficient.

The *pressure distribution* on a flap gate increases hydrostatically from the upstream water surface toward the channel bottom, and turns sharply to the gate crest where atmospheric pressure prevails. The common assumption of hydrostatic pressure on the gate is on the safe side for structural design, but overestimates the real pressure distribution. Based on detailed pressure readings on flap gates, Raemy and Hager (1998) related the hydrodynamic pressure force to the hydrostatic pressure force. This force ratio is close to 1, provided the gate angle δ is small, and decreases with increasing gate angle.

The design configuration involves a water level control for crest elevation, such that the approach flow depth h_o equals the gate length L. The ratio of moments $\mu = M_d/M_s$ with M_d and M_s as the moments on the axis of rotation (gate crest) due to the hydrodynamic and hydrostatic pressure distributions, respectively, varies as

$$\mu = 1 - \frac{1}{4}\tan\delta. \tag{4.6}$$

The conventional assumption involving hydrostatic pressure distribution is based on $\mu = 1$, and the term (1/4) tanδ corresponds to the hydrodynamic pressure effect.

With known coefficient μ, the *discharge coefficient* C_d can be determined using the momentum equation. For $0 < \delta < 30°$, C_d increases slightly with δ [rad] as

$$C_d = 0.60[1 + 0.23\delta - 0.16\delta^2]. \tag{4.7}$$

The design of the Hinged Flap Gate involves the *equation of moments*, with the dynamic moment M_d due to the hydrodynamic pressure force on the gate as disturbing moment, and the sum of moments $M_g + M_w$ due to the gate weight G and the counterweight W as restoring moment (Fig. 4.12). The equilibrium condition requires

$$M_d = \frac{1}{2}GL\sin\delta + eW\cos(\varepsilon + \delta) \tag{4.8}$$

with L as gate length, e distance from pivot to the center of gravity of the counterweight system, and ε as angle of the counterweight relative to the normal of the gate.

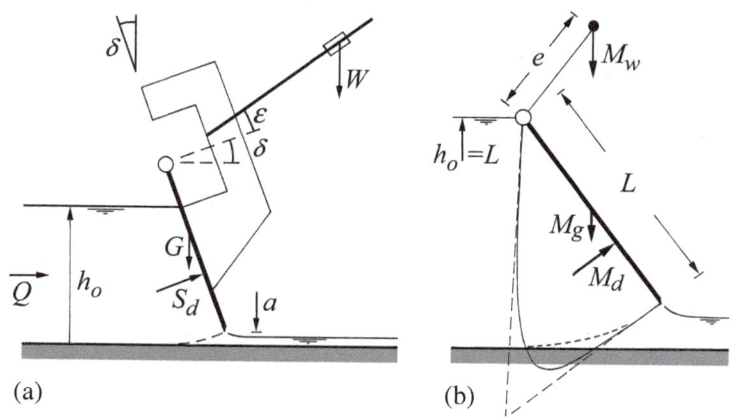

Fig. 4.12 Hinged Flap Gate (**a**) notation, (**b**) equilibrium of moments on pivot

The static moment on the gate is $M_s = (1/2)\rho gbL^2(2L/3)\cos\delta$ with ρ as fluid density. With $M_d = \mu M_s$ and for $L = h_o$, Eqs. (4.6) and (4.8) yield

$$(1/3)\rho gbL^3 - eW\cos\varepsilon = [(1/2)GL - eW\sin\varepsilon + (1/12)\rho gbL^3]\tan\delta. \tag{4.9}$$

To satisfy the *equilibrium requirement* for any angle δ, two conditions result from Eq. (4.9):

$$\text{(1) Restoring moment} \quad eW/(\rho gbL^3) = (3\cos\varepsilon)^{-1}, \tag{4.10}$$

$$\text{(2) Angle } \varepsilon \quad \sin\varepsilon = (eW)^{-1}[(1/2)GL + (1/12)\rho gbL^3]. \tag{4.11}$$

Eliminating the term (eW) gives with the relative gate weight $\gamma = G/(\rho gbL^2)$ the two conditions (Raemy and Hager 1998)

$$\frac{eW}{\rho gbL^3} \cong 0.344\left[1 + \frac{1}{3}\gamma\right], \tag{4.12}$$

$$\tan\varepsilon = \frac{1}{4} + \frac{3}{2}\gamma. \tag{4.13}$$

γ is the ratio of weights of the gate and the reference water volume, and $\gamma << 1$ is typical. According to Eq. (4.12) the restoring moment has to be at least $eW = 0.35(\rho gbL^3)$, and the angle ϵ has a minimum of 14° ($\tan\varepsilon = 0.25$). Omitting the hydrodynamic pressure effect would not at all produce this leading term. Typically, (e/L) is about 3, such that $W = (1/10)(\rho gbL^2)$ and the weight W is 10% of the reference water volume (ρgbL^2). If steel is used, the counterweight volume is thus only about 1% of the typical water volume bL^2.

4.5.3 *Performance Characteristics*

In the laboratory model, the weight of the aluminium gate was $G = 46$ N, the weight of the supporting construction 27.7 N and the counter balance was 106.6 N, thus $W = 134.3$ N. A distance $e = 860$ mm for $h_o = 415$ mm was adjusted, with $\varepsilon = 18.5°$ as the required angle for $Q = 0$ ($\delta = 0$). Increasing the discharge had practically no effect on the approach flow depth, and extreme level variations were confined to ±1% for $0 < \delta < 45°$. The Hinged Flap Gate could be manually moved out of the equilibrium position, and the gate returned immediately to the equilibrium position, normally with an aperiodic oscillation. *Gate stability* increases with increasing discharge. To inhibit oscillations for small discharges ($\delta < 10°$) and surface wave phenomena for large discharge ($\delta > 45°$) *arrest devices* have to limit operation out of the operational range $0 \leq \delta < 45°$.

The *capacity* of the Hinged Flap Gate expressed as a non-dimensional discharge is

$$\frac{Q}{(gb^2L^3)^{1/2}} = 0.90(1 - \cos\delta). \tag{4.14}$$

For $\delta_M = 45°$ this gives a maximum (subscript *M*) discharge $Q_M = 0.27(gb^2L^3)^{1/2}$.

The *freeboard f* required for a completely blocked Hinged Flap Gate ($\delta = 0$) involves gate overflow. The overflow depth h_W measured from the channel bottom is

$$\frac{h_W}{L} = 1 + \left[(1 - \cos\delta)\frac{C_d}{C_W}\right]^{2/3}. \tag{4.15}$$

This gives $h_W = 1.6L$ for $\delta = 45°$, and the freeboard f must be at least $f \geq (h_W–L)$. This relatively large depth can be significantly reduced by allowing for a lateral overflow upstream from the gate.

Figure 4.13 shows gate flows for various angles δ and a proper level control without the formation of surface waves. Figure 4.14 relates to details of flows including flow contraction downstream from the gate, and stagnation flow upstream from the

(a)
(b)
(c)

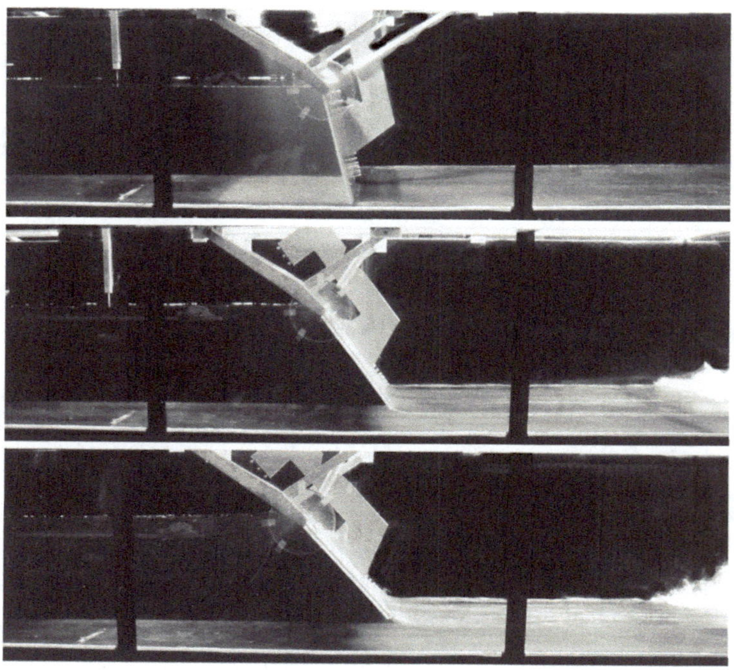

Fig. 4.13 Hinged Flap Gate for $\delta =$ (**a**) 13°, (**b**) 31°, (**c**) 45°

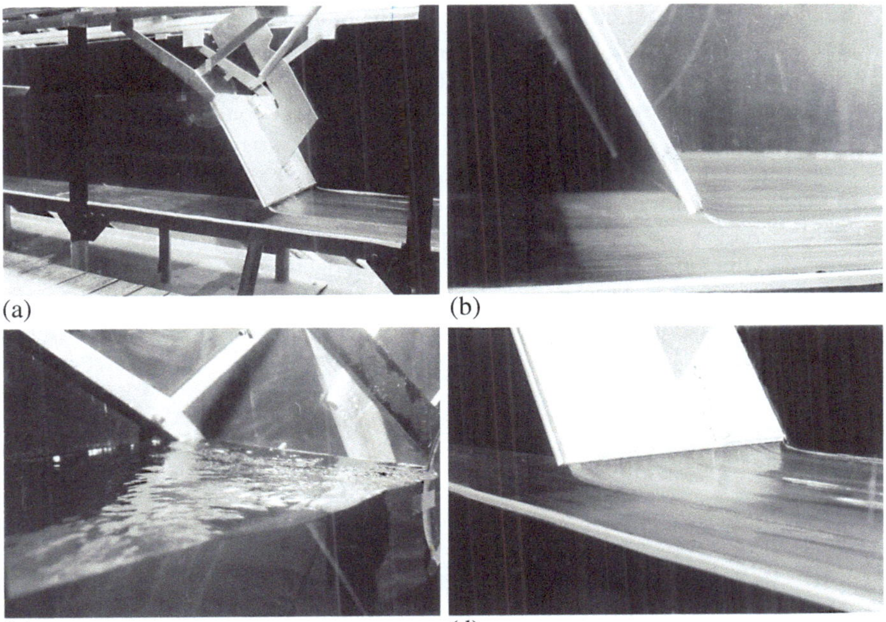
(a) (b) (c) (d)

Fig. 4.14 Details of flow for Hinged Flap Gate (**a**) downstream view, (**b**) jet contraction, (**c**) upstream view onto gate crest, (**d**) gate performance and excellent tailwater flow conditions (Raemy and Hager 1998)

gate. The *Hinged Flap Gate* was also tested for $\varepsilon = 0$ as was previously suggested by various designers. Observations indicated a poor hydraulic performance.

4.6 Discharge Control

A storm water drainage basin is often widely subdivided in space. Also, *security* from service breakdown is often of foremost concern in urban agglomerations. Therefore, a trend towards automatic monitoring and control exists. The data collection and regulating devices in a storm water drainage basin must remain functionally effective with reasonable maintenance, cost and effort.

A changeover from formerly mechanically-regulated to controlled storm water drainage basins must be planned. Discharge and water level probes must be built and the recording of the current data for the drainage basin installed. To evaluate particular events and to decide on the operational patterns, observations by expert personnel are necessary.

Drainage basins critical with regard to *deposition* or *clogging* are connected with the treatment station to register the relevant data and to report danger conditions. In a further step, arrangements would be made for the control commands emanating from the station to be executed at the drainage basins. Process computers are employed to facilitate decisions for analysis of past events and predictions. The optimum results are then executed. Naturally, hydro-meteorological events have significance, they must be detected early and their impact on the sewer network is correspondingly simulated.

This semi-futuristic *management of sewer systems* has also limitations, especially for failure of individual members of the system. Problems may arise in the modelling of the parts with difficult flow processes as their adaptation from the conventional to the future *management* takes years. In planning a sewage system today, one should be ready to provide for its automatic control. An overview of the problems of discharge control and storage management in drainage systems is given by Schilling (1990).

Notation

a	[m]	outflow depth
b	[m]	channel width
C_d	[−]	coefficient of discharge
C_W	[−]	overflow discharge coefficient
D_s	[m]	diameter of screw
e	[m]	distance from gravity center
f	[m]	freeboard
F_a	[m^2]	area of outflow section
F_d	[m^2]	area of diaphragm
g	[ms^{-2}]	gravitational acceleration
G	[N]	gate weight
h_a	[m]	outflow head
h_o	[m]	approach flow depth
h_p	[m]	pressure head
h_W	[m]	overflow depth
H	[m]	energy head
H_d	[m]	variable water level
H_g	[m]	geodetic head difference
H_t	[m]	total head difference
L	[m]	length of gate
L_s	[m]	length of screw
M	[Nm]	moment
Q	[m^{3}s^{-1}]	discharge
Q_a	[m^{3}s^{-1}]	outflow discharge
Q_k	[m^{3}s^{-1}]	critical treatment discharge
Q_M	[m^{3}s^{-1}]	maximum discharge
r	[−]	ratio of radii
R_1	[m]	radius of sewer

R_2	[m]	radius of diaphragm
s	[m]	diaphragm opening, gap
S	[−]	relative diaphragm opening
V_a	[ms^{-1}]	outflow velocity
W	[N]	counterweight
γ	[−]	relative gate weight
δ	[−]	gate angle
ε	[−]	counterweight angle
μ	[−]	moment coefficient
ρ	[kgm^{-3}]	density

References

ATV (1980). Planung und Bau von Abwasserpumpwerken mit kleinen Zuflüssen (Planning and construction of sewage pumping stations for small discharges). ATV-Regelwerk, *Arbeitsblatt* A134. ATV: St. Augustin [in German].

ATV (1993). Richtlinien für die hydraulische Dimensionierung und den Leistungsnachweis von Regenwasser-Entlastungsanlagen in Abwasserkanälen und -leitungen (Guidelines for the hydraulic design and the capacity proof of stormwater drainage structures in sewers). *Arbeitsblatt* A111. Abwassertechnische Vereinigung: St. Augustin [in German].

Borcherding, H., Brombach, H. (1995). Hydraulische Eigenschaften gehäuseloser Abwasser-Rückstauklappen (Hydraulic characteristics of casingless sewage flap gates). *Wasserwirtschaft* 85(4): 200–203 [in German].

Bos, M.G. (1976). Discharge measurement structures. Laboratorium voor Hydraulica en Afvoerhydrologie. *Rapport* 4. Landbouwhogeschool: Wageningen NL.

Brombach, H. (1982a). Drosselstrecken und Wirbeldrosseln an Regenbecken (Throttling pipes and vortex throttles at stromwater basins). *Schweizer Ingenieur und Architekt* 102(33/34): 670–674 [in German].

Brombach, H. (1982b). Abflusssteuerung von Regenwasserbehandlungsanlagen (Discharge control of stormwater treatment stations). *Wasserwirtschaft* 72(2): 44–52 [in German].

Brombach, H. (1987). Eine späte Nutzung des Bernoulli-Effekts: Die Schlauchdrossel (A late application of the Bernoulli effect: The hose throttle). *Wasser und Boden* 39(11): 564–571 [in German].

Burrows, R. (1984). The hydraulic characteristics of hinged flap gates. *Hydraulic design in water resources engineering: Land Drainage.* K.V.H. Smith, D.W. Rycroft, eds. Springer: Berlin.

Dasek, I.V. (1989). Pumpensumpfbemessung in Abwasserpumpwerken (Pump sump design in sewage pumping stations). *Schweizer Ingenieur und Architekt* 109(41): 1107–1112 [in German].

Fass, W. (1970). Vakuum in Abwasser-Pumpstationen (Vacuum in sewage pumping stations). *Gas – Wasserfach* Wasser/Abwasser 111(6): 334–337 [in German].

Hager, W.H. (1987a). Circular gates in circular and U-shaped channels. *Journal of Irrigation and Drainage Engineering* 113(3): 413–419.

Hager, W.H. (1987b). Abfluss im U-Profil (Flow in U-shaped profile). *Korrespondenz Abwasser* 34(5): 468–482 [in German].

Kaul, G. (1986). Mindestabfluss von Drosseleinrichtungen in Mischwasserkanälen (Minimum discharge of throttling devices in combined sewers). *Korrespondenz Abwasser* 33(7): 587–591 [in German].

Lessmeier, H. (1983). Pumpen und kleine Hebewerke für die Abwasserförderung (Pumps and small lift structures for sewage pumping). *Wasser und Boden* 35(10): 452–456 [in German].

Raemy, F., Hager, W.H. (1998). Hydraulic level control by Hinged Flap Gate. Proc. Institution Civil Engineers *Water Maritime & Energy* 130(June): 95–103.

Schilling, W. (1990). *Operationelle Siedlungsentwässerung* (Operational urban drainage). Oldenbourg: München [in German].

Tuttahs, G. (1987a). Gesichtspunkte bei der Bemessung von Abwasserpumpwerken (Points of view for designing sewage pumping stations). *Wasser und Boden* 39(1): 30–34 [in German].

Tuttahs, G. (1987b). Förderanlagen für Abwasser (Pumps for sewage). *Wasser und Boden* 39(5): 246–250 [in German].

Vischer, D. (1979). Die selbsttätige Schlauchdrossel zur Gewährleistung konstanter Beckenausflüsse (Automatical hose throttle to guarantee constant basin outflow discharge). *Wasserwirtschaft* 65(12): 371–375 [in German].

Volkart, P.U., de Vries, F. (1985). Automatic throttle hose – a new flow regulator. *Journal of Irrigation and Drainage Engineering* 111(3): 247–264.

Witschi, R. (1970). Die Schneckenpumpe und ihr Einsatz bei der Abwasserreinigung. (The screw pump and its use for sewage purification) [in German].

Uniform flow is hardly observed because lengths are too short in man-made hydraulic structures. The bottom slope of this connection channel is close to critical, developing am undular hydraulic jump

Chapter 5
Uniform Flow

Abstract The equilibrium condition between driving and retaining forces causes uniform flow. The conditions of uniform flow are discussed in detail first. Then, uniform flow is described with formulae and applied on the partially-filled circular, the egg-shaped and the horseshoe profiles. Approximate expressions are derived for discharge, flow depth, energy head and critical bottom slope allowing for a simple application of uniform flow. In addition, aspects such as air entrainment in steep sewers or choking of sewers from supercritical to pressurized flow are described. The design procedure is finally outlined and illustrated with examples.

5.1 Introduction

Contrary to pressurized flow with a change of cross-section only in the streamwise direction, free surface flows may vary in depth h and with the streamwise coordinate x. *Free surface flows* thus have an additional degree of freedom and are particularly attractive. Free surface waves do not exist in pressurized flows, and backwater phenomena are absent.

The additional freedom degree of free surface flows as compared to pressurized flows complicates the hydraulic approach. By comparing the governing equations between those two classes of flows, free surface flows have the additional term $\partial(V^2/2g)/\partial x$, with V as cross-sectional average velocity, and g as gravitational acceleration. The so-called convective acceleration term accounts for the local change of velocity head between two sections. Accordingly, also steady flow ($\partial/\partial t = 0$) with t as time can be subjected to acceleration that change a certain flow, in contrast to solid bodies. This particular feature of fluid mechanics adds considerably to the mathematical problems that govern fluid flows.

Fundamentally, *steady flows* that are independent of a temporal variation can be classified as either uniform or non-uniform. For an uniform flow all flow parameters are independent of both space and time. As discussed in the following, *uniform flow* corresponds to an idealized condition occuring rarely in nature. This chapter deals with uniform flow exclusively, and Chap. 8 relates to non-uniform flow, where the flow depth may vary with space. Here, the uniform flow is defined first, and the assumptions for uniform flow are presented. Then, methods of computation for

W.H. Hager, *Wastewater Hydraulics*, 2nd ed., DOI 10.1007/978-3-642-11383-3_5,

uniform flow in the standard sewer profiles are outlined. Further, aspects of surface air entrainment and flow choking in large velocity flows are considered. This chapter is concluded with design procedures and is illustrated with computational examples.

5.2 Description of Uniform Flow

Uniform flow (German: Normalabfluss; French: Ecoulement uniforme) can be defined differently, such as the conditions of:

- Constant velocity V,
- Constant flow depth h,
- Surface profile parallel to channel bottom,
- Equilibrium between driving and retaining forces.

All these descriptions are inexact, however, and even wrong for certain cases, because they relate to a certain flow feature only and do not specify the necessary conditions for uniform flow to occur. Because uniform flow is a fundamental flow process, a definite and rather exhaustive description follows.

Uniform flow must satisfy the following *necessary conditions:*

- Bottom slope S_o be constant,
- Wall roughness be uniformly distributed,
- Discharge remains constant both in time and space,
- Cross-section be prismatic,
- Channel axis be straight,
- Air pressure above free surface be constant,
- Fluid be homogeneous, and
- Governing parameters be independent of time.

It is only such a fluid that can – but may not have – uniform flow.

Uniform flow is an *asymptotic process* and is established only after a long flow distance. If neither bottom slope nor roughness, discharge nor cross-section do vary over this reach, an equilibrium condition may result, with a compensation of the gravitational force component in the flow direction with the friction force. Then, a fluid particle is not subjected by any force, or no acceleration according to Newton, and the condition of equilibrium is retained. This means:

- Velocity field and thus the average velocity remain invariable along the channel,
- Cross-sectional area remains constant for a given discharge, and the flow depth does not change locally in a prismatic channel,
- Free surface is straight and parallel to the channel bottom, and
- Energy line is also parallel to both free surface profile and bed profile.

If the numerous assumptions for uniform flow to occur are considered, and the large distances needed for backwater curves are computed from Chap. 8, it can be realized that uniform flow corresponds indeed to an *idealized condition of flow*. Despite the rarity of occurrence, uniform flow has a significant feature, because the ratio of uniform flow depth to other depths alone allows to account for the effective flow characteristics. The uniform flow depth is thus computed as a function of the main flow variables. Apart from its computation, one should not forget about the restrictions of its meaning as a particular flow depth, and the limitations of uniform flow should be kept in mind.

5.3 Uniform Flow Law

Uniform flow in open channels has similarity with pressurized flows without a pressure gradient. Such basic flows were investigated for a long time and are presented in Chap. 2. With V as average cross-sectional velocity, f as friction coefficient, and D as pipe diameter, the friction gradient S_f according to the *equation of Darcy and Weisbach* is

$$S_f = \frac{V^2}{2g}\frac{f}{D}. \tag{5.1}$$

The friction coefficient f is currently determined with the universal *Colebrook and White formula* (Press and Schröder 1966, Daily and Harleman 1966)

$$\frac{1}{\sqrt{f}} = -2 \cdot \log\left(\frac{\varepsilon}{37} + \frac{2.51}{\mathsf{R}\sqrt{f}}\right), \mathsf{R} > 2300. \tag{5.2}$$

Here, $\varepsilon = k_s/D$ is the relative roughness and $\mathsf{R} = VD/\nu$ the Reynolds number with ν as kinematic viscosity (see Sect. 2.2.1).

Equation (5.1) and (5.2) are conventionally extended to arbitrary pressurized flows in non-circular profiles by introducing four times the *hydraulic radius* ($4R_h$) instead of the pipe diameter D. This procedure is also extended to uniform flow in open channels. For flows in the turbulent smooth regime, the effects of *channel shape* (i.e. Marchi 1961, Bock 1966, Marchi and Rubatta 1981, Kazemipour and Apelt 1982) and *Froude number* (e.g. Rouse 1965) are known to exist. To apply the system of Eqs. (5.1) and (5.2), so-called shape parameters ϕ_f have been introduced that are discussed by Sinniger and Hager (1989). Considering the difficulties in determining the wall roughness or the exact discharge, the conventional approach is followed here. The number and the ranges of experiments conducted until today are too small to quantify definite effects of channel shape on the roughness law. It should be noted that the flow in a pressurized pipe produces normally less resistance than an equivalent open channel flow. In general, it may be stated that Eqs. (5.1) and (5.2) apply the better, the closer the flow profile is circular-shaped.

The *uniform flow formula* can thus be written by expanding Eq. (5.1.) as

$$S_f = \frac{V^2}{2g} \cdot \frac{f}{4R_h} \tag{5.3}$$

with f from Eq. (5.2), $\varepsilon = k_s/(4R_h)$ and $\mathsf{R} = V(4R_h)/\nu$. In the same sense all approximations of the general flow equation as introduced in Chap. 2 can be applied when inserting $(4R_h)$ instead of the pipe diameter D.

5.4 Flow Formulae

Depending on the magnitudes of the relative roughness ε and the Reynolds number R three turbulent flow regimes can be distinguished (Chap. 2):

- *Turbulent smooth regime* with the effect of ε much smaller than that of R,
- *Turbulent rough regime* with the effect of ε much larger than the effect of R, and
- *Transition regime with* both effects significant on the flow resistance.

Note the particularity that for a specific channel or sewer, all three regimes may occur, depending on the velocity.

In wastewater hydraulics with narrow ranges of equivalent roughness, diameter and velocities, the transition regime has particular relevance. Relating for example to a sewer of 1 m in diameter for common sewage with $\nu = 1 \times 10^{-6}$ m^{2}s^{-1} at a temperature of 20°C, the equivalent sand roughness height k_s has to be larger than k_{sr} specified in Table 5.1 (Chap. 2). Analogously, the equivalent sand roughness height k_s must be smaller than k_{ss} for turbulent smooth flow to occur (Table 5.1).

For a velocity of $V = 1$ ms^{-1} the flow in concrete pipe with $k_s = 2.5$ mm is thus hydraulically rough, an asbestos pipe with $k_s = 0.1$ mm generates a flow in the transition regime and a glass or new PVC pipe behaves nearly turbulent smooth. Based on the equations presented in Chap. 2, the transitions from one to the other regime may be determined when inserting $4R_h$ instead of the pipe diameter D.

For the hydraulic rough regime an approximation of the Colebrook-White formula (Chaps. 2 and 4) is the *flow formula of Manning-Strickler*

$$V = (1/n)S_o^{1/2}R_h^{2/3}. \tag{5.4}$$

Table 5.1 Limit roughness heights k_{sr} [mm] for turbulent rough (subscript r) and k_{ss} [mm] for turbulent smooth (subscript s) flows as a function of velocity V [ms^{-1}] according to Sinniger and Hager (1989)

V [ms^{-1}]	0.1	0.5	1	2	3	5	10
k_{sr} [mm]	10.5	2.1	1	0.52	0.35	0.21	0.11
k_{ss} [mm]	0.01	0.002	0.0013	–	–	–	–

The quadratic resistance law relative to the bottom slope is characterized by the *roughness coefficient* $1/n$ whose dimension is $[m^{1/3}s^{-1}]$. In a sewer with a constant boundary roughness, the n-value remains thus constant. For a range of relative roughnesses $7 \times 10^{-4} < k_s/(4R_h) < 7 \times 10^{-2}$ Eq. (5.4) applies only provided the absolute roughness satisfies the sufficient condition $k_s > 30\nu\left[g^2S_o^2Q\right]^{-1/5}$ (Hager 1988). Strickler (1923) has also related k_s to n as

$$\frac{(1/n)k_s^{1/6}}{g^{1/2}} = 8.2 \tag{5.5}$$

If this relation is inserted into the previous two conditions for rough turbulent flow, then

$$0.036 < \frac{g^{1/2}}{(1/n)D^{1/6}} < 0.179; \quad (1/n) < 8.2\,g^{1/2}\frac{\left[g^2S_o^2Q\right]^{1/30}}{(30\nu)^{1/6}} \tag{5.6a}$$

For typical values $D^{1/6} \cong 1$ m and $\nu = 10^{-6}$ m^{2}s^{-1}, the *dimensional* conditions for the roughness coefficient $1/n$ are, respectively,

$$18\,m^{1/3}s^{-1} < 1/n < 87\,m^{1/3}s^{-1}; \quad 1/n < 170(S_o^2Q)^{1/30}. \tag{5.6b}$$

Equation (5.6b) demonstrates that the Manning-Strickler formula applies only for n-values between limit values, and that minimum values for both the bottom slope S_o and the discharge Q have to be accounted for. Table 5.2 illustrates the relation between those two parameters for $n = 0.011$ m$^{-1/3}$s. For bottom slopes smaller than 0.1% (0.001) attention should be drawn to the application of the Manning-Strickler formula because of viscous effects. Such small bottom slopes are not relevant in sewers because of minimum velocity requirements (Chap. 3).

Example 5.1 Given a pipe of $D = 0.50$ m, $S_o = 1\%$ and $k_s = 0.1$ mm. Can the Manning-Strickler formula be applied for a discharge $Q = 0.15$ m^{3}s^{-1}?
With a relative sandroughness height $k_s/D = 10^{-4}/0.5 = 2\cdot 10^{-4}$ and $30\nu[g^2S_o^2Q]^{-1/5} = 30\cdot 1.3\cdot 10^{-6}[9.81^2 0.01^2 0.15]^{-1/5} = 1.44\cdot 10^{-4}\ \text{m} > k_s = 10^{-4}$ m, both conditions for turbulent rough flow are not satisfied and Eq. (5.4) should not be applied.

Table 5.2 Minimum discharge Q_{min} as a function of bottom slope S_o for $n = 0.011$ m$^{-1/3}$s for which the Manning Strickler formula still applies

S_o [‰]	0.1	0.5	1	5	10
Q_{min} [Ls^{-1}]	520	21	5	0.2	–

Example 5.2 Demonstrate with Eq. (5.6a) for the numbers given in Example 5.1 that the Manning-Strickler formula cannot be applied.
The roughness coefficient is $1/n = 8.2 \cdot 9.81^{1/2}(10^{-4})^{-1/6} = 119\ \mathrm{m}^{1/3}\mathrm{s}^{-1}$ from Eq. (5.5), and from Eq. (5.6a) this should be smaller than $8.2 \cdot 9.81^{1/2}$ $[9.81^2 0.01^2 0.15]^{1/30}/(30 \cdot 1.3 \cdot 10^{-6})^{-1/6} = 112\ \mathrm{m}^{1/3}\mathrm{s}^{-1}$.

According to Eq. (5.6b) $1/n = 119\ \mathrm{m}^{1/3}\mathrm{s}^{-1} > 170\ (0.01^2 0.15)^{1/30} = 117\,\mathrm{m}^{1/3}\mathrm{s}^{-1}$, and the condition is not satisfied. Eq. (5.4) cannot be applied, therefore.

The previous discussion demonstrates that the Manning-Strickler formula (5.4) is highly *relevant in practice*, because:

- Simplicity in application,
- Explicit solution for velocity V, hydraulic radius R_h and bottom slope S_o,
- Application in wide ranges of relative roughnesses, particularly in the range of commercially fabricated pipes and sewers, and
- Roughness coefficient $1/n$ is known for the available pipe materials.

Further comments on the Manning-Strickler formula follow in Sect. 5.8.2.

5.5 Conditions of Partial Pipe Filling

5.5.1 *Partial Pipe Filling Diagrams*

Conduits can be partially filled or pressurized, in contrast to open channels. Although a full-pipe filling under uniform flow cannot be realized experimentally, i.e. a pipe cannot be filled completely under atmospheric pressure, full conduit flow can be assumed if the flow depth is equal to the pipe diameter. *The pipe filling ratio* (German: Teilfüllung; French: Remplissage de conduite) corresponds to the ratio of flow depth to the pipe diameter and it varies from zero to one. The discharge Q related to the full (subscript v) filling discharge Q_v is thus a function of the filling ratio and cross-sectional shape only. The vertical scaling is the diameter D for circular pipe and the profile height T for other conduit shapes. Analogous relations can also be established for the ratios of velocity V/V_v, cross-sectional area F/F_v and other parameters in terms of the filling ratio $y = h/T$, where V_v, F_v are the quantities for full pipe filling.

Partial filling diagrams are popular because of the simplicity in representation. They can be applied exclusively for uniform flow, because additional effects such as roughness, bottom slope or Reynolds number had to be accounted for in addition. The *discharge equation* based on the Manning-Strickler formula thus reads from Eq. (5.4)

$$Q/Q_v = (F/F_v)(R_h/R_{hv})^{2/3}. \tag{5.7}$$

The two numbers of the right hand side depend exclusively on the *sewer geometry* and are independent of bottom slope and roughness coefficient. If the Colebrook-White formula would be used, this simple relation could not be reproduced.

5.5.2 *Choking of Sewer Flow*

The deviations between uniform flow in an open channel and pressurized pipe flow increase as the cross-sectional shape deviates from the circle. In closed channels additional effects due to the closed profile geometry must be expected. Whereas the first effect can be accounted for with the shape parameter ϕ_f previously discussed, the second effect depends essentially on the pressure conditions of the free surface. This effect is notable for a relatively large filling ratio, because the air access from the atmosphere may be reduced, and free surface flow as in an open channel is not guaranteed. A minute flow perturbation may already suffice to choke the pipe flow. *Choking* (German: Zuschlagen, French: Fermeture brusque) of a pipe means a sudden and abrupt transition from free surface to pressurized flow. The inverse process (German: Aufschlagen, French: Ouverture brusque) has not yet received an English expression, and 'striking' or 'opening' do not really describe the process. A similar flow pattern occurs with air-water flows containing air-pockets in steep sewers. This flow condition is dangerous because of *flow pulsations* and depressions, and should be avoided (5.7).

The choking of a pipe has received scarce attention until now, although it is typical for *overcharged sewer reaches* and may be dangerous for capacity reasons. The effect of approach flow conditions is essential and the outflow from a basin with a sharp-crested pipe inlet, or flows perturbed by upstream bends may behave differently from uniform approach flow, mainly because of shock waves (Chap. 16). Figure 5.1 relates to two examples for which uniform flow may practically not be possible.

For flow depths that are close to the pipe vertex, the *access of air* from a manhole is reduced and the inlet flow is characterized by intermittent opening and closing of the section (Fig. 5.1a). Such flow is typical for outflow from a reservoir, or for subcritical flow from a manhole, for which the inlet acts as a gate with a fractional air access only, such as with intake vortices. Minute under- or over-pressures in the

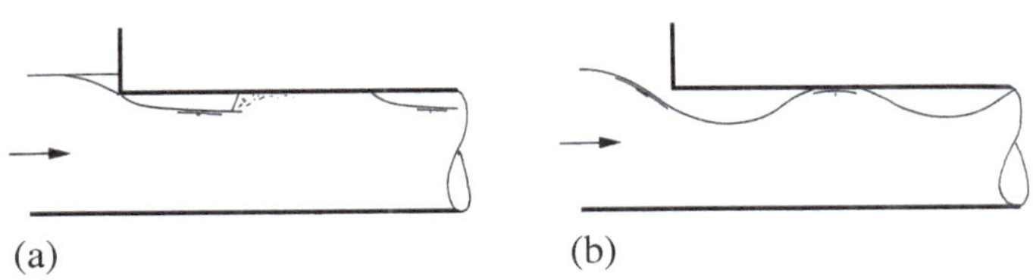

Fig. 5.1 Choking of sewers caused by (**a**) inlet perturbation, (**b**) wave generation

pipe inhibit the formation of uniform flow, and the flow contains *bubbles* or *pockets* instead (5.7). These poor flow conditions can be improved by sufficient air supply or a continuous acceleration of the flow. For sewers, both methods are complicated and the only means to counter choking is reducing the pipe filling, associtated with a loss in discharge capacity.

Figure 5.1b relates to choking for *transitional flows* with a subcritical upstream and a weakly supercritical downstream flow. If such a flow touches the pipe vertex, an undular hydraulic jump (Chap. 7) may cut down the air supply. Then, air pockets are formed and the uniform flow pattern cannot establish.

The choking and opening phenomena of pipe flow can be modelled with the data of Sauerbrey (1969), provided the initial and the terminal flow conditions are uniform, respectively. Choking of a pipe is generated *in* the flow direction, and opening of the pipe *against* the flow direction. Sauerbrey found that choking for $S_o < 0.8\%$ occurs also for pipes that have longitudinal slits in the pipe vertex. Until now, the effect of approach flow conditions previously referred to has not yet been investigated, and the following results apply to Sauerbrey's experimental arrangement exclusively.

With $q_D = Q/(gD^5)^{1/2}$as the relative discharge and S_{oz} as the bottom slope where *choking* (subscript *z*) occurs, the choking condition can be expressed for $S_{oz} < 0.8\%$ as (Hager 1991)

$$S_{oz}\,[\text{‰}] = 20.5\,(q_D - 0.36)\,. \tag{5.8}$$

Figure 5.2a shows that choking does not occur for $q_D < 0.36$, whereas the pipe is always choked for $q_D > 0.70$.

The filling ratio y for choking to occur can also be plotted as a function of S_{oz}. Figure 5.2b shows that pipes with a filling ratio larger than 92% always choke. For

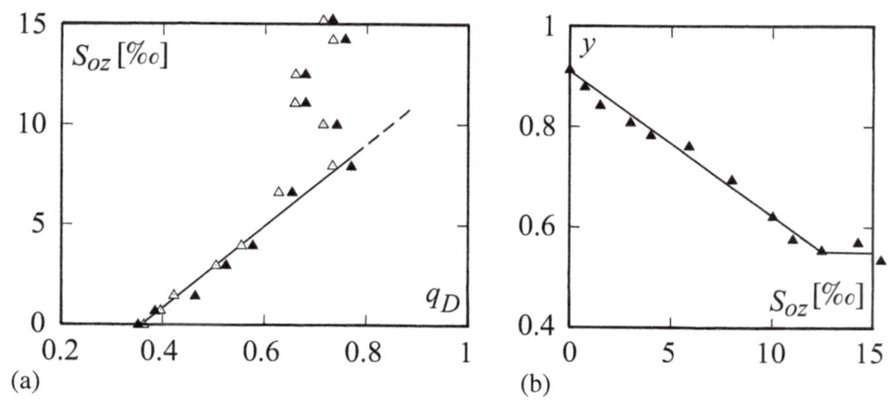

Fig. 5.2 Limit bottom slope S_{oz} [‰] for *pipe choking* to occur according to data of Sauerbrey (1969) as function of (**a**) relative discharge $q_D = Q/(gD^5)^{1/2}$ for (▲) non-aerated and (Δ) aerated pipe with (—) Eq. (5.8), (**b**) corresponding filling ratio $y = h/D$ and (—) Eq. (5.9)

larger S_{oz}, the corresponding filling ratio decreases and the data may be expressed for $y > 0.55$ with $\phi_z = 0.03$ for choking as

$$y = 0.92 - \phi S_{oz}\,[‰] \tag{5.9}$$

For $S_{oz} > 12‰$ the effect of choking seems to reduce. According to Eq. (5.9) a pipe with a bottom slope of $S_o = 12‰$ would choke whenever $y > 0.56$. Sewers with bottom slopes of 1% are typical in hilly regions, and choking occurs for a relatively small filling ratio already. The data of Sauerbrey (1969) should be substantiated to assess the design of sloping sewers.

Figure 5.3 relates to *pipe opening* (subscript a) under uniform flow conditions as a function of the limit bottom slope S_{oa}. According to Fig. 5.3a, a pipe with $S_o > 3.6‰$ opens always if $q_D < 0.50$, whereas pipes with $q_D \geq 0.50$ seem to run always under pressure once they have choked. The corresponding filling ratio y is shown in Fig. 5.3b as a function of S_{oa}, and Eq. (5.9) applies if $\phi_a = 0.05$. Also included is the curve for pipe choking and it may be observed that for a given filling ratio y, the pipe opens for a smaller bottom slope S_o than it chokes. The transitional phenomena depend thus on the variation of discharge, and pipe choking is *not* inversely corresponding to pipe opening.

Additional reasons for flow choking are *flow perturbations*, such as changes in streamwise direction and bottom slope, sewer junctions, drop structures or sewer sedimentation. All these perturbations tend to enhance surface waves and thus reduce the air supply over the flow surface. Self-aerated sewer flow is discussed in Sect. 5.6, and the mechanism of two-phase flow in general in Sect. 5.7. The effects of limited air supply and choking due to shockwaves in bend and junction manholes are outlined in Chap. 16. The hydraulic capacity of these manholes is also discussed.

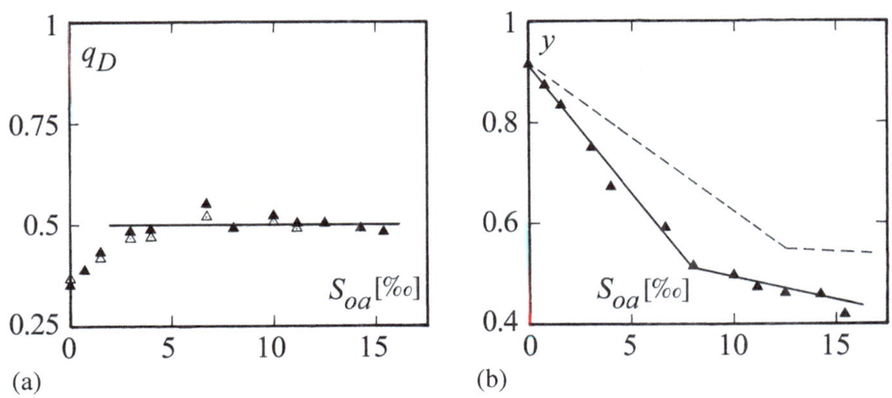

Fig. 5.3 Limit bottom slope S_{oa} [‰] for *pipe opening* to occur according to data of Sauerbrey (1969), as function of (**a**) relative discharge $q_D = Q/(gD^5)^{1/2}$ with (—) $q_D = 0.50$, and (**b**) filling ratio $y = h/D$ with (—) Eq. (5.9), (- - -) Fig. 5.2b. Notation Fig. 5.2

The previous discussion relates to weakly supercritical flows with a Froude number up to about 2. The effect of *air flow* above the free surface is significant because it can perturb uniform flow. This effect of course increases with the filling ratio. Therefore, the flow in highly-filled sewers can be modified as compared to open channels with an atmospheric free surface pressure. Until now, only limited studies are available for circular sewers, and other profiles have not at all received attention. A detailed experimental approach to the transition phenomena of sewers is definitely needed to allow for their rational design.

5.5.3 *Partial Filling of Circular Sewer*

The *geometry* of the circular sewer of diameter D is related to the half center angle δ (Fig. 5.4) as

$$\frac{h}{D} = \frac{1}{2}(1 - \cos\delta) \tag{5.10}$$

$$\frac{P}{D} = \delta \tag{5.11}$$

$$\frac{F}{D^2} = \frac{1}{4}(\delta - \sin\delta\cos\delta) \tag{5.12}$$

Here h is flow depth, P wetted perimeter, and F cross-sectional area. The hydraulic radius is $R_h = F/P$.

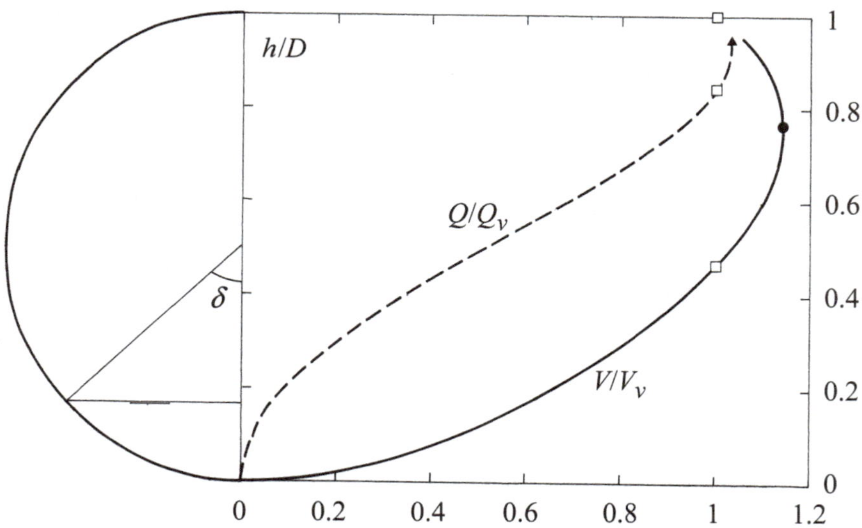

Fig. 5.4 Filling curves for *circular sewer* according to Eq. (5.14), (- - -) discharge Q/Q_v and (—) velocity V/V_v as a function of filling ratio h/D. (□) Full filling, (•) maximum velocity, and (▲) maximum discharge

For full (subscript v) pipe filling with $h_v/D = 1$, i.e. $\delta_v = 180°$ or π, one has $P_v/D = \pi$, $F_v/D^2 = \pi/4$ and $R_h/D = 1/4$. Thus, the discharge in a full-flowing circular pipe is $Q_v = (\pi/4^{5/3})(1/n)S_o^{1/2}D^{8/3}$. The discharge ratio $q_v = Q_N/Q_v$ according to Eq. (5.7) indicates for uniform (subscript N) sewer flow

$$q_v = \left(\frac{\pi}{\delta}\right)^{2/3}\left(\frac{\delta - \sin\delta\cos\delta}{\pi}\right)^{5/3}. \tag{5.13}$$

The relation between relative discharge q_v and filling ratio $y = h/D$ is given in Eq. (5.10). The curve $q_v(y_N)$ has a *maximum* $(y_N;\ q_v)_M = (0.94;\ 1.076)$, i.e. the full-flowing discharge is smaller. This effect is due to the rapid increase of the hydraulic radius close to pipe vertex, with a maximum $(R_h/R_{hv})_M = 1.217$ for $y_N = 0.81$.

Sauerbrey (1969) certainly conducted the most complete analysis on the problem of partial pipe filling. He recommended a filling curve $q_v(y_N)$ up to $y_N = 0.95$ and broke off the curve for $0.95 < y_N \le 1$ because such flows are physically irrelevant. *Stable flows* for such large filling ratios are not possible. Hager (1985) proposed a discharge formula for $q_N = nQ/(S_o^{1/2}D^{8/3})$ that follows his recommendation for $y_N \le 0.5$ and then the data of Sauerbrey (1969) as

$$q_N = \frac{nQ}{S_o^{1/2}D^{8/3}} = \frac{3}{4}y_N^2\left(1 - \frac{7}{12}y_N^2\right), \quad y_N < 0.95. \tag{5.14}$$

The maximum deviations from Sauerbrey's empirical formula are less than 5% for $0.2 < y_N < 0.95$, or less than 3% for $y_N > 0.4$. Equation (5.14) gives always slightly smaller values than according to Sauerbrey and is on the safe side for discharge, therefore.

Equation (5.14) can be explicitly solved for the *filling ratio* $y_N = h_N/D$ as a function of relative discharge q_N

$$y_N = 0.926[1 - (1 - 3.11q_N)^{1/2}]^{1/2}. \tag{5.15a}$$

An approximation of Eq. (5.15a) to ±10% accuracy is

$$y_N = (4/3)q_N^{1/2}. \tag{5.15b}$$

The *cross-sectional area* F is to ±1%

$$F/D^2 = \frac{4}{3}y^{3/2}\left(1 - \frac{y}{4} - \frac{4y^2}{25}\right), \tag{5.16a}$$

and a power function accurate to ±10% is

$$F/D^2 = y^{1.4}. \tag{5.16b}$$

With given discharge and cross-sectional area, the *average velocity* $V = Q/F$ can be determined as a function of filling ratio y. With $\mu_N = q_N/(F_N/D^2) = (Q/F_N)[n/(S_o^{1/2}D^{2/3})]$ as the relative velocity for uniform flow, Eqs. (5.14) and (5.16a) yield to ±5% accuracy for $0.01 < y_N < 0.75$

$$\mu_N = 0.56\, y_N^{1/2}. \tag{5.17}$$

The *hydraulic radius* R_h is from Eq. (3.8)

$$R_h/D = (2/3)\, y\left(1 - \frac{1}{2}y\right), \tag{5.18a}$$

or as a power function accurate to ±10% for $0.05 < y < 0.85$

$$R_h/D = 0.40 y^{0.80}. \tag{5.18b}$$

Equations (5.14), (5.15a), (5.16a), (5.17) and (5.18a) are thus an excellent aid for a simple and accurate prediction of partially-filled sewer flows. All quantities necessary for *design* including discharge, velocity, cross-sectional area and hydraulic radius thus depend explicitly on the filling ratio $y_N = h_N/D$. Table 5.3 gives 'exact' values of these ratios as a fraction of full filling quantities, based on the Manning-Strickler formula.

Example 5.3 Given a circular sewer of bottom slope $S_o = 0.4\%$, $D = 0.70$ m and $1/n = 85\ \text{m}^{1/3}\text{s}^{-1}$. What is the uniform flow depth h_N for a discharge of $Q = 0.46\ \text{m}^3\text{s}^{-1}$?
With $q_N = 0.46/(85 \cdot 0.004^{1/2} 0.70^{8/3}) = 0.222$ as the relative discharge, Eq. (5.15a) gives $y_N = 0.926[1 - (1 - 3.11 \cdot 0.222)^{1/2}]^{1/2} = 0.616$, and thus $h_N = y_N \cdot D = 0.616 \cdot 0.7 = 0.43$ m.

The corresponding cross-section is $F_N/D^2 = (4/3)0.616^{3/2}[1 - 0.616/4 - (4/25)0.616^2] = 0.506$ from Eq. (5.16a), thus $F_N = 0.506 \cdot 0.7^2 = 0.248\ \text{m}^2$. With the uniform velocity $V_N = Q/F_N = 0.46/0.248 = 1.85\ \text{ms}^{-1}$, the energy head becomes $H_N = h_N + V_N^2/(2g) = 0.43 + 1.85^2/19.62 = 0.605$ m. With $\mu_N = 0.56 \cdot 0.616^{1/2} = 0.44$ as the relative velocity from Eq. (5.17) one has $V_N = \mu_N[(S_o^{1/2}D^{2/3})/\text{n}] = 0.44 \cdot 0.004^{1/2} 0.7^{2/3} 85 = 1.86\ \text{ms}^{-1}(+1\%)$.

The effect of a closed sewer on uniform flow may be expressed as the ratio Q_{ex}/Q_{th} between the experimentally (subscript *ex*) determined discharge from Eq. (5.14) and the theoretically (subscript *th*) computed discharge. The deviations are less than 4% for $y_N > 0.50$ and are thus insignificant compared to the errors in estimating the roughness coefficient. It is therefore recommended to drop a separate effect of large filling ratios in the equation for uniform flow and to apply a *simple* filling ratio-discharge equation, such as Eq. (5.14). Maximum deviations are estimated to be below 5%, i.e. in the range of the accuracy of sewer hydraulics.

Table 5.3 Filling curves for circular pipe as a function of filling ratio $y = h/D$. Maximum values are in *italic* (ATV 1988)

h/D	F/F_v	P/P_v	R_h/R_{hv}	B/D	V/V_v	Q/Q_v
0.02	0.0048	0.0903	0.0528	0.2800	0.1592	0.0008
0.04	0.0134	0.1262	0.1047	0.3919	0.2440	0.0033
0.06	0.0245	0.1575	0.1555	0.4750	0.3125	0.0077
0.08	0.0375	0.1826	0.2053	0.5426	0.3717	0.0139
0.10	0.0520	0.2048	0.2541	0.6000	0.4247	0.0221
0.12	0.0680	0.2252	0.3018	0.6499	0.4730	0.0322
0.14	0.0851	0.2441	0.3485	0.6940	0.5175	0.0440
0.16	0.1033	0.2620	0.3942	0.7332	0.5589	0.0577
0.18	0.1224	0.2789	0.4388	0.7684	0.5976	0.0732
0.20	0.1424	0.2952	0.4824	0.8000	0.6340	0.0903
0.22	0.1631	0.3108	0.5248	0.8285	0.6684	0.1090
0.24	0.1845	0.3259	0.5662	0.8542	0.7008	0.1293
0.26	0.2066	0.3406	0.6065	0.8773	0.7316	0.1511
0.28	0.2292	0.3550	0.6457	0.8980	0.7608	0.1744
0.30	0.2523	0.3690	0.6838	0.9165	0.7885	0.1990
0.32	0.2759	0.3828	0.7207	0.9330	0.8149	0.2248
0.34	0.2998	0.3963	0.7565	0.9474	0.8400	0.2518
0.36	0.3241	0.4097	0.7911	0.9600	0.8638	0.2800
0.38	0.3487	0.4229	0.8246	0.9708	0.8865	0.3091
0.40	0.3735	0.4359	0.8569	0.9798	0.9080	0.3392
0.42	0.3986	0.4489	0.8880	0.9871	0.9284	0.3701
0.44	0.4238	0.4617	0.9179	0.9928	0.9478	0.4017
0.46	0.4491	0.4745	0.9465	0.9968	0.9662	0.4340
0.48	0.4745	0.4873	0.9739	0.9992	0.9836	0.4668
0.50	0.5000	0.5000	1.0000	*1.0000*	1.0000	0.5000
0.52	0.5255	0.5127	1.0248	0.9992	1.0154	0.5336
0.54	0.5509	0.5255	1.0483	0.9968	1.0299	0.5674
0.56	0.5762	0.5383	1.0704	0.9928	1.0435	0.6013
0.58	0.6014	0.5511	1.0913	0.9871	1.0561	0.6351
0.60	0.6265	0.5641	1.1106	0.9798	1.0677	0.6689
0.62	0.6513	0.5771	1.1285	0.9708	1.0785	0.7024
0.64	0.6759	0.5903	1.1449	0.9600	1.0883	0.7356
0.66	0.7002	0.6037	1.1599	0.9474	1.0971	0.7682
0.68	0.7241	0.6172	1.1732	0.9330	1.1050	0.8002
0.70	0.7477	0.6310	1.1849	0.9165	1.1119	0.8313
0.72	0.7708	0.6450	1.1950	0.8980	1.1178	0.8616
0.74	0.7934	0.6594	1.2033	0.8773	1.1226	0.8907
0.76	0.8154	0.6741	1.2097	0.8542	1.1264	0.9185
0.78	0.8369	0.6892	1.2143	0.8285	1.1290	0.9448
0.80	0.8576	0.7048	1.2158	0.8000	1.1305	0.9695
0.82	0.8776	0.7211	*1.2171*	0.7684	*1.1306*	*0.9922*
0.84	0.8967	0.7380	1.2150	0.7332		
0.86	0.9149	0.7559	1.2104	0.6940		
0.88	0.9320	0.7748	1.2029	0.6499		
0.90	0.9480	0.7952	1.1921	0.6000		
0.92	0.9625	0.8174	1.1775	0.5426		
0.94	0.9755	0.8425	1.1579	0.4750		
0.96	0.9866	0.8718	1.1316	0.3919		
0.98	0.9952	0.9097	1.0941	0.2800		
1.00	*1.0000*	*1.0000*	1.0000	0.0000	1.0000	*1.0000*

Based on a literature review on *uniform pipe flow* including over 80 references in the past 120 years, Hager (1991) concluded:

- Manning-Strickler's equation relating to full-filling quantities is recommended,
- Full pipe filling is computed with the equation of Colebrook and White,
- Roughness coefficients $1/n$ smaller than 90 $m^{1/3}s^{-1}$ are only relevant,
- Filling diagrams are limited to 95%, and a full filling at zero pressure does not exist,
- Equation (5.14) is a simple and explicit approximation for relative discharge with an accuracy of about ±5%, and
- Transition from partial to full pipe filling is always abrupt, and reference is made to choking.

5.5.4 *Partial Filling of Non-circular Profiles*

Although there is currently a tendency to select circular sewers, large sewer systems or particular cases exist where other sewer profiles are used. Reasons for selecting non-circular sewers are:

- Improved flow features for small filling ratios and increased velocity to reduce sedimentation (Chap. 3), and
- Constructional simplifications for limited height or large sewers.

Egg-shaped sewers are higher than wide and respond to the first item, and they are advantageous in terms of static resistance. *Horseshoe sewers* need less height because of their compact shape. Instead, one could also design parallel circular sewers. *Rectangular sewers* are often selected for large dimensions, typically for widths larger than 1.50 m. They are also found on treatment stations. Besides all standard sewer shapes, other existing sewers are found, and their capacity and flow characteristics have also to be assessed. These profiles should be amenable for a simplified design. Table 5.4 summarizes the main sewer geometries and gives their main dimensions.

The *standardisation* of sewer profiles was introduced by Thormann (1944) toward the end of World War II by retaining 15 profile geometries. He also introduced a so-called reduction curve for large filling ratios that involved an empirical formula for uniform flow. The profiles include three egg shapes, two stretched circular shapes, four horseshoe shapes and three special shapes (Table 5.4). ATV (1988) kept all these profiles as non-standard shapes. The *standard sewer profiles* retained are:

- Circular pipe,
- Egg-shaped sewer 2:3, and
- Horseshoe sewer 2:1.5.

Table 5.4 Basic dimensions of standard (top) and non-standard sewer profiles according to Thormann (1944), or ATV (1988), with B largest width, T height of profile, see also Fig. 5.5

Type	Name	T/B	F_v/B^2	P_v/B	R_{hv}/B
A	Circle	2:2	0.785	3.141	0.250
B	Egg	3:2	1.149	3.965	0.290
C	Horseshoe	1.5:2	0.595	2.801	0.212
a	Elongated circle	2.5:2	1.036	3.642	0.285
b	Stretched circle	3:2	1.286	4.141	0.311
c	Stretched egg	3.5:2	1.373	4.425	0.311
d	Wide egg	2.5:2	0.956	3.515	0.272
e	Shortened egg	2:2	0.775	3.141	0.246
f	Stretched horseshoe	2:2	0.845	3.301	0.256
g	Shortened horseshoe	1.25:2	0.484	2.585	0.187
h	Elongated horseshoe	1.75:2	0.723	3.070	0.235
i	Upset horseshoe	1:2	0.402	2.461	0.163
k	Cap	2.5:2	1.097	3.819	0.288
l	Parabola	2:2	0.752	3.141	0.239
m	Kite	2:2	0.730	3.063	0.238

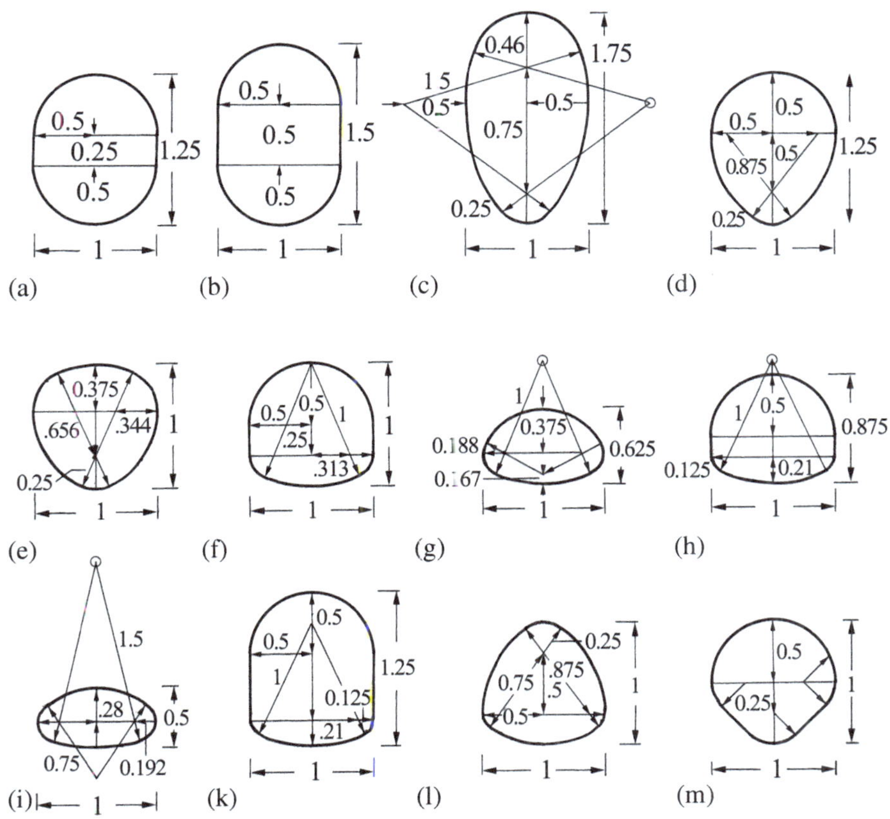

Fig. 5.5 Non-standard sewer profiles with unit width $B = 1$, and fractions for other lengths (ATV 1988)

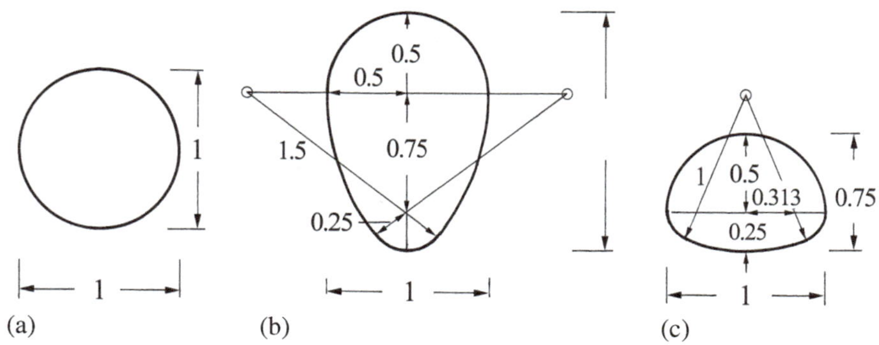

Fig. 5.6 Standard sewer profiles based on unit width $B = 1$. (**a**) Circular sewer, (**b**) egg-shaped sewer 2:3, (**c**) horseshoe sewer 2:1.5

Figure 5.5 shows the non-standard profiles of unit width B and all other dimensions as a fraction of the unit width. The height of the profile is T. Figure 5.6 shows the standard sewer profiles circle, egg and horseshoe, again for a unit width $B = 1$ and the other lengths as fractions. These profiles are subsequently discussed in terms of uniform flow, critical flow (Chap. 6) and sequent depths (Chap. 7). The complete geometries of the non-standard profiles are given by Thormann (1944).

Egg-shaped Sewer 2:3

The egg-shaped profile (German: Eiprofil, French: profil ovoïd) consists of three portions, namely a circular arc of radius $B/4$ up to $h/B = 0.25$, then of two circular arcs of radius $1.5B$ for $0.25 \leq h/B \leq 1$, and for $1 < h/B < 1.5$ of a half circle of radius $B/2$ (Fig. 5.6b). The main dimensions are given in Table 5.4.

Full sewer filling gives thus

$$Q_v = 0.503(1/n)S_o^{1/2}B^{8/3} = 0.171(1/n)S_o^{1/2}T^{8/3}. \tag{5.19}$$

Table 5.5 reproduces the profile geometry and the hydraulic characteristics for *partially filled egg-profiles*, based on the Manning-Strickler formula. The maximum discharge results for 95% filling, and maximum velocity for 86% filling. Full filling occurs also for 57% in terms of velocity and 80% in terms of discharge.

Uniform flow in egg-shaped sewers can be approximated better than ±3% as

$$q_v = Q/Q_v = 1.9\,y_N^2(1 - 0.42\,y_N^2), \quad y_N < 0.95. \tag{5.20}$$

This is similar to Eq. (5.14), with Q_v as given in Eq. (5.19). Equation (5.20) can also be explicitly solved for the filling ratio $y_N = h_N/T$ as

$$y_N = 1.09\left[1 - (1 - 0.884q_v)^{1/2}\right]^{1/2}. \tag{5.21}$$

Table 5.5 Partially filled egg-shaped sewer 2:3 as a function of vertex radius $r = B/2$ with $T = 3r$. Maximum values in *Italic*

h/r [–]	h/T [%]	F/r^2 [–]	P/r [–]	R_h/r [–]	V/V_v [%]	Q/Q_v [%]
3	*100*	*4.593*	*7.930*	0.579	100	100
2.95	98.33	4.573	7.296	0.627	105.5	105.0
2.9	96.67	4.536	7.029	0.645	107.5	106.2
2.85	95.0	4.488	6.822	0.658	108.9	*106.4*
2.8	93.33	4.430	6.643	0.667	109.9	106.0
2.75	91.67	4.367	6.485	0.673	110.5	105.1
2.7	90.0	4.299	6.339	0.678	111.1	104.0
2.65	88.33	4.225	6.204	0.681	111.4	102.5
2.6	86.67	4.147	6.075	0.683	111.6	100.8
2.55	85.0	4.066	5.955	*0.683*	*111.6*	98.8
2.5	83.33	3.980	5.836	0.682	111.5	96.6
2.4	80.0	3.802	5.612	0.677	111.0	91.9
2.3	76.67	3.614	5.398	0.669	110.1	86.6
2.2	73.33	3.422	5.191	0.659	109.0	81.2
2.1	70.0	3.223	4.989	0.646	107.6	75.5
2	66.67	3.023	4.788	0.631	105.9	69.7
1.9	63.33	2.823	4.588	0.615	104.1	64.0
1.8	60.0	2.624	4.388	0.598	102.2	58.4
1.7	56.67	2.426	4.187	0.579	100.0	52.8
1.6	53.33	2.231	3.985	0.560	97.8	47.5
1.5	50	2.037	3.784	0.538	95.2	42.2
1.4	46.67	1.847	3.580	0.516	92.6	37.2
1.3	43.33	1.662	3.375	0.492	89.7	32.5
1.2	40.0	1.481	3.169	0.468	86.8	28.0
1.1	36.67	1.306	2.960	0.441	83.4	23.7
1.0	33.33	1.136	2.749	0.413	79.8	19.7
0.9	30.0	0.974	2.536	0.384	76.1	16.1
0.8	26.67	0.820	2.319	0.354	72.0	12.9
0.7	23.33	0.675	2.099	0.322	67.6	9.9
0.6	20.0	0.538	1.875	0.287	62.6	7.3
0.5	16.67	0.414	1.647	0.251	57.3	5.2
0.45	15	0.366	1.531	0.233	54.5	4.3
0.4	13.33	0.300	1.413	0.212	51.2	3.3
0.35	11.67	0.248	1.294	0.192	47.9	2.6
0.3	10.0	0.199	1.174	0.170	44.2	1.9
0.25	8.33	0.154	1.051	0.147	40.1	1.3
0.2	6.67	0.112	0.927	0.121	35.2	0.86
0.15	5.0	0.074	0.795	0.093	29.5	0.48
0.1	3.33	0.041	0.644	0.064	23.0	0.21
0.05	1.67	0.015	0.451	0.033	14.8	0.05
0	0	0	0	0	0	0

Note that both Eqs. (5.20) and (5.21) apply only if $y_N < 0.95$, and that full-filling is determined from either Eq. (5.19) or based on Chap. 3.

The *cross-sectional area* related to the vertex radius $r = T/3$ can be expressed in analogy to Eq. (5.16) better than 2% as

$$F/r^2 = 6.25y^{3/2}\left[1 - 0.15y - 0.10y^4\right]. \tag{5.22}$$

For full filling, this gives 4.68 (+ 2.1%) instead of 4.593 (Table 5.5).

Example 5.4 Given an egg-shaped profile 120/180 with $S_o = 1.5\%$ and $1/n = 80\ \mathrm{m^{1/3}s^{-1}}$. Compute the uniform flow depth for a discharge of $Q = 1.2\ \mathrm{m^3s^{-1}}$. With the full filling discharge $Q_v = 0.503 \cdot 80 \cdot 0.015^{1/2} 1.2^{8/3} = 8.01\ \mathrm{m^3s^{-1}}$ from Eq. (5.19) one has $q_v = 1.2/8.01 = 0.15$, such that the filling ratio is $y_N = 1.09[1 - (1 - 0.884 \cdot 0.15)^{1/2}]^{1/2} = 0.285$, i.e. $h_N = y_N T = 0.285 \cdot 1.8 = 0.51$ m.

The sectional area $F_N/r^2 = 6.25 \cdot 0.285^{1.5}(1 - 0.15 \cdot 0.285 - 0.1 \cdot 0.285^4) = 0.91$ follows from Eq. (5.22), thus $F_N = 0.91 \cdot 0.6^2 = 0.327\ \mathrm{m^2}$ with $r = B/2 = 0.60$ m. The uniform flow velocity is $V_N = Q/F_N = 1.2/0.327 = 3.66\ \mathrm{ms^{-1}}$, and uniform energy head $H_N = h_N + V_N^2/(2g) = 0.51 + 3.66^2/19.62 = 1.19$ m.

Horseshoe-profile 2:1.5

Computational full filling of the horseshoe-profile (German: Maulprofil, French: profil aplati) of width $B = (4/3)T$ is

$$Q_v = 0.212(1/n)S_o^{1/2}B^{8/3} = 0.457(1/n)S_o^{1/2}T^{8/3}. \tag{5.23}$$

Table 5.6 relates to the filling characteristics according to ATV (1988).

Table 5.6 Partially filled horseshoe-profile 2:1.5 as a function of vertex radius $r = B/2$ with $T = 1.5r$. Maximum values in *italic*

h/D	F/F_v	P/P_v	R_h/R_{hv}	B/D	V/V_v	Q/Q_v
0.02	0.0058	0.1238	0.0470	0.3451	0.1479	0.0009
0.04	0.0164	0.1753	0.0936	0.4862	0.2275	0.0037
0.06	0.0301	0.2150	0.1399	0.5932	0.2925	0.0088
0.08	0.0462	0.2486	0.1859	0.6823	0.3493	0.0161
0.10	0.0644	0.2783	0.2315	0.7599	0.4007	0.0258
0.12	0.0844	0.3024	0.2791	0.8205	0.4504	0.0380
0.14	0.1057	0.3209	0.3293	0.8627	0.4994	0.0528
0.16	0.1279	0.3367	0.3797	0.8952	0.5460	0.0698
0.18	0.1508	0.3509	0.4297	0.9212	0.5899	0.0889
0.20	0.1743	0.3640	0.4789	0.9422	0.6312	0.1100

Table 5.6 (continued)

h/D	F/F_v	P/P_v	R_h/R_{hv}	B/D	V/V_v	Q/Q_v
0.22	0.1983	0.3763	0.5270	0.9593	0.6701	0.1329
0.24	0.2227	0.3881	0.5739	0.9729	0.7067	0.1574
0.26	0.2474	0.3994	0.6193	0.9835	0.7412	0.1834
0.28	0.2723	0.4105	0.6633	0.9914	0.7737	0.2107
0.30	0.2974	0.4214	0.7058	0.9967	0.8043	0.2392
0.32	0.3226	0.4312	0.7465	0.9995	0.8330	0.2687
0.34	0.3478	0.4428	0.7854	*0.9999*	0.8599	0.2991
0.36	0.3730	0.4536	0.8225	0.9992	0.8850	0.3301
0.38	0.3982	0.4643	0.8577	0.9975	0.9035	0.3618
0.40	0.4234	0.4750	0.8912	0.9950	0.9306	0.3940
0.42	0.4484	0.4858	0.9231	0.9915	0.9512	0.4365
0.44	0.4734	0.4966	0.9532	0.9871	0.9705	0.4594
0.46	0.4982	0.5075	0.9817	0.9818	0.9885	0.4925
0.48	0.5229	0.5185	1.0086	0.9755	1.0054	0.5258
0.50	0.5475	0.5295	1.0340	0.9682	1.0211	0.5590
0.52	0.5718	0.5406	1.0577	0.9600	1.0357	0.5922
0.54	0.5959	0.5518	1.0799	0.9507	1.0492	0.6252
0.56	0.6198	0.5631	1.1006	0.9404	1.0617	0.6580
0.58	0.6434	0.5746	1.1197	0.9229	1.0732	0.6905
0.60	0.6666	0.5862	1.1373	0.9165	1.0837	0.7224
0.62	0.6896	0.5979	1.1533	0.9028	1.0932	0.7539
0.64	0.7122	0.6099	1.1677	0.8879	1.1018	0.7847
0.66	0.7344	0.6221	1.1806	0.8717	1.1093	0.8147
0.68	0.7562	0.6345	1.1918	0.8542	1.1159	0.8438
0.70	0.7775	0.6472	1.2014	0.8352	1.1215	0.8720
0.72	0.7983	0.6601	1.2093	0.8146	1.1261	0.8990
0.74	0.8186	0.6735	1.2155	0.7924	1.1297	0.9248
0.76	0.8383	0.6872	1.2199	0.7684	1.1323	0.9492
0.78	0.8573	0.7014	1.2224	0.7424	1.1337	0.9720
0.80	0.8757	0.7161	*1.2230*	0.7141	*1.1341*	*0.9931*
0.82	0.8934	0.7314	1.2215	0.6834		
0.84	0.9102	0.7475	1.2177	0.6499		
0.86	0.9261	0.7644	1.2116	0.6131		
0.88	0.9411	0.7825	1.2027	0.5724		
0.90	0.9550	0.8020	1.1908	0.5268		
0.92	0.9676	0.8233	1.1753	0.4750		
0.94	0.9789	0.8474	1.1552	0.4146		
0.96	0.9885	0.8757	1.1287	0.3412		
0.98	0.9959	0.9123	1.0916	0.2431		
1.00	*1.0000*	*1.0000*	1.0000	0.0000	1.0000	*1.0000*

Uniform discharge as a function of filling ratio $y_N = h_N/T$ in analogy to Eq. (5.14) is

$$q_v = Q/Q_v = 2.8\, y_N^2(1 - 0.71\, y_N^2), \quad y_N < 0.80. \tag{5.24}$$

If the domain up to $y_N = 0.93$ should be reproduced with a 3% accuracy, the expression is more complicated

$$q_v = 2.8\, y_N^2(1 - 0.8\, y_N^2 + 0.25\, y_N^6), \quad y_N < 0.93. \tag{5.25}$$

Both Eqs. (5.24) and (5.25) are not very exact for high filling ratios.

The *filling ratio* $y_N = h_N/T$ as a function of relative discharge $q_v = Q/Q_v$ is better than 5% approximated with the explicit relation

$$y_N = 0.85[1 - (1 - q_v)^{1/2}]^{1/2}. \tag{5.26}$$

The *cross-sectional* area based on the full-filling area $F_v = 0.595B^2 = 1.058T^2$ is better than 5% with

$$F/F_v = 2y^{3/2}[1 - 0.6y^{3/2} + 0.1y^3]. \tag{5.27}$$

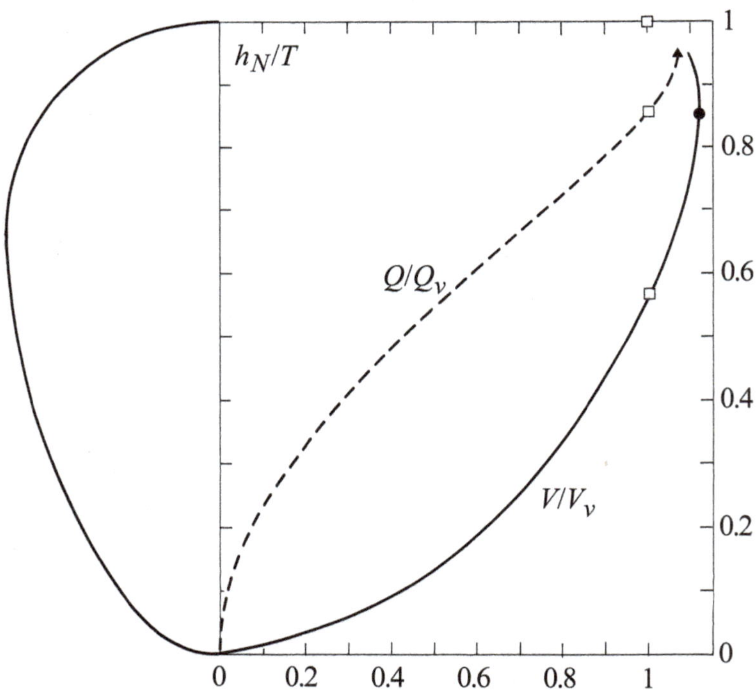

Fig. 5.7 Partial filling diagram for *egg-shaped profile* 2:3 with (□) full filling, (▲) discharge maximum, and (•) velocity maximum. Relative (—) velocity and (- -) discharge as functions of $y_N = h_N/T$

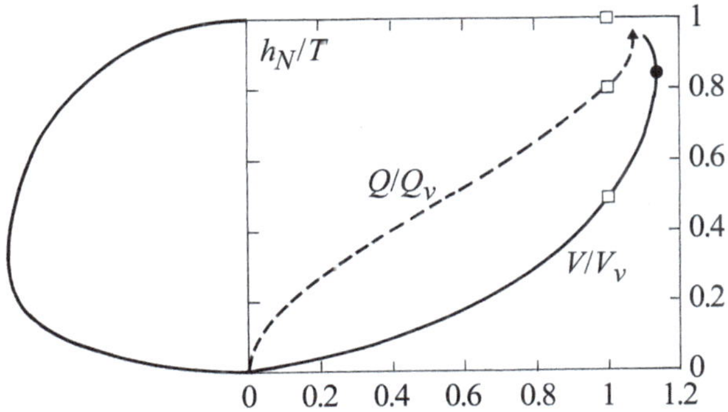

Fig. 5.8 Partially filled *horseshoe-profile* 2:1.5 with (□) full filling, (▲) maximum discharge and (•) maximum velocity. Relative values of (—) velocity and (- - -) discharge as a function of filling ratio $y_N = h_N/T$

Example 5.5 The discharge in a horseshoe profile 150/200 is $Q = 2.2\ \mathrm{m^3s^{-1}}$ for a bottom slope of $S_o = 2\%$ and a roughness coefficient of $1/n = 75\ \mathrm{m^{1/3}s^{-1}}$. What is the uniform flow depth?
With $B = 2.00$ m, the full filling discharge is $Q_v = 0.212 \cdot 75 \cdot 0.02^{1/2} 2^{8/3} = 14.3\ \mathrm{m^3s^{-1}}$ from Eq. (5.23), and $q_v = 2.2/14.3 = 0.154$. From Eq. (5.26) $y_N = 0.85[1-(1-0.154)^{1/2}]^{1/2} = 0.241$, thus $h_N = 0.241 \cdot 1.5 = 0.36$ m.

With $y_N = 0.241$, the relative cross-section of uniform flow is $F_N/F_v = 2 \cdot 0.241^{3/2}(1-0.6 \cdot 0.241^{3/2} + 0.1 \cdot 0.241^3) = 0.22$ from Eq. (5.27), thus $F_N = 0.22 \cdot 0.595 \cdot 2^2 = 0.524\ \mathrm{m^2}$. The uniform velocity is then $V_N = Q/F_N = 2.2/0.542 = 4.2\ \mathrm{ms^{-1}}$, and the uniform energy head obtains finally $H_N = h_N + V_N^2/(2g) = 0.36 + 4.2^2/(19.62) = 1.26$ m.

Summary of Results

The uniform flow filling $y_N = h_N/T$ can be expressed exclusively as a function of relative discharge $q_N = nQ/(S_o^{1/2}T^{8/3})$ with T as the *profile height*. Table 5.7 gives a summary of previous results for the standard sewer profiles for both filling ratio y_N and cross-sectional area F/T^2.

Table 5.7 Uniform filling ratio $y_N = h_N/T$ and cross-sectional area F/T^2 for standard sewer profiles

Profile (1)	Uniform filling ratio (2)	Cross-sectional area (3)
Circle	$y_N = 0.926[1-(1-3.11q_N)^{1/2}]^{1/2}$	$F/D^2 = 1.333y^{3/2}\left[1-0.25y-0.16y^2\right]$
Egg	$y_N = 1.090[1-(1-5.18q_N)^{1/2}]^{1/2}$	$F/T^2 = 0.695y^{3/2}[1-0.15y-0.10y^4]$
Horseshoe	$y_N = 0.850[1-(1-2.19q_N)^{1/2}]^{1/2}$	$F/T^2 = 2.116y^{3/2}[1-0.6y^{3/2}+0.1y^3]$

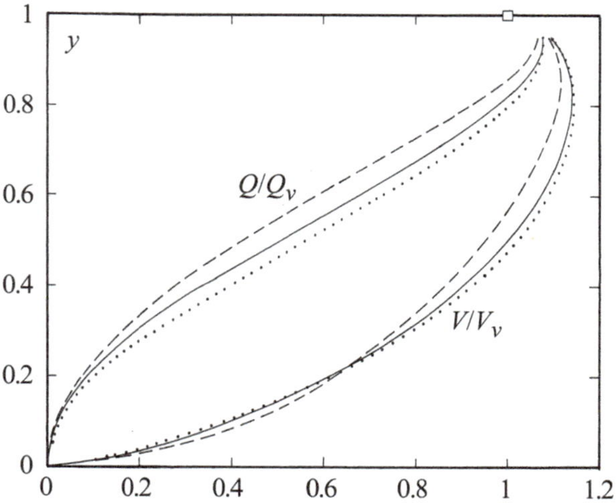

Fig. 5.9 Comparison between filling curves for discharge Q/Q_v and velocity V/V_v in terms of y for (—) circular, (- -) egg-shaped and (. . .) horseshoe profiles based on uniform flow equation by Manning-Strickler. (□) full-filling condition

Figure 5.9 compares the filling curves for the three standard sewer profiles circle, egg and horseshoe. Note the relatively good agreement between the three curves for discharge and velocity, respectively, despite the geometrical differences between the sections considered. As a simplification, one may thus approximate the *non-standard sewer profiles* with one of the standard section, depending on the similarity of the profiles. Typical errors are below ±10%, i.e. acceptable in terms of preliminary design. Compared to errors in the estimation of the roughness coefficient n, such deviations are justifiable provided the capacity of a non-standard sewer has to be estimated.

The agreement of filling curves for three such widely varying profile geometries is astonishing, and deviations of about 10% are relatively small. In pre-projects, this fact is currently not often applied, to simplify the design of a sewer system.

5.5.5 *Uniform Energy Head*

The energy head (German: Energiehöhe; French: Charge) is a significant quantity in open channel flows (Chap. 1). In the following, the energy head H_N for uniform flow in the three standard sewer profiles is considered. It is defined as

$$H_N = h_N + \frac{Q^2}{2gF_N^2}. \tag{5.28}$$

Relating to the profile height T, Eq. (5.28) may also be expressed as

$$Y_N = \frac{H_N}{T} = \frac{h_N}{T} + \frac{Q^2}{2gT^5}\frac{1}{(F_N/T^2)^2}. \tag{5.29}$$

To compute the relative energy head Y_N, the uniform discharge $Q(y_N)$ and the cross-sectional area F_N/T^2 must be known as a function of filling ratio $y_N = h_N/T$. Then, the function $Y_N[Q/(gT^5)^{1/2}]$ may be derived, to directly determine Y_N for given relative discharge for a specified profile.

Circular Profile

Based on Eqs. (5.15) and (5.16), and with $T = D$ and $q_N = nQ/(S_o^{1/2}D^{8/3})$, the filling ratio for small values of q_N is

$$y_N = \frac{2}{\sqrt{3}}q_N^{1/2}. \tag{5.30}$$

Eliminating y_N in Eq. (5.29) and introducing the so-called *roughness characteristic* $\chi = S_o^{1/2}D^{1/6}/(ng^{1/2})$ gives

$$Y_N = \frac{2}{\sqrt{3}}q_N^{1/2}\left[1 + \frac{1}{2}\left(\frac{9}{16}\chi\right)^2\right]. \tag{5.31}$$

For larger q_N this may be modified to

$$Y_N = \frac{2}{\sqrt{3}}q_N^{3/5}\left[1 + \frac{1}{6}\chi^2\right]. \tag{5.32}$$

The relative energy head thus depends essentially on the relative discharge q_N and the roughness characteristic.

Egg-shaped Profile

In analogy to the circular profile one has with $q_N = nQ/\left(S_o^{1/2}B^{8/3}\right)$ for small filling ratios $y_N = h_N/T$

$$y_N = 1.022q_N^{1/2}. \tag{5.33}$$

The expression analogous to Eq. (5.31) is for small filling ratios

$$Y_N = 1.022q_N^{1/2}\left[1 + \frac{1}{8}\chi^2\right] \tag{5.34}$$

and a generalized relation of $\pm 5\%$ accuracy is

$$Y_N = 1.20q_N^{0.55}\left[1 + \frac{1}{8}\chi^2\right]. \tag{5.35}$$

Equation (5.35) is valid for $\chi < 2$.

Horseshoe Profile

In analogy to Eqs. (5.30) and (5.33), and with $q_N = nQ/\left(S_o^{1/2}B^{8/3}\right)$, the uniform filling ratio for small $y_N = h_N/T$ is

$$y_N = 1.305q_N^{1/2} \tag{5.36}$$

and thus for $y_N = H_N/T$

$$Y_N = 1.305q_N^{1/2}\left[1 + \frac{1}{6}\chi^2\right]. \tag{5.37}$$

For larger values of q_N, the expansion corresponding to Eqs. (5.32) and (5.35) is to 10% accuracy

$$Y_N = 2.0q_N^{0.60}\left[1 + \frac{1}{6}\chi^2\right]. \tag{5.38}$$

This expression should also not be applied for both large y_N and χ.

Example 5.6 Determine directly the energy head of Example 5.3.
With $q_N = 0.222$ and $\chi = 5 \cdot 0.004^{1/2}0.7^{1/6}/9.81^{1/2} = 1.62$, one has with Eq. (5.32) $Y_N = 1.5 \cdot 0.222^{0.6}[1 + 0.17 \cdot 1.62^2] = 0.879$, corresponding to $H_N = 0.879 \cdot 0.7 = 0.615$ m (+ 2%).

Example 5.7 Determine directly the energy head of Example 5.4.
With $q_v = 0.150$, thus $q_N = 0.075$ and $\chi = 80 \cdot 0.015^{1/2}1.2^{1/6}/9.81^{1/2} = 3.22$ one has $Y_N = 1.2 \cdot 0.075^{0.55}[1 + 0.125 \cdot 3.22^2] = 0.664$ from Eq. (5.35), corresponding to $H_N = 0.664 \cdot 1.8 = 1.20$ m (+1%).

Example 5.8 Determine directly the energy head of Example 5.5.
With $q_v = 0.154$, thus $q_N = 0.033$ and $\chi = 75 \cdot 0.02^{1/2}2^{1/6}/9.81^{1/2} = 3.80$, one has $Y_N = 2 \cdot 0.033^{0.6}[1 + 0.17 \cdot 3.80^2] = 0.875$ from Eq. (5.38), and thus $H_N = 0.875 \cdot 1.5 = 1.31$ m (+ 4%).

5.6 Steeply Sloping Sewer

5.6.1 *Self-Aerated Flow*

Flows in steeply-sloping sewers generate *self aeration* (German: Selbstbelüftung, French: Aération). The phenomena of so-called white water are attributed to flow turbulence that is able to eject fluid particles once incipient aeration is reached. According to Volkart (1980) water drops falling back on the flow are able to entrain air bubbles, and an *air-water mixture flow* (German: Gemischabfluss, French: Melange eau-air) is developed.

Currently, the knowledge on incipient aeration, the local change of air concentration and the air bubble distribution in spillway flows are known (ICOLD 1992, Hager and Schleiss 2009). In closed conduit flow, this information relates exclusively to uniform-aerated flow. The main knowledge is reviewed below and compared with results relating to the rectangular open channel with an unlimited air supply.

5.6.2 *Incipient Aeration*

Figure 5.10 shows supercritical flow in a rectangular channel of constant bottom slope S_o. From the upstream control section at $x = 0$ a *boundary layer* (German: Grenzschicht; French: Couche limite) of thickness δ_g develops and intersects the free surface profile at location $x = x_i$ of incipient (subscript i) aeration. Then, the turbulent flow has reached the free surface and defines the point of incipient aeration (German: Belüftungsanfang; French: Début d'aeration). The velocity head $H_s = V_i^2/(2g)$ relative to the free surface elevation and the equivalent sand roughness height k_s determine the boundary layer thickness $\delta_g(x)$ as (ICOLD 1992)

$$\delta_g/x = 0.021\ (k_s/H_s)^{0.10}. \tag{5.39}$$

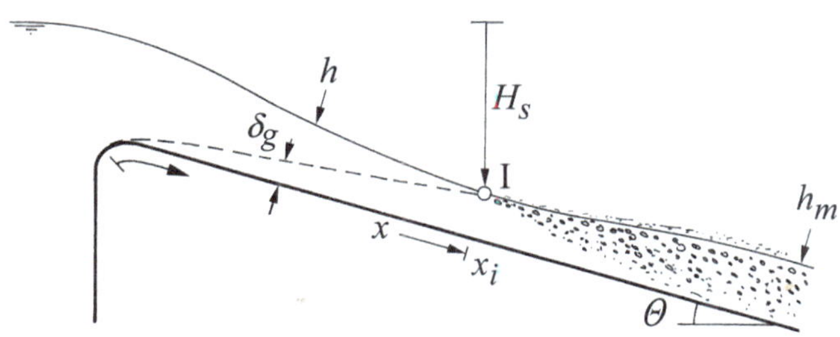

Fig. 5.10 Incipient aeration on steeply sloping channel

Below the boundary layer the flow is fully turbulent and a potential flow is developed above the boundary layer. Water particles sufficiently turbulent may thus be ejected downstream from the air inception point I. The location x_i where $\delta_g = h$ may be determined with a drawdown curve, and the boundary layer development according to Eq. (5.39). A generalized analysis for rectangular channels was provided by Hager and Blaser (1998).

The location of incipient aeration in sewers has not yet been studied. It is recommended to determine x_i with the drawdown curves described in Chap. 8 and to use Eq. (5.39) for the curve $\delta_g(x)$, with H_s as the difference elevation between the critical energy head at the upstream control section and the velocity head at the location of incipient aeration. The approach is iterative because location x_i is initially unknown.

5.6.3 *Uniform-aerated Flow*

The cross-sectional average *air concentration* $\bar{C}$ of an air-water flow depends on the Boussinesq number $\mathsf{B} = V/(gR_h)^{1/2}$, in which the reference velocity involves the hydraulic radius R_h instead of the hydraulic depth F/B_s for the Froude number (Chap. 6), with B_s as surface width. For uniform-aerated flow Volkart (1978) found

$$\bar{C} = 1 - \left[1 + 0.02(\mathsf{B} - 6)^{1.5}\right]^{-1}. \tag{5.40}$$

Incipient aeration in circular sewers is thus defined by $\mathsf{B} = 6$, corresponding to a velocity V larger than $6(gR_h)^{1/2}$. This condition may be expressed explicitly with the roughness coefficient $1/n$ according to Manning-Strickler, bottom slope S_o, sewer diameter D and gravitational acceleration g as (Hager 1985)

$$\chi_i = (1/n)S_o^{1/2}D^{1/6}g^{-1/2} = 8. \tag{5.41}$$

The *roughness characteristic* χ_i for incipient aeration to occur involves a main effect of the roughness coefficient $(1/n)$, a medium effect of bottom slope and only a minor effect of sewer diameter.

Eliminating n with Eq. (5.5) gives $S_{oi} = 0.93(k_s/D)^{1/3}$. For a typical equivalent roughness height $k_s = 0.5$ mm and for $D = 1$ m, the bottom slope for incipient aeration to occur is $S_{oi} = 0.07$, and the corresponding values for minimum (subscript *min*) and maximum (subscript *M*) diameters $D_{min} = 0.30$ m and $D_M = 2$ m are 0.11 and 0.06, respectively. It can thus be stated that whenever the bottom slope is larger than about 5 to 10%, self aeration is likely to occur. It should be noted that Eq. (5.41) is independent of discharge and flow depth.

Mixture (subscript *m*) flow has a total discharge Q_m that is composed of air discharge Q_a and water discharge Q_w. A mixture flow thus needs more area than

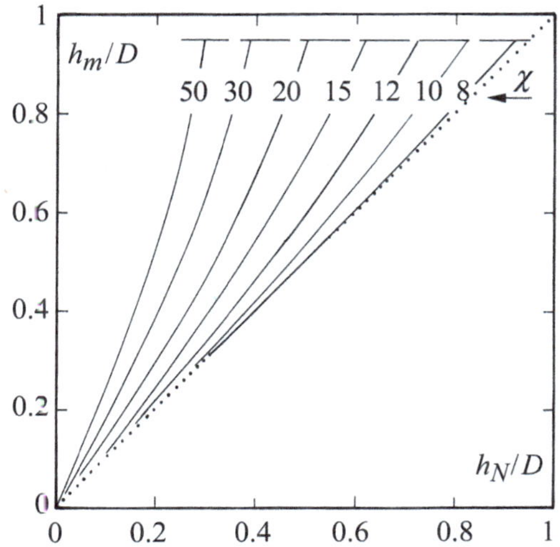

Fig. 5.11 *Uniform-aerated flow* in circular sewer. Ratio between mixture filling h_m/D and pure water filling h_N/D versus roughness characteristics $\chi = S_o^{1/2}D^{1/6}/\left(ng^{1/2}\right)$. (- - -) Maximum sewer filling, (. . .) limit of self-aeration

the corresponding water discharge alone. The *mixture flow depth* h_m is thus always larger than the corresponding pure water depth h. Figure 5.11 shows the mixture filling ratio h_m/D as a function of pure water filling h_N/D for various roughness characteristics $\chi > 8$. The plot is based on Eq. (5.40) for uniform mixture flow, thereby excluding high filling ratios $y_N > 0.95$. From Sect. 5.5.2 a mixture flow chokes prior to $y_N = 0.95$, depending mainly on the bottom slope.

The relation for the mixture flow depth as plotted in Fig. 5.11 can be approximated for $h_m > h_N$ as

$$h_m/D = (1/4)\chi^{2/3}\,(h_N/D)^{10/9}\,. \tag{5.42a}$$

Inserting the roughness characteristics $\chi = S_o^{1/2}D^{1/6}/\left(ng^{1/2}\right)$ gives simply

$$\frac{h_m}{h_N} = \frac{1}{4}\left(\frac{S_o h_N^{1/3}}{n^2 g}\right)^{1/3}\,. \tag{5.42b}$$

The *mixture ratio* thus depends significantly on the roughness coefficient, somewhat on the bottom slope and is almost independent of the uniform water depth.

Example 5.9 Given a discharge of $Q = 1.7$ m^{3}s^{-1}, a diameter $D = 0.9$ m, the roughness coefficient $1/n = 80$ m$^{1/3}$s^{-1} and a bottom slope of $S_o = 40\%$. What is the uniform mixture flow depth h_m?

As $\chi = 80 \cdot 0.4^{1/2} 0.9^{1/6}/9.81^{1/2} = 15.87 > 8$, the uniform flow gets aerated. With $q_N = 1.7/(80 \cdot 0.4^{1/2} 0.9^{8/3}) = 0.044$ the uniform filling ratio is $y_N = 0.248$ from Eq. (5.15).

With $(y_N; \chi) = (0.248; 15.87)$, Fig. 5.11 gives $y_m = 0.34$, and thus $h_m = 0.34 \cdot 0.9 = 0.305$ m as the uniform mixture depth, which is by 37% larger than h_N. From Eq. (5.42a) one therefore may compute $h_m/D = 0.25 \cdot 15.87^{2/3} 0.248^{1.11} = 0.335$ (– 3%).

5.6.4 *Design Procedure*

The design of *aerated sewer flows* can thus easily be integrated in the conventional hydraulic approach. First pure water flow is considered with h_N as the corresponding uniform flow depth, either determined with the Colebrook-White or the Manning-Strickler approach. Then, the parameter χ is determined to check the criteria whether or not the flow is self-aerated. For a roughness characteristics $\chi > 8$, the uniform mixture flow depth h_m is computed, either from Fig. 5.11 or with Eq. (5.42). For all subsequent computations, such as drawdown or backwater curves (Chap. 8), the pure water depth h_N is replaced by the mixture flow depth h_m. The procedure is thus simple because flow dynamics and flow aeration can be split.

Further information on aerated spillway flow that can be transmitted approximately to flow in steep sewers involve the (ICOLD 1992):

- Velocity distribution,
- Distribution of air concentration in a section,
- Effect of air concentration on roughness coefficient, and
- Bottom air concentration to determine the potential of cavitation damage.

In addition backwater and drawdown curves for supercritical spillway flow are discussed with a simplified approach. This information is not yet available for sewers where the air supply is not infinite, as for a chute spillway, and where air currents may modify the mixture flow.

Mixture flows in sewers are currently not fully understood. Questions relating to shockwaves due to curves, junctions or drops have not been investigated from this point of view. Also, dangerous phenomena such as *air pockets* and *flow pulsations* cannot yet be predicted. Based on these current deficiencies, steeply sloping sewers have to be designed liberally. The sewer should be straight in plan view and be sufficiently aerated to inhibit zones of low pressure that may generate flow choking. To obtain clear hydraulic conditions, both the air transport and the mixture flow should be studied in detail. Filling ratios of above 60% or so have to be checked against *choking*.

5.7 Air-Water Flows

5.7.1 *Introduction*

Gas-liquid flows (German: Gas-Flüssigkeitsströmungen; French: Ecoulements à gas-liquide) have received much attention mainly in industrial hydraulics. Such flows typically occur with siphons, heat exchangers, cooling-towers and nuclear engineering. The following parameters are significant in practical applications: Fluid fraction in suspension, void fraction, pressure loss and percentage liquid entrainment.

Hewitt and Hall-Taylor (1970) distinguished six regimes for *horizontal* gas-liquid flows (Fig. 5.12):

① *Stratified flow* (German: Schichtströmung; French: Ecoulement stratifié), where the liquid phase flows below the gas phase, with an almost straight interface,
② *Wave flow* (German: Wellenströmung; French: Ecoulement ondulé), with an undular interface between the top gas and bottom liquid phases,
③ *Slug flow* (German: Pulsationsströmung; French: Ecoulement en bouchon), with a wavy surface that reaches to the pipe vertex and thus separates the gas phase in singular cells,
④ *Plug flow* (German: Blasenströmung; French: Ecoulements en bulles), with gas bubbles and pockets that are distributed mainly over the top portion of the pipe,
⑤ *Bubbly flow* (German: Tropfenströmung; French: Ecoulement en gouttes), with an almost uniform distribution of gas bubbles in the liquid phase, and

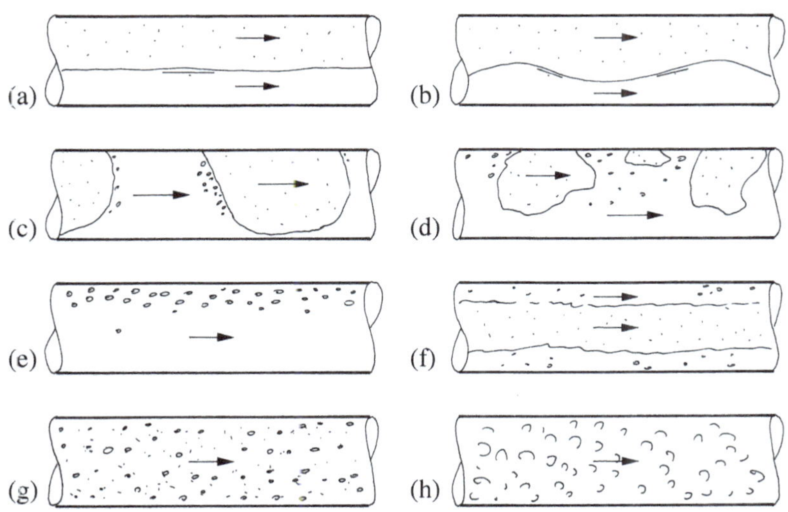

Fig. 5.12 Regimes for gas-liquid flows in a nearly horizontal pipe, details see main text

⑥ *Annular flow* (German: Ringströmung; French: Ecoulement annulaire), with a large gas portion that pushes the liquid phase to the wall region, and a fluid ring developing around the gas phase.

In addition, there are:

⑦ *Spray flow* (German: Sprayströmung; French: Ecoulement en spray), with an almost uniform mixture of both phases, and

⑧ *Froth flow* (German: Schaumströmung; French: Ecoulement en mousse), with a froth flow structure.

The various regimes may be identified in a *flow diagram* containing discharges of air V_a and water V_w in [kg/m^{2}s]. Figure 5.13a shows four main lines, namely the transitions from ① stratified to ④ plug flows, and from ② wavy to ③ slug flows on the one hand, and from ④ plug and ③ slug flows to ⑦ spray (homogeneous flow) on the other hand. For large air velocity V_a one has always ⑥ annular flow.

Regimes ② and ⑥ are also summarized as *separated flows* (German: Abgelöste Strömung; French: Ecoulement séparé) and regimes ③ and ④ as *intermittent flows* (German: Aussetzende Strömung; French: Ecoulement intermittent). An alternative classification to Fig. 5.13 containing modern data presented Weismann (1983). Flow diagrams for vertical pipes and counter-current flows for two phases are also available.

For *co-current flows* with both phases in the same flow direction, five hydrodynamical instabilities may occur, namely the instability of:

- *Kelvin-Helmholtz* due to the interaction of both phases propagating with different velocities. The instability is described with the densimetric Froude number that accounts for buoyancy and inertia,

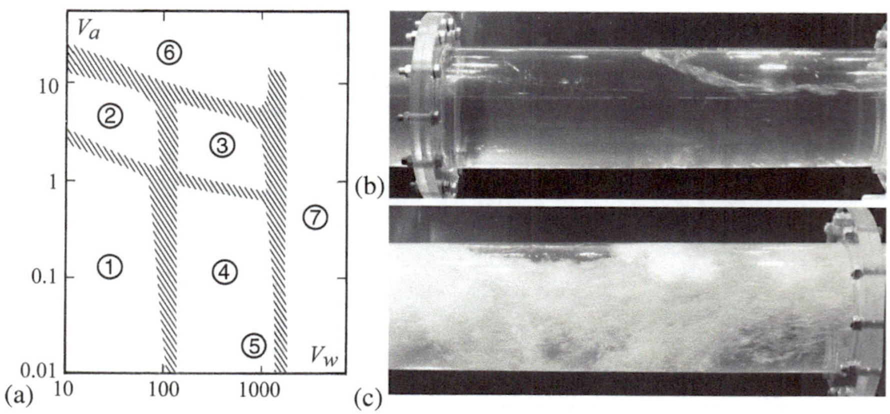

Fig. 5.13 (**a**) Flow diagram for air-water flow in horizontal pipe with $D = 50$ mm. Air discharge V_a and water discharge V_w in [kg/m^{2}s], notation in text. (**b**) Slug flow, (**c**) Annular flow

- *Tollmien-Schlichting* as the transition between laminar and turbulent flows that is described by the Reynolds number,
- *Rayleigh-Taylor* due to density differences of the two phases,
- *Rayleigh-Bénard* due to a phase differences in temperature or concentration, and
- *Marangoni* due to surface tension effects.

This list of possible instabilities of two-phase flows demonstrates the complexity of physical interactions with these flow phenomena. In the following, some important relations for *air-water flows* are presented.

5.7.2 Empirical Correlations

The *head loss* of two-phase flows can be modelled based on the experiments of Lockhart and Martinelli (1949). Two parameters are introduced

$$\Phi_g = \left(\frac{\mathrm{d}p/\mathrm{d}x)_{gw}}{\mathrm{d}p/\mathrm{d}x)_g}\right)^{1/2}, \quad \Phi_w = \left(\frac{\mathrm{d}p/\mathrm{d}x)_{gw}}{\mathrm{d}p/\mathrm{d}x)_w}\right)^{1/2}, \text{ respectively, and,} \tag{5.43}$$

$$X = \left(\frac{\mathrm{d}p/\mathrm{d}x)_w}{\mathrm{d}p/\mathrm{d}x)_g}\right)^{1/2}. \tag{5.44}$$

Here, $(\mathrm{d}p/\mathrm{d}x)_g$ and $(\mathrm{d}p/\mathrm{d}x)_w$ relate to pressure gradients of the gas phase (subscript g) and the water phase (subscript w) *alone*, i.e. considered as a one-phase flow, and $(\mathrm{d}p/\mathrm{d}x)_{gw}$ refers to the *gas-water mixture* (subscript gw). Although current observations have resulted in modifications of the classical Lockhart-Martinelli data, their approach is generally applied with the symmetric expressions

$$\Phi_g = 1 + 4X^{+0.7}, \tag{5.45}$$

$$\Phi_w = 1 + 4X^{-0.7}. \tag{5.46}$$

Because both Φ_g and Φ_w are always larger than 1, the *mixture pressure loss* is always larger than the loss of the corresponding one-phase flow. The mixture loss increases with the phase content.

The so-called *fluid holdup* X^{-1} varies with the *void fraction* R_a as

$$\frac{1}{X} = 1 - \frac{\rho_c}{\rho_w}\left(1 - \frac{K_h}{R_a}\right), \tag{5.47}$$

$$K_h = \mathrm{Tanh}[0.2(\mathsf{Z} - 0.5)]. \tag{5.48}$$

Here, $\mathsf{Z} = \mathsf{R}^{1/6}\mathsf{F}^{2/3}\mathsf{Ri}^{-1/4}$ with $\mathsf{R} = VD/\nu$ as Reynolds number, $\mathsf{F} = V/(gD)^{1/2}$ as pipe Froude number and Ri as Richardson number.

5.7.3 *Slug Flow*

Slug flow is significant in nearly horizontal sewers because of *flow choking*. This two-phase flow can also be described as a fluid flow with elongated gas pockets close to the pipe vertex, of which the wave crests propagate much faster than the fluid. The pressure variations between air and fluid flows can be so large as to result in inacceptable conditions for static resistance.

Figure 5.14 shows the *development of a slug* as the transition between stratified and slug flows. Such phenomena often occur close to the inlet reach of a pipe that are subjected with a stratified flow. Figure 5.14a relates to a wave with a low wave tail and a subsequent wave crest. A small surface perturbation moves to the wave crest (Fig. 5.14b) and wave steepening may result in pipe choking that cuts the air transport (Fig. 5.14c). The fluid phase is thus accelerated by the gas phase to result in a surface depression behind the wave that is below the approach fluid level (Fig. 5.14d). Taitel and Dukler (1977) introduced a numerical model to predict this process. The flow depths in front and upstream of the wave crest are of particular relevance.

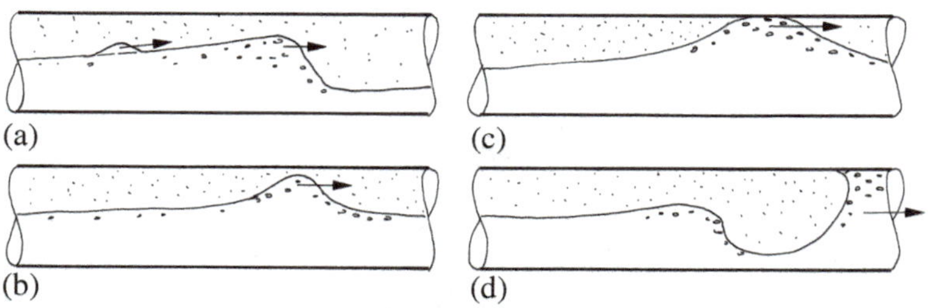

Fig. 5.14 Mechanism of slug flow development according to Taitel and Dukler (1977), details in text

5.7.4 *Wave Instability*

The theory of two-dimensional wave instability along the interface of a two-phase flow containing air (subscript a) and water (subscript w) according to *Kelvin-Helmholtz* refers to the densimetric Froude number (Fig. 5.15)

$$\mathsf{F}_\Delta = \frac{V_a - V_w}{\left[g(\rho_w - \rho_a)\left(\frac{h_a}{\rho_a} + \frac{h_w}{\rho_w}\right)\right]^{1/2}}. \tag{5.49}$$

Here V is the average phase velocity, h the average phase depth and ρ the phase density. A small wave becomes unstable provided $\mathsf{F}_\Delta \geq 1$ and the wave amplitude increases without limits. The experiments of Wallis and Dobson (1973) demonstrate that this instability occurs for $\mathsf{F}_\Delta = 0.5$ already. Physically, this instability may be

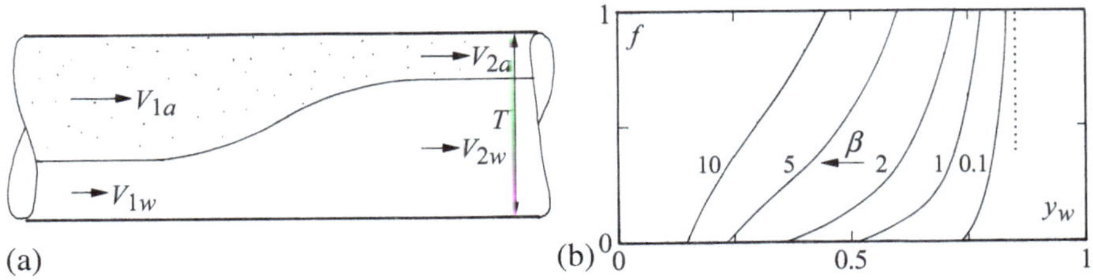

Fig. 5.15 (**a**) Wave instability of a wavy two-phase flow (adopted from Gardner 1979), (**b**) Densimetric Froude number ratio $f_d = \mathsf{F}_\Delta/\mathsf{F}_d$ as a function of liquid filling ratio $y_w = h_w/D$ for various phase portions $\beta = Q_a/Q_w$

explained as the dominant *suction effect* of the gas phase to the pipe vertex compared with the gravity effect of the fluid phase.

In an arbitrary profile the *densimetric Froude* number is (Kubie 1979)

$$\mathsf{F}_\Delta = \frac{V_a - V_w}{\left[g(\rho_w - \rho_a)\left(\frac{F_w}{\rho_w B_i} + \frac{F_a}{\rho_a B_i}\right)\right]^{1/2}} \tag{5.50}$$

where F is the phase cross-sectional area, and B_i the interface width. In circular channels, the geometry is from Eq. (5.16) $F/D^2 = y^{1.4}$ and $B_i/D = 2\ (y - y^2)^{1/2}$. Inserting in Eq. (5.50) gives (Hager 1995)

$$f_d = \mathsf{F}_\Delta/\mathsf{F}_d = \frac{\beta/(\pi/4 - y^{1.4}) - 1/y^{1.4}}{\left[\frac{y^{1.4}}{2(y-y^2)^{1/2}} + \frac{\pi/4 - y^{1.4}}{2(y-y^2)^{1/2}} R\right]^{1/2}}. \tag{5.51}$$

Here $\mathsf{F}_d = Q_w/(g'D^5)^{1/2}$ is *densimetric pipe Froude number* with $g' = [(\rho_w - \rho_a)/\rho_w]g$ as the reduced gravity constant, $\beta = Q_a/Q_w$ the discharge ratio and $R = \rho_w/\rho_a$ the phase density ratio. Figure 5.15b shows the ratio f_d as a function of $y_w = h_w/D$ for various values of β and $R = 1000/1.4 = 714$. For given F_d and β, the limit fluid filling y_w can thus be determined for *incipient wave instability* to occur, i.e. for $\mathsf{F}_\Delta = 0.50$. It should be noted that for any β, the pipe chokes whenever $y_w >$ 0.85.

Example 5.10 Given $Q = 2.1$ m^{3}s^{-1}, $D = 0.80$ m and $\beta = 2$. What is the limit fluid filling ratio?
With $g' = [(1000 - 1.4)/1000]9.81 = 9.80$ ms^{-2} the densimetric pipe Froude number is $\mathsf{F}_d = 2.1/(9.80 \cdot 0.8^5)^{1/2} = 1.17$ and thus $f_d = 0.5/1.17 = 0.43$. For $\beta = 2$, Fig. 5.15b yields $y_w = 0.62$. Whenever the flow depth is larger than $h_w = 0.62 \cdot 0.8 = 0.50$ m, the pipe would choke, therefore.

Figure 5.15b indicates that the limit value y_w decreases with increasing β. Clearly, for a large air discharge the velocity head over a wave crest is large and the pressure head in the gas phase reduces such that the fluid wave expands to the pipe vertex. For equal phase velocities $V_a = V_w$, the discharge ratio is $\beta = (\pi/4)y^{-1.4}-1$ and $f = 0$. Then the condition for y_w is determining and reads from Eq. (5.51)

$$y_w = \left(\frac{\pi/4}{1+\beta}\right)^{0.70}. \quad (5.52)$$

Mishima and Ishii (1980) analyzed the effect of streamline curvature on wave instability. Based on a nonlinear wave analysis, the critical value of $\mathsf{F}_\Delta = 0.5$ was confirmed.

Ruder et al. (1989) determined the necessary condition for wave instability. The flow as shown in Fig. 5.16a contains a surge of which the downstream portion has a flow depth h_1 and a velocity V_1. In Sect. 2 the flow has parallel streamlines and one may note similarity with a hydraulic jump (Chap. 7). Neglecting air entrainment and introducing a moving coordinate system of celerity c the *stability condition* is

$$\mathsf{F}_r = \frac{c - V_1}{(gD)^{1/2}} > 1. \quad (5.53)$$

Accordingly, the Froude number based on the relative velocity $c - V_1$ and the pipe diameter D has to be larger than one. A second stability condition was related to the tail of a slug wave. The necessary conditions for slug flow to occur are shown in Fig. 5.16b. For $y_1 > 0.56$ it relates to the front criteria, and for smaller filling ratios the propagation of the wave end is affected. On the left of the transition curve one has bubbly flow, and slug flow on the right of the transition curve. The stability curves based on Lin and Hanratty (1986) are also included.

According to Fan et al. (1992) the surface profile of a slug tail may be approximated with a so-called *Benjamin bubble*. An existing slug flow can be removed from

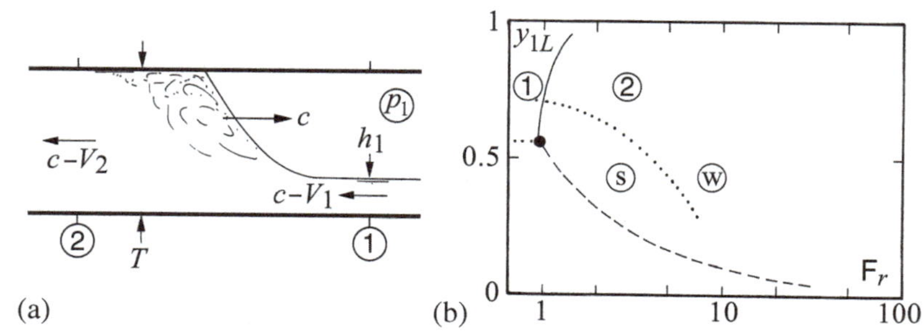

Fig. 5.16 (**a**) Slug flow in pipe flow, notation. (**b**) Limit filling ratio $y_{1L} = h_{1L}/D$ in circular pipe as a function of relative Froude number $\mathsf{F}_r = (c - V_1)/(gD)^{1/2}$ for transition between 1 plug flow to 2 slug flow according to Ruder et al. (1989), (•) Benjamin bubble, and (...) stability limit between s stable and w growing slugs waves

a slightly sloping pipe provided the mixture discharge Q_m is decreased below

$$Q_m = 0.54(\pi D^2/4)\,(gD)^{1/2}. \tag{5.54}$$

For smaller discharge *stratified* flow appears.

5.7.5 *Benjamin Bubble*

Benjamin (1968) defined a *gravity current* as the intrusion of a heavier fluid into a lighter fluid by neglecting the effects of viscosity and entrainment along the interface. Figure 5.17a relates to a density current as occurs for instance in final settling tanks. The current consists of a head at the front, the body and a tail upstream.

Figure 5.17b relates to the *emptying process* of a *rectangular duct* of height T originally filled with a heavy fluid. If the downstream cover is removed, a negative surge propagates into the duct of which the tail has a height h_2 and a velocity V_2. Because point P is a stagnation point, the Bernoulli equation gives

$$V_2^2/(2g) = T - h_2. \tag{5.55}$$

Assuming atmospheric pressure above the bubble, one has from the energy relation far upstream (subscript 2) and far downstream (subscript 1) from point P relative to the duct

$$\frac{p_1}{\rho g} + \frac{V_1^2}{2g} = H_2 = 0. \tag{5.56}$$

The pressure head upstream from the stagnation point is thus located by $-V_1^2/(2g)$ below the cover of the duct. Assuming hydrostatic pressure and uniform velocity distributions, the momentum equation yields

$$p_1 T + \frac{1}{2}\rho g T^2 + \rho V_1^2 T = \frac{1}{2}\rho g h_2^2 + \rho V_2^2 h^2. \tag{5.57}$$

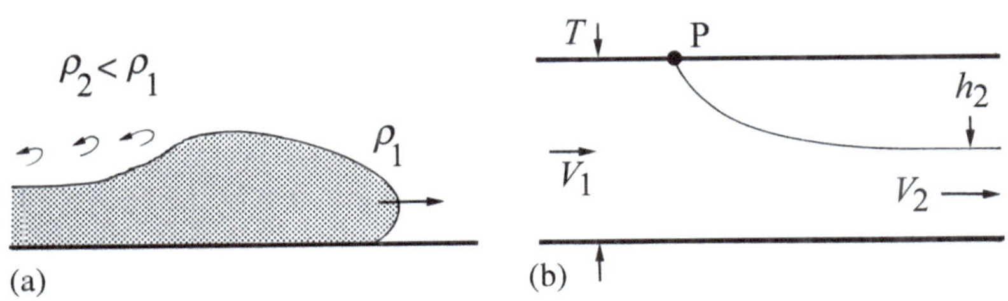

Fig. 5.17 (**a**) Density current on a slightly sloping bottom, (**b**) Benjamin bubble in a rectangular duct

Eliminating p_1 with Eq. (5.56) and using the continuity equation $V_1T = V_2h_2$ gives

$$V_2^2 = \frac{g(T^2 - h_2^2)T}{(2T - h_2)h_2}. \tag{5.58}$$

Eliminating further V_2^2 from Eqs. (5.55) and (5.57) gives two solutions, namely the trivial solutions $y = h_2/T = 1$ and the second physically relevant solution $y = 1/2$. Inserting the latter in Eqs. (5.56) and (5.55) gives $V_1/(gT)^{1/2} = 1/2$ and $V_2/(gT)^{1/2} = 2^{1/2}$, with a transition from pressurized to supercritical duct flow between the stagnation point. For $y > 0.653$ Benjamin found a subcritical downstream flow, and the flow is supercritical for $0.5 < y < 0.653$. For *circular pipes* of diameter D, the corresponding values are $V_2/(gD)^{1/2} = 0.935$ and $V_1/(gD)^{1/2} = 0.542$, and the downstream Froude number is $\mathsf{F}_2 = 1.328$ instead of $2^{1/2}$ (Benjamin 1968).

The *propagation velocity* c of surges in pipes (Fig. 5.18) was analyzed by Blind (1956). Based on the momentum equation his result was (Fig. 5.18c)

$$C = \frac{c}{(gh_1)^{1/2}} = \left[\varphi + \frac{3}{2}Y_h + \frac{1}{2\varphi}Y_2^2\right]^{1/2}, \tag{5.59}$$

where $\phi = F_1/B_mh_1$ and $Y_h = \Delta h/h_1$. If the average surface width B_m is approximated with B_1, then ϕ can be related to y_1. Figure 5.18c indicates only a small effect of y_1 on C and an approximation is (Hager 1995)

$$C = 0.85\,(1 + Y_h). \tag{5.60}$$

The relative propagation velocity C thus increases linearly with the relative slug height.

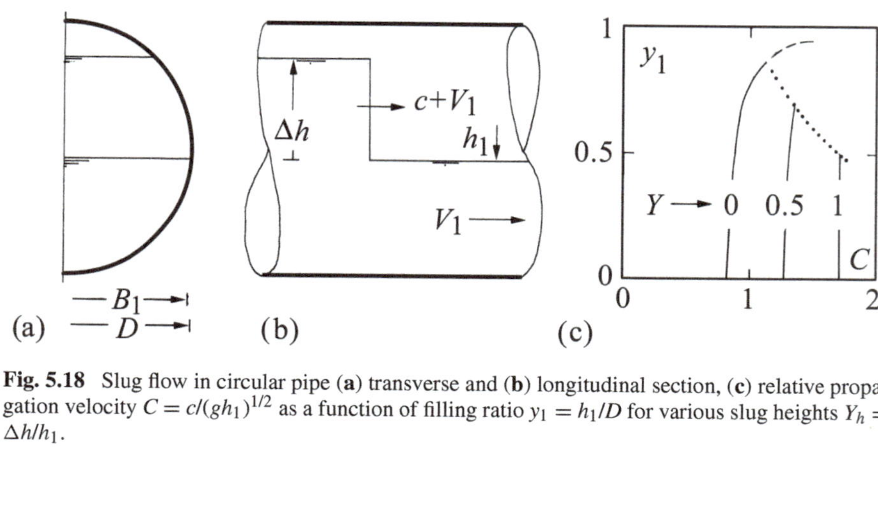

Fig. 5.18 Slug flow in circular pipe (**a**) transverse and (**b**) longitudinal section, (**c**) relative propagation velocity $C = c/(gh_1)^{1/2}$ as a function of filling ratio $y_1 = h_1/D$ for various slug heights $Y_h = \Delta h/h_1$.

5.8 Design of Sewers

5.8.1 *Design Principle*

Closed sewer channels (German: Geschlossener Kanal; French: Canal découvert) as typically used in sewage engineering are available as prefabricated pipes up to about interal diameters of $D = 2$ m. Diameters have standard dimensions such as $D = 0.25, 0.30, 0.40$ m etc. The design results in a computationally *required diameter* and a larger diameter is selected, depending on the accuracy of the design and on constructional simplifications. Often, the selection of a larger sewer diameter has little effect on cost but there may be problems with the minimum required velocity for small discharges. For diameters larger than 1.5 or 2 m sewers are often rectangular in shape and fabricated with *in-situ* concrete. A distinction is thus needed between:

- Computationally *required diameter* based on a hydraulic design, and
- *Selection of diameter* and sewer material based on criteria such as sewage, construction and availability.

According to Sect. 5.5, sewers filled over about 90% have a larger capacity than under full filling. This phenomenon is not accounted for in design, however, because:

- Additional capacity is 'only' about 5%,
- Stable flow condition is required with security against pipe choking,
- Large filling ratios depend significantly on inlet and outlet conditions that are not included for uniform flow, and
- Air flow above water flow should not be influenced to provoke subpressure.

Normally, the computed *full filling discharge* Q_V is used for design, and ATV (1988) recommends only 90% of Q_V. If the computed capacity is larger, the next larger diameter should be selected. This recommendation is fully supported here, provided no detailed hydraulic design is made, including backwater effects and local hydraulic phenomena. This approach has the following *advantages*:

- For both partial and full sewer filling, the design is based on the same formula,
- Filling curves can be applied in a optimal manner. First, full filling is considered for capacity, and partial filling is used for all remaining flow conditions, and
- Filling ratio is independent of capacity design and one has to account only for parameters describing full filling condition (Chaps. 2 and 3).

5.8.2 *Design Procedure*

The full filling condition of a sewer can be determined either with the Colebrook-White or the Manning-Strickler approaches. The latter is restricted to the turbulent rough regime as described in Chap. 2, or in Sect. 5.3. The Manning-Strickler approach has definite advantages, particularly because the diameter required can be explicitly determined. With D as the sewer diameter and B as the width of the egg-shaped and the horseshoe standard profiles, the following *design relations* apply

$$\text{Circular profile} \qquad D = 1.55\left(\frac{nQ}{S_o}\right)^{3/8}, \tag{5.61}$$

$$\text{Egg-shaped profile 2:3} \quad B = 1.294\left(\frac{nQ}{S_o}\right)^{3/8}, \tag{5.62}$$

$$\text{Horseshoe profile 2:1.5} \quad B = 1.791\left(\frac{nQ}{S_o}\right)^{3/8}. \tag{5.63}$$

The relative discharge sewer capacity is thus $q_N = \left[nQ/\left(S_o^{1/2}B^{8/3}\right)\right]$, and the generalized design formula (subscript D) is

$$q_N = \frac{nQ}{S_o^{1/2}B^{8/3}} < q_D. \tag{5.64}$$

The parameter q_D has then the following rounded off numbers

Circular profile $q_D = 1.55$,
Egg-shaped profile 2:3 $q_D = 1.30$,
Horseshoe profile 2:1.5 $q_D = 1.80$.

It is advantageous to use [m, s] dimensions, i.e. discharge Q in [m^3s^{-1}], roughness coefficient ($1/n$) in [$m^{1/3}s^{-1}$] and width B in [m]. The bottom slope S_o is expressed in absolute numbers, and neither in [%] nor in [‰].

Usual flow conditions in sewers are in the turbulent rough or in the turbulent transition regime. The rough regime has always larger losses than the corresponding flow in the transition regime, and the diameter design is on the safe side, therefore. To estimate the effect of viscosity ν on the diameter D of a circular sewer, Hager and Schwalt (1992) have advanced the explicit diameter relation

$$D^{*-5/2} = -\frac{\pi}{\sqrt{2}}\log\left[\left(0.361k_s^*\right)^{15/16} + \left(3.434\nu^*\right)^{15/16}\right], \tag{5.65}$$

with the notation used in Chap. 2

$$D^* = D/D_o, \quad k_s^* = k_s/D_o, \quad \nu^* = \nu D_o/Q \tag{5.66}$$

where the reference diameter D_o is

$$D_o = \left[Q^2/(gS_o)\right]^{1/5}. \tag{5.67}$$

With known parameters (ν, k_s, Q, S_o) one may compute D_o first, then determine k_s^* and ν^* from Eq. (5.66), introduce them in Eq. (5.65) to solve for D^*. The diameter according to Eq. (5.65) deviates always less than $\pm 1\%$ from the mathematically exact equation based on Colebrook-White.

The deviations of the velocity according to Manning-Strickler from the Colebrook-White approach are less than 5% provided $k_s^* < 30\nu^*$. The Manning-Strickler equation should only be applied for relative roughnesses in the range $7 \cdot 10^{-4} < k_s/D < 7 \cdot 10^{-2}$, as discussed in Sect. 5.4. Note that Chap. 5 relates exclusively to *uniform flow* and the effects of up- and downstream perturbations have to be accounted for separately.

A *modified design procedure* based on the Manning-Strickler equation involves thus for every sewer reach five computational steps:

1. Compilation of basic parameters:

   - Discharge Q
   - Bottom slope S_o
   - Equivalent sand roughness height k_s and
   - Kinematic viscosity ν

   The kinematic viscosity of ordinary sewage is $\nu = 1.3 \cdot 10^{-6}$ m^{2}s^{-1} (Table 3.1) for an ambient temperature of 12°C (ATV 1988). The equivalent sand roughness height is given in Chaps. 2 and 3.

2. Determination of flow regime:

   - Computation of reference diameter D_o from Eq. (5.67),
   - Calculation of $k_s^* = k_s/D_o$ and $\nu^* = \nu D_o/Q$,
   - If $k_s^* > 30\nu^*$ the flow is in the turbulent rough regime, and if $7 \cdot 10^{-4} < k_s/D < 7 \cdot 10^{-2}$, the Manning-Strickler formula applies,
   - If these two conditions are invalidated, the design follows Eq. (5.65) according to Colebrook-White.

3. Diameter design:

   - Depending on the standard profile, Eqs. (5.61) to (5.63) are applied,
   - If the Manning-Strickler equation does not apply, then Eq. (5.65) for circular sewers is used and $D = D_o D^*$.

4. Effect of self-aeration:

   - Computation of roughness characteristic $\chi = S_o^{1/2} D^{1/6} / \left(ng^{1/2}\right)$,
   - If $\chi < 8$, self-aeration is absent and the diameter previously determined applies,
   - If $\chi \geq 8$, self-aeration occurs and the diameter has to be increased by the factor h_m/h_N from Eq. (5.42).

5. Choking security:

   - Depending mainly on χ a choking security can be introduced. For $\chi > 5$, say, the maximum filling ratio should be 80%, and the filling ratio should be below 70% for $\chi > 10$. Definite numbers are not yet available, and effects of shockwaves (Chap. 16) have also to be accounted for.

Contrary to the design method based on the Colebrook-White formula, the *modified approach* has the disadvantage that two computational approaches may apply, depending on the parameter combinations. The advantages of the modified approach are additional information, such as the flow regime and simplicity when determining conditions of partial filling.

Example 5.11 What is the sewer diameter for a discharge $Q = 2\ \mathrm{m^3s^{-1}}$ and a bottom slope $S_o = 0.1\%$ for a precast concrete pipe?

1. $Q = 2\ \mathrm{m^3s^{-1}}$, $S_o = 0.001$, $k_s = 2.5 \cdot 10^{-4}$ m, $\nu = 1.3 \cdot 10^{-6}\ \mathrm{m^2s^{-1}}$.
2. $D_o = [2^2/(9.81 \cdot 0.001)]^{1/5} = 3.33$ m, $k_s^* = 2.5 \cdot 10^{-4}/3.33 = 7.5 \cdot 10^{-5}$, $\nu^* = 1.3 \cdot 10^{-6} \cdot 3.33/2 = 2.16 \cdot 10^{-6}$, $30\nu^* = 6.5 \cdot 10^{-5}$. Therefore, $k_s^* < 30\nu^*$ and the flow is *not* in the turbulent rough regime.
3. With $k_s^* = 7.5 \cdot 10^{-5}$ and $\nu^* = 2.16 \cdot 10^{-6}$ from 2, the relative sewer diameter is $D^{*-5/2} = -2.22\log[(2.71 \cdot 10^{-5})^{15/16} + (7.42 \cdot 10^{-6})^{15/16}] = 9.35$ from Eq. (5.65), thus $D^* = 9.25^{-0.40} = 0.411$ and $D = 0.411 \cdot 3.33 = 1.37$ m. Selected is $D = 1.50$ m.
4. With parameters $1/n$ and D as determined below in 5, one has $\chi = 102 \cdot 0.001^{1/2} 1.29^{1/6} 9.81^{-1/2} = 1.07 < 8$, and *no* self-aeration occurs.
5. No special addition for choking security is provided.

If the Manning-Strickler formula would have been used, the roughness coefficient is $1/n = 8.2 \cdot 9.81^{1/2}(2.5 \cdot 10^{-4})^{-1/6} = 102\ \mathrm{m^{1/3}s^{-1}}$ according to Eq. (5.5) and thus the diameter would be $D = 1.55[2/(102 \cdot 0.001^{1/2})]^{3/8} = 1.29$ m (–5%). The effect of viscosity follows also from the extremely large $1/n$ value.

Example 5.12 Design the sloping sewer for the discharge $Q = 1.2\ \mathrm{m^3s^{-1}}$ if $S_o = 14\%$, for $1/n = 85\ \mathrm{m^{1/3}s^{-1}}$.

1. $Q = 1.2\ \mathrm{m^3s^{-1}}$, $S_o = 0.14$, $\nu = 1.3 \cdot 10^{-6}\ \mathrm{m^2s^{-1}}$, $1/n = 85\ \mathrm{m^{1/3}s^{-1}}$.
2. $D_o = [1.2^2 9.81^{-1} 0.14^{-1})]^{1/5} = 1.01$ m,
   $k_s = (8.2 \cdot 9.81^{1/2} 85^{-1})^6 = 0.76 \cdot 10^{-3}$ m according to Eq. (5.5), and thus $k_s^* = 0.75 \cdot 10^{-3}$ and $\nu^* = 1.3 \cdot 10^{-6} 1.01/1.2 = 1.09 \cdot 10^{-6}$. Because $k_s^* = 0.75 \cdot 10^{-3} > 3.28 \cdot 10^{-5} = 30\nu^*$, the present flow *is* in the turbulent rough regime and the equation of Manning-Strickler can be applied.

3. Diameter $D = 1.55\,(1.2/85 \cdot 0.14^{1/2})^{3/8} = 0.45$ m from Eq. (5.61).
4. $\chi = 85 \cdot 0.14^{1/2} 0.45^{1/6} 9.81^{-1/2} = 8.9 > 8$, and self aeration occurs. From Fig. 5.11 for $h_m/D = 0.85$, the pure water filling is $h_N/D = 0.80$, thus $h_m/h_N = 1.06$. Increasing the diameter by this factor gives $1.06 \cdot 0.45 = 0.48$ m, selected is $D = 0.50$ m.
5. The security against choking is provided because of a larger diameter.

Apart from a design of sewer capacity, one has often to determine the uniform flow depths for smaller discharges. This is straightforward with the indications given in Sect. 5.5.

Notation

B	[m]	profile width
B	[–]	Boussinesq number
c	[ms^{-1}]	celerity
C	[–]	$= c/(gh_1)^{1/2}$
$\bar{C}$	[–]	mean air concentration
D	[m]	diameter
D_o	[m]	reference diameter
D^*	[–]	relative diameter
f	[–]	resistance coefficient
f_d	[–]	$= \mathsf{F}_\Delta/\mathsf{F}_d$
F	[m^2]	cross-sectional area
F	[–]	pipe Froude number
F_d	[–]	densimetric pipe Froude number
F_Δ	[–]	densimetric Froude number
F_r	[–]	$= (c - V_1)/(gD)^{1/2}$
g	[ms^{-2}]	gravitational acceleration
g'	[ms^{-2}]	reduced gravitational acceleration
h	[m]	flow depth
H	[m]	energy head
H_s	[m]	velocity head
k_s	[m]	equivalent sand roughness height
k_s^*	[–]	relative sand roughness
K_h	[–]	parameter
$1/n$	[m$^{1/3}$s^{-1}]	roughness coefficient
p	[Nm^{-2}]	pressure
P	[m]	wetted perimeter
q_B	[–]	relative design discharge

q_D	[–]	discharge referred to $(gD^5)^{1/2}$
q_N	[–]	discharge referred to $(1/n)S_o{}^{1/2}D^{8/3}$
q_v	[–]	discharge referred to full filling
Q	$[m^3s^{-1}]$	discharge
Q_m	$[m^3s^{-1}]$	mixture discharge
r	[m]	vertex radius
R	[–]	phase density ratio
R	[–]	Reynolds number
Ri	[–]	Richardson number
R_a	[–]	void fraction
R_h	[m]	hydraulic radius
S_f	[–]	friction slope
S_o	[–]	bottom slope
t	[s]	time
T	[m]	profile height
V	$[ms^{-1}]$	average velocity
x	[m]	streamwise coordinate
y	[–]	filling ratio
Y_h	[n]	relative surge height
Z	[–]	void number
β	[–]	discharge ratio
δ	[–]	half central angle
δ_g	[m]	boundary layer thickness
ε	[–]	relative roughness
ϕ	[–]	aspect ratio
ν	$[m^2s^{-1}]$	kinematic viscosity
ν^*	[–]	relative viscosity
χ	[–]	roughness characteristics
X	[–]	pressure gradient ratio
ρ	$[kgm^{-3}]$	density
ϕ_f	[–]	shape factor
ϕ	[–]	coefficient
Φ	[–]	pressure loss ratio

Subscripts

a	opening, air
D	design
g	gas
i	incipient aeration
m	air-water mixture
min	minimum
M	maximum

N	uniform flow
r	rough
s	smooth
v	full filling
w	water
z	choking
1	downstream surge portion
2	pressurized surge portion

References

Abwassertechnische Vereinigung ATV (1988). Richtlinien für die hydraulische Dimensionierung von Abwasserkanälen und -leitungen (Guidelines for the hydraulic design of sewers). Regelwerk Abwasser *Arbeitsheft* A110. ATV: St. Augustin [in German].

Benjamin, T.B. (1968). Gravity currents and related phenomena. *Journal Fluid Mechanics* 31(2): 209–248.

Blind, H. (1956). Nichtstationäre Strömungen im Unterwasserstollen (Unsteady flows in tunnels). *Dissertation*. Technische Hochschule Karlsruhe. Springer: Berlin [in German].

Bock, J. (1966). Einfluss der Querschnittsform auf die Widerstandsbeiwerte offener Gerinne (Shape effect on open channel resistance coefficients). *Technischer Bericht* 2, O. Kirschmer, ed. Institut für Hydromechanik und Wasserbau, TH Darmstadt: Darmstadt [in German].

Daily, J.W., Harleman, D.R.F. (1966). *Fluid dynamics*. Addison-Wesley: Reading MA.

Fan, Z., Jepson, W.P., Hanratty, T.J. (1992). A model for stationary slugs. *International Journal of Multiphase Flow* 18(4): 477–494.

Gardner, G.C. (1979). Onset of slugging in horizontal ducts. *International Journal of Multiphase Flow* 5: 201–209.

Hager, W.H. (1985). Abflusseigenschaften in offenen Kanälen (Flow features of open channels). *Schweizer Ingenieur und Architekt* 103(13): 252–264 [in German].

Hager, W.H. (1988). Abflussformeln für turbulente Strömungen (Flow formulae for turbulent flow). *Wasserwirtschaft* 78(2): 79–84 [in German].

Hager, W.H. (1991). Teilfüllung in geschlossenen Kanälen (Partial filling of closed conduit flow). *Gas-Wasserfach* Wasser/Abwasser 132(10): 558–564; 132(11): 641–647 [in German].

Hager, W.H. (1995). Zuschlagen von teilgefüllten Rohren (Transition of partially filled to pressurized pipe flows). gwf-Wasser/Abwasser 136(4): 200–210 [in German].

Hager, W.H., Blaser, F. (1998). Drawdown curve and incipient aeration for chute flow. *Canadian Journal of Civil Engineering* 25(3): 467–473.

Hager, W.H., Schleiss, A.J. (2009). *Constructions hydrauliques – Ecoulements stationnaires (Hydraulic structures – steady flows)*, ed. 2. Presses Polytechniques et Universitaires Romandes: Lausanne [in French].

Hager, W.H., Schwalt, M. (1992). Explizite Fliessformel für turbulente Rohrströmung (Explicit flow formula for turbulent pipe flow). *3R-International* 31(1/2): 18–21 [in German].

Hewitt, G.F., Hall-Taylor, N.S. (1970). *Annular two-phase flow*. Pergamon: Elmsford.

ICOLD (1992). Spillways and bottom outlets: Shockwaves and air entrainment in chutes. *Bulletin* 81. Commission Internationale des Grands Barrages: Paris.

Kazemipour, A.K., Apelt, C.I. (1982). New data on shape effect in smooth rectangular channels. *Journal of Hydraulic Research* 20(3): 225–233.

Kubie, J. (1979). The presence of slug flow in horizontal two-phase flow. *International Journal of Multiphase Flow* 5(4): 327–339.

Lin, P.Y., Hanratty, T.J. (1986). Prediction of the initiation of slugs with linear stability theory. *International Journal of Multiphase Flow* 12(1): 79–98.

Lockhart, R.W., Martinelli, R.C. (1949). Proposed correlation of data for isothermal two-phase, two-component flow in pipes. *Chemical Engineering Progress* 45(1): 49–58.

Marchi, E. (1961). Il moto uniforme delle correnti liquide nei condotti chiusi e aperti. (Uniform flow in closed and open channels). *L'Energia Elettrica* 38(4): 289–301; 38(5): 393–413 [in Italian].

Marchi, E., Rubatta, A. (1981). *Meccanica dei fluidi* (Fluid mechanics). UTET: Torino [in Italian].

Mishima, K., Ishii, M. (1980). Theoretical prediction of onset of horizontal slug flow. *Journal Fluids Engineering* 102(12): 441–445.

Press, H., Schröder, R. (1966). *Hydromechanik im Wasserbau*. Ernst & Sohn: Berlin [in German].

Rouse, H. (1965). Critical analysis of open-channel resistance. Proc. ASCE *Journal of the Hydraulics Division* 91(HY4): 1–25; 91(HY6): 247–248; 92(HY2): 387–409; 92(HY4): 154; 92(HY5): 204–206.

Ruder, Z., Hanratty, P.J., Hanratty, T.J. (1989). Necessary condition for the existence of stable slugs. *International Journal of Multiphase Flow* 15: 209–226.

Sauerbrey, M. (1969). Abfluss in Entwässerungsleitungen unter besonderer Berücksichtigung der Fliessvorgänge in teilgefüllten Rohren (Flow in drainage pipes with particular reference to partially filled conduits). *Wasser und Abwasser in Forschung und Praxis* 1. Erich Schmidt: Bielefeld [in German].

Sinniger, R.O., Hager, W.H. (1989). *Constructions hydrauliques – Ecoulements stationnaires* (Hydraulic structures: Steady flows). Presses Polytechniques et Universitaires Romandes: Lausanne [in French].

Strickler, A. (1923). Beiträge zur Frage der Geschwindigkeitsformel und der Rauhigkeitszahlen für Ströme, Kanäle und geschlossene Leitungen (Contributions to the question of flow roughness coefficients for rivers, channels and conduits). *Mitteilung* 16. Amt für Wasserwirtschaft: Bern, Switzerland [in German].

Taitel, Y., Dukler, A.E. (1977). A model for slug flow frequency during gas-liquid flow in horizontal and near horizontal pipes. *International Journal of Multiphase Flow* 3: 585–596.

Thormann, E. (1944). Füllungskurven von Entwässerungsleitungen (Filling curves of drainage pipes). *Gesundheits-Ingenieur* 67(2): 35–47 [in German].

Volkart, P. (1978). Hydraulische Bemessung teilgefüllter Steilleitungen (Hydraulic design of partially filled steep sewers). *Gas – Wasser – Abwasser* 58(11): 658–667 [in German].

Volkart, P.U. (1980). The mechanism of air bubble entrainment in self-aerated flow. *International Journal of Multiphase Flow* 6: 411–423.

Wallis, G.B., Dobson, J.E. (1973). The onset of slugging in horizontal stratified air-water flow. *International Journal of Multiphase Flow* 1: 173–193.

Weisman, J. (1983). Two-phase flow patterns. *Handbook of fluids in motion*: 409–425. N.P. Cheremisinoff, R. Gupta, eds. Ann Arbor Science: Houston.

Chapter 6
Critical Flow

Abstract Critical flow occurs when a set of conditions are fulfilled. These are enumerated in this chapter. Along with these, the critical flow is hydraulically described and the requirements for critical flow are explained in detail. Critical flow is then computed for the three standard cross-sections, namely, the circular, the egg-shaped and the horseshoe sections. The critical energy head and the critical slope are also determined in terms of basic flow parameters.

As an application of the critical flow principle, the transition from a mild to a steep sewer reach is described. This relevant problem can, for all practical purposes, be presented with the aid of simplified relations.

6.1 Introduction

Next to uniform flow, the critical flow (German: kritischer Abfluss; French: Ecoulement critique) represents another special flow state with reference to which a given channel flow is described. While the uniform flow forms a fixed 'static criterion' by which the frictional forces are balanced with the driving force, the critical flow may be considered a *dynamic* criterion. Critical flow prevails if the velocity of flow corresponds exactly to the celerity of an elementary gravity wave.

Like uniform flow, critical flow also depends on the cross-sectional channel geometry. In the following, two closely related problems are investigated:

- Definitions and description of critical flow, and
- Practical evaluation of critical flow.

First, the *energy equation* is used to investigate the critical flow condition and then – as an application – the flows in the three standardized sections, namely, the circular, the egg-shaped and the horseshoe sewers, are studied. These three sections are already mentioned in Chap. 5. Besides the critical depth, the critical slope is also investigated. General summaries on the critical depth are given by Chow (1959), Henderson (1966) and Naudascher (1987). The results can in particular be applied to the backwater and drawdown curves (Chap. 8) as also to discharge measuring

W.H. Hager, *Wastewater Hydraulics*, 2nd ed., DOI 10.1007/978-3-642-11383-3_6,

devices (Chaps. 10, 11, 12 and 13). As a further application, the transition of flow from a mild to a steep sewer reach is finally investigated.

6.2 Description of Critical Flow

The *energy head* H_* measured from the bed of a channel is (Chap. 1)

$$H_* = h + \frac{Q^2}{2gF^2} \tag{6.1}$$

where h is the flow depth, F the cross-sectional area of the flow, g the acceleration due to gravity and Q the discharge. The cross-sectional area F in an *open channel* is a function of the water depth h and the longitudinal coordinate x. Dependence on x occurs for non-prismatic channels in which the cross-section varies with every position x. If, however, the cross-sectional shape does not change, the channel is referred to as *prismatic* for which $F = F(h)$. In contrast, for pressurized flows in a closed conduit, the cross-section may vary only with x. It is repeated that free surface flow can occur both in open and closed channels. For closed channels both free-surface flow *or* pressurized flow are possible.

Consider for the moment a channel of constant discharge Q so that the energy head H_* at station $x = x_*$ is represented by a function solely of the flow depth h. The cross-sectional area at this station is $F(x = x_*) = F_*$. It is useful to pursue further the relation between H_* and h.

All channels are characterized by a cross-section F that increases with increasing flow depth h. A continuous cross-section in which the relation $F_*(h)$ does not exhibit any kink gives $\mathrm{d}F_*/\mathrm{d}h > 0$. In the mathematical analysis of a continuous function, the determination of the *extreme values* is often relevant. The first two derivatives of Eq. (6.1), with $F = F_*(h)$, are

$$\frac{\mathrm{d}H_*}{\mathrm{d}h} = 1 - \frac{Q^2}{gF_*^3}\frac{\mathrm{d}F_*}{\mathrm{d}h}, \tag{6.2}$$

$$\frac{\mathrm{d}^2H_*}{\mathrm{d}h^2} = \frac{3Q^2}{gF_*^4}\left(\frac{\mathrm{d}F_*}{\mathrm{d}h}\right)^2 - \frac{Q^2}{gF_*^3}\frac{\mathrm{d}^2F_*}{\mathrm{d}h^2}. \tag{6.3}$$

The function $H_*(h)$ possesses an extreme for $\mathrm{d}H_*/\mathrm{d}h = 0$. The extreme corresponds to a maximum value if $\mathrm{d}^2H_*/\mathrm{d}h^2 < 0$ and a minimum value if $\mathrm{d}^2H_*/\mathrm{d}h^2 > 0$. The dimensionless function $1 - (\mathrm{d}H_*/\mathrm{d}h)$ is denoted as the square of the *Froude number* (German: Froudezahl; French: Nombre de Froude), so named after the Englishman William Froude (1810–1879). Therefore

$$\mathsf{F}^2 = \frac{Q^2}{gF_*^3}\frac{\mathrm{d}F_*}{\mathrm{d}h}. \tag{6.4}$$

Consequently, an extreme value of the energy head H_* follows from Eq. (6.2) for the Froude number $\mathsf{F} = 1$. No useful information can be obtained by setting $\mathsf{F} = 1$ in Eq. (6.3). If, however, the second term on the right side of Eq. (6.2) is replaced by the term of the left side of Eq. (6.4) and the resulting equation is derived with respect to h, it follows

$$\frac{d^2H_*}{dh^2} = -2\mathsf{F}\frac{d\mathsf{F}}{dh}. \tag{6.5}$$

For constant discharge, the Froude number always decreases with increasing depth of flow, i.e. $d\mathsf{F}/dh < 0$. Therefore, the extreme value of the energy head H_* ($\mathsf{F} = 1$) is a *minimum value*. Accordingly, the first definition of the critical flow condition is: Free-surface flow is said to be critical if the energy head, measured from the channel bed, is minimum.

Equation (6.1) can be solved for discharge Q, thus

$$Q = F[2g(H_* - h)]^{1/2}. \tag{6.6}$$

If a constant energy head H_* is assumed, a maximum value of the discharge can be found for a flow depth $h = h_c$ for $\mathsf{F} = 1$. Equation (6.4) then gives the relation between the discharge and the depth of flow. The second definition of the critical (subscript c) flow condition is: Free-surface flow under a given energy head H_* carries the maximum discharge Q_c for critical flow.

It can further be shown that for the critical flow condition:

- Velocity of flow equals the celerity $(gh)^{1/2}$ of an elementary gravity wave, and
- Specific force S, according to Eq. (1.11), assumes also a minimum value.

It may readily be accepted that the critical flow condition decisively influences all channel flows. Since the Froude number of a pressurized conduit flow is practically nil, this characteristic number may be considered a true characteristic of all *free-surface flows*. Further special features of the critical flow state are highlighted in the following section.

6.3 Features of Critical Flow

6.3.1 *Critical Depth*

Consider here a *rectangular channel* of constant width b. This is in order that the calculations do not become too involved while restricting ourselves only to the presentation of the principles of critical flow. If now a constant discharge Q is considered, the discharge per unit width $q = Q/b$ is similarly a constant quantity. With $F = bh$ as the cross-sectional area of flow, Eq. (6.1) gives

$$H_* = h + \frac{q^2}{2gh^2}. \tag{6.7}$$

With $F = bh$ and $\mathrm{d}F/\mathrm{d}h = b$, Eq. (6.4) gives for the Froude number in the rectangular channel

$$\mathsf{F}^2 = \frac{q^2}{gh^3}. \tag{6.8}$$

The pre-requisite for critical flow (subscript c) to occur is $\mathsf{F} = 1$. The accompanying critical depth h_c is, therefore,

$$h_c = (q^2/g)^{1/3}, \tag{6.9}$$

and the corresponding *critical energy head*, following Eqs. (6.7) and (6.9), is given by

$$H_{*c} = h_c\left[1 + \frac{1}{2}\frac{q^2}{gh_c^3}\right] = \frac{3}{2}h_c. \tag{6.10}$$

If the critical flow depth is eliminated between the Eqs. (6.9) and (6.10), the following relation between the *critical discharge* q_c and the critical energy head H_* is

$$q_c = g^{1/2}\left[\frac{2}{3}H_*\right]^{3/2}. \tag{6.11}$$

Equations (6.7), (6.8), (6.9), (6.10) and (6.11) are exclusively valid for the rectangular cross-section. Here the relationships are simple and all the critical parameters can be solved explicitly.

6.3.2 *Influence of Bottom Geometry*

There exists an entire series of relations that are peculiar to the critical flow condition. It is, however, not known which of them are necessary as requirements for critical flow to occur. In order to examine this issue, consider a variable but continuously changing bed geometry $z(x)$ (Fig. 6.1). The origin O of the coordinate system is arbitrarily chosen.

The energy head H of a flow in a rectangular channel is given by (Chap. 1)

$$H = z + h + \frac{q^2}{2gh^2} = z + H_*. \tag{6.12}$$

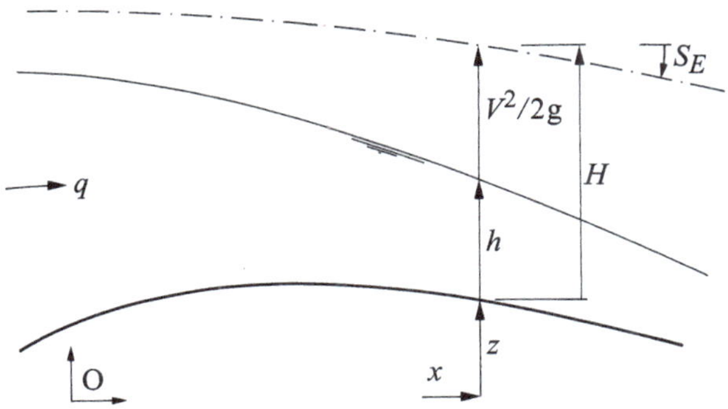

Fig. 6.1 Definition sketch of flow over a continuously curved channel bed

First, the condition is investigated which must be satisfied for critical flow to occur in a frictionless fluid, i.e. a flow for which the energy slope S_E – unlike that shown in Fig. 6.1 – is zero. Then, all derivatives of H with respect to x are identically zero. The first two derivatives of Eq. (6.12) with respect to the streamwise direction x are

$$\frac{dH}{dx} = \frac{dz}{dx} + \frac{dh}{dx}\left(1 - \frac{q^2}{gh^3}\right) = 0, \tag{6.13}$$

$$\frac{d^2H}{dx^2} = \frac{d^2z}{dx^2} + \frac{d^2h}{dx^2}\left(1 - \frac{q^2}{gh^3}\right) + \frac{3q^2}{gh^4}\left(\frac{dh}{dx}\right)^2 = 0. \tag{6.14}$$

For the critical flow condition Eq. (6.8), it follows from Eqs. (6.13) and (6.14) that:

- Critical flow $q^2/(gh^3) = 1$ can be established only for an *extreme value* $dz/dx = 0$ of the bottom geometry, and
- Water surface slope dh/dx at the critical point equals

$$\frac{dh_c}{dx} = \pm\left(-\frac{h_c}{3}\frac{d^2z_c}{dx^2}\right)^{1/2}. \tag{6.15}$$

The resulting slope is directly proportional to the square-root of the product of the bottom profile curvature and the flow depth at the critical point. Because only one real quantity is physically relevant, the term under the radical sign must be positive, i.e. $d^2z_c/dx^2 < 0$. Consequently, critical flow in a prismatic channel can take place only at a *local maxima* of the bottom profile.

If the friction losses are taken into consideration, $dH/dx = -S_E$ and $d^2H/dx^2 = -dS_E/dx$ instead of $dH/dx = 0$ and $d^2H/dx^2 = 0$. The right hand sides of Eqs. (6.13) and (6.14) change correspondingly, and it follows:

- Critical flow occurs at the location where the channel bottom slope $S_o = -\,\mathrm{d}z/\mathrm{d}x$ equals the energy line slope S_E. The condition $S_o = S_{Ec}$ is, in analogy to Chap. 5, denoted *pseudo-uniform flow*.
- Since the energy line curvature $\mathrm{d}S_E/\mathrm{d}x$ is normally much smaller than the curvature of the bottom slope, the critical point lies not at the crest, but slightly downstream of it.

6.3.3 *Influence of Cross-sectional Geometry*

A second case of practical interest represents a flow, without losses at first, in a horizontal channel with uniformly varying cross-sectional geometry. For the sake of simplicity consider again a *rectangular channel* of variable width $B = B(x)$. The energy head relative to the channel bottom is then

$$H = h + \frac{Q^2}{2gB^2h^2}. \tag{6.16}$$

With $\mathrm{d}H/\mathrm{d}x = \mathrm{d}^2H/\mathrm{d}x^2 = 0$, the first two derivatives of Eq. (6.16) are

$$\frac{\mathrm{d}H}{\mathrm{d}x} = \frac{\mathrm{d}h}{\mathrm{d}x}\left(1 - \frac{Q^2}{gB^2h^3}\right) - \frac{Q^2}{gB^3h^2}\frac{\mathrm{d}B}{\mathrm{d}x} = 0, \tag{6.17}$$

$$\begin{aligned}\frac{\mathrm{d}^2H}{\mathrm{d}x^2} &= \frac{\mathrm{d}^2h}{\mathrm{d}x^2}\left(1 - \frac{Q^2}{gB^2h^3}\right) - \frac{Q^2}{gB^3h^2}\frac{\mathrm{d}^2B}{\mathrm{d}x^2} + \frac{3Q^2}{gB^4h^2}\left(\frac{\mathrm{d}B}{\mathrm{d}x}\right)^2\\ &\quad + \frac{4Q^2}{gB^3h^3}\frac{\mathrm{d}B}{\mathrm{d}x}\frac{\mathrm{d}h}{\mathrm{d}x} + \frac{3Q^2}{gB^2h^4}\left(\frac{\mathrm{d}h}{\mathrm{d}x}\right)^2 = 0.\end{aligned} \tag{6.18}$$

If the Froude number $\mathsf{F}^2 = Q^2/(gB^2h^3)$ is introduced in these equations, it follows

$$\frac{\mathrm{d}H}{\mathrm{d}x} = \frac{\mathrm{d}h}{\mathrm{d}x}(1 - \mathsf{F}^2) - \frac{h}{B}\mathsf{F}^2\frac{\mathrm{d}B}{\mathrm{d}x} = 0, \tag{6.19}$$

$$\begin{aligned}\frac{\mathrm{d}^2H}{\mathrm{d}x^2} &= \frac{\mathrm{d}^2h}{\mathrm{d}x^2}(1 - \mathsf{F}^2) - \frac{h}{B}\mathsf{F}^2\frac{\mathrm{d}^2B}{\mathrm{d}x^2} + \frac{3h}{B^2}\mathsf{F}^2\left(\frac{\mathrm{d}B}{\mathrm{d}x}\right)^2\\ &\quad + \frac{4}{B}\mathsf{F}^2\frac{\mathrm{d}B}{\mathrm{d}x}\frac{\mathrm{d}h}{\mathrm{d}x} + \frac{3}{h}\mathsf{F}^2\left(\frac{\mathrm{d}h}{\mathrm{d}x}\right)^2 = 0.\end{aligned} \tag{6.20}$$

Continuing the investigation of critical flow for $\mathsf{F} = 1$, one obtains from Eq. (6.19) the condition $(h/B)(\mathrm{d}B/\mathrm{d}x) = 0$. In general, this condition is only satisfied if $\mathrm{d}B/\mathrm{d}x = 0$ since neither $h = 0$ nor $B^{-1} = 0$. The critical flow is then established at an *extreme location of the cross-sectional geometry*. Whether this location is at a contraction ($\mathrm{d}^2B/\mathrm{d}x^2 > 0$) or expansion ($\mathrm{d}^2B/\mathrm{d}x^2 < 0$) is fixed by Eq. (6.20). For $\mathsf{F} = 1$, this equation simplifies to

$$-\frac{h_c}{B}\frac{\mathrm{d}^2B}{\mathrm{d}x^2}+\frac{3}{h_c}\left(\frac{\mathrm{d}h}{\mathrm{d}x}\right)_c^2=0, \tag{6.21}$$

from which one obtains

$$\left(\frac{\mathrm{d}h}{\mathrm{d}x}\right)_c=\pm\left(\frac{1}{3}\frac{h_c^2}{B}\frac{\mathrm{d}^2B}{\mathrm{d}x^2}\right)^{1/2}. \tag{6.22}$$

This solution is physically relevant only if $\mathrm{d}^2B/\mathrm{d}x^2 > 0$. The extreme value therefore corresponds to a *contraction of the cross-section*.

If the potential flow assumption is discarded and the energy line slope is $\mathrm{d}H/\mathrm{d}x = -S_E$, the location of the critical flow is determined with the relation $(h/B)(\mathrm{d}B/\mathrm{d}x) = S_E$; the location corresponds to a point slightly downstream of the geometrical constriction. In principle this does not change the simplified derivation.

6.3.4 *Discussion of Results*

In view of the preceding analysis of the energy equation, it is evident that for critical flow to occur, at least two requirements must be satisfied:

- *Necessary condition*, that the channel has either a curved bottom profile or a variable cross-sectional geometry. Critical flow can then occur only at the top of a channel hump or at the narrowest section of a channel constriction.
- *Sufficient condition*, that the Froude number must be $\mathsf{F} = 1$, in addition. Accordingly no downstream submergence may exist as otherwise the flow is drowned.

The necessary condition can be extended to arbitrary steady flows for which the energy head can be written as

$$H = z + h + \frac{Q^2}{2gF^2}. \tag{6.23}$$

For flow in the turbulent rough regime, the energy gradient obtained from the Manning-Strickler formula (Chap. 5) is

$$\frac{\mathrm{d}H}{\mathrm{d}x} = -S_E = -\frac{n^2Q^2}{F^2R_h^{4/3}}. \tag{6.24}$$

For a flow with:

- Variable bottom profile $z(x)$,
- Spatially-varied discharge $Q(x)$,
- Non-prismatic cross-sectional geometry $F(x)$ and $R_h(x)$, and
- Variable wall roughness $n(x)$

one therefore obtains

$$\frac{\mathrm{d}z}{\mathrm{d}x}+\frac{\mathrm{d}h}{\mathrm{d}x}(1-\mathsf{F}^2)+\frac{Q}{gF^2}\frac{\mathrm{d}Q}{\mathrm{d}x}-\frac{Q^2}{gF^3}\frac{\partial F}{\partial x}=-\frac{n^2Q^2}{F^2R_h^{4/3}}. \tag{6.25}$$

For a Froude number $\mathsf{F}=1$, it follows for the location of the critical point

$$S_o-S_E-S_Q-S_F=0 \tag{6.26}$$

where

S_o = – dz/dx is bottom slope,

S_E = is energy line slope, from Eq. (6.24), due to wall friction,

S_Q = $(F/Q)[(\mathrm{d}Q/\mathrm{d}x)(\mathrm{d}F/\mathrm{d}x)]$ is the slope due to the local variation of discharge, and

S_F = $(\partial F/\partial x)/(\partial F/\partial h)$ is slope due to the local variation in cross-sectional geometry.

As one, two or even three of the total four possible mechanisms for critical flow are absent, Eq. (6.26) gets correspondingly simplified. Two cases, one with $S_o\neq 0$ and the other with $S_F\neq 0$ were already discussed under Sects. 6.3.2 and 6.3.3, respectively.

Hager (1985) demonstrated that critical flow can establish for a local variation in:

- Channel bottom profile,
- Resistance characteristics,
- Discharge, or
- Cross-sectional geometry.

In particular, critical flow can *not* occur in a channel if:

- Channel bottom has constant slope $S_o\neq S_c$,
- Channel wall possesses uniform roughness,
- Discharge remains spatially constant, and
- Cross-sectional geometry does not change.

The *necessary conditions* for critical flow are therefore diametrically opposed to those for the uniform flow. If such a critical flow does appear, the flow structure must undergo changes, it calls for a dynamic flow against which a static, invariable flow, the uniform flow, is barely rendered possible. One sole exception is the flow in a channel in which the bottom slope S_o is exactly equal to the critical slope S_c, as discussed in Sect. 6.4.5.

Besides these necessary requirements, the *sufficient condition* for critical flow must also be satisfied. This can be expressed with the Froude number $\mathsf{F}=1$; but under no circumstances may this be the only one. The statement that critical flow takes place only for the condition $\mathsf{F}=1$ is definitely incomplete, therefore.

As is explained in detail in Chap. 8 on backwater and drawdown curves, the flows with $\mathsf{F} < 1$ are denoted *subcritical* or tranquil flows, and those with $\mathsf{F} > 1$ are called *supercritical* or shooting flows. Here the notations sub- and supercritical are preferred because of the more precise description of the flow condition. Until now, the condition $\mathsf{F} = 1$ alone was considered to represent critical flow. The condition $\mathsf{F} = 1$ is, however, specifically important for the transition from subcritical to supercritical flow. With H as the energy head as defined in Eq. (6.23), S_E as the slope of the energy line as given in Eq. (6.24) and assuming a prismatic channel ($\partial F/\partial x \equiv 0$) with constant bed slope ($\mathrm{d}S_o/\mathrm{d}x \equiv 0$) and discharge ($\partial Q/\partial x \equiv 0$), the equation for the water surface profile is

$$\frac{\mathrm{d}H}{\mathrm{d}x} = -S_o + \frac{\mathrm{d}h}{\mathrm{d}x}\left(1 - \frac{Q^2}{gF^3}\frac{\mathrm{d}F}{\mathrm{d}h}\right) = -S_E. \tag{6.27}$$

Solving for the *water surface slope* gives

$$\frac{\mathrm{d}h}{\mathrm{d}x} = \frac{S_o - S_E}{1 - \dfrac{Q^2}{gF^3}\dfrac{\mathrm{d}F}{\mathrm{d}h}} = \frac{S_o - S_E}{1 - \mathsf{F}^2}. \tag{6.28}$$

It can be seen from Eq. (6.28) that for critical flow ($\mathsf{F} = 1$), two cases may arise:

1. Either $S_o \neq S_E$, as a condition different from uniform flow, and the water surface becomes vertical ($\mathrm{d}x/\mathrm{d}h = 0$), or
2. The condition $S_o = S_E$ exists and result is the undefined expression $\mathrm{d}h/\mathrm{d}x = 0/0$.

For hydrostatic pressure distribution, one of the basic assumptions in hydraulics (Chap. 1) and also for Eq. (6.28), the vertical water surface is physically irrelevant. It is apparent therefore that critical flow ($\mathsf{F} = 1$) and pseudo-uniform flow ($S_o = S_E$) prevail *simultaneously*. This, however, corresponds exactly to the generalized condition of Sect. 6.3.2 for a flow with losses. Remember at this point the computational assumptions for a flow with $\mathrm{d}Q/\mathrm{d}x = 0$ in prismatic channels.

The water surface slope $(\mathrm{d}h/\mathrm{d}x)_c$ at the critical point $x = x_c$ can be calculated using L'Hopital's rule. If Eq. (6.28) is differentiated with respect to x, a slightly modified expression, vis-à-vis Eq. (6.15), results for the loss term. Important in this connection is the recognition that the flow depth increases for the + sign, and for – sign the flow surface drops in the positive direction x. Because converging streamlines involve no additional losses, critical flow applies only at the *transition* from sub- to supercritical flow. For the reverse case, a hydraulic jump takes place (Chap. 7). The hydraulic jump is accompanied by a high degree of energy dissipation and the assumptions of pseudo-potential flow are then no longer valid.

Because Eq. (6.15) for a variable bottom profile as well as Eq. (6.22) for a variable cross-sectional geometry show that the critical water surface slope $(\mathrm{d}h/\mathrm{d}y)_c$ is finite, the critical flow must pass through a distinctive cross-section. Infinitesimal distances upstream and downstream of this cross-section result in sub- and supercritical flow conditions, respectively. One therefore speaks also of a *critical cross-section* or, for the one-dimensional representation of Fig. 6.1, of a so-called *critical point* $x = x_c$. At the critical point the following features are established:

- Critical flow depth $h = h_c$,
- Critical velocity $V = V_c$,
- Critical discharge $Q = Q_c$ and
- Critical energy head $H = H_c$.

These conditions are again diametrically opposite to those of uniform flow: Critical flow occurs in an *unique* cross-section whereas uniform flow describes an asymptotic condition that requires an infinitely long channel to establish, except for $S_o = S_E$.

In summary, the critical flow condition can be described as:

- By *critical flow*, one understands the condition in which either for a given discharge Q, a minimum energy head H_c results, or for a given energy head H, a maximum discharge Q_c is obtained.
- Critical flow generally occurs as the *Froude number* is $\mathsf{F} = 1$, and the channel geometry must vary locally thereby excluding downstream submergence.
- The *required* local changes in the flow characteristics are, either individually or in combination, a bottom elevation, a roughness reduction, an increase or decrease in discharge, or a contraction of the cross-section. Figure 6.2 shows the five basic cases where critical flow may establish, and the remaining cases for which the true extreme value lies along the energy line $H(x)$ but does *not* result in a minimum energy head.
- The *sufficient condition* for critical flow requires a Froude number $\mathsf{F} = 1$ at the critical point; accordingly, critical flow can occur only for supercritical downstream flow.
- Critical flow can only exist as a *transitional condition* from subcritical ($\mathsf{F} < 1$) to supercritical ($\mathsf{F} > 1$) flow. As a result, critical flow occurs in a so-called critical cross-section. The corresponding other parameters at this location are the critical point x_c, the critical depth h_c, the critical velocity V_c, and the critical energy head H_c.
- The single-valued relation between the critical depth h_c and the *critical discharge* Q at the critical point is the condition $\mathsf{F} = 1$. Critical discharge therefore is influenced neither by the bottom slope nor by channel roughness but solely depends on cross-sectional shape.
- Since the critical discharge is related to the cross-sectional parameter, as in Eq. (6.11), the corresponding discharge $Q = Q_c$ is obtained as a function of cross-sectional geometry alone. Critical flow is often applied for *discharge measurement* (Chaps. 10, 11, 12 and 13).

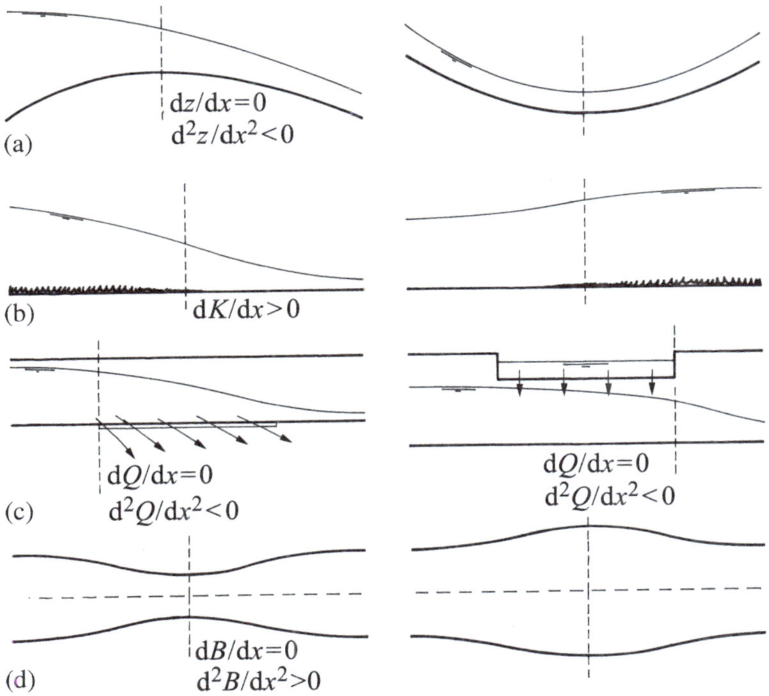

Fig. 6.2 Flow configurations satisfying the sufficient conditions for critical flow (with mathematical relations) and those with only an extreme in the energy head (without text). (- - -) Location of extreme energy head. Roughness coefficient $K = 1/n$

6.3.5 *Significance of Froude Number*

Open channel flows that have a certain minimum scale are governed by the Froude similarity law. If both the discharge Q and the flow depth h are too small, *viscous effects* become dominant with an additional effect of Reynolds number. Also, for curved free surface flows such as is typical close to the critical cross-section, *capillary effects* may play a role with an additional influence of the Weber number. These features are typical in hydraulic modelling with scale models and can only be controlled if the model scale is such that so-called *scale effects* remain insignificant. The theory of hydraulic modelling is complex, and reference is made to standard texts such as those of Ivicsics (1975), Kobus (1980, 1984), Novak and Cabelka (1981), Sharp (1981), Chadwick and Morfett (1996), and Miller (1994).

For models with a free surface flow and for which scale effects are insignificant, the Froude number is the main characteristic. Knowing the *Froude number* allows to draw basic conclusions on the main flow features. These include:

- $0 < \mathsf{F} \leq 0.3$ as *nearly pressurized flows* with $\mathsf{F}^2 < 10^{-1}$. Such flows are practically unaffected by free surface effects and the assumption of a rigid lid placed on the flow does not substantially alter the flow pattern.
- $0.3 < \mathsf{F} \leq 0.7$ as *subcritical type flows*, with typical backwater and drawdown effects. Here, free surface effects are obvious because of the formation of a surface gradient.
- $0.7 < \mathsf{F} \leq 1.5$ as *transitional flows* with a relatively unstable flow pattern that may be easily disturbed, particularly for decelerating flows. Examples are undular hydraulic jumps (Chap. 7) and breakdown of shockwaves due to low Froude number.
- $1.5 < \mathsf{F} \leq 3$ as *supercritical type flow*, with a stable flow pattern comparable to the corresponding subcritical flow. For those flows, the effect of shockwaves and hydraulic jumps is relatively small and these flows may be controlled by appropriate hydraulic means (Chap. 16).
- $3 < \mathsf{F}$ as *hypercritical flows*, with $\mathsf{F}^2 > 10^1$. These flows contain practically only kinetic energy and are extremely stable, provided flow disturbances are weak. If such a flow is significantly perturbed, large damage may result due to enormous hydraulic forces that are dissipated, however.

The classification of free surface flows is somewhat arbitrary because the limit Froude numbers as specified are intended to give an order of magnitude. It is noteworthy that by knowing the Froude number at a particular location the main features of flow are evident. This property has not yet been adopted until now by many designers, and the Froude number at a particular sewer location is currently often not determined. The next section is intended to improve this situation by introducing simple expressions for F such that an immediate application is provided.

6.4 Computation of Critical Flow

6.4.1 *Computational Principles*

In the context of critical flow there are two quantities to calculate:

(1) Froude number F from Eq. (6.4), and
(2) Relation between discharge Q and flow depth h_c for a Froude number $\mathsf{F} = 1$.

The first relation enables the determination of the Froude number and thus the character of flow ($\mathsf{F} < 1$ subcritical or tranquil, $\mathsf{F} > 1$ supercritical or shooting flow). Further, the magnitude of the number F indicates, in the context of the hydraulic jump, either air entrainment or the formation of shock waves. The Froude number may generally be considered to be the most important parameter of supercritical flows.

The critical flow depth h_c is often required for comparison with the prevailing flow depth. For example, in the computations of backwater and drawdown curves the parameter h/h_c is closely related to the Froude number. According to Eq. (6.4), the Froude number can also be described as

$$\mathsf{F} = Q/Q_c \tag{6.29}$$

where

$$Q_c = [gF^3/(\mathrm{d}F/\mathrm{d}h)]^{1/2}, \tag{6.30}$$

in which the subscript «*» has now been dropped. The parameter Q_c may be interpreted as the critical discharge; as the discharge Q equals the critical discharge Q_c, the Froude number is $\mathsf{F} = 1$. From Eq. (6.30) it is clear that Q_c depends only on the *local cross-sectional geometry*. The critical discharge is not influenced by characteristic elements of the channel such as the bottom slope S_o or the wall roughness. The critical discharge is determined exclusively by the function $F(h)$ of the cross-sectional geometry at the critical point $x = x_c$.

Besides these two parameters, the *critical energy head* H_c is often required. For a known critical flow depth h_c and cross-sectional area F_c, the critical velocity equals $V_c = Q_c/F_c$, and

$$H_c = h_c + V_c^2/(2g). \tag{6.31}$$

The ratio H_c/h_c of the critical energy head and the critical depth depends solely on the cross-sectional shape. In a rectangular channel, Eq. (6.10) gives $H_c/h_c = 3/2$.

Example 6.1 Consider once again the prismatic rectangular channel of width b and cross-sectional area $F = bh$. The derivative of this function with respect to the flow depth, synonymous with the free surface width, is $\mathrm{d}F/\mathrm{d}h = b$. Setting these in Eq. (6.30) yields Eq. (6.9) for the critical discharge.

In the following sections approximate expressions for the Froude number F as well the critical discharge Q_c and the critical energy head H_c are derived for the three standard cross-sections of Chap. 5, namely, the circular, the egg-shaped and the horse-shoe sections.

6.4.2 *Circular Section*

From Chap. 5 a close approximation of the cross-sectional area of a circular section is

$$F/D^2 = \frac{4}{3}y^{3/2}\left[1 - \frac{1}{4}y - \frac{4}{25}y^2\right] \tag{6.32}$$

where D is diameter and $y = h/D$ filling ratio. In the entire range of $0 < y < 1$, the above expression gives a maximum deviation of less than 1.3% from the exact values.

The derivative of F with respect to the filling ratio y is

$$\frac{1}{D}\frac{\mathrm{d}F}{\mathrm{d}h} = 2y^{1/2}\left[1 - \frac{5}{12}y - \frac{28}{75}y^2\right]. \tag{6.33}$$

Substitution of Eqs. (6.32) and (6.33) into Eq. (6.4) yields (Hager 1990)

$$\mathsf{F} = \frac{Q}{(gD^5)^{1/2}}\frac{3}{4}\left(\frac{3}{2}\right)^{1/2} y^{-2}\frac{[1-(5/12)y-(28/75)y^2]^{1/2}}{[1-(1/4)y-(4/25)y^2]^{3/2}}. \tag{6.34}$$

The expression $f_K = \mathsf{F}/[Q/(gD^5)^{1/2}]$ therefore depends only on the filling ratio y. Values of the expression $f_K y^2$ calculated from Eq. (6.34) for $0 < y < 1$ vary between 0.919 and 0.929 (Table 6.1). For the usual range $0.3 < y < 0.95$ of filling ratios in sewers, the values of $f_K y^2$ differ from unity by less than 3.4%. The quantity $f_K y^2$ can therefore be set approximately equal to unity and consequently the *Froude number* is (Hager 1990)

$$\mathsf{F} = \frac{Q}{\sqrt{gDh^4}}. \tag{6.35}$$

The Froude number for the circular cross-section is therefore proportional to the discharge Q, the square of the flow depth h but merely to the square root of the diameter D. Equation (6.35) is extremely simple and enables explicit determination of all the parameters.

Setting $\mathsf{F} = 1$ in Eq. (6.35) and solving for the *critical depth* h_c, one gets the simple expression

$$h_c = [Q/(gD)^{1/2}]^{1/2}. \tag{6.36}$$

According to Eq. (6.36), h_c depends only on the fourth root of the diameter D. In the range $0.2 < h_c/D < 0.91$, the deviations of the values of h_c, calculated from Eq. (6.36), from its exact values are less than 4%. The exact expression for h_c is not derived here at all, as it is too complex for practical purposes. Table 6.2 presents values of the exact (subscript e) filling ratio $y_e = (h_c/D)_e$ versus the approximations (subscript a) $y_a = (h_c/D)_a$ calculated from Eq. (6.36).

Table 6.1 Values of the expression $f_K y^2$ from Eq. (6.34) in relation to the filling ratio y

y	0	0.1	0.2	0.3	0.4	0.5	0.6
$f_K y^2$	0.919	0.935	0.952	0.970	0.988	1.006	1.022
y	0.65	0.7	0.75	0.8	0.9	0.95	1
$f_K y^2$	1.028	1.032	1.034	0.031	1.006	0.977	0.929

Table 6.2 Critical filling ratio y according to geometrically exact (subscript e) and approximate (subscript a) relations with the relative discharge $q_D = Q/(gD^5)^{1/2}$

q_D	0	0.1	0.2	0.3	0.4	0.5
y_e	0	0.3216	0.4522	0.5509	0.6329	0.7043
y_a	0	0.3162	0.4472	0.5477	0.6325	0.7071
q_D	0.6	0.7	0.8	0.9	0.95	1
y_e	0.7684	0.8280	0.8867	0.9543	1.0061	∞
y_a	0.7746	0.8367	0.8944	0.9487	0.9747	1

Apart from the critical flow depth, the *critical energy head* H_c is also of practical interest. The following expression for the relative critical energy head $Y_c = H_c/D$ developed by Hager (1990), gives values, in the range $0.1 < Q/(gD^5)^{1/2} < 0.75$, whose deviations from their exact values are less than 4%

$$\frac{H_c}{D} = \frac{5}{3}q_D^{3/5} = \frac{5}{3}\left[\frac{Q}{(gD^5)^{1/2}}\right]^{3/5}. \tag{6.37}$$

Equation (6.37) is numerically evaluated in Table 6.3.

For less than 10% filling ratio, the following more complex relation can be used

$$Y_c = \frac{H_c}{D} = \frac{41}{32}q_D^{1/2}\left[1+\frac{1}{9}q_D^{1/2}\right], \quad y_c < 0.1. \tag{6.38}$$

Equation (6.37) represents a power function and is therefore convenient for algebraic calculations. Elimination of the relative discharge $Q/(gD^5)^{1/2}$ between Eqs. (6.36) and (6.37) provides finally a relation between the critical flow depth h_c and the critical energy head H_c as

$$\frac{h_c}{D} = \frac{2}{3}\left(\frac{H_c}{D}\right)^{5/6}. \tag{6.39}$$

Figure 6.3 shows graphically the relation between the relative discharge $q_D = Q/(gD^5)^{1/2}$ and the critical filling ratio h_c/D as also the critical energy head ratio H_c/D.

The available methods for calculating critical flows deter many users by the formidable nature of the expressions. The system of equations presented here may be considered simple and user-friendly. It remains to be hoped that these will be used, because the critical flow determines important observations as compared

Table 6.3 Critical energy head $Y_c = H_c/D$ in relation to relative discharge $q_D = Q/(gD^5)^{1/2}$ from Eq. (6.37)

q_D	0	0.1	0.2	0.3	0.4	0.5	0.6	0.7	0.8	0.9	1
Y_c	0	0.419	0.635	0.809	0.962	1.100	1.227	1.346	1.458	1.565	1.667

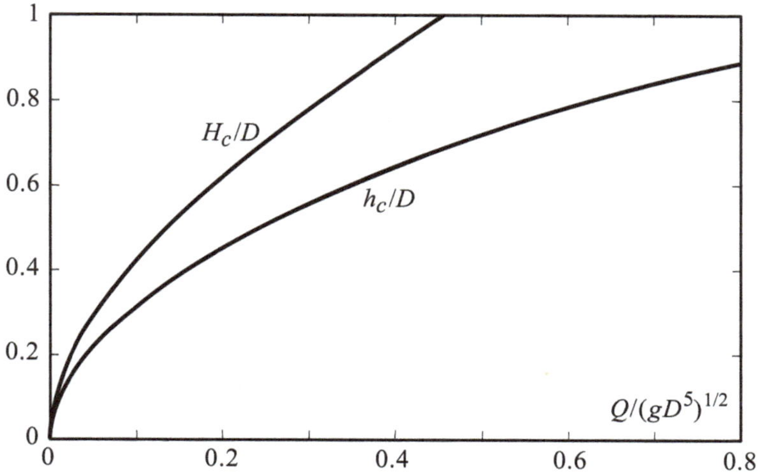

Fig. 6.3 Critical parameters in circular section

with effective discharge patterns. The approximate expressions recommended here, namely, Eq. (6.35) for the Froude number, and Eqs. (6.37) and (6.39) for the critical energy head, are comparable to the corresponding expressions for the rectangular cross-section.

Example 6.2 Calculate the critical flow parameters for a discharge $Q = 0.8\ \mathrm{m^3s^{-1}}$ in a circular pipe of diameter $D = 0.9$ m.
With $q_D = Q/(gD^5)^{1/2} = 0.8/(9.81 \cdot 0.9^5)^{1/2} = 0.332$, one obtains from Eq. (6.36) for $y_c = h_c/D = q_D^{1/2} = 0.332^{1/2} = 0.577$, and for the critical flow depth $h_c = 0.577 \cdot 0.9 = 0.519$ m.

The critical energy head $Y_c = 1.667 \cdot 0.332^{3/5} = 0.860$ follows from Eq. (6.37). Therefore, $H_c = 0.860 \cdot 0.9 = 0.774$ m. Using Eq. (6.39) one calculates $y_c = (2/3)0.860^{5/6} = 0.588$, whereas from Eq. (6.36) the 2% smaller value of 0.577 is obtained.

6.4.3 *Egg-shaped Section*

From Sect. 5.5.4, the approximate cross-sectional area function for the egg-shaped section 2:3 is

$$F/T^2 = \frac{25}{36} y^{3/2}[1 - 0.15y - 0.10y^4], \tag{6.40}$$

where $T = 3r$ is the height of the section from invert to soffit and $y = h/T$ is the filling ratio. The derivative of cross-sectional area F with respect to flow depth h equals

$$\frac{d(F/T^2)}{d(h/T)} = \frac{75}{72}y^{1/2}\left[1 - \frac{1}{4}y - \frac{11}{30}y^4\right]. \tag{6.41}$$

Combining now Eq. (6.41) with Eq. (6.4) one gets for the *Froude number*

$$\mathsf{F} = \frac{Q}{\sqrt{gT^5}} \frac{\left[\frac{75}{72}y^{1/2}(1 - 0.25y - 0.367y^4)\right]^{1/2}}{\left[\frac{25}{36}y^{3/2}(1 - 0.15y - 0.10y^4)\right]^{3/2}}. \tag{6.42}$$

In analogy to Sect. 6.4.2, the factor f_E is defined as the zeroth order expression

$$f_E = \frac{\mathsf{F}}{Q/(gT^5)^{1/2}} = \frac{36}{25}\left(\frac{3}{2}\right)^{1/2} y^{-2}. \tag{6.43}$$

The higher order expression left out in Eq. (6.43) varies between 0.953 for $y = 1$ and 1.056 for $y = 0.68$. For $0 < y < 0.95$ however the values always lie between 1 and 1.056. Multiplying therefore the constant of Eq. (6.43) by the average value 1.02, one obtains the approximate expression

$$\frac{\mathsf{F}}{Q/(gT^5)^{1/2}} = \frac{9}{5}y^{-2}. \tag{6.44}$$

Values of the *Froude number* in egg-shaped sections calculated from Eq. (6.44) within the range $0 < y < 0.95$ deviate from their exact values by less than 3%, and are therefore sufficiently accurate for practical computations. Solving this relation for F, it follows

$$\mathsf{F} = \frac{9}{5}\frac{Q}{\sqrt{gTh^4}}. \tag{6.45}$$

Equation (6.45) resembles closely Eq. (6.35), with a quadratic relationship between F and the flow depth h, but only a weak relation between F and the height T of the cross-section.

The *critical flow depth* h_c is calculated explicitly from Eq. (6.45) as

$$h_c = 1.34\,[Q/(gT)^{1/2}]^{1/2}. \tag{6.46}$$

In the entire range of possible values of the filling ratio, namely $0 < y < 0.95$, Eq. (6.46) yields values of h_c that agree within ±2% their exact values.

For the *critical energy head* $Y_c = H_c/T$, the corresponding quantity to the circular section is

$$Y_c = y_c + \left(\frac{Q}{\sqrt{gTD^4}}\right)^2 \frac{1}{2(F_c/D^2)^2}. \tag{6.47}$$

If Q is eliminated between Eqs. (6.46) and (6.47), an approximate relation between Y_c and y_c is

$$Y_c = \frac{4}{3} y_c \left[1 + 0.15 y_c^2\right]. \tag{6.48}$$

For $0 < y_c < 0.95$ the deviations are less than ±1%. A cruder approximation with deviations of the order of ±7% is given by the linear expression

$$Y_c = \sqrt{2} y_c. \tag{6.49}$$

An explicit relation between *critical discharge* Q and critical energy head H_c is obtained by eliminating h_c between the Eqs. (6.46) and (6.49) as

$$Q = 0.278 \sqrt{gT} H_c^2. \tag{6.50}$$

Critical discharge, therefore, depends essentially on the critical energy head and only weakly on cross-sectional geometry. The analogy with the circular section is noteworthy.

Example 6.3 Determine the critical flow characteristics for an egg-shaped section 2:3 of height $T = 1.8$ m and a discharge of $Q = 2$ m^{3}s^{-1}.
The critical flow depth is $h_c = 1.34[2/(9.81 \cdot 1.8)^{1/2}]^{1/2} = 0.924$ m from Eq. (6.46). The cross-sectional area of flow corresponding to the critical filling ratio $y_c = h_c/T = 0.924/1.8 = 0.514$ is obtained from Eq. (6.40) as $F_c/T^2 = 0.234$. From this, $F_c = 0.234 \cdot 1.8^2 = 0.759$ m^2. The critical energy head is thus $H_c = 0.924 + 4/(2 \cdot 9.81 \cdot 0.759^2) = 1.278$ m. From Eq. (6.48), $Y_c = 1.33 \cdot 0.514[1 + 0.15 \cdot 0.514^2] = 0.711$, corresponding to $H_c = 0.711 \cdot 1.8 = 1.279$ m. The discharge calculated from Eq. (6.50) is 1.91 m^{3}s^{-1} and about 4.5% less than the given value of $Q = 2$ m^{3}s^{-1}.

6.4.4 *Horseshoe Section*

From Sect. 5.5.4, the approximate cross-sectional area function for the horseshoe section 2:1.5 is

$$F/T^2 = 2.11 y^{3/2}[1 - 0.6 y^{3/2} + 0.1 y^3], \tag{6.51}$$

where T is the height of the cross-section from invert to soffit and $y = h/T$ is the filling ratio. Values of cross-sectional areas calculated from Eq. (6.51) differ from their geometrically exact values by less than ±3.6%. The derivative of Eq. (6.51) is

$$\frac{\mathrm{d}(F/T^2)}{\mathrm{d}(h/T)} = 3.17 y^{1/2} [1 - 1.2 y^{3/2} + 0.3 y^3]. \tag{6.52}$$

The *Froude number* is now calculated from Eq. (6.4) as

$$\mathsf{F} = \frac{0.58}{y^2}\frac{Q}{(gT^5)^{1/2}}\frac{(1-1.2y^{3/2}+0.3y^3)^{1/2}}{(1-0.6y^{3/2}+0.1y^3)^{3/2}}. \tag{6.53}$$

The expression $f_H = \mathsf{F}/[(0.58/y^2)Q/(gT^5)^{1/2}]$ varies, for values of $y < 0.95$, between 1.0 and 1.12. Taking the average value, one obtains

$$\frac{\mathsf{F}}{Q/(gT^5)^{1/2}} = 0.62y^{-2}. \tag{6.54}$$

The approximate *Froude number* for the horseshoe section is then given by

$$\mathsf{F} = 0.62\frac{Q}{\sqrt{gTh^4}}. \tag{6.55}$$

Equation (6.55) is analogous to Eq. (6.35) for the circular, and Eq. (6.45) for the egg-shaped sections. Setting $\mathsf{F} = 1$ in Eq. (6.55) the *critical flow depth* in the horseshoe section is

$$h_c = 0.787(Q/\sqrt{gT})^{1/2}. \tag{6.56}$$

The *critical energy head* in the horseshoe section further follows with ±1% accuracy

$$Y_c = 1.30y_c\left(1+\frac{5}{8}y_c^{5/2}\right). \tag{6.57}$$

The following expression for the critical energy head has a larger range of deviations (±10%) from the geometrically exact values, but is of power form

$$Y_c = 1.8y_c^{6/5}. \tag{6.58}$$

Analogous to Eq. (6.50), a convenient explicit relation between Q and H_c is

$$\frac{Q}{\sqrt{gT^5}} = \frac{3}{5}Y_c. \tag{6.59}$$

Example 6.4 Given a horseshoe section of height $T = 1.5$ m. Calculate the critical characteristics for a discharge of $Q = 2\ \mathrm{m^3s^{-1}}$.
Critical depth from Eq. (6.56) is $h_c = 0.787(2/(9.81\cdot 1.5)^{1/2})^{1/2} = 0.568$ m. The critical filling ratio is then $y_c = 0.568/1.5 = 0.379$. For the area of the critical cross-section, we get $F_c/T^2 = 0.426$, and $F_c = 0.426\cdot 1.5^2 = 0.96\ \mathrm{m^2}$. The critical velocity obtains $V_c = Q/F_c = 2.08\ \mathrm{ms^{-1}}$, and the critical energy head is $H_c = 0.568 + 2.08^2/19.62 = 0.789$ m. From Eq. (6.57) one calculates $H_c = 0.780$ m (–1.2%), and from Eq. (6.58) $H_c = 0.842$ m (+6.8%).

6.4.5 *Critical Slope*

Besides the critical flow depth h_c and the critical energy head H_c, a fictitious channel bottom slope S_c can introduced. The *critical slope* S_c is that slope for which critical flow is exactly established under uniform flow conditions. If the actual bottom slope $S_o < S_c$, subcritical uniform flow prevails, whereas for $S_o > S_c$ supercritical uniform flow occurs.

According to the conclusions of Chap. 5, the uniform flow law of Manning and Strickler often represents a remarkable approximation of the real flow situation. If the energy line slope S_E equals the channel bed slope S_o and if n denotes Manning's roughness coefficient, F the cross-sectional area of flow and $R_h = F/P$ the hydraulic radius with P as the wetted perimeter, the bottom slope S_o required to ensure *uniform flow* is

$$S_o = \frac{n^2 Q^2}{F^2 R_h^{4/3}}. \tag{6.60}$$

The critical slope can now be calculated if the discharge Q, for example, is replaced by the critical discharge obtained from Eq. (6.4) by setting $\mathsf{F} = 1$. This gives

$$S_c = \frac{gF}{\mathrm{d}F/\mathrm{d}h} \cdot \frac{n^2}{(F/P)^{4/3}}. \tag{6.61}$$

The right side of Eq. (6.61) depends exclusively on the n-value, on the flow depth and on the cross-sectional geometry. If in Eq. (6.61) one sets $F = T^2(F/T^2)$, $P = T(P/T)$ and $h = T(h/T)$, the generalized relationship for critical slope in terms of the preceding dimensionless relative values is obtained as

$$j_c = \frac{T^{1/3}}{n^2 g} S_c = \frac{1}{[\mathrm{d}(F/T^2)/\mathrm{d}(h/T)]} \cdot \frac{(P/T)^{4/3}}{(F/T^2)^{1/3}}. \tag{6.62}$$

The expression for j_c therefore depends only on cross-sectional geometry and on the filling ratio $y = h/T$. For a circular cross-section, the cross-sectional height T corresponds to the diameter D of the conduit.

Circular Section

The wetted perimeter of a circular section is

$$P/D = \arccos(1 - 2y), \tag{6.63}$$

and from Eq. (3.8) the hydraulic radius is

$$R_h/D = (2/3)y[1 - (1/2)y]. \tag{6.64}$$

Using these, j_c can be determined from Eq. (6.62). For the range $y_c < 0.9$, j_c can be approximated, with ±5% accuracy, by the expression

$$j_c = \frac{[3/(2y_c)]^{1/3}}{1 - 0.87y_c}. \tag{6.65}$$

This relation can simplify the determination of the critical slope $j_c = [D^{1/3}/(n^2 g)]S_c$ in the circular section ($T = D$) for the usual circumstances considered here. As can be seen, S_c decreases sharply with decreasing n-value but only slowly with increasing diameter. For small filling ratios, namely $y_c < 0.2$, the empirical formula $S_c = (gn^2)[3/(2h_c)]^{1/3}$ indicates that for small part-full flows, the critical slope is independent of the conduit diameter D.

Example 6.5 How large is the critical slope in Example 6.2 for $n^{-1} = 85$ $m^{1/3}s^{-1}$?
With $Q = 0.8$ m^3s^{-1}, $D = 0.9$ m and the critical depth $h_c = 0.519$ m, one gets $y = y_c = 0.519/0.9 = 0.577$, thus $j_c = [3/(2 \cdot 0.577)]^{1/3}/(1 - 0.87 \cdot 0.577) = 2.76$ from Eq. (6.65). Therefore, with $T = D$, the critical slope is $S_c = gn^2D^{-1/3}j_c = 9.81/(85^2 \cdot 0.9^{1/3})2.76 = 0.40\%$.

With $q_{Nc} = nQ/(S_c^{1/2}D^{8/3}) = 0.2$ as the normalized critical discharge, the relevant critical depth is $y_c = 0.575$, and $h_c = 0.517$ m (−0.3%).

How large is the critical slope for a discharge of $Q = 0.1$ m^3s^{-1}?

For $Q = 0.1$ m^3s^{-1}, from Eq. (6.36), $h_c = [0.1/(9.81 \cdot 0.9)^{1/2}]^{1/2} = 0.183$ m, and $y_c = 0.183/0.9 = 0.204$. With that and Eq. (6.65), one has $j_c = [3/(2 \cdot 0.204)]^{1/3}/(1-0.87 \cdot 0.204) = 2.36$. This gives $S_c = gn^2D^{-1/3}j_c = 9.81/(85^2 0.9^{1/3})2.36 = 0.33\%$.

With $y_c = y_N = 0.204$, $q_N = 0.030$, and $n^{-1}S_o^{1/2}D^{8/3} = Q/q_N = 0.1/0.030 = 3.28$ m^3s^{-1}. Therefore $S_o = [3.28/(85 \cdot 0.9^{8/3})]^2 = 0.262\%$. The deviation of 0.06% is due to the various approximations involved.

Egg-shaped Section

For the egg-shaped section, the wetted perimeter P as a function of the filling ratio $y = h/T$, is

$$P/T = 0.693[\arccos(1 - 2y)]^{5/4}, \quad y < 0.95. \tag{6.66}$$

For $0.05 < y < 0.9$, the deviations from the exact expression are smaller than ±3%. The *hydraulic radius* R_h for $y < 0.85$ can be approximated, with ±9% accuracy, by the relation

$$R_h/T = 0.29y^{3/4}. \tag{6.67}$$

Substitution of Eqs. (6.40), (6.41) and (6.67) in Eq. (6.62) yields, for the relative critical slope, an equation analogous to Eq. (6.65)

$$j_c = \frac{[4/(3y_c)]^{1/3}}{1 - 0.87y_c^{1/2}}, \; y_c < 0.95. \tag{6.68}$$

Example 6.6 How large is the critical slope in Example 6.3 for $n^{-1} = 75$ $m^{1/3}s^{-1}$?
With $Q = 2$ m^3s^{-1}, $T = 1.8$ m, and critical depth $h_c = 0.924$ m, $y_c = 0.514$. Substitution in Eq. (6.68) gives $j_c = [4/(3 \cdot 0.514)]^{1/3}/(1 - 0.87 \cdot 0.514^{1/2}) = 3.65$, and $S_o = g/(n^{-2}T^{1/3})j_c = 9.81/(75^2 \cdot 1.8^{1/3})3.65 = 0.52\%$.

From Eq. (5.20) with $y = y_c = 0.514$ the relative discharge is $q_v = 0.446$. Since the width is $B = 1.2$ m, $Q_v = Q/q_v = 2/0.446 = 4.48$ $m^{1/3}s^{-1}$, from Eq. (5.42) $S_o = 1.294^{16/3}[Q_v/(n^{-1}B^{8/3})]^2 = 3.95[4.48/(75 \cdot 1.2^{8/3})]^2 = 0.53\%$, which is 0.01% larger than the value previously calculated.

Horseshoe Section

The wetted perimeter P of the horseshoe section is given as a function of the filling ratio $y = h/T$ to ±5% accuracy by the expression

$$P/T = 0.10\,[\arccos(1 - 2y)]^{4/5}, \tag{6.69}$$

in which the exponent is inversed as compared to Eq. (6.66). The hydraulic radius is approximated, with ±6% accuracy, as

$$R_h/T = 0.65y(1 - 0.6y^3). \tag{6.70}$$

Using Eqs. (6.51), (6.52) and (6.59) in Eq. (6.62) and proceeding in the manner of developing Eq. (6.68), a ±10% accurate expression for the critical slope j_c is

$$j_c = \frac{[4/(3y_c)]^{1/3}}{1 - y_c^{3/2}}, \; y_c < 0.85. \tag{6.71}$$

6.4.6 *Summary of Results*

To provide a better overview, the computational results pertaining to the critical flow parameters for the circular, the egg-shaped and the horseshoe cross-sections are summarized in Table 6.4.

- The *Froude number* F can be determined from the relation $f = \mathsf{F}/\mathsf{F}_o$ where the reference number is $\mathsf{F}_o = Q/(gTh^4)^{1/2}$. The values of the factor f for the three cross-sections lie between 0.62 and 1.8.

Table 6.4 Summary of critical flow parameters

Parameter (Normalized)	Horseshoe section	Circular section	Egg-shaped section
Froude number $\frac{F}{Q/(gTh^4)^{1/2}}$	0.62	1.0	1.8
critical flow depth $y_c = h_c/(Q/(gT)^{1/2})^{1/2}$	0.787	1.0	1.34
critical energy head $H_c/T = f_1(q_D)$	$(1.36)q_L^{3/5}$	$(5/3)q_D^{3/5}$	$(0.527)q_D^{1/2}$
$H_c/T = f_2(y_c)$	$1.8y_c^{6/5}$	$(3y_c/2)^{6/5}$	$\sqrt{2}y_c$
critical slope $j_c = (K^2T^{1/3}/g)S_c$	$[4/(3y_c)]^{1/3}/(1 - y_c^{3/2}$	$[3/(2y_c)]^{1/3}/(1 - 0.87y_c)$	$[4/(3y_c)]^{1/3}/(1 - 0.87y_c^{1/2})$

- The *critical flow depth* h_c can be expressed as $y_f = h_c/h_{co}$ with the reference measure $h_{co} = (Q/(gT)^{1/2})^{1/2}$. The numerical values of y_f for the three cross-sections correspond to the square root of the preceding series of f-values and lie between 0.787 and 1.34.
- The *critical energy head* $Y_c = H_c/T$ can be expressed in terms of either the discharge Q or the critical flow depth $y_c = h_c/T$. It is surprising that the exponent for both the circular and the horseshoe sections are identical.
- The *normalized critical slope* j_c can be expressed, with a reasonable degree of approximation, by a power formula alone over the entire range of filling ratios y_c. The basic exponent of y_c in the formula is –1/3 in all the three cases; for $y_c \rightarrow 1$, the value of j_c is large.

Lastly, it should be remembered that the cross-sectional height T for the circular section corresponds to the pipe diameter D. Also, the relations presented in Table 6.4 represent approximate expressions with an accuracy of about ±5%. With these expressions the *practical calculation* of the critical flow parameters is simple and straightforward. In any case, the argument of large computational effort with the expressions presented herein can not be sustained anymore.

6.5 Transition from Mild to Steep Sewer Reaches

6.5.1 *Computational Assumptions*

If the approach flow to a manhole is subcritical and the outflow is supercritical, critical flow occurs in the manhole itself. The slope of the approach (subscript o) flow channel S_{oo} is therefore smaller and the downstream (subscript u) channel slope S_{ou} larger than the critical slope. For a manhole which has to pass a large discharge Q, the transition from subcritical to supercritical flow is examined here in more detail than in Sect. 6.4. In particular, it is not sufficient to determine only the critical depth h_c, but the *water surface profile* in the neighbourhood of the critical point must also be predicted. The downstream sewer must be correctly designed so that a transition from free-surface to pressurized flow owing to a reduced diameter, as discussed in Sect. 7.4, is avoided (Hager, 1987).

Because in the downstream sewer a velocity higher than in the approach flow sewer prevails, the diameter D_u in the downstream reach compared to the diameter D_o of the upstream sewer is reduced. Usually, a linear transition profile $D(x)$ between the two sewers is chosen as

$$D = D_o - \theta x. \tag{6.72}$$

The origin of co-ordinates $x = 0$ is set at the end of the approach flow sewer (Fig. 6.4) and the x-direction coincides with its direction. The contraction angle is $\theta = (D_o - D_u)/L_u$ where L_u is the length of the transition reach.

To prevent flow separation from the channel bottom, the transition curve from the upstream slope S_{oo} to the downstream slope S_{ou} is continuous. The simplest bottom transition profile $z(x)$ is a circular arc of radius R_u. For small differences in the up- and the downstream bottom slopes, the circular arc is approximated by the *parabolic profile*

$$z = -x^2/(2R_u). \tag{6.73}$$

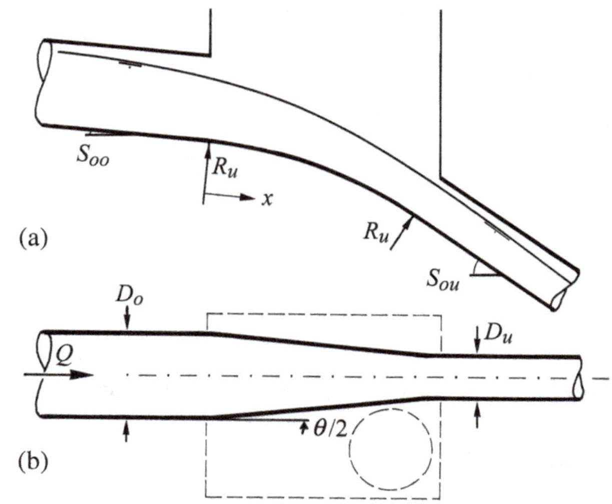

Fig. 6.4 (**a**) Longitudinal section and (**b**) plan of control manhole with change from sub- to supercritical flow

The cross-section of the channel transition in the control manhole is usually U-shaped. The U-shape is formed by a rectangular section set on top of a semicircular bottom portion, and its cross-sectional area F can be approximated for filling ratio $y = h/D < 1.2$ by

$$F/D^2 = \frac{4}{3}y^{3/2}\left(1 - \frac{1}{3}y\right), \tag{6.74}$$

where the diameter D varies with x in accordance with Eq. (6.72).

In a gradual channel contraction the *energy loss* is only on account of wall friction. In usual sewers, the friction slope S_f is given by Eq. (6.24). Since a control manhole is hydraulically a short structure (Chap. 8), the change in the friction slope S_f along the manhole length is small and its influence on the free surface profile even smaller. While $S_{oo} < S_f < S_{ou}$ holds for the slopes, the water depth relations are $h_o < h_{oN}$ at the upstream end of the manhole and $h_u > h_{uN}$ at its downstream end. In order that the calculations are simple, a *constant energy slope* $S_E = S_{fm} = S_{oo}$ is assumed. If the manhole is also tilted by this angle S_{oo}, the energy line becomes horizontal such that $\mathrm{d}H/\mathrm{d}x = 0$. Figure 6.5 illustrates the hydraulic substitute system. In this figure are shown the critical depth h_c, the corresponding critical energy head H_c as also the location of the critical point x_c.

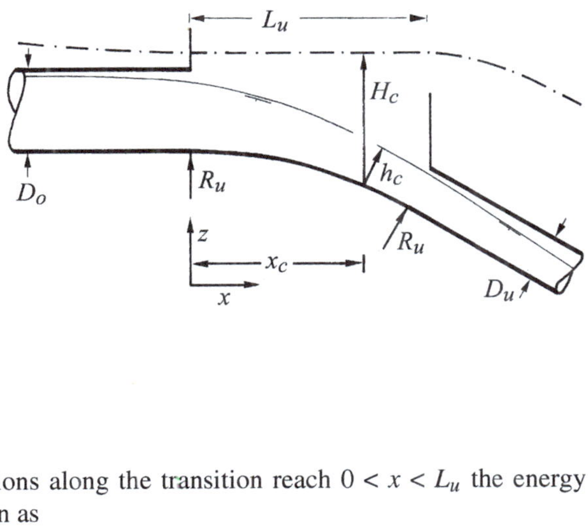

Fig. 6.5 Hydraulically equivalent system with (—) flow surface, (- · · -) energy line

6.5.2 *Critical Point*

With the preceding assumptions along the transition reach $0 < x < L_u$ the energy equation (6.23) can be written as

$$H = -\frac{x^2}{2R_u} + h + \frac{Q^2}{2gF^2}, \tag{6.75}$$

$$\frac{\mathrm{d}H}{\mathrm{d}x} = 0. \tag{6.76}$$

The *position of the critical point* is determined by taking the derivative of Eq. (6.75) under the assumption of constant discharge and setting it, in accordance with Eq. (6.76), equal to zero, thus

$$\frac{\mathrm{d}H}{\mathrm{d}x} = -\frac{x}{R_u} + \frac{\mathrm{d}h}{\mathrm{d}x} - \frac{Q^2}{gF^3}\left(\frac{\partial F}{\partial x} + \frac{\partial F}{\partial h}\frac{\mathrm{d}h}{\mathrm{d}x}\right) = 0. \tag{6.77}$$

Rearranging terms gives

$$-\left[\frac{x}{R_u} + \frac{Q^2}{gF^3}\frac{\partial F}{\partial x}\right] + \frac{\mathrm{d}h}{\mathrm{d}x}\left[1 - \frac{Q^2}{gF^3}\frac{\partial F}{\partial h}\right] = 0. \tag{6.78}$$

Since the expression in the second bracket is identical to $(1 - \mathsf{F}^2)$ and vanishes at the critical point x_c, it follows

$$\frac{x_c}{R_u} + \frac{Q^2}{gF^3}\frac{\partial F}{\partial x} = 0. \tag{6.79}$$

Setting the critical condition $\mathsf{F} = 1$ from Eq. (6.4) gives $Q^2/gF^3 = (\partial F/\partial h)^{-1}$. If this and the derivatives $\partial F/\partial x$ and $\partial F/\partial h$ obtained from Eq. (6.74) are substituted into Eq. (6.79), one gets with $y_c = h_c/D_c$

$$\frac{x_c}{R_u} = -\frac{\partial F/\partial x}{\partial F/\partial h} = \frac{\theta}{3}y_c\left[1 + \frac{1}{3}y_c\right]\left[1 - \frac{5}{9}y_c\right]^{-1}. \tag{6.80}$$

The location of the critical point $x_c/(R_u\theta)$ therefore depends only on the filling ratio y_c at the critical point. Dimensionless parameters scaled with the approach flow diameter D_o are introduced as

$$X = \theta x/D_o, y_o = h/D_o, \rho_u = R_u\theta^2/D_o \tag{6.81}$$

Noting the identities $D_c = D_o - \theta x_c = D_o(1 - X_c)$ and $y_c = h_c/D_c = h_c/[D_o(1 - X_c)] = y_{oc}(1 - X_c)$, where $y_{oc} = h_c/D_o$, y_{oc} can be represented as a function of only x_c and ρ_u (Fig. 6.6b). With the relation $y_{oc} = y_{oc}(X_c, \rho_u)$, the parameter h_c/D_c is eliminated, and a relation $y_{oc} = y_{oc}(q_D, \rho_u)$ with $q_D = Q/(gD_o^5)^{1/2}$ follows (Fig. 6.6a).

Example 6.7 Given $D_o = 1.5$ m, $D_u = 0.8$ m, $R_u = 4$ m, $L_u = 3$ m, $Q = 4$ m^{3}s^{-1}. Where is the critical point x_c?
With $\theta = (D_o - D_u)/L_u = (1.5 - 0.8)/3 = 0.233$, $\rho_u = 4 \cdot 0.233^2/1.5 = 0.145$ and $Q/(gD_o^5)^{1/2} = 4/(9.81 \cdot 1.5^5)^{1/2} = 0.463$, Fig. 6.6a) gives $y_{oc} =$

0.726, and Fig. 6.6b) indicates $X_c = 0.087$. Calculating back for dimensional quantities, $h_c = 0.726 \cdot 1.5 = 1.09$ m and $x_c = 0.087 \cdot 1.5/0.233 = 0.56$ m.

As a check, the critical depth in a circular section can be determined with $D_c = D_o - \theta x_c = 1.5 - 0.233 \cdot 0.56 = 1.37$ m. From Eq. (6.36) it then follows $h_c = [4/(9.81 \cdot 1.37)^{1/2}]^{1/2} = 1.05$ m, a value insignificantly smaller than the above result of 1.09 m.

For a particular value of ρ_u the function $y_{oc}(q_D)$ breaks off as the maximum value q_{DM} is reached. A higher discharge through this manhole cannot be realized. This maximum (subscript M) relative discharge can be approximated as

$$q_{DM} = 0.14\rho_u^{-0.8}. \tag{6.82}$$

Equation (6.82) in turn yields $Q_M = 0.14g^{1/2}D_o^{3.3}/(\theta^2 R_u)^{0.8}$. The maximum discharge Q_M thus depends largely on the diameter D_o and the contraction angle θ.

Once the distance x_c and the critical depth h_c are known, the critical energy head can be calculated, practically independent of ρ_u and related to the relative discharge $q_D = Q/(gD_o^5)^{1/2}$ as

$$Y_{oc} = H_c/D_o = 1.28q_D^{1/2}\left[1 + \frac{1}{4}q_D^{1/2}\right]. \tag{6.83}$$

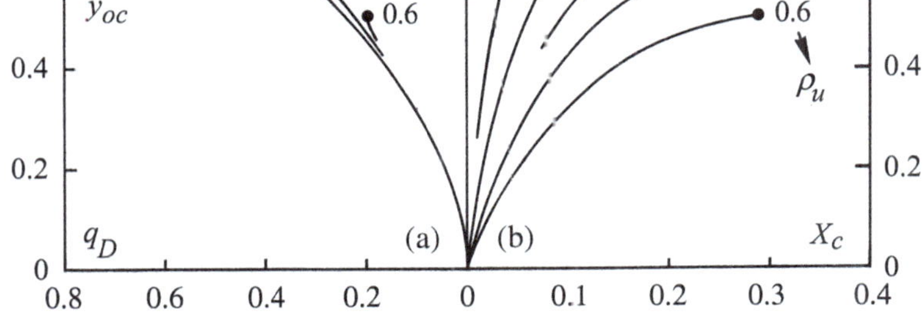

Fig. 6.6 Relations at the critical cross-section. Critical depth $y_{oc} = h_c/D_o$ as a function of parameters $\rho_u = \theta^2 R_u/D_o$ and (**a**) relative discharge $q_D = Q/(gD_o^5)^{1/2}$ as well as (**b**) critical distance $X_c = x_c\theta/D_o$. (•) Maximum value

Example 6.8 According to Example 6.7, $D_c = 1.37$ m. Equation (6.74) then gives $F_c = (4/3)(1.37^2 \cdot 1.09^3)^{1/2}[1 - 1.09/(3 \cdot 1.37)] = 1.305$ m^2. Accordingly, the critical velocity is $V_c = 4/1.305 = 3.06$ ms^{-1}. Then, the critical energy head is $H_c = 1.09 + 3.06^2/19.62 = 1.57$ m. The energy head at the manhole inlet is $H_o = H_c - x_c^2/2R_u = 1.57 - 0.56^2/(2 \cdot 4) = 1.53$ m.
With the relative discharge $q_D = 4/(9.81 \cdot 1.5^5)^{1/2} = 0.463$, Eq. (6.83) gives $H_c/D_o = 1.28 \cdot 0.463^{1/2}(1 + 0.25 \cdot 0.463^{1/2}) = 1.02$, which in turn gives also $H_c = 1.02 \cdot 1.5 = 1.53$ m.

6.5.3 *Free Surface Profile*

Consideration of Eq. (6.83) in the system of Eqs. (6.75) and (6.76) implicitly yields for the *free surface profile*

$$1.28q_D^{1/2}\left[1 + \frac{1}{4}q_D^{1/2}\right] = \frac{X^2}{2\rho_u} + y_o + \frac{(9/32)q_D^2}{(1-X)y_o^3\left[1 - \dfrac{y_o/3}{1-X}\right]^2}. \qquad (6.84)$$

The free surface profile $y_o(X)$, with $y_o = h/D_o$, depends only on q_D and ρ_u. Figure 6.7 shows $y_o(X)$ for different relative discharges $q_D = 0.1$, 0.2, 0.3 and 0.4. In these profiles, the critical points are marked by bold circles.

For a given discharge Q and manhole geometry (D_o, R_u, θ) the parameters q_D and ρ_u are first calculated and then the free surface is determined. In particular, the

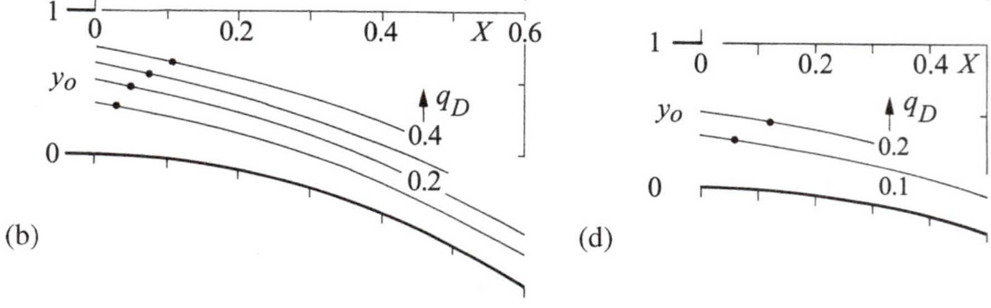

Fig. 6.7 Dimensionless free surface profiles $y_o(X)$ as a function of the relative discharge $q_D = Q/(gD_o^5)^{1/2}$ for $\rho_u = \theta^2 R_u/D_o =$ **(a)** 0.1, **(b)** 0.2, **(c)** 0.3, **(d)** 0.4. (•) Critical point (Hager 1987)

flow depth at the downstream manhole end is determined from Fig. 6.7, based on the smallest diameter that still guarantees free surface flow. For neighbouring values of ρ_u, the values of y_o read from the curves of Fig. 6.7 are interpolated to find an approximate value of y_o corresponding to the known value of ρ_u. A more correct value of y_o is then determined by solving for Eq. (6.84) and considering that there are two solutions, one for $\mathsf{F} < 1$ and the other for $\mathsf{F} > 1$.

Example 6.9 Given: upstream $D_o = 1.5$ m, $S_{oo} = 0.2\%$, $n_o^{-1} = 85$ m$^{1/3}$s^{-1}; downstream $D_u = 0.8$ m, $S_{ou} = 20\%$, $n_u = n_o$; manhole geometry $L_u = 3$ m, $R_u = 15$ m. What is the free surface profile for a discharge of $Q = 3$ m^{3}s^{-1}?
For uniform flow $h_{No} = 1.11$ m and $h_{Nu} = 0.38$ m. For critical flow, $h_{co} = 0.90$ m and $h_{cu} \cong D = 0.8$ m. For uniform flow, therefore, $\mathsf{F}_o < 1$ and $\mathsf{F}_u > 1$. With $\theta = (D_o - D_u)/L_u = 0.7/3 = 0.233$ the maximum discharge according to Eq. (6.82) is $Q_M = 0.14 \cdot 9.81^{0.5} 1.5^{3.3} 0.233^{-1.6} 15^{-0.8} = 1.97$ m^{3}s^{-1} $< Q = 3$ m^{3}s^{-1}.

For a given upstream diameter D_o, the maximum discharge can only be increased by reducing either θ or R_u. Both of these parameters are coupled with the downstream bottom slope S_{ou}. Since $S_{ou} = -\mathrm{d}z/\mathrm{d}x$ at $x = L_u$, it follows $S_{ou} = L_u/R_u$, and the contraction angle is $\theta = (D_o - D_u)/(S_{ou} R_u)$.

In Example 6.9 therefore, the downstream diameter D_u is too small for the bottom slope S_{ou} prescribed. The required *minimum downstream diameter* D_{um} can be obtained by setting the preceding expression for θ in Eq. (6.82), thus

$$D_{um} = D_o - S_{ou}(R_u D_o)^{1/2}(0.14/q_D)^{5/8}. \quad (6.85)$$

Therefore, the required minimum diameter is larger for smaller downstream bottom slope and larger relative discharge $q_D = Q/(gD_o^5)^{1/2}$.

Example 6.10 Consider again the control manhole of Example 6.9. From the data given, $q_D = 3/(9.81 \cdot 1.5^5)^{1/2} = 0.348$ the minimum downstream diameter is, from Eq. (6.85), $D_{um} = 1.5 - 0.2(15 \cdot 1.5)^{1/2}(0.14/0.348)^{5/8} = 0.96$ m, chosen $D_u = 1.1$ m. Then $\theta = (1.5 - 1.1)/(0.2 \cdot 15) = 0.133$ and $\rho_u = 0.133^2 15/1.5 = 0.178$.
With a manhole length $L_u = S_{ou} R_u = 3$ m, the downstream location obtains $X_u = \theta L_u/D_o = 0.133 \cdot 3/1.5 = 0.267$. For $q_D = 0.35$, Fig. 6.7a) gives $y_o = 0.48$, and for $\rho_u = 0.2$, Fig. 6.7b) gives $y_o = 0.58$. Interpolating for $\rho_u = 0.178$, we get $y_o = 0.56$. The corresponding value is $h_u = y_o D_o = 0.56 \cdot 1.5 = 0.84$ m. From Fig. 6.6a), $y_{oc} = 0.61$, and from Fig. 6.6b), $X_c = 0.07$. Correspondingly then, $x_c = X_c D_o/\theta = 0.07 \cdot 1.5/0.133 = 0.79$ m.

Once the location of the critical point is determined, the flow depths h_o and h_u at the manhole entry and exit sections, respectively, can either be obtained from Fig. 6.7 or calculated from Eq. (6.84). Surface profiles up- and downstream from the manhole can then be continued in the conventional manner using the backwater and drawdown curves (Chap. 8) for reaches which do not include a change in the flow type.

This section may have shown how the calculation of a flow transition is possible by using simplifying assumptions. Not considered are *shockwaves* (Chap. 16) originating from the converging transition profile to the downstream prismatic sewer. These shockwaves give rise to surface disturbances. Currently no experimental investigation on the height of these waves appears to exist. Also to be examined is the phenomenon of *air entrainment* in steep conduits (Sect. 5.6). The surface of air-entrained water flow is higher than calculated by the previous procedure.

Finally, reference is made, once again, to Eq. (6.85) that allows a preliminary design. The parameter R_u of this equation is a proper variable which, however, has a lower limit as to prevent the separation of the flow from the channel bottom.

6.5.4 *Experimental Verification*

Flows from a mild to a steep bottom are of general interest, mainly in dam engineering and irrigation works. The configuration as discussed for sewer hydraulics is particular in so far as the increase of bottom slope is combined with a decrease of width. Experiments on such flows have not yet been conducted. The following is a summary of some tests performed in a *rectangular* channel of constant width. A summary of results is available (Hager 1995).

Hasumi considered in 1931 an almost horizontal rectangular channel up to the transition point $x = 0$, and a steep downstream channel of either $\alpha = 45°$ or $60°$. The transition elements between the two channel reaches were either abrupt, or rounded with $R_u = 200$ mm, and the channel was $b = 402$ mm wide. Figure 6.8 shows a definition sketch with N as the flow depth measured perpendicularly to the bottom profile.

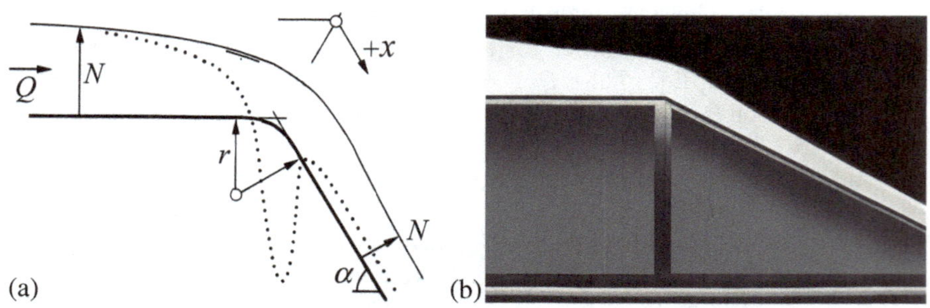

Fig. 6.8 (**a**) Definition of flow over a bottom drawdown element, (—) free surface profile, (...) pressure head profile, (**b**) typical flow configuration

The *free surface profile* $N(x)$, or if normalized with the critical flow depth $h_c = [Q^2/(gb^2)]^{1/3}$ as $X_c = x/h_c$ and $y_c = N/h_c$, can be expressed as

$$\frac{y - y_a}{y_o - y_a} = \tanh\left[\sigma(X_a - X)\right]. \tag{6.86}$$

Here, $(X_a; y_a) = (-0.75; +0.75)$ are the coordinates of the virtual origin of the surface profile, and $y_o = y(X \to \infty) = 1.10$ is the maximum upstream flow depth. The parameter $\sigma = 0.40$ defines the steepness of the curve $y(X)$. Equation (6.86) is valid for $-4 \le X \le +4$ and includes all observations of Hasumi for $\alpha = 45°$ and $60°$ for both transition geometries, and various discharges.

The *pressure head distribution* is nearly hydrostatic upstream from the draw-down point, experiences a minimum value close to the bottom kink and is parallel to the steep bottom in the downstream reach. Of particular relevance is the *minimum pressure* p_m or $p_m/(\rho g h_c) = -1.03$, which is practically independent of discharge and angle α for sharp-cornered transitions. The locations of minimum pressure is at $x_m = (1/3)h_c$, thus slightly downstream of the bottom kink.

The minimum pressure head $P_m = p_m/(\rho g h_c)$ for the *rounded* transition geometry varies essentially with the relative radius of curvature $r = h_c/R_u$ and angle α. Based on limited data, one may approximate

$$P_m = 0.75 r^{0.75} \tan\alpha. \tag{6.87}$$

Typically, the sub-pressure head is also of the order of the critical flow depth, therefore. Note that flows over kinks may *separate* from the bottom geometry if aerated, for which the flow structure changes dramatically.

The kink flow depth h_c is comparable to the end depth of an overfall (Chap. 11). For a rounded kink, the end section is located at the beginning of the curved bottom profile and $h_e = 0.72h_c$, almost independent of α and r. For a sharp-crested kink, the corresponding end depth ratio is $h_e/h_c = 0.65$.

Westernacher (1965) considered a *drop structure* in a 0.60 m wide rectangular channel of bottom radii $R_u = 0.20$ m, or 0.30 m with $\alpha = 52.5°$ ($\tan\alpha = 1.50$). The approach flow channel was horizontal. If x_o is the location where flow depth is h_o (Fig. 6.9), then the relative discharge $q_o = Q/[b(gR_u{}^3)^{1/2}]$ is related to $X_o = x_o/R_u$ as

$$h_o/R_u = 0.90(0.078 + X_o)^{0.09} q_o^{0.605}. \tag{6.88}$$

This relation applies exclusively for the design considered, and for $-2 \le X_o \le 0$ and $q_o < 0.60$. The end depth ratio $y_e = h_e/h_o$ is

$$y_e = 0.715(r/h_c)^{0.09}. \tag{6.89}$$

The *minimum pressure head* $p_m/(\rho g)$ occurs at 40° deflection from the horizontal channel and is

$$p_m/(\rho g R_u) = -0.8(h_c/R_u)^2. \tag{6.90}$$

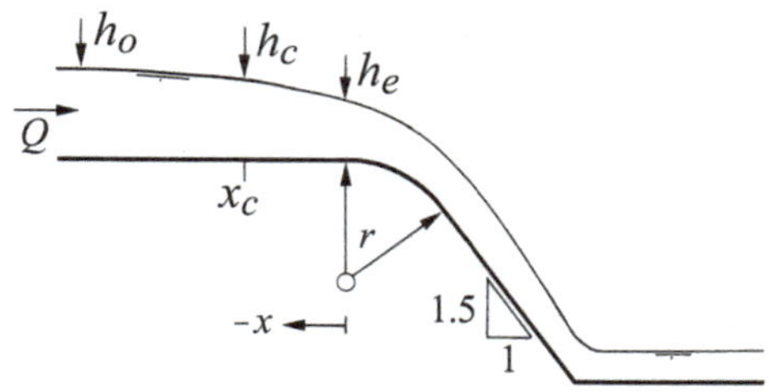

Fig. 6.9 Drop structure of Westernacher (1965)

Weyermuller and Mostapha (1976) considered the abrupt transition from a horizontal approach flow reach to a sloping downstream channel, with bottom slopes between 1 and 100% ($\alpha = 45°$). Contrary to Fig. 6.8 the free surface profiles may not be generalized. The end depth ratio $y_e = h_e/h_c$ was found to vary with both the aspect ratio b/h_c and the downstream bottom slope, but the channel used was so small that scale effects might have occurred. The problem of change from mild to steep bottom slopes needs obviously further experimentation. Castro-Orgaz and Hager (2009) contributed numerically to this problem using the extended Boussinesq equations (Chap. 1), thereby resulting in excellent agreement between observations and predictions.

Notation

b	[m]	channel width
B	[m]	width of channel
D	[m]	conduit diameter
f_E	[–]	relative Froude number in egg-shaped section
f_K	[–]	relative Froude number in circular section
f_H	[–]	relative Froude number in horseshoe section
F	[m^2]	cross-sectional area
F	[–]	Froude number
g	[ms^{-2}]	gravitational acceleration
h	[m]	flow depth
h_c	[m]	critical depth
H	[m]	energy head
H_*	[m]	energy head relative to channel bottom
L_u	[m]	length of transition
$1/n$	[m$^{1/3}$s^{-1}]	Manning's roughness coefficient
N	[m]	length of normal
p	[Nm^{-2}]	pressure
P	[m]	wetted perimeter
q	[m^{2}s^{-1}]	discharge per unit width

q_c	[m^2s^{-1}]	critical discharge per unit width
q_D	[–]	discharge normalized with respect to diameter
Q	[m^3s^{-1}]	discharge
Q_c	[m^3s^{-1}]	critical discharge
r	[–]	curvature ratio
R_h	[m]	hydraulic radius
R_u	[m]	radius of transition curve
S_c	[–]	critical slope
S_E	[–]	slope of energy line
S_F	[–]	slope due to variable cross-section
S_Q	[–]	slope due to spatially variable discharge
S_c	[–]	relative critical slope
S_o	[–]	channel bottom slope
T	[m]	height of cross–section from invert to soffit
V	[ms^{-1}]	average velocity
x	[m]	streamwise coordinate
X	[–]	relative streamwise coordinate
x_a	[–]	$= -0.75h_c$
X_c	[–]	$= x/h_c$
y	[–]	filling ratio
y_a	[–]	$= +0.75h_c$
y_c	[–]	$= N/h_c$
y_{oc}	[–]	critical depth normalized with respect to D_o
y_o	[–]	$= 1.10$
Y_c	[–]	relative critical energy head
z	[m]	vertical height of channel bed above datum
α	[–]	bottom angle
θ	[–]	angle of contraction
ρ	[kgm^{-3}]	density
ρ_u	[–]	relative channel bottom radius
σ	[–]	shape parameter

Subscripts

a	approximate
c	critical
e	geometrically exact
m	minimum
N	uniform flow
o	inlet, inflow
u	outlet, outflow
$*$	relative to channel bottom

References

Castro-Orgaz, O., Hager, W.H. (2009). Curved-streamline transitional flow from mild to steep slopes. *Journal of Hydraulic Research* 47(5): 574–584.

Chadwick, A., Morfett, J. (1996). *Hydraulics in civil and environmental engineering*. Spon: London.

Chow, V.T. (1959). *Open channel hydraulics*. McGraw-Hill: New York.

Hager, W.H. (1985). Critical flow condition in open channel hydraulics. *Acta Mechanica* 54: 157–179.

Hager, W.H. (1987). Übergang von Flach- auf Steilstrecke in Kanalisationen (Transition from mild to steep sewer). *Gas-Wasser-Abwasser* 67(7): 420–426 [in German].

Hager, W.H. (1990). Froudezahl im Kreisprofil (Froude number in circular conduit). *Korrespondenz Abwasser* 37(7): 789–791 [in German].

Hager, W.H. (1995). Übergang von Flach- auf Steilstrecken (Transition from mild to steep channels). *Wasser und Boden* 47(9): 20–24.

Henderson, F.M. (1966). *Open channel flow*. MacMillan: New York.

Ivicsics, L. (1975). *Hydraulic models*. Research Institute for Water Resources Development: Budapest.

Kobus, H. (1980). *Hydraulic modelling*. Pitman: Boston.

Kobus, H., ed. (1984). Symposium on *Scale effects in modelling hydraulic structures*. Technische Akademie: Esslingen, Germany.

Miller, D., ed. (1994). *Discharge measurement structures*. IAHR Hydraulic Structures Design Manual 8. Balkema: Rotterdam, The Netherlands.

Naudascher, E. (1987). *Hydraulik der Gerinne und Gerinnebauwerke* (Hydraulics of channels and channel structures). Springer: Wien, New York [in German].

Novak, P., Cabelka, J. (1981). *Models in hydraulic engineering*. Pitman: Boston.

Sharp, J.J. (1981). *Hydraulic modelling*. Butterworths: London.

Westernacher, A. (1965). Abflussbestimmung an ausgerundeten Abstürzen mit Fliesswechsel (Discharge determination at rounded drops involving transcritical flow). *Dissertation*. TU Karlsruhe, Germany [in German].

Weyermuller, R.G., Mostapha, M.G. (1976). Flow at grade-break from mild to steep slopes. *Journal of the Hydraulics Division* ASCE 102(HY10): 1439–1448; 103(HY8): 946–947; 103(HY9): 1110–1111; 104(HY2): 307–308.

Outlet from a stormwater basin into receiving water. The quality of these flows reflects often the performance of the entire scheme. It is also one of the rare locations where the population has eye-contact with a sewage system

Chapter 7
Hydraulic Jump and Stilling Basins

Abstract For the change of flow type from supercritical to subcritical, a hydraulic jump forms. The phenomenon is first described and then the sequent depths are determined using the specific force principle. The profiles include the rectangular channel and the three standard cross-sections, namely, the circular, the egg-shaped and the horseshoe sections. Choking of the circular section on account of a hydraulic jump is also introduced.

In the second part, relevant outlet structures are presented. The difference between the hydraulic jump and the stilling basin is explained. Subsequently a number of standard basin types are described and finally cases are enumerated which can lead to unplanned energy dissipation.

7.1 Introduction

Hydraulic flows usually run continuously, i.e. all parameters, including bottom slope, roughness, flow depth or discharge vary continuously with the streamwise and the transverse coordinates. Then, the equations of continuum mechanics may successfully be applied since the elementary considerations of either Euler or Navier-Stokes are based on these suppositions. In other words, all the elements participating in the fluid motion can be represented by *continuous functions* possessing, in particular, continuous derivatives at every point in the domain of their definition.

Beside such domains, other zones exist, which possess discontinuities comparable to water falls or breaking waves. Here the flow continuity is not guaranteed. Similarly, in channel hydraulics such domains of discontinuity exist also – although confined locally – which cannot be described with the usual differential equations because differentials do not exist. The differential forms of the equations of continuity and of motion are then replaced by integral equations or, for sudden changes in one-dimensional parameters, by integral relationships over the discontinuous region. The hydraulic jump (Deutsch; Wassersprung or Wechselsprung; French: Ressaut hydraulique), which is considered in detail in this chapter, is the classical phenomenon in which the above mentioned discontinuity occurs. This phenomenon is

W.H. Hager, *Wastewater Hydraulics*, 2nd ed., DOI 10.1007/978-3-642-11383-3_7,

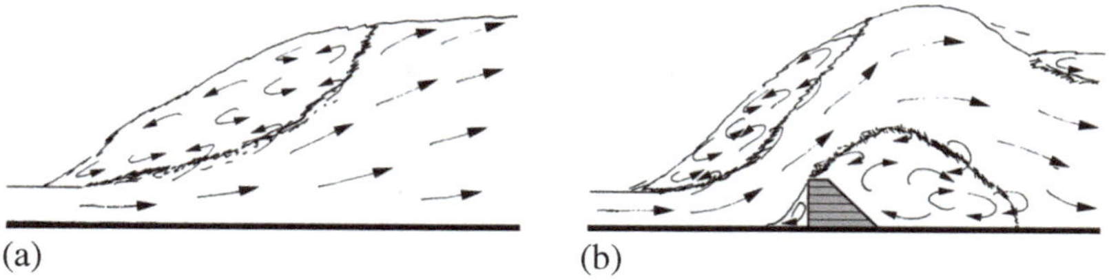

Fig. 7.1 Shear surfaces for (**a**) classical hydraulic jump, (**b**) hydraulic jump with appurtenance (stilling basin)

discussed together with stilling basins. By the term *hydraulic jump* one understands the transition from supercritical to subcritical flow accompanied by considerable local turbulence production, associated with an energy dissipation (Fig. 7.1a). The hydraulic jump can, for example, occur in a channel whose slope becomes flatter, or for a supercritical flow in a closed channel impinging suddenly on a pool of standing water downstream. Depending on the conditions in the up- and downstream reaches, the hydraulic jump shifts along the channel to the stabilizing position, corresponding to the equations of motion.

In contrast, the *stilling basin* (German: Tosbecken; French: Bassin amortisseur) is a standard structure containing the hydraulic jump under all conditions of flow (Fig. 7.1b). By appropriate design, the lengths of the jump and consequently of the structure are minimized. The purpose of a stilling basin is therefore to dissipate energy. By energy dissipation one understands, in the preceding sense, the conversion of mechanical energy principally into thermal energy. The conversion phenomenon is accompanied by turbulence production, that is, the production of large scale vortex formations, which finally die out through viscosity, transforming into heat. In the following, as in the previous chapter, first the nature of the phenomenon is explained, then follows the derivation of the basic equations and the computational procedure is finally introduced.

7.2 Phenomenon of Hydraulic Jump

According to Chap. 6 open channel flows are either subcritical and supercritical, depending on the *Froude number*

$$\mathsf{F} = \frac{Q}{(gF^3)^{1/2}} \left(\frac{\mathrm{d}F}{\mathrm{d}h}\right)^{1/2} \tag{7.1}$$

being larger or smaller than $\mathsf{F} = 1$. Herein, Q is discharge, g acceleration due to gravity, F cross-sectional area, h flow depth and $(\mathrm{d}F/\mathrm{d}h)$ the derivative of the flow area with respect to the water depth corresponding to the water surface width.

According to Chap. 1 three equations form the foundation of hydromechanics. These are the continuity equation, the energy and the momentum equations which ensure the conservations of mass, energy and momentum, respectively. For the sake of simplicity, consider first a *prismatic channel of rectangular cross-section* of width b. From Eq. (7.1) the Froude number in the rectangular channel is $\mathsf{F} = q/(gh^3)^{1/2}$. With $q = Q/b$ as the discharge per unit width, the energy head H and the specific force S relative to the channel bed are

$$H = h + \frac{q^2}{2gh^2}, \tag{7.2}$$

$$S = \frac{h^2}{2} + \frac{q^2}{gh}. \tag{7.3}$$

If these two relations are derived with respect to the flow depth h, then

$$\frac{\mathrm{d}H}{\mathrm{d}h} = 1 - \frac{q^2}{gh^3} = 1 - \mathsf{F}^2, \tag{7.4}$$

$$\frac{\mathrm{d}S}{\mathrm{d}h} = h - \frac{q^2}{gh^2} = h(1 - \mathsf{F}^2), \tag{7.5}$$

and thus

$$h \cdot \frac{\mathrm{d}H}{\mathrm{d}h} = \frac{\mathrm{d}S}{\mathrm{d}h}. \tag{7.6}$$

The proportionality between the changes of energy head $\mathrm{d}H/\mathrm{d}h$ and of specific force $\mathrm{d}S/\mathrm{d}h$ holds for all cross-sections. It therefore follows, as a rule, that the specific force S remains constant if the energy head H does not change, and vice-versa. This assertion is, however, valid only if $\mathsf{F} = 1$ (critical flow) for a transition from sub- to supercritical flow.

If one forms *finite differences* instead of differentials, and considers two cross-sections with flow depths $(h + \Delta h)$ and h, the difference in the energy head $\Delta H = H(h) - H(h + \Delta h)$ is

$$\frac{\Delta H}{h} = \frac{\Delta h}{h}\left[1 - \frac{q^2}{gh^3}\frac{(1 + \Delta h/2\,h)}{(1 + \Delta h/h)^2}\right] \tag{7.7}$$

or, by retaining terms up to the order of $(\Delta h/h)^2$,

$$\frac{\Delta H}{h} \cong \frac{\Delta h}{h}\left[1 - \mathsf{F}^2\left(1 - \frac{3}{2}\frac{\Delta h}{h}\right)\right]. \tag{7.8}$$

For the specific force, if follows correspondingly

$$\frac{\Delta S}{h^2} \cong \frac{\Delta h}{h}\left[1 + \frac{\Delta h}{2\,h} - \mathsf{F}^2\left(1 - \frac{\Delta h}{h}\right)\right]. \tag{7.9}$$

These differences in ΔH and ΔS result by considering terms up to the quadratic order $(\Delta h/h)^2$. If linear terms are only retained, as previously derived, the result is again $h\Delta H = \Delta S$. This relation therefore applies only to small changes of Δh, for example, an infinitesimal change in the flow depth. If, however, the change is large and the second order terms become significant, the difference between the conservations of energy head and specific force also becomes larger.

Eliminating F^2 between Eqs. (7.8) and (7.9), one finds

$$\frac{\Delta S}{h^2} = \left(1 + \frac{\Delta h}{2h}\right)\frac{\Delta H}{h}. \tag{7.10}$$

From this relation the following important *principles* can be derived:

(1) For constant energy head ($\Delta H = 0$), the specific force automatically becomes constant.
(2) For constant specific force ($\Delta S = 0$), however, the energy head needs not necessarily be constant.
(3) Since the energy head can only decrease in the flow direction, the flow depth in case 2 must increase ($\Delta h > 0$) so that the specific force remains constant.

Case (1) applies only for the transition from sub- to supercritical flow whereas case (3) corresponds to the transition from super- to subcritical flow. It is quite clear then that an *energy dissipation* $\Delta H > 0$ results.

7.3 Computation of Hydraulic Jump

7.3.1 *Basic Equation*

Before the characteristics of hydraulic jumps are introduced, the essential computational features are derived. These are the so-called *sequent flow depths* (German: Konjugierte Wassertiefen; French: Hauteurs conjuguées) corresponding to the flow depths up- and downstream of a jump (Fig. 7.2). From this, further relationships can be derived regarding the mechanical energy losses in a hydraulic jump.

Since for hydraulic jumps significant energy loss takes place, the momentum principle must be applied (Chap. 1). For the special case of a *prismatic channel*

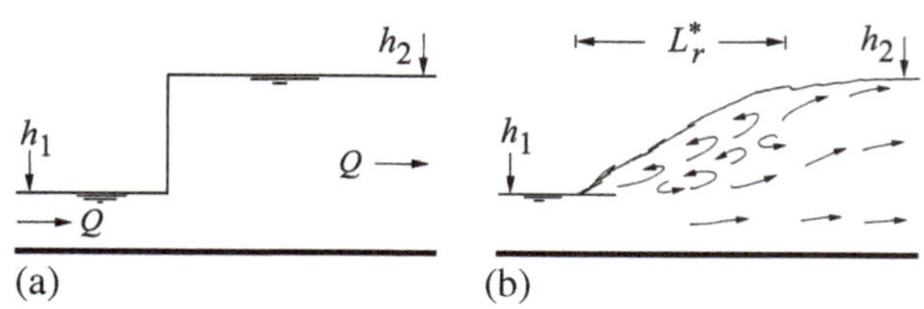

Fig. 7.2 Hydraulic jump (**a**) theoretical abstraction, (**b**) schematic in rectangular channel

for which the friction resistance is compensated for by the component of weight in the flow direction, the specific force principle becomes elementary. According to Sect. 1.3, the specific force principle gives $S_1 - S_2 = 0$. In the following, the subscripts «1» and «2» refer, respectively, to the cross-sections up- and downstream of the hydraulic jump. The specific force S of an one-dimensional flow is according to Chap. 1

$$S = z_s F + \frac{Q^2}{gF} \tag{7.11}$$

where z_s is the vertical distance of the centroid of the cross-sectional area from the water surface. The application of the specific force principle under the assumption of constant discharge then yields

$$z_{s1} F_1 + \frac{Q^2}{gF_1} = z_{s2} F_2 + \frac{Q^2}{gF_2}, \tag{7.12}$$

where both z_s and the cross-sectional area F are functions of only cross-sectional shape and the flow depth. As in Chaps. 5 and 6, the quantities z_s and F are related to the flowing full quantities with the parameters $Z_s = z_s/T$ and $\Phi = F/D^2$ to represent the filling ratio and the area characteristic, respectively. Here T is the maximum height from invert to soffit of the cross-section. It then follows

$$Z_{s1}\Phi_1 + \frac{Q^2}{gT^5}\Phi_1^{-1} = Z_{s2}\Phi_2 + \frac{Q^2}{gT^5}\Phi_2^{-1}. \tag{7.13}$$

Because for a particular cross-sectional geometry both Z_s and Φ depend only on the filling ratio, Eq. (7.13) gives a relation between $y_1 = h_1/T$, $y_2 = h_2/T$ and the dimensionless discharge $Q^2/(gT^5)$. In the following, this relationship is derived for the three standard sections, namely, the circular, the egg-shaped and the horseshoe sections. First, the rectangular cross-section is considered because relationships are particularly simple.

7.3.2 *Rectangular Cross-section*

If b is taken as the width of the rectangular channel, the centroid of the wetted cross-section lies at half height of the water depth, i.e. $z_s = h/2$. Then, according to Eq. (7.12), the *specific force law* can be written as

$$\left(\frac{bh^2}{2} + \frac{Q^2}{gbh}\right)_1 = \left(\frac{bh^2}{2} + \frac{Q^2}{gbh}\right)_2 \tag{7.14}$$

or, after division by $(bh_1^2/2)$,

$$1+\frac{2(Q/b)^2}{gh_1^3}=\left(\frac{h_2}{h_1}\right)^2+\frac{2(Q/b)^2}{gh_2h_1^2}. \tag{7.15}$$

If the approach flow Froude number $\mathsf{F}_1 = Q/(gb^2h_1^3)^{1/2}$ is introduced and $Y^* = h_2/h_1$ is the *ratio of sequent depths* in the rectangular channel (marked by an asterisk), Eq. (7.15) reduces to

$$1+2\mathsf{F}_1^2 = Y^{*2}+2\mathsf{F}_1^2Y^{*-1}. \tag{7.16}$$

If the trivial solution $Y^* = 1$ is disregarded, the physically relevant solution ($Y^* > 0$) of Eq. (7.16) is

$$Y^* = \frac{1}{2}\left[(1+8\mathsf{F}_1^2)^{1/2}-1\right]. \tag{7.17}$$

The above relation was first reported by the French hydraulician Jean-Baptiste Charles Joseph Bélanger (1790–1874) in 1838 (Hager 1990a). Because the hydraulic jump in the rectangular channel is by far the most investigated case, and since the neglect of channel bed slope and wall resistance may form the basis of consideration of all hydraulic jumps, the hydraulic jump in this idealized rectangular channel is designated after Rajaratnam (1967) *classical hydraulic jump.* Equation (7.17) is shown in Fig. 7.3a. For large values of F_1, a close approximation of Eq. (7.17) for the sequent flow depth is

$$Y^* = \sqrt{2}\mathsf{F}_1-(1/2). \tag{7.18}$$

For $\mathsf{F}_1 > 2.5$, the differences in Y^* calculated from Eqs. (7.17) and (7.18) are less than 1%. As is shown below, Eq. (7.17) is used for $\mathsf{F}_1 > 2$. It can be observed from Eqs. (7.17) and (7.18) that:

- The ratio of the sequent flow depths Y^* is proportional to the Froude number F_1. For constant approach flow depth h_1, the sequent depth h_2^* varies linearly with discharge per unit width Q/b.
- For a rectangular section, no effect of cross-sectional shape exists because only the two parameters Y^* and F_1 are involved.

Apart from the relation $Y^*(\mathsf{F}_1)$, the *mechanical energy loss* $\Delta H = H_1 - H_2$ can be determined. The *efficiency* of a hydraulic jump

$$\eta = \frac{H_1-H_2}{H_1} \tag{7.19}$$

as the relative energy dissipation with respect to the energy head of the approach flow is for the classical hydraulic jump

$$\eta^* = 1-\frac{Y^*\left[1+\mathsf{F}_1^2/(2Y^{*3})\right]}{1+(1/2)\mathsf{F}_1^2}. \tag{7.20}$$

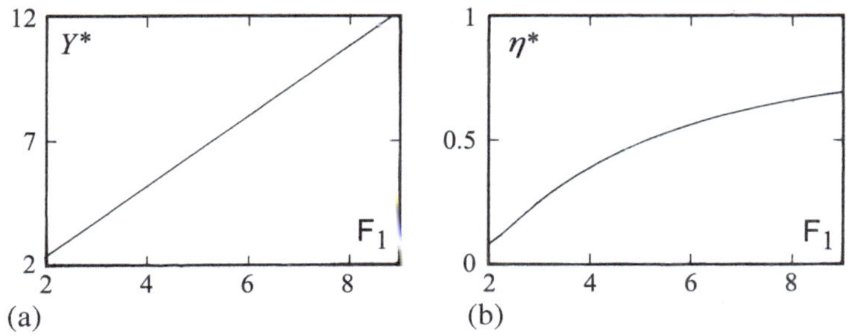

Fig. 7.3 Classical hydraulic jump, (**a**) sequent depth ratio Y^* and (**b**) efficiency η^* in relation to the approach flow Froude number F_1

The parameter Y^* can be expressed as a function of F_1. Substituting Eq. (7.18) into Eq. (7.20) gives, after neglecting the correction term in F_1^{-1}, the relation $\eta^* = 1 - 2\sqrt{2}\mathsf{F}_1^{-1}$. This relation can be further expanded to

$$\eta^* = \left(1 - \frac{3}{2\mathsf{F}_1}\right)^2. \tag{7.21}$$

For $\mathsf{F}_1 > 2.5$, the values of η^* calculated from Eq. (7.21) and from the exact expression of Eq. (7.20) differ by less than 9%. For $\mathsf{F}_1 > 3.5$ the deviations are always less than 1%. The function $\eta^*(\mathsf{F}_1)$ is represented graphically in Fig. 7.3b.

It may be pointed out here that all expressions derived in Sect. 7.3.2 are strictly valid only for the classical hydraulic jump. These are conditional on a prismatic channel of rectangular cross-section with a horizontal bed and a perfectly smooth surface. Also, in order that the effect of viscosity is suppressed, the discharge per unit width must at least be 0.1 m^2s^{-1} (Hager and Bremen 1989).

On the basis of experiments the following additional results apply (Hager 1992). The *roller length* L_r^* (German: Rollerlänge; French: Longueur de rouleau) i.e. the distance from the toe of the jump to the stagnation point at the free surface (Fig. 7.2b) is

$$L_r^*/h_2^* = 4.3, \tag{7.22}$$

whereas the *jump length* L_j^* (German: Wassersprunglänge; French: Longueur du ressaut) is

$$L_j^*/h_2^* = 6.0. \tag{7.23}$$

At the end of the hydraulic jump, the turbulent fluctuations are so reduced that no further bed protection is necessary: A stilling basin required for the classical hydraulic jump possesses therefore the basin length $L_b = L_j^*$. Further characteristics of the internal flow mechanics in a classical jump are given by Rajaratnam

(1967) or Hager (1992). Castro-Orgaz and Hager (2009) considered the roller of the classical hydraulic jump by using both the momentum and the moment of momentum equations. The predicted roller profile was demonstrated to compare excellently with observations, even if effects of streamline curvature, density variation and air concentration effects are neglected.

7.3.3 *Circular Section*

Direct Hydraulic Jump

The term z_sF appearing in Eq. (7.12) can also be interpreted as the pressure force P_s, normalized by the specific weight (ρg), due to the hydrostatic pressure distribution. For example, Hörler (1967) gives

$$P_s/(\rho g T^3) = \frac{1}{8}(\sin\delta - \frac{1}{3}\sin^3\delta - \delta\cos\delta) \tag{7.24}$$

with δ as half central angle. The pressure distribution has the shape of a half, lengthwise horseshoe. For the circular section $T = D$. With the expressions

$$F/T^2 = \frac{1}{4}(\delta - \sin\delta\cos\delta) \tag{7.25}$$

for cross-sectional area and

$$y = \frac{h}{T} = \frac{1}{2}(1 - \cos\delta) \tag{7.26}$$

for the filling ratio, Eq. (7.12) yields for the specific force

$$\frac{S}{T^3} = \frac{P_s}{\rho g T^3} + \frac{Q^2}{gT^5(F/T^2)}. \tag{7.27}$$

The ratio $[(P_s/(\rho g T^3))/(F/T^2)]$ corresponds to the relative height of the centroid, namely $Z_s = z_s/T$. Substituting Eqs. (7.25) and (7.27) in Eq. (7.13) yields

$$\frac{Q^2}{gT^5} = (Z_{s1}\Phi_1 - Z_{s2}\Phi_2)(\Phi_2^{-1} - \Phi_1^{-1})^{-1}. \tag{7.28}$$

The relation between the sequent depths $y_1 = h_1/T$ and $y_2 = h_2/T$ for varying relative discharge is graphically shown in Fig. 7.4. Therein, the condition $h_1 = h_2$ is plotted as a dotted curve. A comparison of measurements with computations indicates general agreement (Hörler 1967). It may be remarked, in passing, that the measured data of h_2 are always smaller than those calculated. The differences can be explained with the effect of viscosity included in the Reynolds number.

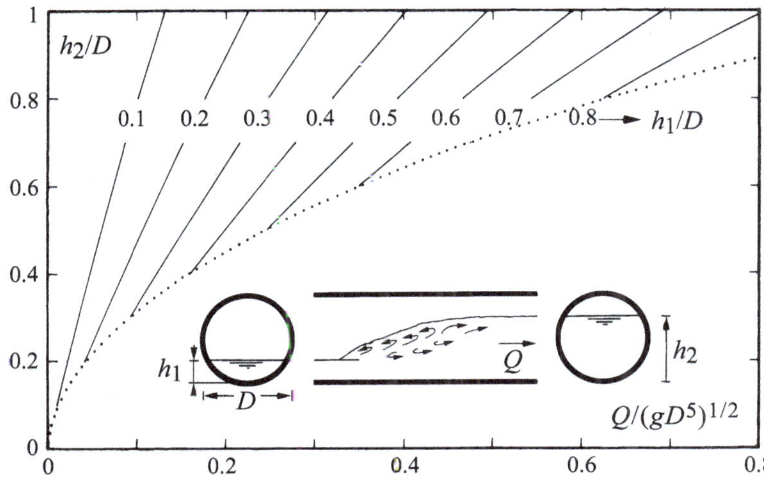

Fig. 7.4 Sequent depths for circular cross-section with (···) relation $h_2 = h_1$. For egg-shaped and horseshoe sections substitute $Q/(gB^2T^3)^{1/2}$ for $Q/(gD^5)^{1/2}$

An *approximate expression* for the sequent depths is obtained from the similarity of the curves in Fig. 7.4 as (Hager 1990c)

$$\frac{y_2 - y_1}{1 - y_1} = \left(\frac{q_D - y_1^2}{q_o - y_1^2}\right)^{0.95} \quad \text{for} \quad y_1 < 0.7, \tag{7.29}$$

where

$$q_o = q_D(y_2 = 1) = \frac{3}{4}y_1^{3/4}\left[1 + \frac{4}{9}y_1^2\right]. \tag{7.30}$$

For known values of y_1 and $q_D = Q/(gD^5)^{1/2}$, q_o is first determined from Eq. (7.30) and y_2 is calculated explicitly from Eq. (7.29). It remains to be remarked that values of h_2 calculated after taking into account the effect of friction are always slightly smaller than the corresponding values obtained experimentally. The sequent flow depths obtained from Eq. (7.29) may be considered an upper bound, therefore.

To avoid complex expressions, simplifications for cross-sectional area F and static pressure force P_s can be introduced as $F/D^2 = y^{1.5}$ and $P_s/(\rho g D^3) = 0.5y^{2.5}$, where $y = h/D$. Inserting in Eq. (7.12) and accounting for the approach (subscript 1) flow Froude number $\mathsf{F}_1 = Q/(gDh_1^4)^{1/2}$ gives for the *sequent depth ratio* $Y = h_2/h_1$ (Stahl and Hager 1998)

$$1 + 2\mathsf{F}_1^2 = Y^{2.5} + 2\mathsf{F}_1^2 Y^{-1.5}. \tag{7.31}$$

The asymptotic solution of Eq. (7.31) for $\mathsf{F}_1 \rightarrow \infty$ is $Y = (2\mathsf{F}_1^2)^{0.4}$, and an approximation within ±2% for $2 < \mathsf{F}_1 < 10$ reads $Y = 1.16\mathsf{F}_1^{0.85}$. Based on experiments conducted in a circular pipe of diameter 240 mm, the data were approximated as

$$Y = \mathsf{F}_1^{0.90}. \tag{7.32}$$

This is slightly smaller than from Eq. (7.31) and accounts thus for the usual viscous effects. In addition, Eq. (7.32) yields the correct result $Y(\mathsf{F}_1 = 1) = 1$ for transitional flow.

Undular Hydraulic Jump

Undular hydraulic jumps generated in closed conduits and open channels look similar. These flow phenomena may be described as a steady free surface wave formation, which are known to be highly sensitive to small perturbations. Reducing for instance the tailwater level by a small amount may shift the jump by multiples of the tailwater depth into the downstream direction (Chanson and Montes 1995, Montes and Chanson 1998, Ohtsu et al. 1997).

Each perturbation of a supercritical flow causes shockwaves. These surface waves may be observed also for undular hydraulic jumps. Depending on the approach flow Froude number $\mathsf{F}_o = V_o/(gh_o)^{1/2}$, four jump types may be distinguished in rectangular channels (Reinauer and Hager 1995):

(1) *Type A* corresponds to an almost plane and smooth surface wave practically free of shock waves, provided $1 \leq \mathsf{F}_o \leq 1.20$,
(2) *Type B* as a spatial free surface wave with shocks across the tunnel, if $1.20 < \mathsf{F}_o < 1.28$,
(3) *Type C* also with a three-dimensional free surface pattern plus a surface roller extending mainly over the first wave crest, if $1.28 \leq \mathsf{F}_o \leq 1.36$, and
(4) *Type D* again as a spatial wave formation and with a breaking central wave portion, if $1.36 < \mathsf{F}_o < 1.60$.

Figure 7.5 shows a definition sketch of the undular hydraulic jump in a circular conduit of diameter D. The approach flow conditions are defined by the approach flow depth h_o and discharge Q, such that the approach flow filling is $y_o = h_o/D$ and the approach flow Froude number $\mathsf{F}_o = Q/(gDh_o^4)^{1/2}$. The successive wave maxima (subscript M) along the conduit axis are h_{1M}, h_{2M}, etc., whereas the axial wave minima (subscript m) are h_{1m}, h_{2m}, etc. The lengths between the wave crests are L_1, L_2, etc., and wave extrema along the side walls are t_{1M}, t_{2M}, etc., respectively. Normally, the first three waves of an undular hydraulic jump were observed because of a wave diffusion process damping further waves considerably.

The types of hydraulic jumps previously developed for the undular jump in rectangular channels was adopted also for the circular conduit. For Froude numbers up to $\mathsf{F}_o = 2$, and filling ratios $0.30 < y_o < 0.70$, the following was observed. Type

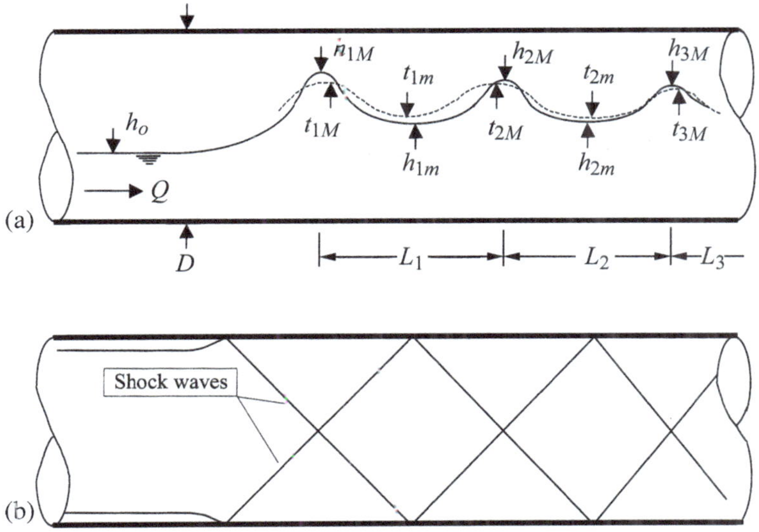

Fig. 7.5 Definition sketch of undular hydraulic jump in circular conduit (**a**) streamwise section, (**b**) plan

A-Jump occurred for $1 \leq \mathsf{F}_o \leq 1.50$, Type B-Jump was generated for small filling ratios $0.30 < y_o < 0.45$, whereas both Type C and Type D-Jumps followed the curve $y_o = 1.5 - 0.60\mathsf{F}_o$.

According to Eq. (7.32), the ratio of sequent depths of a direct jump in a conduit depends only on F_o or F_1, respectively, as $Y(\mathsf{F}_1)$. Further, the sequent depth ratio $Y^*(\mathsf{F}_1)$ of hydraulic jumps in the rectangular channel is given by Eq. (7.17). The first wave maximum Y_{1M} was demonstrated to be contained within these two extreme values. For a large wave height, the conduit chokes.

The *axial wave extrema* (subscript e) $Z_{ie} = h_{ie}/D$, the axial wave length development between two waves, and the wall wave characteristics for wave i may be expressed, based on the test data, as (Gargano and Hager 2002)

$$Z_{iM} = \alpha \mathsf{F}_o y_o - 0.10, \tag{7.33}$$

$$Z_{im} = \beta \mathsf{F}_o y_o - 0.10, \tag{7.34}$$

$$L_i/D = 9.5(y_o/\mathsf{F}_o), \tag{7.35}$$

$$t_{iM}/D = \alpha \mathsf{F}_o y_o - \gamma, \text{ and} \tag{7.36}$$

$$t_{im}/D = 0.94\mathsf{F}_o y_o - 0.06. \tag{7.37}$$

Herein, the coefficients are $\alpha = 1.20$, 1.15 and 1.10, $\beta = 0.965$, 0.975 and 1.00, and $\gamma = 0.15$, 0.125 and 0.10 for $i = 1, 2, 3$, respectively. It is observed that the wave peaks of the undular hydraulic jump decrease in the direction of flow, indicating wave damping. Note also that the upstream wave heights are larger in the axis as along the wall, and both become identical beyond the third wave peak. The above relations hold for $0.50 < \mathsf{F}_o y_o < 1$.

Axial wave profiles $Y(X)$ with $Y = (h - h_{1m})/(h_{1M} - h_{1m})$ and $X = (x - x_{1m})/L_1$ are shown in Fig. 7.6 including various jump types. The streamwise coordinate x is measured from the first wave crest. Note the average wave profile that may be

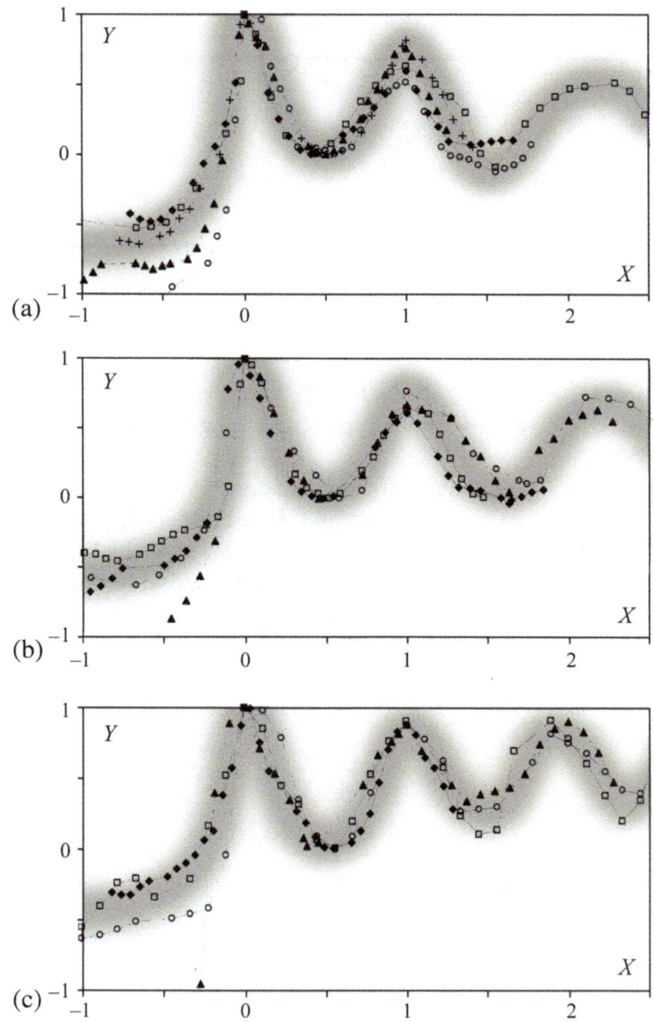

Fig. 7.6 Normalized axial wave profiles $Y(X)$ for selected runs of Series (**a**) A, (**b**) B, and (**c**) C (Gargano and Hager 2002)

drawn particularly over the first wave, whereas scatter increases further downstream. These profiles are close to the cnoidal wave profile, i.e. involving a sharp and narrow wave crest and a rounded wave trough. Further observations indicate that the asymptotic tailwater flow depth corresponds closely to the tailwater depth h_2 according to Eq. (7.32).

The first wave crest and wave trough correspond to the extreme surface elevations governing the *hydraulic design* of closed conduit flow. The governing equation for conduit choking is, therefore, from Eq. (7.33) for $i = 1$

$$Z_{1M} = 1.20\mathsf{F}_o y_o - 0.10. \tag{7.38}$$

Choking is inhibited provided that $Z_{1M} = h_{1M}/D < 1$, or $\mathsf{F}_o y_o < 0.92$. The product $\mathsf{C}_o = \mathsf{F}_o y_o$ was identified by Stahl and Hager (1998) as approach flow choking number related to the hydraulic jump formation in circular conduits. For direct hydraulic jumps with $\mathsf{F}_o > 2$, choking occurs if $\mathsf{C}_o > 1$. For undular hydraulic jumps, choking may be avoided provided that

$$\mathsf{C}_o < 0.90 \quad \text{for} \quad 1 < \mathsf{F}_o < 2. \tag{7.39}$$

The discharge capacity (subscript C) of conduit flow to assure free surface flow is therefore

$$Q_C/(gD^3 h_o^2)^{1/2} < 0.90 \quad \text{for} \quad 1 < \mathsf{F}_o < 2. \tag{7.40}$$

Figures 7.7 and 7.8 relate to photos of typical undular hydraulic jumps.

Features of Hydraulic Jumps

The *energy dissipation* of hydraulic jumps in circular pipes is significantly smaller than in the corresponding rectangular channel. One would never use a dissipator in a conduit, therefore. Also, the jump stability is much lower than in the standard rectangular channel.

Hydraulic jumps may have different appearances, depending both on the approach flow filling y_1 and the approach flow Froude number F_1. For transitional flow up to about $\mathsf{F}_1 = 1.5$, the jump is undular, and the first undulations break for $1.5 < \mathsf{F}_1 < 2$ approximately. For $\mathsf{F}_1 > 2$, say, the undulations disappear and the *direct hydraulic jump* is formed:

- Either with lateral wings and a concentrated surface jet including two asymmetric separation zones and a bottom recirculation (Hager 1992), provided $y_1 < 1/3$,
- Or as an almost plane jump comparable with the classical hydraulic jump (7.3.2) for $y_1 \geq 1/3$, say.

(a)

(b)

(c)

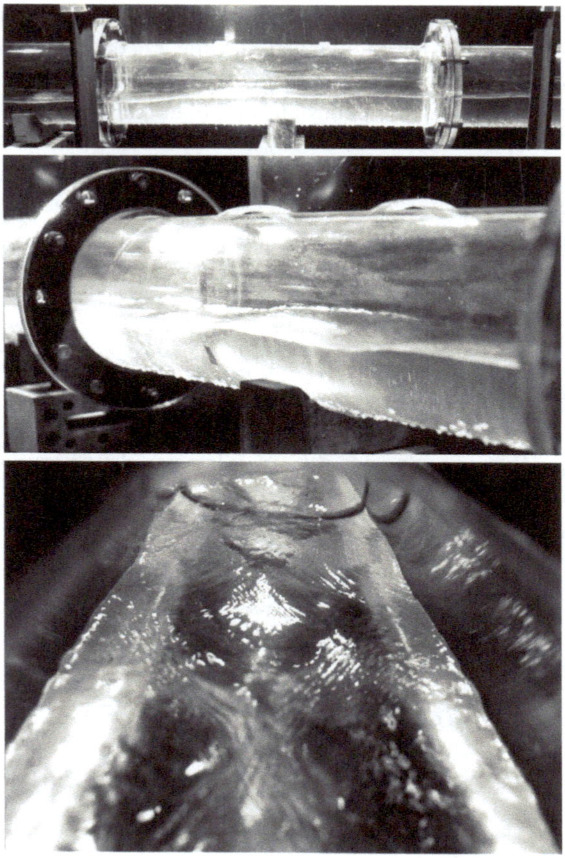

Fig. 7.7 Undular hydraulic jump for $y_o = 0.43$ and $\mathsf{F}_o = 1.47$ (**a**) overall side view, (**b**) free surface profile and (**c**) tailwater view (Gargano and Hager 2002)

Typical lengths of a hydraulic jump in a circular pipe are:

(1) *Recirculation length* L_R measured from the upstream end of the longer wing to the surface stagnation point, with the opposite wing always significantly shorter. The relative length $\lambda_R = L_R/h_2$ varies essentially with F_1 as $\lambda_R = 2\mathsf{F}_1^{1/2}$.
(2) *Aeration length* L_a measured also from the upstream end of the longer wing to the section where air 'clouds' (not single bubbles) have left the flow. The relative aeration length $\lambda_a = L_a/h_2$ is an index for jump length varying as $\lambda_a = 4\mathsf{F}_1^{1/2} = 2\lambda_R$.

Figure 7.9 refers to the (a) undular and (b) direct hydraulic jump with a free surface downstream flow, (c) hydraulic jump with a flow recirculation, and (d) transitional hydraulic jump from free surface to pressurized downstream flow. Note the extent of the aeration length for the latter two jumps.

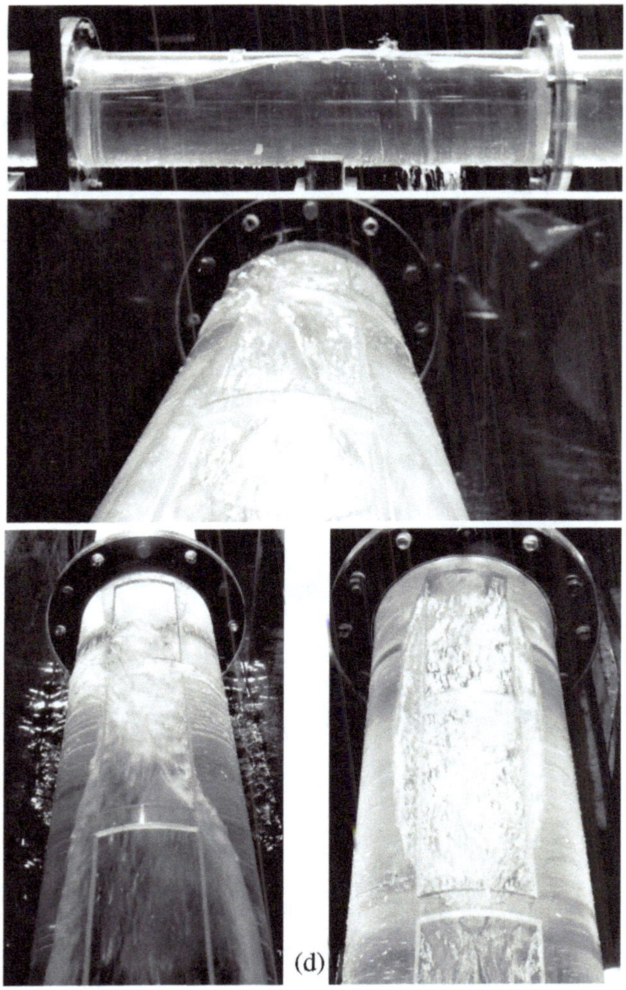

(a) (b) (c) (d)

Fig. 7.8 Undular hydraulic jump for $y_o = 0.62$ and $\mathsf{F}_o = 1.50$ involving choking flow (**a**) overall side view, (**b**) choking detail with shockwaves (**c**) overall tailwater view and (**d**) top view (Gargano and Hager 2002). Note the soffit openings for access of test instrumentation

Top views of these jumps are shown in Fig. 7.10. For $y_1 \geq 1/3$, the jump is similar to the classical hydraulic jump, involving a straight front, a surface roller and a bottom forward flow (Fig. 7.10a). For $y_1 < 1/3$, however, the surface width across the jump increases significantly from the upstream to the downstream sections, and the forward flow is concentrated to a portion of the free surface. At the opposite side, the flow recirculates and exhibits the typical *wing-shaped front* (Fig. 7.10b). For large F_1, the downstream flow gets pressurized with definite surface zones of forward and backward flows (Fig. 7.10c). Such flow is dangerous in terms of pressure fluctuations and breakdown of the free surface condition. So-called *flow choking* is discussed below.

(a)
(b)
(c)
(d)

Fig. 7.9 Side views of hydraulic jumps with F_1 = (**a**) 1.1, (**b**) 2.3, (**c**) 4.1, (**d**) 6.5

Figure 7.11 shows side views relating to Fig. 7.10. Note the significant increase of flow depth across a small length, which may eventually lead to choking. Also, from the abrupt change of 'black' into 'white' water, associated with significant air entrainment, air detrainment can be observed. *Air entrainment* occurs mainly along the mixing layers of the forward and backward flow zones, and air detrainment results basically from the significant turbulence level in a hydraulic jump. Some results on air entrainment are presented below.

Currently no study appears to exist in which the effect of *bottom slope* is considered. Based on similar studies in rectangular ducts, the influence of bottom slope is negligible for bottom slopes smaller than about 5% (Hager 1992).

7.3.4 *Egg-shaped and Horseshoe Sections*

The standard egg-shaped and horseshoe sections are both composed of three cross-sectional parts. Because the computation of the pressure force for the flow in either of these two sections leads to involved expressions, a simpler procedure is presented. Note also that no experimental study was conducted until today.

(a)
(b)
(c)
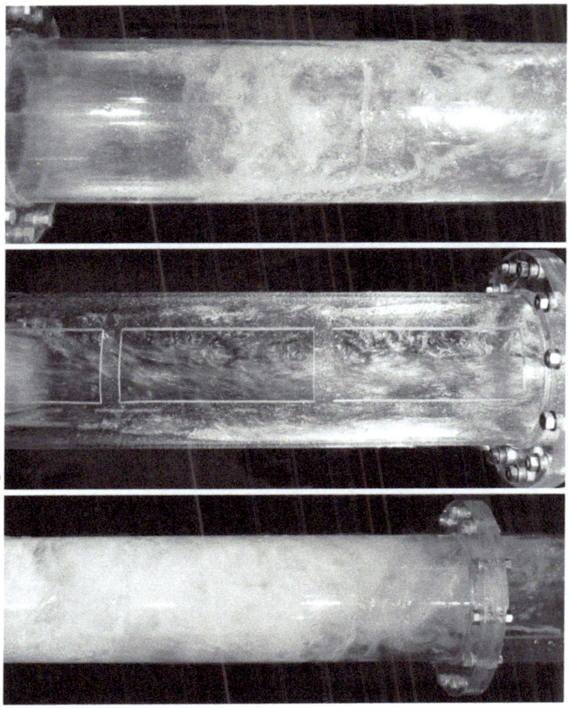

Fig. 7.10 Top views on hydraulic jumps with $\mathsf{F}_1 =$ (**a**) 2.3, (**b**) 4.1, (**c**) 6.5

As shown in Fig. 7.12, both the egg-shaped and the horseshoe cross-sections can be closely approximated by *ellipses* of eccentricity 1:2 as

$$\left(\frac{\bar{x}}{B}\right)^2 + \left(\frac{\bar{y}}{T}\right)^2 = \left(\frac{1}{2}\right)^2, \tag{7.41}$$

where B and T are, respectively, width and height of the cross-section. For a given filling ratio $y = h/T$, the cross-sectional area F is

$$\frac{F}{BT} = \frac{1}{4}\left[\arccos(1-2y) - 2(1-2y)(y-y^2)^{1/2}\right]. \tag{7.42}$$

For $B = T = D$, Eq. (7.42) reduces to Eq. (7.25) for the circular section.

The pressure force in channels of elliptical cross-section is (Hjelmfeldt 1967)

$$\frac{P_s}{\rho g B T^2} = \frac{1}{8}(2y-1)\left[\arccos(1-2y) - 2(1-2y)(y-y^2)^{1/2}\right] + \frac{2}{3}(y-y^2)^{3/2} \tag{7.43}$$

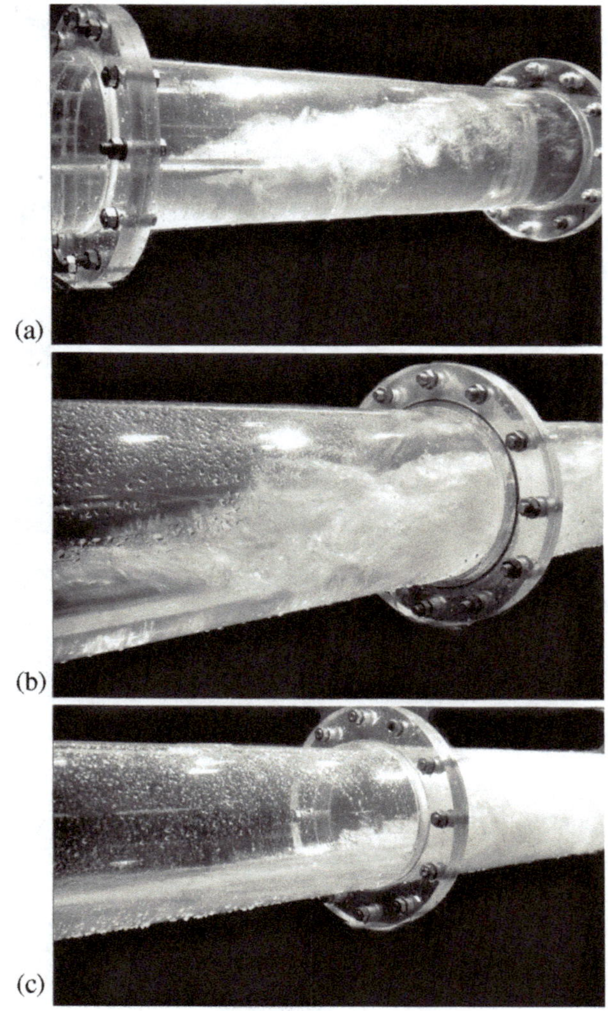

Fig. 7.11 Side views of jumps shown in Fig. 7.10

Remarkable of this result is the agreement of the right-hand side with that of Eq. (7.24) for the circular section. Curves of filling ratio y versus the static pressure force $P_s(y)$ and the cross-sectional area $F(y)$ are therefore identical for the circular, the egg-shaped and the horseshoe cross-sections. Equation (7.29) or Fig. 7.4 then hold also for the egg-shaped and the horseshoe sections, provided the relative discharge is $q_D = Q/(gB^2T^3)^{1/2}$ instead of $q_D = Q/(gD^5)^{1/2}$ for the circular section. This proposed approximation yields an extremely simple method to determine the sequent flow depths for all three standard cross-sections.

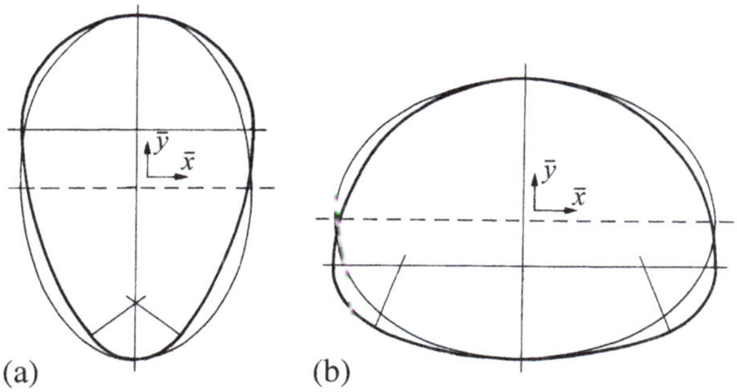

Fig. 7.12 (**a**) Egg-shaped section 2:3, (**b**) horseshoe section 2:1.5 and (—) approximating ellipses, in accordance with Eq. (7.41)

Example 7.1 Calculate the sequent depth in an egg-shaped section 120:180 for $h_1 = 50$ cm and $Q = 1500\ \text{Ls}^{-1}$.
With the values $B = 1.2$ m and $T = 1.8$ m, the relative discharge is $q_D = 1.5/(9.81\cdot1.2^2 1.8^3)^{1/2} = 0.165$. With $y_1 = h_1/T = 0.5/1.8 = 0.278$, the values $q_o = 0.297$ and $(q_D - y_1^2)/(q_o - y_1^2) = (0.165 - 0.278^2)/(0.297 - 0.278^2) = 0.40$ are obtained from Eqs. (7.30) and (7.29), respectively. One therefore calculates from $(y_2 - y_1)/(1 - y_1) = 0.40^{0.95} = 0.419$ the value $y_2 = 0.581$, resulting in $h_2 = 0.581\cdot1.8 = 1.05$ m.

7.4 Hydraulic Jump of Choked Circular Sections

7.4.1 *Introduction*

Although free surface flow should take place in sewers by principle, the flow in reaches can become pressurized due to downstream submergence, large discharge or conduit damage. This condition, strictly speaking, is not a design feature but considered in this section as it represents a dangerous flow case. The description here is a summary from a literature survey by Hager (1989a).

The switching phenomenon from free-surface to pressurized flow is designated by the term *surcharge* or choking and the opposite as *evacuation*. Both these phenomena do not represent continuous flows but correspond always to abrupt flow transitions. Even for uniform flow, careful experiments have failed to produce the so-called full flow condition, corresponding to free pipe full flow without surcharge (Chap. 5).

There may be several causes for the occurrence of abrupt surcharging and evacuation in channels whose cross-sections are not open at the top. Some of these originate from an insufficient flow ventilation, wave formation due to flow disturbances at intakes, curved or contracted pipe reaches in subcritical and especially in supercritical flows or, the formation of local low-pressure zones and hydraulic

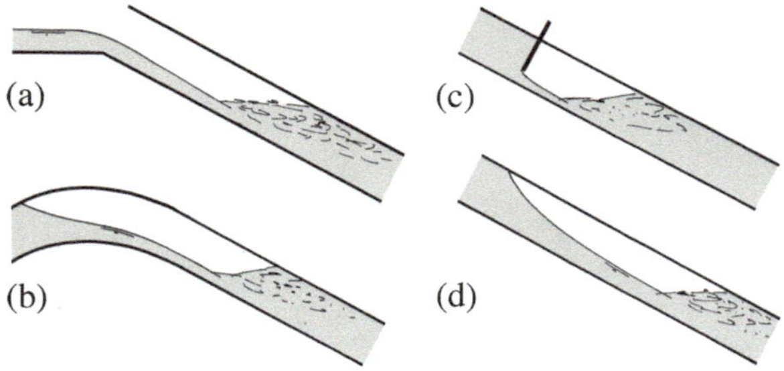

Fig. 7.13 Development of a hydraulic jump in circular sewer due to (**a**) change in bottom slope, (**b**) siphon, (**c**) sluice gate and (**d**) air pocket

jumps caused by downstream submergence. This last aspect shall receive special attention. Currently, the abrupt surcharge and evacuation phenomena of sewers are not entirely understood (Chap. 5). It is hoped that research may lead to useful results in near future.

Figure 7.13a shows an accelerated flow due to a *change of bottom slope* from mild to steep, associated with an abrupt transition to pressurized flow as it impinges the downstream flow. Figure 7.13b shows a typical siphon flow with a drawdown profile, then free-surface flow with change of flow type and finally choking due to pressurized downstream flow. Figure 7.13c relates to the drawdown profile downstream of a sluice gate, whereas Fig. 7.13d shows an air pocket in a pressurized conduit flow. In all these cases air is entrained in the flow due to the presence of a hydraulic jump. Consequently, an *air-water mixture* prevails in the hydraulic jump. Depending on the condition of the flow, the air bubbles are either transported along with the flow, or the bubbles return upstream due to their buoyancy. The hydraulic jump in closed conduits depends particularly on the availability of air.

The advantages of air-entrainment such as conduit ventilation or reduction of cavitation damage on sewers often outweigh the following disadvantages:

- Possibility of pulsating slug flow,
- Reduction in discharge due to two-phase flow, and
- Uncontrolled air outblow, also referred to geysering.

In sewer design, therefore, *free surface flow* is in general required and sufficient ventilation is attempted under all circumstances. In the following, the hydraulic jump in circular sewers from free surface to pressurized flow conditions is discussed with a particular reference to the sequent flow depths, the efficiency of energy dissipation and the jump length.

7.4.2 *Sequent Flow Depths*

In 1943 Kalinske and Robertson presented a first relation between the up- and downstream flow depths of a hydraulic jump in a *circular* sewer. With $y = h/D$ as the

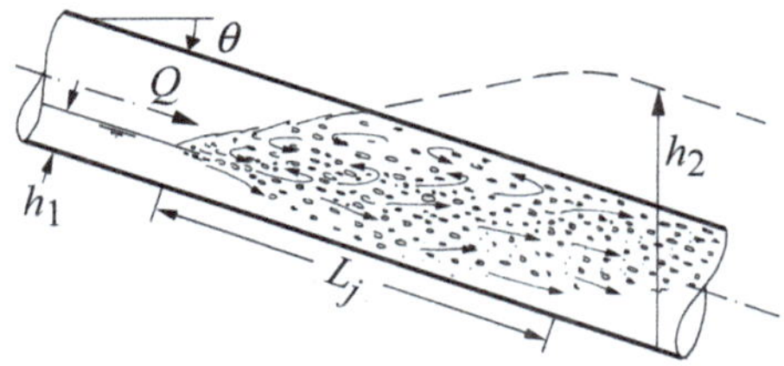

Fig. 7.14 Definition sketch for the hydraulic jump in a sloping conduit with pressurized downstream flow

filling ratio, P_s the static pressure force, $S_o = \arcsin\theta$ the sewer bottom slope, and $\beta_a = Q_a/Q$ the ratio of air (subscript a) to water discharge (Fig. 7.14), the relation for the flow depths h_1 and h_2 up- and downstream of a jump is

$$P_{s1}/(\rho g) + \rho Q V_1 + \frac{\pi}{4} D^2 L_j S_o = P_{s2}/(\rho g) + \rho(1+\beta_a) Q V_2. \tag{7.44}$$

If the relations for the jump length and the air discharge specified below are introduced, Fig. 7.15 results from the solution of Eq. (7.44), in analogy to Fig. 7.4. For the relative discharge $q_D = Q/(gD^5)^{1/2}$ and known upstream sewer filling $y_1 = h_1/D$, the sequent downstream filling $y_2 = h_2/D$ can be determined for various bed slopes S_o. According to Fig. 7.15, for constant discharge q_D, the flow depth h_2 decreases with

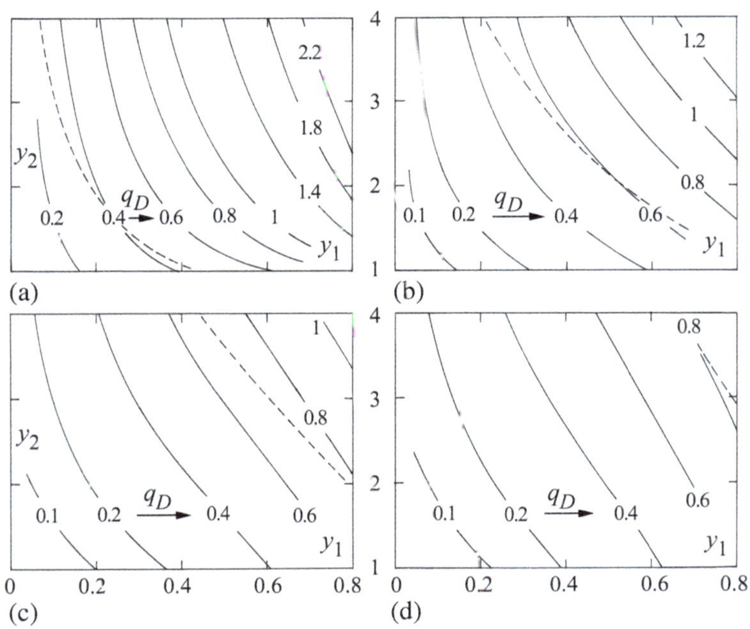

Fig. 7.15 Sequent depths $y_1 = h_1/D$ and $y_2 = h_2/D$ in relation to relative discharge $q_D = Q/(gD^5)^{1/2}$ for bottom slopes $S_o =$ (**a**) 0, (**b**) 10%, (**c**) 20% and (**d**) 30%. (. . .) y_{1L} from Eq. (7.45) (Hager 1989a)

increasing upstream flow depth h_1. The preceding results apply only for $y_2 > 1$ with an abrupt surcharge or choking condition. For free-surface conduit flows the procedure of Sect. 7.3 is relevant.

For all values of $y_1 = h_1/D$ there exists a Froude number below which the hydraulic jump entrains a quantity of air of which only a part is transported downstream. The remaining quantity of entrained air is periodically recirculated upstream by the surface roller. This *limit value* y_{1L} (subscript L) of the inflow flow depth is shown in Fig. 7.15 as dashed line. The following relation expressing the dependence of y_{1L} on F_1 holds for bottom slopes $S_o < 0.4$

$$y_{1L} = \frac{1}{2}\left[\frac{3}{4} + S_o^{0.7}\right]\mathsf{F}_1^{0.6}. \tag{7.45}$$

For $y_1 < y_{1L}$ the air entrainment coefficient β_a and thus Fig. 7.15 are only approximately valid.

Regarding the *lengths of jumps*, one must differentiate between the:

- Roller length L_r,
- Jump length L_j, and
- Ventilation length L_a.

In closed rectangular channels Haindl (see e.g. Hager 1989a) found for the *roller length*

$$\lambda_r = L_r/h_1 = 5.75(\mathsf{F}_1 - 2), \quad \mathsf{F}_1 < 10. \tag{7.46}$$

This *ventilation length* may be approximated with the dimensional relation $L_a[\mathrm{m}] = 10Q\ [\mathrm{m^3s^{-1}}]/b\ [\mathrm{m}]$.

Measurements of the *jump length* by Kalinske and Robertson (1943) resulted in the empirical expression (Hager 1989a)

$$\lambda_j = L_j/h_1 = 1.9[2\exp(1.5y_1) + \exp(-10S_o) - 1](\mathsf{F}_1 - 1). \tag{7.47}$$

This determines the distance between the toe of the jump and the point where the jump profile meets the conduit soffit (Fig. 7.14). In comparison to Eq. (7.23) for the hydraulic jump in the horizontal U-shaped channel, λ_j depends additionally on the bottom slope S_o. The influences of the approach flow Froude number F_1 and the approach flow filling ratio y_1 are comparable to the effect of y^2 (Fig. 7.15).

7.4.3 *Air Entrainment*

Kalinske and Robertson (1943) obtained for the *specific air entrainment*

$$\beta_a = 0.0066(\mathsf{F}_1 - 1)^{1.4}. \tag{7.48}$$

in which the Froude number F_1 for the circular section is given by Eq. (7.1), or approximately by $\mathsf{F}_1 = Q/(gDh_1^4)^{1/2}$, as given in Eq. (6.35).

Further studies on *air bubble transport* have been conducted by Falvey (1980). The ratio of air transport to air entrainment of a pressurized flow in a rectangular conduit has been thoroughly investigated by Ahmed et al. (1984). Studies on the air transport downstream of a hydraulic jump have been likewise summarized by Hager (1989a).

7.4.4 *Choking Criterion*

The sequent depth ratio $Y(\mathsf{F}_1)$ as given in Eq. (7.32) is valid for free surface flows but needs adjustment for choking flows due to the presence of a compressed *air-water mixture*. Based on experimental observations, the sequent depth ratio for flows with $h_2/D > 0.9$ can be expressed as $Y = \mathsf{F}_1$ instead of Eq. (7.32). For choking flow (subscript C) the downstream pressure head h_2 is larger than the sewer diameter D. *Incipient choking* results for $h_2/D = 1$, and the corresponding choking discharge is with $Y_C = \mathsf{F}_1$

$$\frac{Q_C}{(gD^3h_1^2)^{1/2}} = 1. \tag{7.49}$$

Accordingly, the maximum discharge for free surface flow to occur downstream of an hydraulic jump in a sewer varies significantly with the sewer diameter D, and linearly with the approach flow depth h_1.

In Chap. 17, the so-called *choking number* is introduced, based on these considerations for hydraulic jumps in circular conduits. The choking number is an important characteristic for supercritical sewer flow.

7.5 Hydraulic Jump in U-shaped Channel

The U-shaped channel possesses an important application in sewer systems as *transitional cross-section* in manholes and in changes from circular to rectangular cross-sections. For U-shaped channels few measurements exist.

The *U-shaped section* is formed of a semi-circular portion in the lower half ($y \le 1/2$) surmounted by a rectangular section for $y > 1/2$ with D as channel width. Here only the upper portion is considered. With $y = h/D$, the cross-sectional area F and the static pressure force P_s are

$$F/D^2 = y + \frac{\pi}{8} - \frac{1}{2}, \tag{7.50}$$

$$P_s/(\rho g D^3) = \frac{1}{2}\left(y - \frac{1}{2}\right)^2 + \frac{\pi}{8}\left(y - \frac{1}{2}\right) + \frac{1}{12}. \tag{7.51}$$

As for the circular cross-section, it follows, approximately, for $y < 1$ (Hager, 1987)

$$F/D^2 = \frac{4}{3}y^{3/2}\left(1-\frac{1}{3}y\right), \tag{7.52}$$

$$P_s/(\rho g D^3) = \frac{8}{15}y^{5/2}\left[1-\frac{1}{4}y\right]. \tag{7.53}$$

For the nearly horizontal prismatic channel the specific force balance Eq. (7.13) gives for the ratio of *sequent depths*

$$\frac{8}{15}y_1^{5/2}\left(1-\frac{1}{4}y_1\right)+\frac{Q^2/(gD^5)}{\frac{4}{3}y_1^{3/2}\left(1-\frac{1}{3}y_1\right)} = \frac{8}{15}y_2^{5/2}\left(1-\frac{1}{4}y_2\right)+\frac{Q^2/(gD^5)}{\frac{4}{3}y_2^{3/2}\left(1-\frac{1}{3}y_2\right)} \tag{7.54}$$

Solving for the dimensionless discharge $q_D = Q/(gD^5)^{1/2}$, Fig. 7.16 is obtained similar to Fig. 7.4 for the circular sewer. An approximation analogous to Eq. (7.29) may be written for the U-shaped channel as

$$\frac{y_2-y_1}{1-y_1} = \frac{q_D-y_1^2}{q_o-y_1^2}, \quad \text{with } q_o = 0.87y_1^{0.85}. \tag{7.55}$$

For a given discharge Q and for a particular value of approach filling ratio y_1, the experimentally determined sequent depth y_2 is always smaller than calculated. This effect is attributed to the influence of wall friction. The deviations between

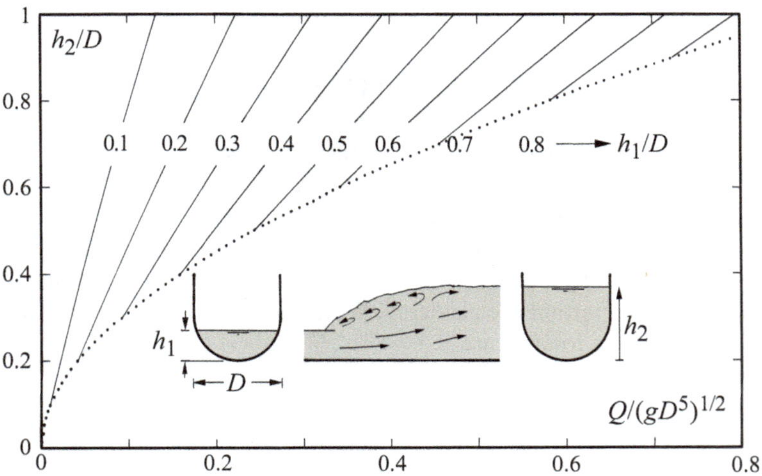

Fig. 7.16 Sequent depths in U-shaped channel, (...) $h_1 = h_2$

observations and predictions decrease as the scale of the structure increases. Usually Eq. (7.55) is sufficient for practical purposes.

For the *jump length* L_j, no systematic relation with neither the Froude number nor the filling ratio exists. A rough estimate obtained by Hager (1987) is

$$L_j/h_2 = 6. \tag{7.56}$$

In comparison with the rectangular channel, the jump in U-shaped sewer is always slightly longer.

The hydraulic jump in U-shaped sewers can be quite different in appearance from that in the rectangular channel. For small values of y, the approach flow width is smaller than the cross-sectional width D (Fig. 7.17). Throughout the hydraulic jump a lateral expansion takes place. Because the approach flow is concentrated to the sewer center it can not abruptly expand laterally but remains as a compact central jet. The forward flow is lifted from the bottom as a surface jet by the lateral mass of water. The axial section involves thus a so-called *bottom roller* below the surface jet (Fig. 7.17b). This feature is opposite to that of hydraulic jumps in the rectangular channel and typical for all sections widening with the depth of flow (such as for the trapezoidal section). In plan (Fig. 7.17a) two *lateral recirculation zones* appear near the surface jet forming two wedge-shaped wings in the inflow zone. The jump length L_j refers to the axial intersection point of the two wings (shown dashed in Fig. 7.17).

Figure 7.17 is only schematic in so far as the hydraulic jump is highly turbulent and symmetrical only in exceptional cases. Often *longitudinal oscillations* are observed and a stable asymmetric jump may occur in extreme cases (Fig. 7.18). Therefore, and on account of further indications (Hager 1992), the hydraulic jump in all channels of non-rectangular sections with a width increase as the flow depth increases are generally less efficient in terms of stability, compactness and wave

Fig. 7.17 Schematic flow pattern for small approach flow filling ratio y_1, (**a**) surface, (**b**) streamwise section

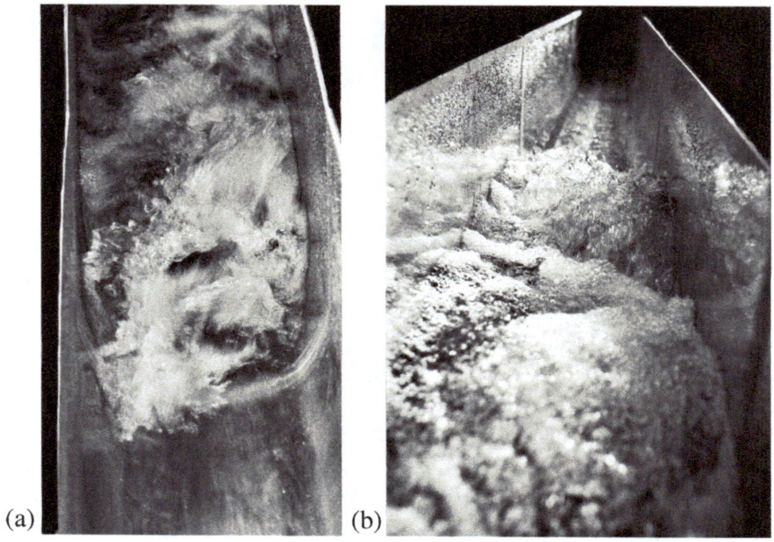

Fig. 7.18 Typical hydraulic jumps in U-shaped channels for $y_1 = 0.13$ with $\mathsf{F}_1 =$ (**a**) 4.6 and (**b**) 6.5

damping, as compared to the classical hydraulic jump. Hydraulic jumps in U-shaped or circular sewers are therefore not recommended for energy dissipation.

The *spatial nature* of the hydraulic jump in an U-shaped channel is most pronounced for $y << 1/2$ because the widening of the supercritical jet is then large. As y_1 approaches the value of 0.5 this influence gets reduced and the hydraulic jump approaches that in the rectangular channel (Hager 1989b).

7.6 Outlet Structures

7.6.1 *Introduction*

Hydraulic jumps in sewer systems occur on account of the channel geometry and the conditions of flow without regard to energy dissipation. A significant energy dissipation requires a *stilling basin* (German: Tosbecken; French: Ouvrage de dissipation), i.e. an hydraulic structure in which energy is efficiently dissipated by means of a hydraulic jump. The feature of an hydraulic jump is its potential of energy dissipation, i.e. the conversion of mechanical energy into primarily heat and sound energies. Since downstream of the stilling basin the energy balance is negative with regard to mechanical energy, one often speaks of an *energy loss* (Chap. 2).

A large energy dissipation in sewers is rare as the discharges are often relatively small in the first place and secondly, large velocities of flow do not occur.

For drop and vortex manholes, special provision is therefore made for energy dissipation (Chap. 15). Energy dissipation in connection with outlets such as downstream from storm water basins or sewage treatment plants is considered in the following. Based on a summary of Hager (1990b), some of the important types of structures are summarized below.

The design of stilling basin has developed its own particular discipline and a number of *standard basin types* have been proposed. Every basin type is more or less exactly described. For example, the USBR-Impact Basin (7.6.4) has not only been subject to laboratory tests but, in addition, was also thoroughly investigated in field installations. At present, even two modifications of this most successful outlet structure exist; by observing the design specifications accurately, this basin can be adopted without further tests.

All stilling basins for sewer outlets have a *rectangular section*. The cross-section therefore either diverges in the flow direction so that a transition section for the downstream width is not necessary, or the stilling basin remains prismatic and adaption to the downstream channel is ensured with a transition reach. In principle the second type of structure is preferred because the flow in an expansion has a strong tendency to *boundary layer separation*. This may lead to poor efficiency combined with the occurrence of dangerous bed erosion or scour. In the following the expanding stilling basins developed by Smith (1955/1988) and the mentioned USBR-Basin VI (Peterka, 1958) are presented. The class of prismatic stilling basins is represented by the basin of Vollmer (1972, 1975).

7.6.2 *Dissipation Mechanisms*

By energy dissipation one understands, in the wider sense, the decrease in mechanical energy flux of a flow due to *fluid viscosity*. The dissipative work is confined principally to the flow boundaries where it stimulates the formation of complex vortex flow structures. These again can penetrate deeper inside the flow and produce highly turbulent flow. Both the production of turbulence and its conversion into heat through dissipative work proceed simultaneously and continue indefinitely.

In the narrow sense, energy dissipation is understood to mean the local concentrated development of macroscopic vortices and their equally concentrated decay. To induce the production of high turbulence, *shear surfaces* are developed close to the boundary surfaces. These develop either unaided as in the classical hydraulic jump (Fig. 7.1a) involving large surface flow separation such as surface or bottom rollers, or by inducing additional separation surfaces by providing *appurtenances* such as sills or blocks (Fig. 7.1b).

Flow *separation zones* can be created by providing a number of selected elements. Since there are no limits to imagination, numerous types of structural elements have been developed, not all of which have led to efficient dissipators. If the separation zone is too large, for example, through an abrupt widening of the channel, along with the development of shear surfaces, insufficient *jet mixing*

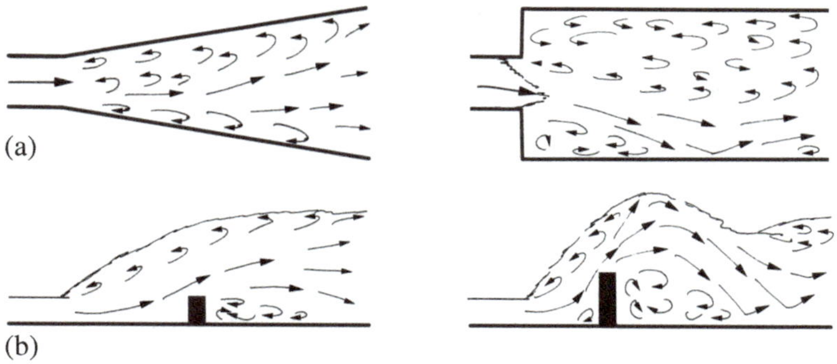

Fig. 7.19 'Good' (*left*) and inadequate (*right*) energy dissipation for (**a**) expanding basin and (**b**) basin containing a baffle sill

occurs. Then, the approach flow jet can be carried downstream without increase of its width and effective jet diffusion does not take place. Figure 7.19 shows two situations. In one of these the energy dissipation is well dispersed while in the other the dissipation remains insufficient. In Fig. 7.19a, the mixing of the incoming jet with the downstream flow is good as long as the opening angle remains small. If, however, the widening of the channel is abrupt and additional stabilizing elements are not provided, jet mixing is absent and the energy dissipation is ineffective.

The *effect of a sill* on the hydraulic jump can be different. For a sill height of the order of the approach flow depth the jet is only moderately deflected and the mixing with the downstream water is good. If the sill is too high, however, the flow develops into a plunging jet and the downstream region can suffer large scale scour.

By *appropriate energy dissipation*, one understands the following (Vischer and Hager 1995):

- High efficiency, i.e., small downstream depth necessary,
- Short length of jump,
- No cavitation damage,
- No development of tailwater waves,
- Low spray development,
- Low dynamic stress, and
- No scour damage.

This list could be extended, but it may be emphasized that *all* desirable objectives can never be met successfully. The *adequate stilling basin* is characterized by a sufficient tailwater depth of the order required without appurtenances. Its efficiency then is not a maximum but the rest of the requirements can be controlled with technique and reasonable expense. In the context of stilling basins provided in sewer systems the requirements for energy dissipation are lower than in hydraulic structures, where both discharge and approach flow velocity can be much larger. For details, the specialized literature on stilling basins should be consulted (Hager 1992).

7.6.3 *Stilling Basin of Smith*

The original arrangement of Smith dates back to 1955. The modified stilling basin of Smith (1988) has a *gradual expansion angle* θ which is related to the approach flow Froude number F_o and the expansion ratio $\beta = b_1/b_o$ as (Fig. 7.20)

$$\tan\theta = \frac{(\beta - 1)^{1/3}}{4.5 + 2\mathsf{F}_o}. \tag{7.57}$$

The entire structural arrangement consists of three parts, namely:

- Prismatic *inlet* channel in the form of either a rectangular or a circular channel of width b_o and depth of flow h_o from which F_o can be calculated; flows with a small approach Froude number $1 < \mathsf{F}_o < 3$ are suitable for this basin. The slope of the inlet channel should be smaller than the critical slope S_c, and the length $L_p = D/2 = b_o/2$ is measured from the outlet section to the beginning of the divergent channel.

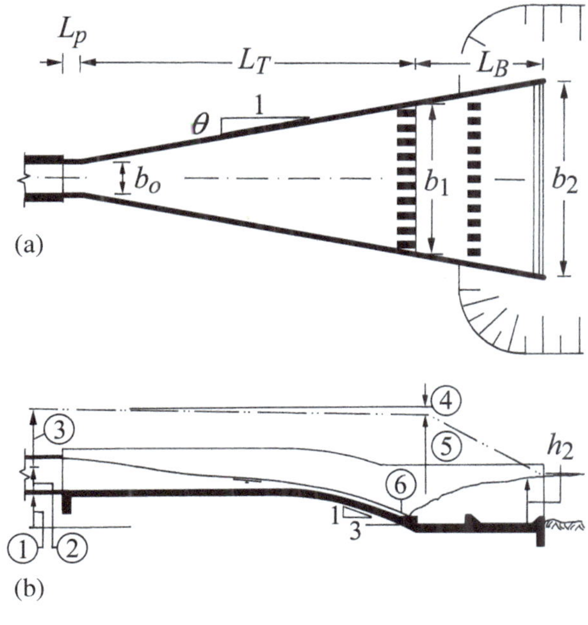

Fig. 7.20 Smith's stilling basin (1988). (**a**) plan and (**b**) side view. ① Δz, ② h_o, ③ $V_o^2/(2g)$, ④ Δz_E, ⑤ H, ⑥ h_1

- *Transition reach* of length $L_T = (\beta - 1)/(2\tan\theta)$ and of width b_1 at the stilling basin beginning given by the dimensional relation

$$b_1[\mathrm{m}] = 1.1Q^{1/2}\,[\mathrm{m}^3\mathrm{s}^{-1}]. \tag{7.58}$$

By limiting the angle θ, the diffuser flow does not separate.

- *Basin* of length L_B containing a row of chute and baffle blocks, respectively, with

$$L_B/h_2^* = 2.7. \tag{7.59}$$

Here h_2^* is the sequent depth of the corresponding hydraulic jump from Eq. (7.17).

Attention should be paid further to the following points:

- The *upstream flow depth* h_1 at the toe of the jump is obtained by a backwater curve, starting from the outlet section. For conduit outlets with curved streamlines only about 50% of the static head is effective. For the local head loss between the outlet section and the toe of the jump, it is about $\Delta H = 0.15V_o^2/(2g)$.
- The height, width and the distance between the *blocks* are equal to either h_1 or $h_2^*/8$, the row of baffle blocks is set at about $L_B/3$ downstream from the beginning of the basin.
- The *floor* of the stilling basin is set about $d_2 = 0.9h_2^*$ below the downstream water surface. The difference Δz between the invert of the approach flow channel and the basin floor should at least correspond to the width of the inlet channel, i.e., $\Delta z \cong b_o$. The transition from the nearly horizontal inlet to the 1:3 inclined chute is smoothly curved.

Figure 7.21 shows two typical views of the stilling basins during model tests.

Example 7.2 Given (after Smith): $Q = 22.7\ \mathrm{m^3s^{-1}}$ for $D = 1.83$ m. Level of conduit outlet = 100.00 m, level of downstream water surface = 101.50 m. Assumption: $\Delta z = 0.90$ m, corresponding to basin floor level = 99.10 m.

The mean conduit velocity is $V_o = Q/A_o = 22.7/(0.785\cdot 1.83^2) = 8.63\ \mathrm{ms^{-1}}$, thus $\mathsf{F}_o = 8.63/(9.81\cdot 1.83)^{1/2} = 2.04$. The pressure head is then 50% of the conduit diameter. The energy head H_o under the assumption of half hydrostatic pressure head is $H_o = 0.5\cdot 1.83 + 8.63^2/19.62 = 4.72$ m, thus $H_1 = H_o + \Delta z = 0.15\cdot 8.63^2/19.62 = 4.72 + 0.90 - 0.57 = 5.05$ m.

The width b_1 from Eq. (7.58) is 5.24 m such that $Q/b_1 = 4.33\ \mathrm{m^2s^{-1}}$. The Froude number at entry to the stilling basin is then calculated as $\mathsf{F}_1 = Q/(gb_1^2h_1^3)^{1/2} = 4.47$. The sequent depth h_2 calculated from Eq. (7.17) is $h_2 = 2.69$ m, thus $d_2 = 0.90\cdot h_2 = 2.42$ m. With the downstream water surface at level 101.50 m, the level of the stilling basin floor is set at

101.50 – 2.42 = 99.08 m, which agrees sufficiently with the assumed value of 99.10 m. With $\beta = b_1/b_o = 5.24/1.83 = 2.86$, the divergence angle from Eq. (7.57) is $\tan\theta = 0.143$, thus $L_T = (b_1 - b_o)/(2\theta) = 11.9$ m. Further, the basin length is $L_B = 3d_2 = 7.26$ m. The exit width of the basin is calculated as $b_2 = b_1 + 2\tan\theta \cdot L_B = 7.32$ m. With that the mean velocity at the basin exit is $V_2 = Q/(b_2 d_2) = 1.28$ ms^{-1}. The total basin length amounts to about $L = 20$ m.

Fig. 7.21 Stilling basin of Smith in operation (a) side view, (b) detail of flow over blocks

7.6.4 *USBR Stilling Basin*

The stilling basin of the United States, Bureau of Reclamation USBR VI (Bradley and Peterka 1957) can be considered special as it does not require *any tailwater* at all for its hydraulic performance. The basin, whose essential feature is a *hanging baffle wall*, may be used for approach flow velocities $V_c < 10$ ms^{-1} and discharges $Q < 13$ m^{3}s^{-1}. For larger discharges multiple units of parallel stilling basins can be employed.

The design parameter corresponds to the *discharge Q* from which the width b of a stilling basin unit is obtained, with ±10% accuracy, as

$$b = 3.3(Q^2/g)^{1/5}. \tag{7.60}$$

On account of the impact baffle wall (Fig. 7.22), *tailwater* is indeed not required. Presence of tailwater does not disturb the basin performance until the tailwater level

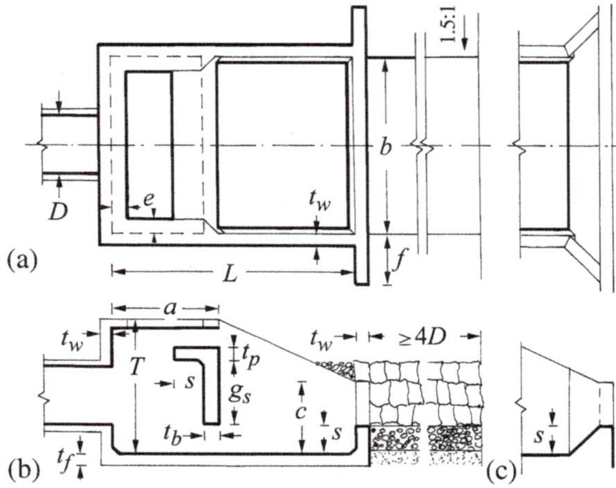

Fig. 7.22 USBR impact basin (**a**) plan, (**b**) longitudinal section, (**c**) alternative design for end sill (Bradley and Peterka 1957)

reaches the level of the axis of the approach flow conduit. Still higher tailwater levels have to be avoided because of unsuitable jet deflection.

The inlet conduit of diameter D can be placed at a maximum inclination of 15° from the horizontal and it can also be a rectangular open channel. The *dimensions* of the box structure, in terms of the stilling basin width b, are:

Total height	$T = (3/4)b$
Total length	$L = (4/3)b$
Height of sill	$s = (1/6)b$
Distance of baffle wall	$a = (7/12)b$
Height of downstream masonry wing	$c = (2/5)b$

Further, $e = (1/10)T$ for the box thickness, $f = (1/4)b$ for the length of the masonry wing and $g_s = (3/8)b$ for the baffle wall height. Finally, $t_w = t_f = t_b \cong (1/12)b$ and $t_p \cong 0.20$ m.

A *riprap* covers the channel bed over a minimum length of (4D) to prevent downstream erosion. The block size d_B [m] should at least be $d_B = V_b/2$, where V_b [ms^{-1}] is the bed velocity at basin exit. It may approximately be taken as $V_b = V_2 = Q/(bh_2)$. This completes the design of USBR Basin VI. Figure 7.23 shows a model basin under a test.

7.6.5 *Stilling Basin of Vollmer*

Another stilling basin working on the principle of counter-current flow has been developed by Vollmer (1972). It consists of a triangular baffle body guiding the inflow on to a semicircular baffle sill (Fig. 7.24).

The essential dimension of the basin is the *diameter D* of the approach flow channel. The high velocity jet first splits and then is turned around at the baffle

Fig. 7.23 Model of USBR Basin VI under test conditions (a) side view, (b) impact flow

sill to combine downstream of the baffle. Small discharges are led directly into the tailwater if the baffle element is set at a small height above the basin floor. For maximum discharge, the water flows both under and over the baffle sill in the second chamber in which the desired pool level is ensured by the provision of an end sill. The basin can also be covered. Then, however, it is necessary to ventilate the basin by providing the basin cover with a ventilating shaft (Vollmer 1975).

7.7 Remarks on Energy Dissipation

In connection with energy dissipation, distinction must be made between:

- *Hydraulic jump*, on the one hand, as the basis for all supercritical flows with a sudden transition to subcritical flow, and

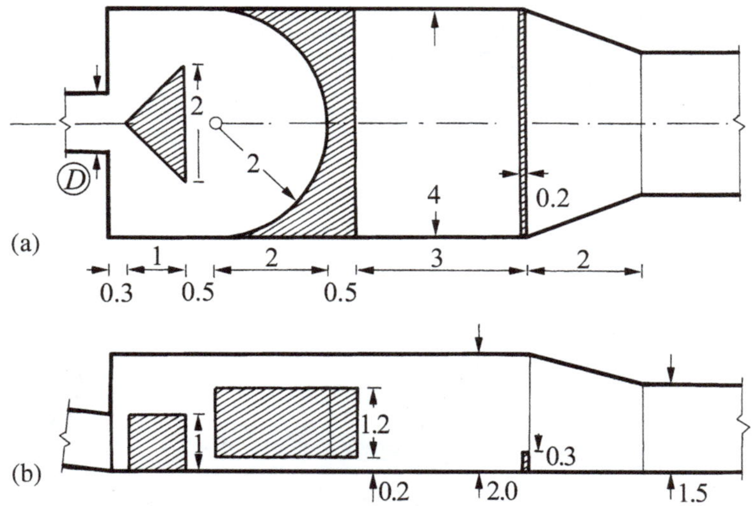

Fig. 7.24 Reverse flow basin of Vollmer (1972) (**a**) plan, (**b**) longitudinal section. Numerical quantities correspond to multiples of the approach flow diameter D

- *Stilling basing*, on the other hand, as a clearly defined location of energy dissipation. In sewer systems, one does usually not provide stilling basins, except for well-defined locations such as at outlets. Nevertheless, hydraulic jumps can occur in flat areas which have to be located in a detailed hydraulic calculation. Figure 7.25 shows typical situations in sewers with hydraulic jumps. Normally there are configurations that are more difficult than those considered in this chapter.

Figure 7.25a shows a longitudinal section of a *drop manhole* (Chap. 15), with the flow in the downstream channel and a highly turbulent mass of water impinging on it. The energy dissipation in this structure can be considerable.

Figure 7.25b shows the entrance of a *vortex drop manhole* (Chap. 15). In this structure the energy dissipation occurs on the one hand by wall friction along the spiral-shaped inlet and the shaft structure, and the stilling chamber at the exit on the other hand. For hydraulic overload, a hydraulic jump can appear in the entrance section for supercritical approach flow. This situation should be avoided because the presence of an adequate air core ensures flow stability.

A short *drop* is represented in Fig. 7.25c. If water backs up in the downstream channel, the energy dissipation is efficient. For insufficient tailwater depth, shock waves in the outflow channel are formed and choking of the downstream sewer may occur. For relatively smaller drop heights, the flow is directed over a so-called negative step (Hager 1992).

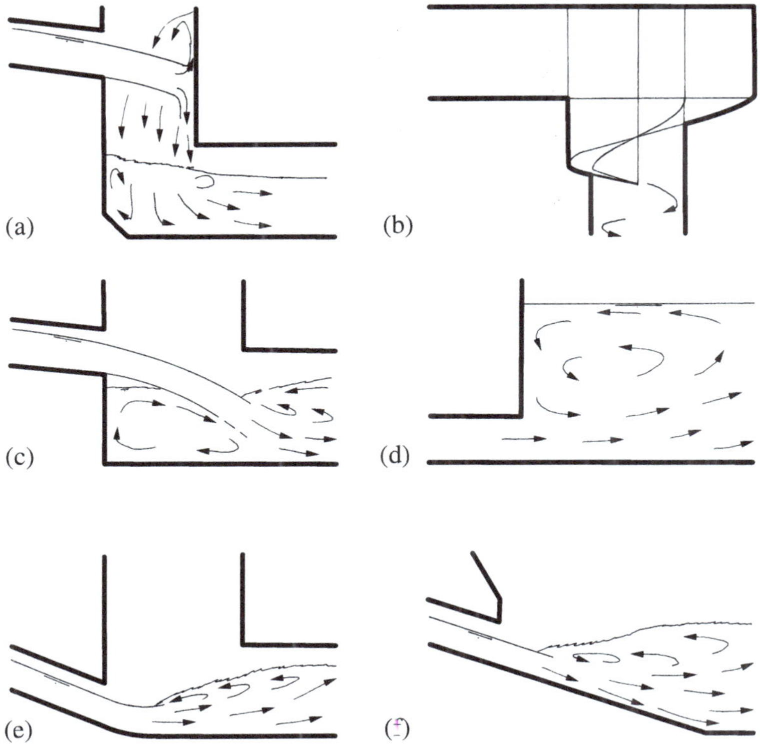

Fig. 7.25 Hydraulic jump in sewer systems, for details see main text

For pressurized flow into a *basin* diffusion phenomena occur, that are responsible for energy dissipation. In the situation shown in Fig. 7.25d, the flow is along the bed as a *turbulent wall jet* whose behaviour was investigated by Rajaratnam (1976).

A type of energy dissipation that occurs frequently is located at the *transition of slopes* from steep to flat sewer reaches (Fig. 7.25e). Depending on the discharge and the relative magnitudes of the two slopes, a hydraulic jump may form either on the steep portion itself or, more frequently, in the downstream sewer reach. If ventilation is not adequate, dangerous low pressure phenomena may develop.

Figure 7.25f shows the last example of energy dissipation in sewer systems as an *outlet channel* discharging into a receiving stream. Here the hydraulic jump often lies on an inclined plane generating a so-called *B-jump*. The relevant literature on hydraulic jumps in rectangular channels is compiled by Hager (1992).

Besides the preceding examples of locations, where hydraulic may jumps occur, the following situations in sewers should also be mentioned, because the occurrence of a hydraulic jump is there usually not considered at the design level. According to Fig. 7.26 transition elements need to be specially investigated for the possibility of occurrence of hydraulic jumps. Figure 7.26a shows the *change of bottom slope* from

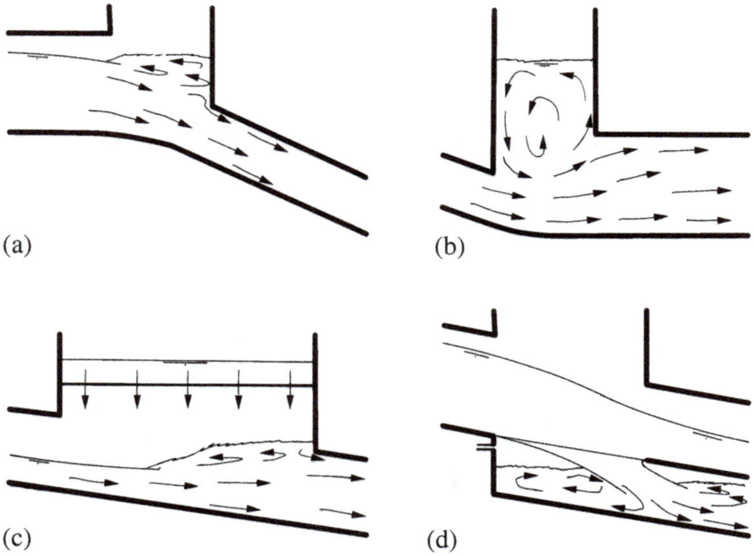

Fig. 7.26 Examples of energy dissipation usually not considered in design

a mild to a steep sewer reach with a corresponding diameter reduction (Chap. 6). If this reduction is too large, the conduit of the steep reach throttles the discharge and the choked flow backs up at the manhole outlet. As a result a hydraulic jump occurs in the manhole which can lead to large scale heading up of water.

A situation opposite to Fig. 7.25a is represented in Fig. 7.26b. For a transition from a steep to a flat reach, the outflow conduit chokes due to *downstream submergence* and gets pressurized. The hydraulic jump is then pushed upstream first in the manhole region, and eventually in the steep sewer reach itself. The control manhole remains under water and generates additional energy dissipation (Chap. 14).

For *lateral discharges* such as in a sideweir or a junction manhole, hydraulic jumps may occur in the collector channel (Fig. 7.26c). If this situation is not carefully designed, the downstream sewer may get pressurized and the flow conditions are drastically altered.

Finally, Fig. 7.26d represents a *bottom opening* over which a supercritical inflow is split up into two parts (Chap. 20). A hydraulic jump may occur in both the outflow and downstream branches due to inadequate ventilation or due to insufficient bottom slope. This again causes an alteration of the design flow conditions.

In principle all sewer reaches with a supercritical flow should be considered with care. In this connection two types of flow should receive close attention:

- For relatively small Froude numbers $1 < \mathsf{F} < 2$ the supercritical flow is not fully developed and often generates standing *surface waves* which can cause choking at the wave crests. These standing waves are also referred to as undular hydraulic jump. In the rectangular channel the *undular hydraulic jump* (German: Ondulierender Wassersprung; French: Ressaut hydraulique ondulé) has a profile which may be considered to be composed of solitary and cnoidal waves (Figs. 7.27 and 7.28). The maximum height h_M of a solitary wave is

$$h_M/h_o = \mathsf{F}_o^2, \qquad \mathsf{F}_o < \sqrt{2},$$

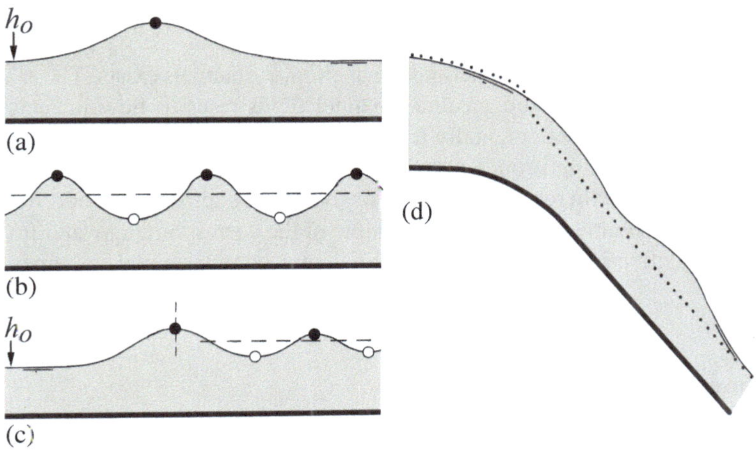

Fig. 7.27 (**a**) Solitary wave, (**b**) cnoidal waves, (**c**) undular hydraulic jump composed of solitary and cnoidal waves and (**d**) transition from flat to steep sewer reaches. Flow surface for (- -) hydrostatic and (—) hydrodynamic pressure distributions

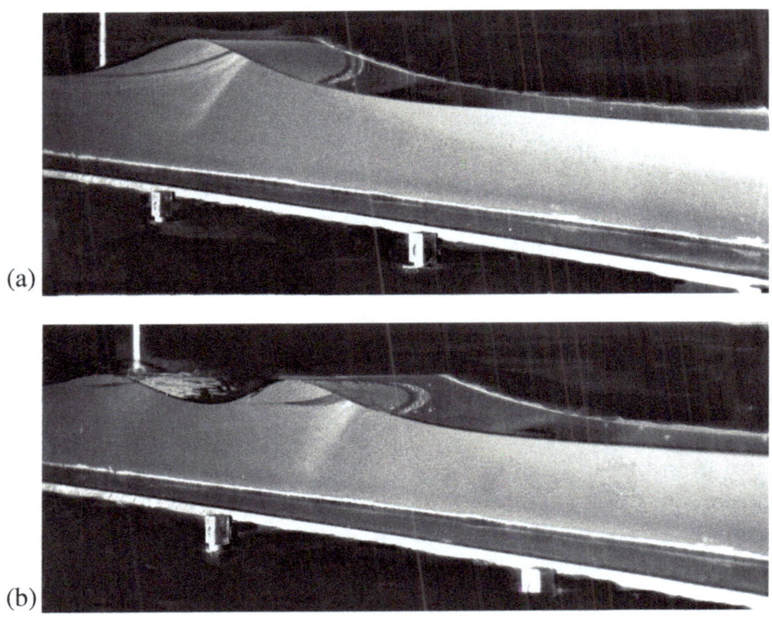

Fig. 7.28 (**a**) Solitary and (**b**) cnoidal waves in rectangular channel

where h_o is the approach flow depth and F_o the Froude number in the approach flow channel (Chap. 1). For $\mathsf{F}_o > \sqrt{2}$, the solitary wave has a broken crest (Reinauer und Hager 1995).

- For Froude numbers $\mathsf{F}_o > 2$, similarly, standing waves in the form of *shock waves* may occur. Every disturbance such as by bottom elements, directional changes or expansions calls for such wave formation. The wave height depends significantly on the approach flow Froude number (Chap. 16).

Currently, no study of these phenomena in circular channels exists. The relatively few works in the area of supercritical channel flows refer to flow in rectangular channels (Chap. 16). Fundamentally the danger from uncontrolled flow conditions increases with increasing Froude number F. For large values of F, a filling ratio of 0.5 may be adopted, thereby accounting for the surface air entrainment. *Air-water mixtures* need a volume larger than the volume of the corresponding water flow. The energy line is often informative and should always be determined for supercritical flows.

Notation

a	[m]	distance of baffle sill
b	[m]	width
b_o	[m]	width at inlet section
B	[m]	width of cross-section
c	[m]	downstream height of wing wall
d_B	[m]	diameter of riprap block
d_2	[m]	downstream height
D	[m]	diameter
e	[m]	thickness of box
f	[m]	length of wing wall
F	[m²]	cross-sectional area
F	[–]	Froude number
g	[ms⁻²]	acceleration of gravity
g_s	[m]	height of baffle sill
h	[m]	flow depth
h_M	[m]	maximum depth
H	[m]	energy head
L	[m]	total length
L_a	[m]	ventilation length
L_B	[m]	basin length
L_P	[m]	length of inlet reach
L_i	[m]	length of hydraulic jump
L_r	[m]	length of roller
L_R	[m]	recirculation length

L_T	[m]	length of transition
P_s	[N]	static pressure force
q	[m^{2}s^{-1}]	discharge per unit width
q_D	[–]	discharge normalized to diameter
q_o	[–]	reference discharge
Q	[m^{3}s^{-1}]	discharge
Q_a	[m^{3}s^{-1}]	air flow discharge
s	[m]	sill height
S	[m^2]	specific force
S_o	[–]	channel bottom slope
t	[m]	wall thickness
T	[m]	height of cross-section
V	[ms^{-1}]	average velocity
V_b	[ms^{-1}]	bottom velocity
V_o	[ms^{-1}]	approach velocity
y	[–]	filling ratio
Y	[–]	ratio of sequent depths
z_s	[m]	distance of centroid from free surface
Z_s	[–]	normalized centroid distance
β	[–]	expansion ratio
β_a	[–]	air-entrainment coefficient
δ	[–]	half central angle
Δz	[m]	height difference
ΔH	[m]	loss of energy head
λ_a	[–]	relative aeration length
λ_j	[–]	relative length of hydraulic jump
λ_R	[–]	relative length of recirculation zone
λ_r	[–]	relative length of roller
Φ	[–]	area function
η	[–]	efficiency
ρ	[kgm^{-3}]	density
Θ	[–]	inclination of channel bottom
θ	[–]	angle of expansion

Subscripts

1	hydraulic jump upstream
2	hydraulic jump downstream
a	air
b	block
B	basin
C	choking

E	energy
f	bed, bottom
L	boundary condition
M	maximum
o	inlet
w	wall
*	classical hydraulic jump

References

Ahmed, A.A., Ervine, D.A., McKeogh, E.J. (1984). The process of aeration in closed conduit hydraulic structures. Symposium on *Scale effects in modelling hydraulic structures* 4.13: 1–11, H. Kobus, ed. Technische Akademie: Esslingen, Germany.

Bradley, J.N., Peterka, A.J. (1957). The hydraulic design of stilling basins: Small basins for pipe or open channel outlets – no tailwater required. Proc. ASCE *Journal of the Hydraulics Division* 83(HY5 Paper 1406): 1–17.

Castro-Orgaz, O., Hager, W.H. (2009). Classical hydraulic jump: Basic flow features. *Journal of Hydraulic Research* 47(6): 744–754.

Chanson, H., Montes, J.S. (1995). Characteristics of undular hydraulic jumps: Experimental apparatus and flow patterns. *Journal of Hydraulic Engineering* 121(2): 129–144.

Falvey, H.T. (1980). Air-water-flow in hydraulic structures. *Engineering Monograph* 41, US Department of Interior. Water and Power Resources Service: Denver CO.

Gargano, R., Hager, W.H. (2002). Undular hydraulic jumps in circular conduits. *Journal of Hydraulic Engineering* 128(11): 1008–1013.

Hager, W.H. (1987). Abfluss im U-Profil (Flow in U-shaped profile). *Korrespondenz Abwasser* 34(5): 468–482 [in German].

Hager, W.H. (1989a). Wassersprung im geschlossenen Kanal (Hydraulic jump in closed conduit). *3R-International* 28(10): 674–679 [in German].

Hager, W.H. (1989b). Hydraulic jump in U-shaped channel. *Journal of Hydraulic Engineering* 115(5): 667–675.

Hager, W.H. (1990a). Geschichte des Wassersprunges (History of hydraulic jump). *Schweizer Ingenieur und Architekt* 108(25): 728–735 [in German].

Hager, W.H. (1990b). Energiedissipation an Auslassbauwerken (Energy dissipation of small outlet works). *Gas – Wasser – Abwasser* 70(2): 123–130 [in German].

Hager, W.H. (1990c). Basiswerte der Kanalisationshydraulik (Basic characteristics of sewer hydraulics). *Gas – Wasser – Abwasser* 70(11): 785–787 [in German].

Hager, W.H. (1992). *Energy dissipators and hydraulic jump*. Kluwer Academic Publishers: Dordrecht-Boston-London.

Hager, W.H., Bremen, R. (1989). Classical hydraulic jump: Sequent depths. *Journal of Hydraulic Research* 27(5): 565–585.

Hjelmfeldt, A.T. (1967). Flow in elliptical channels. *Water Power* 19(10): 429–431.

Hörler, A. (1967). Gefällswechsel in der Kanalisationstechnik bei Kreisprofilen (Change of bottom slope in sewer hydraulics for circular profiles). *Schweiz. Zeitschrift für Hydrologie* 29(2): 387–426 [in German].

Kalinske, A.A., Robertson, J.M. (1943). Closed conduit flow. *Transactions ASCE* 108: 1435–1447; 1513–1516.

Montes, J.S., Chanson, H. (1998). Characteristics of undular hydraulic jumps: Experiments and analysis. *Journal of Hydraulic Engineering* 124(2): 192–205.

Ohtsu, I., Yasuda, Y., Gotoh, H. (1997). Discussion to Characteristics of undular hydraulic jumps: Experimental apparatus and flow patterns. *Journal of Hydraulic Engineering* 123(2): 161–162.

Peterka, A.J. (1958). Hydraulic design of stilling basins and energy dissipators. *Engineering Monograph* 25. US Department of Interior. Bureau of Reclamation: Denver CO.

Rajaratnam, N. (1967). Hydraulic jumps. *Advances in Hydroscience* 4: 197–280. V.T. Chow, ed. Academic Press: New York.

Rajaratnam, N. (1976). Turbulent jets. *Developments in Water Science* 5. V.T. Chow, ed. Elsevier: Amsterdam.

Reinauer, R., Hager, W.H. (1995). Non-breaking undular hydraulic jump. *Journal of Hydraulic Research* 33(5): 683–698; 34(2): 279–287; 34(4): 567–573.

Smith, C.D. (1988). Outlet structure design for conduits and tunnels. *Journal of Waterway, Port, Coastal and Ocean Engineering* 114(4): 503–515.

Stahl, H., Hager, W.H. (1998). Hydraulic jump in circular pipe. *Canadian Journal of Civil Engineering* 26: 368–373.

Vischer, D.L., Hager, W.H. (1995). Energy dissipators. *IAHR Hydraulic Structures Design Manual* 9. Balkema: Rotterdam.

Vollmer, E. (1972). Ein Beitrag zur Energieumwandlung durch Gegenstrom-Tosbecken (A contribution to energy dissipation by counter-current stilling basins). *Mitteilung* 21. Institut für Wasserbau, Universität Stuttgart: Stuttgart [in German].

Vollmer, E. (1975). Energieumwandlung und Leistungsfähigkeit des Gegenstrom-Tosbeckens (Energy dissipation and efficiency of counter-current stilling basin). *Mitteilung* 35. Institut für Wasserbau, Universität Stuttgart: Stuttgart [in German].

(a)

(b)

Backwater curves relate the flows across elements of (**a**) sewers, (**b**) wastewater treatment stations. A rapid check of the main flow characteristics is then possible

Chapter 8
Backwater Curves

Abstract The transition curves of a nearly uniform flow state or, in general, the free surface profiles in open channels are designated backwater curves. They provide, for steady flow, the most general surface profile under a given discharge for specified boundary conditions. Usually, their computation is lengthy but they are required if backwater and drawdown effects have to be determined.

Chapter 8 presents a simplified method to determine backwater curves. It is based on a differential equation which can be solved exactly. Therefore, no convergence criterion with regard to the length of the calculation step is needed. The procedure is detailed by describing the computational scheme and explained with examples. Apart from the circular section, the egg-shaped and the horseshoe sections as well as the rectangular section are also considered. Finally, the effect of the non-hydrostatic pressure distribution on the discharge is discussed.

8.1 Introduction

By the term backwater curve (German: Stau- und Senkungskurve; French: Courbe de remous) one understands an idealized free surface profile obtained as *transition* curves between two or more channel sections. Figure 8.1 shows a long upstream – and an even longer downstream reach of a sewer. In both reaches uniform flow prevails after sufficient distance, provided the conditions of Chap. 5 are satisfied. Backwater curves are simple aids for the characterization of the mean flow characteristics. They provide no information over the cross-section, on the internal flow conditions, separation zones or over the velocity distribution.

Figure 8.1a shows a flow decelerating in the flow direction x. Also indicated are the profiles of uniform and critical flows. Since neither the discharge nor the cross-section changes, the function $h_c(x)$ remains constant. The uniform flow depth h_N however changes from reach to reach following a change in the bottom slope S_o.

For subcritical flow ($\mathsf{F} < 1$), the velocity V is, according to definition, smaller than the wave celerity c. If at a particular location $x = x_o$, the flow is disturbed by a sill, for example, the disturbance is propagated with the velocity (Chow 1959, Press and Schröder 1966)

W.H. Hager, *Wastewater Hydraulics*, 2nd ed., DOI 10.1007/978-3-642-11383-3_8,

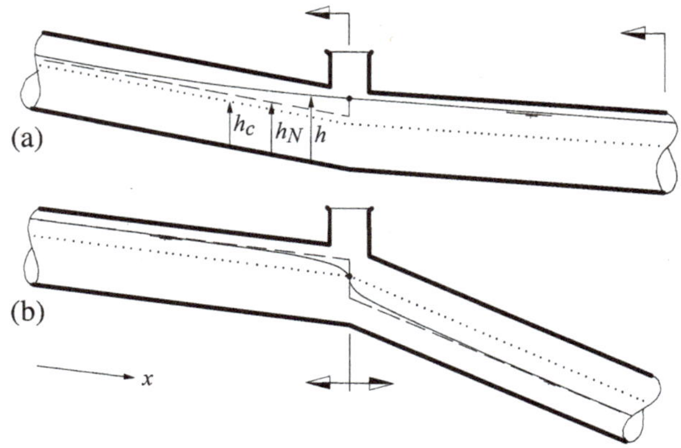

Fig. 8.1 Examples of backwater curves with indications of direction of calculation for (**a**) retarded and (**b**) accelerated flow. (—) Free surface profile $h(x)$, (- - -) uniform flow depth h_N, (. . .) critical depth h_c

$$\frac{\mathrm{d}x}{\mathrm{d}t} = V \pm c = c(\mathsf{F} \pm 1). \tag{8.1}$$

The disturbance that is propagated upstream is a backwater effect while the downstream disturbance dies off quickly. For *supercritical flow* ($\mathsf{F} > 1$) all disturbances travel only downstream since the flow velocity is larger than the propagation speed of a disturbance. This knowledge, which was already formulated in the 18th century by the French mathematicians Lagrange and Laplace, has decisive impact on the *computational procedure* (Fundamental Rule 1):

> For subcritical flows ($\mathsf{F} < 1$), the direction of calculation is always opposite to the flow direction; for supercritical flows ($\mathsf{F} > 1$), the calculation proceeds in the direction of flow.

With regard to Fig. 8.1a, the calculation of continuous subcritical flow proceeds from downstream to upstream whereas in Fig. 8.1b, critical flow is the starting point of the computation. From there, the calculations proceed in both upstream and downstream directions. Only then the requirements under the preceding Fundamental Rule 1 can be fulfilled. The location where critical flow occurs is also called control point or *control section* (German: Kontrollpunkt; French: Point de contrôle), because at this cross-section the flow, both upstream and downstream, is "constrained" through the critical flow state. Backwater curves in prismatic channels have always the tendency to approach the uniform flow state. In Fig. 8.1a uniform flow prevails in the downstream reach. Upstream, the uniform depth is smaller, and the depth decreases in the direction opposite to the flow direction.

In Fig. 8.1b, calculations start at the break of slope where $\mathsf{F} = 1$. Consequently uniform flow is asymptotically reached both in the upstream and downstream reaches. Therefore, Fundamental Rule 2 states:

> The uniform flow depth h_N is asymptotically approached, and cannot be intersected by the water surface profile $h(x)$.

The considerations discussed in this section are mostly qualitative in nature. In the following sections fundamental principles of quantitative calculations are presented. The equations of backwater curves are derived first. Subsequently, the backwater curves are classified and finally, solutions of the backwater equation for selected profiles are presented.

8.2 General Equation of Backwater Curves

Backwater curves are transition profiles. Before the basic backwater equation is derived the *basic underlying assumptions* are enumerated:

- One-dimensional flow, without variations in the cross-section,
- Uniform velocity distribution, and hydrostatic pressure distribution, i.e. the influence of streamline curvature is negligible,
- Flow is homogeneous and continuous, i.e. all parameters have well defined derivatives, and
- Flow is steady and discharge is spatially constant.

According to Chap. 1 both the energy and the momentum principles can be applied to continuous flows. Figure 8.2 shows a continuous flow with Q as discharge, $z = z(x)$ as the bottom geometry and $h = h(x)$ as the free surface profile. Further, $S_o = S_o(x) = -\mathrm{d}z/\mathrm{d}x$ represents the channel bottom slope and $S_e = -\mathrm{d}H/\mathrm{d}x$ the energy line slope.

The *energy head* H (German: Energiehöhe; French: Charge) referred to an arbitrary but fixed coordinate system (x, z) is

$$H = z + h + \frac{V^2}{2g} \tag{8.2}$$

where $V = Q/F$ is the average cross-sectional velocity, and F the cross-sectional area. According to Fig. 8.2, the energy head decreases in the direction of flow, and the energy loss is Δz_e. Expressed across an elementary distance Δx, it follows

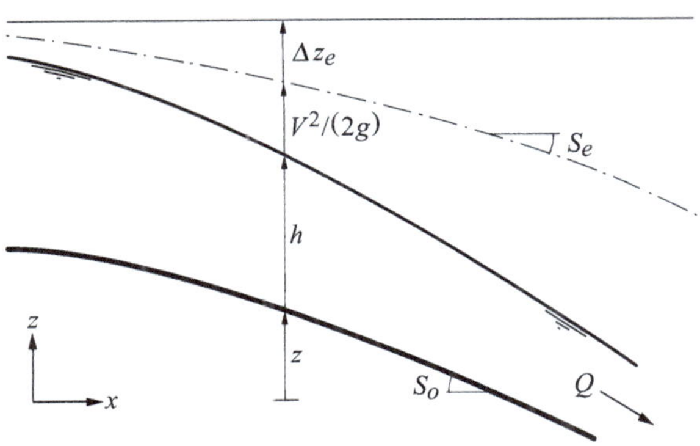

Fig. 8.2 Continuously variable flow, notation

that $\Delta H/\Delta x = -\Delta z_e/\Delta x = -S_e$ or $\mathrm{d}H/\mathrm{d}x = -S_e$, when changing to infinitesimal quantities. From Eq. (8.2), the change in the energy head can be expressed as

$$\frac{\mathrm{d}H}{\mathrm{d}x} = \frac{\mathrm{d}z}{\mathrm{d}x} + \frac{\mathrm{d}h}{\mathrm{d}x} + \frac{1}{2g}\frac{\mathrm{d}}{\mathrm{d}x}\left(\frac{Q^2}{F^2}\right) = -S_e. \tag{8.3}$$

Considering the identity $\mathrm{d}z/\mathrm{d}x = -S_o$ and noting that $\mathrm{d}/\mathrm{d}x(Q^2/F^2) = -(2Q^2/F^3)(\mathrm{d}F/\mathrm{d}x)$ for constant discharge

$$\frac{\mathrm{d}h}{\mathrm{d}x} - \frac{Q^2}{gF^3}\frac{\mathrm{d}F}{\mathrm{d}x} = S_o - S_e. \tag{8.4}$$

This relation is still unsuitable for further calculation because the cross-sectional area F and the flow depth h appear unknown. The cross-sectional area F is a function of the flow depth h and it can also vary with the streamwise coordinate x. This relatively rare case occurs in sewers at changes of cross-section, such as in a manhole. In general, therefore, the cross-sectional area varies as $F = F[x, h(x)]$. The total derivative of a function of two variables is, according to the rules of infinitesimal calculus

$$\frac{\mathrm{d}F}{\mathrm{d}x} = \frac{\partial F}{\partial x} + \frac{\partial F}{\mathrm{d}h}\cdot\frac{\mathrm{d}h}{\mathrm{d}x} \tag{8.5}$$

where $\partial F/\partial x$ is the partial derivative of F with respect to x. If Eq. (8.5) is substituted into Eq. (8.4), the *generalized equation of gradually-varied flow* (German: Allgemeine Gleichung der Stau- und Senkungskurven; French: Equation généralisée des courbes de remous) is obtained as (Chow 1959)

$$\frac{\mathrm{d}h}{\mathrm{d}x} = \frac{S_o - S_e + S_F}{1 - \mathsf{F}^2}. \tag{8.6}$$

Here, three slopes may be distinguished:

- S_o is the locally variable bottom slope which usually is known,
- S_e is the energy line slope accounting for all energy losses, and
- S_F is the additional term on account of a non-prismatic channel geometry, given by

$$S_F = \frac{Q^2}{gF^3}\frac{\partial F}{\partial x}. \tag{8.7}$$

This influence is proportional to the square of the velocity. The expression on the right of Eq. (8.7) shows that the free surface slope increases or decreases whether the cross-section expands ($\partial F/\partial x > 0$) or contracts ($\partial F/\partial x < 0$).

The Froude number F is defined in Chap. 6 as

$$\mathsf{F}^2 = \frac{Q^2}{gF^3}\frac{\partial F}{\partial h}. \tag{8.8}$$

In practice, backwater curves are seldom calculated using Eq. (8.6) because the mathematical formulation of the cross-sectional function is complex. Also, it is still not clear which additional losses are to be included in the term S_e on account of cross-sectional variations. For a non-prismatic channel, calculations are simplified by considering reaches of constant cross-section. For sufficiently small reaches this procedure is sufficiently accurate. In sewers, such reaches are concentrated at manholes.

8.3 Backwater Curves in Prismatic Channels

In a sewer reach the cross-section is invariable, and $\partial F/\partial x = 0$. Also, the energy line slope S_e is equal to the friction slope S_f, as summarized in Chap. 2. Equation (8.6) then simplifies to (Ranga Raju 1990, Townson 1991, Liggett 1994)

$$\frac{\mathrm{d}h}{\mathrm{d}x} = \frac{S_o - S_f}{1 - \mathsf{F}^2}. \tag{8.9}$$

Usually, the bed slope S_o remains reachwise constant – it is a stipulation in sewer design practice that a control manhole be provided at every change of bottom slope – and the right hand side of Eq. (8.9) depends only the flow depth h. The general solution of Eq. (8.9) can thus be written as

$$x + \text{const} = \int \left(\frac{1 - \mathsf{F}^2}{S_o - S_f}\right)\mathrm{d}h. \tag{8.10}$$

Further calculation from Eq. (8.10) in its present form is not directly accessible because the relation between the Froude number F, the friction slope S_f and the flow depth h is still missing.

For any arbitrary cross-section, three fundamental facts can be stated from Eq. (8.9) (see also Sect. 6.3.4):

- For $S_o = S_f$, $\mathrm{d}h/\mathrm{d}x = 0$ corresponds to an extreme flow depth,
- For $\mathsf{F} = 1$, the free surface slope is indefinitely large as $\mathrm{d}x/\mathrm{d}h = 0$, and
- For $S_o - S_f = 0$ and simultaneously $1 - \mathsf{F}^2 = 0$, the free surface slope is not defined.

Consider first the case $S_o = S_f$ but $1 - \mathsf{F}^2 \neq 0$. The second derivative of Eq. (8.9) is

$$-2\mathsf{F}\frac{\mathrm{d}\mathsf{F}}{\mathrm{d}x}\frac{\mathrm{d}h}{\mathrm{d}x}+\left(1-\mathsf{F}^2\right)\frac{\mathrm{d}^2h}{\mathrm{d}x^2}=-\frac{\mathrm{d}S_f}{\mathrm{d}x}. \tag{8.11}$$

Because the functions $\mathsf{F}=\mathsf{F}(x)$ and $S_f=S_f(x)$ depend only on the flow depth $h(x)$, one can write $\mathsf{F}=\mathsf{F}[h(x)]$ and $S_f=S_f[h(x)]$. Equation (8.11) can also be written as

$$-2\mathsf{F}\frac{\mathrm{d}\mathsf{F}}{\mathrm{d}h}\left(\frac{\mathrm{d}h}{\mathrm{d}x}\right)^2+\left(1-\mathsf{F}^2\right)\frac{\mathrm{d}^2h}{\mathrm{d}x^2}=-\frac{\mathrm{d}S_f}{\mathrm{d}h}\frac{\mathrm{d}h}{\mathrm{d}x}. \tag{8.12}$$

Because both functions $\mathrm{d}\mathsf{F}/\mathrm{d}h$ and $\mathrm{d}S_f/\mathrm{d}h$ are different from zero, the result is $\mathrm{d}^2h/\mathrm{d}x^2=0$ for $\mathrm{d}h/\mathrm{d}x=0$. One can show that all higher derivatives of the function $h(x)$ are also identically zero. Fundamental Rule 3 is:

For all flow conditions other than the critical flow, uniform flow ($S_o=S_f$) is approached asymptotically.

Regarding the second case, $\mathsf{F}=1$ for $S_o\neq S_f$, it can only be said that it is not physically significant. In fact the free surface cannot become perpendicular to the channel bottom by assuming hydrostatic pressure distribution. Nevertheless this can be considered as a transition curve without proper physical meaning. It facilitates the representation of the general solution of free surface flows (Chaps. 17, 18 and 19).

The third case, with $S_o=S_f$ and $\mathsf{F}=1$ simultaneously, has been thoroughly investigated in Chap. 6. If both these conditions are introduced in Eq. (8.9), $\mathrm{d}h/\mathrm{d}x$ must not equal zero. According to Chap. 6, this condition is called *pseudo-uniform flow*. Two values for the free surface slope $(\mathrm{d}h/\mathrm{d}x)_c$ normally result, such as given by Eq. (6.15) and Eq. (6.22).

The backwater curves discussed here refer to a special case by assuming *reachwise constant basic parameters* such as discharge, cross-sectional geometry, roughness or bottom slope. This simplification is physically inadmissible for flows separating from the sewer bottom. For changes of slope from a mild to a steep reach a separation zone is formed downstream of the slope break (Fig. 8.3a), and in the region of the break point itself otherwise (Fig. 8.3b).

Related to the backwater curves is the local flow in the *neighbourhood of the point of slope break.* Because the exact transition geometry is usually unknown, the detailed flow pattern in the region of the slope break is separately treated. Therefore, the *computational procedure* for such control sections is:

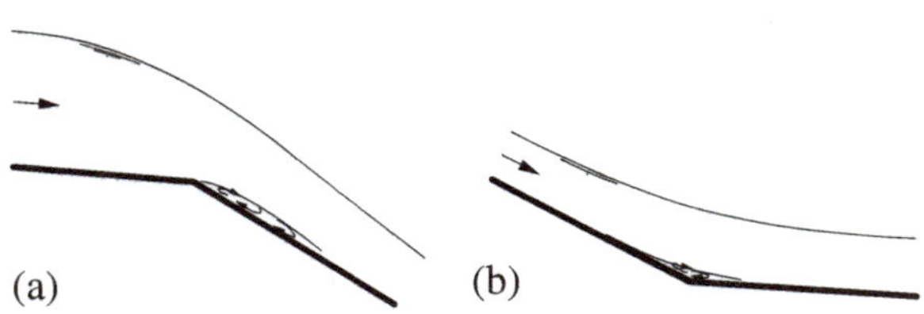

Fig. 8.3 Break in bottom slope and the accompanying flow separation zones from the channel bottom (**a**) flat to steep, and (**b**) steep to flat reaches

- Assume a vertical free surface ($dx/dh = 0$, $\mathsf{F} = 1$).
- Backwater curves for transcritical flow ($0.8 < \mathsf{F} < 1.2$) are excluded. The curves are physically significant only outside of the region of critical flow.
- Transcritical flow may be treated with a local analysis, ultimately connected to the backwater curves (Sect. 6.5).

A *conduit diameter* changes in a control manhole. Reducing the diameter results in a local *contraction*, whereas increasing the diameter corresponds to a local *expansion*. These elements involve additional losses but are not accounted for in conventional backwater calculations. The reason for this simplification is the difference in the flow behaviour in the manholes, and the reaches between two manholes. A manhole can be regarded as a *local flow element* in which the magnitude of the representative velocity head $V^2/(2g)$ relative to the flow depth h plays a dominant role beside the manhole geometry. The flow in a control manhole is thus mainly determined by the *Froude number*.

The flow between two manholes behaves quite differently. Here, the Froude number has relatively small influence, the flow in the region of $\mathsf{F}\sim 1$ being of course excluded. Then, for small and large values of the Froude number F, the following approximation is obtained in place of Eq. (8.9)

$$\zeta_F \frac{dh}{dx} = S_o[1 - (S_f/S_o)], \tag{8.13}$$

where $\zeta_F = 1$ for $\mathsf{F} << 1$ and $\zeta_F = -\mathsf{F}^2$ for $\mathsf{F} >> 1$. From this representation the dominant effect of the friction slope term S_f in relation to the channel bed slope S_o is realized. The equation of Manning and Strickler with n as the roughness coefficient and R_h the hydraulic radius can be used to give

$$S_f/S_o = \frac{n^2 V^2}{R_h^{4/3} S_o} = \frac{V^2}{gh_m} \cdot \frac{n^2 g h_m}{R_h^{4/3} S_o} = \mathsf{F}^2 \chi. \tag{8.14}$$

For the backwater curve then, apart from the square of the Froude number $\mathsf{F}^2 = V^2/(gh_m)$, the *friction characteristic* $\chi = n^2 g h_m/(R_h^{4/3} S_o)$ as introduced in Chap. 5 is significant. It alone determines the equilibrium condition for uniform flow.

For backwater curves with a length scale $X = S_o x/h_N$ one refers to locally lengthened reaches, because a free surface change Δh becomes significant only over a considerable distance Δx. In contrast, a manhole with a typical scale x/h_N is by S_o:1 shorter and therefore referred to as a *local* element. For backwater calculations the manhole influence becomes usually negligible. This concept reflects the exactness of the conventional approach for transition curves. Backwater curves refer to flows between two manholes, and flow details in the manhole itself are excluded (Chap. 14). Therefore, Fundamental Rule 4 holds as follows:

Backwater curves answer the questions regarding the general flow configuration, local flow conditions are not covered, however.

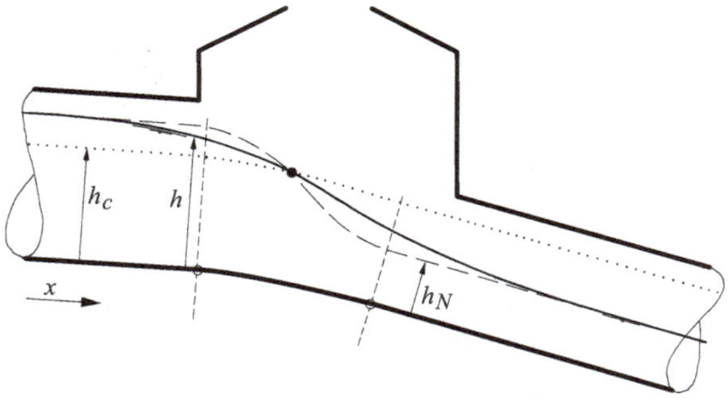

Fig. 8.4 Transition from mild to steep sewer reaches with diameter reduction with (—) flow surface $h(x)$, (. . .) critical depth profile $h_c(x)$, (- - -) uniform depth profile $h_N(x)$. (•) critical point

For supercritical flow, the flow depth of the upstream sewer reach is used as the boundary condition. In contrast, for subcritical flow, the downstream flow depth is taken as the boundary depth for the upstream sewer. Usually, the change of flow depth resulting from the manhole remains unconsidered.

Figure 8.4 shows the change of flow due to a change in the bottom slope. There, as suggested in Sect. 6.5, a continuously curved bottom geometry as also a reduction in the cross-section are provided. The critical point is located at the section where the curve $f_1 = S_o - S_f = 0$ intersects the curve $f_2 = 1 - \mathsf{F}^2 = 0$. In the region of the control manhole the procedure introduced in Chap. 6 is adopted. Outside of the manhole region, conventional backwater curves can be used. As is evident from Fig. 8.4, the water surface and consequently the streamlines in the transition zone are curved. A detailed discussion on the transition from mild to steep sewer reaches is presented in Sect. 6.5.

Every *cross-sectional geometry* requires a different equation for the backwater curve. In the following, two types of cross-sections are considered, namely, the rectangular section and the circular section. In general, relatively complex expressions result. Because the method previously described corresponds to a hydraulic simplification, an approximate procedure is adopted. For practical purposes, the results are reasonably accurate.

8.4 Backwater Curves in Circular Sewers

8.4.1 *Special Solution*

Hager (1990) presented an approximate procedure for the calculation of backwater curves. Based on a literature review it was shown that few computational models for the circular sewer exist and that experimental studies were not even conducted.

Based on Eq. (8.9) one can recognize that the uniform flow and the critical flow conditions have fundamental effect on backwater curves. It is evident from Chaps. 5 and 6 that even these two basic conditions can only be investigated by detailed calculations. If approximations for the uniform and the critical flows are considered,

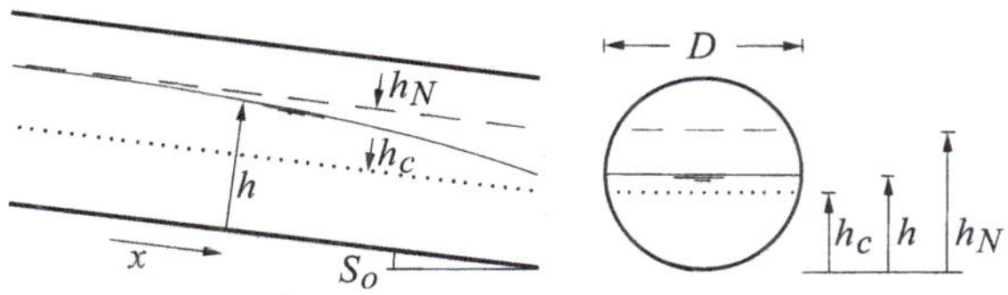

Fig. 8.5 Schematic surface profiles in a circular sewer (**a**) streamwise section, (**b**) cross-section

a simple method can be derived. For *uniform flow* (subscript *N*) Eq. (5.16) gives with $y_N = h_N/D$ (Fig. 8.5)

$$\frac{nQ}{S_o^{1/2}D^{8/3}} = \frac{3}{4}y_N^2\left(1 - \frac{7}{12}y_N^2\right), \quad y_N < 0.95. \tag{8.15}$$

The friction slope is thus

$$S_f = \frac{n^2Q^2}{D^{8/3}}\left[\frac{3}{4}y^2\left(1 - \frac{7}{12}y^2\right)\right]^{-2}. \tag{8.16}$$

From Eq. (6.36) the *critical flow* (subscript *c*) can be approximated as

$$\frac{Q}{(gD^5)^{1/2}} = y_c^2. \tag{8.17}$$

With the dimensionless parameters based on the uniform depth

$$X = S_o x/h_N, \quad Y = h/h_N \text{ and } Y_c = h_c/h_N \tag{8.18}$$

the *gradually-varied flow equation for the circular sewer* is (Hager 1990)

$$\frac{\mathrm{d}Y}{\mathrm{d}X} = \frac{1 - \dfrac{y_N^4\left(1 - \frac{7}{12}y_N^2\right)^2}{y^4\left(1 - \frac{7}{12}y^2\right)^2}}{1 - (Y_c/Y)^4}. \tag{8.19}$$

The length coordinate x is normalized with the length h_N/S_o, whereas the flow depth h is normalized with the uniform flow depth h_N.

According to Eq. (8.19) the free surface slope d*Y*/d*X* depends on four independent parameters, namely, *X*, *Y*, Y_c and $y_N = h_N/D$. A solution in the form of a family of curves does not exist. For small filling ratios $y_N < 0.3$, however, the influence of $(7/12)y^2$ as compared to unity is negligible. Then, the simplification $y_N/y = h_N/h = Y^{-1}$ yields, in place of Eq. (8.19), the three parameter expression

$$\frac{\mathrm{d}Y}{\mathrm{d}X} = \frac{1 - Y^{-4}}{1 - (Y_c/Y)^4}. \tag{8.20}$$

Equation (8.20) is the equation of Tolkmitt (1892) for backwater curves in channels of parabolic cross-section. Its general solution according to Forchheimer (1914) and Tolkmitt (1907) is

$$X = Y - \frac{1}{4}\left(1 - Y_c^4\right)\left[\ln\left|\frac{Y+1}{Y-1}\right| + 2\arctan Y\right] + C \tag{8.21}$$

where C is the constant of integration. Usually, the value of C involves an asymptotic *boundary condition* (subscript r):

- Backwater curves lie 1% above the uniform depth, corresponding to $Y_r(X = X_r) = 1.01$, and
- Drawdown curves lie 1% below the uniform depth, corresponding to $Y_r(X = X_r) = 0.99$.

Figure 8.6 shows the general solution $Y(X)$ for various values of Y_c. Marked in this figure are the uniform depth by a dashed line and the critical depth by a dotted line. The two curves intersect at the singular (subscript s) point $(X_s;Y_s) = (0;1)$ of Eq. (8.20). Also shown by a bold line is the curve $Y_c = 1$. Before applying Fig. 8.6, the general solution of backwater curves in circular channels is presented.

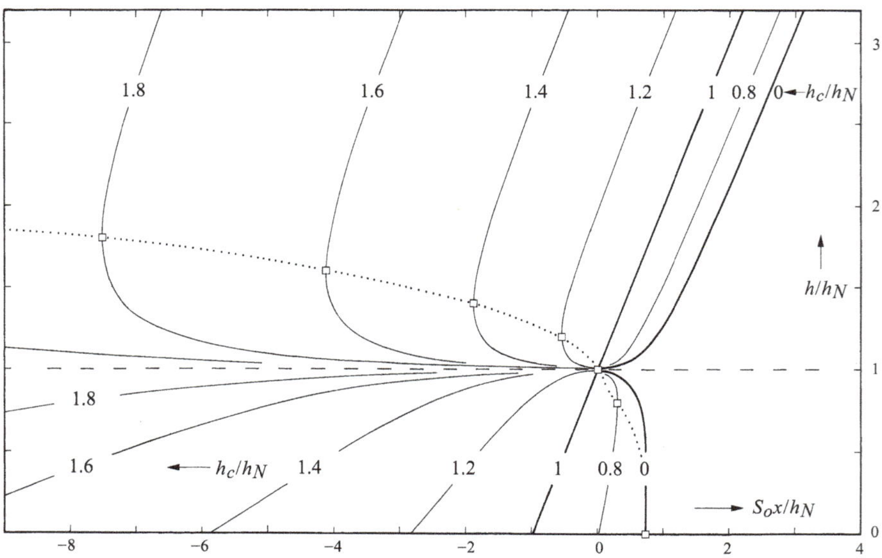

Fig. 8.6 Backwater curves $Y(X)$ in circular sewer for various $Y_c = h_c/h_N$ and $Y_c << 1$ with $X = S_ox/h_N$ and $Y = h/h_N$. The transformation $X \rightarrow X^*$ generalizes this solution for arbitrary values of $y_N < 0.95$. (- - -) Uniform flow, (. . .) critical flow

8.4.2 *General Solution*

The general solution of Eq. (8.19) involves arbitrary values of y_N with $0 < y_N < 0.9$. In order to investigate the effect of the parameter y_N, Eq. (8.19) is solved numerically for various values of y_N. The resulting curves $Y(X)$ for all $y_N > 0$ behave similar to those for the special case $y_N = 0$. These curves are vertical (marked by small circles as they intersect with the critical depth curve, and approach the uniform depth $Y = 1$ asymptotically. To account for the *form parameter* y_N, the length coordinate X of Eq. (8.18) is transformed as

$$X^* = \lambda X \tag{8.22}$$

where the operator λ serves as the *transformation parameter* for arbitrary values of y_N. By optimising the results λ is found as

$$\lambda = \left(1 - 1.1y_N^2\right)^{1/2}, \quad y_N < 0.9. \tag{8.23}$$

Accordingly, from Eq. (8.21), the approximate solution of backwater curves in a circular sewer is with the boundary condition $(X_r^*; Y_r)$

$$X^* - X_r^* = Y - Y_r - \frac{1}{4}\left(1 - Y_c^4\right)\left[\ln\left|\frac{Y-1}{Y-1}\cdot\frac{Y_r-1}{Y_r+1}\right| + 2\arctan Y - 2\arctan Y_r\right]. \tag{8.24}$$

As mentioned in Sect. 8.4.1, the generalized boundary conditions are $(X_r^*;Y_r) = (0;1.01)$ for backwater curves and $(X_r^*;Y_r) = (0;0.99)$ for drawdown curves. Figure 8.6 can therefore be used directly as the general solution with the transformation $X \rightarrow X^* = \lambda X$. Backwater curves for circular sewers can be approximately predicted and *explicitly* calculated. Transition curves are thus determined simply and without integration for detailed hydraulic calculations of sewer systems.

8.4.3 *Backwater Length and Drawdown Length*

Before computational examples are presented certain special cases are examined. The *backwater length* (German: Staulänge; French: Longueur de remous) is the distance between a sewer section with subcritical flow ($h > h_N > h_c$) and the section with uniform flow ($h = h_N$). The backwater length (subscript o) can also be interpreted as the flow distance with a backwater effect. Inserting for the cross-section considered the value $(X_o^*;Y_o)$ and for the uniform flow section (0;1.01), it follows from Eq. (8.24)

$$X_o^* = Y_o - 1.01 + \frac{1}{4}\left(1 - Y_c^4\right)\left[\ln\left(201\frac{Y_o-1}{Y_o+1}\right) + 2(0.785 - \arctan Y_o)\right]. \tag{8.25}$$

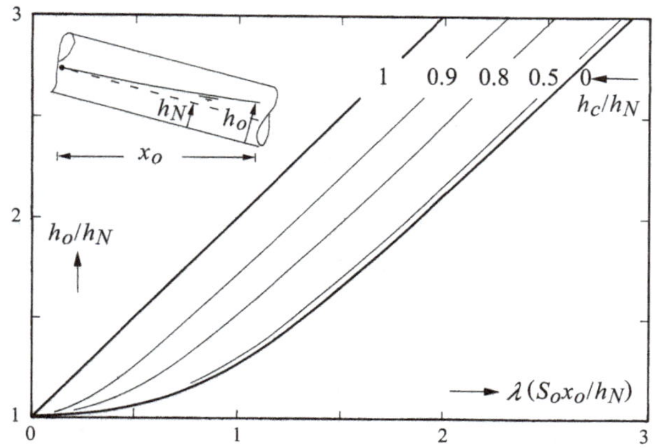

Fig. 8.7 Backwater length λX_o as a function of flow depth ratio h_o/h_N for various values of h_c/h_N

Figure 8.7 shows the solution of Eq. (8.25) and corresponds to the top right hand corner of Fig. 8.6.

Example 8.1 Given a conduit of diameter $D = 1250$ mm with a bottom slope $S_o = 0.5‰$ and a discharge $Q = 1050\ \mathrm{ls}^{-1}$. What is the backwater length for a downstream depth of 1.10 m? The roughness coefficient is $n^{-1} = 90\ \mathrm{m}^{1/3}\mathrm{s}^{-1}$.

1. Depth of uniform flow $y_N = 0.76$ from Eq. (8.15), thus $h_N = 0.76 \cdot 1.25 = 0.95$ m.
2. Depth of critical flow $y_c = 0.438$ from Eq. (8.17), thus $h_c = 0.438 \cdot 1.25 = 0.55$ m.
3. The Froude number is smaller than one, because $h_c/h_N = 0.55/0.95 = 0.58 < 1$.
4. With $y_N = 0.76$, $\lambda = (1-1.1 \cdot 0.76^2)^{1/2} = 0.604$ from Eq. (8.23).
5. With $Y_o = h_o/h_N = 1.10/0.95 = 1.158$ and $Y_c = h_c/h_N = 0.55/0.95 = 0.58$ one reads from Fig. 8.7 $\lambda X_o = 0.75$; Eq. (8.25) yields $X_o^* = \lambda X_o = 0.714$, and thus $X_o = 0.714/0.604 = 1.182$ and $x_o = X_o h_N/S_o = 1.182 \cdot 0.95/0.0005 = 2245$ m. The result is a backwater length of over 2 km.

The *drawdown length* (German: Absenkungslänge; French: longueur de chute) is the distance from an arbitrary cross-section to the critical point. For the drawdown length a distinction is made between subcritical and supercritical flow.

The *drawdown length for* $\mathsf{F} < 1$ is located upstream from the critical section and calculated from Eq. (8.24). Using the boundary conditions $(X^*, Y) = (0;0.99)$ and $Y_o = Y_c < 1$ results in

$$X_o^* = Y_c - 0.99 + \frac{1}{4}\left(1 - Y_c^4\right)\left[\ln\left(199\frac{1-Y_c}{1+Y_c}\right) + 2(0.785 - \arctan Y_c)\right]. \quad (8.26)$$

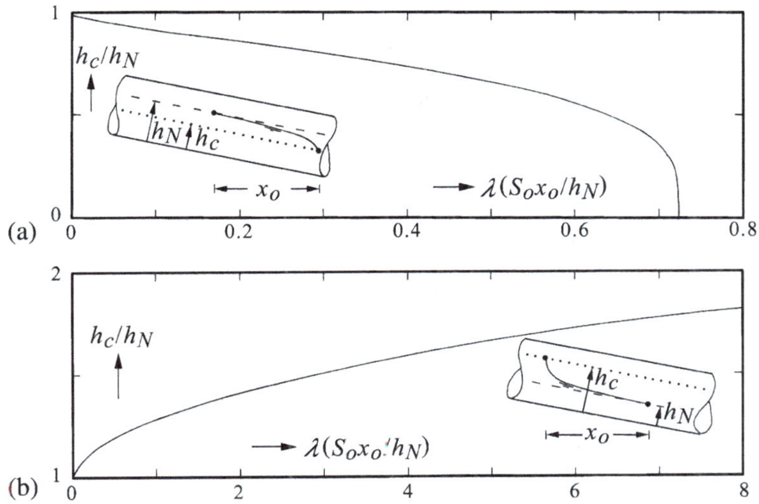

Fig. 8.8 Drawdown length $X_o^* = \lambda X_o$ as a function of h_c/h_N for (**a**) F < 1, (**b**) F > 1

This equation for $X_o^*(Y_c)$ is valid up to the maximum value $\lambda X_o = 0.72$ (Fig. 8.8). Equation (8.26) can be approximated by the simpler relation

$$X_o^* = (Y_c - 1) + \left(1 - Y_c^4\right)[1.72 - Y_c]. \quad (8.27)$$

Example 8.2 How long is the drawdown length for Example 8.1?

1. Uniform depth $h_N = 0.95$ m.
2. Critical depth $h_c = 0.55$ m.
3. Froude number F < 1.
4. Value of $\lambda = 0.604$.
5. With $Y_c = h_c/h_N = 0.58$, one reads from Fig. 8.8a $X_o^* = 0.58$. The value calculated from Eq. (8.26) is $X_o^* = 0.574$, whereas Eq. (8.27) gives $X_o^* = 0.591$ (+3%). The drawdown length therefore amounts to $x_o = 0.591 \cdot 0.95/(0.0005 \cdot 0.604) = 1860$ m. From the critical section about 2 km upstream distance is required to obtain uniform flow.

The *drawdown length for supercritical flow* (F > 1) involves the boundary condition $Y_o = Y_c > 1$ and the uniform flow condition $(X_o^*;Y) = (0;1.01)$. The relation represented in Fig. 8.8b is

$$X_o^* = Y_c - 1.01 + \frac{1}{4}\left(1 - Y_c^4\right)\left[\ln\left(201\frac{Y_c - 1}{Y_c + 1}\right) + 2(0.785 - \arctan Y_c)\right]. \quad (8.28)$$

For $1 < Y_c < 1.6$, this can be approximated as

$$X_o^* = 10(Y_c - 1)^{1.8}. \tag{8.29}$$

Example 8.3 How long is the drawdown length for a sewer of $D = 800$ mm diameter with a discharge of $Q = 1.5\ \mathrm{m^3s^{-1}}$, a bottom slope of $S_o = 1.7\%$ and a roughness coefficient of $n^{-1} = 75\ \mathrm{m^{1/3}s^{-1}}$?

1. Uniform depth $y_N = 0.74$, i.e. $h_N = 0.59$ m.
2. Critical depth $y_c = 0.915$, corresponding to $h_c = 0.73$ m.
3. Froude number larger than one, because $Y_c = h_c/h_N = 1.237 > 1$.
4. Value of $\lambda = 0.631$.
5. With $Y_c = 1.237$ Eq. (8.29) yields $X_o^* = 0.751$, Eq. (8.28) gives $X_o^* = 0.728$ (-3%) from which $x_o = X_o^* h_N/(\lambda S_o) = 0.728 \cdot 0.59/(0.631 \cdot 0.017) = 40$ m.

Thus, uniform flow is attained 40 m downstream of the critical section.

8.5 Classification of Backwater Curves

Originally, transition curves were determined almost exclusively in rectangular channels of width b. As a special case the very wide channel with $h/b << 1$ was investigated. The governing differential equation is similar to Eq. (8.20) with exponents 3 instead of 4. For this simplest type of backwater curve, Chow (1959) introduced a profile classification. The five *types of backwater curves* are:

- H-Curves for horizontal channel,
- M-Curves for channels of mild slope,
- C-Curves for channels of critical slope,
- S-Curves for channels of steep slope, and
- A-Curves for channels of adverse slope.

Further, the curves are divided in three classes:

- Class 1 $h > h_N$ and $h > h_c$, as backwater curves ($Y > 1$) for subcritical flow ($Y > Y_c$),
- Class 2 $h_N > h > h_c$, as drawdown curves for subcritical flow, and
- Class 3 $h < h_N$ and $h < h_c$, as drawdown curves ($Y < 1$) for supercritical flow ($Y < Y_c$).

Thus a total of 15 curve types are defined. Of these, H1 and A1-curves are not physically relevant in sewer hydraulics. From a practical point of view the H-curves and the A-curves are excluded therefore, and only the M-, the C- and the S-curves

Fig. 8.9 Classification of backwater curves in prismatic channels (extended from Chow 1959) with (- - -) uniform depth, (. . .) critical depth and (—) surface profiles for M-curves and S-curves

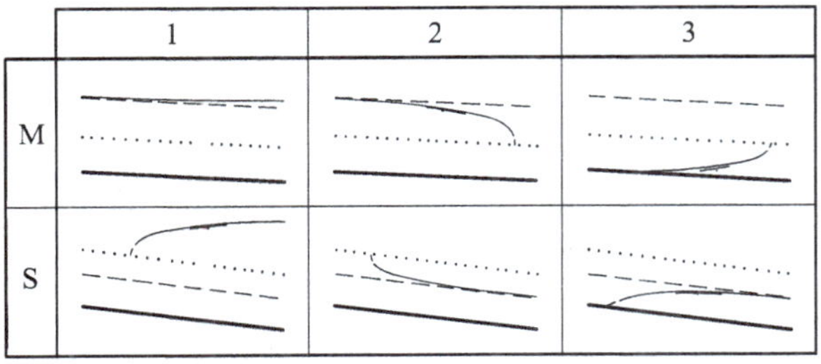

remain. Since flows in the neighbourhood of $\mathsf{F} = 1$ are excluded, only the M-curves and the S-curves are discussed in the following.

If both the numerator and the denominator of the right hand side expression of Eq. (8.20) are multiplied by Y^4 the free surface slope is $(\mathrm{d}Y/\mathrm{d}X)_0 = Y_c^4$ for $Y = 0$. Therefore, the free surface slope is positive for flows with a small flow depth. For large flow depths $Y^{-1} \rightarrow 0$ with a finite value of Y_c gives, in turn, $(\mathrm{d}Y/\mathrm{d}X)_\infty = 1$, corresponding to $\mathrm{d}h/\mathrm{d}x = S_o$. The water surface is then horizontal.

In contrast to other backwater curves, no inflexion point exists for Eq. (8.20). All curves therefore are continuously curved either upwards or downwards. On the basis of this information the relevant types of curves can be drawn. The M- and S-curves are shown in Fig. 8.9.

The M1-*curve* (Fig. 8.10a, b) is certainly the most important transition curve in natural streams, typically upstream of reservoirs or lakes. For submerged sewers, in particular in storage channels, M1-curves can be observed. Such curves are also common in channels connecting two basins.

M2-*curves* prevail wherever a flow accelerates in a sewer of mild bottom slope. Such cases occur for abrupt widening of a channel (Fig. 8.10c), or for a break in the bottom slope (Fig. 8.10d) with supercritical flow downstream of the transition.

M3-*curves* are seldom in practice, as they correspond to supercritical flows on small bottom slopes usually terminating with a hydraulic jump. These transition curves occur downstream of sluice gates or at a break of slopes from steep to mild reaches (Fig. 8.10e and f).

S1-*curves* start with a hydraulic jump and approach the horizontal asymptotically. These flows are supercritical as M3-curves and are a transition from super- to subcritical flow (Fig. 8.10g, h). The hydraulic jump itself is not a part of the S1-curve (Chap. 7). S1-curves prevail in channels with a large backwater.

S2-*curves* for supercritical flows are similar to M2-curves. They start close to critical flow and attain asymptotically uniform flow. Examples of appearance are again at abrupt widening of channels (Fig. 8.10i) or for breaks in the bottom slope (Fig. 8.10j).

Finally, S3-*curves* are found in decelerating supercritical flows, for example, downstream of reductions in the bottom slope (Fig. 8.10k) with a supercritical uniform flow, and downstream of sluice gates (Fig. 8.10l).

Before calculating backwater curves one should consider a sketch of the flow profile, based on the classification of flows. The possible combinations of transition

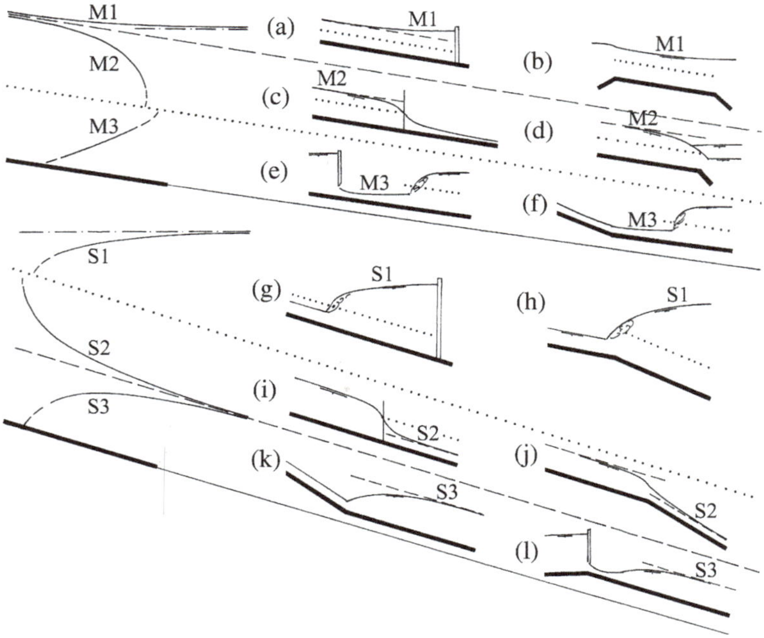

Fig. 8.10 Examples of backwater curves, type (*left*) and typical occurrences at constant bottom slopes, as also at changes of bottom slope (from Chow 1959)

curves for a *break in the bottom slope* of a prismatic channel are discussed. Figure 8.11 is self explanatory and needs no further clarification. Note that the free surface close to the critical section cannot be determined exactly with the foregoing equations. The flow at a change of bottom slope is separately analyzed in Chap. 6.

Hydraulic jumps likewise cannot be determined with the foregoing relations because the extra energy dissipation is not considered. The location of the hydraulic

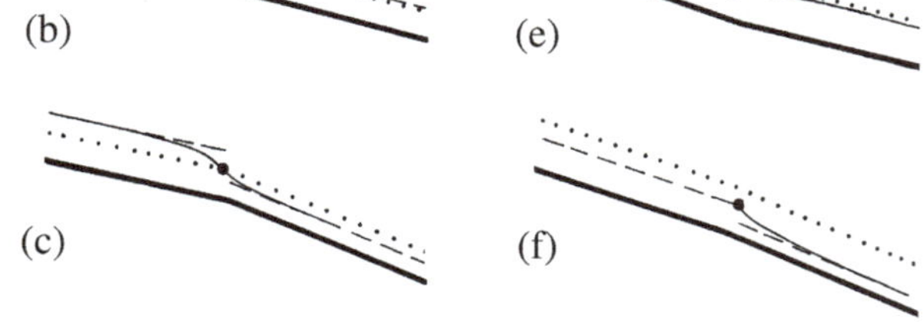

Fig. 8.11 Possible surface profiles for increasing or decreasing bottom slope with (- - -) uniform depth, (. . .) critical depth, (—) flow depth

jump is determined with a special investigation, using backwater calculations for the gradually-varied flow portions. Examples in Sect. 8.6.4 illustrate the computational procedures.

8.6 Computation of Backwater Curves

8.6.1 *Computational Scheme*

Backwater curves can be considered transitions between different uniform flow reaches. Each *change* of the uniform depth due to changes of the:

- Bottom slope,
- Roughness,
- Discharge, or
- Cross-section

causes a deviation from equilibrium and results in a backwater curve. Because these changes call for a control manhole, the important conclusion with regard to the hydraulic design of a sewer system is:

> The basic parameters of backwater curves, the uniform depth h_N and the critical depth h_c, do not change between two adjacent manholes.

This fact simplifies the calculation of backwater curves in sewers. These are therefore carried *reachwise*, with constant values of h_N and h_c along each reach.

The calculation along a sewer reach proceeds as follows:

1. Compilation of *basic parameters:*
   - Discharge Q
   - Diameter D
   - Bottom slope S_o
   - Roughness coefficient n
   - Boundary condition $(x_r; h_r)$
2. Calculation of *uniform depth*:
   Dimensionless discharge $q_N = nQ/(S_o^{1/2}D^{8/3})$, with the corresponding filling ratio according to Sect. 5.5.3

$$y_N = \left(\frac{2}{7}\left(3 - \sqrt{9 - 28q_N}\right)\right)^{1/2}. \tag{8.30}$$

   The uniform depth is $h_N = y_N \cdot D$.
3. Calculation of *critical depth*:
   Relative discharge $q_D = Q/(gD^5)^{1/2}$, with the relative critical depth

$$y_c = h_c/D = [Q/(gD^5)^{1/2}]^{1/2}. \tag{8.31}$$

Accordingly, the critical depth is $h_c = y_c \cdot D$.

4. Calculation of ratio $Y_c = h_c/h_N$. For $Y_c < 1$, the uniform flow is subcritical, otherwise supercritical. For $h_r/h_c > 1$, the flow is subcritical and the direction of calculation is *opposite* to the flow direction. For $h_r/h_c < 1$, the flow is supercritical and the calculation proceeds *in* the flow direction.
5. Sketch of flow with *flow classification*. The sketch contains the boundary conditions, and critical sections are marked at places where the ratio h_c/h_N changes from less than one to larger than one. For changes from $h_c/h_N > 1$ to $h_c/h_N < 1$, a *hydraulic jump* is generated.
6. Calculation of *backwater curve* according to Eq. (8.24)

$$X^* - X_r^* = \mathrm{f}(Y, Y_r, Y_c). \tag{8.32}$$

   On computers, sub-routines are used. The distance Δx for the method proposed is not subject to any restrictions since the solution is not determined with a step-by-step integration but is analytically available.
7. For reachwise changes from $h/h_c < 1$ to $h/h_c > 1$, the sequent depth profile h_2 is determined according to Chap. 7, and related to the subcritical flow. The hydraulic jump occurs at the intersection of the free surface profile $h(x)$ with the sequent depth profile $h_2(x)$. This is explained with an example below.
8. Once the backwater curve is calculated, the *results* are suitably presented. The scales are S_ox/h_N for the streamwise direction, and h/h_N for the direction perpendicular to it. The scales S_ox/D and h/D may also be chosen. This second representation is more advantageous for multiple discharges in a sewer.

The preceding overview clearly indicates that a suitable computational scheme can be worked out although the determination of the backwater curves is indeed a lengthy process. In contrast to the traditional method, only one more function must be explicitly evaluated or a diagram has to be read in the proposed method. It avoids numerical integration requiring sufficiently small steps to satisfy convergence criteria. Therefore, the calculation of backwater curves is elementary.

8.6.2 *Computations for a* Single *Sewer Reach*

In this section step-by-step calculations of backwater curves are presented. The possible six profile types are treated with examples, following the computational scheme presented in Sect. 8.6.1.

Example 8.4 M1-curve
Given a channel with $D = 0.4$ m, $n = (1/85)$ m$^{-1/3}$s and $S_o = 10.5‰$. How large is the flow depth at a distance $\Delta x = 45$ m, if it is at the manhole outlet $h_r = 0.35$ m and the discharge is $Q = 50$ ls^{-1}?

1. $Q = 0.05\ \mathrm{m^3s^{-1}}$, $D = 0.40$ m, $S_o = 0.0015$ and $n = (1/85)\ \mathrm{m^{-1/3}s}$.
2. $y_N = 0.528$, thus $h_N = 0.21$ m.
3. $y_c = 0.397$, thus $h_c = 0.16$ m.
4. $Y_c = h_c/h_N = 0.752 < 1$, consequently uniform flow is subcritical.
5. Boundary condition $h_r = 0.35$ m from which $Y_r = 1.657$, i.e. $h_r > h_N > h_c$ and the flow is similar to that in Fig. 8.10a. Because $h_r/h_c = 2.204 > 1$, the flow is subcritical and the computation proceeds *against* the direction of flow.
6. Boundary condition $(X_r^*; Y_r) = (0; 1.657)$, critical depth $Y_c = 0.752$, transformation parameter $\lambda = 0.833$ according to Eq. (8.23). The backwater curve is

$$\lambda X = Y - 1.657 - 0.170\left[\ln\left(0.247\frac{Y-1}{Y+1}\right) + 2(\arctan Y - 1.028)\right].$$

7. Hydraulic jump does not occur.
8. The equation is evaluated in Table 8.1, resulting in a backwater length of about 200 m. The solution can also be read from Fig. 8.6, with a reduced accuracy.

Table 8.1 Evaluation of Eq. (8.24) in Example 8.4

h [m]	0.35	0.33	0.31	0.29	0.27	0.25	0.23	0.213
Y	1.657	1.564	1.469	1.374	1.280	1.185	1.090	1.01
X	0	−0.125	−0.257	−0.397	−0.546	−0.721	−0.956	−1.477
x [m]	0	−17.58	−36.18	−55.79	−76.85	−101.40	−134.50	−207.80

Example 8.5 M2-curve

Given a circular sewer of diameter $D = 0.90$ m and bottom slope $S_o = 1.9‰$. Determine the free surface profile for $Q = 210\ \mathrm{ls^{-1}}$ and a boundary flow depth $h_r = 0.270$ m, for an equivalent sand roughness $k_s = 0.75$ mm.

From Eq. (5.5) follows $n = (1/8.2)k_s^{1/6}g^{-1/2} = (1/8.2)\cdot 0.00075^{-1/6}\ 9.81^{-1/2} = (1/85)\ \mathrm{m^{-1/3}s}$. One may therefore use the formula of Manning and Strickler.

1. $Q = 0.21\ \mathrm{m^3s^{-1}}$, $D = 0.90$ m, $S_o = 0.0019$ and $n = (1/85)\ \mathrm{m^{-1/3}s}$.
2. With $q_N = 0.075$ follow $y_N = 0.327$, and $h_N = 0.294$ m.
3. From $Q/(gD^5)^{1/2} = 0.087$ one obtains $y_c = 0.295$ and $h_c = 0.266$ m.
4. Because $Y_c = y_c/y_N = 0.902 < 1$, the uniform flow is subcritical.

5. Boundary condition $Y_r = h_r/h_N = 0.270/0.294 = 0.918$. This yields an M2-curve (Fig. 8.10d). With $h_r/h_c = 1.015 > 1$, calculations proceed *opposite* to the flow direction.
6. Boundary condition $(X_r^*;Y_r) = (0;0.918)$, the transformation parameter is $\lambda = 0.939$, and the drawdown curve is

$$X^* = Y - 0.918 - 0.085\left[\ln\left(0.043\frac{1+Y}{1-Y}\right) + 2\arctan Y - 1.485\right].$$

7. No hydraulic jump occurs.
8. The solution is given in Table 8.2.

Table 8.2 Numerical representation of M2-curve of Example 8.5

h [m]	0.270	0.275	0.280	0.285	0.290	0.291
Y	0.918	0.935	0.952	0.969	0.98	0.99
X	0	−0.005	−0.017	−0.041	−0.098	−0.123
x [m]	0	−0.84	−2.66	−6.38	−15.23	−19.00

Example 8.6 M3-curve
Given a circular sewer of diameter $D = 1.50$ m and discharge $Q = 1750$ ls^{-1}. The bottom slope is 0.7‰ and the equivalent roughness $k_s = 0.5$ mm. Determine the transition profile downstream of the sluice gate with a boundary flow depth $h_r = 0.30$ m.

The conversion of k_s into Manning's n yields $n = (1/91)$ m$^{-1/3}$s, and the uniform flow formula of Manning und Strickler may be applied for the calculations.

1. $Q = 1.75$ m^{3}s^{-1}, $D = 1.50$ m, $S_o = 0.0007$ and $n = (1/91)$ m$^{-1/3}$s.
2. With $q_N = 0.247$ one obtains $y_N = 0.666$, and $h_N = 1.00$ m.
3. With $Q/(gD^5)^{1/2} = 0.203$ follows $y_c = 0.45$ and $h_c = 0.675$ m.
4. $Y_c = h_c/h_N = 0.675$, the uniform flow is therefore subcritical.
5. With $Y_r = h_r/h_N = 0.30$ and $h_r/h_c = 0.444 < 1$ the flow is *supercritical* and the calculation proceeds *in* the flow direction. The free surface profile is an M3-curve as illustrated in Fig. 8.10e.
6. The boundary condition is $(X_r^*;Y_r) = (0;0.30)$, and the transformation parameter $\lambda = 0.716$. The equation of the backwater curve reads

$$X^* = Y - 0.30 - 0.198\left[\ln\left(0.538\frac{1+Y}{1-Y}\right) + 2(\arctan Y - 0.583)\right].$$

7. Table 8.3 shows the result. It is evident that at a distance of about 105 m downstream from the sluice gate a hydraulic jump occurs, because the uniform depth is $h_N = 1.0$ m $> h_c$.

Table 8.3 M3-curve of Example 8.6

h [m]	0.30	0.35	0.40	0.45	0.50	0.55	0.60	0.65
Y	0.30	0.35	0.40	0.45	0.50	0.55	0.60	0.65
X	0	0.014	0.027	0.040	0.051	0.061	0.069	0.073
x [m]	0	19.8	38.9	56.95	73.4	87.6	98.55	104.95

Example 8.7 S1-curve

Given a sewer of diameter $D = 1.25$ m discharging into a basin. The discharge is $Q = 500$ ls^{-1}, bottom slope $S_o = 1.2\%$ and $n = (1/80)$ m$^{-1/3}$s. Trace the backwater curve for a basin water level that is 1.5 m above the bed level of the channel at the entry into the basin.

Because the sewer outlet is submerged, a sewer portion is pressurized. Therefore, the point at which the sewer flow changes from free-surface to pressurized flow must be determined. The cross-sectional area remains constant in the pressurized portion and the Froude number is $\mathsf{F} = 0$. The slope of the pressure head curve is then given by Eq. (8.11) as

$$\frac{\mathrm{d}p(\rho g)}{\mathrm{d}x} = S_o - S_f. \tag{8.33}$$

Since both S_o and S_f remain constant and the level difference between the basin water surface and the conduit soffit amounts to $\Delta p/(\rho g) = 0.25$ m, the length L_D of the pressurized conduit reach is $L_D = \Delta p/(\rho g)/(S_o - S_f)$. With $Q = 0.5$ m^{3}s^{-1}, $n = 0.0125$ m$^{-1/3}$s, $F = \pi D^2/4 = 1.227$ m^2 and a hydraulic radius $R_h = D/4 = 0.313$ m follows $S_f = n^2Q^2/(F^2R_h^{4/3}) = [0.5/(80\cdot 1.227\cdot 0.313^{2/3})]^2 = 0.122‰$. With this, $L_D = 0.25/(0.012 - 0.000122) = 21.05$ m. The boundary condition is, therefore, $(x_r;h_r) = (-21.05$ m$;1.25$ m$)$.

1. $Q = 0.50$ m^{3}s^{-1}, $D = 1.25$ m, $S_o = 0.012$ and $n = 0.0125$ m$^{-1/3}$s.
2. With $q_N = 0.031$, one gets $y_N = 0.207$, and $h_N = 0.259$ m.
3. From $Q/(gD^5)^{1/2} = 0.091$ follows $y_c = 0.302$, and $h_c = 0.378$ m.
4. $Y_c = h_c/h_N = 1.459$, the uniform flow is supercritical.
5. With $Y_r = h_r/h_N = 1.25/0.259 = 4.826 > 1$, the surface profile is a backwater curve; because $h_r/h_c = 1.25/0.378 = 3.307 > 1$, the flow is *subcritical*, and the calculations proceed in the direction *against* the flow direction.

6. Boundary condition $(X_r^*;Y_r) = (-0.952;4.826)$, transformation parameter $\lambda = 0.976$. The equation of the backwater curve is

$$X^* = -0.952 + Y - 4.826 + 0.883\left[\ln\left(0.657\frac{Y+1}{Y-1}\right) + 2(\arctan Y - 2.733)\right].$$

7. The backwater curve is numerically reproduced in Table 8.4. A hydraulic jump occurs ultimately at $x = -86.4$ m.

Table 8.4 Backwater curve of Example 8.7

h [m]	1.25	1.15	1.05	0.95	0.85	0.75	0.65	0.55	0.45
Y	4.826	4.440	4.054	3.668	3.282	2.896	2.510	2.124	1.737
X	−0.975	−1.367	−1.759	−2.148	−2.534	−2.913	−3.282	−3.626	−3.909
x [m]	−21.05	−29.50	−37.95	−46.35	−54.70	−62.90	−70.85	−78.25	−84.36

Example 8.8 S2-curve

How does the drawdown curve look like for the channel flow of Example 8.7, with a boundary flow depth of $h_r = 0.35$ m?

1. $Q = 0.50\ m^3s^{-1}$, $D = 1.25$ m, $S_o = 0.012$ and $n = 0.0125\ m^{-1/3}s$.
2. $y_N = 0.207$ and $h_N = 0.259$ m.
3. $y_c = 0.302$ and $h_c = 0.378$ m.
4. $Y_c = 1.459 > 1$, thus the uniform flow is supercritical.
5. $Y_r = 0.35/0.259 = 1.351 > 1$, the surface profile is a drawdown curve, and with $h_r/h_c = 0.926 < 1$ is calculated *in* the direction of flow.
6. Boundary condition $(X_r^*;Y_r) = (0;1.351)$, transformation parameter $\lambda = 0.976$. Equation of the drawdown curve

$$X^* = Y - 1.351 - 0.883\left[\ln\left(0.149\frac{Y+1}{Y-1}\right) + 2\arctan Y - 1.867)\right].$$

7. Table 8.5 shows the free surface profile $h(x)$. Downstream of the coordinate $x = 53.26$ m uniform flow prevails.

Table 8.5 Drawdown curve of Example 8.8

h [m]	0.35	0.33	0.31	0.29	0.27	0.262
Y	1.351	1.274	1.197	1.120	1.042	1.01
X	0	0.064	0.197	0.476	1.236	2.468
x [m]	0	1.37	4.25	10.28	26.68	53.26

Example 8.9 S3-curve
Determine the backwater curve for Example 8.7 with a boundary flow depth $h_r = 0.20$ m.

1. $Q = 0.50$ m^{3}s^{-1}, $D = 1.25$ m, $S_o = 0.012$ and $n = 0.0125$ m$^{-1/3}$s.
2. $y_N = 0.207$ and $h_N = 0.259$ m.
3. $y_c = 0.302$ and $h_c = 0.378$ m.
4. $Y_c = 1.459 > 1$, the uniform flow is supercritical.
5. $Y_r = 0.20/0.259 = 0.772 < 1$, the surface profile is a backwater curve and with $h_r/h_c = 0.529 < 1$ must be calculated *in* the flow direction,
6. The boundary condition is $(X_r^*;Y_r) = (0;0.772)$; with the transformation parameter $\lambda = 0.976$, the equation of the backwater curve is

$$X^* = Y - 0.772 - 0.883\left[\ln\left(0.129\frac{1+Y}{1-Y}\right) + 2\arctan Y - 1.315\right].$$

7. The solution is summarized in Table 8.6. Figure 8.10k) shows the surface profile. Uniform flow prevails in the reach beyond a distance of about 73 m from the boundary point.

Table 8.6 Backwater curve of Example 8.9

h [m]	0.20	0.21	0.22	0.23	0.24	0.25	0.256
Y	0.772	0.811	0.849	0.888	0.927	0.965	0.99
X	0	0.271	0.578	0.944	1.422	2.192	3.378
x [m]	0	5.85	12.45	20.35	30.70	47.30	72.90

It is evident from these examples that the backwater and the drawdown curves can be determined in a straightforward and simple way. In comparison to the traditional method, the proposed integral solution has the facility of arbitrary selection of the *length increment* Δx because the complete analytical solution is available. Since the flow surface along a reach between adjacent manholes either monotonically increases or monotonically decreases, intermediate values can even be discarded and the water depths are determined only at the manholes. Considerable computational time is saved, and even large sewer systems can be analyzed within a reasonable time.

8.6.3 *Computations for* Two *Sewer Reaches*

Before applying the calculation method to an entire sewer system, typical problems of two consecutive manhole reaches are discussed. Consider two reaches of

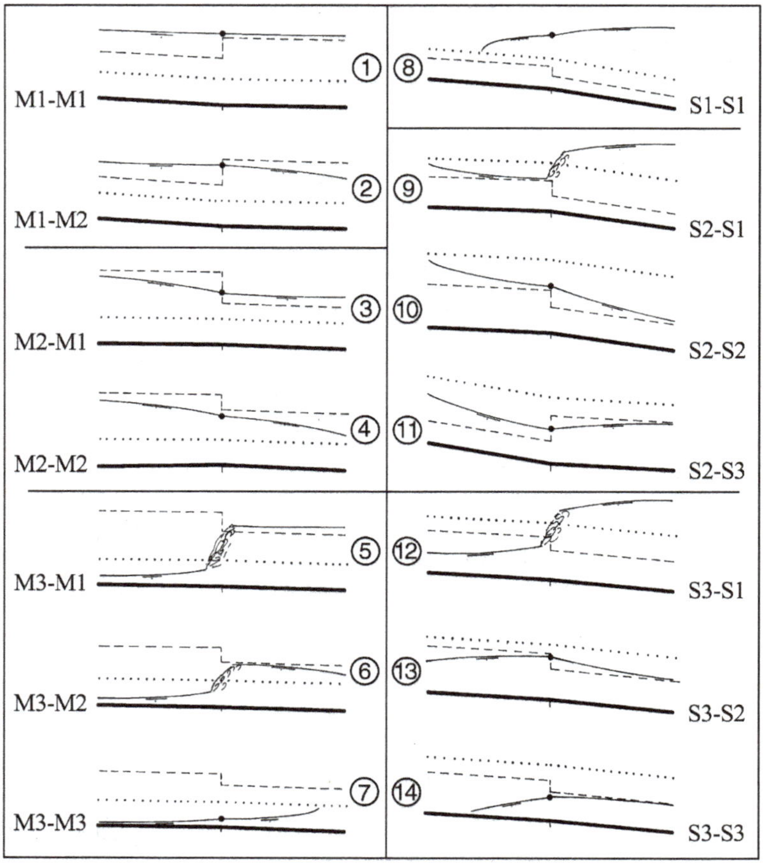

Fig. 8.12 Combination of two M-curves (*left*), or two S-curves (*right*). Configurations M1–M3, M2–M3 as also S1–S2, S1–S3 are physically irrelevant

identical sewer diameter carrying a discharge, and the possible flow conditions are simulated by a *change of bottom slope*. Starting from the six flow types as shown in Fig. 8.10, cases of interest are shown in Fig. 8.12. Combinations of M-curves and S-curves with one another are also shown. Of the possible nine combinations, the cases M1–M3 and M2–M3 as also S1–S2 and S1–S3 are physically irrelevant because no jumps from sub- to supercritical flows are possible. In contrast, *hydraulic jumps* occur in the combinations M3–M1 and M3–M2 as also for S2–S1 and S3–S1. These four cases require caution. For the combinations M3–M2 and S2–S1 the flow surface along the reach between adjacent manholes does not monotonically increase or decrease, respectively.

The direction of calculation proceeds for $h > h_c$ *opposite* to, and for $h < h_c$ *with* the direction of flow. The direction of calculation progresses, therefore, against the

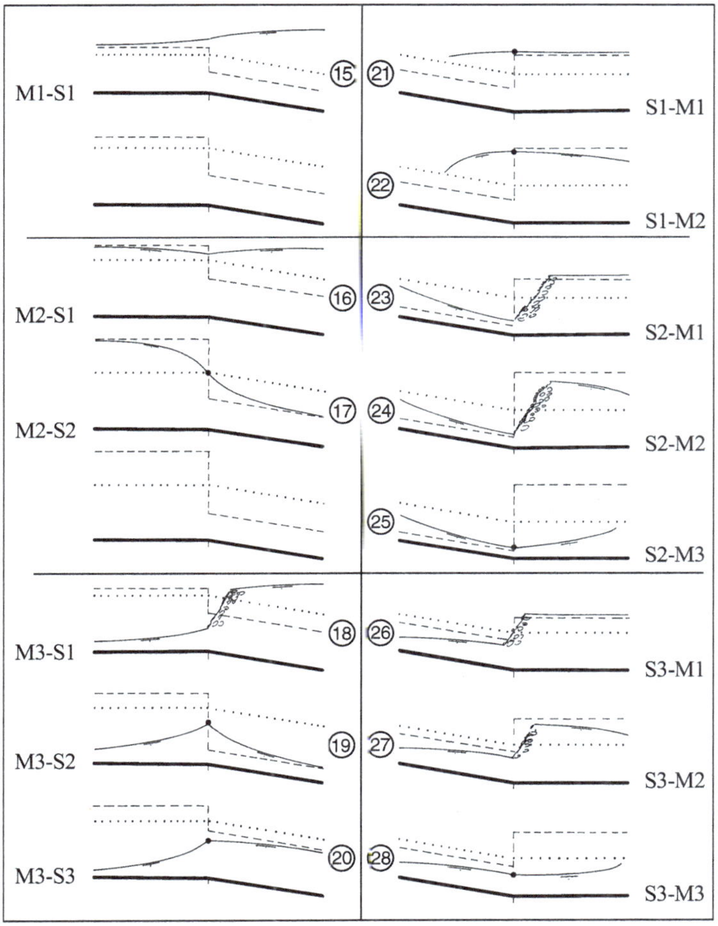

Fig. 8.13 Combination of M-curves *and* S-curves (*left*), and S-curves *and* M-curves (*right*)

flow direction for the M1, M2 and S1 curves, and with the flow direction for the M3, S2 and S3 curves.

For *M- and S-curves*, the combinations M1–S2, M1–S3, M2–S3 as also S1–M3 are physically impossible (Fig. 8.13). For all combinations with M1 and S1-curves the calculations proceed in the direction opposite to the flow, or the flow condition changes abruptly with a hydraulic jump. For combinations involving M3 or S3-curves, the calculations proceed in the flow direction, or a hydraulic jump abruptly changes the flow. M2- and S2-curves require particular attention at certain locations with a *change of direction* of calculation. A specific example is the M2–S1 combination with a flow change at the break point in the bottom slope. This case is important for the selection of appropriate boundary values because of a flow control (Chap. 6).

Figures 8.12 and 8.13 can be extended to arbitrary systems of sewers because the total system can be divided into a number of *two reaches*. In that case, with

$Y_r = h_r/h_N$ boundary value
$Y_c = h_c/h_N$ flow dynamics of uniform flow

as also with the ratio of the two values

$Y_r/Y_c = h_r/h_c$ flow dynamics of boundary value

the curve type is first determined and subsequently the flow is qualitatively sketched. Table 8.7 reproduces the characteristics of the 28 possible cases. This information can conveniently be built into a computer program to control the calculation branch. A scheme can be programmed containing the *flow type number* at each manhole. Thus the information can be represented in a compact form and neuralgic sewer reaches are immediately detected. In the following, differentiation is made between sewer networks of constant and reachwise variable conduit diameter.

Table 8.7 Characteristics of combinations of two sewer reaches

Type	1	2	3	4	5	6	7
Y_r	>1/>1	>1/<1	<1/>1	<1/<1	<1/>1	<1/<1	<1/<1
Y_c	<1/<1	<1/<1	<1/<1	<1/<1	<1/<1	<1/<1	<1/<1
Y_r/Y_c	>1/>1	>1/>1	>1/>1	>1/>1	<1/>1	<1/>1	<1/<1
Type	8	9	10	11	12	13	14
Y_r	>1/>1	>1/>1	>1/>1	>1/<1	<1/>1	<1/>1	<1/<1
Y_c	>1/>1	>1/>1	>1/>1	>1/>1	>1/>1	>1/>1	>1/>1
Y_r/Y_c	>1/>1	<1/>1	<1/<1	<1/<1	<1/>1	<1/<1	<1/<1
Type	15	16	17	18	19	20	21
Y_r	>1/>1	<1/>1	<1/>1	<1/>1	<1/>1	<1/<1	>1/>1
Y_c	<1/>1	<1/>1	<1/>1	<1/>1	<1/>1	<1/>1	>1/<1
Y_r/Y_c	>1/>1	>1/>1	>1/<1	<1/>1	<1/<1	<1/<1	>1/>1
Type	22	23	24	25	26	27	28
Y_r	>1/<1	>1/>1	>1/<1	>1/<1	<1/>1	<1/<1	<1/<1
Y_c	>1/<1	>1/<1	>1/<1	>1/<1	>1/<1	>1/<1	>1/<1
Y_r/Y_c	>1/>1	<1/>1	<1/>1	<1/<1	<1/>1	<1/>1	<1/<1

8.6.4 *Sewer Networks with* Constant *Diameter*

Because sewers are of standardized size, a change of diameter does not necessarily occur at every manhole. Sewer manholes should have a *maximum spacing* of a hundred times the diameter (Chap. 14). In reaches where the bottom slope is nearly constant without lateral inflow, or if the bottom slope increases and thus the effect of a lateral inflow is nearly compensated, no change in the conduit diameter is necessary.

Example 8.10 Given a sewer consisting of four reaches of equal diameter $D = 0.70$ m, and a bottom slope of successive reaches decreasing continuously in the direction of flow. The control point lies at the starting point $x = 0$. The roughness coefficient of the sewer is $n = (1/80)\ \mathrm{m}^{-1/3}\mathrm{s}$. Determine the surface profile of the sewer.

Reach 1. With $h_r = h_c$, thus $Y_r = Y_c = 1.285$ at section $x = 0$ one gets at the cross-section $x = 30$ m, or $X = 1.141$ with $\lambda = 0.919$, the relative flow depth $Y = Y_N$, because the drawdown length according to Eq. (8.28) is $X = 1.102$, i.e. $x = 29$ m. Therefore, the manhole flow depth is $h_{12} = h_N = 0.263$ m, i.e. the flow depth at the end point for the calculation of reach ②.

Reach 2. With $Y_r = 0.263/0.38 = 0.692$, $Y_c = 1.029$ and $\lambda = 0.822$ as end value, and for a reach length $\Delta x = 40$ m, one gets $X = 0.005 \cdot 40/0.380 = 0.526$; these give, from Eq. (8.24), the dimensionless flow depth $Y = Y_N$, because the backwater length is only $X = 0.507$. Hence $h_{23} = h_N = 0.380$ m. This is the flow depth at the end point for the calculations of reach ③.

Reach 3. With $Y_r = 0.380/0.466 = 0.815$, and from $Y_c = 0.414/0.466 = 0.888 < 1$ as also $\lambda = 0.716$, the flow is subcritical. The maximum length for the M3-curve amounts to $X(Y = Y_c) = 0.019$, or $x = 2.576$ m. The hydraulic jump must therefore occur between reaches ② and ③.

Reach 4. Nearly uniform flow prevails in reach ④ because the flow depth in the manhole amounts to $h_{34} = h_{N4} = 0.54$ m. It follows for the backwater curve of *reach* ③ $Y_r = 0.54/0.466 = 1\ 159$, $Y_c = 0.414/0.466 = 0.888$ and $\lambda = 0.716$. Station $\bar{x} = 70$ m corresponds to $x = -40$ m or $X = -0.215$. With this the flow depth is $Y = 1.070$, thus $h = 0.499$ m. Upstream of this station in *reach* 2 the flow depth is $h_{23} = h_{N2} = 0.38\ \mathrm{m} < h_c$. The sequent flow depth corresponding to $y_1 = 0.38/0.7 = 0.543$ and $q_o = 0.537$ calculated from Eq. (7.32) equals $y_2 = 0.579$. This gives $h_2 = 0.405$ m. Because the sequent depth is $h_2 = 0.405$ m, resulting from inflow depth $h_1 = 0.38$ m, is smaller than the downstream flow depth $h_{23} = 0.499$ m, the hydraulic jump takes place further upstream.

Reach 2. The sequent depth to the uniform flow depth $h_{12} = 0.38$ m is with $y_1 = 0.376$ and $q_o = 0.383$ equal to $h_2 = 0.438$ m.

To find the position of the hydraulic jump between manholes ②–③ a backwater curve with $h_o = 0.499$ m is calculated. It is based on $Y_r = 0.499/0.38 = 1.313$, $Y_c = 1.029$ and $\lambda = 0.822$. With a distance of $x = -40$ m, corresponding to $X = -0.526$, the curve is longer than the maximum possible reach length with a continuous subcritical flow. A detailed calculation shows that the hydraulic jump takes place at about $x = 50$ m, i.e. somewhat downstream of the second manhole. Fig. 8.14 shows the resulting surface profile.

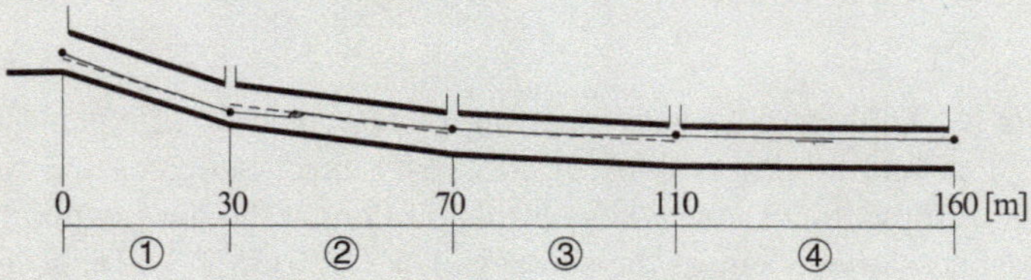

Fig. 8.14 Surface profile of Example 8.10

This example shows how involved the calculations are for a hydraulic jump in a previously unknown reach. For Example 8.10 with Table 8.8, however, it could be guessed in advance that on account of the change from $Y_c > 1$ in the reaches 1 and 2 to $Y_c < 1$ in the downstream reaches, problems of this type might arise. It is proposed that in a reach with $Y_c > 1$, simultaneously with h_N and h_c the value of y_2 be also ascertained. This additional information simplifies the approach from the start of the calculations. Further, the jump can be located more exactly and unnecessary calculations are avoided.

Table 8.8 Characteristics for Example 8.10 (Fig. 8.14)

Reach	①	②	③	④
Q [m^{3}s^{-1}]	0.30	0.40	0.45	0.45
S_o [−]	0.01	0.005	0.0035	0.0025
h_N [m]	0.263	0.380	0.466	0.540
h_c [m]	0.338	0.391	0.414	0.414

Table 8.9 Characteristics for Example 8.11

Reach	①	②	③	④
Q [m^{3}s^{-1}]	0.30	0.40	0.45	0.45
S_o [−]	0.010	0.005	0.0035	0.0025
h_N [m]	0.263	0.380	0.466	0.540
h_c [m]	0.338	0.391	0.414	0.414
h_2 [m]	0.438	0.405	–	–

Example 8.11 The sewer of Example 8.10 has now the control $h_r = h_c$ at the downstream end of manhole ④–⑤. The surface profile is to be recalculated.
Reach 4. With $Y_r = Y_c = 0.414/0.54 = 0.767$ and $\lambda = 0.588$ as also $\Delta x = -50$ m, corresponding to $X = -0.231$, one obtains $Y = 0.956$, and $h_{34} = 0.516$ m.

Reach 3. With $Y_r = 0.516/0.466 = 1.107$, $Y_c = 0.414/0.466 = 0.888$ and $\lambda = 0.716$ as also $\Delta x = -40$ m, corresponding to $X = -0.300$, one obtains $Y = 1.024$ and $h_{23} = 0.477$ m $> h_2 = 0.405$ m. In reach ② subcritical flow prevails throughout.

Reach 2. With $Y_r = 0.477/0.380 = 1.255$, $Y_c = 1.029$ and $\lambda = 0.822$ and $\Delta x = -40$ m, corresponding to $X = -0.526$, no continuous subcritical flow is possible from *downstream* to reach ②. The surface profile is represented in Table 8.10.

Table 8.10 Example 8.11, *Reach* 2, subcritical (*top*) and supercritical (*bottom*) flow

$\bar{x}$ [m]	30	35	40	45	50	55	60	65	70
x [m]					−20	−15	−10	−5	0
X [−]					−0.263	−0.197	−0.132	−0.066	0
Y [−]					–	1.055	1.130	1.195	1.255
h [m]					–	0.401	0.429	0.454	0.477
$\bar{x}$ [m]	30	35	40	45	50	55	60	65	70
x [m]	0	5	10	15	20	25	30	35	40
X [−]	0	0.066	0.132	0.197	0.263	0.329	0.395	0.461	0.526
Y [−]	0.692	0.739	0.784	0.829	0.871	0.912	0.949	0.977	>0.99
h [m]	0.263	0.281	0.298	0.315	0.331	0.347	0.361	0.371	0.377
q_o [−]	0.382	0.405	0.427	0.449	0.470	0.491	0.510	0.524	0.532
y_2 [−]	0.826	0.781	0.741	0.704	0.672	0.641	0.615	0.597	0.586
h_2 [m]	0.578	0.547	0.519	0.493	0.470	0.449	0.430	0.418	0.410

Reach 2. In the *upstream* reach the parameters are as in Example 8.10, namely, $h_{12} = 0.263$ m, $Y_r = 0.692$, $Y_c = 1.029$ and $\lambda = 0.822$. Table 8.10 reproduces the surface profile. Similarly, q_o is calculated from Eq. (7.32) and y_2 from Eq. (7.31) while $h_2 = y_2 \cdot D$.

From Table 8.10a one notes that supercritical flow extends in the extreme case until upstream of $x = 55$ m. In Table 8.10b containing both $h(x)$ for supercritical flow and $h_2(x)$ for the sequent depth, one notices the agreement of h_2 with h of Table 8.10a at $x = 60$ m. The hydraulic jump is located slightly downstream of the location computed for Example 8.10.

This example shows the considerable *computational effort* resulting from a change of flow type. Once the position of the jump is computed, further characteristics of the jump can be determined from the relations developed in Chap. 7.

As against the conventional procedure, the method proposed here for the *explicit* calculation of the surface profile is time efficient in operation.

In a standard *computer solution* all sewer reaches are analyzed from downstream to upstream (in Example 8.11 the analysis upstream of $x = 55$ m is not possible). The results are than retrieved from storage and compared with the corresponding surface profile for supercritical flow calculated in the downstream direction. A sewer system is therefore schematically treated as follows:

1. Estimation of sewer diameter D, or consideration of the existing system based on uniform flow with Q, S_o and n, or k_s and ν as basic values,
2. Calculation of uniform flow depths subject to the condition $h_N \leq 0.8D$,
3. Calculation of critical depths h_c,
4. Marking of all changes from $\mathsf{F}_N < 1$ to $\mathsf{F}_N > 1$ as potential control points,
5. Boundary values h_r at reservoirs or receiving water surface levels, pump sumps or other fixed locations,
6. Computation of *subcritical* flows up to the point where the calculations break down; storage of these results,
7. Calculation of all *supercritical* flows, including sequent depths and comparison with the results of step 6. If the position of the hydraulic jump is fixed, the calculations break down.
8. Presentation of solutions and discussion of results. Comparison with the original rough estimates and adoption of conduit diameters or drops.
9. Subsequent calculation and presentation of final design.

8.6.5 *Sewer Network with* Variable *Diameter*

Generally, 80% filling ratio should be adopted for sewers not too steep, i.e., one may design for 'full flow' condition. This simple procedure yields satisfactory results if diameter changes are smaller than around 20%. For changes from steep to flat sewer reaches, associated with a large change in diameter the hydraulic problems are also small, except for reaches with a large tailwater depth (types 21, 23 and 26 of Fig. 8.13), the conduits tend to get rather overdesigned. However, for a change from *flat to steep sewer reaches* inadmissible conditions can prevail, especially in the entrance region of the downstream conduit (Chap. 6). While passing through the critical depth at the change of slope the flow cannot accelerate to the downstream uniform flow. It therefore impinges on the downstream wall of the manhole at the entrance to the steeply sloping conduit (Fig. 8.15a). This causes pressurized flow in the downstream and submergence in the upstream sewer reaches. In such cases, only a detailed hydraulic design as explained in Chap. 6 gives information on the selection of the downstream diameter.

The conventional computation of backwater curves is based on a vertical free surface profile at the critical point. In the upstream portion, the free surface curves upwards while in contrast the downstream water surface curves downward.

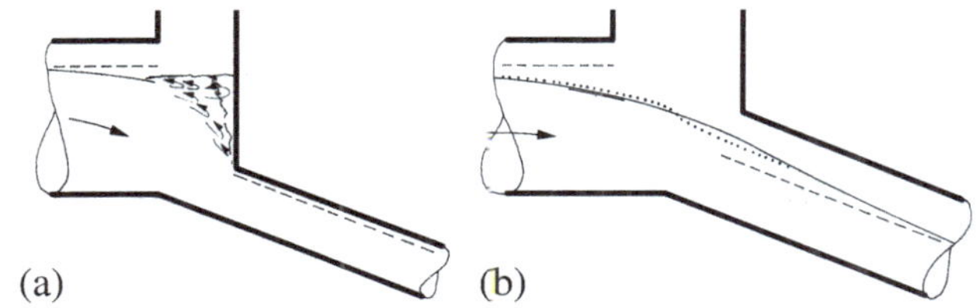

Fig. 8.15 Change of bottom slope from flat to steep reaches. (**a**) Surcharging of the downstream sewer due to insufficient development of bottom transition, (**b**) free surface profile (...) conventionally calculated, (—) measured, (- - -) uniform flow depths

According to this computational model, the critical point is a *point of inflection* (Fig. 8.15b). Measurements in the region of critical points show:

- No vertical free surface,
- No inflection point, and
- Continuous transition from sub- to supercritical flow.

As already mentioned in 8.3, the *conventional calculation method* of backwater curves may not be applicable in the region of critical flow. This region can be defined as

$$0.9 < \frac{h}{h_c} < 1.1, \qquad \text{provided} \left| \frac{h_c}{h_N} - 1 \right| > 0.1. \tag{8.34}$$

If the difference between h_c and h_N is smaller than $\pm 0.1 h_N$, the *validity range* of the conventional calculation becomes questionable. Because nearly critical flows give rise to standing waves, the flows in the neighbourhood are called *undular hydraulic jumps* (7.3.2). Figure 8.16 shows two typical examples: (a) Transition from sub- to supercritical flows with F_N not too far away from the critical value $\mathsf{F} = 1$; (b) undular hydraulic jump with F_N in the downstream channel close to unity. Both cases are manifestations of the nonlinear phenomenon in which both the wave height and the wave length vary with location.

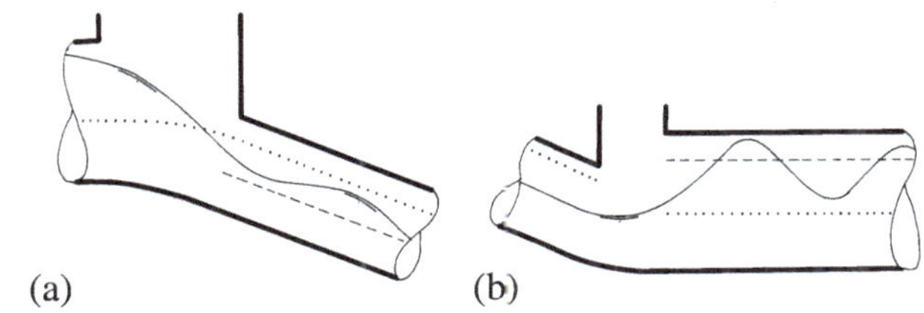

Fig. 8.16 Surface profile for transcritical flow and variable conduit diameter. Transition from (**a**) sub- to supercritical, (**b**) super- to subcritical flows with (- - -) uniform flow depth h_N, (...) critical depth h_c

Example 8.12 Given a sewer (Fig. 8.17) consisting of three reaches ① to ③ with hydraulic characteristics as given in Table 8.11. Determine the surface profile if the approach flow is subcritical and there is no tailwater submergence.

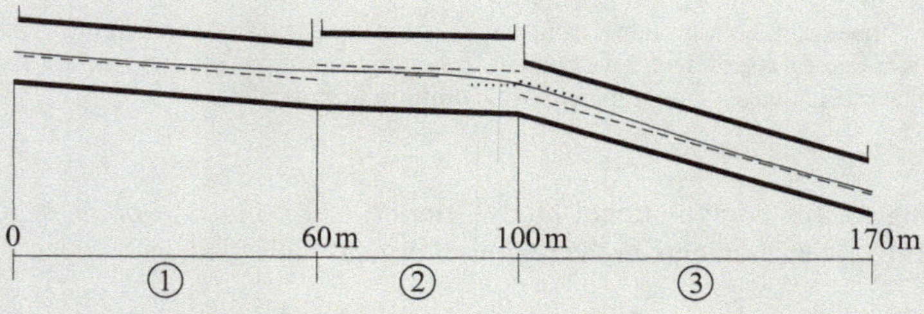

Fig. 8.17 Surface profile for Example 8.12

Table 8.11 Hydraulic characteristics of the three reaches of Fig. 8.17

Reach	①	②	③
Q [m^{3}s^{-1}]	1.00	1.30	1.30
D [m]	1.25	1.50	1.00
S_o [−]	0.0025	0.001	0.029
$1/n$ [m$^{1/3}$s^{-1}]	85	70	85
h_N [m]	0.552	0.860	0.360
h_c [m]	0.534	0.582	0.644

According to Table 8.11 the flow changes from sub- to supercritical at $x = 100$ m (manhole ②/③). The calculations start from there. In contrast to the previous examples the flow depth h would be determined at prescribed locations x. Therefore Eq. (8.24) is solved iteratively to obtain Y for given values of X. Further, shifted coordinates $\bar{x}$ relative to the origin of calculations are introduced. They start at every manhole.

Table 8.12 Surface profiles for Example 8.12 (Fig. 8.17)

h [m]	0.608	0.645	0.685	0.726	0.692	0.664	0.644	0.458	0.419	0.369
Y	1.101	1.169	1.241	0.844	0.805	0.772	1.789	1.272	1.164	1.025
X	−0.272	−0.181	−0.0906	−0.0465	−0.0233	−0.0116	0	0.806	1.611	5.639
x [m]	0	20	40	60	80	90	100	110	120	170
$\bar{x}$ [m]	−60	−40	−20	0/−40	−20	−10	0	+10	+20	+70

The drawdown curve is based on $Y_c = 1.789$, $Y_r = Y_c$, $\lambda = 0.926$ and yields an S2-curve. At $x = 170$ m (manhole ③/④) $h = 0.369$ m is obtained. This corresponds to nearly uniform flow.

The drawdown curve of reach ② based on $Y_c = 0.677$, $Y_r = Y_c$, $\lambda = 0.799$ corresponds to a M2-curve. At $x = 60$ m the boundary depth is $h_r = 0.726$ m and serves for reach ①.

In reach ①, the backwater curve of type M1 is based on the values $Y_c = h_c/h_N = 0.967$, $Y_r = h_r/h_N = 0.726/0.552 = 1.315$ as also $\lambda = 0.886$. At $x = 0$, the beginning of the sewer, the solution is $h = 0.608$ m. For the reaches ① – ② the flow type is ① (Fig. 8.12) and for the reaches ②–③ it is type 17 (Fig. 8.13).

The *final conclusions* of this section are:

- Backwater curves refer only to *gradually-varied* flow,
- Geometric and hydraulic changes within manhole region should be *small*,
- Flows *close* to critical flows are not amenable with conventional backwater curves,
- *Shock waves* can develop in supercritical flows, and
- A *manhole* may, as a local element, be neglected without causing appreciable error.

8.7 Backwater Curves in Egg-shaped and Horseshoe Sewers

8.7.1 *Introduction*

In principle, the governing equations for sewers of circular section can be transformed to egg-shaped and horseshoe sewers. In doing so, the following two problems arise, however:

- Every cross-sectional type leads to a special differential equation requiring separate attention,
- Besides the diameter D of the circular profile, several other cross-sectional parameters may appear, causing additional difficulty in presenting the solutions.

Although backwater curves in arbitrary prismatic water courses can be determined analogous to circular sewers, they are less relevant from a practical point of view. Formidable as the analytical effort, little is their *practical relevance*, because:

- Non-circular sewers seldom occur in practice,
- Backwater curves are similar for all profiles,
- Calculation time is not commensurate with the practical requirements.

These criticisms shall not mean that transition curves in channels of non-circular channels are irrelevant. Rather, for these relatively seldom cases, an *approximate procedure* can be employed, derived directly from the conventional procedure developed in 8.4. In the following the modifications for egg-shaped and the horseshoe sewers are developed.

8.7.2 *Method of Equivalent Cross-section*

Both the standard egg-shaped and the standard horseshoe cross-sections can be considered as distorted circular sections. The backwater curves in these two cross-sections may be regarded also as slightly distorted in comparison to the circular sewer (Fig. 8.18). Backwater curves are *transition curves* between various uniform flow depths for which the critical depth is to be taken into consideration:

- The *uniform depth* in the non-circular sewer is reproduced as exactly as possible, and
- The corresponding *critical depth* is to be taken into account.

These requirements are satisfied with the so-called *equivalent cross-section* if both the uniform depth h_N and the critical depth h_c are calculated in the actual channel, and the flow depth h is computed in the equivalent circular sewer. The equivalent cross-section can be derived from the effective cross-section in the following different ways:

- Identical velocity for uniform flow,
- Equal cross-section at full flow, or
- Equal height of cross-section.

Since for both the standard egg-shaped and horseshoe cross-sections, the eccentricities are relatively small, the last two procedures involve only small deviations. It is recommended that the criterion of *equal cross-sectional area at full flow* be taken as the basis of equivalence. Then, the procedure even becomes independent of

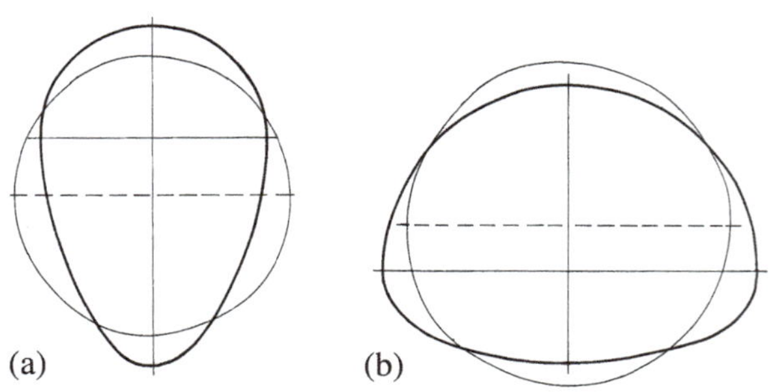

Fig. 8.18 Actual cross-section and corresponding equivalent cross-section for (**a**) egg-shaped and (**b**) horseshoe sewer

discharge and furnishes sufficient accuracy. The fictitious diameter D_f is of course different to the commercially available diameter D. The *computational procedure* then is:

1. Calculation of uniform flow depth h_N in the actual cross-section,
2. Calculation of critical depth h_c in the actual cross-section,
3. Statement of boundary value $h(x = x_r) = h_r$,
4. Classification of backwater curve in the equivalent cross-section,
5. Calculation of backwater curve in the equivalent cross-section, and
6. Transfer of result to original sewer.

The backwater curves for the egg-shaped and horseshoe sections could be exactly derived, the effort is out of proportion to their practical relevance, however.

Example 8.13 Given an egg-shaped section 600/900 with a bottom slope $S_o = 0.15\%$, a coefficient of roughness $n = (1/85)\ \mathrm{m}^{-1/3}\mathrm{s}$ and a discharge $Q = 0.25\ \mathrm{m}^3\mathrm{s}^{-1}$. Determine the backwater curve and the flow depth at the upstream manhole located −70 m upstream from the origin, where critical flow prevails.

1. *Uniform flow* according to Eq. (5.21), with $Q_v = 0.503 \cdot 85 \cdot 0.0015^{1/2} 0.6^{8/3} = 0.424\ \mathrm{m}^3\mathrm{s}^{-1}$, thus $q_v = 0.25/0.424 = 0.59$, and

$$y_N = 1.09[1 - (1 - 0.884 q_v)^{1/2}]^{1/2}, \tag{8.35}$$

or $y_N = 1.09[1 - (1 - 0.844 \cdot 0.59)^{1/2}]^{1/2} = 0.605$, and $h_N = 0.605 \cdot 0.90 = 0.544$ m.
2. *Critical flow* from Eq. (6.46)

$$h_c = 1.34[Q/(gT)^{1/2}]^{1/2}, \tag{8.36}$$

or $h_c = 1.34[0.25/(9.81 \cdot 0.90)^{1/2}]^{1/2} = 0.389$ m. It follows then $h_N > h_c$ and the uniform flow is *subcritical*.
3. *Boundary value* at $x_r = 0$ is $h_r = h_c$, thus $Y_r(X_r = 0) = Y_c = 0.389/0.544 = 0.715$.
4. *Classification*. The surface profile is an M2-curve (drawdown curve). The equivalent diameter is $D_e = [(4/\pi)1.149]^{1/2}B = 1.21B = 1.21 \cdot 0.60\ \mathrm{m} = 0.726$ m (Table 5.3).
5. *Calculation of drawdown curve*. The drawdown length X_0^* is from Eq. (8.27)

$$X_0^* = (Y_c - 1) + \left(1 - Y_c^4\right)[1.72 - Y_c]. \tag{8.37}$$

With $Y_c = h_c/h_N = 0.389/0.544 = 0.715$ one has $X_0^* = (0.715-1) + (1-0.715^4)\,[1.72-0.715] = 0.457$. Equation (8.23) gives

$$\lambda = (1 - 1.1y_N^2)^{1/2}, \tag{8.38}$$

or $\lambda = [1-1.1(0.544/0.726)^2]^{1/2} = 0.618$. Converting the quantity X_o^* to $X_o = X_o^*/\lambda = 0.457/0.618 = 0.739$, thus $x_o = X_o h_N/S_o = 0.739{\cdot}0.544/0.0015 = 268$ m. The drawdown curve extends beyond the upstream manhole.

The drawdown curve profile is given by Eq. (8.24). With the boundary value $(X_r^*; Y_r^*) = (\lambda X_r; Y_r) = (6.18{\cdot}0; 0.715) = (0; 0.715)$, Eq. (8.24) yields

$$X^* = Y - 0.715 - \frac{1}{4}\left(1 - 0.715^4\right)[\ln\left|\frac{Y+1}{Y-1}\frac{0.715-1}{0.715+1}\right| \tag{8.39)$_1$}$$
$$+ 2\arctan Y - 2\arctan(0.715)].$$

Simplification gives

$$X^* = Y - 0.486 - 0.185\left[\ln\left|-0.166\frac{Y+1}{Y-1}\right| + 2\arctan Y\right]. \qquad (8.39)_2$$

Table 8.13 shows the results and it can be seen that at $x = -70$ m the flow depth is $h = 0.506$ m.

Table 8.13 Calculations for Example 8.13

x	[m]	0	−1.10	−5.80	−17.0	−40.95	−56.90	−79.80
X	[−]	0	−0.003	−0.016	−0.047	−0.113	−0.157	−0.220
X^*	[−]	0	−0.002	−0.010	−0.29	−0.070	−0.097	−0.136
Y	[−]	0.715	0.75	0.80	0.85	0.90	0.92	0.94
h	[m]	0.389	0.408	0.435	0.462	0.490	0.500	0.511

Lakshmana Rao and Sridharan (1971) have studied backwater curves for a number of cross-sectional shapes. It is worth noting that differences between the individual curves are small. The recommended procedure is sufficiently accurate for sewer hydraulics, therefore.

8.8 Backwater Curves in Rectangular Channels

8.8.1 *Introduction*

In sewage engineering, channels of rectangular cross-section are frequent apart from circular conduits. For large size, 1.5–2.0 m *in-situ* concrete sewers, they connect channels in sewage treatment plants or outlet channels discharging to receiving waters. Because the rectangular section differs widely from a circular sewer and corresponds to an elementary cross-sectional form, it is treated here in some detail.

From the developments in Sect. 8.8.2, the backwater curves in rectangular channels can also be represented with the coordinates X, Y and Y_c as defined in Eq. (8.18). In the present case, the additional parameter $y_b = h_N/b$ would also appear. In view of this four parameter representation of the governing equation, the general solution gets more involved. Because a practice-friendly method of calculation for the backwater curves is desired, the approximate procedure neglects the explicit influence of the channel width b. This is similar to the method derived in Sect. 8.4.2 and the calculation follows the procedure of Sect. 8.6.1.

8.8.2 *Equation of Free Surface Profile*

The rectangular section is geometrically described with the flow depth h and the channel width b. Then, the cross-sectional area is $F = bh$, the wetted perimeter $P = b + 2h$ and the hydraulic radius $R_h = F/P = bh/(b + 2h)$.

Assuming that the flow formula of Manning and Strickler be valid for the gradually-varied flow, the *friction slope* obtains

$$S_f = \frac{n^2Q^2}{F^2R_h^{4/3}}. \tag{8.40}$$

Here Q is discharge and n is Manning's roughness coefficient (Chap. 2). If the parameters F and P are introduced

$$S_f = \left(\frac{nQ}{bh}\right)^2 \left(\frac{b+2h}{bh}\right)^{4/3}. \tag{8.41}$$

For uniform flow $S_o = S_f$, the relation for the bottom slope is

$$S_o = \left(\frac{nQ}{bh_N}\right)^2 \left(\frac{b+2h_N}{bh_N}\right)^{4/3}. \tag{8.42}$$

Eliminating the discharge Q between Eqs. (8.41) and (8.42), the ratio between the friction and the bottom slopes is

$$S_f/S_o = \left(\frac{b+2h}{b+2h_N}\right)^{4/3} \left(\frac{h_N}{h}\right)^{10/3}. \tag{8.43}$$

The Froude number in the rectangular channel is (Chap. 6)

$$\mathsf{F} = \frac{Q}{(gb^2h^3)^{1/2}} \tag{8.44}$$

and the critical depth is defined as $h_c^3 = Q^2/(gb^2)$. The equation of *gradually-varied flow in the rectangular channel* obtains from Eq. (8.9)

$$\frac{1}{S_o}\frac{\mathrm{d}h}{\mathrm{d}x} = \frac{1 - \left(\frac{b+2h}{b+2h_N}\right)^{4/3}\left(\frac{h_N}{h}\right)^{10/3}}{1 - \left(\frac{h_c}{h}\right)^3}. \tag{8.45}$$

If the three dimensionless parameters X, Y and Y_c defined in Eq. (8.18) as well as the parameter $y_b = h_N/b$ are introduced in Eq. (8.45), the *dimensionless* equation for the gradually-varied flow in a rectangular channel is (Hager 1981)

$$\frac{\mathrm{d}Y}{\mathrm{d}X} = \frac{1 - \left(\frac{1+2y_bY}{1+2y_b}\right)^{4/3} Y^{-10/3}}{1-(Y_c/Y)^3}. \tag{8.46}$$

As already mentioned in Sect. 8.8.1, the solution $Y(X)$ depends on the two parameters Y_c and y_b. The form parameter y_b can take values in the range $0 < y_b < \infty$.

The *classical treatment* of backwater curves refers to the case $y_b = 0$, i.e. to a *very wide channel* (Rouse and Ince 1957, Muller 1972). For this case, Eq. (8.46) simplifies to an equation similar to Eq. (8.20) for the circular channel as

$$\frac{\mathrm{d}Y}{\mathrm{d}X} = \frac{1 - Y^{-10/3}}{1-(Y_c/Y)^3} \tag{8.47}$$

For the other extreme case $y_b^{-1} = 0$, i.e. an *extremely narrow rectangular channel*, Eq. (8.46) reduces for not too small flow depths to

$$\frac{\mathrm{d}Y}{\mathrm{d}X} = \frac{1 - Y^{-7/3}}{1-(Y_c/Y)^3}. \tag{8.48}$$

For the entire range of cross-sectional shapes, a generalized relationship is (Bakhmeteff 1932)

$$\frac{\mathrm{d}Y}{\mathrm{d}X} = \frac{1 - Y^{N}}{1-(Y_c/Y)^M}. \tag{8.49}$$

where the exponents M and N are characteristics of the respective cross-sectional shapes. This equation has been integrated by Chow (1959) in tabular form with M and N as variables. For the integration, the exponents are assumed invariant with the flow depth Y. This proves right for the rectangular, the triangular, and approximately for the circular channel shapes, but leads to incorrect results, for example, for the trapezoidal channel or for channels of complex cross-sectional shape. The disadvantage of this widely accepted *method of Chow* is the need to interpolate from tabulated values, in contrast to the explicitly determined solution presented here. Therefore, the method of Chow as also that of Henderson (1966) is not considered further. As a variant remains the numerical solution of the differential equation.

Although Eq. (8.47) has been solved analytically by Gill (1976) – requiring five lines simply to write down the equation and the specification of parameters – nothing can be said about its calculation for practical application. In deriving the present approximate method, the exponents in the numerator of Eq. (8.46) are simplified to entire numbers. Further, the expression $[(1+2y_bY)/Y]^{1/3}$ is developed through a series expansion about the uniform flow condition $Y=1$, and reduces to $(1+2y_b)^{1/3}[1-(1/3)(Y-1)/(1+2y_b)]$. With this, the *modified* equation of backwater curves in rectangular channels is

$$\frac{\mathrm{d}Y}{\mathrm{d}X}=\frac{Y^3-\left[\frac{1+2y_bY}{1+2y_b}\right]\left[1-\frac{Y-1}{3(1+2y_b)}\right]}{Y^3-Y_c^3}. \tag{8.50}$$

Although a closed form solution of this relation is possible, the equation contains four parameters X, Y, Y_c and Y_b and is not suitable for practical purposes. From a numerical analysis (Hager 1981) the influence of the shape parameter y_b on the surface profile was demonstrated to be small.

For typical values of y_b between 0.1 and 10 a representative average value is $y_b=1$. For flows in the neighbourhood of uniform flows $(0<Y<3)$ Eq. (8.50) simplifies to

$$\frac{\mathrm{d}Y}{\mathrm{d}X}=\frac{Y^3-\frac{1}{3}(1+2Y)\left[1-\frac{Y-1}{9}\right]}{Y^3-Y_c^3}=\frac{(Y-1)(27Y^2+29Y-10)}{27(Y^3-Y_c^3)}. \tag{8.51}$$

The solution of this equation is shown in Fig. 8.19 for the range of values $-9<X<4$ and $0<Y<3$. One recognizes the similarity between this figure, and Fig. 8.6 for the circular sewer. Accordingly, the foregoing procedure can also be employed analogously for computations in the rectangular channel. In contrast to Eq. (8.24), however, the solution of Eq. (8.51) is not obtained explicitly.

To present a simpler analytical solution of Eq. (8.46), the term in bracket must be modified. Setting $y_b=1$ in Eq. (8.46) the resulting term $[(1+2Y)/Y]^{1/3}$ causes a maximum deviation of about 10% in the range $0.5<Y<5$. In the region of uniform flow $(Y\sim 1)$ the approximate relation in analogy to Eq. (8.20) is

$$\frac{\mathrm{d}Y}{\mathrm{d}X}=\frac{1-Y^{-3}}{1-(Y_c/Y)^3}. \tag{8.52}$$

The integration of Eq. (8.52) is simple and discussed in the following section.

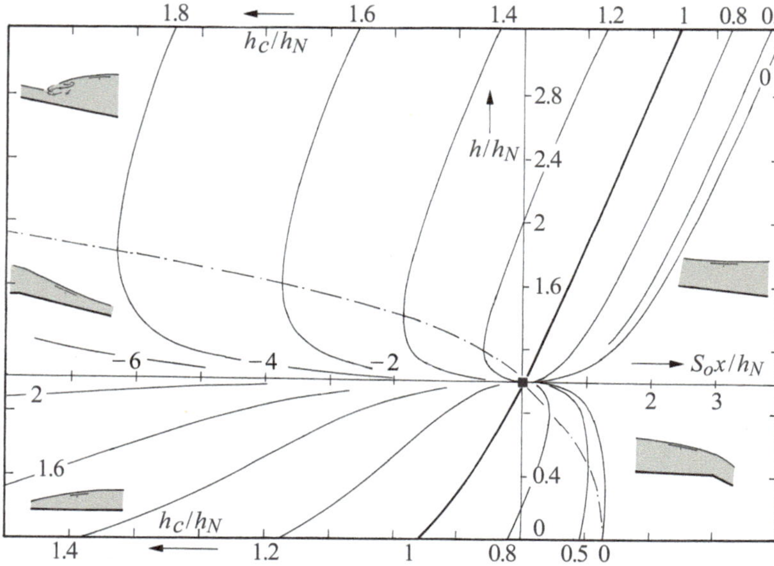

Fig. 8.19 Solution of Eq. (8.51) for gradually-varied flows in rectangular channels, after Hager (1981)

8.8.3 *Approximate Solution*

The integral solution of the differential equation (8.52) has already been reported in 1860 by the Frenchman Jacques Antoine Bresse (1822–1883) (Hager 1990). With the boundary values $Y(X = X_r) = Y_r$, this integral solution is

$$X - X_r = Y - Y_r - (1 - Y_c^3)\left\{\frac{1}{6}\ln\left[\left(\frac{1 - Y_r}{1 - Y}\right)^2\left(\frac{1 + Y + Y^2}{1 + Y_r + Y_r^2}\right)\right]\right.$$
$$\left. + \frac{1}{\sqrt{3}}\arctan\left(\frac{1 + 2Y}{\sqrt{3}}\right) - \frac{1}{\sqrt{3}}\arctan\left(\frac{1 + 2Y_r}{\sqrt{3}}\right)\right\}. \quad (8.53)$$

A relation analogous to Eq. (8.24) is now available allowing to explicitly determine backwater curves in the rectangular channel. The *computational procedure* is analogous to that for the circular sewer detailed in 8.6. It is therefore explained with a representative example.

Example 8.14 Determine the surface profile in a 1.2 m wide 120 m long reach of a rectangular channel carrying a discharge $Q = 3.1\ \mathrm{m^3s^{-1}}$ at a bottom slope $S_o = 3\%$ and a coefficient of roughness $n = (1/90)\ \mathrm{m^{-1/3}s}$. Critical flow prevails at the beginning of the channel and the downstream flow depth is $h_u = 1.2$ m.

1. *Uniform flow.* $y_b = h_N/b$ obtains by iteratively solving Eq. (8.42)

$$\frac{n^2 Q^2}{S_o b^{16/3}} = \frac{y_b^{10/3}}{(1+2y_b)^{4/3}} \tag{8.54}$$

giving $y_b = 0.351$ and $h_N = 0.351 \cdot 1.2$ m $= 0.421$ m.
The value y_b follows also from the explicit solution of the approximate equation proposed by Sinniger und Hager (1989)

$$y_b = \left[\left(\frac{nQ}{S_o^{1/2} b^{8/3}}\right)^{-3/5} - \frac{2}{3}\right]^{-1}. \tag{8.55}$$

With $nQ/(S_o^{1/2}b^{8/3}) = 3.1/(90 \cdot 0.03^{1/2} 1.2^{8/3}) = 0.122$, one obtains for $y_b = [0.122^{-0.6} - 0.666]^{-1} = 0.3495$, which is about 0.5% smaller than the previous value.

2. *Critical depth,* according to Chap. 5, is

$$h_c = [Q^2/(gb^2)]^{1/3}. \tag{8.56}$$

Introducing numerical values yields $h_c = [3.1^2/(9.81 \cdot 1.2^2)]^{1/3} = 0.879$ m. Accordingly, uniform flow is supercritical because $h_N < h_c$ and downstream flow is subcritical because $h_u > h_c$. Therefore, supercritical flow develops and is followed by a hydraulic jump.

3. *Boundary values*
upstream end $h_{ro} = h_c = 0.879$ m with positive direction of calculation
downstream end $h_{ru} = h_u = 1.20$ m with negative direction of calculation.

4. *Classification*
upstream reach: S2-curve (drawdown to uniform flow) downstream reach: S1-curve (subcritical flow in steep channel).

5. *Calculation* of backwater curves. With $X_r(Y_r = Y_c = 0.879/0.421 = 2.088) = 0$, Eq. (8.53) reduces to

$$X = Y - 2.088 - (1 - 2.088^3)$$
$$\times \left\{\frac{1}{6}\ln\left[\frac{(1-2.088)^2}{(1-Y)^2}\frac{(1+Y+Y^2)}{(1+2.088+2.088^2)}\right] + \frac{1}{\sqrt{3}}\arctan\left(\frac{1+2Y}{\sqrt{3}}\right) - \frac{1}{\sqrt{3}}\arctan\left(\frac{1+2\cdot 2.088}{\sqrt{3}}\right)\right\}$$
$$= Y - 2.088 + 8.103\left\{\frac{1}{6}\ln\left[0.159\frac{(1+Y+Y^2)}{(1-Y)^2}\right] + \frac{1}{\sqrt{3}}\arctan\left(\frac{1+2Y}{\sqrt{3}}\right) - 0.720\right\}$$
$$= Y - 7.922 + 1.351 \cdot \ln\left[0.159\frac{1+Y+Y^2)}{(1-Y)^2}\right] + 4.677\arctan\left(\frac{1+2Y}{\sqrt{3}}\right).$$

For given values Y the relative distance X can be calculated. The drawdown curve is obtained with $h = Yh_N$ and $x = Xh_N/S_o$, and $h(x)$ is given (Table 8.14). The drawdown length is 132.70 m.

Table 8.14 Drawdown curve from upstream boundary, Example 8.14

x [m]	0	0.135	1.24	4.32	11.3	17.26	28.50
X [−]	0	0.010	0.089	0.308	0.805	1.266	2.031
Y [−]	2.088	2.0	1.8	1.6	1.4	1.3	1.2
h [m]	0.879	0.842	0.758	0.674	0.589	0.547	0.505
x [m]	37.0	49.95	57.40	67.30	81.67	106.9	132.70
X [−]	2.638	3.559	4.091	4.796	5.82	7.62	9.456
Y [−]	1.15	1.1	1.08	1.06	1.04	1.02	1.01
h [m]	0.484	0.463	0.455	0.446	0.438	0.429	0.425

For the transition curve in the *downstream* reach the boundary condition is $x_r(h_r = h_u) = 120$ m. The corresponding non-dimensional parameter is $X_r = S_o x_r/h_N$ $(Y_r = h_u/h_N = 1.20/0.421 = 2.85) = 0.03{\cdot}120/0.421 = 8.551$. Introducing these values together with $Y_c = 2.088$ in Eq. (8.53) gives

$$X - 8.551 = Y - 2.85 - (1 - 2.088^3) \times \left\{\frac{1}{6}\ln\left[\frac{(1-2.85)^2}{(1-Y)^2}\frac{(1+Y+Y^2)}{(1+2.85+2.85^2)}\right] + \frac{1}{\sqrt{3}}\arctan\left(\frac{1+2Y}{\sqrt{3}}\right) - \frac{1}{\sqrt{3}}\arctan\left(\frac{1+2\cdot 2.85}{\sqrt{3}}\right)\right\}$$

$$= Y - 2.85 + 8.103\left\{\frac{1}{6}\ln\left[0.286\frac{(1+Y+Y^2)}{(1-Y)^2}\right] + \frac{1}{\sqrt{3}}\arctan\left(\frac{1+2Y}{\sqrt{3}}\right) - 0.761\right\}$$

or solving for $X(Y)$

$$X = Y - 0.465 + 1.351 \cdot \ln\left[0.286\frac{1+Y+Y^2}{(1-Y)^2} + 4.678\arctan\left(\frac{1+2Y}{\sqrt{3}}\right)\right].$$

The corresponding solutions $Y(X)$ as also $h(x)$ are presented in Table 8.15. Subcritical flow is possible only in the lower portion of the reach up to $x = 115.85$ m. Further upstream supercritical flow prevails throughout.

Table 8.15 Backwater curve from downstream end upwards, Example 8.14

x [m]	120.0	117.3	116.3	115.85
X [−]	8.55	8.36	8.285	8.254
Y [−]	2.85	2.50	2.3	2.088
h [m]	1.20	1.05	0.97	0.879
F_2 [−]	0.627	0.767	0.863	1.0
h_1^* [m]	0.622	0.729	0.795	0.879

Similarly, the sequent depths h_1^*, conjugate to the depths $h = h_2$ are calculated from Eq. (7.14) which can also be written with $\mathsf{F}_2 = Q/(gb^2h_2^3)^{1/2}$ as

$$h_1^*/h_2 = \frac{1}{2}\left[\left(1+8\mathsf{F}_2^2\right)^{1/2} - 1\right]. \tag{8.57}$$

Equation (8.57) is analogous to Eq. (7.17), but Eq. (7.18) may not be used. For small values $\mathsf{F}_2 < 0.25$, the following relation holds

$$h_1^*/h_2 = 2\mathsf{F}_2^2\left[1 - 2\mathsf{F}_2^2\right]. \tag{8.58}$$

A comparison of Tables 8.14 and 8.15 shows that no hydraulic jump occurs in the reach on account of the small downstream water depth h_u.

Example 8.15 Calculate the surface profile for the channel of Example 8.14 if the downstream flow depth is $h_u = 1.6$ m!

1. *Uniform flow depth* $h_N = 0.421$ m.
2. *Critical depth* $h_c = 0.879$ m, thus $Y_c = 2.088$.
3. *Boundary value* $h_r = h_u = 1.6$ m $> h_c$. Calculation must therefore proceed *opposite* to the flow direction.
4. *Classification* S1-curve as in Example 8.14.
5. *Calculation* of backwater curve
   With $X_r(Y_r = h_r/h_N = 1.60/0.421 = 3.80) = 120{\cdot}0.03/0.421 = 8.551$, the relation for $X(Y)$ is

$$X = 8.551 + Y - 3.8 - (1 - 2.088^3) \times \left\{\frac{1}{6}\ln\left[\frac{(1-3.8)^2}{(1-Y)^2}\frac{(1+Y+Y^2)}{(1+3.8+3.8^2)}\right] + \frac{1}{\sqrt{3}}\arctan\left(\frac{1+2Y}{\sqrt{3}}\right) - \frac{1}{\sqrt{3}}\arctan\left(\frac{1+2\cdot 3.8}{\sqrt{3}}\right)\right\}$$

$$= Y - 1.667 + 1.351\cdot\ln\left[0.407\frac{1+Y+Y^2}{(1-Y)^2} + 4.678\arctan\left(\frac{1+2Y}{\sqrt{3}}\right)\right]$$

   The backwater curve $h(x)$ is presented in Table 8.16. Upstream of $x = 105.6$ m only supercritical approach flow is possible.
6. Comparison of Tables 8.14 and 8.16 indicates as location of the hydraulic jump $x = 120$ m, i.e. at the end of the reach.

Table 8.16 Backwater curve from downstream end upwards, Example 8.15

x [m]	120.0	116.5	111.2	107.1	105.6
X [−]	8.55	8.30	7.924	7.633	7.529
Y [−]	3.80	3.5	3.0	2.5	2.088
h [m]	1.60	1.475	1.265	1.053	0.879
F_2 [−]	0.408	0.460	0.580	0.763	1
h_1^* [m]	0.421	0.473	0.582	0.726	0.879

Backwater curves for channels of *standard cross-sections* can be expressed in closed form. On the one hand integration is avoided and on the other, the solutions are not subject to any condition of convergence. Consequently the backwater curves can be determined economically in terms of time even for large calculation domains, provided a suitable computer program is developed.

The computational scheme can be executed step-by-step in which the uniform flow and the critical flow form the basic conditions. For a change of flow type, the procedure is indeed retained in principle, but additional problems arise with regard to the location of *hydraulic jumps*. A systematic calculation of surface profiles for both sub- and supercritical flows as also the corresponding determination of the sequent depth profile pay off indeed for small catchment areas. The stability of the position of a hydraulic jump should be tested by varying the downstream boundary condition. If the inflow Froude number is in the range $1 < \mathsf{F} < 1.7$, the hydraulic jump is unstable and undular.

Notation

b	[m]	channel width
c	$[\mathrm{ms}^{1}]$	wave celerity
D	[m]	conduit diameter
F	[−]	Froude number
F	$[\mathrm{m}^2]$	cross-sectional area
g	$[\mathrm{ms}^{-2}]$	acceleration due to gravity
h	[m]	water depth
h_c	[m]	critical depth
h_m	[m]	mean flow depth
h_N	[m]	uniform depth
H	[m]	energy head
k_s	[mm]	equivalent sand roughness height
L_D	[m]	length of pressurized flow
n	$[\mathrm{m}^{-1/3}\mathrm{s}]$	Manning's roughness coefficient
$p/(\rho g)$	[m]	pressure head

q_N	[−]	discharge relative to uniform flow
Q	[m^3s^{-1}]	discharge
R_h	[m]	hydraulic radius
S_e	[−]	slope of energy line
S_f	[−]	wall friction gradient
S_F	[−]	slope due to cross-sectional variation
S_o	[−]	bottom slope
t	[s]	time
V	[ms^{-1}]	velocity
x	[m]	longitudinal coordinate
X	[−]	dimensionless longitudinal coordinate
X^*	[−]	transformed longitudinal coordinate
y	[−]	filling ratio
y_b	[−]	form parameter
y_N	[−]	filling ratio for uniform flow
Y	[−]	dimensionless flow depth
Y_c	[−]	dimensionless critical depth
z	[m]	height of channel bottom above datum
λ	[−]	transformation parameter
χ	[−]	friction characteristic
ζ_F	[−]	coefficient

Subscripts

c	critical
e	equivalent
f	fictitious
N	uniform flow
o	backwater and drawdown length
r	boundary value
u	downstream

References

Bakhmeteff, B.A. (1932). *Hydraulics of open channels*. McGraw Hill: New York.

Chow, V.T. (1959). *Open channel hydraulics*. McGraw Hill: New York.

Forchheimer, P. (1914). *Hydraulik* (Hydraulics), ed. 1. Teubner: Leipzig, Berlin [in German].

Gill, M.A. (1976). Discussion to Numerical errors in water profile computation. Proc. ASCE, *Journal of Hydraulics Division* 102(HY9): 1405–1407.

Hager, W.H. (1981). Stau- und Senkungskurven im Kanalbau (Backwater and drawdown curves in channels). *Gas – Wasser – Abwasser* 61(5): 157–167; 61(11): 398 [in German].

Hager, W.H. (1990). Stau- und Senkungskurven im Kreisprofil (Backwater and drawdown curves in circular profile). *Gas – Wasser – Abwasser* 70(6): 422–430; 70(7): 520–523 [in German].

Henderson, F.M. (1966). *Open channel flow*. MacMillan: London.

Lakshmana Rao, N.S., Sridharan, K. (1971). Effect of channel shape on gradually varied flow profiles. Proc. ASCE, *Journal of Hydraulics Division* 97(HY1): 55–64; 97(HY9): 1562–1565; 98(HY4): 712.

Liggett, J.A. (1994). *Fluid mechanics*. McGraw-Hill: New York.

Müller, R. (1972).Geschlossene Berechnung von Stau- und Senkungslinien (Closed form solution for backwater curves). *Mitteilung* 9, P.G. Franke, ed. Institut für Hydraulik und Gewässerkunde, Technische Hochschule: München [in German].

Press, H., Schröder, R. (1966). *Hydromechanik im Wasserbau* (Hydromechanics in hydraulic structures). Ernst & Sohn: Berlin [in German].

Ranga Raju, K.G. (1990). *Flow through open channels*. Tata McGraw-Hill: New Delhi, India.

Rouse, H., Ince, S. (1957). *History of hydraulics*. Iowa Institute of Hydraulic Research, State University of Iowa. Edwards Brothers: Ann Arbor.

Sinniger, R.O., Hager, W.H. (1989). *Constructions hydrauliques – écoulements stationnaires* (Hydraulic structures – steady flows). Presses Polytechniques Romandes: Lausanne [in French].

Tolkmitt, G. (1892). Stauwerke (Backwater structures). *Handbuch der Ingenieurwissenschaften* 3(1/1), ed. 3. Engelmann: Leipzig [in German].

Tolkmitt, G. (1907). *Grundlagen der Wasserbaukunst* (Basics of hydraulic structures). Ernst & Sohn: Berlin [in German].

Townson, J.M. (1991). *Free-surface hydraulics*. Unwin-Hyman: London, U.K.

Chapter 9
Pipe Culverts – Throttling Pipes – Inverted Siphons

Abstract *Culverts* are characterized hydraulically by various flow conditions. They are discussed first and then analyzed with a generalized diagram to present a simple design method. The so-called simple culvert is also considered, and the computational approach is compared with model tests.

Throttling pipes perform correctly only under pressurized flow. The principle of pipe filling is described and two conditions of flow are highlighted. Examples illustrate the computational procedure.

Inverted siphons can be arranged in various ways thereby complicating a general design method. The most important features regarding the intake structure, the throttling pipe and the outlet structure are discussed. The design is also illustrated with a typical example. Thus, the three structures in which either free surface or pressurized flow may appear are described with a comparatively simple hydraulic approach.

9.1 Introduction

Whereas combined sewers are generally designed for free surface flow, various structures can have both either free surface or *pressurized flow* (German: Druckabfluss; French: Ecoulement en charge). The culvert is one of the more common of these structures and it can be found whenever an embankment has to be crossed. Less known and not so much applied is the *throttling pipe* (German: Drosselrohr; French: Conduite d'étranglement) by which the discharge from a storage basin or a side overflow to the treatment station can be limited. The *inverted siphon* (German: Düker; French: Aqueduc-siphon) is a typical structure to underpass streets, railways or other structures by a sewer, because the overpass cannot be realized due to space limitations, aesthetic reasons or economical constraints.

All three structures have in common a complicated hydraulic design procedure and the proper definition of the effective *flow condition.* Often pressurized flow is assumed to occur in culverts but the pipe is too short and free surface flow establishes instead. The consequences of such an erroneous design can be fatal because a free surface flow often needs more head than pressurized flow, and the

W.H. Hager, *Wastewater Hydraulics*, 2nd ed., DOI 10.1007/978-3-642-11383-3_9,

embankment may be seriously eroded by overflow. To exclude such basic design errors, the detailed hydraulic design for *underpass structures* is important. In this chapter, the geometry of these structures is simplified to allow for a generalized hydraulic approach. After a description of the relevant flow types, their basic elements are discussed and design equations are proposed. Computational examples illustrate the design procedure.

9.2 Culvert

9.2.1 *Description*

A culvert (German: Durchlass; French: Ponceau) is an *underpass structure* for water flow. It is characterized by a dual flow pattern: Either free surface flow, or pressurized flow. This particular characteristics makes a culvert a flexible design element, provided the effective flow condition is considered. The consequences of an erroneous assumption of the flow condition can have significant impact, however, with large damages in urban areas. The main flow types of a culvert are presented and the hydraulic design equations are specified.

Culverts can have a wide variety of geometries. To introduce a generalized design, *simplifications* are necessary. Usually, the cross-sectional culvert shape is circular, but it can be also egg-shaped and rectangular. A prismatic circular pipe of diameter D, constant equivalent sand roughness height k_s and constant bottom slope S_o is considered. The approach flow velocity is often relatively small, or of the same order as the outlet velocity, and the energy heads at the intake (subscript o) and outlet (subscript u) sections are then relevant instead of the flow depths. The intake can be rounded with a radius r_d and the length of culvert is L_d (Fig. 9.1).

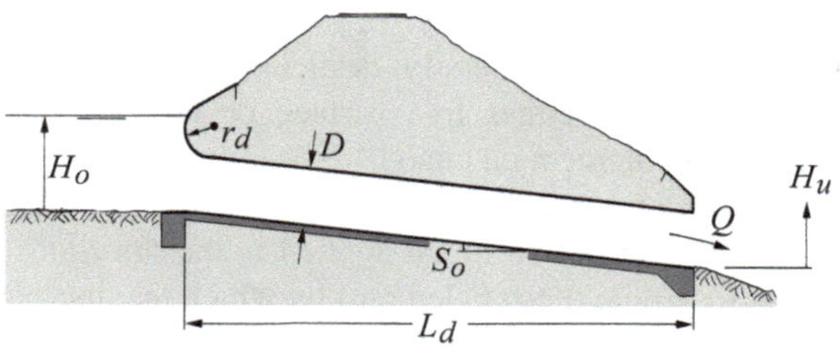

Fig. 9.1 Scheme of culvert, longitudinal section

9.2.2 *Conditions of Flow*

Figure 9.2 shows the main *flow conditions* of a culvert (Chow 1959). Basically, one may distinguish between free surface flow and pressurized flow. Further, there are

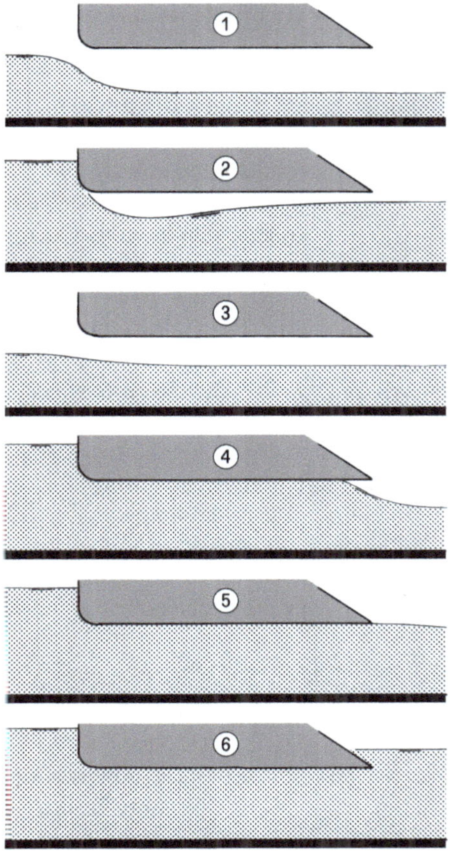

Fig. 9.2 Basic flow types in culvert, for details see main text

flows with an upstream control at the intake section, and with a downstream control at the outlet section. The normalizing length is thus the diameter D.

For small approach flow level $H_o/D < 1.2$ and a bottom slope $S_o > S_c$ with S_c as the critical bottom slope (Chap. 6), *critical flow* can establish provided submergence from the tailwater is absent. Flow type ① (Fig. 9.2) is thus characterized by a relation between the energy head H_o and the discharge Q. Note that air has free access to the flow from both culvert sides.

For larger approach flow level $H_o/D > 1.2$ to 1.5 and free culvert flow, *gated flow* appears. Then, the upstream section is sealing the culvert against air flow, except for air flow by vortices with an air core. The air is normally provided by the downstream access. Flow type ② can thus be compared with a gate, and the hydraulic control is at the intake section. The culvert length has no effect on the discharge, except for a reduction of air discharge due to the formation of shockwaves and two-phase air-water flow. The transition from flow types ① to ② is unsteady mainly due to the formation of *intake vortices* adding to surface fluctuations, flow instability and noise generation. These disadvantages of hydraulic structures (Sinniger and Hager

1989) are not significant for sewers because of relatively small discharges. Clogging is significant, however, and any sort of anti-vortex elements can have a deteriorate effect.

For $H_o/D < 1.2$ and a small bottom slope $S_o < S_c$ the culvert flow is subcritical, and its control is at the outlet section. In general, backwater curves (Chap. 8) had to be computed along the culvert. For a long culvert and a relatively small outflow depth, *uniform flow* can establish close to the inlet, as referred to by flow type ③. Then, a direct relation between the energy head H_o and the discharge is available.

For $h_c/D < H_u/D < 1$ and small bottom slope $S_o < S_c$ the hydraulic control is downstream. Flow type ④ relates to a *partial culvert submergence*. For $H_u/D > 1$ and small bottom slope $S_o < S_c$ *pressurized flow* occurs indicated with flow types ⑤ and ⑥. For large bottom slope $S_o > S_c$ where gated culvert flow may establish, the pressurized condition has to be verified. Alternatively, one could also have flow type ② in the upper culvert portion and flow type ⑤ in the lower culvert portion, with a hydraulic jump separating the two reaches.

This description of culvert flows is due to Chow (1959) and Henderson (1966). It refers to the so-called *long culvert* with a relative culvert length L_d/D of at least 10–20. These culverts have a good throttling effect desired for sewers. For short culverts, complex flows can appear that depend significantly on the exact inlet geometry. They are excluded in the following.

9.2.3 *Generalized Flow Diagram*

The various flow types are now described with hydraulic design equations. The relation between critical energy head $H_c < (1.2\text{–}1.4)D$ and discharge Q follows from Eq. (6.37). Flow type ① is thus characterized by

$$\frac{H_{oc}}{D} = \frac{5}{3}\left[\frac{Q}{\left(gD^5\right)^{1/2}}\right]^{3/5}. \tag{9.1}$$

Uniform flow in a circular pipe is described in Sect. 5.5.3. With $\chi = S_o^{1/2}D^{1/6}/(ng^{1/2})$ as the roughness characteristic, *flow type* ③ may be described from Eq. (5.31) as

$$\frac{H_{oN}}{D} = \frac{2}{\sqrt{3}}\left(\frac{nQ}{S_o^{1/2}D^{8/3}}\right)^{1/2}\left[1+\left(\frac{9}{16}\chi\right)^2\right]. \tag{9.2}$$

Note that the maximum filling ratio should be less than 85%. According to Hager and Wanoschek (1986), the flow is critical whenever $\chi > 2$.

Gated culvert flow can be described by a generalized Bernoulli equation (Chap. 1). *Flow type* ② thus follows with C_d as the discharge coefficient from

$$Q = C_d(\pi/4)D^2[2g(H_o - C_dD)]^{1/2}. \tag{9.3}$$

The discharge coefficient depends mainly on the relative inlet rounding $\eta_d = r_d/D$ as

$$C_d = 0.96[1 + 0.5\exp(-15\eta_d)]^{-1}. \tag{9.4}$$

The discharge of a certain culvert depends thus exclusively on the approach flow head H_o. Note the discharge increase for relatively small rounding $\eta_d \geq 0.15$, such that the head losses can be reduced with a comparatively small effort. In sewer hydraulics, inlets are often sharp-crested mainly because of simplicity in design. To inhibit cavitation damage, larger intakes have to be designed for a minimum pressure at the pipe invert.

Pressurized flow (subscript p) is modelled with the generalized Bernoulli equation (Chap. 2). With a difference of heads between the up- and downstream sections $H_d = H_o + S_oL_d - H_u$ the discharge is

$$Q_p = (\pi/4)D^2[2gH_d/(1 + \Sigma\xi)]^{1/2}. \tag{9.5}$$

Here $\Sigma\xi$ is the sum of all losses along the culvert, including intake and friction. Note that the outlet loss is already included. For a straight culvert with a sufficiently rounded inlet geometry $\eta_d > 1/6$, only wall friction according to Chap. 2 has to be accounted for. If the culvert is hydraulically rough, then $\Sigma\xi = \xi_f = 2\cdot4^{4/3}[n^2gL_d/D^{4/3}]$ with $1/n$ as the roughness coefficient according to Manning and Strickler.

Equations (9.1) to (9.5) involve a number of geometric and hydraulic parameters. If the discharge is related to the reference discharge $D^2(gH_o)^{1/2}$, and the approach flow head to the reference head $(H_o + D)$, a generalized flow diagram is obtained (Sinniger and Hager 1989). Figure 9.3 shows the relative discharge $Q/[D^2(gH_o)^{1/2}]$ versus the normalized head $H_o/(H_o + D) \leq 1$. On the lower left, *uniform flow* appears as a function of χ, with full pipe flow indicated with the dotted curve. The limiting curve to the lower right corresponds to *critical flow*. A culvert is full for conditions of about (0.55; 0.55) and *gated flow* starts for larger heads. Depending on the inlet curvature $\eta_d = r_d/D$ various curves start from the full flow point. The largest discharge capacity is given for $\eta_d \geq 1/6$, as previously discussed. Pressurized flow cannot be included in Fig. 9.3 because of the different governing parameters.

Example 9.1 Given a culvert of length $L_d = 20$ m, bottom slope $S_o = 1\%$ and roughness coefficient $1/n = 70\ \text{m}^{1/3}\text{s}^{-1}$. Its entrance is rounded with $r_d = 0.2$ m, and the downstream head is $H_u = 0.60$ m. What is the discharge for an approach flow energy head $H_o = 2.5$ m and a diameter $D = 1.5$ m?
With $\eta_d = 0.2/1.5 = 0.13$ and $H_o/(H_o + D) = 2.5/(1.5 + 2.5) = 0.625$, Fig. 9.3 yields $Q/[D^2(gH_o)^{1/2}] = 0.675$, thus $Q = 0.675[1.5^2(9.81\cdot2.5)^{1/2}] =$

$7.5\ m^3s^{-1}$. It could be increased by about 5% when rounding the inlet with a radius $r_d = 0.25$ m.

With $H_d = 2.5 + 0.01 \cdot 20 - 0.6 = 2.1$ m and $\Sigma\xi = \xi_f = 2 \cdot 4^{4/3} 9.81 \cdot 20/(70^2 1.5^{4/3}) = 0.30$ as the total loss coefficient, the discharge for pressurized flow according to Eq. (9.5) would be $Q_p = (\pi/4)1.5^2[19.62 \cdot 2.1/(1 + 0.3)]^{1/2} = 9.95\ m^3s^{-1} > Q$. The pressurized discharge is not physically determining because gated flow is relevant. The result is thus $Q = 7.5\ m^3s^{-1}$.

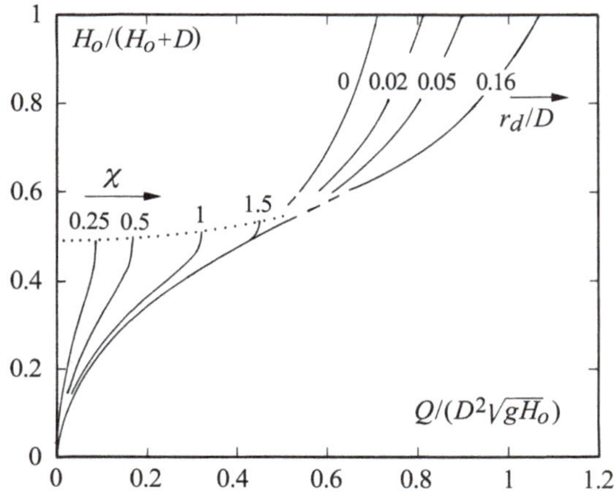

Fig. 9.3 Generalized discharge diagram for free culvert outflow with roughness characteristic $\chi = S_o^{1/2}D^{1/6}/(ng^{1/2})$ and $\eta_d = r_d/D$ as relative inlet rounding. (...) full pipe for uniform flow, (- - -) transition between free surface and gated flow types

9.2.4 *Design Equations*

A disadvantage of Fig. 9.3 is the combined appearance of the two design parameters Q and H_o. A modified approach for the capacity design of a culvert, i.e. the determination of the *culvert diameter D*, is needed. Normally, a culvert is designed for a relative head $H_o/D > 1$. The transition between free surface and pressurized flow is given for $H_o/D < 1.2$ as (Fig. 9.3)

$$Q_{cM} = 0.61\left(gD^5\right)^{1/2}. \tag{9.6}$$

If free surface flow is required in a culvert, then the maximum (subscript M) discharge depends essentially on the culvert diameter D, and Eq. (9.6) refers to the *maximum critical discharge*.

Gated flow with $H_o/D > 1.2$ can be approximated by

$$D/H_o = \delta(\bar{q})^{1/2}\left[1 + \frac{1}{8}(\bar{q})^{1/2}\right]^2 \tag{9.7}$$

with $\bar{q} = Q/(gH_o^5)^{1/2}$ as the pipe Froude number related to the approach flow head H_o, and $\delta = 1.05$ for a well-rounded, and $\delta = 1.2$ for a sharp-crested inlet geometry. For small $\bar{q}$, the diameter depends thus essentially on H_o and not so much on discharge. The effect of inlet rounding on discharge can be up to 15%.

For *pressurized flow* an average total head loss of about $\Sigma\xi = 1/3$ may be assumed, and the diameter is given as

$$D = \left[Q/\left(gH_d\right)^{1/2}\right]^{1/2}. \tag{9.8}$$

The main effect has thus the discharge, and the difference head has a relatively small effect. The resulting diameter of Eq. (9.8) should be checked with Eq. (9.5).

Example 9.2 Consider the culvert of Example 9.1 with $L_d = 20$ m, $S_o =$ 1%, $1/n = 70$ m$^{1/3}$s^{-1}, $r_d = 0.20$ m and $H_u = 0.60$ m. What is the minimum diameter for diversion of a design discharge of $Q_D = 5$ m^{3}s^{-1}?

1. *Critical discharge* with Eq. (9.6) is $D_c = [(5/0.61)^2/9.81]^{1/5} = 1.47$ m. Then $H_{oc}/D = (5/3)[5/(9.81\cdot1.47^5)^{1/2}]^{3/5} = 1.24$ from Eq. (9.1). This is in the transition between critical and gated flow types.
2. *Pressurized discharge* is with $H_d = 1.47 + 0.01\cdot20 - 0.60 = 1.07$ m equal to $D_p = [5/(\pi/4)(1.3/19.62\cdot1.07)^{1/2}]^{1/2} = 1.25$ m $< D_c$ from Eq. (9.5). Critical flow is thus determining.
3. *Uniform discharge* is characterized by $\chi = 70\cdot0.01^{1/2}1.5^{1/6}/9.81^{1/2} =$ $2.4 > 2$, and thus critical flow is also determining.
4. *Gated discharge* with $H_o = 1.24\cdot1.47$ m $= 1.82$ m according to 1. gives $\bar{q} = 5/(9.81\cdot1.82^5)^{1/2} = 0.36$, thus from Eq. (9.7) $D/H_o = 1.05\cdot0.60(1 + 0.60/8)^2 = 0.73$ and $D_g = 0.73\cdot1.82 = 1.33$ m.

This example results thus in the transition between critical and gated flow types. To obtain a definite flow structure, the diameter should be $D = 1.60$ m, such that $H_o/D = 1.09$ and $H_o = 1.74$ m, for which $[H_o/(H_o + D);$ $Q/D^2(gH_o)^{1/2}] = [0.52;0.47]$, corresponding definitely to flow type ①.

9.2.5 *Simple Culvert Structure*

Flow Types

The simple culvert has a specific geometry that can be described as follows:

- Upstream basin with a negligible velocity of approach,
- Sharp-crested intake geometry at a vertical intake wall,

- Straight circular barrel of definite roughness and constant bottom slope, and
- Free outlet into the atmosphere.

The length of the culvert is at least 10 times the culvert diameter. Depending on the bottom slope S_o, the flow in the culvert may be critical or of gate flow type for intake control ($S_o > S_c$), or uniform flow may prevail for $S_o < S_c$ (Fig. 9.4).

Critical flow is described with Eq. (9.1). Introducing the pipe Froude number $\mathsf{F}_D = Q/(gD^5)^{1/2}$, this may equally be expressed with $Y_c = H_c/D$ as

$$Y_c = \frac{5}{3}\mathsf{F}_D^{3/5}. \tag{9.9}$$

Uniform flow follows essentially Eq. (9.2). With $Y_N = H_N/D$ and $\mathsf{F}_N = \mathsf{F}_D/\chi$ in which $\chi = S_o^{1/2}D^{1/6}/(ng^{1/2})$ is the roughness characteristic a simple approximation is

$$Y_N = 2^{1/2}\mathsf{F}_N^{5/9}\left(1 + \frac{1}{6}\chi^2\right). \tag{9.10}$$

It can be demonstrated that uniform flow is subcritical for $\chi < 1$, and supercritical for $\chi \geq 1$. The latter condition is not relevant here because of inlet control.

Gated flow (subscript g) is described with Eq. (9.3). With $Y_g = H_g/D$ as the relative intake energy head, and $C_d = 0.64$ for sharp-crested intake flow, Eq. (9.3) may be written as

$$\mathsf{F}_D = 1.11C_d\left[Y_g - C_d\right]^{1/2}. \tag{9.11}$$

The discharge thus varies essentially with the square root of Y_g.

Pressurized flow (subscript p) follows the generalized Bernoulli equation (Fig. 9.4)

$$z_p + H_p = (1 + \xi_e + \xi_f)\frac{V_p^2}{2g} + \frac{D}{2} \tag{9.12}$$

with z_p as the elevation difference between inlet and outlet, H_p the upstream head, $\xi_e = 0.50$ as intake loss coefficient, ξ_f as friction loss coefficient, and V_p as pipe flow

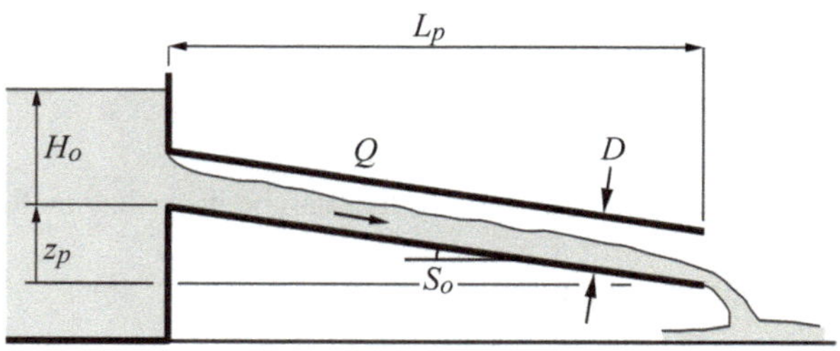

Fig. 9.4 Culvert with free outlet into atmosphere, notation

velocity. The pressure at the outlet section is assumed to be (1/2)D. For turbulent smooth flow, the friction coefficient is $f = 0.2\ \mathsf{R}^{0.2}$ with $\mathsf{R} = V_pD/\nu$ as the Reynolds number, and for turbulent rough flow $f = 2{\cdot}4^{4/3}(gn^2/D^{1/3})$ with g as the gravitational acceleration and $1/n$ the Manning roughness coefficient. Introducing k_s as the equivalent sand roughness height, this may also be expressed as $f = 0.19(k_s/D)^{1/3}$ from Sect. 2.2.3. The internal pressure at the pipe outlet (Chap. 11) is neglected because of the complexity of parameter definition. For long culverts with a relatively large bottom slope, this simplification is acceptable. Equation (9.12) thus reads with $Z_p = z_p/D = S_o(L_p/D) = S_o\lambda_p$ with $\lambda_p = L_p/D$ as relative pipe length

$$\mathsf{F}_D = 1.11\left[\left(S_o\lambda_p + Y_p - 0.50\right) / \left(1.5 + f\lambda_p\right)\right]^{1/2}. \tag{9.13}$$

Figure 9.5 shows the relative upstream energy head Y as a function of pipe Froude number F_D including all four flow regimes. The curves with intake control cross those with pipe control at the *full flow point* $(Y;\mathsf{F}_D)_f = (1.2;0.53)$. A friction coefficient of $f = 0.014$ was assumed. The effect of bottom slope on the discharge relation is seen to be significant for pressurized flow.

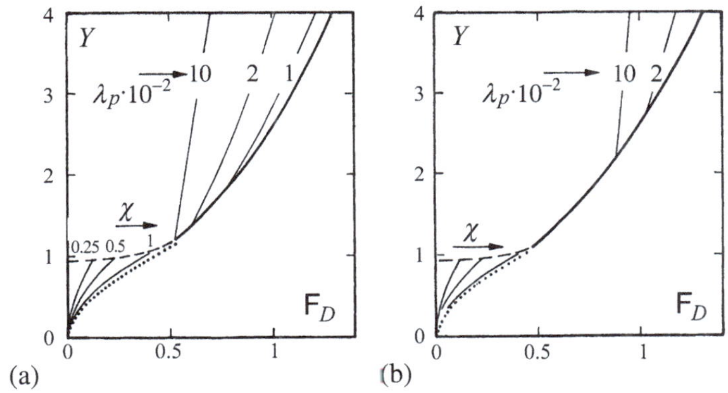

Fig. 9.5 Generalized culvert flow diagram with (...) critical flow, (- - -) uniform flow, (—) gated flow and (—) pressurized flow; (- - -) choking limit. S_o = (**a**) 0.003, (**b**) 0.010 (Hager and Del Giudice 1998)

Transition Types

Depending on the bottom slope S_o, either pressurized or gated flow may establish. The transition (subscript t) between these two flow regimes depends on S_o, λ_p and f, and is from Eqs. (9.11) and (9.13) with $C_d = 0.64$

$$Y_t = \frac{(S_o + 0.26f)\lambda_p - 0.11}{0.41f\lambda_p - 0.39}. \tag{9.14}$$

For $Y > Y_t$ (>1.2) the flow is pressurized and gated flow occurs otherwise. An increase of bottom slope increases Y_t and results in *blowout* of pressurized flow.

This approach was verified with experiments conducted with a plexiglass culvert pipe of internal diameter $D = 0.100$ m. Pipe lengths between 20 and 60 diameters were considered, for bottom slopes between 0.3% and 3.2%. For large bottom slope the transition from $Y < 1.2$ to $Y > 1.2$ involves a *hydraulic jump* in the pipe. For any given upstream head, gated flow was developed in the pipe and the transition to pressurized flow occurs due to blocking of the outlet section, associated with a blowout of air at the pipe inlet. Once the latter flow occurred, it was stable and the transition back to gated flow developed only by addition of air downstream of the inlet section.

Two types of *flow transitions* can be identified in a steep culvert pipe with $S_o > S_c$: (1) Supercritical upstream flow followed by a hydraulic jump, and air supply by intake vortices combined with deaeration by the roller of the jump into the downstream reach; and (2) Hydraulic jump that chokes and seals the pipe against air entrainment from the outlet section. Type (1) occurred up to $\mathsf{F}_D = 1.05$, and type (2) started at about $\mathsf{F}_D = 0.95$, with an overlap of 0.1 in F_D.

Photographs

To illustrate the complex flow phenomena, photos are presented to describe the features observed. These refer to a culvert with a bottom slope of $S_o = 1\%$ and a length of $40D$. Figure 9.6 provides lateral views to the culvert pipe. For $\mathsf{F}_D = 0.51$ the free surface flow is supercritical and involves significant undulations, due to an approach flow to the hydraulic jump with $\mathsf{F}_1 = 1.25$. For $\mathsf{F}_D = 0.63$ the sequent depth of the supercritical approach flow is larger than D, and the pipe chokes from the downstream end, i.e. transition type (2) is developed. For $\mathsf{F}_D = 1.27$, corresponding to $Y = 4.1$, the pipe flow gets pressurized with a complete blowout of air from the conduit.

Figure 9.7 shows top views onto the culvert inlet for $\mathsf{F}_D = 0.63$, with the flow contraction downstream of the inlet (Fig. 9.7a), and the hydraulic jump sealing the air layer against the pipe outlet.

(a)
(b)
(c)

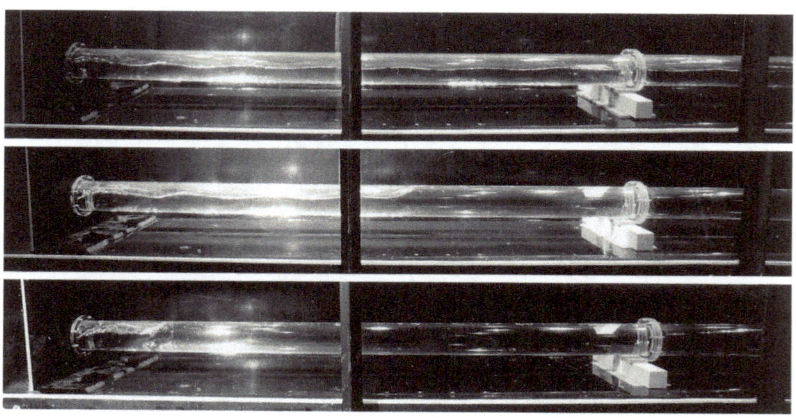

Fig. 9.6 Culvert flow with $\mathsf{F}_D =$ (**a**) 0.51, (**b**) 0.63, (**c**) 1.27. Flow direction from left to right

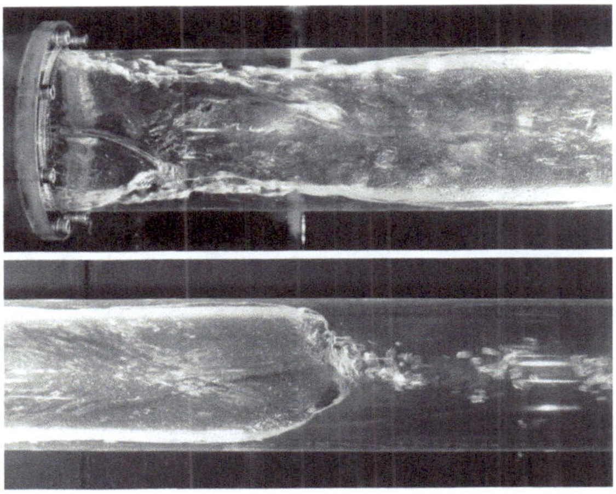

(a)

(b)

Fig. 9.7 Top views on culvert pipe flow for $\mathsf{F}_D = 0.63$. (**a**) Inlet detail, (**b**) hydraulic jump

The choking phenomenon of a culvert pipe is normally abrupt, associated with water hammering in a two-phase flow. Figure 9.8 relates to sequences of choking. If the free surface width reduces locally to below 40 or 50% of the pipe diameter choking may be initiated (Fig. 9.8a). Starting close to the culvert outlet, the downstream pipe portion gets pressurized (Fig. 9.8b) and the hydraulic jump moves upstream until the equilibrium position is reached (Fig. 9.8c). Choking always occurs provided the friction slope is larger than the bottom slope, and the pipe has a sufficient length for flow development.

Length Characteristics

For $H/D > 1.2$ or so, a long culvert pipe may contain a flow with three distinctly different *flow types*:

(1) Free surface flow,
(2) Upstream free surface flow followed by a hydraulic jump, and
(3) Downstream free surface flow.

Combinations of the latter two types of flow are also possible, with the presence of air pockets. The lengths of the upstream and downstream air cavities L_u and L_d are particularly interesting because they allow to located the transition sections. The results below refer to bottom slopes between 0.3 and 1.0% in a smooth culvert pipe for pipe lengths between 20 and 60 diameters.

The *upstream length* L_u, or the length ratio $\lambda_u = L_u/D$ as a function of pipe Froude number is almost constant and equal to $\mathsf{F}_D = 0.60$. A pipe thus gets pressurized for $S_o < 1\%$ provided $\mathsf{F}_D > 0.60$ ($\pm 10\%$).

The transition between critical and gated flows was previously determined to $\mathsf{F}_D = 0.53$, corresponding to a lower limit of F_D for the appearance of flow type (3).

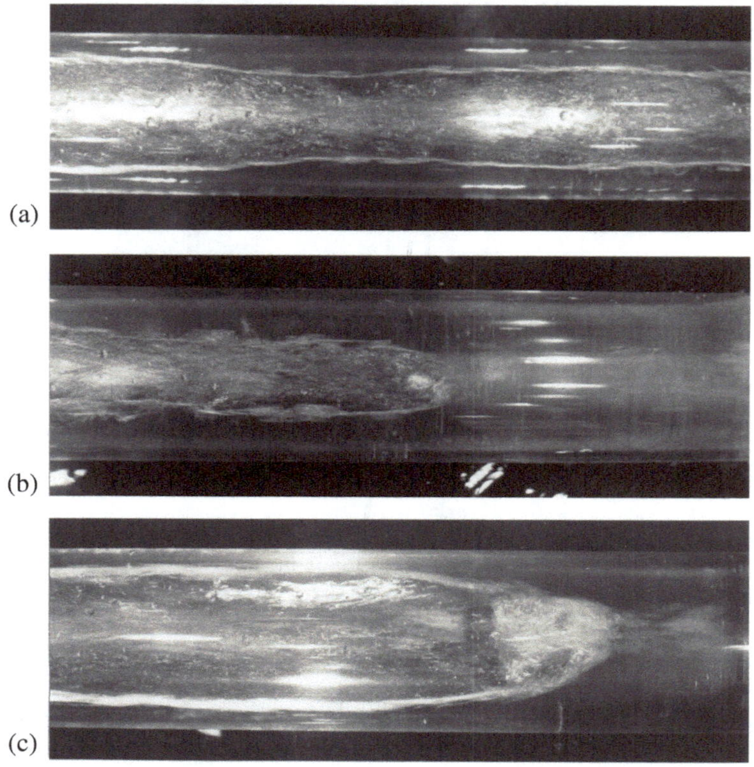

Fig. 9.8 Incipient choking of culvert pipe (**a**) free surface flow close to choking, (**b**) incipient choking close to pipe outlet, (**c**) transition from free surface to pressurized culvert flow

The corresponding *downstream length* $\lambda_d = L_d/D$ varies with F_D as

$$\lambda_d = 0.10\mathsf{F}_D^{-4.5},\ \mathsf{F}_D > 0.50. \tag{9.15}$$

For $0.53 < \mathsf{F}_D < 0.60$, the data scatter because of minute details at the inlet and outlet geometries. For $\mathsf{F}_D \geq 0.60$, the trend is definite, however. Pressurized outlet flow has established if $\mathsf{F}_D > 1.1$, as demonstrated in Chap. 11.

9.3 Throttling Pipe

9.3.1 *Description*

The throttling pipe (German: Drosselstrecke; French: Vanne de conduite) is a sort of culvert that should limit the discharge to a design (subscript D) discharge $Q = Q_D$. Ideally, a throttling pipe should result in a free flow for $Q < Q_D$, and divert a maximum discharge Q_D. The hydraulic performance of a throttling pipe is

$\eta_D = (Q\text{–}Q_D)/Q_D$. Among a variety of other devices reviewed in Chap. 4, the throttling pipe thus limits the discharge by a hydraulic control.

The throttling pipe is usually connected with a diversion structure, such as a storage basin or a sideweir in a combined sewer system. Because rainwater storage basins can have a significant variation of basin level, ATV (1993) recommends throttling pipes exclusively for *rainwater overflow structures* such as sideweirs (Chap. 18) or bottom openings (Chap. 20). Figure 9.9 shows a typical arrangement with a sewer sideweir, including the inlet, the throttling pipe and the outlet. The inlet is normally sharp-crested at the pipe vertex, has a vertical upstream wall and an U-shaped converging profile along the bottom. The minimum diameter should be $D = 0.20$ m (ATV 1993) and the upper limit is roughly 0.50 m. Often, diameters should be smaller to obtain a sufficient *throttling effect*, but non-clogging requirements would not be satisfied (Chap. 3).

Depending on the *relative length* L_d/D, reference to a short ($L_d/D < 10$) and a long throttling pipe is made. Because the throttling effect involves mainly friction for pressurized flow, short throttling pipes are not recommended. A minimum length is $L_d/D = 20$ (ATV 1993) and the maximum length of $L_d = 100$ m should be respected mainly for purposes of maintenance and cleaning (Munz 1977). A relatively long throttling pipe is thus required.

The *bottom slope* S_o of the throttling pipe is an important design parameter. For large bottom slopes $S_o > S_c$ the critical and gated flow types normally occur, and the pipe has an inlet control. Then, the pipe cannot add to the throttling effect, and is useless. For small bottom slope, either uniform or pressurized flow types occur, and the throttle can significantly add to the reduction of discharge. Throttling pipes should therefore be:

- Small in diameter,
- Long relative to the diameter, and
- Almost horizontal to generate pressurized flow.

A throttling pipe has a usual boundary roughness to inhibit clogging by protruding elements. The design discharge for the throttling effect to start is the *critical treatment discharge* Q_K. Additional requirements are (ATV 1993):

- *Throttling ratio* between throttling discharge under maximum approach flow discharge Q_D and critical approach flow Q_K should be less than 1.2 (Table 9.1), and

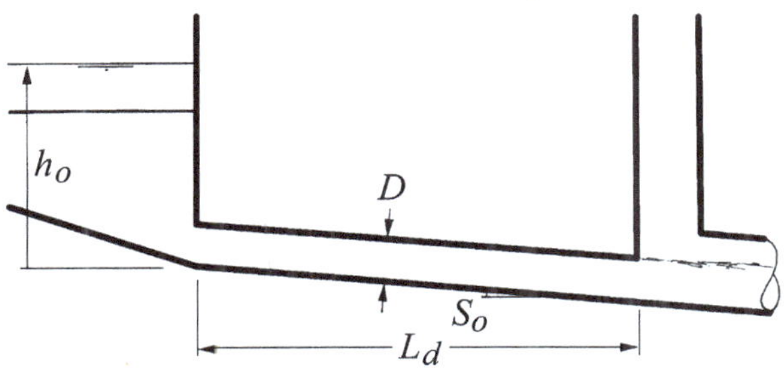

Fig. 9.9 Throttling pipe downstream of sewer sideweir, notation

Table 9.1 Discharges Q_i for various diameters D_i for Example 9.6

D_i	ξ_f	$\Sigma\xi$	A_i	Q_i	V_i
[m]	[–]	[–]	[m^2]	[m^{3}s^{-1}]	(m/s)
(1)	(2)	(3)	(4)	(5)	(6)
0.3	5.58	7.6	0.071	0.117	1.14
0.4	3.8	5.8	0.126	0.237	1.88
0.5	2.82	4.8	0.196	0.406	2.07
0.6	2.21	4.2	0.283	0.627	2.22
0.7	1.8	3.8	0.385	0.896	2.33
0.8	1.5	3.5	0.503	1.22	2.49

- *Minimum velocity* for dry weather flow should be at least V_s [ms^{-1}] $= 0.50 + 0.55D$ [m] from Chap. 3. This condition defines the minimum required bottom slope S_o.

A throttling pipe can thus be hydraulically designed.

9.3.2 *Hydraulic Design*

Inlet Loss

The transition form a sewer sideweir to a throttling pipe is guided along the lower profile portion, with an abrupt inlet (subscript e) from the vertical end wall to the pipe vertex. With h_u as the downstream flow depth of the sideweir (Chap. 18) of diameter $D_u = D$, the *inlet loss* can be estimated. The loss coefficient $\xi_e = \Delta H_e/[V^2/(2g)]$ with ΔH_e as the entrance loss height and V as the pipe velocity depends on the relative approach flow depth $y_u = h_u/D$ and the entrance angles α_1 and α_2 (Fig. 9.10a). For the limit case $y_u \rightarrow 1$ the loss coefficient tends to $\xi_e = 0$, and the upper limit coefficient is reached as $y_u >> 1$. For a sharp-crested intake from a tank, the loss coefficient is $\xi_e = 0.50$ (Chap. 2).

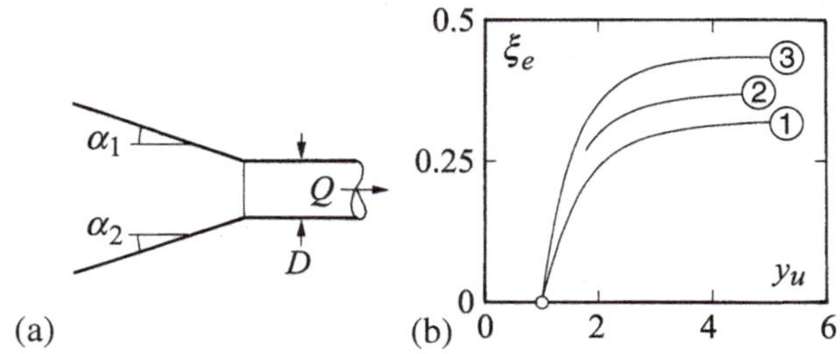

Fig. 9.10 (**a**) Inlet geometry from sewer sideweir to throttling pipe. (**b**) Loss coefficient ξ_e according to Kallwass (1967) as a function of approach filling $y_u = h_u/D$ with ① $\alpha_1 = \alpha_2 = 13.5°$ ② $\alpha_1 = 0, \alpha_2 = 27°$, ③ $\alpha_1 = 0, \alpha_2 = 13.5°$

The effect of *approach flow depth* on the head loss coefficient can be estimated. Sinniger and Hager (1989) suggested

$$\xi_e = \xi_{e\infty} 2\sin^2(\alpha_o/2)\left[1 - y_u^{-1}\right]^{2-(\alpha_o/\pi)}. \tag{9.16}$$

Here, the basic value $\xi_{e\infty}$ varies with the approach flow angles α_1 and α_2 (Fig. 9.10b). Typically, the approach flow direction relative to the pipe axis is $\alpha_o = 45°$, such that

$$\xi_{e45} = \xi_{e\infty}\left[1 - y_u^{-1}\right]^{1.5}. \tag{9.17}$$

This expression deviates from the observations of Kallwass (1967) because the approach flow angles are not contained in Eq. (9.17). However, because the effect of values y_u close to one is insignificant for sideweirs with a high crest (Chap. 18), its effect on ξ_e is negligible, and one may approximate $\xi_e = \xi_{e\infty}$. In the average, a design coefficient of $\xi_e \cong 0.40$ may be adopted.

Self-priming

For free flow from a throttling pipe, air may advance to the inlet from the downstream end. As for culverts, throttling pipes with an approach flow depth larger than about $h_u/D > 1.2$ generate two flow types, namely gated flow and pressurized flow. The transition between the two has been examined by Li and Patterson (1956), and Kallwass (1967). *Self-priming* flow occurs by (Fig. 9.11):

(a) Divergent pipe flow,
(b) Hydraulic jump, or
(c) Standing waves.

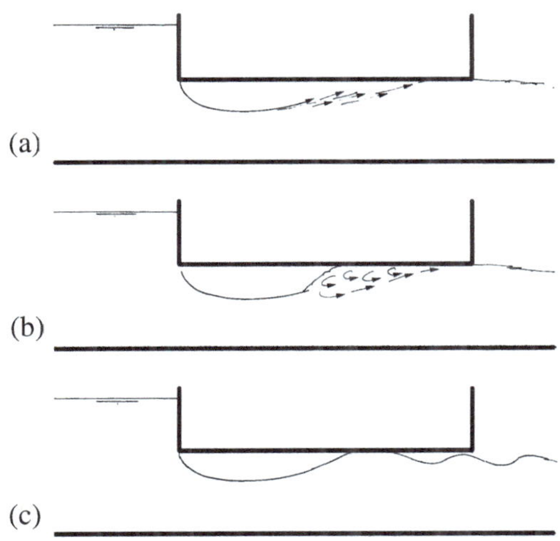

Fig. 9.11 Flow types in a throttling pipe according to Li and Patterson (1956). (**a**) Diverging flow, (**b**) hydraulic jump, (**c**) standing waves

For small bottom slope S_o a hydraulic jump occurs in a long pipe ($L_d/D > 20$). For horizontal pipes, self-priming does even not occur for $L_d/D < 8$. For larger bottom slope and pipes longer than $35D$, pressurized flow may occur by shockwaves crossing the pipe vertex. If the pipe is still too short, or if the slope is too large, self-priming may be generated by surface waves for a large approach flow head of the order $H_o/D = 10$.

For a given throttling pipe and an approach flow head larger than $H_o/D = 1.2$ to 1.5, pressurized flow has a *larger discharge capacity* than gated flow. This results from the sub-atmospheric pressure distribution downstream from the inlet section. For a value $D^{1/3}/(n^2 g) \cong 750$ Li and Patterson (1956) found:

- Stable gated flow for $Q/(gD^5)^{1/2} < 0.6$, and
- Stable pressurized flow for $Q/(gD^5)^{1/2} > 1.6$.

Between these two limit values, which may further depend on the bottom slope, Blaisdell – a discusser of Li and Patterson (1956) – found an unstable transition reach that should be avoided. The slope corresponding to this transition (subscript t) region can be approximated as

$$S_t\,[\%] = \frac{1}{80}\left[\frac{L_d}{D} + 20\right]. \tag{9.18}$$

For a slope of say $S_o = 0.35\%$, a length of $L_d/D = 8$ is sufficient for pressurized flow, whereas the length has to be $45D$ for a bottom slope of 0.8%. These limits refer to the arrangement of Li and Patterson and may be subject to significant variation for slight geometrical modifications. The approach of Kallwass (1967) is questionable due to the assumption of a constant roughness coefficient.

Outflow Features

The *outflow* from a throttling pipe can be either guided by the following U-shaped profile in the downstream manhole (Fig. 9.12), or it can discharge on an outflow

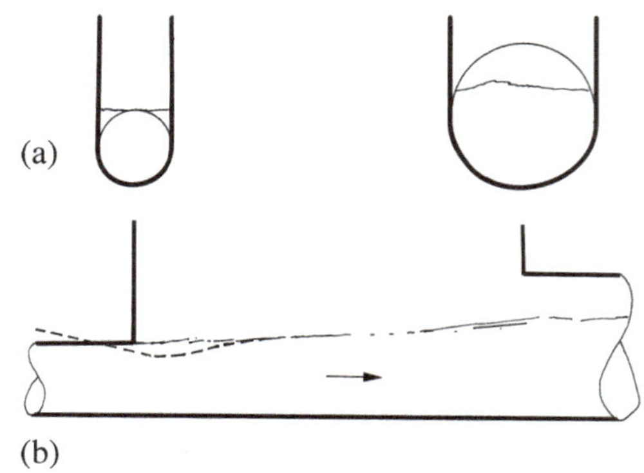

Fig. 9.12 Pressurized outflow from throttling pipe into manhole with (- - -) pressure head line, (—) free surface. (**a**) Transverse sections at manhole inlet and outlet, (**b**) longitudinal section

plane, or it can even discharge with a bottom drop. In the latter two unusual cases, the pressure distribution is no more close to hydrostatic and the pressure line does not coincide with the free surface. These two latter cases are not relevant in the present context because the outflow from a throttling pipe is discharged in a standard manhole with 100% benches (Chap. 14). In addition, no outlet loss has to be considered because the velocity remains practically constant.

A *free outlet* should be provided in any case to separate hydraulically the throttling pipe from the manhole flow. Also, if deposits are prevented, maintenance is not required after each rainfall event.

9.3.3 *Discharge Equations*

Transition Characteristics

For *gated flow* in the throttling pipe, and a sharp-crested inlet configuration ($\eta_d = 0$) the discharge coefficient is $C_d = 0.64$ from Eq. (9.4). For *pressurized flow* with a downstream energy head $H_d = H_o + S_o L_d - D$, $\Sigma\xi = \xi_e + \xi_f$ for inlet and friction losses, and $(\pi/4)C_d = 0.50$, the relative discharge $q_d = Q/(gD^5)^{1/2}$ follows from Eq. (9.5). The results are thus

$$\text{Gated flow:} \qquad q_d = 0.71\,(Y_o - 0.64)^{1/2}, \tag{9.19}$$

$$\text{Pressurized flow:} \qquad q_d = 0.94\left[\frac{Y_o + j_d - 1}{1 + 9 j_d \chi_d^{-2}}\right]^{1/2}. \tag{9.20}$$

Here $Y_o = H_o/D$ is the relative approach flow head, $j_d = S_o L_d/D$ the relative pipe slope and $\chi_d = S_o^{1/2} D^{1/6}/(ng^{1/2})$ the roughness characteristics. Equation (9.20) involves four parameters and cannot be plotted on a single diagram. Because the bottom slope is normally of the order of 0.3–0.5%, the parameter j_d has an order of 0.1–0.2, and one may approximate Eq. (9.20) with $R_d = gL_d n^2/D^{4/3}$ as

$$q_d = 0.94\left[\frac{Y_c - 0.9}{1 + 9R_d}\right]^{1/2}. \tag{9.21}$$

Eliminating q_d from Eqs. (9.19) and (9.21) gives the *transition roughness characteristic*

$$R_d^* = \frac{1}{9}\left[1.75\frac{Y_o - 0.90}{Y_o - 0.64} - 1\right]. \tag{9.22}$$

For $R_d < R_d^*$, i.e. for a short throttling pipe, the governing flow type is gated flow, and for $R_d > R_d^*$ pressurized flow results.

Example 9.3 Given a throttling pipe of length $L_d = 20$ m, diameter $D = 0.30$ m, bottom slope $S_o = 1\%$ with a roughness coefficient $1/n = 90\ \mathrm{m^{1/3}s^{-1}}$. What is the discharge for a head $H_o = 1.0$ m?
The relative approach flow head is $Y_o = 1/0.3 = 3.33$ and the transition roughness characteristics is $R_d^* = (1/9)[1.75(3.33–0.90)/(3.33–0.64)–1] = 0.065$. The effective value is $R_d = 9.81\cdot 20/(90^2 0.3^{4/3}) = 0.12 > R_d^*$ such that pressurized flow occurs. With $j_d = 0.01\cdot 20/0.3 = 0.67$ and $\chi_d = 90\cdot 0.01^{1/2} 0.3^{1/6}/9.81^{1/2} = 2.35$ the relative discharge is $q_d = 0.94[(3.33 + 0.67–1)/(1 + 9\cdot 0.67/2.35^2)]^{1/2} = 1.12$ from Eq. (9.20), or approximately $q_d = 0.94(3.33–0.90)/(1 + 9\cdot 0.12)]^{1/2} = 1.02$ from Eq. (9.21) (–10%), and thus $Q_d = 1.12(9.81\cdot 0.3^5)^{1/2} = 0.173\ \mathrm{m^3s^{-1}}$.
For gated flow the result would be $q_d = 0.71(3.33–0.64)^{1/2} = 1.16$ from Eq. (9.19), thus slightly larger than according to Eq. (9.20).

The main purpose of throttling pipes is to generate *pressurized flow*. This can be reached by increasing the parameter R_d, including:

- Increasing the length L_d,
- Increasing the roughness coefficient n, or
- Decreasing the diameter D.

In practise, only the length is increased, of which the effect on R_d is linear, or the diameter is decreased with a more significant effect on R_d. As mentioned previously, the roughness coefficient should not be increased because of poor performance with sewage, and possible clogging of the throttling pipe.

The *effect of R_d* on the relative discharge q_d is rather small. A maximum (subscript *M*) value is $R_{dM} = gL_{dM}n^2/D_m^{4/3}$ with $L_{dM} = 100$ m and a minimum (subscript *m*) throttle diameter of $D_m = 0.20$ m, thus with $1/n = 85\ \mathrm{m^{1/3}s^{-1}}$ one has for $R_{dM} = 9.81\cdot 100\cdot 0.012^2/0.2^{4/3} = 1.20$. Then, the denominator of Eq. (9.21) tends to $[1 + 9R_d]^{1/2} \rightarrow 3R_d^{1/2}$, such that the minimum discharge tends to the expression $q_{dm} = 0.31[(Y_o–0.9)/R_d]^{1/2}$, or

$$Q_m = 0.31\frac{D^{8/3}}{nL_d^{1/2}}(H_o - 0.9D)^{1/2}. \tag{9.23}$$

Equation (9.23) demonstrates the significant effect of pipe diameter D, with an increase of 50% in diameter D resulting in a three times larger discharge Q_m. The length effect is comparatively small.

Example 9.4 What is the discharge in Example 9.3 for a throttling pipe diameter $D = 0.25$ m and a length $L_d = 80$ m?
With $R_d = 9.81 \cdot 80/(90^2 0.25^{4/3}) = 0.615 > R_d^*$ one has also pressurized flow, with $q_d = 0.94[(4 + 3.2-1)/(1 + 9 \cdot 3.2/2.28^2)]^{1/2} = 0.915$ and thus a discharge of $Q = 0.915(9.81 \cdot 0.25^5)^{1/2} = 0.090$ m^{3}s^{-1}. The discharge is reduced to 52%. Minimum discharge is $Q_m = 0.31(0.25^{8/3}/0.011 \cdot 80^{1/2})$ $[1-0.9 \cdot 0.25]^{1/2} = 0.069$ m^{3}s^{-1} (–25%) according to Eq. (9.23).

Excess Discharge Limit

The overall design includes the condition that the maximum pipe discharge Q_M is only 20% larger than the design discharge Q_D (9.3.1), i.e. $Q_M/Q_D \leq 1.2$. The maximum increase of upstream flow depth $\Delta Y_o = Y_{oM} - Y_{oD}$ is from Eqs. (9.19) and (9.20)

$$\text{Gated flow} \qquad \Delta Y_o = 0.44 Y_{oD} - 0.28, \tag{9.24}$$

$$\text{Pressurized flow} \qquad \Delta Y_o = 0.44[Y_{oD} + j_d - 1]. \tag{9.25}$$

For an equal relative approach flow depth Y_{oD} and with $j_d = 0.2$ both characteristics $\Delta Y_o(Y_{oD})$ are almost identical. From the throttling process, no significant difference between gated and pressurized flow can thus be established.

Example 9.5 How much can the approach flow head increase in Example 9.3 to increase the design discharge by 20%?
With $Y_{oD} = 3.33$ and $j_d = 0.67$, Eq. (9.25) gives for pressurized flow $\Delta Y_o = 0.44(3.33 + 0.67-1) = 1.32$, thus $\Delta H_o = 1.32 \cdot 0.30 = 0.40$ m. Then, discharge is $q_{dM} = 0.94(3.33 + 1 \cdot 32 + 0.67-1)/(1 + 9 \cdot 0.67/2.35^2)]^{1/2} = 1.35$ m^{3}s^{-1}, thus by 20% larger than $q_{dD} = 1.12$.

Design Constraints

The *throttling pipe* is influenced by a number of parameters, namely the:

- Design discharge Q_{dD},
- Length of pipe L_d,

- Pipe diameter D,
- Roughness coefficient n,
- Approach flow head H_o and
- Bottom slope S_o.

In addition, the maximum discharge is $Q_{dM} = 1.2Q_{dD}$. The *design discharge* determines the pipe geometry, whereas the maximum depth has to satisfy the discharge condition. The problem of throttling pipe is thus complex when adding a sideweir (Chap. 18).

For small discharges of about $Q \leq 0.020$ m^{3}s^{-1}, and a minimum pipe diameter of $D = 0.20$ m, the dimensionless discharge is $q_d = 0.36$ with a corresponding approach flow depth $Y_o = 0.9$ for gated flow. For a modified roughness characteristics $R_d = 0.5$ the corresponding relative approach flow depth is $Y_o = 1.1$ for pressurized flow. Therefore, a throttling pipe is unable to control small discharges because of the large increase of Y_o for maximum discharge. Throttling pipes should have a relative approach flow depth of at least 3 for which $q_d = 1$ and the *minimum discharge* amounts to $Q_{dD} = 0.050$ m^{3}s^{-1} for a diameter of 0.20 m. For smaller design discharges, a vortex throttle as introduced in Chap. 4 is recommended. It seems that the actual design guidelines have not been critically reviewed by the wastewater authorities. There is a conflict between the throttling pipe as a hydraulic device and the ratio between throttling discharges for maximum and design discharges. Design throttling discharges of 0.050 m^{3}s^{-1} as previously determined are comparatively so large that they apply to only very large catchment areas. It is not clear on which bases the 20% addition of discharge for maximum approach flow has been advanced, nor the effect of a larger discharge ratio on a treatment station of, say 1.5. If throttling pipes are to be used in the future, this problem has to be attacked scientifically and verified with prototype observations. The author is also convinced that numerous throttling pipes do not perform as prescribed by the authorities, or that their performance has never been tested. With the improved methods of sewer control, such unsatisfactory conditions may be detected and possibilities should be available to improve a certain state.

9.4 Inverted Siphon

9.4.1 *Description of Structure*

An inverted siphon (German: Düker; French: Aqueduc-syphon) is a pressurized duct to underpass another structure (Fig. 9.13). It is particularly prone to *sedimentation* (Muth 1974). Because the discharge ratio of the night minimum and the rainfall maximum are different by a factor of typically 20 or more, a single siphon pipe in a combined sewer system is inadequate. Therefore a *series of parallel siphon pipes* with gradated diameters is selected to satisfy both capacity and minimum velocity requirements. Often, a rainwater outlet is located upstream from an inverted siphon such that the structure has to be designed for the maximum dry weather discharge.

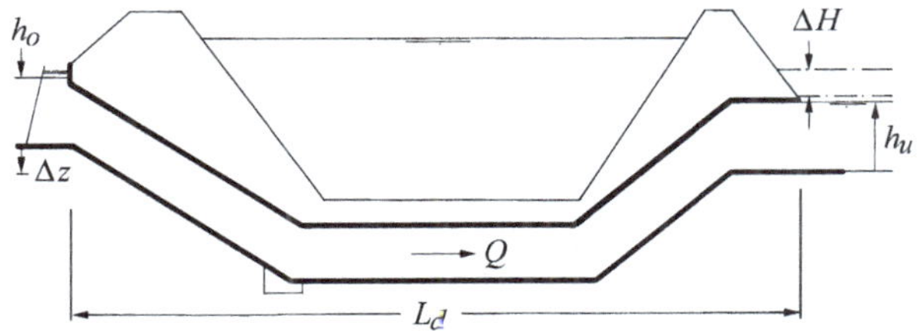

Fig. 9.13 Longitudinal section across an inverted siphon

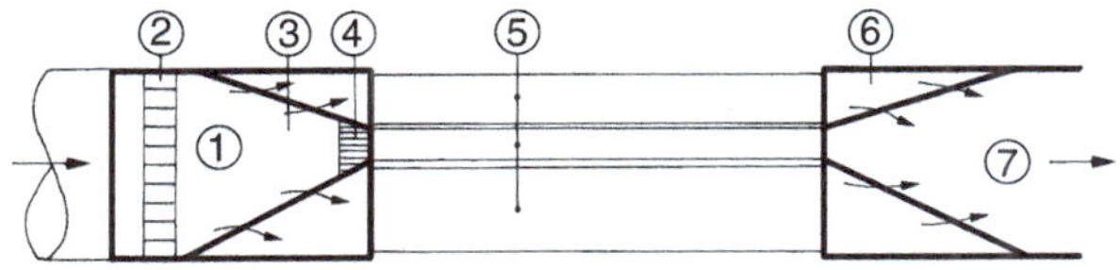

Fig. 9.14 Components of inverted siphon ① intake structure, ② trash rack, ③ sideweir, ④ fine trash rack, ⑤ siphon pipes, ⑥ siphon outlet, ⑦ outlet works

The inverted siphon consists of the inlet structure, the pipe series and the outlet structure. The inlet structure contains sideweirs of staggered crest elevations. The approach flow discharge is thus divided on the various pipes to satisfy the flow conditions. A *trash rack* of a 0.05–0.10 m free passage width located upstream from the sideweirs prevents clogging of the siphon pipes. The area of the trash rack can be enlarged with a box design accessible also during flood conditions (Fig. 9.14). The siphon pipe with the minimum diameter should have a velocity of at least 0.60 ms^{-1} for the discharge during the night minimum flow. The *minimum diameter* should be 0.30 m for combined sewer systems (SIA 1981) and 0.25 m for a separated sewer system.

The *siphon pipes* are usually horizontal. The transition from the inlet should be at least 1:3 sloping, and 1:6–1:1 for the ascending transition. At the lowest point, the pipe should be connected with a pump sump and a valve for maintenance. The siphon profile is often circular, and it can be rectangular for larger structures.

The *outlet works* of the inverted siphon, and the downstream sewer should involve minimum energy losses. As for the inlet, the outlet is staggered to prevent a backflow into the siphon pipes. For security reasons, inverted siphons are often protected with a downstream rack.

9.4.2 *Hydraulic Design*

For given minimum discharge Q_m, dry weather discharge Q_T and critical treatment discharge Q_K, the uniform and critical flow depths are determined in the approach flow and the downstream sewers (SIA 1981). For supercritical approach flow, a sufficient submergence at the inlet should move the hydraulic jump into the upstream sewer. Head losses involve the following elements:

- Trash rack loss ξ_R (Sect. 2.3.8),
- Inlet loss ξ_e (Sect. 2.3.4),
- Bend loss ξ_k (Sect. 2.3.2),
- Expansion loss ξ_E (Sect. 2.3.3), and
- Outlet loss (Sect. 2.3.3) equal to zero.

Typical design values are $\xi_R = 0.5$, $\xi_e = 0.4$, $\xi_k = 4{\cdot}0.15 = 0.6$, $\xi_E = 0.5$ and thus a total of $\Sigma\xi = 2 + \xi_f$ with $\xi_f = 2{\cdot}4^{4/3}[gL_dn^2/D^{4/3}]$ for friction losses. An inverted siphon results in additional losses of about twice the velocity head. The energy equation thus gives

$$h_o + \Delta_z + \frac{V_o^2}{2g} = h_u + \frac{V_u^2}{2g}(1 + \Sigma\xi) \tag{9.26}$$

with h as flow depth, Δz as elevation difference between the inlet and outlet and V as the cross-sectional velocity. Approximately, one may assume an almost constant velocity for subcritical flow, i.e. $V_o = V_u = V_d$. The discharge of the i-th pipe is then with $\Delta h = h_o + \Delta z - h_u$

$$Q_i = A_iV_i = A_i\left[\frac{2g\Delta h}{\sum\xi}\right]^{1/2}. \tag{9.27}$$

First, Eq. (9.27) is applied to the critical treatment discharge $Q_K = n_TQ_T$ by assuming diameters of $D = 0.30$ m, 0.40 m up to $D = 1.0$ m. Then, diameter combinations are selected with a total capacity that is by 10–20% larger than required. Because the ratio of critical to minimum discharges is normally large, an inverted siphon structure is composed of a series of pipes to satisfy both aspects of discharge capacity and minimum required velocity.

Example 9.6 Given $Q_T = 0.67$ m^{3}s^{-1}, $Q_K = 3Q_T = 2$ m^{3}s^{-1} in the approach flow sewer with a diameter $D_o = 1.25$ m, $S_{oo} = 0.3\%$, and $1/n_o = 85$ m$^{1/3}$s^{-1}, and a downstream sewer with $D_u = 1.25$ m, $S_{ou} = 0.35\%$, $n_u = n_o$ and $\Delta z = 0.85$ m. The inverted siphon has a roughness coefficient $1/n = 90$ m$^{1/3}$s^{-1}, its length is $L_d = 65$ m and the acceptable water level difference is $h_o - h_u = 1.05$ m. What are the siphon pipe dimensions resulting in a minimum velocity of $V_u = 0.6$ ms^{-1}?
With $\xi_f = 12.7{\cdot}9.81{\cdot}65/(85^2D^{4/3}) = 1.12D^{-4/3}$ and $A_i = (\pi/4)D_i^2$, discharge Q_i is determined as a function of D_i (Table 9.1). Possible combinations are:

① $D_1 = 0.40$ m, $D_2 = 0.60$ m, $D_3 = 0.80$ m;
② $D_1 = 0.30$ m, $D_2 = 0.70$ m, $D_3 = 0.80$ m;

③ $D_1 = 0.50$ m, $D_2 = 0.60$ m, $D_3 = 0.80$ m; and
④ $D_1 = 0.50$ m, $D_2 = 0.60$ m, $D_3 = 0.70$ m.

Selected is combination ① with $Q_K = 0.237 + 0.627 + 1.22 = 2.08$ m^{3}s^{-1} > 2 m^{3}s^{-1}.

Shrestha and De Vries (1991) described an interactive computer program for designing inverted siphons by accounting also for the effect of sediment.

Notation

A	[m^2]	cross-sectional area
C_d	[–]	discharge coefficient
D	[m]	diameter
f	[–]	friction coefficient
F_1	[–]	approach Froude number
F_D	[–]	pipe Froude number
F_N	[–]	Froude number for uniform flow
h	[m]	flow depth
H	[m]	energy head
j_d	[–]	relative slope
k_s	[m]	equivalent sand roughness height
L_d	[m]	length of siphon
L_p	[m]	culvert length
n_T	[–]	multiple of dry weather discharge
$1/n$	[m$^{1/3}$s^{-1}]	roughness coefficient
$\bar{q}$	[–]	discharge related to energy head
q_d	[–]	discharge related to diameter
Q	[m^{3}s^{-1}]	discharge
Q_T	[m^{3}s^{-1}]	dry weather discharge
r_d	[m]	rounding radius
R_d	[–]	roughness characteristic
R_d^*	[–]	transition value of R_d
R	[–]	pipe Reynolds number
S_c	[–]	critical slope
S_o	[–]	bottom slope
V	[ms^{-1}]	average velocity
V_s	[ms^{-1}]	required velocity
y	[–]	relative flow depth
Y_c	[–]	relative critical energy
Y_o	[–]	energy head related to D
z_p	[m]	elevation difference

Z_p	[–]	$= z_p/D$
α	[–]	approach angle
δ	[–]	coefficient
Δz	[m]	elevation difference
ξ	[–]	head loss coefficient
η_d	[–]	relative curvature
η_D	[–]	throttling performance
λ_d	[–]	relative length
λ_p	[–]	relative culvert length
λ_u	[–]	relative upstream length
ν	[m^2s^{-1}]	kinematic viscosity
χ_d	[–]	roughness characteristic
ξ_e	[–]	asymptotic loss
ξ_E	[–]	expansion loss
ξ_f	[–]	friction loss coefficient
ξ_k	[–]	bend loss coefficient
ξ_R	[–]	trash rack loss

Subscripts

c	critical
d	difference pressure
D	design
e	entrance
f	friction
g	gated
i	multiple value
K	critical treatment
m	minimum
M	maximum
N	uniform
o	upstream
p	pressurized
t	transition
u	downstream

References

ATV (1993). Richtlinien für die hydraulische Dimensionierung und den Leistungsnachweis von Regenwasser-Entlastungsanlagen in Abwasserkanälen und -leitungen (Guidelines for hydraulic design and performance of combined sewers). *Arbeitsblatt* A111. ATV: St. Augustin [in German].

Chow, V.T. (1959). *Open channel hydraulics*. McGraw Hill: New York.

Hager, W.H., Wanoschek, R. (1986). Die Hydraulik des Durchlasses (The hydraulics of culverts). *Wasserwirtschaft* 76(5): 197–202 [in German].

Hager, W.H., Del Giudice, G. (1998). Generalized culvert design diagram. *Journal of Irrigation and Drainage Engineering* 124(5): 271–274.

Henderson, F.M. (1966). *Open channel flow*. MacMillan: New York.

Kallwass, G.J. (1967). Der Füllzustand beim Abfluss in Kreisdurchlässen (Filling condition for flows in circular culverts). *Wasserwirtschaft* 57(10): 367–370 [in German].

Li, W.-H., Patterson, C.C. (1956). Free outlets and self-priming action of culverts. Proc. ASCE, *Journal of Hydraulics Division* 82(HY3), Paper 1009: 1–22; 82(HY5) Paper 1131: 5–8; 83(HY1) Paper 1177: 23–40; 83(HY4) Paper 1348: 3–5.

Munz, W. (1977). Berechnung der Drosselstrecke von Regenüberläufen und Regenbecken (Computation of throttling pipe for overflows in combined sewers). *Gas – Wasser – Abwasser* 57(12): 869–875 [in German].

Muth, W. (1974). Düker (Inverted siphons). *Wasser und Boden* 26(5): 141–147 [in German].

Shrestha, P., De Vries, J.J. (1991). Interactive computer-aided design of inverted syphons. *Journal of Irrigation and Drainage Engineering* 117(2): 233–254.

SIA (1981). Sonderbauwerke der Kanalisationstechnik (Special structures in sewers). *SIA Dokumentation* 40. SIA: Zürich [in German].

Sinniger, R.O., Hager, W.H. (1989). *Constructions hydrauliques* (Hydraulic structures). Presses Polytechniques Romandes: Lausanne [in French].

Chapter 10
Overfalls

Abstract Overfalls allow to accurately determine discharge on the one hand, and may be used to submerge a channel on the other hand. This important control structure has a wide range of application, and many overfall shapes have been suggested until today. The purpose of this chapter is to introduce the main overfall types and to highlight their particular advantages.

For *sharp-crested overflow structures*, the standard rectangular and triangular weirs are described. For discharge measurement, both accuracy and application limits are specified. In addition, general guidelines are presented regarding their installation in sewers or channels. The submerged weir flow is also described.

For larger discharges, the overflow structure has to be sufficiently massive to satisfy static considerations. Particular attention is directed to the *cylindrical weir* and the *broad-crested weir* used for overflow depths of up to about 2 m.

10.1 Introduction

Overfalls (German: Überfall; French: Déversoir) are structures that induce a backwater effect up to the weir crest, and have a significant discharge capacity for larger flow depths. This overfall structure has thus a perfect *hydraulic control characteristics*. In addition, it is known to be a precise *discharge measurement structure*, provided the conditions of observation are well defined.

An overfall structure corresponds to a bottom obstruction. The weir, i.e. the body of the overfall structure can be designed in a wide variety of geometrical shapes (Fig. 10.1). Regarding the *weir section*, one may distinguish between:

- rectangular,
- triangular, or
- circular shapes.

In the *longitudinal section*, a weir may be

- sharp-crested,
- broad-crested, or
- round-crested.

W.H. Hager, *Wastewater Hydraulics*, 2nd ed., DOI 10.1007/978-3-642-11383-3_10,

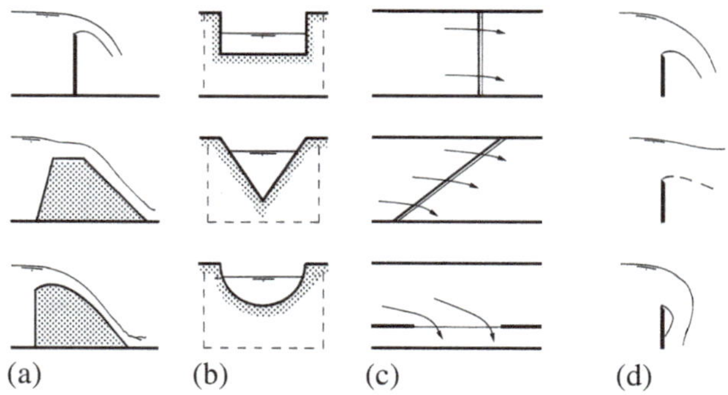

Fig. 10.1 Geometric and hydraulic arrangements of overflow structures (**a**) longitudinal section, (**b**) transverse section, (**c**) plan and (**d**) jet shapes including free and submerged overflow, and adherent nappe

It may also have a so-called 'standard crest' for discharges much larger than are typical in sewer hydraulics, or it can even be polygonal-crested.

The *plan* of a weir can be:

- prismatic,
- oblique,
- convergent, or a
- sideweir.

This compilation of weir geometries demonstrates the diversity of overfall structures. In addition, weir flows can be either free or submerged, and the overfall jet can even cling to the downstream weir face. Around the last turn of century, a large number of overfall structures have been analyzed, and the Frenchman Henri Bazin (1898) has specially contributed to the knowledge of weir flows. His main interests were the possibility of discharge measurement with a suitable overflow structure, and the definition of the so-called standard weir. Rehbock (1929) defined the basic geometrical arrangement of the discharge measurement structure and specified the discharge coefficient. A further development was due to the Americans Kindsvater and Carter (1957), who expanded the concept of *substitute overfall depth* previously introduced by Rehbock. Accordingly, the effects of viscosity and surface tension can be included by an overall approach. Finally, the careful observations of the Englishman White (1977) can be mentioned, who has given a precise definition for the location of surface measurement.

Not all arrangements of overfall structures are relevant in sewer hydraulics. Whereas large capacities are required for dam structures, the discharges in wastewater hydraulics are small or medium, but the flow contains solid matter with a potential to clog. In the following, some typical structures relating to sewer hydraulics are discussed. Here, the presentation is limited to overfalls with a weir crest perpendicular to the approach flow direction, whereas the *oblique weir* (Fig. 10.1c) or *labyrinth weirs*, or even *proportional weirs* are not detailed.

Sideweirs with an oblique approach flow direction are discussed below in Chaps. 17 and 18. In addition, the overflow jet can be either free or submerged, but flows with an adherent jet are excluded (Fig. 10.1d). The end overfall as a particular flow arrangement involving a brink is described in Chap. 11.

The *crest shapes* considered are sharp-crested, broad-crest or round-crested. Due to the simple finish, only the rectangular and the triangular cross-sections are relevant. Weirs have been described extensively in the past, such as by Lakshmana Rao (1975), Bos (1976), Ackers et al. (1978), and Herschy (1985).

10.2 Sharp-Crested Overfalls

10.2.1 *Sharp-Crested Rectangular Weir*

The *rectangular sharp-crested weir* (German: Scharfkantiger Rechteck-Überfall; French: Déversoir rectangulaire en mince paroi) can either be installed in a rectangular prismatic channel as a standard arrangement, or into any channel shape as a non-standard overflow arrangement. Rectangular weirs inserted in a circular sewer are thus non-standard, and the information given below can only be regarded as approximative for this case.

A standard sharp-crested weir in a rectangular channel has a vertical weir plate of small surface roughness. The crest thickness is 2.0 mm, with a sharp upstream edge and a 45° chamfered downstream side. Depending on the weir width b relative to width B of the approach flow channel, one may distinguish between the:

- Contracted-rectangular weir, and
- Full-width rectangular weir.

Figure 10.2 shows the fully contracted rectangular standard weir that is subject to the following conditions (Bos 1976):

- Approach flow width $B - b > 4h$,
- Weir height $h/w < 0.5$,
- Overflow depth $h/b < 0.5$.

To prevent scale effects, both the overfall width b and the weir height w should be at least 0.30 m. The ratio of overfall width b and approach flow width B is $\beta = b/B$.

To prevent an *adherent overflow jet*, the overflow depth should be at least 30 mm, or better 50 mm. Then, the effects of viscosity and surface tension are nearly negligible for water overflow. To extend the application of discharge measurement, the maximum overflow depth can be up to twice the weir height, with $w \geq 0.10$ m. Then, the weir is partially contracted, for which a minimum width $b = 0.15$ m and a minimum tailwater level of 0.05 m below the crest are required. The accuracy of

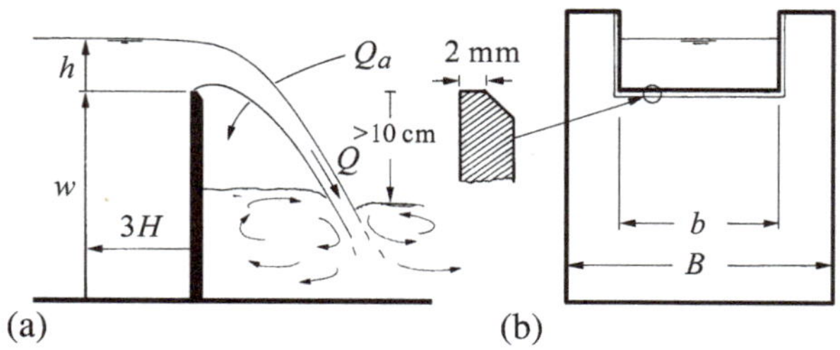

Fig. 10.2 Fully-contracted standard sharp-crested rectangular weir with (**a**) longitudinal section where Q is water discharge and Q_a air discharge, (**b**) cross-section and detail of weir crest geometry

discharge measurement depends significantly on the conditions of weir installation. Often, space limitations, hydraulic conditions or accessibility to the weir decide the *performance* of a weir. For mobile discharge measurement, weirs are unsuitable, as described in Chap. 13.

The *overflow equation* of a rectangular weir is

$$Q = C_d b_e (2gh_e^3)^{1/2} \tag{10.1}$$

with C_d as discharge coefficient, $b_e = b + \Delta b$ the effective (subscript e) overflow width and $h_e = h + \Delta h$ the effective overflow height. According to Kindsvater and Carter (1957) the width correction depends on the width ratio $\beta = b/B$ as

- $\Delta b = +3$ mm for $\beta < 0.8$,
- $\Delta b = -1$ mm for $\beta = 1$.

The overflow height correction is $\Delta h = 1$ mm independent of β.

The *discharge coefficient* C_d varies linearly with the relative overflow depth h/w as

$$C_d = 0.392 + 0.050(h/w + 0.2)\beta^{2.5} \tag{10.2}$$

For a prismatic weir ($\beta = 1$), one has

$$C_d = 0.402 + 0.050(h/w). \tag{10.3}$$

According to Rehbock (1929), the discharge coefficient is

$$C_d = 0.4023 + 0.054(h/w). \tag{10.4}$$

and Ackers et al. (1978) have forwarded probably the most accurate expression as

$$C_d = 0.3988 + 0.060(h/w). \tag{10.5}$$

The *location* of overflow depth reading should be between three and five times (h) upstream from the weir crest. The *accuracy* of a rectangular sharp-crested weir is between 1 and 2% in laboratory, and 5–10% in field conditions, depending on the overall conditions.

Example 10.1 What is the discharge over a prismatic weir if its width is $b =$ 0.50 m and the weir height $w = 0.30$ m for an overflow depth of $h = 0.125$ m? With $\beta = 1$ and $\Delta h = 0.001$ m, the effective overflow depth is $h_e = 0.126$ m, and $b_e = 0.499$ m. With $h/w = 0.125/0.300 = 0.417$ one has $C_d = 0.423$ from Eq. (10.3), $C_d = 0.425$ from Eq. (10.4) and $C_d = 0.424$ from Eq. (10.5). The differences in C_d are thus ±0.25%. With $C_d = 0.424$, the discharge is $Q =$ $0.424 \cdot 0.499(19.62 \cdot 0.126^3)^{1/2} = 0.0419\ \text{m}^3\text{s}^{-1}$.

Example 10.2 How changes discharge in Example 10.1 for an approach flow channel of $B = 1.0$ m width?
With a width ratio $\beta = 0.5/1 = 0.5$, the discharge coefficient is $C_d =$ $0.392 + 0.050(0.417 + 0.2)0.5^{2.5} = 0.397$ from Eq. (10.2) and $Q = 0.397 \cdot$ $0.503(19.62 \cdot 0.126^3)^{1/2} = 0.0396\ \text{m}^3\text{s}^{-1}$. The discharge is thus by 6% reduced, due to the effect of weir constriction.

Figure 10.3 shows lateral views of standard full-width weir flow in a rectangular channel. Note that the *lower nappe* rises beyond the weir crest to a maximum. The stability of the nappe is significantly increased for fully-aerated flow because nappe pulsations may be generated otherwise.

For small overflow depths below 50 mm, two *alternative flow patterns* may occur. Figure 10.4a shows the pattern with a free jet and reveals capillary waves on the upper nappe due to surface tension. Figure 10.4b refers to the adherent nappe, with

(a)  (b) 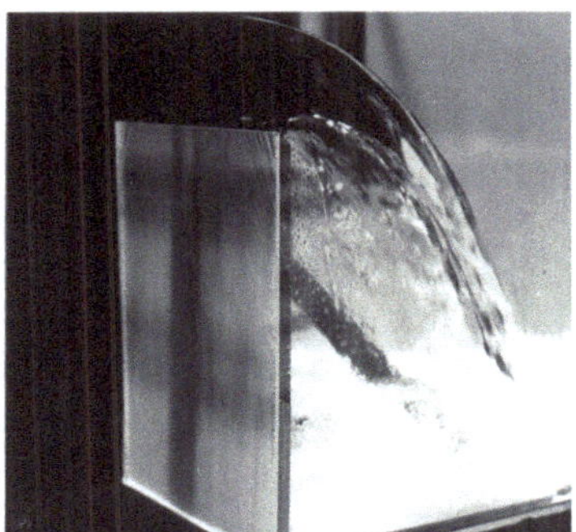

Fig. 10.3 Standard sharp-crested weir in a full width rectangular channel. (**a**) Detail of crest flow, (**b**) overflow jet

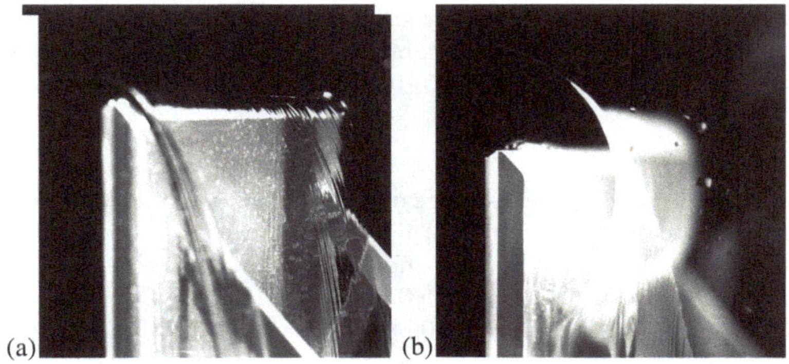

Fig. 10.4 Small overflow-depth weir flow (**a**) capillary waves, (**b**) adherent nappe flow

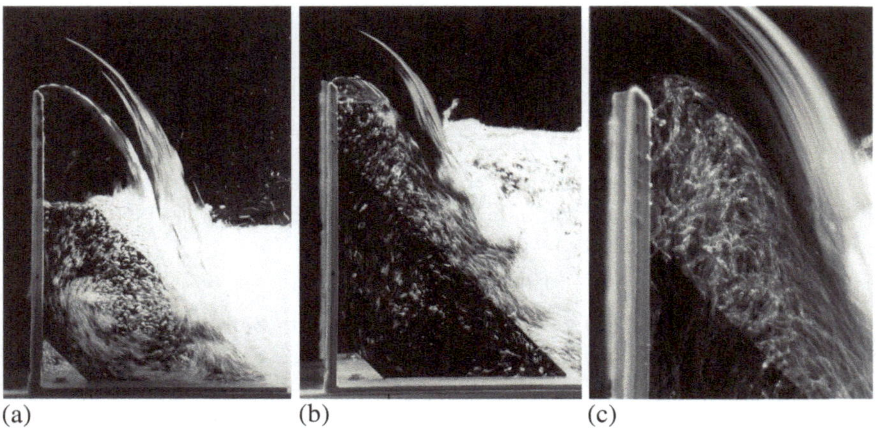

Fig. 10.5 Effect of submergence on overflow pattern. (**a**) Small, (**b**) large and (**c**) definite submergence of weir flow

a vorticity zone downstream of the weir crest. This flow does definitely not follow the discharge equation (10.1) because of significant streamline curvature effects.

Figure 10.5 relates to various *submergence* degrees. In Fig. 10.5a the downstream elevation is much below the weir crest, and no submergence effect has to be expected. In Fig. 10.5b the height of the air cavity below the lower nappe is smaller than the overflow depth, such that a definite effect of submergence follows. Note also the significant elevation difference between the downstream levels on either nappe. In Fig. 10.5c the lower nappe is not aerated at all and a sub-pressure effect increases the discharge as compared to fully-aerated weir flow. Such conditions must be excluded for discharge measurement.

Figure 10.6 shows the conventional arrangement for discharge measurement in a rectangular laboratory channel. The channel is made of smooth boundary material, such as glass or PVC, is rectangular, straight and normally horizontal. Upstream, a flow straightener improves the approach flow and surface waves are reduced with a floater. The overflow depth is measured with a point gage of reading accuracy

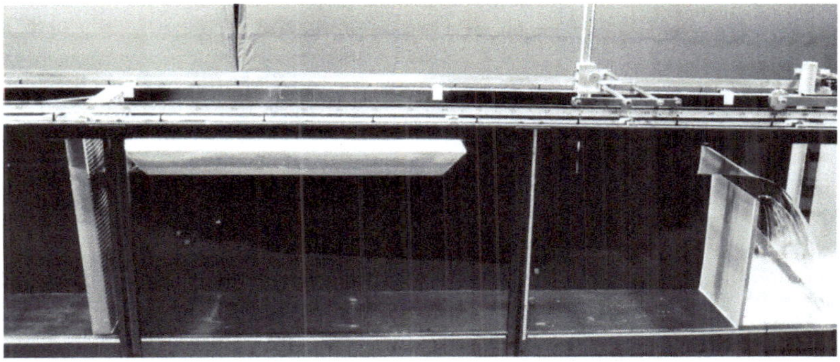

Fig. 10.6 Typical arrangement of discharge measurement in a laboratory channel

of at least 0.2 mm. The weir plate is positioned vertically, is either of metal sheet or PVC, and has the standard-crest shape previously described. The overflow jet is fully aerated with an air supply pipe.

10.2.2 *Sharp-Crested Triangular Weir*

For small overflow depth of $h < 0.10$ m, the sensitivity of the rectangular weir is relatively small. The triangular weir (German: Dreiecküberfall; French: Déversoir triangulaire) then yields precise results. It is made of metal sheet with a symmetrical notch of angle α that is vertically inserted in a rectangular channel of width B (Fig. 10.7). The tailwater should be at least 50 mm below the crest for a fully-aerated nappe. Further, the weir height w should be at least as high as the overflow depth h. The *standard triangular weir* has a sharp crest of crest thickness $e = 1$ mm and is chamfered by 60° (Fig. 10.7b).

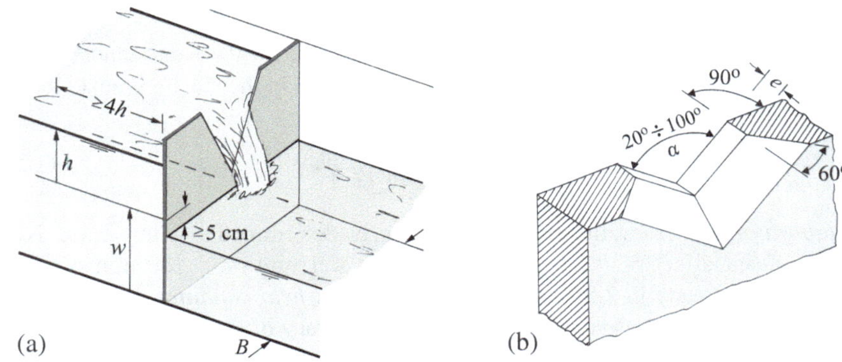

Fig. 10.7 Triangular standard weir. (**a**) Overall view, (**b**) crest detail

Table 10.1 Characteristics of partially-contracted (PCW) and fully-contracted (FCW) triangular weirs (Bos 1976)

Parameter	PCW	FCW
h/w	<1.2	$\leq$0.4
h/B	<0.4	$\leq$0.2
h>50 mm	<600 mm	<380 mm
w	>100 mm	$\geq$450 mm
B	$\geq$600 mm	$\geq$900 mm

The triangular weir was introduced by the Englishman James Thomson in 1859, and reference is also made to the *Thomson weir*. Bos (1976) made a distinction between the partially- and the fully-contracted weir flow (Table 10.1). For precise discharge measurement, the fully-contracted weir should be used.

The *discharge equation* of the triangular weir for free flow is

$$Q = \frac{8}{15} C_d \tan(\alpha/2)(2gH^5)^{1/2}. \tag{10.6}$$

Here, C_d is the discharge coefficient, α the notch angle, and $H = h + Q^2/[2gB^2 (h + w)^2]$ the approach flow energy head relative to the weir crest.

Based on a literature review including a large quantity of available data, Hager (1990) determined for water of usual temperature and notch angles $14° \leq \alpha \leq 100°$

$$C_d = \frac{1}{\sqrt{3}}\left[1 + \left(\frac{h^2 \tan(\alpha/2)}{3B(h+w)}\right)^2\right]\left[1 + \frac{0.66}{h^{3/2}\tan(\alpha/2)}\right]. \tag{10.7}$$

Here, $3^{-1/2}$ is a basic overflow value, the first term in square brackets accounts for the velocity of approach, and the second term includes the effects of viscosity and surface tension for small overflow depths h of water. The flow depth in the dimensional term of the second bracket is in [cm]. The discharge is expressed as

$$Q = \frac{8}{15} C_d \tan(\alpha/2)(2gh^5)^{1/2}, \tag{10.8}$$

involving h instead of H in Eq. (10.6).

Provided the approach flow velocity is small, and if $\alpha = 90°$ as the standard notch angle, then Eqs. (10.7) and (10.8) simplify with h [cm] in the square bracket to

$$Q = 0.308\left[1 + \frac{0.66}{h^{3/2}}\right](2gh^5)^{1/2}. \tag{10.9}$$

For *laboratory accuracy*, the discharge can thus be determined to within $\pm 1\%$. Note the significant effect of h on Q. A small error of 1 mm on a 100 mm overflow depth results already in 2.5% error in discharge. For *field conditions*, the accuracy is typically $\pm 5\%$, and usual conditions give an accuracy of less than $\pm 10\%$.

For *submerged weir overflow*, the ratio $S = h_u/h_o$ is significant (Fig. 10.8). According to Hager (1990) the ratio of discharge for submerged flow Q_s and free

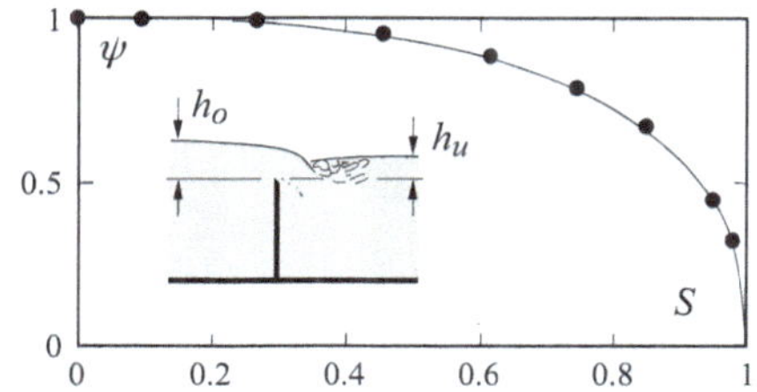

Fig. 10.8 Submerged triangular weir: Definition sketch and Eq. (10.10) compared with own observations for the submergence factor $\psi = Q_s/Q$ versus relative weir submergence $S = h_u/h_o$

discharge Q from Eq. (10.9) follows essentially American measurements in the 1940s as

$$Q_s/Q = (1 - S^{2.5})^{0.385}. \tag{10.10}$$

Precise discharge measurement should be based on *free overflow* because the definition of the downstream flow depth h_u involves a considerable degree of uncertainty.

Figure 10.9 shows side views of free and submerged flows over a triangular weir. As the submergence S increases, the elevation between the two levels decreases and the air entrainment rate is reduced. Note how the air bubbles may be used as tracers

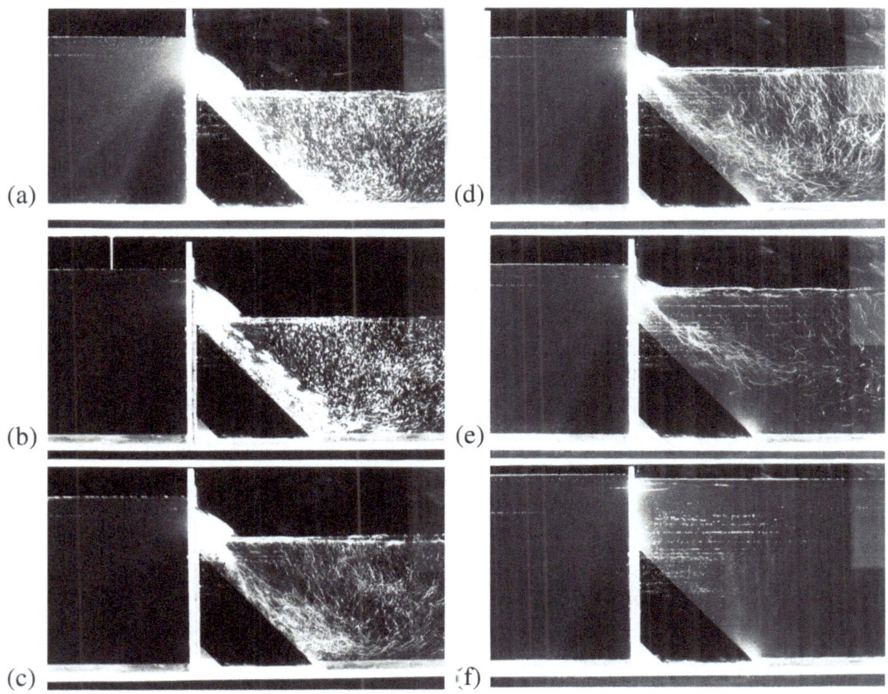

Fig. 10.9 Side views of triangular weir flow for various submergences $S =$ (**a**) 0, (**b**) 0.10, (**c**) 0.27, (**d**) 0.45, (**e**) 0.61 and (**f**) 0.95

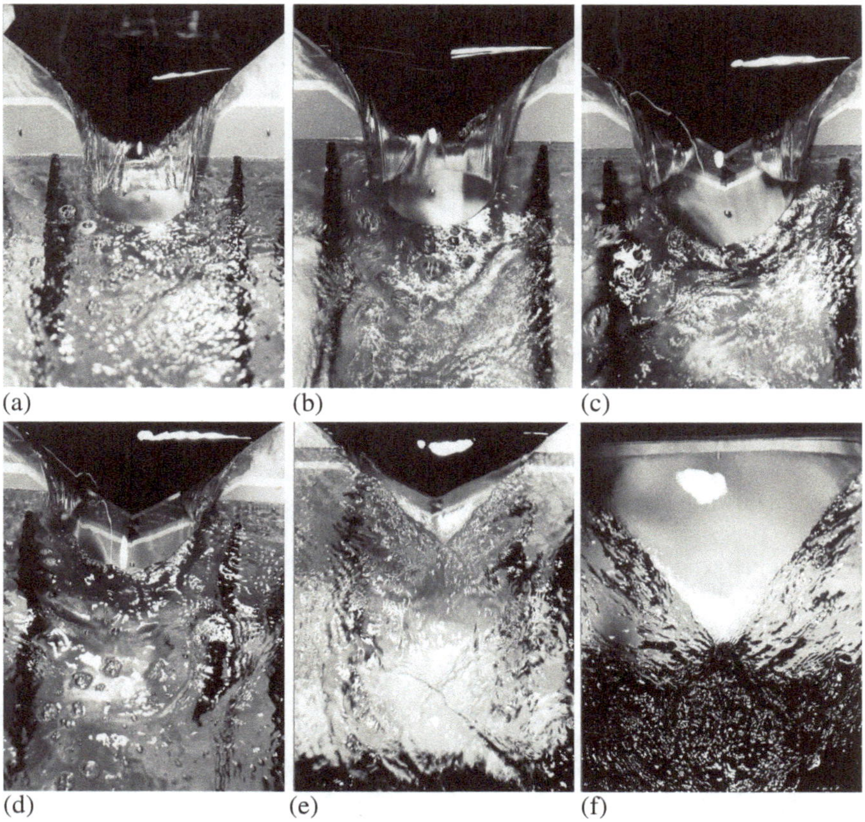

Fig. 10.10 Plans of overfall flows for S = (**a**) 0, (**b**) 0.10, (**c**) 0.27, (**d**) 0.45, (**e**) 0.61, (**f**) 0.75. Flow direction from top to bottom

of the overflow jet. For Fig. 10.9a–d a plunging jet is discharged into the tailwater, which transforms into a surface jet with typical surface waves in Fig. 10.9e. In Fig. 10.9f the flow is deeply submerged with an almost static flow pattern.

Figure 10.10 shows downstream views for various weir submergences. Note the transparency of the overflow jet for $S = 0.27$ as compared with $S = 0$. Then, the capillary waves on the overflow jet are practically suppressed. The transition from the plunging to the surface jet is seen to occur at about $S = 0.61$, with a sharp triangular front from the V-notch weir to the tailwater.

10.3 Broad-Crested Weir

Figure 10.11a shows a definition sketch of the broad-crested weir (German: Breitkroniger Überfall; French: Déversoir à seuil épais), with w as the weir height, h as the approach flow overflow depth, $H = h + Q^2/[2gb^2(h + w)^2]$ as the

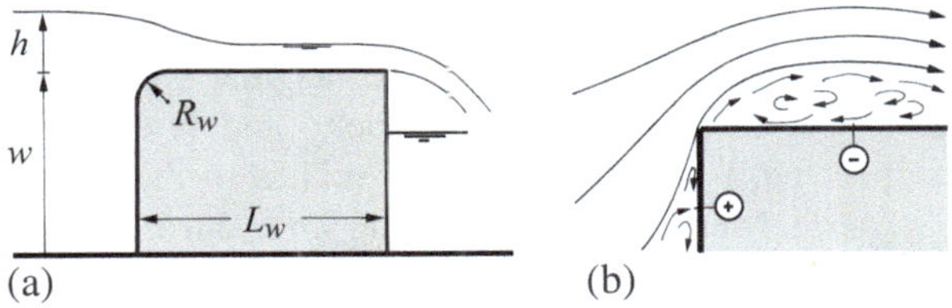

Fig. 10.11 Broad-crested weir (**a**) definition sketch, (**b**) detail of flow separation from upstream crest for $R_w = 0$

corresponding energy head, L_w as the length of weir crest and R_w as the upstream crest radius. For the *standard broad-crested weir* both the up- and downstream weir faces are vertical, the crest surface is horizontal and both crests are sharp. Figure 10.11b shows the *separation pattern* of the flow from a sharp upstream weir crest, with a depression beyond the upstream crest and a overpressure zone along the upstream face.

For broad-crested weirs with sharp crests ($R_w = 0$) the *free surface profile* depends on the relative weir length $\zeta_w = H/L_w$. Figure 10.12 relates and defines various crest length ratios, including the long weir, the broad-crested weir, the short weir and the practically sharp-crested weir, as discussed in Sect. 10.2.

Discharge Q is related to the approach flow energy head H and the corresponding discharge coefficient C_D to suppress the effect of approach flow velocity as

$$Q = C_D b (2gH^3)^{1/2}. \tag{10.11}$$

For overflow depths larges than about 50 mm, the effects of viscosity and surface tension are small, and may be dropped. Then, the *discharge coefficient* of the standard geometry broad-crested weir is (Hager 1994)

$$C_D = 0.326(1 + C_w)\left[\frac{1 + (9/7)\zeta_w^4}{1 + \zeta_w^4}\right] C_B C_R. \tag{10.12}$$

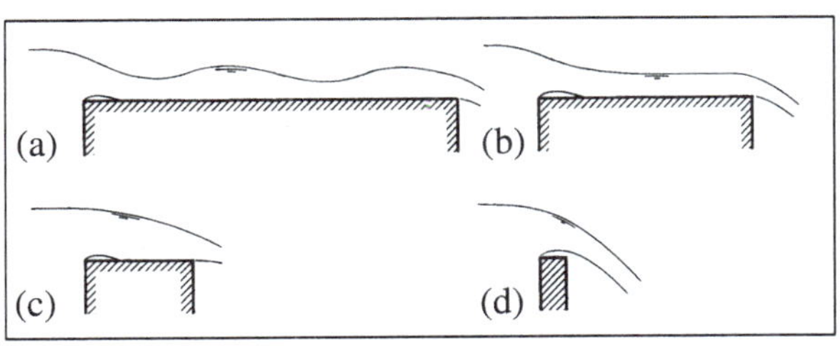

Fig. 10.12 Flow patterns for broad-crested weir with sharp corners and vertical faces, as a function of relative weir length $\zeta_w = H/L_w$. (**a**) $0 < \zeta_w < 0.1$, (**b**) $0.1 \le \zeta_w < 0.35$, (**c**) $0.35 \le \zeta_w < 1.5$ and (**d**) $1.5 \le \zeta_w$

The basic discharge coefficient 0.326 is considerably smaller than the corresponding value 0.42 (+29%) for the sharp-crested rectangular weir. The parameter C_w accounts for the effect of approach flow velocity and is equal to $C_w = 0$ for $H/w < 1/2$. The next term in square brackets relates to the relative weir length $\zeta_w = H/L_w$ and varies between 1 for the very long weir, and $9/7 = 1.29$ for $\zeta_w > 1.5$. The next coefficient C_B refers to the plan contraction, from a wide approach flow channel to the weir channel. For the usual prismatic channel $C_B = 1$, whereas for constrictions $\beta = b/B$ between 0.25 and 0.75, one has (Hager 1994)

$$C_B = 1 - \frac{0.133}{1+\zeta_w^4}. \tag{10.13}$$

The last coefficient C_R refers to the upstream *crest rounding* that depends on the relative crest radius $\rho_w = 3R_w/w$ as

$$C_R = 1 + 0.1[\rho_w \exp(1-\rho_w)]^{1/2}. \tag{10.14}$$

For $\rho_w \cong 0.3$ results thus an increase of discharge of about 10%, associated with a considerable pressure reduction close to the upstream crest (Fig. 10.11b).

Usually, one has $C_w = 0$ and $C_B = C_R = 1$, and the *discharge equation* for the standard broad-crested weir is

$$Q = 0.326\left[\frac{1+(9/7)\zeta_w^4}{1+\zeta_w^4}\right] b(2gH^3)^{1/2}. \tag{10.15}$$

Under the conditions:

- Minimum flow depth $h = 50$–75 mm,
- Minimum weir width $b = 0.30$ m,
- Minimum weir height $w = 0.15$ m,
- Usual weir length $0.08 \le h/L_w \le 0.85$, and
- Small approach flow velocity $0.18 \le h/(h+w) \le 0.60$

the standard broad-crested weir has an accuracy of discharge measurement under laboratory conditions of about $\pm$(2 to 3)%, thus inferior to the sharp-crested weir.

Broad-crested weirs are insensitive to tailwater submergence. The submergence factor $\psi = Q_s/Q$ seems to depend mainly on the relative overflow height $\zeta_w = H/L_w$. With $\sigma_L = H_L/H$ as the modular limit (subscript L), for which the discharge follows Eq. (10.15) for free overflow, one may state (Hager 1994)

$$\sigma_L = 0.85 \tanh(L_w/H). \tag{10.16}$$

Consequently, a sharp-crested weir with $L_w/H = 0$ has a value $\sigma_L = 0$, and it may not be submerged at all. For $L_w/H = 1$ (short-crested weir) the modular limit is $\sigma_L = 0.65$ from Eq. (10.16), and a long weir with $L_w/H = 10$ has $\sigma_L = 0.05$. Broad-crested weirs are thus highly insensitive to submergence. Equation (10.15) may be

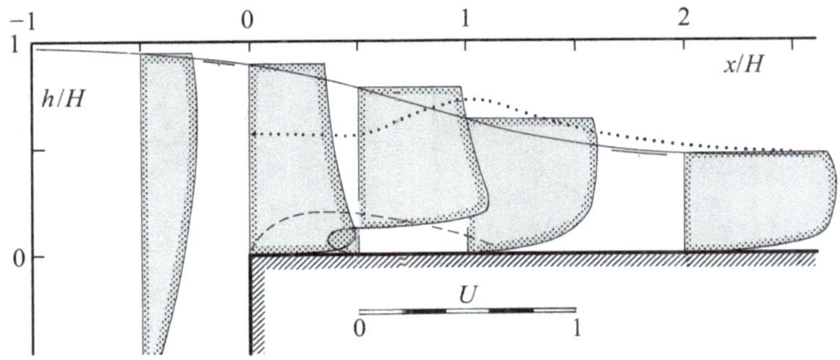

Fig. 10.13 Main flow characteristics of standard broad-crested weir ($0.10 \le H/L_w \le 0.40$) including (—) free surface profile, (- - -) separation profile, (· · ·) bottom pressure profile, and velocity distributions $U(Z)$

applied for $\sigma < \sigma_L$, but a discharge measurement under submerged flow conditions is not recommended.

The main *flow characteristics* of the standard broad-crested weir ($0.10 \le H/L_w \le 0.40$) are shown in Fig. 10.13 (Hager and Schwalt 1992):

- *Free surface profile* h/H as a function of the streamwise coordinate $X = x/H$,
- *Bottom separation zone* downstream from the crest, with a maximum height $0.20H$ at location $0.44H$,
- *Bottom pressure profile*, starting at $0.565H$, with a maximum of $0.73H$ at $x = 1.05H$,
- *Velocity profiles* $U(Z)$ with $U = u/(2gH)^{1/2}$ as the horizontal velocity component and $Z = z/H$ as the vertical coordinate, at locations $X = -0.5$, 0, 0.5, 1 and 2.

Figure 10.14 relates to typical photographs of flow over broad-crested weirs. For $\zeta_w = 0.06$ the surface is undular with almost 2D free surface waves, and for $0.1 < \zeta_w < 0.4$ the central flow portion has a nearly constant flow depth of $0.46H$. Often, it is assumed that the flow is critical in the usual sense, but this is definitely not true, because of *flow separation* from the upstream crest, and a flow over a separated crest zone.

Figure 10.15 refers to *submerged flow* over a broad-crested weir. With $y_u = h_u/H$ as the submergence ratio and h_u the downstream elevation above the weir crest, one may note six distinctly different flow types:

- *Plunging jet* for $y_u < 0.45$.
- *Straight surface flow* for $y_u \cong 0.52$, with a nearly horizontal free surface.
- *Surface jet* for $y_u > 0.60$ with supercritical flow over the weir surface, and a hydraulic jump close to its downstream crest.
- *Surface wave* with a hydraulic jump on the weir surface. For $y_u \cong 0.78$, the jump front is close to the end of bottom separation, and the overflow gets submerged only now.

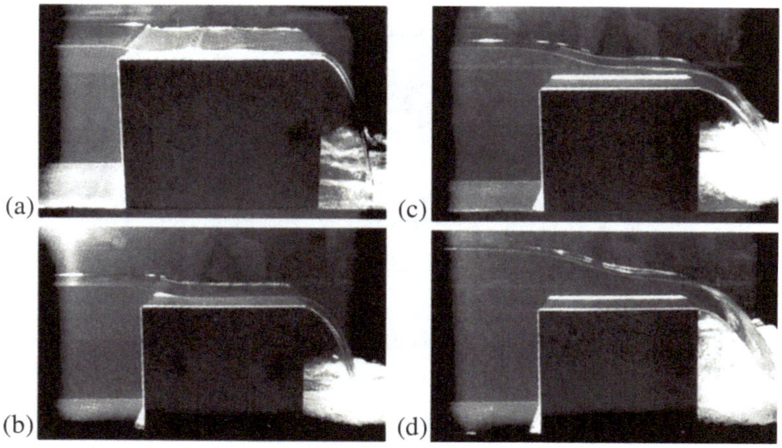

Fig. 10.14 Side views of standard broad-crested weir flow for $\zeta_w =$ (**a**) 0.06, (**b**) 0.13, (**c**) 0.27, (**d**) 0.39

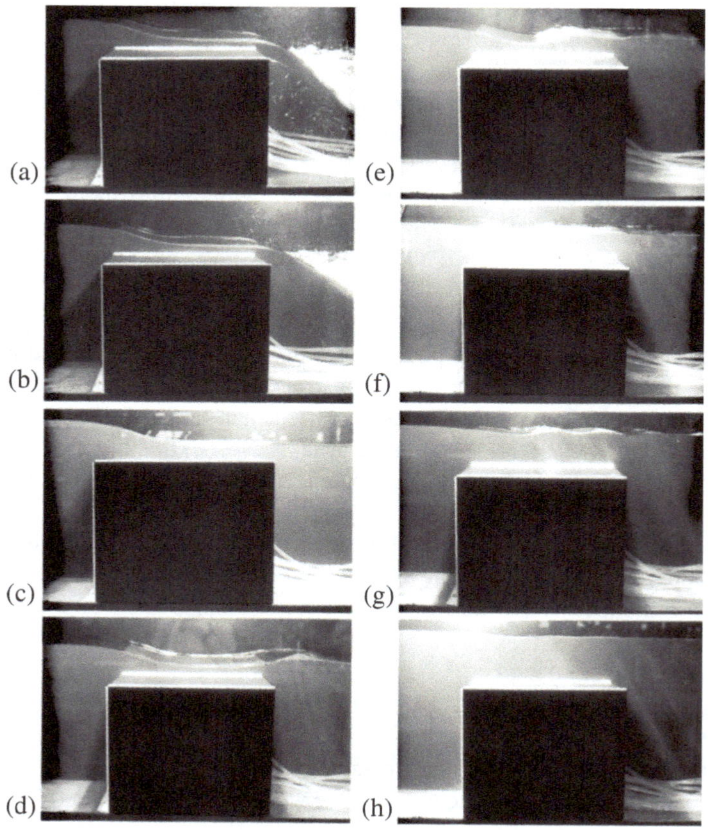

Fig. 10.15 Submerged flow over broad-crested weir for $\zeta_w =$ (**a**) 0.03, (**b**) 0.20, (**c**) 0.52, (**d**) 0.68, (**e**) 0.78, (**f**) 0.85, (**g**) 0.90, (**h**) 0.95. Note that the weir is submerged only for $\zeta_w > 0.78$

- *Undular surface flow* for $y_u \cong 0.85$, and
- *Surface jet* for $y_u \cong 0.95$, with an almost horizontal free surface.

Further information on broad-crested weir flow is available from Hager and Schwalt (1994). Note particularly the undular wave pattern for small relative overflow ratio. The flow over *embankments* with a symmetrical side slope of 1 (vertical): 2 (horizontal) has been investigated by Fritz and Hager (1998). Their results refer to the flow pattern downstream of the crest for both free and submerged flows. The discharge equation (10.15) was generalized, and the modular limit relation detailed in terms of the governing flow conditions. Also, the two-dimensional velocity distributions for the four flow types were generally analyzed, based on laboratory observations. It was found that the velocity distribution for the plunging jet formation may be considered inverse to that of the surface jet regime. The modular limit was again found to be high as compared with other weir type structures.

10.4 Cylindrical Weir

A cylindrical weir (German: Zylinderüberfall; French: Déversoir cylindrique) is an overflow structure involving vertical up- and downstream walls and a crest of half-cylindrical shape (Fig. 10.16a). The crest can simply be finished with a halved commercial pipe made of steel or concrete. To aerate the nappe, the downstream wall has an offset, as shown in Fig. 10.16b.

The *discharge Q* over a cylindrical weir is influenced by the weir height w, the overflow approach flow depth h, and the crest radius R_K. With $H = h + Q^2/[2gF_o^2]$ as the approach flow energy head with F_o as the approach cross-sectional area, one may write

$$Q = C_d b(2gH^3)^{1/2}. \tag{10.17}$$

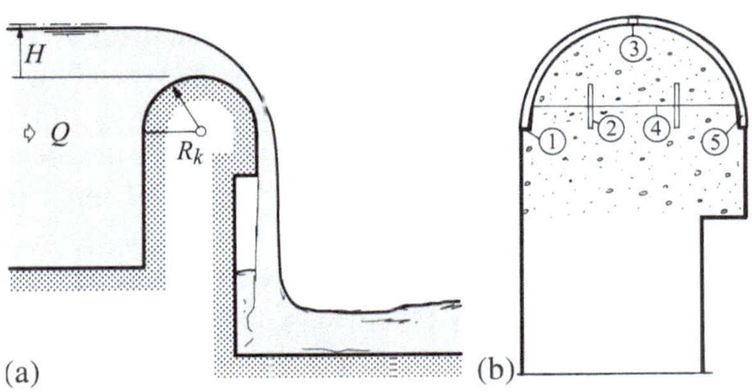

Fig. 10.16 Cylindrical weir (**a**) notation, (**b**) detail of possible crest finish with ① support and sealing, ② bolt, ③ deaeration hole during concrete placement, ④ constructional joint, ⑤ half pipe

Based on a literature review, Hager (1992) studied the *discharge coefficient* C_d as a function of the relative crest curvature $\rho_K = H/R_K$. Rouvé and Indlekofer (1974) distinguished between adherent and free nappe flows. To ensure a definite flow pattern, the latter flow condition should result, whereas a relative overflow depth $H/R_K \leq 1.5$ guarantees crest pressures above the atmospheric pressure. Then, effects of cavitation are avoided.

For overflow depths $H \geq 70$ mm, effects of scale are inhibited, and the discharge coefficient C_d is (Hager 1992)

$$C_d = 0.374\left(1 + \frac{3\rho_K}{11 + 2.5\rho_K}\right). \tag{10.18}$$

For shallow water flows ($\rho_K \to 0$), the minimum discharge coefficient is $C_d = 0.374$, i.e. between the broad-crested and the sharp-crested weir types. For maximum overflow conditions ($\rho_K = 1.5$) the discharge coefficient can be increased by +30% to $C_d = 0.49$. This effect is due to streamline curvature over the weir crest.

Currently, no indication regarding the *limit submergence* of cylindrical weirs is available. Because the weir crest is relatively short, the submergence effect on discharge may start as the downstream level reaches the weir crest. Thus, Eq. (10.18) is valid only for downstream elevations lower than the weir crest elevation.

10.5 Comparison of Weirs

The weir types reviewed here, and the standard weirs as used for large discharges in hydraulic structures are often applied in practice. For discharge measurement both in the laboratory and in the field, weirs with a *sharp crest* are normally used. Up to about 40 Ls^{-1}, the triangular sharp-crested weir is recommended, whereas sharp-crested fully-aerated weirs are employed for discharges larger than about 20 Ls^{-1}. The accuracies of these two standard weirs are about $\pm 1\%$ under laboratory conditions, and between 2 and 5% for field structures, depending on the site conditions. Normally, such precision is accurate enough for the purposes of sewer hydraulics. Due to static limitations and the limits of application, the maximum overflow depth for the rectangular sharp-crested weir is about 1 m. For an overflow depth of 0.60 m, one has a discharge of 800 Ls^{-1} per meter, indicating the typical limit of discharge.

Broad-crested and round-crested weirs are applied from about 0.50 m overflow depth. An upper limit of overflow depth is roughly 1.5 m for the broad-crested weir, corresponding to 3 m^3s^{-1} discharge per meter crest length. If commercially finished pipes of 2 m diameter are considered as the upper limit of cylindrical weirs, the maximum overflow depth can be 3 m, with a discharge capacity of about 11 m^3s^{-1} per meter of crest length. This can be regarded as a lower limit of head on standard-shaped weirs, such as according to Creager or the Waterways Experiment Station (WES). The *type of weir* shape depends thus significantly on the discharge, and the weirs discussed previously can be used for typical applications in wastewater hydraulics.

For discharge measurement, it is recommended that *free flow conditions* be imposed. Then, the discharge can be measured with a single depth reading typically 3–5 times the overflow depth upstream from the weir crest. Note that the modular limit depends only on the crest geometry. For sharp-crested weirs the downstream surface elevation has to be below the crest elevation, whereas the modular limit can be as high as 70–80% for broad-crested weirs.

For structures where the crest elevation may be subject to later adjustment, a sharp-crested weir is recommended that is mounted on a concrete wall. The crest elevation should be secured against uncontrolled changes, because such an adjustment has to be based on a *detailed hydraulic design* involving both the up- and downstream reaches of the structure (ATV 1978).

Notation

b	[m]	weir width
B	[m]	approach width
C_B	[–]	constriction coefficient
C_d	[–]	discharge coefficient
C_D	[–]	discharge coefficient based on H
C_R	[–]	rounding coefficient
C_w	[–]	effect of approach flow velocity
e	[m]	crest thickness
F	[m^2]	cross-sectional area
g	[ms^{-2}]	gravitational acceleration
h	[m]	overflow depth
H	[m]	overflow energy head
L_w	[m]	weir length
Q	[m^{3}s^{-1}]	discharge
Q_a	[m^{3}s^{-1}]	air discharge
R_K	[m]	crest radius
R_w	[m]	rounding radius
S	[–]	submergence ratio
u	[ms^{-1}]	streamwise velocity component
U	[–]	$= u/(2gH)^{1/2}$ relative velocity
w	[m]	weir height
x	[m]	streamwise coordinate
X	[–]	$= x/H$
y_u	[–]	$= h_u/H$ downstream depth ratio
z	[m]	vertical coordinate
Z	[–]	normalized vertical coordinate
α	[–]	notch angle
β	[–]	width ratio
ρ_K	[–]	rounding radius of cylindrical weir

ρ_w [–] rounding radius of weir crest
σ_L [–] modular limit
ψ [–] submergence factor
ζ_w [–] $= H/L_w$ relative weir length

Subscripts

e effective
L limit
o upstream
s submerged
u downstream

References

Abwassertechnische Vereinigung (1978). Bauwerke der Ortsentwässerung (Structures of urban drainage). *Arbeitsblatt* A241. ATV: St. Augustin [in German].

Ackers, P., White, W.R., Perkins, J.A., Harrison, A.J.M. (1978). *Weirs and flumes for flow measurement*. J. Wiley & Sons: Chichester, New York and Toronto.

Bazin, H. (1898). *Expériences nouvelles sur l'écoulement en déversoir* (New experiments on weir flow). Dunod: Paris [in French].

Bos, M.G. (1976). Discharge measurement structures. *Rapport* 4. Laboratorium voor Hydraulica en Afvoerhydrologie. Landbouwhogeschool: Wageningen NL.

Fritz, H.O., Hager, W.H. (1998). Hydraulics of embankments weirs. *Journal of Hydraulic Engineering* 124(9): 963–971.

Hager, W.H. (1990). Scharfkantiger Dreiecküberfall (Sharp-crested V-notch weir). *Wasser, Energie, Luft* 82(1/2): 9–14 [in German].

Hager, W.H. (1992). Abfluss über Zylinderwehr (Flow over cylindrical weir). *Wasser und Boden* 44(1): 9–14 [in German].

Hager, W.H. (1994). Breitkroniger Überfall (Broad-crested weir). *Wasser, Energie, Luft* 86(11/12): 363–369 [in German].

Hager, W.H., Schwalt, M. (1994). Broad-crested weir. *Journal of Irrigation and Drainage Engineering* 120(1): 13–26.

Herschy, R.W. (1985). *Streamflow measurement*. Elsevier: Amsterdam, London, New York.

Kindsvater, C.E., Carter, R.W. (1957). Discharge characteristics of rectangular thin-plate weirs. *Journal of the Hydraulics Division* ASCE 83(HY6) Paper 1453: 1–36; 1958, 84(HY2) Paper 1616: 93–100; 1958, 84(HY3) Paper 1690: 21–30; 1958, 84(HY6) Paper 1856: 39–41; 1959, 85(HY3): 45–49.

Lakshmana Rao, N.S. (1975). Theory of weirs. *Advances in Hydroscience* 10: 309–406. Academic Press: New York.

Rehbock, T. (1929). Wassermessung mit scharfkantigen Überfallwehren (Discharge measurement with sharp-crested weirs). *Zeitschrift VdI* 73(24): 817–823 [in German].

Rouvé, G., Indlekofer, H. (1974). Abfluss über geradlinige Wehre mit halbkreisförmigem Überfallprofil (Flow over straight-lined weirs with a half-circular shaped overflow crest). *Bauingenieur* 49(7): 250–256 [in German].

White, W.R. (1977). Thin plate weirs. *Proc. Institution Civil Engineers* 63: 255–269.

Overfalls are significant structures in wastewater hydraulics, both for discharge measurement and hydraulic control between up- and downstream reaches

Chapter 11
End Overfall

Abstract End overfalls can often be employed as a substitute structure for discharge measurement at locations previously not having been considered. In this chapter the current results are reviewed for the rectangular- and the circular-shaped end overfalls. Answers are given to questions such as: What is the discharge for a certain end depth, and Is the end depth a sufficient parameter for discharge determination?

In addition, nappe geometries originating from ducted outflow into the atmosphere are presented. Further, nappe differences between laterally guided and discharging freely into the atmosphere are given. Effects of roughness are discussed, and the transitions from pressurized to free surface flows in a pipe are reviewed.

11.1 Introduction

Despite the discharge in a sewer should often be known, the additional cost for an accurate discharge measurement structure as discussed in Chap. 10 is too large. The procedure of discharge measurement is complicated in *circular sewers* because of access limitations and working conditions with sewage. In such cases, an end overfall (German: Endüberfall; French: Déversoir terminal) or even another structure such as presented in Chap. 13 may be an economical compromise.

An end overfall is based on the fluid efflux at a pipe outlet into the atmosphere. The flow over an end overfall is normally free and it can be found in existing sewer systems more often than assumed. Typically, these weir type structures may be located at junction manholes, drop manholes, intake or outlet structures or at side channels. Normally these structures have not been designed for discharge measurement but they can be modified with a small effort.

In the following, two configurations are presented, namely the end overfall in a:

- Rectangular channel, and
- Circular pipe.

End overfalls in *rectangular channels* have received attention. The relation between the end depth ratio and the approach flow Froude number is predicted with the

W.H. Hager, *Wastewater Hydraulics*, 2nd ed., DOI 10.1007/978-3-642-11383-3_11,

momentum equation. The effects of streamline curvature cannot be neglected. Also, the effects of bottom slope and boundary roughness have been investigated.

End overfalls in *circular pipes* as typically encountered in sewer hydraulics received limited attention. The literature available was reviewed by Hager (1993a, b). The following is a summary of the relevant knowledge for sewer hydraulics. The end overfall in trapezoidal channels is irrelevant in the present context, however.

11.2 Rectangular Channel

11.2.1 *Flow Description*

Figure 11.1 shows the end overfall with the main parameters. These are the uniform flow depth h_o, the bottom slope S_o, and the end depth h_e (subscript *e*). The American Hunter Rouse (1906–1996), a former professor of hydraulics at the University of Iowa, and one of the most prominent hydraulicians of the twentieth century, can be considered the father of the end overfall. By application of the *momentum equation* and neglect of friction and bottom slopes he determined the ratio between the vertical jet thickness t_∞ and the uniform depth h_o for subcritical approach flow as $t_{e\infty}/h_o = 2/3$. At the end section the end depth ratio was measured to $h_e/h_o = 0.715$ (Rouse 1936). The deviation between the two numbers is related to the residual jet pressure (Fig. 11.1).

Later, Rouse (1943) investigated the effect of the approach flow Froude number $\mathsf{F}_o = Q/(gb^2h_o^3)^{1/2}$ and found for the asymptotic jet thickness

$$t_\infty/h_o = \frac{2\mathsf{F}_o^2}{1+2\mathsf{F}_o^2}. \tag{11.1}$$

For subcritical approach flow with a transition to supercritical flow shortly upstream from the end section, the approach flow Froude number is $\mathsf{F}_o = 1$, and thus $t_\infty/h_o = 2/3$ as previously discussed. For the other extreme $\mathsf{F}_o \to \infty$, the result is $t_\infty/h_o = 1$, i.e. the jet thickness remains equal to the approach flow depth. Rouse (1943) also determined experimentally the lower nappe geometry $Z_u(X)$ with

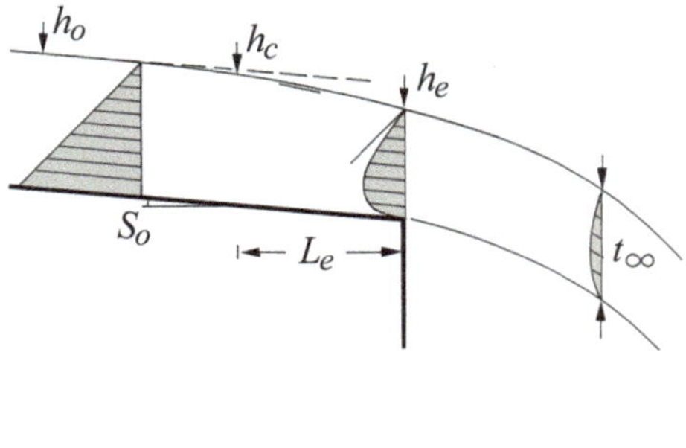

Fig. 11.1 End overfall, notation and typical pressure distributions for subcritical approach flow

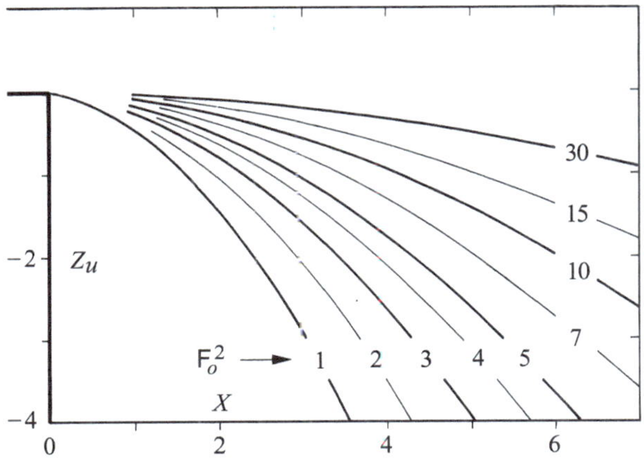

Fig. 11.2 Lower nappe geometry $Z_u(X)$ versus the approach flow Froude numbers F_o according to Rouse (1943)

$Z_u = z_u/h_o$ and $X = x/h_o$. Figure 11.2 shows a generalized diagram. A discussion of these results follows.

Delleur et al. (1956) investigated the effects of bottom slope S_o and wall roughness on the end depth ratio $T_e = h_e/h_o$. The wall roughness is characterized with the critical slope S_c as introduced in Chap. 6. For $S_o/S_c < -5$ the experiments indicated a nearly constant ratio $T_e = 0.75$ (Fig. 11.3b), and the ratio decreases to $T_e = 0.715$ for $S_o/S_c \rightarrow 0$. For large bottom slope such as $S_o/S_c = 10$, the value of $T_e = 0.45$ is approached.

Figure 11.3(a) shows the distance L_e from the end section to the section with a nearly hydrostatic pressure distribution (Fig. 11.1), where $h_c = [Q^2/(gb^2)]^{1/3}$. Note that L_e/h_c varies with the critical slope, and the location of critical depth varies with discharge. The location of observation for a given pipe thus varies with discharge.

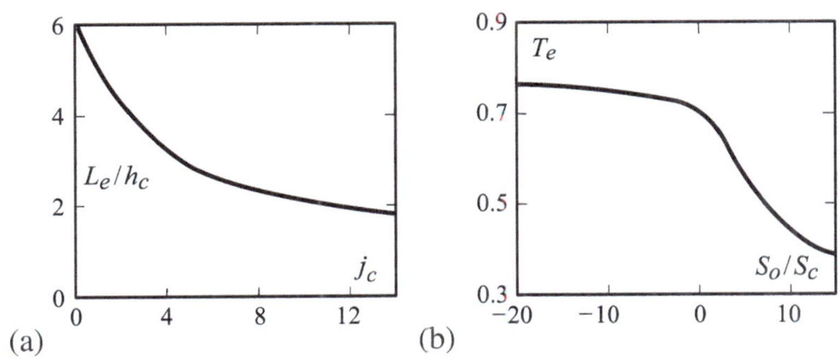

Fig. 11.3 (**a**) Distance L_e/h_c of critical depth section from end section as a function of $j_c = (S_o - S_c) \times 10^3$ according to Carstens and Carter (1955), (**b**) end depth ratio T_e as a function of slope ratio S_o/S_c according to Delleur et al. (1956)

Rajaratnam and Muralidhar (1964a) considered a *laterally unconfined* free overfall. Contrary to end overfalls with a lateral guidance, the flow then transversally expands or contracts, depending mainly on the effect of surface tension. With the *Weber number* $\mathsf{W} = V_c/[\sigma/(\rho h_c)]^{1/2}$ a jet issued from a rectangular channel contracts if $\mathsf{W} < 16$, and expands laterally otherwise. With $\mathsf{D} = [q^2/(gz^3)]^{1/3} = h_c/z$ as the *drop number* where z is the vertical coordinate measured positively downwards from the end section, the upper nappe geometry along the jet axis is for $\mathsf{F}_o = 1$

$$\mathsf{D}_o = h_c/z_o = 204(x/h_c)^{-6.56}, \quad 1.8 < x/h_c < 10. \tag{11.2}$$

The end depth ratio was $T_e = 0.705$, as compared to $T_e = 0.715$ according to Rouse (1936).

The *effect of roughness* was systematically analyzed by Rajaratnam et al. (1976). An end overfall with a subcritical approach flow can be considered hydraulically smooth provided $k_s/h_c < 0.10$, where k_s is the equivalent sand roughness height. For $k_s/h_c > 0.4$, the end depth ratio is related to the rough (subscript r) regime whereas $h_{er}/h_e = 0.80$ in the smooth regime. For discharge measurement, a *smooth channel* is recommended.

Hager (1983) computed the end depth ratio as

$$T_e = \frac{\mathsf{F}_o^2}{0.4 + \mathsf{F}_o^2}, \tag{11.3}$$

thus $T_e = 0.714$ for $\mathsf{F}_o = 1$.

11.2.2 *Free Surface Profile*

The *free surface profile* $T(X)$ with $T = t/h_o$ corresponds to a portion of a *solitary wave* (Chap. 1) and follows from Eq. (1.29) by setting $z = 0$, by assuming that the friction losses are compensated for by the bottom slope. The solution for the upstream portion ($X \leq 0$) and subcritical approach flow is

$$X = \frac{2}{\sqrt{3}}\left[(1 - T_e)^{-1/2} - (1 - T)^{-1/2}\right], \tag{11.4}$$

whereas for $\mathsf{F}_o > 1$

$$X = -2\left[\frac{\mathsf{F}_o^2}{3\left(\mathsf{F}_o^2 - 1\right)}\right]^{1/2}\left[\operatorname{Arctanh}\left(\frac{\mathsf{F}_o^2 - T}{\mathsf{F}_o^2 - 1}\right)^{1/2} - \operatorname{Arctanh}\left(\frac{\mathsf{F}_o^2 - T_e}{\mathsf{F}_o^2 - 1}\right)^{1/2}\right]. \tag{11.5}$$

Equations (11.4) and (11.5) compare well with the observations of Rouse (1943), provided T_e is expressed with Eq. (11.3).

For the downstream portion $X > 0$, a simple expression may also be derived. The momentum equation applied in the horizontal direction indicates an almost *constant vertical jet thickness* t equal to the end depth h_e. The lower nappe profile $Z_u(X)$ corresponds with $\varepsilon = (T_e/\mathsf{F}_o)^2$ to the parabola (Hager 1993a)

$$X = \varepsilon^{-1}\left[\left(Z_o'^2 - 2\varepsilon Z_u\right)^{1/2} - Z_o'\right]. \tag{11.6}$$

where

$$Z_o'^2 = 2\left(1 - T_e\mathsf{F}_o^{-2}\right)(1 - T_e)^2 \tag{11.7}$$

is the slope of the lower nappe profile $Z_u(X)$ at the end section. The nappe profile can be expressed with the generalized coordinates

$$\overline{X} = \left(\varepsilon/Z_o'\right)X \quad \text{and} \quad \overline{Z}_u = \left(\varepsilon/Z_o'^2\right)Z_u \tag{11.8}$$

as

$$\overline{X} = \left(1 - 2\overline{Z}_u\right)^{1/2} - 1. \tag{11.9}$$

This formulation involves thus a two-parameter set, instead of the three parameter set in Fig. 11.2. Transforming to the original parameters gives for the *lower nappe profile*

$$\frac{1.77\mathsf{F}_o^2 X}{\left[\left(\mathsf{F}_o^2 + 0.4\right)\left(\mathsf{F}_o^2 - 0.6\right)\right]^{1/2}} = \left[1 - \left(2.5^2\mathsf{F}_o^2\right)\frac{\mathsf{F}_o^2 + 0.4}{\mathsf{F}_o^2 - 0.6}Z_u\right]^{1/2} - 1. \tag{11.10}$$

Equation (11.10) determines the lower nappe profile $Z_u(X)$, whereas the upper nappe profile is given by $Z_o(X) = Z_u(X) + T_e$.

A number of recent studies have considered the end overfall, including those of Marchi (1992), Montes (1992), and Khan and Steffler (1996). Ferreri and Ferro (1990) studied the *laterally unconfined* end overfall for subcritical approach flow. They recommended $20h_c$ as the minimum length for the approach flow channel to reduce upstream perturbations. Contrary to the laterally-guided end overfall, the jet from their configuration expands laterally, as shown in Fig. 11.4.

The end depth ratio is influenced by both the aspect ratio b/h_c and the relative drop height d/h_c. A minimum channel width of $b = 0.20$ m is recommended. Non-aerated overfalls that are not recommended for design were studied by Christodoulou (1985).

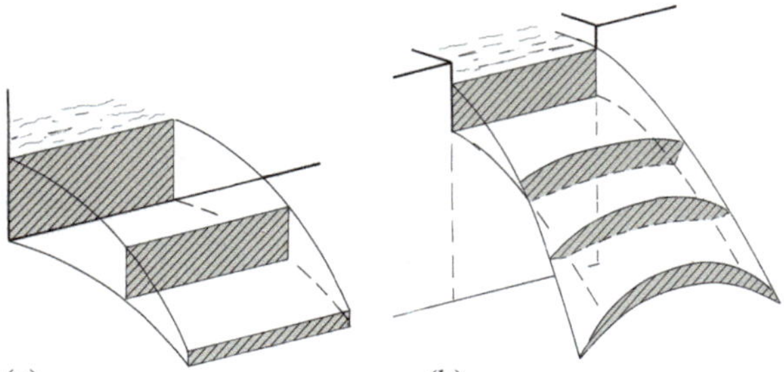

Fig. 11.4 Jet geometry of end overfall (**a**) laterally confined, (**b**) laterally-unguided flow (Ferreri and Ferro 1990)

11.2.3 *Discharge Equation*

For a certain channel, involving the roughness coefficient n of the Manning-Strickler equation, bottom slope S_o and channel width b, the *uniform flow* equation reads (Chap. 5)

$$S_o = S_f = \left(\frac{nQ}{bh_o}\right)^2 \left[\frac{b + 2h_o}{bh_o}\right]^{4/3}. \tag{11.11}$$

With the parameters

$$\Phi = S_o h_e^{1/3}/(n^2 g), \quad \zeta = 1 + 2(h_e/b)T_e^{-1}, \tag{11.12}$$

and after elimination of F_o with Eq. (11.3), one may express Eq. (11.11) as (Hager 1993a)

$$\Phi = \frac{2}{5}\frac{\left[T_e + 2(h_e/b)\right]^{4/3}}{1 - T_e}. \tag{11.13}$$

For a given channel geometry and observed end depth h_e, the values of Φ and thus of $h_o = h_e/T_e$ may be computed (Fig. 11.5). For $\tau = h_e/b > 0.5$, one may explicitly compute T_e with

$$T_e = \frac{2.5\Phi - (2\tau)^{4/3}}{2.5\Phi + 1.68\tau^{1/3}}. \tag{11.14}$$

The *discharge* Q follows then from Eq. (11.3)

$$\frac{Q^2}{gb^2 h_e^3} = \frac{2}{5T_e^2(1 - T_e)}. \tag{11.15}$$

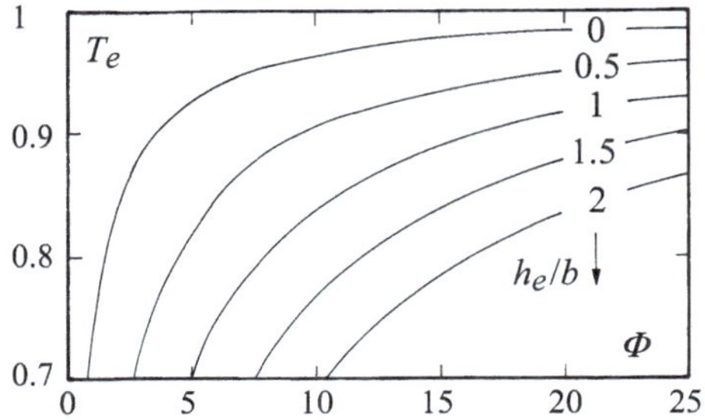

Fig. 11.5 End depth ratio $T_e = h_e/h_o$ for given channel characteristics $\Phi = S_o h_e^{1/3}/(n^2 g)$ and end depth aspect ratio h_e/b

Example 11.1 Given a rectangular channel with $1/n = 85$ m$^{1/3}$s^{-1}, bottom slope $S_o = 2\%$ and width $b = 0.90$ m. How large is discharge Q for an end depth of $h_e = 0.32$ m?
With $\Phi = 85^2 0.02 \cdot 0.32^{1/3}/9.81 = 10.07$ and $\tau = h_e/b = 0.32/0.90 = 0.356$, Fig. 11.5 gives $T_e = 0.924$, i.e. $Q^2/(gb^2h_e^3) = 2/[5 \cdot 0.924^2(1-0.924)] = 6.16$ from Eq. (11.15), and $Q = 6.16^{1/2}(9.81 \cdot 0.9^2 0.32^3)^{1/2} = 1.27$ m^{3}s^{-1}.
Equation (11.14) yields for the aspect ratio $\tau = 0.356$ an end depth ratio $T_e = [2.5 \cdot 10.07-(2 \cdot 0.356)^{4/3}]/[2.5 \cdot 10.07+1.68 \cdot 0.356^{1/3}] = 0.931$ (+1%), such that the discharge is by +4% too large.

Equation (11.15) is highly sensitive to errors in the prediction of T_e because $\Delta Q/Q = (1/5)[(3T_e-2)/(1-T_e)](\Delta T_e/T_e)$ If for example $T_e = 0.9$ and $\Delta T_e/T_e = 1\%$, one had for $\Delta Q/Q = 1.4\%$. Such an error would originate from an incorrect estimation of the roughness coefficient n, or the bottom slope S_o. On the other hand, an error in the end depth reading results in $\Delta Q/Q = (1/5)\Delta T_e/(1-T_e)^2$. Using the same numbers as above, one had $\Delta Q/Q = 18\%$. The error in $\Delta Q/Q$ thus increases significantly as the end depth ratio tends to $T_e = 1$, i.e. for large approach flow Froude numbers. It is recommended that Froude numbers larger than 5 be not considered because of this *reading sensitivity* and the presence of *shockwaves* that make surface readings almost impossible.

11.3 Circular Pipe

11.3.1 *Flow Description*

The end overfall in circular, partially-filled pipes was introduced around 1920. Smith (1962) transferred the knowledge from the rectangular to the circular conduit. Figure 11.6(a) defines the flow configuration, with D as pipe diameter, h_o as approach flow depth, h_e as end depth and S_o as bottom slope. For subcritical approach flow, the pipe chokes for an end depth ratio $y_e = h_e/D = 0.56$. This is to say that the pipe generates free surface flow for smaller, and pressurized flow for larger filling ratio at the end section.

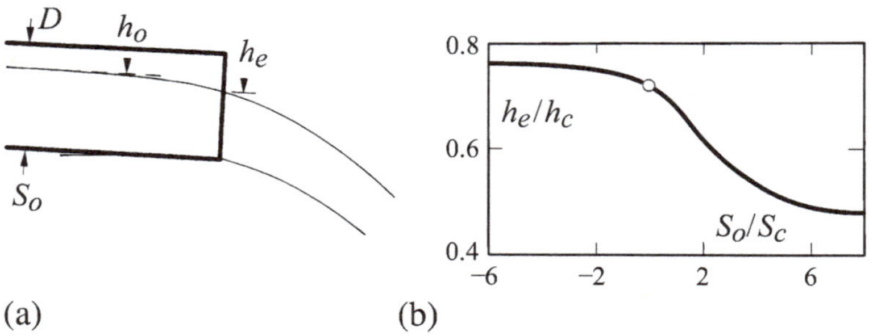

Fig. 11.6 Pipe outflow for a bottom slope S_o. (**a**) Notation, (**b**) end depth ratio h_e/h_c versus S_o/S_c

By discussing Smith's contribution, Rajaratnam and Muralidhar suggested a minimum approach flow length of $L_o = 12H_o$ for a rounded sewer intake, with H_o as the approach flow energy head. The end depth ratio h_e/h_c was plotted also as a function of the slope ratio S_o/S_c, with S_c as the critical slope (Chap. 6). From Fig. 11.6(b) the extreme values are $h_e/h_c = 0.76$ for large negative bottom slope, and $h_e/h_c = 0.48$ for large positive bottom slope. Accordingly, the slope effect is small for adversely sloping pipes, whereas it is large for pipes with a positive slope. The latter configuration is not relevant because the approach flow Froude number then is large, and the measuring accuracy therefore small from the above discussion pertaining to rectangular channels.

Blaisdell, a second discusser of Smith (1962), investigated the *transition* from pressurized to free surface flows. He established a relation between the length L_a measured from the point of free surface flow to the end section, and the *pipe Froude number* $q_D = Q/(gD^5)^{1/2}$. According to Fig. 11.7c) a horizontal conduit has free surface flow for $q_D < 0.43$, and pressurized flow for $q_D > 0.91$. For intermediate values, the flows are as shown in Fig. 11.7(a) and (b).

Rajaratnam and Muralidhar (1964b) determined the end depth ratio $h_e/h_c = 0.725$ for a horizontal, sufficiently long approach flow pipe. The *discharge-end depth relation* is

$$\frac{Q}{\left(gD^5\right)^{1/2}} = 1.54\,(h_e/D)^{1.84}\,. \tag{11.16}$$

Sloping pipes have to be accounted for with Fig. 11.6 to include the effect of friction slope. Until today, the sloping pipe has not yet been thoroughly investigated. A recent contribution is given below.

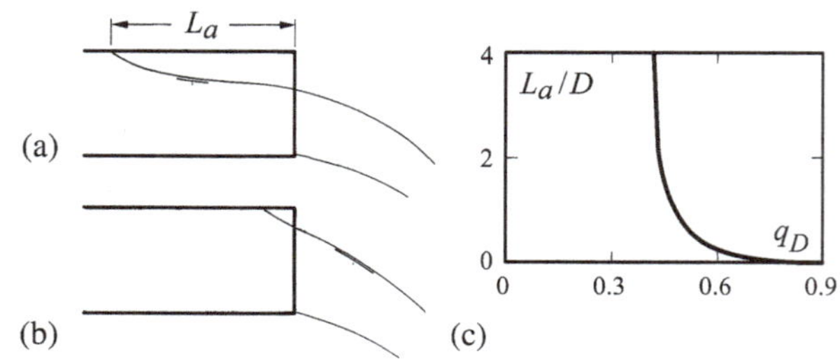

Fig. 11.7 End overfall in a *horizontal* conduit. Transition from (**a**) pressurized to free surface flows, (**b**) free surface to pressurized flows, (**c**) relative length of free surface flow L_a/D as a function of q_D

Example 11.2 Determine discharge Q of a horizontal pipe ($S_o = 0$) of diameter $D = 1.25$ m for an end depth $h_e = 0.42$ m!
For $S_o = 0$, Fig. 11.6(b) gives $h_e/h_c = 0.72$. From Eq. (6.36) the critical depth in a pipe is $h_c = [Q/(gD)^{1/2}]^{1/2}$, such that $Q = h_c{}^2(gD)^{1/2} = (h_e/0.72)^2(gD)^{1/2} = (0.42/0.72)^2(9.81\cdot 1.25)^{1/2} = 1.20\ \text{m}^3\text{s}^{-1}$. From Eq. (11.16) follows $Q = (9.81\cdot 1.25^5)^{1/2}1.54(0.42/1.25)^{1.84} = 1.13\ \text{m}^3\text{s}^{-1}$ if $\mathsf{F}_o = 1$. The difference of 6% may originate from the reading accuracy of Fig. 11.6(b).

Example 11.3 What is the discharge for Example 11.2 for a bottom slope of 1%?
Approximately, the end depth ratio is $h_e/h_c = 0.6$, such that $h_o = 0.42/0.6 = 0.70$ m. Thus, the approach filling ratio is $y_o = h_o/D = 0.7/1.25 = 0.56$, and $R_h/D = 0.27$ from Eq. (6.64), corresponding to $R_h = 0.33$ m as well as $F/D^2 = 0.45$, corresponding to $F = 0.71\ \text{m}^2$. Then, the critical discharge is $Q = h_o{}^2(gD)^{1/2} = 0.70^2(9.81\cdot 1.25)^{1/2} = 1.72\ \text{m}^3\text{s}^{-1}$ and the approach flow velocity $V_o = 1.72/0.71 = 2.42\ \text{ms}^{-1}$. With $1/n = 85\ \text{m}^{1/3}\text{s}^{-1}$ as roughness coefficient, critical slope is $S_c = n^2V_c{}^2/R_{ho}{}^{4/3} = 0.012^2 2.42^2/0.33^{4/3} = 0.35\%$ and thus $S_o/S_c = 1/0.35 = 2.87$.

For $S_o/S_c = 2.87$, the end depth ratio is $h_e/h_o = 0.58$ from Eq. 11.6(b) and a second iteration yields $h_o = 0.42/0.58 = 0.725$ m. With $y_o = 0.725/1.25 = 0.58$ one has $R_{ho} = 0.34$ m and $F_o = 0.74\ \text{m}^2$, thus $Q_c = h_o{}^2(gD)^{1/2} = 0.725^2(9.81\cdot 1.25)^{1/2} = 1.84\ \text{m}^3\text{s}^{-1}$ and $V_o = 1.84/0.74 = 2.49\ \text{ms}^{-1}$, and $S_c = 0.012^2 2.49^2/0.34^{4/3} = 0.36\%$, or $S_o/S_c = 1/0.36 = 2.76$, corresponding to the estimation.

For $h_o = 0.725$ m and $S_o = 1\%$, uniform discharge is $nQ/(S_o{}^{1/2}D^{8/3}) = 0.75\cdot 0.58^2(1 - 0.583\cdot 0.58^2) = 0.203$ from Eq. (5.14) and the discharge $Q = 0.203\cdot 85\cdot 0.01^{1/2}1.25^{8/3} = 3.13\ \text{m}^3\text{s}^{-1}$, thus almost three times as large as from Example 11.2.

11.3.2 *Effect of Approach Flow Froude Number*

Clausnitzer and Hager (1997) investigated the effects of the approach flow Froude number F_o and the approach flow filling ratio $y_o = h_o/D$. With the approximations for cross-sectional area

$$F = \left(Dh^3\right)^{1/2}, \tag{11.17}$$

and static pressure force

$$P/(\rho g) = \frac{1}{2}\left(Dh^5\right)^{1/2}, \tag{11.18}$$

and if neglecting the residual jet pressure at the end section, the *momentum equation* yields

$$\frac{1}{2}\left(Dh_o^5\right)^{1/2}+\frac{Q^2}{g\left(Dh_o^3\right)^{1/2}}=\frac{Q^2}{g\left(Dh_e^3\right)^{1/2}}. \tag{11.19}$$

With the approach flow Froude number $\mathsf{F}_o = Q/(gDh_o^4)^{1/2}$ according to Eq. (6.35), Eq. (11.19) may be solved for the end depth ratio $Y_e = h_e/h_o$ as

$$Y_e=\left(\frac{2\mathsf{F}_o^2}{1+2\mathsf{F}_o^2}\right)^{2/3}. \tag{11.20}$$

For the limit cases $\mathsf{F}_o = 1$, and $\mathsf{F}_o \to \infty$, the results are $Y_e(1) = (2/3)^{2/3} = 0.763$ and $Y_e(\infty) = 1$. The currently accepted standard value for $Y_e = 0.725$ according to Rajaratnam and Muralidhar (1964b) is in reasonable agreement with the simplified approach.

For $\mathsf{F}_o > 2$ the free surface decreases continuously from h_o to h_e, and a standing surface undulation may set up for smaller Froude numbers. The length L_e of the drawdown curve varies significantly with F_o and may be approximated as

$$L_e/h_o = 5 + 0.90\mathsf{F}_o. \tag{11.21}$$

Example 11.4 Reconsider Example 11.3.

Uniform flow in a circular pipe follows from Eq. (5.14). The approach flow Froude number may be expressed with y_o as $\mathsf{F}_o = Q/(gDh_o^4)^{1/2} = Q/[(gD^5)^{1/2}y_o^2] = (S_o^{1/2}D^{1/6})/(ng^{1/2})](3/4)[1 - (7/12)y_o^2]$. Equation (11.20) may be written as $h_e/D = y_o[2\mathsf{F}_o^2/(1 + 2\mathsf{F}_o^2)]^{2/3}$. With $\kappa = (S_o^{1/2}D^{1/6})/(ng^{1/2})$ as the roughness characteristic (Chap. 5), one has

$$h_e/D = y_o\left[1+\frac{8/9}{\kappa^2\left(1-(7/12)y_o^2\right)^2}\right]^{-2/3} \tag{11.22}$$

For given values of h_e/D and κ, one may solve for y_o, therefore.

With $n = 0.012\ \mathrm{m}^{-1/3}\mathrm{s}$, $D = 1.25$ m, $S_o = 0.01$, the roughness characteristic is $\kappa = 0.01^{1/2}1.25^{1/6}/(0.012\cdot 9.81^{1/2}) = 2.76$ and the solution of Eq. (11.22) for $h_e/D = 0.336$ is $h_o/D = 0.366$, and thus $h_o = 0.458$ m. Accordingly, $\mathsf{F}_o = (3/4)\kappa[1-(7/12)y_o^2] = 0.75\cdot 2.76[1 - 0.58\cdot 0.366^2)] = 1.91$ and $Y_e = 0.918$, as is also verified with the previous numbers.

Discharge $Q = \left(gDh_o^4\right)^{1/2}\mathsf{F}_o = (9.81\cdot 1.25\cdot 0.458^4)^{1/2}1.91 = 1.40\ \mathrm{m^3s^{-1}}$ is much smaller than in Example 11.3. The differences must be attributed to the inexact definition of the slope effect in Fig. 11.6b). The length of the drawdown curve amounts to $L_e/h_o = 5 + 0.9\mathsf{F}_o = 5 + 0.9\cdot 1.91 = 6.72$ from Eq. (11.21) and $L_e = 6.72\cdot 0.46 = 3.10$ m.

11.3.3 *Jet Geometry*

Figure 11.8 defines the relevant parameters both up- and downstream from the end section. With known values of h_e, h_o and L_e, the dimensionless coordinates are $X_e = x/L_e$ and $Z_e = (h{-}h_e)/(h_o{-}h_e)$, where x is the coordinate parallel to the approach flow bottom, measured from the end section.

Fig. 11.8 Definition plot for jet geometry in circular pipe

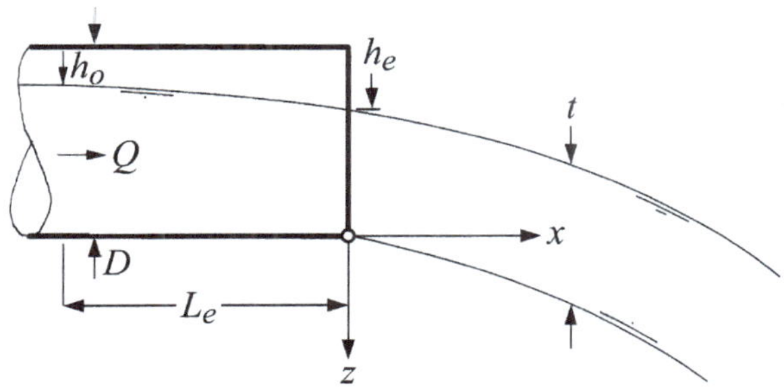

Generalized Drawdown Curve

Experiments were conducted for filling ratios $y_o < 90\%$ and approach flow Froude numbers F_o from 1 to 8. Figure 11.9 indicates that the drawdown curve along the pipe axis up to the end section follows the parabola (Clausnitzer and Hager 1997)

$$Z_e = 1 - (1 + X_e)^2 . \tag{11.23}$$

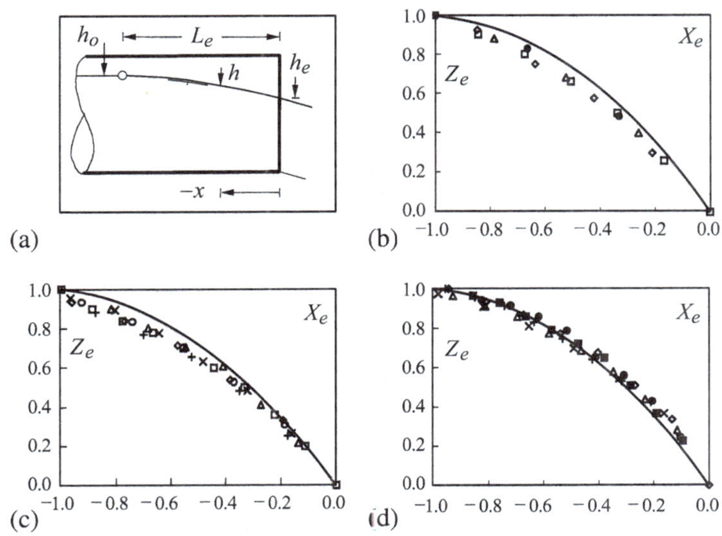

Fig. 11.9 Drawdown curve up to end section, (**a**) definition plot. Curve $Z_e(X_e)$ for $y_o =$ (**b**) 0.2, (**c**) 0.4, (**d**) 0.6 and $\mathsf{F}_o =$ (×) 0.8, (○) 1.3, (+) 1.9, (●) 2.5, (■) 3.8, (Δ) 5.2, (◇) 6.1, (□) 8.0; (—) Eq. (11.23)

Lower Nappe Trajectory

For hypercritical flow ($\mathsf{F}_o > 3$) for which $1-\mathsf{F}_o^2 \to -\mathsf{F}_o^2$ the governing streamwise coordinate in a rectangular channel is $(x/h_o)\mathsf{F}_o^{-1}$. Due to shape effects, the governing streamwise coordinate in a *circular pipe* is $X = (x/h_o)\mathsf{F}_o^{-0.8}$, and the vertical coordinate is $Z = z/h_o$. For filling ratios $0.2 < y_o < 0.9$ and $\mathsf{F}_o \leq 8$, the axial lower nappe trajectory $Z(X)$ may be expressed as (Clausnitzer and Hager 1997)

$$Z = \frac{1}{3}X + \frac{1}{4}X^2. \tag{11.24}$$

Figure 11.10 shows that the data follow perfectly Eq. (11.24). Note that the axial lower nappe trajectory is a parabola, but that the nappe slope at the end section is $dZ/dX = 1/3$, in disagreement with the conventional assumption of a tangent towards the approach bottom ($dZ/dX = 0$). Also, the lower nappe profiles of the rectangular and circular cross-sections deviate, particularly for $\mathsf{F}_o = 1$. This is attributed again to the *cross-sectional shape effect*.

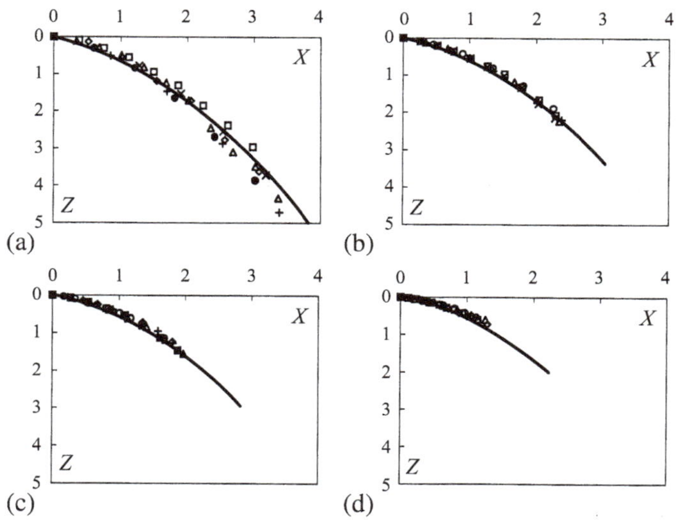

Fig. 11.10 Lower nappe trajectory $Z(X)$ for $y_o =$ (**a**) 0.2, (**b**) 0.4, (**c**) 0.6, (**d**) 1.0. (—) Eq. (11.24), notation Fig. 11.9

Upper Nappe Trajectory

The jet thickness $T(X)$ with $T = t/h_e$ (Fig. 11.8) can be expressed as

$$T = 1 + 0.06X. \tag{11.25}$$

(a) (b) (c) (d)
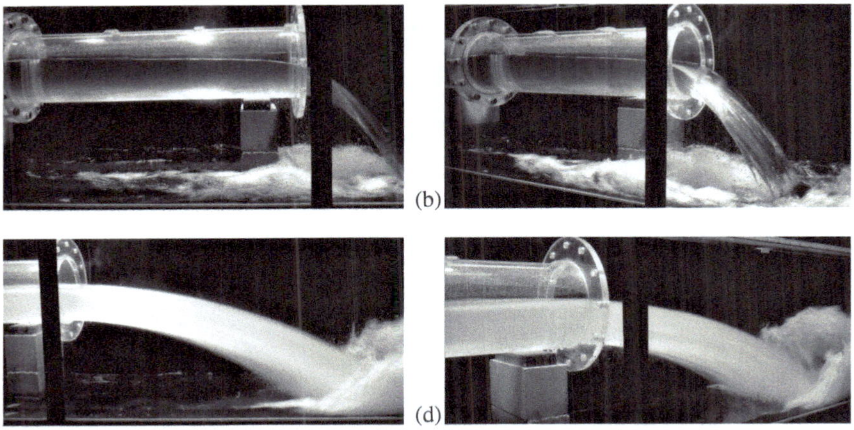

Fig. 11.11 Lateral views of pipe outflows for $y_o = 60\%$ and $\mathsf{F}_o =$ (**a**) 1, (**b**) 2, (**c**) 4, (**d**) 8

Whereas the nappe thickness remains essentially constant for the rectangular cross-section, it increases for jets issued by a circular pipe.

Figure 11.11 shows photographs of the *outflow jet* from a partially-filled pipe with $y_o = 60\%$. For $\mathsf{F}_o = 1$ the jet is strongly deflected by gravity and a large discharge portion is recirculated upon impinging on the tailwater bottom. This is attributed to the large impact angle. The condition $\mathsf{F}_o = 2$ corresponds to typical pipe outflow with an impact angle much smaller than for $\mathsf{F}_o = 1$, and thus less jet recirculation. For $\mathsf{F}_o = 4$, the jet dispersion on the model tested had an effect on the lower nappe trajectory, and the jet became fully aerated for $\mathsf{F}_o = 8$. Here, recirculation is negligible because of the small impact angle.

11.3.4 *Submergence Effects*

Figure 11.12(a) shows a definition plot relative to the submergence of pipe outflows under supercritical approach flow. The pipe is defined to be submerged (in terms of the modular limit) if the shock fronts from the pipe outflow are located at the end section. For a slightly larger downstream depth, a hydraulic jump occurs in the pipe. The downstream flow of the pipe outlet is characterized by a *standing wave* (subscript w) of axial wave height h_w. The determining parameters are h_{ow} as the lateral flow depth at modular limit conditions and h_{uw} as the downstream flow depth. All depths are related to the pipe invert elevation at the end section.

The lateral flow depth ratio $Y_{ow} = h_{ow}/h_o$ may be expressed as (Fig. 11.12b)

$$Y_{ow} = 1 + 0.05(\mathsf{F}_o - 1). \tag{11.26}$$

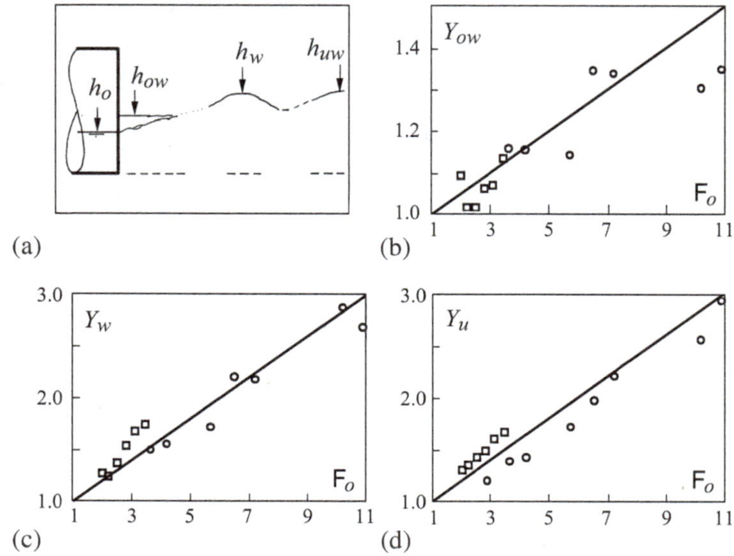

Fig. 11.12 Modular limit of pipe outflow. (**a**) Definition sketch, (**b**) lateral flow depth ratio $Y_{ow} = h_{ow}/h_o$, (**c**) wave height ratio $Y_w = h_w/h_o$ and (**d**) tailwater depth ratio $Y_u = h_{uw}/h_o$ as functions of F_o for $y_o =$ (○) 0.2 and (□) 0.6. (—) Average experimental curves

The *maximum wave height* $Y_w = h_w/h_o$ and the downstream flow depth $Y_u = h_{uw}/h_o$ are almost identical, namely (Fig. 11.12c, d)

$$Y_w = Y_u = 1 + 0.2(\mathsf{F}_o - 1). \tag{11.27}$$

Depending on the approach flow Froude number F_o, the downstream water elevation may be significantly lifted above the approach flow depth. A conservative design would require $Y_{ow} = 1$, i.e. the maximum lateral flow depth be as high as the pipe flow depth.

11.3.5 *Egg-Shaped Sewer*

The geometry of jet trajectories issued by egg-shaped sewers has been determined by Biggiero (1963). The lower (subscript 1) and upper (subscript 2) nappe profiles $Z_1(X)$ and $Z_2(X)$ with $X = x/h_o$ and $Z = z/h_o$ can be approximated as the parabolas

$$Z_1 = -\theta_1 - A_1 X^2, \tag{11.28}$$

$$Z_2 = T_e - \theta_2 X - A_2 X^2. \tag{11.29}$$

Provided $1 \leq \mathsf{F}_o \leq 3.6$, the experimentally determined coefficients θ_i, A_i and T_e depend on the approach flow Froude number F_o and may be approximated as (Hager 1993b)

$$A_1 = 0.28\mathsf{F}_o^{-1}, A_2 = (1/3)\mathsf{F}_o^{-1.4}; \tag{11.30}$$

$$\theta_1 = 0.05(3.7 - \mathsf{F}_o), \quad \theta_2 = 1.70\,\theta_1 \quad \text{and} \tag{11.31}$$

$$T_e = 1 - 0.25\mathsf{F}_o^{-5/3}. \tag{11.32}$$

Figure 11.13 shows the lower jet trajectories $Z_1(X)$ for $1 \leq \mathsf{F}_o \leq 5$ resulting in significant differences when compared with jets issued from a rectangular channel (Fig. 11.2).

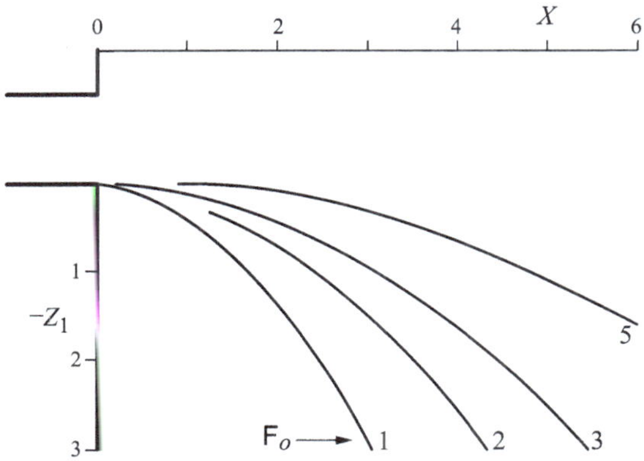

Fig. 11.13 Axial lower jet trajectory $Z_1(X)$ for various F_o issued by the egg-shaped sewer (Biggiero 1963)

11.4 Cavity Outflow

11.4.1 *Outflow Features of Pipes*

Wallis et al. (1977) have distinguished four *outflow types* for circular pipes:

- Free surface flow (Fig. 11.6a),
- Bubble flow (Fig. 11.7a),
- Bubble washout flow (Fig. 11.7b), and
- Pressurized pipe flow.

Free surface flow in the outflow vicinity is either critical for subcritical approach flow, or supercritical (Sect. 11.3.2), whereas bubble flow has a surface stagnation point at distance L_a from the end section. Bubble washout is reached if $L_a/D \leq 1$. Figure 11.14 relates to these four flow types by the pipe Froude number

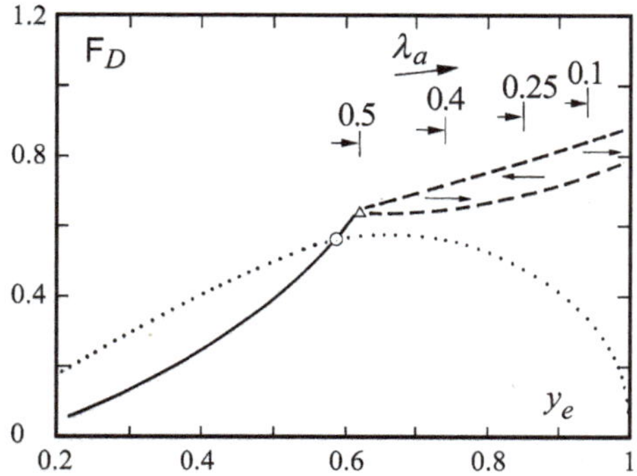

Fig. 11.14 Outflow diagram $F_D(y_e)$ according to Wallis et al. (1977) with $y_e = h_e/D$ and $F_D = (4/\pi)[Q/(gD^5)^{1/2}]$. (—) Critical flow for $S_o = 0$, (...) bubble flow with (○) bubble development, and (Δ) bubble washout depending on $\lambda_a = L_a/D$. (- - -) Oscillating outflow close to pressurized flow

$F_D = 4Q/[\pi(gD^5)^{1/2}]$. To exclude effects of surface tension, the pipe diameter should be at least $D = 0.10$ m. Because results relating to two-phase pipe flow (Chap. 5) often relate to chemical engineering, the pipe diameter used for experiments is normally smaller.

Example 11.5 What is the flow type in Example 11.2?
With $y_e = 0.42/1.25 = 0.336$ and $F_D = (4/\pi)[1.19/(9.81 \cdot 1.25^5)^{1/2}] = 0.277$, Fig. 11.14 indicates a location between critical flow and bubble flow. Figure 11.6(a) relates to the critical flow type.

Example 11.6 For which discharge starts the oscillating pressurized flow type?
According to Fig. 11.14, the transition between bubble washout and pressurized flow is reached for $F_D = 0.64$, corresponding to $Q = 0.64(\pi/4)(9.91 \cdot 1.25^5)^{1/2} = 2.75\ \text{m}^3\text{s}^{-1}$.

11.4.2 *Description of Cavity Outflow*

Figure 11.15(b) shows a sketch of cavity outflow from a circular pipe, with the stagnation point S as the transition from pressurized to free surface flows, the end depth h_e, the jet thickness $t_t(x_t)$, and the vertical coordinate z_t measured from the end section positively downwards. The problem to be solved involves the end depth ratio, the distance to the stagnation point and the free surfaces both up- and downstream from the end section. Also, the conditions of cavity outflow to occur are relevant.

Figure 11.15(a) relates to a similar but simpler problem of the so-called Benjamin bubble (5.7.5). Such flow may result in a horizontal pipe for a frictionless fluid, provided several conditions are satisfied. For free downstream flow, the asymptotic

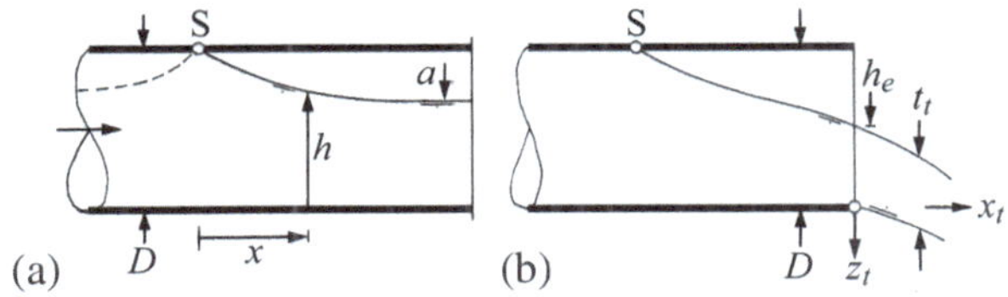

Fig. 11.15 (**a**) Cavity flow in long circular pipe, (**b**) cavity outflow close to end section

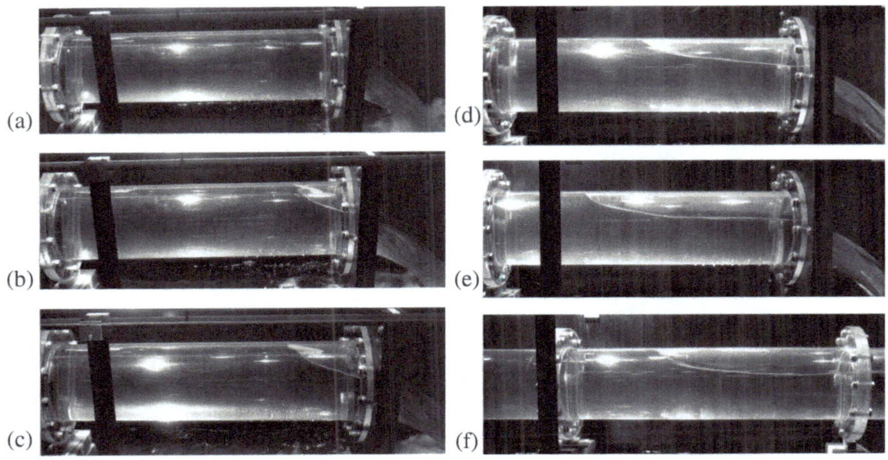

Fig. 11.16 Cavity outflow from circular pipe for f = (**a**) 0.80, (**b**) 0.73, (**c**) 0.70, (**d**) 0.66, (**e**) 0.64, (**f**) 0.586

flow depth a is almost 58% of the pipe diameter, and potential flow is generated. Although Benjamin described such flow in connection with gravity currents, Cola (1967, 1970) made significant contributions to cavity flow, and one should rather refer to as the *Cola bubble* than to the Benjamin bubble, therefore.

Figure 11.16 shows typical outflow geometries for various values of the *pipe Froude number* defined as $f = V/(gD)^{1/2}$ with $V = Q/(\pi D^2/4)$ as the full pipe average velocity. Both the locations and the outflow patterns depend only on f, with a lower limit $f = 0.544$ for the transition between free surface and cavity flows, whereas the transition between cavity and pressurized flows occurs for $f = 1.15$. The effect of bottom slope on the outflow pattern is small for S_o confined between $\pm 1\%$ (Hager 1998).

For $f = 0.80$ the cavity is nearly washed out from the pipe, with a transition length $X_t = x_t/D = 0.60$ to the stagnation point (Fig. 11.16a). For $f = 0.73$ comparable flow is generated (Fig. 11.16b), but the inflexion point of the free surface profile is shifted near to the end section, resulting in a cavity length of $X_t = 1.1$. This corresponds roughly to the limit between *bubble washout* and *bubble flow* as defined by Wallis et al. (1977). The cavity length increases now significantly with a decrease of f, such as to $X_t = 1.8$ for $f = 0.66$ (Fig. 11.16d) or $X_t = 3.0$ for $f = 0.64$ (Fig. 11.16e). The cavity resembles now a Cola bubble, except for the downstream

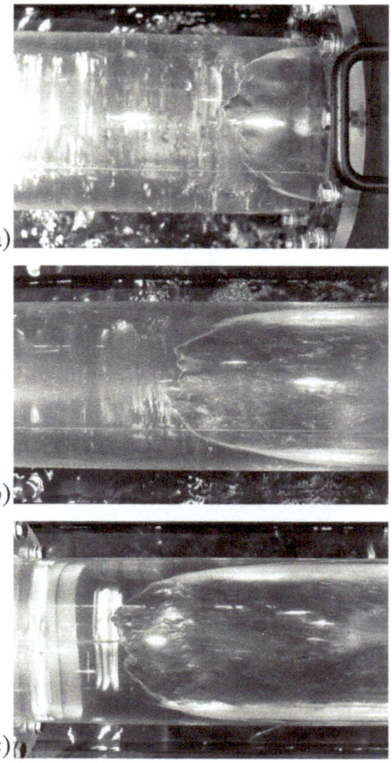

Fig. 11.17 Cavity top views close to stagnation point for f= (**a**) 0.73, (**b**) 0.66, (**c**) 0.586

end of the outflow portion. For $f = 0.586$ the cavity length is $X_t = 6.8$ (Fig. 11.16f), and a lower experimental value of $f = 0.548$ was attained. This value depends on details of flow generation, with values between 0.51 to 0.64 being quoted in the literature.

Figure 11.17 refers to top views on the cavities and it seems that those for larger f have a smaller radius of curvature, i.e. are slimmer than those for smaller f. The latter are definitely less stable and may oscillate easily by $\pm 0.25D$.

Profiles for a typical cavity outflow for f between 0.60 to 0.70 are shown in Fig. 11.18. The cavity angle at the stagnation point is 33° (±3°), and the constant depth ratio is 0.625 ± 0.02. For such cavity flow, there is a tendency for *wave formation* with an amplitude of about $\pm 0.02D$. This tendency increases with decreasing f, and a transition to free surface pipe flow is always accompanied with downstream wave formation.

Figure 11.19 relates to a sequence of images between *cavity flow* and *slug flow*, i.e. during a slight reduction of f close to the lower limit discharge for cavity flow. First, the waves downstream of the stagnation point break (Fig. 11.19a), and an air pocket moving against the flow into the low pressure zone (Fig. 11.19b) upstream from the stagnation point (see below). Once an isolated cavity is formed (Fig. 11.19c) air is entrained by the hydraulic jump and this cavity experiences a sub-pressure. Therefore, cavity flow develops another breaking wave (Fig. 11.19d)

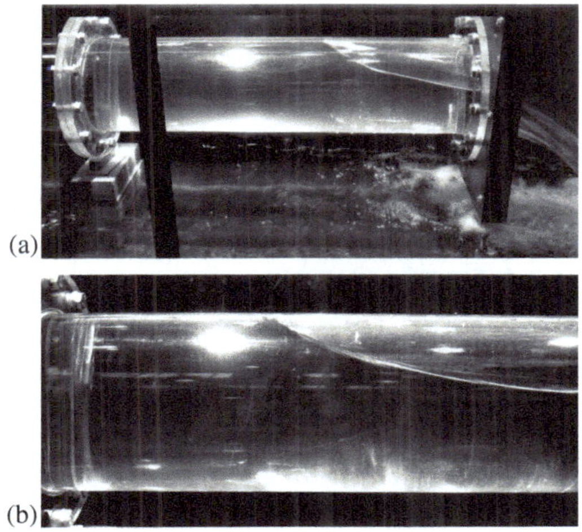

Fig. 11.18 Side views of cavities for f = (a) 0.66, (b) 0.586

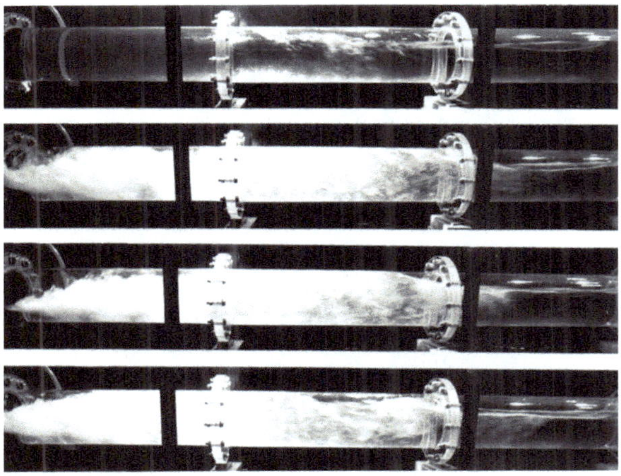

Fig. 11.19 Transition from cavity to slug flow

and a further air pocket moves upstream. The following relates exclusively to cavity flow, whereas air pockets are described in Chap. 5.

11.4.3 *Cavity Shape*

The cavity shape can be demonstrated, at least in the rectangular channel, to be a portion of the *solitary wave profile* (Chap. 1 and Sect. 11.2.2). For the circular pipe, the free surface profile $T(X)$ with $T = t/a$ as the height above the invert relative to the asymptotic flow depth a and $X = x/a$ is somewhat more involved. From energy considerations the maximum discharge results for $a/D = 2/3$, a value close to observations.

Based on the generalized energy Eq. (1.29) allowing for streamline curvature effects, and accounting for the difference between pressure head h and flow depth t in a free surface flow of appreciable *streamline curvature*, the profile $T(\chi)$ obtains

$$T = \frac{2}{3}\left[1 + 0.5\exp\left(-\frac{3}{2}\chi\right)\right], \tag{11.33}$$

with $\chi = \bar{x}/D$ where $\bar{x} = 0$ is located at the stagnation point S. Clearly, $T(\chi = 0) = 1$ as required, and a 1% deviation from $T = 2/3$ is reached for $\chi = 2.6$. The asymptotic flow depth $a/D = 2/3$ is thus reached almost at $3D$ downstream of the stagnation point. Cavities with $f < 0.64$ can thus approach a surface profile close to Eq. (11.33).

A comparison of observations with Eq. (11.33) indicates small but not systematic deviations, with an asymptotic flow depth $T = 0.625$ rather than 2/3, as assumed previously. The cavity slope at the stagnation point is practically constant $\mathrm{d}T/\mathrm{d}\chi = -0.5$, independent of f. The corresponding asymptotic Froude number for $a/D = 2/3$ is $\mathsf{F} = 1.23$, with undular surface waves being typical for this value.

Cavity and nappe profiles for various pipe Froude numbers are shown in Fig. 11.20. The value $f = 1.16$ corresponds to the upper discharge limit for cavity flow, with the pipe just fully pressurized, and the end depth equal to D (Fig. 11.20a). For $f = 0.79$ and 0.70, the end depth ratio is significantly reduced and cavity lengths

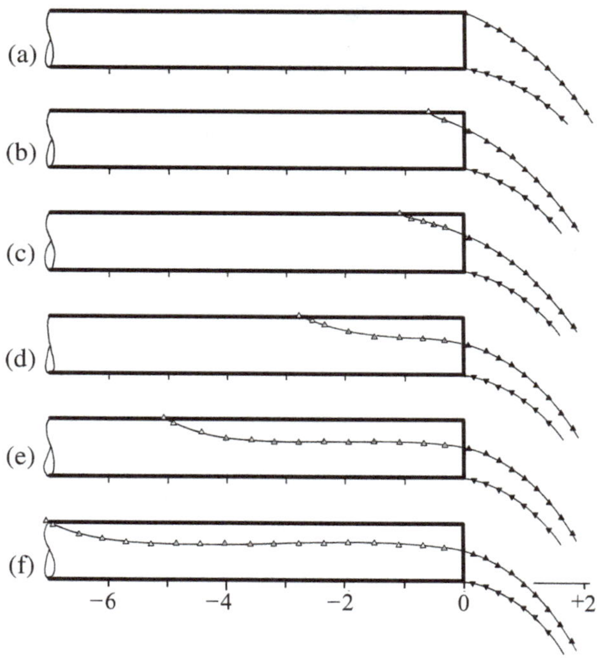

Fig. 11.20 Cavity and nappe profiles for $f =$ (**a**) 1.16, (**b**) 0.79, (**c**) 0.70, (**d**) 0.64, (**e**) 0.63, (**f**) 0.60

are close to D (Fig. 11.20b and c). Typical *cavity flows* are shown in Fig. 11.20(d) and (e) for $f = 0.64$ and $f = 0.63$, with a fully-developed horizontal flow portion. For $f = 0.60$, the cavity length is $7D$ and a single surface wave may be seen. The flow is still relatively stable, and wave breaking did not occur.

11.4.4 *End Depth Ratio*

The end depth ratio $y_e = h_e/D$ (Fig. 11.15b) may be determined with the momentum equation provided the pressure conditions are correctly modelled. For cavity outflow, the *residual pressure* at the end section is negligible, whereas for bubble washout, both the pressure and velocity distributions are not as simply determined. With σ_p as a *pressure coefficient* the momentum equation can be written as (Hager 1998)

$$\frac{\rho Q^2}{(\pi/4)D^2} + \frac{\rho g \pi D^3}{8} - \sigma_p \frac{\rho V^2}{2}\frac{\pi D^2}{4} = \frac{\rho Q^2}{(\pi/4)Dh_e}. \quad (11.34)$$

Solving for y_e in terms of f gives

$$y_e = \frac{2f^2}{1 + (2 - \sigma_p)f^2}. \quad (11.35)$$

The solution for cavity outflow involves $\sigma_p = 1$. Based on experimental data, an overall coefficient $\sigma_p = 2/3$ can be adopted. Equation (11.35) applies for $0.544 < y_e < 1$, i.e. for $0.65 < f < 1.20$. Figure 11.21a shows $y_e(f)$, indicating substantial agreement between prediction and experimental data. These can also be approximated as

$$y_e = \frac{5}{6}f, \quad 0.5 \le y_e \le 1. \quad (11.36)$$

Accordingly, the end depth h_e increases linearly with the pipe velocity V.

The *distance* of the end section from the stagnation point $X_t = x_t/D$ varies also with f. Fig. 11.21b shows that the function

$$X_t = 0.09(f - f_t)^{-3/2}, \quad 0.6 \le f < 1.2 \quad (11.37)$$

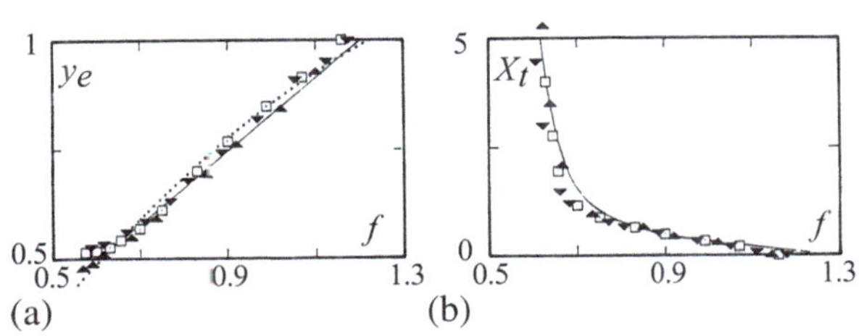

Fig. 11.21 (**a**) End depth ratio $y_e(f)$ for bottom slopes S_o = (▼) −1%, (□) 0, (▲) +1% and (···) Eq. (11.35), (—) Eq. (11.36). (**b**) Relative distance of stagnation point $X_t(f)$, (—) Eq. (11.37)

reproduces the data well, provided the fictitious transition pipe Froude number is $f_t = 0.55$. Note the sharp increase of relative length for $f < 0.70$, i.e. the transition from bubble washout to cavity outflow.

11.4.5 *Nappe Trajectories*

The lower and upper nappes of outflow jets from circular pipes are discussed in Sect. 11.3.2. The end depth ratio $Y_e = h_e/h_o$ as a function of F_o is given by Eq. (11.20). For the present flow configuration all lengths were normalized with the end depth h_e instead of the approach flow depth h_o. With $Z_e = z_t/h_e$ as the vertical distance of the lower nappe profile from the outlet invert, and $X_e = (x_t/h_e)\mathsf{F}_o^{-0.8}$ where $\mathsf{F}_e = Q/(gDh_e{}^4)^{1/2}$ is the end depth Froude number, one has (Hager 1998)

$$Z_e = \frac{1}{3}X_e + \frac{1}{4}X_e^2, \quad f < 0.79. \tag{11.38}$$

Equation (11.38) is identical with Eq. (11.24) except of the normalisation. For $f > 0.79$ the washout as a modified outflow causes deviations not contained in Eq. (11.38).

The *axial nappe thickness* $Z_t(X_e)$ where $Z_t = t_t/h_e$ follows the relation

$$Z_t = 1 - 0.265\,\nu X_e. \tag{11.39}$$

Either $\nu = 1$ for cavity outflow ($f \leq 0.70$), or $\nu = 2$ for bubble washout ($f > 0.70$). Accordingly, the axial nappe thickness decreases considerably for cavity flows (Fig. 11.20).

11.5 Velocity Distribution

The axial velocity distribution $v(z)$ with z as the vertical coordinate measured upwards from the pipe invert depends significantly on f, again. Figure 11.22 shows $\mu(Z)$ with $\mu = v/(gD)^{1/2}$ as the normalized horizontal velocity component and

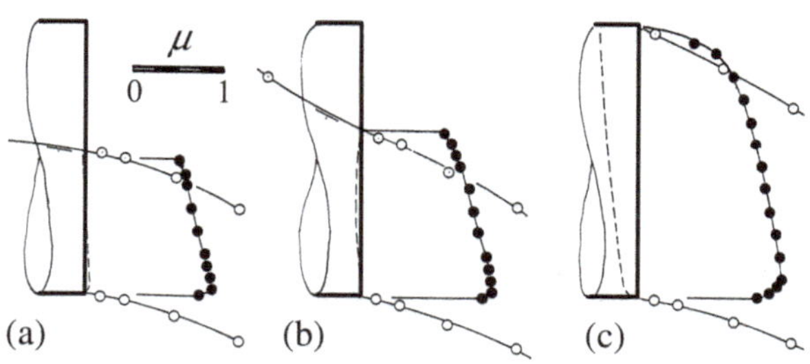

Fig. 11.22 Axial velocity distribution $\mu(Z)$ at end section for $f =$ (**a**) 0.63, (**b**) 0.75, (**c**) 1.16. (•) Local velocity, (○) trajectory geometry, (- - -) pressure distribution

$Z = z/D$. For $f = 0.63$ the velocity increases almost linearly, from the upper to the lower nappe, and the pressure head is close to zero indicating atmospheric internal jet pressure (Fig. 11.22a). For $f = 0.75$ a slight negative pressure occurs across the end section (Fig. 11.22b). For $f = 1.16$, the velocity is practically equal to zero at the pipe vertex, associated with a significant sub-pressure of $-0.50(\rho gD)$. The velocity increases significantly towards the lower pipe boundary with a decrease of the internal pressure (Fig. 11.22c).

Notation

a	[m]	asymptotic pipe flow depth
A	[–]	initial nappe curvature
b	[m]	width
d	[m]	drop height
D	[m]	diameter
D	[–]	drop number
f	[–]	pipe Froude number
f_t	[–]	transition pipe Froude number
F	[m^2]	cross-sectional area
F	[–]	Froude number
g	[ms^{-2}]	gravitational acceleration
h	[m]	flow depth
h_e	[m]	end depth
H	[m]	energy head
j_c	[–]	relative slope
k_s	[m]	equivalent sand roughness height
L_a	[m]	length of free surface
L_e	[m]	distance to critical section
L_o	[m]	approach distance
$1/n$	[m$^{1/3}$s^{-1}]	Manning's roughness coefficient
P	[N]	static pressure force
q	[m^{2}s^{-1}]	unit discharge
q_D	[–]	Pipe Froude number
Q	[m^{3}s^{-1}]	discharge
R_h	[m]	hydraulic radius
S_c	[–]	critical slope
S_f	[–]	friction slope
S_o	[–]	bottom slope
t	[m]	vertical jet thickness
T	[–]	$= t/h_e$, t/a
T_e	[–]	end depth ratio
v	[ms^{-1}]	horizontal velocity component
V	[ms^{-1}]	velocity
W	[–]	Weber number

x	[m]	streamwise coordinate
$\bar{x}$	[m]	location measured from stagnation point
X	[–]	relative longitudinal coordinate
X_e	[–]	$= x/L_e$ or $(x_t/h_e)\mathsf{F}_e^{-0.8}$
y	[–]	filling ratio
y_e	[–]	$= h_e/D$
Y	[–]	$= h/h_o$ depth ratio
Y_o	[–]	$= h_{ow}/h_o$
Y_u	[–]	$= h_{uw}/h_o$
Y_w	[–]	$= h_w/h_o$
z	[m]	vertical coordinate
Z	[–]	dimensionless vertical coordinate
Z_e	[–]	$= (h-h_e)/(h_o-h_e)$ or z_t/h_e
Z_o'	[–]	boundary slope of lower nappe profile
Z_t	[–]	$= t_t/h_e$
ϵ	[–]	relative Froude number
θ	[–]	boundary jet slope
κ	[–]	roughness characteristic
λ_a	[–]	relative length
μ	[–]	normalized velocity component
ν	[–]	trajectory parameter
ρ	[kgm^{-3}]	density
Φ	[–]	roughness parameter
σ	[Nm^{-1}]	surface tension
σ_p	[–]	pressure coefficient
τ	[–]	end depth aspect ratio
χ	[–]	normalized coordinate
ζ	[–]	shape parameter

Subscripts

c	critical
D	related to $(gD^5)^{1/2}$
e	end section
o	approach, upper
t	outflow jet
u	lower
v	full filling
w	standing wave
1	lower nappe
2	upper nappe

References

Biggiero, V. (1963). Sul tracciamento dei profili delle vene liquide (On the nappe geometry of liquid jets). *8 Convegno di Idraulica* Pisa A11: 1–19 [in Italian].

Carstens, M.R., Carter, R.W. (1955). Discussion to Hydraulics of the free overfall, by A. Fathy and M. Shaarawi. Proceedings ASCE *Journal of the Hydraulics Division* 81(Paper 719): 18–28.

Christodoulou, G.C. (1985). Brink depth in nonaerated overfalls. *Journal of Irrigation and Drainage Engineering* 111(4): 395–403.

Clausnitzer, B., Hager, W.H. (1997). Outflow characteristics from circular pipe. *Journal of Irrigation and Drainage Engineering* 123(10): 914–917.

Cola, R. (1967). Esame teorico e sperimentale dei fenomeni ondosi di vuotamento di una condotta a sezione circolare (Theoretical and experimental study of undular waves in a circular pipe). *Atti dell' Istituto Veneto di Scienze, Lettere ed Arti* 125: 257–294 [in Italian].

Cola, R. (1970). Sul moto permamente in prossimità del sbocco di una condotta a sezione circolare (On permanent motion close to a circular pipe outlet). *L'Acqua* 48(3): 70–79 [in Italian].

Delleur, J.W., Dooge, J.C.I., Gent, K.W. (1955). Influence of slope and roughness on the free overfall. Proceedings ASCE *Journal of the Hydraulics Division* 82 (HY4) (Paper 1038): 30–35.

Ferreri, G.B., Ferro, V. (1990). Efflusso non guidato di una corrente lenta da un salto di fondo in canali a sezione rettangolare (Non-guided outflow of a subcritical current over a drop in a rectangular channel). XXII *Convegno di Idraulica e Costruzioni Idrauliche* Cosenza 1: 195–213 [in Italian].

Hager, W.H. (1983). Hydraulics of plane free overfall. *Journal of Hydraulic Engineering* 109(12): 1683–1697; 110(12): 1887–1888.

Hager, W.H. (1993a). Abflussverhältnisse beim Endüberfall (Flow features of end overfall). *Österreichische Wasserwirtschaft* 45(1/2): 36–44 [in German].

Hager, W.H. (1993b). Ausfluss aus Rohren (Outflow from pipes). *Korrespondenz Abwasser* 40(2): 184–186 [in German].

Hager, W.H. (1998). Cavity outflow from a nearly horizontal pipe. *International Journal of Multiphase Flow* 25: 349–364.

Khan, A.A., Steffler, P.M. (1996). Modelling overfalls using vertically averaged and moment equations. *Journal of Hydraulic Engineering* 122(7): 397–402.

Marchi, E. (1992). On the free overfall. *Journal of Hydraulic Research* 31(6): 777–796.

Montes, J.S. (1992). A potential flow solution for the free overfall. Proceedigs Institution of Civil Engineers *Water Maritime & Energy* 96: 259–266; 112: 81–87.

Rajaratnam, N., Muralidhar, D. (1964a). Unconfined free overfall. *Journal of Irrigation and Power* 21(1): 73–89.

Rajaratnam, N., Muralidhar, D. (1964b). End depth for circular channels. Proceedings ASCE *Journal of the Hydraulics Division* 90 (HY2): 99–119; 90(HY5): 261–270; 90(HY6): 293–297; 91(HY3): 281–283; 92(HY1): 81.

Rajaratnam, N., Muralidhar, D., Beltaos, S. (1976). Roughness effects on rectangular free overfall. Proceedings ASCE *Journal of the Hydraulics Division* 102(HY5): 599–614; 103(HY3): 337–338.

Rouse, H. (1936). Discharge characteristics of the free overfall. *Civil Engineering* 6(4): 257–260.

Rouse, H. (1943). Discussion to Energy loss at the base of a free overfall, by W.L. Moore. *Transactions ASCE* 108: 1343–1392.

Smith, C.D. (1962). Brink depth for a circular channel. Proceedings ASCE *Journal of the Hydraulics Division* 88(HY6): 125–134; 89(HY2): 203–210; 89(HY3): 389–405; 89(HY4): 249–258; 89(HY6): 253–256; 90(HY1): 259.

Wallis, G.B., Crowley, C.J., Hagi, Y. (1977). Conditions for a pipe to run full when discharging liquid into a space filled with gas. *Journal of Fluids Engineering* 99(6): 405–413; 100(3): 136.

Chapter 12
Venturi Flume

Abstract Due to solid matter contained in sewage, the discharge in sewer technology is usually measured with a Venturi flume. This flume corresponds to a locally constricted channel normally without a bottom inset to force a transition between sub- and supercritical flows. Then, deposition of material in the approach flow channel can be often prevented, and this design is superior to a weir structure.

In this chapter, flumes of *long* and *short* structural length are presented. Based on an analysis of advantages and disadvantages, three flume types are retained and recommended for design. These are characterized by simplicity, economy and hydraulic performance. A fourth type of flume is mentioned, because of its large discharge range for a limited approach flow depth.

The design equation of a Venturi flume depends significantly on the effects of velocity of approach, constriction geometry and streamline curvature. For sufficiently large structures, effects of scale may be suppressed. A comparison with the discharge equations indicates a reasonable accuracy, and Venturi flumes of standard geometry may be designed with a rational approach.

12.1 Introduction

A Venturi flume (German: Venturikanal; French: Canal Venturi) is a discharge measurement structure involving a *local constriction*. A wide variety of flume geometries has been proposed, and the most important designs are discussed in the following.

In *longitudinal section* flumes either have a constant bottom slope, or the bottom has a local hump or sill, or even a weir type structure (Fig. 12.1a). For a relatively large tailwater level, the upstream level is thus artificially increased to reduce submergence effects. This bottom sill cannot be recommended because the significant advantage of the Venturi flume – corresponding to a flow without bottom constriction and thus unlimited sediment transport – is removed. If the flow is close to submergence, a bottom drop downstream of the flume is much more efficient (Fig. 12.1b). Therefore, a Venturi flume can be considered essentially as an *horizontal channel*, for which a simplified hydraulic approach is amenable.

W.H. Hager, *Wastewater Hydraulics*, 2nd ed., DOI 10.1007/978-3-642-11383-3_12,

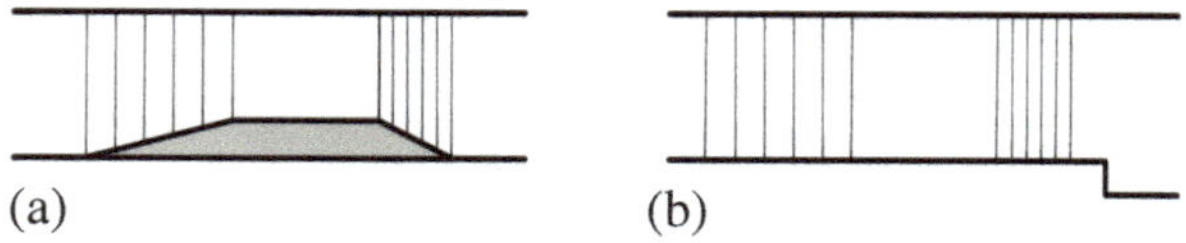

Fig. 12.1 Longitudinal section of Venturi flume with (**a**) bottom inset (not recommended) and (**b**) downstream bottom drop

The *flume geometry* at the constricted flume section is often trapezoidal, whereas the rectangular cross-section corresponds to the standard section. For a large ratio between the maximum and minimum discharges, the Venturi flume has a trapezoidal-constricted section with a more complex inset. The *standard section* of a Venturi flume is rectangular because of finish accuracy, simplicity and cost.

Venturi flumes are normally inserted in open channels, such as of rectangular or U-shaped shape. Palmer and Bowlus (1936) have suggested a design for the U-shaped profile that has been modified and improved by Wells and Gotaas (1958). This particular flume is presented in Sect. 12.2.6.

In a *rectangular approach flow channel*, the constriction may be obtained (Fig. 12.2):

- Either by constriction of a rectangular section,
- Or by constriction of the bottom width and keeping the top width constant. The constricted section is then of trapezoidal shape.

Flume configurations in *plan* can have a great geometrical variety. To standardize the flume geometries, two basic types are discussed (Fig. 12.3):

- Polygonal-shaped flume, and
- Circular-arc shaped flume.

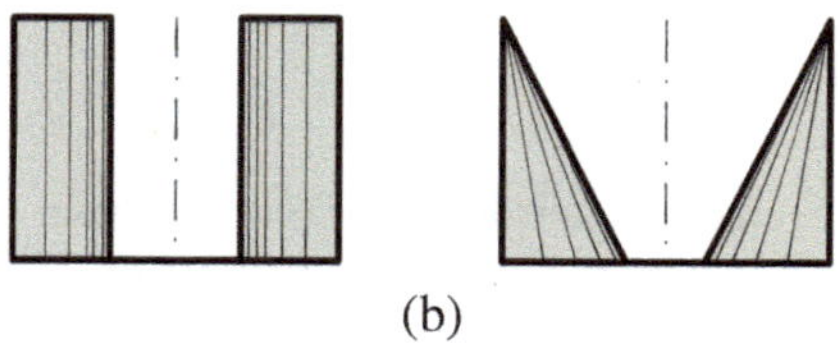

Fig. 12.2 Constricted flume cross-section for rectangular approach flow channel (**a**) rectangular and (**b**) trapezoidal constriction geometry

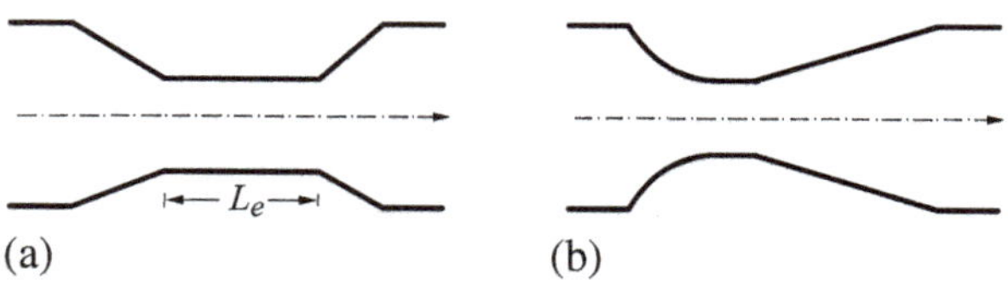

Fig. 12.3 Plan of Venturi flumes in a rectangular channel with (**a**) polygonal and (**b**) arc-shaped constriction inlet

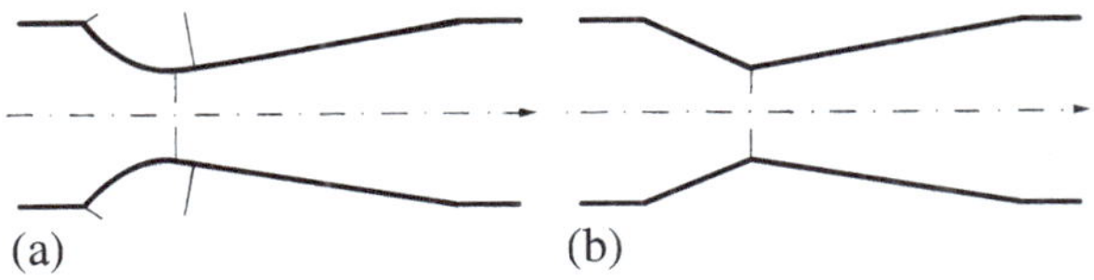

Fig. 12.4 (**a**) Khafagi long-throated flume, and (**b**) Cut-throat flume with a short throat

The polygonal-shaped flume can be simpler in finish and less expensive due to plan shaped elements, but flow separation from the walls make a hydraulic design more complex. The Venturi flume with an arc-shaped inlet configuration *without* a prismatic constriction reach looks elegant, performs well and is referred to as *Khafagi flume*. The Egyptian Anwar Khafagi (1912–1972) made his PhD-thesis at ETH Zurich in 1942.

Depending on the relative constriction length L_e/h_o and the inlet geometry, reference is made to short ($L_e/h_o < 1$) and long flumes. The distinction is mainly hydraulic, because the flow in a short flume with an abruptly varied intake geometry separates from the side walls, whereas flow separation is prevented in long and rounded flumes. Figure 12.4 shows two extremes, namely the *Khafagi flume* as a typical long-throated structure, and the *Cut-throat flume* (Sect. 12.3) as a typical short-throated structure.

In Sect. 12.2, the hydraulic design of the long-throated Venturi flume is presented, including the effects of streamline curvature. Then, in Sect. 12.3, the short-throated flume is discussed. Conclusions and design recommendations are given in Sect. 12.4.

12.2 Long-Throated Flume

12.2.1 *Discharge Equation*

The cross-sectional area F of a symmetrical trapezoidal profile of side slope m (horizontal): 1 (vertical) and base width b is

$$F = bh + mh^2. \tag{12.1}$$

The free surface width is $B_s = \partial F/\partial h = b + 2mh$. Critical (subscript c) discharge is thus defined based on the Froude number with $\mathsf{F} = 1$ (Chap. 6) or

$$Q_c = \frac{\left[g\left(b_c h_c + m_c h_c^2\right)^3\right]^{1/2}}{(b_c + 2m_c h_c)^{1/2}}. \tag{12.2}$$

Further, the critical energy head H_c is

$$H_c = h_c + \frac{Q_c^2}{2gF_c^2} \tag{12.3}$$

or, when eliminating Q_c with Eq. (12.2),

$$H_c = h_c + \frac{b_c h_c + m_c h_c^2}{2(b_c + 2m_c h_c)}. \tag{12.4}$$

With $y_c = m_c h_c / b_c$ as the relative filling and $Y = m_c H_c / b_c$ as the relative energy head, the dimensionless discharge $q = [m_c^3/(g b_c^5)]^{1/2} Q_c$ can be expressed with Eqs. (12.2) and (12.4) as

$$q = \frac{[y(1+y)]^{3/2}}{(1+2y)^{1/2}}, \tag{12.5}$$

$$Y = y + \frac{y(1+y)}{2(1+2y)}. \tag{12.6}$$

The relation between critical discharge and critical energy head involves thus relative filling y. For small values of y one has $q = y^{3/2}$, $Y = (3/2)y$, and therefore $y = (2/3)Y$ and $q = [(2/3)Y]^{3/2}$. By transforming these relations back to dimensional quantities, they are independent of m_c as for the rectangular profile with $m_c = 0$ (Chap. 6). Higher order terms in Y can be added by Taylor series, and the *critical discharge* in a symmetric trapezoidal profile is

$$q = [(2/3)Y]^{3/2}\,[1 + 0.70Y]. \tag{12.7}$$

For $Y \leq 2$, Eq. (12.7) deviates less than 1% from the mathematically exact expression. For $Y > 2$, an additional quadratic term Y^2 had to be added to the square bracket of Eq. (12.7). However, other effects discussed below then become significant.

Example 12.1 Compute the critical discharge for a trapezoidal Venturi flume with $m_o = 0$ (approach flow channel), $m_c = 2$ (critical section), for $b_o = 1.2$ m, $b_c = 0.50$ m and an approach flow depth $h_o = 0.87$ m!
Assuming first that $H_o \approx h_o = 0.87$ m, the relative critical energy head is $Y = m_c H_c / b_c = 2 \cdot 0.87/0.5 = 3.48$, thus $q = [(2/3)3.48]^{3/2}$ $[1 + 0.70 \cdot 3.48] = 12.14$ from Eq. (12.7) and $Q_c = q/[m_c^3/(g b_c^5)]^{1/2} = 12.14$ $(9.81 \cdot 0.5^5)^{1/2}\, 2^{-3/2} = 2.38$ m^{3}s^{-1}.

The approach flow energy head is $H = h_o + Q^2/(2g b_o^2 h_o^2) = 0.87 + 2.38^2/$ $(19.62 \cdot 1.2^2 0.87^2) = 1.13$ m, thus much larger than previously assumed. The effective critical discharge is determined in Example 12.2.

Both the critical flow depth h_c and the critical energy head H_c are *no directly measurable parameters*. At the critical section, i.e. at the flume constriction, the free surface is considerably sloped and the error in discharge is significant because of the inexact knowledge of the exact critical section. Therefore, the flow depth is measured in the approach flow channel of practically constant flow depth h_o. Further, for the usual bottom slopes of some pars pro mille, one may assume a compensation with the friction slope, and the energy head remains *essentially constant*. Accordingly, the approach flow energy head H_o is equal to the critical energy head H_c (Fig. 12.5). This concept of discharge measurement is absolutely identical with that of weirs.

In analogy to overflow structures (Chap. 10), the section of *depth observation* should be located about one to two times the approach flow depth upstream from the flume entrance. For a rectangular approach flow channel, conservation of energy head requires

$$H_o = h_o + \frac{Q^2}{2gb_o^2h_o^2} = H_c. \tag{12.8}$$

With the approach (subscript o) flow filling ratio $y_o = m_c h_o/b_c$, and the width ratio $\beta = b_o/b_c$, Eq. (12.8) may be written as

$$Y = y_o + \frac{q^2}{2\beta^2 y_o^2}. \tag{12.9}$$

Eliminating q with Eq. (12.7) yields further

$$Y = y_o + \frac{4Y^3}{27}\left[(1+0.7Y)/(\beta y_o)\right]^2. \tag{12.10}$$

Because the effect of y_o in the square bracket is small, one may approximate y_o with Y, and

$$Y = y_o + \frac{4Y}{27}\left[(1+0.7y_o)/\beta\right]^2. \tag{12.11}$$

Solving for Y and developing under the condition $(1+0.7y_o)/\beta << 1$ gives

$$Y = y_o\left[1 - \frac{4}{27\beta^2}(1+0.7y_o)^2\right]^{-1} \cong y_o\left[1 + \frac{4}{27\beta^2}(1+0.7y_o)^2\right]. \tag{12.12}$$

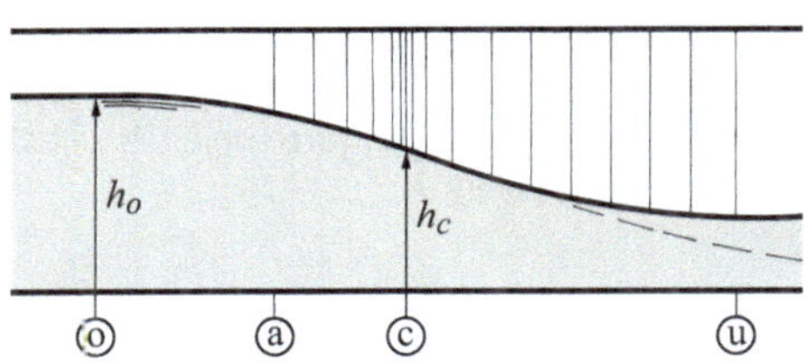

Fig. 12.5 Typical sections along Venturi flume including upstream (o), start (a), critical (c) and downstream (u) sections

Inserting this in Eq. (12.7), the explicit *discharge equation* is expressed as a function of flume geometry and measurable approach flow depth

$$\frac{Q}{m_c^{-3/2} g^{1/2} b_c^{5/2}} = [(2/3)y_o]^{3/2}[1+0.7y_o]\left[1+\frac{2}{9\beta^2}(1+0.7y_o)^2\right]. \quad (12.13)$$

Therefore, the discharge in the *symmetrical trapezoidal Venturi flume* with a rectangular approach flow channel depends essentially on the filling ratio $y_o = m_c h_o / b_c$ and the width ratio $\beta^{-1} = b_c / b_o$. Equation (12.13) is accurate provided the last term is smaller than 0.2.

Example 12.2 Recompute Example 12.1 using Eq. (12.13).
With $y_o = m_c h_o / b_c = 2 \cdot 0.87/0.50 = 3.48$ and $\beta = 1.2/0.5 = 2.4$, the relative discharge is $q = [(2/3)3.48]^{3/2}[1+0.70 \cdot 3.48]\ [1+(2/9 \cdot 2.4^2)(1+0.7 \cdot 3.48)^2] = 18.64$. The effect of the last term is considerable as it increases q by +53%. Its value is larger than 0.2 and Eq. (12.12) cannot be applied.

With a discharge $Q = 1.46 \cdot 2.38 = 3.47\ \mathrm{m^3s^{-1}}$, the approach flow velocity is $V_o = 3.47/(1.2 \cdot 0.87) = 3.32\ \mathrm{ms^{-1}}$, thus $H_o = 0.87 + 3.32^2/19.62 = 1.43$ m and thus $Y = 2 \cdot 1.43/0.50 = 5.73$. Inserting in the basic equation (12.7) gives $q = [(2/3)5.73]^{3/2}[1+0.70 \cdot 5.73] = 37.4$, thus three times more than in Example 12.1. The contraction of this Venturi flume with such a large approach flow velocity is too strong, and significant surface waves are set up in the approach flow channel. The observation of h_o is thus complicated and the accuracy in discharge poor. The *upper limit* of approach flow Froude number is discussed in the next section.

In the *rectangular flume* the terms with y_o drop and the discharge is given from Eq. (12.13) as

$$\frac{Q}{(gb_c^5)^{1/2}} = [(2/3)(h_o/b_c)]^{3/2}\left[1+(2/9)(b_c/b_o)^2\right]. \quad (12.14)$$

12.2.2 *Discussion of Result*

In the general Eq. (12.13) for critical discharge in a trapezoidal Venturi flume, two additional effects compared to the conventional discharge equations appear:

- Approach flow depth ratio $m_c h_o / b_c$, or the observation parameter $m_c h_o / b_o$, and
- Constriction rate $\beta^{-1} = b_c / b_o$.

From Chap. 6, the dimensionless discharge may be interpreted as the Froude number in the trapezoidal channel. The present approach thus is based on *Froude similitude.*

Typically, the constriction rate is around $\beta^{-1}= 0.4$. Then, the right hand side of Eq. (12.13) may be approximated with the power function

$$\frac{Q}{\left(gb_c^5/m_c^3\right)^{1/2}} = 0.95y_o^{1.73}. \tag{12.15}$$

For other values of β, corresponding expressions could be derived. Compared to Eq. (12.13), the dimensional unity of Eq. (12.15) is no more guaranteed because of the exponent 1.73, provided that the $Q(h_o)$ relation is introduced. Such equations have often been proposed for practical purposes.

As mentioned, the present model involves exclusively gravity and inertia forces. If the flume length quantities are smaller than some limit dimensions, other effects can become dominant, such as:

- *Viscosity* characterized by Reynolds number $\mathsf{R}_c = V_c(4R_{hc})/\nu$ where R_h is the hydraulic radius and ν the kinematic viscosity,
- *Surface tension* characterized by the Weber number $\mathsf{W}_c = V_c/[\sigma/(\rho h_c)]^{1/2}$ where σ is surface tension and ρ fluid density, and
- *Boundary roughness* k_s/R_{hc} with k_s as the equivalent sand roughness height.

All these effects are negligible, provided that the:

- Fluid is almost water,
- Channel widths have a minimum of $b_o = 0.30$ m and $b_c = 0.10$ m, and
- Flow depth in the approach flow channel is at least $h_o = 50$ mm.

So-called *scale effects* can thus be suppressed with models not too small in size. Note that the scale between prototype and model is not significant, because *minimum lengths* specified previously have to be retained in a hydraulic model. A detailed discussion on scale effects give Ackers et al. (1978), Kobus (1984) and Miller (1994).

As mentioned in Chap. 13, the *Froude number* in the approach flow channel has to be limited because the transition from sub- to supercritical flow in the Venturi flume is weak, otherwise. If the constriction rate is too large, *undular waves* develop in the approach flow channel and accurate surface measurement becomes impossible. According to Chap. 13, the maximum approach flow Froude number should be $\mathsf{F}_o = Q/(gb_o^2h_o^3)^{1/2} = 0.5$. In Example 12.2, an unrealistically high value of $\mathsf{F}_o = 1.14$ resulted.

12.2.3 *Effect of Streamline Curvature*

The effect of streamline curvature has not been addressed so far, because the analysis was based on an one-dimensional flow with an uniform velocity and an hydrostatic pressure distribution. These basic assumptions of hydraulics are satisfied as long as free surface slope and curvature are small. According to Chap. 1, these effects increase in the horizontal channel with $(\mathrm{d}h/\mathrm{d}x)^2$ and $h(\mathrm{d}^2h/\mathrm{d}x^2)$, respectively. The free surface curvature is thus related to the quantity flow depth h, and the *shallow water approximation* used so far requires that this term be typically smaller than 10^{-1}.

Starting computations with the *generalized energy equation* Eq. (1.29) instead of Eq. (12.3), the energy head in a horizontal rectangular channel is

$$H = h + \frac{Q^2}{2gb^2h^2}\left(1 + \frac{2hh'' - h'^2}{3}\right). \tag{12.16}$$

For *critical flow* to occur, the sufficient condition requires $\mathrm{d}H/\mathrm{d}h = (\mathrm{d}H/\mathrm{d}x)/(\mathrm{d}h/\mathrm{d}x) = 0$. Differentiating Eq. (12.16) gives (Hager 1985)

$$\frac{Q^2}{gb_2^2h_c^3} = 1 + \frac{2hh'' - h'^2 - h^2\,h'''/h'}{3}. \tag{12.17}$$

Compared to the one-dimensional approach presented in Sect. 12.2.1, additional terms appear:

- $h'^2 = (\mathrm{d}h/\mathrm{d}x)^2$ square of free surface slope,
- $hh'' = h(\mathrm{d}^2h/\mathrm{d}x^2)$ relative free surface curvature, and
- $h^2h'''/h' = h^2(\mathrm{d}^3h/\mathrm{d}x^3)/(\mathrm{d}h/\mathrm{d}x)$ relative change of free surface curvature.

These terms can be determined with the conventional energy equation by a *singular point analysis*. Consider the lowest order of approximation of Eq. (12.16) for energy head

$$H = h + \frac{Q^2}{2gb^2\,h^2}. \tag{12.18}$$

For potential flow over a horizontal bottom, or approximately for a flow in which the friction slope is compensated for by the bottom slope, the energy head is *constant* along the flume, and one has $\mathrm{d}H/\mathrm{d}x = \mathrm{d}^2H/\mathrm{d}x^2 = \mathrm{d}^3H/\mathrm{d}x^3 = 0$. For constant discharge, successive differentiation of Eq. (12.18) gives

$$H' = h'\left(1 - \frac{Q^2}{gb^2\,h^3}\right) - \frac{Q^2b'}{gb^3\,h^2} = 0, \tag{12.19}$$

$$H'' = h''\left[1 - \frac{Q^2}{gb^2\,h^3}\right] + \frac{3Q^2\,(b'h + bh')^2}{gb^4\,h^4} - \frac{Q^2\,(2b'h + b''h)}{gb^3\,h^3} = 0, \tag{12.20}$$

$$H''' = h''' \left[1 - \frac{Q^2}{gb^2 h^3} \right] - \frac{12Q^2 \left(b'h + bh'\right)^3}{gb^5 h^5} + \frac{9Q^2 \left(b'h + bh'\right) \left(b''h + 2b'h' + bh''\right)}{gb^4 h^4}$$
$$- \frac{Q^2 \left(b'''h + 3b''h' + 3b'h''\right)}{gb^3 h^3} = 0. \tag{12.21}$$

To lowest approximation the critical discharge is defined as $Q^2/(gb^2h^3) = 1$, or $\mathsf{F} = 1$. According to Eq. (12.19), the location of critical flow is then $b' = 0$, i.e. at the constricted channel section. Inserting $b' = 0$ in Eqs. (12.20) and (12.21) yields

$$\frac{3}{h} h'^2 - \frac{b''h}{b} = 0, \tag{12.22}$$

$$-\frac{12(bh')^3}{b^3 h^2} + \frac{9bh' \left(b''h + bh''\right)}{b^2 h} - \frac{b'''h + 3b''h'}{b} = 0. \tag{12.23}$$

Equation (12.22) valid at the critical point $x = x_c$ indicates for the square of the free surface slope

$$h'^2 = \frac{b''h^2}{3b}, \tag{12.24}$$

For a physical solution, b'' must be positive with the extreme $b' = 0$ corresponding to a minimum of the width function $b(x)$, with $b' = 0$ and $b'' > 0$. *Transitional flow* develops at a *local width constriction* as has already been explained in Chap. 6. Equation (12.23) further demonstrates that

$$hh'' = \frac{4}{3} h'^2 - \frac{2}{3} (h^2/b) b'' + \frac{1}{9} b'''/bh' \tag{12.25}$$

Inserting the expression for h'^2 from Eq. (12.24) and considering the *circular-arc intake geometry* of which $b''' = 0$, then

$$hh'' = -(2/9) h^2 b''/b. \tag{12.26}$$

Higher order derivatives can be obtained in this manner as (Hager 1985)

$$h^2 h'''/h' = -(5/3) h^2 b''/b. \tag{12.27}$$

The effect of streamline curvature can be accounted for by the parameter $u = h^2b''/b$, i.e. the square of critical flow depth divided by the critical width and the radius of curvature at the critical section. Inserting in Eqs. (12.16) and (12.17) gives

$$H = h + \frac{Q^2}{2gb^2 h^2} \left[1 - \frac{7u}{27} \right], \tag{12.28}$$

$$\frac{Q^2}{gb^2 h^3} = \left[1 - \frac{8u}{27}\right]^{-1}, \quad (12.29)$$

and thus

$$H = \frac{3h}{2}\left[1 - \frac{5u}{27}\right]. \quad (12.30)$$

As mentioned in Sect. 12.1, the observational quantity is not the critical depth h_c but the *critical energy head* H_c. Introducing the curvature parameter $U = H^2 b_c''/b_c$ instead of u their relation is

$$u = \frac{4U}{9}\left[1 + \frac{1}{6}U\right], \quad (12.31)$$

and Eq. (12.30) may be written as

$$H = (3/2)h\left[1 - \frac{20}{243}U\right]. \quad (12.32)$$

Inserting Eqs. (12.31) and (12.32) in Eq. (12.29) gives finally (Hager 1985)

$$Q = (2/3)^{3/2} b_c (gH_c^3)^{1/2}[1 + (14/243)U]. \quad (12.33)$$

For *parallel streamlines* ($U = 0$), Eq. (12.33) simplifies to Eq. (12.7). For a specific energy head H_c, the discharge conventionally computed is thus always too small. It should be noted that only first order terms have been retained in Eq. (12.33), and terms in U^2 are neglected. The result is thus valid up to a maximum of $U = 1$.

Example 12.3 Given a rectangular Venturi flume of widths $b_o = 1.50$ m, $b_c = 0.60$ m and rounding radius $R_c = 1/b_c'' = 2.0$ m. How large is the discharge for an approach flow depth $h_o = 0.28$ m?
Assuming a velocity head of $V_o^2/(2g) = 0.05$ m, then $H_o = 0.33$ m and $U = 0.33^2/(2 \cdot 0.60) = 0.09$. The discharge is accordingly with Eq. (12.33) $Q = (2/3)^{3/2}\ 0.60(9.81 \cdot 0.33^3)^{1/2}[1 + (14/243)0.09] = 0.195\ \mathrm{m^3s^{-1}}$. For the estimated energy head of $H_o = 0.33$ m, the corresponding approach flow depth is $h_o = 0.31$ m, and the assumption in $V_o^2/(2g)$ was too large.

For the second iteration, consider $H_o = 0.29$ m, the curvature parameter is then $U = 0.29^2/(2 \cdot 0.60) = 0.07$, and $Q = (2/3)^{3/2}\ 0.60(9.81 \cdot 0.29^3)^{1/2}$ $[1 + (14/243)0.07] = 0.160\ \mathrm{m^3s^{-1}}$. Then, $h_o = 0.28$ m, as required.

In this example, the effect of streamline curvature is negligible because $1 + (14/243)0.07 = 1.004$ only.

For the *trapezoidal flume*, the computations are much more complex, because of the additional parameters $s = mh/b$ and $t = m''h/b''$. According to Hager (1985) the effect of t is insignificant, and the main correction factor is $[1 + U/(6S)] = [1 + H_c/6m_cR_c)]$ with R_c as the curvature of the flume section. For $m_cH_c/b_c > 1$ and $H_c^2/(b_cR_c) < 1$, the *generalized discharge equation* instead of Eq. (12.7) reads

$$q = [(2/3)Y]^{3/2}[1 + 0.70Y]\left[1 + H_c/(6m_cR_c)\right]. \qquad (12.34)$$

This result applies to weakly two-dimensional potential flow.

12.2.4 *Submerged Flow*

For *free flow* with a transition from the sub- to the supercritical condition, the discharge Q of a certain flume depends only on the approach flow depth h_o. For *submerged flow*, the approach flow depth h_o varies not only with discharge but also with the downstream flow depth h_u. The transition between free and submerged flows is referred to as the *modular limit* (subscript L) σ_L (German: Grenzeinstau; French: Limite de submersion). The modular limit is an important characteristic of discharge measurement structures.

The following *types of free surface flow* may be observed in Venturi flumes as a function of downstream flow depth (Fig. 12.6):

- Supercritical flow with shockwave formation (Chap. 16),
- Supercritical flow in downstream flume expansion, followed by a direct hydraulic jump (Chap. 7),
- Undular downstream flow (Chap. 7), and
- Deeply submerged flow with a surface jet (Chap. 10).

Both cases (a) and (b) of Fig. 12.6 are typical features of free flume flow, and direct hydraulic jumps are a definite index for the free flow pattern. The transition from

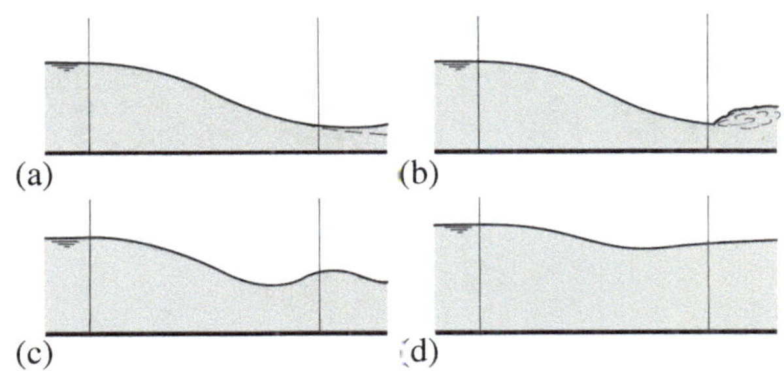

Fig. 12.6 Flow in Venturi flume subjected with an increasing downstream flow depth. (**a**) Supercritical flow with shockwave formation, (**b**) direct hydraulic jump, (**c**) undular hydraulic jump, and (**d**) submerged flow with a surface recirculation

the direct to the undular hydraulic jump occurs for approach flow Froude numbers between 1 and 2, or so, and the latter cannot be regarded as an index for either a free or a submerged flume flow. This transition is complicated by the development of jumps in an expanding channel portion, and a more detailed information on the flume hydraulics is needed.

The *modular limit* is often expressed as a percentage, with 70% indicating for example that the structure gets submerged as the downstream flow depth h_u increases over 70% of the approach flow depth h_o. For $h_u/h_o < 70\%$, the flow is free, and submerged otherwise. For Venturi flumes with a sloping bottom, or even a bottom sill, these flow depths are related to the bottom elevation at the critical section. For usual Venturi flumes, the modular limit is often as high as 70 – 80% and thus extremely high as compared to sharp-crested structures. The modular limits of standard broad-crested weirs (Chap. 10) and Venturi flumes are comparable. Once the downstream flow depth is larger than the modular limit depth, the discharge is extremely sensitive to h_u, and the *accuracy* of a flume gets poor. It is recommended that Venturi flumes be used exclusively for *free flow conditions*, therefore. If submergence is a problem, the entire structure can be lifted, as shown in Fig. 12.1b.

12.2.5 *Comparison with Observations*

Hager (1993) reviewed the current knowledge on Venturi flumes. Although flumes were introduced at the end of World War I in the USA, and Italians used flumes extensively already prior to World War II, the first proposition of a flume type structure is due to Khafagi (1942). A *Khafagi flume* has the following features (Fig. 12.7):

- Circular-arc intake geometry,
- Rectangular cross-section,
- Constricted section without a prismatic reach, and
- Diffusor splay of approximately 1:10.

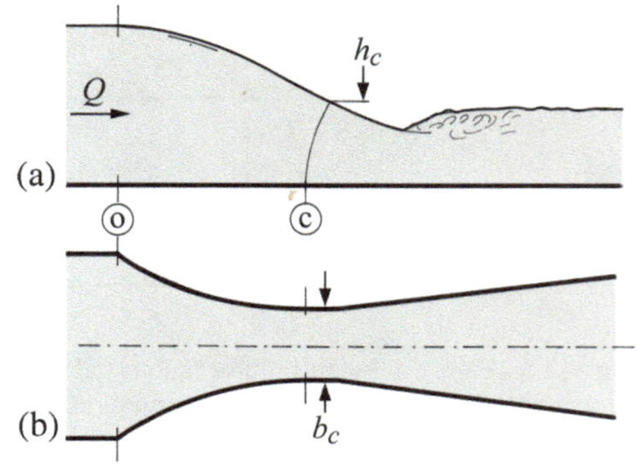

Fig. 12.7 Khafagi flume (**a**) longitudinal section with the *critical length* and (**b**) plan (Khafagi 1942)

Although a computational procedure is given, that includes approximately an effect of streamline curvature, no experimental data are available. For $U<0.5$, the design equations published are in agreement with Eq. (12.33), as do also those published by Blau (1960).

Barczewski and Juraschek (1983) have analyzed a model series of Khafagi flumes and found an overall effect of the relative flow depth h_o/b_c, that can be regarded as an index of streamline curvature. In the domain of application, Eq. (12.33) compares well with their data.

Robinson and Chamberlain (1960) introduced the *trapezoidal Venturi flume* with a constant side slope m and plan transition sections. For $m = 3^{1/2}$, 1 and $3^{-1/2}$, and critical widths of $b_c = 0$, 50 and 100 mm 'good' agreement with the conventional approach resulted. This flume looks not really economic, and it has been designed for irrigation rather than sewage techniques. The modular limit is between 80 and 90%.

Bos and Reinink (1981) have developed the previous flume. The side slope was 1:1 and flow separations are suppressed if the convergence angle is smaller than 1:2. It was further verified that the conventional approach applies provided that the flow is free and the approach flow depths are relatively small. The modular limit was between 85 and 90%, but this design is again spacious and has a small bottom drop in addition.

The study of Skogerboe and Hyatt (1967a) refers mainly to *submerged*, long-throated Venturi flumes with a plan bottom and a polygonal plan (Fig. 12.8). This flume has three locations of depth reading, namely one in the converging and one in the diverging reaches, and one to observe the minimum flow depth h_m. For the channel with a critical width $b_c = 0.305$ m (1 foot), the *discharge-head equation* in [ft; s] units is

$$Q = 2.87h_a^{1.525}. \tag{12.35}$$

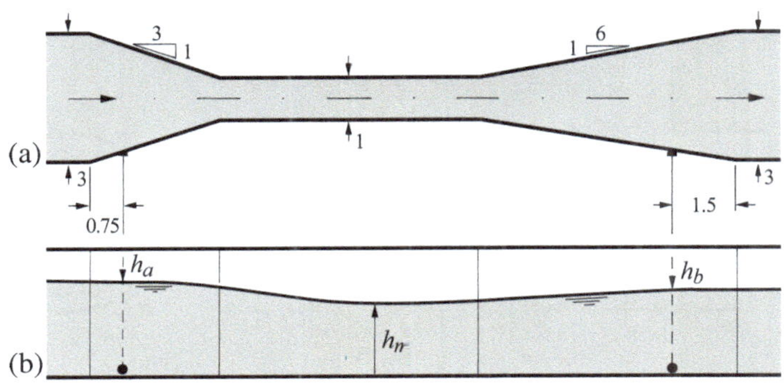

Fig. 12.8 Long-throated rectangular Venturi flume of Skogerboe and Hyatt (1967a). (**a**) Plan, (**b**) section, numbers in feet (1 ft = 0.305 m), (•) observational locations

Compared to Eq. (12.33), Eq. (12.35) does not account for any other effect. The modular limit S_L is about 88%, and submerged discharge Q_s can be included with the *submergence factor* $\psi = Q_s/Q$. With $S = h_u/h_o$ as the depth ratio, the data can be expressed as

$$\psi = 1 - (S - S_L)^{1.5}. \tag{12.36}$$

One may state as a conclusion of this review that nothing speaks against long-throated flumes, except for their spacious and thus *uneconomic design*. Based on economic reasons, a long-throated flume is practically always less effective than short-throated flumes. An exception is possibly the Khafagi flume as a transition between the too extreme designs.

12.2.6 *Venturi Flume in Manhole*

A special type of Venturi flume was introduced by Palmer and Bowlus (1936). The *portable device* can be inserted at the downstream side of a manhole with an U-shaped section of diameter D. The constricted section has a length $L_e = D$ and a trapezoidal geometry of side slope $m = 2$ (Fig. 12.9). The invert is increased by a bottom sill of height $s_e/D = 1/3$ (Wells and Gotaas 1958), and the optimum base width is $b_e = D/3$. The upstream ramp has a slope 1:3 and the downstream drop is abrupt. The observational location of the approach flow depth h_o should be between $(1/2)D$ and $(3/2)D$ upstream from the ramp.

A comparison between observed and conventionally computed discharges (see also Chap. 13) starts at 0.97 for $(h_o - s_e)/D \rightarrow 0$, is equal to 1.0 for $(h_o - s_e)/D = 0.75$ and increases then rapidly to 1.1 for $(h_o - s_e)/D \cong 0.92$. In the main domain of the trapezoidal section, i.e. for $0.1 < (h_o - s_e)/D < 0.85$, the *discharge coefficient* varies only within ±3%.

The *modular limit* varies slightly with the approach flow depth h_o, with the upper limit of 85%. Wells and Gotaas (1958) have thus significantly contributed to discharge measurement in sewers. Figure 12.10 has been taken from Komiya et al. (1981) who report further results. It is not clear whether this last development is

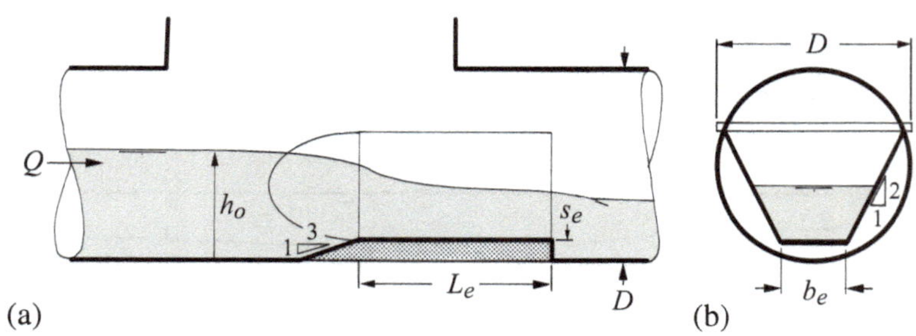

Fig. 12.9 Palmer-Bowlus flume, modified by Wells and Gotaas (1958) (**a**) longitudinal section, (**b**) transverse section

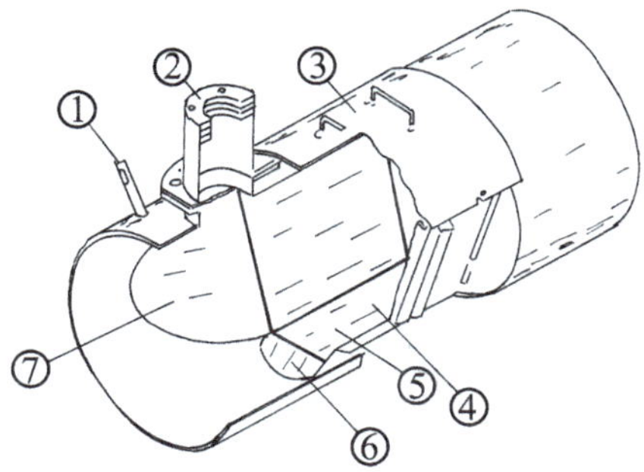

Fig. 12.10 Palmer-Bowlus flume modified by Komiya et al. (1981). ① Suspension, ② transmission for depth evaluation, ③ cover, ④ bottom inset, ⑤ constriction, ⑥ bottom ramp, ⑦ side wings

currently still in use, or if a further advance on this design was made. Note also that portable Venturi-type flumes are also described in Chap. 13, with a special reference to their performance in the wastewater environment.

12.3 Short Venturi Flume

The first short Venturi flume, referred to also as *Cut-throat Flume* has been introduced only in 1967 by Skogerboe and Hyatt (1967b). The plan flume geometry was inserted in a rectangular channel, with a contraction ratio of 1:3, and a diverging ratio of 1:6. A prismatic constriction reach is missing, however. The flume can be applied for both free and submerged flow conditions, yet two depth readings are necessary. The approach flow depth h_a is observed at (1/3) of the contracted reach, and the downstream flow depth h_b at (1/6) from the flume end (Fig. 12.11). The major *advantages* of the configuration are:

- Simple geometry,
- Similarity between model and prototype, and
- Observation within the flume length.

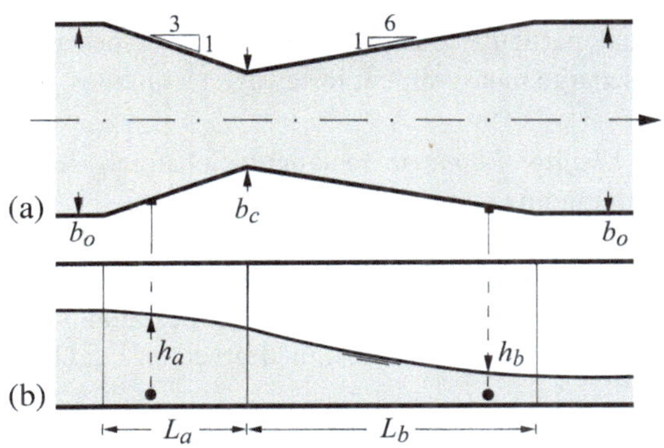

Fig. 12.11 Cut-throat Flume according to Keller (1984). (**a**) Plan, (**b**) longitudinal section with locations of observation. (•) Observational positions

The Cut-throat Flume was developed by Keller (1984) using a width ratio of $b_c/b_o = 0.52$. The *discharge-head equation* for free flow can be expressed based on his data as (Hager 1993)

$$Q = (2/3)^{3/2}\left(gb_c^2h_a^3\right)^{1/2}\left[1 + 0.25h_a/b_c\right]. \tag{12.37}$$

This is similar to Eq. (12.33), with a drawdown effect on the flow depth h_a. Deviations of the data from Eq. (12.37) are less than ±5%. The *modular limit* S_L varies mainly with the relative flow depth h_a/b_c as

$$S_L = (2/3)(b_c/h_a)^{1/3}. \tag{12.38}$$

Equation (12.38) is valid for $0.5 < S_L < 0.95$ and shows definitely that shallow water flows are much less-sensitive to submergence than are flows with a significant streamline curvature. Therefore, submerged flows should not be used for discharge measurement.

Example 12.4 Given a Cut-throat Flume with $b_o = 1.0$ m, thus $b_c = 0.52$ m, $h_a = 0.47$ m and $h_b = 0.32$ m. What is the discharge?
The modular limit is $S_L = (2/3)(0.52/0.47)^{1/3} = 0.69$ from Eq. (12.38), thus $h_{bL} = 0.69 \cdot 0.47 = 0.33$ m. With $h_b = 0.32$ m < 0.33 m, free flow is just established.

The discharge is $Q = (2/3)^{3/2}(9.81 \cdot 0.52^2 0.47^3)^{1/2}[1 + 0.25(0.47/0.52)] = 0.35$ m^{3}s^{-1} from Eq. (12.37). The upstream Froude number is $\mathsf{F}_a = Q/(gb_a^2h_a^3)^{1/2} = 0.35/(9.81 \cdot 0.84^2 0.52^3)^{1/2} = 0.35 < 0.5$, i.e. surface waves are of no concern.

12.4 Design Recommendations

Currently, there seem to be no reasons to favour long-throated flumes, except for local particularities. Therefore, three types of flumes can be recommended for discharge measurement in sewage techniques:

- *Khafagi Flume* in rectangular channel, designed according to Eq. (12.33) by including streamline curvature effects,
- *Cut-throat Flume* as a simple and compact structure including the observational locations, and
- *Palmer-Bowlus Flume* as a trapezoidal structure that may be inserted in a manhole. It can be designed based on Eq. (12.34).

Until today, the *trapezoidal Venturi flume* in a rectangular approach channel with a Khafagi inlet design has not yet received systematic attention. This structure has

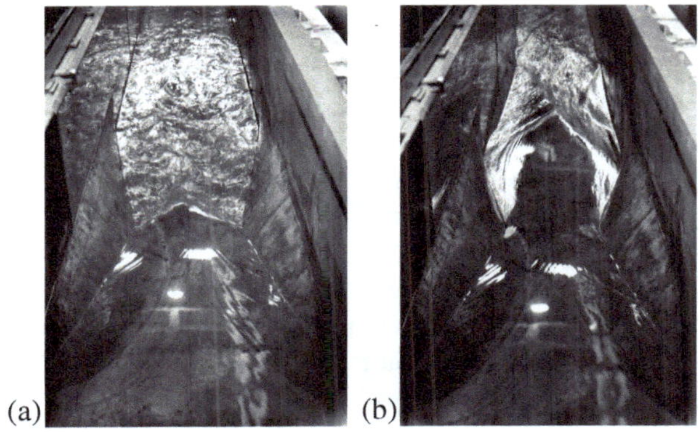

Fig. 12.12 Trapezoidal Venturi flume as a laboratory model. (**a**) Free flow with a direct hydraulic jump (VAW 26784), (**b**) formation of shock-waves for supercritical downstream flow (VAW 26782)

been analyzed by Hager (1989) and a prototype has performed well over 10 years on a large Swiss sewage treatment station. This design can be of interest if the ratio of maximum to minimum discharges is large and the loss of head is limited. Figure 12.12 shows a model that was designed with Eq. (12.34).

With these indications, Venturi flumes can be designed for *free flow* to at least $\pm 3\%$ accuracy, provided small flow depths and discharges are excluded. Indications on the modular limit are available, which is with typically 80% high. Thus, the recommended flumes are amenable to a hydraulic design, if they have a *standard finish*. Submerged flow adds considerably to the inaccuracy and should be prevented.

Notation

b	[m]	channel width
D	[m]	pipe diameter
F	[m^2]	cross-sectional area
F	[−]	Froude number
g	[ms^{-2}]	gravitational acceleration
h	[m]	flow depth
H	[m]	energy head
k_s	[m]	equivalent sand roughness height
L_e	[m]	length of construction
m	[−]	side slope of trapezoidal section
$1/n$	[m$^{1/3}$s^{-1}]	Manning's roughness coefficient
q	[−]	critical unit discharge
Q	[m^{3}s^{-1}]	discharge
R_c	[m]	radius of curvature at constriction
R_h	[m]	hydraulic radius

R	[−]	Reynolds number
s	[−]	$= mh/b$ relative depth in trapezoidal channel
s_e	[m]	drop height
S	[−]	$= h_u/h_o$ submergence height ratio
S_c	[−]	critical slope
S_o	[−]	bottom slope
S_L	[−]	modular limit
t	[−]	$= m''h/b''$ relative curvature in trapezoidal channel
u	[−]	curvature parameter related to h
U	[−]	curvature parameter related to H
V	[ms^{-1}]	cross-sectional velocity
W	[−]	Weber number
x	[m]	streamwise coordinate
y	[−]	$= m_c h_c/b_c$ relative critical flow depth
y_o	[−]	$= m_c h_o/b_c$ relative approach flow depth
Y	[−]	$= m_c H_c/b_c$ relative critical head
z	[m]	bottom coordinate
β	[−]	$= b_o/b_c$ width ratio
ψ	[−]	submergence factor
ν	[m^{2}s^{-1}]	kinematic viscosity
ρ	[kgm^{-3}]	fluid density
σ	[Nm^{-1}]	surface tension

Subscripts

a	inlet
b	outlet
c	critical
e	inset
N	uniform
o	upstream
s	submerged
u	downstream

References

Ackers, P., White, W.R., Perkins, J.A., Harrison, A.J.M. (1978). *Weirs and flumes for flow measurement.* John Wiley & Sons: Chichester, New York.

Barczewski, B., Juraschek, M. (1983). Ermittlung der Abflussbeziehung von Venturikanälen (Determination of discharge equation of Venturi flumes). *Wasserwirtschaft* 73(5): 149–154 [in German].

Blau, E. (1960). Die modellmässige Untersuchung von Venturikanälen verschiedener Grössen und Form (Model studies on Venturi flumes of different sizes and shape). *Veröffentlichung* 8. Forschungsanstalt für Schiffahrt, Wasser- und Grundbau. Akademie-Verlag: Berlin [in German].

Bos, M.G., Reinink, Y. (1981). Required head loss over long-throated flumes. *Journal of Irrigation and Drainage Division* ASCE 107(IR1): 87–102.

Hager, W.H. (1985). Critical flow conditions in open channel hydraulics. *Acta Mechanica* 54: 157–179.

Hager, W.H. (1989). Venturikanäle (Venturi flumes). *Gas – Wasser – Abwasser* 69(7): 389–395 [in German].

Hager, W.H. (1993). Venturikanäle langer und kurzer Bauweise (Venturi flumes of long and short finish). *Gas – Wasser – Abwasser* 73(9): 733–744 [in German].

Keller, R.J. (1984). Cut-throat flume characteristics. *Journal of Hydraulic Engineering* 110(9): 1248–1263; 112(11): 1105–1107.

Khafagi, A. (1942). Der Venturikanal (The Venturi flume). Versuchsanstalt für Wasserbau, *Mitteilung* 1. Leemann: Zürich [in German].

Kobus, H. ed. (1984). *Symposium on scale effects in modelling hydraulic structures*. Technische Akademie: Esslingen, Germany.

Komiya, K., Utsumi, H., Satori, T. (1981). Flow test on a 500 mm Palmer-Bowlus flume. *Flow – Its measurement and control in science and industry* 2: 613–619.

Miller, D.S. ed. (1994). *Discharge characteristics*. IAHR Hydraulic Structures Design Manual 8. Balkema: Rotterdam.

Palmer, H.K., Bowlus, F.D. (1936). Adaption of the Venturi flumes to flow measurements in conduits. *Trans. ASCE* 101: 1195–1239.

Robinson, A.R., Chamberlain, A.R. (1960). Trapezoidal flumes for open-channel flow measurement. *Trans. American Society of Agricultural Engineers* 3: 120–124.

Skogerboe, G.V., Hyatt, M.L. (1967a). Analysis of submergence in flow measuring flumes. *Journal of the Hydraulics Division* ASCE 93(HY4): 183–200; 1968 94(HY3): 774–794; 94(HY6): 1530–1531.

Skogerboe, G.V., Hyatt, M.L. (1967b). Rectangular cut-throat flow measuring flumes. *Journal of Irrigation and Drainage Division* ASCE 93(IR4): 1–13; 1968 94(IR3): 357–362; 94(IR4): 527–530; 1969 95(IR3): 433–439.

Wells, E.A., Gotaas, H.B. (1958). Design of Venturi flumes in circular conduits. *Trans. ASCE* 123: 749–775.

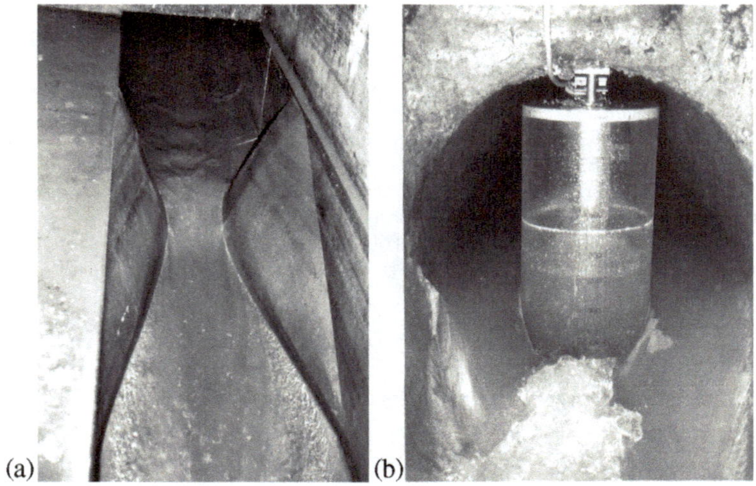

(**a**) Venturi flume and (**b**) Mobile Venturi flume in stormwater sewers

Chapter 13
Mobile Discharge Measurement

Abstract The discharge of a sewer has to be determined often over a short period only. This can serve as design basis, and an accurate discharge measurement is required only later. In this chapter, the so-called mobile discharge measurement for rectangular, trapezoidal and circular profiles is introduced. Starting from the simplest configuration of a circular cylinder in a rectangular channel, the discharge-head curves, the modular limits, the submergence effect and practical questions are discussed for a number of designs. At optimal sites, the discharge can then be determined up to ±5% accuracy.

In addition, the thin-plate Venturi flume is presented that is proposed for larger channels or rectangular sewers of width up to 3 m, with an access by a standard manhole. Mobile discharge measurement with overflow structures is not recommended, because the mounting precision, the security and accuracy are demonstrated to be unsatisfactory.

13.1 Introduction

Because the *discharge* (German: Durchfluss; French: Débit) of a sewer or a sewage channel of a treatment station is a design quantity, its value is of significance. On the one hand, processes can be controlled with the discharge, and the flow discharged to a receiving water of a combined sewer system depends highly on the discharge on the other hand. In addition, the discharge characteristic is an important design basis for systems that have to be refurbished. Another problem that depends significantly on the discharge is the determination of *parasite water* in a sewer system, typically due to sewer leakage if the groundwater level is high. In the future, the quality of numerical analyses of a sewer system has to be evaluated, based on the knowledge of discharge at selected locations.

Currently, the discharge is evaluated by two distinctly different methods:

- *Permanent* discharge measurement at an observational location, with a recording of data and basis for process control, and

W.H. Hager, *Wastewater Hydraulics*, 2nd ed., DOI 10.1007/978-3-642-11383-3_13,

- *Mobile* discharge measurement normally not planned previously, at various locations of a system to derive important decisions for design, refurbishing and adaptation.

In the first case, a *discharge measurement* structure with all the components necessary is available. The second case is normally improvised, not really equipped and allows only an estimation of discharge. In this chapter, the latter method is improved with selected devices that allow for a more precise determination of discharge at almost arbitrary locations of a sewer system.

The mobile Venturi flume (German: Mobiler Venturikanal; French: Canal Venturi mobile) is considered in this chapter, whereas Chap. 12 relates to the design of conventional Venturi flumes. The mobile Venturi flume has performed well in practice. Overfall and other mobile discharge measurement structures are also considered, and their advantages are highlighted.

13.2 Mobile Venturi Flume

13.2.1 *Principle of Measurement*

The mobile Venturi flume corresponds to an *inverted* conventional flume (Chap. 12). Normally, the constriction elements are located at the channel walls, whereas these are concentrated axially of a prismatic channel. The latter have a rectangular, trapezoidal or circular cross-sectional shape due to simplicity and accuracy in finish.

A *Venturi body* allows to determine discharge by a single depth reading. This is particularly suited for sewage containing solid matter, and methods involving the measurement of either pressure or velocity are not needed. The geometry of the Venturi body can be arbitrary, as long as the transition from sub- to supercritical flow occurs (Chap. 6). In practice, the shape of the body is carefully selected, however, to satisfy conditions of exact finish. To prevent large-scale separation from the body, that depend normally on fluid viscosity, the Venturi body is rounded and has a definite geometry. A particular shape is of course circular, and the *circular cylinder* is usually taken as the standard Venturi body. Figure 13.1 shows a rectangular prismatic channel with a standard Venturi body, once in conventional and once in the mobile arrangement.

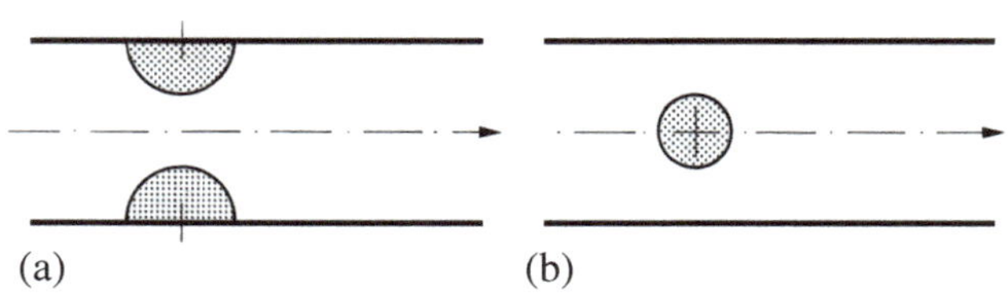

Fig. 13.1 Venturi flume in (**a**) conventional and (**b**) mobile arrangement

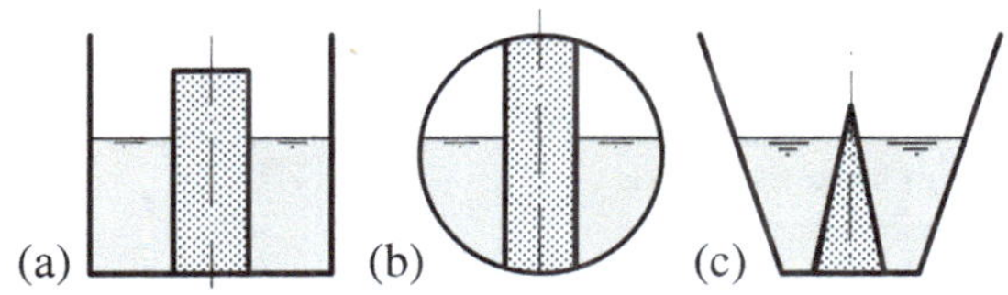

Fig. 13.2 Section through mobile Venturi flumes of (**a**) rectangular, (**b**) circular and (**c**) trapezoidal cross-section

The circular-shaped Venturi corresponds to an overall optimum. A *circular cone* may be better relative to the discharge-head curve of trapezoidal channels. Its geometry involves an additional parameter (Fig. 13.2). The *measurement principle* of the mobile and conventional Venturi flumes is analogue. Transitional flow is forced with a local channel constriction. Of interest are the discharge-head curve, and the modular limit.

13.2.2 *Mobile Venturi Flume in Rectangular Channel*

Free Flow

Diskin (1963) first described a mobile arrangement for discharge measurement by using Venturi bodies similar to bridge piers instead of circular cylinders. He determined discharge-head curves for various pier geometries and found a modular limit of about 80% (Chap. 10). For raw sewage, solid matter such as paper or cloth wind around the body and the discharge measurement is affected.

Hager (1985a) introduced the *circular cylinder* mounted on the lateral walls of a channel. A motor set on the axis of the cylinder turns the body such that solid matter is transported downstream, and the discharge measurement remains unaffected. The hydraulics of this device are influenced by streamline curvature effects (Chap. 12), with details analyzed by Hager (1985b).

Ueberl and Hager (1994) tested the standard Venturi body in a rectangular channel (Fig. 13.3). For free flow, *shockwaves* are set up in the downstream channel with the typical shape as discussed in Chap. 16. These are a definite index for supercritical flow, i.e. for free Venturi flow. Compared to the original determination of the approach flow depth h_o, the energy head H is measured in the cylinder. The Venturi body is thus provided with holes of about 5 mm diameter arranged along the front of the cylinder, with an interdistance of about 50 mm. If this line of holes is directed against the flow direction within a maximum deviation of about ±5, the flow depth in the cylinder is equal to the *stagnation head*, i.e. the energy head H of the flow. In the following, the discharge Q of such a flume is determined as a function of cylinder diameter D_V and rectangular channel width B.

According to the conventional critical (subscript k) flow theory, the *critical discharge* is (Chap. 12)

$$Q_k = (B - D_V)[g(2H/3)^3]^{1/2}. \tag{13.1}$$

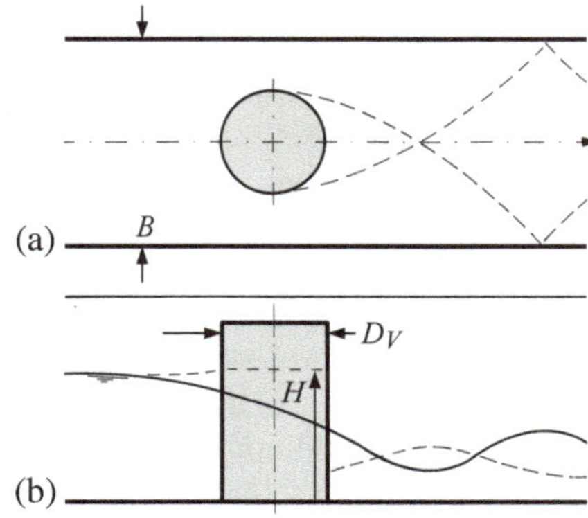

Fig. 13.3 Measurement arrangement for mobile Venturi flume, extended with line of holes to observe stagnation depth H in cylinder (**a**) plan, (**b**) longitudinal section

This expression has to be corrected for effects of viscosity, surface tension and streamline curvature. The first two effects can be removed by choosing sufficiently large flume and cylinder dimensions such as $B \geq 0.30$ m, $(B - D_V) \geq 0.10$ m and $H \geq 0.10$ m. Then, the discharge follows the *Froude similitude* and the effective discharge Q can be expressed as $Q = qQ_k$, where q is the streamline curvature effect that depends exclusively on the curvature parameter $U = 2H^2/[D_V(B–D_V)]$, as derived in Chap. 12. To the first approximation, $q = 1+(14/243)U$ from Eq. (12.33); Ueberl and Hager (1994) expanded this result for large U to

$$q = 1 + \frac{(14/243)U}{1 + (1/7)(1 - \delta)U}. \tag{13.2}$$

Here $\delta = D_V/B < 1$ is the rate of constriction. For large values of U, the curvature effect tends to $q_\infty = q(U >> 1) = 1 + 0.4/(1 - \delta)$. For small δ, this value q_∞ is thus smaller than for larger δ. In the latter case, the streamline curvature is increased, and the pressure along the cylinder reduces such that the discharge is increased. The *optimum* constriction degree (German: Verbauungsgrad; French: Degré d'obstruction) is $\delta = 0.40$, with relevant values in practice of $0.3 < \delta < 0.6$, depending on modular limit and capacity. Then, Eq. (13.2) simplifies for $U < 5$ to

$$q = 1 + \frac{0.058U}{1 + 0.08U}. \tag{13.3}$$

The degree of constriction thus has no direct influence on discharge.

Figure 13.4 shows the *effect of asymmetry* of a Venturi body on discharge, with $q_z = Q/Q_a$ where Q_a corresponds to the axial (subscript a) position. The transverse normalized coordinate $Y_z = 2y_z/(B - D_V)$ varies between –1 and +1, corresponding to the body at either side wall at the limits. According to Fig. 13.4, the deviations in discharge are less than $\pm 1\%$ for $|Y_z| < 0.5$, and the exact axial positioning of a Venturi body is *not* important. In addition, a body located close to a wall has a larger discharge at otherwise identical flow conditions. This feature simplifies greatly the

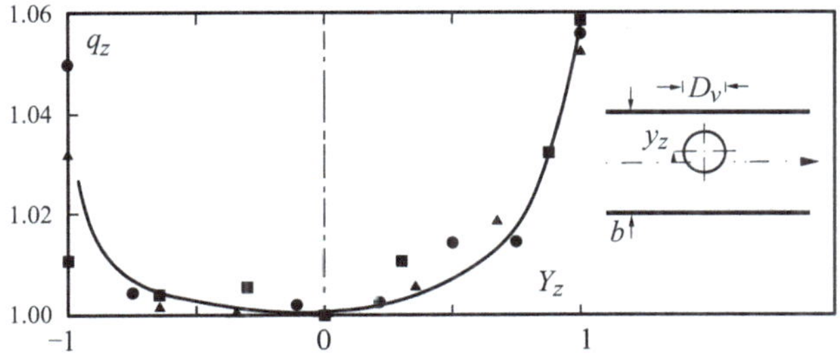

Fig. 13.4 Effect of body asymmetry Y_z on discharge correction q_z for constriction rate $\delta =$ (■) 0.32, (▲) 0.40, (•) 0.52

practical use of mobile Venturi bodies, because the discharge may often not be fully stopped. The exact positioning would also be complicated by sediment on the sewer bottom, or poor optical conditions that make such a procedure almost impossible.

Example 13.1 Given a rectangular channel of width $B = 1.2$ m. What is the discharge for an energy head of $H = 0.57$ m for a Venturi body of diameter $D_V = 0.60$ m?
With a constriction rate of $\delta = 0.6/1.2 = 0.50$ and a curvature parameter of $U = 2{\cdot}0.57^2/[(1.2–0.6)0.6] = 1.81$ the discharge correction due to streamline curvature is $q = 1 + 0.058{\cdot}1.81/(1 + 0.08{\cdot}1.81) = 1.09$ from Eq. (13.3). From Eq. (13.1) $Q_k = (1.2–0.6)[9.81(2{\cdot}0.57/3)^3]^{1/2} = 0.44$ m^{3}s^{-1}, such that $Q = Q_k q = 0.44{\cdot}1.09 = 0.48$ m^{3}s^{-1}.

Modular Limit

The modular limit (German: Grenzeinstau; French: Submersion limite) corresponds to a downstream flow depth $h_u = h_L$ for which the approach flow is just not yet submerged. The depth ratio $y_L = h_L/H$ varies with the constriction rate δ and the curvature parameter U, the latter effect being negligible if $0.2 < \delta < 0.7$. According to Ueberl and Hager (1994) the modular limit of a standard mobile Venturi flume is then

$$y_L = 0.84 - 0.35\delta. \tag{13.4}$$

In the usual domain of constriction rates δ, the downstream depth can thus reach up to 60–75% of the energy head, and for the optimum constriction rate $\delta = 0.4$, the modular limit is $y_L = 0.70$. Compared to short overflow structures the mobile Venturi flume is thus *insensitive* to submergence. Compared to the conventional flume (Chap. 12), the limit is somewhat reduced for a mobile Venturi flume.

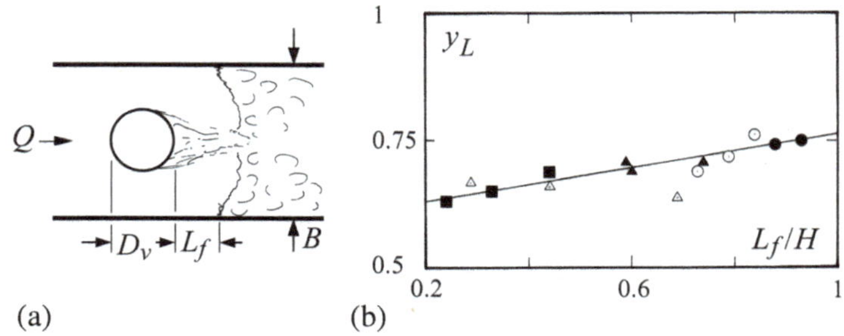

Fig. 13.5 Front distance L_f of jump toe from cylinder end (**a**) geometrical configuration, (**b**) observations for δ = (•) 0.26, (○) 0.32, (▲) 0.40, (△) 0.52, (■) 0.60

The distance L_f of the *front* of the hydraulic jump from the downstream cylinder end (Fig. 13.5a), or the relative length $\lambda = L_f/D_V$ is a *visual index* of modular limit. According to Fig. 13.5b)

$$y_L = 0.6 + (1/6)(L_f/H), \tag{13.5}$$

and thus after elimination of y_L with Eq. (13.4)

$$L_f/H = 1.5(1 - 1.5\delta). \tag{13.6}$$

For $0.2 < \delta < 2/3$ the front of the hydraulic jump is located always downstream of the cylinder end. If the front is along the cylinder, the flow is submerged, and Eq. (13.1) does not apply.

Example 13.2 What is the limit downstream depth for free flow in Example 13.1?

With $\delta = 0.50$, one has from Eq. (13.4) $y_L = 0.84–0.35·0.5 = 0.67$. The limit downstream flow depth is thus $h_L = 0.67·0.57 = 0.38$ m.

Assuming that the toe of the hydraulic jump is just where the shock fronts have reached the side walls (Fig. 13.3a), one may compute the corresponding downstream flow. Up to the toe (subscript 1), the energy head is nearly constant. For H = 0.57 m and Q = 0.480 m^{3}s^{-1}, the downstream flow depth is computed to h_1 = 0.13 m for which H_1 = 0.137 + 0.48/(19.62·1.2^{2}0.137^2) = 0.571 m. For an average approach flow velocity V_1 = 0.48/(1.2·0.137) = 2.92 ms^{-1}, the approach flow Froude number is $\mathsf{F}_1 = V_1/(gh_1)^{1/2}$ = 2.92/(9.81·0.137)$^{1/2}$ = 2.52 and thus the ratio of sequent depths $Y = 2^{1/2}\mathsf{F}_1–(1/2)$ = 1.41·2.52–0.5 = 3.06 from Eq. (7.18), and $h_2 = Yh_1$ = 3.06·0.137 = 0.42 m. The downstream energy head is H_2 = 0.42 + 0.48^2/(19.62·1.2^{2}0.42^2) = 0.466 m. For this case, one would have $h_u/H = h_2/H$ = 0.42/0.57 = 0.74 and thus practically the same as for the modular limit.

The *submergence effect* can be accounted for by the parameter $y_u = (h_u - h_L)/(H - h_L)$, and the submergence rate is (Chap. 10)

$$\psi = Q_s/Q. \tag{13.7}$$

For $\psi = 1$, the submerged (subscript s) discharge is equal to the free discharge, and complete submergence occurs for $\psi = 0$, i.e. $h_u = H$. From experiments, no effects of both U and δ on ψ have been observed. For $y_u < 0.6$, the submergence is $\psi = 1 - (y_u/1.8)^2$. To attain the asymptotic value $\psi(y_u = 1) = 0$ this has to be generalized as (Ueberl and Hager 1994)

$$\psi = \left[1 - (y_u/1.8)^2\right]\left[1 - y_u^{10}\right]. \tag{13.8}$$

For $y_u = 0.5$, the discharge reduction is thus only 10%. For precise discharge measurement, only the *free flow* condition is recommended, with a laboratory accuracy of about ±1.5% (Fig. 13.6). For submerged flow, the accuracy is lower than ±5%.

Example 13.3 What is discharge in Example 13.1 for a downstream depth $h_u = 0.50$ m?
With $h_u = 0.50$ m, and $H = 0.57$ m, $h_L = 0.38$ m (Example 13.2) one has $y_u = (0.5–0.38)/(0.57–0.38) = 0.63$, such that the submergence rate is $\psi = [1–(0.63/1.8)^2][1–0.63^{10}] = 0.88 \cdot 0.99 = 0.87$ from Eq. (13.8), and the discharge $Q = 0.87 \cdot 0.48 = 0.417$ m^{3}s^{-1}.

Further details on the mobile Venturi flume are given in Sect. 13.3.

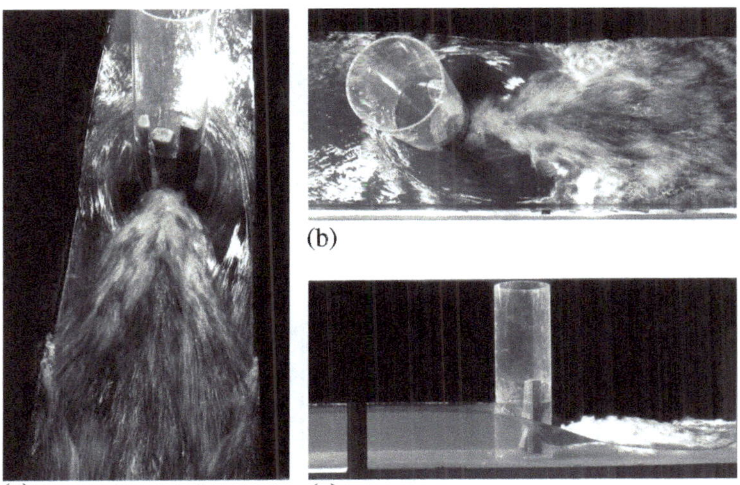

(a) (b) (c)

Fig. 13.6 Views of mobile Venturi flume in rectangular laboratory channel (**a**) against flow direction, (**b**) top view and (**c**) side view

13.2.3 *Mobile Venturi with Circular Cone*

Often, a mobile flume in a rectangular channel requires a significant head difference between the up- and downstream water elevations for free flow. For larger discharges, a *trapezoidal channel* with a circular cylinder may be better suited. In sewage technology, the rectangular channel is a standard open channel, however. To improve the discharge-head relation for a wide-range of discharges, one may use a *circular cone* instead of a circular cylinder, therefore. Figure 13.7 shows both arrangements.

Hager (1986) expanded the analysis of streamline curvature to rectangular channels with a cone. The *discharge correction* due to streamline curvature is

$$q = 1 + \frac{\alpha U}{1 + \alpha U} \tag{13.9}$$

with $\alpha = 1/27 = 0.037$ for relative cone inclinations $S_K = mH/(B-D_V) > 0.5$, where m (horizontal): 1 (vertical) is the cone slope. The critical discharge based on the conventional approach for a trapezoidal profile is from Eq. (12.7)

$$\left[\frac{m^3}{g(B-D_V)^5}\right]^{1/2} Q = \left[(2/3)mH/(B-D_V)\right]^{3/2}\left[1 + 0.70\frac{mH}{B-D_V}\right]. \tag{13.10}$$

Here, D_V is the base diameter of the cone. If the approach flow depth h_o instead of the energy head H is measured, then refer to Sect. 12.2.1.

Example 13.4 Given a rectangular channel of width $B = 0.90$ m, in which a cone is inserted of base diameter $D_V = 0.55$ m and side slope $m = 0.50$ (opening angle 26.5°). What is the discharge for an energy head $H = 0.42$ m? The parameter is $S_K = mH/(B-D_V) = 0.5{\cdot}0.42/(0.9\text{–}0.55) = 0.60$, thus $[0.5^3/9.81(0.9\text{–}0.55)^5]^{1/2}Q_k = 1.56Q_k = [(2/3)0.6]^{3/2}[1 + 0.7{\cdot}0.6] = 0.359$ according to Eq. (13.10), corresponding to $Q_k = 0.23$ m^{3}s^{-1}.

The effect of streamline curvature is with $U = 2H^2/[(B-D_V)D_V] = 2{\cdot}0.42^2/[(0.9\text{–}0.55)0.55] = 1.83$ from Eq. (13.9) equal to $q = 1 + 0.037{\cdot}1.83/(1 + 0.037{\cdot}1.83) = 1.063$, i.e. $Q = 1.063Q_k = 0.245$ m^{3}s^{-1}.

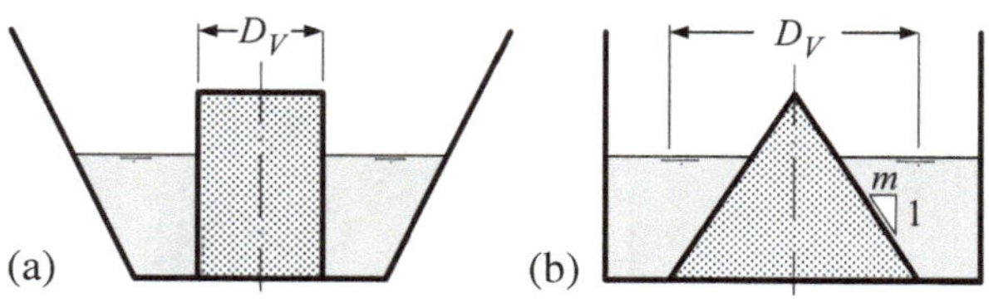

Fig. 13.7 Mobile Venturi flume, (**a**) circular cylinder in trapezoidal channel, (**b**) circular cone in rectangular channel

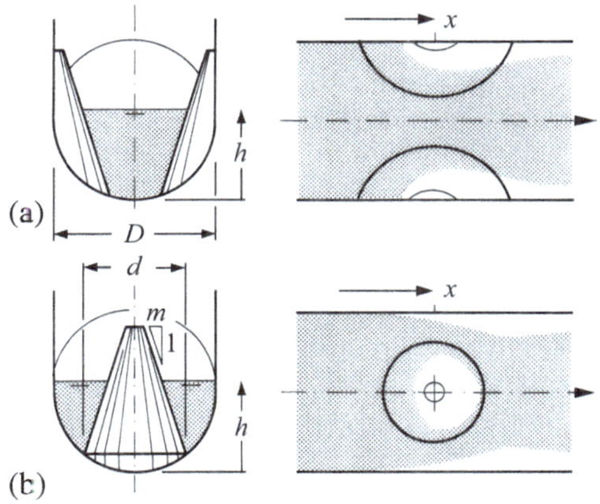

Fig. 13.8 Venturi flume in U-shaped profile (**a**) modified Palmer-Bowlus flume, (**b**) central arrangement with a circular cone

Until today, no experiences are available on cones with pressure taps. Therefore, the approach flow depth h_o is measured and the energy head H computed with an estimated discharge Q. The effective discharge is then determined iteratively. The submergence characteristics have so far also not been evaluated.

Hager and Züllig (1987) applied this Venturi flume for practical purposes. Figure 13.8 shows a modified version of the Palmer-Bowlus flume (Chap. 12) in a *U-shaped channel*. Mounted on a base that is also the bearing of the cone, a compact structure was developed. The device is inserted with a horizontal strut into the manhole. A battery driven motor on the top of the cone rotates the Venturi body, either permanently or intermittently to inhibit clogging by solid matter. The Venturi body can be completely immersed without a major effect on the discharge measurement (Fig. 13.9). Suspended particles such as silt pass the device easily because of the *horseshoe vortex* generated at the pier front, and granular material settling upstream from the device without an effect on the discharge measurement.

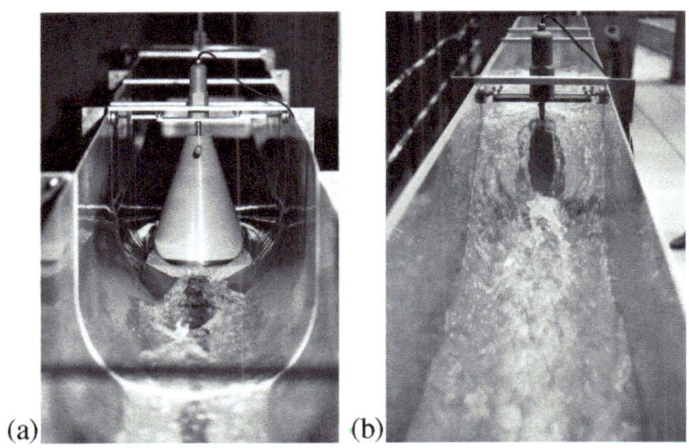

Fig. 13.9 Flow in a mobile Venturi flume of cone shape for (**a**) small and (**b**) large discharge

Fig. 13.10 Modified mobile Venturi flume for discharge measurement in U-shaped manhole channel (Hager and Züllig 1987)

A significant disadvantage of the *standard manhole* is the geometrical inaccuracy of the U-shaped section. Because the manhole banks (Chap. 14) are usually hand made, the cross-sectional geometry of the U-shaped profile may be accurate to less than 1 cm. As the discharge depends linearly on the width $(B-D_V)$, the accuracy of the device may be poor. The proposal of Hager and Züllig (1987) refers to a mobile flume inserted in a *prefabricated* U-shaped channel of which the dimensions are accurate to the mm. Thus, the geometry of the flume is well defined. The cone is inserted in a piece of channel, the motor is mounted on the strut and the flow depth is controlled with an ultrasonic gage for surface detection. The data can be treated electronically, for example, it can be connected with a sampler. Data loggers can be used for reproduction of the hydrograph as a basis for management of a sewer reach. The fixation and sealing of the device is made with selected inflatable hoses at its upstream end (Fig. 13.10).

13.3 Mobile Venturi Flume in Circular Pipe

13.3.1 *Basic Device*

Hager (1988) introduced the mobile flume in the circular pipe. Based on the inaccuracies of U-shaped manhole profiles (Sect. 13.2.3), the device is inserted in the circular sewer at the manhole outlet. The commercially finished sewer has an accuracy in diameter D of about 1 mm. The Venturi body is a vertical circular cylinder of diameter d, of which the dimensions are more accurate than for cones. Figure 13.11 shows the basic arrangement at a manhole *outlet*.

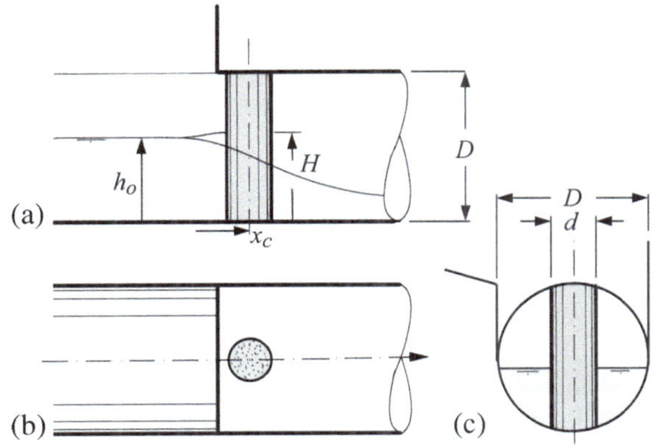

Fig. 13.11 Arrangement of mobile Venturi flume *downstream* of a standard manhole (**a**) longitudinal section, (**b**) plan, (**c**) transverse section

The cross-sectional shape of a cylinder at the critical section involves the difference between a partially-filled flow in a pipe and the cylinder. With $\delta = d/D$ as the constriction rate of the cylinder, and with $y = h/D$ as the filling ratio, one has

$$F/D^2 = (4/3)y^{3/2}[1 - (1/4)y - (4/25)y^2] - \delta[y - (1/12)\delta^2]. \tag{13.11}$$

Critical flow is conventionally defined as (Chap. 6)

$$\frac{Q^2}{gD^5} = \frac{(F/D^2)^3}{\mathrm{d}(F/D^2)/\mathrm{d}y} \tag{13.12}$$

and the corresponding energy head ratio $Y = H/D$ is

$$Y = y + (1/2)\frac{F/D^2}{\mathrm{d}(F/D^2)/\mathrm{d}y}, \tag{13.13}$$

with the dimensionless surface width

$$\mathrm{d}(F/D^2)/\mathrm{d}y = y^{1/2}[2 - (5/6)y - (3/4)y^2] - \delta. \tag{13.14}$$

For a specific filling $y < 1$ and a known constriction rate δ, the discharge can thus be defined as a function of y and δ. Further, the filling ratio y can be eliminated between Eqs. (13.12) and (13.13) to obtain a relation $Q(H)$ for the device (Fig. 13.12).

An approximation for the *discharge relation* is with the exponent $\sigma = 0.525 - 0.36\delta$

$$Q_k/(gD^5)^{1/2} = (Y/1.45)^{1/\sigma}. \tag{13.15}$$

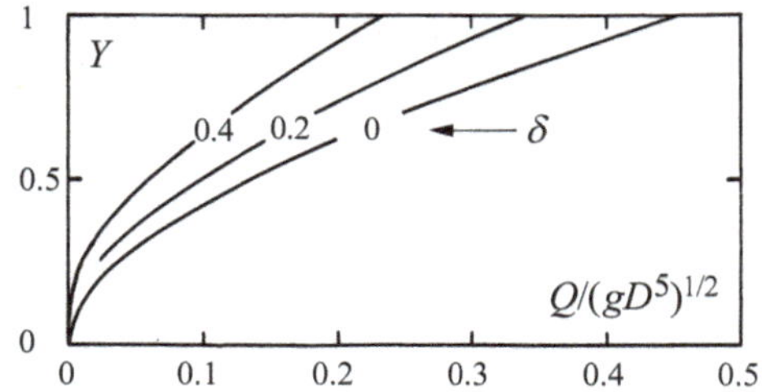

Fig. 13.12 Relative energy head $Y = H/D$ versus dimensionless discharge $Q/(gD^5)^{1/2}$ for variable constriction rate $\delta = d/D$

The *curvature effect* varies for this device exclusively with y/δ^2. Based on tests with $Y < 1$ and for constriction rates $0.25 < \delta < 0.35$, it can be expressed as (Hager 1988)

$$q = 0.985 + 0.205Y. \tag{13.16}$$

The modular limit is nearly 80%, independent of Y.

Example 13.5 Given a pipe of diameter $D = 0.70$ m with a mobile Venturi body of diameter $d = 0.20$ m. The observed energy head is $H = 0.27$ m. What is the discharge?
The constriction rate is $\delta = 0.2/0.7 = 0.286$ and the relative energy head $Y = 0.27/0.70 = 0.386$. With $\sigma = 0.525–0.360 \cdot 0.286 = 0.422$ or $1/\sigma = 2.37$, the discharge according to the conventional approach is $Q_k/(gD^5)^{1/2} = (0.386/1.45)^{2.37} = 0.0434$, thus $Q_k = 0.0434(9.81 \cdot 0.7^5)^{1/2} = 0.0558$ m^{3}s^{-1}. If Eqs. (13.11) to (13.14) are used, then $Q_k = 0.0479(9.81 \cdot 0.7^5)^{1/2} = 0.0615$ m^{3}s^{-1} (+10%). The effect of streamline curvature is $q = 0.985 + 0.205 \cdot 0.386 = 1.064$, thus $Q = 1.064 \cdot 0.0615 = 0.0654$ m^{3}s^{-1}.

The *ideal constriction rate* of the mobile Venturi flume inserted in a pipe is $\delta = 0.30$, i.e. smaller than in the rectangular channel. For $0.25 < \delta < 0.35$, the *discharge capacity* of the device is about $Q/(gD^5)^{1/2} = 0.35$. Figure 13.13 shows various views of the device. Because commercially-finished pipes are available only for specific diameters, such as $d = 0.15$ m and $D = 0.50$ m, the discharge rating scale can be directly mounted on the cylinder. Thus, the discharge has to be determined only once, and may be read simply for application. From the photographs, the flows with a small discharge are seen to be very smooth. For larger discharges, such as under a rainfall discharge, turbulence in the approach flow channel is increased but no larger flow perturbations both in the up-and downstream portions of the cylinder are noted. The effect of *bottom slope* was not evaluated, but it should be smaller than 1% to prevent the formation of hydraulic jumps close to the cylinder. This method of discharge measurement is thus not suited for steeper sewers.

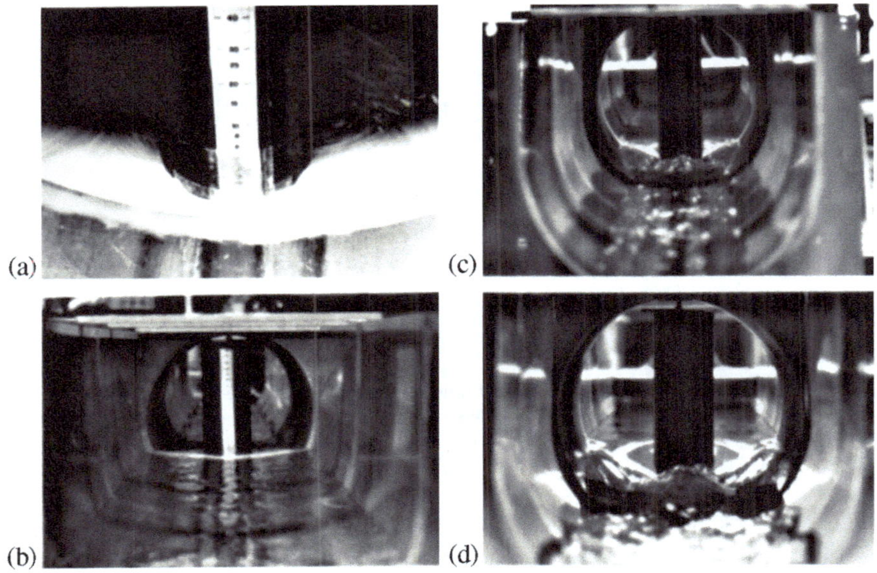

Fig. 13.13 Views on Venturi body in circular pipe

A problem with the permanent installation of a mobile flume is *solid matter*, mainly paper or cloth obstructing the critical section. Before further studies on the long-time observation in prototype sewers are being finished, only the *mobile* measurement of discharge is recommended. A further paper to this topic was presented by Samani et al. (1991).

13.3.2 *Optimized Design*

Kohler and Hager (1997) introduced the mobile flume mounted *upstream* from a sewer manhole. Thus the geometries of both the sewer and the *plexiglass pipe* are accurate to at least ±1 mm if commercially fabricated. Figure 13.14 shows a sketch of the arrangement, also referred to as the *Pipe Flume*.

The original arrangement of mobile discharge measurement (13.3.1) has two serious disadvantages that are removed with the optimized design:

- *Stagnation zone* upstream from the cylinder is relatively turbulent, and the hydraulic head fluctuation on the device may be as large as ±5%,
- Effect of *pipe slope* on the location of depth reading is not negligible, as for all flumes.

The optimized device is provided with a series of holes drilled axially into the cylinder, with an interdistance of about $0.3d$ through which water enters the pipe. The

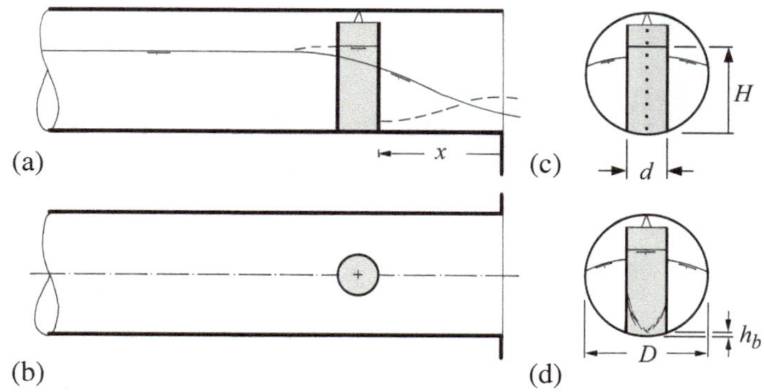

Fig. 13.14 Circular cylinder as mobile flume in a circular pipe *upstream* from manhole. (**a**) Longitudinal section, (**b**) plan, (**c**) view from upstream with holes arranged vertically, (**d**) downstream view

filling level H of the cylinder corresponds exactly to the stagnation depth and can be measured with a reduced surface fluctuation of typically ±0.5 mm, excluding the effect of pipe slope. The depth reading can be either made visually from the manhole through the Plexiglass, or with a pressure cell mounted on the bottom of the device and connected to a data logger.

For *positioning*, the device is shifted from the manhole into the upstream sewer and fixed with a clam-type mechanism. The device should be secured with a rope to the manhole against loss under storm water conditions. No problems of sealing have been experienced because both the pipe and the bottom of the device are accurate in shape.

Because the positioning of a Pipe Flume may be complicated by poor access conditions, its axis can deviate relative to the vertical by the angle η (Fig. 13.15a). The *effect of distortion* η on the corresponding head H_η, or the normalized head H_η/H, was determined as a function of relative discharge $q = Q/(gD^5)^{1/2}$. Note that

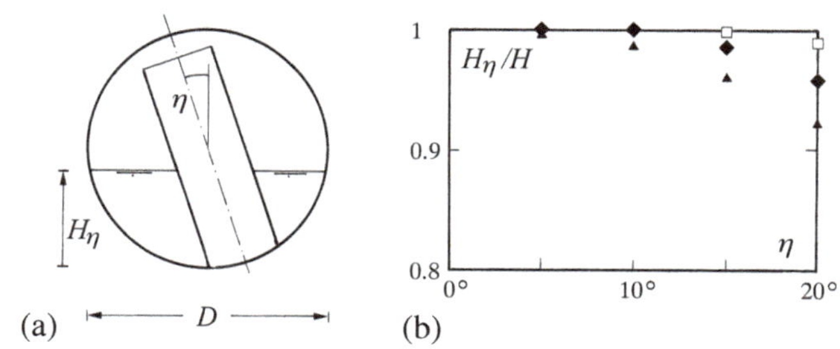

Fig. 13.15 Effect of distortion (**a**) definition sketch, (**b**) normalized head H_η/H as a function of distortion angle η for $q\times10^{-2}$= (▲) 0.24, (♦) 1.47, (□) 19.6

the cylindrically-shaped bottom of the device ensures no longitudinal distortion. Figure 13.15b) shows that H_η/H decreases both as the distortion angle η and the relative discharge increase. For a distortion angle smaller than $\pm 10°$, the accuracy is better than $\pm 1\%$. Such a large distortion angle can be visually detected while mounting the device, provided an auxiliary axial line is added on the downstream cylinder face.

The *cylinder position* may be (Fig. 13.14):

- Either at the downstream end of a pipe outlet into the atmosphere, such as for end overfalls (Chap. 11),
- Or in a sewer outlet to a manhole with a laterally guided flow in a U-shaped profile.

With $X = x/D$ where x is the streamwise coordinate measured from the end section, the effect of outlet is negligible if $X \geq 2.5$. Therefore, the device may be positioned at any location except too close to the end section of a pipe discharging into the atmosphere. In the following, the arrangement upstream from a *standard manhole* is considered.

The *effect of bottom slope* of a circular sewer on the stagnation depth was determined up to $S_o = 3\%$. No effect of pipe slope was detected and the optimized Pipe Flume is highly advantageous in this regard. The pipe slope is limited by the presence of hydraulic jumps. As the constriction rate δ increases, the jump migrates upstream. Bottom slopes S_o larger than 3% are too large for accurate discharge measurement, because of the *stability* of the approach flow, and therefore should be excluded.

The *modular limit* of the Pipe Flume decreases with the filling ratio $Y = H/D$. In all cases, the modular limit $y_L = h_L/H$ was larger than 60% with h_L as the limiting downstream flow depth. This modular limit is comparable to short-throated Venturi flumes (Chap. 12).

The *discharge-head equation* of a Pipe Flume was derived in Sect. 13.3.1. Computations get rather involved because of the effects of cross-sectional geometry and streamline curvature. The relative discharge $q = Q/(gD^5)^{1/2}$ depends on the relative head in the cylinder $Y = H/D$ and the relative crest height $Y_b = h_b/D$ (Fig. 13.14d). The simplest approach is the power function

$$q = M(Y - Y_b)^r, \tag{13.17}$$

where M is a coefficient of proportionality, and r an exponent. For $r = 2$ the section is rectangular, for $r = 2.5$ triangular and for $r = 3$ parabolic. From a detailed analysis, the exponent $r = 2.5$ was retained, and the corrections were applied to the M value.

The *effect of constriction* $M_1(\delta)$ is

$$M_1 = 0.611 - 0.560\delta \tag{13.18}$$

with the width effect being linear. The effect of bottom slope was insignificant, as previously noted. The *effect of pipe filling* $M_2(Y)$ or of streamline curvature is

$$M_2 = 0.64\ln(3.62/Y). \tag{13.19}$$

The generalized discharge-head equation for Pipe Flumes discharging into manholes with high benches is with Eq. (13.17) if using $M = M_1 \cdot M_2$ (Kohler and Hager 1997)

$$q = 0.64(0.611 - 0.56\delta)\cdot\ln(3.62/Y)(Y - Y_b)^{2.5}. \tag{13.20}$$

Here, the overflow height $Y_b = Y_h + Y_\delta$ can be related to the capillary height $Y_h = 3\,[\text{mm}]/D$ plus the constriction height $Y_\delta = (1/2)\,[1-(1-\delta^2)^{1/2}] \cong (1/4)\delta^2$. The relative discharge depends thus exclusively on the pipe filling Y and the constriction rate δ. For $Y > 0.3$, all data are within $\pm 5\%$ of the prediction according to Eq. (13.20). This accuracy can easily be obtained in prototype structures.

Example 13.6 Determine the discharge of a Pipe Flume of sewer diameter $D = 0.80$ m and a cylinder diameter $d = 0.25$ m for a stagnation head $H = 0.63$ m.
With $Y = 0.63/0.80 = 0.788$ and $Y_b = 0.003/0.80 + 0.25(0.25/0.80)^2 = 0.028$, the discharge for a constriction rate $\delta = 0.25/0.80 = 0.313$ is from Eq. (13.20) $q = 0.64(0.611–0.56\cdot 0.313)\cdot\ln(3.62/0.788)\cdot(0.788–0.028)^{2.5} = 0.64\cdot 0.436\cdot 1.525\cdot 0.504 = 0.214$, or $Q = q(gD^5)^{1/2} = 0.214(9.81\cdot 0.80^5)^{1/2} = 0.384\ \text{m}^3\text{s}^{-1}$.

Figure 13.16 refers to an unsubmerged Pipe Flume located upstream of the end section with $Y = 0.32$. From the side view, the flow appears smooth and drops at the end section into a pool. Tailwater and plan views are also shown.

To analyze the *effect of clogging* sand was added to the approach flow. No upstream deposits were found because of a self-cleaning mechanism. In front of the pier-shaped body, a horseshoe vortex develops and scours the material away, as with bridge piers. Neither material deposits in front of the cylinder modifying the approach flow, nor does material enter the axially positioned pressure holes. Material is transported around the cylinder into the tailwater, without an effect on discharge measurement.

Field observations in sewers with heavily-polluted sewage showed that *clogging* of the Pipe Flume occurred mainly due to paper and cloth. The Pipe Flume should be checked against clogging at least once a day for permanent measurement. The Pipe Flume was also inserted into a drainage pipe, where fine sand and silt were contained in the flow. Clogging was again no problem and the discharge measurement remained unaffected over a period of a month.

Fig. 13.16 Photographs of pipe flume for $Y = 0.32$ (**a**) side view, (**b**) downstream view, (**c**) plan

13.4 Mobile Discharge Measurement with Lateral Constriction

13.4.1 *Plate Venturi*

Balloffet (1955) introduced simple elements for mobile discharge measurement in rectangular channels. Figure 13.17 shows arrangements with, and without a streamwise extension that can be simply inserted, and the discharge-head curves determined. Balloffet has not detailed information on the modular limit and submergence curves.

Because the arrangements shown in Figs. 13.17b and c correspond basically to a Cut-throat Flume (Chap. 12) the element shown in Fig. 13.17a is of particular interest due to simplicity and compactness. It can be used as a *mobile device* in rectangular sewers of larger width, such as $B \geq 2$ m, with an access by a man-hole. Contrary to short-throated Venturi flumes with a guided flow along the walls, flow separation from the so-called *Plate-Venturi* (German: Platten-Venturi; French: Venturi à plaque) is forced (Fig. 13.18). If both arrangements should result in

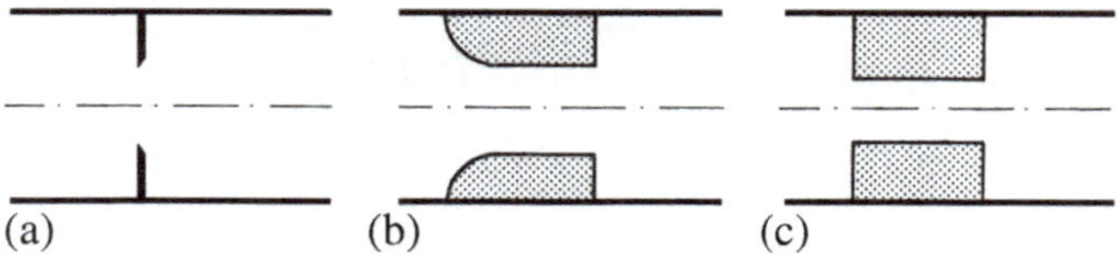

Fig. 13.17 Constriction elements for mobile discharge measurement (Balloffet 1955)

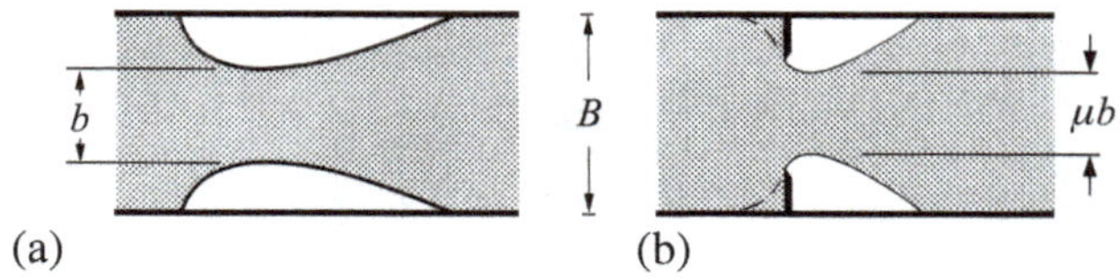

Fig. 13.18 Discharge measurement with (**a**) short-throated flume, (**b**) Plate Venturi flume

identical flow contraction, then the make of Fig. 13.18b needs less wide elements than does that of Fig. 13.18a.

13.4.2 *Discharge Equation*

According to Hager (1988), the discharge of a Plate Venturi can be determined with Eq. (13.1) and the effect of curvature is accounted for by parameter U as introduced in Sect. 13.2.2. The radius of streamline curvature is now no more given by the flume geometry but by the separation geometry of the flow from the Plate Venturi. Streamline curvature can be demonstrated to correlate with the shallowness H/b of the flow, where b is the contracted flow width downstream of the thin-crested plate element (Fig. 13.18b). With $\delta_B = b/B$ as constriction factor and $W = H/b$ as the normalized energy head, the *discharge equation* is to ±3% accuracy

$$Q = bg^{1/2}\left[(2/3)H\right]^{3/2}\left[0.828 + 0.057\left(\delta_B + 2\delta_B^4\right)\right]\left[1 + \frac{W^2}{3 + 5W^2}\right]. \quad (13.21)$$

Here the first term corresponds to the conventional critical discharge from Eq. (13.1), the second term incorporates the effect of constriction and the third term is due to the velocity of approach. Equation (13.21) was experimentally verified for $0.6 \leq \delta_B \leq 0.8$ and $H/b < 2$.

The flow is free as long as the shock fronts in the downstream portion of the constriction reach the side walls, and the roller return flow from a hydraulic jump does not migrate to the separation zone. Based on a simplified approach, a relation between the submergence ratio $S = h_u/h_o$ and the approach flow Froude number

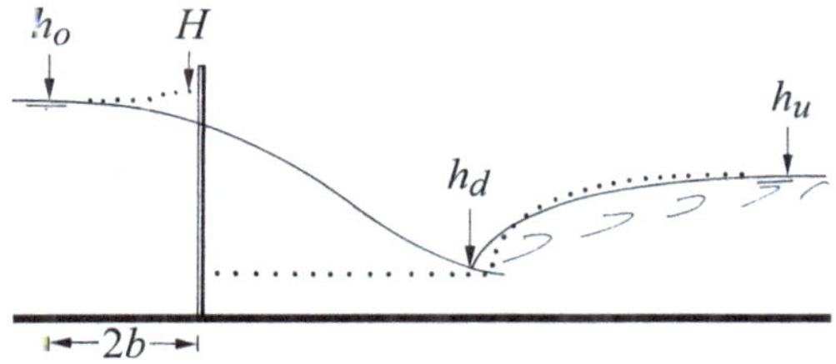

Fig. 13.19 Determination of modular limit with free surface profiles along (—) channel axis, (. . .) channel walls

$\mathsf{F}_o = Q/(gB^2h_o^3)^{1/2}$ was derived (Fig. 13.19), resulting in a *modular limit* (subscript L)

$$S_L = \left[\sin\left(\mathsf{F}_o \cdot 90°\right)\right]^{1/2}. \tag{13.22}$$

Submerged flow across a Plate Venturi is not recommend, as for all other discharge measurement structures.

Example 13.7 Given a Plate Venturi in a channel of $B = 2.7$ m width with a contracted width $b = 1.8$ m. What is the discharge for an energy head $H = 0.96$ m?
With $\delta_B = 1.8/2.7 = 0.67$ and $W = 0.96/1.8 = 0.53$ the constriction coefficient is $[0.828 + 0.057(0.67 + 2{\cdot}0.67^4)] = 0.889$ from Eq. (13.21), and the effect of energy head is $[1 + 0.53^2/(3 + 5{\cdot}0.53^2)] = 1.064$, such that the discharge is $Q = 1.8{\cdot}9.81^{1/2}[(2/3)0.96]^{3/2}0.889{\cdot}1.064 = 2.73\ \text{m}^3\text{s}^{-1}$.
With $H = 0.96$ m, the approach flow depth is $h_o = 0.90$ m and therefore $\mathsf{F}_o = Q/(gB^2h_o^3)^{1/2} = 2.73/(9.81{\cdot}2.7^2 0.90^3)^{1/2} = 0.38$. The modular limit is $S_L = [\sin(0.38{\cdot}90°)]^{1/2} = [\sin 34.2°]^{1/2} = 0.75$ from Eq. (13.22), such that $h_L = 0.75{\cdot}0.90 = 0.675$ m.

Figure 13.20 shows flow configurations of a Plate Venturi, including free flow (Fig. 13.20a), and the modular limit condition (Fig. 13.20b), both in plan and downstream views. Note the differences of flow between the jet expansion reach immediately downstream of the constriction section and the hydraulic jump region further downstream. The hydraulic jump fronts are arc-shaped with small 'tongues' at either channel wall. Once flow from the hydraulic jump enters these dead-water zones downstream of the constriction, the supercritical flow structure abruptly breaks down.

13.4.3 *Practical Aspects*

Normally, the cover of a manhole has a diameter of 0.60 m, and a linear element of about 0.55 m can just be lowered to the sewer. The fixation of a Venturi element uses

(a)

(b)

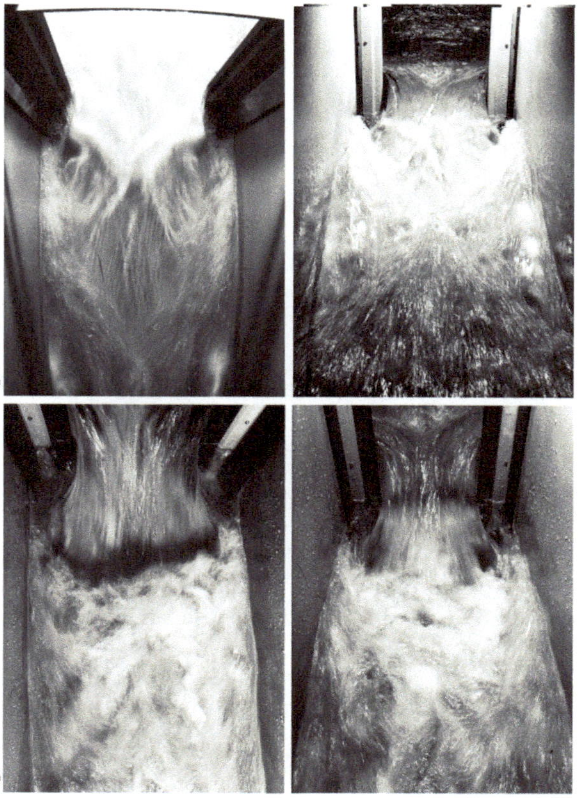

Fig. 13.20 Plan (*left*) and downstream view (*right*) of flow across Plate Venturi for (**a**) free flow and (**b**) modular limit flow

a strut which is welded into a 90° corner of a plate element, whose length is smaller than 0.40 m. The maximum constriction rate $\delta_B = b/B$ should be 0.75 because the *stability* of approach flow is not guaranteed, otherwise, and the maximum channel width should be $B = 3$ m. For larger widths the Venturi elements have to be entered from larger openings than a standard manhole cover.

The *plate element* consists of a steel plate with a standard sharp crest (Chap. 10), on which a perpendicular fixation plate is attached and the structure is reinforced with triangular elements that are attached to the strut (Fig. 13.21). The strut can be varied in height and is fixed to bottom and ceiling of the rectangular duct. The elements are sealed with the walls. They can of course also be inserted in an open rectangular channel with slight modifications.

The *energy head H* could be measured in the corner between a side wall and the Plate Venturi (Fig. 13.19). However, due to considerable vortex generation, the fluctuations of the corner flow depth can be significant. Therefore, the energy head is determined from the approach flow depth h_o measured axially at a location $2b$

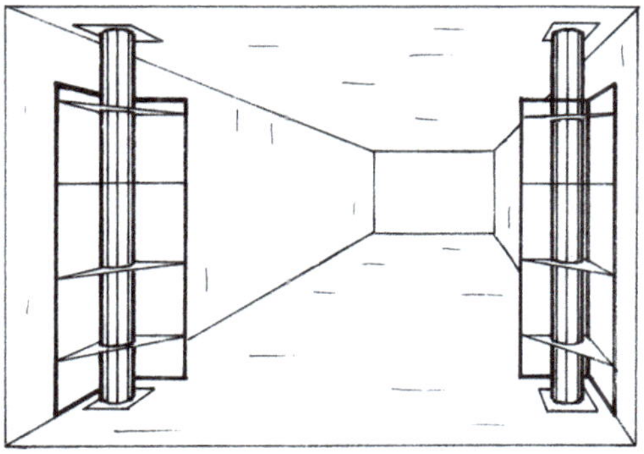

Fig. 13.21 Sketch of mobile discharge measurement in a rectangular sewer

upstream form the contracted section. The transformation of the $Q(H)$ to the $Q(h_o)$ relation can be made based on the relative values $w = h_o/b$ and $q^* = Q/(gb^5)^{1/2}$ as (Hager 1988)

$$W = w + \frac{1}{2}\left(\frac{q^*\delta_B}{w}\right)^2, \tag{13.23}$$

$$q^* = [(2/3)W]^{3/2}\left[0.828 + 0.057\left(\delta B + 2\delta_B^4\right)\right]\left[1 + \frac{W^2}{3 + 5W^2}\right]. \tag{13.24}$$

Figure 13.22 shows a typical relation $q^*(w)$ for various constriction rates δ_B.

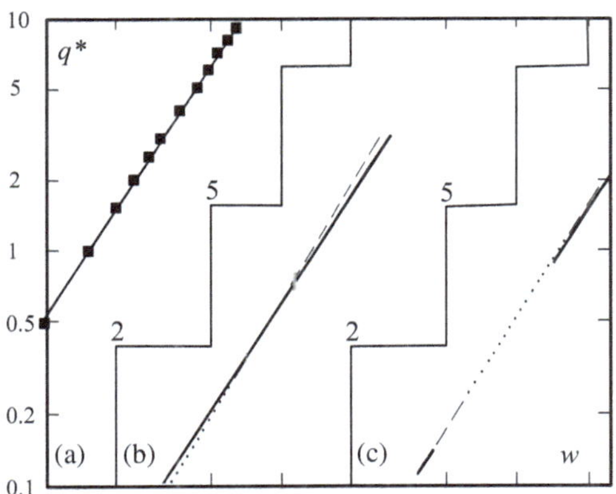

Fig. 13.22 Relative discharge $q^* = Q/(gb^5)^{1/2}$ as a function of relative approach flow depth $w = h_o/b$ for various constriction rates $\delta_B =$ (**a**) 0.20, (**b**) 0.66, (**c**) 0.33

13.5 Mobile Discharge Measurement with Weirs

Because conventional Venturi flumes cannot be inserted as a mobile discharge measurement structure into rectangular or U-shaped channels, there have been some investigations with mobile weirs. In all applications, a sharp-crested weir of triangular or rectangular section has been used (Chap. 10). Such *thin-plate weirs* are inserted normally at the inflow section of a manhole fixed by a simple mechanism and sealed against seepage. Because the U-shaped manhole profile is rather inaccurate in shape, problems may arise in exact weir positioning, definition of crest elevation above the sewer bottom and adjustment of the weir against the horizontal.

Because the mobile weir induces a significant reduction of the approach flow velocity, *solid matter* deposits in the upstream sewer reach, and a long-time installation suffers from disadvantages that have been discussed for mobile flumes (13.3). Therefore, mobile weirs can be used in raw sewage conditions only over a strictly limited time period.

A significant *disadvantage* of mobile weirs is the determination of the overflow head. Normally, the height is determined in the approach flow sewer at about two diameters upstream from the manhole, with extremely poor access conditions. For sewer diameters below $D = 0.50$ m, the overflow depths are often so small that the error in surface reading gives large errors in discharge. Based on these disadvantages the discharge determination by mobile weirs cannot be recommended as a serious design basis. Weir positioning, determination of weir geometry and measurement of weir crest elevation and of the approach flow depth are so uncertain that large errors may result allowing at best an estimation of discharge. Compared to mobile weirs, the mobile flumes are much simpler to mount and the discharge can be determined simpler and more accurate. These conclusions are in agreement with those of Saitenmacher (1967) who compared the mobile V-notch weir with a conventional Venturi flume.

Notation

b	[m]	contracted channel width
B	[m]	approach channel width
d	[m]	cylinder diameter in circular sewer
D	[m]	sewer diameter
D_V	[m]	cylinder diameter
F	[m^2]	cross-sectional area
F	[–]	Froude number
g	[ms^{-2}]	gravitational acceleration
h	[m]	flow depth
h_c	[m]	critical depth

h_L	[m]	limit depth
h_u	[m]	downstream depth
H	[m]	energy head
H_η	[m]	energy head with distortion
L_f	[m]	distance of jump toe from cylinder end
m	[–]	transverse slope
M	[–]	coefficient of proportionality
$1/n$	[$m^{1/3}s^{-1}$]	Manning's roughness coefficient
q	[–]	critical unit discharge
$q*$	[–]	$= Q/(gb^5)^{1/2}$ relative discharge
Q	[m^3s^{-1}]	discharge
Q_a	[m^3s^{-1}]	discharge for symmetric arrangement
Q_k	[m^3s^{-1}]	critical discharge by conventional approach
Q_s	[m^3s^{-1}]	submerged discharge
r	[–]	exponent
R_h	[m]	hydraulic radius
S	[–]	submergence ratio
S_L	[–]	modular limit
S_K	[–]	cone parameter
S_o	[–]	bottom slope
u	[–]	curvature parameter
U	[–]	curvature parameter related to H
V	[ms^{-1}]	cross-sectional average velocity
w	[–]	$= h_o/b$ relative flow depth
W	[–]	$= H/b$ relative energy head
x	[m]	streamwise coordinate
X	[–]	$= x/D$ normalized coordinate
y	[–]	relative pipe filling
y_L	[–]	modular limit filling
y_o	[–]	approach filling ratio
y_u	[–]	relative submergence
y_z	[m]	transverse coordinate
Y	[–]	relative critical energy head
Y_b	[–]	relative crest height
Y_z	[–]	relative transverse position
Y_δ	[–]	relative constriction height
δ	[–]	constriction rate of Mobile Venturi Flume
δ_B	[–]	constriction rate of Plate Venturi
η	[–]	distortion angle
λ	[–]	relative toe distance
ψ	[–]	submergence effect
σ	[–]	auxiliary parameter

Subscripts

L	limit
o	upstream
s	submerged
u	downstream

References

Balloffet, A. (1955). Critical flow meters. *Journal of the Hydraulics Division* ASCE 81(HY4, Paper 743): 1–31.

Diskin, M.H. (1963). Temporary flow measurement in sewers and drains. *Journal of Hydraulics Division* ASCE 89(HY4): 141–159; 90(HY2): 383–387; 90(HY6): 241–247.

Hager, W.H. (1985a). Der «mobile» Venturikanal (Mobile Venturi flume). *Gas – Wasser – Abwasser* 65(11): 684–691 [in German].

Hager, W.H. (1985b). Modified Venturi channel. *Journal of Irrigation and Drainage Engineering* 111(1): 19–35.

Hager, W.H. (1986). Modified, trapezoidal Venturi channel. *Journal of Irrigation and Drainage Engineering* 112(3): 225–241.

Hager, W.H. (1988). Venturi flume of minimum space requirements. *Journal of Irrigation and Drainage Engineering* 114(2): 226–243; 115(5): 913.

Hager, W.H., Züllig, H. (1987). Der modifizierte, mobile Venturikanal zur Anwendung in der Kanalisationstechnik (Modified mobile Venturi flume for application in sewer technology). *Korrespondenz Abwasser* 34(5): 460–467 [in German].

Kohler A., Hager, W.H. (1997). Mobile flume for pipe flow. *Journal of Irrigation and Drainage Engineering* 123(1): 19–23.

Saitenmacher, L. (1967). Die Abwassermengenmessung im Kanalisationsnetz (Discharge measurement in sewers). *Wissenschaftliche Zeitschrift TU Dresden* 16(1): 153–159 [in German].

Samani, Z., Jorat, S., Yousaf, M. (1991). Hydraulic characteristics of circular flume. *Journal of Irrigation and Drainage Engineering* 117(4): 558–566.

Ueberl, J., Hager, W.H. (1994). Mobiler Venturikanal im Rechteckprofil (Mobile Venturi flume in rectangular channel). *Gas – Wasser – Abwasser* 74(9): 761–768 [in German].

Chapter 14
Standard Manhole

Abstract Each change of a flow parameter or of the sewer geometry requires a manhole. If no changes occur, a manhole is needed after about 100 sewer diameters for inspection and rehabilitation purposes. In this chapter, the latter type of standard manhole is described, whereas the following two chapters refer to special manholes.

The recommended standard manhole is characterized by *benches* up to the pipe vertex, to obtain completely guided flow across the manhole. The manhole flow for both free surface and pressurized flows is described and loss coefficients in terms of manhole geometry are specified for pressurized manhole flow. The choking features of manhole flow are also considered.

14.1 Introduction

Manholes (German: Schacht; French: Puits) allow for inspection of a sewer. In addition, they serve for:

- Maintenance and rehabilitation,
- Reconstruction of damaged sewers,
- Inspection of larger sewers,
- Design of special manholes,
- Aeration and deaeration of flow, and
- Emergency overflow during clogging.

Although the latter condition should be excluded by a sound static and hydraulic design, conditions particularly during floods can cause uncontrolled overflow. This points to an highly unsatisfactory performance or finish of a sewer system, while other weaknesses such as flood overflows remain often unrecognized. To inhibit uncontrolled *sewer overflow* out of a manhole, the structure must satisfy conditions of statics, construction as well as of hydraulics and sewer technology. The latter conditions are specified.

W.H. Hager, *Wastewater Hydraulics*, 2nd ed., DOI 10.1007/978-3-642-11383-3_14,

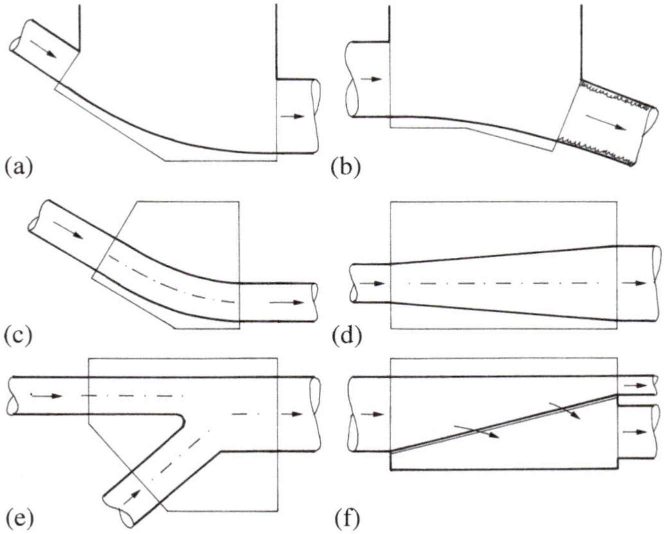

Fig. 14.1 Manholes due to changes of a sewer parameter

Sewers are characterized by *prismatic* reaches between two manholes. All geometrical and hydraulic changes of a sewer are concentrated at the manhole. These include (Fig. 14.1):

- Bottom slope,
- Wall roughness,
- Direction of sewer,
- Cross-sectional shape, or
- Discharge.

Changes of *bottom slope* involve an increase or a decrease (Fig. 14.1a) from a first to a second reach, as described in Chap. 6.

Usually, the *sewer roughness* remains constant, but changes may occur if a new sewer portion is connected to an existing reach (Fig. 14.1b). From hydraulics, this is simply accounted for by backwater curves as described in Chap. 8.

Changes of *sewer direction* (Fig. 14.1c) are met often. Because sewers have to be straight in plan, the angular change is concentrated in the manhole. These are described in Chap. 16, given the similarity with junction manholes.

The *sewer geometry* may change either in diameter or in shape. The standard is based on a through-bottom profile, and an increase or decrease at the sewer vertex. Hydraulically, there are no consequences and the resulting flow disturbance is small (Fig. 14.1d). Typically, such changes are found in sideweirs (Chap. 18) with a transition from a large approach flow pipe to a small throttle pipe (Chap. 9), as shown in Fig. 14.1f). The head losses for subcritical flow are small, whereas supercritical flow may generate unpleasant shock waves.

The discharge may vary *spatially*. In junction manholes (Fig. 14.1e) it increases, whereas it decreases in separating manholes (Chaps. 18, 20). Sewer pipes of buildings are usually not considered separately in sewer hydraulics, except for backwater effects on buildings due to a submerged sewer main (Speerli and Volkart 1991). Combinations of the preceding special manholes are also possible but have not received particular attention in terms of hydraulics.

This chapter is restricted to *standard manholes* with a change from the pipe to the U-shaped profile. In a standard manhole, bottom slope, roughness, direction, diameter and discharge thus remain constant, except for a local diameter expansion. Further, results are currently available only for *pressurized manhole flow*, because the abrupt transition from free surface to pressurized flow at the entrance to the downstream sewer reach has not yet been analyzed. The following gives also a discussion of pipe choking.

14.2 Choking at Sewer Entrance

Figure 14.2 shows flow choking at manhole outlets for sub- and supercritical flows with a filling ratio in excess of 50%.

For *subcritical flow*, the manhole outlet corresponds to a constriction as described in Chaps. 2 or 9. Upstream from the outlet section, the free surface elevation is increased due to a backwater effect, an effect that is modest up to 100% pipe filling. The hydraulic conditions are comparable to a culvert intake (Chap. 9). For flow depths above the sewer vertex, two problems are notable:

- Vortex formation in manhole, and
- Air entrainment into downstream sewer.

For *supercritical flow* a spatial flow pattern is set up in the downstream sewer, comparable to expanding flow (Sect. 16.3.4). Axially, the free surface is drawn down and laterally, the flow piles up just upstream from the intake section. Behind the intake section, the flow separates from the sewer associated with a drawdown of the surface profile. Because the main flow expands further downstream and impacts the sewer sidewalls, a standing wave patterns is generated as is typical for

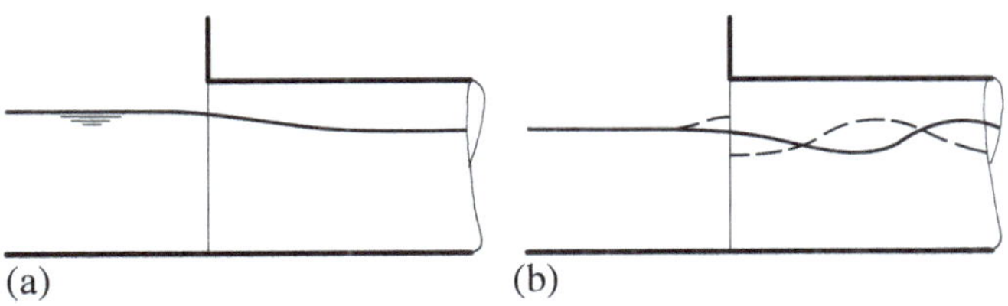

Fig. 14.2 Choking of manhole outlet for (**a**) subcritical and (**b**) supercritical flows. (–) Axial and (- – -) lateral free surface profiles

supercritical channel flows (Chap. 16). Usually, the first wave maximum is larger, and the first wave minimum smaller than all the following wave extrema. The shock angle depends essentially on the local Froude number.

The pattern of wave maxima in the channel axis and wave minima at the channel sides, and vice-versa, is typical for supercritical flows subject to any perturbation. Contrary to subcritical flows, supercritical flows are characterized by a spatially undular free surface. *Choking* downstream of manholes may thus be generated by:

- Touching of sewer vertex by a wave maximum,
- Cut-down of air transport, which can be followed by,
- Breakdown of supercritical flow structure,
- Development of moving hydraulic jump,
- Stabilization of hydraulic jump upstream from the manhole, and
- Transition from free surface to pressurized sewer flow.

Currently, this phenomenon has not yet been systematically investigated. It may occur if supercritical flows have a relatively large filling ratio. Therefore, supercritical flows are prone to choking and the filling ratio has to be limited sufficiently below the sewer vertex. For special manholes as shown in Fig. 14.1, this effect is amplified, and the resulting flow configurations are more complex. The inflow to the sewer downstream of a manhole is then even more prone to choking, if the filling ratio is too large.

14.3 Pressurized Manhole Flow

14.3.1 *Unsuitable Manhole Design*

Manhole flows with a filling ratio around 100% are transitional, they are modelled with an appropriate loss coefficient. Standard manholes are characterized by many geometrical parameters. The plan can be rectangular- or circular-shaped, the conduits can simply be added to the manhole (Fig. 14.3b) or are extended with an *U-shaped manhole* profile (Fig. 14.3a). The latter design is much improved as compared to unguided manhole flow, from hygienic, operating and hydraulic points of view. In the deadwater zones rotten sewage deposits add to conditions that are below standards of modern sewer systems. These zones also increase hydraulic losses; the additional cost for a manhole with an U-shaped through-flow channel is thus more than justified. Accordingly, the studies on manhole losses of Sangster et al. (1959), Hare (1981) or Black and Piggott (1983) are insignificant. The first contribution to *guided manhole flow* is due to Ackers (1959). He found that additional losses due to free surface flow are small, but they may significantly increase for pressurized manhole flow.

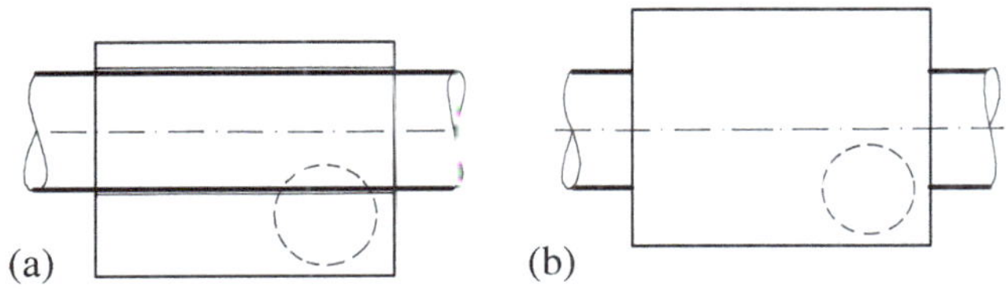

Fig. 14.3 Plan of rectangular manholes with (**a**) through-flow channel with benches, (**b**) no through-flow channel (not recommended for design)

14.3.2 *Results of Liebmann*

Liebmann (1970) refers exclusively to *pressurized* manhole flow. The total head loss includes the deceleration upstream from the manhole, the entry to it, the formation of vortices in the manhole, the outlet and the acceleration in the downstream sewer. His experiments are based on the standard manhole geometry, with a manhole (subscript s) diameter $D_s = 3D$ and with benches of $0.5D$. Also, benches of relative height of 75, 100, and 125% have been considered (Fig. 14.4). The bottom slope was 0.3%, and the sewer diameter $D = 0.30$ m.

If the additional loss ΔH_s (Chap. 2) is plotted as a function of relative filling $y_s = h_s/D$ with h_s as the manhole flow depth, similar plots result for both increasing discharge and submergence (Fig. 14.5). Up to $y_s = 0.5$, the loss is $\Delta H_s = 0$ because the approach flow is completely guided in the U-shaped manhole profile. Increasing y_s results in a significant increase of ΔH_s, particularly for pressurized manhole flow ($y_s > 1$). A maximum value is reached for $y_s \cong 1.5$, and ΔH_s decreases as y_s is further increased to tend to the limit value for 100% benches.

The *manhole flow pattern* in the standard manhole with 50% benches can be described as follows (Fig. 14.5a): For about $y_s = 60\%$, the benches get covered with water and the manhole flow has a rotational component. For $y_s = 70\%$, the free surface starts to oscillate and the amplitude increases as y_s increases. For $y_s = 105\%$ both the manhole inlet and outlet are covered with water, with air being entrained into the downstream sewer. The surface oscillations further increase and the vortices expand over the entire manhole.

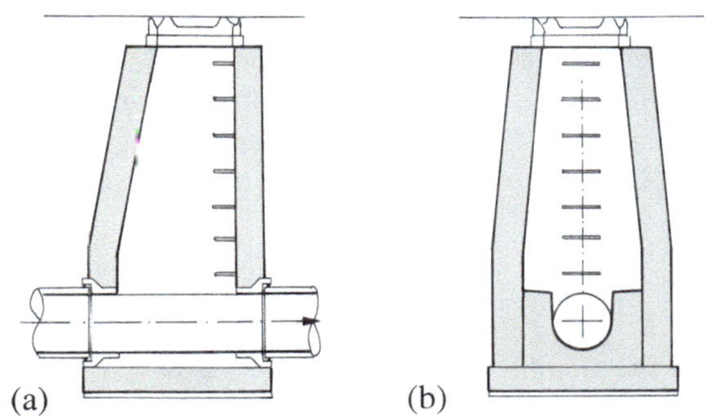

Fig. 14.4 Standard manhole up to $D = 0.50$ m according to ATV (1978). (**a**) Longitudinal section, (**b**) transverse section with 100% benches

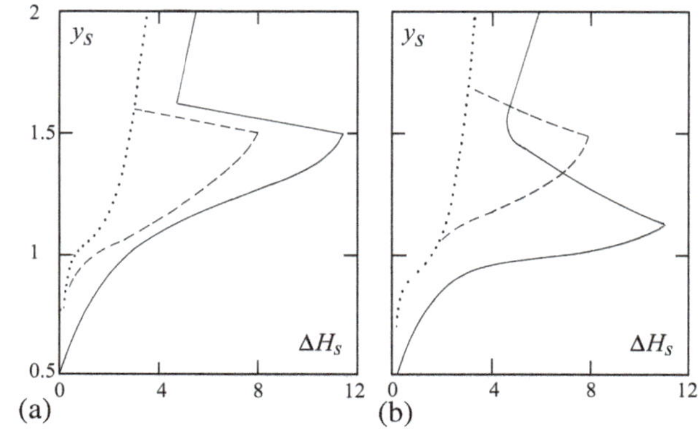

Fig. 14.5 Additional energy loss ΔH_s [cm] as a function of manhole filling $y_s = h_s/D$ for (**a**) variable submergence, and (**b**) variable discharge. Bench height t_b/D = (—) 50%, (- - -) 75%, (· · ·) 100 and 125% (Liebmann 1970).

At about $y_s = 120\%$, the manhole oscillations are abruptly stopped, and the two main vortices with a vertical axis located at the manhole sides are transformed into a single vortex of vertical axis. This main vortex accidentally turns clockwise or anticlockwise over the guided main flow, and involves a significant increase of loss due to *large scale vorticity*.

For $y_s = 150\%$, the single vortex disappears and *pulsations* are set up. Due to large pressure variations, the reading of average values is impossible. Above about 160% filling, the free surface in the manhole is reported to become quiet again, with two vortices over the benches. A detailed description was also provided by Matsushita (1984).

Figure 14.5 reveals that the effect of benches is significant for the head losses across a manhole. Minimum losses result for the *manhole with standard benches* of 100% height, i.e. up to the sewer vertex. This design yields thus advantages in terms of hydraulics, safety, maintenance and hygiene. The additional cost is more than justified.

The maximum *head loss coefficient* $\xi_{sM} = \Delta H_s/[V^2/2g]$ with V as the sewer velocity for the standard manhole is $\xi_{sM} = 0.86$, independent of the approach flow Reynolds number $\mathsf{R} = VD\nu^{-1}$ with ν as kinematic viscosity. For the manhole with 100% benches, the loss coefficient is about 0.1 for $\mathsf{R} = 3 \times 10^5$, increases to the maximum $\xi_{sM} = 0.17$ for $\mathsf{R} = 4 \times 10^5$ and decreases again to the original value for larger R. Therefore, the reduction relative to the manhole with 50% benches involves a factor of 5. For design purposes, one may assume $\xi_s = 1/6$, i.e.

$$\Delta H_s = (1/6)[V^2/(2g)]. \tag{14.1}$$

The significant reduction of ξ_s can be explained with the absence of large scale vortices. If the flow is thus sufficiently guided across the manhole, then energy losses

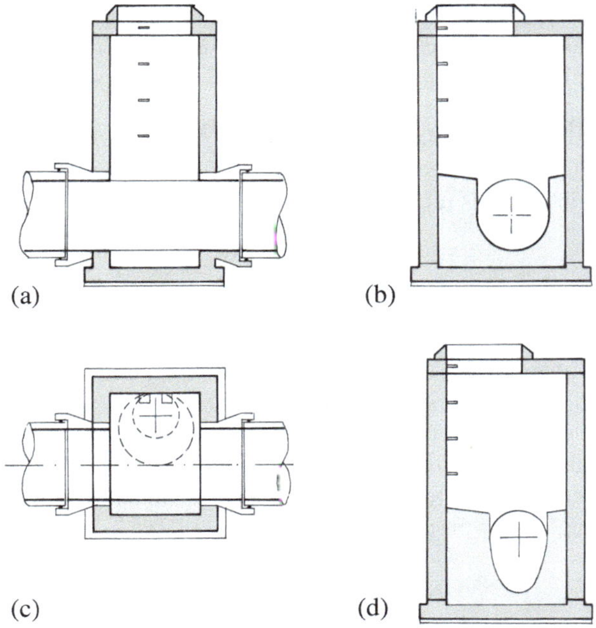

Fig. 14.6 Standard manhole for diameters in excess of $D = 0.50$ m according to ATV (1978) for a circular sewer (**a**) longitudinal section, (**b**) transverse section. Egg-shaped sewer manhole with (**c**) plan and (**d**) transverse section

remain small. Similar techniques are well known with diffusors that are improved with guide vanes (Fig. 14.6). In addition, for manholes with 100% benches, no unpleasant pressure pulsations were identified. These results have been confirmed by Gothe and Valentin (1992).

14.3.3 *Results of Lindvall and Marsalek*

Lindvall (1984) noted also strong manhole oscillations for 50% benches whereas these were practically removed with 100% benches for relative manhole diameters between $D_s/D = 1.7$ and 4.1. Contrary to Liebmann, Lindvall identified a weak maximum in ξ_s for about $y_s = 150\%$, with $\xi_s = 0.20$ for $D_s/D = 3$. For $1 < y_s < 5$, the value according to Eq. (14.1) seems to be reliable. The *effect of manhole diameter* can be specified as

$$\Delta H_s = \frac{1}{12}(D_s/D - 1)\left[V^2/(2g)\right]. \tag{14.2}$$

This reduces to Eq. (14.1) for $D_s/D = 3$.

Marsalek (1984) investigated manholes of circular and square plans, and benches of 0, 50 and 100% relative height. Although the values of ξ_s were slightly larger in the square than in the circular manhole, Eq. (14.2) gives results of about the same order.

According to Dick and Marsalek (1985) and ASCE (1982), the manhole head loss for a large filling depends only on the *manhole geometry* and the relative bench height. Further indications relate to changes in direction (Chap. 16).

14.3.4 *Further Results*

Johnston and Volker (1990) analyzed the standard manhole as a particular case of junction manholes. They related the energy loss to the pipe Froude number $\mathsf{F}_D = V/(gD)^{1/2}$. This effect has to be investigated with additional observations.

Pedersen and Mark (1990) demonstrated that the loss coefficient ξ_s of a manhole structure is essentially composed of the losses of the manhole inlet and outlet. For 100% bench height, they proposed instead of Eq. (14.2)

$$\Delta H_s/[V^2/(2g)] = 0.025(D_s/D). \tag{14.3}$$

This seems to apply for extremely large submergence as well as for $D_s/D < 4$, and may be considered as the lower limit of ξ_s.

Example 14.1 Given a standard manhole of a diameter $D_s = 2$ m, with a sewer of diameter $D = 0.70$ m. What is the upstream energy head H_o for a downstream pressure height of $h_p = 0.35$ m above the sewer vertex, for a discharge of $Q = 0.81$ m^{3}s^{-1}?
With $V_o = Q/(\pi D^2/4) = 0.81/(3.14 \cdot 0.7^2/4) = 2.1$ ms^{-1}, the velocity head is $V_o^2/2g = 0.23$ m, thus $H_u = 0.35$ m $+ 0.23$ m $= 0.58$ m above vertex. For a diameter ratio $D_s/D = 2/0.7 = 2.86$ follows $\Delta H_s = (1/12)(2.86-1)0.23$ m $= 0.04$ m according to Eq. (14.2), and thus $H_o = H_u + \Delta H_s = 0.58$ m $+ 0.04$ m $= 0.62$ m.

Notation

D	[m]	sewer diameter
D_s	[m]	manhole diameter
F_D	[–]	pipe Froude number
g	[ms^{-2}]	gravitational acceleration
h_p	[m]	pressure head

h_s	[m]	manhole flow depth
R	[–]	Reynolds number
t_b	[m]	height of benches
V	[ms^{-1}]	approach velocity
y_s	[–]	manhole filling ratio
ΔH_s	[m]	additional energy loss
ξ_s	[–]	manhole loss coefficient
ν	[m^{2}s^{-1}]	kinematic viscosity

Subscripts

m	minimum
M	maximum
o	approach
u	downstream

References

Abwassertechnische Vereinigung ATV (1978). Bauwerke der Ortsentwässerung (Structures of urban drainage). *Arbeitsblatt* A241. ATV: St. Augustin [in German].

Ackers, P. (1959). An investigation of head losses at sewer manholes. *Civil Engineering and Public Works Review* 54(7/8): 882–884; 54(9): 1033–1036.

ASCE (1982). Design and construction of urban storm water management systems. *ASCE Manuals and Reports of Engineering Practise* 77. American Society of Civil Engineers: New York, updated in 2007.

Black, R.G., Piggott, T.L. (1983). Head losses at two pipe stormwater junction chambers. *Second National Conference on Local Government Engineering* Brisbane: 219–223.

Dick, T.M., Marsalek, J. (1985). Manhole head losses in drainage hydraulics. 21 *IAHR Congress* Melbourne 6: 123–131.

Gothe, E., Valentin, F. (1992). Schachtverluste bei Überstau (Manhole losses due to pressurized sewer flow). *Korrespondenz Abwasser* 39(4): 470–478 [in German].

Hare, C. (1981). Magnitude of hydraulic losses at junctions in piped drainage systems. *Conference on Hydraulics in Civil Engineering* Sydney: 54–59; also in *Civil Engineering Transactions* The Institution of Engineers, Australia 25(1983): 71–76.

Johnston, A.J., Volker, R.E. (1990). Head losses at junction boxes. *Journal of Hydraulic Engineering* 116(3): 326–341; 117(10): 1413–1415.

Liebmann, H. (1970). Der Einfluss von Einsteigschächten auf den Abflussvorgang in Abwasserkanälen (Effect of manholes on sewer flow). *Wasser und Abwasser in Forschung und Praxis* 2. Erich Schmidt: Bielefeld [in German].

Lindvall, G. (1984). Head losses at surcharged manholes with a main pipe and a 90° lateral. Third Int. Conf. on *Urban Storm Drainage* Göteborg 1: 137–146.

Marsalek, J. (1984). Head losses at sewer junction manholes. *Journal of Hydraulic Engineering* 110(8): 1150–1154.

Matsushita, F. (1984). The lost head characteristics of the various stands: The hydraulics of the stands of the open pipeline system (II). *Trans. Japan Society of Irrigation and Drainage Engineering* 111: 85–94.

Pedersen, F.B., Mark, O. (1990). Head losses in storm sewer manholes: Submerged jet theory. *Journal of Hydraulic Engineering* 116(11): 1317–1328; 118(5): 814–816.

Sangster, W.M., Wood, H.W., Smerdon, E.T., Bossy, H.G. (1959). Pressure changes at open junctions in conduits. Proc. ASCE *Journal of the Hydraulics Division* 85(HY6): 13–42; 85(HY10): 157; 85(HY11): 153; 86(HY5): 117.

Speerli, J., Volkart, P. (1991). Rückstau in Hausanschlüsse an die Kanalisation (Submergence to building connections from sewers). *Mitteilung* 111. Versuchsanstalt für Wasserbau, Hydrologie und Glaziologie, ETH Zürich: Zürich [in German].

Chapter 15
Fall Manholes

Abstract Fall manholes either are drop manholes or vortex drops. Both types are described and their relative domains of application are discussed. Differences in their hydraulic characteristics are thereby outlined.

For *drop manholes* with a maximum drop height of 7 to 10 m, particular attention must be paid to the jet trajectory geometry and the outflow conditions. Problems with air and pulsations are also addressed.

For *vortex drops*, the approach flow conditions are important. The designs of the intake and outlet structures are particularly analyzed, and questions relating to the air-water flow in the shaft are discussed. All computations required for design are illustrated by examples.

15.1 Introduction

In steep catchment areas, a sewer system can be designed by two fundamentally different approaches:

- Either with a *steeply sloping sewer* that follows the local topography,
- Or the elevation difference is overcome with a *vertical shaft* of nearly horizontal approach and downstream sewer reaches.

The first solution applies to topographies of uniform slope up to about 50% that is connected to a downstream sewer with supercritical flow. Then, no energy dissipation is needed. Such steeply sloping sewers are addressed in Chap. 5.

Fall manholes apply to a step-type topography, with nearly horizontal reaches both up- and downstream. Then, the energy dissipation structure can be integrated into the fall manhole with a subcritical downstream flow.

Depending on the elevation difference from the approach flow to the outlet sections, and the ratio to the sewer diameter, one can distinguish between the:

- *Drop manhole* for elevation differences up to about 7 m, and the
- *Vortex drop* for elevation differences larger than about 5 m.

W.H. Hager, *Wastewater Hydraulics*, 2nd ed., DOI 10.1007/978-3-642-11383-3_15,

Fig. 15.1 Fall manhole with an S-shaped invert, modified according to ATV (1991), with high benches (not recommended)

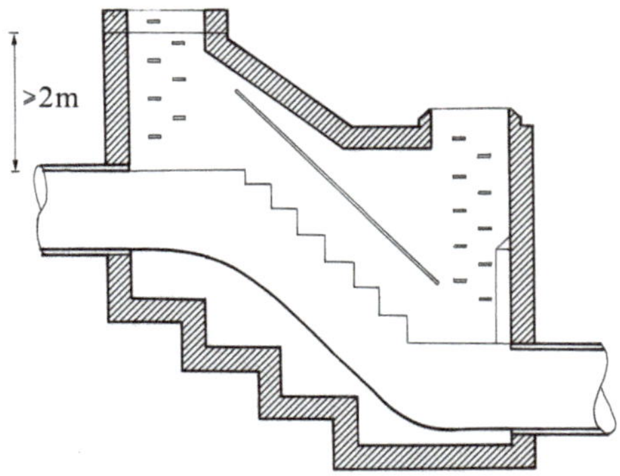

The manhole with an S-shaped invert geometry cannot be considered as a fall manhole, because no energy dissipation is involved. It could also not be designed with the indications of Chap. 5, because of its relative structural shortness. Because the manhole shown in Fig. 15.1 is prone to flow problems in the downstream sewer, and no experiments have been conducted relative to the optimum invert geometry and choking of the downstream pipe, this rather costly structure is not recommended for design.

In the following, the drop manhole and the vortex drop manhole are discussed. Both so-called *special manhole structures* involve hydraulic problems at the intake, in the shaft and at the outlet. Both structures are prone to air-water flow problems and simple hydraulic conditions have to be created. Both the drop manhole and the vortex manhole have been tested and the designs to be considered have proven effective for sewers.

15.2 Drop Manhole

15.2.1 *Manhole Setup*

For smaller differences of elevation, from the intake to the outlet sections, of up to 7 m, or 10 m in extreme cases, a drop manhole (German: Absturzschacht; French: Puits de chute) can be arranged. The current design follows SIA (1980) and consists of the parts as shown in Fig. 15.2.

The approach flow can be either sub- or supercritical. The *dry-weather flow* has a separate fall pipe with a minimum diameter of 0.30 m. Then, the discharge is not too much transformed into spray, and the conditions of hygiene, noise and odour are roughly satisfied.

For *rain-weather flow* the discharge is directed over the impact wall to the opposite manhole wall down to the outlet. The jet component rising up the manhole wall is deflected with an impact nose to avoid water rising up to the access cover.

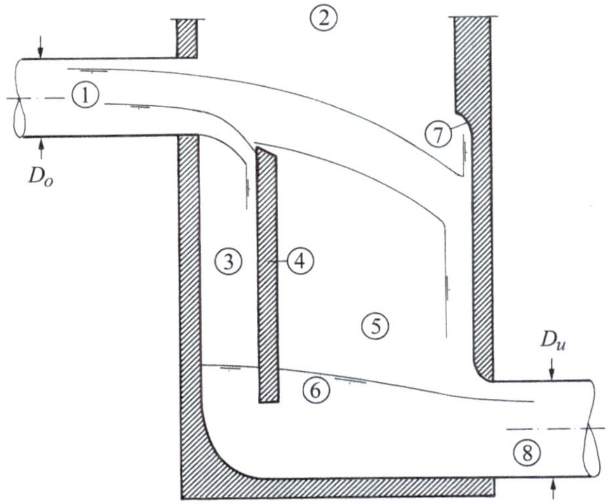

Fig. 15.2 Drop manhole according to SIA (1980) with ① approach sewer, ② access, ③ dry-weather drop, ④ impact wall, ⑤ drop chamber, ⑥ water cushion, ⑦ impact nose, ⑧ outlet

Accordingly, the sewage falls nearly vertically to the manhole bottom. The manhole invert may be strengthened with resistive material, such as granite plates to inhibit abrasion. The effect of this configuration is that the drop manhole adds to the energy dissipation, and the downstream flow has a much reduced velocity.

Drop manholes are restricted in height because of:

- Generation of poor hydraulic conditions due to the air-water flow,
- Pulsations set up by the two-phase flow,
- Poor energy dissipation in the manhole,
- Absence of dissipation chamber for introduction of dissipated flow into the downstream sewer, and
- Noise level that is too high in urban areas.

The hydraulic design of a drop manhole involves at least the dry-weather (subscript K) and the rain (subscript R) discharges Q_K and Q_R. It includes the approach flow sewer, the vertical dry-weather pipe, the shaft for rain-weather discharge and the manhole outlet. An aeration pipe guarantees pressures in the shaft structure close to atmospheric pressure. The main aspects of drop manholes are discussed below.

15.2.2 *Approach Flow Sewer*

The approach (subscript o) flow sewer is characterized by the bottom slope S_{oo}, the Manning roughness coefficient $1/n$ and the diameter D_o. The uniform depth h_{No} and the critical depth h_{co} are determined with Eqs. (5.15) and (6.36) for circular sewers as

$$y_N = 0.926[1 - (1 - 3.11q_N)^{1/2}]^{1/2} \quad (15.1)$$

$$h_{co} = [Q/(gD_o)^{1/2}]^{1/2} \quad (15.2)$$

with $y_N = h_{No}/D_o$ and $q_N = nQ/(S_{oo}^{1/2}D_o^{8/3})$. The Froude number of the approach flow is according to Eq. (6.35)

$$\mathsf{F}_o = Q/(gD_oh_o^4)^{1/2}. \quad (15.3)$$

Example 15.1 Given an approach flow sewer with a bottom slope of $S_{oo} =$ 1.2%, roughness coefficient $n = 0.012$ sm$^{-1/3}$ and diameter $D_o = 1.20$ m. What are the hydraulic approach flow conditions for $Q_K = 0.16$ m^{3}s^{-1} and $Q_R = 2.45$ m^{3}s^{-1}?

With the relative discharges $q_{NK} = 0.012 \cdot 0.16/(0.012^{1/2}1.2^{8/3}) = 0.0106$ and $q_{NR} = 0.012 \cdot 2.45/(0.012^{1/2}1.2^{8/3}) = 0.162$ the filling ratios are, respectively, $y_{NK} = 0.926[1–(1–3.11 \cdot 0.0106)^{1/2}]^{1/2} = 0.119$, and $y_{NR} = 0.503$ for uniform flow. The uniform depths are thus $h_{NK} = 1.2 \cdot 0.119$ m $= 0.143$ m, and $h_{NR} =$ $0.503 \cdot 1.2$ m $= 0.604$ m.

The critical depths are $h_{cK} = [0.16/(9.81 \cdot 1.2)^{1/2}]^{1/2} = 0.216$ m and $h_{cR} = [2.45/(9.81 \cdot 1.2)^{1/2}]^{1/2} = 0.845$ m. The Froude numbers for uniform flow are $\mathsf{F}_{oK} = 0.16/(9.81 \cdot 1.2 \cdot 0.143^4)^{1/2} = 2.28$, and $\mathsf{F}_{oR} =$ $2.45/(9.81 \cdot 1.2 \cdot 0.604^4)^{1/2} = 1.96$, respectively. Both approach flows are supercritical.

The approach flow is assumed to be uniform. To obtain stable flow, uniform flow conditions are required and the *minimum distance* to the upstream manhole should be $20D_o$. For flows containing strong disturbances such as shock waves (Chap. 16), the approach flow distance has to be increased.

15.2.3 *Jet Geometry*

The crest of the *impact* wall has to be set such that the complete dry-weather flow is deflected into the vertical dry-weather pipe, and the lower trajectory of the rain-weather flow does not touch the crest (Fig. 15.2). The following refers to the trajectory geometry of circular sewers. The distance of the dry-weather separating wall to the upstream manhole wall should be at least 0.50 m.

Figure 15.3 shows the so-called *end overfall* as described in Chap. 11. The reference length is the undisturbed approach flow depth h_o equal to about the uniform depth, and the end depth h_e is located at the end section ($x = 0$). The vertical jet thickness is t and z is the vertical coordinate.

According to Chap. 11, the geometry of the *lower jet trajectory* issued by a circular, partially-filled pipe can be described with the parameters $X = (x/h_o)\mathsf{F}_o^{-0.8}$

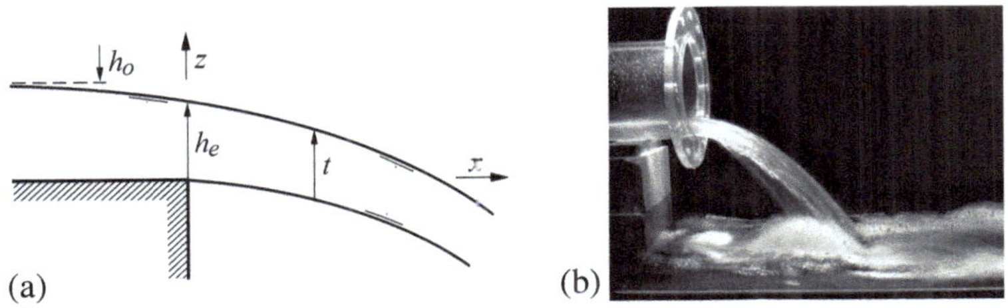

Fig. 15.3 End overfall (**a**) notation, (**b**) photograph of jet

and $Z = z/h_o$. All data for filling ratios $y < 0.9$ and approach flow Froude numbers $0.8 < \mathsf{F}_o < 8$ follow the curve

$$Z = -\frac{1}{3}X - \frac{1}{4}X^2. \quad (15.4)$$

The *axial jet thickness* t, or the thickness ratio $T = t/h_e$ varies with the nondimensional position X as

$$T = 1 + 0.06X. \quad (15.5)$$

Accordingly, the jet thickness increases linearly with increasing distance, starting from the end depth h_e. This is in contrast to plane jets issued by a rectangular duct for which the jet thickness remains practically constant.

The end depth ratio $Y_e = h_e/h_o$ varies essentially with F_o. According to Chap. 11, the result can be expressed as

$$Y_e = \left(\frac{2\mathsf{F}_o^2}{1 + 2\mathsf{F}_o^2}\right)^{2/3}. \quad (15.6)$$

For critical approach flow $\mathsf{F}_o = 1$, the end depth ratio is thus $Y_e = 0.763$, and $Y_e \to 1$ for a large approach flow Froude number. Further results are presented in Chap. 11.

Example 15.2 Determine the crest elevation of the impact wall for the numbers as given in Example 15.1, provided the wall distance from the end section is 0.5 m.

With $Q_K = 0.16\ \mathrm{m^3s^{-1}}$ and $h_{NK} = 0.143$ m one has $\mathsf{F}_{oK} = 2.28$. The equation of the lower jet trajectory is from Eq. (15.4)

$$z_K = -0.172x - 0.468x^2.$$

At location $x = 0.50$ m, one has $z_K = -0.20$ m. The nappe thickness thus is $t_K = [1+0.06 \cdot 0.5/(0.143 \cdot 2.28^{0.8})]0.135 = 0.15$ m from Eq. (15.5), with $h_e/h_o = [2 \cdot 2.28^2/(1+2 \cdot 2.28^2]^{2/3} = 0.941$ or $h_e = 0.941 \cdot 0.143 = 0.135$ m. Thus, the location of the upper nappe is $t_K + z_K = 0.15{-}0.20 = -0.05$ m below the invert of the end section.

With $Q_R = 2.45$ m^{3}s^{-1} and $h_{NR} = 0.604$ m the approach flow Froude number is $\mathsf{F}_{oR} = 1.96$. Thus, the equation of the lower trajectory is

$$z_R = -0.195x - 0.141x^2.$$

and $z_R = -0.13$ m at location $x = 0.50$ m. With $t_K + z_K = -0.05$ m and $z_R = -0.13$ m, the two jets overlap, and the position has to be redesigned for acceptable flow conditions.

Inspection and maintenance require *minimum dimensions* for the drop manhole including (Fig. 15.4a):

- Manhole width $B_s \geq 1$ m,
- Impact width $L_p \geq 0.50$ m, and drop chamber $L_A \geq 1$ m,
- Minimum manhole length $L_s = 2$ m, and
- Dismantling impact wall for dry-weather drop width smaller than 0.6 m.

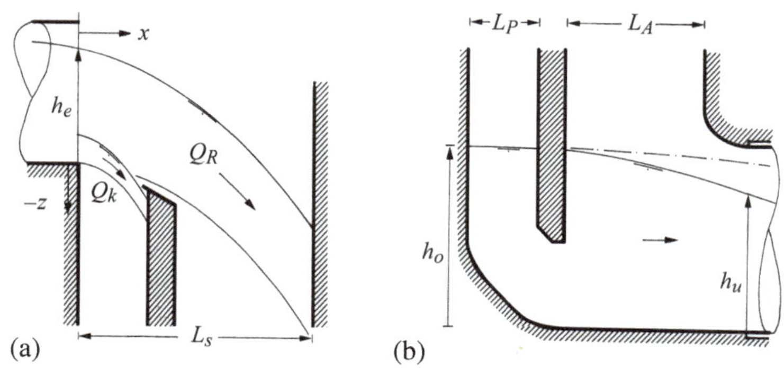

Fig. 15.4 Drop manhole with (**a**) approach flow and (**b**) outlet flow

15.2.4 *Drop Shaft*

The flow in the *manhole shaft* has not been investigated so far. Preliminary experiments have demonstrated an intensive air-water mixture flow that is prone to pulsations. To reduce dangerous vibrations, the drop shaft should be sufficiently aerated from the cover, and the downstream sewer should be sufficiently large for air evacuation. From experience, a number of drop shafts are known for a poor performance, a fact attributed mainly to insufficient shaft dimensions. The above specified dimensions correspond to an absolute minimum, therefore. Also, drop shafts higher than 7 m can work differently than described here, and should only be designed based on a hydraulic model study. Rajaratnam et al. (1997) conducted some experiments for a drop shaft with a downward curved inlet sewer without a dry weather pipe.

15.2.5 *Manhole Outlet*

A drop manhole outlet may be hydraulically modelled as a side channel (Fig. 15.2). To limit abrasion and deposits, the upstream end of this side channel should be provided with a rounded invert. To generate free surface flow in the downstream sewer, its diameter has to be sufficiently large. The distribution of lateral discharge to the side channel is almost uniform, and the friction slope is nearly compensated for by the bottom slope. The momentum flux in the U-shaped profile of diameter D is (Hager 1987)

$$S = \frac{8}{15}D^3 y^{5/2}\left(1 - \frac{1}{4}y\right) + \frac{Q^2}{\frac{4}{3}gD^2 y^{3/2}\left(1 - \frac{1}{3}y\right)}. \qquad (15.7)$$

Here $y = h/D$ is the relative filling, Q is the discharge and g the gravitational acceleration. For a prismatic U-shaped channel of constant diameter D_u, the momentum equation between the upstream dead-end (subscript o) and the outlet (subscript u) sections is

$$\frac{8}{15}y_o^{5/2}\left(1 - \frac{1}{4}y_o\right) = \frac{8}{15}y_u^{5/2}\left(1 - \frac{1}{4}y_u\right) + \frac{Q^2}{\frac{4}{3}gD_u^5 y_u^{3/2}\left(1 - \frac{1}{3}y_u\right)}. \qquad (15.8)$$

Figure 15.5 shows the relation between the two filling ratios y_o and y_u with the relative discharge $Q/(gD_u^5)^{1/2}$. For $Q = 0$, the trivial solution is $y_o = y_u$, and $y_o > y_u$ otherwise.

Equation (15.8) applies to both sub- and supercritical flows at the manhole outlet section. In the first case, the side channel is submerged from the downstream sewer, and computations proceed against the direction of flow (Chap. 8). In the second case, *critical flow* occurs at the outlet section (Chap. 19). Along the side channel, the flow is thus subcritical, and supercritical flow prevails in the downstream sewer. The Froude number in the U-shaped profile is (Hager 1987)

$$\mathsf{F}^2 = \frac{Q^2}{gD^5}\frac{2y^{1/2}\left(1 - \frac{5}{9}y\right)}{\left[\frac{4}{3}y^{3/2}\left(1 - \frac{1}{3}y\right)\right]^3}. \qquad (15.9)$$

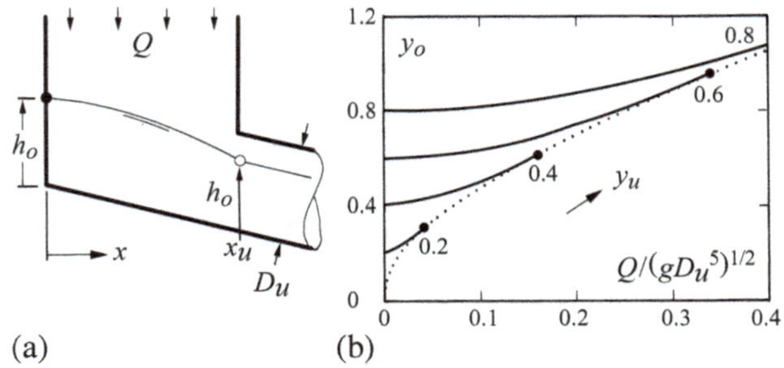

Fig. 15.5 Outlet of drop manhole (**a**) definition sketch, (**b**) upstream filling ratio $y_o = h_o/D_u$ versus $y_u = h_u/D_u$ and relative discharge $Q/(gD_u^5)^{1/2}$ according to Eq. (15.8). ($\cdot\ \cdot\ \cdot$) Eq. (15.10) for critical flow at outlet section, with $y_u = y_c$

Eliminating Q in Eq. (15.8) and imposing $\mathsf{F} = 1$ results in an equation for y_o as a function of $y_u = y_c$ as

$$y_o^{5/2}\left(1-\frac{1}{4}y_o\right) = y_u^{5/2}\left[\left(1-\frac{1}{4}y_u\right)+\frac{5}{3}\frac{\left(1-\frac{1}{3}y_u\right)^2}{1-\frac{5}{9}y_u}\right]. \tag{15.10}$$

Figure 15.5 contains also Eq. (15.10) as the dotted line.

The term in parentheses of Eq. (15.10) varies for $0 < y_u < 1$ only between 2.67 and 2.41, and an average is 2.5. Then, Eq. (15.10) simplifies to

$$y_o^{5/2}\left(1-\frac{1}{4}y_o\right) = 2.5y_u^{5/2}. \tag{15.11}$$

Equation (15.9) can be simplified for critical flow ($\mathsf{F} = 1$) to

$$\frac{Q}{(gD^5)^{1/2}} = \frac{8}{9}y_c^{1.9}, \tag{15.12}$$

such that Eqs. (15.11) and (15.12) may be combined to a relation $y_o(Q)$ as

$$y_o^{5/2}\left(1-\frac{1}{4}y_o\right) = 2.92\left[Q/(gD_u^5)^{1/2}\right]^{1.315}. \tag{15.13}$$

A simple expression of ±10% accuracy is finally $Q/(gD_u^5)^{1/2} = 0.380y_o^{1.88}$, or

$$y_o = \frac{5}{3}\left(\frac{Q}{(gD_u^5)^{1/2}}\right)^{0.5}. \tag{15.14}$$

The flow depth at the upstream dead-end of a drop manhole varies thus with the square root of discharge and with only the forth root of the diameter. It is noted that $[Q/(gD_u^5)^{1/2}]^{1/2} = h_{cu}$ is the critical depth of the downstream reach, and that

$$h_o = (5/3)h_{cu}. \quad (15.15)$$

The upstream depth is thus about 170% of the critical depth, a result that is nearly identical with that for rectangular side channels, where $h_o = 3^{1/2}h_c$. For given discharge Q the upstream filling ratio y_o can be predicted as a function of relative discharge, provided there is a transition from sub- to supercritical flow.

Example 15.3 Determine the upstream flow depth for the numbers given in Example 15.1 for a downstream sewer with $D_u = 1.0$ m and a bottom slope of $S_{ou} = 2\%$.

With $S_{ou} = 2\%$, the downstream flow is supercritical, and critical flow is forced at the outlet section. With $Q/(gD_u^5)^{1/2} = 2.45/(9.81 \cdot 1^5)^{1/2} = 0.782$ for rain-weather discharge, one has $y_c = [(9/8)0.782]^{1/1.9} = 0.935$ from Eq. (15.12), or $h_c = 0.935$ m. The relevant solution of Eq. (15.13) is $y_o = 1.67$, in agreement with Eq. (15.15).

Considering the flow turbulence and the air entrainment due to the plunging jet, a diameter of $D_u = 1$ m is insufficient for adequate discharge capacity. For $D_u = 1.25$ m, one has $Q/(gD_u^5)^{1/2} = 0.45$, so that $h_c = 0.87$ m and $h_o = 1.44$ m. The downstream sewer has to be sufficiently large to inhibit *choking*. A diameter reduction may follow further downstream.

The *aeration* and *deaeration* of drop manholes has not been analyzed until today. Depending on the discharge and the manhole geometry, air discharges of up to 50% of the water discharges can result. The air supply pipes have to be designed sufficiently large to secure adequate aeration and deaeration of the flow. Granata et al. (2009, 2010) have added considerably to the understanding of shaft flow, and preliminary results are available. More results are expected relative to the optimum design of shafts including particular bottom shapes to improve the shaft outflow and to avoid choking flow.

15.3 Vortex Drop

15.3.1 *Limits of Application*

Contrary to a drop manhole, a vortex drop (German: Wirbelfallschacht; French: Puits à vortex) involves a significant *energy dissipation* by wall friction. This favourable effect is due to a superposition of rotational with translatory flow components causing a helicoidal flow across the shaft. Further, the water moves essentially along the shaft walls, whereas the air moves in a central *air core* with a pressure slightly above the atmosphere. Thus, the shaft flow is separated into a fluid and a gas phase, resulting in a stable annular flow pattern (Chap. 5).

Vortex drops apply, provided:

- Elevation difference between inlet and outlet is at least 5–10 m, depending on the shaft diameter,
- Approach flow is either stably subcritical ($\mathsf{F}_o < 0.7$) or stably supercritical ($\mathsf{F}_o > 1.5$), and
- Steeply sloping sewer is no economical solution.

The outflow of a steep sewer is supercritical, whereas the outflow of a vortex drop is generally subcritical. This difference in concept may be determining in the selection of the optimum structure.

The design of a vortex drop depends essentially on the approach flow conditions. The computation involves three items:

- Intake structure,
- Vertical shaft, and
- Outlet structure.

15.3.2 *Intake Structure*

The approach flow channel is rectangular of width b, bottom slope S_o and roughness coefficient according to Manning n. The diameter of the vertical shaft (subscript s) is D_s. The purpose of the intake structure is to transform a rectilinear approach flow to a vertical shaft flow by the *tangential intake.*

Figure 15.6a shows the intake structure for *subcritical approach flow* according to Drioli (1969). Here, a is the distance of the shaft center to the outer approach wall, c is the minimum width of the intake spiral, e the eccentricity, ΔR the radius of the shaft intake and s the minimum width of the guide wall. The geometry can be described as

$$a = R + \Delta R + \frac{1}{2}b + c + s, \tag{15.16}$$

$$e = \frac{1}{7}(b + s). \tag{15.17}$$

The radii of the intake structure are

$$R_4 = R + \Delta R + c + e, \tag{15.18}$$

$$R_3 = R_4 + e, \tag{15.19}$$

$$R_2 = R_4 + 3e, \tag{15.20}$$

$$R_1 = R_4 + 5e. \tag{15.21}$$

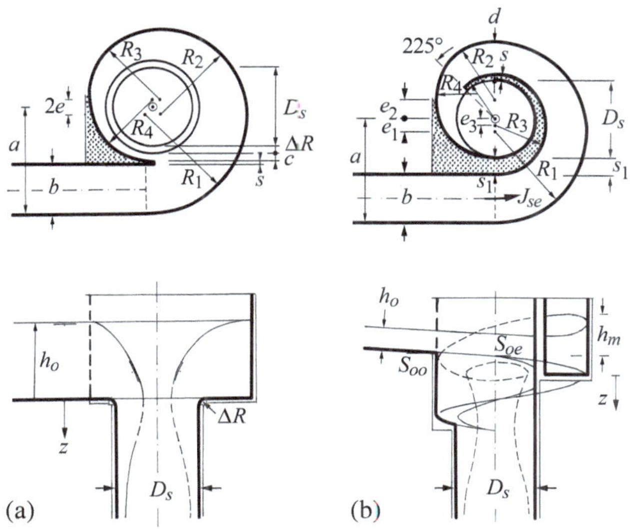

Fig. 15.6 Geometry of vortex drop for (**a**) subcritical and (**b**) supercritical approach flow with plan (*top*) and section (*bottom*)

Reasonable dimensions result for $0.8 < D_s/a < 1$, and $\Delta R > D_s/6$. The thickness of the guide wall is based on structural requirements.

For *supercritical approach flow* the geometry of the intake structure can be described with $R = (1/2)D_s$ as the shaft radius as (Kellenberger 1988)

$$R_1 = (a + R + s + d)/2, \qquad e_1 = a - R_1; \tag{15.22}$$

$$R_2 = (2R + s + d)/2, \qquad e_2 = R + s + d - R_2; \tag{15.23}$$

$$R_3 = (a + R + s - b)/2, \qquad e_3 = a - b - R_3; \tag{15.24}$$

$$R_4 = R + s, \qquad s_1 = a - b - R \tag{15.25}$$

The conditions to satisfy are

$$(R + s + d) \leq a \leq (3R + s), \tag{15.26}$$

$$0.8R \leq b \leq 2R, \tag{15.27}$$

$$0.8R \leq d \leq 2R. \tag{15.28}$$

The transverse bottom geometry is horizontal, and the streamwise bottom slope S_{oe} equal or larger than the approach flow bottom slope S_{oo} Values between 10 and 20%

are an optimum, and $S_{oe} \leq 30\%$. The thickness of the guide wall should again satisfy structural requirements, and favourable are small values of both b/R and d/R.

15.3.3 *Design of Intake*

For *subcritical approach* flow, the relation between discharge Q and approach flow depth h_o is described with the parameters

$$h^* = aR/b, \quad Q^* = (gaR^5/b)^{1/2}. \tag{15.29}$$

According to Hager and Kellenberger (1987)

$$Q/Q^* = \sqrt{2}h_o/h^*. \tag{15.30}$$

The design discharge is equal to the maximum (subscript *M*) discharge Q_M with

$$Q_M = 4R^3(5\,g/b)^{1/2}. \tag{15.31}$$

For $Q > Q_M$, the vertical shaft chokes and the air transport is not secured.

For *supercritical approach* flow, the design discharge Q_M is related to the shaft diameter D_s as (Kellenberger 1988)

$$Q_M = [g(D_s/1.25)^5]^{1/2}. \tag{15.32}$$

The shaft capacity for both sub- and supercritical approach flow is thus similar.

Compared to vortex drops with a subcritical approach flow and a steadily decreasing flow depth along the intake spiral, the supercritical approach flow generates shock waves due to the strong flow deflection at the intake structure, with a *main standing wave* of maximum height h_M. According to Hager (1990), its height is

$$\frac{h_M}{R_1} = \left[\frac{2^{1/2}Q}{(gbh_oR_1^3)^{1/2}} - \frac{1}{2}S_{oe}\right](1.1 + 0.15\mathsf{F}_o). \tag{15.33}$$

Here, R_1 is the radius from Eq. (15.22), h_o the approach flow depth, $\mathsf{F}_o = V_o/(gh_o)^{1/2}$ the Froude number in the rectangular approach flow channel and S_{oe} the bottom slope of the intake structure (Fig. 15.7). The angle α_M with the maximum flow depth is (Fig. 15.7)

$$\alpha_M/\mathsf{F}_o = 75°(h_o/R_1)^{1/2}. \tag{15.34}$$

Figure 15.8 shows two experiments with typical configurations of the free surface. In Fig. 15.8a the free surface has one maximum, whereas two maxima are

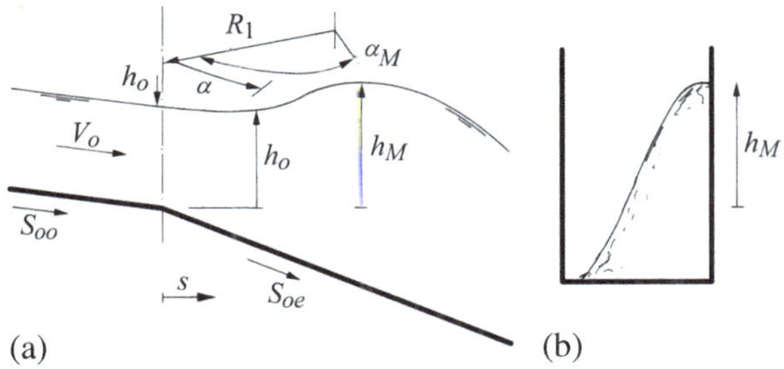

Fig. 15.7 Standing wave along outer wall of the vortex chamber (**a**) side view, (**b**) section of maximum flow depth

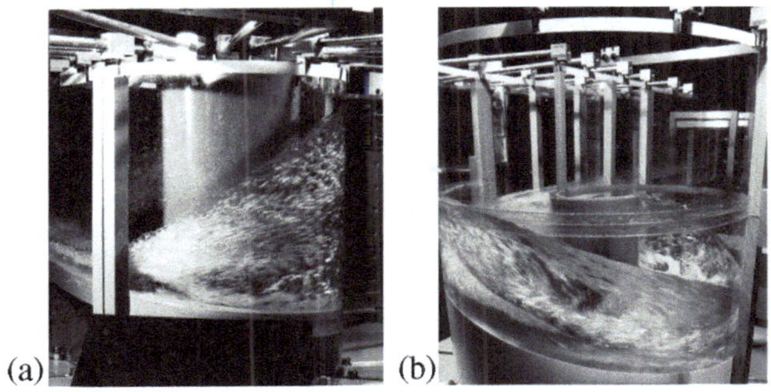

Fig. 15.8 Side views of intake flow to drop shaft for $\mathsf{F}_o =$ (**a**) 5.7, (**b**) 1.8

identified in Fig. 15.8b. Kellenberger's (1988) design inhibits *choking* of the supercritical approach flow resulting eventually in a hydraulic jump in the approach flow channel or in the intake reach, resulting in a completely modified approach flow to the vortex shaft.

Example 15.4 Given an approach flow channel of width $b = 1.2$ m with a bottom slope $S_{oo} = 0.1\%$ and $n = 0.011\ \mathrm{sm}^{-1/3}$. What is the resulting intake structure for a discharge of $Q = 2.5\ \mathrm{m^3s^{-1}}$?
For uniform approach flow with $R_{ho} = bh_o/(b+2h_o)$ one has (Chap. 5)

$$Q = (1/n)S_{oo}^{1/2}bh_oRh_o^{2/3}, \quad \text{or}$$

$$\frac{nQ}{S_{oo}^{1/2}b^{8/3}} = \frac{y_o^{5/3}}{(1+2y_o)^{2/3}},$$

where $y_o = h_{No}/b$ and $nQ/(S_{oo}^{1/2}b^{8/3}) = 0.011 \cdot 2.5/(0.001^{1/2}1.2^{8/3}) = 0.54$, for which $y_o = 1.1$, and $h_{No} = 1.1 \cdot 1.2 = 1.32$ m. The corresponding Froude number is $\mathsf{F}_{No} = Q/(gb^2h_o^3)^{1/2} = 2.5/(9.81 \cdot 1.2^2 1.32^3)^{1/2} = 0.44 < 0.7$. The approach flow is thus stably subcritical.

From Eq. (15.31) the shaft radius R is

$$R = \left[\frac{Q_M/4}{(5g/b)^{1/2}}\right]^{1/3},$$

thus $R = [(2.5/4)/(5 \cdot 9.81/1.2)^{1/2}]^{1/3} = 0.46$ m, and chosen $D_s = 2R = 1$ m. From structural considerations, $s = 0.15$ m, $c = 0.10$ m, $\Delta R = 0.10$ m and $a = 0.5 + 0.1 + 0.6 + 0.1 + 0.15 = 1.45$ m. The requirement $0.8 < D_s/a < 1$ is almost satisfied.

With $h^* = aR/b = 1.45 \cdot 0.5/1.2 = 0.60$ m and $Q^* = (9.81 \cdot 1.45 \cdot 0.5^5/1.2)^{1/2} = 0.61$ m^{3}s^{-1} according to Eq. (15.29), the approach flow depth for $Q = 2.5$ m^{3}s^{-1} is $h_o = (Q/Q^*)(h^*/2^{1/2}) = (2.5/0.61)(0.6/2^{1/2}) = 1.74$ m. Compared to the uniform flow depth $h_{No} = 1.32$ m, the significant *backwater effect* is inacceptable. Either, the approach flow bottom slope has to be reduced, or the shaft diameter must be increased.

If $D_s = 1.20$ m instead of $D_s = 1$ m is assumed, $a = 0.6 + 0.1 + 0.6 + 0.1 + 0.15 = 1.55$ m, with $D_s/a = 0.78 \cong 0.8$. Further, with $h^* = 1.55 \cdot 0.6/1.2 = 0.78$ m, $Q^* = (9.81 \cdot 1.55 \cdot 0.6^5/1.2)^{1/2} = 0.99$ m^{3}s^{-1}, thus $h_o = (2.5/0.99)(0.78/2^{1/2}) = 1.39$ m, and the backwater effect is much reduced.

The geometry of the vortex drop can be described with $R = D_s/2 = 0.60$ m, $a = 1.55$ m, $b = 1.2$ m, $e = (1.2 + 0.15)/7 = 0.20$ m, $R_4 = 1.0$ m, $R_3 = 1.2$ m, $R_2 = 1.6$ m and $R_1 = 2.0$ m.

Example 15.5 Given an approach flow channel of width $b = 1.20$ m, roughness coefficient $n = 0.011$ sm$^{-1/3}$ and $S_{oo} = 3\%$. What is the geometry of the intake structure for $Q = 2$ m^{3}s^{-1}?
With $nQ/(S_{oo}^{1/2}b^{8/3}) = 0.011 \cdot 2/(0.03^{1/2}1.2^{8/3}) = 0.079$, the uniform depth filling is $y_{No} = 0.26$ from indications of Example 15.4, and $h_{No} = 0.31$ m. The corresponding Froude number is $\mathsf{F}_{No} = 2/(9.81 \cdot 1.2^2 0.31^3)^{1/2} = 3.08 > 1.5$, and the approach flow is stably *supercritical.*

The shaft diameter is $D_s = 1.25(Q_M^2/g)^{1/5} = 1.05$ m from Eq. (15.32), and $D_s = 1.20$ m is selected. With $s = 0.10$ m, and $d = 0.9$ m, Eq. (15.26) indicates $(0.6 + 0.1 + 0.9)$ m $\leq a \leq (1.8 + 0.1)$ m, thus $a = 1.9$ m. Conditions (15.27), (15.28) are satisfied. The vortex chamber geometry is $R_1 = (1.9 + 0.6 + 0.1 + 0.9)/2 = 1.75$ m, $R_2 = (2 \cdot 0.6 + 0.1 + 0.9)/2 = 1.10$ m, $R_3 = (1.9 + 0.6 + 0.1 - 1.2)/2 = 0.70$ m and $R_4 = (0.6 + 0.1) = 0.70$ m. In addition, $e_1 = (1.9 - 1.75) = 0.15$ m, $e_2 = (0.6 + 0.1 + 0.9 - 1.1) = 0.5$ m, $e_3 = (1.9 - 1.2 - 0.70) = 0$ and $s_1 = (1.9 - 1.2 - 0.6) = 0.10$ m.

The characteristics of the standing wave for $S_{oe} = 0.1$ are $h_M/R_1 = [2^{1/2}2/(9.81 \cdot 1.2 \cdot 0.31 \cdot 1.75^3)^{1/2} – 0.5 \cdot 0.1](1.1 + 0.15 \cdot 3.08) = 0.92$, thus $h_M = 0.92 \cdot 1.75\ \mathrm{m} = 1.61\ \mathrm{m}$ from Eq. (15.33) at $\alpha_M = 75(0.31/1.75)^{1/2}3.08 = 97°$.

Jain (1984) introduced the so-called *tangential vortex-inlet*, with a straight-walled contraction upstream from the shaft pipe. This simplified approach configuration seems to be suited for smaller discharges. The *vertical slot vortex drop*, still another version, was introduced by Quick (1990). A review on various vortex drop systems provided Jain and Ettema (1987). The inlet flow features of tangential vortex intakes were further investigated by Yu and Lee (2009).

Vortex drops can also be used as a junction structure for various sewers. Volkart's (1984) design refers to two sewers, and Bruschin and Mouchet's (1985) design to even four sewer branches that are combined in a common drop shaft. Normally, the sewer inlet elevations are different, at least for more than two branches.

15.3.4 *Vertical Shaft*

The vertical shaft (German: Vertikalschacht; French: Puits vertical) of diameter D_s has a *smooth boundary* to increase the flow stability. The latter is supported by the central air core with a slightly higher pressure than the wall pressure. Pulsations of flow due to insufficient aeration are strictly to be inhibited. Ideally, the water flow along the shaft boundary is separated from the annular air flow.

Although the shaft surface is smooth, roughness effects along the shaft have to be considered. A drawdown curve analogous to open channel flow develops from the intake structure with a decreasing thickness of annular flow along the shaft to reach finally equilibrium between the driving and the retarding forces (Fig. 15.6). The scaling parameters are the determining height $z^* = V^{*2}/(2g)$ and the end velocity V^* with

$$z^* = \frac{(1/n)^{6/5}}{2g}\left(\frac{Q}{\pi D_s}\right)^{4/5}, \quad V^* = (1/n)^{3/5}\left(\frac{Q}{\pi D_s}\right)^{2/5}. \tag{15.35}$$

The shaft velocity V as a function of depth z below the intake is (Kellenberger 1988)

$$(V/V^*)^2 = \tanh(z/z^*). \tag{15.36}$$

in which V^* is the *uniform shaft velocity* that establishes provided the shaft length L_s is larger than the limit (subscript L) depth $z_L = 3z^*$, or

$$z_L = \frac{3}{2}\frac{(1/n)^{6/5}}{g}\left(\frac{Q}{\pi D_s}\right)^{4/5}. \tag{15.37}$$

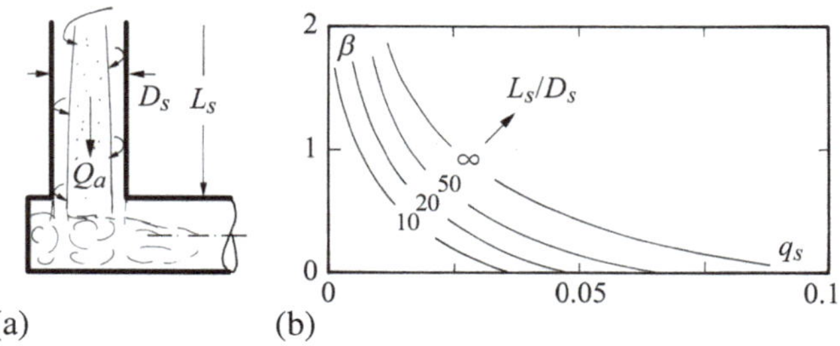

Fig. 15.9 Air entrainment of vortex drops (**a**) notation, (**b**) air entrainment coefficient β as a function of relative water discharge q_s for various relative shaft lengths L_s/D_s

The *shaft efficiency* η of a vortex drop is $\eta = \Delta H/H_a$ with $\Delta H = H_a - H(z)$ and $H_a = H_o + L_s$ as the energy head related to the shaft outlet (Fig. 15.9a). The shaft efficiency specifies the excess energy that has to be dissipated by the dissipation chamber.

Example 15.6 Compute the shaft efficiency for Example 15.4 with a shaft length $L_s = 18$ m.
With $1/n = 90$ m$^{1/3}$s^{-1}, $D_s = 1.20$ m and $Q = 2.5$ m^{3}s^{-1} the scaling parameters are $z^* = (90^{6/5}/19.62)(2.5/\pi \cdot 1.2)^{4/5} = 8.12$ m and $V^* = (2gz^*)^{1/2} = (19.62 \cdot 8.12)^{1/2} = 12.62$ ms^{-1}. The limit shaft height is thus $z_L = 3z^* = 3 \cdot 8.12$ m $= 24.36$ m$ > L_s$. Accordingly, the equilibrium velocity has not established at the shaft outlet. The outlet velocity V_s is with $(V_s/V^*)^2 = \tanh(18/8.12) = \tanh(2.22) = 0.977$ from Eq. (15.36), thus $V_s = 0.977^{1/2} 12.62$ ms^{-1} $= 12.50$ ms^{-1}.

With an approach flow depth of $h_o = h_{No} = 1.32$ m and $V_{No} = 2.5/(1.2 \cdot 1.32) = 1.58$ ms^{-1}, the approach flow energy head is $H_o = 1.32 + 1.58^2/19.62 = 1.45$ m. The outlet energy head is $H_a = H_o + L_s = 1.45 + 18 = 19.45$ m. The shaft efficiency is $\eta = [19.45–12.5^2/(19.62)]/19.45 = 0.60$, i.e. 60% of the approach flow energy are dissipated along the shaft.

According to Eq. (15.35), a small shaft outlet velocity and thus a large efficiency can be reached by a large shaft roughness and a small shaft diameter. This contradicts the requirement of stable two-phase flow, however. An optimum vortex drop involves thus an energy dissipation by both the vortex shaft and the dissipation chamber.

The *air demand* (German: Luftbedarf; French: Besoin d'air) of annular shaft flow depends significantly on the relative shaft length L_s/D_s and the relative discharge $q_s = nQ/(\pi D_s^{8/3})$. The air (subscript a) discharge $Q_a = \beta Q$ with β as the air entrainment coefficient can be determined with Fig. 15.9b, based on Hager and Kellenberger (1987). Approximately, one obtains

$$\beta = \left(\frac{q_e}{q_s}\right)^{1/2} - 1 \qquad (15.38)$$

with the equilibrium (subscript e) discharge

$$q_e = 0.018(L_s/D_s)^{1/3}. \qquad (15.39)$$

By differentiation of Eq. (15.38), the determining water discharge obtains $Q_M = (1/2)^2 q_e (1/n)\pi D_s^{8/3}$, corresponding to $q_M = (1/4)q_e$. The maximum (subscript M) air discharge is then

$$Q_{aM} = Q. \qquad (15.40)$$

The air discharge is thus always smaller than the determining water discharge Q_M. Compared to other drop structures, a vortex drop has a relatively small air discharge, indicating a relatively low degree of turbulence generation and an overall smooth flow. Details of flow have been observed by Farroni et al. (1988).

Example 15.7 Determine the air discharge Q_a for Example 15.6.
With $D_s = 1.20$ m, $L_s = 18$ m, $Q = 2.5$ m^{3}s^{-1} and $n = 0.011$ sm$^{-1/3}$ one has $q_s = 0.011 \cdot 2.5/(\pi \cdot 1.2^{8/3})$=0.0055. Further, with $L_s/D_s = 18/1.2 = 15$ one has with Eq. (15.39) $q_e = 0.018 \cdot 15^{1/3} = 0.044$, thus from Eq. (15.38) for $\beta = (0.044/0.0055)^{1/2} - 1 = 1.83$ and $Q_a = \beta Q = 1.83 \cdot 2.5$ m^{3}s^{-1} $= 4.6$ m^{3}s^{-1}. For this example, the air discharge is significant due to a small water discharge.

With a typical air velocity $V_a = 25$ ms^{-1} in the air supply system, the diameter required is large with $D_a = (4Q_a/\pi V_a)^{1/2} = 0.8$ m. The determining diameter does not necessarily coincide with the diameter needed for maximum discharge (Fig. 15.9).

The maximum air discharge occurs for a water discharge $Q_M = 0.5 \cdot 0.044 \cdot 90\pi \cdot 1.2^{8/3}$ m^{3}s^{-1} $= 10.1$ m^{3}s^{-1} with $Q_{aM} = 1 \cdot 10.1$ m^{3}s^{-1} $= 10.1$ m^{3}s^{-1} according to Eq. (15.40). If such a large discharge may occur, the design consequences have to be accounted for.

15.3.5 *Dissipation Chamber*

The purpose of the dissipation chamber is to transform the remaining excess energy into heat such that the downstream flow is *subcritical* without velocity concentrations. In addition, the flow should be de-aerated and the vertical flow

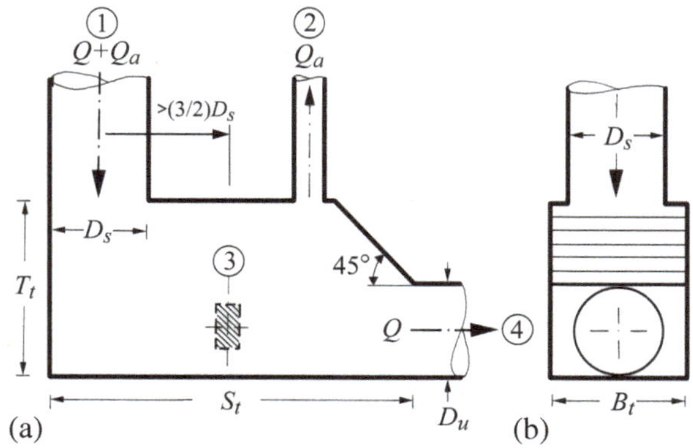

Fig. 15.10 Dissipation chamber of vortex drop (schematic) (**a**) longitudinal and (**b**) transverse section (Hager and Kellenberger 1987)

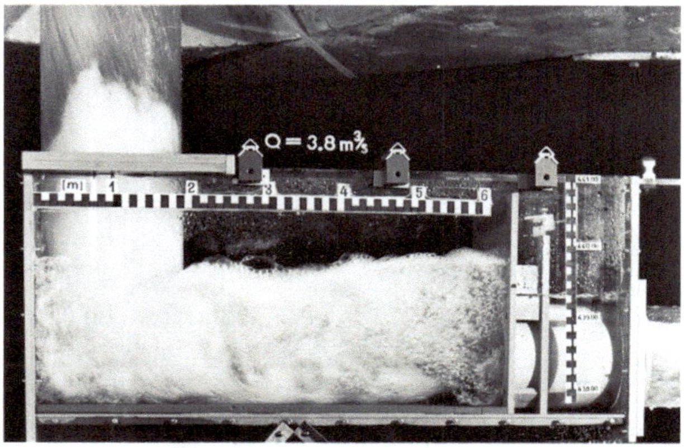

Fig. 15.11 Simplified version of dissipation chamber in a hydraulic model

be carried over into a horizontal flow. Figure 15.10 shows the stilling chamber (German: Toskammer; French: Chambre de dissipation) that was tested experimentally (Fig. 15.11). Its principal dimensions are:

- Length $S_t \cong 4\bar{D}$,
- Width $B_t = (1–1.2)\bar{D}$,
- Height $T_t \cong 2\bar{D}$,

where $\bar{D}$ is the larger value of either the shaft diameter D_s or the downstream diameter D_u. For a reasonable performance, a rather *large dissipation chamber* is recommended. If the chamber is too small, submergence into the vortex shaft, breakdown of air circulation, generation of pulsations or difficulties in access for maintenance can result.

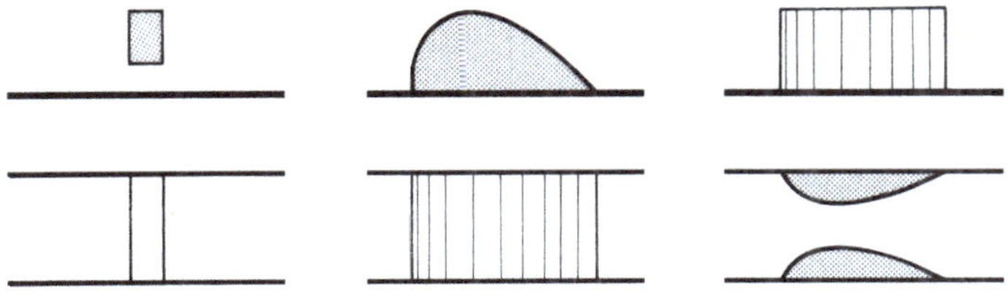

Fig. 15.12 Elements for generation of water cushion in a dissipation chamber (**a**) transverse sill, (**b**) round-crested weir, (**c**) Venturi flume, section (up), plan (below)

The vortex shaft should not protrude into the stilling chamber, and its ceiling should be sloping by 45° into the downstream sewer. For larger structures, a closed *aeration system* is recommended, with design air velocities of 30 ms^{-1}, but maximum air velocities below 50 ms^{-1}. A sufficient flow aeration and a good flow de-aeration are key aspects of a sound chamber design. Further details on the design of the dissipation chamber were presented by Kellenberger and Volkart (1986), and Balah and Bramley (1989).

The *efficiency* of a dissipation chamber can be as high as 80%. Appurtenances such as shown in Fig. 15.12 can slightly add, but their main purpose is to protect the floor by a limited water cushion. Sills can result in an under- or overflow, with the submergence depending on the discharge. Weirs create a permanent water cushion with the danger of upstream sedimentation. Venturi flumes (Chap. 12) combine both effects and allow inspection, but they are more costly. The constricted section should be located at least $(3/2)D_s$ downstream from the axis of the vortex shaft for security reasons (Fig. 15.10a).

Notation

a	[m]	distance
b	[m]	width
B_s	[m]	width of manhole
B_t	[m]	width of dissipation chamber
c	[m]	minimum width of intake spiral
d	[m]	width of intake spiral after 180°
D	[m]	sewer diameter
D_s	[m]	shaft diameter
$\bar{D}$	[m]	determining diameter
e	[m]	eccentricity
F	[–]	Froude number
g	[ms^{-2}]	gravitational acceleration
h	[m]	flow depth
h_e	[m]	end depth

H	[m]	energy head
L_s	[m]	length of shaft
n	[sm$^{-1/3}$]	Manning's roughness coefficient
q_e	[–]	dimensionless discharge
q_N	[–]	$= nQ/(S_o^{1/2}D^{8/3})$ dimensionless uniform discharge
q_s	[–]	$= nQ/(\pi D^{8/3})$ dimensionless shaft discharge
Q	[m^{3}s^{-1}]	discharge
Q^*	[m^{3}s^{-1}]	normalizing discharge
R	[m]	shaft radius
s	[m]	thickness of guide wall
S	[m^3]	relative momentum flux
S_c	[–]	critical slope
S_o	[–]	bottom slope
S_t	[m]	length of dissipation chamber
t	[m]	vertical jet thickness
T_e	[–]	relative end depth
T_t	[m]	height of stilling chamber
V_o	[ms^{-1}]	approach velocity
V^*	[ms^{-1}]	normalizing velocity
x	[m]	longitudinal coordinate
X	[–]	$= (x/h_o)\mathsf{F}_o^{-0.8}$
y	[–]	filling ratio
z	[m]	vertical coordinate
z^*	[m]	scaling depth
Z	[–]	$= z/h_o$
α	[–]	angle of location
β	[–]	air entrainment coefficient
η	[–]	efficiency of dissipation chamber
ΔH	[m]	energy head loss
ΔR	[m]	radius of curvature

Subscripts

a	air
c	critical
K	dry-weather
L	limit
m	minimum
M	maximum
N	uniform
o	approach
R	rain-weather
u	downstream

References

ATV (1991). Bauwerke der Ortsentwässerung (Structures of urban drainage). *Arbeitsblatt* A241. Abwassertechnische Vereinigung: St. Augustin [in German].

Balah, M.I.A., Bramley, M.E. (1989). Standard stilling basin design for use with medium-head vortex drop shafts. *Proc. Institution of Civil Engineers* 86(1): 91–107.

Bruschin, J., Mouchet, P.-L. (1985). Ouvrage de jonction de quatre collecteurs (Junction structure with four sewers). *Ingénieurs et Architectes Suisse* 111(6): 88–91 [in French].

Drioli, C. (1969). Installazioni con pozzo di scarico a vortice (Installations with a vortex drop). *L'Energia Elettrica* 46(2): 81–102; 46(6): 399–409 [in Italian].

Farroni, A., Ianetta, S., Remedia, G. (1988). Rilievi del campo cinematico di uno scaricatore a vortice (Velocity field in a vortex drop). 21 *Convegno di Idraulica e Costruzioni Idrauliche* L'Aquila B(04): 527–537 [in Italian].

Granata, F., de Marinis, G., Gargano, R., Hager, W.H. (2009). Energy loss in circular drop manholes. 33rd *IAHR Congress*, Vancouver, Canada, (CD-Rom).

Granata, F., de Marinis, G., Gargano, R., Hager, W.H. (2010). Hydraulics of circular drop manholes. *Journal of Irrigation and Drainage Engineering* posted ahead of print 16 July 2010.

Hager, W.H. (1987). Abfluss im U-Profil (Flow in U-shaped channel). *Korrespondenz Abwasser* 34(5): 468–482 [in German].

Hager, W.H. (1990). Vortex drop inlet for supercritical approaching flow. *Journal of Hydraulic Engineering* 116(8): 1048–1054.

Hager, W.H., Kellenberger, M.H. (1987). Die Dimensionierung des Wirbelfallschachtes (The design of the vortex drop). *gwf - Wasser/Abwasser* 128(11): 585–590 [in German].

Jain, S.C. (1984). Tangential vortex-inlet. *Journal of Hydraulic Engineering* 110(12): 1693–1699.

Jain, S.C., Ettema, R. (1987). Vortex-flow intakes. *IAHR Hydraulic Structures Design Manual* 1: 125–137, J. Knauss, ed. Balkema: Rotterdam.

Kellenberger, M., Volkart, P. (1986). Der Wirbelfallschacht in Kanalisationsnetzen (Vortex drop in sewer systems). *Schweizer Ingenieur und Architekt* 104(16): 364–371 [in German].

Kellenberger, M. (1988). Wirbelfallschächte in der Kanalisationstechnik (Vortex drops in sewers). *Mitteilung* 98. Versuchsanstalt für Wasserbau, Hydrologie und Glaziologie, ETH: Zürich [in German].

Quick, M.C. (1990). Analysis of spiral vortex and vertical slot vortex drop shafts. *Journal of Hydraulic Engineering* 116(3): 309–325; 118(1): 100–107.

Rajaratnam, N., Mainali, A., Hsung, C.Y. (1997). Observations on flow in vertical drop shaft in urban drainage systems. *Journal of Environmental Engineering* 123(5): 486–491.

SIA (1980). Sonderbauwerke der Kanalisationstechnik (Special structures in sewer techniques). *SIA-Dokumentation* 40. Schweizerischer Ingenieur- und Architektenverein: Zürich [in German].

Volkart, P. (1984). Vereinigungs- und Wirbelfallschacht kombiniert (Combination of junction and vortex drop structures). *Schweizer Ingenieur und Architekt* 102(11): 190–195 [in German].

Yu, D., Lee, J.H.W. (2009). Hydraulics of tangential vortex intake for urban drainage. *Journal of Hydraulic Engineering* 135(3): 164–174.

Chapter 16
Special Manholes

Abstract Special manholes can generate significant losses that lead to poor hydraulic flow conditions. This chapter demonstrates that junctions behave as pressurized conduits for *subcritical* flow in terms of energy losses. Accordingly, the well known loss coefficients can be applied. Starting from specified downstream conditions and loss coefficients, the flows in the upstream junction branches can be determined.

Supercritical junction flow receives particular attention. Following a general discussion of flow phenomena in Sect. 16.3, applications including the channel expansion, the channel bend and the channel junction are described for the rectangular channel. Methods to reduce shockwaves and to improve flow uniformity are also given.

Section 16.4 refers to the through-flow, the bend and the junction manholes. Although there are still various flow features to be explained, the main aspects of junction manholes are currently known and a design procedure is available.

16.1 Introduction

Apart from the standard manhole and the fall manhole (Chaps. 14 and 15), junction manholes (German: Vereinigungsschacht; French: Puits à jonction) are often integrated in sewer systems. The bend manhole and the through-flow manhole are particular cases of the junction manhole, and all are referred to as special manholes. Because those systems are widely branched, and the sewage should be collected to a central treatment station, junction manholes are a basic design element. As long as the velocity is small, problems in terms of flow dynamics are small except for inadmissible submergence. For larger flow velocities particularly at poorly designed junctions, dynamic effects can reduce the efficiency of the sewer system dramatically. For supercritical approach flow, the following *phenomena* can occur:

- Shockwave development, with origin in a junction manhole and generation of cross waves into the downstream sewer,
- Choking of downstream sewer intake due to large shocks,
- Breakdown of air transport into downstream sewer,

W.H. Hager, *Wastewater Hydraulics*, 2nd ed., DOI 10.1007/978-3-642-11383-3_16,

- Breakdown of supercritical flow structure resulting in a transition to pressurized flow in one or two junction branches, and
- Submergence effect, particularly into building connections.

To inhibit these phenomena, a special manhole must be hydraulically designed. As for any structure, a distinction between sub- and supercritical approach flows is important. According to Sect. 16.2, subcritical junction flow can be treated by a one-dimensional approach. Supercritical junction flow is more complex and needs a two-dimensional treatment. To summarize the most important features, Sect. 16.3 gives an introduction to supercritical free surface flows in rectangular channels. In Sect. 16.4, the main manhole types are then treated, based on recent hydraulic experimentation.

Loss coefficients for supercritical flows are insignificant but it is important to verify that a supercritical flow can be maintained across any structure. The various design concepts for sub- and supercritical flows are stressed in the following.

16.2 Subcritical Flow

16.2.1 *Principle of Computation*

A through-flow manhole or a bend manhole as particular cases of the junction manhole can be designed currently based on several assumptions. On the one hand, the manhole geometry has to be simple to exclude large separation zones. As described in Chapt. 2, the particular case where $\mathsf{F} \rightarrow 0$ corresponds to pressurized flow, and the deviations from this asymptotic case increase as the Froude number increases on the other hand.

Using the momentum equation to determine the energy loss across a junction needs knowledge of the wall pressure distribution and the momentum fluxes. The wall pressures can be estimated for *sharp-crested* wall geometries, whereas difficulties arise for rounded wall geometries, particularly due to *unstable separation* phenomena. Figure 16.1 shows junction manholes with a polygonal and a rounded wall geometry. For the sharp-crested manhole, the separation points coincide with the points of wall deflection whereas the separation of flow in Fig. 16.1b varies with

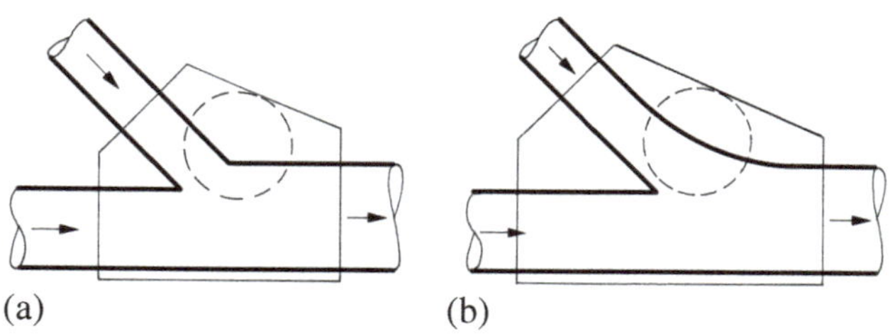

Fig. 16.1 Junction manhole with (**a**) sharp-crested and (**b**) rounded wall geometry

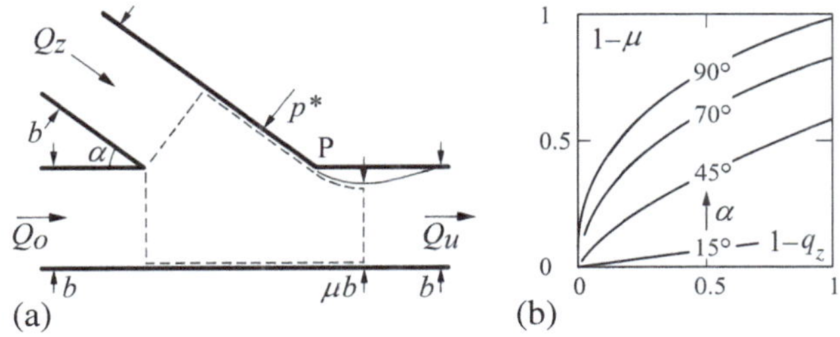

Fig. 16.2 (**a**) Computational model for two-dimensional junction flow, (**b**) contraction coefficient $\mu(q_z)$ according to Eq. (16.6)

discharges in both the through and the lateral branches. Therefore, the wall reactions are difficult to determine, and the common calculation of junctions refers to the sharp-crested design.

According to Hager (1987), the loss coefficient of a junction of combining angle α can be computed for plane ducted flow, thereby excluding the free surface effect and considering asymptotically a flow of zero Froude number. Assuming that the wall friction slope is compensated for by the bottom slope, one may practically assume a potential flow for the contracting flow portion (Fig. 16.2a). Due to the sharp-crested wall deflection, the contracted cross-section (subscript e) is located somewhat downstream from the intake point P. With

$$H = p + \frac{Q^2}{2gb^2} \tag{16.1}$$

as the energy head, and with $E = Q \cdot H$ as the hydraulic energy, the asymptotic case $\mathsf{F} = 0$ requires from *energy conservation*

$$\left(p_o + \frac{Q_o^2}{2gb_o^2}\right) Q_o + \left(p_z + \frac{Q_z^2}{2gb_z^2}\right) Q_z = \left(p_e + \frac{Q_e^2}{2gb_e^2}\right) Q_e. \tag{16.2}$$

Subscripts o, z and u refer, respectively, to the upstream, the lateral and the downstream branches and p denotes the pressure head. With $b_e = \mu b_u$ as the contracted width, and $Q_e = Q_u$, as well as for a constant branch width junction $b_o = b_z = b_u = b$, the *momentum equation* requires

$$p_o b + \frac{Q_o^2}{gb} + p_z b \cos\alpha + \frac{Q_z^2 \cos\alpha}{gb} = p_u b + \frac{Q_u^2}{g\mu b} + p^* b \cos\alpha. \tag{16.3}$$

The pressure head p^* (Fig. 16.2a) corresponds to the average of wall pressure between the boundary values p_z and p_e, i.e. with n_p as pressure coefficient to be determined

$$p^* = \frac{p_z + n_p p_e}{1 + n_p}, \tag{16.4}$$

The extremes of n_p are $p^*(0) = p_z$ and $p^*(\infty) = p_e$.

Vischer (1958) demonstrated that the pressures p_o and p_z have to be identical to satisfy pressure continuity across the junction point. Further, continuity requires

$$Q_o + Q_z = Q_u. \tag{16.5}$$

Thus the pressure heads in Eqs. (16.2) and (16.4) can be eliminated to solve for the contraction coefficient μ as a function of discharge ratio $q_z = Q_o/Q_u$, junction angle α and pressure coefficient n_p. Further, the following conditions have to be satisfied:

- For $q_z = 1$, i.e. no lateral discharge, the contraction coefficient be $\mu(q_z = 1) = 1$,
- For $\alpha = 0$, i.e. parallel branches, the contraction coefficient be $\mu(\alpha = 0) = 1$.

Thus, the coefficient must be $n_p = 1/2$, and the *contraction coefficient* (Hager 1987)

$$\mu^{-1} = \left(1 + \left[(1 - q_z)(2 - q_z)(1 - 2/3\cos\alpha - 1/3\cos^2\alpha) + 1/9\cos^2\alpha\right]^{1/2}\right)$$
$$(1 + 1/3\cos\alpha)^{-1}. \tag{16.6}$$

Figure 16.2b shows the coefficient $(1 - \mu)$ as a function of $(1 - q_z)$ for various junction angles α. The experimental data confirm this approach particularly for small values of α and q_z, whereas deviations are notable for $\alpha \to 90°$ and for $q_z \to 1$.

The *energy loss* results from flow expansion in the downstream branch between the sections of widths μb, and b (Fig. 16.2a). The momentum equation indicates

$$p_e - p_u = \frac{Q_u^2}{gb}\left(1 - \mu^{-1}\right) \tag{16.7}$$

and the generalized energy equation (Chap. 2) requires

$$p_u - p_e = \frac{Q_u^2}{2gb^2}\left[\mu^{-2} - q_z^3 - (1 - q_z)^3\right]. \tag{16.8}$$

By elimination of p_e from Eqs. (16.7) and (16.8) results

$$p_o - p_u = \frac{Q_u^2}{2gb^2}\left[(\mu^{-1} - 1)^2 + 3q(1 - q_z)\right]. \tag{16.9}$$

Introducing *energy loss coefficients* referred to the downstream velocity $V_u = Q_u/(b\times 1)$

$$\xi_o = \frac{H_o - H_u}{Q_u^2/(2gb^2)}, \quad \xi_z = \frac{H_z - H_u}{Q_u^2/(2gb^2)} \tag{16.10}$$

gives

$$\xi_o = (\mu^{-1} - 1)^2 - 1 + 3q_z - 2q_z^2, \quad (16.11)$$

$$\xi_z = (\mu^{-1} - 1)^2 + q_z - 2q_z^2. \quad (16.12)$$

Upon comparing these results with classical data collected for pipe junctions, such as those of Vischer (1958) or Favre, excellent agreement is noted for $\alpha < 70°$. The deviations for $\alpha \to 90°$ or $q_z \to 1$ originate from the assumption that the lateral branch has the angle (α) upon entering the control volume (Fig. 16.2a). However, due to the dominant main branch flow, the determining lateral direction is $\gamma = \sigma\alpha$ where $\sigma < 1$. Based on an optimization process, the coefficient is $\sigma = 8/9$. Figure 16.2b can be applied provided α is substituted by the adjusted angle α/σ, i.e. if $\alpha = 70°$ as computational angle then 79° be used.

The previous computation demonstrates the finding of Chap. 2 that the *energy loss* coefficients determined for pressurized pipe flow can be used also for free surface flow, provided the maximum Froude number across a structure is not too close to 1. Normally, the upper limit is $\mathsf{F}_u < 0.7$. Therefore, the large body of experimental data relating to pressurized flow can be directly transferred to subcritical open channel flows.

16.2.2 *Sub- and Transcritical Flows*

Kumar Gurram et al. (1997) presented an experimental approach to simple channel junctions with a subcritical approach flow. Figure 16.3 shows the geometry and the resulting flow pattern. To apply the *momentum equation* the wall pressure along the lateral branch wall must be known. Due to the flow acceleration from the lateral branch toward the downstream branch, the wall pressure decreases in the flow direction. A *pressure coefficient* $\eta = (P_u - P_o)\sin\alpha/[(1/2)\rho gbhL^2]$ between the downstream (subscript u) and upstream (subscript o) wall pressures was introduced, with α as the junction angle and h_z as the lateral flow depth. Conventionally, one assumes $\eta = 0$ but the experiments indicated that

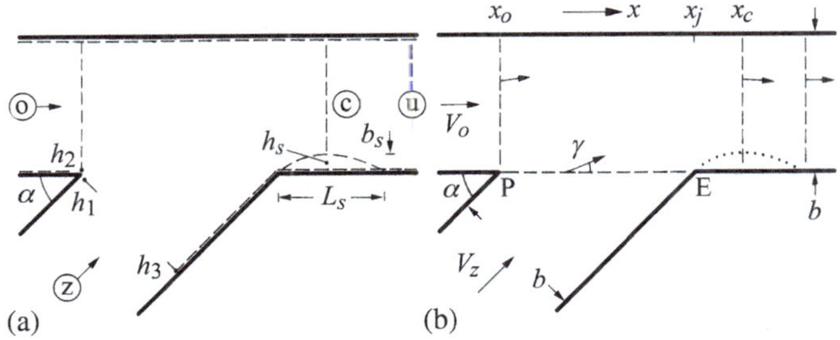

Fig. 16.3 Channel junction, notation for (**a**) flow depths and (**b**) velocities. (– –) Control volumes and profile sections

$$\eta = \cos\alpha \tag{16.13}$$

independent of the Froude number and the discharge distribution.

The flow depths at either side of the junction point P (Fig. 16.3) were compared, and it was found that $h_1/h_2 = 1.0$ ($\pm 2.5\%$) for any flow configuration. Also, the flow depths h_1 and h_3 at the opposite side of the branch channel were compared with the result

$$h_3/h_1 = 1 - 0.09q_z\mathsf{F}_u. \tag{16.14}$$

As shown below, the parameter $q_z\mathsf{F}_u$ is approximately equal to F_z, and the depth ratio h_3/h_1 decreases with increasing Froude number in the lateral branch.

The *separation zone* starts at point E (Fig. 16.3b), has a maximum width b_s at the contracted section (subscript c) and a length L_s. All lengths depend on the downstream Froude number $\mathsf{F}_u = Q_u/(gb^2h_u^3)^{1/2}$. The *width ratio* b_s/b is

$$\frac{b_s}{b} = \frac{1}{2}\left(\mathsf{F}_u - \frac{2}{3}\right)^2 + 0.45\,q_z^{1/2}\left(\frac{\alpha}{90°}\right) \tag{16.15}$$

and the *length ratio*

$$\frac{L_s}{b} = 3.8\sin^2\alpha\left(1 - \frac{1}{2}\mathsf{F}_u\right)q_z^{1/2}; \qquad \mathsf{F}_u < 1; \tag{16.16}$$

$$\frac{L_s}{b} = 0.26\left(1 + \frac{3\alpha}{90°}\right)q_z^{1/2}. \qquad \mathsf{F}_u = 1. \tag{16.17}$$

The length L_s for transcritical flow is much smaller than for subcritical downstream flow. This length reduction is due to a considerable pressure gradient in the vicinity of the *separation zone*. The minimum flow depth h_s in the separation zone is almost independent of the junction angle, and given by

$$\frac{h_s}{h_u} = 1 - 0.6q_z^{1/2}\mathsf{F}_u^{5/2}. \tag{16.18}$$

This indicates a strong effect of the downstream Froude number; for $\mathsf{F}_u < 0.5$, there is practically no sink effect. The shape of the separation bubble was also determined. Figure 16.4 shows typical wall depth and velocity distributions for junctions of various junction angles.

Figure 16.4 shows that the streamlines from the lateral branch have a tendency to turn more towards the downstream direction than according to the junction angle α. The lateral inflow velocity increases from points P to E (Fig. 16.3), as is typical for subcritical curve flows. Far upstream of the junction, the *lateral momentum* is $M_z = \rho g Q_z V_z \cos\alpha$. Toward the junction, both flows in the upstream and lateral branches are accelerated because of constriction. In addition, the lateral direction turns from α to γ (Fig. 16.3b) at the lateral inflow section. Accordingly, the lateral momentum is modified to $M_z = \tau[\rho g Q_z V_z \cos\alpha]$, with

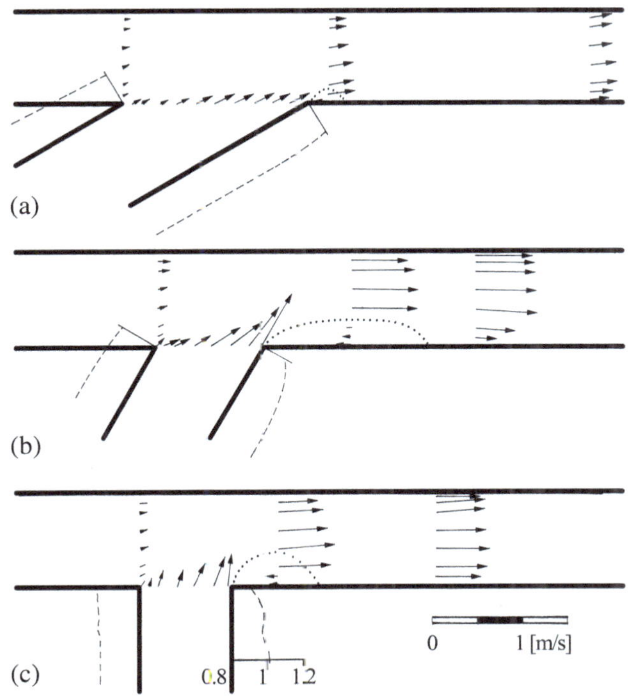

Fig. 16.4 Velocity distribution in channel junction with (– –) lateral wall flow depth and (· ·) separation zone. (F_u,α) = (**a**) (0.25;30°), (**b**) (0.50;60°); (**c**) (1.00;90°), always for $q_z = Q_z/Q_u = 0.75$

$$\tau = \frac{\cos(\sigma\alpha)}{\cos(\alpha)} q_z^{-1}. \qquad (16.19)$$

The effect of the tailwater Froude number on the lateral momentum transfer is thus insignificant, and the coefficient is $\sigma = 0.85$ as previously established for junction flow with a small Froude number. The effect of τ is large for large junction angles α and small $q_z = Q_z/Q_u$, i.e. a large discharge from the upstream branch. Assuming that: (1) the friction slope is compensated for by the bottom slope, (2) nearly one-dimensional flows occur in the upstream and lateral branches, (3) the lateral momentum is determined as before, and (4) the flow depths h_o and h_z are equal, the *momentum equation* requires

$$\frac{bh_o^2}{2} + \frac{Q_o^2}{gbh_o} + \frac{Q_z^2 \cos\alpha}{gbh_z} \cdot \frac{Q_u \cos(\sigma\alpha)}{Q_z \cos\alpha} = \frac{bh_u^2}{2} + \frac{Q_u^2}{gbh_u}. \qquad (16.20)$$

The depth ratio $Y = h_z/h_u$ satisfies then the equation

$$Y^3 - \left(1 + 2\mathsf{F}_u^2\right) Y + 2\,\mathsf{F}_u^2 \left[(1 - q_z)^2 + q_z \cos(\sigma\alpha)\right] = 0. \qquad (16.21)$$

It depends on the downstream Froude number F_u, the discharge ratio q_z and the junction angle. For Y close to unity, Eq. (16.21) can be solved explicitly to yield

$$Y = 1 + \frac{q_z \mathsf{F}_u^2 \left[2 - q_z - \cos(\sigma\alpha)\right]}{1 - \mathsf{F}_u^2}. \tag{16.22}$$

Because q_z and $\cos(\sigma\alpha)$ are smaller than 1, Y is larger than 1, and the effect of the free surface previously omitted is contained in the denominator $(1 - \mathsf{F}_u^2)$. For $\mathsf{F}_u < 0.3$ this effect is negligible because F_u^2 is then one order smaller than the term 1. For $\mathsf{F}_u > 0.7$, Eq. (16.22) ceases to apply because of second order terms that have been neglected, and recourse to Eq. (16.21) has to be made. The term $q_z\mathsf{F}_u$ is essentially equal to Q_zV_u and corresponds to the *momentum flux* of the lateral branch.

The effect of the lateral branch disappears for two asymptotic cases:

(1) No lateral discharge ($q_z = 0$) and $Y = 1$ for any junction angle α and downstream Froude number,
(2) No discharge at all ($\mathsf{F}_u = 0$) and $Y = 1$ for any values of α and q_z.

The *backwater effect* can be controlled with three methods:

(1) Small downstream Froude number, $\mathsf{F}_u < 0.5$,
(2) Discharge ratio not too close to the value $q_M = 1-(1/2)\cos(\sigma\alpha)$, for which the term $q_z[2 - q_z - \cos(\sigma\alpha)]$ has the maximum (subscript M)

$$Y_M = 1 + \frac{\left[1 - (1/2)\cos(\sigma\alpha)\right]\mathsf{F}_u^2}{1 - \mathsf{F}_u^2}. \tag{16.23}$$

Junction angles up to $\alpha = 45°$ are thus not really significant because of their small effect on Y, and
(3) Small junction angle α.

This approach was compared with observations and the agreement was satisfactory for $\alpha < 90°$, and $\mathsf{F}_u < 1$. Figure 16.5 shows side views to a 90° channel junction with transitional flows and various discharge ratios $q_z = Q_z/Q_u$.

The *separation zone* is relevant in junction flow because it determines the backwater effect and the entrainment characteristics. From Figure 16.6 one may hardly see a flow separation for $q_z = 0$. Instead, a series of surface waves are generated downstream of the junction point E. For $q_z = 0.5$, the constriction effect and wall return flow are noted. For $q_z = 1$, the sink created by the lateral flow is significant, and the constriction at the critical section amounts to nearly 50%. Additional work on junction flow include laboratory tests of Hsu et al. (1998a,b), Weber et al. (2001), and Chiapponi and Longo (2008), and numerical tests by Huang et al. (2002) and Shabayek et al. (2002). More information on the numerical modeling of these flows may be expected in the future.

Fig. 16.5 Transitional flow in 90° junction for $q_z =$ (**a**) 0, (**b**) 0.5, (**c**) 1

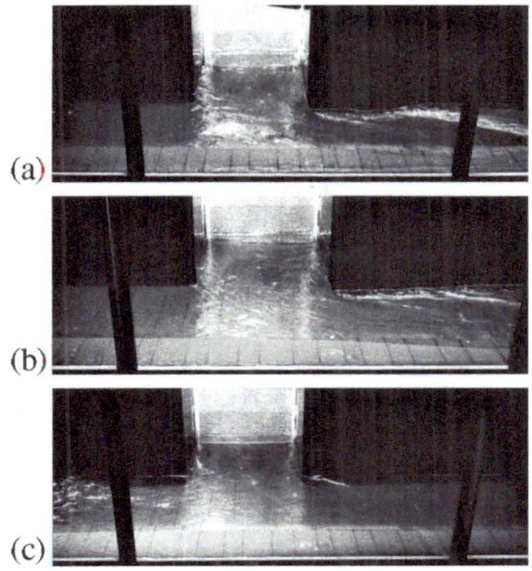

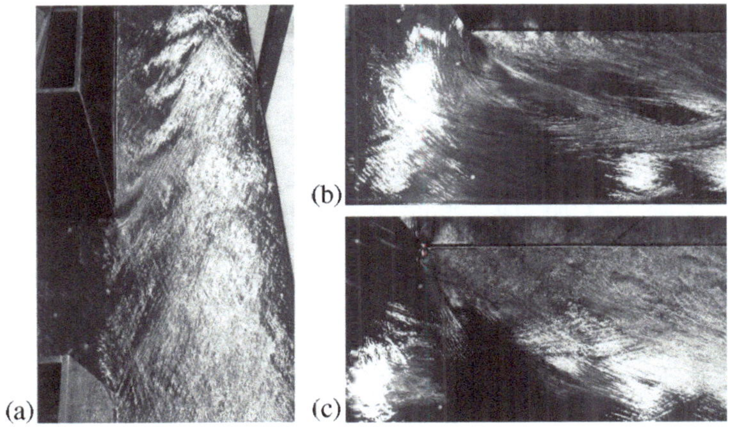

Fig. 16.6 Separation zone for 90° junction and $q_z =$ (**a**) 0, (**b**) 0.5, (**c**) 1

16.2.3 *Loss Coefficients*

The loss coefficients specified in Sect. 2.3.5 refer to sharp-crested junctions. In the following the *round-crested junction* is considered due to its wide-spread application, and to estimate the effect of rounding a wall deflection.

The first source of loss coefficients is again Idel'cik (1979) for a radius $R_z = b_z$. If $q_z = Q_z/Q_u$ is the discharge ratio of lateral to downstream branches, then the loss coefficients ξ_o relative to the approach and ξ_z relative to the lateral branches can be specified as functions of q_z for various ratios of cross-sectional areas $m = F_z/F_u$. Figure 16.7 indicates insignificant losses for $q_z \cong 0.5$, but these increase for $q_z \to 1$ in the lateral branch associated with a decrease in the upstream branch. In addition, the energy distribution becomes poor for small m.

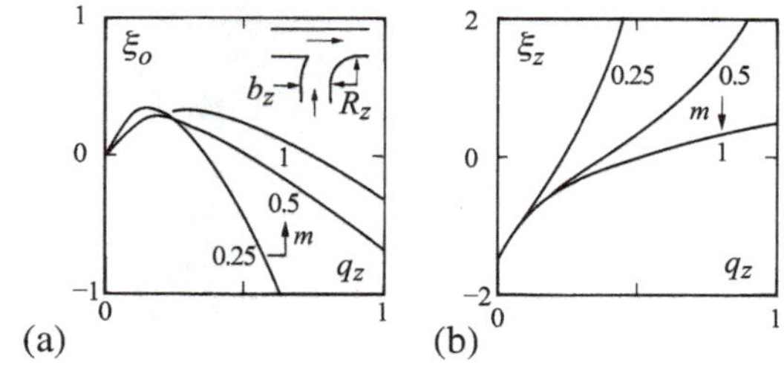

Fig. 16.7 Energy loss coefficients in rounded junction geometry for $\alpha = 90°$ and cross-sectional ratio $m = F_z/F_u$ with $R_z = b_z$ for (**a**) upstream branch ξ_o, and (**b**) lateral branch ξ_z for discharge ratios $q_z = Q_z/Q_u$ and $0.7 < F_o/F_u < 1.1$ (Idel'cik 1979)

Upon comparing the values with these corresponding to sharp-crested junctions, a significant reduction of the loss coefficients is noted. Rounding any structure is *hydraulically efficient* for subcritical flow, therefore. For supercritical approach flow, this statement must be modified.

Figure 16.8 shows the effect of the curvature radius R_z on the loss coefficients of a 90° junction. As from Fig. 2.10, a significant reduction of loss may be observed as the rounding increases. The analogue results also for smaller junction angles.

In addition to this information for pressurized junctions, some results refer to *junction manholes*. Apart from the works of Sangster et al. (1959), Townsend and Prins' (1978) observations have to be mentioned, which refer all to unguided manhole flow (Chap. 14). The data for 45° junction flow indicate all relative small loss coefficients.

According to Lindvall (1984) the submergence of a manhole has practically no effect on the energy loss. His data refer to the 90° junction manhole with a relative radius $R_z/D = 1$ and 100% bench height. For a completely submerged manhole

$$\xi_o = 0.475[3.3 - (D_z/D_u - 0.42)^2]\Omega + 0.024\,\delta_s, \qquad (16.24)$$

$$\xi_z = 0.07 + 0.133(\delta_s + 10)\Omega - 0.575\Omega^{3.5}. \qquad (16.25)$$

Here $\Omega = [1 - (Q_o/Q_u)^2(D_u/D_o)^2]$ is a velocity coefficient and $\delta_s = D_s/D_u$ is the ratio of manhole to tailwater sewer diameters.

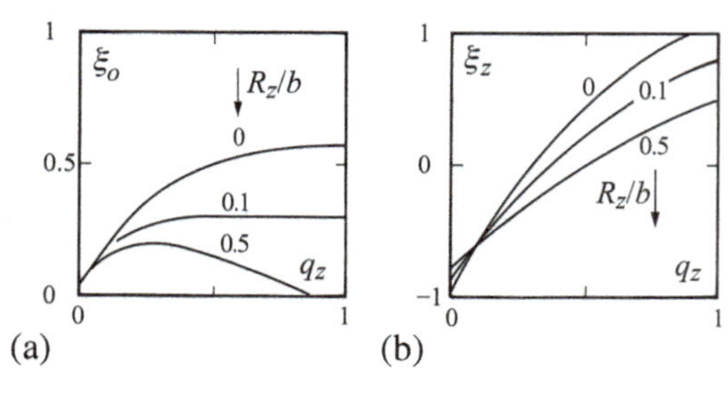

Fig. 16.8 Loss coefficient for a 90° junction and various *relative* roundings R_z/b for equal cross-sections in all branches (Ito and Imai 1973). (**a**) Through branch $\xi_o(q_z)$ and (**b**) lateral branch $\xi_z(q_z)$ with $q_z = Q_z/Q_u$

Whether such complex relations are justified for a limited data basis is questionable. The *maxima* for both ξ_o and ξ_z result for $q_z \to 1$. Approximately, the effect of δ_s can be neglected. With $\delta_s = 3$, $D_u/D_o = 1$ and $0.42 < D_z/D_u \leq 1$, Eqs. (16.24) and (16.25) read with $\Omega = 1-(Q_o/Q_u)^2$

$$\xi_o = 0.475[3.3 - (D_z/D_u - 0.42)^2][1 - (Q_o/Q_u)^2] + 0.07, \quad (16.26)$$

$$\xi_z = 1.73[1 - (Q_o/Q_u)^2] - 0.575[1 - (Q_o/Q_u)^2]^{3.5} + 0.07. \quad (16.27)$$

Although these values may seem relatively large, they apply to completely submerged manhole flow. Figure 16.9 shows Eqs. (16.26) and (16.27) indicating the small effect of the relative lateral diameter. Compared to the previous indications, both ξ_o and ξ_z are always positive, and the suction effect seems to be absent.

Marsalek (1987) investigated *counter flow junctions* with a 90° deflection in the manhole (Fig. 16.10a). According to observations, the effect of manhole submergence is insignificant, although a major effect for standard manholes follows from Chap. 14. For a 100% bench height, the same results ξ_o and ξ_z as functions of

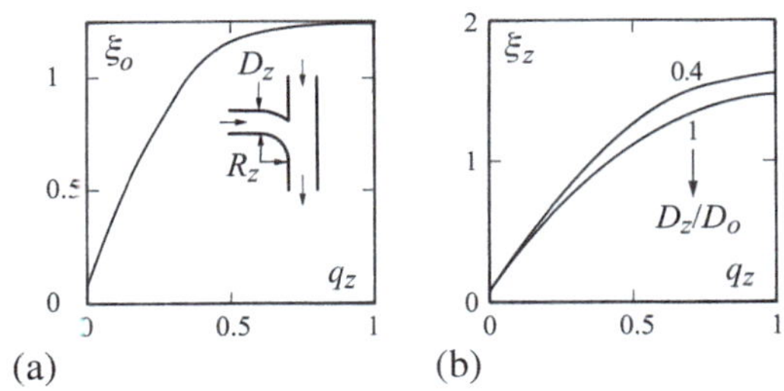

Fig. 16.9 Loss coefficients in the submerged junction manhole with 100% benches for $\alpha = 90°$ and $R_z/D_z = 1$. (**a**) $\xi_o(q_z)$ and (**b**) $\xi_z(q_z)$ for various diameter ratios D_z/D_o, and $q_z = Q_z/Q_u$ (Lindvall 1987)

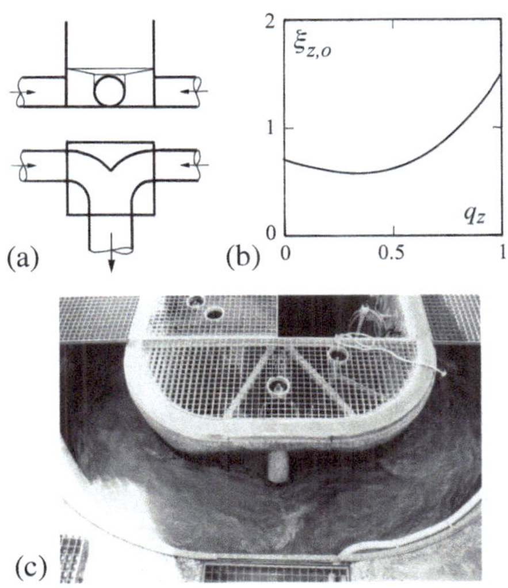

Fig. 16.10 Counter-flow junction (**a**) geometry in section and plan, (**b**) loss coefficients ξ_o or ξ_z as functions of $q_z = Q_z/Q_u$ (Marsalek 1987). (**c**) Y-junction structure in a treatment station

$q_z = Q_z/Q_u$ apply (Fig. 16.10b). Approximately, $\xi = 1$ for all discharges, such that one velocity head of the downstream flow is dissipated across the manhole.

16.2.4 *Computation of Free Surface Profiles*

For given values ξ_o in the through branch and ξ_z in the lateral branch, the generalized energy head equations read, respectively

$$H_o + \Delta z_o = H_u + \xi_o[V_u^2/2g] + \Delta z_{fo}, \tag{16.28}$$

$$H_z + \Delta z_z = H_u + \xi_z[V_u^2/2g] + \Delta z_{fz} \tag{16.29}$$

with H as energy head relative to the sewer bottom, Δz as the elevation difference, V as the cross-sectional velocity and Δz_f as the friction loss height. Depending on whether free surface or pressurized flows occur, the static pressure head is equal to the *flow depth h*, or the *pressure head* h_p.

For both pressurized flow or subcritical flow, the computational direction is opposed to the flow direction. Accordingly, all parameters at the downstream section (subscript u) are known from *backwater calculations*, and the flow depth or pressure heads in both the upstream (subscript o) and lateral (subscript z) branches have to be determined. The distribution of discharge is not always known, and additional conditions have to be satisfied. It should be noted that the kinetic energy is always much smaller than the static energy head, and simplifications in the kinetic energy term are admissible. Neglecting the variation of cross-sectional area, as previously introduced in Chaps. 2 and 10, allows for an insight into the governing flow structure. Thus, if $V_o = Q_o/F_u$ and $V_z = Q_z/F_u$ is assumed, Eqs. (16.28) and (16.29) yield

$$h_o + Q_o^2/(2gF_u^2) + \Delta z_o = h_u + (1 + \xi_o)Q_u^2/(2gF_u^2) + \Delta z_{fo}, \tag{16.30}$$

$$h_z + Q_z^2/(2gF_u^2) + \Delta z_z = h_u + (1 + \xi_z)Q_u^2/(2gF_u^2) + \Delta z_{fz}. \tag{16.31}$$

For junction manholes *without* a bottom drop, the friction slope is nearly compensated for by the bottom slope, and $\Delta z \cong \Delta z_f$. Thus

$$h_o - h_u = \frac{(1 + \xi_o)Q_u^2 - Q_o^2}{2gF_u^2}, \tag{16.32}$$

$$h_z - h_u = \frac{(1 + \xi_z)Q_u^2 - Q_z^2}{2gF_u^2}. \tag{16.33}$$

The water level differences $\Delta h_i = h_i - h_u$ increase with increasing difference in discharge $\Delta Q_i = Q_u - Q_i$, and with large loss coefficients ξ_i. For a negative loss coefficient $\xi_i = (Q_i/Q_u)^2 - 1$, the resulting level difference is $\Delta h = 0$. Eqs. (16.32) and (16.33) apply for subcritical open channel *and* for pressurized flows. For a given discharge distribution, they can be determined explicitly.

Often, the upstream water levels are identical, and $h_o = h_z$. The condition to be satisfied is then $\xi_o - \xi_z = q_o - q_z$, i.e. the difference between the loss coefficients is equal to the difference of the relative discharges. With $q_o = 1 - q_z$, one has also $\xi_o - \xi_z = 1 - 2q_z$. For $q_z = 0$ the result is $\xi_o = 1 + \xi_z$, whereas for $q_z = 1$ the result is $\xi_o = \xi_z - 1$ and $\xi_o = \xi_z$ if $q_z = 1/2$.

Example 16.1 Given a rounded junction of equal diameter $D = 0.80$ m and downstream parameters $Q_u = 0.75$ m^{3}s^{-1}, $h_u = 0.64$ m, bottom slope 0.2% and roughness value $n = 0.011$ sm$^{-1/3}$. What are the flow depths for a discharge distribution of $Q_z/Q_u = 1/3$?
From Fig. 16.7, the loss coefficients are $\xi_o = 0.1$ and $\xi_z = -0.12$ ($m = 1$). For a filling ratio $y_u = h_u/D = 0.64/0.80 = 0.8$, the cross-section is $F_u/D^2 = (4/3)0.8^{3/2}[1 - 0.25 \cdot 0.80 - 0.16 \cdot 0.8^2]$ according to Eq. (5.16), thus $F_u = 0.67 \cdot 0.8^2 = 0.43$ m^2. Further, with the discharges $Q_o = (2/3)0.75$ m^{3}s^{-1} = 0.50 m^{3}s^{-1} and $Q_z = 0.25$ m^{3}s^{-1}, the differences in flow depths are, respectively, $h_o - h_u = [(1+0.1)0.75^2 - 0.5^2]/(19.62 \cdot 0.43^2) = 0.10$ m from Eq. (16.32), and $h_z - h_u = [(1-0.12)0.75^2 - 0.25^2]/(19.62 \cdot 0.43^2) = 0.12$ m, i.e. $h_o = 0.74$ m and $h_z = 0.76$ m.

Example 16.2 What are the effects of wall friction and bottom slope in Example 16.1?
The combined effect may be described with the total slope $J = S_o - S_f$. With a hydraulic radius of $R_{hu} = 0.244$ m and the friction slopes $S_{fo} = [n(Q_o + Q_u)/2]^2/[F_u^2 R_{hu}^{4/3}] = [0.011(0.5 + 0.75)/2]^2/[0.43^2 0.244^{4/3}] = 0.17\%$, and $S_{fz} = [(Q_z + Q_u)/(Q_o + Q_u)]^2 S_{fo} = 0.11\%$, respectively, one has $J_o = 0.20 - 0.17 = 0.03\%$, and $J_z = 0.20 - 0.11 = 0.09\%$. For a determining length of two manhole lengths $L_s = 2.5$ m, i.e. $L = 5$ m, the difference heights are $\Delta z_{do} = 0.0003 \cdot 5 = 0.0015$ m and $\Delta z_{dz} = 0.0009 \cdot 5 = 0.0045$ m, respectively. These heights are much smaller than those due to the additional losses. For junctions with approximately uniform flow, the determining equations are thus Eqs. (16.32) and (16.33).

16.2.5 *Bottom Drop*

According to Eqs. (16.32) and (16.33) the elevation differences $h_o - h_u$, and $h_z - h_u$ increase significantly with $(1 + \xi_i)[Q_u^2/(2gF_u^2)]$, corresponding to $(1 + \xi_i)[V_u^2/2g]$. To inhibit a large submergence into the approach branches, the junction manhole may

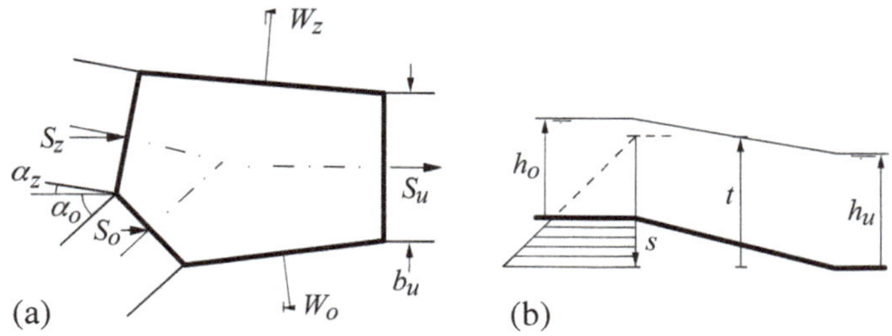

Fig. 16.11 Junction manhole with a *bottom drop* (**a**) plan with forces in the downstream direction, (**b**) section with pressure distribution on the drop

be extended with a bottom drop (German: Sohlabsturz; French: Chute de fond). The drop height is adjusted to the design discharge, such that *uniform flow* establishes in all branches. Then the submergence effect is absent (Fig. 16.11).

The required drop height can be determined with the momentum equation. To simplify computations, a *substitute junction* structure (subscript E) of rectangular cross-sections is considered, with

$$h_E = h, \quad \text{and} \quad F_E = F. \tag{16.34}$$

The width of a branch is thus $b_E = F_E/h_E = F/h$. Note that the energy heads of the effective and the substitute configurations are identical. Assuming again that the friction losses are compensated for by the bottom slope, the governing equation is (Hager 1982)

$$[b_o h_o^2/2 + V_o Q_o/g]\cos\alpha_o - [b_u h_u^2/2 + V_u Q_u/g] + W + Z + B = 0. \tag{16.35}$$

Here V is velocity, α_o the manhole branch angle, W the wall reaction force, Z the lateral momentum, and B the bottom reaction, all in the direction of the downstream channel. These forces are determined below.

The average pressure height on the drop is $t = (1/2)(h_o + s + h_u)$, the average drop width is $(1/2)(b_o\cos\,\alpha_o + b_u)$ and the *bottom reaction B* for a drop height s per unit width $s(t - s/2)$, thus from Fig. 16.11b

$$B = (1/2)s(b_o\cos\alpha_o + b_u)(1/2)(h_o + h_u). \tag{16.36}$$

The *wall reaction* depends on the determining width $(b_u - b_o\cos\,\alpha_o)$ times the average of the flow depths squared $h_m^2 = (1/2)(h_o^2 + h_u^2)$, thus

$$W = (1/2)(b_u - b_o\cos\alpha_o)(h_o^2 + h_u^2). \tag{16.37}$$

For a prismatic junction structure, the wall reaction drops out.

Figure 16.12 shows four different configurations of lateral flow: (a) submerged, that is excluded in the following, (b) plunging jet, (c) free jet, and (d) impact on

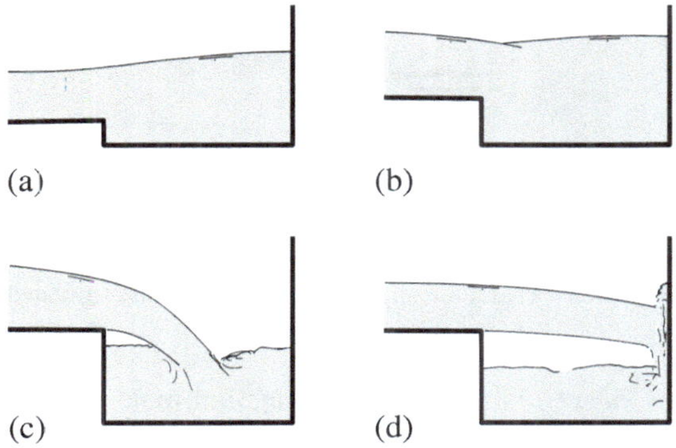

Fig. 16.12 Lateral inflow configurations into a junction manhole, for details see text

opposite wall. For cases (b) and (c), the lateral momentum is compensated for by the wall reaction. Further, for cases (c) and (d), the internal jet pressure is nearly zero, and no static jet pressure has to be included.

The *lateral momentum* is thus $Z = \varepsilon V_z Q_z/g$ with ε as the momentum coefficient ($0 < \varepsilon \leq 1$). For cases (b) and (c) results $\varepsilon = 1$, whereas case (d) may be described with $\varepsilon < 1$. The large number of parameters can be reduced by defining the following ratios

$$Y_o = h_o/h_u,\ Y_z = h_z/h_u,\ \beta_o = b_o/b_u,\ \beta_z = b_z/b_u$$
$$S_o = s_o/h_u,\ S_z = s_z/h_u,\ q_o = Q_o/Q_u,\ q_z = Q_z/Q_u. \tag{16.38}$$

Further, with the velocities $V_o = Q_o/(b_o h_o)$, and $V_z = Q_z/(b_z h_z)$, the downstream Froude number $\mathsf{F}_u = Q_u/(gb_u^2 h_u^3)^{1/2}$ and the continuity equation $q_z = 1 - q_o$, the governing equation is (Hager 1982)

$$S_o = 1 - Y_o + \frac{4\mathsf{F}_u^2\left[1 - \frac{q_o^2 \cos\alpha_o}{\beta_o Y_o} - \frac{\varepsilon(1-q_o)^2 \cos\alpha_z}{\beta_z Y_z}\right]}{(1 + Y_o)(1 + \beta_o \cos\alpha_o)}. \tag{16.39}$$

The relative drop height S_o depends on nine parameters. For $\mathsf{F}_u = 0$, the correct expression $s_o + h_o = h_u$ results from Eq. (16.39), in agreement with Vischer (1958) for the pressurized junction. Currently, no experiments on the junction manhole with a bottom drop are available. The momentum coefficient ε has only a small effect on the result, with the *maximum drop height* resulting for $\varepsilon = 0$. This case can be described with a hypercritical flow or a large drop height. Note that Eq. (16.39) is relevant only for $S_o > 0$.

Minimum energy losses result for small branch angles α_o and α_z. For subcritical flow, energy losses decrease with rounded walls. A hydraulic effective, but structurally more involved junction structure is shown in Fig. 16.13.

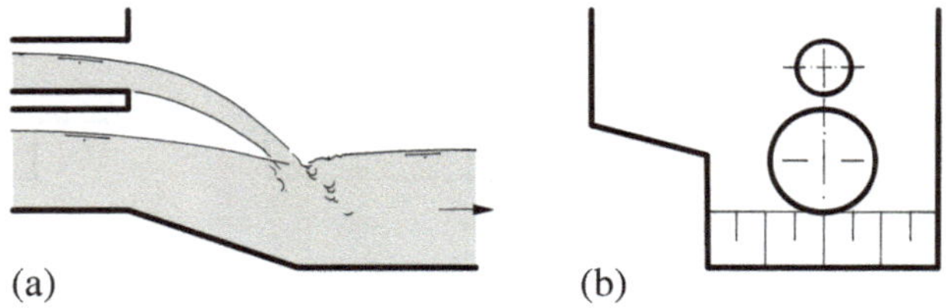

Fig. 16.13 Hydraulic effective junction structure (**a**) longitudinal, (**b**) transverse sections

Example 16.3 Given a junction manhole with $Q_o = 2.5$ m^{3}s^{-1}, $Q_z = 1.7$ m^{3}s^{-1}, $\alpha_o = 10°$, $\alpha_z = 35°$, $S_{oo} = 0.5\%$, $S_{oz} = 0.7\%$, $S_{ou} = 0.3\%$ and $1/n =$ 85 m$^{1/3}$s^{-1}. The lateral branch elevation is 1 m above the outlet elevation. Design the junction structure (after Hager 1982).

Table 16.1 summarizes the governing parameters. The diameter is given by $D \geq 1.55(nQ/S_o^{1/2})^{3/8}$, the uniform flow characteristics are given in Chap. 5. The substitute widths are $b_o = 1.04$ m, $b_z = 0.84$ m and $b_u = 1.34$ m. With $h_{co} = 0.86$ m, $h_{cz} = 0.75$ m and $h_{cu} = 1.05$ m, the flows in the upstream and lateral branches are supercritical, whereas the flow in the downstream branch is subcritical, involving a hydraulic jump.

Table 16.1 Numbers of Example 16.3

Branch	Q [m^{3}s^{-1}]	S_o [%]	D [m]	Q_v [m^{3}s^{-1}]	V_v [ms^{-1}]	h_N [m]	V_N [ms^{-1}]
upstream	2.5	0.5	1.25	3.4	2.77	0.81	2.96
lateral	1.7	0.7	1.00	2.2	2.80	0.67	3.02
downstream	4.2	0.3	1.60	5.1	2.53	1.15	2.72

The dimensionless parameters are $Y_o = h_o/h_u = 0.70$, $Y_z = h_z/h_u = 0.58$, $\beta_o = b_o/b_u = 0.78$, $\beta_z = b_z/b_u = 0.63$, $q_o = 0.60$, $\cos\alpha_o = 0.98$, $\cos\alpha_z = 0.82$ and $\varepsilon = 1$. The so-called *junction parameter* $A_s = 1-q_o^2\cos\alpha_o/(\beta_o Y_o)-\varepsilon(1-q_o)^2\cos\alpha_z/(\beta_z Y_z) = 1-0.65-0.36 = -0.01$ indicates practically no dynamic effect. With $\mathsf{F}_u^2 = Q_u^2/(gb_u^2h_u^3) = 0.66$ and $B_s = 4\mathsf{F}_u^2A_s/(1+Y_o)(1+\beta_o\cos\alpha_o) = 4\cdot 0.66(-0.01)/(1.7\cdot 1.6) = -0.01$, Eq. (16.39) gives $S_o = 1-Y_o+B_s = 0.29$, i.e. the drop height is $s_o = S_o h_u = 0.33$ m (Fig. 16.14).

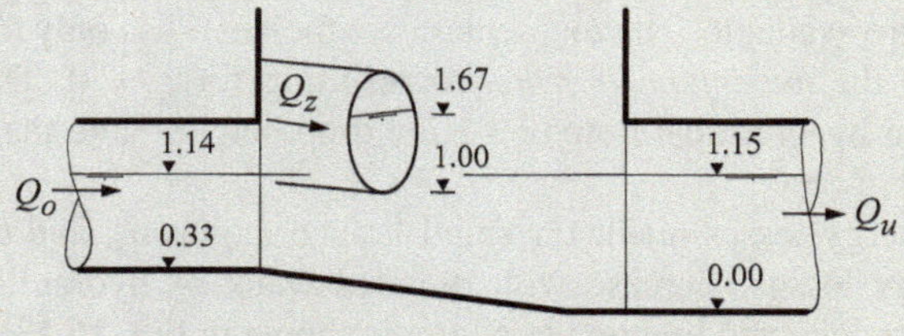

Fig. 16.14 Section across junction manhole for Example 16.3

If $s_z > 1.5$ m, then $\varepsilon = 0$ and $s_o = 0.70$ m from Eq. (16.39). The lateral branch results in a drawdown of the free surface. The minimum loss configuration $\alpha_o = \alpha_z = 0$ would give $s_o = 0.25$ m.

16.3 Supercritical Flow

16.3.1 *Flow Phenomenon*

Supercritical channel flow (German: Schiessender Kanalabfluss; French: Ecoulement torrentiel) is distinctly different from subcritical flow. Whereas the latter can be approximated often as a one-dimensional flow, supercritical flow is fundamentally *two-dimensional*. The 2D dynamic equations of a steady flow with velocity components u and v in the streamwise direction x and the transverse direction y, respectively, read with Fig. 16.15 (Chaudhry 1993)

$$\frac{u}{g}\frac{\partial u}{\partial x} + \frac{v}{g}\frac{\partial u}{\partial y} + \frac{\partial h}{\partial x} = S_{ox} - S_{fx}, \tag{16.40}$$

$$\frac{u}{g}\frac{\partial v}{\partial x} + \frac{v}{g}\frac{\partial v}{\partial y} + \frac{\partial h}{\partial y} = S_{oy} - S_{fy}. \tag{16.41}$$

Here, S_{ox} and S_{oy} are the bottom slopes in the x- and y-directions. Further, the two-dimensional *continuity equation* is

$$\frac{\partial(uh)}{\partial x} + \frac{\partial(vh)}{\partial y} = 0. \tag{16.42}$$

With the Manning equation, the friction slopes are

$$S_{fx} = \frac{u(u^2+v^2)^{1/2}}{(1/n)^2 h^{4/3}}, \quad S_{fy} = \frac{v(u^2+v^2)^{1/2}}{(1/n)^2 h^{4/3}}. \tag{16.43}$$

Imposing the bottom geometry and three boundary conditions for the unknowns u, v, h as functions of x and y, this system of partial differential equations may be numerically solved (Wehausen and Laitone 1960, Liggett 1994).

Provided one has nearly a *potential flow*, the source term $(S_o - S_f)$ may be neglected and the condition for *irrotational flow* is satisfied

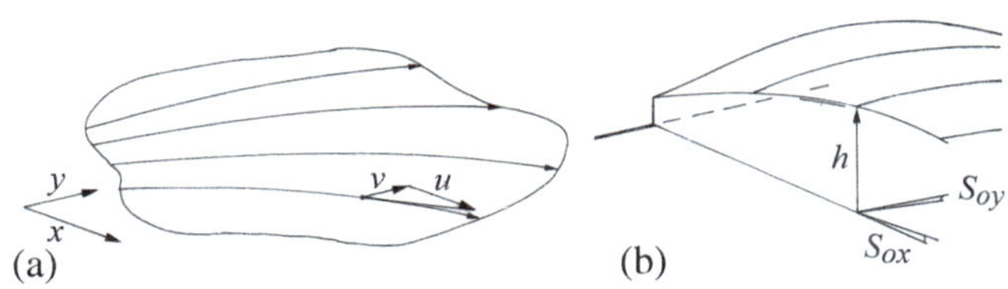

Fig. 16.15 Two dimensional shallow water flow, (**a**) definition of velocity components, (**b**) section across flow

$$\frac{\partial u}{\partial y} = \frac{\partial v}{\partial x}. \tag{16.44}$$

Inserting in Eq. (16.40) gives (Wehausen and Laitone 1960)

$$\frac{u}{g}\frac{\partial u}{\partial x} + \frac{v}{g}\frac{\partial v}{\partial x} + \frac{\partial h}{\partial x} = \frac{1}{2g}\frac{\partial(u^2)}{\partial x} + \frac{1}{2g}\frac{\partial(v^2)}{\partial x} + \frac{\partial h}{\partial x} = 0. \tag{16.45}$$

This can also be expressed as

$$\frac{\partial}{\partial x}\left[\frac{u^2+v^2}{2g} + h\right] = 0, \quad \text{resp.} \ \frac{V^2}{2g} + h = H. \tag{16.46}$$

Here $V^2 = u^2 + v^2$ is the square of the absolute velocity and H is the integration constant, equal to the energy head. The identical result follows also from Eq. (16.41). The energy head relative to the bottom is everywhere equal to $H = h_o + V_o^2/(2g)$ as at the approach flow section.

Equations (16.40), (16.41), (16.42), and (16.43) cannot immediately be applied because a relation between u, v and h as a function of (x,y) is not generally available. The system of partial differential equations can be transformed to a system of *ordinary* differential equations by application of the *method of characteristics*. According to Abbott (1966) let $y_\pm(x)$ be curved coordinates defined as

$$\left[\frac{\mathrm{d}y}{\mathrm{d}x}\right]_\pm = \frac{-uv \pm c[V^2 - c^2]^{1/2}}{c^2 - u^2} \tag{16.47}$$

along which the velocity gradients vary as

$$\left[\frac{\mathrm{d}v}{\mathrm{d}u}\right]_\pm = \frac{-uv \pm c[V^2 - c^2]^{1/2}}{v^2 - c^2}. \tag{16.48}$$

Here $c = (gh)^{1/2}$ is the *elementary wave celerity* and F the Froude number of two-dimensional flow defined as

$$\mathsf{F}^2 = (u^2 + v^2)/c^2 = (V/c)^2. \tag{16.49}$$

It can be demonstrated that the positive characteristic curve $(\mathrm{d}y/\mathrm{d}x)_+$ is at each point perpendicular to the curve $(\mathrm{d}v/\mathrm{d}u)_-$, and vice-versa. The characteristic curves may be determined by a numerical procedure resulting in a relation between the characteristic angle and the local Froude number. Details are omitted here because of extensive works (Wehausen and Laitone 1960, Chaudhry 1993, Liggett 1994).

From Eqs. (16.47), real solutions of the characteristic equations exist only if $(V^2 - c^2)^{1/2}/c = (\mathsf{F}^2 - 1)^{1/2} \geq 0$, i.e. provided $\mathsf{F} > 1$ and the governing system of equations is hyperbolic. *Standing surface waves* exist only for supercritical flow, therefore. For subcritical flows, the governing system of equations is elliptic whose solutions are boundary value problems, such as for overflow or underflow problems. The major conclusions from this short introduction in two-dimensional free surface flows are:

- Subcritical shallow-water flows may be described with a *one-dimensional approach*, because the transverse variation of free surface elevation is normally insignificant.
- Supercritical shallow-water flows are sensitive to perturbations, and these result in standing surface waves. The transverse free surface profiles are therefore not horizontal, and a *two-dimensional approach* is indispensable.

In the following, various basic geometries are considered in which supercritical sewer flow may exist. Currently, all experiments have been conducted on a plane bottom geometry, and conduits have not yet received attention. Two aspects are of major importance:

- Definition of main flow pattern, involving the surface extrema, and
- Choking of flow, corresponding to a breakdown of the supercritical flow structure.

The first aspect is needed for *freeboard considerations* to inhibit lateral overflow of sewage. The second aspect delimits supercritical from the subcritical flow. Wherever the latter flow appears, a sewer may get pressurized, with dramatic submergence effects in the approach flow reaches. The elements to be discussed include the abrupt wall deflection, the channel contraction, the channel expansion, the channel curve, and the channel junction in the rectangular channel. Further, methods to reduce shockwaves are outlined, and typical conclusions relating to the junction and bend manholes are given. Finally, sewer manholes are considered.

16.3.2 *Abrupt Wall Deflection*

Description

The *basic phenomenon* of supercritical flow may be easily explained in a horizontal wide and smooth rectangular channel. Figure 16.16 shows a non-perturbed approach flow of depth h_1, velocity V_1 along the laterally bounded channel. At point P, the flow is abruptly deflected by the angle $+\theta$. Downstream from the deflection point P, the flow follows the new direction, and one may distinguish between the *non-perturbed flow region* ① and the *perturbed flow region* ②. These two regions are separated by a line, originating at P under the angle β_s. This separation line is referred to as shock front, and the corresponding angle is the *shock-angle* β_s (German: Stosswinkel; French: Angle de choc). The velocity in the perturbed region is V_2, and the flow depth h_2.

(a)
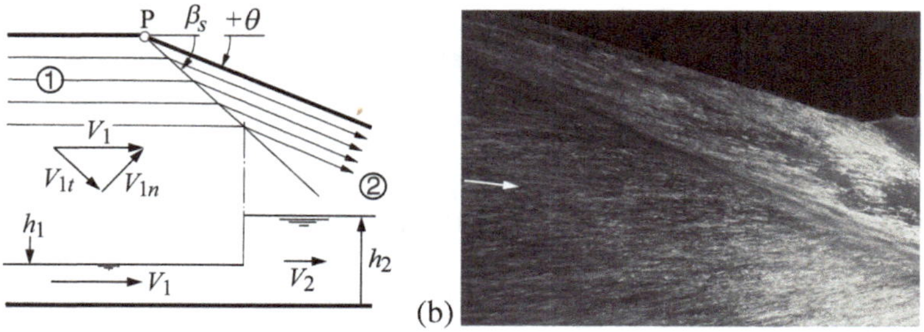

(b)

Fig. 16.16 (**a**) Plan (*top*) and section (*bottom*) of abrupt wall deflection, (**b**) flow downstream of deflection point P. Note the slight curvature of the shock front because of energy dissipation

Shock Relations

Because of the abrupt change of flow depth across a shockfront, energy dissipation cannot be neglected. The main information on the flow features of a *shockwave*, i.e. the relation of parameters in regions ① and ②, must thus be determined by appying the *momentum equation*. In analogy to the Borda-expansion (Chap. 2) or the classical hydraulic jump (Chap. 7), the flow is governed by energy losses that are previously unknown. Assuming hydrostatic pressure and uniform velocity distributions, three relations may be deduced (Chow 1959)

$$\frac{h_2}{h_1} = \frac{1}{2}\left[(1 + 8\mathsf{F}_1^2 \sin^2\beta_s)^{1/2} - 1\right], \tag{16.50}$$

$$\frac{h_2}{h_1} = \frac{\tan\beta_s}{\tan(\beta_s - \theta)}, \tag{16.51}$$

$$\mathsf{F}_2^2 = Y_s^{-1}\left[\mathsf{F}_1^2 - \frac{1}{2Y_s}(Y_s - 1)(Y_s + 1)^2\right]. \tag{16.52}$$

Here $\mathsf{F}_1 = V_1/(gh_1)^{1/2}$ is the approach flow Froude number, and $Y_s = h_2/h_1$ the ratio of flow depths. For given approach flow parameters h_1, V_1 and θ, the unknowns h_2, β_s and $\mathsf{F}_2 = V_2/(gh_2)^{1/2}$ may thus be implicitly determined. For $\mathsf{F}_1 \sin\beta_s > 1$, approximations are (Hager 1992)

$$Y_s = \sqrt{2}\mathsf{F}_1 \sin\beta_s - \frac{1}{2}, \tag{16.53}$$

$$\beta_s - \theta = 1.06\mathsf{F}_1^{-1}. \tag{16.54}$$

For $\beta_s < 45°$ and $\mathsf{F}_1 > 2$, Eq. (16.54) yields deviations of less than 2° from the exact result. Equation (16.53) can be interpreted as the generalization of the sequent depth ratio of a classical hydraulic jump ($\beta_s = 90°$), as presented in Chap. 7. A shockwave is thus a type of hydraulic jump, yet without a surface roller and with a supercritical downstream flow. It can also be demonstrated from Eq. (16.52) that

$$\mathsf{F}_2 = \mathsf{F}_1/(1 + \mathsf{F}_1\theta/\sqrt{2}) > 1. \tag{16.55}$$

Shockwaves due to an abrupt wall deflection can be classified according to Ippen and Harleman (1956) in analogy to hydraulic jumps as:

- *Undular shockwaves* for $1 < Y_s < 2$, and
- *Direct shockwaves* with a sharp front for $Y_s \geq 2$.

Equations (16.53), (16.54), and (16.55) have been experimentally verified for $Y_s > 2$. Figure 16.16b) shows a typical direct shock.

Example 16.4 Given a wide rectangular channel of bottom slope 7%, discharge per unit width $q_z = 15$ m^{2}s^{-1} and roughness coefficient $n = 0.011$ sm$^{-1/3}$. What is the downstream flow downstream of an abrupt wall deflection of $\theta = 4°$, provided uniform flow prevails in the approach flow channel (after Hager 1992)?
The uniform flow depth in the wide rectangular channel ($R_h = h$) is $h_1 = h_N = [nq/(S_o^{1/2})]^{3/5} = 0.75$ m, of velocity $V_1 = q_z/h_1 = 20$ ms^{-1} and $\mathsf{F}_1 = V_1/(gh_1)^{1/2} = 20/(9.81 \cdot 0.75)^{1/2} = 7.37 > 2$, resulting in a direct shock wave with a sharp front.

From Eq. (16.54) the shock angle is $\beta_s = 4° + (180/\pi)1.06/7.37 = 12.2°$, and $Y_s = 1.41 \cdot 7.37\sin(12.2°) - 0.5 = 1.7$ from Eq. (16.53). Further, $\mathsf{F}_2 = 7.37/(1 + 7.37 \cdot 4 \cdot \pi/1.41 \cdot 180°) = 5.4$. The result is thus $h_2 = Y_s h_1 = 1.28$ m, $\beta_s = 12.2°$ for the shock angle and $\mathsf{F}_2 = 5.4$.

Eliminating β_s from Eqs. (16.53) and (16.54), the depth ratio obtains (Hager 1992)

$$Y_s = 1 + \sqrt{2}\mathsf{F}_1\theta. \tag{16.56}$$

The wave height is thus influenced only by the so-called *shock number* $\mathsf{S}_1 = \theta\, \mathsf{F}_1$. The effects of F_1 and θ are interchangable, and a constant shock number yields absolutely the identical effect. It should be noted that the shock number influences also the ratio of angles $\beta_s/\theta = 1 + 1.06\mathsf{S}_1^{-1}$ from Eq. (16.54), as well as the ratio of Froude numbers $\mathsf{F}_2/\mathsf{F}_1 = 1 + \mathsf{S}_1/\sqrt{2}$ from Eq. (16.55). Further, with the downstream Froude number $\mathsf{F}_2 = V_2/(gh_2)^{1/2}$, the downstream velocity is $V_2 = \mathsf{F}_2(gh_2)^{1/2} = \mathsf{F}_1(1 + \mathsf{S}_1/\sqrt{2})(gh_1)^{1/2}Y_s^{1/2} = V_1$ when inserting $Y_s^{1/2} = 1 + (1/2)\sqrt{2}\mathsf{S}_1 + O(\mathsf{S}_1^2)$. The results of this analysis are:

- Relative *wave height* Y_s depends exclusively on the shock number S_1,
- Effects of wall deflection angle θ and approach flow Froude number F_1 are interchangable,
- Downstream velocity V_2 is almost equal to the approach flow velocity V_1.

These important features can be extended to all shock waves. A shock wave thus involves a relatively large surface perturbation yet without a significant change of the velocity, with the associated hydraulic characteristics across a shock front are small.

Shock Surface

Schwalt and Hager (1992) investigated the deflection process along an abruptly deflected wall. With $X = x/(h_1\mathsf{F}_1)$ and $Y = y/h_1$ as the dimensionless coordinates along and perpendicular to the deflected wall, with origin at the deflection point P, a generalized free surface $G = (h - h_1)/(h_2 - h_1)$ may be defined (Fig. 16.17). For $X < 4$, the flow depth is steadily increasing towards the wall, whereas the surface has a maximum of about $G = 1.2$ for larger X. The analysis is limited to $X \leq 6$ for shock numbers $\mathsf{S}_1 < 1$.

Of particular relevance is the *wall profile* (subscript w) of the shock wave $G_w = G(X, Y = 0)$, where $G_w = (h_w - h_1)/(h_2 - h_1)$. Figure 16.18 relates to the function $\gamma_w = (h_w - h_1)/(h_M - h_1)$ as a function of X, with the maximum (subscript M) wall flow depth at position $X_M = 1.75$

$$\frac{h_M}{h_1} = 1 + \sqrt{2}\mathsf{S}_1\left(1 + \frac{1}{4}\mathsf{S}_1\right). \tag{16.57}$$

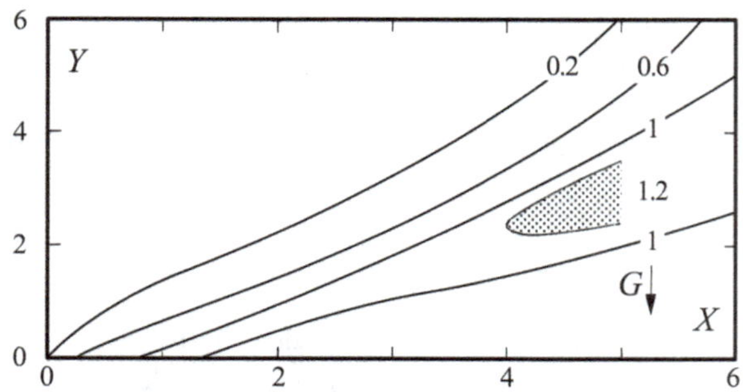

Fig. 16.17 Universal shock surface $G(X,Y)$ for $\mathsf{S}_1 = \theta\,\mathsf{F}_1 < 1$

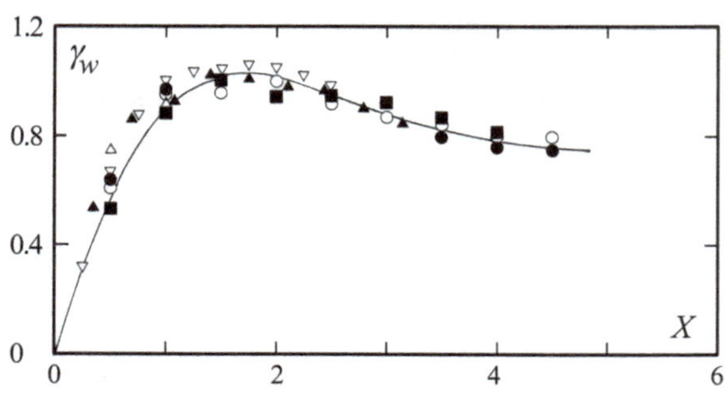

Fig. 16.18 Generalized wall profile $\gamma_w(X)$ due to an abrupt wall deflection for various combinations of approach Froude numbers F_1 and deflection angles θ (Schwalt and Hager 1992)

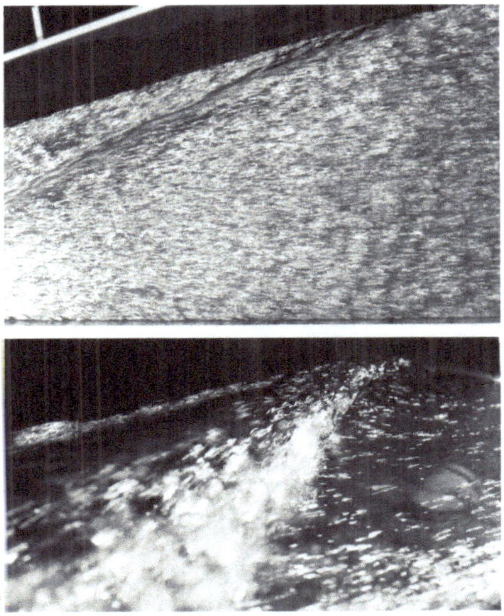

Fig. 16.19 Photographs of shockwaves with $\mathsf{F}_1 = 4$, $\theta = 0.2$ and $h_o = 50$ mm (**a**) plan view, (**b**) view from tailwater. Direction of flow from right to left

Comparing Eqs. (16.56) and (16.57) reveals a second term that is added to the analysis, based on the *shallow-water theory*. For $\mathsf{S}_1 < 1$, this term is relatively small, however. Figure 16.18 shows a rapid increase of the wall profile to the maximum flow depth, and an almost constant downstream flow depth. Note that the velocity in the complete domain of Fig. 16.17 remains nearly constant.

Figure 16.19 refers to photographs with shockwaves due to an abrupt wall deflection. The *shock front* is clearly visible as a separation line between the undisturbed and disturbed flow regions. In Fig. 16.19b the steep front is obvious, and the front is close to breaking. Also, air entrainment into the flow is noted.

Air Entrainment

Up to now, the approach flow to the wall deflection was considered a pure water flow, defined by the approach flow depth h_1 and the approach flow Froude number F_1. Often, the approach flow is aerated and the effect of air entrainment (German: Lufteintrag; French: Entraînement d'air) on the shockwave development has to be determined. The cross-sectional air concentration C can be expressed as the ratio of air discharge Q_a to the total discharge $(Q_a + Q)$ with Q as water discharge, i.e.

$$C = \frac{Q_a}{Q_a - Q} = \frac{Q_a}{Q_m} \tag{16.58}$$

with Q_m as the mixture (subscript m) discharge. The *mixture flow depth* h_m is thus

$$h_m = h/(1 - C). \tag{16.59}$$

Because $C < 1$ for a mixture flow, the mixture flow depth is always larger than the pure water depth h, i.e. the mixture flow needs more depth than a water flow. The following aims at predicting the additional *freeboard* required for mixture flow.

From momentum considerations, is can be demonstrated that the mixture momentum $\rho_m Q_m V_m$ is equal to the pure water momentum ρQV, because the increase in discharge is offset by the reduction of density provided the slip velocity between water and air bubbles is assumed zero, and $V_m = V$.

Applying Eqs. (16.50), (16.51), and (16.52) to mixture flow, it can be demonstrated that the results are identical with Eqs. (16.53), (16.54), and (16.55), provided the shock number $\mathsf{S} = \theta\mathsf{F}$ is replaced by the *mixture shock number* $\mathsf{S}_m = \theta\mathsf{F}_m$. For $\mathsf{S}_m < 0.5$, Eq. (16.56) is valid and for larger shock numbers up to $\mathsf{S}_m < 1.5$ a second order term in S_m has to be added (Reinauer and Hager 1996)

$$Y_m = \left(1 + \frac{1}{\sqrt{2}}\mathsf{S}_m\right)^2. \tag{16.60}$$

The *shock angle* β_s has been predicted in Eq. (16.54). If observations are compared with the prediction, a systematic deviation is noted. The corrected result is

$$\beta_m/\theta = 1 + \frac{0.75}{\mathsf{S}_m}. \tag{16.61}$$

Detailed experiments thus indicate a reduction of the constant from 1.06 to 0.75. Equations (16.60) and (16.61) apply for both pure water and mixture flows, with $\mathsf{F}_m = V_m/(gh_m)^{1/2}$ as the *mixture Froude number*.

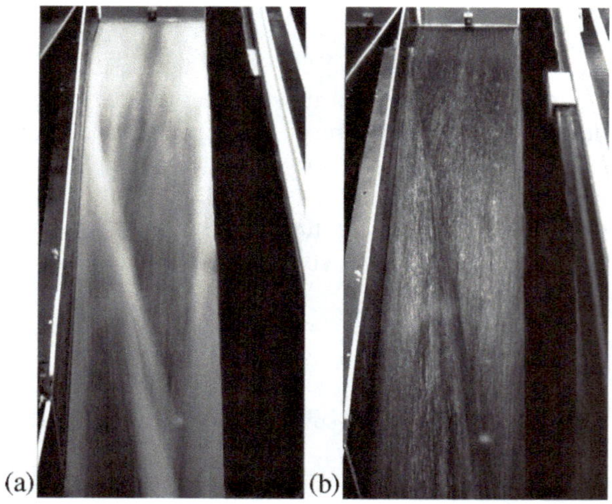

Fig. 16.20 Shockwave (**a**) *with* and (**b**) *without* upstream entrained flow, $\mathsf{S}_{om} = 6.17$

Figure 16.20 compares shock waves of pure water and mixture flows. Whereas the front of the pure water flow is sharply defined, the front of the mixture flow is more diffused, with an accumulation of air along the front. This is due to the overpressure slightly upstream of the front and an *underpressure* at the top of the wave. The air bubbles directed towards the front are thus lifted due to increased buoyancy and may then escape into the atmosphere, such that the air concentration beyond the shock front is considerably reduced. The shock front is thus perfectly traced by air bubbles and shocks can easily be made visible by adding air to the approach flow.

Abruptly Deflected Channel

An application of the abrupt wall deflection is the *mitre bend* of small deflection angle (Fig. 16.21). The wall is positively deflected at point A, and negatively deflected at point B. Positive deflections, i.e. a deflection into the flow, result in a positive surface elevation and negative deflections in a surface depression. The latter cannot be described with the previous theory based on the shallow-water equations because of significant streamline curvature effects and a *wall separation* process.

The undisturbed flow region upstream of points ACB is contrasted to the disturbed flow in regions ACE and BCD, with streamlines parallel to the walls. In the region CDFE, the streamlines are again deflected, depending on the approach flow conditions h_1 and F_1. In this simplified consideration, one may identify all triangular regions with a certain flow depth and a specific velocity, subject to the effects of pressure distribution and viscosity.

Point C is located at the crossing of two shock fronts and it delimitates two regions of disturbed downstream flow. By interference the regions EFG and ACE are identical if effects of non-hydrostatic pressure distribution and viscosity are neglected. Inversion is the condition that applies to region CDFE: Although the streamlines are deflected by the angle $+2\theta$ from the original direction, the original flow parameters h_1 and F_1 are re-established.

Figure 16.21 is typical for supercritical flow downstream from a geometrical perturbation. The flow pattern involving successive minima and maxima along a wall that are connected by shock fronts has received the name *crosswave*. Clearly, a one-dimensional computation is unable to predict such a surface geometry. In particular, it would be impossible to determine the freeboard required in open channels, and the transition features to pressurized flow in ducted profiles.

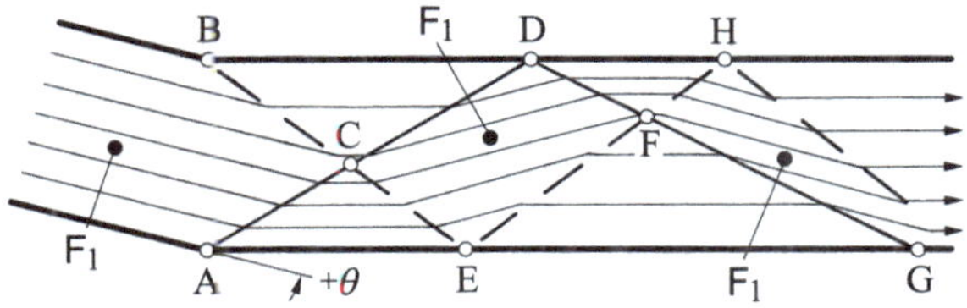

Fig. 16.21 Wall deflection in channel of finite width and abruptly deflected boundaries. Streamlines and points of reflection, (•) zones of constant F_1

Shockwave Treatment

Each *flow perturbation* of any basic parameter, including:

- Bottom geometry,
- Cross-sectional geometry,
- Boundary roughness, and
- Discharge variation

has a direct reaction on the generation of a shockwave. Besides the definition of the shockwave geometry, the *shockwave reduction* deserves attention. A thorough design on any hydraulic structure thus includes a downstream flow without flow concentrations and with a structural economy. If these items are not attained, an intermediate stilling basin would be appropriate to reduce the disadvantages of high-speed flows. From the hydraulic point of view, such a design would be forced or even poor, and solutions to be suggested below aim at an *entirely supercritical flow* across any hydraulic element. Such solutions are advantageous from both the hydraulic and the structural views.

Currently, a number of *methods for shockwave reduction* are available, but these are based on assumptions such as:

- Neglect of boundary layer effects,
- Hydrostatic pressure distribution,
- Omission of bottom and friction slopes, and
- Investigation for design flow only.

Such simplifications are ruled out provided the design is based on either:

- Serious two-dimensional numerical modelling of supercritical flow, or
- Flow analysis by sufficiently large physical models to inhibit scale effects.

Currently, *scale models* are an economic and reliable approach to solve these problems, and numerical models have received less attention.

In the following, supercritical flows of typical structures relating to sewer hydraulics are discussed, such as constrictions, expansions, bends and junctions. First, the flow patterns of the *untreated channel* are presented, and structural means to reduce shockwaves are then outlined. The indications refer normally to rectangular channels, because U-shaped and circular channels have not yet received a systematic attention. Some results pertinent to sewer structures follow in Sect. 16.4. Few results were available until the 1990s; this lack was particularly due to the modeling of supercritical channel flow. Whereas most of the studies were based on a poor approach flow whose surface was perturbed by significant shockwaves prior to enter the test reach, the present Author developed with his colleagues the so-called *jet-box* allowing for an almost perfect approach flow absolutely free of shockwaves and velocity concentrations.

16.3.3 *Channel Contraction*

Conventional Design

A *channel contraction* (German: Kanalverengung; French: Contraction de canal) can be geometrically designed either funnel-, fan- or nozzle-shaped. The following is restricted to polygonal-funnels, because curved contraction walls are not a standard in sewer hydraulics (Fig. 16.22).

A hydraulically sound designed contraction is characterized by a monotonically rising surface in the constricted portion, and a downstream channel practically free of surface waves. In the approach flow region ①, the flow depth is h_1 and the velocity V_1, such that the approach flow Froude number is $\mathsf{F}_1 = V_1/(gh_1)^{1/2}$. In analogy to the abrupt wall deflection (Sect. 16.3.2), the *origin of perturbation* is the contraction angle θ at point A. In region ②, the characteristic parameters are thus h_2 and according to Eq. (16.55) $V_2 = V_1$. Point D generates negative shockwaves due to the negative wall deflection, and the downstream channel has the typical configuration with crosswaves. The flow uniformity is thus poor in the untreated channel.

Ippen and Dawson (1951) suggested a design in which the contraction angle θ is selected such that the positive shock front is directed to the origin of the negative shock front, and points C and D in Fig. 16.22a coincide (Fig. 16.22b). According to the *wave interference principle*, the downstream waves should then be eliminated. A contraction structure so designed has three regions, namely the approach flow region ①, the contraction region ②, and the downstream region ③, with the corresponding flow depths h_1, h_2 and h_3. Applying successively the relations derived in Sect. 16.3.2, the design contraction angle θ (< 10°) can be given as (Hager 1992)

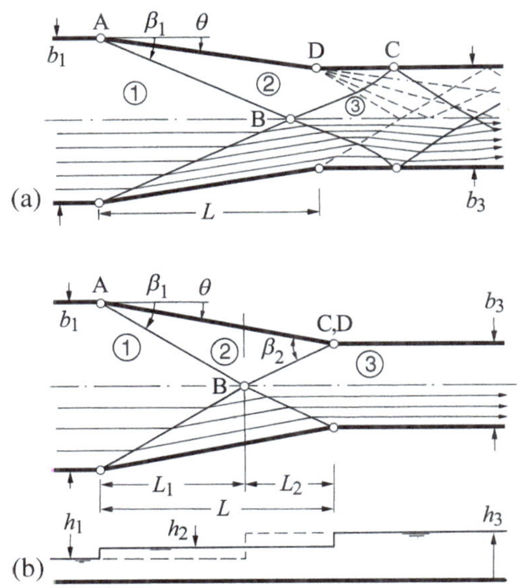

Fig. 16.22 (**a**) Schematic flow in polygonal contraction and (**b**) design in plan (*top*) and section (*bottom*) with (–) wall and (- -) axial profiles (Ippen and Dawson 1951)

$$\arctan\theta = \left[\frac{b_1}{b_3} - 1\right]\frac{1}{2\mathsf{F}_1}. \quad (16.62)$$

To inhibit *choking* (German: Strömungszusammenbruch; French: Effondrement de l'écoulement) of downstream flow, the condition $\mathsf{F}_1 > 2$ must be satisfied. Equation (16.62) can be used to demonstrate a serious disadvantage of all flows designed with the interference principle. A minor deviation from the design conditions leads to serious shockwaves, i.e. the *design flexibility* is extremely narrow. In addition, Reinauer and Hager (1998) have demonstrated experimentally that the wave interference principle as applied in optics cannot be transmitted to hydraulics, mainly because of the nonlinear wave characteristics. Whereas waves in optics have practically no transverse extension, the wave lengths in hydraulics are finite and have the order of a flow depth.

Novel Design

Reinauer and Hager (1998) investigated experimentally supercritical flow in a channel contraction, for bottom slopes up to 30°. Figure 16.23 shows a definition sketch in the horizontal channel without any bottom elements. These have been proposed for chute contractions with a large discharge and are not relevant for sewer contractions.

At the contraction point A a shockwave of shock angle β_1 is generated due to the abrupt wall deflection of angle θ. The shock is propagated to point B in the channel axis and reflected to the wall again at point C. At the contraction end point E, a negative wave forms, resulting in a complicated wave pattern in the downstream channel. In the side view the wall (subscript w) and the axial (subscript a) surface profiles are relevant. Three shockwaves can be distinguished: *Wave* 1 of height h_1 and *wave* 3 of height h_3 along the wall, and the axial *wave* 2 of height h_2. Downstream of wave 1, the flow depth h_p in the wall region is nearly constant up to the height h_e at the

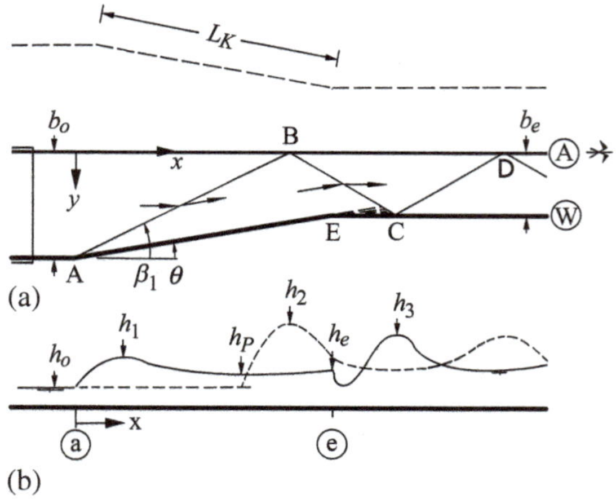

Fig. 16.23 Horizontal channel contraction (**a**) plan (**b**) section with surface profiles along (–) wall and (- -) channel axis

contraction end. The velocity across the contraction is essentially constant, as for a flow behind an abrupt wall deflection (Sect. 16.3.2).

The shock waves in a channel contraction depend mainly on the:

- Approach flow Froude number $\mathsf{F}_o = V_o/(gh_o)^{1/2}$ with $V_o = Q/(b_o h_o)$,
- Width ratio $\omega = b_e/b_o$,
- Wall deflection angle θ, and
- Bottom slope S_o.

Effects of approach flow depth, i.e. *scale effects* are inhibited provided the minimum approach flow depth is $h_o = 50$ mm. For waves 1 and 2, the width ratio is insignificant, such that the approach flow shock number $\mathsf{S}_o = \theta\ \mathsf{F}_o$ is the governing parameter. Shockwaves for $\mathsf{S}_o < 0.5$ are weak, and they become overforced for $\mathsf{S}_o > 2$.

A shockwave may be described with the coordinates $(x;h)$ of location and wave maximum. With the bottom angle α in [deg] and the contraction angle θ in [rad] the relative heights of waves $1-3$ may be expressed as

$$Y_1 = h_1/h_o = \left(1 + \frac{1}{\sqrt{2}}\mathsf{S}_o\right)^2, \tag{16.63}$$

$$Y_2 = h_2/h_o = (1 + \sqrt{2}\mathsf{S}_o)^2, \tag{16.64}$$

$$Y_3 = h_3/h_o = \omega^{-1} + 1.8\mathsf{S}_o^{1/2} - 0.2\,\alpha^{0.6}. \tag{16.65}$$

Therefore, the contraction angle and the bottom slope have an effect only on wave 3. The locations of wave maxima are more complex because they involve a drawdown of the free surface profile along the sloping channel. The application limits of Eqs. (16.63), (16.64), and (16.65) are $0.2 < \omega < 1$, $\alpha < 45°$, and $\mathsf{S}_o < 2$ (Reinauer and Hager 1998).

Equation (16.63) gives a height of the shock wave identical with that of an abrupt wall deflection. For the maximum value $\mathsf{S}_o = 2$ recommended, $Y_1 \cong 6$ resulting in a significant wave height. Wave 2 with $Y_2 \cong 15$ is much higher, however. For a typical parameter combination $\omega = 0.5$, and $\alpha = 10°$, the result for wave 3 is $Y_3 = 3.75$. The important conclusion from this study is that the *effect of bottom slope* decreases the wave height, and a conservative approach assuming $\alpha = 0$ gives results on the safe side. Typically, a shock number of $\mathsf{S}_o = 1$ should be selected for design, and the optimum contraction rate is between $0.5 < \omega < 0.80$. For $\omega < 0.5$, the flow tends to be overforced, and contractions with $\omega > 0.8$ are often not justified because of additional cost. The design parameter for the maximum wave height is the *maximum discharge* Q_M. All other discharges $Q < Q_M$ result in smaller waves.

Choking flow in contractions is a significant design limitation. If the approach flow Froude number F_o or the contraction rate ω are too small, then the supercritical flow structure breaks down, associated with a *hydraulic jump* in the contraction of

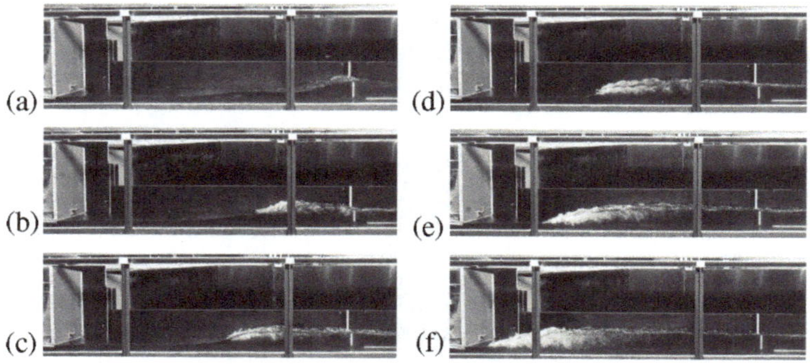

Fig. 16.24 Choking flow in a channel contraction for $\mathsf{F}_o = 2.40$

which the toe is located in the approach flow channel. Figure 16.24 shows a typical sequence between the limiting downstream flow and a stable hydraulic jump in the contraction.

Starting from a supercritical flow, F_o can be reduced up to the occurrence of *incipient choking*, with $\mathsf{F}_o = \mathsf{F}_o^-$. Then, the flow depth at point E is equal to the critical depth, and the hydraulic jump establishes, with a transition from supercritical approach flow to subcritical contraction and supercritical downstream flow. A choked contraction has a much larger flow depth than the maximum wave height. Therefore, choking of any hydraulic structure that was designed for supercritical flow is dangerous, because the flow structure assumed is significantly different from the choked flow pattern. Once a structure has a choked flow, a Froude number F_o^+ much larger than F_o^- is needed to *blow out* the hydraulic jump. For a given approach flow Froude number in an almost horizontal channel, the limit (subscript L) contraction ratio is

$$\omega_z = \mathsf{F}_o \left(\frac{3}{2 + \mathsf{F}_o^2} \right)^{3/2}. \qquad (16.66)$$

A contraction flow chokes if $\omega < \omega_z$. For bottom slopes $\alpha > 5°$ choking for $\omega > 0.5$ is unlikely to occur (Reinauer and Hager 1998).

16.3.4 *Channel Expansion*

A channel expansion (German: Kanalerweiterung; French: Expansion de canal) can be geometrically arranged in various shapes. Mazumder and Hager (1993) stated that only the abrupt expansion and the so-called reversed expansion are relevant. Whereas the latter is applied in hydraulic structures for larger discharges, sewer hydraulics involve normally *abrupt expansions* (Fig. 16.25). The following refers to a rectangular channel in which the friction slope is nearly compensated for by the bottom slope. The approach (subscript o) flow depth is h_o and the approach flow

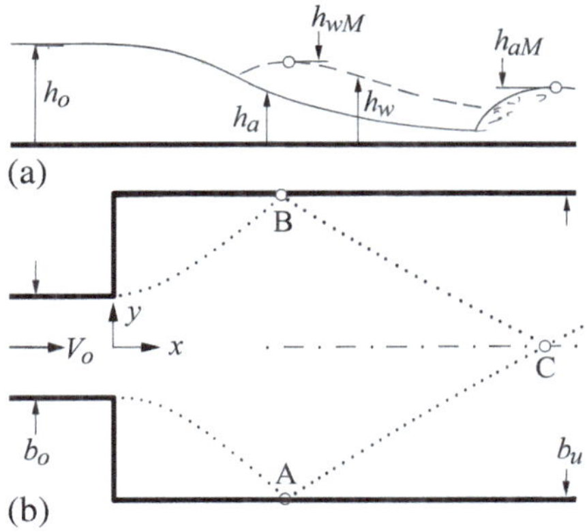

Fig. 16.25 Abrupt channel expansion. (**a**) Section with axial surface profile $h_a(x)$ and wall profile $h_w(x)$. (**b**) Plan with (· · ·) profiles of shock fronts

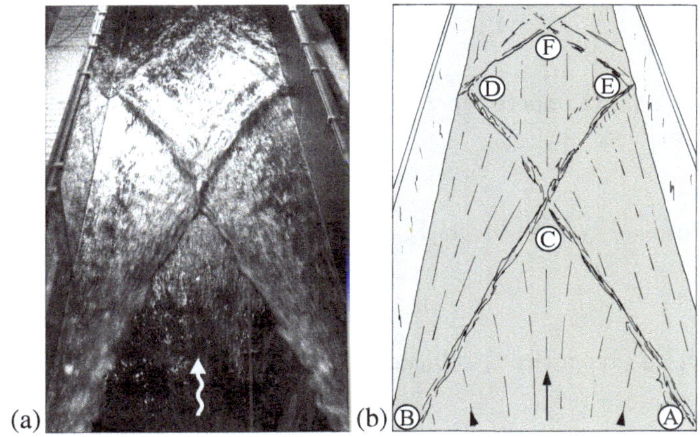

Fig. 16.26 Free surface downstream of abrupt channel expansion with $b_u/b_o = 3$, $\mathsf{F}_o = 2$ and $h_o = 96$ mm. (**a**) Photo, (**b**) schematic plot (Hager and Mazumder 1992)

Froude number $\mathsf{F}_o = V_o/(gh_o)^{1/2}$. The approach flow width b_o increases abruptly to the downstream (subscript u) width b_u at the expansion section.

Following an expansion, the free surface draws down in the approach flow channel (Fig. 16.25a). At the expansion section, the flow has nearly a rectangular cross-section, with vertical sides, such that its pressure distribution is non-hydrostatic. This jet expands over the downstream width b_u and delimits dead-water zones in either corner. At the impinging points A and B of the forward flow fronts, *shockwaves* are generated similar to those due to a wall deflection. Contrary to the latter, the approach flow to the shockwaves is already perturbed due to flow expansion, and the relation between the maximum (subscript M) wall wave height h_{wM} and the approach flow depth h_o is complicated. Downstream from the impact points A and B, the waves are reflected and the various flow zones typical for supercritical flows are again visible (Fig. 16.26). In analogy to Fig. 16.21, region ACB is influenced by the approach flow only, resulting in the *maximum flow depth* at points A and B. In region ACE, the average flow depth is larger than upstream from point A, with an axial maximum flow depth h_{aM} at point C. Downstream, the typical crosswave pattern appears, with successive wave minima and maxima along both the channel axis and walls of which all are smaller than upstream.

The *maximum flow depths* that are determining for freeboard design occur at the first wave both along the channel axis (subscript a) and the channel walls (subscript w). An analysis of these maxima is thus more relevant than information on the complete flow surface. Figure 16.27 refers to the axial profile $Y_a(X)$ with $Y_a = h_a/h_o$ and the wall profile $Y_w(X)$ with $Y_w = h_w/h_o$ in which the dimensionless streamwise coordinate is stretched with F_o to $X = x/(b_o\mathsf{F}_o)$ for various expansion ratios $\beta_e = b_u/b_o \geq 1$. The axial profile is seen to be independent of β_e and F_o and may be expressed up to the crossing point C as

$$Y_a = 0.2 + 0.8\exp(-X^2). \tag{16.67}$$

The *wall profile* $Y_w(X)$ can be represented as

$$Y_w = Y_{wM}\,\tau\exp(1-\tau) \tag{16.68}$$

with the *maximum wall flow depth* $Y_{wM} = h_{wM}/h_o$ for $1.8 < \beta_e < 6$ as

$$Y_{wM} = 1.27\beta_e^{-0.4}. \tag{16.69}$$

The normalized streamwise coordinate

$$\tau = \frac{X - X_m}{X_M - X_m} \tag{16.70}$$

depends on the locations of the *wave extrema*

$$X_m = (1/6)(\beta_e - 1), \quad X_M = 0.52\,\beta_e^{0.86}. \tag{16.71}$$

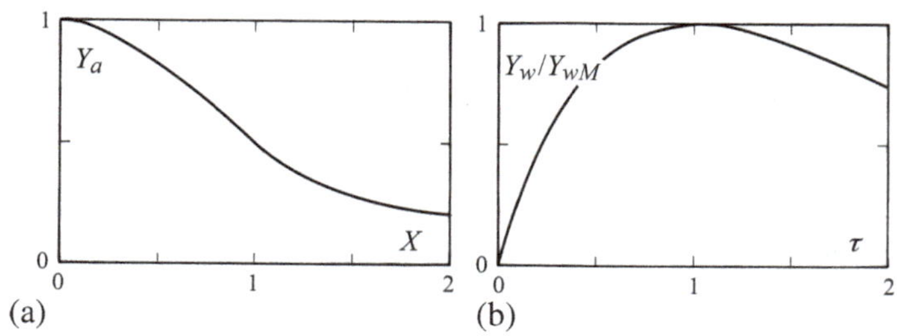

Fig. 16.27 (**a**) Axial surface profile, and (**b**) wall surface profile for *abrupt* channel expansion for $1 < \mathsf{F}_o < 10$ and $1.8 < \beta_e < 6$ (Hager and Mazumder 1992)

The location X_m refers to the transition between the dead-water zone and the wave beginning with the minimum (subscript m) flow depth $h_{wM} = 0$, and X_M is the point of maximum wave height (points A and B of Fig. 16.26).

Example 16.5 Given a rectangular channel of approach flow velocity $V_o = 6\ \mathrm{ms}^{-1}$ and approach flow depth $h_o = 0.60$ m. Describe the downstream flow pattern for an abrupt width increase from $b_o = 0.8$ m to $b_u = 3.0$ m.
With $V_o = 6\,\mathrm{ms}^{-1}$ and $h_o = 0.60$ m, the approach flow Froude number is $\mathsf{F}_o = 6/(9.81 \cdot 0.6)^{1/2} = 2.50$, and the width ratio is $\beta_e = 3.0/0.8 = 3.75$. The locations of extreme wave heights are $X_m = (1/6)(3.75–1) = 0.46$, and $X_M = 0.52 \cdot 3.75^{0.86} = 1.62$, according to Eqs. (16.71). The normalized length coordinate is thus $\tau = (X - 0.46)/1.16$. The maximum wall wave height is $Y_{wM} = 1.27 \cdot 3.75^{-0.4} = 0.75$ from Eq. (16.69), and the equation of the wall profile is

$$Y_w = 0.75[(X - 0.46)/1.16]\exp[1 - (X - 0.46)/1.16] \qquad (16.72)$$

from Eq. (16.68). The results are thus $x_M = b_o\mathsf{F}_oX_M = 0.8 \cdot 1.62 \cdot 2.5 = 3.2$ m, and the maximum wall flow depth is $h_{wM} = 0.75 \cdot 0.60 = 0.45$ m.

Further indications refer to the flow surface, the shock fronts, the transverse surface profiles and the velocity profiles. The dead-water zone downstream from the expansion section can be suppressed by linear insets up to point x_m according to Eq. (16.71). Then, the abrupt expansion is replaced by a *gradual expansion*, without a potential for sedimentation of the sewage.

The internal flow characteristics of expanding flows up to the point of shockwave development has been investigated by Hager and Yasuda (1997). Based on Eqs. (16.40), (16.41), (16.42), and (16.43) the 2D shallow water equations were demonstrated to tend asymptotically to the 1D unsteady flow equations, and a flow analogy between *steady two-dimensional* and *unsteady one-dimensional flows* was established.

16.3.5 *Channel Bend*

Weak Bend Flow

A channel bend (German: Kanalkurve; French: Canal courbé) of deviation ζ involves a supercritical flow similar to a wall deflection with a positive shockwave along the outer and a negative shockwave along the inner wall, respectively. Further downstream the wave maxima and minima alternate and typical *crosswaves* are generated. Knapp (1951) discussed such flow and set up a design procedure.

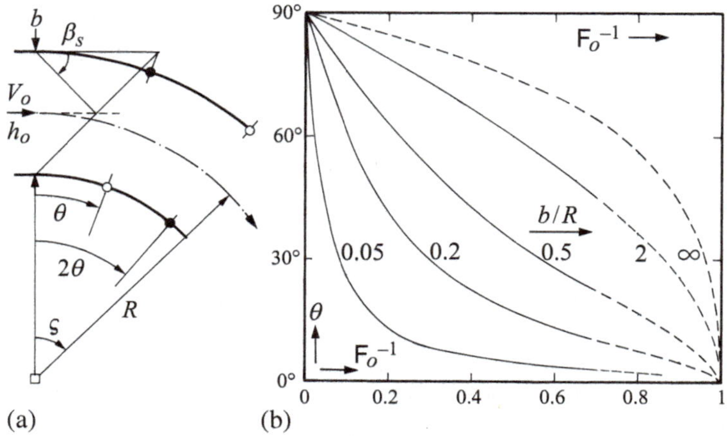

Fig. 16.28 Supercritical flow in a channel bend (**a**) schematic flow process, (**b**) angle β_s versus approach flow Froude number F_o and relative bend curvature *b*/*R* (Hager 1992)

The flow pattern in a rectangular horizontal channel containing a frictionless fluid can be simplified as follows. Extreme wave heights occur at locations with angles θ, 2θ and so on (Fig. 16.28). At the end of the bend, this oscillating surface pattern does not stop but continues into the tailwater channel.

The elementary shockwave angle is $\beta_s = \arcsin(\mathsf{F}_o^{-1})$. Knapp (1951) related this angle to the shockfront angle θ and to the relative centerline bend curvature *b*/*R* as

$$\theta = \arctan\left[\frac{b/R}{(1+2b/R)\tan\beta_s}\right]. \tag{16.73}$$

Figure 16.28b indicates that θ increases as both F_o and *b*/*R* increase. For small values of F_o, the curves are dashed because of the *undular* surface pattern. For larger approach flow Froude numbers resulting in direct shockwaves provided *b*/*R* < 1/2, Eq. (16.73) can be written as

$$\tan\theta = \frac{b/R}{1+(1/2)(b/R)}\sqrt{\mathsf{F}_o^2-1} \cong (b/R)\mathsf{F}_o. \tag{16.74}$$

The *extreme wave heights* (subscript *e*) may be obtained with the energy principle, and Figure 16.29 shows the result of Knapp (1951). For $(b/2R)\mathsf{F}_o^2 < 1$ the extreme values $y_e = h_e/h_o$ may be approximated as

$$y_e = \left[1 \pm \frac{1}{2}(b/R)\mathsf{F}_o^2\right]^2, \tag{16.75}$$

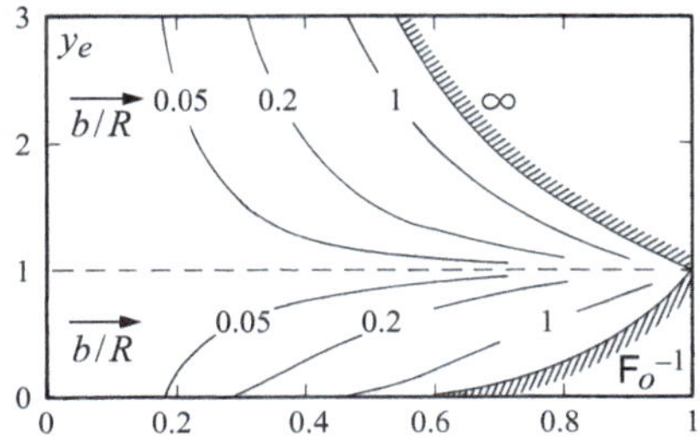

Fig. 16.29 Curve flow in rectangular channel, extreme flow depths $y_e = h_e/h_o$ versus $F_o = V_o/(gh_o)^{1/2}$ and relative centerline curvature b/R (Hager 1992)

with the +sign relating to the wave maximum and the –sign to the wave minimum, respectively. In analogy to the shock number $S = \theta F$ of an abrupt wall deflection, one may define the *bend number* as $B = (b/R)^{1/2}F$. The effects of the square root of relative bend radius $(b/R)^{1/2}$, and approach flow Froude number are thus contained in the bend number. Compared to the abrupt wall deflection, the effect of b/R is smaller. Note that the effect of the approach flow Froude number F_o is significant and shockwaves are high for medium values of F_o.

Example 16.6 Given a rectangular channel with an approach flow velocity $V_o = 7$ ms^{-1}, $b_o = 1.2$ m and $h_o = 0.50$ m. How high should be the sidewalls for a deflection angle of $\zeta = 35°$, if the centerline radius is $R = 8$ m?
With $V_o = 7$ ms^{-1} and $h_o = 0.50$ m, the approach flow Froude number is $F_o = 7/(9.81 \cdot 0.50)^{1/2} = 3.16$, and the relative centerline curvature amounts to $b/R = 1.2/8 = 0.15 < 0.50$. Accordingly, the term $(1/2)(b/R)$ $F_o^2 = 0.5 \cdot 0.15 \cdot 3.16^2 = 0.75 < 1$, and $y_e = (1 \pm 0.75)^2$, thus $y_M = 3.06$ and $y_m = 0.06$, and the maximum wall flow depth is $h_M = 3.06 \cdot 0.5 = 1.53$ m and the minimum flow depth is $h_m = 0.06 \cdot 0.5 = 0.03$ m. From Fig. 16.29 one finds for $F_o^{-1} = 3.16^{-1} = 0.32$ values $y_M = 2.3$ and $y_m = 0.15$, i.e. values closer to 1 than according to the approximations. Further, from Eq. (16.74), $\tan\theta = 0.15 \cdot 3.16 = 0.47$, and $\theta = 25°$. The first wave extrema are thus located within the curved channel portion.

The effects of bottom slope and transverse slope have not yet been systematically investigated. Results for conduit bends follow below. Figure 16.30 shows photographs of flows in horseshoe profiles. The transverse free surface slope can be considerable, and *choking* of such tunnels for large discharges is possible. Then a complex air-water flow may set up, with a hydraulic jump in the upstream portion, and pressurized flow in the tailwater portion. The limit conditions for choking flow in curved channels should be analyzed, therefore.

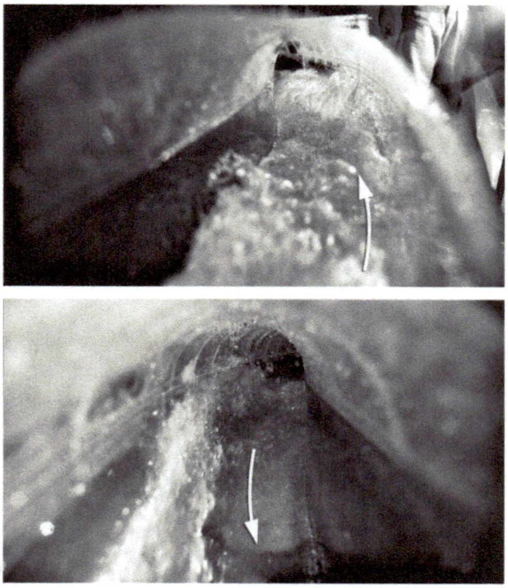

(a)

(b)

Fig. 16.30 Flow in closed conduit of large bottom slope (**a**) upstream view, (**b**) downstream view

Strong Bend Flow

Reinauer and Hager (1997) expanded the approach of Knapp (1951). Figure 16.31 shows flows in bends of rectangular channels in which large wall shockwaves may be observed along the outer channel wall, whereas the flow depth along the inner channel wall may reduce to zero due to the centrifugal force.

The *velocity fields* of bend flow reveal interesting flow features. From detailed observations, one may conclude that the (Reinauer and Hager 1997):

- *Tangential* velocity component remains almost constant along the curve, with a velocity identical to the approach flow velocity for $\mathsf{F}_o > 3$.
- Velocity distribution in the *vertical* direction is also practically constant, except for the boundary layers along the bottom and the walls.

Figure 16.32 shows a definition plot for bend flow in a rectangular channel, with the wave maxima (subscript *M*) and wave minima (subscript *m*) at location θ, 2θ, ... The approach flow velocity is V_o, the approach flow depth h_o, such that $\mathsf{F}_o = V_o/(gh_o)^{1/2}$ is the approach flow Froude number. With R_a as the centerline radius of curvature, or b/R_a as the relative bend curvature, the significant parameter for supercritical bend flow is the *bend number* $\mathsf{B} = (b/R_a)^{1/2}\mathsf{F}$, with $\mathsf{F} = \mathsf{F}_o$. Based on the existing observations and own experiments, the *maximum wave height* $Y_M = h_M/h_o$ depends exclusively on the bend number as (Reinauer and Hager 1997)

$$Y_M = (1 + 0.4\mathsf{B}^2)^2,\ \mathsf{B} \le 1.5, \tag{16.76}$$

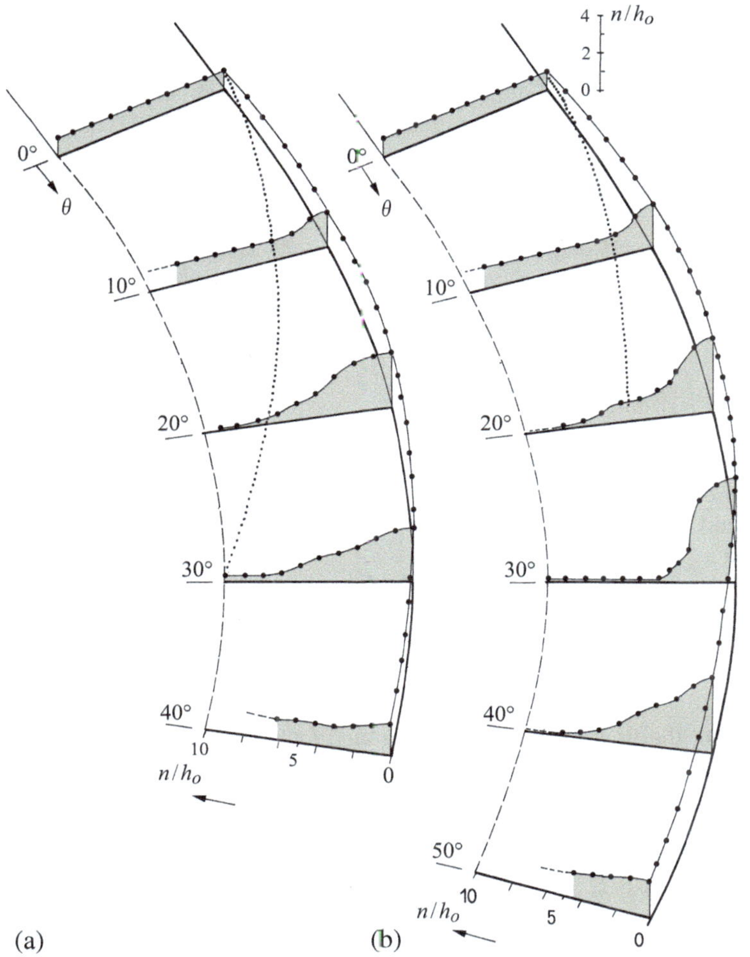

Fig. 16.31 Surface of supercritical bend flow for $\mathsf{F}_o =$ (**a**) 4, (**b**) 6. (· · ·) Shock front

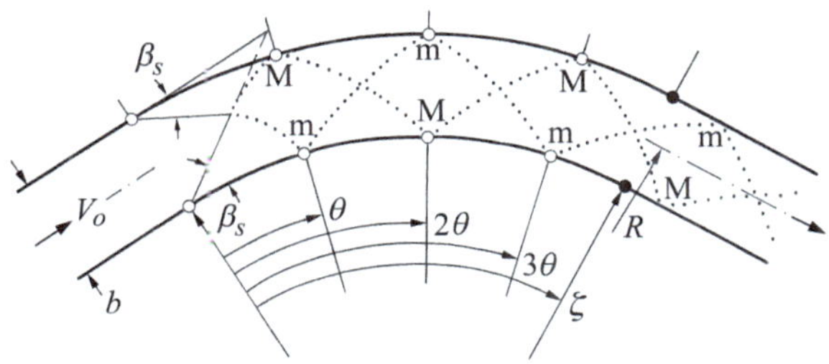

Fig. 16.32 Definition plot for supercritical bend flow in plan view

$$Y_M = (1 + 0.6\mathsf{B})^2, \mathsf{B} > 1.5. \tag{16.77}$$

For relatively small B, Eq. (16.76) is an approximation of Knapp's approach and the present predictions are almost identical. For large values of B, the effect of the bend number is linear, and Eq. (16.77) has to be applied. The change of Eq. (16.76), and (16.77) could not be attributed to a physical feature, such as flow *separation* from the inner bend wall, or wave breaking.

The *location* $\tan\theta_M$ of the wave maximum is governed by the parameter $(b/R_a)\mathsf{F}$ instead of $(b/R_a)^{1/2}\mathsf{F}$, in agreement with Eq. (16.74). The experimental results are

$$\tan\theta_M = (b/R_a)\mathsf{F}, \quad \text{for} \quad (b/R_a)\mathsf{F} \leq 0.35, \tag{16.78}$$

$$\tan\theta_M = 0.6[(b/R_a)\mathsf{F}]^{1/2}, \quad \text{for} \quad (b/R_a)\mathsf{F} > 0.35. \tag{16.79}$$

Accordingly, Ippen and Knapp's approach applies only to flows of relatively small bend number.

The location and height of the *wave minimum* (subscript m) can also be predicted as

$$Y_m = (1 - 0.5\mathsf{B}^2)^2, \tag{16.80}$$

$$\tan\theta_m = \sqrt{2}(b/R_a)\mathsf{F}. \tag{16.81}$$

For $Y_m < 0$, the flow is completely separated from the wall; Reinauer and Hager (1997) predicted the geometry of the separated flow region. Figure 16.33 shows plan views of bend flows and Fig. 16.34 refers to upstream views. In the latter case the bend flow is seen to occupy only a portion of the channel width, and the design may be highly *uneconomic*. Choking was not a problem in the rectangular channel, but it may be dangerous in pipes. The main results are presented below.

The *wave profiles* along the *inner* and outer bend walls have been further analyzed. Figure 16.35 shows the normalized profile $\tau_M(\theta/\theta_M)$ along the outer wall,

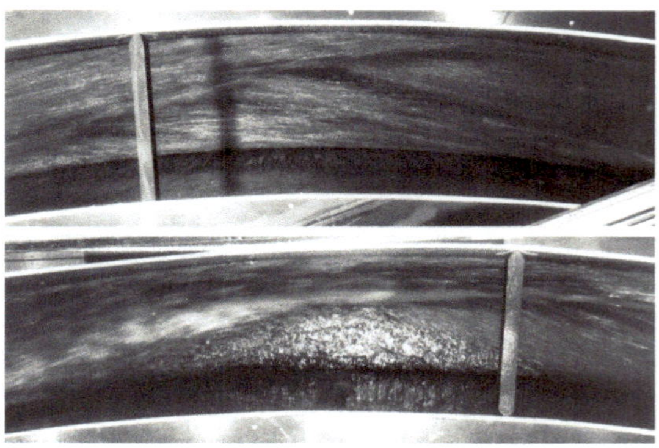

(a) (b)

Fig. 16.33 Plan view for bend flow with $b/R_a = 0.07$ and $\mathsf{F}_o =$ (**a**) 4, (**b**) 8

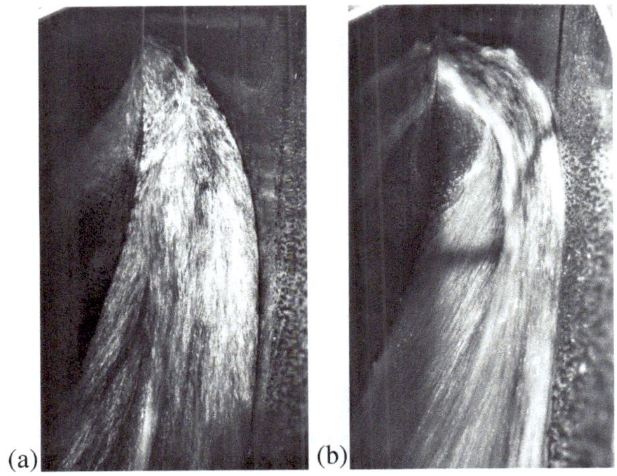

Fig. 16.34 Upstream views of bend flow for $b/R_a = 0.144$ and $\mathsf{F}_o =$ (**a**) 4, (**b**) 8

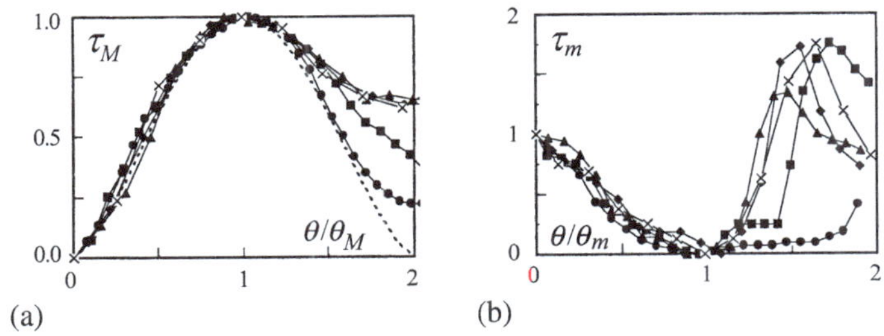

Fig. 16.35 Generalized wall flow profiles along (**a**) outer and (**b**) inner bend walls for $b/R_a = 0.07$ and $\mathsf{F}_o =$ (×) 2.5, (♦) 3, (▲) 4, (■) 6, (•) 8

with $\tau_M = (h - h_o)/(h_M - h_o)$, and similarity is noted up to $\theta/\theta_M = 1.25$. For larger deflection the experimental curves spread, depending on F_o. For θ up to about 50° the *outer wave profile* follows

$$\tau_M = [\sin(\theta/\theta_M)]^{1.5}. \qquad (16.82)$$

The profile along the *inner* bend wall $\tau_m(\theta/\theta_m)$ corresponds to the profile along the outer wall. With $\tau_m = (h - h_o)/(h_m - h_o)$, a single curve may be plotted up to $\theta/\theta_m = 1.2$. For smaller values of F_o, the wave trough is followed by a wave crest, whereas flow separation from the inner wall results for larger F_o. Equation (16.82) is verified for relative curvatures $b/R_a = 0.144$ and 0.31 in Fig. 16.36a.

The effect of *partial bend flow* with a deflection angle ζ smaller than θ_M is shown in Fig. 16.36b). At the bend end, the rise of the surface profile according to Eq. (16.82) stops almost abruptly. Then, the maximum flow depth is predicted from Eq. (16.82) by setting $\theta = \zeta$.

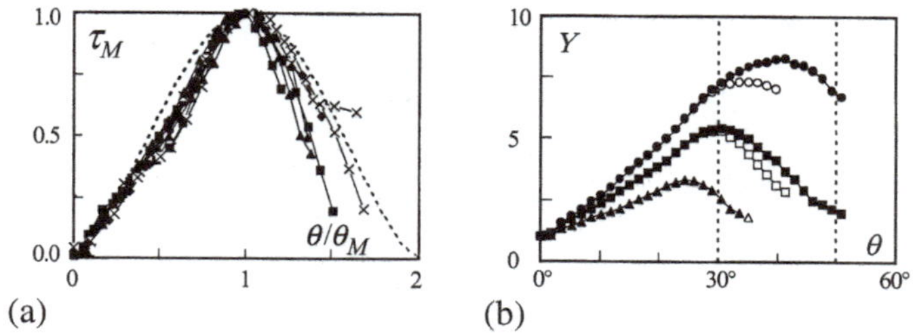

Fig. 16.36 (**a**) Outer wall profile $\tau_M(\theta/\theta_M)$ for $b/R_a = 0.144$ and 0.31, (–) Eq. (16.82), (**b**) partial bend flow $Y(\theta)$ for $\theta = 30°$ (*light*) and $\theta = 50°$ (*solid*). Notation Fig. 16.35

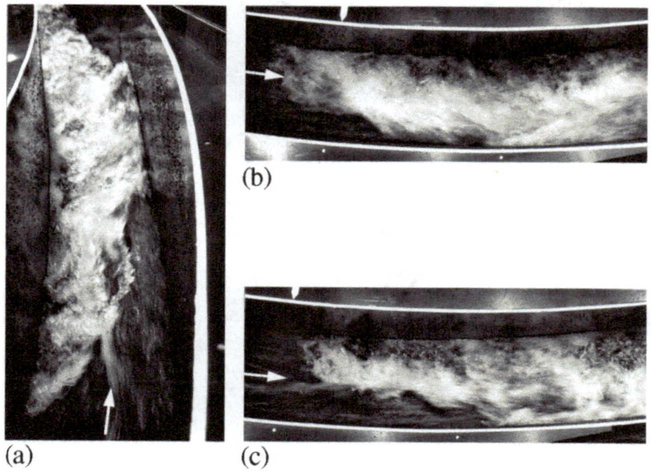

Fig. 16.37 Choked bend flow with forward flow along the outer wall and recirculated flow along the inner bend wall. (**a**) Upstream view, (**b**) and (**c**) plan views

Figure 16.37 shows submerged bend flow with a supercritical approach flow. The forward flow concentrates along the outer wall, and a large *recirculation zone* is observed along the inner bend wall. Such flows are highly unstable and have to be avoided because of the poor hydraulic performance. Currently, such flows have not yet received systematic treatment, however.

Tunnel Bend Flow

Bend flows may choke in a tunnel if its capacity is too small to discharge free surface flow. These flows were analyzed experimentally by Gisonni and Hager (1999) relative to bottom outlets. Whereas a bottom outlet is normally deflected by an angle of 45°, a sewer usually has a 90° bend. In the following, only results for a 45° deflection angle are presented, which apply approximately also to the 90°-deflected sewer bend.

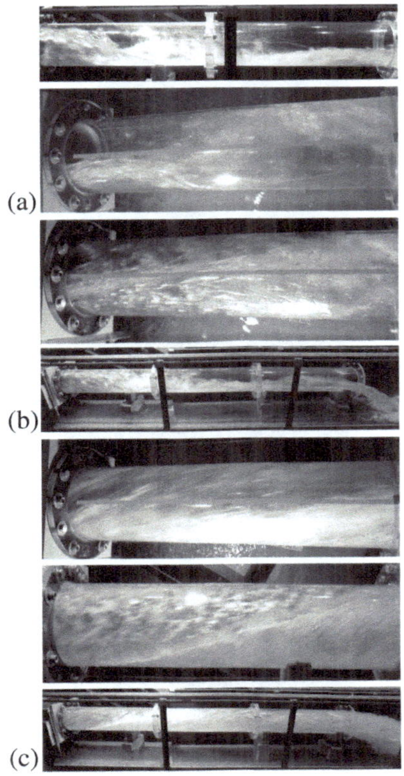

Fig. 16.38 Flow downstream of 45°-deflected tunnel bend for $y_o = 0.24$ and $\mathsf{F}_o =$ (**a**) 5.2, (**b**) 5.8, (**c**) 11.8 with general view (top), plan view on tailwater pipe (center or bottom), and side view (bottom of c)

Figure 16.38 relates to the straight tailwater reach downstream of the bend, for an approach flow filling ratio of $y_o = h_c/D = 0.24$ and a range of approach flow Froude numbers $\mathsf{F}_o = Q/(gD^4h_o)^{1/2}$ (Chap. 6), where D is the tunnel diameter and h_o the approach flow depth. The relative tunnel radius was $R/D = 3$. For $\mathsf{F}_o = 5.2$ (Fig. 16.38a), so called stratified tunnel flow was observed, for which the air phase is above the water flow. As $\mathsf{F}_o = 5.8$ is reached (Fig. 16.38b), the transition from stratified to annular flow occurs just downstream of the tunnel bend. Upon further increasing to $\mathsf{F}_o = 11.8$ (Fig. 16.38c), the entire tailwater reach was in the annular flow regime (Chap. 5).

Figure 16.39 relates to a higher filling ratio of $y_o = 0.61$, for which the flow structure is similar as for $y_o = 0.24$ except that the transition Froude number between stratified and annular flows is reduced. Figure 16.39a shows the transverse surface profile consisting mainly of black water along the outer, and a white water front along the inner tailwater sides. If either y_o or F_o are increased, then the pipe chokes resulting in pressurized two-phase flow. The transition occurs for $\mathsf{F}_o \cong 2.75$ (Fig. 16.39b), and (Fig. 16.39c) shows annular flow for $\mathsf{F}_o = 4.2$ including a long vortex spanning from the bottom of the bend outlet toward the top of the pipe end. The following relates to the main hydraulic features of these flows.

(a)

(b)

(c)

Fig. 16.39 Tunnel flow downstream of 45°-deflected bend for $y_o = 0.61$ and $\mathsf{F}_o =$ (**a**) 2.0, (**b**) 2.75, (**c**) 4.2

Figure 16.40 is a definition plot for the flow considered, including the approach flow water discharge Q and the approach flow air discharge Q_a, the distance d_1 from the bend end to the first wave maximum h_M and the height h_2 of the second wave maximum. As demonstrated in Chap. 7, the sequent depth ratio of flow in a straight

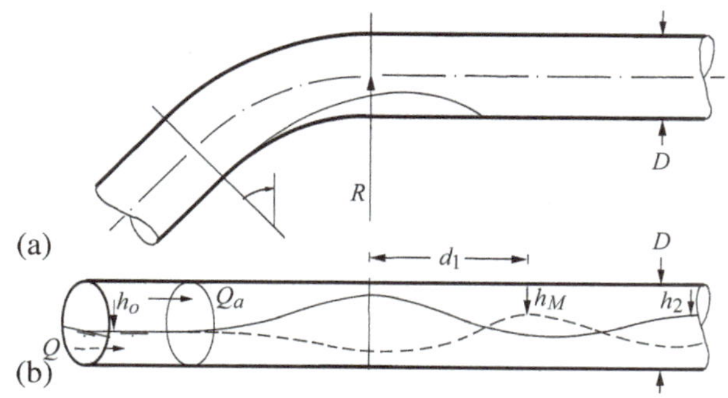

Fig. 16.40 Definition plot for supercritical tunnel bend flow with (**a**) plan, (**b**) streamwise section

circular pipe is $Y = \mathsf{F}_1^{0.90}$, with subscript 1 relating to the supercritical approach flow. Choking from free surface to pressurized pipe flow occurs nominally for $h_2 = D$. Inserting this condition in the previous sequent depth relation leads to the so called *choking number*

$$\mathsf{C} = \mathsf{F}_1/(D/h_1) = Q/\left(gD^3h_1^2\right)^{1/2}. \tag{16.83}$$

Note that all three governing parameters discharge Q, pipe diameter D and approach flow depth h_1 have an almost linear effect on C. The choking number so defined includes both the filling ratio of the approach flow and its dynamics as expressed by the approach flow Froude number.

A tunnel bend generates shockwaves for supercritical flow whose effect reduces the capacity as compared to a straight tunnel reach. The choking number of a tunnel bend is defined from Eq. (16.83) as $\mathsf{C}_o = \mathsf{F}_o/(D/h_o)$ and may be considered the relevant parameter for these flows. The relative wave maximum $Y_M = (h_M/h_o)(h_o/D)^{2/3}$ in the tailwater reach (Fig. 16.40b) may be expressed as (Gisonni and Hager 1999)

$$Y_M = 1.10 \tanh(1.4\mathsf{C}_o), \quad \mathsf{C}_o > 0.80. \tag{16.84}$$

For $\mathsf{C}_o < 0.80$, the tunnel flow was subcritical because a hydraulic jump formed at the bend inlet. For $\mathsf{C}_o > 1$, the maximum relative wave height is almost constant at $Y_M = 1.1$ indicating $h_M = 1.1(h_oD^2)^{1/3}$, corresponding to a maximum filling ratio of $h_M/D = 1.1(h_o/D)^{1/3}$. The distance d_1 from the bend end to wave 1 may be expressed as $d_1/D = 4.8(\mathsf{C}_o - 0.8)^{1/2}$. For incipient choking, the first wave was located some 2 pipe diameters downstream from the tunnel end. Wave 2 downstream of the maximum wave is always less high, i.e. $h_2 < h_M$, and further waves were hardly visible. No effect of a small streamwise bottom slope on this and the following relations was noted.

If annular flow is established downstream of the bend tunnel, there is a transition to stratified flow at a length d_a once the rotational component is insufficient. The relative distance was found to be

$$d_a/D = 0.85(\mathsf{C}_o - 1)^2, \quad \mathsf{C}_o > 1. \tag{16.85}$$

Accordingly, the tunnel just issues a stratified flow for $\mathsf{C}_o = 1$, such that this is the condition for stratified tunnel bend flow.

The distance of the first wave crest from the tunnel outlet was further measured, resulting in

$$d_1/D = 4.8(\mathsf{C}_o - 0.8)^{1/2}, \quad \mathsf{C}_o > 0.8. \tag{16.86}$$

For incipient choking, the first wave was located about 2 pipe diameters downstream of the bend end, with a corresponding filling ratio of $h_M/D = 0.97(h_o/D)^{1/3}$.

The air discharge Q_a across the tunnel bend was also measured. This discharge was observed to increase with C_o and to decrease with the filling ratio, because the cross-sectional area for air flow reduces. Accordingly, the usual air-water discharge ratio as presented in Chap. 5 was generalized to $B = (Q_a/Q)/(1-y_o)^2$. The tests indicated that B increases essentially with C_o if $\mathsf{C}_o > 1$. For lower values of C_o there is also air flow but mainly due to free surface entrainment, as in open channel flow. The data were fitted to

$$B = 3\ \tanh\left[3(\mathsf{C}_o - 1)\right], \quad \mathsf{C}_o > 1. \tag{16.87}$$

indicating a maximum of $Q_a/Q = 3(1-y_o)^2$ for $\mathsf{C}_o > 1$, providing typically values of $Q_a/Q = 1-2$. This is a comparatively large air discharge ratio asking for a separate air supply across the upstream manholes. It was also observed that cover plates as introduced in Sect. 16.3.7 for bend manholes do not at all improve the flow pattern across a bend tunnel.

16.3.6 *Channel Junction*

Flow Description

Figure 16.41 refers to a channel junction (German: Kanalvereinigung; French: Jonction de canal) with a *straight* through (subscript o) and a lateral (subscript z) branch under the junction angle δ. Further, the widths of the upstream and downstream branches are equal to $b_o = b_u$, whereas $b_z \leq b_o$. Because the flow close to the junction point P is only considered, the effects of bottom slope and wall friction are neglected. The general case of a junction structure with three different widths b_o, b_z, and b_u, and two different branch angles δ_o and δ_z has not yet been analyzed.

Regarding the branch discharges, three cases can be discussed:

- Discharge $Q_o > 0$ and no lateral discharge,
- Discharge $Q_z > 0$ and no upstream discharge, and
- Discharges $Q_o > 0$ in the upstream, and $Q_z > 0$ in the lateral branch.

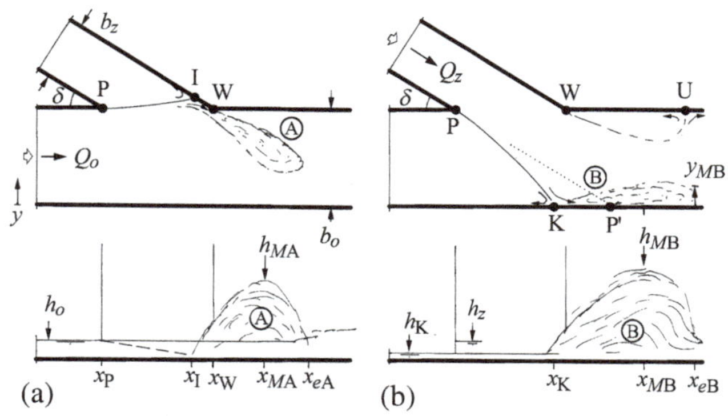

Fig. 16.41 Channel junction with (a) upstream discharge only, (b) lateral discharge only

For junctions with $Q_z = 0$, the governing parameters are (h_o, F_o). Downstream from the junction point P, the flow may laterally expand and impinge on the lateral branch with a stagnation point I (Fig. 16.41a). Depending on the junction angle δ, a small flow portion is directed into the lateral branch but the main discharge portion continues in the downstream branch. By flow impact on the wall between points I and W, a standing wave is formed, referred to as *wall wave* A with a maximum flow depth $h_{M\text{A}}$ at location $x_{M\text{A}}$. The end (subscript e) of the wave is at location $x_{e\text{A}}$.

For junctions with $Q_o = 0$, i.e. lateral discharge only, the flow expands from the junction point P and impinges on the opposite wall, with a stagnation point K. Depending on the junction angle δ, a smaller or larger discharge portion flows into the upstream branch and forms a recirculation zone. Downstream from point K, a standing wave is formed, referred to as *wall wave* B. Its maximum flow depth $h_{M\text{B}}$ is at location $x_{M\text{B}}$, and its end is at $x_{e\text{B}}$. Opposite from wave B, and starting at point W a recirculation zone is established with a stagnation point U in the downstream branch.

Both waves A and B can be either compact or wall-type waves (Fig. 16.42). A *compact wave* is continuous and smooth with a nearly hydrostatic pressure distribution. These waves are typical for weak junction flow. A *wall-type wave* rises up the wall, is narrow compared to the approach flow width and results from overforced flow. Waves A and B may be either pure water waves or highly aerated two-phase flows with a considerable spray development, depending on the Froude number and the absolute velocity. Supercritical junction flows can thus be highly spatial and then tend to *flow concentrations* in the downstream branch, combined with large separation regions.

A channel junction with discharges in both the *upstream and lateral branches* can be defined with the parameters (h_o, F_o) and (h_z, F_z), respectively (Fig. 16.43). Normally, wave A is suppressed by the lateral discharge but *wave C* is formed, starting from the junction point. Wave C is directly related to the merging of two supercritical flows of different direction and different flow depth. It can be interpreted as a vertical movement of two horizontal flows representing the only possibility of momentum exchange between two high-velocity flows. Wave C has a maximum depth $h_{M\text{C}}$ at location $x_{M\text{C}}$, and the shock front has an angle θ relative to

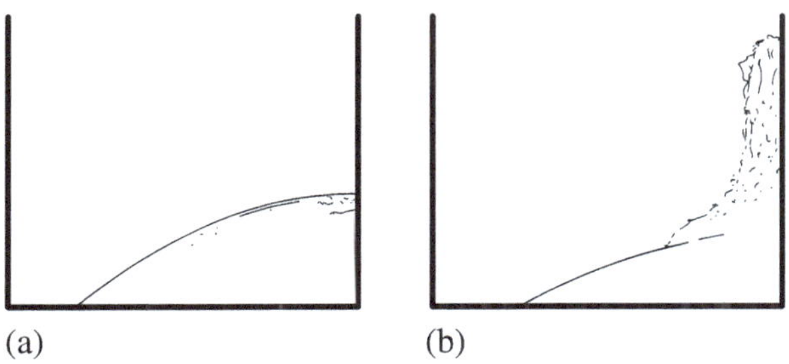

Fig. 16.42 Wall wave types (**a**) compact and (**b**) wall-type waves

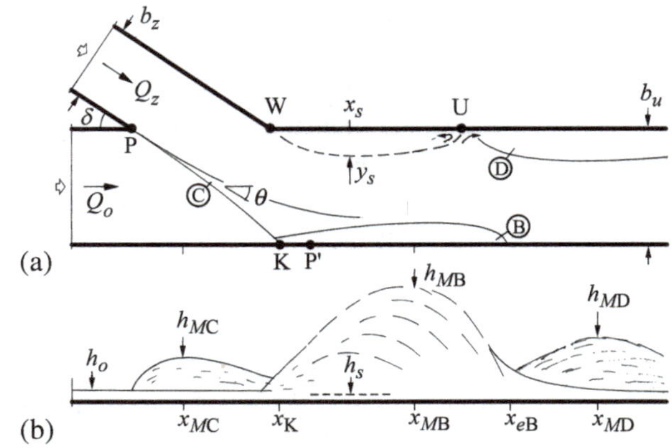

Fig. 16.43 Simple channel junction with supercritical approach flows, (**a**) plan, (**b**) section with wave structure and (- - -) recirculation zone

the downstream direction. Note the analogy between the abrupt wall deflection and the junction structure regarding wave C.

The *shock front* of wave C impinges the opposite wall at point K, generating wave B downstream from it, with h_{MB} as maximum wave height at location x_{MB}. On the opposite wall, the flow separates (subscript s) between point W and the stagnation point U, with the maximum width y_s at location x_s. Wave B is reflected into the downstream branch, to generate *wave D*. Depending on the ratio of wave amplitudes h_{MD}/h_{MB}, the undular flow character is rapidly or slowly dampened in the downstream branch.

Based on the previous description of general junction flow, a complex flow pattern is always established. It is influenced by a large number of parameters such as the ratio of branch flow depths h_o/h_z, the branch Froude numbers F_o and F_z, the width ratio b_o/b_z and the junction angle δ. In addition, various *junction geometries* could be considered, with rounded corners, bottom drops and changes in the boundary roughness. In the following, some results originating from systematic experimentation relating to the *simple junction* as shown in Fig. 16.43 are presented.

Limit Conditions

Incipient choking (German: Grenzzustand; French: Commencement d'effondrement) corresponds to the transition between super- and subcritical junction flows. Depending on whether incipient choking appears in one or even in two branches of a junction, supercritical flow is maintained in one or even no branch. Knowledge of the limit (subscript L) conditions is essential because choked junction flow behaves completely different from the supercritical junction flow described previously.

Figure 16.44 shows the stages of flow choking in one branch alone. Both conditions are generated provided the flow at the stagnation point K for the through branch, and at the stagnation point I recirculate. The limit condition depends thus

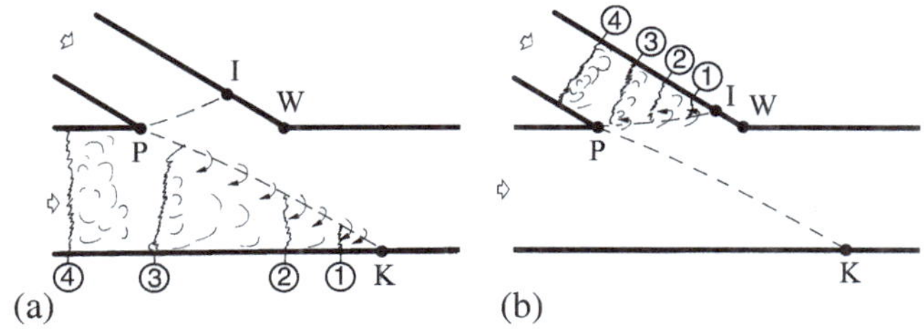

Fig. 16.44 Limit conditions in channel junction for (**a**) through and (**b**) lateral branch, with phases from incipient choking to stable hydraulic jump

on the exact flow structure and the junction geometry. Figure 16.44 relates to the four main phases of choking, from incipient choking to the stable hydraulic jump in one branch, and supercritical flow in the other branch. If the Froude number is further reduced, choking appears also in the latter branch, resulting in fully choked junction flow.

The experimental data regarding the limit (subscript L) condition of *incipient choking* may be expressed as (Schwalt 1993)

$$\mathsf{F}_{Lo} = 2 + (\delta/120°)\mathsf{F}_z Y^{-1}, \tag{16.88}$$

$$\mathsf{F}_{Lz} = 2 + (1/8)(\mathsf{F}_o + 5)Y. \tag{16.89}$$

Both limit Froude numbers thus depend linearly on the Froude number of the neighbour branch and on the flow depth ratio $Y = h_o/h_z$. F_{Lo} varies with the junction angle δ in addition, whereas F_{Lz} does not. An absolute lower limit for both F_{Lo} and F_{Lz} is $\mathsf{F} = 2$, because supercritical approach flow for $\mathsf{F} < 2$ is always choked. These flows typically generate undular hydraulic jumps (Chap. 7). Prior to the design of a junction structure, one has thus to verify both conditions $\mathsf{F}_o > \mathsf{F}_{Lo}$ and $\mathsf{F}_z > \mathsf{F}_{Lz}$.

Wave Geometry

The main waves of a junction subjected to supercritical flow are waves B, C and D (Fig. 16.43). Usually, wave A cannot be observed if both branches operate, and wave D and all successive downstream waves are smaller than the *determining wave* B. Particular attention receive thus waves B and C.

Each wave can be defined by its beginning (subscript a), maximum (subscript M), and end (subscript e). In addition, the width y_{MB} of a wall wave at the maximum location can be defined. Further, the main characteristics of the separation (subscript s) zone are of interest. With x as the streamwise coordinate, y as the transverse coordinate measured from the wall opposite of the lateral branch, and h as the flow depth, these parameters generally vary with the basic parameters $\mathsf{F}_o = V_o/(gh_o)^{1/2}$, $\mathsf{F}_z = V_z/(gh_z)^{1/2}$, $Y = h_o/h_z$, and $\beta = b_z/b_o$. Based on systematic experimentation, Schwalt (1993) found for

Wave C

$$\tan\theta = \frac{x_{MC} - x_P}{b_o - y_{Mc}} = (1/2)\delta(\mathsf{F}_z/\mathsf{F}_o)^{2/3}Y^{-1}, \quad (16.90)$$

$$L_{MC} = \frac{x_{MC} - x_P}{\bar{h}\cos\theta} = 4.3(f - 2), \quad (16.91)$$

$$Z_{MC} = \frac{h_{MC}}{\bar{h}} - 1 = 1 + 0.25(f\sin\delta)^2. \quad (16.92)$$

Here, x_P is the location of junction point P, θ the shock angle, $f = 2\mathsf{F}_o\mathsf{F}_z/(\mathsf{F}_o + \mathsf{F}_z)$ the determining Froude number, $\bar{h} = (h_o h_z)^{1/2}$ the determining approach flow depth, and $\bar{b} = (b_o b_z)^{1/2}$ the determining approach flow width. With $\Delta = 0.55 + 0.05\delta°$, Schwalt (1993) found for the characteristics of

Wave B

$$L_{aB} = \frac{x_{aB} - x_P}{b_o} = (1/\Delta)\left(\mathsf{F}_o/\mathsf{F}_z^{1/3}\right)Y^{2/3}, \quad (16.93)$$

$$L_{MB} = \frac{x_{MB} - x_P}{b_o} = 1.65L_{aB}, \quad (16.94)$$

$$L_{eB} = \frac{x_{eB} - x_P}{(\bar{b}\bar{h})^{1/2}} = 1.35\cos\delta\left[\mathsf{F}_o(\mathsf{F}_z Y)^{1/3} + 5\right], \quad (16.95)$$

$$B_{MB} = \frac{y_{MB}}{\bar{h}} = 0.3(0.1 + \sin\delta)\mathsf{F}_o\mathsf{F}_z^{1/3}, \quad (16.96)$$

$$Z_{MB} = \frac{h_{MB}}{\bar{h}} - 1 = 0.25f^2. \quad (16.97)$$

Wave D starts at location (Schwalt and Hager 1995)

$$L_{aD} = \frac{x_{aD} - x_W}{b_o} = (2.5/\Delta)^{3/2}\left[(\mathsf{F}_o/\mathsf{F}_z^{1/3})Y^{2/3} - 0.7\right] \quad (16.98)$$

and the *recirculation zone* (Fig. 16.45) may be described as

$$L_s = \frac{x_s - x_W}{(\bar{b}\bar{h})^{1/2}} = 1.35\cos\delta(f - 3), \quad (16.99)$$

$$L_{es} = \frac{x_{es} - x_W}{(\bar{b}\bar{h})^{1/2}} = 2.5L_s, \quad (16.100)$$

$$B_s = \frac{y_s}{\bar{b}} = 6.4(\cos\delta)^{1/2}\left[\mathsf{F}_z^{-1}Y^{1/3} - 0.05\right]. \quad (16.101)$$

These results apply for $\delta > 15°$, and wave C in particular behaves differently for smaller junction angles, which are irrelevant in sewer applications, however. The upper limit of application is $\delta < 70°$. According to experiments, junctions with a right-angled branch do always choke. Provided a junction should have a thoroughly supercritical flow, the *limit junction angle* should be about 60°. Typical junction angles are thus $30° < \delta < 60°$.

(a)

(b)

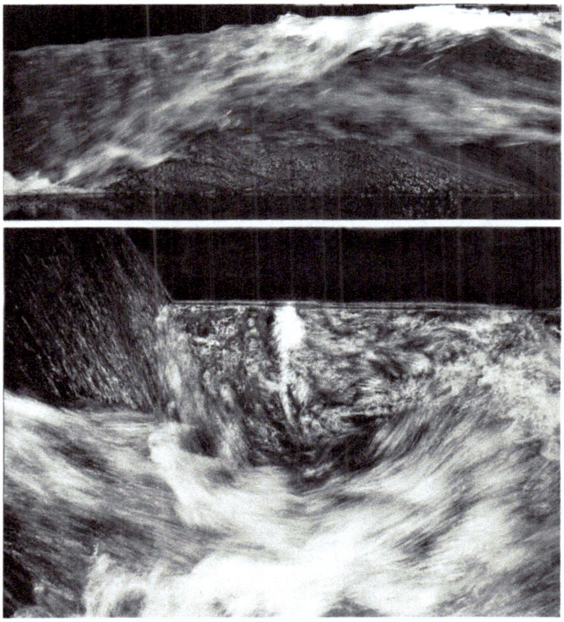

Fig. 16.45 Recirculation zone (bottom) and wave B (top) for junction angles $\delta =$ (**a**) 30°, (**b**) 60°

The transition (subscript t) between compact and wall-type waves B occurs approximately for

$$\mathsf{F}_{ot} = 5(\sin\delta)^{1/2}(\mathsf{F}_z - 2)^{1/3}Y^{-1/3}. \quad (16.102)$$

With these indications, the main features of supercritical junction flow are given.

Example 16.7 The approach flow conditions of a channel junction are $Q_o = 10\ \mathrm{m^3s^{-1}}$, $h_o = 0.8$ m, $b_o = 1.5$ m and $Q_z = 4\ \mathrm{m^3s^{-1}}$, $h_z = 0.45$ m, $b_z = 1$ m. Describe the flow structure for a junction angle of $\delta = 35°$!

Approach branches $\mathsf{F}_o = 10/(9.81 \cdot 1.5^2 0.8^3)^{1/2} = 2.98$,
$\mathsf{F}_z = 4/(9.81 \cdot 1.0^2 0.45^3)^{1/2} = 4.23$, $Y = 0.8/0.45 = 1.78$,
$\bar{b} = 1.225$ m, $\bar{h} = 0.60$ m,
$(\overline{bh})^{1/2} = 0.86$ m, $f = 2 \cdot 2.98 \cdot 4.23/(2.98 + 4.23) = 3.50$.

Limit conditions $\mathsf{F}_{Lo} = 2 + (35/120)4.23/1.78 = 2.69 < 2.98$,
$\mathsf{F}_{Lz} = 2 + (1/8)(2.98 + 5)1.78 = 3.78 < 4.23$ according to Eqs. (16.88) and (16.89), choking does not occur.

Wave C	$\tan\theta = (1/2)35°(\pi/180°)(4.23/2.98)^{2/3}/1.78 = 0.22$,
	thus $\theta = 12.1°$,
	$x_{MC}–x_P = 0.60 \cdot \cos 12.1° 4.3(3.50-2) = 3.78$ m
	$h_{MC} = [2+0.25(3.50 \cdot \sin 35°)^2]\ 0.60 = 1.80$ m.
Wave B	With $\Delta = 0.55+0.05 \cdot 35° = 2.30$,
	$x_{aB}–x_P = [(1/2.30)(2.98/4.23^{1/3})1.78^{2/3}]1.5 = 1.77$ m
	$x_{MB}–x_P = 1.65 \cdot 1.77 = 2.9$ m
	$x_{eB}–x_P = 1.35\cos 35°[2.98(4.23 \cdot 1.78)^{1/3}+5]0.86 =$
	10.3 m, $y_{MB} = 0.3(0.1+\sin 35°)\ 2.98 \cdot 4.23^{1/3} 0.6 = 0.59$ m
	$h_{MB} = [1+0.25 \cdot 3.5^2]0.60 = 2.44$ m
Recirculation zone	$x_s–x_W = 0.86 \cdot 1.35\cos 35°(3.5–3) = 0.5$ m
	$x_{es}–x_W = 2.5 \cdot 0.5 = 1.25$ m
	$y_s = 6.4 \cdot 1.22(\cos 35°)^{1/2}[4.23^{-1} 1.78^{1/3}–0.05] = 1.67$ m.

This indicates a relatively high wave C, and a slightly larger wave B. The width of the recirculation zone is with $y_s > b_o$ physically impossible. With $\mathsf{F}_{ot} = 5(\sin 35°)^{1/2}(4.23–2)^{1/3}/1.78^{1/3} = 4.08 > \mathsf{F}_o$, wave B is compact.

16.3.7 *Methods of Shockwave Reduction*

The preceding discussion has demonstrated that perturbations of a supercritical flow increase significantly with the approach Froude number F_o. Because the effects of viscosity and wall roughness on the decay of wave amplitudes are small, and disadvantages such as flow concentrations, surface aeration and local erosion may occur, the *reduction of shockwaves* is of significance. Basically, shockwaves may be reduced with various methods, as illustrated in Fig. 16.46 (Vischer and Hager 1994):

- *Transverse bottom slope* S_{oy} to compensate the centrifugal acceleration of a particle moving on a streamline radius R, where

$$S_{oy} = \frac{V_o^2}{gR}. \tag{16.103}$$

- *Multiple vanes* to reduce the total channel width b in various sub-sections of reduced maximum wave height, an approach impractical in sewers due to clogging, however.

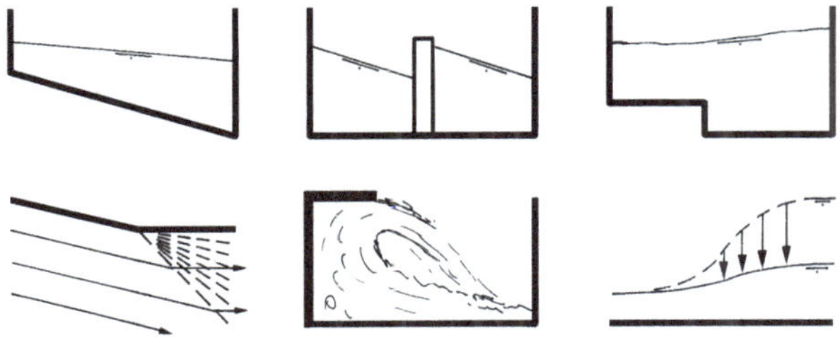

Fig. 16.46 Possibilities of shockwave reduction, for details see main text

- *Transverse step* to guide a flow with a wall minimum depth resulting in a more horizontal surface profile. This approach is unsuitable mainly because of abrasion.
- *Cover plate* to restrict the height of a wall wave.
- Application of *wave interference* techniques based on the superposition of two identical waves of different sign.

Elements such as guide walls, transverse steps and application of wave interference might be justified in hydraulic structures, but they are not a design in sewer techniques. Therefore, only two methods of shockwave reduction in sewers are available:

- Reduction of shock number, or
- Application of cover plates.

Shockwaves increase often linearly with the governing shock number. Accordingly, shockwaves can be reduced in the design phase by either limiting the approach flow Froude number, or by limiting the *perturbation intensity*, such as the deflection angle, the relative curvature or the junction angle. Reducing a shock number is thus a direct *design method* that adds to the performance of a sewer.

Because the Froude number of a sewer flow is $\mathsf{F} = Q/(gDh^4)^{1/2}$ and the discharge is a design variable, F reduces only by increasing either the diameter D or, more significantly, the flow depth h. If considerations refer to a long sewer reach where uniform (subscript N) flow has nearly established, one has from Eq. (5.15) for $y_N < 0.8$

$$y_N = 1.15 q_N^{1/2}(1 + q_N)^{1/2}. \tag{16.104}$$

Eliminating y_N, F_N can be expressed with the relative discharge $q_N = nQ/(S_o^{1/2}D^{8/3})$ as

$$\mathsf{F}_N = \frac{Q/(gD^5)^{1/2}}{1.32 q_N(1 + q_N)} = \frac{0.76\chi}{1 + q_N}. \tag{16.105}$$

Here, $\chi = S_o^{1/2}D^{1/6}/(ng^{1/2})$ is the roughness characteristics according to Sect. 5.6. If an average design value $y_N = 0.50$ is considered, one has $q_N = 0.15$, and $\mathsf{F}_N = (2/3)\chi$ from Eq. (16.104). The Froude number for uniform flow is thus small for *small bottom slope*, because the diameter effect is insignificant and roughness cannot really be influenced for commercial pipe material.

The *perturbation intensity* of a flow includes:

- Deflection angle θ for contractions,
- Expansion ratio $\beta = b_u/b_o$ for expansions,
- Relative centerline curvature b/R for curves, and
- Junction angle δ for junction structures.

The shockwave height depends on the product of approach flow Froude number and the perturbation intensity. From constraints in space and cost, the shock number can often not be limited to a value of, say, below 1. Then, the only method to reduce shockwaves is *wave treatment* by flow forcing. Based on a fundamental analysis of the problem in channel junctions, Schwalt (1993) considered the following geometrical arrangements for *active shock control:*

- *Bottom vane* to separate the branch flows, resulting in a modification of the approach flow direction to wave B,
- *Bottom step* to reduce the transverse surface slope, and
- *Cover plate* to limit the wall wave height.

Both, the bottom vane and the bottom step (Fig. 16.46) have a passive effect on the shockwave but these elements are not flexible enough for variable approach flow conditions. They are designed for a specific condition but either are ineffective or even perform poorer than the basic design. In addition, the elements are prone to sediment deposition or abrasion, such that they cannot be recommended for sewers.

The only reliable, efficient and simple method to reduce shockwaves in sewer manholes is the *cover plate* (German: Deckplatte; French: Plaque de couverture). It is an element that inhibits the rising of wall waves actively and it is mounted on the benches of a manhole. Its prime action is to reduce the risk of choking flow in sewers and thus the breakdown of supercritical pipe flow. The cover plate can be inserted in junction and bend manholes, as described below. Disadvantages of cover plates include corrosion and a certain clogging potential. An alternative is presented in Sect. 16.5.

Figure 16.47a shows typical supercritical flow in a bend manhole, with a submerged downstream sewer due to shockwave choking. Because larger shocks are wall waves (Fig. 16.42b), they can often be reduced with a suitable *cover plate*. According to Fig. 16.47b the plate can be made tip up to easen access and maintenance. The element can also be added to locations where shockwaves have

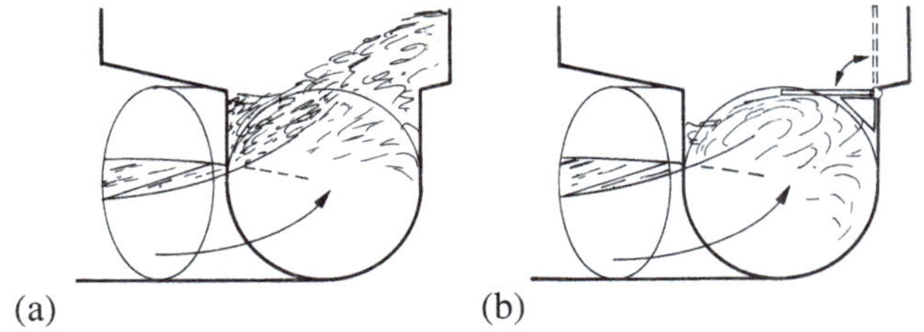

Fig. 16.47 Supercritical approach flow to bend manhole (**a**) untreated and (**b**) treated flow

Fig. 16.48 Waves B and C at junction manholes extended with cover plate. (**a**) section, (**b**) plan

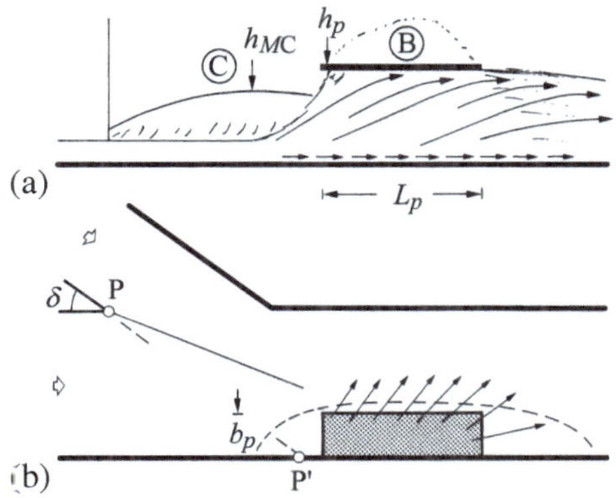

caused damages. Contrary to most other elements, the plate is subjected with an overpressure and vibrations and cavitation damage are no concern.

According to Sect. 16.3.6 wave B is always higher than wave C. Figure 16.48 relates to the cover plate of length L_d, width b_d and height h_d. To exclude overflow of wave C, the height h_d should always be larger than the height h_{MC} of wave C. The position of the cover plate must be such that there is no overflow from the through branch, and a sufficient length guarantees a horizontal free surface towards the downstream branch, because a wall wave is only partially reduced otherwise (Fig. 16.49).

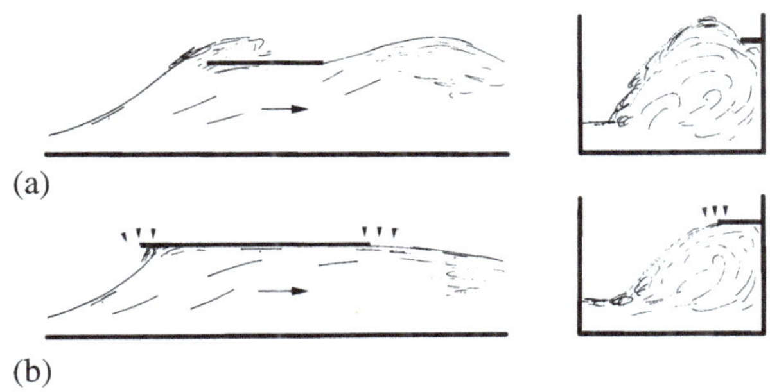

Fig. 16.49 Adaptation of cover plate in a junction manhole. (**a**) Poor and (**b**) appropriate design

Schwalt (1993) conducted experiments in a rectangular channel and found:

- *Cover plates* are suitable for junction angles up to about 45°. For larger angles, waves C and B have nearly the same height according to Eqs. (16.92) and (16.97), and the element becomes inefficient.
- For larger junction angles, a *combination* of a bottom drop and a cover plate may be suitable. The drop height has the order of the lateral flow depth. This design needs further tests.

The following results refer to junction angles δ between 15° and 45°. The major *design elements* of a cover plate (subscript p) are the position of beginning (subscript a), the end position (subscript e), the width and the height above the bottom. According to Schwalt (1993) the beginning of the cover plate x_{ap} is at (Figs. 16.43 and 16.48)

$$\frac{x_{ap} - x_{P'}}{b_o} = \frac{0.27}{\sin\delta} Y^{3/4} \mathsf{F}_o \mathsf{F}_z^{-1/3} - 1.5 \tag{16.106}$$

and its end x_{ep} should satisfy the condition

$$\frac{x_{ep} - x_{P'}}{(\overline{bh})^{1/2}} = \frac{1.15}{\sin\delta} Y^{1/2} \mathsf{F}_o \mathsf{F}_z^{-1/3} - 2.3. \tag{16.107}$$

Both relations are similar and contain the determining Froude number $\mathsf{F}_o/\mathsf{F}_z^{1/3}$. The width of the cover plate is $b_p = h_o + h_z$ and its height h_p above the invert should be

$$\frac{h_p}{h_z} = 2 + 0.67 \sin\delta \mathsf{F}_z (\mathsf{F}_o Y)^{1/3}. \tag{16.108}$$

The Froude number F_z of the lateral branch has a significant effect on h_p. Extreme wall waves occur for $2 < (\mathsf{F}_o/\mathsf{F}_z^{1/3})Y^{1/3} < 7$, and a cover plate yields *reductions* up to 40%. Equation (16.108) also demonstrates the practical limits of the recommended element. For large lateral flow depths, the position of the cover plate is always large and the freeboard required is significant. Figure 16.50 refers to identical flows without, and with a cover plate and illustrates a considerable improvement of flow.

Example 16.8 Design a cover plate for Example 16.7.
With $\mathsf{F}_o = 2.98$, $\mathsf{F}_z = 4.23$, $Y = 1.78$, $h_o = 0.80$ m, $h_z = 0.45$ m, $(\overline{bh})^{1/2} = 0.86$ m, Eqs. (16.106) and (16.107) give for start and end of the cover plate $x_{ap} - x_{P'} = [(0.27/\sin 35°)1.78^{0.75} 2.98 \cdot 4.23^{-1/3} - 1.5]1.5 = -0.24$ m and $x_{ep} - x_{P'} = [(1.15/\sin 35°)1.78^{1/2} 2.98 \cdot 4.23^{-1/3} - 2.3]0.86 = 2.26$ m. Further, $b_p = 0.8 + 0.45 = 1.25$ m, $h_p = [2 + 0.67 \sin 35° 4.23 (2.98 \cdot 1.78)^{1/3}]0.45 = 2.2$ m

are width and height, such that the cover plate length is $L_p = x_{ep} - x_{ap} = 2.26 + 0.24 = 2.5$ m, i.e. practically equal to the downstream width.

Due to the large filling ratio, the cover plate height is slightly below $h_{MB} = 2.44$ m. Because $\mathsf{F}_o(Y/\mathsf{F}_z)^{1/3} = 2.25$ the wave height cannot be much influenced. With $\mathsf{F}_{ot} = 5(\sin 35°)^{0.5}(4.23-2)^{0.33}1.78^{-0.33} = 4.08$ according to Eq. (16.102), wave B is of transitional type.

When applying these results to junction manholes, the Froude number of the U–shaped profile must be determined. Often, the required length of the cover plate is larger than the manhole length, downstream from point P′. The downstream sewer is then considered as a part of the cover plate (Fig. 16.51). According to Eq. (16.108) the required downstream diameter can be quite large, particularly for larger junction angles δ and large Froude number F_z. Then, the only design possibilities are

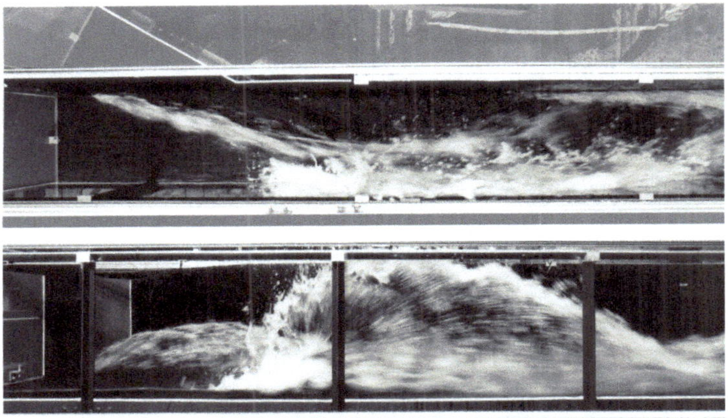

(a)

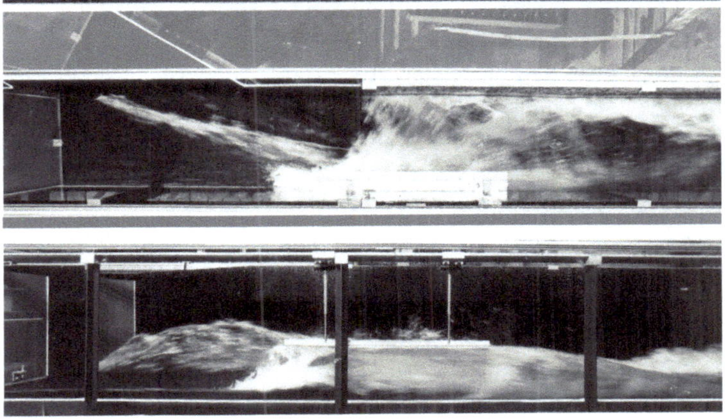

(b)

Fig. 16.50 Junction flow with $h_o = h_z = 40$ mm, $\mathsf{F}_o = \mathsf{F}_z = 8$ and $\delta = 30°$ (**a**) *without* and (**b**) *with* cover plate in plan (*top*) and section (*bottom*)

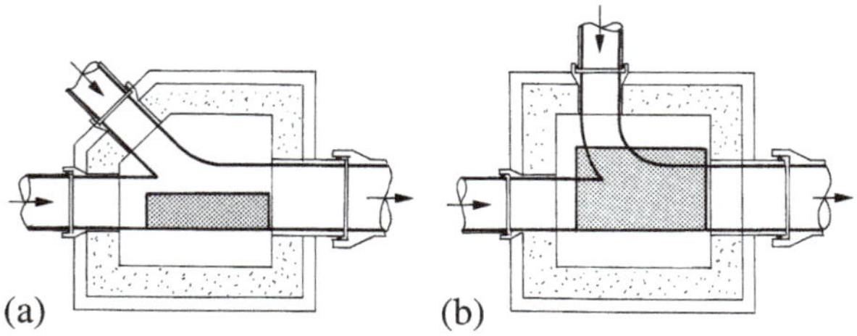

Fig. 16.51 Cover plate in junction manhole for (**a**) 45° junction and straight lateral channel walls, (**b**) 90° junction with curved lateral inlet (not recommended)

either a reduction of the shock number, or an increase of the downstream diameter. Until today, no *experiences* on cover plates are available. If supercritical flow has to be maintained across a manhole, bottom drops as mentioned in Sect. 16.2.4 should not be applied. Fig. 16.51a refers to a 45° junction manhole with a direct transition to the downstream sewer, whereas Fig. 16.51b presents an unrealistic design with a 90° branch channel, for which the junction flow almost certainly will choke. A serious hydraulic design of junction manholes is essential to inhibit submergence and pulsations, which at the limit may generate manhole overflow (Chaps. 5 and 7).

16.4 Bend Manhole

16.4.1 *Introduction*

Each change of sewer direction both in plan and in the section requires a manhole. In the following, the bend manhole (German: Krümmerschacht; French: Puits à coude) is described for plan direction changes. Changes of bottom slope are often so small that no additional losses have to be considered for subcritical flow. Transitional flows in these structures are described in Chaps. 6 and 7.

The following discussion relates to sub- and supercritical flows separately, as for junction manholes. In Sect. 16.4.4 the cover plate as recommended for junction manholes is introduced. For *subcritical flow*, the results relating to pipe flow apply, in analogy to other structures. For *supercritical flow* in bend manholes, a systematic experimental program was conducted to investigate the shockwave effects, and their consequences on air entrainment and choking of the downstream sewer.

16.4.2 *Subcritical Flow*

As for junction manholes, only some information on bend manholes is currently available. For subcritical flow, the knowledge on the *head loss coefficient* in terms of manhole geometry is essential to determine the backwater effect in the upstream sewer.

Table 16.2 Head loss coefficients ξ_k in bend manhole for geometries of Fig. 16.52 (Dick and Marsalek 1985)

Manhole Type	(a)			(b)			(c)	(d)
Angle of deviation	22.5°	45°	90°	30°	60°	90°	90°	90°
ξ_k	0.3	0.6	1.85	0.45	0.9	1.6	1.1	0.55

Dick and Marsalek (1985) investigated the manhole configuration with 100% benches (Chap. 14). In the average, this design has a loss coefficient ξ_k = 0.05–0.10, whereas bend manholes with 50% benches have a ξ_k value of about 0.10–0.20. The poor configuration without benches at all involves ξ_k values of about 0.20–0.30.

The bend (subscript k) manhole has four major designs, namely with 0, 50% and 100% *benches*, and a configuration with a locally expanding U-shaped profile. Table 16.2 summarizes some experimental observations relative to the head loss coefficient $\xi_k = \Delta H/[V_o^2/2g]$ for pressurized manhole flow. The effect of manhole filling was not as significant as for the simple through-flow manhole (Chap. 14). One may note a considerable increase of ξ_k with the deviation angle δ_k, and a reduction of ξ_k as the bench height increases. Compared to the standard design case (c) of Fig. 16.52, case (a) has almost the double loss, and case (b) a loss of 150%, whereas case (d) involves a loss reduction to some 50%. The latter design is costly and not recommended, however.

Marsalek and Greck (1988) considered bend manholes of quadratic plan with identical diameters of the up- and downstream sewers. For larger relative submergence $S_s = s_s/D_o > 1.8$, no *submergence effect* was noted such that the head loss varies only with the manhole geometry. For the configuration corresponding to Fig. 16.52c, the head loss coefficient is also $\xi_k = 1.1$.

Contrary to the previous observations, Johnston and Volker (1990) related their results to the upstream pipe Froude number $\mathsf{F}_o = V_o/(gD_o)^{1/2}$. In the discussion of the results, their data were represented for $\delta = 90°$ and $D_s/D_o = 4$ as

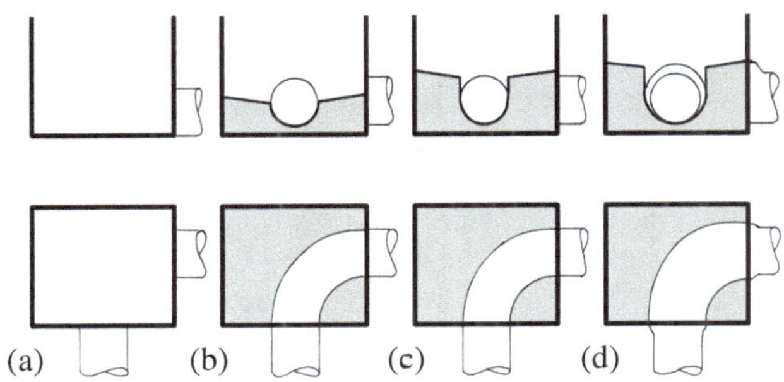

Fig. 16.52 90° bend manholes, geometry in transverse section (*top*) and plan (*bottom*)

$$\xi_k = C_k \mathsf{F}_o^{-1}. \tag{16.109}$$

The coefficient of proportionality C_k varies mainly with the submergence ratio S_s and the manhole geometry. For 100% benches, the results are $C_k = 0.3$ for $S_s = s_s/D_o \cong 1.4$, and $C_k = 0.22$ for $S_s \cong 5$. Therefore, ξ_k decreases with increasing submergence, in analogy to the standard manhole (Chap. 14).

Example 16.9 Given a bend manhole of deflection angle $\delta_k = 45°$. What is the approach flow depth for a downstream sewer with $D_u = 0.70$ m, $h_u = 0.52$ m and $Q = 0.3$ m^{3}s^{-1} for 100% benches?
With a downstream filling ratio $y_u = 0.52/0.70 = 0.74$, the cross-sectional area is $F_u/D_u^2 = y_u^{1.4} = 0.66$ from Eq. (5.16)$_2$, thus $F_u = 0.66 \cdot 0.7^2 = 0.32$ m^2 and $V_u = 0.3/0.32 = 0.93$ ms^{-1}. According to Table 16.2, case b) has nearly a loss coefficient of $\xi_k = 0.7$, and for case c) one may admit $\xi_k = 0.7(1.1/1.6) = 0.48$ for $\delta_k = 90°$, i.e. $\Delta H_k = 0.48 \cdot 0.93^2/19.62 = 0.021$ m. Because $\mathsf{F}_u = Q/(gD_u h_u^4)^{1/2} = 0.3/(9.81 \cdot 0.70 \cdot 0.52^4)^{1/2} = 0.42 < 0.5$, one may set $h_o = h_u + \Delta H_k = 0.54$ m, because the bottom slope compensates approximately the friction slope.

Partially-filled flow in bend manholes with $\delta_k = 45°$ and $90°$ was investigated in terms of head losses and transverse free surface slope. Because both the upstream and downstream sewers were circular pipes with an U-shaped manhole of 100% benches, the head loss ΔH_b refers only to the manhole length, excluding the head-loss in the downstream sewer due to flow realignement. The *headloss coefficient* $\xi_k = \Delta H_b / \left[V_d^2/(2g)\right]$ related to downstream (subscript *d*) velocity is

$$\xi_k = C_k y_d^{-1},\ y_d < 0.90 \tag{16.110}$$

with $y_d = h_d/D$ as the downstream sewer filling, $C_k = 0.07$ for $\delta_k = 45°$, and $C_k = 0.085$ for $\delta_k = 90°$. Accordingly, the headloss coefficient decreases as the filling ratio increases. This effect may be explained with the deviation of the circular cross-sectional shape as y_d decreases. The range of Froude numbers tested was $0.30 < \mathsf{F}_o < 0.70$.

The *maximum superelevation* Δh_M between the outer and inner walls of a bend manhole depends on the ratio of centrifugal to gravitational accelerations. With R_a as the axial radius of curvature, D the width of the U-shaped profile and V_o as the approach flow velocity, the potential vortex flow theory gives $\Delta h_M = DV_o^2/(gR_a)$. By plotting $\Phi = (\Delta h_M/D)[gR_a/V_o^2]$ as a bend superelevation coefficient against the approach flow filling ratio y_o yields $\Phi = 0.80$ for both $\delta_k = 45°$ and 90° up to $y_o = 0.60$, and then an increase of Φ to 1.20 for 45°, and a decrease of $\Phi =$ to 0.40 for 90°, for $y_o = 1$.

Velocity profiles for two typical flows in 90°-manhole bends are shown in Fig. 16.53. One notes an essentially uniform velocity distribution both in the streamwise and the radial directions, except for the boundary layers. A detailed data analysis indicates no real velocity increase towards the curvature center, as would follow from the potential vortex theory. One may also note from the free surface plot that the maximum flow depth is located at ≅ 60°, whereas the minimum is at

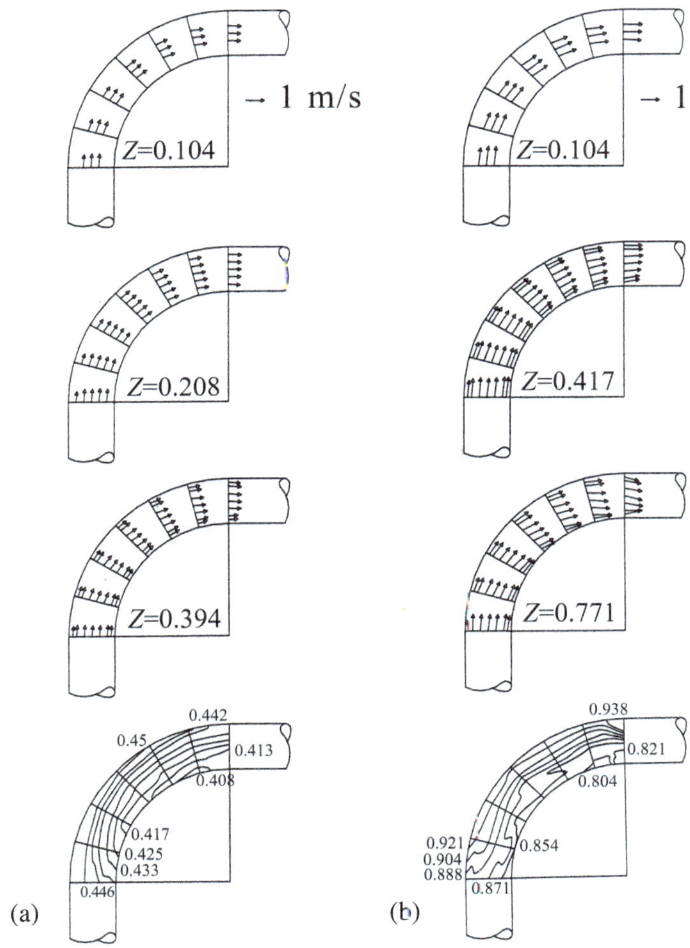

Fig. 16.53 Velocity field for various layers $Z = z/D$ in bend manhole for subcritical flow (**a**) $y_o = 0.45$, $\mathsf{F}_o = 0.41$ and (**b**) $y_o = 0.86$, $\mathsf{F}_o = 0.56$. Flow surface at bottom

75° downstream from the approach flow section. The two flows behave essentially similar, yet with a more complex free surface topography due to impact on the downstream manhole wall for $y_o = 0.86$.

Figure 16.54 shows various typical bend flows as occur in *sewage treatment stations*. In Fig. 16.54a the Froude number is estimated to 0.40 with a typical standing surface wave pattern. The flow in a 90° bend may dissipate a considerable amount of energy, with a transition from supercritical to subcritical flow (Fig. 16.54b). Pump cost can thus be notable. Such flows have not yet been considered with laboratory modelling. The flow in Fig. 16.54c has an even larger approach flow Froude number due to a small drop structure, resulting in an oblique direct *hydraulic jump*.

Bend manholes are typical structures of a sewer system. Figure 16.55a refers to a 60° bend manhole. Figure 16.55b shows a bend manhole during low discharge

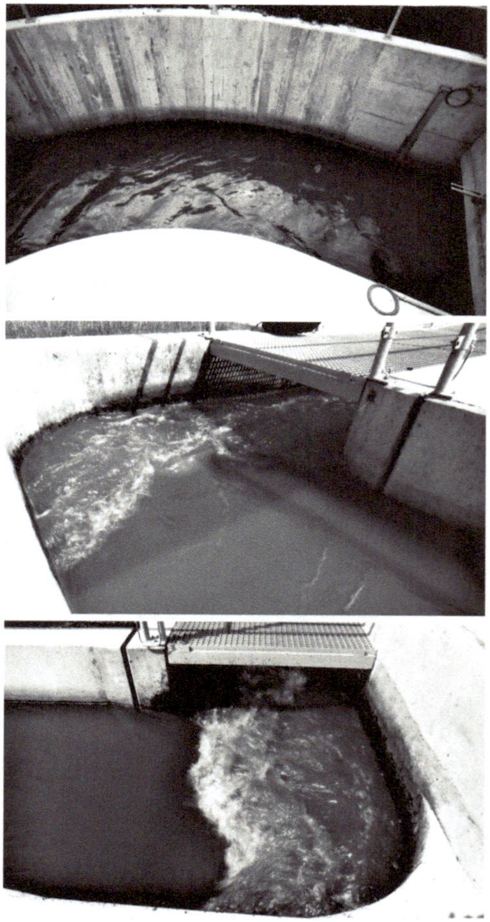

(a) (b) (c)

Fig. 16.54 Bend flow in sewage treatment stations

Fig. 16.55 Bend manholes looking (**a**) through the manhole shaft, (**b**) from benches

(a)

(b)

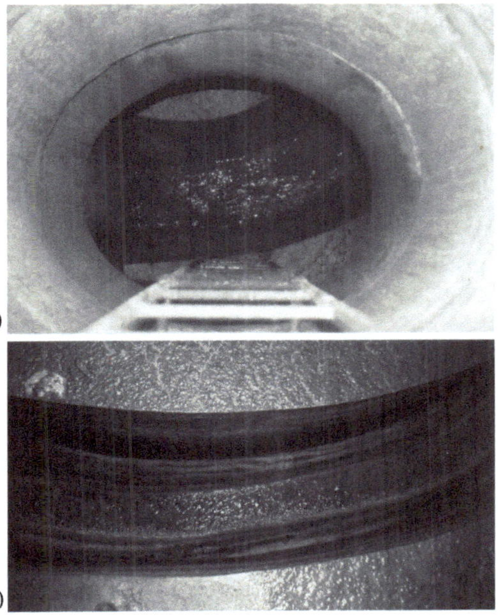

condition, with a considerable sludge layer on the benches. These conditions may produce a slippery walkway with a potential danger for accidents.

16.4.3 *Supercritical Flow*

Supercritical flow in bend manholes is similar to the flows discussed in Sect. 16.3.5. The difference is mainly due to the downstream sewer limiting the flow in its vertical extension. At the downstream manhole end, the flow may impinge onto the manhole wall and hydraulic *impact jumps* may develop. In contrast, the flow in an open channel bend has no such limitations. The following is a summary of current knowledge on bend manholes, involving also the bend cover to increase the discharge capacity and to improve the flow pattern of the downstream sewer.

Christodoulou (1991) analyzed supercritical flow in bend manholes. Figure 16.56 shows the geometry considered, with angles $\delta_k = 0$ and 90°. The head loss ΔH_s was related to the approach flow velocity V_o. The variables are the manhole diameter ratio $\delta_s = D_s/D$, the relative drop height $Z_s = \Delta z_s/D$, the bend deflection angle δ_k, the approach flow bottom slope S_o and the drop number $\mathsf{D} = V_o/(g\Delta z_s)^{1/2}$. From observations no effect of S_o on the loss coefficient $\xi_k = \Delta H_s/\ [V_o^2/(2g)]$ was deduced. For $\xi_k = 90°$ the data follow

$$\xi_k = 0.2 + 2.3\mathsf{D}^{-2}. \tag{16.111}$$

Fig. 16.56 Supercritical flow in bend manhole (**a**) plan, (**b**) section

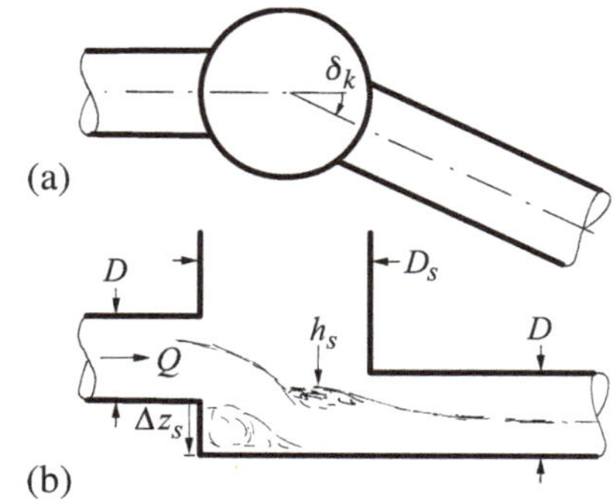

The first term 0.2 can be identified as the loss due to the manhole, and the second is the loss due to the impact (subscript I) on the manhole, with $\Delta H_I = 1.15\Delta z_s$.

With h_s as the flow depth in the manhole downstream from the drop (Fig. 16.56), and $y_s = h_s/\Delta z_s$, the data for $\delta_k = 90°$ may be expressed as

$$y_s = 0.85\left(S_o^{-1} - S_{om}\right)^{0.3} \mathsf{D}^{1.5} \tag{16.112}$$

with $S_o < 10\%$ as the bottom slope in [%] and $S_{om} = 0.15\%$ as the minimum value. For $\delta_k = 0°$, the constant 0.85 in Eq. (16.112) has to be replaced with the constant 1. To inhibit choking of the downstream sewer, the condition $h_s/D < 1$ must be satisfied, and *submergence* into the approach flow sewer is suppressed for $y_s < 1$.

Example 16.10 Given an approach flow sewer with $S_o = 6\%$, $D = 1.0$ m, $1/n = 90$ m$^{1/3}$s^{-1} and $Q = 1.2$ m^{3}s^{-1} discharging into a bend manhole with $\delta_k = 45°$. What is the height of the drop?

For uniform approach flow the dimensionless discharge is $q_N = 0.011 \cdot 1.2/0.06^{1/2} = 0.054$, thus according to Eq. (5.15) $y_N = 0.926[1-(1-3.11 \cdot 0.054)^{1/2}]^{1/2} = 0.276$ and $h_N = 0.276 \cdot 1.0 = 0.28$ m. Further, with $F_o/D_o^2 = y_N^{1.4} = 0.276^{1.4} = 0.165$ from Eq. $(5.18)_2$, the cross-sectional area is $F_o = 0.165 \cdot 1 = 0.165$ m^2 such that $V_o = Q_o/F_o = 1.2/0.165 = 7.3$ ms^{-1}.

Starting from Eq. (16.112), the resulting flow depth in the manhole is $y_s = 0.9(6^{-1} - 0.15)^{0.3}[V_o/(g\Delta z_s)^{1/2}]^{1.5} = 0.94/\Delta z_s^{0.75}$. If submergence has to be inhibited, i.e. $y_s < 1$, then $\Delta z_s = (0.94)^{-4/3} = 1.09$ m, corresponding to $\mathsf{D} = 7.3/(9.81 \cdot 1.09)^{1/2} = 2.23$. If the manhole should flow free, then $h_s/D = 0.9(6^{-1} - 0.15)^{0.3}[V_o/(g\Delta z_s)^{1/2}]^{1.5}(\Delta z_s/D) \leq 1$, with the solution $\Delta z_s = 1.07^{1/4} = 1.02$ m. The drop height should thus be about 1 m with the manhole so long that the jet does not impact the downstream manhole wall (Chap. 11).

Del Giudice et al. (2000) considered bend manholes with a supercritical approach flow for deflection angles 45° and 90°. The effect of bottom slope can be neglected provided $S_o < 2\%$. The bend manholes tested had a relative curvature $\rho_a = D/R_a = 1/3$, with R_a as the radius of the bend axis. The benches were 150% of the diameter because of space limitations with the lateral manhole extension.

Untreated Manhole Flow

Figure 16.57 relates to the 90° bend manhole with an approach flow Froude number $\mathsf{F}_o = 2.18$ and an approach flow filling $y_o = 0.35$. A view from the approach flow sewer shows a significant but smooth increase of the outer wall profile and a

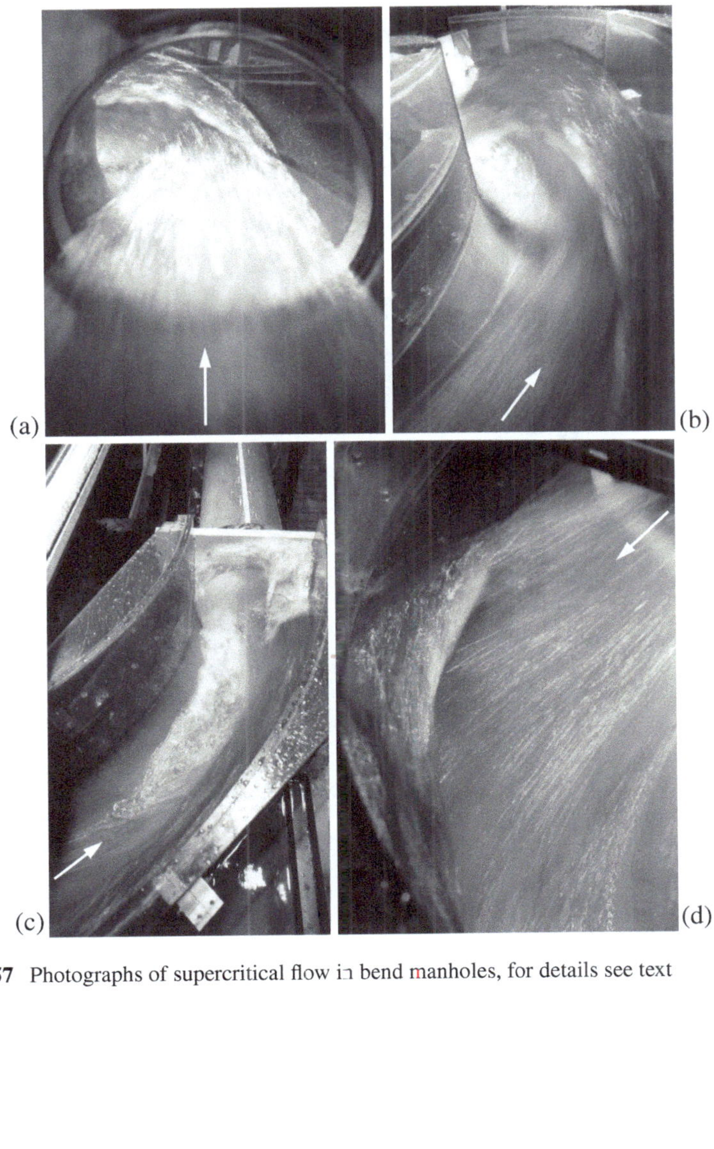

Fig. 16.57 Photographs of supercritical flow in bend manholes, for details see text

decrease of the inner wall profile (Fig. 16.57a). The complete free surface along the manhole is shown in Fig. 16.57b, with the shockwave clearly separated and nearly vertical along the curved channel. At the end of the manhole, the flow impinges on the manhole endwall, with a swell of maximum height z_s at the outer wall (Fig. 16.57c). A swell portion falls back on the supercritical flow which may choke if the *recirculation zone* is too long. The capacity of the manhole is thus governed by the features of the swell. For large discharges, the recirculation can even have an influence on the inner wall profile, as is seen in the photograph. The curved shock front as seen towards the manhole upstream end is also shown in Fig. 16.57d. Again, the free surface gradients are seen to be immense as compared to usual hydraulic configurations.

Figure 16.58a and b relate to a complete overview, including the approach pipe flow, the outer wall wave and impingement onto the manhole end wall. Figure 16.58c shows the flow in the downstream sewer with the swell on the left side, contraction of flow due to the pipe inlet and formation of wave 1 on the inner side of the downstream sewer. Wave 2 on the outer side can be seen in its beginning. In all cases tested, with always a free pipe outflow, these waves were never a problem, and choking never occurred in the downstream pipe due to wave formation.

The *velocity field* pertaining to the flows presented in Figs. 16.57 and 16.58 is shown in Fig. 16.59 for various levels $Z = z/D$ above the invert, with z as the

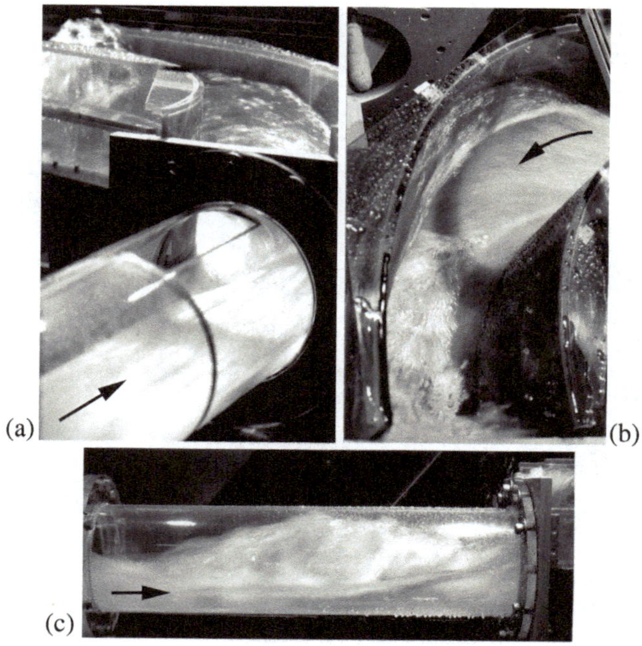

Fig. 16.58 Manhole flow for supercritical approach flow, for details see text

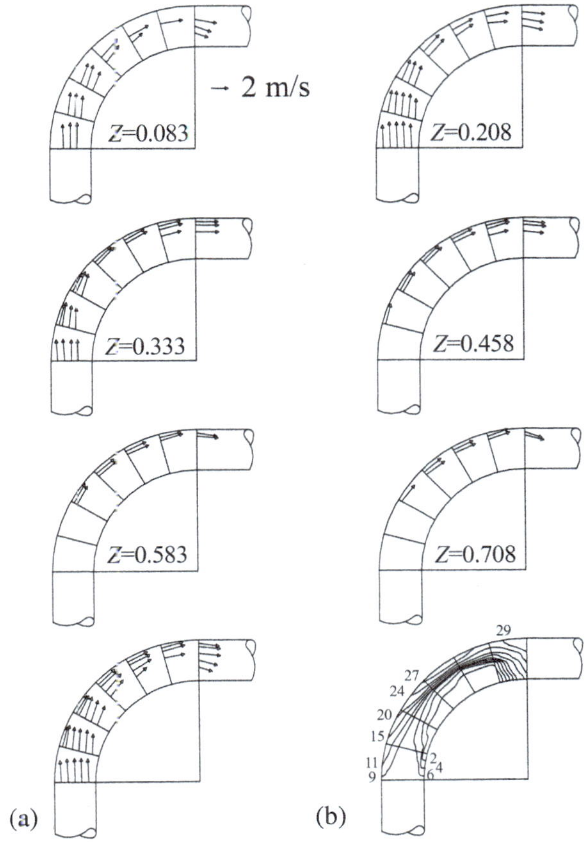

Fig. 16.59 Velocity field pertaining to experimental run shown in Figs. 16.57 and 16.58. Bottom: average velocities (*left*) and free surface (*right*)

vertical coordinate. Comments as for subcritical bend flow apply: (1) In a certain height above the invert, the variation of velocity is small, except close to the walls and in the recirculation region; (2) The transverse velocity has a slight tendency to increase towards the center of curvature: (3) Except for the bottom layer, the velocity decreases with height; (4) The effect of shock wave may not really be detected in the velocity field but it clearly follows from the free surface plot. Accordingly, supercritical flows across bend manholes can be approximated with an almost *constant velocity* equal to the average approach flow velocity, and the particular hydraulic concern is the *free surface* because of extreme wave formation.

The *extreme flow depths* have a maximum (subscript M) flow depth h_M and location θ_M at the outer wall, and a minimum (subscript m) flow depth h_m at location θ_m, with θ as the angle measured along the manhole from the intake section (Fig. 16.60). Based on Sect. 16.3.5, the data for the extreme wave elevations $Z_M = Y_M^{1/2} - 1$

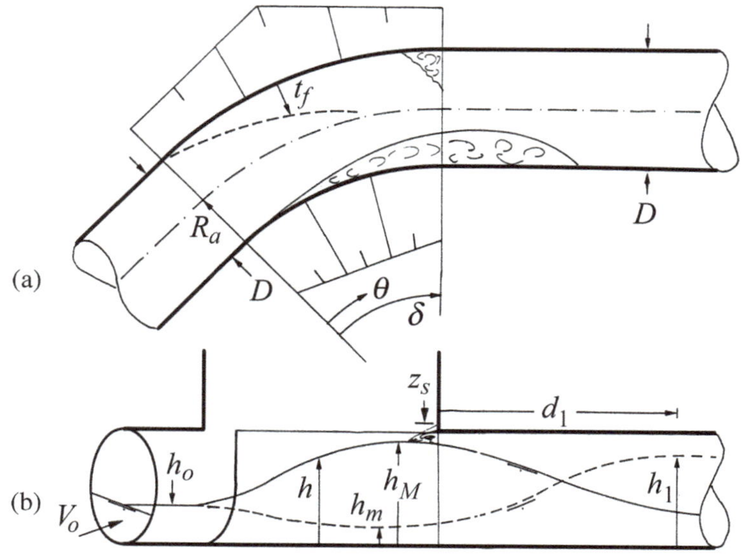

Fig. 16.60 Definition plot for supercritical flow across bend manhole (**a**) plan, (**b**) section

and $Z_m = Y_m^{1/2} - 1$ may be expressed with the approach flow bend number $\mathsf{B}_o = \rho_\alpha^{1/2}\mathsf{F}_o$ as

$$Z_M = Z_m = 0.50\mathsf{B}_o^2. \tag{16.113}$$

The difference in Z_M between the rectangular and the U-shaped channels is obviously a shape effect. Note that $\mathsf{F}_o = V_o/(gh_o)^{1/2}$, because the upper portion of the U-shaped channel is rectangular.

The locations of extreme wave heights vary with $\rho_a\mathsf{F}_o$, as for the rectangular channel bend. The results for $\mathsf{B}_o < 1.50$ are

$$\tan\theta_M = 2.8(\rho_a\mathsf{F}_o)^2, \tag{16.114}$$

$$\tan\theta_m = \sqrt{2}(\rho_a\mathsf{F}_o). \tag{16.115}$$

The *wave profile* along the outer bend wall can also be described with Eq. (16.82). The height of the swell $Z_s = z_s/D$ is equal to

$$Z_s = \sigma_s\mathsf{B}_o^2, \quad \mathsf{B}_o < 1.5 \tag{16.116}$$

with $\sigma_s = 0.80$ for $\delta_k = 45°$ and $\sigma_s = 0.50$ for $\delta_k = 90°$. The swell may be higher or smaller than the maximum wave height, depending on its location, therefore.

Waves 1 *and* 2 in the downstream sewer can be described by their distances d_1 and d_2 from the manhole end, and their respective heights h_1 and h_2. The details are given by Del Giudice et al. (2000). As already mentioned, these waves have never had an effect on the flow in the bend manhole, and are not of design concern, therefore.

The *discharge capacity* (subscript *C*) of a bend manhole is reached when Z_s has the order of 1.5. Then, the impact of the supercritical flow onto the manhole end wall is so strong that the recirculating flow may choke, and a hydraulic jump forms in the manhole. For $\delta_k = 45°$, the absolute maximum (subscript *L*) approach flow filling was $y_{oL} = 0.70$, whereas it amounted to only $y_{oL} = 0.55$ for $\delta_k = 90°$. For $y_o > y_{oL}$, the manhole flow chokes, with a hydraulic jump moving into the upstream sewer. Sewers with an approach flow filling ratio larger than either 55%, or 70%, respectively, choke and supercritical flow may not be maintained. This finding is important in terms of manhole choking, breakdown of supercritical flow, upsurging of manhole water level up to blasting of manhole covers. The conventional sewer design involving the full flow concept cannot be maintained for supercritical flow, therefore. A novel design of such configurations is presented below.

For $0.20 < y_o < y_{oL}$ the capacity Froude number F_{oC} can be expressed as

$$\mathsf{F}_{oC} = 3\sin\delta_k(1 - y_o) + y_o. \tag{16.117}$$

This transitional condition corresponds to a lower bound of the approach flow Froude number, below which a supercritical manhole flow cannot be maintained. The details of this analysis are also presented by Del Giudice et al. (2000), and it can be stated here that supercritical flow breaks down for $\mathsf{F}_o < 1.50$. Then, an *undular hydraulic jump* may establish first in the downstream pipe, and eventually move upstream when decreasing F_o toward 1. As a general consequence, approach flows with $0.75 < \mathsf{F}_o < 1.50$ should be inhibited because of the formation of a transitional flow character. Further results relating to the energy loss across a bend manhole for supercritical flow, and the shockfront development are not discussed here (Del Giudice et al. 2000).

Example 16.11 Given a bend manhole with $\delta_k = 90°$ and $D = 0.60$ m, approach flow bottom slope $S_o = 3\%$ and roughness coefficient $1/n = 85$ $\text{m}^{1/3}\text{s}^{-1}$. Determine the bend flow characteristics for a discharge of $Q = 0.90$ m^3s^{-1}.

Assuming uniform approach flow gives with Eq. (5.15) $h_o/D = 0.65$, thus $h_o = 0.65 \cdot 0.60 = 0.39$ m, $V_o = 4.65$ ms^{-1}, $\mathsf{F}_o = 4.65/(9.81 \cdot 0.39)^{1/2} = 2.38$ and $\mathsf{B}_o = 2.38(1/3)^{1/2} = 1.37$. Therefore, $h_M/h_o = (1{+}0.50 \cdot 1.37^2)^2 = 3.77$, $\tan\theta_M = 2.8(2.38/3)^2 = 1.76$ and $Z_s = 0.5 \cdot 1.37^2 = 0.94$ from Eqs. (16.113), (16.114) and (16.116). The characteristics of the bend flow are, therefore, $h_M = 3.77 \cdot 0.39 = 1.47$ m, $\theta_M = 60°$ and $z_s = 0.94 \cdot 0.60 = 0.56$ m.

16.4.4 *Shockwave Reduction*

Abrupt Wall Deflection

As discussed in Sect. 16.3.7, wall waves may effectively be reduced with the *cover plate*. Schwalt (1993) considered the *mitre bend* for deflection angles $\delta = 30°$ and $60°$. Because bend manholes are short, his experimental indications are relevant. For $20° < \delta < 60°$, the beginning x_{ad} of the cover plate is at location x_P, i.e. at the extension of the lateral wall to the opposite side (Fig. 16.48b). For a junction where the through branch has no discharge, the position of beginning is $x_{ad} = x_P$, and the cover plate starts at the junction point P.

The end of cover plate is, independent of δ,

$$\frac{x_{ed} - x_{P'}}{b_z} = 0.6\mathsf{F}_z^{1/2}. \tag{16.118}$$

The *height* h_d of the cover plate varies with the deflection angle δ, the approach flow Froude number F_z and the width ratio b_z/b_d as

$$h_d/h_z = 2 + 0.23 \sin\delta \mathsf{F}_z(b_z/b_d)^{1/2}. \tag{16.119}$$

The *width* of the cover plate should be equal to $b_d = 1.5h_z$ for $\delta = 30°$ and $b_d = 2h_z$ for $\delta = 60°$. Figure 16.61 shows the flow deflection for $\delta = 30°$ with and without a cover plate. The effect of the plate is significant, and the downstream flow is much improved, mainly as regards uniformity and shockwave damping. To further improve the flow, Schwalt suggested a second cover plate on the opposite side.

Example 16.12 Given an abrupt channel deflection with $\delta = 30°$, $\mathsf{F}_z = 6$, $h_z = 0.4$ m and $b_z = b_u = 0.9$ m. What is the geometry of the cover plate required?
With $x = 0$ at point P′, the plate starts at $x_{ap} = x_{P'} = 0$ and its width is $b_p = 1.5h_z = 0.6$ m. With Eq. (16.118) the end of the cover plate is at $x_{ep}-x_{P'} = 0.6 \cdot 6^{1/2}0.9 = 1.32$ m. The cover plate height is from Eq. (16.119) $h_p = [2+0.23\sin30°6(0.9/0.6)^{1/2}]0.4 = 1.15$ m. The plate dimensions are thus $L_p = 1.32$ m, $b_p = 0.6$ m, and $h_p = 1.15$ m. The downstream channel should be at least $1.15/0.9 = 1.25$ m high, therefore. Because the wave height in the channel without a cover plate would be $h_{MB}/h_z = 1+(\sin\delta/4)\mathsf{F}z^2 = 5.5$, i.e. $h_{MB} = 5.5 \cdot 0.4 = 2.2$ m, the reduction of wave height is with more than 40% significant.

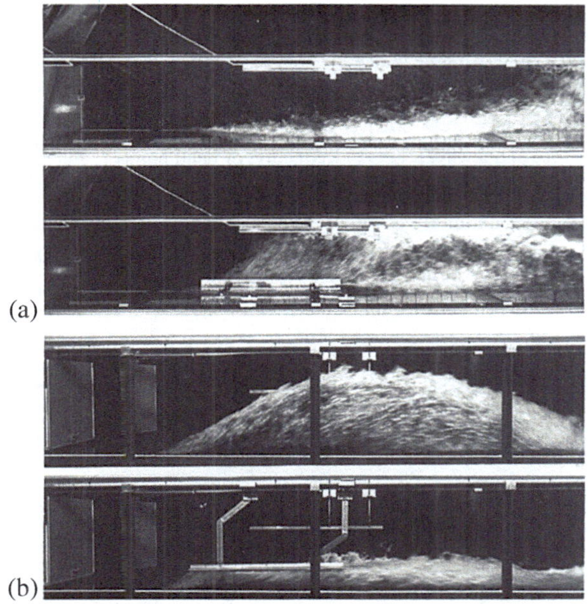

Fig. 16.61 Deflection flow (**a**) *without* and (**b**) *with* cover plate in plan (*top*) and section (*bottom*)

Bend Manhole

Shockwaves in a bend manhole with an U-shaped profile may get high, for a large bend number B_o and a typical design approach flow filling of $y_o \cong 0.50$. Also, the supercritical flow in a bend manhole may break down if $y_o > y_{oL}$, as described previously. Out of the two methods to reduce shockwaves, i.e. (1) Reduction of bend number, and (2) *Cover plate*, only the latter method can be often applied because of limitations in the upstream sewer. Figure 16.62 shows a definition sketch of the

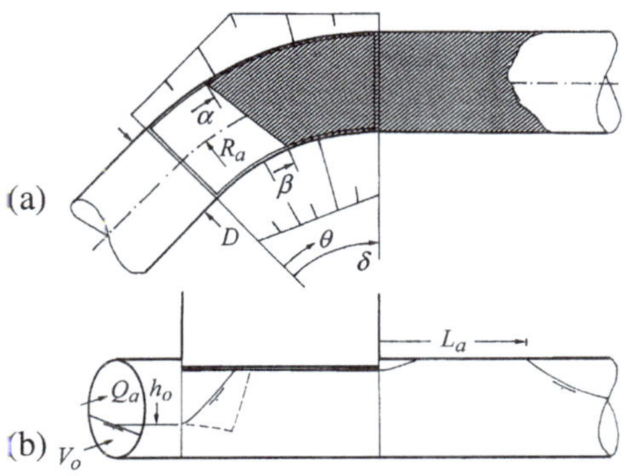

Fig. 16.62 Definition plot for cover plate in bend manhole (**a**) plan, (**b**) section

bend manhole with a cover (subscript c), including the angles α_c and β_c where the flow attaches the cover, the air discharge Q_a and the downstream length L_a measured from the manhole end where free surface flow is reestablished.

The *optimum height* of the cover plate above the manhole invert was found to be 0.90 D, providing both a large flow section below the cover, and a sufficiently large space above it for aeration of the downstream sewer. The cover *impinging angles* vary essentially with $y_o = h_o/D$ and $\mathsf{F}_o = V_o/(gh_o)^{1/2}$ as (Del Giudice et al. 2000)

$$\alpha_c y_o = 10(\mathsf{F}_o - 1)^{-1/3}, \tag{16.120}$$

$$\beta_c y_o = 8.5[1 + 2\exp(-(\mathsf{F}_o - 1)^2)]. \tag{16.121}$$

If α_c and β_c are smaller than the manhole deflection angle δ_k, the latter has no effect on flow reattachment. To simplify the design, the complete manhole can be covered. The cover plate should be secured against *uplift pressure*. A conservative design involves the maximum pressure over the cover cross-sectional area $A_c = (\delta_k - \alpha_c)R_aD$. Using the expression for the maximum wave height h_M from Eq. (16.113) gives $P_c = \rho g(h_M - h_o)A_c$ for the maximum pressure force. A cover should be easily removable from the U-shaped profile for maintenance and sewer inspection. The cover may not be suited for shockwave control in all cases of practice.

More practical information has to be collected for the optimum arrangement of the cover plate, and its potential for clogging a sewer. Figure 16.63 shows photos from the lab model for $\delta_k = 45°$ and $y_o = 0.49$, $\mathsf{F}_o = 3.40$. It can be seen that the transition from free surface to covered flow occurs within a short reach, and that the flow in the tailwater pipe may become annular.

The *air supply* Q_a required for a flow without underpressure in the downstream sewer was also determined. The air supply depends mainly on the approach flow Froude number F_o, the water discharge Q and the approach flow filling ratio y_o because of the limited air cross-sectional area above the air-water mixture flow. From experiments, the relative air discharge $B = (Q_a/Q)y_o^{4/3}$ is close to zero for $\mathsf{F}_o < 2$, increases steeply with F_o and remains almost constant at $B = 0.24$ for $\mathsf{F}_o > 4$. The

(a)
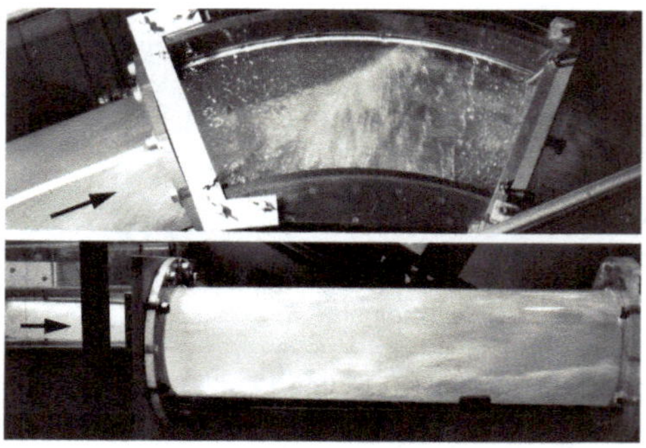
(b)

Fig. 16.63 Flow below manhole cover. (**a**) Plan view with flow from left, (**b**) side view of tailwater pipe including in flow direction cover (left) and annular pipe flow (right) experimental data follow the relation (Del Giudice et al. 2000)

$$B = 0.24[\tanh(\mathsf{F}_o - 2)^2]^2. \tag{16.122}$$

Depending on y_o and F_o, a bend cover may induce either free surface or full-pipe air-water flow in the downstream sewer. Because of limited discharge, a fully choked flow under the cover with a hydraulic jump in the upstream sewer was never attained for both $\delta_K = 45°$ and $90°$. The *maximum approach flow Froude number* in the experiments conducted was $\mathsf{F}_{oM} = 1 + 1.3y_o^{-1}$, and this value may give a guidance of the upper limit conditions for the bend manhole with the cover plate. For $y_o = 0.50$, say, the maximum approach flow Froude number would thus be $\mathsf{F}_{oM} = 3.60$.

The increase of *discharge capacity* due to the presence of the bend cover is typically 60% for $y_o > 0.30$ and slightly less for smaller values of y_o. Accordingly, the bend cover can be regarded as an effective device to increase discharge, to improve the flow in the downstream sewer, to inhibit flow choking and the transition to pressurized sewer flow. Because pressure on the cover is always positive, cavitation is of no concern.

For both large values of y_o and F_o, the downstream sewer may run full with an *air-water mixture flow*. The length L_a of the mixture flow measured from the manhole end is (Del Giudice et al. 2000)

$$L_a/D = 2y_o(\mathsf{F}_o - 1)^2. \tag{16.123}$$

For all flows tested up to $\mathsf{F}_o = 5$ such a transition to full-pipe mixture flow exhibited no problems, provided that sufficient air supply discharge Q_a is available. Accordingly, the flow downstream of the bend manhole was free of vibrations, pulsating mixture flow, and large air pockets.

Clearly, the conditions tested in the experimental setup involved exclusively free downstream conditions, i.e. the downstream sewer should never be submerged. A *submerged bend manhole* could perform adversely because of the breakdown of the air transport into the downstream sewer. Conditions for downstream submergence have not yet been investigated.

The cover plate was suggested for practical application by ATV (1996).

Example 16.13 What are the characteristics of a covered bend manhole assuming the conditions of Example 16.11?
The cover angles are from Eq. (16.120) $\alpha_c = 10/[0.65(2.38-1)^{1/3}] = 14°$ and from Eq. (16.121) $\beta_c = (8.5/0.65)[1+2\exp(-(2.38-1)^2)] = 17°$. Using $h_M = 1.37$ m as the maximum pressure head, the pressure force on the cover is $P_c = \rho g(h_M - h_o)(\delta_k - \alpha_c)(\pi/180°)R_aD = 1.40$ t. The air discharge is

$Q_a = 0.24[\tanh(\mathsf{F}_o - 2)^2]^2 Q/y_o^{4.3} = 0.24[\tanh 0.34^2]^2 0.90/0.65^{4/3} = 0.008$ m^3s^{-1} from Eq. (16.122), i.e. no additional air is required. The length of mixture flow is $L_a = 2h_o(\mathsf{F}_o - 1)^2 = 2 \cdot 0.39(2.38 - 1)^2 = 1.49$ m from Eq. (16.123). The limit Froude number for the bend manhole is $\mathsf{F}_{oM} = 1 + 1.3/0.65 = 3 > \mathsf{F}_o = 2.38$ and the design is not prone to breakdown of flow.

16.5 Definite Manhole Design

16.5.1 *Introduction*

Supercritical manhole flow is governed by either *shockwaves* generated at each flow discontinuity, or *hydraulic jumps*, if the discharge capacity is too small to convey a fully supercritical flow. Whereas shockwaves involve mainly a medium increase of the flow depth beyond a shock front, a hydraulic jump may result in the collapse of the supercritical flow regime and a backwater effect. The latter may be considered a serious problem for a sewer because of an abrupt change from free surface to pressurized two-phase flow. This choking phenomenon is accompanied further with water hammer, a decrease of the discharge capacity finally resulting in so called *geysering* of wastewater out from the manhole onto public space (Fig. 16.64). Sewer breakdown must be avoided in any case (ATV 1996, 2000).

The following intends to present the definite recommendation for the through-flow, the bend and the junction manholes, based on extensive hydraulic modeling at VAW, ETH Zurich. This research was conducted after it had been observed that the cover plate does not satisfy all requirements, mainly relating to sewer access and the potential of clogging, as already mentioned in Sect. 16.4. The purpose of this sub-chapter is to present a method that is simple in design and allows for a straightforward determination of all pertinent parameters.

Fig. 16.64 Geysering of manhole in a combined sewer (Hager and Gisonni 2005)

16.5.2 *Through-Flow Manhole*

A *through-flow manhole* is the simplest sewer manhole arrangement for control and maintenance purposes (Chap. 14). The manhole of U-shaped flow profile and length L is connected to equal up- and downstream sewers of diameter D. Figure 16.65 shows a definition sketch involving the approach flow depth h_o and velocity V_o. For $y_o = h_o/D \leq 0.50$, the flow remains entirely in a circular-shaped pipe, whereas the flow abruptly expands at the manhole inlet for $y_o > 0.50$, forming a side depression (Sect. 16.3.4) which is followed by a shockwave of height h_i shortly downstream because of flow impact onto the side walls. Whereas this phenomenon is relatively small, a more dramatic change occurs at the manhole outlet because of flow impact onto the upper portion of the circular profile, resulting in a shaft (subscript s) flow depth h_s. Depending on its height relative to D, the flow may either continue as a supercritical flow, or it breaks down due to the formation of an *impact hydraulic jump*. Choking then results at the manhole outlet because the jump formation and the breakdown of the air transport from the up- to the downstream sewer reaches (Fig. 16.66). If the discharge increases fast, the choking phenomenon may initiate even geysering, as previously described.

Given that the U-shaped profile corresponds essentially to a rectangular channel, the determining Froude number is $\mathsf{F}_U = Q/(gD^2h_o^3)^{1/2}$. The relative *shaft outflow depth* was experimentally found to (Gargano and Hager 2002)

$$h_s/h_o = 1 + (1/3)(\mathsf{F}_U y_o)^2. \tag{16.124}$$

Therefore, the relative wave amplitude $[(h_s - h_o)/h_o]$ increases quadratically with $\mathsf{F}_U y_o$, or the ratio $[(h_s - h_o)/D]$ depends exclusively on F_U.

The discharge capacity (subscript C) Q_C of this manhole is of design interest. According to Eq. (16.124) the approach flow filling y_o is relevant. The transition from free surface to pressurized manhole flow may be accounted for by the capacity Froude number $\mathsf{F}_C = Q_C/(gD^5)^{1/2}$. Gargano and Hager (2002) proposed for $0.70 < y_o < 0.75$

$$\mathsf{F}_C = 14.6 - 17.3y_o. \tag{16.125}$$

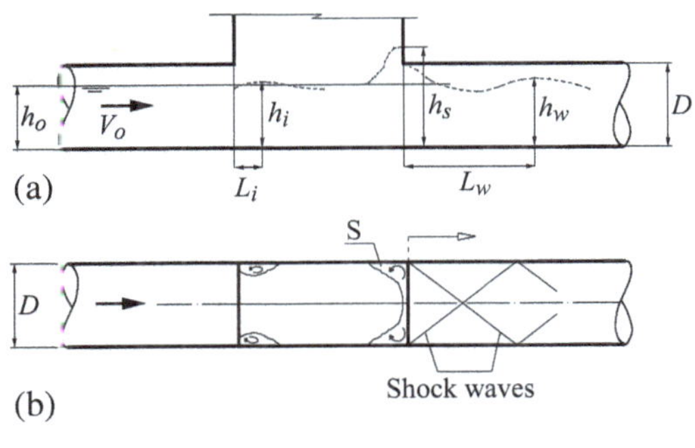

Fig. 16.65 Hydraulics and design of through-flow manhole (**a**) section, (**b**) plan (Hager and Gisonni 2005)

During all tests, no free surface flow resulted if $y_o > 0.75$, but choking never occurred for $y_o < 0.70$. In the average, the choking Froude number amounted to $\mathsf{F}_C = 2$.

The current sewer design practice accounts for the so-called full-flow approach (Chap. 5), involving a relative sewer filling of some 85%, independent of the flow conditions. This condition was originally introduced for nearly uniform flows, which differ significantly from the supercritical flows previously described. The observations presented clearly indicate that the standard design procedure results always in a breakdown of the manhole flow. Accordingly, supercritical flows in through-flow manholes must be limited both in the filling ratio and the discharge capacity, to ensure no change of the flow regime.

16.5.3 *Bend Manhole*

The bend manhole may be often found in the urban infrastructure, given that roads are normally arranged in a rectangular grid. Of particular interest is the 90° bend manhole, but also the 45° deflection angles may be relevant. The average bend radius is usually $R_a = 3D$, with D as the sewer diameter. One might think that the 90° bend manhole is more critical in terms of discharge capacity than the 45° manhole. Figure 16.67 shows a definition sketch involving the approach flow depth h_o and velocity V_o for a deflection angle of $\delta = 45°$. As discussed in Sects. 16.2 and 16.4.3, two shockwaves form along the inner and the outer walls. In the following, only the wave along the *outer wall* of maximum height h_M is considered. Del Giudice et al. (2000) found with $\mathsf{F}_U = Q/(gD^2h_o^3)^{1/2}$ that

$$h_M/h_o = [1 + 0.50(D/R_a)\mathsf{F}_U{}^2]^2. \qquad (16.126)$$

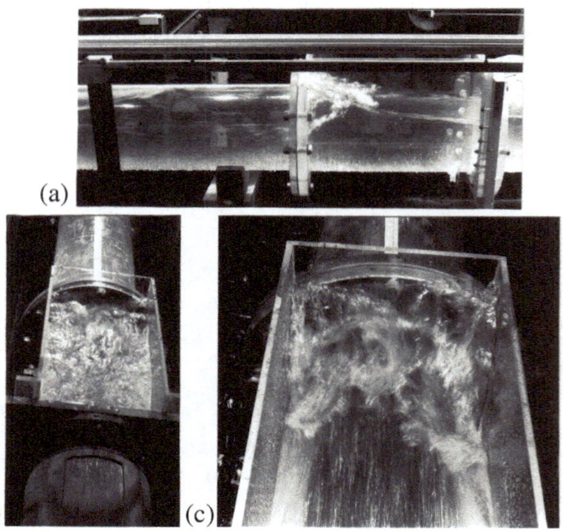
(a) (b) (c)

Fig. 16.66 Choking flow at through-flow manhole outlet for $y_o = 0.75$ and $\mathsf{F}_o = 1.30$, (**a**) section, (**b**) view from upstream, (**c**) impact flow (Hager and Gisonni 2005)

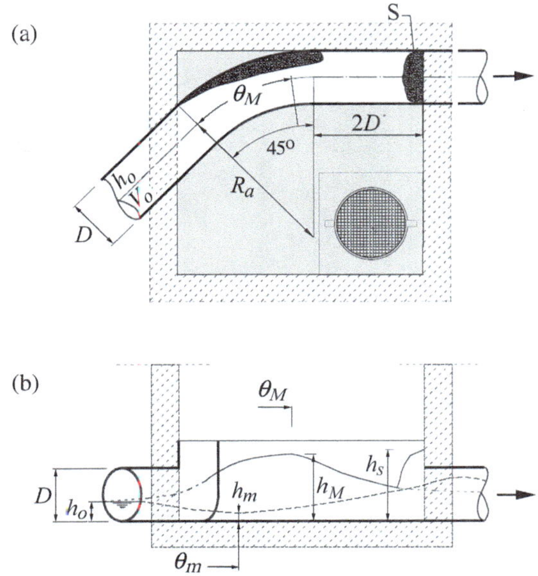

Fig. 16.67 Bend manhole with manhole extension (**a**) plan where S is swell, (**b**) section (Hager and Gisonni 2005)

The angle θ_M of maximum wave location is located between 35° and 55° measured from the manhole inlet (Gisonni and Hager 2002a). The discharge capacity of the 45° bend manhole as compared to a 90° deflection is therefore dramatically reduced due to the presence of the maximum wave at the manhole outlet.

To improve the capacity of this manhole, a straight tailwater *manhole extension* of length $2D$ was added to the structure, as shown in Fig. 16.67. The extension length was fround from detailed hydraulic tests, resulting in a wave maximum upstream of the manhole outlet and a second wave maximum within the tailwater sewer which does not result in flow choking. The manhole extension increases significantly the discharge capacity, which was determined for $y_o < 2/3$ from model tests to (Gisonni and Hager 2002a)

$$\mathsf{F}_C = (3 - 2y_o)y_o^{3/2}. \tag{16.127}$$

The discharge capacity of bend manholes is thus significantly smaller than of the corresponding through-flow manhole, with a maximum of $\mathsf{F}_{CM} = 0.90$ for $y_o = 0.67$, and only $\mathsf{F}_C = 0.80$ for a typical sewer filling of $y_o = 0.60$. Note that in all tests, the flow across the bend manhole choked if the approach flow filling was in excess of 65%, as compared to 75% for the through-flow manhole.

Figure 16.68 shows typical flow features in a bend manhole prior to flow choking. The discharge capacity may be increased if the tailwater sewer diameter D_d is increased thereby using a manhole extension length of $2D_d$ instead of $2D$. No tests were so far conducted to analyze the effect of an increased tailwater diameter on both the approach flow filling and the discharge capacity. Note also that the choking was always related to so-called gate flow type, as described by Hager (1994).

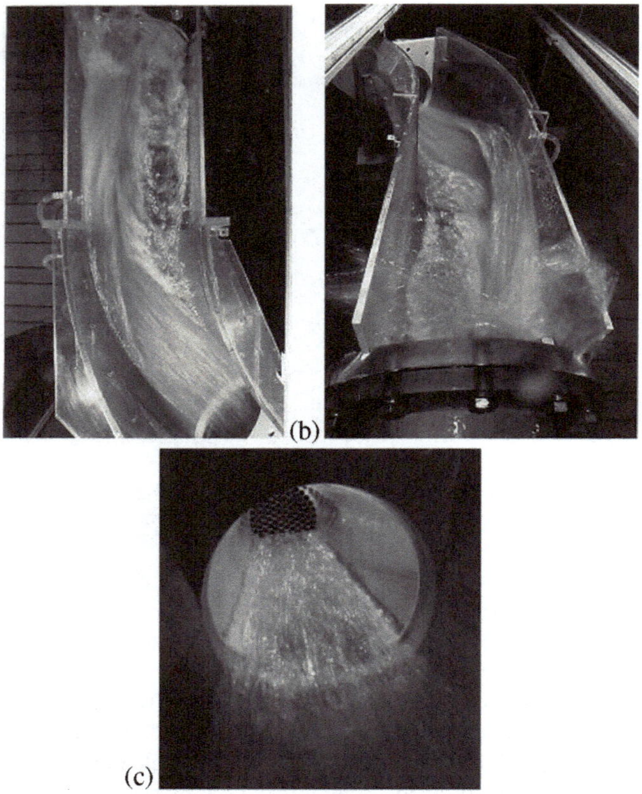

Fig. 16.68 Flow features of bend manhole with a manhole extension (**a**) plan, (**b**) downstream view, (**c**) intake section (Hager and Gisonni 2005)

16.5.4 *Junction Manhole*

A junction manhole may be considered hydraulically intermediate between through-flow and bend manholes. The discharge capacity of the junction manhole is therefore also intermediate. However, the flow structure of junction manholes differs from the other two manhole types. Figure 16.69 shows an equal branch diameter manhole, including the upstream (subscript o) branch of approach flow depth h_o and velocity V_o and the lateral (subscript z) branch with h_z and V_z, respectively. The junction angle between the two branches is δ, with a sharp-crested intersection at the junction point P.

Depending on the branch discharges, four waves may appear, as previously dicussed in Sect. 16.3.6. In the following all flow conditions except for entirely subcritical flow as discussed in Sect. 16.2 are considered. A data analysis of Del Giudice and Hager (2001) and Gisonni and Hager (2002b) indicated that the relevant Froude numbers are $\mathsf{F}_z = Q_z/(gD_zh_z^4)^{1/2}$ for the lateral branch, and

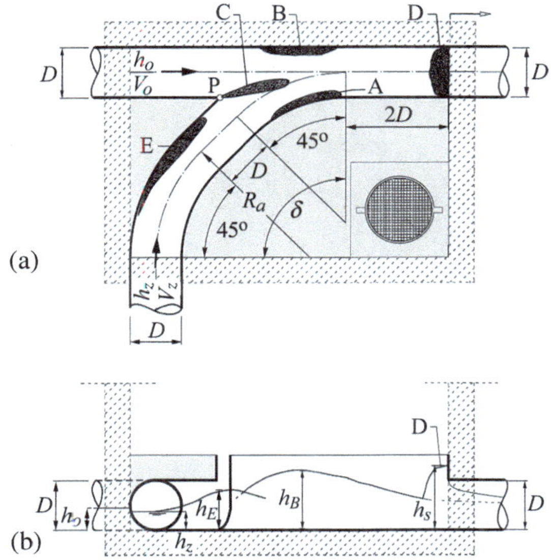

Fig. 16.69 Definition sketch of junction manhole (**a**) plan, (**b**) section (Hager and Gisonni 2005)

$\mathsf{F}_o = Q_o/(gD_oh_o^4)^{1/2}$ for the upstream branch. The height h_B of wave B described in Sect. 16.3.6 for $2 < \mathsf{F}_z < 6$ and both $\delta = 45°$ and $90°$ was found to be independent from F_o

$$h_\mathrm{B} = 1 + (8/7)(\mathsf{F}_z - 1). \tag{16.128}$$

The impact wave height also referred to as the swell height h_S was for $\mathsf{F}_z > 1$

$$h_S = 1 + C_\delta \mathsf{F}_z, \tag{16.129}$$

where $C_\delta = 1$ for $\delta = 45°$ and $C_\delta = 2/3$ for $\delta = 90°$. This height is comparable with h_B as given in Eq. (16.128), influencing then the height of the manhole chamber.

As already noted, two distinctly different phenomena may occur in supercritical manhole flow, namely (1) Choking of manhole outlet due to swell generation associated with an abrupt breakdown of the supercritical manhole flow, and (2) Choking of one or even both branch pipes due to flow blockage of the other branch or poor combining flow conditions in the manhole. A complicated hydraulic jump then may generate submerging either one or even both branch pipes and causing also the breakdown of the supercritical flow structure. In both cases, the breakdown may become so abrupt and strong that manhole geysering results (Sect. 16.5.1). Figure 16.70 shows fully supercritical flow in a junction manhole, whereas Fig. 16.71 relates to choking type 1. The latter results from gate-type flow due to the swell impact onto the manhole outlet wall. Given the complexity of

Fig. 16.70 Typical supercritical flow in junction manhole with waves B and D for $y_o = y_z = 0.27$, $\mathsf{F}_o = 5.95$ and $\mathsf{F}_z = 2.84$. (**a**) upstream view, (**b**) downstream view, (**c**) plan (Hager and Gisonni 2005)

independent parameters, and the variety in flow regimes to be considered, the engineering design of these manholes must be simplified, and retain the main features of flows. An approach is given below.

16.5.5 *Manhole Discharge Capacity*

The previous results indicate that the 45° and the 90° bend and junction manholes are governed by similar flow mechanisms, provided that a manhole extension of length 2*D* is added beyond the lateral branch deflection (Figs. 16.67 and 16.69). Further, both the 45° and the 90° bend and junction manholes behave hydraulically

Fig. 16.71 Choking of junction manhole outlet for $y_o = y_z = 0.34$, $\mathsf{F}_o = 4.19$ and $\mathsf{F}_z = 4.0$ with view from (**a**) upstream, (**b**) downstream (Hager and Gisonni 2005)

similar if an intermediate bend extension of $1D$ is added to the 90° deflection (Fig. 16.69). Then, the bend wave occurring at roughly 45° from the manhole inlet section is allowed to fall, and the flow gets more uniform as compared with an abrupt flow deflection by 90°.

Observations indicated that the maximum discharge capacity of junction manholes designed according to Fig. 16.69 is $\mathsf{F}_C = Q/(gD^5)^{1/2} = 1.4$, as compared to a value of $\mathsf{F}_C = 0.8$ for bend manholes. Therefore, the through-flow branch lessens the effect of the lateral branch flow, and increases the discharge capacity of the junction manhole. A junction manhole thus behaves intermediate to the through-flow ($Q_z = 0$) and the bend manholes ($Q_o = 0$). The capacity Froude numbers and the maximum approach flow filling ratios y_C of the three basic manhole types are given in Table 16.3. This contrasts strongly the current design basis, with the sewer filling independent of flow regime and manhole presence of some 85%, and no limitation of discharge for manholes. Assuming therefore the traditional 'uniform flow concept' along with the full-flow sewer may result in undesirable and dangerous flow conditions for which the sewer was not designed. Note that the conditions expressed in Table 16.3 allow for a basic manhole design, if the indications of Figs. 16.65, 16.67 and 16.69 both in plan and in the streamwise section are considered.

Table 16.3 Capacity Froude numbers F_C and maximum filling ratios y_C for basic manhole types

Manhole Type	Through-flow	Junction	Bend
Discharge F_C	2.0	1.4	0.8
Filling ratio y_C	0.75	0.70	0.65

16.5.6 *Practical Recommendations*

Compared to the currently designed manholes involving circular sewers and an U-shaped manhole through-flow profile placed symmetrically within the structure, the designs shown in Figs. 16.67 and 16.69 are asymmetric. The manhole wall opposite the bend branch is flush with the upstream branch with a vertical-sided through-flow profile, such that a bend wave may not flow onto a bench. The manhole entrance is therefore arranged opposite from this wall, with the manhole platform confined within the bend. Given that both the bend and junction manholes include a manhole extension of $2D$ length, there is enough space for maintenance and sewer control works. Note that this design does not have elements such as the cover plate intruding the flow section, and clogging is of no concern. If the wave along the inner bend side is large, then the benches may also be higher than $1D$ usually adopted. The novel design performed excellently in the laboratory, and designs installed in prototypes have so far not resulted in problems. Long-time tests will indicate whether additional elements should be included for an improved hydraulic performance. The present design is in any case a significant improvement of existing manholes subjected by supercritical flow.

Notation

A_c	[m^2]	cover surface
b	[m]	channel width
B	[−]	relative air discharge
B	[−]	bend number
c	[ms^{-1}]	wave celerity
C_k	[−]	manhole coefficient
d	[−]	location of downstream wave maximum
D	[m]	diameter
D	[−]	drop number
D_s	[m]	manhole diameter
F	[m^2]	cross-sectional area
F	[−]	Froude number
F_D	[−]	pipe Froude number
g	[ms^{-2}]	gravitational acceleration
G	[−]	dimensionless flow depth
h	[m]	flow depth
Δh_M	[m]	maximum superelevation
h_p	[m]	pressure head
h_s	[m]	flow depth in manhole
H	[m]	energy head
J	[−]	total slope
L_a	[m]	mixture length

L_s	[m]	length of manhole
m	[−]	cross-sectional ratio
$1/n$	[$m^{1/3}s^{-1}$]	Manning's roughness coefficient
n_p	[−]	pressure coefficient
P_c	[N]	cover uplift force
$p/(\rho g)$	[m]	pressure head
q_z	[−]	discharge ratio
Q	[m^3s^{-1}]	discharge
Q_a	[m^3s^{-1}]	air discharge
R	[m]	centerline radius of curvature
R	[−]	Reynolds number related to D and V
R_a	[m]	bend radius
R_h	[m]	hydraulic radius
R_z	[m]	rounding radius
s_s	[m]	submergence of manhole
s	[m]	drop height
S	[−]	relative drop height
S	[−]	shock number
S_o	[−]	bottom slope
S_s	[−]	relative manhole submergence
t	[m]	average bottom pressure height
t_b	[m]	height of benches
u	[ms^{-1}]	streamwise velocity component
v	[ms^{-1}]	transverse velocity component
V	[ms^{-1}]	absolute velocity
x	[m]	streamwise coordinate
X	[−]	dimensionless streamwise coordinate
y	[−]	dimensionless transverse coordinate
y_d	[−]	downstream sewer filling
y_s	[−]	filling ratio in manhole
Y	[−]	ratio of flow depths
Y_s	[−]	relative height of shockwave
Δz	[m]	drop height
Z	[m^3]	relative lateral momentum
z_s	[m]	swell height
α	[−]	junction angle
α_c	[−]	outer cover angle
β_c	[−]	inner cover angle
β_i	[−]	ratio of widths
β_e	[−]	expansion ratio
β_s	[−]	shock angle
γ_w	[−]	dimensionless wall profile
δ	[−]	junction angle
δ_k	[−]	bend deflection angle

δ_s	[−]	ratio of manhole diameters
ΔH	[m]	energy loss height
ΔH_s	[m]	additional loss due to manhole
Δz	[m]	bottom difference level
Δz_f	[m]	friction loss
ε	[−]	lateral momentum coefficient
θ	[−]	deflection angle
Φ	[−]	superelevation coefficient
μ	[−]	contraction coefficient
ρ_a	[−]	relative curvature
σ	[−]	coefficient of adaptation
σ_s	[−]	swell coefficient
τ	[−]	transformed location
ξ	[−]	loss coefficient
ξ_k	[−]	loss coefficient of bend manhole
ξ_s	[−]	loss coefficient of manhole
χ	[−]	friction characteristic
ω	[−]	width ratio
ζ	[−]	deflection angle
Ω	[−]	velocity coefficient

Subscripts

a	axial, beginning
b	bend
C	capacity
d	downstream
e	contracted section, end
E	substitute cross-section
k	bend
L	limit
m	minimum
M	maximum
N	uniform
o	upstream, through branch
p	cover plate
s	manhole, swell
t	transition
u	downstream
v	full filling
w	wall
z	lateral branch
1	upstream from shock
2	downstream from shock

References

Abbott, M.B. (1966). *An introduction to the method of characteristics*. American Elsevier: New York.

ATV (1996). ATV-Handbuch *Bau und Betrieb der Kanalisation* (Sewer construction and management), ed.4. W. Ernst & Sohn: Berlin [in German].

ATV (2000). *Bau und Betrieb von Kanalisationen*. Ernst & Sohn: Berlin [in German].

Chaudhry, M.H. (1993). *Open channel flow*. Prentice Hall: Englewood Cliff.

Chiapponi, L., Longo, S. (2008). Experimental study of river junctions. *RiverFlow* 2008 Cesme 3: 2117–2124, M. Altinakar, ed. Kubaba: Izmir.

Chow, V.T. (1959). *Open channel hydraulics*. McGraw Hill: New York.

Christodoulou, G.C. (1991). Drop manholes in supercritical pipelines. *Journal of Irrigation and Drainage Engineering* 117(1): 37–47; 118(5): 832–834.

Del Giudice, G., Gisonni, C., Hager, W.H. (2000). Supercritical flow in bend manhole. *Journal of Irrigation and Drainage Engineering* 126(1): 48–56.

Del Giudice, G., Hager, W.H. (2001). Supercritical flow in 45° junction manhole. *Journal of Irrigation and Drainage Engineering* 127(2): 100–108.

Dick, T.M., Marsalek, J. (1985). Manhole head losses in drainage hydraulics. 21st *IAHR Congress* Melbourne, Seminar A6: 123–131.

Gargano, R., Hager, W.H. (2002). Supercritical flow across combined sewer manhole. *Journal of Hydraulic Engineering* 128(11): 1014–1017.

Gisonni, C., Hager, W.H. (1999). Studying flow at tunnel bends. *Hydropower and Dams* 6(2): 76–79.

Gisonni, C., Hager, W.H. (2002a). Supercritical flow in manholes with a bend extension. *Experiments in Fluids* 32(3): 357–365.

Gisonni, C., Hager, W.H. (2002b). Supercritical flow in the 90° junction manhole. *Urban Water* 4: 363–372.

Hager, W.H. (1982). Die Hydraulik von Vereinigungsbauwerken (Hydraulics of junction structures). *Gas – Wasser – Abwasser* 62(7): 282–288; 63(3): 148–149 [in German].

Hager, W.H. (1987). Discussion to Separation zone at open-channel junctions. *Journal of Hydraulic Engineering* 113(4): 539–543.

Hager, W.H. (1992). Spillways – Shockwaves and air entrainment. ICOLD *Bulletin* 81. International Commission for Large Dams: Paris.

Hager, W.H., Mazumder, S.K. (1992). Supercritical flow at abrupt expansions. Proc. Institution Civil Engineers *Water, Maritime & Energy* 96(9): 153–166.

Hager, W.H. (1994). Impact hydraulic jump. *Journal of Hydraulic Engineering* 120(5): 633–637.

Hager, W.H., Yasuda, Y. (1997). Unconfined expansion of supercritical water flow. *Journal of Engineering Mechanics* 123(5): 451–457.

Hager, W.H., Gisonni, C. (2005). Supercritical flow in sewer manholes. *Journal Hydraulic Research* 43(6): 659–666.

Hsu, C.-C., Wu, F.-S., Lee, W.-J. (1998a). Flow at 90° equal-width open channel junctions. *Journal of Hydraulic Engineering* 124(2): 186–191.

Hsu, C.-C., Lee, W.-J., Chang, C.-H. (1998b). Subcritical open-channel junction flow. *Journal of Hydraulic Engineering* 124(8): 847-855; 126(1): 87–89.

Huang, J., Weber, L.J., Lai, Y.G. (2002). Three-dimensional numerical study of flows in open-channel junctions. *Journal of Hydraulic Engineering* 128(3): 268–280.

Idel'cik, I.E. (1979). *Memento des pertes de charge* (Review of hydraulic losses). 2nd ed. Eyrolles: Paris [in French].

Ippen, A.T., Dawson, J.H. (1951). Design of channel contractions. *Trans. ASCE* 116: 326–346.

Ippen, A.T., Harleman, D.R.F. (1956). Verification of theory for oblique standing waves. *Trans. ASCE* 121: 678–694.

Ito, H., Imai, K. (1973). Energy losses at 90° pipe junctions. Proc. ASCE *Journal of the Hydraulics Division* 99(HY9): 1353–1368; 100(HY8): 1183–1185; 100(HY9): 1281–1283; 100(HY10): 1491–1493; 101(HY6): 772–774.

Johnston, A.J., Volker, R.E. (1990). Head losses at junction boxes. *Journal of Hydraulic Engineering* 116(3): 326–341; 117(10): 1413–1415.

Knapp, R.T. (1951). Design of channel curves for supercritical flow. *Trans. ASCE* 116: 296–325.

Kumar Gurram, S., Karki, K.S., Hager, W.H. (1997). Subcritical junction flow. *Journal of Hydraulic Engineering* 123(5): 447–455.

Liggett, J.A. (1994).*Applied fluid mechanis.* McGraw-Hill: New York.

Lindvall, G. (1984). Head losses at surcharged manholes with a main pipe and a 90° lateral. 3rd Intl. Conf. on *Urban Storm Drainage* Göteborg 1: 137–146.

Lindvall, G. (1987). Head losses at surcharged manholes. 4th Intl. Conf. *Urban Storm Drainage* Lausanne: 140–141.

Marsalek, J. (1987). Head loss at junctions of two opposing lateral sewers. 4th Intl. Conf. *Urban Storm Drainage* Lausanne: 106–111.

Marsalek, J., Greck, B.J. (1988). Head losses at manholes with a 90° bend. *Canadian Journal Civil Engineering* 15: 851–858.

Mazumder, S.K., Hager, W.H. (1993). Supercritical expansion flow in Rouse modified and reversed transitions. *Journal Hydraulic Engineering* 119(2): 201–219.

Reinauer, R., Hager, W.H. (1996). Shockwave in air-water flows. *Intl. Journal Multiphase Flow* 22(6): 1255–1263.

Reinauer, R., Hager, W.H. (1997). Supercritical bend flow. *Journal of Hydraulic Engineering* 123(3): 208–218.

Reinauer, R., Hager, W.H. (1998). Supercritical flow in chute contraction. *Journal of Hydraulic Engineering* 124(1): 55–64.

Sangster, W.M., Wood, H.W., Smerdon, E.T., Bossy, H.G. (1959). Pressure changes at open junctions in conduits. Proc. ASCE *Journal of the Hydraulics Division* 85(HY6): 13–42; 85(HY10): 157; 85(HY11): 153; 86(HY5): 117.

Schwalt, M. (1993). Vereinigung schiessender Abflüsse (Supercritical junction flows). *Dissertation* 10370 ETH Zürich. Appeared also as VAW *Mitteilung* 129, D. Vischer, ed. VAW: Zürich [in German].

Schwalt, M., Hager, W.H. (1992). Shock pattern at abrupt wall deflection. *Environmental Engineering* Water Froum'92, Baltimore ASCE: 231–236.

Schwalt, M., Hager, W.H. (1995). Experiments to supercritical junction flow. *Experiments in Fluids* 18: 429–437.

Shabayek, S., Steffler, P., Hicks, F. (2002). Dynamic model for subcritical combining flows in channel junctions. *Journal of Hydraulic Engineering* 128(9): 821–828.

Townsend, R.D., Prins, J.R. (1978). Performance of model storm sewer junctions. *Journal of the Hydraulics Division* ASCE 104(HY1): 99–104.

Vischer, D. (1958). Die zusätzlichen Verluste bei Stromvereinigungen (The additional losses of junction flows). *Dissertation*, Arbeit 147, Theodor-Rehbock-Flussbau-Laboratorium, TH Karlsruhe: Karlsruhe [in German].

Vischer, D.L., Hager, W.H. (1994). Reduction of shockwaves: A typology. *Journal of Hydropower and Dams* 1(4): 25–29.

Weber, L.J., Schumate, E.D., Mawer, N. (2001). Experiments on flow at a 90° open-channel junction. *Journal of Hydraulic Engineering* 127(5): 340–350.

Wehausen, J.V., Laitone, E.V. (1960). Surface waves. Handbuch der Physik 9: 446–778 *Strömungsmechanik*, S. Flügge, C. Truesdell, eds. Springer: Berlin-Göttingen-Heidelberg.

(a)

(b)

Distribution (**a**) and collector (**b**) channels on Swiss treatment station

Chapter 17
Distribution Channel

Abstract If a discharge should be locally distributed along the length of a channel reach, a distribution channel has to be designed. In sewer hydraulics, sideweirs are often used as the discharging structure, but other types are also possible.

As a supplement to Chap. 18, in which the sewer sideweir is considered, the sideweir in the *rectangular* channel is treated in Chap. 17. The generalized free surface profile is discussed, solutions are presented, and examples illustrate the computational procedure. The pseudo-uniform flow condition is accounted for in particular as a significant aid for designing distribution channels.

In Sect. 17.6, the distribution channel is demonstrated to correspond to a generalisation of a single lateral opening, and the governing substitute system is introduced. The general outflow geometry is then described and examples are given to present the relatively involved solution procedure.

In Sect. 17.7, the channel bifurcation is described as a short distribution channel. The main flow features are presented, and the relation to sideweirs of zero weir height is outlined.

17.1 Introduction

A distribution channel or manifold (German: Verteilkanal; French: Canal de distribution) has lateral openings which discharge fluid over the distribution reach. Such structures may be designed for both free surface or pressurized flow and are applied widely in distributing flow systems. Typical technical applications are found in aeration techniques, cooling water processes, blast-furnaces or in sewage technology. For the latter, *pressurized manifolds* are typically used in sewer recirculation ducts to rivers or estuaries, to enhance the mixing quality of the effluent, or in *open channels*, such as for basin inlets on a sewage treatment station (Fig. 17.1). Usually, a distribution channel should result in an *uniform distribution* of discharge, a requirement that cannot be easily satisfied.

The rectangular cross-section is often used for distribution channels and circular cross-sections can be provided for manifolds as well. A typical question is then: What is the design of the distribution channel for given approach and downstream

W.H. Hager, *Wastewater Hydraulics*, 2nd ed., DOI 10.1007/978-3-642-11383-3_17,

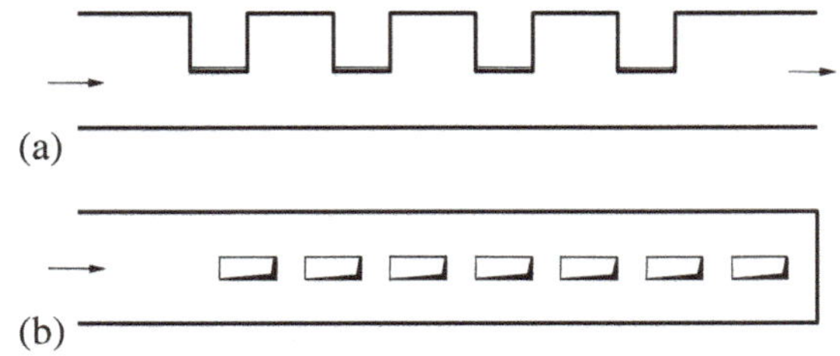

Fig. 17.1 Distribution structure in (**a**) free surface channel, (**b**) ducted channel with a dead-end

flow conditions that provide a specified discharge distribution?, or: How can a uniform distribution be obtained? Both questions cannot be answered by considering only the up- and downstream sections, with the manifold as a black-box, but internal flow features have to be known in addition. This adds to the complexity of distribution channels. In the following, *rectangular* distribution channels as typically used for aeration basins on sewage treatment stations are considered, and indications are given for other manifolds used in sewer hydraulics.

The distributing structure can be principally designed as (Fig. 17.2):

- *Sideweir* as a weir with a lateral approach flow direction,
- *Side opening* as a gate-type structure with a lateral approach flow, and
- *Bottom opening* as an orifice-type structure with a lateral approach flow.

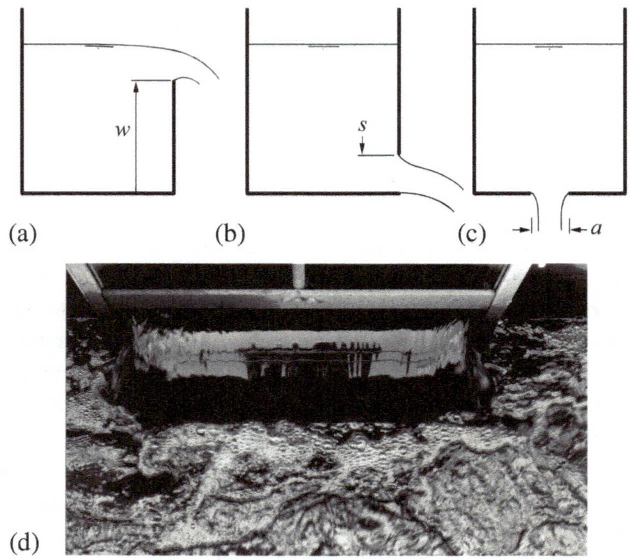

Fig. 17.2 Geometries of lateral outflow structures (**a**) sideweir, (**b**) side opening, (**c**) bottom opening and (**d**) lateral outflow into aeration basin

Normally all three configurations may be characterized with one parameter alone, namely the weir height w for the sideweir, the opening height s of the side opening and the bottom opening width a of the bottom opening. The crest geometry of standard outflow structures is sharp-crested with the crest either horizontal or parallel to the bottom. Circular bottom orifices may be substituted by equivalent rectangular bottom slits of width a and length ΔL.

Contrary to sewer sideweirs, the geometry of sideweirs as used for distribution channels is simple. However, the effects of bottom and friction slopes may play a role due to the streamwise extension of a distribution channel. In addition, manifold distribution has to be investigated for distribution channels. First, the basic equations for *spatially-varied flow* are derived, then the lateral outflow equations are discussed and applied on the distribution channel. Third, the numerical results are presented and generalized solutions are applied for the design of distribution channels. Finally, the hydraulic characteristics of channel bifurcation structures are taken as a particular case of flow distribution.

17.2 Governing Equations

With $E = QH$ as the specific energy with Q as discharge and H as energy head (Chap. 1), the energy conservation over a length element Δx is composed of the increase due to the bottom slope S_o and the decrease due to the energy line slope S_e, times the discharge. Based on the definition

$$E = QH = Q\left[h + \frac{Q^2}{2gF^2}\right], \tag{17.1}$$

energy conservation requires

$$\frac{\mathrm{d}E}{\mathrm{d}x} = (S_o - S_e)Q + \left[p + \frac{U^2}{2g}\right]\frac{\mathrm{d}Q}{\mathrm{d}x}. \tag{17.2}$$

Here, p is the lateral outflow pressure head and U the lateral outflow velocity (Fig. 17.3). Differentiation of Eq. (17.1) and accounting for Eq. (17.2) gives for

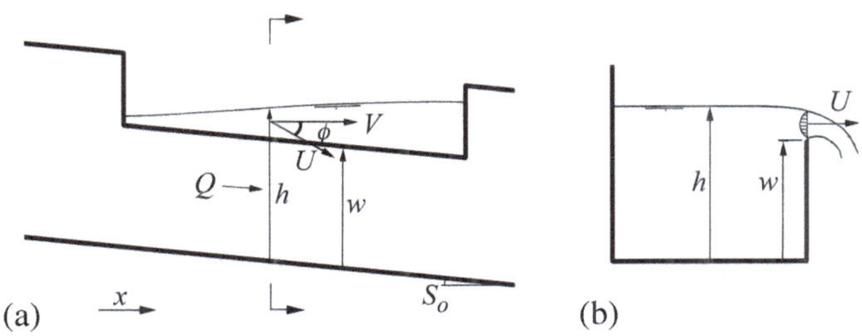

Fig. 17.3 Notation for sideweir (**a**) longitudinal section, (**b**) transverse section

the free surface profile based on *energy conservation* (Yen and Wenzel 1970, Hager 1986a)

$$\frac{\mathrm{d}h}{\mathrm{d}x} = \frac{S_o - S_e + \dfrac{\mathrm{d}Q/\mathrm{d}x}{Q}\left[p - h + \dfrac{U^2 - 3V^2}{2g}\right] + \dfrac{Q^2}{gF^3}\dfrac{\partial F}{\partial x}}{1 - \mathsf{F}^2}. \tag{17.3}$$

Here $\mathsf{F}^2 = [Q^2/(gF^3)](\partial F/\partial h)$ is the square of the local Froude number. Based on *momentum conservation* a similar equation for $h(x)$ may be derived as (Favre 1933)

$$\frac{\mathrm{d}h}{\mathrm{d}x} = \frac{S_o - S_f + \dfrac{\mathrm{d}Q/\mathrm{d}x}{Q}\left[\dfrac{V \cdot U\cos\phi - 2V^2}{g}\right] + \dfrac{Q^2}{gF^3}\dfrac{\partial F}{\partial x}}{1 - \mathsf{F}^2}. \tag{17.4}$$

Contrary to the energy line slope S_e, the friction slope S_f is contained in the momentum approach. Further, $U\cos\phi$ is the streamwise component of lateral outflow velocity and $\partial F/\partial x$ the streamwise change of cross-sectional area. Assuming that $S_e \cong S_f$, the *lateral outflow slope* S_L may be defined from Eqs. (17.3) and (17.4) as

$$S_L = -\left[1 - \frac{U\cos\phi}{V}\right]\frac{Q(\mathrm{d}Q/\mathrm{d}x)}{gF^2}. \tag{17.5}$$

Because the lateral outflow intensity $\mathrm{d}Q/\mathrm{d}x$ is negative for sideweir flow, one has a:

- Positive slope S_L for $V > U\cos\phi$, and
- Negative slope S_L for $V < U\cos\phi$.

This *anomaly* is typical for spatially-varied flows and has been noted in Chaps. 2 and 16. The total energy line slope is thus composed of bottom slope S_o, friction slope S_f and additional slope S_L.

The lateral outflow of outlet structures is normally non-forced, i.e. there are no additional losses and one may assume $V = U\cos\phi$, except for hydraulic jumps. Although detailed analyses of Hager (1987) indicate a slight effect, it is neglected here for simplicity. This statement indicates that the transverse velocity distribution of the streamwise velocity component in *non-forced sideweir flow* is uniform. Forced sideweir flow results typically from baffles inserted in a sideweir to control overflow of float.

The problem may further be simplified from a mathematical point of view when considering the *average friction slope* $S_{fa} = (1/L_v)\int_{Lv} S_f \mathrm{d}x$ over length L_v of the distribution channel, instead of the locally varied function $S_f(x)$. Normally, one may assume that $S_{fa} = (1/2)(S_{fo} + S_{fu})$, i.e. the average (subscript a) of the friction slopes at either boundary section. Then, the so-called total slope $J = S_o - S_{fa}$ is independent of the coordinate x, and Eq. (17.4) may be integrated to yield the *generalized Bernoulli equation*

$$H = H_r + Jx = h + \frac{Q^2}{2gF^2}. \tag{17.6}$$

Here, subscript r refers to the boundary value to be specified below. For typical applications, the bottom slope is nearly compensated for by the friction slope, and one may further assume a *constant energy head* relative to the bottom of the distribution channel. Then, the computation of the free surface simplifies significantly, because

$$H = H_r = h + \frac{Q^2}{2gF^2}. \tag{17.7}$$

Differentiating Eq. (17.7) with respect to x gives

$$\frac{dH}{dx} = \frac{dh}{dx} + \frac{Q(dQ/dx)}{gF^2} - \frac{Q^2(dF/dx)}{gF^3} = 0. \tag{17.8}$$

Eliminating discharge Q from Eqs. (17.7) and (17.8), the *equation of the free surface profile* is

$$\frac{dh}{dx}\left[1 - \frac{2(H-h)}{F}\frac{\partial F}{\partial h}\right] = 2(H-h)\frac{\partial F/\partial x}{F} - \frac{\left[2g(H-h)\right]^{1/2}(dQ/dx)}{gF}. \tag{17.9}$$

This equation may be integrated once the discharge intensity dQ/dx is known and a boundary (subscript r) flow depth $h(x = x_r) = h_r$ is prescribed. The discharge distribution $Q(x)$ can be evaluated then from Eq. (17.7).

Note that the assumptions $S_L = 0$ and $J = 0$ split the complex problem in two equations, one for $h(x)$ and the other for $Q(x)$. Otherwise, a non-linear system of equations had to be solved. This is mathematically a standard problem, but two disadvantages would result:

- The *solution* is complicated by a numerical procedure that does not reveal the basic features, and
- The *number of parameters* is too large for a general solution, and no discussion of the fundamental results is amenable.

Therefore, the simplified approach is followed here in favor of an involved mathematical procedure. Accordingly, the design of distribution channels is straightforward and does not require involved solution techniques. Effects of friction should be included if the bottom slope deviates by more than 1%, say, from the friction slope, and the total head loss due to friction ΔH_f amounts to a significant portion of H_r, say $\Delta H_f > 0.1H_r$. Such conditions were considered for example by Hager (1986b) for pressurized distribution conduits.

17.3 Lateral Outflow

The lateral outflow from a distribution channel contains effects of outflow geometry, approach flow velocity and approach flow direction to the outflow structure. In general, the lateral outflow is different from the conventional outflow from a basin, therefore. For zero approach flow velocity, the two approaches become identical.

According to Hager (1987), the classical derivation of the outflow equation due to De Marchi (1934) is exact only for small Froude numbers. For all other cases, *generalized outflow equations* govern sideweir flow, such as these presented by Hager and Volkart (1986)

$$\frac{dQ}{dx} = -0.6n^*c_k(gH^3)^{1/2}(y-W)^{3/2}\left[\frac{1-W}{3-2y-W}\right]^{1/2}\left[1-(J+\theta)\left(\frac{3(1-y)}{y-W}\right)^{1/2}\right], \tag{17.10}$$

$$\frac{dQ}{dx} = -n^*S\left[\frac{2gH^3}{3(4-3y)}\right]^{1/2}\left[1-(J+\theta)\left(\frac{1-y}{y}\right)^{1/2}\right], \tag{17.11}$$

$$\frac{dQ}{dx} = -0.62A\left[2gH^3\left(\frac{y}{2-y}\right)\right]^{1/2}. \tag{17.12}$$

Equation (17.10) relates to the sideweir (German: Streichwehr; French: Déversoir latéral), Eq. (17.11) to the side opening (German: Seitenöffnung; French: Ouverture latérale), and Eq. (17.12) to the bottom opening (German: Bodenöffnung; French: Ouverture de fond). According to Fig. 17.2, n^* is the number of lateral outflow sides ($n^* = 1$ or 2), c_k is a crest parameter ($c_k = 1$ for sharp-crested outflow), J is the total slope defined in 17.2 and θ is the contraction angle in the plan. Further, with $y = h/H$ as the relative flow depth referred to the energy head H, the geometry of the outflow types can be defined for sideweir, side opening and bottom opening as

$$W = w/H, \quad S = s/H, \quad A = a/H. \tag{17.13}$$

In the following, the sideweir is considered as the main outflow structure for distribution channels, and the other two structures may be computed similarly (Hager 1985, Ramamurthy et al. 1989).

Referring to the rectangular channel of cross-sectional area $F = Bh$ where the width B varies *linearly* with the streamwise coordinate x, i.e.

$$B = b + \theta x. \tag{17.14}$$

where b is the width at the origin $x = 0$, one has $\partial F/\partial x = \theta h$ and $\partial F/\partial h = b + \theta x = B$. With the dimensionless parameters

$$X = kx/b, \quad y = h/H, \quad \Theta = \theta/k, \quad W = w/H. \tag{17.15}$$

where $k = n^* c_k$ is a constant, the *dimensionless equation of the free surface profile* may be written from Eq. (17.9) as

$$\frac{\mathrm{d}y}{\mathrm{d}X} = \frac{2\Theta(1-y) - (\bar{Q}'/k)[2(1-y)]^{1/2}}{(3y-2)(1+\Theta X)}. \tag{17.16}$$

The *dimensionless outflow intensity* $\bar{Q}'/k = (\mathrm{d}Q/\mathrm{d}x)/[k(gH^3)^{1/2}]$ of the sideweir according to Eq. (17.10) is for $J = 0$

$$\frac{\bar{Q}'}{k} = -0.6(y-W)^{3/2}\left[\frac{1-W}{3-2y-W}\right]^{1/2}\left[1-\Theta\left(\frac{3(1-y)}{y-W}\right)^{1/2}\right]. \tag{17.17}$$

Therefore, the dimensionless surface profile $y(X)$ depends exclusively on the parameters Θ and W plus one boundary condition. If the surface profile is determined (Sect. 17.5), the discharge may be computed from Eq. (17.7) as

$$\frac{Q}{(gB^2H^3)^{1/2}} = y\,[2(1-y)]^{1/2}. \tag{17.18}$$

The relative simplicity of the system of governing equations as compared to a rigorous analysis are thus worth mentioning. The *accuracy* of this approach is roughly ±10%, mainly due to the 1D-approach, the outflow equation and the neglect of additional losses.

17.4 Pseudo-Uniform Flow

17.4.1 *Effect of Width Reduction*

Uniform flow as defined in Chap. 5 involves the *equilibrium* between the driving and the retaining forces in a water flow, resulting in a constant flow depth in prismatic channels of constant roughness, discharge and bottom slope. From Eq. (17.16), a constant flow depth ($\mathrm{d}y/\mathrm{d}X = 0$) is obtained under certain conditions, referred to as pseudo-uniform flow with a *pseudo-uniform flow depth* (subscript PN) provided

$$(\bar{Q}'/k)_{PN} = \Theta y_{PN}[2(1-y_{PN})]^{1/2}, \tag{17.19}$$

and the sufficient conditions $y = 2/3$ (critical flow) and $X = 1/\Theta$ (zero channel width) are excluded. Eliminating the discharge intensity with Eq. (17.17), a relation for y_{PN} as a function of the relative weir height $W = w/H$ and the width convergence rate Θ may be obtained. Figure 17.4 shows that real solutions for y_{PN} result only for $\Theta < 0$, and the trivial solution $\Theta = 0$ requires that $y_{PN} = W$, i.e. zero lateral overflow according to Eq. (17.17).

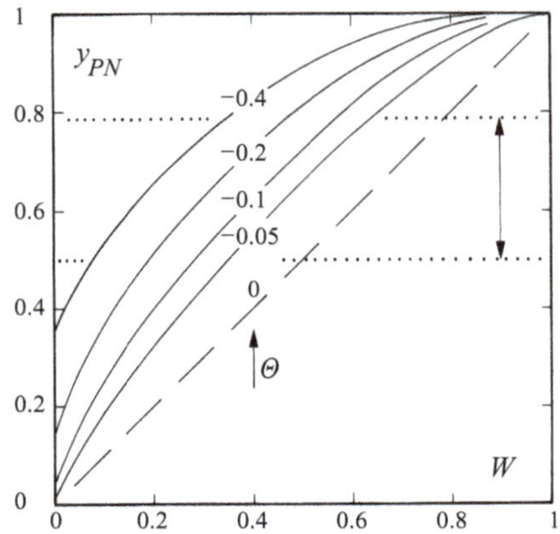

Fig. 17.4 Pseudo-uniform flow depth $y_{PN} = h_{PN}/H$ as function of relative weir height $W = w/H$ and relative contraction angle $\Theta = \theta/k$

For $W < 0.8$, the pseudo-uniform flow depth may be approximated as

$$\frac{y_{PN} - 1.5|\Theta|^{1.5}}{1 - 1.5|\Theta|^{1.5}} = 1.2\, W^{0.8}. \tag{17.20}$$

providing an explicit solution $y_{PN}(\Theta, W)$. For pseudo-uniform flow, a constant flow depth along the lateral outflow reach of $h_{PN} = y_{PN}H$ is established, the velocity $V = V_{PN} = Q/F_{PN}$ remains thus invariable and so does the lateral outflow intensity $\bar{Q}'_{PN}$. Pseudo-uniform flow is a significant *design concept* that results in a fundamental flow condition for spatially-variable discharge and converging channel width.

Inserting the pseudo-uniform flow condition in Eq. (17.7) for two boundary cross-sections of widths B, and $B–\Delta B$, where the discharges are Q, and $Q–\Delta Q$, respectively, the result is (Ramamurthy et al. 1978)

$$H = h_{PN} + \frac{Q^2}{2gB^2h_{PN}^2} = h_{PN} + \frac{(Q - \Delta Q)^2}{2\,g(B - \Delta B)^2 h_{PN}^2}. \tag{17.21}$$

This may be simplified to (Fig. 17.5)

$$\frac{\Delta Q}{Q} = \frac{\Delta B}{B}. \tag{17.22}$$

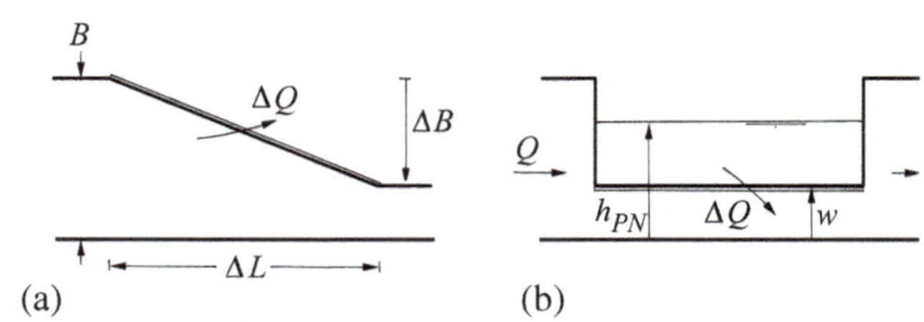

Fig. 17.5 Pseudo-uniform flow in distribution channel of converging section (**a**) plan, (**b**) section

Therefore, the reduction of width ΔB relative to the approach flow width B has to be equal to the reduction of discharge ΔQ relative to the approach flow discharge Q. For the limit case $\Delta Q \to 0$ results a prismatic channel of $\Delta B \to 0$, whereas the plan of the distribution channel is triangular for the other limit $\Delta Q \to Q$. This significant feature provides the *optimum performance* of a distribution channel of constant velocity, where sedimentation of particles is inhibited. Also, constant flow depth requires the minimum freeboard, and the constant outflow intensity yields a minimum lateral outflow length. In addition, a trapezoidal or even a triangular plan of the distribution channel may be supplemented with a diverging collecting channel (Chap. 19) to result in a *compact structural* unit.

Example 17.1 Given a sideweir with $Q_o = 1.2$ m^{3}s^{-1}, $B_o = 1.5$ m, $h_o = 0.6$ m. Design the sideweir for $\Delta Q = 1$ m^{3}s^{-1} and a weir height of $w = 0.4$ m for pseudo-uniform flow.
According to Eq. (17.22) $\Delta B/B = \Delta Q/Q = 1/1.2 = 0.833$, thus $\Delta B = 0.833 \cdot 1.5 = 1.25$ m and $B_u = 0.25$ m.

With $V_o = Q_o/(B_o h_o) = 1.2/(1.5 \cdot 0.6) = 1.33$ ms^{-1} as the approach flow velocity and $H_o = 0.6 + 1.33^2/19.62 = 0.69$ m for the energy head, the sideweir convergence is assumed to $\Theta = \theta \cong -0.1$ with $k = 1$. Further, with $W = 0.4/0.69 = 0.58$, $y_{PN} = 0.8$ from Fig. 17.4, and thus $\bar{Q}'/k = -0.6 \cdot 0.22^{3/2}[0.42/0.82]^{1/2}\ [1 + 0.1(0.6/0.22)^{1/2}] = -0.052$ from Eq. (17.17), corresponding to $Q' = -(9.81 \cdot 0.69^3)^{1/2} 0.052 = -0.093$ m^{2}s^{-1}. The length of sideweir is thus $\Delta L = Q/(-Q') = 1.2/0.093 = 12.95$ m, and $\theta = -1.25/12.95 = -0.097 \cong -0.10$, as previously assumed.

Conventionally, one would obtain with $h_{PN} = 0.8 \cdot 0.69 = 0.55$ m for the lateral discharge $\Delta Q = 0.4 \cdot 19.62^{1/2}(0.55 - 0.40)^{3/2} 12.9 = 1.33$ m^{3}s^{-1}, i.e. 33% of the effective value. The Froude number is $\mathsf{F}_{PN} = 1.33/(9.81 \cdot 0.55)^{1/2} = 0.57$, and a considerable reduction of lateral discharge due to the approach conditions results, therefore.

Example 17.2 What is the overflow depth in Example 17.1?
With $W = 0.4/0.69 = 0.58$ and $\Theta = -0.1$ Fig. 17.4 yields $y_{PN} = 0.80$ and $h_{PN} = 0.55$ m. The overflow depth is thus $h_{PN} - w = 0.55 - 0.4 = 0.15$ m. According to Eq. (17.20), $y_{PN} = 1.5 \cdot 0.1^{1.5} + 1.2 \cdot 0.58^{0.8}(1 - 1.5 \cdot 0.1^{1.5}) = 0.047 + 0.776\,(1 - 0.047) = 0.79\ (-1\%)$ results.

Note that Example 17.1 is overdetermined. Normally, the approach flow discharge, the approach flow width, the approach flow depth and thus H_o are specified, and the weir height as a function of weir length is sought. Stable flow should occur, and Froude numbers between 0.5 and 2, or at least 0.75 and 1.5 are excluded. With $H_{PN} = h_{PN} + Q^2/(2gB^2h_{PN}^2)$ or $y_{PN}^{-1} = 1 + (1/2)\mathsf{F}_{PN}^2$ the domains of y_{PN} should be $y_{PN} > 0.78$ for *subcritical flow* and $y_{PN} < 0.50$ for *supercritical flow*. The domain of relative weir heights is thus $W > 0.4$ for subcritical flow and $W < 0.4$ for supercritical flow (Fig. 17.4). Because the latter is not recommended for design, the minimum relative weir height should be $W = 0.4$.

17.4.2 *Effect of Bottom Elevation*

For a distribution channel with a total slope $J \neq 0$, Eq. (17.21) can be generalized to

$$h_{PN} + \frac{Q^2}{2gB^2h_{PN}^2} = h_u + \frac{(Q-\Delta Q)^2}{2g(B-\Delta B)^2h_u^2} + \Delta z \tag{17.23}$$

where $\Delta z = J\Delta L$ is approximately the elevation difference between the boundary sections (Fig. 17.6a), resulting in

$$\frac{\Delta Q}{Q} = \frac{\Delta B}{B} + \frac{\Delta z}{h_{PN}} - \frac{\Delta B}{B}\frac{\Delta z}{h_{PN}}. \tag{17.24}$$

The flow in distribution channels can thus be uniformized with an *adversely sloping bottom*. The relative bottom height $\Delta z/h_{PN}$ should be limited to 0.5 to inhibit transitional flow. Combining bottom increase and width reduction allows to discharge all flow, $\Delta Q/Q = 1$. In practise this case cannot really be attained due to structural limitations requiring $B_u \geq 0.20$ m for maintenance. An auxiliary bottom outlet enables complete drainage of the distribution channel for inspection purposes.

Example 17.3 What is the lateral discharge in Example 17.1 for $\Delta z/h_{PN} = 0.2$?
With $\Delta B/B = 0.833$ and $\Delta z/h_{PN} = 0.2$, one has $\Delta Q/Q$–$0.833 = 0.2\,(1 - 0.833) = 0.033$, i.e. $0.033 \cdot 1.2 = 0.04$ m^{3}s^{-1} more water is laterally discharged.

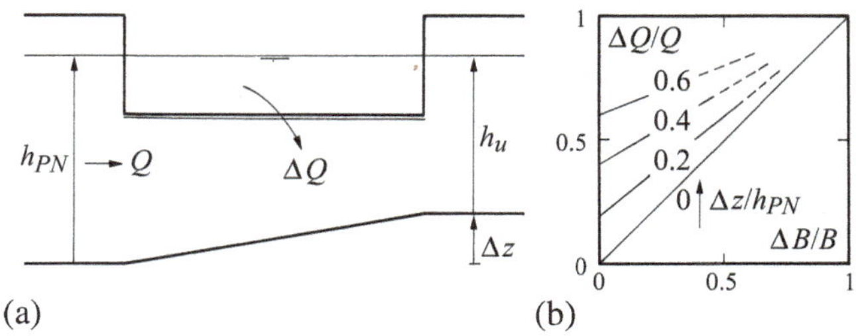

Fig. 17.6 Distribution channel with an increasing bottom profile to improve flow uniformity (**a**) section, (**b**) diagram showing Eq. (17.24). (- - -) Unstable flow

Because the boundary flow depths vary with discharge, the distribution channel is normally *designed* for pseudo-uniform flow. For all other relevant discharges the local surface profile must be determined, based on the following section.

17.5 General Free Surface Profile

17.5.1 *Representation of Solution*

The free surface profile of sideweir flow is prescribed by Eq. (17.16), once a boundary condition is specified. The solution depends significantly on the:

- Sign of the pseudo-uniform flow parameter $\sigma = \Theta y[2(1-y)]^{1/2} - \bar{Q}'/k$, and the
- Local Froude number $\mathsf{F} = [2(1-y)/y]^{-1/2}$.

For $y < y_{PN}$ the sign of σ is opposed to that for $y > y_{PN}$, and so is the free surface slope dy/dX. For $y > 2/3$, the flow in a rectangular channel is subcritical, and supercritical otherwise. Figure 17.7 summarizes the six possible cases of free surface profile patterns, together with the location of the pseudo-uniform depth.

In the *prismatic* distribution channel ($\Theta = 0$), the flow depth increases for $\mathsf{F} < 1$ and decreases for $\mathsf{F} > 1$. This basic concept introduced by De Marchi (1934) is valid also for the *converging* distribution channel provided $\sigma > 0$, i.e. the width decrease is relatively modest. For $\sigma < 0$, corresponding to a strong width convergence, the free surface profile decreases for subcritical and increases for supercritical flow, however. It should be noted that the pseudo-uniform flow has an *equalizing effect*, and the flow depth remains relatively close to the pseudo-uniform depth even for off-design. Also, the profile has never a local minimum nor a local maximum. If the flow depth increases, then it will attain the largest depth at either channel end. This feature is correct for continuous flows and may be invalidated by *hydraulic jumps* (see below).

The governing Eq. (17.16) may be solved *generally* when considering the most general *boundary conditions* (subscript r). For subcritical flow, computations proceed *against* the flow direction and $\mathsf{F}_r = 0$ or $y_r = 1$ is the most extreme downstream

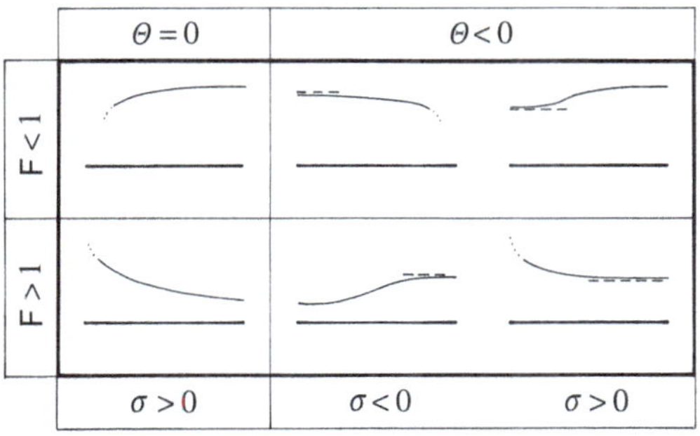

Fig. 17.7 Classification of free surface profiles in sideweir flow with $J = 0$ as a function of Froude number and parameter σ (Hager and Volkart 1986). (- - -) Pseudo-uniform depth, (. . .) transitional flow

condition. All other conditions $y_r < 1$ are located upstream between the stagnation depth $h_r = H$ and the pseudo-uniform flow depth $h_r = h_{PN}$. For subcritical flow with $h_r < h_{PN}$, the most extreme downstream condition can be demonstrated to be the critical depth, and $y_r = 2/3$ ($\mathsf{F}_r = 1$) is the general boundary condition, provided $2/3 < y_r < y_{PN}$.

For supercritical flow, the boundary condition is $y_r = 2/3$ ($\mathsf{F}_r = 1$) for $y_{PN} < y_r < 2/3$, with the computational direction *in* the flow direction (Fig. 17.7). For the last case $W < y_r < y_{PN}$ the most extreme condition is $y_r = W$. As a result, the general boundary conditions are:

- $y_r = 1$ for subcritical, and $y_r = 2/3$ for supercritical flow in the prismatic distribution channel, and
- $y_r = 1$ and $y_r = 2/3$ for subcritical, and $y_r = 2/3$ and $y_r = W$ for supercritical flow in the converging distribution channel.

Figures 17.8 and 17.9 show the *general solution* of the free surface profile $y(X)$ for subcritical ($2/3 < y < 1$) and supercritical ($W < y < 2/3$) flows for channel convergences $\Theta = 0, -0.1, -0.2$ and -0.4. For a given boundary condition y_r and specified values W and Θ, the location X_r can be determined and the surface $y(X)$ can be explicitly obtained. The application of the diagrams is as in Chap. 8 or Chap. 18. They are principally based on the assumption of *constant energy head H* along the distribution channel.

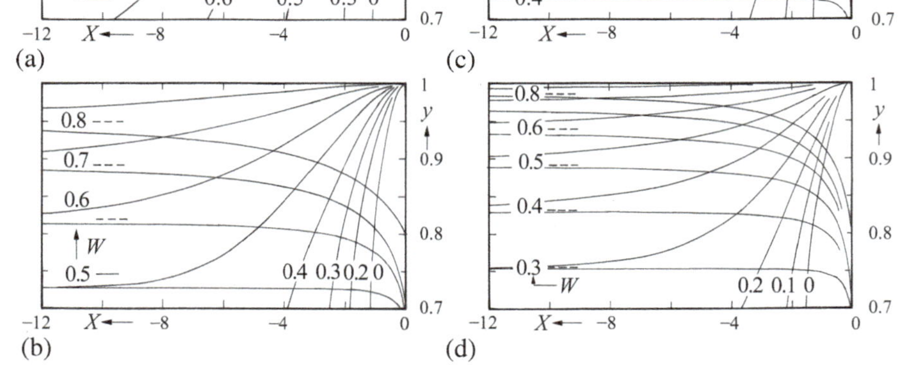

Fig. 17.8 Free surface profiles $y(X)$ in rectangular distribution channel for *subcritical* flow with $\Theta =$ (**a**) 0, (**b**) −0.1, (**c**) −0.2, (**d**) −0.4. (- -) Pseudo-uniform flow depth

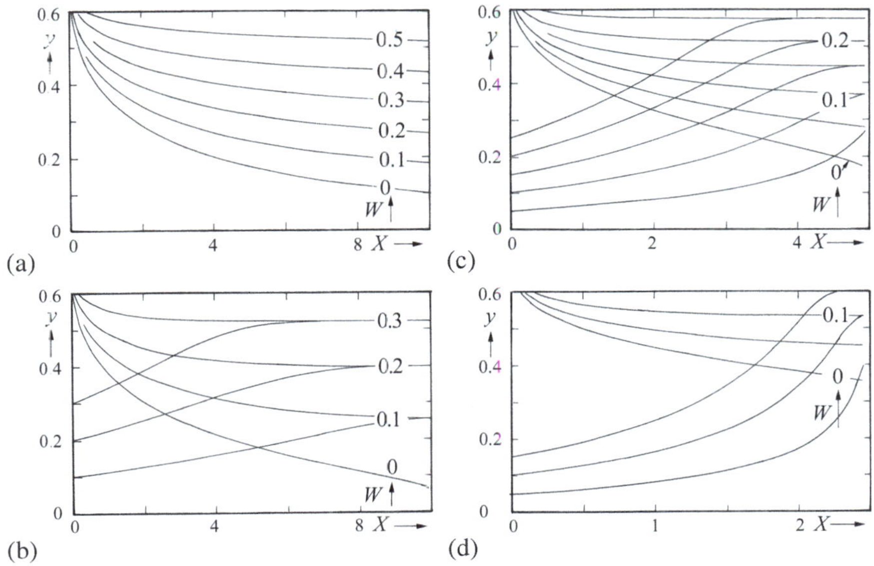

Fig. 17.9 Free surface profiles $y(X)$ in rectangular distribution channel for *supercritical* approach flow with Θ = (**a**) 0, (**b**) –0.1, (**c**) –0.2, (**d**) –0.4

Example 17.4 Given a prismatic one-sided sideweir of width $B = 0.7$ m, length $\Delta L = 4$ m and weir height $w = 0.35$ m. What is the lateral discharge for a downstream flow depth of $h_u = 0.52$ m, provided $Q_u = 0.27$ m^{3}s^{-1}?
With $V_u = 0.27/(0.7 \cdot 0.52) = 0.74$ ms^{-1}, $H_u = 0.52 + 0.74^2/19.62 = 0.55$ m as energy head and thus $\mathsf{F}_u = 0.74/(9.81 \cdot 0.52)^{1/2} = 0.33$. The flow is subcritical and the computation has to proceed *against* the flow direction.

For a sharp-crested weir with $c_k = 1$ one has $k = 1$. Further, with the boundary flow depth $y_r = h_u/H_u = 0.52/0.55 = 0.94$ and the relative weir height $W = 0.35/0.55 = 0.64$ application of Fig. 17.8a yields $X_r = -2.9$. For $\Delta L = 4$ m the dimensionless weir length is $\Delta X = -1 \cdot 4/0.7 = -5.7$ from the definition equation $(17.15)_1$, thus $X_o = X_r + \Delta X = -2.9–5.7 = -8.6$. Moving along the curve with the fictitious value $W = 0.64$, the relative depth $y_o = 0.71$ is reached at the prescribed location $X_o = -8.6$ (Fig. 17.8a), and $h_o = 0.71 \cdot 0.55 = 0.39$ m. The corresponding discharge is $Q_o/(9.81 \cdot 0.7^2 0.55^3)^{1/2} = 0.71[2(1-0.71)]^{1/2} = 0.54$ from Eq. (17.18), and $Q_o = 0.54 \cdot 0.9 = 0.48$ m^{3}s^{-1}. The lateral discharge is finally $\Delta Q = Q_o – Q_u = 0.48–0.27 = 0.21$ m^{3}s^{-1}.

Example 17.5 How does the lateral outflow change in Example 17.4 if the approach flow width is $B_o = 1.1$ m?
With $h_u = 0.52$ m, $Q_u = 0.27$ m^{3}s^{-1}, $b_u = 0.7$ m and $w = 0.35$ m, the energy head is $H_u = 0.52 + 0.27^2/(19.62 \cdot 0.7^2 \cdot 0.52^2) = 0.55$ m, $y_u = 0.52/0.55 =$

0.94 and $W = 0.35/0.55 = 0.64$. Further with $\theta = -0.4/4 = -0.1$, thus $\Theta = -0.1/1 = -0.1$, Fig. 17.8b gives $X_u = -4.1$ for $y_u = 0.94$ and $W = 0.64$. The width at the origin is $b = B-\theta x$ from Eq. (17.14), corresponding to the value $B-(\theta/k)(kx/b)b = B-\Theta Xb$, and thus with $B(X_u) = 0.7$ m for $b = 0.7(1-0.1\cdot 4.1) = 0.41$ m.

The dimensionless overflow length is $\Delta X = -1\cdot 4/0.41 = -9.75$, and the upstream end is located at $X_o = -4.1-9.75 = -13.85$, where the flow has practically reached the pseudo-uniform flow depth $y_{PN} = 0.85$, thus $h_o = 0.85\cdot 0.55 = 0.47$ m. From Eq. (17.14) $B/b = 1+\Theta X = 1+0.1\cdot 13.85 = 2.38$, thus $B = 2.38\cdot 0.41 = 0.98$ m and $Q_o = (9.81\cdot 0.98^2 0.55^3)^{1/2} 0.85[2(1-0.85)]^{1/2} = 0.58$ m^{3}s^{-1}, from which results the outflow $\Delta Q = 0.58 - 0.27 = 0.31$ m^{3}s^{-1}. Compared to Example 17.4, the lateral discharge is by 48% increased.

Example 17.6 The approach flow discharge to a *prismatic* distribution channel of width $B_o = 1.6$ m is $Q_o = 4.7$ m^{3}s^{-1}. What is the length required to discharge $\Delta Q = 0.7$ m^{3}s^{-1} for a weir height of $w = 0.2$ m and $h_o = 0.7$ m?

With the approach flow velocity $V_o = 4.7/(1.6\cdot 0.7) = 4.2$ ms^{-1}, the approach flow energy head is $H_o = 0.7 + 4.2^2/19.62 = 1.6$ m and with $\mathsf{F}_o = 1.60$, the approach flow is weakly *supercritical.* For $W = 0.2/1.6 = 0.125$ and $y_o = 0.7/1.6 = 0.44$ Fig. 17.9a indicates $X_r = 0.75$ as the boundary value. For $Q_u = 4.7-0.7 = 4.0$ m^{3}s^{-1} Eq. (17.18) gives $4/(9.81\cdot 1.6^2\cdot 1.6^3)^{1/2} = 0.4 = y_u[2(1-y_u)]^{1/2}$, with the relevant solution being $y_u = 0.35$. Following in Fig. 17.9a the fictitious curve with $W = 0.125$, one reads $X_u = 2.1$, from where $\Delta X = 2.1 - 0.75 = 1.35$, or $\Delta x = 1.35\cdot 1.6 = 2.16$ m. The structure has thus a length of 2.2 m.

The previous examples indicate the laborious computations even with the diagrams. If the results are compared with the conventional approach, *significant reductions* of the lateral outflow are observed, mainly due to the effects of velocity and direction of the approach flow. The present method does not explicitly include effects of the friction and bottom slopes but assumes a constant energy head. If these additions would be included, the complications were so involved that an application of diagrams is not justified and the solutions had to be obtained by *numerical integration*. Often, the total slope J is almost equal to zero in practise, and Figs. 17.8 and 17.9 reflect the reality at least approximately. The predictions were compared with tests, and the agreement was satisfactory for cases where the Froude number was not too close to unity.

17.5.2 *Similarity Solutions*

As for sideweirs (Chap. 18), similarity solutions provide another basis for the solution of spatially-varied flow. The purpose of this section is to present solutions in which the relative weir height W is not explicitly contained. In terms of W, one may distinguish between there different cases (Fig. 17.10):

- *Case 1* $1 > y > W$, respectively y_{PN}, with y located between the two limit cases $y = 1$ and $y = W$, or $y = y_{PN}$, starting and ending with a horizontal asymptote,
- *Case 2* $1 > y > y_c$, thus starting with a horizontal asymptote and ending vertically for $\mathsf{F} = 1$, and
- *Case 3* $y_c > y > W$, respectively y_{PN}, with a vertical start ($\mathsf{F} = 1$) and ending horizontally at $y = W$, or $y = y_{PN}$.

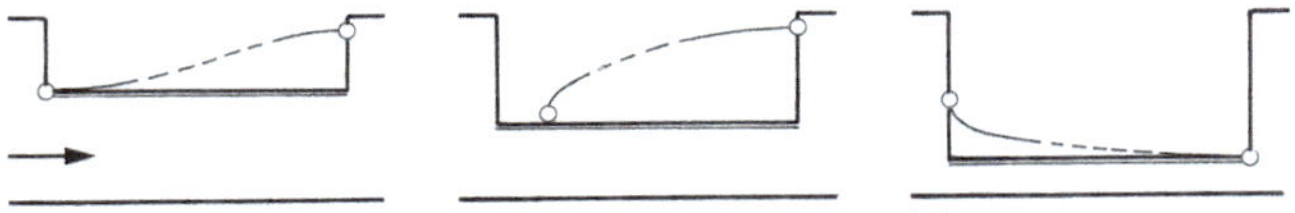

Fig. 17.10 (○) Boundary limits in prismatic distribution channels for (**a**) and (**b**) subcritical, (**c**) supercritical flow

For *case 1* the reference length $X_{-1/2}$ is the length at half overflow depth $X[y = (1+W)/2]$. Based on Fig. 17.8a the result is

$$X_{-1/2} = 3.8 \tan(90° W). \tag{17.25}$$

The normalized free surface profile $Y_- = (Y-W)/(1-W)$ may be related to the normalized coordinate $X_- = X/X_{-1/2}$ as (Fig. 17.11a)

$$Y_- = (5/8)(X_- + 1.8). \tag{17.26}$$

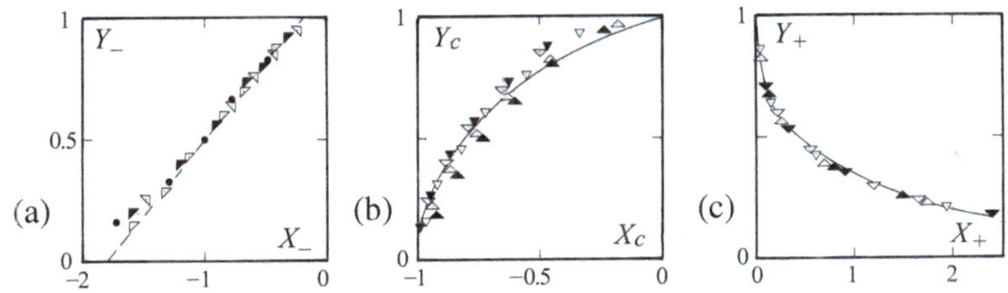

Fig. 17.11 Normalized free surface profiles based on a two-parameter plot for *prismatic* distribution channel for W = (◤) 0.1, (◇) 0.2, (□) 0.3, (◁) 0.4, (▷) 0.5, (▽) 0.6, (•) 0.7, (○) 0.8, (◥) 0.9

This relation is valid for $0.2 < Y_- \leq 0.95$ and comparable to Eq. (18.25), stating that the free surface profile between $y = W$ and $y = 1$ is essentially *linear* (Fig. 17.8a).

Case 2 involves the critical depth $y_c = 2/3$ as the lower limit value, the normalized depth is $Y_c = (y - 2/3)/(1 - 2/3) = 3y - 2$, and the streamwise coordinate $X_c = X/\chi_c$ where χ_c is the length of profile between the boundary values $y = 1$ and $y = 2/3$. Based on Fig. 17.8a

$$\chi_c = 1.1 + 1.9[\tan 90°(1.5\,W)^{1.5}] \tag{17.27}$$

and the resulting profile is for $W < 0.55$ (Fig. 17.11b)

$$Y_c = (1 + X_c)^{0.4}. \tag{17.28}$$

For *case 3* with a supercritical approach flow, $Y_+ = (Y - W)/(2/3 - W)$ as the normalized flow depth is considered a function of $X_{+1/3} = X(Y_+ = 1/3)$, in analogy to $X_{-1/2}$ for subcritical approach flow. Based on Fig. 17.9a, the result is

$$X_{+1/3} = 3.4(1 - 1.2\,W^2), \tag{17.29}$$

and the profile may be expressed as (Fig. 17.11c)

$$Y_+ = 1 - 0.65X_+^{0.3}. \tag{17.30}$$

Equations (17.26), (17.28) and (17.30) include a two-parameter set for *prismatic* distribution channels.

For the *converging* distribution channel, various cases might be distinguished, but only the cases analogous to the prismatic channel are considered here (Fig. 17.12). For *case 1* with $y_{PN} < y < 1$, the reference length is also $X_{-1/2} = X[(1+y_{PN})/2]$, and $X_- = X/X_{-1/2}$, $Y_- = (y - y_{PN})/(1 - y_{PN})$ are the generalized coordinates. *Case 2* refers to $2/3 < y < y_{PN}$ with $X_c = X/\chi_c$ and $Y_c = 3y - 2$, and for *case 3* with $y_{PN} < y < 2/3$ the coordinates are as previously $X_{+1/3} = X[Y_{+1/3} = 1/3]$ with $Y_{+1/3} = (y - y_{PN})/(2/3 - y_{PN})$. For $-0.4 \leq \Theta \leq -0.05$, the normalizing lengths are

$$X_{-1/2} = 2 + 10\,W^{2.5}, \tag{17.31}$$

$$\chi_c = 8.2[1 + 7(-\Theta)^{1.6}]W, \tag{17.32}$$

$$X_{+1/3} = -0.02/(\Theta W), \tag{17.33}$$

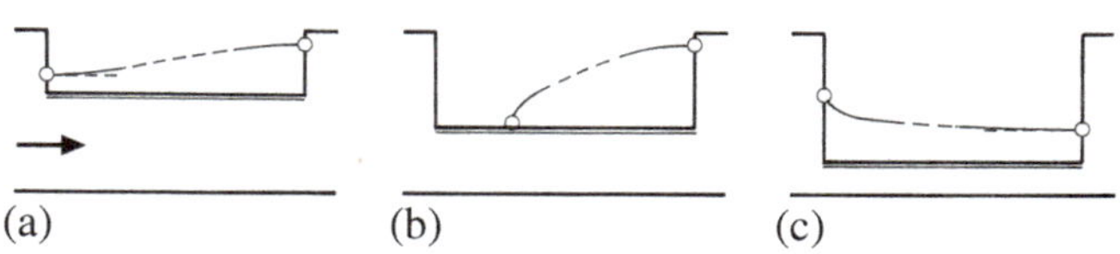

Fig. 17.12 (o) Boundary limits in *converging* distribution channel for (**a**), (**b**) subcritical, (**c**) supercritical flow

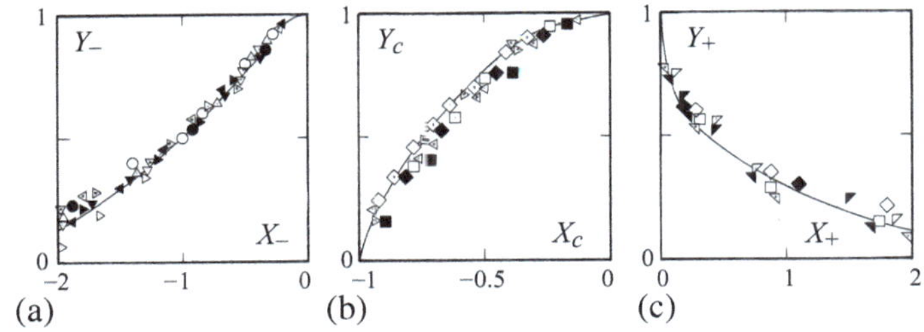

Fig. 17.13 Normalized free surface profiles based on two-parameter plot for *converging* distribution channel and various values of Θ and W, for symbols see Fig. 17.11

and the profile equations read, respectively (Fig. 17.13),

$$Y_- = 1 - [\tanh(-0.74X_-)]^{1.5}, \tag{17.34}$$

$$Y_c = [\sin(1 + X_c)90°]^{0.75}, \tag{17.35}$$

$$Y_+ = 1 - 0.70X_+^{0.35}. \tag{17.36}$$

Note that the scaling $X_{-1/2}$ is independent of Θ. Eqs. (17.34), (17.35), and (17.36) provide relations accounting for the nonprismatic distribution channel with a two-parameter set, with the effects of both Θ and W contained in this formulation.

Example 17.7 Recompute Example 17.5 with the two-parameter formulation. From $H_u = 0.55$ m, $y_u = 0.94$, $y_{PN} = 0.85$, $W = 0.64$ and $\Theta = -0.1$ results case 1. Thus $X_{-1/2} = 2 + 10 \cdot 0.64^{2.5} = 5.28$ from Eq. (17.31) such that $Y_{-u} = (y_u - y_{PN})/(1 - y_{PN}) = (0.94 - 0.85)/(1 - 0.85) = 0.60$, and the adequate solution of Equation (17.34) is $X_{-u} = -(1/0.74)\text{arctanh}(1 - Y_{-u})^{2/3} = -1.35\text{arctanh}(0.54) = -1.35 \cdot 0.49 = -0.67$. The dimensionless length of the distribution channel is $\Delta X = -9.75$ from Example 17.5, thus $\Delta X_- = -9.75/5.28 = -1.85$ and $X_{-o} = -0.67 - 1.85 = -2.52$, from where $Y_{-o} = 1 - [\tanh(0.74 \cdot 2.52)]^{1.5} = 0.069$ according to Eq. (17.34) and $y_o = y_{PN} + (1 - y_{PN})Y_- = 0.85 + (1 - 0.85)0.069 = 0.86$. As a result, y_o is practically equal to y_{PN}, as also found in Example 17.5.

Example 17.8 Recompute also Example 17.6.
For the prismatic distribution channel with supercritical approach flow, the basis is Eq. (17.30). With $W = 0.125$ results $X_{+1/3} = 3.4(1 - 1.2 \cdot 0.125^2) = 3.34$ from Eq. (17.29). For $y_o = 0.44$, Eq. (17.30) gives $Y_{+o} = (0.44 - 0.125)/(0.66 - 0.125) = 0.59$, thus $X_{+o} = [(1 - Y_{+o})/0.65]^{1/0.30} = 0.215$. For

$y_u = 0.35$, one has further $Y_{+u} = (0.35 - 0.125)/(0.66–0.125) = 0.42$ and thus $X_{+u} = 0.685$. Because $\Delta X_+ = 0.685–0.215 = 0.47$, the corresponding length is $\Delta X = 0.47 \cdot 3.34 = 1.57$, compared with 1.35 (–14%) from Example 17.6.

The computation using similarity solutions includes one additional transformation but is purely analytical, compared to a semi-graphical solution of Sect. 17.5.1. Its advantage is the *analytical approach* incorporated in an explicit computation without recourse to numerical integration. In addition, observations on distribution flow use one plot such as Fig. 17.13. In 18.4, comparisons with test data are made for the circular sideweir. Figure 17.14 compares similar flows in prismatic and converging distribution channels.

(a) (b)

Fig. 17.14 Distribution channel with weakly supercritical approach flow in (**a**) prismatic, (**b**) converging sideweir. Note the standing wave pattern because of weakly supercritical approach flow

17.6 Distribution Channel

17.6.1 *Substitute Distribution Channel*

Distribution channels (German: Verteilkanal; French: Canal de distribution) combine *single laterals* to an unit structure. For sideweirs as shown in Fig. 17.15, one would have to compute each opening successively, and consider the intermediate reach with a backwater curve (Chap. 8), a lengthy and involved procedure to be simplified.

For distribution channels with more than roughly five closely spaced laterals, i.e. if the interdistance between two laterals is smaller than about ten times the approach flow depth, or smaller than about the length of a lateral, a *substitute system* (subscript E) may be considered. It is characterized by a single lateral of length L_E that is equal to the sum of all laterals of length L_v equal to the original distribution channel length (Fig. 17.16). The friction losses are computed for length L_v with $S_{fa} = \int_{Lv} S_f \, dx$, whereas the lateral outflow process is referred to length L_E.

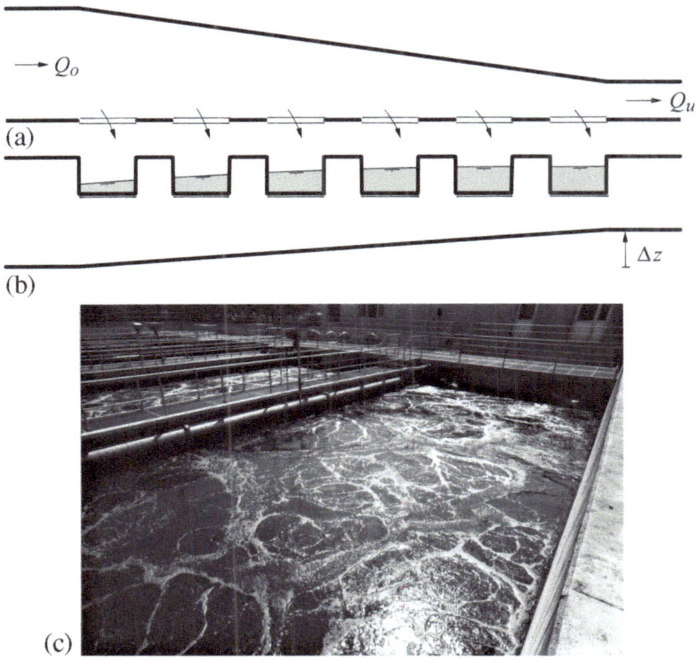

Fig. 17.15 Distribution channel with width reduction and bottom elevation increase (**a**) plan, (**b**) section. (**c**) Aeration basin with a lateral discharge distribution

The substitute system behaves practically as the effective distribution channel, provided the lateral has a minimum length. For smaller lateral length, the lateral outflow angle and thus the lateral discharge depend in addition on the length, and the fundamental assumption of a *continuous* spatially-varied flow is invalidated. The discontinuities at both the beginning and the end of a lateral are then significant,

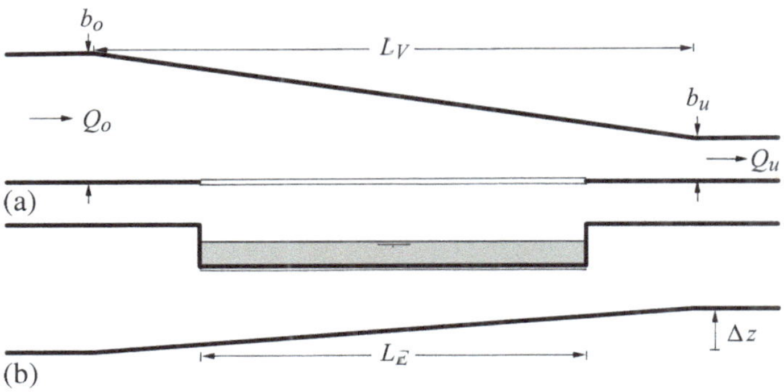

Fig. 17.16 Substitute system of distribution channel (**a**) plan, (**b**) section

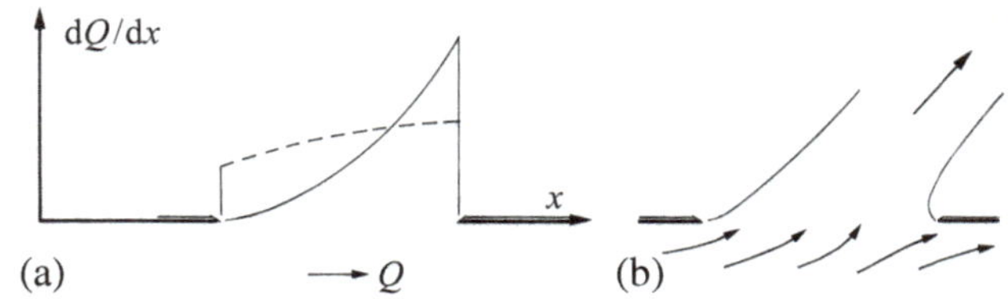

Fig. 17.17 (**a**) Outflow intensity at plane lateral with (—) short and (- - -) long lateral opening, (**b**) approach flow to short lateral (schematic)

and the intermediate reach is too short for being dominant. At the beginning of a sharp-crested lateral, effects of jet contraction reduce the outflow intensity, whereas the oblique direction of approach flow increases the outflow intensity at the end of the lateral, compared to the center reach (Fig. 17.17). For a continuous flow, these local effects are neglected and the outflow intensity is supposed to vary exclusively with the overall geometry of the lateral. Also, for short laterals, the outflow cannot be considered free, and the additional slope J_L is different from zero.

Example 17.9 Given a prismatic channel of width b = 0.50 m with an approach flow discharge of Q_o = 0.45 m^{3}s^{-1}. Distribute the discharge over a length L_v = 25 m.

To achieve an approximately *uniform distribution*, subcritical flow is required. The critical depth is $h_{co} = (Q_o^2/gb^2)^{1/3} = (0.45^2/9.81 \cdot 0.5^2)^{1/3} =$ 0.44 m, thus $H_o > H_{co} = 1.5 \cdot 0.44 = 0.66$ m. These indications determine the overfall length ΔL and the weir height w. Assuming for security reasons $y_o = 0.75$ ($\mathsf{F}_o = 0.82$), gives $H_o^{3/2} = [Q/(gb^2)^{1/2}]/y_o[2(1 - y_o)]^{1/2} =$ $[0.45/(9.81 \cdot 0.5^2)^{1/2}]/0.75[2(1 - 0.75)]^{1/2} = 0.54$ from Eq. (17.18), corresponding to $H_o = 0.667$ m, such that $H_o = 0.69$ m is selected.

The sideweir length depends on the relative weir height, from Eq. (17.17). One computes $\Delta X(0) = 1.1$, $\Delta X(0.2) = 1.6$, $\Delta X(0.4) = 2.8$ and $\Delta X(0.5) =$ 4.2, or $\Delta x(0) = (b/k)\Delta x = (0.5/1)1.1 = 0.55$ m, $\Delta x(0.2) = 0.8$ m, $\Delta x(0.4) =$ 1.4 m and $\Delta x(0.5) = 2.1$ m. Select $\Delta x = 2$ m, with five single openings each 0.40 m long and of $W = 0.5$, corresponding to $w = 0.50 \cdot 0.69 = 0.35$ m. From Fig. 17.8a and $X_u = 0$, one has $X_o = 2/0.5 = 4$ for $W = 0.50$, thus $y_o = 0.70$, generating subcritical flow over the distribution channel.

The center locations of the laterals are $x_1 = -0.20$ m, $x_2 = -0.60$ m, $x_3 = -$ 1.0 m, $x_4 = -1.4$ m, $x_5 = -1.8$ m, thus $X_{1..5} = -0.4, ...-3.6$. The corresponding flow depths are $y_1 = 0.996$, $y_2 = 0.97$, $y_3 = 0.92$, $y_4 = 0.85$, $y_5 = 0.76$ from Fig. 17.8a with discharges $Q_1 = 0.08$ m^{3}s^{-1}, $Q_2 = 0.24$ m^{3}s^{-1}, $Q_3 = 0.37$ m^{3}s^{-1}, $Q_4 = 0.466$ m^{3}s^{-1}, $Q_5 = 0.473$ m^{3}s^{-1}from Eq. (17.18). Difference discharges are $\Delta Q_{12} = 0.16$ m^{3}s^{-1}, $\Delta Q_{23} = 0.13$ m^{3}s^{-1}, $\Delta Q_{34} = 0.10$ m^{3}s^{-1}, $\Delta Q_{45} = 0.01$ m^{3}s^{-1}. This indicates a strong increase of lateral discharges toward the end of the distribution channel.

17.6.2 *Effect of Friction*

Distribution channels may be so long as compared to the approach flow depth that wall friction becomes significant, and the assumption of a constant energy head may not be acceptable. The following example is aimed to determine the *friction losses* for a simplified free surface configuration.

Example 17.10 Compute the friction loss along a distribution channel!
The friction losses can only be approximated because the surface profile $h(x)$ and the local discharge $Q(x)$ are not explicitly known. Assuming a *linear* variation of all parameters, then flow depth $h(x)$, channel width $b(x)$ and average velocity $V(x)$ are, respectively

$$h(x) = h_o + (h_u - h_o)(x/L),$$

$$B(x) = b_o + (b_u - b_o)(x/L),$$

$$V(x) = V_o + (V_u - V_o)(x/L).$$

The relative values $\alpha_o = (h_u - h_o)/h_o$, $\beta_o = (b_u - b_o)/b_o$ and $\gamma_o = (V_u - V_o)/V_o$ plus relative location $X_L = x/L$ are introduced. According to Manning and Strickler, the friction slope S_f for turbulent rough flow is with the hydraulic radius $R_h = h(x) \cdot b(x)/[b(x)+2h(x)]$

$$S_f = \frac{n^2 V^2}{R_h^{4/3}} \tag{17.37}$$

and the average (subscript a) over length L is

$$S_{fa} = \int_0^1 S_f \mathrm{d}X_L. \tag{17.38}$$

Inserting the equations for $h(x)$, $b(x)$ and $V(x)$ in Eq. (17.38) gives

$$\begin{aligned} S_{fa} &= \frac{n^2 V_o^2}{R_{ho}^{4/3}} \int_0^1 (1 + \gamma_o X_L)^2 (R_{ho}/R_h)^{4/3} \mathrm{d}X_L \\ &\cong S_{fo} \int_0^1 (1 + \gamma_o X_L)^2 (R_{ho}/R_h) \mathrm{d}X_L. \end{aligned} \tag{17.39}$$

The ratio of hydraulic radii R_{ho}/R_h is with the aspect ratio $\delta_o = h_o/b_o$

$$\frac{R_{ho}}{R_h} = \frac{1}{1+2\delta_o}\frac{1}{1+\alpha_o X_L} + \frac{2\delta_o}{1+2\delta_o}\frac{1}{1+\beta_o X_L}. \tag{17.40}$$

The result of integration obtains

$$S_{fa}/S_{fo} = \frac{1}{(1+2\delta_o)a_o^3}\left[2a_o^2\gamma_o + (1/2)a_o^2\gamma_o^2 - \alpha_o\gamma_o^2 + (\gamma_o - \alpha_o)^2 \ln|1+a_o|\right]$$
$$+\frac{2\delta_o}{(1+2\delta_o)\beta_o^3}\left[2\beta_o^2\gamma_o + (1/2)\beta_o^2\gamma_o^2 - \beta_o\gamma_o^2 + (\gamma_o - \beta_o)^2 \ln|1+\beta_o|\right]. \tag{17.41}$$

For a prismatic channel ($\beta_o = 0$), this simplifies to

$$S_{fa}/S_{fo} = \frac{1}{(1+2\delta_o)a_o^3}\left[2a_o^2\gamma_o + (1/2)a_o^2\gamma_o^2 - \alpha_o\gamma_o^2 + (\gamma_o - \alpha_o)^2 \ln|1+a_o|\right]$$
$$+\frac{2\delta_o}{1+2\delta_o}\left[1+\gamma_o+(1/3)\gamma_o^2\right]. \tag{17.42}$$

The ratio S_{fa}/S_{fo} depends on the three parameters α_o, γ_o and δ_o, and Fig. 17.18 shows the result for $V_u = 0$, corresponding to $\gamma_o = -1$. It is noted that $S_{fa}/S_{fo} \leq 1/3$, i.e. the average friction slope is always *smaller* than one third of the approach flow friction slope S_{fo}.

For *pseudo-uniform flow* one has nearly $\alpha_o = \gamma_o = 0$, and $\beta_o = (Q_u - Q_o)/Q_o$, thus from Eq. (17.41)

$$S_{fa}/S_{fo} = \frac{2\delta_o}{1+2\delta_o}\frac{\ln|1+\beta_o|}{\beta_o}. \tag{17.43}$$

Fig. 17.18b shows that the ratio S_{fa}/S_{fo} increases with δ_o. In all cases, the ratio is also smaller than 1, and a simple assumption for the average friction slope would be $S_{fa} = (1/2)(S_{fo} + S_{fu})$, as previously introduced.

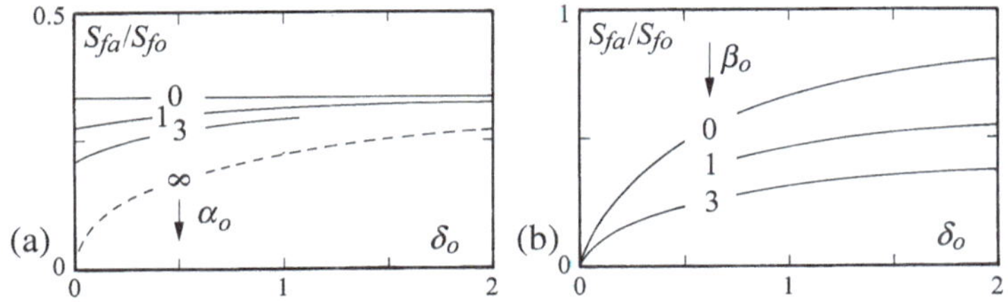

Fig. 17.18 Ratio of average and approach flow friction slopes S_{fa}/S_{fo} as a function of $\delta_o = h_o/b_o$ for (**a**) prismatic dead-end distribution channel and various surface slopes $\alpha_o = (h_u - h_o)/h_o$ and $V_u = 0$, (**b**) pseudo-uniform flow with $\alpha_o = \gamma_o = 0$

17.6.3 *Lateral Outflow Features*

For a long lateral, the *lateral outflow angle* depends only on the opening geometry and the approach flow conditions. For the sideweir, the governing parameters are y and W, thus the relative overflow depth h/w and the local Froude number (Hager 1987). For *short laterals* an effect of the lateral length ΔL adds, which can be accounted for only by including viscosity and surface tension, from where *scale effects* result, as for weirs with a small overflow depth. Figure 17.19 shows a photographic series under identical approach flow conditions in a 0.30 m wide rectangular channel for a single lateral of which the length was reduced progressively from

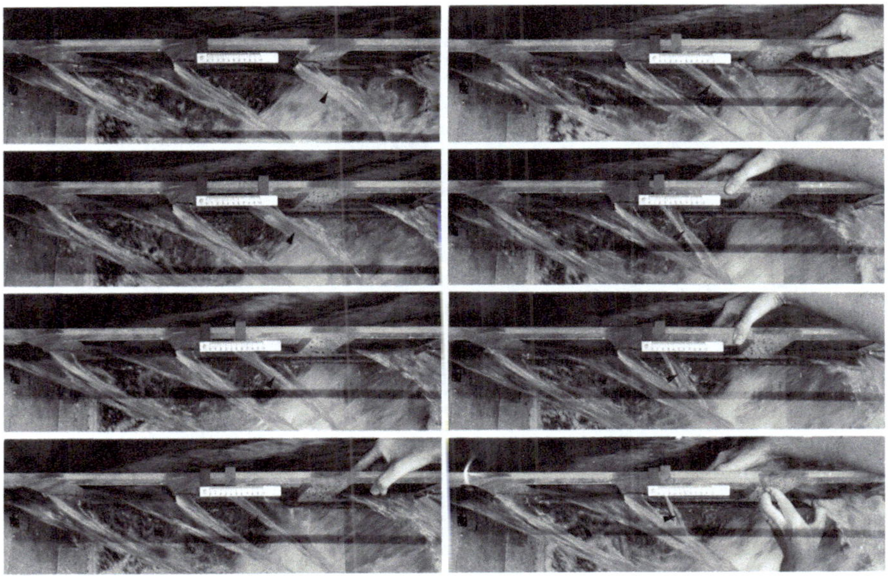

Fig. 17.19 Lateral outflow from distribution channel under otherwise invariable flow conditions for a progressively reduced outflow length ΔL [mm] = 200, 100, 50, 30, 20, 10, 5, 1 (Hager 1984)

0.20 m. Down to about $\Delta L = 0.050$ m, no significant effect is noted, but the outflow angle tends to $\phi = 90°$ beyond this limit. For extremely small length, the outflow leaves the distribution channel perpendicular, independent of the local Froude number and the relative overflow depth. Under such conditions, the outflow is forced, and the present computational approach is invalid.

If a *perforated plate* is mounted at the outflow plane, the lateral outflow directions of the individual holes are close to 90°, and the outflow direction compared to a normal outflow modified. Figure 17.20 shows the effect with two adjacent laterals, and verifies the differences between the 'long' and the 'short' laterals. In the extreme, the outflow angle is $\phi = 90°$, and the additional loss due to Eq. (17.5) is $S_L = -Q(\mathrm{d}Q/\mathrm{d}x)/gF^2$. Then, the present analysis is also no more valid, as noted previously. The minimum length of a lateral required for the application of this model is about one to two times the approach flow depth, because the flow is then nearly continuous.

Distribution channels have received relatively scarce attention. Mention might be made of the research works at the Oklahoma State University (Sweeten et al. 1969, Sweeten and Garton 1970, Uhl and Garton 1972). They refer to rectangular channels with lateral siphons. The effect of oblique approach flow is not significant, however, because velocities are smaller than about 1 ms^{-1}. The effect of friction is to be included because of the significant length of the channels.

Sweeten and Garton (1970) considered the distribution channel with sideweirs because of its simple performance. Based on energy conservation and the friction losses, and a semi-empirical outflow equation, a computational model was set up and compared with observations. These cannot be used to verify the present results because of information missing. Also, the lateral outflow equations are dimensionally incorrect.

(a) (b)

Fig. 17.20 Lateral outflow from distribution channel with manifold sideweirs. Effect of lateral length on the outflow direction

17.6.4 *Pseudo-Uniform Flow*

Pseudo-uniform flow has been developed late as an effective means to improve distribution flow. Mention can be particularly made of the research conducted by

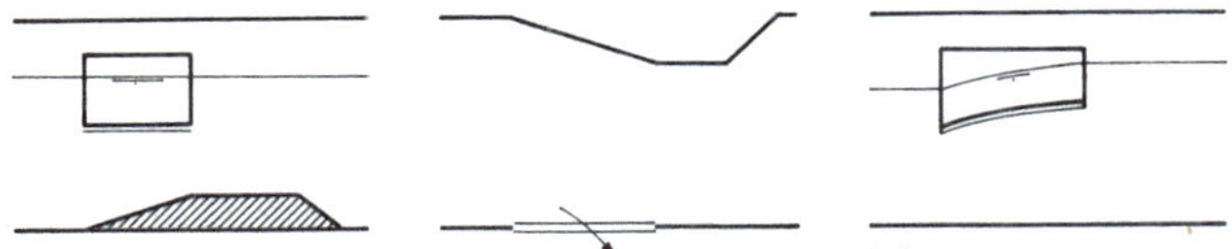

Fig. 17.21 Influence of pseudo-uniform flow by local insets (Ramamurthy et al. 1975)

Prof. A.S. Ramamurthy at Concordia University, Montreal (Canada). Since 1975, he and his collaborators advanced several contributions to this special case of free surface flow. Basically, three different *methods* are available (Fig. 17.21):

- Local bottom inset,
- Local channel constriction, and
- Increase of weir height.

Contrary to later works, a *locally* uniform flow was sought, and the distribution channel was prismatic. Only in 1978, Ramamurthy et al. have proposed a distribution channel as shown in Fig. 17.5 with a pseudo-uniform flow over the entire length of the distribution channel.

Without recourse to pseudo-uniform flow, Chao and Trussell (1980) refered to the distribution channel. They pointed at the variability of the overflow coefficient with the Froude number. Contrary to the present formulation with a locally varied coefficient, their coefficient varies from lateral to lateral. To improve the *outflow uniformity*, they suggested:

- Adaption of outflow geometry, such as weir height,
- Width increase of distribution channel to reduce channel velocity,
- Linear width reduction, without mentioning pseudo-uniform flow,
- Linear width reduction combined with an adaption of weir height, and
- Reduction of overflow lengths by keeping the weir height constant.

For distribution channels as applied in sewage treatment stations, the third method is recommended because of advantages for both design and off-design conditions.

Example 17.11 Given a distribution channel of length $L_v = 15$ m, approach flow width $b_o = 1.2$ m and approach flow discharge $Q_o = 2.1$ m^{3}s^{-1}. Describe the design provided half of the approach flow discharge should be distributed and the minimum weir height is $w = 0.40$ m?
The critical flow depth is $h_{co} = [Q_o^2/(gb_o^2)]^{1/3} = (2.1^2/9.81 \cdot 1.2^2)^{1/3} = 0.68$ m, thus $H_{co} = 1.5h_{co} = 1.02$ m. To obtain a sufficiently small Froude number, the approach flow energy head is set $H_o = 1.10$ m, thus $W = w/H = 0.4/1.10 = 0.36$.

For a constant energy head, the width is reduced according to Eq. (17.22) from $b_o = 1.2$ m to $b_u = 0.60$ m. Figure 17.4 relates y_{PN} to Θ. For example, one has with $W = 0.36$ for $\Theta = -0.05$ a value of $y_{PN} = 0.51$, and $y_{PN}(-0.1) = 0.59$, $y_{PN}(-0.2) = 0.69$, and $y_{PN}(-0.4) = 0.80$. Accordingly, Θ must either be smaller than –0.2, or W is increased. Selected are $W = 0.36$ and $\Theta = -0.3$, thus $y_{PN} = 0.75$. Equation (17.19) gives $\bar{Q}'/k = -0.3 \cdot 0.75[2(1-0.75)]^{1/2} = -0.16$ thus with $k = 1$ follows $Q' = \mathrm{d}Q/\mathrm{d}x = -1 \cdot 0.16(9.81 \cdot 1.1^3)^{1/2} = -0.57$ m^{2}s^{-1} and $\Delta x = -\Delta Q/Q' = 1.05/0.57 = 1.83$ m. With $\Theta = \theta = -0.3$ one has $\Delta L = \Delta B/\Theta = 0.6/0.3 = 2$ m almost as above. With $h_{PN} = 0.75 \cdot 1.1 = 0.83$ m and $w = 0.4$ m the overflow depth is $h_{PN} - w = 0.43$ m.

The distribution channel can be arranged with, say, five laterals each 0.40 m long, and the substitute contraction angle would be $\Theta = -0.6/15 = -0.04$. A second iteration should be made with this value.

The design procedure of distribution channels is rather elaborate. In addition, the effects of friction have to be included in the final design. In the first step, pseudo-uniform flow is assumed for the *design discharge* and other discharges are then accounted for with the generalized procedure according to Sect. 17.5. Finally, the free surface profiles, the lateral discharges and the boundary flow depths are presented as a function of discharge. Numerical techniques may be used for simplification of the computations.

The design of a distribution channel should be based on thoroughly *subcritical flow*. In particular cases, thoroughly supercritical flows may be used as a design basis. Froude numbers in the domain $0.75 < \mathsf{F} < 1.5$ should be excluded with structural arrangements. A design containing *hydraulic jumps* is normally poor, and must be excluded. Such flows are highly sensitive and the computational procedure outlined becomes inaccurate. Figure 17.22 shows the aesthetics of a distribution channel with closely arranged sideweirs.

Fig. 17.22 Distribution channel with sideweirs as outlets

17.7 Channel Bifurcation

17.7.1 *Flow Pattern*

A channel bifurcation (German: Kanalverzweigung; French: Canal à dérivation) is a structure dividing the approach flow discharge into various portions. Such structures can be compared with a distribution channel, but they involve hydraulically *short* elements, such as the channel junction. The counterpart to the distribution channel is the side channel (Chap. 19).

In general, bifurcations are hydraulically more complex than junctions, because the latter have a nearly one-dimensional approach flow, whereas the bifurcation is characterized by large scale separation, such as in diffusors. Therefore, the main body of literature to bifurcations is based on experiments. In what follows the key information is highlighted because a detailed state-of-the-art is given elsewhere (Hager 1991).

Figure 17.23a refers to a bifurcating rectangular channel with δ as bifurcation angle. Figure 17.23b shows the complex secondary currents, with a strong bottom current toward the lateral branch causing a significant sediment transport capacity. Intake channels are thus located in the through branch with a small sediment transport, and not in the lateral branch.

Mock (1960) investigated *subcritical flow* in a channel bifurcation of equal branch widths. With $q_a = Q_a/Q_o$ as the branch discharge ratio and subscripts a, o and u denoting the lateral, the upstream and the downstream branches, respectively, the loss coefficients based on the approach flow velocity V_o are (Chap. 2)

$$\xi_u = \frac{H_o - H_u}{V_o^2/2g} \quad \text{and} \quad \xi_a = \frac{H_o - H_a}{V_o^2/2g}. \tag{17.44}$$

His model observations may be expressed as

$$\xi_a = 1 - (5/4)q_a + 0.725 q_a \left[1 + q_a^2\right] \tan(\delta/2), \tag{17.45}$$

$$\xi_u = 0.45 q_a (q_a - 0.5). \tag{17.46}$$

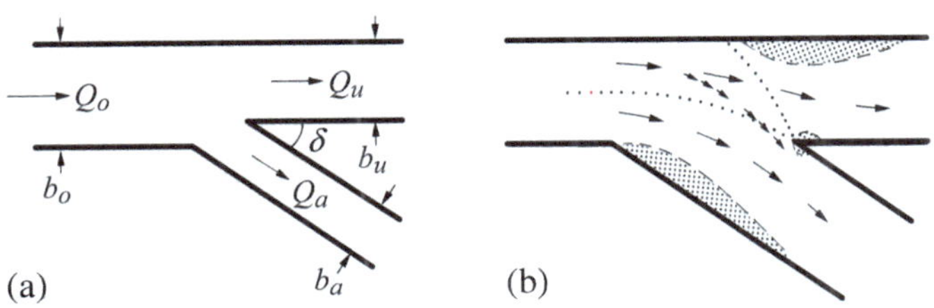

Fig. 17.23 Channel bifurcation (**a**) definition sketch, (**b**) (→) surface and (→→→) bottom currents, (...) corresponding separation lines with separated flow region (*shaded*)

The conclusions are:

- These observations are identical with those for *pressurized flow* and the transfer law for Froude numbers up to about 0.5, or even 0.7, as discussed in 2.4 can be applied.
- Coefficient ξ_u of the through branch is *independent* of the branch angle, in agreement with distribution channels.
- Coefficient ξ_u can be both positive and negative, and the energy head can *increase* or decrease.
- The lateral branch may have significant losses, particularly as $q_a \to 1$, depending significantly on the *bifurcation angle* δ.

The flow pattern can be described with Fig. 17.24. For $q_a = 0$, i.e. no lateral discharge, a strong secondary current governs the flow in the lateral branch. Because the flow depth in the branch channel is almost equal to that of the main branch, all the approach velocity is dissipated and $\xi_a(0) = 1$.

For $q_a > 0$ the flow can be compared with subcritical bend flow. Depending on the bifurcation angle and the discharge ratio, a separation zone is developed either at the inner bend side, or at the through-flow branch in addition. Then, the losses are again significant, as may be seen from Eq. (17.45). If sewage is transported across bifurcations, particles may settle along the separation zones and cause additional maintenance. Sharp-edged geometries are thus not a satisfactory design, although they are often found combined with gates to block either of the branches.

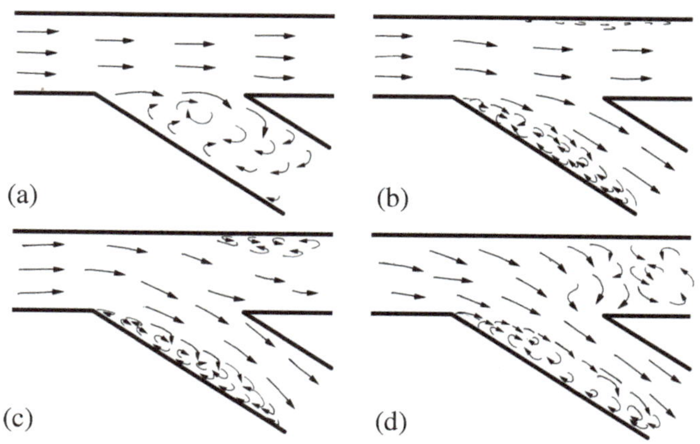

Fig. 17.24 Flow patterns in channel bifurcation for $q_a = Q_a/Q_o =$ (**a**) 0, (**b**) 1/3, (**c**) 2/3, (**d**) 1 (Mock 1960)

17.7.2 *T-Bifurcation*

A T-bifurcation has a straight through branch and a perpendicular lateral of equal or smaller width (Fig. 17.25). For $\mathsf{F}_u < \mathsf{F}_o < 0.75$ and without submergence from the lateral branch into the main branch, the *distribution of discharge* is with $\beta_a = b_a/b_o \leq 1$ as the width ratio (Krishnappa and Seetharamiah 1963)

$$q_a = Q_a/Q_o = (1.55 - 1.45\mathsf{F}_o)\beta_a + 0.16(1 - 2\mathsf{F}_o). \tag{17.47}$$

No relation between the flow depths in the up- and downstream branches was given.

Lakshmana Rao et al. (1968) also considered this case and an alternative equation to Eq. (17.47) is

$$Q_u/Q_o = \tanh[5(0.56 - \mathsf{F}_a)\mathsf{F}_u^{1/2}]. \tag{17.48}$$

Flows with $\mathsf{F}_a > 1/3$ develop a hydraulic jump to be computed with $\mathsf{F}_a = 1/3$ (Fig. 17.26b). The contraction coefficient C_c of the lateral branch (Fig. 17.26c) depends on the downstream Froude number F_u as

$$\frac{1 - C_c}{1 - C_{c\infty}} = \left[\tanh(3\mathsf{F}_u)\right]^{2/3} \tag{17.49}$$

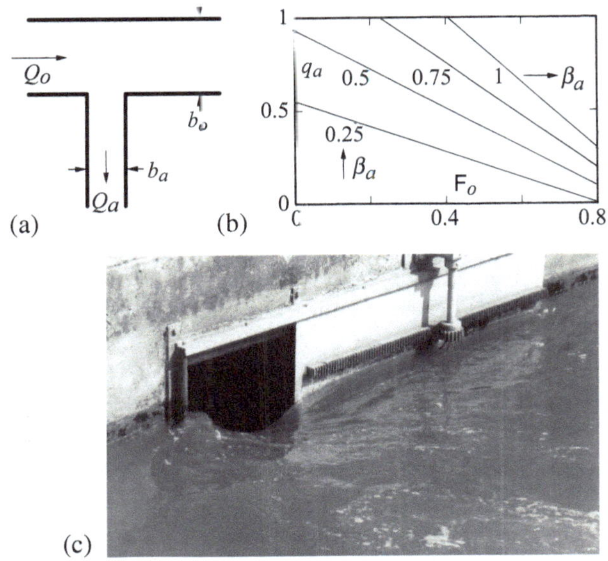

Fig. 17.25 T-bifurcation (**a**) notation, (**b**) discharge ratio $q_a = Q_a/Q_o$ versus approach flow Froude number F_o for supercritical flow in lateral branch ($\mathsf{F}_a > 1$) and width ratio $\beta_a = b_a/b_o$ (Krishnappa and Seetharamiah 1963). (**c**) Inflow to lateral channel with local drawdown effect

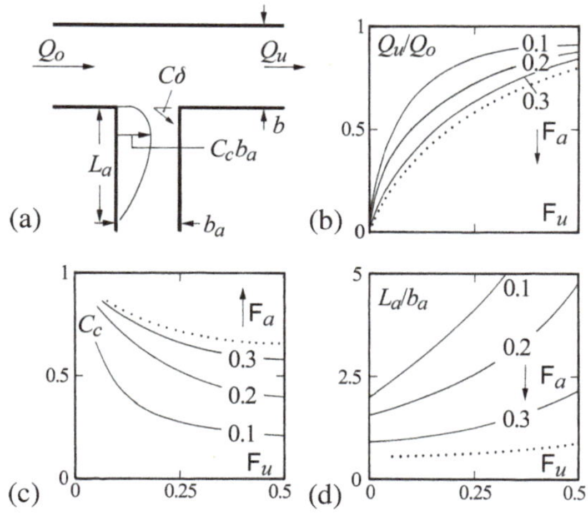

Fig. 17.26 T-bifurcation (**a**) definition plot, (**b**) discharge distribution Q_u/Q_o, (**c**) coefficient of contraction C_c and (**d**) relative length of separated flow L_a/b_a as functions of downstream Froude number $F_u = Q_u/(gb_u^2h_u^3)^{1/2}$ and $F_a = Q_a/(gb_a^2h_a^3)^{1/2}$. (...) Hydraulic jump in lateral branch (Lakshmana Rao et al. 1968)

with $C_{c\infty}$ as contraction coefficient for $F_u >> 0$

$$C_{c\infty} = (2/3)\tanh(3.5F_a). \tag{17.50}$$

The section of *maximum flow contraction* (Fig. 17.26a) is located about $(1/2)b_a$ downstream from the intake section. The separation length depends on both F_u and F_a (Fig. 17.26d). The average bifurcation direction $C\delta$ (Fig. 17.26a) varies with F_u as

$$C\delta = (2/3)(1 - F_u). \tag{17.51}$$

For $Q_u = 0$ ($q_a = 1$) the lateral direction is thus larger than for cases with discharge in the downstream branch.

Example 17.12 Given a channel bifurcation with $b_o = b_u = 1.2$ m and $b_a = 0.80$ m. Determine the main hydraulic characteristics for $Q_o = 0.75$ m^{3}s^{-1}, $Q_u = 0.60$ m^{3}s^{-1}and $F_u = 0.22$.
With $Q_u/Q_o = 0.6/0.75 = 0.8$ and $F_u = 0.22$, Eq. (17.48) gives $F_a = 0.56 - 1/(5F_u^{1/2})\ \text{arctanh}(Q_u/Q_o) = 0.56 - 0.43(1/2)\ln[(1+0.8)/(1-0.8)] = 0.09$. Further, from Eq. (17.50) $C_{c\infty} = (2/3)\tanh(3.5\cdot 0.09) = 0.20$, thus $(1 - C_c)/(1 - C_{c\infty}) = [\tanh(3\cdot 0.22)]^{2/3} = 0.69$ and $C_c = 1 - 0.69(1 - 0.20) = 0.45$ from Eq. (17.49). According to Fig. 17.26c C_c is somewhat smaller and the lateral flow is considerably contracted. With $C = (2/3)(1 - 0.22) = 0.52$ from Eq. (17.51) the average bifurcation angle is equal to $0.52\cdot 90° = 47°$, and the contracted section is located at $0.5\cdot 0.8 = 0.40$ m downstream from the intake section. The length of the separated zone is with $L_a/b_a = 4$ from Fig. 17.26d equal to $L_a = 4\cdot 0.8 = 3.2$ m.

Joliffe (1981) investigated sub- and supercritical flows in partially-filled pipe bifurcations of equal branch diameter and $\delta = 90°$. For *subcritical* downstream flow ($\mathsf{F}_u < 0.6$) the discharge ratio is, independent of bottom slope,

$$Q_a/Q_o = \exp[-(7/3)\mathsf{F}_u], \quad (17.52)$$

and for *supercritical* approach flow ($\mathsf{F}_c > 1$) the discharge ratio is

$$Q_a/Q_o = \mathsf{F}_o/(8.8\mathsf{F}_o - 4). \quad (17.53)$$

Ramamurthy et al. (1990) investigated the effect of *submergence* on T-bifurcations of equal branch width b. Figure 17.27 shows a plot presented in the discussions of their paper relating the relative flow depths $Y = h_o/h_u$ with the discharge ratio $q_a = Q_a/Q_o$ and the downstream Froude number $\mathsf{F}_u = Q_u/(gb^2h_u^3)^{1/2}$. Note that Y is limited to 0.80 because of the appearance of a hydraulic jump, otherwise.

Example 17.13 Given a T-bifurcation of width $b = 1.2$ m with $h_u = 0.8$ m, and $Q_u = 0.95$ m^{3}s^{-1}. What is the upstream depth for $Q_o = 1.2$ m^{3}s^{-1}?
With $1 - q_a = 1\text{–}0.25/1.2 = 0.79$ and $\mathsf{F}_u = 0.95/(9.81 \cdot 1.2^2 \cdot 0.8^3)^{1/2} = 0.35$, Fig. 17.27 gives $Y = 0.95$, thus $h_o = 0.95h_u = 0.95 \cdot 0.8 = 0.76$ m and the upstream Froude number of $\mathsf{F}_o = 1.2/(9.81 \cdot 1.2^2 \cdot 0.76^3)^{1/2} = 0.48$. With $V_u = 0.95/(1.2 \cdot 0.8) = 1$ ms^{-1} and $V_o = 1.2/(1.2 \cdot 0.76) = 1.32$ ms^{-1}, the energy heads are $H_u = 0.8 + 1^2/19.62 = 0.851$ m and $H_o = 0.76 + 1.32^2/19.62 = 0.849$ m. The head loss coefficient is with $\xi_u = (H_o - H_u)/[V_o^2/2g)] = -0.002/[1.32^2/19.62] = -0.023$ almost equal to zero. From Eq. (17.46) one computes $\xi_u = -0.027$.

The *energy losses* of bifurcations have been determined for angles $\delta = 45°$ up to 90° by Peruginelli and Pagliara (1992). Semi-empirical loss coefficients are presented in terms of branch Froude numbers and branch discharge ratio. The three-dimensional flow structure in bifurcating channels was numerically investigated by Neary and Odgaard (1993), and Shettar and Keshava Murthy

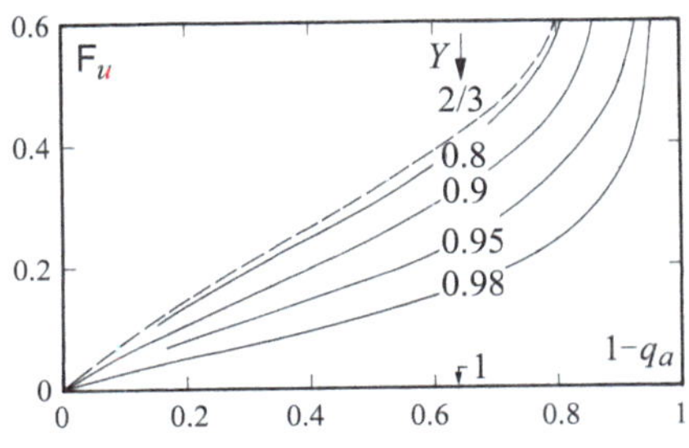

Fig. 17.27 Flow depth ratio $Y = h_o/h_u$ for T-bifurcation as a function of the discharge ratio $1-q_a$ and Froude number F_u in downstream branch

(1996). These studies provide detailed insight into the secondary currents and the predictions are in general agreement with observations.

More complex bifurcation geometries involving a lateral gate or a bottom drop have been investigated by Acuna et al. (1972), Nougaro and Boyer (1974), and Nougaro et al. (1975), respectively. Typical flow patterns are shown in Fig. 17.28.

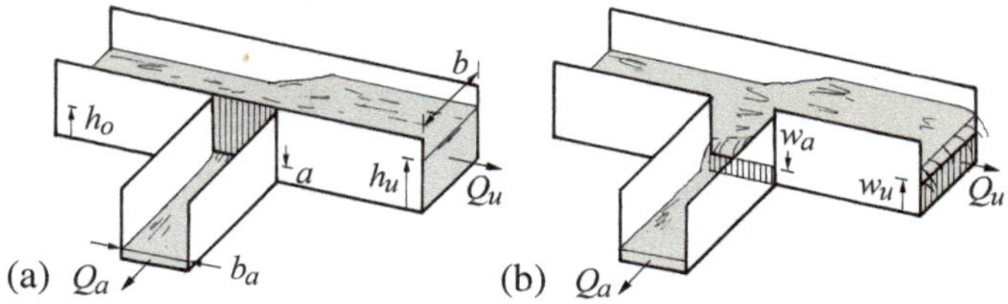

Fig. 17.28 Bifurcation geometries of (**a**) Acuna et al. (1972) (**b**) Nougaro and Boyer (1974), Nougaro et al. (1975)

Notation

a	[m]	bottom opening width
A	[–]	$= a/H$ relative opening width
b	[m]	channel width
B	[m]	variable channel width
c_k	[–]	crest coefficient
C	[–]	bifurcation angle coefficient
C_c	[–]	contraction coefficient
C_d	[–]	discharge coefficient
E	[m^{4}s^{-1}]	energy related to (ρg)
F	[m^2]	cross-sectional area
F	[–]	Froude number
g	[ms^{-2}]	gravitational acceleration
h	[m]	flow depth
H	[m]	energy head
J	[–]	total slope
k	[–]	$= n^* c_k$ constant
L_a	[m]	length of separated branch flow
L_E	[m]	length of substitute system
L_v	[m]	length of distribution channel
ΔL	[m]	overflow length
n^*	[–]	number of lateral outflow sides
$1/n$	[m$^{1/3}$s^{-1}]	Manning's roughness coefficient
p	[m]	pressure head
q	[–]	discharge intensity
q_a	[–]	$= Q_a/Q_o$ discharge ratio
Q	[m^{3}s^{-1}]	discharge

ΔQ	[m^{3}s^{-1}]	lateral discharge
R_h	[m]	hydraulic radius
s	[m]	height of side opening
S	[–]	$= s/H$ relative height
S_e	[–]	energy line gradient
S_f	[–]	friction slope
S_L	[–]	additional slope
S_o	[–]	bottom slope
u	[ms^{-1}]	lateral outflow velocity
V	[ms^{-1}]	channel velocity
w	[m]	weir height
W	[–]	$= w/H$ relative weir height
x	[m]	streamwise coordinate
X	[–]	$= kx/b$ relative location
X_L	[–]	relative position
y	[–]	$= h/H$ relative flow depth
Δz	[m]	bottom height increase
α_o	[–]	flow depth ratio
β_a	[–]	$= b_a/b_o$ width ratio
β_o	[–]	width difference ratio
γ_o	[–]	velocity difference ratio
δ	[–]	bifurcation angle
δ_o	[–]	aspect ratio
ϕ	[–]	lateral outflow angle
θ	[–]	contraction slope
Θ	[–]	$= \theta/k$ relative contraction
λ_s	[–]	relative length
ρ	[kgm^{-3}]	density
σ	[–]	pseudo-uniform flow parameter
χ_c	[–]	maximum length without jump
ξ_a	[–]	loss coefficient of bifurcation branch
ξ_u	[–]	loss coefficient of downstream branch

Subscripts

a	average, lateral branch
c	critical
M	maximum
N	uniform
o	upstream
PN	pseudo-uniform
r	boundary
u	downstream

References

Acuna, E., Aravena, L., Flores, J., Miquel, J., Fuentes, R. (1972). Determinacion del coefficiente de gasto de compuertas laterales con resalto rechazado. 5 *Congreso Latinoamericano de Hidraulica* Lima A1: 1–10 [in Spanish].

Chao, J.-L., Trussell, R.R. (1980). Hydraulic design of flow distribution channels. *Journal of Environmental Engineering Division* ASCE 106(EE2): 321–334; 106(EE6): 1212–1213; 107(EE1): 299–303; 107(EE2): 432–433; 107(EE5): 1109.

De Marchi, G. (1934). Saggio di teoria del funzionamento degli stramazzi laterali (Theoretical knowledge on the functioning of sideweirs). *L'Energia Elettrica* 11(11): 849–860 [in Italian].

Favre, H. (1933). *Contribution à l'étude des courants liquides* (Contribution to the study of liquid flows). Rascher: Zürich [in French].

Hager, W.H. (1984). Some scale effects in distribution channels. Symposium *Scale Effects in Modelling Hydraulic Structures* 2(9): 1–6, H. Kobus, ed. Technische Akademie: Esslingen, Germany.

Hager, W.H. (1985). Bodenöffnung in Entlastungsanlagen von Kanalisationen (Bottom openings of sewer laterals). *Gas-Wasser-Abwasser* 65(1): 15–23 [in German].

Hager, W.H. (1986a). L'écoulement dans des déversoirs latéraux (Flow in sideweirs). *Canadian Journal of Civil Engineering*13(5): 501–509 [in French].

Hager, W.H. (1986b). Flow in distribution conduits. *Proc. Institution of Mechanical Engineers* 200 (A3): 205–213.

Hager, W.H. (1987). Lateral outflow over sideweirs. *Journal of Hydraulic Engineering* 113(4): 491–504; 115(5): 682–688.

Hager, W.H. (1991). Abflussverhältnisse in Kanalverzweigungen (Flow patterns of channel bifurcations). *Korrespondenz Abwasser* 38(10): 1350–1357 [in German].

Hager, W.H., Volkart, P. (1986). Distribution channels. *Journal of Hydraulic Engineering* 112(10): 935–952; 114(2): 235.

Joliffe, I.B. (1981). Accurate pipe junction model for steady and unsteady flows. 2nd Intl. Conf. *Urban Storm Drainage* Urbana: 93–100.

Krishnappa, G., Seetharamiah, K. (1963). A new method of predicting the flow in a 90° branch channel. *La Houille Blanche* 18(7): 775–778.

Lakshmana Rao, N.S., Sridharan, K., Yahia Ali Baig, M. (1968). Experimental study of the division of flow in an open channel. Conf. *Hydraulics and Fluid Mechanics*: 139–142. The Institution of Engineers, Australia: Sydney.

Mock, F.-J. (1960). Strömungsvorgänge and Energieverluste in Verzweigungen von Rechteckgerinnen (Flow patterns and energy losses in bifurcations of rectangular channels). *Mitteilung* 52. Institut für Wasserbau und Wasserwirtschaft, TU Berlin: Berlin [in German].

Neary, V.S., Odgaard, A.J. (1993). Three-dimensional flow structure at open-channel diversions. *Journal of Hydraulic Engineering* 119(11): 1223–1230.

Nougaro, J., Boyer, P. (1974). Sur la séparation des eaux dans les dérivations de canaux à section rectangulaire. *La Houille Blanche* 29(3): 199–203 [in French].

Nougaro, J., Boyer, P., Claria, J. (1975). Comportement d'une dérivation de canaux lorsque les biefs aval sont pourvus de retenues. *La Houille Blanche* 30(4): 267–273 [in French].

Peruginelli, A., Pagliara, S. (1992). Energy loss in dividing flow. *Entropy and energy dissipation in water resources*, V.P. Singh, M. Fiorentino, eds. Kluwer: Dordrecht NL.

Ramamurthy, A.S., Subramanya, K., Carballada, L. (1975). Uniformly discharging outlets for irrigation systems. Proc. 2nd World Congress on Water Resources *Water for Human Needs* 5: 323–326.

Ramamurthy, A.S., Subramanya, K., Carballada, L. (1978). Uniformly discharging lateral weirs. *Journal of Irrigation and Drainage Division* ASCE 104(IR4): 399–412.

Ramamurthy, A.S., Tran, D.M., Carballada, L.B. (1989). Open channel flow through transverse floor outlets. *Journal of Irrigation and Drainage Engineering* 115(2): 248–254; 117(1): 148–151.

Ramamurthy, A.S., Tran, D.M., Carballada, L.B. (1990). Dividing flow in open channels. *Journal of Hydraulic Engineering* 116(3): 449–455; 118(4): 634–637.
Shettar, A.S., Keshava Murthy, K. (1996). A numerical study of division of flow in open channels. *Journal of Hydraulic Research* 34(5): 651–675.
Sweeten, J.M., Garton, J.E. (1970). The hydraulics of an automated furrow irrigation system with rectangular sideweir outlets. *Trans. American Society Agricultural Engineers* 13: 746–751.
Sweeten, J.M., Garton, J.E., Mink, A.L. (1969). Hydraulic roughness of an irrigation channel with decreasing spatially-varied discharge. *Trans. American Society Agricultural Engineers* 12: 466–470.
Uhl, V.W., Garton, J.E. (1972). Semi-portable sheet metal flume for automated irrigation. *Trans. American Society Agricultural Engineers* 15: 256–260.
Yen, B.C., Wenzel, H.G. (1970). Dynamic equations for steady spatially varied flow. *Journal of the Hydraulics Division* ASCE 96(HY3): 801–814.

Sideweir in combined sewer system, view against flow direction, without weir crest of sheet metal.

Chapter 18
Sewer Sideweir

Abstract A combined sewer system depends significantly on the performance of the lateral outlets. If these behave hydraulically poor, uncontrolled overflow may reach the receiving waters. It is thus imperative to direct the attention on the hydraulics of sewer sideweirs.

Two different designs of sewer sideweirs are currently used: First the standard design with a high weir crest, and second the design with a low weir crest. Both designs cannot be considered as weirs with a perpendicular approach flow, and the effects of approach flow velocity and direction as well as the weir geometry have to be accounted for.

Despite of relative complex relations, a simple design procedure is presented answering the various questions relating to the overflow pattern. In addition, the design of the sewer sideweir is generalized to reduce the number of parameters and to standardize future sewer sideweirs. The short sewer sideweir and the sideweir with a throttling pipe are introduced at the end of this chapter, highlighting the effect of the approach flow Froude number.

18.1 Introduction

According to Fahrner et al. (1990) there is currently a confusion regarding the design of sewer sideweirs in *combined sewer systems* (German: Mischsystem; French: Reseau unitaire d'assainissement). Apart from sharp-crested, round-crested and broad-crested weirs (Chap. 10) a variety of particular arrangements have been suggested for which no design bases are available. Currently, a standard structure has not yet been suggested and the hydraulics of sewer sideweirs are incomplete. This defect has to be removed to increase the efficiency and the performance of the sewer sideweir and to assess its environmental performance.

Basically, two types of sewer laterals are currently available, namely the sewer sideweir (German: Streichwehrentlastung; French: Déversoir latéral de décharge) with an *overflow* element, and the leaping weir (German: Springüberfall; French: Ouverture de décharge de fond) with an *orifice* element. As explained in Chap. 20,

W.H. Hager, *Wastewater Hydraulics,* 2nd ed., DOI 10.1007/978-3-642-11383-3_18, the leaping weir is applied for supercritical approach flow only. The sewer sideweir has two typical configurations:

- Either with a high weir to apply a backwater effect on the upstream reach,
- Or with a low weir to allow for supercritical flow along the weir crest.

The sewer sideweir with the *high weir crest* corresponds to the classical sewer lateral because the flow pattern can be modelled with a conventional hydraulic approach. Due to the large flow depth, the flow in the approach flow sewer is subcritical and hydraulic jumps are eventually located away from the sewer outlet. Approach flow sewers with a mild bottom slope can ideally be used as *storage channels* that are emptied after the rainfall has ceased.

The cross-section of the sewer sideweir is U-shaped and the transition between the approach flow reach and the downstream throttling pipe is *linearly converging*. The bottom slope along the sideweir is relatively large to prevent submergence for minimum discharges. The overflow crest can be arranged on either one or two sides, with free flow in the side channel receiving the overflow. Figure 18.1 shows the typical sewer sideweir as currently used in Europe, with a converging plan and a diverging section. The weir crest is always horizontal and should be sharp-crested made of metal sheet to allow for weir height adjustment. The design weir crest level should be fixed to prevent unauthorized personnel from crest level alteration. This was recognized as one of the major causes for a poor performance of sewer sideweirs in the past. Altering a crest elevation can only be made based on a serious hydraulic design. Sewer sideweirs are normally not modified by skimming or baffle walls because they complicate a sound hydraulic design and may lead to problems with clogging.

Sewer sideweirs with a *low weir crest* were designed in the past to reduce the overflow length at locations of limited space. These are shorter than the standard structure, and apply no submergence on the upstream sewer, but they are hydraulically difficult to assess. In general, this structure is not recommended. Recently, a systematic series on short sewer sideweirs was conducted with the results presented in Sect. 18.5.

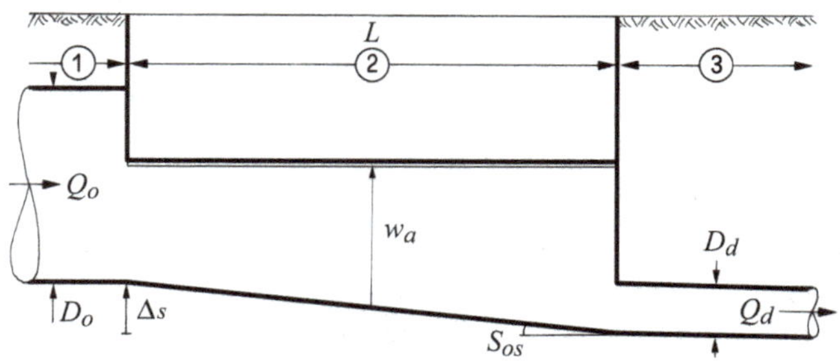

Fig. 18.1 Typical sewer sideweir with *high* weir crest ① approach flow sewer, ② weir overflow reach of length ΔL, ③ throttling pipe

In the following the sewer sideweir with a *high weir crest* is discussed in detail and the computational procedure is illustrated with selected examples. Sewer sideweirs with a low weir crest are also analyzed, mainly because of the many existing structures. In Sect. 18.2, the knowledge is summarized to present a standard design geometry. Because of the many geometrical parameters, this *standardisation* is imperative to quantify its features. It should be noted that the main results refer to *subcritical approach flow*. For supercritical approach flow, leaping weirs are recommended (Chap. 20). The throttling pipe is partly discussed in Sect. 9.3, in Sect. 18.3.3, and particularly in Sect. 18.6.

18.2 Design Basis

18.2.1 *Basic Knowledge*

According to a literature review of Hager (1993), there is currently no thorough work available on sewer sideweirs. Of interest are the papers by Kallwass (1964) who has considerably added to the current geometrical design. Further, Taubmann (1972) has performed experiments on sewer sideweirs without a data analysis, however. Mention might be made also of SIA (1980) who introduced a computational approach, based on the previous findings. In the following, a modified approach is presented that applies to standard sewer sideweirs. For complex configurations such as those in Basel, Switzerland (Siegenthaler 1981), scale modelling is actually the only method for reliable predictions. In addition, an *optimisation* of the structure can be developed to increase the overall safety of the sewer system.

18.2.2 *Description of Standard Structure*

Sewer sideweirs with a high crest and a throttling pipe may be considered as the main *storm water outlet* (German: Regenentlastungsbauwerk; French: Déchargeur de pluie). It is recommended for future designs in combined sewers, because of its advantages (SIA 1980):

- Small excess throttling discharge for maximum approach discharge,
- Security against flooding from receiving water, and
- Utilization of storage capacity of sewer.

The disadvantages of the sewer sideweir include:

- Control of laterals in upstream reach against submergence,
- Control of minimum approach flow velocities to prevent sedimentation.

Sewer sideweirs have at least three design discharges, namely the dry weather discharge Q_T, the design discharge Q_K of the treatment station, and the maximum discharge Q_M during rainfall. Contrary to the proposal of SIA (1980) limiting the maximum approach flow bottom slope to 1%, a *stable subcritical approach flow* ($\mathsf{F}_o < 0.75$) is required. In general, the *two-sided* lateral outflow configuration is recommended, both for reasons of compactness and flow symmetry.

A sewer sideweir consists of three parts (Fig. 18.1):

- *Approach flow sewer* as the transition between a circular pipe of diameter D_o and the U-shaped sideweir profile,
- *Lateral overflow reach* with a horizontal crest and an average (subscript m) weir height w_m, sideweir (subscript s) bottom slope $S_{os} = \Delta s/\Delta L$ with Δs drop height and ΔL overflow length, and
- *Throttling pipe* (subscript d) of diameter D_d, bottom slope S_{od} and length L_d.

The lateral discharge is collected by a side channel (Chap. 19). The maximum flow depth in the side channel should be always lower than the weir crest to prevent submergence.

Depending on the approach flow discharge, a sewer sideweir has a different task:

- *Dry weather discharge* Q_T without sedimentation along the invert of the sideweir,
- *Design discharge* Q_K of the treatment station just prior to overflow, and pressurized flow in the throttling pipe, and
- *Maximum discharge* Q_M during storm water conditions with a separation accuracy of 20%. Then, the discharge in the throttling pipe should be less than 120% of Q_K.

The following refers to sewer sideweirs with a high crest.

18.3 High-Crested Sewer Sideweir

18.3.1 *Approach Sewer*

In the approach flow reach of a sewer sideweir, the overflow start has to be fixed, and the weir height w_o, the drop height Δs and the minimum velocity must be determined. According to ATV (1993) the weir height w_o at the beginning of the sideweir must be between 0.5 and $0.8D_o$. The elevation difference Δs between beginning and end of the sideweir should be at least 0.03 m (Fig. 18.1). It can be fixed such that a nearly constant flow depth h_T develops along the converging cross-section for the dry weather discharge Q_T. For given discharges Q_T, Q_K and Q_M as well as diameters D_o and D_d, all geometrical parameters except for the weir length may thus be determined.

Example 18.1 Given the discharges $Q_T = 0.018$ m^{3}s^{-1}, $Q_K = 0.18$ m^{3}s^{-1} and $Q_M = 2$ m^{3}s^{-1}, an approach flow sewer with $D_o = 1.10$ m and $S_{oo} = 0.4\%$, a throttling pipe with $D_d = 0.35$ m, $S_{od} = 0.5\%$ and $L_d = 35$ m. Design the structure for a roughness coefficient $n = 0.012$ sm$^{-1/3}$(after SIA 1980). Table 18.1 shows the hydraulic characteristics of the approach flow sewer (subscript o) for uniform flow (subscript N) according to Chaps. 5 and 6. Note that the Froude numbers for all three discharges are close to 1.

Assuming a relative weir height of $w_o = D_o/2 = 0.55$ m gives for $Q_K = 0.18$ m^{3}s^{-1} an approach flow velocity $V_{oK} = 0.18/0.475 = 0.38$ ms^{-1}, which is too low according to Chap. 3. However, the values of Table 3.5 are difficult to attain for such conditions. According to SIA (1980), the minimum approach flow velocity to sideweirs may be reduced to

$$V_{mo}(Q_K)\ [\mathrm{ms}^{-1}] = 0.5 - 0.3S_{oo}\ [‰] \tag{18.1}$$

corresponding to $V_{mo} = 0.38$ ms^{-1} for $S_{oo} = 0.4\%$.

The elevation difference between the beginning and the end of the sewer sideweir based on Q_T assumes equal flow depth $h_{oT} = h_{uT}$. According to Eq. (5.14) one has with $y_{Nu} = y_{No}$

$$S_{os}^{1/2} = \frac{nQ/(D_d^{8/3})}{(3/4)y_{Nu}^2\left[1-(7/12)y_{Nu}^2\right]}. \tag{18.2}$$

For $h_{No} = 0.065$ m according to Table 18.1, $y_{Nu} = 0.065/0.35 = 0.186$, thus for $nQ/(D_d^{8/3}) = 0.012 \cdot 0.018/0.35^{8/3} = 0.0035$. The bottom slope along the sideweir is thus $S_{os}^{1/2} = 0.0035/0.0254 = 0.138$, corresponding to $S_{os} = 1.9\%$. The height $\Delta s = S_{os}\Delta L$ is determined once the sideweir length is known.

Table 18.1 Approach flow conditions for Example 18.1 and various discharges

Condition	Discharge [m^{3}s^{-1}]	q_N [–]	Uniform flow depth [m]	Uniform flow velocity [ms^{-1}]	Froude number [–]	Energy head [m]
Dry weather	0.018	0.003	0.065	0.79	1.30	0.097
Treatment design	0.180	0.026	0.207	1.44	1.28	0.311
Maximum	2.000	0.289	0.840	2.59	0.86	1.183

18.3.2 *Overflow Reach*

Overflow Formula

A weir in a straight rectangular channel is influenced by the (Chap. 10):

- Relative overflow depth $(h–w)/h$,
- Absolute overflow depth $(h–w)$,
- Crest geometry,
- Weir geometry, and
- Fluid properties.

For water or common sewage, the fluid has not a particular effect on the overflow process. The *sharp-crested weir* is a standard design provided the nappe is fully aerated. The crest level is sealed to prevent unauthorized modification. Based on a hydraulic design, a sharp-crested weir may easily be adapted to updated flow conditions. Provided overflow depths $(h–w)$ smaller than about 0.050 m are excluded, the effects of viscosity and surface tension are insignificant (Chap. 10), and the weir geometry is the only parameter influencing the overflow discharge Q. Conventionally, the latter is related to $(h–w)$, and the effect of the approach flow velocity has to be considered. However, if the *overflow energy head* $(H–w)$ instead of the overflow depth is accounted for, one may write

$$Q = C_D b(2g)^{1/2}(H - w)^{3/2}. \tag{18.3}$$

The discharge coefficient C_D related to the overflow energy head depends exclusively on the crest geometry and is typically $C_D = 0.402$ for the sharp-crested and $C_D = 0.325$ for the broad-crested weir. Further, C_D varies with the relative crest radius H/R for round-crested weirs (Chap. 10). The energy head is $H = h + Q^2/[2gb^2h^2]$ with g as gravitational acceleration, b as channel width and w as weir height.

Sideweirs are characterized by an *oblique approach flow* toward the weir plane. Compared to weirs with a perpendicular approach flow direction the following effects have to be accounted for (Hager 1987):

- Flow depth,
- Approach flow velocity,
- Lateral outflow angle,
- Sideweir geometry, and
- Weir configuration.

Whereas the flow depth h and the energy head H for a weir with a perpendicular approach flow direction are almost identical, sideweirs may have values of H much larger than h. In addition, sideweirs have rarely a perpendicular approach flow, and the effects of approach flow velocity and direction compensate to a certain degree.

Normally, the upper weir face close to the crest is vertical, except for sideweirs with a low crest. Then, the crest inclination has an additional effect on the outflow process. The overflow process of sideweirs in an U-shaped channel as considered in this section may be approximated with the overflow over a rectangular channel because of the large filling ratio.

Sewer sideweirs have a *converging width* in plan and the streamlines have a certain angular deviation from the crest direction. Because of the increased velocity component perpendicular to the crest line, these sideweirs have a somewhat larger discharge capacity than the sideweir in a prismatic channel.

The effects previously discussed have been analyzed experimentally; Hager et al. (1982) suggested as *generalized overflow equation* (Chap. 17)

$$\frac{\mathrm{d}Q}{\mathrm{d}x} = -0.6n^* c_k (gH^3)^{1/2}(y-W)^{3/2}\left[\frac{1-W}{3-2y-W}\right]^{1/2}\left[1-(\theta+S_{os})\left(\frac{3(1-y)}{y-W}\right)^{1/2}\right]. \tag{18.4}$$

where:

- n^* is the number of lateral outflow sides (either $n^* = 1$ or 2),
- c_k is the crest coefficient ($c_k = 1$ for sharp-crested weir),
- $(gH^3)^{1/2}(y–W)^{3/2} = g^{1/2}(h–w)^{3/2}$ is the determining overflow depth effect,
- $[(1–W)/(3–2y–W)]^{1/2}$ incorporates the effects of approach flow velocity and direction, and
- the term in the second parenthesis includes effects of sideweir convergence θ and bottom slope S_{os}.

Equation (18.4) was determined for the rectangular channel, and can be applied equally for *U-shaped sideweirs*. All lengths are normalized by the energy head H, such as the relative flow depth $y = h/H$ and the relative weir height $W = w/H$. It should be noted that all parameters in Eq. (18.4) may vary with the streamwise coordinate x. Also, Eq. (18.4) looks complicated at first sight, but includes the main factors that have an effect on a rather complex flow process. Simplified versions of the lateral outflow equation are presented below.

There are two *extreme cases* of overflow. First, for a sideweir with a perpendicular approach flow, the velocity tends to zero, i.e. $h = H$ or $y = 1$. Then, both terms in square brackets tend to 1 and Eqs. (18.3) and (18.4) are identical. This extreme case is thus reached for sideweirs with a dead-end, where the average velocity tends to zero.

Second, the case $y \rightarrow W$ or $(h–w) \rightarrow 0$ corresponds to zero overflow discharge, and the sideweir tends to a channel with a constant discharge. The first term in the parenthesis tends to $3^{-1/2}$ and the reduction of overflow is about 40% compared to the first case. The second term in parenthesis tends to $(y–W)^{3/2}$, and thus to zero. This term has in general a secondary effect of about 10%. For converging sideweirs of 10% ($\theta = –0.1$), and for $y = 0.8$ and $W = 0.5$ the term is 1.14 and its variation with the local flow depth $h(x)$ often neglected. For practical purposes relating to standard sewer sideweirs the determining *lateral outflow intensity* is thus

$$\frac{dQ}{dx} = -0.6n^* c_w (gH^3)^{1/2}(y - W)^{3/2}\left[\frac{1 - W}{3 - 2y - W}\right]^{1/2}. \qquad (18.5)$$

Here, the effects of crest geometry and sideweir convergence are contained in the lumped parameter c_w. For the *standard sideweir* with a sharp overflow crest and a nearly prismatic section, $c_w = 1$. For other geometries, Hager et al. (1982) recommended a design basis.

Example 18.2 Determine the lateral outflow intensity dQ/dx for a sideweir with an U-shaped cross-section of diameter $D = 0.7$ m and a flow depth of $h = 0.55$ m, a discharge of $Q = 0.21$ m^{3}s^{-1} and a weir height of $w = 0.40$ m. The approach flow diameter is $D_o = 1.2$ m and the downstream diameter $D_d = 0.25$ m, for a length $\Delta L = 2.8$ m with $\Delta s = 0.05$ m.
With $S_{os} = \Delta s/\Delta L = 0.05/2.8 = 0.018$, and $\theta = (D_d–D_o)/\Delta L = (0.25–1.2)/\ 2.8 = –0.34$ one has $(\theta + S_{os}) = –0.32$. Further, with a cross-sectional area of $F = \pi D^2/8 + (h–D/2)D = 0.39 \cdot 0.7^2 + (0.55–0.35)0.7 = 0.33$ m^2, thus $V = Q/F = 0.21/0.33 = 0.63$ ms^{-1} and $V^2/(2g) = 0.021$ m, the energy head amounts to $H = h + V^2/(2g) = 0.55 + 0.02 = 0.57$ m. The relative values are, respectively, $y = h/H = 0.55/0.57 = 0.965$ and $W = 0.4/0.57 = 0.70$. Inserting in Eq. (18.4) gives $dQ/dx = -1.2 \cdot 1(9.81 \cdot 0.57^3)^{1/2}(0.965 - 0.7)^{3/2}[(1 - 0.7)/(3 - 2 \cdot 0.965 - 0.7)]^{1/2} \times [1 - (-0.32)(3(1 - 0.965)/(0.965 - 0.70))^{1/2}] = -1.2 \cdot 1.348 \cdot 0.136 \cdot 0.90 \cdot 1.20 = -0.238$ m^{2}s^{-1} for the sharp-crested two-sided weir. Thus, because of the significant contraction, the correction factor $c_w = 1.20$ is rather large.

Conventionally, $dQ/dx = -C_d n^*(2g)^{1/2}(h–w)^{3/2}$ with $C_d = 0.42$ as average discharge coefficient. Then, $dQ/dx = –0.42 \cdot 2 \cdot 19.62^{1/2}(0.55–0.40)^{3/2} = –0.216$ m^{2}s^{-1}, thus mainly due to θ a reduction of –10%. In the present case, the effect of approach flow is insignificant because the approach flow Froude number is small with $\mathsf{F} \cong 0.63/(9.81 \cdot 0.55)^{1/2} = 0.27$.

Free Surface Profile

Apart from the equation for the lateral outflow intensity, the free surface profile $h(x)$ must be known to determine the flow features of a sideweir. It can be obtained with the *momentum approach* (Chap. 1) formulated as a differential equation. As in Chap. 17, the friction slope is often compensated for by the bottom slope, and the additional losses due to lateral outflow (Chap. 17) are negligible for unbaffled outflow (Hager and Schleiss 2009). If such baffles were present such as to retain floating matter, then this assumption had to be verified.

Sewer sideweirs with a high overflow crest are characterized by subcritical approach flow, involving a relatively small local variation of the flow depth. To simplify the approach, one may assume that the *free surface profile* $h(x)$ depends only on the boundary flow depths h_o and h_u. This approach thus neglects the internal features of sideweir flow. From Fig. 18.2, one may write

$$h(x) = h_o + (h_u - h_o)(x/\Delta L)^{1/2}, \tag{18.6}$$

with x as the streamwise coordinate measured from the beginning of the sideweir, thus $h(x = 0) = h_o$ and $h(x = \Delta L) = h_u$. Equation (18.6) produces the correct tendency of the free surface profile for both subcritical ($h_o < h_u$) and supercritical flows ($h_o > h_u$) because the profile is parabolically increasing, or decreasing, respectively (Fig. 18.2).

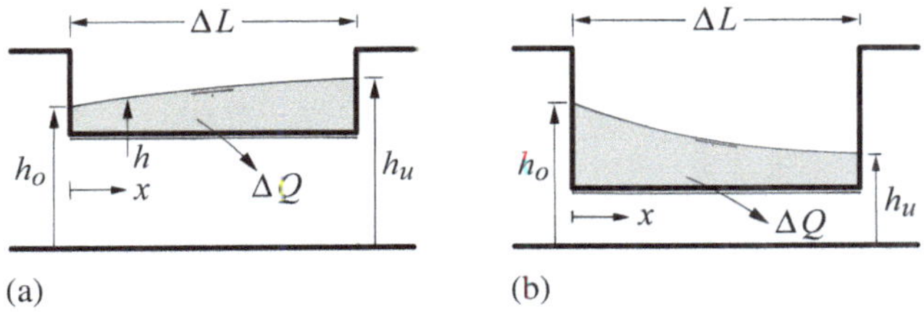

Fig. 18.2 Typical free surface profile $h(x)$ for (**a**) subcritical and (**b**) supercritical flow

The average flow depth h_m over the sideweir length is

$$h_m = h_o + \frac{2}{3}(h_u - h_o), \tag{18.7}$$

i.e. larger than the arithmetic average $(1/2)(h_o + h_u)$.

Assuming essentially a *constant energy head* across the relatively short sideweir, Bernoulli's energy equation reads

$$h_o + \frac{Q_o^2}{2gF_o^2} = h_u + \frac{Q_u^2}{2gF_u^2}. \tag{18.8}$$

As previously described, the difference of increase or decrease of energy head due to gravity and friction is nearly compensated. Further, the downstream velocity head $Q_u^2/(2gF_u^2)$ for subcritical flow is much smaller than the downstream flow depth h_u, and asymptotically drops out for storm water discharge to result in

$$h_o = h_u - \frac{Q_o^2}{2gF^{*2}}. \tag{18.9a}$$

Then $Q_u << Q_o$ and $Q_u^2/(2gF_u^2) \rightarrow 0$. Because computations start at the downstream end for subcritical flow, Eq. (18.9a) for the upstream flow depth is implicit but since

$Q_o^2/(2gF_o^2) < h_o$ the upstream depth is approximated with the known downstream flow depth $h = h_u$, and $F^* = \pi D_o^2/8 + (h_u - D_o/2)D_o$. Note that this is correct as $\mathsf{F}_o \to 0$ $(h_o = h_u)$ and that the first solution can be corrected by a second iteration.

Equation (18.9a) states that $h_u > h_o$. The difference $(h_u - h_o)$ increases as the velocity head increases. Because the nonuniformity of lateral outflow increases with the power (3/2) of the difference flow depth $(h_u - h_o)$, flows with a large difference level should be prevented. The only method available is, as in Chap. 17, the *converging sideweir* with a velocity decrease much smaller than for the prismatic sideweir. Inserting the expression for the cross-section F^* in Eq. (18.9a) yields with $f = Q_o/(gD_o^4 h_u)^{1/2}$ as the relative discharge and $y_u = h_u/D_o$ as the downstream filling ratio

$$\frac{h_o}{h_u} = 1 - \frac{f^2/2}{\left[\frac{\pi}{8} + \left(y_u - \frac{1}{2}\right)\right]^2}. \tag{18.9b}$$

This equation provides an *explicit* solution for h_o in terms of f and y_u.

Overflow Relation

The lateral overflow ΔQ is determined with Eq. (18.5) by inserting average values as

$$\frac{\Delta Q}{\Delta L} = 0.6 n^* c_w g^{1/2} (h_m - w_m)^{3/2} \left[\frac{H_m - w_m}{3H_m - 2h_m - w_m}\right]^{1/2}, \tag{18.10}$$

and when accounting for Eq. (18.7)

$$\frac{\Delta Q}{\Delta L} = \frac{1}{3} 0.6 n^* c_w g^{1/2} (h_o + 2h_u - 3w_m)^{3/2} \left[\frac{H_u - w_m}{9H_u - 2h_o - 4h_u - 3w_m}\right]^{1/2}. \tag{18.11}$$

If the downstream velocity V_u is again assumed to be small, then $h_u = H_u$ and

$$\frac{\Delta Q}{\Delta L} = 0.2 n^* c_w g^{1/2} (h_o + 2h_u - 3w_m)^{3/2} \left[\frac{h_u - w_m}{5h_u - 2h_o - 3w_m}\right]^{1/2}, \tag{18.12}$$

corresponding in non-dimensional terms to

$$\frac{\Delta Q}{g^{1/2} D_o h_u^{3/2}} = 0.2 \left[\frac{h_o}{h_u} + 2 - 3\frac{w_m}{h_u}\right]^{3/2} \left[\frac{1 - \frac{w_m}{h_u}}{5 - 2\frac{h_o}{h_u} - 3\frac{w_m}{h_u}}\right]^{1/2} \frac{n^* c_w \Delta L}{D_o}. \tag{18.13}$$

Normally, the design of a sideweir involves the *length determination* for given approach flow parameters D_o, h_o, Q_o. With

$$F^* = (\pi/8)D_o^2 + (h_o - D_o/2)D_o \cong D_o h_o \cong D_o h_u \tag{18.14}$$

as an approximation for the upstream cross-sectional area in a rectangular substitute cross-section, Eq. (18.9a) gives instead of Eq. (18.9b) simply

$$\frac{h_o}{h_u} = 1 - \frac{(Q_u + \Delta Q)^2}{2gD_o^2 h_u^3}. \tag{18.15}$$

Eliminating h_o/h_u from Eqs. (18.13) and (18.15) gives the relative sideweir length $\lambda_s = n^* c_W \ \Delta L/D_o$ as a function of the relative discharges $q_u = \Delta Q/(gD_o^2 h_u^3)^{1/2}$ and $Q_u/(gD_o^2 h_u^3)^{1/2}$, and of relative weir height w_m/h_u. For storm-weather discharge, $Q_u << \Delta Q$, one may asymptotically consider $Q_u/\Delta Q \to 0$, and omit the second discharge ratio. Figure 18.3 shows that the sideweir length λ_s increases significantly with both the relative weir height and the relative lateral discharge. For $q_u > 0.6$ the assumptions governing this simplified approach are invalid, and the results incorrect.

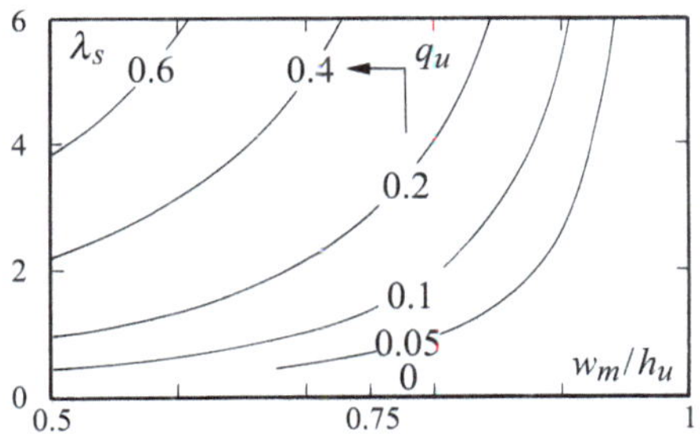

Fig. 18.3 Relative sideweir length $\lambda_s = n^* c_w \Delta L/D_o$ versus w_m/h_u and various lateral outflow intensities $q_u = \Delta Q/(gD_o^2 h_u^3)^{1/2}$ for $Q_u/Q_o \to 0$

Example 18.3 Given a sideweir with $Q_{oM} = 1.2 \ m^3 s^{-1}$, $Q_u = 0.07 \ m^3 s^{-1}$, $D_o = 0.90$ m, $D_d = 0.20$ m and $w_m = 0.5$ m. What is the sideweir length required for a downstream flow depth $h_u = 0.80$ m?
Conventionally, one would determine lateral outflow with a discharge coefficient of $C_d = 0.42$

$$\Delta Q = n^* C_d (2g)^{1/2} (h_u - w_m)^{3/2} \Delta L \tag{18.16}$$

resulting in $\Delta L = 1.13/[0.42 \cdot 19.62^{1/2}(0.8 - 0.5)^{3/2}] = 3.70$ m (Fig. 18.4).

The governing parameters of the *modified approach* are $w_m/h_u = 0.5/0.8 = 0.625$ and $q_u = \Delta Q/(gD_o^2 h_u^3)^{1/2} = 1.13/(9.81 \cdot 0.9^2 0.8^3)^{1/2} = 0.56$, thus

$n^*c_W\Delta L/D_o = 5.76$. With $n^*c_W = 1$ (one-sided outflow) the sideweir length is $\Delta L = 5.76 \cdot 0.9 = 5.18$ m, or with Eq. (18.15) $h_o/h_u = 0.82$, corresponding to $h_o = 0.82 \cdot 0.8 = 0.66$ m. With $f = 1.2/(9.81 \cdot 0.9^4 0.8)^{1/2} = 0.53$ and $y_u = 0.8/0.9 = 0.89$ results for $h_o/h_u = 1 - 0.5 \cdot 0.53^2/(0.39 + 0.89 - 0.5)^2/0.77$ (–5%) from Eq. (18.9b).

With the arithmetic average $\overline{h}_m = (1/2)(h_o + h_u) = (0.66 + 0.80)/2 = 0.73$ m instead of the downstream flow depth h_u, one would compute $\Delta L = 5.5$ m (+6%) from Eq. (18.16), which is significantly larger than $\Delta L = 3.7$ m based on $h_u = 0.8$ m. In both formulations, the effects of approach flow conditions are not contained, and this may be significant in the correct estimation of the sideweir length.

Turning back to Eq. (18.9a), one may also verify the assumption that $h_o \cong h_u$. With $h_o = 0.66$ m the upstream cross-sectional area is $F_o = (\pi/8)0.9^2 + (0.66–0.45)0.9 = 0.51$ m^2 and $V_o = 1.2/0.51 = 2.37$ ms^{-1}, compared to $F^* = (\pi/8)0.9^2 + (0.8–0.45)0.9 = 0.63$ m^2 and $V^* = 1.9$ ms^{-1}. The two velocity heads $V_o^2/(2gh_u) = 0.36$ upstream and $V^{*2}/(2gh_u) = 0.23$ downstream of the sideweir are thus quite different in this example, and a second iteration had to be performed, based on $h_o = 0.66$ m.

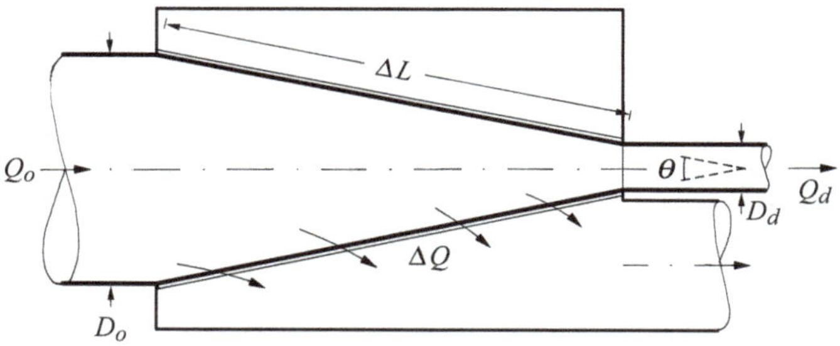

Fig. 18.4 Two-sided high sideweir with throttling pipe

Example 18.4 What is the lateral discharge for a two-sided weir with $h_u = 1.2$ m, $w_m = 0.8$ m and $D_o = 1.5$ m, $D_d = 0.35$ m if its length is $\Delta L = 4.5$ m? With $w_m/h_u = 0.8/1.2 = 0.67$ and $\lambda_s = 2 \cdot 1 \cdot 4.5/1.5 = 6$, Fig. 18.3 gives $q_u \cong 0.5$, thus $\Delta Q = 0.5(9.81 \cdot 1.5^2 1.2^3)^{1/2} = 3.12$ m^{3}s^{-1}. From Eq. (18.15) the ratio of boundary flow depths is $h_o/h_u = 1–(1/2)q_u^2 = 1–0.5 \cdot 0.5^2 = 0.87$, and $h_o = 0.87 \cdot 1.2 = 1.05$ m.

Equation (18.16) gives $\Delta Q = 2 \cdot 0.41 \cdot 19.62^{1/2}(1.2–0.8)^{3/2}4.5 = 4.13$ m^{3}s^{-1}, thus more than +30% in excess of the previously determined lateral discharge. Relating this number to the average overflow depth $\overline{h}_m = (1/2)(1.05 + 1.20) = 1.125$ m, one would have $\Delta Q = 3.03$ m^{3}s^{-1} comparing well with $\Delta Q = 3.1$ m^{3}s^{-1}.

The advantage of Eq. (18.13) is that both design parameters, namely length ΔL and lateral discharge ΔQ can explicitly be solved for. This is further simplified if developing a Taylor series, resulting in

$$q_u = 0.6(1 - W_m)^{3/2}\left[1 - \frac{0.36q_u^2}{(1 - W_m)}\right]\lambda_s, \quad (18.17a)$$

where $W_m = w_m/h_u$ and $q_u = \Delta Q/(gD_o^2h_u^3)^{1/2}$ as previously. For small values of q_u, the square bracket tends to one and Eq. (18.17a) tends to Eq. (18.16). The reduction of overflow discharge depends thus significantly on q_u. Increasing the overflow discharge increases the difference (h_u–h_o) and thus reduces ΔQ. Because h_o and h_u are often of a similar order, and ΔQ is close to the approach flow discharge Q_o, one may approximate $q_u = \Delta Q/(gD_o^2h_u^3)^{1/2} \cong Q_o/(gD_o^2h_u^3)^{1/2} \cong \mathsf{F}_o$ in U-shaped channels. Therefore, the accuracy of the conventional overflow equation (18.16) reduces as the approach flow Froude number increases. Equation (18.16) is strictly correct only for $\mathsf{F}_o \rightarrow 0$. An upper limit of application is typically $\mathsf{F}_o = 0.3$, because the correcting term in Eq. (18.17a) is then $0.36 \cdot 0.3^2/(1-0.5) < 0.1$ for $W_m = 0.5$. Note that Eqs. (18.13), (18.14), (18.15), (18.16), and (18.17a) are based on the assumption of *subcritical approach flow* to the sewer sideweir. Also, an effect of width reduction is not contained because the downstream velocity is neglected.

Equation (18.17a) may also be explicitly solved for q_u as

$$\frac{\Delta Q}{(gD_o^2h_u^3)^{1/2}} = 0.6(1 - W_m)^{3/2}\lambda_s[1 - (0.36(1 - W_m)\lambda_s)^2]. \quad (18.17b)$$

To lowest order, Eq. (18.16) results again, whereas a quadratic term is retained to second order. The term $(1-W_m)\lambda_s$ should be smaller than 0.25. This term disappears as either the overflow depth (h_u–w_m) or the sideweir length ΔL tend to zero. The *application limits* of the conventional weir equation to sideweir flow are thus:

- $q_u \rightarrow 0$ corresponding to a small lateral overflow discharge,
- $\mathsf{F}_o \rightarrow 0$ and the approach flow Froude number has to be small ($\mathsf{F}_o < 0.3$),
- $h_m/w_m \rightarrow 1$ for which the overflow depth is small, or
- $\Delta L/D_o < 1$ stating that the overflow length is small.

These application limits imply that the conventional weir equation does often not apply for sideweir flow. Evidently, all four criteria point to sideweir efficiency, with a designer asking for a short structure and thus for a large overflow depth. Then, the reduction of lateral discharge may become significant. Typical causes are thus an increase of the downstream flow depth h_u over the designed depth, a reduction of the approach flow depth below the designed depth, and either a generation of supercritical approach flow with a hydraulic jump along the overflow reach, or increased discharge to the treatment station. Both consequences can have a strongly adverse impact on the performance of the sewer reach, and should be countered by a sound hydraulic design.

Example 18.5 Re-compute Example 18.4 by applying Eq. (18.17b). Given $W_m = 0.67$, $\lambda_s = 6$, such that the iterative solution of Eq. (18.17a) is $q_u = 0.505$, practically identical with the solution of Eq. (18.13). Applying Eq. (18.17b) yields $0.36(1-W_m)\lambda_s = 0.36 \cdot 0.33 \cdot 6 = 0.72$, that is too large and this relation cannot be used here.

Equation (18.17a) allows a simple determination of the:

- Required overflow length ΔL for given storm water discharge Q_M,
- Lateral discharge ΔQ for all other, smaller approach flow discharges, and
- Weir height w_m required.

Therefore, the design of the sewer sideweir with a high overfall crest can be based on an explicit computation that is governed by energy conservation. Compared to the present design method, no empirical reduction factors are contained but the design is founded on physical facts combined with suitable simplifications.

Method of Uyumaz and Muslu

A simplified method to predict the lateral outflow discharge in *prismatic* circular-shaped sideweirs was presented by Uyumaz and Muslu (1985). Based on the weir equation (18.16) with $h_m = (1/2)(h_e + h_u)$ as the determining flow depth instead of h_u, where h_e is the end depth (Fig. 18.5), the discharge coefficient C_d depends on the:

- Approach flow Froude number F_o,
- Relative weir height w/D, and
- Relative weir length $\Delta L/D$.

Their data refer to the sharp-crested weir with $0.24 < w/D < 0.56$, $1 < \Delta L/D < 3.4$ and $\mathsf{F}_o < 2$. Based on a data reanalysis of Hager (1993), the *average discharge*

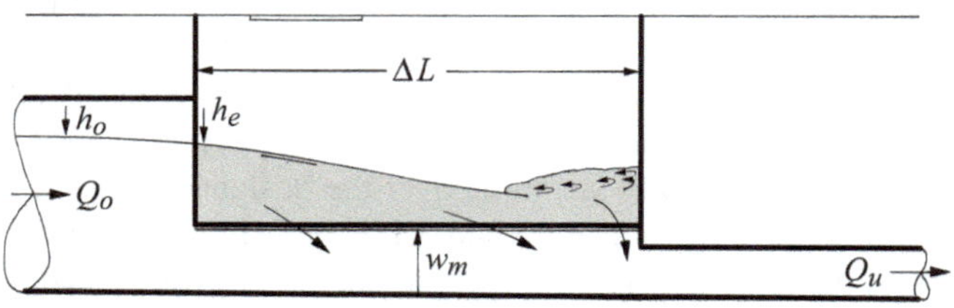

Fig. 18.5 Low-crested sewer sideweir, notation

coefficient C_{dm} can be expressed for the weir with a nearly vertical outflow plane ($h_m/D > 0.4$) as

$$C_{dm} = 0.40 + 0.01(\Delta L/D) - \frac{0.185\mathsf{F}_o^2}{\Delta L/D}. \tag{18.18}$$

This demonstrates a significant effect of F_o and indicates a reduction of C_{dm} as F_o increases. For the conventional weir ($\mathsf{F}_o \rightarrow 0$) the effect of relative length ΔL may be felt by an increase of up to 10%, which is reduced for $\mathsf{F}_o > 0$ due to boundary effects, as discussed in Sect. 17.6. Note that Eq. (18.18) is valid only for short prismatic sideweirs and the effect of $\Delta L/D$ tends to zero for longer weirs. In addition, Eq. (18.18) can be applied by an iterative procedure because the end depth h_e is initially unknown. A detailed discussion has been presented by Naudascher (1992).

18.3.3 *Throttling Pipe*

For both sub- and supercritical approach flows, the throttling (subscript d) pipe (Chap. 9) is a component of the sideweir governing the downstream sewer. For given bottom slope S_{od}, length L_d, diameter D_d and roughness n_d (Fig. 9.4), the relation between the head h_{od} on the throttling pipe and discharge $Q = Q_d$ is established by accounting for the appropriate intake and friction losses. The governing type of pipe flow is also defined. Equations (9.19) and (9.20) refer to gated and pressurized pipe flow, respectively.

If the relations $Q_d(h_{od})$ relative to the throttling pipe and $Q_u(h_u)$ for the downstream sideweir end are determined, respectively, the sideweir flow is computed for various *throttling scenarios*. For subcritical sideweir flow, the direction of computation starts at the downstream sideweir end. For supercritical sideweir flow with a hydraulic jump along the sideweir, the computations start at either end of the structure and proceed towards the jump center, where the relation for the sequent depths links the two profiles. This involved procedure is presented in Sect. 18.4.5 and additional results are given in Sect. 18.5. Recent experimental results relating to the sewer sideweir with a throttling pipe are introduced in Sect. 18.6.

18.4 Low-Crested Sewer Sideweir

18.4.1 *Flow Patterns*

Contrary to high-crested sideweirs, supercritical flow may establish along a sideweir with a low crest despite a subcritical approach flow (Chap. 6). Because the velocity upstream from the throttling pipe is relatively small, a *hydraulic jump* then develops along the overflow reach complicating the hydraulic design considerably. Low-crested sideweirs thus complicate the design.

Figure 18.5 shows a low-crested sewer sideweir. Upstream from the sideweir, the free surface profile has a *drawdown curve* from h_o to h_e, similar to the end overfall (Chap. 11) because of non-hydrostatic pressure distribution. For a subcritical approach flow, the transition from sub- to supercritical flows is slightly upstream from the sideweir beginning. For supercritical approach flow with a smaller drawdown effect the depth ratio tends to $h_e/h_o \rightarrow 1$ as the Froude number is large.

The flow along the sideweir is accelerated from the approach flow sewer up to the toe of the hydraulic jump. Particles close to the sideweir bottom continue with almost the same velocity along the jump, whereas the surface current is significantly influenced by roller flow. Depending on the hydraulic jump location relative to the end section made up by a vertical end wall, it either impinges on it or the jump is separated by a subcritical flow reach from the end wall (Fig. 18.6). The *impact hydraulic jump* (German: Aufprall-Wassersprung; French: Ressaut hydraulique d'impact) deviates considerably from the classical hydraulic jump (Chap. 7). It can also be observed close to regulating gates in power channels that slightly submerge a supercritical approach flow. With $\lambda_g = L_g/L_r^*$ as the ratio of the length L_g between the toe and the end wall, and L_r^* as the roller length, impact hydraulic jumps develop for $\lambda_g < 1.2$ (Hager 1994b). Then, an instability is set up upstream from the end wall with surface waves propagating into the upstream portion. The variation of flow depth at the end wall can be significant, depending on the value of λ_g. To inhibit such *unsteady flow features* the ratio should be $\lambda_g > 1.5$, i.e. there should be a reach of subcritical flow downstream of the end of the hydraulic jump for flow reestablishment.

Sideweirs with a hydraulic jump in the center portion can be treated approximately with a hydraulic approach. Up to today, no systematic observations on hydraulic jumps along a sideweir were conducted, and the lateral outflow mechanism is also unclear. The question is whether such flow can indeed be approximated with the conventional approach. Such computations have been presented by Hager et al. (1982) and the results compared reasonably well with limited observations. In the following, this simplified model is presented. In the first part, continuous flow in the *prismatic sideweir* is considered, and converging sideweirs are analyzed in the second part. It should be clearly stated again that the low-crested sideweir is not recommended for design, but there exist a large number of these structures that have to be reanalyzed in the future.

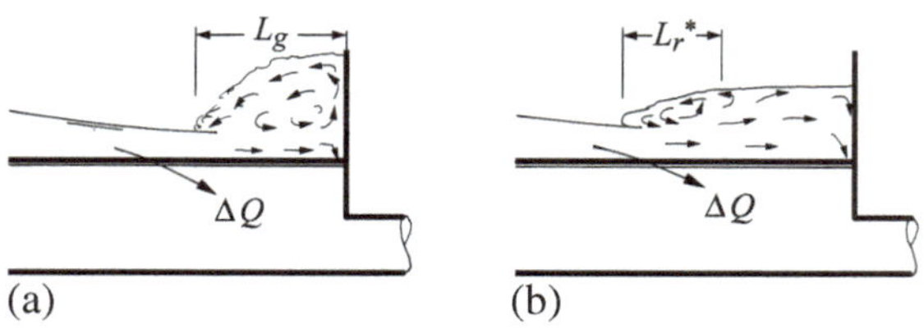

Fig. 18.6 Hydraulic jump along a low-crested sewer sideweir with (**a**) impact jump, (**b**) reestablishment reach between end of jump and end wall

18.4.2 *Prismatic Sideweir*

Although standard sewer sideweirs converge in plan from the approach flow diameter D_o to the throttle diameter D_d, prismatic sideweirs correspond to a special case having received attention mainly due to simplicity. A summary of experimental studies was provided by (Hager 1993). The following *simplifications* are introduced:

- Cross-sectional area F as a function of diameter D and flow depth h is approximated better than 10% up to a filling ratio $y = h/D$ of 85% by the power function

$$F = (h/D)^{1.4} D^2, \tag{18.19}$$

- Energy line slope S_e along the sideweir is compensated for by the bottom slope S_o, such that the energy head H relative to the bottom remains constant,
- Lateral outflow along the sideweir is spatially-varied, and the boundary reaches at the up- and downstream sideweir portions are comparably small,
- Hydraulic jumps are treated as for backwater curves, and the sub- and supercritical flow reaches can be connected based on the sequent depth ratio,
- Lateral outflow induces no additional losses,
- Weir height w is constant, and
- Flow depth $h(x)$ determines both the channel flow and the outflow pattern.

For a *constant energy head H* relative to the sideweir bottom, the relation between the discharge $Q(x)$ and the flow depth $h(x)$ is

$$H = h + \frac{Q^2}{2gF^2}. \tag{18.20}$$

Introducing the dimensionless coordinates

$$X = \frac{x}{(D/H)^{0.6} H}, \quad y = h/H \tag{18.21}$$

as nondimensional location and relative flow depth, and with $q = [\mathrm{d}Q/\mathrm{d}x]/(gH^3)^{1/2}$ as the dimensionless outflow intensity, the equation of the free surface profile obtains by differentiation of Eq. (18.20) subject to $\mathrm{d}H/\mathrm{d}x = 0$, thus with $c = 0.737$ (Hager 1994a)

$$\frac{\mathrm{d}y}{\mathrm{d}X} = -\frac{0.372(1-y)^{1/2} q}{(y-c)y^{0.4}}. \tag{18.22}$$

The lateral outflow intensity is according to Eq. (18.5)

$$q = -0.6n^*\left[(y-W)^3(1-W)/(3-2y-W)\right]^{1/2}. \tag{18.23}$$

As in Chap. 17, n^* is the number of lateral outflow sides ($n^* = 1$ or 2), and $W = w/H$ the relative weir height. Inserting in Eq. (18.22) gives the dimensionless equation of the *free surface profile*

$$\frac{\mathrm{d}y}{n^*\mathrm{d}X} = \frac{0.223\left[(1-y)(y-W)^3(1-W)\right]^{1/2}}{(y-c)y^{0.4}[3-2y-W]^{1/2}}. \tag{18.24}$$

Accordingly, the function $y(X)$ varies only with the weir height W and the critical depth ratio $c = h_c/H = 0.737$. The free surface is *horizontal* ($\mathrm{d}y/\mathrm{d}X = 0$) if (Fig. 18.7):

- $y = 1$, corresponding to the dead-end of the sideweir with $Q = 0$, and
- $y = W$, as the flow depth is equal to the weir height.

The free surface is *vertical* ($\mathrm{d}X/\mathrm{d}y = 0$) for critical flow ($y = c$). Equation (18.24) is thus a generalized backwater equation for sideweirs in a circular channel. To solve it, two generalized boundary conditions must be imposed, as discussed in Chap. 17:

- $X(y = 1) = 0$ for subcritical flow ($c < y < 1$), and
- $X(y = c) = 0$ for supercritical flow ($W < y < c$).

The general solution of Eq. (18.24) as numerically obtained is shown in Fig. 18.8.

Figure 18.8 includes three basic types of flow, namely (Fig. 18.7):

- *Case* 1 for $c < y < 1$ and $W > c$, with an S-shaped profile for subcritical flow,
- *Case* 2 for $c < y < 1$ and $W < c$, with a profile starting vertically and increasing for subcritical flow also, and
- *Case* 3 for $W < y < c$, starting also vertically and decreasing for supercritical flow.

A disadvantage of Fig. 18.8 is the presence of the third parameter W instead of one profile for each case only. However, the profiles look similar, and

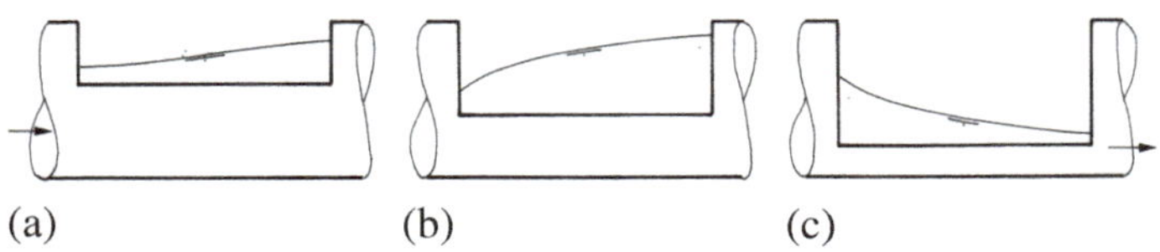

Fig. 18.7 Generalized surface profiles in prismatic sideweir for Case (**a**) 1, (**b**) 2, (**c**) 3

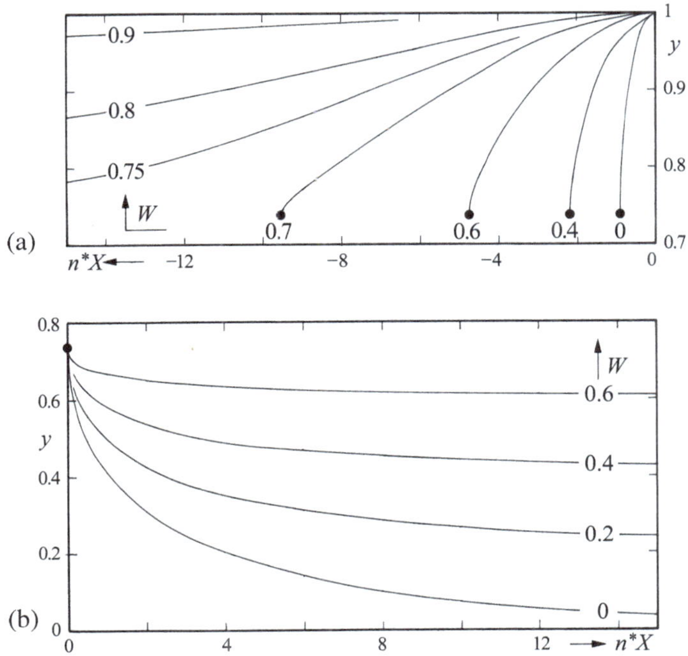

Fig. 18.8 Generalized free surface profile $y(n^*X)$ for various relative weir heights W based on Eq. (18.24) for (**a**) subcritical, (**b**) supercritical sideweir flow in *prismatic* circular channel (Hager 1994a)

approximations can be introduced on the basis of *generalized surface equations* as (Hager 1994a)

$$\frac{y-W}{1-W} = 1.06 + 0.32(1-W)^{1.15}n^*X, \tag{18.25}$$

$$\frac{y-W}{1-W} = 1 - \left[\frac{n^*X}{\exp[1.4\tan(W/1.4)]}\right]^{0.4}, \tag{18.26}$$

$$\frac{y-W}{c-W} = 0.8 - \frac{0.3(n^*X)^{1/2}}{[1.4-2.5(W-0.1)^{2.5}]^{1/2}}. \tag{18.27}$$

The left hand sides of Eqs. (18.25), (18.26), and (18.27) are normalized such that the depths vary between zero and one. Observations were collected for Case 3 only, namely by Sassoli (1963) and Buffoni et al. (1986). Both used circular pipes of diameter $D = 0.20$ m with bottom slopes of 0.1 and 0.2%, and weir lengths of 0.60 m and 1.20 m, respectively. Weir heights varied between 0.050 and 0.133 m, with Sassoli referring to the one-sided ($n^* = 1$), and Buffoni et al. to two-sided ($n^* = 2$) overflow.

Hager (1994a) analyzed the data based on the previous parameters and observed that Eq. (18.27) describes the upstream profile portion well, whereas deviations increase towards the downstream sideweir portion. The profile similarity has been verified and the experiments for *supercritical approach flow* may be represented as

$$\frac{y-W}{c-W} = N\exp\left[\frac{-n^*X/2}{1.4-2.5|\mathrm{W}-0.1|^{2.5}}\right]. \tag{18.28}$$

Here $N = 0.725$ for $n^* = 1$ and $N = 0.80$ for $n^* = 2$. Equation (18.28) is valid exclusively for the supercritical flow reach, thus either to the downstream sideweir end for supercritical downstream flow, or up to the toe of a hydraulic jump for subcritical downstream flow.

Example 18.6 Given a prismatic one-sided sideweir of diameter $D = 0.90$ m, upstream discharge of $Q_o = 1.2$ m^{3}s^{-1}, weir height of $w = 0.40$ m and length of $\Delta L = 3.7$ m. How much water is laterally discharged for an end depth of $h_e = 0.65$ m at the beginning of the sideweir?
With $h_e = 0.65$ m, the energy head at the sideweir beginning is $H_e = 0.65 + 1.2^2/[19.62\cdot 0.9^4(0.65/0.9)^{2.8}] = 0.93$ m from Eqs. (18.19) and (18.20). Further, $W = 0.40/0.93 = 0.43$ and $c = 0.737$. With $y_e = h_e/H = 0.65/0.93 = 0.70$, the boundary location is $X_e = -0.48$ from Eq. (18.28). For $\Delta L = 3.7$ m, that is $\Delta X = 3.7/[(0.9/0.93)^{0.6}0.93] = 4.06$, thus for $X = X_e + \Delta X = -0.48 + 4.06 = 3.58$, Eq. (18.28) gives $(y-W)/(c-W) = 0.17$, and $y = y_u = 0.48$, or $h_u = 0.48\cdot 0.93 = 0.45$ m. The downstream discharge is then $Q_u = F_u[2g(H-h_u)]^{1/2} = (0.45/0.9)^{1.4}0.9^2[19.62(0.93-0.45)]^{1/2} = 0.94$ m^{3}s^{-1} from Eq. (18.20) and thus $\Delta Q = Q_o - Q_u = 1.20-0.94 = 0.26$ m^{3}s^{-1}.

With the conventional overflow formula applied to h_e, and a 5% reduction of C_d to 0.40, one had $\Delta Q = 0.40\cdot 19.62^{1/2}(0.65-0.40)^{3/2}3.6 = 0.81$ m^{3}s^{-1}, a value three times too large! If the computation is based on the initially *unknown* average flow depth $\overline{h} = (1/2)(h_e + h_u) = 0.55$ m then $\Delta Q = 0.37$ m^{3}s^{-1}, which is still +40% too large. For supercritical sideweir flow, the conventional approach is thus always considerably erroneous as compared to observations.

18.4.3 *Converging Sideweir*

For a standard sewer sideweir the plan converges from the approach flow diameter D_o to the throttle pipe diameter D_d, with the convergence angle θ as an additional parameter (Fig. 18.4). The hydraulic approach is further complicated by the explicit appearance of the coordinate x, on the right hand side of Eq. (18.24), requiring to solve a differential equation $\mathrm{d}y/\mathrm{d}X = f_2(X,y)$ instead of an integral $\mathrm{d}y/\mathrm{d}X = f_1(y)$, complicating the presentation of the generalized solution because the boundary condition may vary as a function of X.

Hager et al. (1982) attempted the solution for the U-shaped profile with a cross-sectional area $F/D^2 = (h/D)^{1.5}$. In the present case, the cross-sectional area $F/D^2 = (h/D)^{1.4}$ for the circular channel is preferred, yet the differences between the two approaches are small. The complexities of sideweir flow do not result from such details, however, because the appropriate lateral outflow intensity, the three-dimensional flow pattern and the implementation of relevant boundary conditions finally influence the performance of a computational model. The hydraulic approach presented below is thus a significant *simplification* of the real flow situation. Among steady free surface flows, the most complex originate from those in channels of variable width, variable discharge and eventually containing hydraulic jumps. The only approach to solve these problems is currently the scale model in which the main parameters are varied over wide reaches. A particular study of this type is presented in Sect. 18.5.

For a *linearly converging sideweir*, the diameter varies as

$$D(x) = D_o - \theta x \tag{18.29}$$

with $\theta = (D_o - D_u)/\Delta L$ as the convergence providing a continuous transition between the upstream sewer and the throttling pipe. The cross-section along the sideweir is either U-shaped for $h_o/D_o > 1/2$, or changes from circular in the approach flow portion to U-shape along the sideweir. To apply the procedure presented for prismatic sideweirs in Sect. 18.4.2 Eq. (18.19) is applied in either case. For $h/D > 1$, the cross-sectional area is smaller than for the U-shaped channel but this is insignificant because of the relatively small velocities at the end of the sideweir. The procedure outlined here can be modified once systematic observations to the standard sewer sideweir are available.

The governing equation for the *free surface profiles* can be deduced from Eqs. (18.19), (18.20) and (18.29) as

$$\frac{\mathrm{d}y}{\mathrm{d}X} = -\frac{0.372(1-y)^{1/2}q}{(y-c)y^{0.4}} + \frac{1.2y(1-y)(\mathrm{d}D/\mathrm{d}x)}{3.8(D/H)^{0.4}(y-c)}. \tag{18.30}$$

Compared to Eq. (18.22), the effect of diameter change $\mathrm{d}D/\mathrm{d}x$ is added. Inserting the lateral outflow intensity q from Eq. (18.23) gives

$$\frac{\mathrm{d}y}{n^*\mathrm{d}X} = \frac{0.223\left[(1-y)(y-W)^3(1-W)\right]^{1/2}}{(y-c)y^{0.4}[3-2y-W]^{1/2}} - \frac{0.316\theta}{n^*(D/H)^{0.4}}\frac{y(1-y)}{y-c}. \tag{18.31}$$

The right hand side of Eq. (18.31) consists of two terms, and their difference can be either positive or negative. If the term $(D/H)^{0.4}$ is considered as corrective that can be replaced by the average diameter $D_m = (1/2)(D_o + D_u)$, i.e. $(D/H)^{0.4} \cong (D_m/H)^{0.4}$, the additional parameter $\theta x/D_o$ is removed and the right hand side of Eq. (18.31) remains independent of X, as in Eq. (18.24).

Equation (18.31) allows for the determination of the pseudo-uniform flow depth h_{PN} by setting $dy/dX = 0$, and this special flow type has been referred to in Chap. 17 as *pseudo-uniform flow*. The ratio $y_{PN} = h_{PN}/H$ is determined from

$$\Theta_{PN} = \frac{\sqrt{2}\theta}{n^*(D_m/H)^{0.4}} = \frac{\left[(1-W)(y_{PN}-W)^3\right]^{1/2}}{y_{PN}^{1.4}[(1-y_{PN})(3-2y_{PN}-W)]^{1/2}}. \quad (18.32)$$

Figure 18.9 shows that y_{PN} increases with increasing Θ_{PN} and W. For $y_{PN} > 0.4$, the expression for Θ_{PN} can be approximated as

$$\Theta_{PN} = \frac{(y_{PN}-W)^{3/2}}{(1-y_{PN})^{1/2}}. \quad (18.33)$$

The convergence rate Θ_{PN} for pseudo-uniform flow to occur depends thus significantly on the overflow depth $(y_{PN}-W)$.

Example 18.7 Compute the pseudo-uniform flow depth for Example 18.6, if $D_u = 0.50$ m instead of $D = 0.90$ m.
With $W = 0.43$, $y = 0.70$, and $\theta = (0.9-0.5)/3.7 = 0.11$, and assuming $D \cong D_m = 0.70$ m, thus $(D_m/H)^{0.4} = 0.89$, gives $\Theta_{PN} = 2^{1/2}0.11/(1\cdot 0.89) = 0.17$ and the solution of Eq. (18.32) is $y_{PN} = 0.62$, or $y_{PN} = 0.65$ from Eq. (18.33). The approach flow depth is thus $h_{PN} = 0.62\cdot 0.9 = 0.56$ m, and the approach flow Froude number $\mathsf{F}_o = 1.2/(9.81\cdot 0.9\cdot 0.56^4)^{1/2} = 1.29 > 1$ for which surface undulations are likely to appear. To prevent these conditions, the downstream diameter should be further reduced, such as to $D_u = 0.3$ m.

The corrective term $T = [-0.316\theta/n^*(D_m/H)^{0.4}]\times[y(1-y)/(y-c)]$ in Eq. (18.31) can be split into two terms Θ and $T_y = y(1-y)/(y-c)$. The convergence rate θ has typically the order 10^{-1} and $(D_m/H)^{0.4}$ has order 1, such that Θ is of order 10^{-2}. The

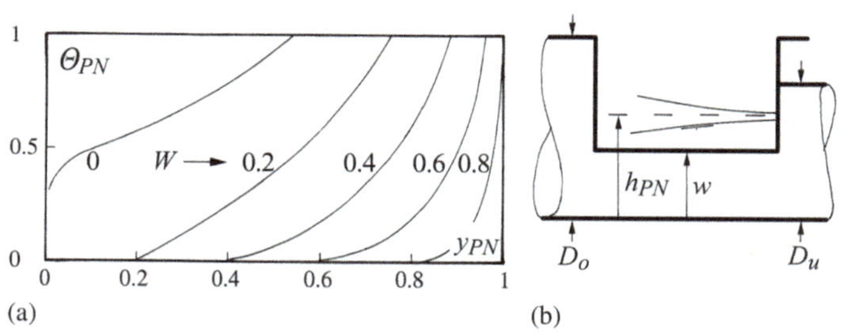

Fig. 18.9 Pseudo-uniform flow (**a**) convergence rate $\Theta_{PN}(y_{PN}, W)$ according to Eq. (18.32), (**b**) schematic surface profiles (—) $h(x)$ and (- - -) h_{PN}

term T_y is equal to zero for $y=0$ and $y=1$ and tends to infinity as $y\to c$. Therefore, the effect of T cannot be neglected in general but be regarded as a correction of the first term, provided transitional flows ($y\cong c$) are excluded. Expressing Eq. (18.31) as

$$\frac{\mathrm{d}y}{n^*\mathrm{d}X}=\frac{0.223\left[(1-y)(y-W)^3(1-W)\right]^{1/2}}{(y-c)y^{0.4}[3-2y-W]^{1/2}}\left[1-\frac{\Theta y^{1.4}(1-y)^{1/2}[3-2y-W]^{1/2}}{\left[(1-W)(y-W)^3\right]^{1/2}}\right] \quad (18.34)$$

and considering the case $y\to W$ as the limit without lateral discharge, than the correction in the square bracket remains small compared to 1, if $(y–W)>0.2$. With $y_m=(1/2)(y_o+y_u)$ and $W_m=(1/2)(W_o+W_u)$ the product P in the square bracket of Eq. (18.34)

$$P=\frac{\sqrt{2}\theta}{n^*(D_m/H)^{0.4}}\frac{y_m^{1.4}(1-y_m)^{1/2}[3-2y_m-W_m]^{1/2}}{[(1-W_m)(y_m-W_m)^3]^{1/2}} \quad (18.35)$$

is independent of X and Eq. (18.31) may be expanded to

$$\frac{\mathrm{d}y}{n^*(1-P)\mathrm{d}X}=\frac{0.223[(1-y)(1-W)(y-W)^3]^{1/2}}{y^{0.4}(y-c)[3-2y-W]^{1/2}}. \quad (18.36)$$

Because all corrections are now included into the streamwise coordinate $n^*(1–P)X$, the solution of the *converging* sideweir is basically identical with the solution of *prismatic* sideweirs. Their solutions Eqs. (18.25), (18.26), and (18.27), or Eq. (18.28), are thus readily applicable provided the transformation $n^*X\to n^*(1–P)X$ is introduced. This approximate method is currently recommended until detailed observations on converging sewer sideweirs are available. Then, the results had to be compared with those of Eq. (18.31) to derive an updated design procedure.

Example 18.8 Re-investigate Example 18.6 for diameters $D_o=0.90$ m and $D_u=0.50$ m.
With the energy head $H=H_e=0.93$ m, $W=0.43$ and $y_e=0.70<c$, the approach flow is slightly supercritical, and the flow depth decreases along the sideweir. Yet, the decrease is smaller as for the corresponding prismatic sideweir. According to Example 18.6, $y_u=0.48$, and $y_{PN}=0.62$ from Example 18.7. The average from Eq. (18.7) is thus $y_m=0.65$.

With $\theta=0.4/3.7=0.11$, $n^*=1$ and $(D_m/H)^{0.4}=0.89$ the corrective term is $P=[(1.41\cdot0.11)/(1\cdot0.89)][0.65^{1.4}0.35^{1/2}1.27^{1/2}0.57^{-1/2}0.22^{-3/2}]=0.175\cdot4.68=0.82$ and *large*. The simplified method is thus only a poor approximation because of the neglect of higher order terms, and cannot be applied.

For $P > 0.5$ the effect of diameter reduction is significant and the influence of the *pseudo-uniform flow* on the free surface profile cannot be neglected. Then, one may approximate the downstream flow depth y_u with the pseudo-uniform depth y_{PN}. Accordingly, the average flow depth is computed from Eq. (18.7) and the lateral outflow determined from Eq. (18.23).

Example 18.9 Re-compute Example 18.8!
Given $D_o = 0.90$ m, $D_u = 0.50$ m, $Q_o = 1.2$ m^{3}s^{-1}, $w = 0.40$ m, $\Delta L = 3.7$ m and $h_e = 0.65$ m. Further, $H = 0.93$ m and $h_{PN} = y_{PN} \cdot H = 0.62 \cdot 0.93$ m $= 0.58$ m.

The average overflow depth from Eq. (18.7) is $h_m = 0.70 + (2/3)(0.58–0.70) = 0.62$ m, corresponding to $y_m = 0.62/0.93 = 0.67$. With $W = 0.43$ the average outflow intensity is $q = –0.6 \cdot 1[0.24^3 0.57/1.23]^{1/2} = –0.048$, thus $\Delta Q/\Delta x = –0.048(9.81 \cdot 0.93^3)^{1/2} = –0.135$ m^{2}s^{-1} and thus $\Delta Q = 0.135 \cdot 3.7 = 0.50$ m^{3}s^{-1}. This corresponds almost to the double value of Example 18.6.

Pseudo-uniform flow corresponds to an important design basis to uniformize the lateral outflow distribution and to obtain a *compact distribution structure*. This concept is currently not often applied, however. Figure 18.10 compares the prismatic with the *converging sideweir* for both subcritical and supercritical flows. In both cases, the width reduction causes a free surface profile that is closer to the horizontal than for the prismatic sideweir. Based on these features, *pseudo-uniform flow* is an important design basis. Provided the boundary flow depth h_r is equal to the pseudo-uniform flow depth h_{PN}, the flow depth, the velocity and the lateral outflow intensity along the sideweir remain constant. This condition can be imposed only for one discharge, however, such as the design discharge, and all other discharges have to be computed with methods that are also explained in Chap. 17. It is highly recommended therefore to include the pseudo-uniform flow concept in all channels with a spatially-decreasing discharge.

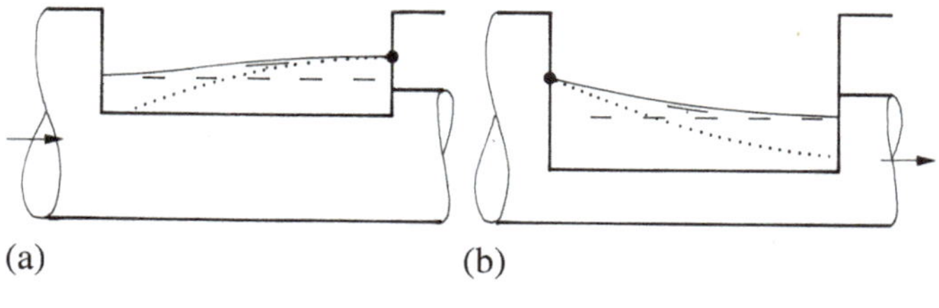

Fig. 18.10 Comparison of free surface profiles in (...) prismatic and (—) converging sideweirs for (**a**) subcritical and (**b**) supercritical flow. (•) boundary flow depth, (- - -) pseudo-uniform flow depth

18.4.4 *Hydraulic Jump in Sideweir*

Hydraulic jumps occur frequently in sideweirs with a low crest elevation, and such flow cannot be compared with hydraulic jumps as discussed in Chap. 7. Figure 18.11 shows differences also with the simplified approach assuming a linear variation of flow depth from the boundary value h_o to h_u, and the modified approach with a decreasing free surface profile in the upper portion, followed by an increasing surface profile in the downstream portion.

Depending on the jump length and its position relative to the downstream sideweir end, the hydraulic jump has a typical appearance, or not. Figure 18.6a shows an *impact hydraulic jump* and Fig. 18.6b the typically developed hydraulic jump (Chap. 7).

Based on the momentum equation (Chap. 1) and by neglecting the lateral outflow component, the *sequent depths* h_1 and h_2 in front and downstream of the jump are

$$h_2/h_1 = \mathsf{F}_1. \tag{18.37}$$

This result is similar to Eq. (7.18) for flows with large Froude numbers in rectangular channels. Equation (18.37) compares reasonably well with the data of Sassoli (1963) for one-sided outflow, and Buffoni et al. (1986) for two-sided outflow from a circular pipe, provided $\mathsf{F}_1 < 3$. Sideweirs with a supercritical approach flow can be roughly analyzed for the location of hydraulic jumps. For jumps which are not impact hydraulic jumps, the length of roller L_r, and the length of jump

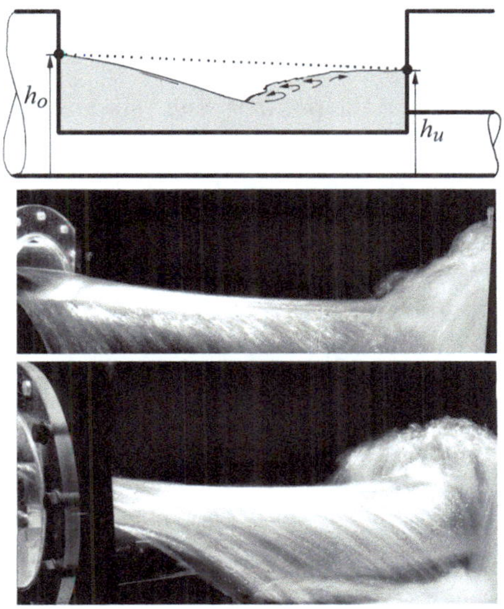

Fig. 18.11 (**a**) (—) Free surface profile of sideweir with a hydraulic jump, (. . .) simplified approach with a linear surface profile, (•) boundary flow depths. (**b**) and (**c**) impact hydraulic jump along sideweir (Hager 1994a)

L_j can be estimated from data collected on U-shaped and rectangular channels (Chap. 7) as

$$L_r \cong 4.5h_2, \tag{18.38}$$

$$L_j \cong 6.0h_2. \tag{18.39}$$

Impact hydraulic jumps are discussed in Sect. 18.4.1. If the length from the toe to the sideweir end is smaller than $1.5L_r$, then the computations have a considerable degree of uncertainty.

18.4.5 *Computational Approach for Converging Sideweir*

In converging sideweirs involving a low weir crest, hydraulic jumps occur frequently. This additional flow complication has to be integrated into the computational procedure. The free surface profiles are thus governed by backwater curves, such as those discussed in Chap. 8. The uniform flow depth and the critical depth have a main influence on the resulting free surface profile. In converging sideweirs, these parameters are the *pseudo-uniform depth* and the *critical depth.* Whereas the backwater effect in usual channels is due to bottom and friction slopes, sideweir flows are modified by the lateral outflow effect. Despite of this fundamental difference the basic computational rules for spatially-varied flow are:

- *Subcritical sideweir flow* is governed by a downstream boundary condition, and the computational direction is against the flow direction, and
- *Supercritical sideweir flow* is governed by an upstream boundary condition, and the computational direction is identical with the flow direction.

Therefore, the portions with sub- and supercritical sideweir flows are separated by the hydraulic jump (Fig. 18.12).

In a generalized method of computation, all *boundary conditions* must be specified, i.e. flow depths and discharges at the upstream (subscript o) or downstream (subscript u) ends of the sideweirs. Then, the thoroughly sub- and supercritical flow regions are computed, followed by a verification whether:

- Thoroughly supercritical flow occurs,
- A hydraulic jump establishes along the outflow reach, or
- Thoroughly subcritical flow forms.

Depending on the computational result, the free surface profile $h(x)$ and thus the discharge distribution $Q(x)$ may be determined. The *computational scheme* can be summarized as follows:

1. Computation of upstream characteristics Q_o, n_o, S_{oo}, D_o, and downstream characteristics Q_u, n_u, S_{ou} and D_u.

(a)

(b)

Fig. 18.12 Sideweir with a two-sided low-crested overflow for $S_{oo} = 0.1\%$ and $Q_o = 10Q_K$ (Taubmann 1972) for (**a**) short and (**b**) long sideweir

2. Uniform flow depths h_{No} and h_{Nu}.
3. Critical depths h_{co} and h_{cu}.
4. Boundary flow depths h_o and h_u, determination of corresponding energy heads H_o and H_u, and relative flow depths $y_o = h_o/H_o$ and $y_u = h_u/H_u$, respectively,
5. Specification of average weir height w_m, and relative weir heights $W_o = w_m/H_o$ and $W_u = w_m/H_u$, respectively, plus weir length ΔL.
6. Computation of parameter Θ from Eq. (18.32) and thus the two pseudo-uniform flow depths h_{PNo} and h_{PNu}.
7. Determination of free surface profile $h(x)$ for *supercritical* flow reach by accounting for pseudo-uniform flow. Simultaneous computation of sequent flow depth profile $h_2(x)$ from Eq. (18.37).
8. Determination of free surface profile $h(x)$ for *subcritical* flow reach and comparison with the profile $h_2(x)$.
9. Definition of the final free surface profile $h(x)$ by including the hydraulic jump, and the discharge distribution $Q(x)$.
10. Discussion of the solution, and possibly advancing an improved flow by modifying the basic parameters.

From this compilation of the computational steps for one approach flow condition, a considerable effort is seen to result. This deficiency and the uncertainties of basic assumptions make sideweirs with a hydraulic jump unattractive. Hager et al. (1982) have also discussed an alternative approach. Two examples are presented for both converging sideweirs without, and with a hydraulic jump.

Example 18.10 Given an approach flow channel with $D_o = 1.25$ m, bottom slope $S_{oo} = 1\%$, $1/n = 85$ m$^{1/3}$s^{-1} with a maximum discharge $Q_o = 4$ m^{3}s^{-1}. The one-sided sideweir is $\Delta L = 6.5$ m long with an average weir height of $w_m = 0.60$ m. The throttling pipe has a diameter of $D_u = 0.30$ m and the downstream flow depth is $h_u = 0.90$ m. Determine the free surface profile $h(x)$ for $Q_u = 0.1$ m^{3}s^{-1}.

1. $Q_o = 4$ m^{3}s^{-1}, $1/n_o = 85$ m$^{1/3}$s^{-1}, $S_{oo} = 1\%$, $D_o = 1.25$ m, $Q_u = 0.1$ m^{3}s^{-1}, $1/n_u = –$, $S_{ou} = –$, $D_u = 0.30$ m.
2. *Uniform flow* $q_{No} = 0.26$, thus $y_{No} = 0.694$ and $h_{No} = 0.87$ m (Chap. 5). For the throttling pipe, no indications are available.
3. *Critical flow* $h_{co} = 1.07$ m and $h_{cu} = 0.24$ m (Chap. 7). The corresponding approach flow Froude number for uniform approach flow is $\mathsf{F}_o = 4/(9.81 \cdot 1.25 \cdot 0.87^4)^{1/2} = 1.51 > 1$.
4. *Boundary conditions* (subscript r) are: upstream $h_{or} = h_{No} = 0.87$ m, $Q_o = 4$ m^{3}s^{-1}, $H_o = 1.87$ m; downstream $h_{ur} = 0.90$ m, $Q_u = 0.10$ m^{3}s^{-1}, $H_u = 0.91$ m. The difference of energy heads $(H_o–H_u)/H_u = 1.05$ points at the appearance of a hydraulic jump.
5. *Sideweir* with $w_m = 0.60$ m, thus $W_o = 0.6/1.87 = 0.32$ and $W_u = 0.6/0.91 = 0.66$, weir length $\Delta L = 6.5$ m.
6. *Pseudo-uniform flow* convergence $\theta = (1.25–0.3)/6.5 = 0.146$, thus with $(D_m/H_o)^{0.4} = (0.775/1.87)^{0.4} = 0.70$ and one-sided outflow ($n^* = 1$), one has $\Theta_o = 1.41 \cdot 0.146/(1{\cdot}0.7) = 0.29$, and $y_{PN} = 0.58$ from Eq. (18.32), thus $h_{PNo} = 0.58 \cdot 1.87 = 1.08$ m $> h_o$. Accordingly, the pseudo uniform flow has no direct effect on the surface profile.

   From the downstream boundary with $W_u = 0.66$ and $(D_m/H_u)^{0.4} = 0.94$, and $\Theta_u = 1.41 \cdot 0.146/(1{\cdot}0.94) = 0.22$, pseudo-uniform flow depth is $y_{PN} = 0.86$ and $h_{PNu} = 0.86 \cdot 0.91 = 0.78$ m.
7. The equation of the free surface profile is from Eq. (18.28)

$$\frac{y(X) - 0.32}{0.737 - 0.32} = 0.725 \exp\left[\frac{-X/2}{1.4 - 2.5(0.32 - 0.1)^{2.5}}\right], \tag{18.40}$$

or

$$y(X) = 0.32 + 0.302 \exp(-0.372X). \tag{18.41}$$

With the boundary condition $h_{No} = 0.87$ m, thus $y_o = 0.87/1.87 = 0.465$, Eq. (18.41) gives $X_o = 2$. Table 18.2 shows the profiles $y(X)$ and $h(x)$. The discharge is, from Eq. (18.20),

$$Q = (h/D)^{1.4} D^2 [2\,g(H - h)]^{1/2}, \tag{18.42}$$

the Froude number F_1 varies with the diameter $D(x)$, the flow depth $h(x)$ and the discharge $Q(x)$ according to Eq. (6.35). From Table 18.2, the sequent flow depth is always smaller than 1.35 m.

8. The *subcritical* flow reach cannot be specified because $h_u = 0.90$ m is always below h_2 from Table 18.2. Thus, the flow is supercritical across the sideweir with an *impact hydraulic jump* at the sideweir end, and a height of almost 1.20 m. Its length is estimated to $L_p \cong L_r/3 = 1.5h_2 = 1.75$ m. The average flow depth from Eq. (18.7) along the hydraulic jump is $h_{mu} = 0.64 + (2/3)(1.2–0.64) = 1.01$ m, and the corresponding lateral outflow discharge based on the conventional approach is $\Delta Q = 0.42 \cdot 19.62^{1/2}(1.01–0.60)^{3/2}1.75 = 0.85$ m^{3}s^{-1}.
9. The effective free surface profile is given in Table 18.2.
10. The lateral outflow $\Delta Q = 4–0.1 = 3.9$ m^{3}s^{-1} cannot be discharged with $h_{ru} = 0.9$ m. The latter value has to be increased to at least $h_{ru} = 1.2$ m.

Table 18.2 Free surface profiles, from *upstream* sideweir end for $H_o = 1.87$ m and $Q_o = 4$ m^{3}s^{-1}

x [m]	0	1	2	3	4	5	6	6.5
D [m]	1.25	1.10	0.96	0.81	0.67	0.52	0.37	0.30
X [-]	2.0	2.37	3.60	4.65	5.96	7.76	10.48	12.42
y [-]	0.464	0.43	0.40	0.37	0.35	0.34	0.33	0.32
h [m]	0.87	0.80	0.75	0.70	0.66	0.63	0.61	0.60
Q [m^{3}s^{-1}]	(4.15)	3.55	3.06	2.56	2.14	1.74	1.37	1.19
F_1 [-]	1.55	1.69	1.77	1.85	1.92	1.94	1.93	1.93
h_2 [m]	1.35	1.35	1.33	1.30	1.27	1.22	1.18	1.16
h_{eff} [m]	0.87	0.80	0.75	0.70	0.66	0.85	1.10	1.20
Q [m^{3}s^{-1}]	4.0	3.55	3.06	2.56	2.14	1.70	1.30	1.00

Example 18.11 Re-compute Example 18.10 for $h_{ru} = 1.3$ m.
Steps 1 to 3, and 5 and 7 are identical as in Example 18.10.

4. *Boundary conditions* at downstream end are $h_{ru} = 1.3$ m, $Q_u = 0.1$ m^{3}s^{-1} and thus $H_u = 1.30$ m. The energy loss is now significantly reduced to $(1.87–1.30)/1.87 = 30\%$.

6. *Pseudo-uniform flow* with $\theta = 0.146$, $(D_m/H)^{0.4} = (0.775/1.3)^{0.4} = 0.81$, $W_m = 0.6/1.3 = 0.46$ and $n^* = 1$, one has $\Theta = 0.25$, thus $y_{PN} = 0.71$, or $h_{PN} = 0.92$ m.
8. From Eq. (18.25) the profile equation reads

$$\frac{y - W}{1 - W} = 1.06 + 0.32(1 - 0.46)^{1.15}1 \cdot X \text{ or} \tag{18.43}$$

$$y(X) = 0.46 + 0.54(1.06 + 0.16X) = 1.03 + 0.085X. \tag{18.44}$$

Table 18.3 shows the free surface profile for *subcritical* flow, based on $X_r = -0.35$ or $x_r = -0.35(0.3/1.30)^{0.6}1.3 = -0.19$ m. The dimensionless coordinates X and $y(X)$ may thus be computed.

9. The effective surface profile is given in Table 18.3. It is characterized by a *hydraulic jump* at location $x = 5.8$ m, i.e. it is deformed because of its short distance up to the end wall. The distribution of discharge $Q(x)$ was obtained from Eq. (18.20) by accounting for the respective energy heads H_o and H_u.
10. The computational procedure is lengthy for sideweirs with a hydraulic jump. For practical application, several downstream depths h_u had to be assumed to vary ΔQ with h_u. One could also determine an optimum structural length because the reach upstream from the hydraulic jump is not really effective (Hörler and Hörler 1973).

Table 18.3 Free surface profile, from *downstream* end $x = 6.5$ m for $H_u = 1.30$ m

x [m]	6.5	6	5	4	3	2	1	0
D [m]	0.30	0.37	0.52	0.67	0.81	0.96	1.10	1.25
X [-]	−0.35	−1.13	−2.25	−3.08	−3.78	−4.34	−4.85	−5.28
y [-]	1.0	0.93	0.84	0.77	0.71	0.66	0.62	0.58
h [m]	1.30	1.21	1.09	1.00	0.92	0.86	0.81	0.75
h_2 [m]	1.16	1.18	1.22	1.27	1.30	1.33	1.35	1.35
h_{eff} [m]	1.30	1.21	0.63	0.66	0.70	0.75	0.80	0.87
Q [m^{3}s^{-1}]	0.10	0.96	1.70	2.14	2.56	3.06	3.55	4.00

18.5 Short Sewer Sideweir

18.5.1 *Introduction*

Due to space shortage in most cities and the increasing maximum discharges due to urbanisation, existing sewer sideweirs are modernized based on the original structural dimensions. To increase the *lateral outflow capacity* for a fixed structural length, the weir height has to be decreased. Then, the non-uniformity of lateral outflow increases and the revised design may be poorer than the original design. The question is thus whether a sideweir with an increased capacity may have an acceptable hydraulic performance. Also, the separation characteristics between undissolved matter and water should not be poor. To study the hydraulic limits of a sideweir, the short sewer sideweir was model-tested in detail (Fig. 18.12). The design is based on the fact that the lateral outflow should currently not be discharged directly to a receiving water but introduced in a *stormwater storage basin* (German: Regenrückhaltebecken; French: Bassin de retenue des eaux de pluie). The sewage is thus treated with a primary settling process and only the basin overflow continues to the receiving water. After a rainfall, the storage basin is emptied to the treatment station, and the main load is thus treated further.

The short sewer sideweir was introduced by Hörler and Hörler (1973). Based on the poor outflow intensity of sideweirs with a supercritical approach flow, and an impact hydraulic jump at the end wall, the late Prof. Arnold Hörler (1903–1995), a Swiss pioneer of sewage technology, advanced the idea of the short sewer sideweir. It consists of only two portions, namely the supercritical drawdown portion and the hydraulic jump, with the intermediate reach of small overflow depth upstream from the hydraulic jump dropped.

Gisonni and Hager (1997) defined the *short sewer sideweir*. Its main features are:

- Length L smaller than three approach flow diameters,
- Weir height w at least half the approach flow diameter,
- Converging and symmetrical plan to improve the outflow uniformity,
- Sharp-crested sideweir to allow for overflow modification,
- Approach flow Froude number smaller than 1.5, and
- Flow stabilisation by the so-called End Plate (see below).

To reduce the number of parameters, in all tests $Q_u = 0$, i.e. the lateral outflow was equal to the approach flow discharge. This is close to the conditions of maximum approach flow discharge, and smaller discharges may be computed as previously outlined.

18.5.2 *End Plate*

For small relative weir height w/D_o a supercritical flow impinges the end wall of the sideweir. Because of the complex spatial flow, an instability is set up that can be described as follows. For typical bottom slopes in the approach flow sewer of the order of 1%, the flow is close to critical accelerating due to the lateral outflow (Fig. 18.13a). The channel flow impinges heavily onto the end wall and two macro-vortices are generated. Because of the impact flow and the lateral confinement by the weir walls a bottom vortex in the clockwise direction is developed, associated with a surface vortex in the counter-clockwise direction due to the pile-up at the end wall.

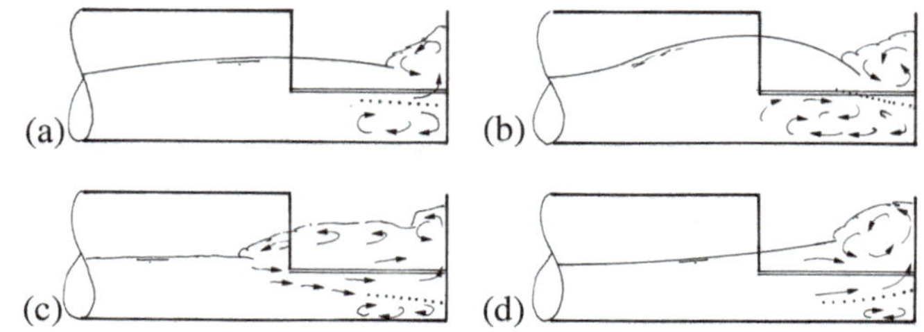

Fig. 18.13 Sequences of dead-end instability, for details see main text

The *shear layer* between the two vortices is unstable and the surface flow plunges, resulting in a backwater effect (Fig. 18.13b). A solitary type-wave is propagated into the upstream sewer which may eventually choke due to surging (Fig. 18.13c). If the filling ratio of the pipe is sufficiently large, the cross-section gets obstructed and a plug-type two-phase flow is set up. Such cases occurred in laboratory pipes already for filling ratios of the order of 60%.

The cycle of an instability may be described as follows: In the beginning, the surface is nearly horizontal, and so is the interface between the plunging and the rising flow cells upstream from the end wall (Fig. 18.13a). Due to the large pressure at the end wall the interface is lowered and the bottom vortex is filled with excess fluid, resulting in a curved free surface and so-called plunging flow, with a swell upstream from the end wall (Fig. 18.13b). As the free surface attains a limit slope, wave breaking at the rising swell portion occurs and a hydraulic jump type choking phenomenon is developed. Due to the free surface return flow, the bottom cell is emptied (Fig. 18.13c) and the surface roller is pushed downstream to fill the surface vortex cell again (Fig. 18.13d). It eventually grows strong enough to initiate the next instability cycle.

The *flow instability* for short, low-crested sideweirs has to be removed for a satisfactory and efficient sideweir design. Otherwise, surges are set up in the upstream sewer that can initiate transitions from free surface to pressurized flow, associated with water hammering and structural vibration (Hamam and McCorquodale 1982). The flow instability is clearly caused by the *unstable vortex interface* between the two vortex cells.

Figure 18.14 shows the so-called *End Plate* stabilizing the vortex interface at the end wall. The End Plate has a length L_p and a width b_p, and is mounted on the end

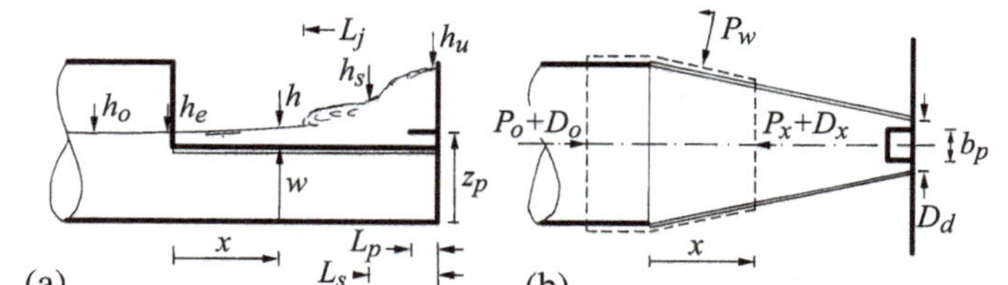

Fig. 18.14 Sewer sideweir with End Plate (**a**) section (**b**) plan with (- - -) control volume

wall. The approach flow Froude number has practically no effect of the plate size, and the main parameter is the approach flow filling ratio $y_o = h_o/D$, provided its elevation from the invert is larger than $z_p = 0.15D_o$. The width of the plate was effective if $b_p \geq (1/2)D_u$ and the section is only partially constricted. The minimum plate length should be $L_p = (1/6)L$, provided the sideweir is short. For $L/D_o > 3$, no End Plate is needed because the instability disappears due to its shortness compared to the sideweir length L. The relative elevation of the End Plate $Z_p = z_p/D_o$ over the sideweir bottom varies with the relative length L/D_o and $y_o = h_o/D_o$ as (Gisonni and Hager 1997)

$$Z_p = 0.38(L/D_o)^{0.75}y_o^{1/3}. \tag{18.45}$$

The elevation of an effective End Plate thus increases as both the length of the sideweir and the approach flow filling ratio increase. The elevation z_p should not be higher than according to Eq. (18.45), nor higher than h_o because the separation of the vortex interface is then ineffective.

Example 18.12 Design an End Plate for a sewer sideweir of discharge $Q_o = 1.5\ \text{m}^3\text{s}^{-1}$, diameters $D_o = 1$ m and $D_d = 0.25$ m with a bottom slope $S_{oo} = 1.6\%$ and $n = 0.011\ \text{sm}^{-1/3}$. The sideweir length is $L = 2.5$ m.
With $q_{No} = nQ_o/(S_{oo}^{1/2}D_o^{8/3}) = 0.011 \cdot 1.5/(0.016^{1/2}1^{8/3} = 0.130$, the uniform flow filling ratio is $y_{No} = 0.926[1-(1-3.11 \cdot 0.13)^{1/2}]^{1/2} = 0.44$ from Eq. $(5.15)_1$, thus $h_{No} = 0.44 \cdot 1 = 0.44$ m, and the uniform approach flow Froude number is $\mathsf{F}_{No} = 1.5/(9.81 \cdot 1 \cdot 0.44^4)^{1/2} = 2.50 > 1$, i.e. stably supercritical.

With $L/D_o = 2.5$ and $y_o = y_{No} = 0.44$, the elevation of the End Plate is $Z_p = 0.38 \cdot 2.5^{0.75}0.44^{0.33} = 0.58$ from Eq. (18.45), thus $z_p = 0.58 \cdot 1 = 0.58$ m, its width is $b_p \geq 0.5 \cdot 0.25 = 0.125$ m and its length $L_p \geq (1/6)2.5 = 0.42$ m. Selected are $z_p = 0.58$ m, $b_p = 0.15$ m, $L_p = 0.45$ m.

Separate experiments with uniform sand of 1 and 2 mm diameter demonstrated a significant effect of the End Plate on the *resuspension of solids*. Without the End Plate, the vortices were unstable and turbulence lifted material over the weir crest. With the End Plate mounted, the sediments remained practically at the bottom and were much less resuspended. The device is simple and effective because it improves the hydraulic performance and the sediment transport across a sideweir. It can easily be added to existing structures.

18.5.3 *Discharge Distribution*

For a conventional weir with a horizontal crest, the discharge is *uniformly* distributed over the length L. If $x = 0$ is the beginning and $x = L$ the end of a sideweir, $Q/Q_o = 1$

for $\overline{X} = x/L = 0$ and $Q/Q_o = 0$ for $\overline{X} = 1$, with $Q/Q_o = 1 - \overline{X}$. Increasing the approach flow Froude number F_o results in a non-uniform discharge distribution $Q(\overline{X})/Q_o$. Based on detailed and systematic experiments, the discharge depends mainly on F_o as

$$Q/Q_o = 1 - \overline{X}^{1+\mathsf{F}_o}. \tag{18.46}$$

For $\mathsf{F}_o \rightarrow 0$, the discharge is uniformly distributed and the degree of *non-uniformity* increases as F_o increases. For larger F_o, the upstream overflow portion is small due to the low overflow depth, and the main overflow is concentrated at the sideweir end. Note that Eq. (18.46) is independent of the relative weir height and the approach filling ratio.

18.5.4 *Free Surface Profile*

Figure 18.14a defines the free surface profile in a short sewer sideweir, with the approach flow reach (subscript o), the upstream section (subscript e), the swell section (subscript s) and the downstream section (subscript u). The axial free surface profile $h(x)$ may be divided into a more or less gradually varied reach between h_e and h_s, and the swell reach of depth h_u and length L_s.

The surface profile can be normalized by accounting for the extreme values h_e and h_u, with the coordinate $Y_p = (h - h_e)/(h_u - h_e)$ varying between 0 and 1. A linear increase would correspond to $Y_p = \overline{X}$. Again, the nonlinearity of the free surface profile increases with the approach Froude number F_o, and Gisonni and Hager (1997) found

$$Y_p = \overline{X}^{1+\mathsf{F}_o}. \tag{18.47}$$

The distributions of flow depth $Y_p(\overline{X})$, and discharge $1 - Q/Q_o$ are thus identical. Short sewer sideweirs of low weir height and a large approach flow Froude number have a *nonuniform* outflow pattern, and should be avoided, therefore. This confirms the current design practise but the limit between acceptable and inacceptable flows has yet to be defined.

The *free surface profile* can approximately be computed by assuming:

- One-dimensional flow with a hydrostatic pressure distribution,
- Compensation of friction slope by bottom slope,
- Discharge distribution according to Eq. (18.46),
- Cross-sectional area approximated by a power formula.

Because of the swell development at the end wall, the additional losses cannot be neglected, and the free surface profile must be determined based on momentum considerations. The solution consists of portions for the supercritical upstream and the subcritical downstream reaches. The two reaches are separated by a hydraulic jump of which the sequent depth ratio was determined to

$$h_2/h_1 = (2\mathsf{F}_1^2)^{0.4} - 0.2. \tag{18.48}$$

The depth ratio increases thus almost linearly with F_1. Differences with the empirical Eq. (18.37) are small for $1 < \mathsf{F}_1 < 5$.

Of particular relevance is the end depth ratio $Y_u = h_u/h_e$ predicted with the momentum equation. With $\Delta = D_d/D_o$ as the diameter ratio, Gisonni and Hager (1997) found

$$Y_u = \left(1 + \frac{4\mathsf{F}_o^2}{1 + \Delta^{1/2}}\right)^{2.5}. \tag{18.49}$$

Equation (18.49) is independent of convergence angle, weir height and sideweir length, depending mainly on the approach flow Froude number F_o. The agreement with observations is excellent up to $\mathsf{F}_o = 3$. The length of the swell and the length of the jump were also measured and correlated to F_o.

Example 18.13 Determine the discharge distribution and the axial surface profile for the sideweir given in Example 18.12.
With $\mathsf{F}_o = 2.5$, the discharge distribution is $Q/Q_o = 1 - \overline{X}^{3.5}$ resulting in $Q/Q_o(0) = 1$, $Q/Q_o(0.2) = 1$, $Q/Q_o(0.4) = 0.96$, $Q/Q_o(0.6) = 0.83$, $Q/Q_o(0.8) = 0.54$, and $Q/Q_o(1) = 0$. Note that more than 50% of the discharge is diverted over the last 20% of sideweir length.

With $D_d = 0.30$ m, the diameter ratio is $\Delta = 0.3/1 = 0.3$ and thus from Eq. (18.49) $Y_u = [1 + 4 \cdot 2.5^2/(1 + 0.3^{1/2})]^{0.4} = 3.12$ or $h_u = 3.12 \cdot 0.44 = 1.37$ m. The nonuniformity of the free surface profile is notable.

With $h_o \cong h_e$ because of a small drawdown effect (Chap. 20), the profile equation is $(h - 0.44)/(1.37 - 0.44) = \overline{X}^{3.5}$ from Eq. (18.47), thus $h(\overline{X}) = 0.93\overline{X}^{3.5} + 0.44$. The resulting depths are $h(0) = 0.44$ m, $h(0.2) = 0.44$ m, $h(0.4) = 0.48$ m, $h(0.6) = 0.60$ m, $h(0.8) = 0.87$ m and $h(1) = 1.37$ m.

18.5.5 *Lateral Discharge*

The lateral outflow from a short sewer sideweir is complicated by spatial currents, and a simple model was introduced to retain the most prominent flow features. For flows with an approach flow depth $(h_o - w) > 0$, the overflow discharge ΔQ is related to $(h_o - w)$, the approach overflow velocity $[2g(h_o - w)]^{1/2}$ and the overflow length ΔL. All remaining parameters are accounted for considering the *average discharge coefficient* C_{da}. From observations, C_{da} varies considerably with F_o as (Gisonni and Hager 1997)

$$C_{da} = 1 + \frac{1}{2}(\Delta L/D_o)\mathsf{F}_o^2. \tag{18.50}$$

For sideweirs without the End Plate, the term (1/2) is reduced to (1/3) due to the large flow nonuniformity. Equation (18.50) is valid for F_o up to 3.

The *lateral discharge* of the short sideweir is thus

$$\Delta Q = C_{da}[2\,g(h_o - w)^3]^{1/2}\Delta L. \tag{18.51}$$

Equation (18.51) cannot be explicitly applied because C_{da} depends on both Q_o and ΔL. For the design discharge, the sideweir length is sought, whereas the sideweir length ΔL is fixed for all other discharges.

Prediction of sideweir length. According to Gisonni and Hager (1997) the relevant parameters are dimensionless length Λ and generalized Froude number Ψ, given as

$$\Lambda = \left(\frac{2D_o}{h_o}\right)^{1/3}\left(\frac{h_o - w}{h_o}\right)\frac{\Delta L}{D_o}, \quad \Psi = \left(\frac{h_o}{2\,D_o}\right)^{1/6}\left(\frac{h_o}{h_o - w}\right)^{1/2}\mathsf{F}_o. \tag{18.52}$$

Inserting in Eq. (18.51) results in

$$\Lambda = \Psi^{-2}\left[(1 + 2\Psi^3)^{1/2} - 1\right]. \tag{18.53}$$

It may be demonstrated that the basic solution with $C_{da} \cong 1$ is valid only if $\Psi < 0.5$. For supercritical approach flow, the average discharge coefficient based on h_o is >> 1.

Example 18.14 What is the weir height required for Example 18.13?
With $\mathsf{F}_o = 2.5$, $\Delta L = 2.5$ m and $D_o = 1$ m, $C_{da} = 1 + 0.5(2.5/1)2.5^2 = 8.8$, and $w = h_o - [(\Delta Q/C_{da}\Delta L)^2/2g]^{1/3} = 0.44 - [(1.5/8.8 \cdot 2.5)^2/19.62]^{1/3} = 0.38$ m.

Example 18.15 What is the weir length for a weir height of $w = 0.35$ m?
With $\mathsf{F}_o = 2.5$, $w = 0.35$ m, $h_o = 0.44$ m and $D_o = 1$ m, one has for $\Psi = [0.44/(2 \cdot 1)]^{1/6}[0.44/(0.44 - 0.35)]^{1/2}2.5 = 4.3$ and thus $\Lambda = 0.63$ from Eq. (18.53). The weir length required is thus $\Delta L = 0.63(0.44/2 \cdot 1)^{1/3}[0.44/(0.44 - 0.35)]1 = 1.86$ m. From Eq. (18.50) the discharge coefficient is $C_{da} = 1 + 0.5(1.86/1)2.5^2 = 6.81$ and thus smaller than in Example 18.14.

Prediction of overflow discharge. Equation (18.51) may also be used to predict the lateral outflow discharge ΔQ, provided $\Delta Q/Q_o \cong 1$. Then, the governing parameters are

$$f = \left(\frac{\Delta L}{D_o}\right)^{1/2} \mathsf{F}_o, \quad \lambda = \left(\frac{h_o - w}{h_o}\right)^{3/2} \left(\frac{\Delta L^3}{h_o D_o^2}\right)^{1/2} \tag{18.54}$$

and the relevant solution of the transformed Eq. (18.51) is

$$f = \frac{1}{2^{1/2}\lambda}[(1 - (1 - 4\lambda^2)^{1/2}] \tag{18.55}$$

The function $\lambda(f)$ has the maximum (subscript M) $\lambda_M = \lambda(f = 2^{1/2}) = 1/2$ and the physical domain of λ is $0 < \lambda < 1/2$. For small values of λ, i.e. for a small relative overflow depth, the asymptotic value $f = 2^{1/2}\lambda$ corresponds to $C_{da} = 1$, and the effect of F_o is small. For $\lambda > 0.25$, the effect of the approach flow Froude number becomes significant and cannot be neglected.

Example 18.16 What is the lateral outflow ΔQ for $\mathsf{F}_o = 2.5$, $D_o = 1$ m, $h_o = 0.44$ m and $w = 0.35$ m, if $\Delta L = 1.5$ m only?
Parameter $\lambda = [(0.44–0.35)/0.44]^{3/2}[1.5^3/(0.44 \cdot 1^2)]^{1/2} = 0.26$ is just at the limit of significance, and Eq. (18.55) yields $f = 0.39$, thus $\mathsf{F}_o = 0.39(1/1.5)^{1/2} = 0.32$ and $Q_o = (gD_o h_o^4)^{1/2}\mathsf{F}_o = (9.81 \cdot 1 \cdot 0.44^4)^{1/2} 0.32 = 0.19$ m^{3}s^{-1}. The corresponding discharge coefficient is from Eq. (18.50) $C_{da} = 1 + 0.5(1.5/1)0.32^2 = 1.08$. A length reduction to $\Delta L = 1.5$ m generates subcritical approach flow, with a significantly reduced lateral outflow.

18.5.6 *Momentum Transfer*

The lateral sideweir outflow involves an *additional loss slope* S_L, as given in Eq. (17.5). The momentum transfer is thus governed by the term $\Phi = [1–(U\cos\phi/V)]$ where U is the lateral outflow velocity, ϕ the lateral outflow angle and V the cross-sectional channel velocity. The controlling parameter of the momentum exchange is the ratio of discharge intensity times the local channel width and local discharge, thus $\Omega = –(\mathrm{d}Q/\mathrm{d}x)B/Q$ (Hager 1987). With Eq. (18.46) for the discharge distribution, thus a given discharge intensity $\mathrm{d}Q/\mathrm{d}x$ follows $\Omega = (D_o/\Delta L)(1 + \mathsf{F}_o)$ $[(1 - \theta\overline{X})/(\overline{X}^{-\mathsf{F}_o} - \overline{X})]$. Introducing the parameter $\omega = [\Omega/(1 + \Omega)]^{\mathsf{F}_o/(1+\mathsf{F}_o)}$ that is confined to $0 < \omega < 1$, it can be further shown that

$$\Phi = \omega. \tag{18.56}$$

The lateral outflow velocity $U\cos\phi$ in the streamwise direction can thus be predicted, once the approach flow Froude number F_o is known, and the sideweir geometry is specified. It can further be concluded that the effect of Φ on the decay of energy head along the sideweir is small, and may practically be neglected. The assumption of a *constant energy head* on which the computations in Sects. 18.3 and 18.4 are based is thus verified. This conclusion is significant from an engineering point of view because a complex hydraulic problem can be attacked with a comparably simple approach.

The *energy loss* of the short sewer sideweir is determined based on the generalized Bernoulli equation (Chap. 1) between the approach flow and downstream sections. Accounting for Eq. (18.49), the headloss coefficient $\xi_s = \Delta H/[V_o^2/2g]$ depends exclusively on F_o and $\Delta = D_d/D_o$. Typically the headloss coefficient varies between −1 and +1, with $\xi_s = 0$ for about $\mathsf{F}_o = 1.5$. For small Froude numbers $\mathsf{F}_o < 1.5$, there is a slight increase of the energy line due to the lateral outflow, whereas one may expect a slight reduction of energy head for $\mathsf{F}_o > 1.5$, mainly due to the hydraulic jump at the end wall. Converging the sideweir plan involves slightly smaller losses than for the prismatic sideweir. These observations are based on detailed experimentation, from which these conclusions have been derived.

18.5.7 *Experimental Observations*

The complex flow patterns of short sewer sideweirs are illustrated by photographs. All documents pertain to the relative lengths $\Delta L/D_o = 4$, 2 and 1, with the End Plate always mounted.

Typical sideweir flow occurs for $\Delta L/D_o = 4$. Figure 18.15 refers to *longitudinal views* for various approach flow Froude numbers F_o and $y_o \cong 0.7$ for $w/D_o = 0.60$. For $\mathsf{F}_o = 0.13$ and 0.30 the free surface is nearly horizontal and the flow is comparable to conventional weir flow. Increasing F_o to 0.75 results in a standing wave pattern close to the sideweir start, and an impact swell at the sideweir end. For supercritical approach flow with $\mathsf{F}_o = 1.25$ the axial flow depth increases gradually up to the swell, with an impact hydraulic jump at the sideweir end, associated with a reversed lateral outflow. For $\mathsf{F}_o = 2.5$ and $y_o = 0.42$ there is no lateral overflow along the first portion of the crest length. Then, the flow depth increases rapidly to a second surface level and impinges again the end wall by generating the swell. Such flow is rather nonuniform, although steady and without pulsations.

Figure 18.16 refers to the corresponding *axial views* in the flow direction. For $\mathsf{F}_o = 0.3$, the flow is tranquil and also for $\mathsf{F}_o = 0.75$ practically no surface waves may be identified except close to the downstream sideweir end. The front portion for the flow with $\mathsf{F}_o = 1.25$ is surprisingly smooth but the swell is now large, though steady. For $\mathsf{F}_o = 2.5$ the front portion looks like an undular hydraulic jump, with the upstream shock fronts crossing the channel to form a central rim and a relatively smooth transition to subcritical flow, impinging on the end wall. *Lateral views* in the downstream direction for this configuration exhibit typical features of sideweir

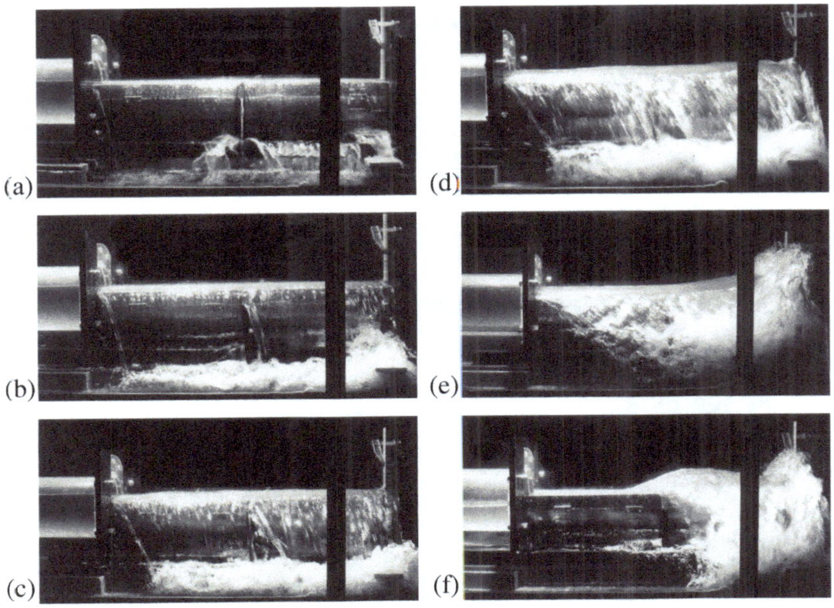

Fig. 18.15 Side views to sideweir flow for $y_o \cong 0.7$, $w/D_o = 0.60$ and $\mathsf{F}_o =$ (**a**) 0.13, (**b**) 0.30, (**c**) 0.50, (**d**) 0.75, (**e**) 1.25, (**f**) 2.50 (Gisonni and Hager 1997)

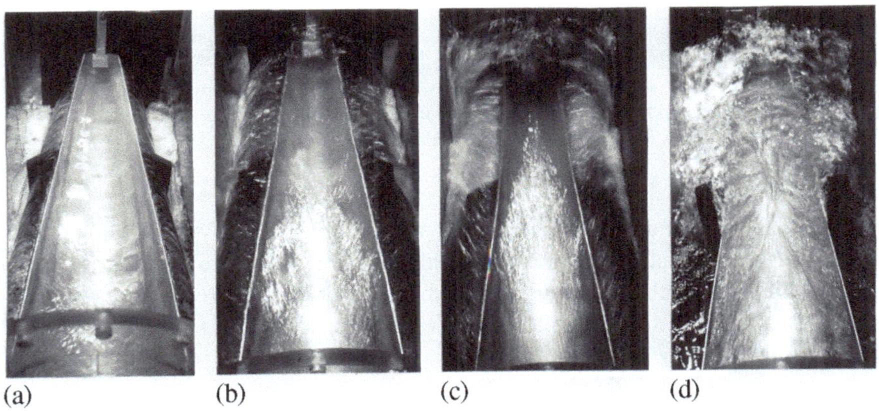

Fig. 18.16 Axial views in the flow direction corresponding to flows shown in Fig. 18.15 for $\mathsf{F}_o =$ (**a**) 0.30, (**b**) 0.75, (**c**) 1.25, (**d**) 2.50

flow (Fig. 18.17). For $\mathsf{F}_o = 0.30$ smooth flow is generated that becomes rougher for $\mathsf{F}_o = 0.75$. The flow in the upstream zone is similar for $\mathsf{F}_o = 1.25$, except for the large swell at the downstream end.

Short sideweirs of lengths $\Delta L/D_o = 2$ and 1 show a quite different flow pattern as compared to these of intermediate length. The flow is essentially made up

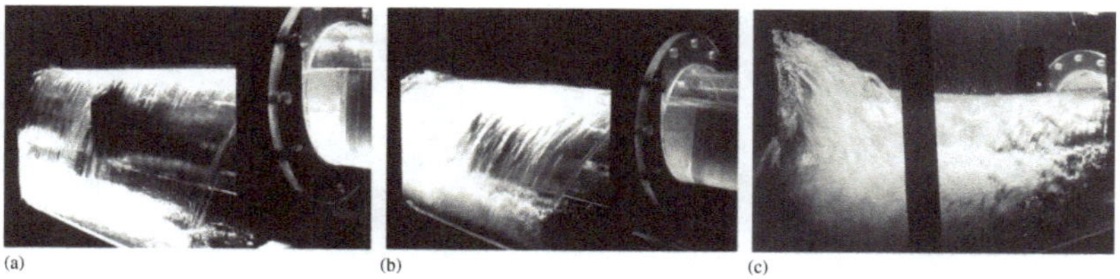

Fig. 18.17 Lateral views corresponding to Fig. 18.15 for F_o = (**a**) 0.30, (**b**) 0.75, (**c**) 1.25

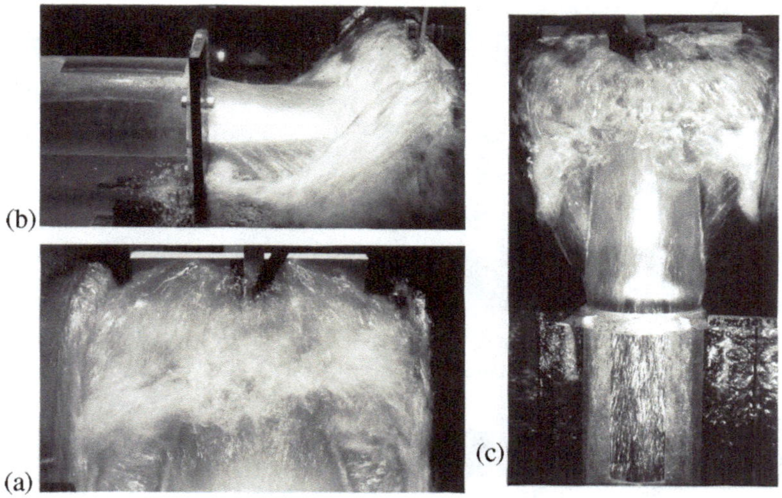

Fig. 18.18 Supercritical flow in short sewer sideweir for $\Delta L/D_o = 2$, $y_o = 0.66$, $w/D_o = 0.4$ and $\mathsf{F}_o = 1.25$. (**a**) Side view, (**b**) upstream view, (**c**) plan view

by the upstream and downstream reaches, without the gradually-varied flow reach. The typical flow features of the short sideweir are illustrated in Fig. 18.18 for $\Delta L/D_o = 2$, for which the flow may be subdivided into a first reach with an essentially horizontal free surface, followed by a second reach with a swell due to the end wall.

18.6 Sewer Sideweir with Throttling Pipe

18.6.1 *Introduction*

The observations resulting for the short sewer sideweir without a throttling pipe as described in Sect. 18.5 were extended to account for the *throttling device* (Del Giudice and Hager 1999). The experiments were conducted with the installations previously used by Gisonni and Hager (1997), involving an approach flow

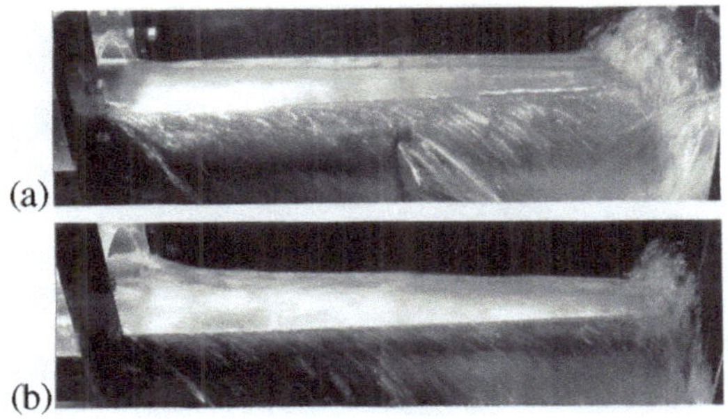

Fig. 18.19 Sideweir with throttling pipe, side (*top*) and upstream (*below*) views of sewer sideweir flow for $w/D_o =$ (**a**) 0.40, (**b**) 0.60

pipe of diameter $D_o = 0.240$ m and a throttling pipe of $D_d = 0.100$ m, a constant bottom slope of $S_o = 0.3\%$ and a sideweir 1.00 m long with two sharp-crested symmetrical overflow reaches. The throttling pipe (subscript p) was of plexiglass to allow for visualization, had lengths $L_p/D_d = 20$, 40 and 60, and was connected flush to the end wall of the sideweir with a sharp-crested intake geometry. The outlet of the throttling pipe discharged into the atmosphere, and thus was never submerged. Sideweirs of relative height $\overline{W} = w/D_c = 0.40$, 0.60 and 0.80 and with two-sided horizontal crests were tested for approach flow Froude numbers F_o between 0.15 and 1.50.

Compared to the sideweir without the throttling pipe, the flow close to the end wall was much *more stable* due to the throttle intake effect. An end plate was *never* required to improve the flow. Figure 18.19 shows typical flows for a medium crest height and almost supercritical approach flow. Note the nearly horizontal surface along the sideweir up to the *hydraulic jump* due to impact onto the end wall. Except for the hydraulic jump, the flow can be regarded as gradually-varied, with the characteristic flow depth for the throttling discharge at the end wall. These photos clearly indicate excellent flow characteristics even for a relatively large approach flow Froude number, particularly because of the converging sideweir plan geometry. For otherwise identical configurations, the *drawdown effect* followed by the hydraulic jump would be larger for the prismatic sideweir, resulting in a much less uniform flow pattern along the sideweir.

Figure 18.20 shows downstream views to the throttling pipe and details of discharge measurement (Del Giudice and Hager 1999). For design discharge, the throttling pipe is normally pressurized as is observed from the figure, or gated flow resulted if the bottom slope was larger than selected here.

18.6.2 *Free Surface Profile*

The free surface profile along a sewer sideweir may increase or decrease, depending on the contraction rate and the approach flow Froude number (Chap. 17). For the present observations with a contraction rate of $(D_o–D_d)/L = 0.14$, and usually subcritical approach flow, the free surface profile either remained constant or *increased*

Fig. 18.20 Throttling pipe downstream of sewer sideweir (**a**) overall view, (**b**) overflow discharge measurement with Mobile Venturi Flume, (**c**) sideweir flow and resulting two-phase flow in throttling pipe (Del Giudice and Hager 1999)

in the flow direction for side overflow, and *decreased* without overflow. Also, the energy head H relative to the approach flow energy head H_o increased over 1 in the upper overflow portion, and decreased below 1 close to the end wall, in agreement with indications in Sect. 18.5. The *energy head* along a converging sewer sideweir may be assumed constant, such that the Bernoulli equation is applicable.

The cross-sectional shape of a converging sewer sideweir is rather involved with a linear diameter variation $D = D_o - [D_o - D_d](x/L)$, where L is the sideweir length, and a cross-sectional area following either the circular, or the U-shaped profile. Because the dynamic effect for subcritical flows is not as significant as the static effect, and to allow for a straightforward approach, consider a *rectangular substitute* sewer sideweir of cross-sectional area $F = Dh$. Further, if the bottom slope is adjusted to the average friction slope, and all additional headlosses due to lateral flow are neglected, the *axial surface profile* follows the Bernoulli equation as

$$h_o + \frac{Q_o^2}{2gD_o^2 h_o^2} = h + \frac{Q^2}{2gD^2\,h^2}. \tag{18.57}$$

Here, subscript o refers to the approach flow section of the sideweir. With the approach flow Froude number $\mathsf{F}_o = Q_o/(gD_o^2 h_o^3)^{1/2}$, this can also be expressed as (Del Giudice and Hager 1999)

$$1+\frac{1}{2}\mathsf{F}_o^2 = y+\frac{1}{2}\mathsf{F}_o^2\left(\frac{\rho}{\delta y}\right)^2 \tag{18.58}$$

where $y = h/h_o$, $\rho = Q/Q_o$ and $\delta = D/D_o$. The first order solution of Eq. (18.58) is

$$y = 1+\frac{\mathsf{F}_o^2[1-(\rho/\delta)^2]}{2[1-(\rho/\delta)^2\mathsf{F}_o^2]}. \tag{18.59}$$

Experiments compare well with Eq. (18.59), except close to $(\rho/\delta)\mathsf{F}_o = 1$.

The free surface remains horizontal for two conditions: (1) $\mathsf{F}_o = 0$, and (2) $\rho/\delta = 1$. The first case is the hydrostatic configuration and not of further interest, therefore. So-called *pseudo-uniform* flow conditions result for $\rho/\delta = 1$, and $y = 1$ can be maintained along the entire sideweir provided

$$\frac{D(x)}{D_o} = \frac{Q(x)}{Q_o}. \tag{18.60}$$

This condition was explored in detail in Chap. 17 and is discussed below. Because the discharge Q decreases with x, pseudo-uniform flow results exclusively for converging sewer sideweirs. In general, the free surface may increase or decrease, as is indicated by Eq. (18.59). The Froude number is $\mathsf{F} = Q/(gD^2h^3)^{1/2} = (\rho/\delta)\mathsf{F}_o y^{-3/2}$ at any section of the sideweir, or nearly $\mathsf{F} = (\rho/\delta)\mathsf{F}_o$ for conditions close to pseudo-uniform flow. Introducing the *pseudo-uniform parameter* $P_u = \rho/\delta = [(Q/Q_o)/(D/D_o)]$ transforms Eq. (18.59) to

$$y = 1+\frac{\mathsf{F}_o^2(1-P_u^2)}{2[1-(P_u\mathsf{F}_o)^2]}. \tag{18.61}$$

This basic solution for converging sideweirs reveals the following information:

(1) For no flow ($\mathsf{F}_o = 0$) the free surface remains horizontal, and deviations of the free surface increase quadratically with the approach flow Froude number;
(2) Depending on the pseudo-uniform parameter P_u, the flow depth may increase or decrease along the sideweir. For $P_u = 1$, independent of all other conditions, the flow depth remains constant, except for $\mathsf{F}_o = 1$;
(3) The effect of the local Froude number is retained, given the previously introduced approximation. Equation (18.61) corresponds to a generalized backwater equation in which the effects of bottom slope and friction are neglected but the effects of spatially-varied discharge and diameter are retained; and
(4) Because all source terms due to gravity, friction and lateral outflow were dropped, Eq. (18.61) is valid only close to the sideweir beginning. Due to impact flow onto the end wall, this formulation breaks down close to the downstream sideweir end.

For *spatially constant discharge* ($\rho = 1$) along the converging sideweir prior to overflow, Eq. (18.59) simplifies to

$$y = 1 + \frac{\mathsf{F}_o^2(1-\delta^2)}{2(\mathsf{F}_o^2-\delta^2)}. \tag{18.62}$$

Because $0 \leq \delta \leq 1$, the surface profile depends essentially on the approach flow Froude number F_o. For $\mathsf{F}_o < 1$ the flow depth decreases in the direction of flow, and increases for $\mathsf{F}_o > 1$. This feature is well known from contracting channels, and Eq. (18.62) reflects the main flow patterns, again close to the origin. Note that for strong contraction, such as $\delta^2 \rightarrow 0$, the asymptotic solution obtains $y = 3/2$, and stagnation flow occurs at the end wall with $h_u = (3/2)h_o$.

18.6.3 *End Depth Ratio*

The ratio of boundary flow depths $Y_u = h_u/h_o$ as also discussed in Sect. 18.5.4 results from the momentum equation. Accounting for the rectangular substitute sideweir instead of the U-shaped profile, the result is similar to Eq. (18.49), as demonstrated by Del Giudice and Hager (1999). The agreement with observations is good for $\mathsf{F}_o > 0.8$, but deviations are notable otherwise.

Using a generalized *energy approach*, one may demonstrate that

$$Y_u = 1 + \beta \mathsf{F}_o^2 \tag{18.63}$$

with $\beta = 2/3$ correlating excellently with the data up to $\mathsf{F}_o = 1$. For given approach flow conditions h_o and F_o, one may thus simply determine the flow depth h_u at the end wall. Note that $\beta = 0.50$ would correspond to potential flow, and that $\beta = 2/3$ accounts for an increase of energy head due to lateral outflow, as discussed previously.

18.6.4 *Discharge Distribution*

Discharge Q along a sideweir at the relative position $X = x/L$ has the features $Q(0) = Q_o$ and $Q(1) = Q_u$. Introducing $q_x = (Q-Q_u)/(Q_o-Q_u)$ as the relative discharge yields the normalization $q_x(0) = 1$ and $q_x(1) = 0$. Gisonni and Hager (1997) considered the case where $Q_u = 0$, resulting in Eq. (18.46). For the general case with a throttle discharge, the previous approach remains valid, and the *spatial discharge* distribution along a sideweir with $\mathsf{F}_o < 1$ is (Del Giudice and Hager 1999)

$$q_x = 1 - X^{1+\mathsf{F}_o}. \tag{18.64}$$

The lateral outflow discharge is coupled in a complex manner with the local flow depth, the local velocity, and the weir height. Equation (18.64) retains the most significant effect of the approach flow Froude number F_o. Increasing F_o thus increases the *non-uniformity* not only of the free surface profile but also of the lateral discharge intensity, whereas decreasing F_o reduces the lateral outflow capacity. The optimum sideweir configuration accounts thus for a *compromise* between performance and discharge capacity.

18.6.5 *Discharge Characteristics*

The lateral discharge ΔQ of a sideweir is related in a complex manner to the approach flow conditions, the free surface profile and the sideweir geometry. To allow for a simple design, simplifications have to be introduced. Based on the discharge equation (18.51), the average (subscript a) discharge coefficient C_{da} was determined for the configuration with the throttling pipe, and Eq. (18.50) was verified for $\mathsf{F}_o < 1$. The lateral discharge for two-sided overflow is, from Eqs. (18.50) and (18.51),

$$\Delta Q = \left(1 + \frac{1}{2}\frac{\Delta L}{D_o}\mathsf{F}_o^2\right)[2g(h_o - w)^3]^{1/2}\Delta L. \tag{18.65}$$

As for the discharge distribution $Q(x)$ the lateral discharge ΔQ thus depends exclusively on the approach flow conditions and on the sideweir geometry.

The ratio of discharges in the throttling pipe Q_p and treatment plant design at zero overflow Q_{pD} determines the throttling performance. According to ATV (1993), the *performance parameter* is limited to $\eta = Q_p/Q_{pD} \geq 1.20$. Figure 18.21a shows $\eta(\mathsf{F}_o)$ for various relative weir heights $\overline{W} = w/D_o$ and the limit (subscript L) value η_L is attained for either $\overline{W}$ enough large, or F_o sufficiently small. The sideweir with $\overline{W} = 0.60$ would satisfy ATV regulations if F_o is restricted to values below 0.50. For $\overline{W} \geq 2/3$, say, these restrictions are satisfied for all F_o, and no difficulties in design

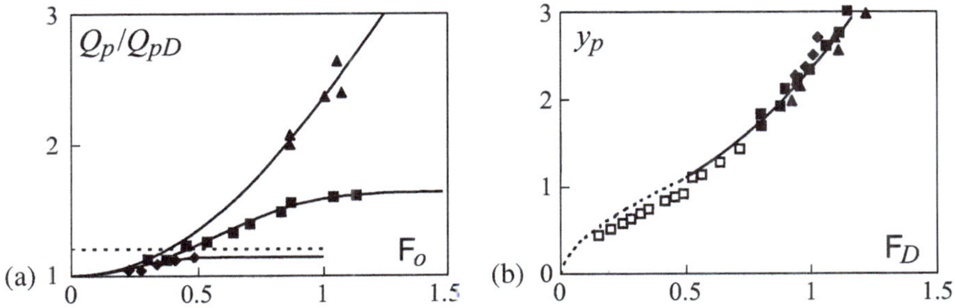

Fig. 18.21 (**a**) Throttling performance $\eta(\mathsf{F}_o)$ for $\overline{W} = w/D_o =$ (▲) 0.40, (■) 0.60, (♦) 0.80, (· · ·) ATV limit $\eta_L = 1.20$. (**b**) Discharge characteristics of throttling pipe $y_p(\mathsf{F}_D)$ with and without lateral overflow, (—) Eq.(18.67)

are to be expected. Del Giudice and Hager (1999) discussed the limit number $\eta_L = 1.20$ in terms of hydraulics, maintenance and sewage treatment. It is not clear why this limit number was advanced, independent of sewer and wastewater characteristics. The limit number η_L should be set by the local water authorities, and ATV regulations may be considered as a standard minimum value. Increasing η_L results in a more economic sideweir design, and the compromise between capacity and performance previously addressed applies of course also here. The *optimum sewer sideweir* has a length of $3 < L/D_o < 6$ (or up to 10) and a diameter ratio D_d/D_o of about 0.1–0.5. Minimum velocity constraints and minimum diameters for combined sewers have to be accounted for (Chap. 3).

The *discharge capacity* of a sewer sideweir is determined with Eq. (18.65). For relatively small approach flow Froude numbers ($\mathsf{F}_o < 0.75$), the effect of the variable discharge coefficient is small. For example, if assuming $\mathsf{F}_o = 0.50$ and $\Delta L/D_o = 4$, then $C_{da} = 1.5$ is a typical number. For capacity considerations, the downstream discharge Q_u may be neglected, and $\Delta Q \cong Q_o$. Inserting in Eq. (18.65) thus gives

$$\frac{Q_o/\Delta L}{(gD_o^3)^{1/2}} = 2\left(y_o - \frac{2}{3}\right)^{3/2} \tag{18.66}$$

for a lower limit weir height of $w/D_o = 2/3$. The discharge capacity thus varies mainly with the approach flow diameter D_o and linearly with the overflow length ΔL. Assuming an upper limit approach flow filling of $y_o = 0.85$ would give $Q_o = \Delta Q = (1/6)(gD_o^3)^{1/2}\Delta L$ for maximum design discharge. This relation can be used for preliminary determination of D_o and ΔL.

18.6.6 *Throttling Discharge Characteristics*

The hydraulic characteristics of a sideweir throttle are similar to a *culvert structure*, as discussed in Chap. 9. In general, four basic flow types may occur for long throttles as considered in the present context, including: (1) Critical flow, (2) Uniform flow, (3) Gated flow, and (4) Pressurized flow. For sideweirs with a weir crest elevation $\overline{W} > 2/3$ and a diameter ratio $D_d/D_o < 1/2$, only flow types ③ and ④ are relevant because the relative intake depth is $h_u/D_d > 1.2$ (Chap. 9). Depending on the bottom slope S_o relative to the critical slope S_c, either *pressurized flow* or *gated flow* occur, associated with an inflow control.

Gated flow can be described with Eq. (9.11) involving the pipe Froude number $\mathsf{F}_D = Q/(gD_d^5)^{1/2}$ and the relative intake pressure head ratio $y_p = h_u/D_d$ as

$$\mathsf{F}_D = 1.11C_d(y_p - C_d)^{1/2}, \tag{18.67}$$

where $C_d = 0.70$ was fitted from the experimental data. The increase of 10% in C_d over the standard discharge coefficient results from the converging intake plan geometry as compared with the standard intake from a large basin. Figure 18.21b shows agreement between predictions and observations for gated flow conditions.

For *critical flow* conditions prior to overflow, the prediction is too low because of the appreciable approach flow velocity effect. Introducing the upstream head instead of the upstream energy head would significantly improve the analysis. The present prediction $h_u = y_p D_d$ can thus be regarded as the downstream energy head instead of the downstream flow depth. Priming of the throttle occurs for $y_p \cong 1.2$, as is also seen from Fig. 18.21b.

Pressurized flow follows the generalized Bernoulli equation and is expressed with (Chap. 9)

$$z_p + H_p = (1 + \xi_e + \xi_f)\frac{V_p^2}{2g} + \sigma_p D_d, \tag{18.68}$$

where $\xi_e \leq 0.5$ is the intake loss coefficient, ξ_f the friction loss coefficient, z_p the elevation difference between intake and outlet, $V_p = Q_p/(\pi D_d^2/4)$ the throttling pipe velocity and $\sigma_p D_d$ the pressure head at the outlet section. For outlet into the atmosphere a minimum of σ_p is 0.50. For outlet into a sewer manhole with a jet guidance one may assume $\sigma_p = 2/3$, and for a jet with a hydrostatic pressure distribution one would have $\sigma_p = 1$. An upper limit of the intake loss coefficient is 0.5, whereas sideweirs with a converging intake arrangement are approximately modelled with $\xi_e = 1/3$ (Sect. 9.3).

Assuming that the friction losses can be approximated with the Manning-Strickler equation (5.4) gives $\xi_f = (2 \cdot 4^{4/3} g n^2 / D_d^{1/3})(L_p/D_d)$, with n as the Manning roughness coefficient. Typically, $1/n = 85\ \mathrm{m}^{1/3}\mathrm{s}^{-1}$ and $D_d = 0.20$ m, such that $\xi_f = 0.030(L_p/D_d)$. With $\lambda_p = L_p/D_d$ as the relative throttle length, and S_o its bottom slope, the discharge equation obtains (Hager and Del Giudice 1998)

$$\mathsf{F}_D = 1.11\left[\left(y_p + S_o\lambda_p - \frac{1}{2}\right) / \left(\frac{4}{3} + 0.03\lambda_p\right)\right]^{1/2}. \tag{18.69}$$

Accordingly, Eqs. (18.67) and (18.69) have a similar dependence $\mathsf{F}_D(y_p)$, with the latter equation involving S_o and λ_p in addition to y_p. The determining discharge results from the equation with the smaller value in Q, for a given throttle geometry. Obviously, the transition (subscript t) condition between gated and pressurized flows is

$$(S_o\lambda_p)_t = C_d^2\left(\frac{4}{3} + 0.03\lambda_p\right)^2 (y_p - C_d) + 0.5 - y_p. \tag{18.70}$$

Pressurized flow prevails for long throttles with a small bottom slope, whereas gated flow is typical for steep throttling pipes with a supercritical flow. The air-water two-phase flow features in throttling pipes have so far not received systematic attention. Also, further experimental work is required to assess the previous findings.

18.6.7 *Design Recommendations*

Based on the information obtained from Sects. 18.5 and 18.6, sewer sideweirs are effective provided the following conditions are satisfied:

- *End Plate* is inserted for very short sideweir with $L/D_o < 3$ to stabilize the vorticity interface close to the downstream end,
- A *compact and economic* overflow structure results for usual sideweir lengths between 3 and 6 (or even 10) approach flow diameters D_o,
- Relative *weir height* should be at least 50% of the approach flow diameter, to inhibit spiral currents, locally supercritical flow and nonuniform overflow,
- Relative *weir height* should be better $w/D_o = 0.67$ to force subcritical approach flow for arbitrary flow conditions,
- Plan of sideweir is *linearly contracting*, typically with $D_d/D_o = 0.10$ to 0.4,
- *Approach flow Froude number* is less than 0.75 to increase overflow uniformity, and
- Sideweir has a symmetric sharp-crested overflow geometry.

Based on a hydraulic design, the main flow features of short sewer sideweirs can be determined. The performance of the novel design has to be tested in prototype structures, mainly with regard to separation characteristics, maintenance and safety.

18.7 Closing Comments

The geometry of a sewer sideweir is currently defined but there is still a significant lack of observations on sideweir flow, particularly in prototype conditions (Hager 1993). In addition, the following comments are appropriate:

- To apply *Froude similitude* the hydraulic models have to be sufficiently large, with a minimum throttle pipe diameter of about 0.10 m. To exclude effects of surface tensions on the overflow features, minimum overflow depths have to be about 0.05 m. To generate gradually-varied flow, a minimum length of ten approach flow pipe diameters is required. A *hydraulic model* of a sideweir is thus quite large.
- The geometry of the sewer sideweir is influenced by a large quantity of *parameters*, including the boundary diameters, the bottom slope, the boundary weir heights, the roughness coefficients and the sideweir length. In addition, the boundary discharges and flow depths are variables and of interest are at least the axial and the lateral free surface profiles, and the corresponding velocities.
- To design sewer sideweirs, a hydraulic approach is required based on systematic observations of sideweirs with both low and high weir crests. For example, the current design guidelines do not account for *pseudo-uniform flow* although it was determined in Chap. 17 as the significant design variable. The

basis of an improved lateral outflow management, and simplifications in design and maintenance involve thus systematic hydraulic experiments. Provided these observations are available, the computational models can be validated to remove deviations in the present computations. The hydraulic design of a sewer sideweir remains a difficult task because of the complex flow pattern and the mixed influences from the up- and downstream sewer reaches.

- Based on these observations, the sewer sideweir with a *low weir height* cannot be recommended for design. The transition from the upstream sewer to the sideweir on the one hand, and the formation of hydraulic jumps along the outflow length on the other hand result in hydraulic conditions that are currently not amenable to prediction. The minimum upstream weir height recommended is $w_o/D_o = 0.5$. Also, the approach flow Froude number should be stably subcritical, such as $\mathsf{F}_o < 0.75$.
- *Experiments* to sideweir flow as described previously are considered significant to increase the efficiency of the combined sewer systems. Before a hydraulic design based on reliable experiments is available, these systems include always a potential risk for large scale pollution. A combined sewer system thus stands and falls mainly with the outfalls from sideweirs.

Notation

b	[m]	channel width
c	[–]	$= 0.737$
c_k	[–]	crest shape parameter
c_W	[–]	overall outflow coefficient
C_d	[–]	discharge coefficient
C_{da}	[–]	average discharge coefficient
C_D	[–]	discharge coefficient based on approach energy head
D	[m]	diameter
D_o	[–]	approach flow diameter
D_d	[–]	diameter of throttle
f	[–]	relative approach Froude number
F	[m^2]	cross-sectional area
F	[–]	Froude number
F_D	[–]	pipe Froude number
g	[ms^{-2}]	gravitational acceleration
h	[m]	flow depth
H	[m]	energy head
k_s	[m]	equivalent roughness height
L	[m]	sideweir length
L_d	[m]	length of throttling pipe
L_j	[m]	length of hydraulic jump
L_p	[–]	length of throttling pipe

L_r	[m]	length of roller
n^*	[–]	number of lateral outflow sides
$1/n$	[m$^{1/3}$s^{-1}]	Manning's roughness coefficient
N	[–]	coefficient
P	[–]	correction coefficient
P_u	[–]	pseudo-uniform parameter
q	[–]	lateral discharge intensity
q_u	[–]	$= \Delta Q/(gD_o^2h_u^3)^{1/2}$ normalized lateral outflow
q_x	[–]	$= (Q-Q_u)/(Q_o-Q_u)$
Q	[m^{3}s^{-1}]	discharge
Q_o	[m^{3}s^{-1}]	approach flow discharge
R	[m]	crest radius
S_o	[–]	bottom slope
V	[ms^{-1}]	velocity
V_p	[ms^{-1}]	pipe velocity
w	[m]	weir height
W	[–]	$= w/H$ relative weir height
$\overline{W}$	[–]	$= w/D_o$
x	[m]	streamwise coordinate
X	[–]	$= x/[(D/H)^{0.6}H]$ dimensionless coordinate
$\overline{X}$	[–]	$= x/L$ coordinate scaled by sideweir length
y	[–]	$= h/H$ relative flow depth, or h/h_o
y_o	[–]	approach filling ratio
y_u	[–]	$= h_u/D_o$
y_p	[–]	$= h_u/D_u$
Y_p	[–]	normalized flow depth
Y_u	[–]	$= h_u/h_e$ end depth ratio
z_p	[m]	elevation of End Plate, or throttling pipe
Z_p	[–]	$= z_p/D_o$
β	[–]	kinetic energy coefficient
δ	[–]	$= D/D_o$
ΔL	[m]	outflow length
ΔQ	[m^{3}s^{-1}]	lateral discharge
Δs	[m]	drop height
$\varDelta$	[–]	$= D_d/D_o$ diameter ratio
η	[–]	performance parameter
θ	[–]	contraction angle
$\varTheta$	[–]	relative contraction angle
λ	[–]	relative sideweir length
λ_g	[–]	relative jump location
λ_p	[–]	relative throttle length
λ_s	[–]	relative length
$\varLambda$	[–]	relative sideweir length

Ψ	[–]	relative approach Froude number
ρ	[–]	$Q(x)/Q_o$
σ_p	[–]	pressure head coefficient
ω	[–]	normalized value of Ω
Ω	[–]	momentum exchange coefficient
Φ	[–]	momentum transfer coefficient
ξ_e	[–]	intake loss coefficient
ξ_f	[–]	friction loss coefficient
ξ_s	[–]	headloss coefficient of sideweir

Subscripts

c	critical flow
d	throttle
D	design
e	end section
K	critical treatment
L	limit
m	average
M	maximum
N	uniform
o	upstream, approach
p	End Plate, pipe
PN	pseudo-uniform
r	boundary
s	sideweir
t	transition
T	dry weather
u	downstream
1	upstream of jump
2	downstream of jump

References

ATV (1993). Richtlinien für die hydraulische Dimensionierung und den Leistungsnachweis von Regenwasser-Entlastungsanlagen in Abwasserkanälen und -leitungen (Guidelines for the hydraulic design and the capacity proof of rainwater outlets in sewers). *Arbeitsblatt* A111. Abwassertechnische Vereinigung: St. Augustin [in German].

Buffoni, F., Sassoli, F., Viti, C. (1986). Ricerca sperimentale sugli sfioratori bilaterali in canali a sezione circolare (Experimental research on two-sided sideweirs in circular channels). 20 *Convegno di Idraulica e Costruzioni Idrauliche* Padova C (2): 679–688 [in Italian].

Del Giudice, G., Hager, W.H. (1999). Sewer sideweir with throttling pipe. *Journal of Irrigation and Drainage Engineering* 125(5): 298–306.

Fahrner, H., Peter, G., Seybold, W. (1990). Problematik der Entlastungsmessung an Überlauf-Bauwerken der Mischkanalisation (Problems of lateral outfalls in combined sewers). *Korrespondenz Abwasser* 37(10): 1175–1188 [in German].

Gisonni, C., Hager, W.H. (1997). Short sewer sideweir. *Journal of Irrigation and Drainage Engineering* 123(5): 354–363.

Hager, W.H. (1987). Lateral outflow over sideweirs. *Journal of Hydraulic Engineering* 113(4): 491–504; 115(5): 684–688.

Hager, W.H. (1993). Streichwehre mit Kreisprofil (Sideweirs with a circular profile). *gwf - Wasser/Abwasser* 134(3): 156–163 [in German].

Hager, W.H. (1994a). Supercritical flow in circular-shaped sideweirs. *Journal of Irrigation and Drainage Engineering* 120(1): 1–12.

Hager, W.H. (1994b). Impact hydraulic jump. *Journal of Hydraulic Engineering* 120(5): 633–637.

Hager, W.H., Hager, K., Weyermann, H. (1982). Die hydraulische Berechnung von Streichwehren in Entlastungsbauwerken der Kanalisationstechnik (Hydraulic design of sewer sideweirs). *Gas – Wasser – Abwasser* 63(7): 309–329 [in German].

Hager, W.H., Del Giudice, G. (1998). Generalized culvert design diagram. *Journal of Irrigation and Drainage Engineering* 124(5): 271–274.

Hager, W.H., Schleiss, A.J. (2009). *Constructions hydrauliques – Ecoulements stationnaires (Hydraulic structures – steady flows),* ed. 2. Presses Polytechniques et Universitaires Romandes: Lausanne [in French].

Hamam, M.A., McCorquodale, J.A. (1982). Transient conditions in the transition from gravity to surcharged sewer flow. *Canadian Journal of Civil Engineering* 9(2): 189–196.

Hörler, A., Hörler, E. (1973). Streichwehre mit niedrigen Überlaufschwellen in kreisförmigen Kanälen (Sideweirs with low crest in circular channels). *gwf – Wasser/Abwasser* 114(2): 579–584 [in German].

Kallwass, G.J. (1964). Beitrag zur hydraulischen Berechnung gedrosselter Regenüberläufe (Contribution to the hydraulic design of throttled sewer sideweirs). *Dissertation* TH Karlsruhe: Karlsruhe [in German].

Naudascher, E. (1992). *Hydraulik der Gerinne und Gerinnebauwerke*. Springer: Wien [in German].

Sassoli, F. (1963). Ricerca sperimentale sugli sfioratori laterali in canale a sezione circolare (Experimental research on sideweirs in circular channels). 8 *Convegno di Idraulica* Pisa A(12): 1–19 [in Italian].

SIA (1980). Sonderbauwerke der Kanalisationstechnik (Special structures of sewer techniques). *SIA-Dokumentation* 40. Schweizerischer Ingenieur- und Architektenverein: Zürich [in German].

Siegenthaler, A. (1981). Hydraulische Modellversuche der Regenauslässe Riehenstrasse Basel (Hydraulic modeling of the sewer sideweir at Riehenstrasse, Basel). *Gas – Wasser – Abwasser* 61(5): 152–156 [in German].

Taubmann, K.-C. (1972). Regenüberläufe (Sewer sideweirs). *Gas-Wasser-Abwasser* 52(10): 297–308 [in German].

Uyumaz, A., Muslu, Y. (1985). Flow over sideweirs in circular channels. *Journal of Hydraulic Engineering* 111(1): 144–160.

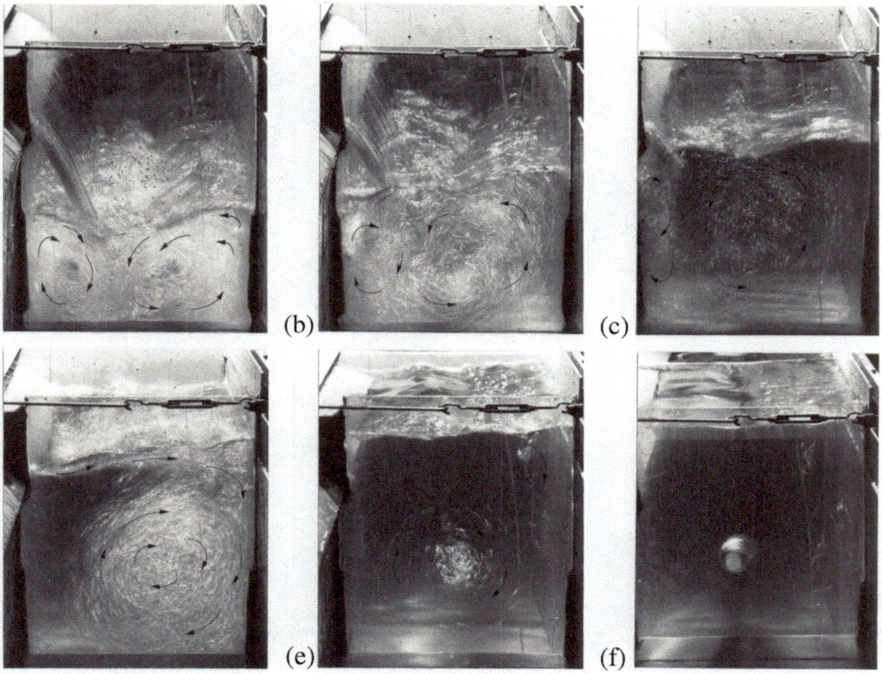

(a) (b) (c) (d) (e) (f)

Formation of tornado vortex along side channel due to increasing tailwater submergence. Development from two-vortex to one- vortex systems following a transition from plunging in (**a–c**), to surface jet in (**d–f**) (*Wasser, Energie, Luft* 82(11/12): 325-330)

Chapter 19
Side Channel

Abstract The discharge increases spatially in a side channel. Because side channels are relatively short structures, the effect of local friction variation may be neglected, and a generalized approach is possible. The differences of flow in the rectangular and the U-shaped side channels are small, such that all computations may be related to a rectangular substitute channel.

Flows with a singular, and a critical point are particularly addressed. The latter case can be accounted for even by including wall friction, to present a generalized design procedure. Thus, lengthy numerical computations are not necessary.

Finally, the spatial flow features of side channels are described by including the rotational flow component. Orders of magnitude for the lateral pile up of flow along the channel walls are given, and the effects of so-called tornado vortices are described.

19.1 Introduction

A side channel (German: Sammelkanal; French: Canal collecteur) is a hydraulic structure collecting lateral flow and directing it into the downstream channel. The gutter is obviously a simple example of a side channel. Each sewer system is also a collector, but one would not refer to a side channel because of the *concentrated* lateral addition of discharge at manholes, whereas the discharge varies *spatially* for a side channel (Fig. 19.1). To apply the theory of spatially-varied flow, the minimum length of a side channel has to be at least one order larger than the flow depth. If both lengths are about the same, reference to a *channel junction* is made (Chap. 16). Discharges from buildings are so small that they are not considered individually in a sewer system.

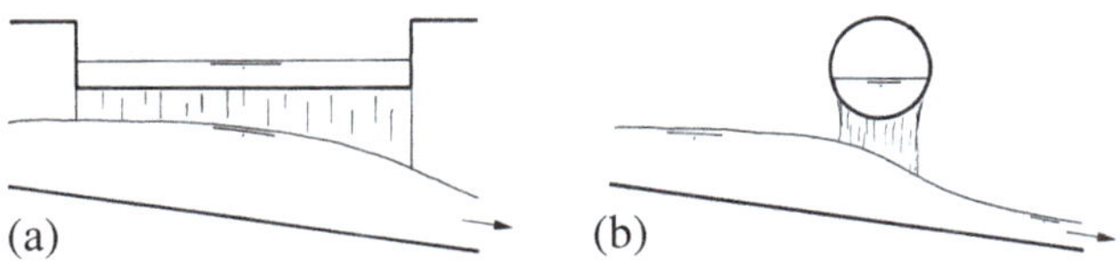

Fig. 19.1 Lateral discharge added to (**a**) side channel and (**b**) channel junction

W.H. Hager, *Wastewater Hydraulics*, 2nd ed., DOI 10.1007/978-3-642-11383-3_19,

A side channel may be computed with a hydraulic approach if the transverse velocity components are neglected. This simplification is not always appropriate, and the consequences are discussed in Sect. 19.4. In the following, the flow with a *linearly increasing discharge* is considered by the conventional hydraulic approach. First the governing equation for the free surface profile is derived, then simplifications are introduced allowing for a generalized solution. The latter is comparable with the solution for the distribution channel (Chap. 17) and can be represented graphically. Particular solutions are then discussed to allow for a simple design of side channels. Some two-dimensional features of side channel flow are also presented.

19.2 Basic Equations

Continuous free surface flows with a spatially-varied discharge involve:

- Either flows with a spatially *de*creasing discharge, such as those in sideweirs or bottom openings (Chaps. 17 and 20),
- Or flows with a spatially *in*creasing discharge, such as those in side channels.

The governing equation for the free surface profile is based on the *momentum equation*, because the lateral flow exerts losses on the flow that cannot be predicted with the energy equation. With $U\cos\phi$ as the velocity component of the lateral flow in the streamwise direction, and thus the lateral momentum $(\rho g)U\cos\phi(\mathrm{d}Q/\mathrm{d}x)$, where ρ is density, g gravitational acceleration and $(\mathrm{d}Q/\mathrm{d}x)$ lateral discharge intensity, the *equation of the free surface profile* is (Chow 1959, Naudascher 1992)

$$\frac{\mathrm{d}h}{\mathrm{d}x} = \frac{S_o - S_f - \left[2 - \frac{U\cos\phi}{V}\right]\frac{QQ'}{gF^2} + \frac{Q^2}{gF^3}\frac{\partial F}{\partial x}}{1 - \mathsf{F}^2}. \tag{19.1}$$

Here h is flow depth, x streamwise coordinate, S_o bottom slope, S_f friction slope, $V = Q/F$ average cross-sectional velocity, $Q' = \mathrm{d}Q/\mathrm{d}x$ discharge intensity, F cross-sectional area in the streamwise direction x, and Q discharge (Fig. 19.2). The term $\mathsf{F}^2 = Q^2(\partial F/\partial h)/(gF^3)$ is the local Froude number, as discussed in Chap. 6. Equation (19.1) is based on the usual assumptions of a one-dimensional flow, i.e. uniform velocity and hydrostatic pressure distributions. A detailed derivation of Eq. (19.1) is presented in Chap. 17.

For $Q' = 0$, i.e. without lateral discharge, Eq. (19.1) reduces to Eq. (6.25), or Eq. (8.6), provided $\partial F/\partial x = 0$, i.e. the backwater equation of a prismatic channel. Equation (19.1) corresponds to the basic equation of steady open channel flow. Often such equations are complicated by correction terms α and β to allow for non-uniform velocity and non-hydrostatic pressure distributions (Chap. 1). These are only approximately known, and their differentials $\mathrm{d}\alpha/\mathrm{d}x$ or $\mathrm{d}\beta/\mathrm{d}x$ are normally unknown. Therefore, Eq. (19.1) is retained.

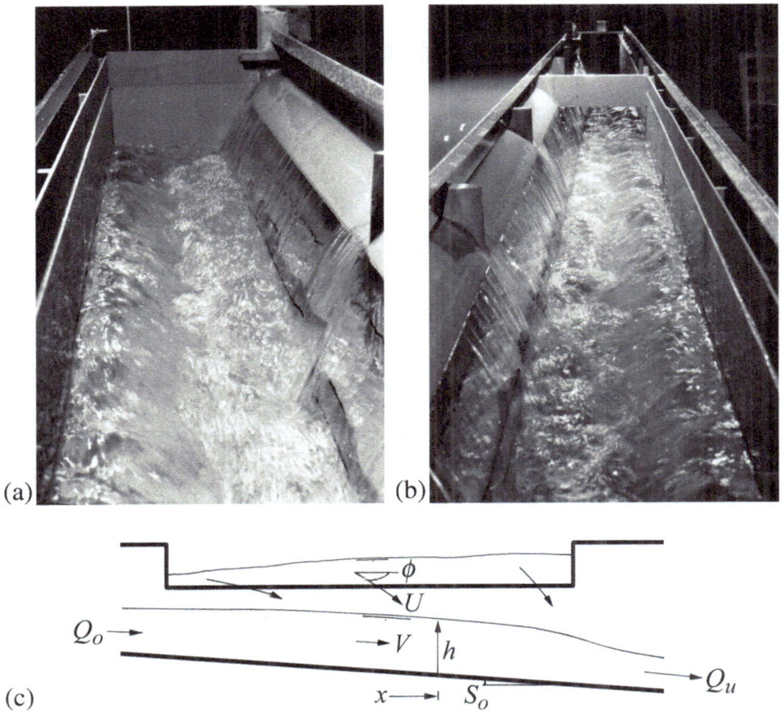

Fig. 19.2 Side channel flow (**a**) against and (**b**) in flow direction. (**c**) Side channel with notation

Equation (19.1) can also be deduced from the *energy conservation principle* as (Hager 1987)

$$H = z + h + \frac{Q^2}{2gF^2}, \quad \frac{dH}{dx} = -\left[S_f + \left(1 - \frac{U\cos\phi}{V}\right)\frac{QQ'}{gF^2}\right], \tag{19.2}$$

where $dz/dx = -S_o$ is the bottom slope. $dH/dx = H' = H'_f + H'_L$ is the energy line gradient with $H'_f = -S_f$ as the friction slope. The *additional slope* due to lateral discharge is

$$H'_L = -\left(1 - \frac{U\cos\phi}{V}\right)\frac{QQ'}{gF^2}. \tag{19.3}$$

Depending on the sign of Q', i.e. positive for lateral inflow and negative for lateral outflow, the sign of H'_L varies (Chap. 17). For distribution channels, the energy head may thus increase along the channel, whereas a lateral outflow reduces the energy content. This anomaly is also reflected by the head loss coefficients that can be negative for the latter flows (Chap. 2).

If all parameters in Eq. (19.1) are known as functions of either x or h, the free surface profile can be determined subject to a boundary (subscript r) condition $h(x = x_r) = h_r$. Depending on the *condition of flow*, the computational direction is against the flow direction for subcritical and in the direction of flow for supercritical flows (Chap. 8). An exception are *transitional flows* from sub- to supercritical. Starting at the critical flow section ($\mathsf{F} = 1$), the computation proceeds against the flow direction for the upstream reach and in the flow direction for the downstream reach. At the critical point ($\mathsf{F} = 1$) two cases are possible:

- Either a *singular* point $x = x_s$ for which both numerator and denominator in Eq. (19.1) are zero and $\mathrm{d}h/\mathrm{d}x$ is previously undefined. A mathematical analysis yields the slope $(\mathrm{d}h/\mathrm{d}x)_s$ at the singular (subscript s) point,
- Or a *critical* point $x = x_c$ for which $\mathsf{F} = 1$ but where the numerator of Eq. (19.1) is different from zero and the free surface slope $(\mathrm{d}h/\mathrm{d}x)_c \rightarrow \infty$ at the critical (subscript c) point. The condition of continuous flow is then locally invalidated.

In practice, both cases are relevant. In the following, the basic equation is simplified to obtain a general solution in the form of a generalized backwater curve. For supercritical flow in the downstream channel, the hydraulic configuration is particularly simple.

19.3 Side Channel of Rectangular Cross-Section

19.3.1 *Equation of Free Surface Profile*

Normally, the lateral discharge of side channels is assumed *linearly increasing*, with a constant discharge intensity $p_s = Q_s/L_s$. Here, Q_s is the lateral discharge and L_s the length of the side channel (subscript s). Also, the momentum component $U\cos\phi$ is neglected and the lateral discharge assumed to enter perpendicularly the side channel. These assumptions are not always true, such as for sideweirs (Chap. 18) with a spatially-varied discharge intensity, or for steeply sloping side channels of which the vertical jet results in an additional momentum component in the streamwise direction. For usual side channels of small bottom slope, the simplifications are justified, however.

The origin $x = 0$ (subscript o) corresponds either to the upstream end for a dead-end side channel, or to the beginning of the lateral discharge portion for an upstream discharge $Q_o > 0$. The downstream (subscript u) discharge is constant $Q_u = Q_o + p_s L_s$ (Fig. 19.3).

For a side channel of bottom slope S_o, the streamwise momentum component due to a vertical lateral jet is $U\cos\phi = [2g(z_o + h_{ü} + S_o x – h)]^{1/2} S_o$, with z_o as the elevation difference between the weir crest and the bottom at $x = 0$, and $h_{ü}$ as the overflow depth. For usual configurations, with a relatively small height $h_{ü}$, a bottom slope S_o of some percents and modest depth z_o, the expression $U\cos\phi/V$ is much smaller

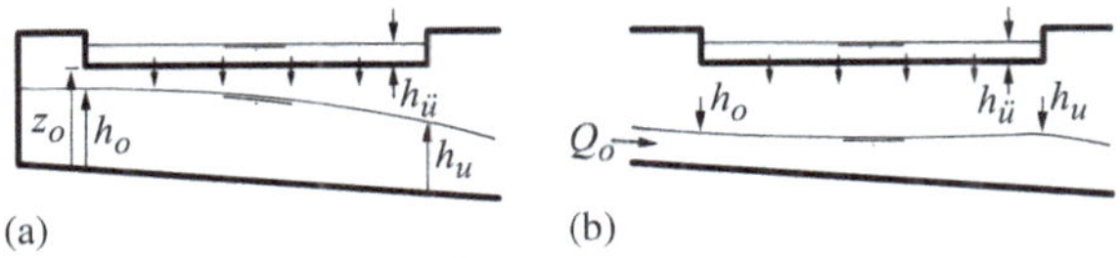

Fig. 19.3 Side channel (**a**) with an upstream dead-end and subcritical flow ($Q_o = 0$), (**b**) with a supercritical approach flow ($Q_o > 0$)

than 1 and may be neglected. Regarding the submergence effect on the discharging channel, this assumption is on the safe side, and the effect of secondary parameters can be dropped.

In analogy to sideweir flow, one might introduce the concept of *pseudo-uniform flow* and thus design a diverging side channel (Fig. 19.4). According to Eq. (19.3) and for $U\cos\phi/V \to 0$ the energy loss slope tends to $H'_L = -QQ'/(gF^2)$. Integrating over the average (subscript m) cross-section F_m along the side channel gives $\Delta H_L = -Q_u^2/(2gF_m^2) \cong V_u^2/(2g)$, indicating that the losses increase with the downstream velocity head. If the prismatic and diverging side channels of equal downstream width b_u are compared, then less energy loss results for the *prismatic side channel* because of smaller local velocity. In practice, an intermediate design involves a compact structure that includes a converging distribution and a diverging collection channel. Then, a rectangular unit structure can be designed with a diagonal separation wall between the distributor and collector portions. Side channels as used in sewage treatment stations are normally prismatic, because the additional cost for increasing the width are too high.

To simplify the approach, the *prismatic side channel* is considered a standard. Further, the friction slope is accounted for by the average over length L_s as $S_{fm} = (1/2(S_{fo} + S_{fu})$, because:

- The length of the side channel is small compared to a sewer reach and the effect of a locally varied friction slope therefore small (Chap. 17), and
- The computational approach would unnecessarily be complicated.

Side channels have either rectangular or U-shaped cross-sections. For design discharge, the U-shaped channel is at least half filled, corresponding to a rectangular channel whose bottom elevation is lifted by $[(1/2)–(\pi/8)]D = 0.107D$, thus by about 10% from the original bottom profile. The *substitute profile* of the U-shaped channel is rectangular resulting in a closed form computation, whereas the upper and lower portions of the U-shaped channel had to be separated for a computational approach.

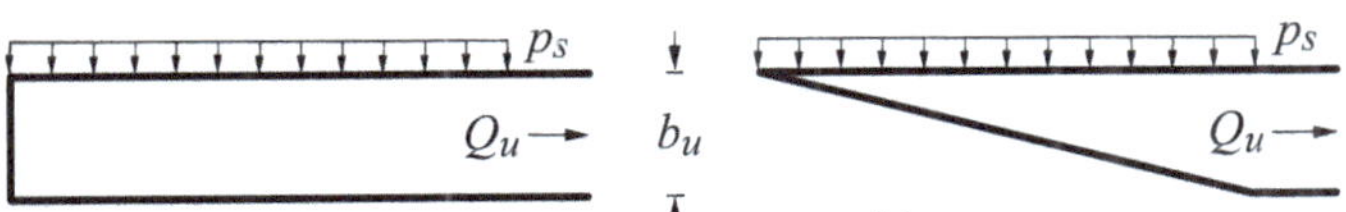

Fig. 19.4 Plans of side channels with (**a**) prismatic and (**b**) triangular diverging geometry

With these simplifications the *free surface profile* can be deduced from Eq. (19.1) as

$$\frac{dh}{dx} = \frac{S_o - S_{fm} - \frac{2p_s^2 x}{gb^2h^2}}{1 - \frac{p_s^2 x^2}{gb^2h^3}}. \tag{19.4}$$

The free surface profile $h(x)$ thus depends only on the bottom slope S_o, the average friction slope S_{fm}, the discharge intensity p_s and the channel width b. With $J = S_o - S_{fm}$ as the substitute side channel slope, as introduced in Chap. 17, the scaling coordinates are

$$x_s = \frac{8p_s^2}{gb^2J^3}, \quad h_s = \frac{4p_s^2}{gb^2J^2}. \tag{19.5}$$

With $X = x/x_s$ and $y = h/h_s$ as normalized coordinates, the *generalized equation of the free surface* is

$$\frac{dy}{dX} = 2\frac{y^3 - Xy}{y^3 - X^2}. \tag{19.6}$$

Equation (19.6) becomes singular, i.e. $dy/dX = 0/0$ provided $X = y = 0$, and $X = y = 1$. The first case corresponds to the asymptotic condition $\mathsf{F} \rightarrow \infty$ and is not further discussed. For the second case, singular flow ($\mathsf{F} = 1$) occurs but this is in contrast to the case $y^3 = X^2$ ($\mathsf{F} = 1$) and $y^3 \neq Xy$, referred to as critical flow. At the *singular point* (subscript s), the free surface slope is $(dy/dX)_s = 1 - 3^{-1/2}$ and the solution of $y(X)$ close to the singular point is, therefore,

$$y = 1 + (1 - 3^{-1/2})(X - 1), \quad |X - 1| << 1. \tag{19.7}$$

The numerical solution has to start at a control point, i.e. either at a critical point or at the singular point. For the latter, Eq. (19.7) is used close to the singularity, because then Eq. (19.6) remains finite further away. Flows with a critical point have been discussed by Hager (1985). The simplified procedure accepts an almost vertical surface profile despite it is physically irrealistic due to the hydrostatic pressure assumption.

19.3.2 *General Classification*

A classification of free surface profiles can be made based on Fig 19.5. A distinction must be made between flows with an increasing or decreasing flow depth of which the centre of curvature is above or below the free surface. The governing parameters are $\chi_s = J/(p_s h/Q)$, $u = (1 + 3^{1/2})\mathsf{F}^2 + (1 - 3^{1/2})$ and $v = (1 - 3^{1/2})\mathsf{F}^2 + (1 + 3^{1/2})$.

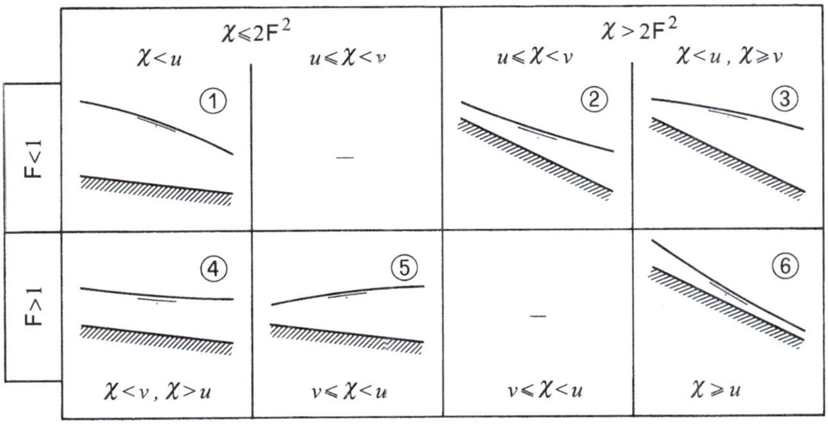

Fig. 19.5 Classification of free surface profiles $y(X)$ with parameters χ, u and v for $\mathsf{F}<1$ and $\mathsf{F}>1$ (Hager and Bremen 1990)

Figure 19.6 shows the general solution of the free surface profile $y(X)$ for $0 \le X \le 5$, and $0 \le y \le 3$ including critical and pseudo-uniform flows.

Domain ① is relevant because it corresponds to dead-end side channels with a supercritical downstream flow, as often encountered in practice. Figure 19.7 shows an enlarged domain of Fig. 19.6. Reaches ② and ③ involve subcritical flow and are submerged from the downstream channel. Such cases are seldom in practise.

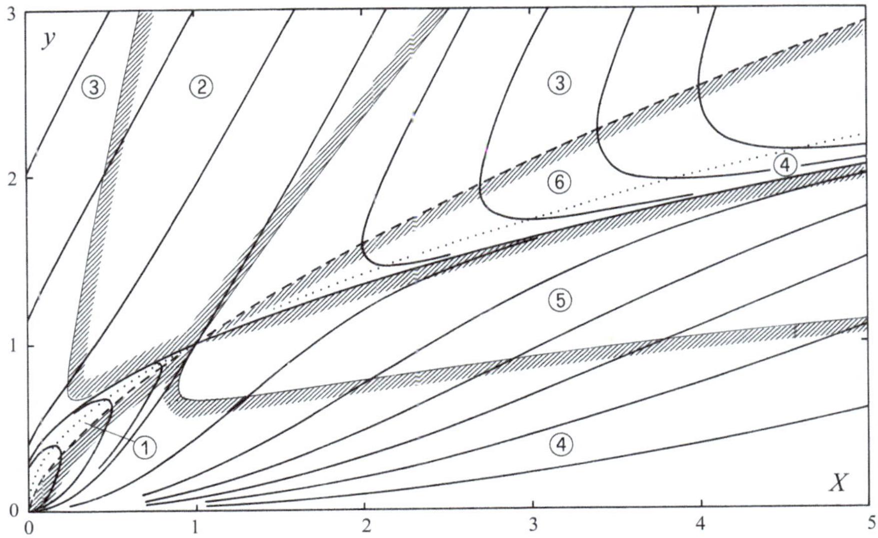

Fig. 19.6 General solution for free surface profile $y(X)$ for side channel in rectangular prismatic channel. Classification according to Fig. 19.5, (- - -) critical depth, (· · ·) pseudo-uniform depth

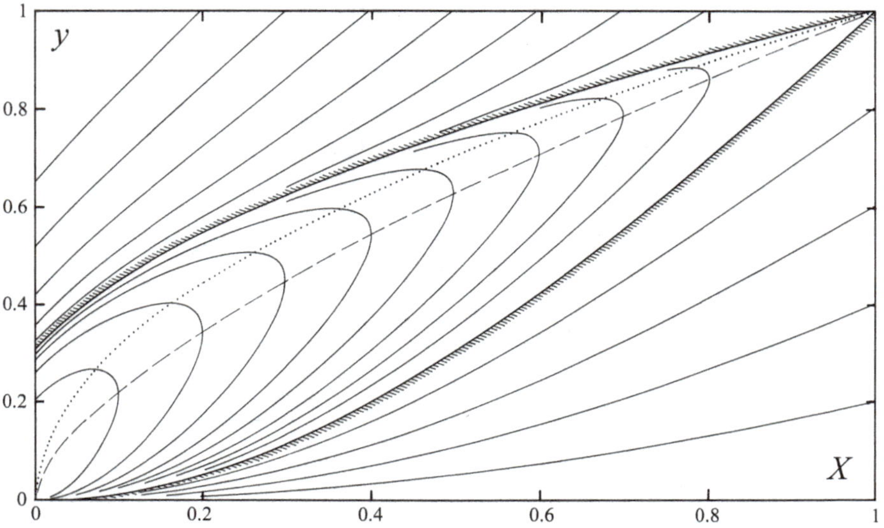

Fig. 19.7 Detail of Fig. 19.6 for free surface profile $y(X)$ for $0 \leq X \leq 1$ and $0 \leq y \leq 1$

Reaches ④ to ⑥ refer to supercritical side channel flows that are also not typical in applications.

For $X < 0.1$, Eq. (19.6) is approximated as $dy/dX = 2$, thus with the boundary condition $y(X = X_u) = y_u$, the solution is $y - y_u = 2(X - X_u)$ or $h(x) - h_u = J(x - x_u)$. For dead-end channels $h(x = 0) = h_o$, the solution is simply $h_o = h_u - Jx_u$. If losses were neglected the solution is thus a horizontal free surface profile close to the dead-end.

Example 19.1 Given a side channel with $Q_s = 0.5\ \text{m}^3\text{s}^{-1}$, $L_s = 4$ m, $Q_o = 0$, $b = 0.7$ m, $S_o = 0.2\%$ und $1/n = 80\,\text{m}^{1/3}\text{s}^{-1}$. Describe the free surface profile.

1. *Uniform flow*. With $q_N = nQ/(S_o^{1/2}b^{8/3}) = 0.362$ the uniform depth is $y_N = h_N/b = 0.795$ from Eq. (8.42), thus $h_N = 0.7 \cdot 0.795 = 0.56$ m.
2. *Critical flow*. With $h_c = [Q^2/(gb^2)]^{1/3} = [0.5^2/(9.81 \cdot 0.7^2)]^{1/3} = 0.373$ m, the uniform flow is subcritical, and the corresponding Froude number is $\mathsf{F}_N = Q/(gb^2h_N^3)^{1/2} = 0.5/(9.81 \cdot 0.7^2 0.56^3)^{1/2} = 0.54 < 1$.
3. *Singular point*. With $p_s = Q_s/L_s = 0.5/4 = 0.125\ \text{m}^2\text{s}^{-1}$ and $J \cong S_o$, Eq. (19.5) gives for $x_s = 8 \cdot 0.125^2/(9.81 \cdot 0.7^2 0.002^3) = 3.25 \times 10^6\,\text{m} > L_s$. Because $J < S_o$, the effective distance x_s is even larger.
4. *Free surface profile*. $X = 4/(3.25 \times 10^6) = 1.23 \times 10^{-6} << 1$ for $x = L_s$ such that an approximation is $h_o = h_u - Jx_u = h_u - JL_s = 0.56 - 0.002 \cdot 4 = 0.55$ m.

For side channels with an approach flow discharge $Q_o > 0$, its upstream end is artificially expanded by the length $L_o = Q_o/p_s$, such that Fig. 19.6 may also be applied.

Example 19.2 Given a side channel of length $L_s = 7$ m and width $b = 1.1$ m of bottom slope $S_o = 3\%$ and roughness coefficient $1/n = 85$ m$^{1/3}$s^{-1}. What is the free surface profile for a lateral discharge of $Q_s = 0.7$ m^{3}s^{-1}, if the approach flow of discharge $Q_o = 0.9$ m^{3}s^{-1} is uniform?

1. *Uniform flow.* $q_{No} = nQ/(S_o^{1/2}b^{8/3}) = 0.012 \cdot 0.9/(0.03^{1/2}1.1^{8/3}) = 0.047$ in the upstream channel, thus $y_{No} = 0.18$ according to Eq. (8.42), and $h_{No} = 0.18 \cdot 1.1 = 0.20$ m. Further, for the downstream channel, $q_{Nu} = nQ_u/(S_o^{1/2}b^{8/3}) = 0.012 \cdot 1.6/(0.03^{1/2}1.1^{8/3}) = 0.084$, thus $y_{Nu} = 0.27$ and $h_{Nu} = 0.295$ m.
2. *Critical flow.* $h_{co} = [Q_o^2/(gb^2)]^{1/3} = [0.9^2/(9.81 \cdot 1.1^2)]^{1/3} = 0.41$ m in the upstream channel, thus $\mathsf{F}_{No} = Q_o/(gb^2h_{No}^3)^{1/2} = 2.92 > 1$. In the downstream channel $h_{cu} = [Q_u^2/(gb^2)]^{1/3} = [1.6^2/(9.81 \cdot 1.1^2)]^{1/3} = 0.6$ m and $\mathsf{F}_{Nu} = 1.6/(9.81 \cdot 1.1^2 \cdot 0.295^3)^{1/2} = 2.90$. Both uniform flows are thus supercritical.
3. *Singular point.* To lowest order one has $S_o \cong J$ and from Eq. (19.5) $x_s = 8(1.6/7)^2/(9.81 \cdot 1.1^2 0.03^3) = 1304$ m $> L_s$, which is irrelevant. The singular flow depth would be $h_s = Jx_s/2 = 0.03 \cdot 1304/2 = 19.55$ m.
4. *Free surface profile.* With the boundary condition $h_r = h_{No} = 0.20$ m one has $y_r = h_r/h_s = 0.20/19.55 = 0.010$. The fictitious side channel origin with a dead end would be located at $x_o = Q_o/p_s = 0.9/(0.7/7) = 9$ m upstream from the lateral beginning, thus $X_r = x_r/x_s = x_o/x_s = 9/1304 = 0.0069$. The boundary values $(X_r;y_r) = (0.0069;\ 0.010)$ are extremely small, and readings from Fig. 19.6 are impossible. The surface profile has a boundary slope $(\mathrm{d}y/\mathrm{d}X)_r \cong 2(0.01^3 - 0.0069 \cdot 0.01)/(0.01^3 - 0.0069^2) = 2.91$ from Eq. (19.6), thus

$$y = y_r + (\mathrm{d}y/\mathrm{d}X)_r \times (X - X_r).$$

For $x_u = x_o + L_s = 16$ m, corresponding to $X_u = 16/1304 = 0.0123$, one has $y_u = 0.01 + 2.91(0.0123 - 0.0069) = 0.0256$, or $h_u = y_u h_s = 0.0256 \cdot 19.55 = 0.5$ m. The profile is of type ④ (Fig. 19.5), and a drawdown curve has to be calculated for the downstream channel, based on the uniform flow depth $h_{Nu} = 0.295$ m (Chap. 8).

This example illustrates the lengthy computational procedure for side channels. In a second approximation, the effect of friction had to be accounted for.

Example 19.3 Determine the friction slope for Example 19.2.
With the Manning formula (Chap. 2)

$$S_f = \frac{n^2 Q^2}{b^2 h^2}\left(\frac{b+2h}{bh}\right)^{4/3}$$

one has $S_{fo} = [0.012\cdot 0.9/(1.1\cdot 0.20)]^2[(1.1+2\cdot 0.20)/(1.1\cdot 0.20)]^{4/3} = 0.03$ at the upstream section of the side channel as previously assumed, and $S_{fu} = [0.012\cdot 1.6/(1.1\cdot 0.5)]^2[(1.1+2\cdot 0.5)/(1.1\cdot 0.50)]^{4/3} = 0.007$ resulting in $S_{fm} = (0.03+0.007)/2 = 0.0185$, or $J = S_o - S_{fm} = 0.03 - 0.0185 = 0.0115$. With $(x_r;h_r) = (9\text{ m};0.20\text{ m})$ one has from the dimensional equation (19.4) $(\mathrm{d}h/\mathrm{d}x)_r = [0.0115\text{–}2\cdot 0.1^2 9/(9.81\cdot 1.1^2 0.2^2)]/[1\text{–}0.1^2 9^2/(9.81\cdot 1.1^2 0.2^3)] = 0.049$. The downstream flow depth is, from a linear approximation as derived in Example 19.2, $h_u = 0.2+0.049\cdot 7 = 0.542$ m, i.e. the original depth was by 10% too small.

By a third approximation that accounts for $h_u = 0.54$ m and thus for a modified J value, one could integrate Eq. (19.4) directly. There is no use of doing so from the beginning, because of the unknown downstream flow depth. An alternative method starts from the basic Eq. (19.1) by expressing S_f with the above equation.

19.3.3 *Transitional Flow*

The *computational direction* of free surface flow follows the rules of generalized backwater curves:

- Subcritical flows have a computational direction *against* the flow direction, and the most downstream condition is critical,
- Supercritical flows have the computational direction *in* the flow direction, and the most upstream condition is again critical.

Side channels with a bottom slope S_o larger than the critical slope S_c (Chap. 6) have a particular flow pattern. Downstream from the lateral discharge increase ($x > L_s$) critical flow cannot exist because all the bottom slope, the roughness coefficient and the channel width remain constant. As the flow is subcritical at the upstream dead-end, and the bottom slope is larger than the critical slope in the downstream channel, transitional flow with $\mathsf{F} < 1$ to $\mathsf{F} > 1$ must occur along the side channel, whose discharge is spatially-varied. According to Chow (1959), the singular point corresponds then to the *control section*, and the flow is subcritical for $x < x_s$, and supercritical for $x > x_s$ (Fig. 19.8).

For $x_s > L_s$ the singular point is located outside the side channel, where Eq. (19.6) ceases to apply. Then, critical flow is *forced*, with the critical section located at the

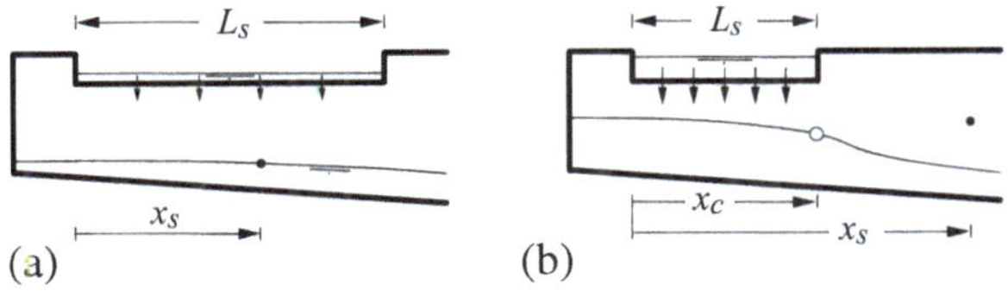

Fig. 19.8 Transitional flow in side channel with (•) singular point located (**a**) inside and (**b**) outside of lateral discharge reach. (○) critical point

side channel end $x_c = L_s$, of flow depth $h = h_c$, (Chap. 6). According to Hager (1985) critical flow is *forced* at the transition from $p_s > 0$ to $p_s = 0$. Often, this condition is assumed to apply but the necessary condition $x_s < L_s$ is not verified. For maximum discharge, the critical flow is normally at the downstream end of the side channel, if $8p_s^2/(gb^2J^3) > L_s$. This condition applies for large lateral discharge intensity per unit width p_s/b, and small total slope J, i.e. conditions for the economical design of a side channel.

Example 19.4 Given a side channel with $b = 0.6$ m, $S_o = 2\%$, $Q_o = 0$, $Q_u = 0.2$ m^{3}s^{-1} and $L_s = 10$ m. What is the surface profile for $1/n = 85$ m$^{1/3}$s^{-1}?

1. *Uniform flow*. With $q_{Nu} = 0.012 \cdot 0.2/(0.02^{1/2}0.6^{8/3}) = 0.065$, one has $y_{Nu} = 0.225$ and $h_{Nu} = 0.225 \cdot 0.6 = 0.135$ m.
2. *Critical flow*. With $h_{cu} = [0.2^2/(9.81 \cdot 0.6^2)]^{1/3} = 0.225$ m the downstream Froude number is $\mathsf{F}_{Nu} = 0.2/(9.81 \cdot 0.6^2 0.135^3)^{1/2} = 2.15$ and the uniform flow is supercritical. The critical downstream slope is $S_{cu} = [0.012 \cdot 0.2/(0.6 \cdot 0.225)]^2[(0.6 + 2 \cdot 0.225)/(0.6 \cdot 0.225)]^{4/3} = 0.005$.
3. *Singular point*. With $J = 0.02 - 0.005 = 0.015$ as a first approximation, and a discharge intensity of $p_s = 0.2/10 = 0.02$ m^{2}s^{-1}, one has from Eqs. (19.5) for the singular point $x_s = 8 \cdot 0.02^2/(9.81 \cdot 0.6^2 0.015^3) = 268$ m and $h_s = 4 \cdot 0.02^2/(9.81 \cdot 0.6^2 0.015^2) = 2.01$ m. Because $\mathsf{F}_u > 1$ and $x_s > L_s$, the control section is forced and $h_u = h_c = 0.225$ m.
4. *Free surface profile*. With boundary values $(x_r;h_r) = (10$ m$;0.225$ m$)$, thus $(X_r;y_r) = (0.037;0.112)$. Fig. 19.7 gives $y_o = 0.10$, and $h_o = 0.1 \cdot 2.01 = 0.20$ m. A more exact solution follows in 19.3.4.

Example 19.5 What is the resulting free surface profile for the minimum discharge $Q_u = 0.02$ m^{3}s^{-1} of the side channel described in Example 19.4?

1. *Uniform flow*. With $q_{Nu} = 0.012 \cdot 0.02/(0.02^{1/2}0.6^{8/3}) = 0.0065$, $y_{Nu} = 0.05$ and $h_{Nu} = 0.05 \cdot 0.6 = 0.03$ m.
2. *Critical flow*. With $h_c = [0.02^2/(9.81 \cdot 0.6^2)]^{1/3} = 0.0485$ m, the downstream Froude number is $\mathsf{F}_u = 0.02/(9.81 \cdot 0.6^2 0.03^3)^{1/2} = 2.05 > 1$, and

the downstream flow is supercritical. The critical slope is $S_{cu}=[0.012\cdot 0.02/(0.6\cdot 0.0485)]^2[(0.6+2\cdot 0.0485)/(0.6\cdot 0.0485)]^{4/3}=0.0045$.

3. *Singular point*. With $J=S_o-S_{cu}=0.02-0.0045=0.0155$ and $p_s=0.02/10=0.002$ m^{2}s^{-1}, Eq. (19.5) give $x_s=8\cdot 0.002^2/(9.81\cdot 0.6^2 0.0155^3)$ $=2.43$ m$<L_s$, and $h_s=Jx_s/2=0.0155\cdot 2.43/2=0.019$ m. For such a small discharge, the control section is inside the lateral overflow length at $x=x_s$. For $0\leq x\leq 2.43$ m, the flow is subcritical, and supercritical for $x>2.43$ m.
4. *Free surface profile*. With $(x_s;h_s)=(2.43$ m;0.019 m$)$ one has $X_u=L_s/x_s$ $=10/2.43=4.11$. From Fig. 19.6, one starts at the singular point (1;1) and follows the singular curve between regions ④ and ⑤ up to $X=4.11$ where $y_u=1.90$, corresponding to $h_u=1.9\cdot 0.019=0.036$ m, where the Froude number is $\mathsf{F}_u=0.02/(9.81\cdot 0.6^2 0.036^3)^{1/2}=1.56$.

Example 19.5 demonstrates the smallness of discharges required to obtain a singular point inside the side channel. For maximum discharges and an economic design, the control of flow is thus always at the *downstream end* of the side channel structure. This important case is treated in the following section.

19.3.4 *Critical Flow*

Often, a side channel has an upstream dead-end. For bottom slopes S_o larger than the critical slope S_{cu}, the control of flow is thus *forced* at the downstream end of the side channel. This case is treated here, with the results derived from Fig. 19.6.

The downstream section (subscript *d*) where critical flow is forced ($\mathsf{F}=1$) serves as the boundary condition. Between the two ends of the side channel the free surface profile has a maximum flow depth (subscript *M*) h_M, and the upstream (subscript *o*) flow depth is h_o, where $Q_o=0$. Figure 19.9 shows $y_o=h_o/h_d$ and $y_M=h_M/h_d$ as functions of $X_d=L_s/x_s$. These follow better than 5% the approximations

$$y_M=X_d^{1/\sqrt{3}},\quad X_d>0.02; \tag{19.8}$$

$$y_o=0.311[\tanh(5X_d)]^{1/\sqrt{3}},\quad X_d>0.05. \tag{19.9}$$

The location of maximum flow depth is $X_M=y_{max}^2$ from Eq. (19.6).

With $i_s=p_s/b$ as discharge load per unit length and unit width of the side channel, one has with $X_d=L_s/x_s=gL_sJ^3/(8i_s^2)$ according to Eq. (19.8)

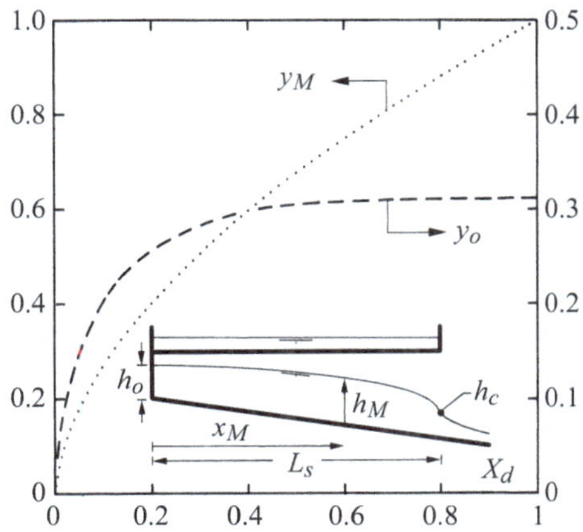

Fig. 19.9 Side-channel with critical flow $h_u = h_c$ at the downstream end, (- - -) y_o and $(\cdots)$ y_M as functions of X_d, notation (*inset*)

$$h_M = 1.2\frac{i_s^{0.845}L_s^{0.577}}{g^{0.423}J^{0.268}}; \quad x_M = \left[\left(\frac{gJ^3}{8i_s^2}\right)L_s\right]^{0.155} L_s. \tag{19.10}$$

The *maximum flow depth* h_M thus depends significantly on i_s, but less on the length L_s. The effect of J is relatively small, such that its estimation must not be very exact. The location of maximum flow depth x_M varies considerably with L_s and somewhat with J, whereas the effect of i_s is small.

The *upstream flow depth* $h_o = y_o h_s$ may be approximated as

$$h_o/h_s = 0.311(5X_d)^{1/\sqrt{3}}, \quad X_d < 0.1; \tag{19.11}$$

$$h_o/h_s = 0.311, \quad X_d > 0.3. \tag{19.12}$$

This can also be expressed as

$$h_o \simeq 0.95\frac{i_s^{0.845}L_s^{0.577}}{g^{0.423}J^{0.268}} = 0.79h_M, \quad X_d < 0.1; \tag{19.13}$$

$$h_o = 1.25\frac{i_s^2}{gJ^2} \simeq 1.25\frac{h_{cu}^3}{L_s^2J^2}, \quad X_d > 0.3. \tag{19.14}$$

For relatively short side channels, the upstream flow depth is always about 80% of the maximum flow depth, whereas for longer side channels it depends significantly on the critical depth $h_{cu}^3 = Q_u^2/(gb^2)$. Here, the effects of i_s and J are significant.

Because of the *decreasing energy line*, the upstream flow elevation is always higher than the elevation at the section of maximum flow depth. To inhibit *submergence* into a lateral channel, its crest must therefore be above the water elevation of the upstream end.

Example 19.6 Re-computation of Example 19.4.

1. *Uniform flow* $h_{Nu} = 0.135$ m.
2. *Critical depth* $h_{cu} = 0.225$ m.
3. *Singular point* $x_s = 268$ m, $h_s = 2.01$ m, corresponding to $X_d = 10/268 = 0.0373$.
4. *Free surface profile* $y_M = 0.0373^{1/\sqrt{3}} = 0.150$ according to Eq. (19.8), thus $h_M = 0.150 \cdot 2.01 = 0.30$ m and for the maximum flow depth, and from Eq. (19.9) for $y_o = 0.311[\tanh(5 \cdot 0.0373)]^{1/\sqrt{3}} = 0.311 \cdot 0.184^{0.577} = 0.117$, $h_o = 0.117 \cdot 2.01$ m $= 0.235$ m as upstream depth.

   The intensity of discharge is $i_s = Q_u/(L_s b) = 0.2/(10 \cdot 0.6) = 0.0333$ ms^{-1}. With Eq. (19.10) the maximum flow depth is determined to $h_M = 1.2 \cdot 0.033^{0.845} 10^{0.577}/(9.81^{0.423} 0.015^{0.268}) = 0.298$ m, as previously, with $x_M = [9.81 \cdot 0.015^3/(8 \cdot 0.033^2)10]^{0.155} 10 = 6$ m as the corresponding location.

   With $X_d = 0.0373 < 0.1$, Eq. (19.13) gives for the upstream flow depth $h_o \simeq 0.79 h_M = 0.79 \cdot 0.30 = 0.237$ m, as previously.

19.3.5 *Comparison With U-Shaped Profile*

The side channel of the U-shaped profile has been considered in Chap. 15. With $\bar{y} = \bar{h}_o/D$ as the upstream filling ratio and for critical flow at the downstream end of the side channel, Eq. (15.15) reads

$$\bar{y}_o^{5/2}\left(1 - \frac{1}{4}\bar{y}_o\right) = 2.92[Q_u^2/(gD^5)]^{2/3}. \tag{19.15}$$

For the rectangular channel, this computation yields with $y_o = h_o/b$ for $J = 0$

$$y_o = \sqrt{3}[Q_u^2/(gb^5)]^{1/3}. \tag{19.16}$$

For U-shaped profiles that are at least half filled, the relation between the flow depths is $\bar{h}_o = h_o + 0.11D$, and $D = b$. For a specific value y_o, one may thus compute the relative discharge $Q_u/(gb^5)^{1/2}$. Table 19.1 compares the results of Eqs. (19.15) and (19.16) for the two profiles, and differences of only about 2% are noted. The concept of substitute profile may thus be applied.

If the right hand side of Eq. (19.16) is substituted by the critical downstream depth $h_{cu} = [Q_u^2/(gb^2)]^{1/3}$, then $h_o/h_{cu} = 3^{1/2}$. This classical result is thus valid exclusively for $J = 0$, i.e. if the friction slope is compensated for by the bottom slope. The energy loss is then $\Delta H = H_o - H_{uc} = h_o - (3/2)h_{uc} = (3^{1/2} - 1.5)h_{uc} = 0.232 h_{uc} = \xi_s[V_u^2/(2g)]$ along the side channel due to lateral discharge, with

Table 19.1 Relative discharge $Q_u/(gb^5)^{1/2}$ according to (**a**) Eq. (19.15) and (**b**) Eq. (19.16) as functions of relative upstream filling y_o

y_o	0.5	0.6	0.7	0.8	0.9	1
a)	0.157	0 203	0.254	0.309	0.367	0.427
b)	0.155	0 204	0.257	0.314	0.375	0.439

$\xi_s = 2 \cdot 0.232 = 0.464$. This head loss coefficient ξ_u is relatively large and comparable with that of an intake (Chap. 2). The assumption of a constant energy head along a side channel is in considerable disagreement with observations, and not appropriate, therefore.

19.4 Practical Aspects of Side Channel Flow

Side channels generate flows of a considerable *spatial* degree. The previous one-dimensional computation reflects the average flow pattern. Depending on the tailwater level and the lateral inflow conditions, *spiral currents* of one or two vortex cells may be superposed the main forward flow. Figure 19.10 shows three typical sections of side channels. In the first case, the lateral inflow is free and a plunging jet with two rotational cells is developed. In the second case, the lateral flow is submerged resulting in a surface jet. In the third case with a trapezoidal section a plunging jet with only one rotational cell develops. The flow depth $h(x)$ as determined in 19.3 relates to the cross-sectional average and no information on the maximum flow depth t_s opposite from the lateral is available. According

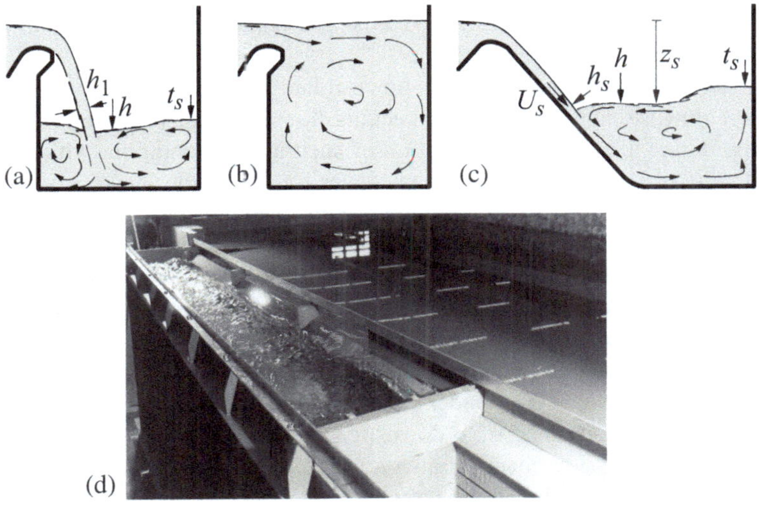

Fig. 19.10 Flow configurations in side channel of rectangular profile for (**a**) plunging jet and (**b**) surface jet. (**c**) Plunging jet into a trapezoidal channel. (**d**) Submerged side channel flow

to Bremen and Hager (1990) the computed value h corresponds nearly to the minimum cross-sectional depth. For plunging jets, it is close to the impact point (Fig. 19.10a, c), and the effect of spiral flow is not contained.

The flow in a side channel may be approximated as a *superposition* of primary and secondary flows. The primary flow is defined in Sect. 19.3, and the secondary flow is composed of either one or two vortices with axes along the side channel. The vortices depend significantly on its geometry and arrangement. Currently scarce information is available, such as of Gedeon (1962), Marangoni (1963) with particular reference to treatment stations, USBR (1938), Viparelli (1952), and Hager and Bremen (1990). Based on Japanese model observations for a trapezoidal channel of side slope 1(H):0.6(V), the local boundary flow depth ratio is (Fig. 19.10c)

$$\frac{t_s}{h} = 1 + 5.5\left[\frac{p_s^2}{ghP_h}(z_s/h)^{1/2}\right]^{1/2}. \tag{19.17}$$

Here $P_h = b + 2h$ is the wetted perimeter and z_s the elevation difference between the lateral and the side channel flows. In the rectangular side channel Hager and Bremen (1990) found

$$\frac{t_s}{h} = 1 + \gamma_s \frac{p_s U_s}{gh^2} \tag{19.18}$$

with $\gamma_s \cong 1$ as a factor of proportionality, that can increase up to $\gamma_s = 1.5$. Further, U_s is the lateral inflow velocity and $h = h(x)$ is the local flow depth according to the one-dimensional computation. According to Eqs. (19.17) and (19.18) the wall flow depth ratio (t_s/h–1) increases with the lateral discharge intensity p_s, the lateral jet velocity U_s and decreases as h increases. The parameter t_s is important for the definition of the *freeboard* required.

Side channels discharge often in a downstream sewer. For large discharges, or for a large downstream submergence, peculiar flow conditions may develop in side channels. Figure 19.11d shows a section of such a flow condition, with so-called gated flow (Chap. 9) in the downstream sewer for which the intake is submerged and the sewer is aerated from the downstream manhole. For a sufficiently high flow depth in the side channel, the lateral flow may be submerged, with a surface jet as shown in Fig. 19.10b. Consequently, a *single line vortex* is generated as a secondary current along the side channel axis. Because the tangential velocity component increases toward the vortex core, and the energy is almost constant for a potential vortex in the outer region, the pressure in the vortex core may drop below the atmospheric pressure. The under-pressure along the vortex core may become so strong that air is entrained from downstream. As a result, a winding line vortex develops along such side channels also referred to as the *Tornado vortex* (Hager 1990). According to the Helmholtz vortex theorem, the upstream end of the vortex is at the dead-end wall.

The presence of a Tornado vortex disturbs the primary flow of a side channel and of the tailwater reach considerably. In addition, the cross-sectional flow area is constricted because of *annular two-phase flow*, and the submergence to the overflow section increases. Further, flow pulsations are generated that propagate into the downstream region adding to possible choking of the sewer flow. Therefore, flow conditions with a Tornado vortex have to be prevented, and the downstream sewer of a side channel has to be sufficiently large to guarantee stable free surface flow along the overflow portion. Figure 19.12 shows a side channel with a partial Tornado vortex along the collecting reach.

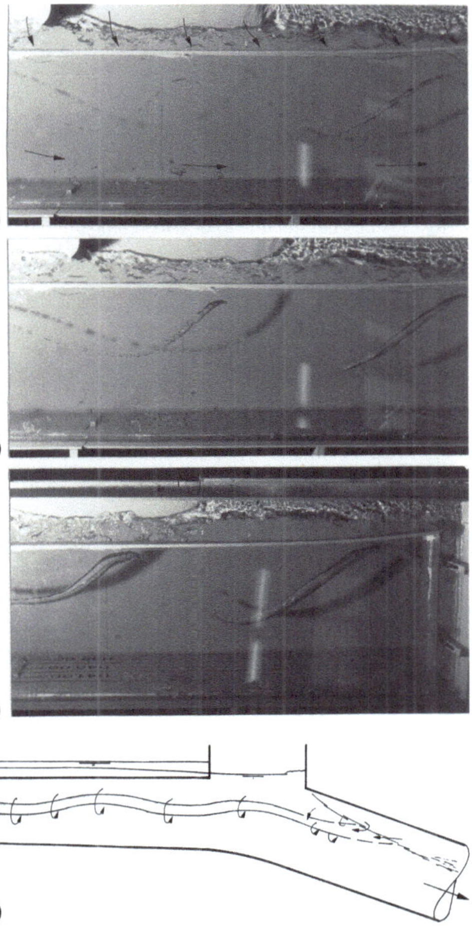

(a) (b) (c) (d)

Fig. 19.11 (**a–c**) Stages of tornado vortices with increasing lateral discharge. (**d**) Tornado vortex in side channel due to hydraulic overload

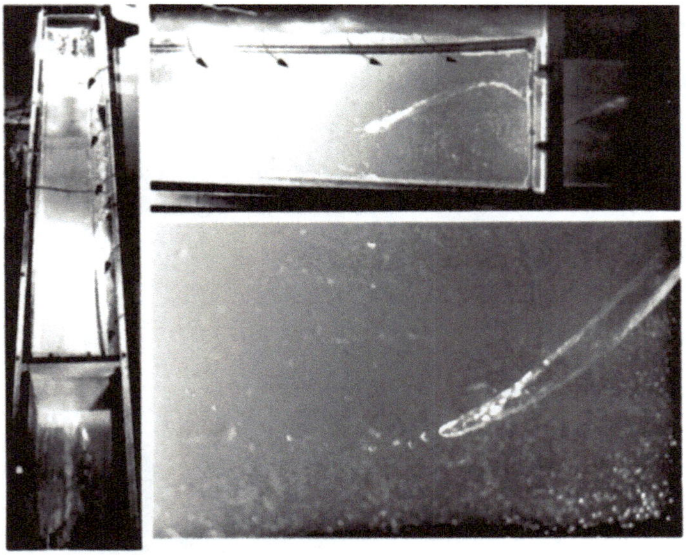

Fig. 19.12 Tornado vortex in side channel due to downstream submergence

Notation

b	[m]	channel width
D	[m]	diameter
F	[m^2]	cross-sectional area
F	[–]	Froude number
g	[ms^{-2}	gravitational acceleration
h	[m]	flow depth
$\bar{h}$	[m]	flow depth in U-shaped channel
h_s	[m]	flow depth at singular point
$h_{ü}$	[m]	overflow depth
H	[m]	energy head
H_f'	[–]	loss gradient due to friction
H_L'	[–]	loss gradient due to lateral discharge
i_s	[ms^{-1}]	$= p_s/b$ lateral discharge load
L_s	[m]	lateral length
$1/n$	[m$^{1/3}$s^{-1}]	Manning's roughness coefficient
p_s	[m^{2}s^{-1}]	lateral discharge intensity
P_h	[m]	wetted perimeter
Q	[m^{3}s^{-1}]	discharge
Q_s	[m^{3}s^{-1}]	lateral discharge

Q'	[m^{2}s^{-1}]	lateral discharge intensity
q_N	[–]	uniform discharge parameter
S_f	[–]	friction slope
S_o	[–]	bottom slope
t_s	[m]	side flow depth
u	[–]	flow characteristic
U	[ms^{-1}]	lateral velocity
v	[–]	flow characteristic
V	[ms^{-1}]	channel velocity
x	[m]	streamwise coordinate
x_s	[m]	location of singular point
X	[–]	$= x/x_s$
y	[–]	$= h/h_s$
y_N	[–]	filling ratio for uniform flow
z	[m]	vertical coordinate
z_o	[m]	upstream overflow height
z_s	[m]	difference of water elevations
γ_s	[–]	factor of proportionality
ΔH	[m]	energy head loss
ϕ	[–]	lateral approach direction
ρ	[kgm^{-3}]	density
χ	[–]	$= J/(p_s h/Q) = J\ (V/i_s)$ relative slope
ξ_s	[–]	loss coefficient

Indices

c	critical
d	forced critical
f	friction
m	average
M	maximum
N	uniform flow
o	upstream
r	boundary
s	singular
u	downstream

References

Bremen, R., Hager, W.H. (1990). Experiments in side-channel spillways. *Journal of Hydraulic Engineering* 115(5): 617–635.

Chow, V.T. (1959). *Open channel hydraulics*. McGraw-Hill: New York.

Gedeon, D. (1962). Discussion to Side-channel spillway design. *Journal of the Hydraulics Division* ASCE 88(HY6): 227–231.
Hager, W.H. (1985). Trapezoidal side-channel spillways. *Canadian Journal of Civil Engineering* 12: 774–781.
Hager, W.H. (1987). Lateral outflow over side weirs. *Journal of Hydraulic Engineering* 113(4): 491–504; 1989, 115(5): 684–688.
Hager, W.H. (1990). Tornado-Wirbel im Wasserbau (Tornado vortex in hydraulic engineering). *Wasser, Energie, Luft* 82(11/12): 325–330 [in German].
Hager, W.H., Bremen, R. (1990). Flow features in side-channel spillways. 22 *Convegno di Idraulica e Costruzioni Idrauliche* Cosenza 4: 91–105.
Marangoni, C. (1963). Un particulare tipo di sfioratore costruito all'estremità della derivazione "Castelletto-Nervesa" (A particular overflow type designed at the end of Castelletto-Nervesa diversion). 8 *Convegno di Idraulica* Pisa C(4): 1–10 [in Italian].
Naudascher, E. (1992). *Hydraulik der Gerinne und Gerinnebauwerke* (Hydraulics of channels and channel structures). Springer: Wien [in German].
USBR (1938). Model studies of spillways. *Boulder Canyon Project*: Final Reports. Bulletin 1 Part VI, Hydraulic Investigations. United States Bureau of Reclamation USBR: Denver, Colorado.
Viparelli, C. (1952). Sul proporzionamento dei colletori a servizio di scarichi di superficie (On the design of collector channels for free surface overflow). *L'Energia Elettrica* 29(6): 341–353 [in Italian].

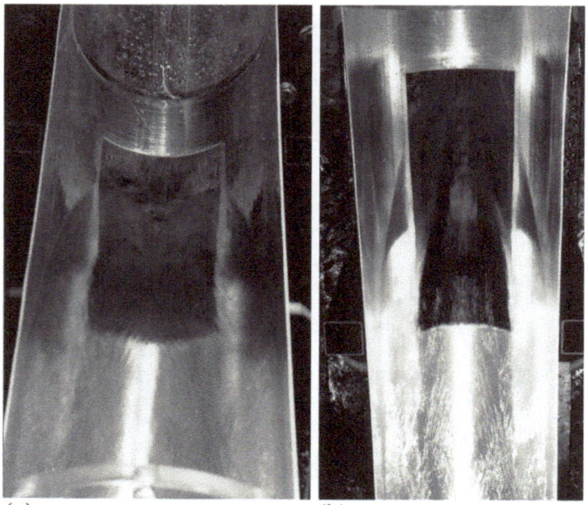

(a) (b)

Bottom opening with full discharge into the lateral (**a**) small, (**b**) limit discharge (flow direction from *bottom* to *top*)

Chapter 20
Bottom Opening

Abstract Bottom openings are overflow structures used for *supercritical* approach flow. The concept was introduced 35 years ago. This chapter introduces a simplified structure tested in detailed model observations.

The approach is based on a prismatic U-shaped profile with a rectangular slot along the bottom. The slot geometry is defined in terms of the approach flow conditions and it is demonstrated that the excess discharge to the treatment station is within ATV regulations for arbitrary approach flow discharges, if the jet discharged from the bottom opening is fully aerated. The analysis includes expressions for the end depth ratio, the discharge characteristics, the excess treatment discharge, the downstream flow features, and the free surface profile. In addition principles for the design of a bottom opening are specified. The novel structure is simple in operation, effective in hydraulic separation and may be regarded as the analogue to sewer sideweirs for supercritical approach flow.

20.1 Introduction

Two types of outlet structures are available in combined sewers. For subcritical approach flow, *the sewer sideweir* with a high weir crest and a throttling pipe is recommended, as described in Chap. 18. This concept is not suited for supercritical approach flow because of the occurrence of hydraulic jumps and the corresponding non-uniform lateral outflow distribution.

The bottom opening or *leaping weir* (German: Springüberfall; French: Ouverture de fond) was developed for supercritical approach flow. A different approach was presented by Biggiero (1969). Hager (1985) summarized the hydraulic knowledge on bottom openings. The lateral discharge is deflected through an opening in the bottom of a prismatic U-shaped channel, whose width and length depend on the approach flow characteristics at critical treatment and maximum discharges. For the limit discharge to the treatment station, all flow is discharged across the bottom opening. For the maximum approach flow discharge during the design rainfall, the bottom opening should receive a limit surplus discharge. The bottom opening corresponds thus to a simple hydraulic discharge separator, of which the

W.H. Hager, *Wastewater Hydraulics*, 2nd ed., DOI 10.1007/978-3-642-11383-3_20,

opening geometry can be easily modified whenever the discharges in the upstream or downstream reaches are adjusted to updated design conditions.

Bottom openings are well known as Tyrol weirs in hydraulic structures, such as in Alpine intakes. The cross-section is then rectangular, and the channel bottom consists of a rack preventing gravel and rock from being discharged into the bottom opening, and thus towards the intake structure (Brunella et al. 2002, Hager and Minor 2003). In sewer hydraulics, no systematic experiments on bottom openings were available, due to the considerable number of parameters and the geometry of the structure (Hager 1992). Oliveto et al. (1997) thus conducted systematic experiments on the sewer bottom opening, with a summary of the results following below. In the future the effect of submergence on bottom openings has to be defined because the previously mentioned results are based on free bottom outflow. From other indications, the *effect of submergence* is known to be significant and the general recommendation given in Chap. 19 for side channels can only be repeated here.

20.2 Computational Assumptions

Taubmann (1972) standardized previous designs of bottom openings such that a hydraulic approach is amenable. Figure 20.1 shows a scheme of a bottom opening structure with the approach flow sewer, the bottom opening discharging to the treatment station, and the storm water outlet that is eventually connected with a storage basin. To reduce *flow pulsations* and to define definite jet flow, the outflow from the bottom opening is fully aerated, such that the pressure on the jet is atmospheric. To ensure stable approach flow, Taubmann introduced the following requirements (SIA 1980):

- Minimum approach flow Froude number $\mathsf{F}_o = 1.5$;
- Constant sewer diameter D_o and no laterals over the approach flow length of at least $10D_o$,
- Constant bottom slope S_o of at least 1%, and
- Straight approach flow sewer to avoid shock waves.

Then, the approach flow (subscript o) is nearly uniform.

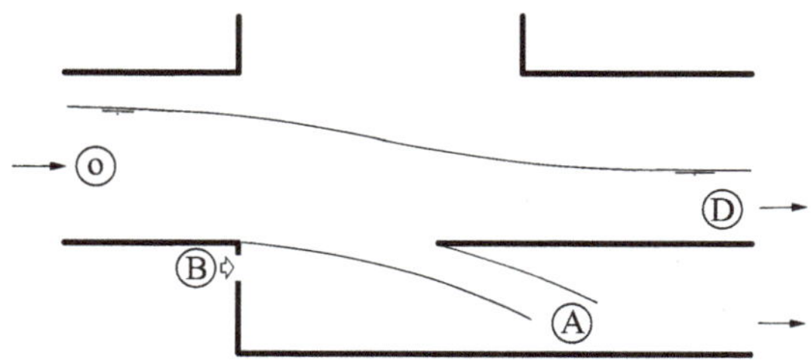

Fig. 20.1 Bottom opening without throttling pipe (schematic) with Z approach flow sewer, A discharge to treatment station and D downstream sewer, eventually connected with a storage basin. B jet aeration

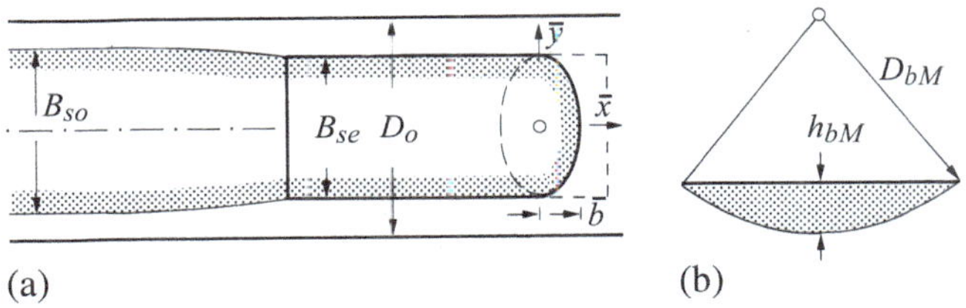

Fig. 20.2 Bottom opening according to Taubmann (1972) with separation sheet in (**a**) plan and (**b**) section

The bottom opening introduced by Taubmann (1972) has the following features (Fig. 20.2):

- *Approach flow sewer* of diameter D_o extends into the downstream reach to reduce shock waves due to a diameter reduction,
- *Bottom opening* involves a rectangular opening to separate the treatment discharge from the outlet discharge,
- *Separation sheet* has an elliptic geometry at its downstream end,
- *Modification* of lateral outflow capacity by adapting the metal sheet geometry, and
- Sewer discharging into the treatment station can be extended with a *throttling pipe*.

In addition, the opening length should at least be 0.50 m and the downstream sewer be straight and free of any constriction to prevent submergence effects on the bottom jet.

The hydraulic design involves at least two discharges, namely:

- *Treatment discharge* Q_K for which all discharge from the approach flow sewer continues to the downstream sewer (Z to A in Fig. 20.1), and
- *Maximum discharge* Q_M for which the treatment discharge has to be limited.

The treatment discharge defines the geometry of the bottom opening and one has to verify that the discharge to the treatment station is limited for the maximum discharge. The hydraulic design refers to three different sections, namely the approach flow section (subscript *o*), the end section (subscript *e*) at the outlet of the approach flow sewer to the bottom opening, and the downstream section (subscript *d*).

Hager (1992) developed Taubmann's approach by presenting equations for the length and width of the bottom opening, and the separation characteristics under maximum discharge conditions. However, Taubmann never verified his approach with experiments in a sewer bottom opening, and his results have thus never been compared with observations. Therefore, Oliveto et al. (1997) conducted experiments on a bottom opening that are presented below.

20.3 End Depth Ratio

Figure 20.3a shows a definition sketch of the approach flow portion of a bottom opening with h as flow depth, Q discharge and subscripts o, e and d referring to the approach flow, the end and the downstream sections, respectively. The end section is located at the beginning of the bottom opening and reference is made to the end section of a pipe outlet, as described in Chap. 11.

The cross-sectional area F and the hydraulic pressure force P of free surface flow in a circular pipe or in an U-shaped channel of diameter D can be approximated as ($\pm 10\%$)

$$F = (Dh^3)^{1/2}, \tag{20.1}$$

$$P/(\rho g) = \frac{1}{2}(Dh^5)^{1/2}. \tag{20.2}$$

Here, ρ is fluid density and g gravitational acceleration. For pipe outflow into the atmosphere, it is commonly assumed that the internal jet pressure at the end section may be neglected. For a bottom opening of width b, or relative width $\beta = b/D \leq 1$ one may assume that $P_e/(\rho g) = [(D - b)h_e^5]^{1/2}$. For the two particular cases $\beta = 0$ and $\beta = 1$ the basic results are retained.

The momentum equation applied between the approach flow (subscript o) and the end (subscript e) sections provides the *end depth ratio* $Y_e = h_e/h_o$. It may be expressed for an uniform velocity distribution as (Fig. 20.3a)

$$\frac{1}{2}(\rho g)(Dh_o^5)^{1/2} + \frac{\rho Q^2}{(Dh_o^3)^{1/2}} = \frac{1}{2}(\rho g)\left[(D-b)h_e^5\right]^{1/2} + \frac{\rho Q^2}{(Dh_e^3)^{1/2}}. \tag{20.3}$$

Introducing the Froude number of circular pipe flow (Chap. 6)

$$\mathsf{F} = \frac{Q}{(gDh^4)^{1/2}}, \tag{20.4}$$

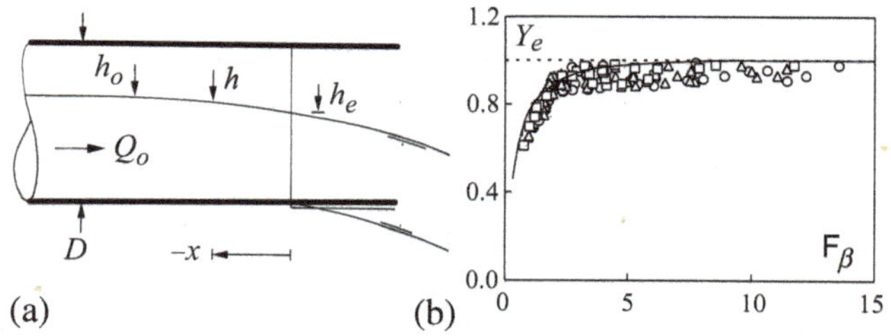

Fig. 20.3 End depth ratio (**a**) definition of parameters, (**b**) $Y_e(\mathsf{F}_\beta)$ for β = (○) 40%, (△) 60%, (□) 80%, and (—) Eq. (20.6)

the governing equation is

$$1 + 2\mathsf{F}_o^2 = (1-\beta)^{1/2} Y_e^{5/2} + 2\mathsf{F}_o^2 Y_e^{-3/2}. \tag{20.5}$$

For $\beta = 1$ (efflux into the atmosphere) and for the lower limit approach flow Froude number $\mathsf{F}_o = 1$, Eq. (20.5) results in $Y_e = (2/3)^{2/3} = 0.763$, comparing reasonably well with observations (Chap. 11). For $\beta = 0$ (no opening at all), the trivial solution is $Y_e = 1$, i.e. no drawdown effect due to the bottom opening. The effect of decreasing β is to increase Y_e, therefore. Figure 20.3b shows the end depth ratio $Y_e = h_e/h_o$ as a function of the *combined Froude number* $\mathsf{F}_\beta = \beta^{-1/2}\mathsf{F}_o$ as

$$Y_e = \left(\frac{2\mathsf{F}_\beta^2}{1 + 2\mathsf{F}_\beta^2}\right)^{2/3} \tag{20.6}$$

Note the nonlinear effect of β on F_β, and thus also on the end depth ratio. Equation (20.6) is located slightly above the test data which must be attributed to difficulties in determining h_o, and to simplifications in the computational approach.

Example 20.1 Given a sewer of $D = 0.90$ m, $S_o = 6\%$, $1/n = 85$ m$^{1/3}$s^{-1} and $Q = 1.2$ m^{3}s^{-1}. What is the drawdown of uniform flow depth for $\beta = 60\%$? With $q_N = nQ/(S_o^{1/2}D^{8/3}) = 0.012 \cdot 1.2/(0.06^{1/2}0.90^{8/3}) = 0.0763$, Eq. (5.15)$_1$ yields $y_N = 0.33$ and thus $h_N = 0.30$ m. The corresponding uniform flow Froude number is $\mathsf{F}_N = 1.2/(9.81\ \ 0.9 \cdot 0.3^4)^{1/2} = 4.5$. With $\beta = 0.60$, the combined Froude number is $\mathsf{F}_\beta = 0.6^{-1/2}4.5 = 5.8$, and Eq. (20.6) gives $Y_e = [2 \cdot 5.8^2/(1 + 2 \cdot 5.8^2)]^{2/3} = 0.99$. The drawdown from uniform flow is thus 1%, to $h_e = 0.99 \cdot 0.30 = 0.297$ m.

20.4 Discharge Characteristics

The discharge ΔQ across a bottom opening is related in a complex way to the approach flow conditions, the outlet geometry, shock waves generated by the supercritical flow, resulting in a highly-spatial flow pattern. To introduce a simple model, the *lateral discharge* ΔQ is computed with an average (subscript *a*) *discharge coefficient* C_{da} as

$$\Delta Q = C_{da}(2gh_o)^{1/2}\, b\Delta L \tag{20.7}$$

with ΔL as the average jet length at the bottom elevation between the axis and the lateral boundary, and $\Delta L = L$ for a bottom opening with a downstream discharge $Q_d > 0$. The cross-sectional area of lateral discharge is $b\Delta L$, and the determining approach flow velocity perpendicular to the slot is $(2gh_o)^{1/2}$.

For a conventional sharp-crested orifice subjected by a very small approach flow velocity ($\mathsf{F}_o \to 0$), the discharge coefficient is theoretically $\pi/(\pi+2) = 0.611$. Increasing the approach flow Froude number reduces C_{da} due to streamline curvature and a reduction of flow depth along the bottom opening. Also, for relatively short openings, boundary effects reduce the outflow, as for sideweirs (17.6). Streamline curvature and shape effects in a circular pipe may be accounted for with the pipe filling $y_o = h_o/D$, where h_o is the approach flow depth. The average discharge coefficient C_{da} thus reduces as F_o, y_o and $D/\Delta L$ increase, such that $\Phi = y_o^{1/2}(D/\Delta L)\mathsf{F}_o$ is the governing reduction parameter. The experimental data may be expressed to ±5% as (Oliveto et al. 1997)

$$C_{da} = 0.61 - 0.10\beta^{1/2}\Phi. \tag{20.8}$$

The lateral discharge across a bottom opening is then

$$\Delta Q = 0.61(2gh_o)^{1/2}b\Delta L - 0.14\left(\frac{b^3}{Dh_o^2}\right)^{1/2} Q_o \tag{20.9}$$

The conventional expression for orifice flow is therefore extended by a term proportional to the aspect ratio and the approach flow discharge.

The *length* ΔL required for a bottom opening can be determined from the condition that for $Q = Q_K$ all discharge flows across the bottom opening to the treatment station. According to Taubmann (1972) the *width* of the bottom opening should be equal to the approach flow width B_s, i.e. $\beta = 2(y_o - y_o^2)^{1/2}$ from elementary geometry. Because the fraction of discharge Q_K to the treatment station is small compared with the maximum approach flow discharge Q_M, the filling ratio $y_o = h_o/D$ for treatment discharge is usually below 30%. For $0.10 < y_o < 0.35$, the result is then simply (Oliveto et al. 1997)

$$\Delta L = h_o\mathsf{F}_o. \tag{20.10}$$

The required length of the bottom opening increases thus linearly with the approach flow depth h_o and Froude number F_o.

Example 20.2 Given the bottom opening of Example 20.1 for a treatment discharge of $Q_K = 0.2\ \mathrm{m^3s^{-1}}$. What is the geometry of the bottom opening? With $q_N = nQ/(S_o^{1/2}D^{8/3}) = 0.012 \cdot 0.2/(0.06^{1/2}0.90^{8/3}) = 0.013$ for uniform approach flow, one has $y_o = 0.131$ from Eq. (5.15) and $h_{oK} = 0.131 \cdot 0.9 = 0.118$ m. The approach flow Froude number is then $\mathsf{F}_o = 0.2/(9.81 \cdot 0.9 \cdot 0.118^4)^{1/2} = 4.83$ from Eq. (20.4). The approach flow width is with $\beta = 2(0.131 - 0.131^2)^{1/2} = 0.67$ equal to $b = 0.67 \cdot 0.9 = 0.60$ m. The length of the bottom opening is $\Delta L = 0.118 \cdot 4.83 = 0.57$ m, selected is a quadratic opening 0.60 m × 0.60 m.

20.5 Excess Treatment Discharge

According to ATV (1994) a bottom opening should have a *separation accuracy* M (German: Trennschärfe; French: Limite de séparation) of less than 20%, i.e. the treatment discharge under maximum approach flow condition should be less than 120% of Q_K. With subscript M for maximum approach flow conditions, the requirement is thus $M = \Delta Q_M/Q_K \leq 1.2$. This condition is difficult to satisfy because the maximum discharge can easily be ten times larger than the treatment discharge. The latter is often the double or a multiple of the dry weather discharge. For example, one can have $Q_K = 0.1\ \text{m}^3\text{s}^{-1}$ and $Q_M = 1.5\ \text{m}^3\text{s}^{-1}$. Then, the maximum discharge to the treatment station should be smaller than $0.12\ \text{m}^3\text{s}^{-1}$.

For a sufficiently long and straight approach flow sewer, the approach flow is uniform and the corresponding Froude number is approximately (Oliveto et al. 1997)

$$\mathsf{F}_M = \frac{Q_M}{(gDh_M^4)^{1/2}} = \left(\frac{3}{4}\right)^2 \frac{S_o^{1/2} D^{1/6}}{ng^{1/2}}. \tag{20.11}$$

F_M is thus independent of discharge and varies mainly with the bottom slope S_o and the roughness coefficient n. The approach flow Froude numbers for the treatment and the maximum discharges are thus identical.

The separation accuracy can be expressed exclusively as a function of discharge ratio $q = Q_M/Q_K$ as (Oliveto et al. 1997)

$$M = 2^{3/2} q^{1/4}[0.61 - 0.20\beta^{-1/2} q^{1/4}]. \tag{20.12}$$

The function $M(q)$ has the maximum value of $M_M = 1.32\beta^{1/2}$ for $q = 5.4\beta^2$. For example $M_M(\beta = 0.6) = 1.02$, and this function decreases below $M = 1$ for $q > 2$. The separation accuracy is thus always smaller than 1.02 for $\beta = 0.6$ which may be considered as a lower bound for β. For an upper limit of $\beta = 0.80$ the maximum is $M_M = 1.18$ for $q = 3.5$. Therefore, within the practical limits of β between 0.60 (or even 0.50) and 0.80, the condition for the separation accuracy is *always* satisfied, and the bottom opening as described here corresponds to an excellent hydraulic discharge separator for sewers. Note that the basic requirements are stable supercritical approach flow, a sufficiently long approach flow sewer resulting in uniform flow, a fully aerated jet from the bottom opening and absence of tailwater submergence.

Example 20.3 Determine the separation accuracy for the conditions described in Example 20.1 and Example 20.2.

Treatment discharge $Q_K = 0.2\ \text{m}^3\text{s}^{-1}$, $\mathsf{F}_K = 4.83$, $\Delta L = 0.57$ m, $b = 0.60$ m.

Maximum discharge $Q_M = 1.2\ \text{m}^3\text{s}^{-1}$, $\mathsf{F}_M = 4.50$, $h_M = 0.30$ m.

With $\Delta Q_M = [0.61(19.62 \cdot 0.3)^{1/2} 0.6 \cdot 0.57 - 0.14(0.6^3/0.9 \cdot 0.3^2)^{1.2}]1.2 = 0.51 - 0.28 = 0.23$ m^{3}s^{-1} from Eq. (20.9), the separation accuracy is $M = \Delta Q_M/Q_K = 0.23/0.2 = 1.15 < 1.2$. From Eq. (20.12) results directly $M = 2.82 \cdot 6^{1/4}[0.61 - 0.2 \cdot 0.67^{-1/2} 6^{1/4}] = 1.01$ for $q = 1.2/0.2 = 6$, thus slightly less due to simplifications in the derivation of Eq. (20.12).

20.6 Downstream Flow

The momentum equation can also be applied across the bottom opening to predict the ratio $Y_d = h_d/h_o$ of the boundary flow depths. With $U\cos\phi$ as the resulting outflow velocity component in the streamwise direction, and assuming hydrostatic pressure and uniform velocity distributions (Fig. 20.4)

$$\frac{1}{2}(\rho g)(Dh_o^5)^{1/2} + \frac{\rho Q_o^2}{(Dh_o^3)^{1/2}} = \rho U \cos\phi \Delta Q + \frac{1}{2}(\rho g)(Dh_d^5)^{1/2} + \frac{\rho Q_d^2}{(Dh_d^3)^{1/2}}. \quad (20.13)$$

With $R = Q_d/Q_o < 1$ as the discharge ratio follows

$$1 + 2\mathsf{F}_o^2 = \frac{2U\cos\phi\Delta Q}{g(Dh_o^5)^{1/2}} + Y_d^{5/2} + 2R^2\mathsf{F}_o^2 Y_d^{-3/2}. \quad (20.14)$$

The *lateral outflow component* is estimated with the Bernoulli equation resulting in $2U\cos\phi\Delta Q/[g(Dh_o^5)^{1/2}] = 2(1-R)\mathsf{F}_o^2$ (Oliveto et al. 1997). Inserting in Eq. (20.14) gives

$$\mathsf{F}_o^2 = \frac{Y_d^{3/2}(1 - Y_d^{5/2})}{2R(R - Y_d^{3/2})}. \quad (20.15)$$

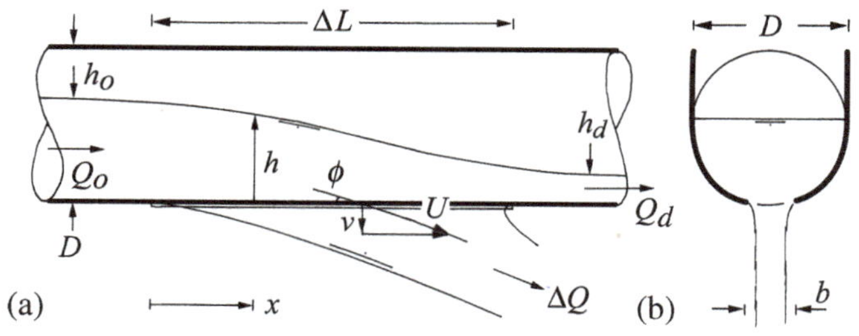

Fig. 20.4 Definition of variables in bottom opening (**a**) streamwise section, (**b**) transverse section

An approximation of Eq. (20.15) is

$$Y_d R^{-2/3} = \left[1 + \frac{1}{3R\mathsf{F}_o^2}\right]^{-2/3}. \tag{20.16}$$

Figure 20.5a shows the relation $Y_q(\mathsf{F}_q)$ where $Y_q = R^{-2/3} Y_d$ and $\mathsf{F}_q = R^{1/2}\mathsf{F}_o$ as suggested by Eq. (20.16), and excellent agreement with the data is noted. The boundary depth ratio is thus always smaller than 1 and increases as both F_o and R increase.

Example 20.4 Determine the downstream flow depth for Example 20.3.
With $R = Q_d/Q_o = (Q_o - \Delta Q)/Q_o = (1.2 - 0.23)/1.2 = 0.81$, $\mathsf{F}_o = 4.5$, Eq. (20.16) gives $Y_d = 0.81^{2/3}[1 + 1/(3 \cdot 0.81 \cdot 4.5^2)]^{-2/3} = 0.86$, thus $h_d = 0.86 \cdot 0.30 = 0.26$ m. The corresponding downstream Froude number is then $\mathsf{F}_d = Q_d/(gDh_d^4)^{1/2} = 0.97/(9.81 \cdot 0.90 \cdot 0.26^4)^{1/2} = 4.83$.

To inhibit tailwater submergence, the sequent depth is determined. From Eq. (7.32), and with $y_1 = 0.26/0.90 = 0.29$, one has as reference discharge $q_o = (3/4)0.29^{0.75}[1 + (4/9)0.29^2] = 0.31$ and therefore for dimensionless discharge $q_D = Q/(gD^5)^{1/2} = 0.97/(9.81 \cdot 0.9^5)^{1/2} = 0.40$. Thus, this results in $(y_2 - y_1)/(1 - y_1) = [(0.40 - 0.29^2)/(0.31 - 0.29^2)]^{0.95} = 1.375$, and further $y_2 = 0.29 + (1 - 0.29)1.375$ or $y_2 = 1.27$. For free surface flow ($y_2 < 1$), there is no danger of submergence for the bottom opening.

The *energy loss* of resulting from a bottom opening can be simply predicted with known downstream flow depth. Assuming that the friction slope is compensated for by the bottom slope, and if the loss due to the lateral discharge is accounted for as $\Delta H = \xi_d[V_o^2/(2g)]$, the generalized Bernoulli equation reads

$$H_o = h_o + \frac{V_o^2}{2g} = h_d + \frac{V_d^2}{2g} + \xi_d \frac{V_o^2}{2g}. \tag{20.17}$$

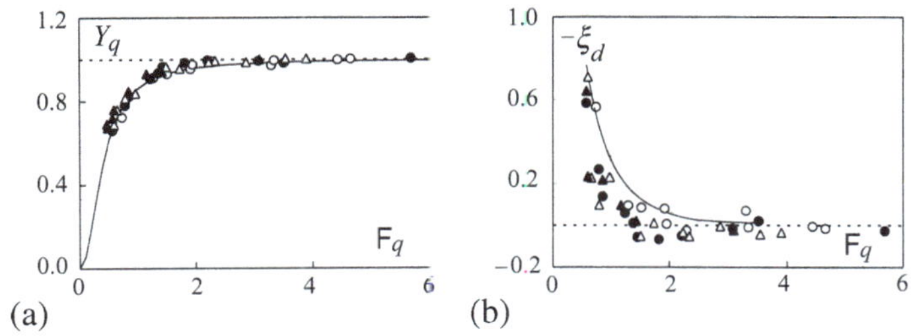

Fig. 20.5 (**a**) Tailwater depth ratio Y_q as a function of F_q according to (—) Eq. (20.16), (**b**) loss coefficient $\xi_d(\mathsf{F}_q)$ with (—) Eq. (20.18)

Substituting for h_d Eq. (20.16), ξ_d may be expressed as a function of the discharge ratio $R = Q_d/Q_o$ and the approach flow Froude number F_o. A simple approximation for the *loss coefficient* ξ_d is (Fig. 20.5b)

$$\xi_d = -\frac{1}{3\mathsf{F}_q^2} = -\frac{1}{3R\mathsf{F}_o^2}. \tag{20.18}$$

For $\mathsf{F}_q > 2$, the losses due to lateral discharge (Chap. 17) are negligible because $\xi_d \to 0$, and for $\mathsf{F}_q < 2$ the effect is also small because the energy loss $\Delta H/h_o = \xi_d[Q_o^2/(2gDh_o^4)] = -1/(6R)$ is independent of F_o. For all practical purposes the head loss due to lateral outflow is thus small, though negative. The friction losses are nearly compensated for by the bottom slope for uniform approach flow, and one may indeed assume a *constant energy head* across the bottom opening. The conventional approach for spatially-varied flow as originally introduced by De Marchi in 1934 (Chow 1959) corresponds thus to a powerful engineering method.

Example 20.5 Determine the head loss of the bottom opening described in Example 20.4.

The cross-sectional area of the U-shaped profile is $F \cong (Dh^3)^{1/2}$ and the hydraulic radius is for $h/D < 0.7$ equal to $R_h/D = 0.42(h/D)^{0.75}$. The approach flow velocity is, therefore, $V_o = Q_o/(Dh_o^3)^{1/2} = 1.2/(0.9 \cdot 0.3^3)^{1/2} = 7.7\ \mathrm{ms}^{-1}$, thus $H_o = h_o + V_o^2/(2g) = 0.3 + 7.7^2/19.62 = 3.32$ m. Further, downstream velocity is $V_d = 0.97/(0.9 \cdot 0.26^3)^{1/2} = 7.72\ \mathrm{ms}^{-1}$ and $H_d = 0.26 + 7.72^2 = 3.30$ m, thus $\Delta H = H_o - H_d = 3.32 - 3.30 = +0.02$ m. According to Eq. (20.18) $\xi_d = -1/(3 \cdot 0.82 \cdot 4.5^2) = -0.02$ as the loss coefficient, thus $\Delta H = -0.02 \cdot 7.7^2/19.62 = -0.06$ m.

The average (subscript a) friction slope according to Manning and Strickler is $S_{fa} = n^2 V_a^2/R_{ha}^{4/3}$. With $V_a = (1/2)(V_o + V_d) = 7.71\ \mathrm{ms}^{-1}$ and $R_{ha} = (1/2)(R_{ho} + R_{hd}) = 0.5(0.134 + 0.165) = 0.150$ m one has therefore $S_{fa} = (0.012 \cdot 7.71)^2/0.15^{4/3} = 0.107$. The energy loss over the bottom opening is $\Delta H_f = 0.107 \cdot 0.6 = 0.064$ m, and the energy increase due to the bottom slope is $S_o \Delta L = 0.06 \cdot 0.6 = 0.036$ m. The effect of energy change is thus $\Delta H = 0.036 - 0.064 - 0.02 = -0.048$ m, or practically equal to zero for the large approach flow energy head. The assumption of a constant energy head is justified.

20.7 Surface Profile

The surface profile across a bottom opening is of practical significance to determine the location of a hydraulic jump. Such flow has to be prevented because the previous approach is based on supercritical flow along the entire bottom opening. The free surface profile across a bottom opening is strongly influenced by streamline curvature, and a conventional hydraulic approach is erroneous. No attempt has been made so far to compute this profile with a generalized approach, involving the *Boussinesq equations* (Chap. 1), and a semi-empirical procedure is presented below.

For a pipe flow discharging into the atmosphere, the lower and upper nappe profiles are functions of the dimensionless streamwise coordinate $X_s = (x/h_o)\mathsf{F}_o^{-0.8}$ (Chap. 11). This case corresponds to an asymptotic bottom opening for $\beta = 1$. Figure 20.6 shows experimental data for the upper jet trajectory $Y_s = h/h_e$ as a function of X_s for various values of $\beta = b/D$ collected by Oliveto et al. (1997). The data for $\beta = 60\%$ and 80% are nearly identical but the data for $\beta = 40\%$ are definitely higher. Bottom openings with $\beta < 50\%$ are impractical because of the potential to clog, with the corresponding approach flow filling ratio smaller than 7%. For $\beta \geq 0.50$ the generalized free surface profile reads

$$Y_s = 1 - \gamma X_s^{1.5}, \tag{20.19}$$

where $\gamma = 0.54$ for all discharge across the bottom opening ($Q_d = 0$), and $\gamma = 0.40$ for a downstream discharge ($Q_d > 0$). Also included in Fig. 20.6 is Eq. (11.25) for $\beta = 1$, in agreement with the end overfall for $Q_d = 0$.

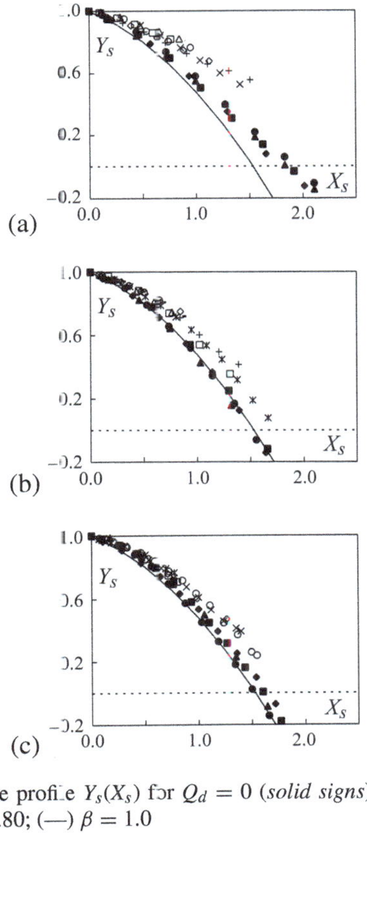

Fig. 20.6 Axial free surface profile $Y_s(X_s)$ for $Q_d = 0$ (*solid signs*) and $Q_d > 0$ (*light signs*) for $\beta =$ (**a**) 0.40, (**b**) 0.60, (**c**) 0.80; (—) $\beta = 1.0$

Example 20.6 Determine the free surface profile for the bottom opening as given in Example 20.1.

With $\mathsf{F}_o = 4.50$, $h_o = 0.30$ m and $h_e = 0.297$ m, the profile equation is $Y_s = 1 - 0.4X_s^{1.5}$. Table 20.1 gives the coordinates of the upper jet profile.

It is seen that the downstream flow depth is slightly below the depth $h_d = 0.26$ m as determined in Example 20.4. This is due to the minimum location of Eq. (20.19) for $\beta = 100\%$. The differences are normally within some percents, as are the data plotted in Fig. 20.6. Note also that the profiles apply to a jet within lateral walls, because these data were collected within a channel. For jets issued into the free atmosphere, slightly different profiles particularly at the sides would be expected.

Table 20.1 Free surface profile for Example 20.6

x	[m]	0.1	0.1	0.2	0.3	0.4	0.5	0.6
X_s	[−]	0	0.1	0.2	0.3	0.4	0.5	0.6
Y_s	[−]	1	0.987	0.964	0.934	0.900	0.859	0.814
h	[m]	0.297	0.293	0.286	0.277	0.267	0.255	0.242

Figures 20.7, 20.8, and 20.9 refer to *photographs* of bottom opening flows with $L/D = 1.67$, $\beta = 60\%$ and an approach flow filling ratio of nearly 60%. Figure 20.7 relates to side views. For $\mathsf{F}_o = 1$ all discharge flows through the bottom opening, whereas some flow is directed to the downstream sewer for $\mathsf{F}_o = 2$ and 3. Note how the jet angle relative to the horizontal decreases as the approach flow increases. Also the space between the side channel and the bottom opening is relatively large to prevent submergence onto the lateral jet.

(a) (b) (c)

Fig. 20.7 Side views at bottom opening for $\mathsf{F}_o =$ (**a**) 1, (**b**) 2, (**c**) 3

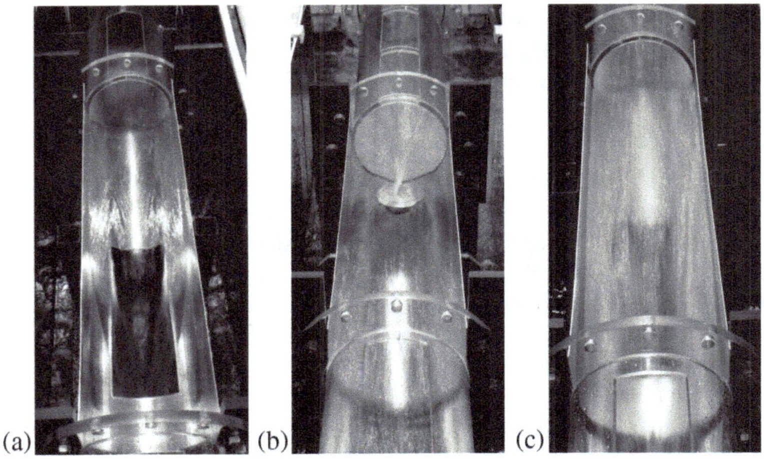

Fig. 20.8 Top views on bottom opening for F_o = (a) 1, (b) 2, (c) 3

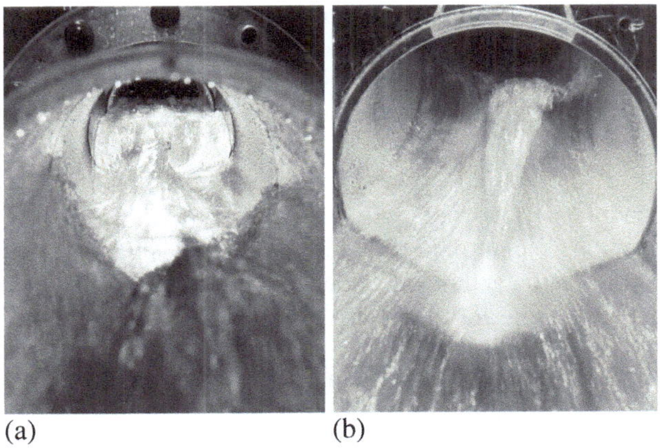

Fig. 20.9 Development of axial shockwave at the downstream end of the bottom opening. Views (a) in and (b) against the flow direction

Figure 20.8 shows top views of the corresponding flows. For $\mathsf{F}_o = 2$, the transverse flow contraction along the bottom opening becomes evident, associated with the formation of shock waves beyond the end of the bottom opening. For $\mathsf{F}_o = 1$ these shocks are missing because of the absence of downstream flow. For $\mathsf{F}_o = 3$, one may hardly recognize the flow contraction because of the high approach flow Froude number.

Figure 20.9 refers to a flow with $\mathsf{F}_o = 3$ and shows details at the end of the bottom opening. Because of the abrupt end of the opening the flow is disturbed and a *central shockwave* is developed. Its height is of the order of the approach flow depth. The central shock expands to the sides of the sewer and generates the well known wave reflections, as is typical for disturbed supercritical flows. Abrupt transitions from the free surface to pressurized flow have never occurred, due to the reduction of the average flow depth, from the approach flow to the downstream sewer. It would be unwise, however, to design a bottom opening with a converging U-shaped profile.

20.8 Design Principles

Bottom openings in U-shaped profiles correspond to a simple hydraulic separation structure for combined sewers. It can be recommended for design provided the following *conditions* are satisfied:

1. The approach flow is stably supercritical, with a minimum approach flow Froude number of about 1.50.
2. The approach flow sewer has a minimum length of about 30 diameters, such that uniform flow has established at the end section.
3. The prismatic U-shaped profile is connected accurately with the upstream and downstream sewers to suppress the formation of shockwaves.
4. The opening geometry is rectangular of length ΔL and width b, with a minimum relative width $\beta = b/D$ of 50%.
5. The opening width b is equal to the approach flow width of the treatment discharge. The bottom opening should be made of metal sheet to allow for a later adaptation.
6. The length of the bottom opening varies essentially with the approach flow depth h_K and the corresponding Froude number.
7. Bottom openings should only be applied if the potential for clogging is small. A periodic inspection and maintenance of the structure is important.
8. The sewers continuing for the treatment station and the outlet structure should not submerge a bottom opening.
9. The lateral jet generated from the bottom opening has to be sufficiently aerated to guarantee a fully-aerated outflow.
10. The access to the structure has to be guaranteed for inspection.

If these conditions are satisfied, the bottom opening is a simple and effective separation structure with an excellent hydraulic performance. It can be considered as a structure analogous to the sewer sideweir, provided that the approach flow is supercritical.

Notation

B_s	[m]	free surface width
C_c	[–]	contraction coefficient
C_{da}	[–]	average discharge coefficient
D	[m]	sewer diameter
D_s	[m]	manhole diameter

F	[m^2]	cross-sectional area
F	[–]	Froude number
F_β	[–]	$= \beta^{-1/2}\mathsf{F}_o$
F_q	[–]	$= R^{1/2}\mathsf{F}_o$
g	[ms^{-2}]	gravitational acceleration
h	[m]	flow depth
h_e	[m]	end depth
H	[m]	energy head
L	[m]	length of bottom opening
$1/n$	[m$^{1/3}$s^{-1}]	roughness coefficient
P	[N]	pressure force
Q	[m^{3}s^{-1}]	discharge
R	[–]	$= Q_d/Q_o$ discharge ratio
S_f	[–]	friction slope
S_o	[–]	bottom slope
U	[ms^{-1}]	lateral outflow velocity
V	[ms^{-1}]	channel velocity
x	[m]	streamwise coordinate
X_s	[–]	$= (x/h_o)\mathsf{F}_o^{-0.8}$ dimensionless coordinate
y	[–]	$= h/D$ filling ratio
Y	[–]	height ratio
Y_q	[–]	$= R^{-2/3}Y_d$
Y_d	[–]	$= h_d/h_o$
ΔH_s	[m]	additional energy loss
ΔQ	[m^{3}s^{-1}]	lateral discharge
β	[–]	$= b/D$ width ratio
γ	[–]	profile parameter
Φ	[–]	$= y_o^{1/2}(D/\Delta L)\mathsf{F}_o$
ρ	[kgm^3]	density
ϕ	[–]	outflow angle
ξ_d	[–]	loss coefficient

Subscripts

a	average
d	downstream
e	end section
K	treatment station
L	lateral outflow
M	maximum
o	approach
s	surface profile

References

ATV (1978). Bauwerke der Ortsentwässerung (Structures of urban drainage). *Arbeitsblatt* A241. Abwassertechnische Vereinigung: St. Augustin [in German].

ATV (1994). Richtlinien für die hydraulische Dimensionierung und den Leistungsnachweis von Regenwasser-Entlastungsanlagen in Abwasserkanälen und -leitungen (Guidelines for the hydraulic design and the capacity proof of storm water outlets in sewers). *Arbeitsblatt* A111. Abwassertechnische Vereinigung: St. Augustin [in German].

Biggiero, V. (1969). Scaricatori di piena per fognature (Outlets for combined sewers). *Ingegneri* 10(11/12): 1–36 [in Italian].

Brunella, S., Hager, W.H., Minor, H.-E. (2002). Hydraulics of bottom rack intakes. *Journal of Hydraulic Engineering* 129(1): 2–10.

Chow, V.T. (1959). *Open channel hydraulics*. McGraw-Hill: New York.

Hager, W.H. (1985). Bodenöffnungen in Entlastungsanlagen von Kanalisationen (Bottom openings as discharge structures for sewers). *Gas-Wasser-Abwasser* 65(1): 15–23 [in German].

Hager, W.H. (1992). Vereinfachte Berechnung von Springüberfällen (Simplified computation of leaping weirs). *Gas-Wasser-Abwasser* 72(7): 469–475 [in German].

Hager, W.H., Minor, H.-E. (2003). Hydraulic design of bottom rack intakes. 30 *IAHR Congress* Thessaloniki D: 495–502.

Oliveto, G., Biggiero, V., Hager, W.H. (1997). Bottom outlet for sewers. *Journal of Irrigation and Drainage Engineering* 123(4): 246–252.

SIA (1980). Sonderbauwerke der Kanalisationstechnik (Special structures of sewer techniques). *SIA-Dokumentation* 40. Schweizerischer Ingenieur- und Architektenverein: Zürich [in German].

Taubmann, K.-C. (1972). Regenüberläufe (Overflows for combined sewer systems). *Gas-Wasser-Abwasser* 52(10): 297–308 [in German].

Appendix: Short History of Wastewater Hydraulics

Introduction

Sewage has been a significant problem for all civilizations mainly because of health aspects and drinking water supply. Ancient cities such as Babylon, Athens or Rome provided a sewer system to divert sewage. The *cloaca maxima* of ancient Rome was covered and followed the course of a small river to the sea. The Romans installed sewers also in Paris, Cologne and Bath, among other cities, which have been partly in use even during the 19th century. These sewers were poorly constructed and required considerable maintenance, which was often secured by slaves. Frontinus (40–103) provided a detailed look at water supply and sewer systems of ancient Rome.

Until the end of the middle ages, sewage was no topic even for large cities, which must have had terrible conditions of living in terms of cleanliness, hygiene and odour. Conditions definitely improved only once roads were paved and designed for sewage discharge. Diseases including cholera urged the city of London in 1848 to introduce the *Public Health Act*, a health authority. Such commissions were founded later in other countries marking the start of an orderly development of water supply and sewage drainage. Sewers as a closed profile were being used instead of open channels, and the toilet was developed. In Germany, the English engineer William Lindley (1808–1900) was asked to design the sewer system of Hamburg after the Great Fire of 1842. Other cities like Frankfurt, Stettin or Danzig followed by 1860 in the aftermath of cholera epidemics. Detailed reports on the sewerage systems of cities of Danzig (Wiebe 1865), Berlin (Hobrecht 1884) or Zurich (Weyl 1903) contain a large number of tables and details of special manholes. By the end of the 19th century, the relation between water quality and improvement of life quality was definitely realized (Olshausen 1899). Conditions on rural sites did not really improve, however, until the mid-20th century.

By the turn of the century, institutions dealing with water quality were founded in Berlin, Paris or later in Zurich and other cities. Water was no longer of private concern but controlled by specialized governmental agencies. The water qualities of the USA have received particular attention and cities such as New York, Chicago and Boston have had among the most modern sewers by 1900. W.P. Gerhard

W.H. Hager, *Wastewater Hydraulics*, 2nd ed., DOI 10.1007/978-3-642-11383-3, presented his 'Drainage works for the American building' by 1897 already. Sewage technology, i.e. the art of sewage drainage and treatment, was developed only in the 20th century. What was obvious by the turn to the 21st century for the developed countries has not been introduced worldwide, however. More than half of the total world population are estimated to have insufficient water quality and quantity, and it will require decades to satisfy this basic human need.

Early Developments

Georges BECHMANN's (1848–1927) book (1905) is one of the first of its kind and reflects the French development. It includes 28 chapters subdivided into hydrology, agricultural hydraulics, and urban hydraulics, the latter of which is described here. After an historical account on the French developments relating particularly to the efforts of Marie-Francois-Eugène BELGRAND (1810–1878) and Alfred DURAND-CLAYE (1842–1888), both water supply and sewage treatment are described. Bechmann's approach is descriptive, however, leaving the engineer without analytical design criteria. Note that France and England were the leading nations in wastewater technology in the 19th century (Fig. A1).

Certainly the most comprehensive work on sewage technology was published by August FRÜHLING (1846–1910) in 1910. As a professor at the Dresden Technical University, he gave a state-of-the-art of German practice around 1900. Over 700 pages on details of sewer system planning, construction, maintenance and cost are presented in part 1. Part 2 describes sewage treatment including self-purification, sedimentation basins, and biological treatment. The book contains numerous sketches and references, including a short history on sewers and sewage treatment.

Details on the minimum sewer slope, on the advantages of the egg-shaped sewer as compared to the circular section, on building inlets, on construction methods, on toilets and on optimum siphon shapes are discussed among many other items. Uniform flow formulae of Weisbach, Darcy and Bazin, Ganguillet and Kutter, and Knauff are listed, and the so-called short Kutter formula with an appropriate roughness coefficient was recommended for design purposes. The flow in partly-full

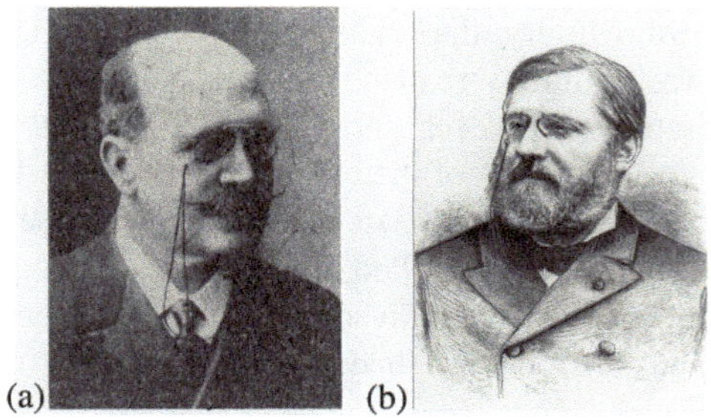

Fig. A1 Portraits of (**a**) Georges Bechmann (1848–1927) (*La Technique Sanitaire et Municipale* 22(5): 97), and (**b**) Alfred Durand-Claye (1842–1888) (Durand-Claye 1890)

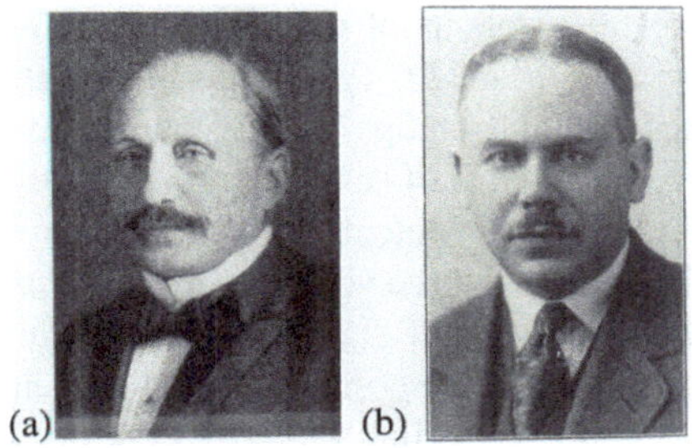

Fig. A2 Portraits of (**a**) August Frühling (1846–1910) (*Zentralblatt der Bauverwaltung* 30(64): 427), (**b**) G.M. Fair (1894–1970) (*Water Works Engineering* 88(5): 515)

sewers is computed by tables. A wide variety of sewer shapes was introduced, but their selection is not based on rational assumptions. Also, sideweirs as lateral outlet works of combined sewers are discussed, yet without any hydraulic approach. Frühling's book clearly marks the German leadership in wastewater technology from around 1900 (Fig. A2).

Special chapters are devoted to construction materials and static resistance of sewers due to road traffic. Construction methods involve both the open ditch and tunnelling in cities. Another chapter describes fabrication methods of sewer pipes and their laying in roads. Of particular interest here is the chapter on special manholes such as drops, junctions, bends and lateral outlets. Frühling recommended manholes with 50% benches for hygienic and working reasons. Optimum manhole geometries were described, including covers and access ladders. Even methods of flushing a sewer are highlighted, with a large variety of operational procedures. Additional chapters are devoted to building drainage, culverts and outlets into receiving waters. This immense amount of information gives a picture of the engineer Frühling. It is noteworthy that about 100 years ago, sewage engineering was not made by a computational approach, but by using books such as those discussed, adapting the optimum of various designs presented, and using experience from previous designs. From the present point of view, classic books thus do not favour a particular design, nor are standard structures recommended. The art of engineering was by far more important than a detailed design.

Karl Imhoff (1876–1965) published in (1907) his first *Taschenbuch*, a highly successful engineering guide comprising 20 pages and 16 tables. He recommended graphical solutions for sewer problems because this was considered sufficient in terms of accuracy and significance of results. Imhoff suggested that the Large Kutter formula be used for uniform flow computations, thereby designing sewers for full-flowing conditions. Imhoff's Taschenbuch has had many expansions, and over ten editions. The third edition published in 1922 comprised already 56 pages and 16 tables. Additional chapters as compared to the first edition include the determination of design discharge, the design of treatment stations including the so-called *Emscherbrunnen* and industrial treatment plants (Imhoff 1928).

Imhoff served also as an editor of *Stadtentwässerung* for the advance of German sewage technology (Brix et al. 1934). In the first volume the sewer schemes and treatment plants of most German cities are compiled. Volume 2 includes chapters on city drainage, sewage purification, receiving waters, and chemical analyses of sewage and sludge. Both volumes provide excellent overviews on the German technical state of sewage drainage and treatment in the early 1930s.

Eddy and Metcalf published the most popular US book in sewage engineering. The original three volumes of *American Sewerage Practice* was released in 1914 by Leonard Metcalf (1870–1926) and Harrison P. Eddy (1870–1937). Their *Sewerage and Sewage Disposal* was published in 1922 as a single volume textbook, with a second edition in 1930. A recent book *Wastewater engineering* including collection, treatment and disposal was edited by Clark and Ungersma (Metcalf and Eddy 1972). It includes 16 chapters on: (1) Historical developments, (2) Sewage flow rates, (3) Hydraulics of sewers, (4) Design of sewers, (5) Sewer appurtenances and special structures, (6) Pumping stations, (7) Wastewater characteristics, (8) Physical unit operations, (9) Chemical unit processes, (10) Biological unit processes, (11) to (13) Design of wastewater treatment facilities, (14) Advanced wastewater treatment, (15) Water-pollution control, and (16) Wastewater treatment studies. The almost 800 pages book also contains a number of appendices. The book is directed to design and numerous figures illustrate typical standard structures. The third edition of 1991 was significantly enlarged and contains the latest advancements. The book of Fair and Geyer (1954) may be considered complimentary to these of Metcalf and Eddy.

Armin Schoklitsch (1888–1969) in his *Wasserbau* (1930) presented an overview of water engineering including a summary on the knowledge of wastewater treatment. The circular sewer is considered as the standard section, whereas the egg-shaped sewer is recommended for conditions critical against sewage deposition. Its advantage is relatively small compared to the circular sewer, however, and not generally justified. The Small Kutter velocity formula is also recommended for design and filling curves for 19 different standardized sewer sections are provided. Manholes have to be installed at transitional locations. A sewer should be located at least 1.5 m below ground level. Recommended minimum and maximum velocities are 0.5 ms^{-1} and 6 ms^{-1}, because of deposition and abrasion, respectively. A sewer should be designed for just full-flow conditions. Schoklitsch's book includes also special manholes that have always at least 50% benches. Of particular importance are side overflows in combined sewers that may be extended with stormwater detention basins. Figure A3 shows an outlet with a baffle wall and a broad-crested weir. Junction manholes should be arranged with small intake angles. Also, the

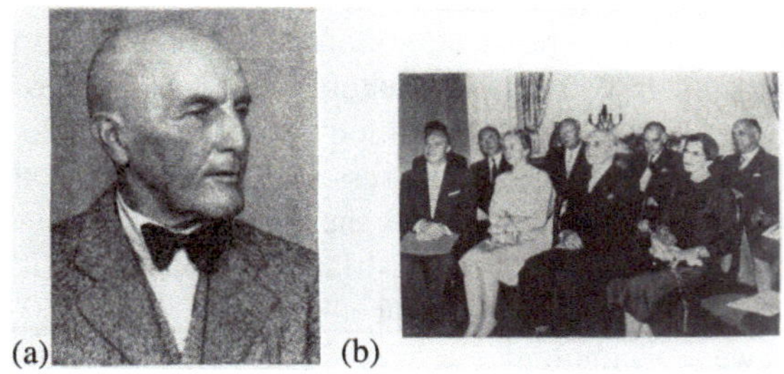

Fig. A3 Karl Imhoff (1876–1965) (**a**) around 1950 (*Engineering News-Record* 159(October 24): 24), (**b**) Award community (*Civil Engineering* 27(12): 884)

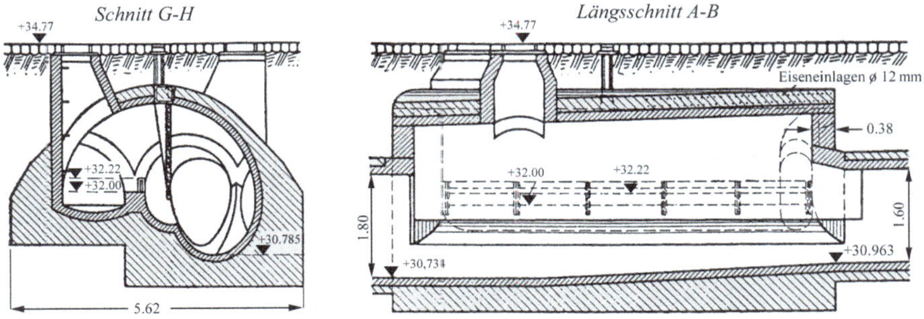

Fig. A4 Sideweir from combined sewer (Schoklitsch 1930)

downstream sewer should not submerge any of the branch sewers. A simple design rule requires all branch vertexes at equal elevation. Drop manholes can be designed either with cascades or almost as currently recommended (Chap. 15). Other works refer to inverted siphons and pumping plants (Fig. A4).

Wilhelm Geissler (1875–1937), a professor of sewage treatment at the University of Dresden presented an updated version of Frühling's contribution. His book (1933) is subdivided in sewers and sewage treatment, the latter of which is not further discussed here. On 200 pages, the fundamentals of sewers are described including details on sewer design and execution. Junction manholes, drop manholes, sideweirs, and sewer outlets to rivers receive special attention. A hydraulic design is often not available, and one does not really know why certain design assumptions are made in this and not the other way.

Erich Thormann, a head of drainage systems of the city of Berlin during World War II, aimed to standardize sewer sections. In his 1944 paper, the last of a paper series, 15 sections were discussed both from the hydraulic and the static points of view. His name is still common in Germany for his proposal to modify the upper branch of velocity and discharge filling curves. His proposal should account for the effect of air flow above water flow although the approach was never verified experimentally.

Chapter 7 on special structures introduces the standard manhole with 100% benches. Bend manholes are stated to need careful construction to produce acceptable flow conditions. A centerline radius of three sewer diameters was recommended. However, a hydraulic design for bends, junctions, drops and outlets is missing. Also, the Froude number as the basic hydraulic characteristic is introduced but not used.

Modern Developments

Arnold Hörler (1903–1995), a notable Swiss engineer and professor at ETH Zurich, wrote a widely used guide to sewer design (Hörler 1966). The booklet includes the *hydrologic* data of Switzerland with procedures for the design discharge, the hydraulic design bases including recommendation of the Strickler flow formula, critical flow, hydraulic jumps in sewers, backwater curves, sideweirs, bottom openings, special manholes and drop manholes, static requirements for sewers, and the design of rainwater basins for combined sewers. Hörler's guideline is still popular because of its conciseness and relevance (Fig. A5a).

Peter Ackers (*1924) of the UK wrote outstanding papers on sewage engineering, including a paper on the temporal modification of sewers (Ackers and Holmes 1964). The main results were: (1) Sliming of sewers varies considerably with location and hydraulic conditions and an equilibrium slime layer is established within a short time, (2) For high velocity regions, the slime thickness is small, and (3) Roughness values are recommended for sewers with a developed slime layer. Ackers' contributions to sewer sideweirs are also notable. His 1957 paper is directed mainly to the hydraulic design, including recommendations regarding the so-called dip-plates, and tapering effects on the free surface profile. Ackers (1967) realized that a sideweir cannot be used for both hydraulic discharge regulation *and* separation of undissolved matter and sewage. Ackers et al. (1967) evaluated various sideweir designs and recommended the weir with a high crest as the overall optimum in terms of discharge capacity and environmental performance (Fig. A5b).

The Swiss Society of Engineers and Architects (SIA) contributed a manual on sewer design (SIA 1977). The guideline is composed of six chapters including: (1) Terms, (2) Project bases, (3) Hydraulic and structural designs, (4) Material, (5) Execution, (6) Supply and guarantee. Appendices relating to standard sewer structures complete the manual. SIA (1985) contributed to sewer hydraulics with an additional documentation. It includes junction structures, drop manholes, sideweirs and bottom openings, as well as inverted siphons. Most of the recommendations are reviewed in the present work, and basic assumptions were verified in experimental models. The documentation can be regarded as one of the first of its kind, because a number of problems are solved with a hydraulic design.

Ben Chie Yen (1935–2001) was a professor of civil engineering at the University of Illinois, Urbana IL, after having submitted a PhD thesis to the University of Iowa in 1965, and having been a research associate at Princeton University from 1960 to 1964 (Fig. A6a). Yen had worked in general hydraulics with a particular devotion to wastewater hydraulics, notably with his book Urban stormwater hydraulics and hydrology (1982a). He was certainly influenced by his countryman Ven Te Chow (1919–1981), a master in hydraulics and hydrology and also a professor at

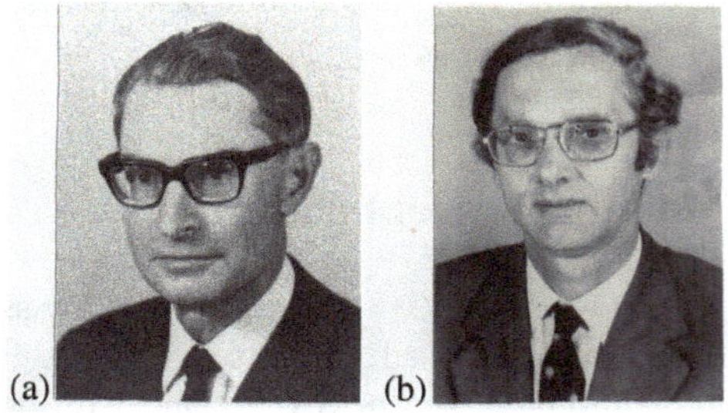

Fig. A5 (**a**) Arnold Hörler (1903–1995) (*Schweizerische Wasserwirtschaft* **66**(1/2): 92), (**b**) Peter Ackers (*1924) around 1960 (private communication)

(a)

(b)

Fig. A6 (**a**) Ben Chie Yen (1935–2001) (Marsalek (2002), (**b**) Nallamuthu Rajaratnam (*1935) ~ 1970 (private communication)

University of Illinois. At the same time, Yen (1982b) presented his Urban stormwater quality book. During the 22nd IAHR Congress held in Lausanne, Yen (1987) edited another conference on Topics in urban drainage hydraulics and hydrology. Another similar conference was held in Italy (Cao et al. 1993). The hydraulics community met Ben during the 2001 Beijing IAHR Congress for the last time, because he passed away soon after, leaving a great gap (Chiu 2002).

Nallamuthu Rajaratnam (*1935) made his Master degree in 1958 at University of Madras, India, and his PhD thesis at the Indian Institute of Science, Bangalore, India (Fig. A6b). From 1963, Rajaratnam was associated to the University of Alberta, Edmonton, Canada, first as a post-doctoral Fellow, from 1967 as an Associate Professor and from 1971 as professor. He has mainly conducted research on hydraulic jumps and energy dissipators from 1965 to 1980, then continued with problems in jet flow, erosion and scour until around 1990, and finally contributed significantly to stepped spillways and wastewater hydraulics, notably by analyzing drop manholes and jet diffusion. He was also active in the design of fishways. He was awarded the ASCE Hydraulic Structures Medal in 1998, and the ASCE Hunter Rouse Hydraulic Engineering Lecture Award in 1999. Rajaratnam, though retired for years, continues in research and particularly loves to collaborate with young people.

One of the first textbooks in wastewater hydraulics was presented by Benefield et al. (1984). This 200 pages book treats the topics: (1) Flow in pipes, (2) Multiport diffuser outfalls, (3) Flow in open channels, (4) Flow measurement and hydraulic control points, (5) Pumps, and (6) Design examples. This book has been successfully used in the US. Typical structures of sewers are not treated, however. Appendices include several computer programs.

A more recent book was published by Casey (1992), a professor of University College, Dublin (Ireland). Totalling more than 270 pages, this work includes chapters on: (1) Fluid properties, (2) Fluid flow, (3) Steady flow in pipes, (4) Flow in pipe manifolds, (5) Steady flow in pipe networks, (6) Unsteady flow in pipes, (7) Steady flow in open channels, (8) Open channel flow measurement structures, (9) Dimensional analysis, (10) Unsteady flow in open channels, and (11) Pumping installations. Computer algorithms are available in an appendix. This book provides an overview on water supply and wastewater hydraulics, without going much into the details of civil engineering structures. It can be regarded as a useful text for environmental engineers.

The American Society of Civil Engineers (ASCE 1992) has published a noteworthy 700 pages book on sewers with a large international distribution. The book comprises sections on: (1) Urban stormwater management, (2) Financial, legal and regulatory concerns, (3) Surveys and investigations, (4) Design concept and master planning, (5) Hydrology and water quality, (6) Storm drainage hydraulics, (7) Computer modelling, (8) Drainage conveyances, (9) Special structures, (10) Combined sewer systems, (11) Stormwater impoundments, (12) Stormwater managements, (13) Materials, (14) Structural requirements, (15) Contract documents, and (16) Construction surveys. The chapter on sewer hydraulics is comparable to an undergraduate text book, with a limited account for sewer profiles. The governing formulae for uniform flow, for example, include those of Manning-Strickler and Hazen-Williams without recourse to the friction coefficient of Colebrook-White. Chapter 9 on special structures has an excellent section on erosion and sedimentation. In addition, various standard energy dissipators and drop structures are reviewed. Several structures such as sideweirs or side channels are not considered, however. Hydraulic guidelines of special structures have thus not really been included in the engineering design procedures. The ASCE book has a significant international impact.

Paoletti (1997) edited a notable 900 pages book on the Italian practice of sewer systems. The book comprises twenty one chapters: (1) Introduction, (2) Legal aspects, (3) Design criteria, (4) Basic information, (5) Design of dry weather discharge, (6) Storm rainfall, (7) Sewage quality, (8) Maximum stormwater discharge, (9) Mathematical models for urban drainage, (10) Control of lateral outfalls, (11) Types and materials for sewers, (12) Hydraulic design of sewers, (13) Statics of sewers, (14) Standard sewer structures, (15) Tunnels and culverts, (16) Outlet works, (17) Stormwater outlets, (18) Storage basins, (19) Energy dissipators, (20) Pumping stations, and (21) Discharge measurement in sewers. This manual has updated design information and can be regarded as a useful support for sewage engineers as it takes a definite step into the solution of urban drainage problems of the 21st century.

The *Abwassertechnische Vereinigung* (ATV) of German wastewater authorities, the current Deutsche Wasser- und Abwasservereinigung DWA, has published an impressive handbook series, including ATV (1996). The 660 pages book is subdivided into ten chapters, including: (1) Submission of sewerage design, (2) Design bases, (3) Construction methods, (4) Management and maintenance, (5) Building drainage, (6) Public toilets, (7) Pumping works, (8) Quality engineering, (9) Financing of sewers, and (10) Cost of sewers. As for the ASCE-Manual, the hydraulic design bases are presented to a minimum. The ATV handbook contains a large number of standard manholes, including bend, junction, and drop manholes. ATV guidelines are currently used as a basis for the European guidelines on sewage treatment. ATV includes also a number of subsections that elaborate guidelines to all topics of sewage drainage and wastewater treatment. ATV had 1998 its 50th anniversary, and a special series to the history of sewage technology was published. The contributions of Imhoff (1998), Sickert (1998), and Kieslinger (1998) are of special relevance in the present context.

References

Ackers, P. (1957). A theoretical consideration of sideweirs as storm water overflows. *Proc. Institution of Civil Engineers* 6: 250–269; 7: 180–184.

Ackers, P. (1967). The hydraulics of storm overflows. *Surface water and storm sewage*: 27–44. J. Pickford, ed. Loughborough University: Loughborough UK.

Ackers, P., Holmes, D.W. (1964). Effects of use on the hydraulic resistance of drainage conduits. *Proc. Institution of Civil Engineers* 28: 339–360; 34: 219–230.

Ackers, P., Brewer, A.J., Birkbeck, A.E., Gameson, A.L.H. (1967). *Storm overflow performance studies using crude sewage.* Storm sewage overflows: 63–77. Institution of Civil Engineers: London UK.

ASCE (1992). *Design and construction of urban stormwater management systems.* ASCE Manual and Report of Engineering Practise 77. ASCE: New York.

ATV (1996). *Bau und Betrieb der Kanalisation* (Construction und management of the sewerage system). Ernst & Sohn: Berlin [in German].

Bechmann, G. (1905). *Hydraulique agricole et urbaine* (Agricultural and urban hydraulics). Béranger: Paris [in French].

Benefield, L.D., Judkins jr., J.F., Parr, A.D. (1984). *Treatment plant hydraulics for environmental engineers.* Prentice-Hall: Englewood Cliffs NJ.

Brix, J., Imhoff, K., Weldert, R. (1934). *Die Stadtentwässerung in Deutschland* (City drainage in Germany). G. Fischer: Jena [in German].

Cao, C., Yen, B.C., Benedini, M., eds. (1993). *Urban storm drainage.* Proc. US-Italy bilateral seminar. Water Resources Publications: Highlands Ranch CO.

Casey, T.J. (1992). *Water and wastewater engineering hydraulics.* Oxford University Press: Oxford NY.

Chiu, C.-L. (2002). Ben Chie Yen (1935–2001): A distinguished leader in hydraulic engineering and hydrology. *Journal of Hydraulic Engineering* 128(7): 652–655.

Durand-Claye, A. (1890). *Hydraulique agricole et génie rural.* Octave Doin: Paris [in French].

Fair, G.M., Geyer, J.C. (1954). *Water supply and wastewater disposal.* J. Wiley & Sons: New York.

Frühling, A. (1910). *Die Entwässerung der Städte* (Drainage of cities). Der Wasserbau, in Handbuch der Ingenieurwissenschaften 4, ed. 4. Engelmann: Leipzig [in German].

Geissler, W. (1933). *Kanalisation und Abwasserreinigung* (Sewers and wastewater purification). Springer: Berlin [in German].

Hobrecht, J. (1884). *Die Canalisation von Berlin* (The sewerage system of Berlin). Ernst & Korn: Berlin [in German].

Hörler, A. (1966). *Kanalisation* (Canalisation). Separate print of *Ingenieur-Handbuch* 2 [in German].

Imhoff, K. (1907). *Taschenbuch für Kanalisations-Ingenieure* (Pocket-guide for sewerage engineers). R. Oldenbourg: München und Berlin [in German].

Imhoff, K. (1928). *Taschenbuch der Stadtentwässerung* (Pocket-guide for city drainage). R. Oldenbourg: Berlin [in German].

Imhoff, K.R. (1998). Geschichte der Abwasserentsorgung (History of sewage treatment). *Korrespondenz Abwasser* 45(1): 32–38 [in German].

Kieslinger, R.M. (1998). 50 Jahre ATV (50 years of ATV). *Korrespondenz Abwasser* 45(5): 824–834; 45(6): 1062–1069 [in German].

Marsalek, J. (2002). Prof. Ben Chie Yen. *IAHR Newsletter* 40(1): 9.

Metcalf & Eddy (1972). *Wastewater engineering.* McGraw-Hill: New York.

Olshausen, J. (1899). Ein halbes Jahrhundert der Sanierung 1850–1900 (Half a century of sanitation). *Gesundheits-Ingenieur* 22(11): 175–177; 22(12): 194–197; 22(13): 213–214 [in German].

Paoletti, A., ed. (1997). *Sistemi di fognatura* (Sewer systems). Hoepli: Milano [in Italian].

SIA (1977). *Kanalisation* (Sewerage). SIA *Norm* 190. SIA: Zurich, Switzerland [in German].

SIA (1985). *Sonderbauwerke der Kanalisationstechnik* 1 (Special sewer structures 1). Documentation 40. SIA: Zurich, Switzerland [in German].

Sickert, E. (1998). Kanalisation im Wandel der Zeit (Sewers in the change of time). *Korrespondenz Abwasser* 45(2): 220–246 [in German].

Schoklitsch, A. (1930). *Der Wasserbau* (Hydraulic structures). Springer: Wien [in German]. English translation by S. Shulits in 1937.

Thormann, E. (1944). Füllhöhenkurven von Entwässerungsleitungen (Filling curves for drainage pipes). *Gesundheitsingenieur* 67(2): 35–47 [in German].

Weyl, T. (1903). *Die Assanierung von Zürich* (The restoration of Zurich). Engelmann: Leipzig [in German].

Wiebe, E. (1865). *Die Reinigung und Entwässerung der Stadt Danzig* (The purification and drainage of the city of Danzig). Ernst & Korn: Berlin [in German].

Yen, B.C., ed. (1982a). *Urban stormwater hydraulics and hydrology*. Proc. 2nd Intl. Conference. Water Resources Publications: Littleton CO.

Yen, B.C., ed. (1982b). Urban stormwater quality, management and planning. Proc. 2nd Intl. Conference. Water Resources Publications: Littleton CO.

Yen, B.C., ed. (1987). Topics in urban drainage hydraulics and hydrology. Proc. 6th Intl. Conf. *Urban storm drainage*, Lausanne. IAHR: Delft NL.

Author Index

Note: Citations in text are written standard, those in references *italic*.

Züllig, H., 363–364, *378*

Subject Index

T

Kolibri 360 GmbH
Ferdinand-Jühlke-Straße 7
99095 Erfurt, Deutschland
produktsicherheit@kolibri360.de

Lucien F. Trueb
Günther Ramm · Peter Wenzig

Electrifying the Wristwatch

Lucien F. Trueb
Günther Ramm · Peter Wenzig

Electrifying the Wristwatch

Schiffer Publishing Ltd
4880 Lower Valley Road • Atglen, PA 19310

edited by Christian Pfeiffer-Belli

Library of Congress Control Number: 2013937222

Photography: Photo Design Rohner, Munich; Günther Ramm, Brunswick; Peter Wenzig, Munich Manufacturer Photos

Type set in Optima LT Std

ISBN: 978-0-7643-4304-9
Printed in China

Published by Schiffer Publishing, Ltd.
4880 Lower Valley Road
Atglen, PA 19310
Phone: (610) 593-1777; Fax: (610) 593-2002
E-mail: Info@schifferbooks.com

In Europe, Schiffer books are distributed by
Bushwood Books
6 Marksbury Ave.
Kew Gardens
Surrey TW9 4JF England
Phone: 44 (0) 20 8392 8585; Fax: 44 (0) 20 8392 9876
E-mail: info@bushwoodbooks.co.uk
Website: www.bushwoodbooks.co.uk

Contents

9. QUARTZ MOVEMENTS FROM GERMANY

10. QUARTZ MOVEMENTS FROM FRANCE

11. QUARTZ MOVEMENTS FROM BRITAIN

12. QUARTZ MOVEMENTS FROM JAPAN

Foreword

The Authors

The aim of this book is to retrace, in words and pictures, the technical revolution that fundamentally changed and decimated the watch industry – mostly in the Old World – from about 1970 to 1985. What emerged from the ruins was a totally different industry with new products and new centers of gravity, mostly in Asia.

Until the early 1950s, wristwatches and pocket watches necessarily had mechanical movements. They were manufactured in Switzerland, France, Germany, the USSR, the United States, Japan, and Hong Kong. Half a century later, the proportion of mechanical watches has shrunk to one half of a percent (the luxury types generate huge sales though, being very expensive). The quartz wristwatch reigns supreme, mostly with analog displays, i.e. with hands. Sports watches, some fashion watches and the very cheapest timepieces sold in the Third World countries are equipped with a digital liquid crystal display.

One has to keep in mind that the geography of the watch industry too was totally changed by the quartz revolution. Nowadays, some 96 percent of the global watch production comes from Asia, China being the uncontested leader. Europe (where Switzerland is the dominating force) has become nearly irrelevant in quantitative terms, the total production being of the order of 30 million watches a year versus 1.2 billion made in Asia. However, Switzerland wins out by far in terms of cash flow. Central Europe (i.e. Switzerland, with Germany far behind, France being practically out) is specialized in luxury watches, the industry having regrouped at the tip of the price pyramid after the quartz crisis. In the United States, once a top rate horological power, mass production of watches is long gone. A few master watchmakers still create small series of beautifully crafted mechanical timepieces, but watches with American brand names are mostly manufactured in the Far East. The exceptions are two well known brands that belong to the Swiss and the Japanese, respectively.

What is commonly called the quartz revolution led to new classes of watches that are much more accurate and much less costly than the absolutely best mechanical chronometers, the technology of which was fully established around 1800. In many respects, the limits imposed by physics had already been reached at that time. This is the way things stayed for 150 years. The watch was practically the last industrially made product to make the transition from a purely mechanical device to an electromechanical device: this happened in the 1950s. Ten years later, electronics gained a foothold with the tuning fork watch: it was controlled by a single transistor. Finally, the integrated circuit made quartz wristwatches technically possible in the late sixties; they became affordable in the early 1970s. The general trends to electrification and electronics had started much earlier, but enormous efforts had to be made in order to "squeeze" existing clock movements into a wristwatch case.

We chose 1950 to 1985 as the focal point of our research on electromechanical and electronic watches, but one must be aware of the fact that these new products were the crowning achievement of a long process that started with big clocks. The earliest electromechanical pendulum clock was built in 1841 while the first quartz clock was put in operation in 1929. All the early developments in the field of electromechanical and electronic wristwatches followed a straightforward evolutionary pattern. In some cases, mass produced truly novel timepieces actually hit the marketplace, but disappeared after a relatively short time. The classical example was the tuning fork watch that had a product life of just 17 years.

In view of those historical facts, we decided to split the book: Part I, with the chapters 1 to 7, covers electrical clocks and watches up to the very first quartz wristwatch movement prototypes of 1967. Part II, with the chapters 8 to 15, covers the worldwide development of quartz wristwatch calibers up to 1990. At that time, the technology was almost totally standardized: there have been no significant further developments in the past twenty years. Even technically fascinating niche products, such as thermocompensated and radio-controlled movements, have now been manufactured on an industrial base for more than two decades.

In the pioneering days, many variants of the quartz wristwatch movement were put forward. At the end of the day, only one of them survived for good – well over a billion pieces of it are produced year after year, with either analog or digital displays. Combining a silver oxide battery with a photolithographic 32 kHz quartz resonator, a 15-step binary CMOS dividing circuit (often with digital setting) and a two pole Lavet-type stepping motor (or a digital liquid crystal display) resulted in a winner that near-monopolized the field. And no alternative is in view.

The future means space and atoms. GPS-receivers for the wrist have been available for some time, so are watches controlled by the atomic time transmitted by satellites. The rubidium clock with laser diode optical pumping has been miniaturized to the point that it fits small GPS receivers and could even be squeezed into a wristwatch case. The problem is the energy supply: the smallest rubidium clock still needs a thousand times more power than a standard quartz movement. Tremendous efforts must still be made in order to achieve a battery life of more than a day. Inductive charging of watch batteries is a possibility: people are already used to recharging their smartphones and tablet computers on a pad overnight.

Amusing problems came up when the integration of a cell phone into a watch was tried. Those projects all failed for the simple reason that the power requirements of such a device are simply too high, quite aside from their unpractical size. The issue could have been easily predicted: the phone was not integrated with the watch, but the watch went into the phone. Nowadays, millions of people walk around who do not wear a watch: instead, they just check their cell phone when they want to know the time: the watch went into the phone. This is a kind of renaissance of the pocket watch.

Our aim was to retrace the evolutionary and revolutionary steps that led to the present day technical concept of the quartz wristwatch – a process subject to Darwinian selection with a fair share of failures and dead ends. As already stated, we purposely limited our horizon to the early nineties, even though it was necessary to go slightly beyond this limit in a few instances. Setting a lower limit on the other hand could not be justified both historically and technically.

As far as quartz was concerned, we had to start with Horton and Morrison and the year 1927 in order to open the road to the quartz wristwatch, the intermediate steps being quartz wall clocks, table clocks, marine chronometers, deck watches, and pocket chronometers. The wristwatch had not been just waiting for quartz: its electrification started with the contact controlled and later transistor controlled electromechanical watch. With some hindsight, this development proceeded in an absolutely logical, yet unforeseeable way. The electromechanical watch first without, then with transistor as well as the transistorized tuning fork watch were logical steps on the way to the quartz wristwatch that made its commercial debut in 1969-1970.

By the early eighties, quartz technology had been optimized nearly to the limits imposed by physics. However, thanks to ingenious tricks, the power consumption of the movements was further reduced to a small fraction of a microampere. From then on, the technical design had to be adapted to the imperatives of fully automated production and assembly. This led to dramatic price reductions. In the early seventies, a quartz movement cost several hundred dollars. Forty years later, the functionally similar article can be bought in China for 40 cents – including a short-lived battery with very little silver oxide.

We have tried hard to describe in great detail the early electromechanical and electronic movements from the important and less important horological centers all over the world. This made a lot of travelling necessary in order to visit surviving plants and interview veterans of the wristwatch electrification and of the quartz revolution in Switzerland, Germany, France, the Czech Republic, Belarus, Japan and the United States. As to South Korea, we only succeeded in the last minute to obtain pertinent information on the early phases of their watch industry that produced vast quantities of digital LCD watches in the eighties and nineties.

We have knowingly left out Hong Kong, mainland China, and Taiwan. First of all, nothing really original ever came from those quarters, even though enormous efforts were made to industrialize the know-how created in North America, Europe, and Japan on a huge scale, which permitted spectacular price reductions. Another point is that the watch industry in Hong Kong and China consists of an extremely complex convolution of hundreds of companies that change names, and are bought and sold in rather obscure ways. They do not compete with each other in terms of technology (which is always the same), but strictly on price. A fraction of a cent per movement can make or break a big sale – and may threaten the very existence of the company.

Researching the sometimes tortuous way that led from the electromechanical movement to the quartz watch via the tuning fork watch was not without its frustrations. In quite a few instances no movements or documents survived. Japanese companies, however, have an excellent sense of history – at least the ones that are still in the watch business. Quite the opposite is true for American companies: their once promising diversification into the watch business ended as a flop which nobody really wants to remember.

Needless to say that the many interviews and talks with old and new friends, as well as plant and museum visits in Switzerland, Germany, France, Eastern Europe, and Japan were a revealing and fun experience both socially and technically. We also learned that time is much more than a physical dimension that man learned to measure ever more accurately. On a house in the Black Forest we discovered an unforgettable piece of wisdom. The inscription said "Time is the very best teacher. Unfortunately it has the bad habit of killing all its pupils."

Lucien F. Trueb
Günther Ramm
Peter Wenzig

Preface

Estelle Fallet

The electronic wristwatch is part of our everyday life. Most of us take its precision, dependability and long service life for granted. For many years now, 99.5 percent of the watches manufactured all over the world have belonged to this type. But the enormous amounts of highly creative work that went into the development of this entirely new kind of watch are almost forgotten. Remarkably enough, two research laboratories in Switzerland and in Japan both had functioning prototypes in the late summer of 1967. They had worked in total independence, hardly being aware of each other's existence. Other labs in Switzerland, Japan Germany, France, the United States, and the Soviet Union were not far behind. The present book retraces in great detail the technical and human history of those developments that bore fruit despite totally different cultural environments. The basic recipe for success was a total mastery of the most advanced electrical and electronic technologies. The old art of watchmaking was left far behind.

Less than half a century elapsed between the earliest R&D efforts for a quartz frequency standard (1920/1930) and commercial "atomic" clocks based on an extremely stable oscillation of the cesium atom (1970/1975). This exemplifies the steady acceleration of technological progress one observes in the discipline of time measurement and timekeeping since the 19th century. Parallel to it, miniaturization made enormous strides, reducing the volume of quartz movements from cubic meters to cubic centimeters.

The present book retraces the research that was conducted in European, American, and Japanese laboratories – amazing success stories but also painful dead ends that led to the downfall of once powerful companies. The confrontation of new devices with market forces resulted in a truly Darwinian selection: at the end of the day, only one type of quartz movement was left, which all the manufacturers adopted. This design was fully established in 1980 and has not basically changed ever since. Just minor adjustments had to be made in order to adapt it to ever more sophisticated, fully automated assembly lines. The evolution lies in the realm of new applications, such as GPS and mobile telephones. The authors tell a tale of modern times, more particularly of the linkage of the once totally separate worlds of watchmakers and electronic engineers. Nearly all of the purely electronic companies that tried to diversify into watches failed abysmally. On the other hand, traditional watch manufacturers successfully adapted the new technologies to their product line – but not without severe hardships.

Furthermore, the scientists and engineers responsible for research and technical development had to keep in mind that the new products would have to be mass produced on automated assembly lines at very low cost. Mankind's strive for an immutable, universal, and reproducible frequency and time standard succeeded so well that having the exact time within a second a day is within anybody's reach anywhere in the world. Thermo-compensated or radio-controlled watches are very reasonably priced too and do much better.

The authors of the present book took great pains in carefully defining the physics and the materials science aspects of the quartz watch in everyday language. Their book has thus become an essential, interdisciplinary tool. Countless human interest stories and anecdotes retrace in a pleasant way the very special circumstances that led to the conquest of near-absolute precision at everybody's wrist. A single research scientist or engineer could never have achieved this. Both in Switzerland and in Japan, big teams of highly specialized and motivated members were put together in the early sixties to achieve the common goal. Long before the first prototype started to tick, many important issues pertaining to industrial production and automation had to be addressed, even though their impact would only be felt in the following decades. The "quartz revolution" saw the emergence of a few pioneer brands such as Seiko, Lip, Hamilton, Bulova, Longines, Girard-Perregaux, and Omega: they were soon followed by the whole gamut of watch brands dominating the horological scene in Switzerland and all over the world.

The very first electrical pendulum clock was built in 1841 and combined the watchmaker's art with the new tenets of electrical engineering. In the following decades, the balance wheel-controlled, battery-powered table clock and alarm clock appeared. The first electromechanical wristwatches were launched in the early 1950s and at last put electricity on people's wrist. It took another ten years to replace the slow moving, balance wheel by the hum of a transistorized tuning fork. The integrated circuit with hundreds of transistors on a single silicon chip led the way to the quartz wristwatch.

A few milestones are worth recalling: in 1948, the Neuchâtel observatory purchased one of the very first quartz clocks built in Switzerland. Less than two decades later, in 1967, prototype quartz wristwatch movements developed by the Centre Electronique Horloger (CEH) in Neuchâtel obtained the first ten ranks at the Observatory's chronometer contest. Technically quite different prototypes sent in by Seiko obtained the next ranks. Further development led to the Astron, world's first commercial quartz wristwatch; it was launched in Tokyo at the end of 1969. On the Swiss side, the CEH prototypes evolved into the Beta 21, small series of which were available in 1970.

In collaboration with Ebauches SA, the International Watch Museum in La Chaux-de-Fonds organized a special exhibit as early as 1978 under the name of Chronatome. The theme was electrical and electronic watchmaking, the history of which was still being written. Since that time, many articles, but only a precious few books, on the same subject were published. We had to wait until 2011 for the comprehensive, profusely illustrated overall view of the watch electrification process written by Lucien Trueb. He was assisted by Günther Ramm and Peter Wenzig, the foremost collectors of electronic watches in Europe who supplied many of the pictures. The reader of *Electrifying the Wristwatch* is thus provided with an extremely thorough, brand to brand history of the electromechanical and electronic watch from 1967 to the early nineties. But past achievements naturally lead to new perspectives; physicists and engineers are presently working at miniaturizing several variants of the atomic clock without sacrificing too much of the millionth of a millionth of a second per day that is routinely available with standard clocks of that type.

Estelle Fallet

Estelle Fallet is the Conservator of the watch, enamel, jewelry and miniatures collection of the Art and History Museum in Geneva.

PART I

130 Years of Electrical Clocks and Watches: 1840-1970

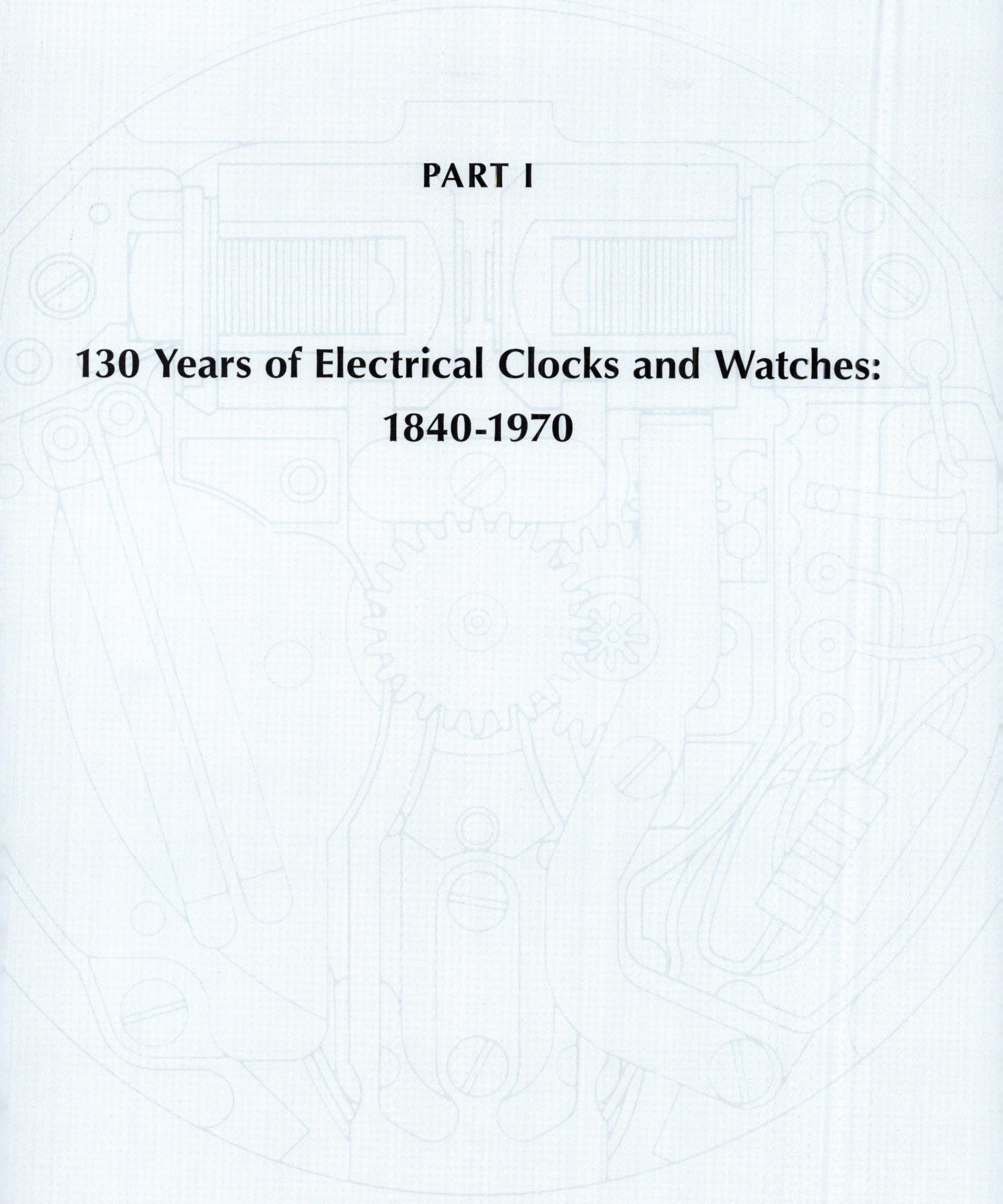

1. The First Electrical Clocks

One must strictly differentiate between the terms "electrical" and "electronic." Regrettably, various definitions exist for the latter. In the latter part of the 19th century, a kind of electronic technology existed. It was based on *passive* (i.e. non-amplifying) components that actually belonged to classical electrical technology. They included coils, capacitors (the Leyden jar being an "antique" prototype), resistors (that are given by any non-superconducting electrical conductor), solid state rectifiers, as well as many kinds of switches and relays.

In the early days of wireless telegraphy, the semiconducting properties of iron oxide and certain minerals such as pyrite and galena were put to good use. But "real" electronics started for good in 1906, with the invention of the electronic tube by the American physicist Lee de Forest (1873-1961). Just like the transistor that followed four decades later, the electronic tube was an active component. Its amplifying and very fast switching properties started the electronic age.

In this first chapter, we limit ourselves to electrical clocks without active electronic components. The first timekeepers of this kind were "electrified" mechanical clocks. Their energy source was not the traditional weight or mainspring, but a battery that powered an electro-mechanical device.

The challenge of developing the very first electrical clocks consisted in using the electromagnetic force for periodically supplying enough energy to the pendulum (or balance wheel) so that it would constantly oscillate with the same amplitude and entrain the entire movement. This was equivalent to a paradigm change, as the pendulum now acted as a motor. There are reports that in 1685, the Polish mathematician Adam Kochanski designed a magnetic clock. Such a device would never work as long as only permanent magnets were available. What was needed was a magnet that could be switched on and off. Hans Christian Oersted (1777-1851) rose to that challenge in 1820 by designing a wire coil generating a magnetic field when it carried an electrical current.

Alexander Bain

In 1840, the Scottish watchmaker Alexander Bain (1811-1877) figured out a way to use the electromagnet in a clock. He was granted a patent for this revolutionary innovation in the following year. Bain first attached an arc-shaped permanent magnet to the pendulum of his clock. At the end of its travel, the magnet would swing into the bore of a coil of insulated copper wire. The coil was energized just then, which set up a repulsive magnetic force. Thus, the pendulum was given a push at each period or half-period of the pendulum (depending on the system). The coil was controlled by a contact attached to the rod of the pendulum.

As mentioned above, the clockwork was now driven by the pendulum acting both as an oscillator and a motor. It was not far fetched to transfer this concept to the balance wheel of clocks and watches. By periodically applying a magnetic force, the balance wheel became an electrical motor – a balance wheel motor. Starting from the 1920s it would be extensively used in miniature clocks, car clocks, and alarm clocks, as well as in the electromechanical wristwatches of the 1950s and 1960s.

The balance wheel motor was also invented by Alexander Bain, who patented it in 1847. There are several variants, as the magnetic force of a fixed coil can act either on a magnet or a piece of soft iron on the balance wheel. Conversely, a coil mounted on the balance wheel (i.e. a moving coil) can interact with stationary magnets or pieces of soft iron. It is not known whether Bain actually built a clock equipped with a balance wheel motor. Yet it is remarkable that Bain managed to overcome the "central dogma" of the watchmaker, i.e. the belief that the pendulum or balance wheel should swing as freely as possible so that it would oscillate with a constant frequency. For the classical watchmaker, using the oscillator as a motor was heresy. But Bain demonstrated in a convincing manner that this kind of "abuse" does not necessarily compromise the daily rate of a timekeeper.

Matthias Hipp

Matthias Hipp (1813-1893) was another pioneer of the electromechanical pendulum clock. This ingenious watchmaker of German origin spent most of his life in Switzerland. With A. de Peyer, he was the co-founder of an important company that would later be called Favag in Neuchâtel. It is now part of the Ascom Group. Hipp started with the production of telegraph and telephone equipment. In the electrical clock he built in 1860, a piece of soft iron was attached to the bottom of the pendulum rod. It swung back and forth over the pole shoe of an electromagnet that was switched on by a system of cams and contacts activated by the pendulum rod.

1.1. Electromagnetic regulator. Alexander Bain, London, around 1845

1.2. Wall clock with electromagnetic torsion pendulum. Fabrik elektrischer Uhren und Apparate, Hanau, around 1885

1.3. Rocking motor clock. Möller, Berlin, around 1910

1.4. Clock with electromagnetic balance wheel. Eureka, London, 1910

The ingenious part of it was that the pendulum swung freely most of the time, the electromagnet being switched on only after the pendulum had reached minimum amplitude. In this case, the pendulum was attracted by the magnet until it had nearly reached the lowest point of its trajectory. The magnet was switched off just then, so that the pendulum swung freely to the other end of its trajectory and then swung back.

Maurice Favre-Bulle

Maurice Favre-Bulle (1870-1954) was trained as a watchmaker at the school of horology in Besançon from 1885 to 1888. With his brother, he took over the very successful Favre-Heinrich watch factory that his father had founded. Bulle moved from Besançon to Paris in 1907 and kept perfecting various inventions. After World War I broke out, he was ordered by the military to join the mechanical laboratory of the university, where he designed pilot watches, radio gear, and speedometers. After the war was over, he founded the Bulle-Clock & Cie., which was renamed Compagnie Générale des Appareils Horo-Electriques in 1920. Maurice Favre-Bulle remained in charge of this very successful company until he died in 1952. Bulle had no children, so his company was liquidated in 1955.

Marcel Moulin (1881-1914) was the first scientist in Favre-Bulle's employ. This brilliant physicist became a professor at the University of Besançon when he was just 29 years old and was also put in charge of the French Institute of Chronometry. This is when he started working with Maurice Favre-Bulle and designed the electrical part of the Bulle-Clock. A prototype worked in early 1914, but when World War I broke out, reserve officer Moulin was immediately called to arms. He was killed in action in September 1914 at the head of his regiment while storming enemy lines on the Marne River.

Right after the end of the war, Maurice Favre-Bulle hired Marius Lavet (1891-1980), another very brilliant inventor. Lavet's first assignment was writing patents concerning the Bulle-Clock on the base of Moulin's notebooks. This was a rather frustrating job as those patents were the intellectual property of Favre-Bulle and Moulin's widow. Lavet had studied electrical engineering at the Ecole Supérieure d'Electricité where he had obtained his degree in 1914. He designed the commercial version of the Bulle-Clock and set up production in Boulogne-Billancourt, a Paris suburb. The Bulle-Clock was an immediate success in France and abroad, particularly in Great Britain: production continued until 1970.

The most important part of the Bulle-Clock was a hollow, coil that was attached to the end of the pendulum rod and acted as a bob. This coil freely swung through a three-pole, arc-shaped permanent magnet. Its north pole was in the middle, both ends were south poles. A system of contacts was actuated by the pendulum rod in such a manner that the coil was switched on shortly before it reached its lowest position. Now the coil's north pole was strongly repelled by the central north pole of the permanent magnet. The pendulum thus swung all the way to its turning point, swung back, reached the other turning point and returned to the central north pole where it was again accelerated, etc.

Thus, the pendulum was subjected to a repulsive magnetic force once per cycle. The unusual location of the poles on the magnetic rod had the advantage that the induction current in the coil minimized sparking at the contacts. A driving pawl attached to the pendulum rod set in motion a geared wheel and pushed it forwards by one tooth for each cycle. A second, stationary pawl prevented the ratchet wheel from turning back. The motion of the ratchet wheel was transferred by gears to a conventional clock movement. The Bulle-Clock was energized by 1.5-volt Leclanché dry cells that had to be replaced every five years.

The clock was protected by a glass case with a height of 38 cm. Another, smaller type with a height of 20 cm was also available. It was called the Bulle-Clockette and was characterized by a U-shaped permanent magnet. Strong cobalt steel magnets were introduced in the 1930s. In consequence, a 5 cm long magnet and a much smaller coil that was set in the middle of the pendulum rod now sufficed.

Léon Hatot – ATO

Léon Hatot (1883-1953) was an incredibly gifted and many-sided individual. He is cited as a watchmaker, goldsmith, inventor, engineer, artist and entrepreneur. He was amazingly successful in every respect.

Born in Besançon, Léon Hatot was admitted to the city's school of horology when he was only 12 years old, having passed the admission exam with flying colors. After graduation, he held various jobs and opened his own store in 1905. Selling jewelry and watches that he had designed and made himself proved to be very profitable. In 1911, he moved to Paris, where most provincials fail. But Hatot who started out as an engraver, was soon discovered by Bredillard, a major manufacturer of luxury jewelry and watches. This led to many orders from leading Paris jewelers such as Cartier, Van Cleef & Arpels, Boucheron etc.

Aside from the workshops where Hatot and his employees made thousands of pieces of jewelry, he set up a research lab where he experimented with electrical devices. In 1919, he invented the famous ATO electromagnetic pendulum clock. His most important associate was Marius Lavet (1894-1980), whom he had hired away from the Bulle-Clock company. In 1923, Lavet rose to the post of CEO. He was granted a patent for the motor that bears his name in 1937.

Today, well over a billion Lavet motors are produced each year. Every quartz watch with a conventional, analog display is actuated by a Lavet motor. In 1940, Lavet obtained a very broad patent for the transistor control of electromechanical clocks and watches. The licenses on this patent alone yielded royalty payments of up to 2 million dollars per year. After Léon Hatot passed away, Lavet stayed in charge of the company until his retirement in 1962. The Léon Hatot brand was acquired by the Swatch Group in 1999 and now stands for technically and artistically original, high end jewelry and watches

Production of the ATO electromechanical pendulum clock began in 1920; three different types were available at that time. They were equipped with a half second pendulum for individual clocks and master clocks, a quarter second pendulum for table clocks and secondary clocks, and a 2/5 second pendulum for table clocks and wall clocks. The three clock types mostly differed by the contact mechanisms and the pendulum suspension. In 1929 the half second version was adopted by the French National Railways SNCF as their standard time base. This clock was extremely sturdy and reliable, its daily rate was +0.4 to +1.8 seconds.

The basic principle of the ATO clock was similar to that of the Bulle-Clock. The major difference was that now the coil was fixed, while the pendulum bob was an arc-shaped, cobalt steel magnet; one end oscillated back and forth through the hollow, ironless coil. A system of contacts actuated by the pendulum rod saw to it that the coil was energized just at the moment when the pendulum bob was reaching the end point of its travel inside the coil. This caused the bob to be strongly repelled. What seemed to be a second coil on the other side was just a piece of brass looking like a coil: its function was mostly esthetic, as it restored the symmetry. But it actually had an electrical function, as the magnetic rod travelling through it induced eddy currents that dampened excessive deflections of the

1.5. Electromagnetic pendulum clock. M. Hipp, Neuchâtel, around 1870

1.6. Electromagnetic pendulum clock, Favre-Bulle. Compagnie Générale des Appareils Horo-Electriques, Paris, around 1921

pendulum, thus further improving the daily rate of the clock. The movement was actuated by a driving pawl attached to the pendulum rod.

In 1924, Hatot started selling foreign licenses for his clock. Haller & Benzing located in Schwenningen was the first licensee in Germany. This company went bankrupt in 1929, but the license was taken over by the Hamburg-Amerikanische Uhrenfabrik HAU. A year later, HAU was bought by Junghans who remained the German licensee until World War II broke out. The Junghans-Hatot cooperation was officially reinstated as early as 1949.

The electromagnetic pendulum clocks invented by Bain, Hipp, Favre-Bulle and Hatot were all based on the interaction of a permanent magnet with a coil that was periodically energized. The current flow was controlled by the pendulum itself. As this clock was in great demand, it triggered the development of many similar designs, such as the clocks by Bomeyer, Lemoine, Féry, Gregory-Lippman, Favarger, Brillié Frères, Joly, Siemens-Schuckert, and Irk.

Electrical Winding

Winding the mainspring of a conventional mechanical clock with an electrical motor was an obvious concept. The electrification of clocks with weight drives was quite easy too. A movement of that type required long life batteries or a connection to the grid as well as efficient actuators.

Louis Breguet (1804-1883), a grandson of the watchmaking genius Abraham-Louis Breguet (1747-1823), is often cited as the inventor of the electrically wound clock. The first prototype clock of this kind was actually built in 1856 and was promptly imitated by numerous watchmakers. But good quality batteries only became available around 1880, while the industrial production of electrically wound clocks started around the beginning of the 20th century. Thus, electromagnetic pendulum clocks preceded them by two generations.

In the early models, the mainspring was electrically wound every hour, but this rate kept accelerating until 30 seconds were reached. This meant that only a small mainspring was needed, and it was almost fully wound all the time. This had a positive effect on the isochronism. The winding actuator usually was an asynchronous motor, alternatives were eddy current motors and synchronous motors. Reduction gears were, of course, needed between the motor and the mainspring.

Electrically wound clocks were widely used as car clocks as long as quartz clocks were more expensive. Table clocks and wall clocks of this type were manufactured well into the 1960s. Junghans was the most important producer in Germany; in Switzerland Schild & Co, and Patek Philippe were major players, while Seth Thomas was big in the United States.

The most remarkable electrical winding mechanism for weight driven clocks was found in the precision pendulum clocks made by Sigmund Riefler (1847-1912). It required no energy storage in a spring and came very close to the ideal of a freely swinging pendulum as it did not need controlling electrical contacts. The drive mechanism was limited to a coil with an iron core that periodically lifted a rocking lever. The other end of the lever lifted another lever weighing about 10 gram in one stroke. As it fell back, the 10 gram lever actuated a gear-shaped ratchet wheel and interrupted the current flow in the coil. The ratchet wheel was connected to a gear that actuated the Graham escapement. The pendulum of the Riefler clock was made of Invar and was nearly immune to temperature variations.

The entire Riefler mechanism was protected by a hermetically closed glass bell jar that was evacuated with a manual pump. The pendulum frequency was adjusted by varying the pressure in the bell jar. Similar mechanisms based on lifting a weight with an electromagnet in order to actuate a clock movement were designed by Favarger in Neuchâtel, Normal-Zeit in Berlin, Cohen in Munich, Wagner in Wiesbaden, and others.

A rather weird electrothermal-pneumatic winding mechanism was developed by Junghans from 1928 to 1930 and was named "Elektronom." It was based on patents of the Swiss inventor Martin Fischer. The energy source of this clock was a special type of electrical bulb (a so-called compressor lamp) that heated air in a glass jar. The air expanded and drove a piston, the motion of which actuated a system of ratchets and wheels that wound the mainspring of a clock. Fischer produced such clocks in his own plant in Zurich and used the brand name "Pneuora" for them.

The "Elektronom" served as a master clock for up to six secondary clocks. The latter were equipped with their own compressor lamp that was switched on and off by the master clock. There also was a fully pneumatic version, the secondary clocks receiving pressure pulses through hoses that were called "hollow cables." Each of those pulses advanced the secondary clocks by one minute.

Another version used an electrically heated bellows that contained a highly volatile solvent such as ether or acetone. The bellows expanded when it was heated, thus winding the mainspring. A prototype was built in 1928 but never functioned properly.

Martin Fischer also developed very interesting and successful electromechanical timekeeping systems. A generator actuated by the weight driven master clock generated very strong minute pulses. The secondary clocks thus needed no external power source. Fischer's company was called Magneta and was bought out by Landis & Gyr in Zug in 1919. The clock business continued under the name Inducta, but was sold to Wittwer-Bär in Gwatt and Glattbrugg in 1961 (see Chapter 6).

Timekeeping Systems

From the Middle Ages all the way to the Industrial Revolution, man was quite content with the time signals provided by clocks mounted in church towers and city halls. They were very visible; their bells could be heard within several hundred yards or even a couple of miles, depending on the wind direction. The daily rate of such clocks was anywhere between bad and disastrous, but they were periodically set to the infallibly correct sundial time. Nobody really worried about the fact that each town or village thus had a different local time, depending on the longitude: mobility meant speeds of a few miles per hour. The concepts of legal time and time zones were totally unknown.

Things changed radically with the coming of railroads and industrialized production in the 19th century. A standardized time and perfectly synchronous clock systems became a necessity. This meant developing accurate master clocks to which many secondary clocks could be connected, particularly in railroad stations and within large factories or businesses. Synchronizing clocks mechanically was quite impractical: electrical systems were indispensable. In the mid-19th century, master clocks appeared that were still mechanical, but generated an electrical pulse every minute (or even every second) that controlled the hands of many secondary clocks. The pioneers of this new concept were the aforementioned Alexander Bain in Great Britain and August Steinheil (1801-1870) in Germany.

At about the same time, telegraph wires were laid just about everywhere, mostly by railroad companies and postal administrations. Those wires could of course be used for the transmission of time signals. The latter were generated by precision pendulum clocks, which themselves were set to astronomical time, as measured by observing the passage of well known stars by means of a meridian telescope. A good example is the Neuchâtel Observatory in Switzerland that was founded in 1858. Its first director Adolph Hirsch arranged to have the

1.7. Teaching model of an ATO master clock. Junghans, Schramberg, around 1950

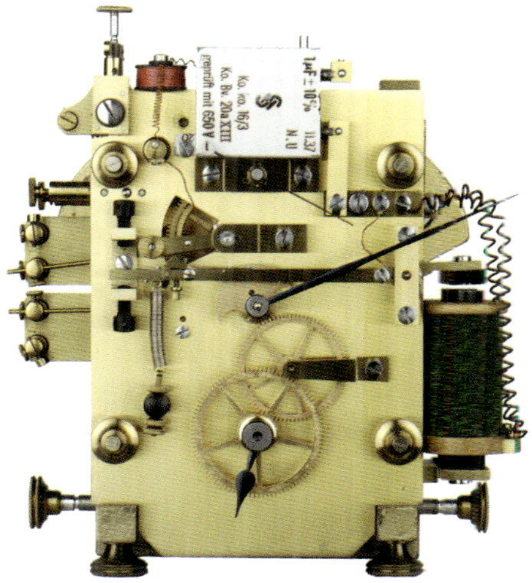

1.8. Riefler precision pendulum clock No. 556 with gravity escapement, Munich/Nesselwang, 1935

1.9. Electro-pneumatic movement "Elektronom." Junghans, Schramberg, around 1930

observatory connected to the Federal telegraph system, so that astronomically defined, precision time signals could be transmitted to the major centers of the Swiss watch industry.

However, the system was often disturbed by storms, as the telegraph wires were out in the open and slung from pole to pole. Major cities such as Paris and Vienna thus installed pneumatic systems for controlling their public clocks. This caused another nightmare, as the pipes and hoses inevitably leaked. Also, the costs of installing and maintaining the network were prohibitive.

The problem was eventually solved by underground electrical wiring.

1.10. Demonstration model of a motor-wound movement. Mauthe, Schwenningen, around 1930

Radio-Controlled Clocks

Hatot the Pioneer

Transmitting time signals by radio made them accessible to anybody having a receiver. This was achieved for the first time in Paris where a radio transmitter was installed on top of the Eiffel Tower in 1912. At twelve noon and midnight, watches could now be set to the Paris Observatory time, where the signals were generated and then transmitted by wire to the Eiffel Tower. In 1928, Léon Hatot and his partner Marius Lavet designed the world's first radio-controlled clock. It was called "Radiola" and synchronized itself automatically with broadcast time signals.

The Eiffel Tower time signal was rhythmically musical: the receiver thus needed a special earpiece, the resonant frequency of its membrane corresponding to the rhythm of the time signal. The vibration amplitude of the membrane was then high enough to close a contact that switched on the synchronization circuit of an ATO clock. A small motor set the hour and minute-hands to the 12 o'clock position by means of heart-pieces, as used in the return mechanism of mechanical chronographs. The system worked, but few people bought it, as the range of the Eiffel Tower transmitter was only 250 kilometers or 150 miles at that time.

1.11. Master clock. Synchronome. London, around 1930

Junghans was interested in Hatot's system and tried to improve it, but nothing commercial came of it. Radio-controlled clocks and watches made a real breakthrough only after the German time signal transmitter DCF77 in Mainflingen near Frankfort began broadcasting coded time signals in 1979. It operates on long waves (77.5 kHz) and has a range of about 2000 kilometers or roughly 1300 miles. In 1985, the German companies Kundo and Junghans developed table clocks with receivers for DCF77. Five years later, Junghans launched Mega 1, the world's first radio-controlled wristwatch.

The technology had been developed by Wolfgang Hilberg at the Darmstadt technical university and was much more sophisticated than Hatot's system. More particularly, DCF77 transmits a full "time telegram" every minute. with year, date, weekday, hour, minute, and second. Leap years, the winter time/summer time changes, and leap seconds are taken into account automatically. Such a system could not be sufficiently miniaturized before the advent of advanced microprocessors.

The CEH Prangins Clock

Many quartz clocks were developed as a step towards the ultimate goal of a quartz wristwatch. But the Centre Electronique Horloger (CEH) in Neuchâtel (see Chapter 7) developed a radio-controlled clock without a quartz movement in 1966, i.e. just a year before the quartz wristwatch prototypes Beta 1 and Beta 2 started working. The lab needed something interesting for the CEH shareholders who started to be impatient. They could not be informed of the progress that had actually been made, otherwise leaks would have been inevitable.

It is quite clear that the CEH scientists and engineers thought of a radio-controlled wristwatch as a promising new development, but this kind of product did not correspond to the Swiss watch culture. Switzerland had always produced timepieces that could be used anywhere in the world (and even in space). A watch that only gave the exact time within a thousand miles of a foreign transmitter sounded like heresy. It is a sad fact that the radio-controlled wristwatch was left to the Germans, the Japanese, and finally the Chinese.

The CEH ventured into radio control only once, developing several versions of the so-called Prangins Clock. The name was derived from the town of Prangins near Nyon on Lake Geneva, where a Swiss time signal transmitter started operating in 1965 (it was closed at the end of 2011). Prangins broadcasted time signals on long waves (75 kHz); they consisted originally of a continuous tone, that was interrupted for 0.1 second every second. There was a double dip at the full minute, a triple dip at the full hour and a quadruple dip at midnight. Those signals could be received in Europe and North Africa all the way to the North Cape. Adequate reception required a field strength of more than 30 microvolt per meter.

The project of a non-quartz time signal clock was financed by Ebauches SA, Omega, Portescap, Complications SA, and Flückiger & Cie. Jean Fellrath was in charge and four different versions were built. The last one was designed for industrial production. This clock had an internal time base consisting of a multivibrator that was constantly synchronized with the Prangins frequency. If this signal could not be received, the clock continued running for four minutes without losing or

1.12. Time signal receiver SM-400. Silver-Marshall, Chicago IL, around 1930

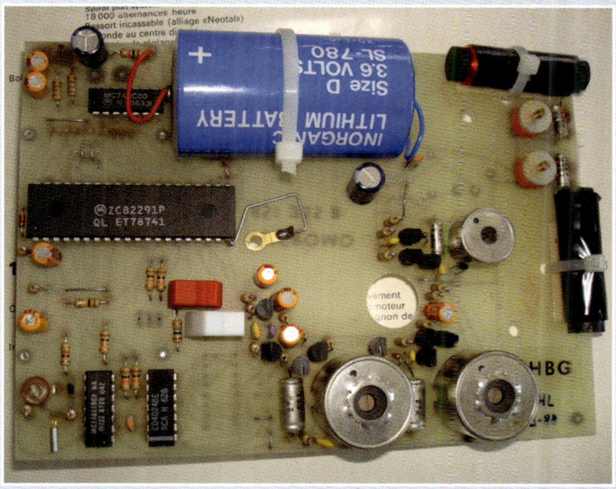

1.13. Time signal receiver for the transmitter HBG in Prangins

gaining time. It would have been much simpler to work with a quartz movement, but the cost of such calibers was prohibitive in 1966.

By 1967, 20 Prangins clock prototypes had been built and were tested in a wide range of distances from the transmitter. This made it apparent that local topographic conditions and the materials used for construction of the houses where the tests were made strongly influenced reception quality. Furthermore, long wave broadcasts often interfered with time signal reception. If the daily rate of the clock was expected to be 1 second, reception had to be good for at least two hours a day. The design was reworked in 1969 by improving the antenna and adding interference filters.

In the end, the venture was unsuccessful. Only Complications SA (i.e. Piaget) ordered 1000 Prangins clocks; Portescap (that had been chosen as the subcontractor) built a few hundred and then interest waned. By that time, good quality quartz clocks and alarm clocks were available and they cost a lot less than the Prangins clock. Junghans in Germany was first to realize that the way to proceed was to synchronize a quartz clock with radio time signals. On the base of Wolfgang Hilberg's patents, Junghans succeeded in producing radio-controlled clocks that were quite successful.

On the base of this know-how, the German companiès Junghans and Ruhla, as well as Citizen and Casio in Japan, developed radio-controlled wristwatches that set themselves once every 24 hours to radio time signals. If it gets out of range, such a watch continues working as a normal quartz watch that returns to radio-control as soon as it can receive a time signal.

The La Chaux-de-Fonds-based companies Imhof and Montremo were among the early manufacturers of Prangins radio-controlled quartz clocks. The receiver modules were supplied by the Neuchâtel Observatory where such systems were built in large series for synchronizing church clocks and railway clocks with the Prangins time signals. Later on, Montremo started producing its own receiver modules.

Electrical Clocks and Alarm Clocks

In the 1930s, many companies developed balance wheel motors for electrical clocks. Early pioneers were Junghans and Henker & Co. in Germany, Léon Hatot and Maurice Favre-Bulle in France, Citizen in Japan, as well as Friedlie Frères and Buèche-Risse in Switzerland.

Four different types of contact-controlled clock movements with a balance wheel motor appeared in the 1910s and 1920s; production continued well into the 1950s. In the system with fixed coils and pieces of soft iron mounted on the balance wheel or, in the case of a moving coil on the balance wheel and fixed pieces of soft iron, the iron-magnet interaction was based on the attracting magnetic force, the iron being attracted to the coil as long as the current flow was maintained. Conversely, a piece of soft iron mounted on the balance was attracted by a fixed coil. Current flow in the coil was controlled by cams on the balance wheel axle acting on long-lived contacts.

Quite a few manufacturers of electromechanical clocks chose to place permanent magnets on the balance wheel: this was the fixed coil type using repulsive magnetic forces. On the other hand, the moving coil type was also a possibility, but required a rather complicated, very light balance wheel construction. A cam-controlled contact device switched on the coil each time it approached one or several fixed magnets. In this case also, energy was supplied to the balance wheel by repulsive magnetic forces.

The Eureka table clock was first built in London around 1910 and became legendary. It is characterized by a gigantic balance wheel bearing a coil and rotating on ball bearings. This coil interacted with fixed pieces of soft iron. Eureka-type clocks were produced by several manufacturers, e.g. Leroy in Paris in the 1920s. In Germany, Palmtag was still in business with Eureka clocks in the 1950s.

All four types of balance wheel motors were used in clocks and alarm clocks between the war and well into the 1960s. A great variety of such clocks were manufactured in France, Germany, Japan, and Switzerland. They established the technological base of the electromechanical wristwatches of the 1950s and 1960s. In later days, contact control was replaced by transistor control, but this did not prevent electromechanical clocks from being eradicated as soon as the much more accurate quartz movements approached their price.

Clocks with Synchronous Motors

It was soon realized that time signals could not only be transmitted via telegraph wires, but also through the electrical grid. Such grids expanded very rapidly in the second third of the 19th century, after Werner von Siemens (1816-1892) and Charles Wheatstone (1802-1875) had invented the electrical generator. After battling short lived, low power and costly batteries for half a century, mankind at last had a dependable source of electrical energy that allowed high power applications such as light bulbs and motors. A very simple method of synchronizing clocks was developed in the United States where big power systems were established in the 1920s. They supplied alternating current with a very stable frequency of 60 Hz.

The system of generators (that were synchronized to a small fraction of a Hertz) was a kind of master clock that could supply frequency and thus time signals to an unlimited number of secondary clocks. The latter were extremely simple, and consisted of a small synchronous motor and the gears required for hour, minute, and seconds-hands. Such clocks were very accurate and did not require resetting as long as the power was not interrupted and the frequency remained at exactly 60 Hz. However, normal synchronous motors did not restart on their own after a power failure: this had to be done manually.

Clocks equipped with self starting synchronous motors restarted on their own when power was restored, but displayed the wrong time. A mechanical power reserve in the clock solved that problem, but cost extra. In Germany, clocks with synchronous motors were not feasible for a long time, as there was a big variety in local grids with different voltages and frequencies. But in 1935, the herculean task of interconnecting all those grids and standardizing voltage and frequency to 220 volts and 50 Hertz respectively, was completed.

Electromechanical Pocket Watches

Pocket watch movements are not much smaller than clock movements. Thus, making electrical pocket watches was relatively easy once the balance wheel motor became available. But there were no batteries small enough to fit into a watch case. Electrical pocket watches thus needed a separate battery case and a cable – which was thoroughly unpractical. Even though prototypes were built, it does not seem that the electrical pocket watch was ever produced in series. Two such prototypes are exhibited at the Nederlands Goud-, Zilver en Klokkenmuseum in Schoonhoven. They are dated 1910 and 1920, respectively.

One of the watches in the Schoonhoven museum was equipped with two electromagnets that pulled an iron anchor to the pole pieces whenever the magnets were energized. The current was then switched off, the anchor was pulled back by a spring, which caused the magnets to be switched on again, etc. A pawl was attached to the anchor and pushed on a wheeled gear. This motion was used to wind the mainspring of a conventional watch movement. There are reports about another electric pocket watch built in France by Huguenard and Bonneuil in 1926. On the other hand, wristwatch size clock movements with a balance wheel motor were produced in large series in the 1950s, 1960s, and 1970s. The earliest ones were developed in the United States, in France and in Germany; Switzerland and Japan followed soon afterwards.

1.14. Model of a movement with synchronous motor. Mauthe, Schwenningen, 1935-1939

1.15. Synchronous motor for frequency measurement and frequency divider for the first serially built quartz clocks of PTR (later Rohde & Schwarz) for Kienzle, Schwenningen, 1930s.

2. Fixed Coil Wristwatch Movements

We have seen in Chapter 1 that in electromechanical clocks the pendulum or the balance wheel served both as an oscillator and a motor. In other words, both types of oscillators supplied energy to the movement. Quite naturally, this technical concept was transferred to wristwatches. In a widely used design, a ruby roller pin fitted on the balance staff disc moved the fork of a specially shaped anchor back and forth. The pallets of the anchor alternatively pushed or blocked the teeth of a ratchet wheel and moved it forward one tooth at a time.

Forerunners of a Technical Revolution

Electromechanical wristwatches of the fixed coil and moving coil types were launched commercially in the 1950s. They were followed by the tuning fork watch in the 1960s. At that time, nobody would have even dreamed that these new watches would lead in a straight line to the quartz Revolution.

We owe the fixed coil electromechanical watch to the American Elgin Company that was managed by John G. Shennan, and to the French watch producer LIP in Besançon that was owned and managed by Fred Lip. A loose cooperation was established between the two companies, but this led to quite different products. Another approach was taken by Hamilton Watch (also in the USA), whose technical director was John A. van Hoorn. Hamilton worked with the small German watch manufacturer Helmut Epperlein in order to develop a moving coil balance wheel motor.

Wristwatches with a balance wheel motor were decorated with arrows and called themselves "electronic." As a matter of fact, those "electronics" were limited to a battery and a diode for suppressing the switching spark (the cheaper alternative was a high impedance resistor). Batteries, diodes and resistors are passive components and should be called "electronic" only in the context of at least one transistor – a truly active component.

The coil – fixed or moving – of a balance wheel motor had to be switched on and off just at the right moment, so that the repulsive force between it and the magnets was applied with the best possible yield. This was usually achieved with a system of cams on the balance staff, actuating a springy contact lamella. Inevitably, a small spark formed whenever the contact was broken. As this happened several times a second or ten thousands of times per hour, it caused the electro erosion or "burning" of any contact material one could imagine. All kinds of schemes were tried to make the contacts last several years, but a true solution had to await the coming of affordable transistors. Then the transistor switched the coil on and off, the mechanical switch being only used to control the emitter. Using inductivity control of the transistor, one could even do entirely without the switch.

Both the contact-controlled and the transistor-controlled electromechanical movement with a balance wheel motor were made either with a moving coil on the balance wheel and fixed magnets, or with a fixed coil and moving magnets on the balance wheel. In both cases the coil-magnet interaction generated a repulsive magnetic force. The concept of the Eureka clock with a moving coil interacting with a fixed piece of soft iron was also used in certain electromechanical calibers for wristwatches. In that case the attractive force between iron and magnet was put to good use.

In 1953, the appearance of electromechanical watches with balance wheel motors caused the Swiss electrical engineer Max Hetzel, then employed by Bulova in Bienne, to invent the (truly) electronic tuning fork watch. It was much more accurate and reliable than the watches with a balance wheel motor. No watchmaker would ever have used an electronically excited tuning fork both as the oscillator and motor of a watch movement. Louis-Clément Breguet, a grandson of Abram-Louis Breguet had actually thought up such a scheme in the 19th century, but this had been thoroughly forgotten; Hetzel (and the Swiss patent office) were not aware of it. Furthermore, Breguet had to excite his tuning fork mechanically, while Hetzel used a simple but effective transistor circuit to achieve this goal.

Hetzel's work led to a spectacular innovation: the Bulova tuning fork wristwatch Accutron that was protected by a very broad, almost perfect patent. For seven years, Bulova was very profitable with its electronic watch and refused to sell patent licenses. By so doing, the company unknowingly clouded its own future. Its big competitors had been convinced that horology was shifting to electronics, so they had to find something even better than the transistor-controlled tuning fork that they were not allowed to manufacture. The obvious alternative was quartz.

Thus, the Centre Electronique Horloger in Neuchâtel (Switzerland) and Seiko in Suwa (Japan) working in total secrecy both developed wristwatch-size quartz watches, the first prototypes of which started ticking in late summer 1967, within weeks of each other. Just a few years later, the quartz wristwatch caused the disappearance of both electromechanical and tuning fork watches.

Elgin

The National Watch Company was founded in 1864 in Elgin, about 25 miles west of Chicago. Next to Waltham in Massachusetts, Elgin was one of the pioneers of partly automated, machine made precision watch parts. The company grew very quickly: in the 1920s it employed 4600 workers. The manufacturing complex in Elgin then was the world's biggest watch factory.

When the United States entered World War II, Elgin switched to fabricating armament-related items such as fuses for bombs and artillery shells. By so doing, it lost the know-how of efficient and modern watchmaking. After 1945, it could not keep up with the Swiss, who had systematically continued to develop micromechanical technologies. Furthermore, thanks to the so-called "Geleitscheine" issued by the Nazi authorities, 7.4 million good quality Swiss watches were exported to the United States during each war year, which corresponded to 75% of the pre-war demand.

The massive imports from Switzerland were the death knell for the entire American watch industry. American consumers became used to the fact that a good quality watch was "Swiss made." It is almost funny that American authorities forced the Swiss industry to print or engrave this declaration of origin on both the dial and the movement. The aim was to protect the domestic watch industry, but this backfired badly: "Swiss made" became a symbol of superior quality – and still is.

Caliber 722

Elgin had already started to develop an electromechanical wristwatch in the 1930s, but the war halted this deployment. The project was reactivated in 1946 and received a strong boost from the cooperation with LIP. On March 19, 1952, Elgin president John G. Shennan gave a press conference at the Blackstone Hotel in Chicago, where he showed the media a round, gold cased watch with the electromechanical caliber 722 (as it was called later). The watch was exhibited in a glass case with a 7 foot long, red arrow pointing to it.

The astonishingly small movement measured just 18 mm x 15 mm with a height of 4 mm. Once it was cased, enough space remained for a kidney-shaped battery. In the early models, the hands were set conventionally by means of a crown on the right side of the case at 3 o'clock. Shennan limited himself to showing the watch and explaining its mechanism: the movement was not shown to the journalists. A month later, a US patent was solicited for the movement: it was granted six years later.

The first sentence in an Elgin ad published in the journal *The Watch Word* is quite interesting: *"A laboratory model of the electronic wrist watch of tomorrow, a timepiece that carries its own source of power and runs initially more than a year before its tiny energy capsule needs replacing, was revealed to the world by our Company on March 19."*

Caliber 725: A Failure

It soon became apparent that the Elgin caliber 722 was too small and could not be produced industrially

2.1. Elgin caliber 722

2.2. Elgin caliber 725 with kidney-shaped battery

2.3. Elgin caliber 725 with round batteries

2.4. Elgin caliber 910

with the necessary quality. Development was then centered on the slightly larger caliber 725 with a diameter of 25 mm and a linear coil with an iron core. Energy was supplied by two button cells of the mercury oxide type in parallel. As batteries never have exactly the same voltage, a small current always flowed between the cells and reduced their lifetime. This problem was solved by going to just one, larger size battery.

Replacing the battery of the prototypes was quite a ritual: the crown on the case bottom had to be removed and the watchglass lifted off: then, the movement could be taken out of the case. Only a qualified watchmaker was capable of performing this feat. The case of the commercial version was more customer-friendly, as the battery could now be replaced after unscrewing a lid in the case-bottom with a coin.

Despite its larger size, Caliber 725 got into endless difficulties: a movement with a completely new design just measuring 11 lines was probably much too ambitious. It took years to fabricate 6000 watches equipped with this movement. They were called "Lord Elgin" and were sold in Chicago, that city serving as a test market; a spare battery was included in the price. About 90 percent of those watches were returned, as they did not even stand up to mild sports, or stopped working for other reasons. Furthermore, the contacts burned out easily, even though a diode was connected in parallel with the contacts. The returns were scrapped – to the present-day despair of collectors, as Lord Elgin watches are quite rare.

It is not clear whether the improved caliber 910 ever was produced in series. In any case, the electromechanical watch was an extremely expensive venture for Elgin. Actually the company never recovered and went bankrupt in 1966.

Fred Lip: Trigger of an Unintended Revolution

The LIP Company in France produced an electromechanical table clock with a balance wheel motor in the 1930s. Rumors have it that it was developed in cooperation with the Ericsson Company in Sweden. Such clocks were not really revolutionary, as the balance wheel motor had already been invented and patented in 1847 by Alexander Bain (see Chapter 1). The oscillator was the classical balance wheel/hairspring combination, but a piece of soft iron was attached to the balance wheel and was attracted to a fixed coil whenever the latter conducted current.

The LIP-Ericsson clock was not terribly exciting. Before and after World War II, large numbers of clocks and alarm clocks with a balance wheel motor (in many variants) were produced in Germany, France, and Switzerland. The technology was quite well known, but miniaturization to wristwatch size proved to be extremely difficult. This explains why seven years passed between the presentation of the electromechanical LIP watch prototype and the start of industrial production.

The construction chosen by LIP (and others) was characterized by at least one coil mounted on the bottom plate under the balance wheel or just next to it. A piece of soft iron was attached to the balance wheel. A magnetic field was created whenever the coil was connected to the battery, attracting the iron on the balance wheel and giving it a mechanical impulse. As mentioned above, a cam on the balance staff switched on the coil for a short time just before the iron on the balance wheel started to move towards the coil. This caused the balance wheel to oscillate constantly, thus driving the movement.

A specially shaped anchor or a ratchet transformed the oscillations of the balance wheel into the continuous motion of the anchor wheel. This motion was transferred to the watch hands with a conventional set of wheels.

When Northern France was invaded by Nazi Germany in 1941, Fred Lip, the owner of the LIP Company, moved to the south of the country where he joined the Résistance and became an officer. When Besançon was liberated by the First French Army in 1945, Lip reoccupied his plant at the head of his men. This was totally bloodless, as the German forces had evacuated the city just in time. But the plant was in a desolate condition, having been converted to the production of fuses for bombs and shells. Everything had to be reconditioned or replaced in order to revert to watch production.

Three years later, things worked reasonably well and watches were in great demand. But sooner or later the market would be saturated and the LIP brand had to prove its innovative force. So Fred Lip gave his engineers an exciting challenge: miniaturize the old electromechanical table clock movement to wristwatch size. The first electrical LIP-Caliber would be called R27.

It was eventually learned that Elgin in the USA had a similar project. This led to a loose cooperation between LIP and the Americans. LIP supplied a semi-automatic assembly chain, while Elgin transferred electrotechnical know-how to Besançon. Some 20 LIP engineers visited the Elgin plant during that time.

Caliber R27

Paul Dargier de Saint Vaulry led the team of about 15 engineers and craftsmen that worked on the R27 project. In 1950, a strongly miniaturized version of the clock's balance wheel motor worked satisfactorily. It was still too big for a wristwatch, but by 1952 the

2.5. Dial for Elgin caliber 725

2.6. LIP caliber R27, dial side

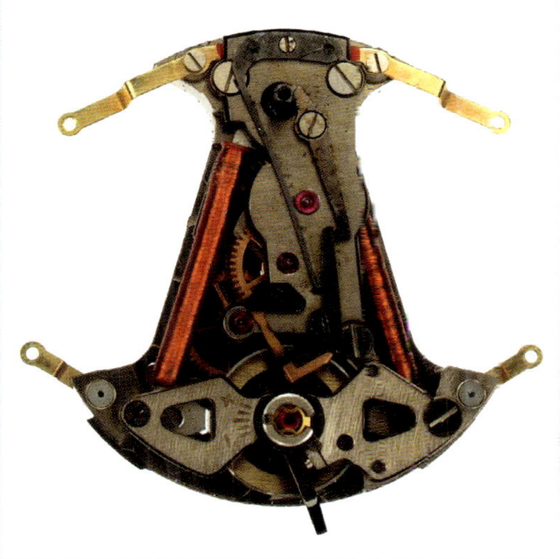

2.7. LIP caliber R27 bridge side

2.8. LIP caliber R148

2.9. Landeron caliber 4751

desired specifications had been reached. The balance wheel with its iron fitting oscillated between the poles of a split electromagnet that received an electrical pulse once per period.

The best geometry for the electromagnet was found in 1954: it consisted of two iron-cored coils forming the long sides of an isosceles triangle with an apex angle of 30 degrees. Two bean-shaped batteries ("Piles haricot") were located between the coils and the edge of the plate, thus assuring optimal use of the available space. Unfortunately, batteries with such an odd shape leaked invariably, badly corroding the movement; some of them even exploded due to excessive hydrogen pressure. Only ten prototypes with bean-shaped batteries were built, none of them was preserved for posterity. It soon became obvious that one had to switch to round batteries.

The very first prototype worked in early 1952; it obtained a chronometer certificate at the Besançon observatory on 5 March 1952. Exactly two weeks later the media were informed about the new watch movements by Fred Lip in Paris and John Shennan in Chicago. Soon afterwards, Lip was invited to give a talk at the prestigious "Académie des Sciences" in Paris. The theme evidently was the electric watch of the future. Lip had brought along one of the R27 prototypes, but according to the records of the session, the watch was not in working order. We know it was a 12-line movement with a sweep seconds-hand and a frequency of 2.5 Hz (18,000 vibrations per hour).

The industrial production of the R27 caliber was much more difficult than originally thought and only started in December 1958. The disadvantages were obvious to the designers: with two batteries, the volume was excessive and production costs were horrendous. But 85 percent of the R27 movements tested at the observatory received a chronometer certificate.

Caliber R148/R184

Jean-Georges Laviolette and his team immediately started to design a much-improved electromechanical three-hand movement. It was called R148; the diameter was 11½ lines, the height 5.8 mm, the frequency 2.5 Hz (18,000 vibrations per hour). The R148 was launched in the United States in 1962; France had to wait until 1963. It resembled the Elgin caliber 725 in some respects, but it was much more reliable. It only needed one coil and one battery.

Just the electrical part was made in-house; for the conventional mechanical part, LIP adopted the Fontainemelon caliber FHF 55. The balance could be stopped mechanically in order to obtain a perfect synchronization with the radio time signal. The variant with a date disc was designated as R148 Dato. The R148 was a huge success, particularly in the United States; Benrus, Belforte, Elgin, and Waltham used it in their watches. Major customers in Switzerland were Electra, Marvin, Nivada, Universal Genève, and Vulcain.

Starting out with caliber 148, Jean Pommier developed the 11½ lines caliber R184 "Datolip" with sweep seconds-hand and date disc; it replaced caliber R148. Production started in 1964 and lasted until 1967. In the beginning, energy was supplied by a 1.35-volt mercury oxide battery; it was eventually replaced by a 1.55-volt silver oxide cell.

Ebauches SA

The management of Ebauches SA in Neuchâtel was rather disturbed by the presentation of the LIP and Elgin electromechanical movements. A team of top caliber designers led by André Beyner and René Besson was given the assignment to develop something similar. They came up with the 11-line caliber L4750. A prototype supposedly worked in the lab as early as 1953, but the final version was presented to the media in 1960; production started at the end of that year.

Caliber L4750/4751

The Ebauches SA electromechanical caliber L4750 was exhibited at the Basel watch show of 1961. Its most massive part was a donut-shaped, two-part fixed coil. Each half consisted of 3000 turns of enamelled copper wire with a diameter of 0.03 mm. That coil defined the outer edge of the movement. The magnetic field it produced was conducted by horizontal pole shoes to a cross-shaped piece of soft iron on the balance wheel that oscillated with a frequency of 2.5 Hz (18,000 vibrations per hour).

A double cam on the balance wheel pressed together two gold plated contact lamellae so that the coil was energized in the right moment. Two of the arms of the cross-shaped iron piece on the balance wheel were then aligned with the pole shoes of the electromagnet: this gave the balance wheel a mechanical impulse at every vibration, so it would in principle oscillate continuously.

As mentioned above, the mechanically actuated contacts were the weak point of all electromechanical watch movements. But they were indispensable: thanks to them, the coils were energized at the right time, so the balance wheel motor would get its mechanical impulse. There was no alternative: of course the contact lamellae had to be as light as possible and resist sparking in the long run. Sparking was minimized in the case of caliber L4750, as the coil current flowed through two pairs of springs that were pressed against lateral contact studs by the cams which themselves were insulated and carried no current.

The ESA caliber L4750 (11½ lines x 3.7 mm) with sweep seconds-hand was manufactured by Hahn in Le Landeron (this explains the L) and delivered to customers in kit form. However, the coil molded in white plastic and the balance wheel motor were delivered complete.

A mercury oxide cell powered the first model. The battery could easily be replaced by unscrewing the lid in the watch bottom. Some 30 watch brands actually used the L4750 in their watches. A special model was called Mickey Mouse watch, as two circular appendices of the watchcase had been provided to house the battery and a spare.

Other watch brands squeezed the battery and a spare into special compartments in the bracelet, between the horns. Carrying a spare was practical, as battery life was only 8 to 12 months. A special model (the adapter version 4750/4751) was equipped with a Leclanché nickel-cadmium rechargeable battery located in the bottom of the case. Charging current was supplied via a miniature plug in a second crown at 9 o'clock. The charging unit simply consisted of a 1.5-volt flashlight battery with a cable. Another version was clamped between the two crowns for recharging, which needed to be done every six months. However, it was soon realized that such small nickel-cadmium batteries only survived a few charge-discharge cycles, particularly when they were charged at the wrong time. Thus, the rechargeable version of the L4750 was soon discontinued and taken from the market.

The height of the next version with the caliber number L4751 was 4.35 mm. The extra volume was used to accommodate a bigger battery that lasted about two years. Power could be switched off by pulling the crown. This also stopped the balance wheel for synchronization with radio time signals. Production of the L4751 ran only for six months and was stopped after the LIP R148 caliber had been launched, as it was flatter and less costly.

Shanghai Clock & Watch Parts

In the mid-1970s, the Chinese ministry of light industry decided that all four technological phases of electromechanical and electronic movements should be understood locally and needed to be developed in parallel in order to be competitive with the West. The Shanghai Clock & Watch Parts Factory was ordered to develop a contact controlled electromechanical movement. Starting point was an existing 5 Hz caliber (36,000 vibrations per hour). The battery had to find a place in the bracelet between the horns. It is not known whether SD1 was ever produced; probably not, as much superior quartz movements both of the analog and digital display type were already available.

3. Moving Coil Wristwatch Movements

The electromechanical movement with a moving coil was sometimes called "electrodynamic." The basic principle was similar to the one with a fixed coil, but now the coil was mounted on the balance wheel where it was moving. A cam on the balance staff actuating flexible contacts switched the coil on and off. The advantage of this design was that the coil was very light, which allowed a higher balance wheel frequency. The most important calibers of this type are listed below:

Hamilton 500/500A, 505
Standard Time Vantage 130, 130E etc.
Ricoh 555
Epperlein Electric 100
Laco-Durowe Electromat
Laco Electric 860/861 (Timex M 67)
Laco Electric 870/871 (Timex M 84), 880/881 (M 85)
Timex Lady Electric 900 (Timex M 82)
Timex M 40/41
Timex Lady Electric M 69/71
UMF Ruhla Electric 25/26
Porta Elechron PUW 1000
Porta Ladychron PUW 3000

Hamilton

After World War II and four years of war-related fabrication of fuses for shells and bombs, Hamilton, in Reading (Pennsylvania), could finally revert to its old core business: manufacturing watches. But radically new products were needed in order to achieve recognition as a watch producer. Research director G. Luckey suggested in 1946 that Hamilton should develop an electrical wristwatch. This project was approved, but progress was disappointingly slow.

It was a complete surprise when Elgin president John G. Shennan showed the prototype of an electromechanical wristwatch at a press conference in Chicago on 19 March 1952 (see Chapter 2). Suddenly, the development of a similar Hamilton product received the highest priority. Three weeks later, John A. van Hoorn, then Hamilton's technical director, circulated a confidential report in which he reviewed possible ways of developing an electrical wristwatch. This report was discussed at length in a meeting, the protocol of which has been preserved. Remarkably enough, even the feasibility of a tuning fork watch was mentioned. But the idea was dropped in view of the technical difficulties that were expected from such a revolutionary design. Conversely, developing an electrodynamic movement with a moving coil and fixed magnets seemed to be state of the art. Hamilton was granted several patents in this field between 1956 and 1957.

PM-1 and EM-1

In July 1952, work began on the prototype PM-1 (Permanent Magnet Model 1) in the Hamilton watch laboratory. A coil wound on the balance wheel interacted with permanent magnets that were mounted on the main plate of the movement. The prototype worked on 23 February 1953. As there was no room for the battery in the case, it was integrated into the bracelet between the horns.

In October 1952, work was started on another prototype called EM-1 (Electromagnetic Model 1) with a ring-shaped coil similar to the one that was used in the Ebauches caliber L4750. This actually was a miniaturized version of the "Koehler Clock," of which Hamilton had built six prototypes between 1950 and 1952. A US-Patent was applied for on 15 October 1951.

It was soon obvious that the Koehler alternative had a much higher power consumption than PM-1, so this project was abandoned. A hand made prototype of the PM-1 was available in early 1956. This was the final design for all practical purposes: Hamilton started the industrial production of their caliber 500 (as it was now called) in December 1956.

Caliber 500

On 3 January 1957, the Hamilton "Electric" with the caliber 500, was launched with much fanfare. It was designated as the "Watch of the future" and "most important horological innovation for the last 447 years." This referred to the invention of the torsion pendulum or balance wheel with hairspring by Christiaan Huygens in 1510.

It is interesting that for the electromechanical wristwatch there is both a technical priority and a commercial priority – which is also the case with the quartz wristwatch. Elgin and LIP have the technical priority, as their prototypes 722 and R27 were shown to the media on 19 March 1952 in Chicago and Paris, respectively. On the other hand, Hamilton definitely has the commercial priority, since the "Electric" was launched on 3 January 1957.

Furthermore, it was a great idea to ask the extremely creative Richard Arbib to design cases for the Hamilton

Electric. Arbib's masterpiece was a shield-shaped, asymmetrical case that was combined with a very simple dial and dot-shaped indexes. This was a truly revolutionary watch that was recognized from afar: it was called "Ventura" and was very successful.

The Hamilton 500 was a 12½-line caliber with a height of 6.55 mm; the balance wheel oscillated with a frequency of 2.5 Hz (18,000 vibrations per hour). The stationary magnets were made of platinum-cobalt; the flat, very light coil was integrated into the balance wheel. Hamilton was granted a total of 45 patents for caliber 500 and later models.

In 1957, Hamilton started working with the German Company Epperlein for reasons that are not very clear. Hamilton was well ahead technically with its own design, but possibly Epperlein had accumulated important practical know-how for fabricating flat coils that could be mounted on a balance wheel. It is quite clear that the priority of the electrodynamic wristwatch with a moving coil belongs to Hamilton and not to Epperlein.

Many Variants

The product life of the Hamilton Electric was 12 years. During that time, many improvements were made. Prototype 501 never went into production; the same applies to caliber 502 that used the balance wheel of caliber 500 but the contact system of caliber 505. Starting with 1959, caliber 500A replaced caliber 500. It was characterized by a different battery holder and an eccentric adjustment.

Caliber 505 was launched in 1961; the contact system now was much sturdier, the ratchet wheel doubling up as a contact wheel. The variants 505C and 507 with date disc never made it into production. Between 1966 and 1968, a shaped caliber especially designed for ladies' watches (Hamilton Ladies Electric 1) went through about ten prototype stages, but was not produced either. Caliber 510 was designed as a slimmer variant of caliber 505, but it, too, never went into production. The balance wheel was much too small and the movement was powered by two batteries with an insufficient capacity. Production was started, but the project was cancelled almost immediately as it became obvious that this caliber was unreliable.

The slim version of caliber 505 was the "Thin Electric," caliber 511 with a height of 5.3 mm. The contacts were made of platinum and were very long-lived. But due to the catalytic effect of platinum, they soon were coated with amorphous carbon. This problem was solved by inverting the polarity of the whole movement. Eleven prototypes were built in 1966: one of them ran for 26 years without interruption with the same set of contacts. Yet, the project was cancelled:

3.1. Hamilton caliber 500

3.2. Hamilton caliber 500A

3.3. Hamilton caliber 505

3.4. Hamilton-Ricoh caliber 555

the movement was not slim enough, the commercial potential was insufficient.

As early as the 1950s, Hamilton considered manufacturing an electrodynamic movement that was small enough for ladies' watches. Of course the problems that had to be solved with the calibers 500 and 505 had the first priority. But in the 1960s, design work on a 6¾ x 8 lines movement began in earnest.

Ten prototypes were ready for extended tests in 1968, but at that time, development work on the Pulsar was in full swing. Obviously, electrodynamic movements would soon be obsolete; production of the Hamilton Electric was therefore stopped in 1969. Total production is not known, but in the first year alone, about 42,000 watches of this kind were sold.

Aside from the calibers 500 and 505 that had been developed in-house, Hamilton also equipped watches with the Ebauches SA calibers Dynotron 9154 or 9158: they were called Hamilton Electronic. Even quartz-synchronized Dynotrons were experimented with. But in 1970, Time Computer, a company under the umbrella of the HMW holding (Hamilton Watch also belonged to HMW), launched the revolutionary LED quartz watch "Pulsar." It was technologically more advanced than any other watch of the time (see Chapter 7).

Vantage and Hamilton-Ricoh

In the 1960s, the American watch industry started suffering rather badly from the Swiss and Japanese competition that offered good quality watches at attractive prices. At that time, many consumers bought watches strictly on the basis of their price, technical design and looks were not the most relevant criteria. This explains the amazing success of Timex, which frontally attacked the Hamilton Electric with the Laco-Timex models.

Hamilton had the reputation of producing only high quality watches. Therefore, the company could not afford selling low-priced watches under its old brand name. For this reason, a subsidiary called Standard Time Company was founded. It assembled watches consisting of imported, Swiss made parts in St. Croix (Virgin Islands). Major brands were Vantage and Township.

There even was an electrodynamic model equipped with the caliber 130E that was practically identical with the Hamilton caliber 505. The electrical part was actually supplied by Hamilton, the rest was imported from Switzerland and Japan. Caliber 130E was used in simple, chrome plated brass cases that were distributed under the name "Vantage Electric;" production continued until 1969.

In order to gain a foothold in the important Japanese market, Hamilton signed a joint venture agreement

3.5. Epperlein prototype

3.6. Balance with coil of the Epperlein prototype

3.7. Epperlein preliminary series

3.8. Epperlein caliber 100

with Ricoh in 1962. Caliber Hamilton-Ricoh 555E was produced in Japan, it was practically identical to the Vantage caliber 130E. It was used in asymmetrical, cushion-shaped watches that were only moderately successful. Production stopped in 1969.

Epperlein

Between 1952 and 1956, Helmut Epperlein, the owner of Uhren-Werk Ersingen (UWERSI) in Kampfelbach-Ersingen near Pforzheim, developed an electrodynamic watch movement that looked interesting to Hamilton. This led to a cooperation contract, Hamilton profiting from the patented Epperlein flat coil construction and gold plated contact spring system. On the other hand, Epperlein could access the Hamilton technology and started to produce the Epperlein Electric 100 in 1958; it was very similar to the Hamilton 500. The gold plated movement had a diameter of 11½ lines, the balance wheel frequency was 2.5 Hz (18,000 vibrations per hour). Battery life was at least one year.

Epperlein produced just a few hundred watches with his caliber 100. In 1959, he replaced it with a much improved version that was also sold under the name Epperlein 100. Both the contact springs and the balance wheel had been reworked. The production run was about 5000 pieces, a third of which was returned under warranty. Epperlein also developed the model "Electric Piccolo" in 1960. This was the first electric ladies' watch, but it never went into production. Epperlein went bankrupt in 1971.

The development work done by Epperlein and Hamilton was very carefully followed in the USSR. Epperlein's prototypes were copied meticulously in the 1950s in Moscow, but there was no production. Wisely, the Soviet watch industry did not get involved with contact controlled systems and concentrated on transistor control, putting on the market an excellent copy of the Junghans caliber 600.

Laco-Timex

Laco-Durowe has the signal merit of having produced transistor-controlled electrodynamic wristwatch movements under the name "Electromat" no later than 1956. However, they never went beyond the prototype stage or the test series (see Chapter 4). In 1959, Laco-Durowe was bought out by the American Timex (at that time it was called U.S. Time Corp.). Timex actually bought the Durowe Company; Laco was a brand that Timex inherited in the course of that transaction, but never used.

Back to Contact Control

Due to excessive costs, the Management of U.S. Time decided against marketing the transistor-controlled Electromat. It was replaced by the less costly but technically inferior contact control of the balance wheel motor oscillating at the rate of 3 Hz (21'600 vibrations per hour). The basic layout of the Electromat movement was left unchanged, but a new, vibration-damped construction was developed for the contact springs that were made of a gold-silver-palladium alloy. Those contacts lasted up to five years before replacement was necessary. The "Laco-Timex-Electric" (Laco 861, Timex M67) was launched in 1961 and was produced by the millions.

Several variants were produced in small series for extended tests. Their main characteristic was a special, banana-shaped battery that had been developed in collaboration with Varta. This was not a very elegant solution, but was meant to take advantage of the space formerly occupied by the electronic part. Furthermore it became apparent quite soon that batteries that are not perfectly circular like the button cell are prone to leakage – with disastrous effects. Changing the battery was a rather complicated procedure as it involved removing the crystal and lifting the movement out of the case. Another characteristic was the tail of the regulator: it was eliminated in later models.

The stem of the setting mechanism ran through the bottom of the case. Pulling the crown interrupted the current flow from the battery: this feature was used to synchronize the watch with a radio time signal. The individually numbered prototypes used for tests still were designed for the banana-shaped battery, but power was supplied by a much more reliable button cell. The space for it was obtained by bulging out the case bottom.

M67: A Very Reliable Caliber

After the Laco 860 test series (without a date disc), Laco caliber 861 (Timex M67) was launched in 1961. It was the first serially produced moving coil movement with a balance motor oscillating at 3 Hz (21'600 vibrations per hour). The diameter was 13½ lines, the height was 6.0 mm. The bottom plate was laid out for the banana-shaped battery of the early models, but was later modified for a button cell.

The coil mounted on the balance wheel swung back and forth through the field of permanent magnets mounted on the plate and was connected to the battery by a cam controlling the contact springs in both directions. The energized coil thus turned into an electromagnet that was repelled by the permanent magnets. In this manner, the balance wheel received enough energy to keep the movement in motion.

The gear train was, of course, not subjected to the force of a mainspring and advanced step by step, the seconds-hand jumping quite precisely. Caliber 961 stayed in production for several years; watches equipped with it are easy to recognize as they have no crown, the hands being set from the case bottom, which already had been the case with the Laco prototypes II and III. But in 1965, Timex sold the Durowe Company to the Swiss Ebauches SA, keeping just the Laco brand until 1988. However, this brand name was never used during that time, causing Timex to lose its legal rights. The Lacher family eventually recovered their old brand name.

3.9. Laco caliber 860 prototype

Electrifying the Ladies' Wrist

The Laco calibers 870/871 (Timex M84) and 880/881 (M85) were derived from caliber 861; production started in 1964. The calibers 880/881 were equipped with an unusual date disc and a rather special actuating mechanism for displaying the AM and PM hours. In the morning, a black dot appeared under the date; after twelve noon, a similar dot appeared on top of the date. This was achieved by a slight jump of the date disc at 12 o'clock.

3.10. Laco caliber 861 – Timex caliber M67

The "tin plate calibers" M40/41/42, measuring 9¾ x 12¾ lines, were launched in 1970. There was no brass bottom plate: it was replaced by stamped and bent steel plate. The moving coil balance wheel still oscillated with a frequency of 3 Hz (21'600 vibrations per hour); M41 displayed the date, M42 featured day and date discs.

The "Electric Dynabeat" calibers M253, M254 (with date) and M255 (with day/date) used the M40 caliber, but the balance wheel oscillated with a frequency of 4 Hz (28,800 vibrations per hour). The contacts were protected by a diode. A second hour-hand could be individually set in the "Time Zone" model. Production continued until 1982 and was of the order of several million pieces.

3.11. Laco caliber 900 Lady Electric – Timex caliber M82

Caliber Laco 900 (Timex M82) "Lady Electric" was the world's first electrical movement for ladies' watches. It was launched in 1965 and production ran until 1971. Quite an original slogan was used in newspaper and magazine ads. It said "Electrify your wife. $50." The movement was 6¾ x 8 lines, the height 6.0 mm. The technology was very similar to the one developed for the 860 family of calibers. M82 was replaced by M69/71 (M71 displayed the date) and was introduced in 1972. It was another tinplate caliber; with a height of 5.8 mm it was slightly slimmer than its predecessor and featured a sweep seconds-hand.

3.12. Timex caliber M69

Caliber M57 "Electro-Alarm" launched in 1974 was rather exotic. This actually was caliber M69 with an electric alarm: the latter function was powered by a separate battery. Only a few prototypes were built, the alarm mechanism being similar to the one invented by Spadini for his Nepro brand of alarm clocks and watches. As mentioned above, Durowe was bought by Ebauches SA and produced ETA watch movements for duty free export into the European Union countries.

Porta (PUW)

The PUW Company (PUW stands for Pforzheimer Uhren-Rohwerkefabriken) was founded in 1934 by Rudolf Wehner (1899-1986), a watchmaker from Dresden (Saxony) and two partners. Their ambition was to produce top quality ebauches for the German watch industry. As Switzerland did not allow the export of the required precision machinery, the latter was bought by straw men in Switzerland, taken apart, brought across the border in the course of many car trips and reassembled in Pforzheim. Once production had started, the quality of the PUW ebauches was practically identical to that of Swiss made ones. PUW also produced watches under the "Porta" brand.

The PUW plant was totally destroyed during World War II and had to be rebuilt under very difficult circumstances. As PUW's competitors had electrical movements in their catalogs in the early 1960s, PUW had to act. In 1966, it launched caliber 1000 (12½ lines by 5.9 mm). Variants were the calibers 1001 (with date and fast date corrector) and 1002 (with day/date). They were moving coil movements with a balance wheel frequency of 4 Hz (28,800 vibrations per hour) and were used in watches with the brand name "Porta Elechron." In 1969, PUW launched a transistorized version of caliber 1000 (see Chapter 4).

This caliber was used by Helbros in their "Luna Lectric" model, the first wrist watch with lunar time. It was advertised as "World's first Lunar Watch." It displayed lunar hours, tenths of lunar hours and lunar milli-hours. The watch was cased in 14 K yellow gold. It was issued in honor of the first lunar landing on 20/21 July 1969; only a very small number of pieces was actually produced. The American astronomer Kenneth L. Franklin designed the watch that was available commercially in 1971. Other brands that used the PUW 1000 caliber were Croton, Gruen, Fricoma and Feuvrier.

The smaller, 6¾ x 8 lines 4 Hz caliber Porta Ladychron PUW 3000 was launched in 1972. Its height was 6.0 mm including the battery. This turned out to be a very sturdy caliber, the double-action contact springs lasted 8 years. This caliber was used by quite a few brands: Elgin, Fleurier, Gruen, Helvetia, Invicta, Sandoz, Solvil & Titus, Pallas, and Waltham.

3.13. Timex caliber M40

3.14. Timex caliber M71

3.15. PUW caliber 1000

3.16. PUW caliber 3000 Ladychron

UMF Ruhla

By the end of the 19th century, the small metalworking company that had been founded in 1862 by the Thiel brothers in Ruhla (Thuringia) had developed into an important producer of watches and machinery. It was confiscated by the Soviet occupation forces after the end of World War II and was USSR property until 1952. It was then turned over to the German Democratic Republic (GDR) and was state property until the Berlin wall fell in 1989.

In 1960, Ruhla developed an interesting contact controlled electrodynamic caliber with a diameter of 12 lines and a height of 6.3 mm with date disc. At that time, this was hi-tech and was instrumental in convincing the GDR authorities that they should maintain a watch industry in East Germany. The USSR had requested that the watch plants in Ruhla and Glashütte be closed, so they could sell Russian watches in the GDR.

The 13-jewel Ruhla electrodynamic movement was given the designation UMF cal. 25, the balance wheel motor oscillated with a frequency of 2.5 Hz. The caliber was shown at the Leipzig Fair in September 1962; watches with the caliber 25 were available in the stores at the end of that year. Two models went into production: UMF 25/82 with a sweep seconds-hand and the self-starting UMF 25/12. The caliber also was used in a steering wheel clock for the Wartburg 353 car. In 1965, limited special editions of a picture clock and of a gold plated globe watch became available. Production of caliber 25 stopped in 1967.

Caliber 25 was replaced in that same year by caliber UMF Electric 26 with 13 jewels. UMF 26-02, a special model with only 7 jewels was developed for export to the United States, where duty increases rapidly with the number of rubies. The oscillation frequency of this 12-line caliber was 4 Hz (28'800 vibrations per hour). Production started in 1969 and ended in 1975.

The UMF-Ruhla caliber 26 served as a starting point for the development of a small alarm clock with the "Minimat" brand name (cal. 26-09). The alarm time was set by turning the crystal. An electromechanical buzzer was switched on at the set alarm time. Sales of the Minimat started in 1969.

Watches with the electrodynamic caliber UMF 26 as well as its predecessor UMF 25, did not rate as very dependable. Production never exceeded 50,000 pieces a year and was stopped in 1975. Watches and clocks with the movements UMF 25 and 26 were sold to Champion Electric, Ermi, Gsell, Lafayette, Targa Electric and Worldtime.

3.17. UMF Ruhla caliber 25, bridge side

3.18. UMF Ruhla caliber 25, dial side

3.19. UMF Ruhla caliber 26, bridge side

3.20. UMF-Ruhla alarm clock type 26-09

4. Transition to Electronics

It was already mentioned in Chapters 2 and 3 that the electromechanical movements of both the fixed coil and moving coil types had one inherent disadvantage: sparking eroded the contacts that switched the coil(s) on and off. The transistor brought the ultimate solution: the emitter-collector circuit now carried the coil current. The gate current required for switching was so small that no sparking occurred. There was an even better alternative: the gate voltage could be generated inductively and no material contacts were necessary. This was actually the preferred scheme.

Small Clocks and Alarm Clocks

As soon as transistors became affordable, clocks, alarm clocks, and wristwatches with a transistor-controlled balance wheel motor appeared in the marketplace. In terms of clocks, this chapter is limited to Derby Swissonic (once a well-known brand owned by the Ebauches-Group), the transistor-pioneer Léon Hatot, and the legendary Mouvement de Paris (MP).

Derby Swissonic

In the early 1960s, Ali Schneiter from Ebauches SA in Neuchâtel, developed a transistor-controlled electromechanical movement for clocks and alarm clocks. A piece of soft iron was mounted on the balance wheel and was periodically attracted by a fixed coil. This movement was available with or without a date disc and an alarm. Time and alarm time were set with the same knob; power was supplied by a single AA-type battery. This movement was mostly used inside the Ebauches group, but was also sold to Tissot and Bulova.

"Derby" was chosen as the brand name: it was adopted from the Derby Timer Company in La Chaux-de-Fonds that had been acquired by the Ebauches group in 1964. Derby originally manufactured pedometers and parking meters. Many clock models also bore the designation "Swissonic." The latter brand name had been bought from the Centre Electronique Horloger CEH in Neuchâtel and had to be used somehow; otherwise it would soon have become public domain. This was the first time that Ebauches SA had a complete watch product in its catalog. Using an existing brand minimized the costs of marketing.

The cases were designed by Gaston Vuilleumier. Round plastic cases with a diameter of 88 mm and elliptical cases (95 x 65 x 60 mm) were available in several bright colors. An elliptical case milled out of massive aluminum was particularly successful. Cylinder-, spindle-, and cushion-shaped aluminum cases were also available. The Derby movement was used in wall clocks and table clocks, as well as in desk clocks with world and solar time displays. The picture clocks were reproductions of old etchings showing a church tower with a functioning clock. The biggest derby clock was a 276 cm x 90 cm wall panel with a world map; the clock movement was from Favag and was powered by 220 V AC.

Léon Hatot

The potential importance of transistor control in clocks and watches was understood at a very early time by the Hatot engineers Lavet and Dietsch in Paris. In 1953, they were granted basic patents for transistor circuits that eliminated sparking in movements with a balance wheel motor. The main goal of this work was a further improvement of the ATO pendulum clock. But the electromechanical table clocks manufactured by Junghans in Germany in the early 1960s were also covered by the Hatot patents. This explains the Junghans brand names Atomat, Atovox, Atophon, and Atolux; the latter had a solar cell power supply (see the section on Junghans).

In the last years of his life, Léon Hatot developed a transistor-controlled electromechanical wristwatch. But he died in 1953, before finishing the prototype. Lavet and Dietsch took over and showed the watch at a session of the French Chronometer Society in 1956. It was never produced in series.

Mouvement de Paris

The "Mouvement de Paris" (MP) was invented in Paris in 1960 and was designed for small clocks. In 1964, the technical division of the company moved to Besançon. The MP was based on a transistorized balance wheel motor with a fixed coil. However, the "escapement" that moved the gears was very special and was quite different from other constructions. The balance wheel staff bore two lifting arms spaced 180 degrees apart and bent upwards and downwards, respectively. They acted alternatively on a horizontal plastic wheel with 15 prismatic blades. As the balance wheel swung forwards, it moved one of the blades by

12 degrees. On the backwards swing, the same blade was moved by another 12 degrees. The bladed wheel thus rotated by 24 degrees for each oscillation of the balance wheel.

The bladed wheel arbor was prevented from slipping backwards by a ratchet mechanism. The electronic part just consisted of a transistor and a capacitor. Magnets on the balance wheel generated a current pulse in the control loop of the coil that switched on the transistor. This energized the coil, generating a magnetic field that pushed away the closest magnet on the balance wheel. This type of balance wheel motor had to be started manually with a small lever. The MP was available with two hands, three hands, and an optional date disc. Major MP components were purchased from external vendors: JAZ supplied the gear train, Trophy the electronics, Vedette the balance wheel cage with a shock absorber: the same company assembled the movement. The compact version had the designation MPC. Most of the French manufacturers of clocks and alarm clocks adopted the MP; it disappeared in 1970, as competitively priced quartz movements became available.

Wristwatches

Bifora

In 1971, Bifora in Schwäbisch-Gmünd to the east of Stuttgart started the production of two transistorized electrodynamic calibers: B10 ($8\frac{1}{4}$ lines, for ladies' watches) and B11 (13 lines, for men's watches). They were based on the contact controlled electrodynamic calibers B8 ($8\frac{1}{4}$ lines) and B9 (13 lines) with a 5 Hz balance wheel (36,000 A/h) and a height of 5.9 mm; those latter calibers never went into production.

The same, rather exotic, horseshoe-shaped, 5 Hz balance wheel with two hairsprings connected in series was used for the B10 and B11 calibers. A similar design had already been used for the B8/B9 calibers. All four calibers had the same height and were equipped with 6 jewels. The day/date mechanism and the sweep seconds-hand were exactly alike. A mercury oxide battery with a life of about 3 years powered those movements.

Two coils were mounted on the balance wheel of the Bifora calibers B10/B11; they were energized via contact cams and a transistor. Four permanent magnets were mounted on the bottom plate. An integrated circuit with 6 transistors and 9 resistors controlled the amplitude of the balance wheel. The rate was ± 2 minutes per month.

4.1. Bifora caliber B10

4.2. Bifora caliber B11

4.3. Citizen caliber 0802 Cosmotron

4.4. Citizen caliber 7803A Cosmotron

Bulova

The tuning fork watch Accutron by Bulova was launched in 1960 and generated big headlines as well as big profits. But it was rather expensive: the brand needed some sort of an electronic watch that was significantly less costly. This gap was filled with electromechanical calibers with a balance wheel motor that were supplied by external vendors.

In 1967, Bulova started buying Citizen's 12½-line electrodynamic calibers X-8 0811 and 4830, calling them Caravelle 12 TC and 12 OTC, respectively. They were used in men's wristwatches and pocket watches. A year later, Bulova purchased the Citizen caliber 0820 that was called 12 OUC. Also in 1968, Bulova purchased the Citizen caliber 0830 and gave it the designation 12 OUCD. In 1970, LIP sold Bulova the first series of its transistorized caliber R 50 – some 70,000 pieces; it was called Caravelle 7 OT. This 7¾-line caliber was used in ladies' watches. Starting with 1975, Bulova also purchased the ESA Dynotron caliber 9157 (with date) and 9158 (with day/date): they were renamed 13 UECD and 13 UECB, respectively.

Citizen

Citizen managers were well aware of the importance of the transistor for the electrical watch. The Tokorozawa research laboratory started developing a transistor controlled movement with fixed coil and moving magnets in 1962. A working prototype (on the base of an ATO-license) with two complementary silicon transistors was ticking in 1966; it was exhibited at the 1967 Osaka Fair.

X-8 Cosmotron

Citizen watch model X-8 Cosmotron with caliber 0802 Chronotron Chronometer (12½ lines x 5.5 mm, 25 jewels, sweep seconds-hand) was marketed in 1967. The shock-protected balance wheel motor consisted of four platinum-cobalt magnets on the balance wheel and two fixed coils, the frequency was 3 Hz (21'600 vibrations per hour). The next model was 0811 Cosmotron with the same specifications, and 0820/0830 Cosmotron Chronomaster (Bulova Caravelle OUC, 1968).

The Cosmotron Chronomaster was followed by the caliber 0840/0880 Cosmotron (0880 was the export version) and caliber 0884/0888 Cosmotron Chronometer (3 Hz). Caliber 4801 Cosmotron Electronic had a height of 6.4 mm and also displayed the date. The dimensions were the same but the balance wheel frequency of caliber 4830 (1970) was only 2.5 Hz (18,000 vibrations per hour), it was sold to Bulova as Caravelle OTC. Caliber 4840 displayed the date, the balance wheel frequency was 3 Hz (21,800 vibrations per hour). It was mentioned above that Bulova adopted two of the X-8 calibers and used them in their Caravelle watch models. Citizen supplied the balance wheel motor, Bulova manufactured the other parts in its own facilities.

The Cosmotron Special (cal. 7800, 7801, 7802, 7803A, 7804A) launched in 1972 was particularly accurate, thanks to the unusually high frequency of the balance wheel (5 Hz, 36,000 vibrations per hour). The seconds-hand could be set to zero with a pushpiece and maintained there, for synchronization with a radio time signal. The more economical caliber 7804 did not have the stop function.

Citizen Electronic IC 12

The Citizen Electronic IC 12 of 1970 used the three-hand caliber 5800 with 8 magnets and 2 fixed coils. This 3-hand ladies' caliber measured 6¾ x 8 lines, the height was 5.0 mm. The balance wheel motor oscillated with the unusually high frequency of 6 Hz (43'200 A/h). This was the first time that movement parts were made by precision injection molding of a blue plastic. 5820, 5830 and 5835 were two-hand calibers, caliber 5810 had a sweep seconds-hand and 17 jewels.

In 1971, a Citizen-Bulova joint venture launched the tuning fork watch Hi-Sonic. Despite of that, Citizen continued developing and producing the Cosmotron calibers with a balance wheel frequency of 6 Hz. Thus, caliber 7801 was launched in 1972: it was followed by the variants 7802, 7803, 7803A (with sweep seconds-hand and day/date, respectively); the calibers 7804 and 7804A had both a sweep seconds-hand and a date disc.

Synchronization with Quartz

Due to rapid advance of quartz movements, Citizen launched an electromechanical movement with quartz synchronization in 1973. Many parts were made by precision injection molding of plastic.

The beam-shaped 16'384 Hz, flexural mode quartz resonator had a length of 15.8 mm and was suspended in its capsule by springy wires. An integrated circuit supplied by Toshiba divided the quartz resonator frequency ten times by two in order to obtain 16 Hz. A trimmer capacitor was provided for fine adjustment of the quartz frequency.

The balance wheel, with a diameter of just 7.8 mm, was not used as an oscillator, but solely as a motor that was equipped with four pairs of magnets. Two flat coils that interacted with the magnets were mounted on a lateral protrusion of the baseplate. Synchronization was achieved by means of the oscillation amplitude of the 8-Hz balance wheel, the phase of which was compared in each cycle with the phase of the divided quartz signal. Deviations were compensated by adjusting the current through the coil.

The basic caliber 8810 (11½ lines, a height of 5.65 mm and a frequency of 5 Hz) with sweep seconds-hand also had a day/date display. The variant 8810B E.F.A. (Extra fine accuracy) consisted of hand checked movements the rate of which was guaranteed to be 10 seconds per month. Caliber 8811B launched in 1973 was nearly identical with the basic caliber 8810.

Ebauches SA

In 1967, Ebauches SA launched their caliber 9150 "Dynotron" or "Swissonic 10" (12½ lines x 5.50 mm) with a balance wheel frequency of 3 Hz (21'600 vibrations per hour). It was based on a patent by Hatot in Paris, "Licence Hatot" was engraved on the baseplate (see the section Léon Hatot). Caliber 9150 was equipped with a sweep seconds-hand and a date disc. It was replaced by caliber 9154 in 1970; the dimensions had hardly changed (the height was now 5.55 mm), but the balance wheel frequency had been increased to 4 Hz (28'800 vibrations per hour). In 1975, Ebauches launched the calibers 9157 with date and 9158 with day/date; both had a diameter of 13 lines.

Dynotron – A Success Story

Electronic watches with a transistor-controlled balance wheel motor of the Dynotron type were significantly more accurate than mainspring powered mechanical watches. There were two main reasons for this: first and foremost, the nonlinear characteristic of the mainspring was eliminated; furthermore the movement ran with much lower bearing pressures. Another advantage was that such watches ran for a year or two, regardless of whether they were worn or not. They could even be switched off by pulling the crown, thus extending the mercury oxide battery life. Tens of thousands of Dynotrons received a chronometer certificate.

The balance wheel of all transistorized electromechanical movements of Ebauches SA had two levels and consisted of aluminum alloy discs on the inner side, on which two barium ferrite magnets were glued. They were connected on the outer side by kidney-shaped soft magnetic iron platelets. The two-sided flat coil was mounted on the baseplate. Whenever the magnets reached the control section of the coil (1350 windings of enamel insulated 18 micrometer copper wire), a voltage pulse was

4.5. Citizen caliber 4801 Cosmotron Electronic

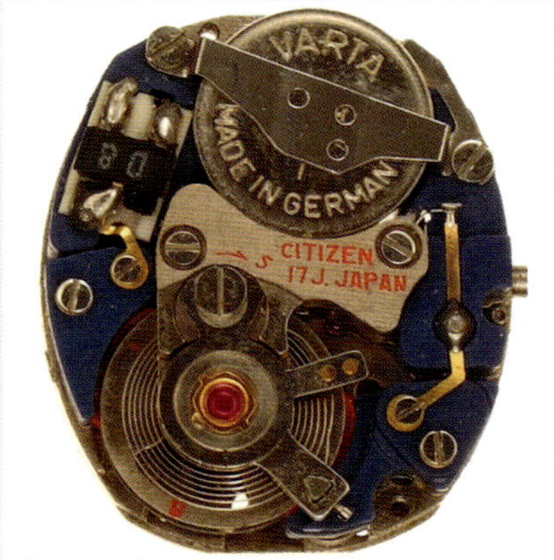

4.6. Citizen caliber IC 12 caliber 5800

4.7. ESA caliber 9150 Dynotron

4.8. ESA caliber 9154 Dynotron

generated that switched on the silicon transistor. Now current flowed through the working part of the coil (3350 windings), the resulting magnetic field repelling the magnets and thus the whole balance wheel. As the balance wheel swung back, a voltage pulse of the opposite polarity was generated that blocked the transistor.

The balance wheel, thus, was supplied with energy once per cycle. Its back and forth motion actuated a specially shaped anchor that advanced the escape wheel one tooth at the time. In later Dynotron models, the anchor was replaced by a simple pin on the balance wheel. The balance wheel's amplitude was stabilized with a simple circuit consisting of a diode, a resistor and a capacitor.

Electronic Chronometers

The electromechanical watches, first contact-controlled, then transistor-controlled, had a product life of 15 years. During that time, Ebauches SA produced millions of them. Several new technologies and much know-how had to be acquired, that later helped the transition to quartz, e.g. working with electronic parts, precision molding of plastic, and the fabrication of odd-shaped, small magnets.

The success of transistor-controlled electromechanical watches was mainly due to their attractive price and good performance. They did not cost more than a regular automatic watch, yet they could be certified as chronometers quite easily. Some fifty different watch brands used the Dynotron movements.

The mechanical part of the ESA 9150 "Dynotron" calibers and its variants was supplied by the Ebauches plant in Le Landeron: it had already gained valuable experience with the contact controlled electromechanical caliber L4750. Components of the balance wheel motor came from Ebauches group companies in La Chaux-de-Fonds and Le Locle, the electronic module was made in Porrentruy, A. Schild in Grenchen supplied the gear train. Derby assembled the movement in temporary facilities until 1971 Then, the complete Dynotron fabrication was transferred to Marin where construction of the Ebauches Electroniques (EEM) complex had just been completed.

Disc and Drum Displays

As mentioned above, there were several different Dynotron calibers. In the original caliber 9150, the balance wheel motor had a frequency of 3 Hz, which was increased to 4 Hz for the caliber 9154 of 1970. The ladies' watch caliber ESA 9190/LIP R50 (7½ lines, 3 Hz) was developed jointly in 1975 by Fontainemelon (FHF) and LIP.

4.9. ESA caliber 9158 Dynotron

4.10. ESA caliber 9175

4.11. ESA caliber 9156

4.12. ESA caliber 9200 Frontronic

Another ladies' watch caliber ESA 9200 Frontronic (6¾ x 8 lines x 4.6 mm) was launched in 1976. It was developed by Ebauches SA in Marin, but Fontainemelon (FHF) was responsible for the production. The balance wheel's frequency was 4 Hz. Longines used it in its watches and called it L7102.

This caliber was extremely sensitive to external magnetic fields. The mainplate had to be made of non-magnetic and, therefore, expensive brass. As brass is constantly recycled, it is unavoidable that steel chips contaminate the melts. This is why the cheapest types of brass contain significant amounts of iron and are weakly ferromagnetic: this is easy to detect with a strong magnet that attracts the material.

Caliber 9159 (1975) with jumping mechanical-digital display was particularly interesting (14 lines, date). It was derived from the 13-line calibers 9157 with date and the variant 9158 with Day/Date. Caliber 9176 was quite unique: the display consisted of 12 mm diameter drums with the numerals 0 to 9 as they were used in old car odometers. Those drums were moved by the balance wheel motor. Hours and minutes were digitally displayed, a co-rotating segment that made one turn per minute gave an approximation of the second. This caliber was cased by JAZ, but was not very successful. Watches with the drum-type display are quite rare and are much sought by collectors.

Junghans

Junghans gave its first transistor controlled caliber with a balance wheel motor the designation J 100; the dimensions were 30.30 mm x 6.88 mm. A test series of a thousand pieces was produced in 1961. However, the performance was not up to the specifications, whenever the caliber was subjected to motion: it was acceptable as long as it was stationary. For this reason, most of the J 100 calibers were used in table clocks, particularly in the Atolux clocks with solar cells.

In 1967, almost simultaneously with the Ebauches caliber Dynotron, Junghans launched its fixed coil caliber 600.00; it too needed an Hatot license. It was called Junghans Electronic and measured 13½ lines x 5.7 mm; the oscillation frequency was 3 Hz (21'600 A/h). This was Germany's first industrially produced transistorized wristwatch movement. It was used in the Atochron watch collection; within a year, about 10,000 pieces of those three-hand watches with a sweep-seconds hand were sold.

Between 1968 and 1970, Junghans produced 30,000 pieces of the calibers 600.10 (without date disc) and 600.12 (with date disc). Starting with 1970, the balance wheel amplitude of those calibers was

4.13. Junghans caliber J 100

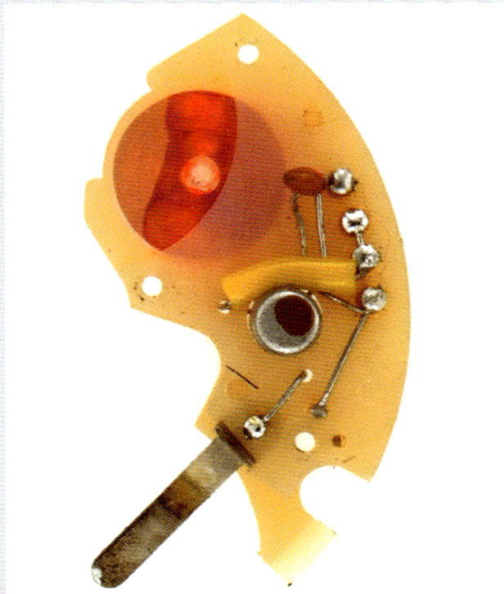

4.14. Junghans electronic module of the caliber J 100

4.15. Junghans caliber J 600

4.16. Laco Electromat

stabilized by means of a simple electronic circuit. The basic specifications were the same as those of caliber 600.00. Until 1975, some 100,000 pieces of the chronometer version with yellow gold, white gold or gold plated case were sold. A ladies' version was also developed by Junghans, but this project did not progress beyond the prototype stage.

Laco-Durowe

Lacher & Co. was founded in the mid-twenties in Pforzheim. In the early days, ebauches were bought in Switzerland, assembled by Lacher and cased using German exterior parts. Those watches were sold under the Laco brand name. As Switzerland severely restricted the export of prime quality ebauches, Laco started producing their own ebauches under the Durowe brand in 1933.

A Transistor Pioneer

The quality of Durowe ebauches with lever escapement was quite good, thanks to the semi-legal import of Swiss precision machinery. Those calibers were not only used by Laco, but were sold to other watch manufacturers in the Pforzheim area. Everything went well, but just a few weeks before the end of World War II, the Laco-Durowe plant was totally destroyed by allied bombs. Rebuilding was slow and difficult; the new plant finally started operating in 1949.

There is little doubt that everybody in Pforzheim was aware of Fred Lip's 1952 talk at the French Academy in Paris where he showed his first electromechanical wristwatch. One may safely assume that Laco managers also knew about Elgin's numerous patents covering the electromechanical watch. The weak point of contact control – contact erosion due to sparking – was soon recognized in Pforzheim. The problem could only be solved by going from contact control to transistor control.

Laco began developing transistor-controlled electrodynamic movements in 1956. In that same year, William B. Shockley, John Bardeen, and Walter Brattain received the Nobel Prize in Physics for the invention of the transistor back in 1947.

Prototypes

Aside from the transistor, the availability of very thin, enameled copper wire with a diameter of 0.02 to 0.03 mm for winding the coil on the balance wheel, and of high quality magnetic materials such as platinum-cobalt or ferrites for the fixed magnets (samarium cobalt had not yet been discovered) were very critical. Three different prototypes were developed by Laco engineers between 1956 and 1958. All of them were called "Laco Electromat."

Prototype I (1956) used a germanium transistor as a switch. Two button cells were needed for powering the movement, they were located in little compartments in the bracelet, next to the horns; their life was about a year. The Fixoflex bracelet was attached to those compartments. The hands were set conventionally with a crown at 3 o'clock. In later prototypes, the setting crown was hidden in the back of the case.

Prototype II (30 mm x 6 mm) was shown at the 1958 Hannover Fair. Switching was now done with a pnp-type silicon transistor combined with an npn germanium transistor. Combining pnp and npn transistors, i.e. creating a complementary pair, considerably reduced the temperature dependence of the circuit.

Silicon transistors cost 32 Deutschmarks apiece at that time, about 10% of the monthly wages of an unskilled worker. The seconds-hand jumped by a full second at a time. The electronics were sealed in a plastic ring that surrounded the mechanical part. The balance wheel motor found its place in an opening in that ring. The hands were set with a flat crown in the case bottom. Like in Prototype I, the batteries were located in small compartments in the bracelet.

Test Series

Prototype III (1958/1959) was also equipped with a complementary pair of silicon and germanium transistors. A single button-type battery was now sufficient, it was located inside the movement. The fixed permanent magnets consisted of barium ferrite. Hamilton had launched the electrodynamic "Electric" (with a moving coil) in 1958, while LIP just started the production of its electromechanical caliber R27 (with a fixed coil). Both were contact-controlled, Laco was way ahead of them with its transistor control.

The Laco prototypes were the world's first truly electronic wristwatches. They were called Electromat; Prototype III was ready for industrial production in 1959. Actually, several test series were produced with a total of some 1000 pieces. Unfortunately, this was the end. The watch was ahead of its time; transistors would have made the watch unaffordable to most potential buyers.

In February 1959, Laco was sold to the United States Time Corp. (which later became Timex and the development of the Electromat was stopped. It was replaced by a contact-controlled, electrodynamic caliber with a moving coil. Prototypes were ready for tests in 1959 and led to the industrially produced Laco Electric caliber 861 (the test series model was designated as caliber 860, see Chapter 3).

LIP

The transistor controlled electromechanical Ladies' watch caliber LIP Electronic Rex 50, RE 50 or R 50 was shown to the media in 1970. Its development had taken seven years and had involved the cooperation of the LIP shareholder Ebauches SA; Paul George led the development team. An ATO license was necessary for the electronic part. The caliber was equipped with a fixed coil with a height of just 0.45 mm that protruded into the spacing of the two-part balance wheel. Two magnets were attached to each side of the balance wheel; they generated voltage pulses for switching the transistor.

The LIP R50 caliber with 14 jewels (ESA 9190 for Ebauches SA) had a diameter of 7¾ lines, the balance wheel oscillated with a frequency of 3 Hz (21'600 vibrations per hour). The variant R51 was equipped with a date disc, R53 had a sweep seconds-hand. Production was concentrated at the Ebauches SA Fleurier plant. The first batch of 70,000 pieces was delivered to Bulova USA where it received the designation Caravelle 7 OTC. It was also sold to Gruen where it was named Gruen 80; other R50 customers were Waltham and Favor.

Porta (PUW)

PUW's watch brand Porta transistorized the contact controlled "Porta Elechron" caliber and gave it the designation 2000. The current for the coil on the balance wheel was controlled by a circuit consisting of three transistors, three resistors and a capacitor. One end of the coil winding contacted the hairspring while the other end was connected to the electronic module via an insulated extra hairspring.

The Porta-Elechron concept led to the calibers 2501 and 2502 (12½ lines x 5.90 mm). The electronic module was replaced by an integrated circuit (TBA 840 by ITT). The coil was mounted on the balance wheel and interacted with three fixed magnets on the baseplate. The disc on the balance staff carried a roller pin that engaged the teeth of a wheel, the pinion of which moved the intermediate wheel. The ladies' watch caliber 3000 (6¾ x 8 lines x 4.85 mm) was launched in 1972 and still was contact controlled.

Industrial production started in 1973; some 30,000 movements of the Porta Elechron type were produced within the next three years. They worked

4.17. LIP caliber R50

4.18. PUW caliber 2000

4.19. PUW caliber 2500

4.20. Seiko caliber 31B

remarkably well and were sold to Elgin, Fleurier, Gruen, Hamilton, Helvetia, Invicta, Pallas-Ormo, Sandoz, Solvil & Titus, and Waltham; despite of that, they generated little profit for PUW.

Seiko

One must keep in mind that Seiko is not a "homogenous" brand, as watches are made by two different Seiko companies. On the one hand, the old Seikosha (founded in 1892) now is known as Seiko Epson. On the other hand, Daini Seikosha (founded in 1937) is now Seiko Instruments. Both companies are controlled by the Hattori family; their original purpose was the development and production of mechanical watches. From the end of the 1960s and well into the 1970s, the production program also included electromechanical watches.

Both Seiko companies bypassed the tuning fork watch, as Bulova refused to sell them a license for the Accutron. This certainly accelerated the coming of quartz wristwatches: they were developed both by Seiko Epson in Suwa and Daini Seikosha in Tokyo.

In principle, the two companies of the Hattori group were competitors, but coordination meetings were organized, so the engineers and physicists of the quartz development teams could exchange basic information: each side knew approximately what the other side was doing.

Now, Seiko Watch Corp. is responsible for the entire watch program of the group. Development and production mandates are given to Seiko Epson and Seiko Instruments while Seiko Watch is responsible for worldwide marketing and distribution of Seiko watch products.

Seiko Epson in Suwa

Seiko Epson in Suwa developed and produced several transistorized electromechanical calibers. The first of them – caliber 31A – was launched in 1968. It was quite different from the American and European constructions, as it was equipped with an anchor motor and not with the usual balance wheel motor. A similar motor had been used in the Beta 1 prototype of the Centre Electronique Horloger (CEH).

The magnetic circuit consisted of three serially connected coils with a common soft iron core. The inner coil was horizontal, the two outer coils made an angle of 45 degrees. The anchor was located between the poles and carried four permanent magnets.

Cams on the balance wheel staff switched the transistors controlling the coils, two of which receiving current pulses with alternating polarity. Thus, the lever was attracted alternatively to either magnetic pole. This motion was transferred in the usual manner to the balance wheel, oscillating at 3 Hz (21,000 vibrations per hour).

The escape wheel was alternatively advanced one tooth at the time by the lever pallets, thus keeping the movement going. As the balance wheel swung back, a spring blocked a sawtooth wheel mounted on the balance wheel staff, preventing the escape wheel from turning backwards.

Caliber 31A with 15 jewels was used in a three-hand watch. The next models 3102A and 3102B launched in 1969 were provided with a date disc. The diameter was 29 mm (13 lines), the height being 4.8 mm; the date disc added another 0.9 mm. The balance wheel frequency was kept at 3 Hz (21,600 vibrations per hour).

The 12-line (26.9 mm) three-hand caliber 37EL with day/date display became available in December 1970. The oscillation frequency of the balance wheel had been reduced to 2.5 Hz (18,000 vibrations per hour). The battery life was indicated to be 20 months, the monthly rate was between minus 30 seconds and plus 60 seconds.

The 29 mm x 5.5 mm calibers 3702A (with date) and 3703B (29 mm x 6.5 mm with day/date) were launched in 1971. They were characterized by an unusually high balance wheel frequency of 6 Hz (43,200 vibrations per hour) and 15 jewels. Battery life was 20 months.

Daini Seikosha (Seiko Instruments)

In 1971, Daini Seikosha started producing the first transistorized electromechanical caliber of its 3300 family; the oscillation frequency was 3 Hz (21,600 vibrations per hour). A bar magnet interacted with a fixed coil mounted on the lever at the rate of the balance wheel. Cams on the balance wheel staff controlled the transistor. A diode protected the latter from the voltage pulses that were generated whenever the coil was switched off.

The balance wheel was not encumbered by coils or magnets while the transistor avoided sparking. The wheel train was kept in motion by the escape wheel. The diameter of caliber 3300 was 28.3 mm (12½ lines), the height was 5.55 mm. Caliber 3302 was provided with a date disc, caliber 3303 displayed day and date, the dimensions were the same. This caliber was used in the "Le Gant" brand watches of the American department house chain Montgomery Ward. Stellaris and Westclox also were important customers.

4.21. Seiko caliber 3702A

4.22. Seiko caliber 3303A

4.23. Seiko caliber 0702A

4.24. Lutch caliber 3045

4.25. Timex caliber M87 / Laco caliber 882 Transistor

The 17.2 mm x 5.2 mm ladies' watch caliber 32 ELC became available in 1972. The electronic circuit consisted of two transistors, a capacitor and a resistor.

In the following year (1973), Seiko Epson made a breakthrough with quartz calibers. Yet, Daini Seikosha launched several new transistorized electromechanical calibers 0702A (with date) and 0703A (with day/date). The balance wheel frequency was 4 Hz (28,800 vibrations per hour), the dimensions were 26 mm x 5.4 mm. The caliber pair 0722A (with date) and 0723A (with day/date) had the same technical specifications as 0702A and 0703A, respectively.

Movements of this kind were much cheaper than quartz movements and significantly more accurate than mechanical calibers. Therefore they kept occupying a niche until quartz movements beat them in terms of price and completely swept them from the marketplace. This happened in the early 1980s.

Tegrov

The transistorized electrodynamic 13½-line caliber Junghans 600 was meticulously copied in the USSR (illegaly of course, but that was the rule). Watches with this caliber were sold in the USSR under the brand names Tegrov, Lutch 3045, Slava, Poljot, Raketa, Wostok, Zara and Tchaika.

Timex M87

For Timex too, replacing contact control by transistor control was big progress. The Laco Electromat of 1956 had shown the way: the transistor eliminated sparking-induced erosion of the contacts: delicate and expensive repairs were no longer necessary. However, the price of transistors was simply too high to allow an industrial production of such movements.

In the 1960s, the price of transistors dropped dramatically. There was no reason for not using them even in the low-priced Laco-Timex movements. In 1969, caliber 882 (M87, 13½ Lines, three hands and date) was developed on the base of calibers M85 by replacing the mechanical contact by a transistor circuit. It consisted of two complementary silicon transistors (pnp and npn), 4 resistors, and one capacitor. Pulling the crown interrupted the current flow to the circuit. Variant 872 (M86) did not have a date disc.

The earliest version of caliber 882 (M87) used a hybrid thick film circuit on a ceramic substrate. Conductive and resistive pastes were applied by silk screen printing. A capacitor, four resistors, and two transistors were soldered on. In the later M87 version, uncased planar transistors without wire connectors were used. This was an important step towards integration that was fully achieved in the circuits of quartz calibers.

The 3 Hz electrodynamic balance wheel used the same oscillating coil and the same system of magnets as caliber Model M85. The circuit was triggered inductively by the voltage pulses appearing in the coil as it swung over the fixed magnets.

An important characteristic of the electronically controlled balance wheel motor was its self-starting capability. When the balance wheel was not in motion, the electronic circuit generated strong starter pulses. The circuit was connected to the moving coil by means of a second hairspring.

Xiang Yang

In 1967, at the height of the Cultural Revolution, a small plant called Shanghai Metal Watch Bands Factory, which had only very basic and simple mechanical equipment in its workshop, amazingly produced the first balance wheel motor electrodynamic watch in China. This was a transistorized 5 Hz (36,000 vibrations per hour) high frequency movement. A year later, the project was transferred to the Xiang Yang Shanghai Ballpoint Pen Plant.

Before long, a test series of China's first balance wheel electronic watch SD2 was successfully completed. It was a turning point for China's watch manufacturing, as it marked the shift from the era of mechanical time pieces to the age of electronic watches. The watch developed in Shanghai was called "Facing the Sun" and was put into commercial production around 1976. It is not known how long production lasted, but several dial and case variants are known.

5. Wristwatches with a Tuning Fork

Aside from a purely mechanical precursor that had been developed in the 19th century, the electronic tuning fork watch was absolutely unique and had only one inventor: Max Hetzel (1921-2005). He put down his ideas in writing no later than April 1952.

The history of the tuning fork watch is rather complicated, among other factors because the technical concept was so attractive: numerous watch manufacturers wanted to produce it. But Hetzel's patent was ironclad to the point that he himself was not able to circumvent it for years. Bulova who owned the original Swiss patent and many others based on it, refused for a long time to sell licenses.

The most important milestones in the history of the tuning fork watch, from the basic idea of 1952 all the way to the end of the product life in 1977 were as follows:

1952: Max Hetzel, an electrical engineer working for Bulova in Bienne (Switzerland) invents the tuning fork watch. He is granted a Swiss patent in December 1953. According to Swiss law, the patent belongs to Bulova.

1960: Bulova launches the "Accutron" tuning fork watch on the base of Hetzel's concept.

1963: Max Hetzel, now in the employ of the Centre Electronique Horloger (CEH) in Neuchâtel, develops a tuning fork watch with a massive, ring-shaped resonator. It infringes upon Hetzels own patent of 1953 and cannot be produced industrially.

1963: Heinz Waldburger develops tuning fork watch movements with H- or figure 8-shaped resonators with a near-zero position error. The project is cancelled in 1966.

1964: The First Moscow Watch Factory launches the "Kamertoni", an illegal copy of the Accutron.

1967: Ebauches SA develops the tuning fork caliber 9162/9164 "Mosaba" with a greatly reduced position error. The patent license is bought from Bulova in 1968. The product is launched under the designation "Swissonic 100."

1971: A Citizen Watch/Bulova joint venture is licensed to produce the tuning fork watch "Hi-Sonic" in Japan.

1973: Omega launches the tuning fork watch "Megasonic" designed by Max Hetzel, finally circumventing his own patents.

1973: An Accutron timer is part of an instrument designed for measuring the seismic activity of the Moon. It is brought to the Moon by the Apollo 17 crew and left there.

1977: After only 17 years, the production of tuning fork watches is stopped everywhere. The quartz watch is more accurate and significantly cheaper than mechanical, electromechanical and any other type of electronic timekeeper.

Hetzel's Accutron

Arde Bulova, son of Bulova Watch founder Joseph Bulova and company president, was very concerned after hearing about the first electromechanical watches that were shown to the media in 1952 by the American Elgin and the French LIP. He was afraid that a technical revolution was brewing and that it might gravely affect his company. He flew to Switzerland, where Bulova had been producing most of its watch movements for about 30 years.

The Bulova Company had started in Bienne by assembling purchased ebauches. In 1950, Bulova built a brand new plant in order to be able to design and produce its own calibers: Bulova then became a true manufacturer. This is where Arde Bulova met the young electrical engineer Max Hetzel: the company had hired him in 1950, mostly for automating the precision machines used in the production department. Arde Bulova asked Hetzel to write a detailed technical report on the new Elgin and LIP calibers.

A Report for the Big Boss

Hetzel wrote his report in April 1952 and was anything but enthusiastic about the new technology. Obviously, the electromechanical watch was only marginally superior to the classical mechanical watch. Nevertheless, it had two advantages: there was no mainspring with its inherent nonlinearity as it unfolds, which affects the isochronism of the watch. Furthermore, the battery-powered watch ran for a full year, regardless of whether it was worn or not.

Hetzel added that it should be possible to design a much better watch with an electronically excited tuning fork acting both as an oscillator and a motor. The old concept of the torsion pendulum with balance wheel and hairspring was out in Hetzel's perspective: the tuning fork was a mechanical oscillator with a much higher resonance frequency and thus would be more accurate. A transistor was indispensable for exciting the resonance frequency of the tuning fork. No watchmaker in his right mind would have thought up such a concept.

Hetzel was not a watchmaker, but an electrical engineer: just a few years before he had obtained his master's degree at the Swiss Federal Institute of Technology (ETH) in Zurich. The subject of his thesis (which was rewarded with a silver medal) had been the properties of electronically excited tuning forks and flat springs. Arde Bulova was greatly pleased with the idea and asked Hetzel to have the concept of the tuning fork watch patented. Hetzel thus wrote an "iron clad" patent that meant a lot of trouble for competitors. First of all Hetzel's patent just could not be circumvented. Furthermore, the Bulova Company did not sell any patent licenses until 1968.

Hetzel calculated that increasing the frequency of the resonator from a few Hertz in the classical balance wheel to several hundred Hertz in the tuning fork would improve the daily rate of a watch by an order of magnitude.

A Patent and its Consequences

Hetzel did not know about Louis-Clément Breguet's clock with a mechanically excited tuning fork: he only became aware of it long after filing his patent. Louis-Clément was a grandson of the watchmaking genius Abraham-Louis Breguet. Hetzel knew about the inherent advantages of the tuning fork since the time of his thesis, i.e. only a marginal temperature dependence of the resonance frequency and a great purity of the sound, i.e. very weak harmonics.

According to Hetzel's notebooks, the following chronology could be reconstructed: in May 1952, he received more documents concerning the electromechanical Elgin watch as well as a point-transistor made by Western Electric. In June, he obtained a picture of the LIP movement caliber R27 with a fixed coil and magnets on the balance wheel.

In March 1953, Bulova New York sent Hetzel three transistors of the type CK 722. Three months later, Hetzel excited a 200 Hz tuning fork with a transistor circuit that he had built on a wooden board: a true breadboard design. On 19 June 1953, Hetzel applied for a patent at the Swiss Agency for Intellectual Property in Bern. The patent "Electrical Watch" was granted on 31 December 1953. It bore the Swiss patent number 312290.

The Excitation of the Tuning Fork

Hetzel designed a simple but effective circuit for the excitation of the tuning fork. Cylindrical (later conical) nickel-cobalt permanent magnets were attached to the end of the fork prongs; they were surrounded by soft

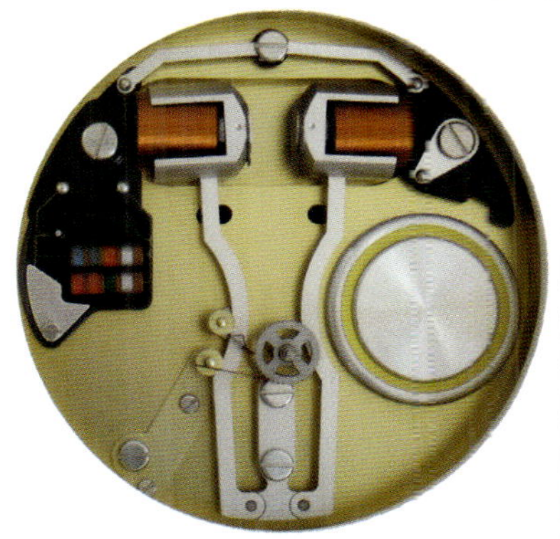

5.1. 10 x enlarged model of Bulova Accutron 214 for demonstration and instruction purposes

5.2. Bulova Accutron caliber 214

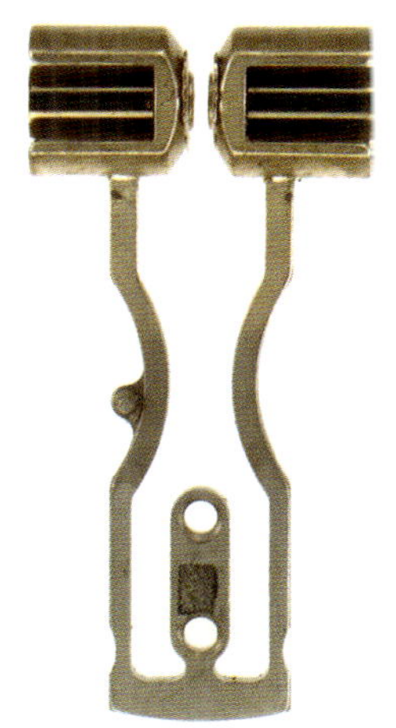

5.3. Tuning fork of the Bulova Accutron caliber 214

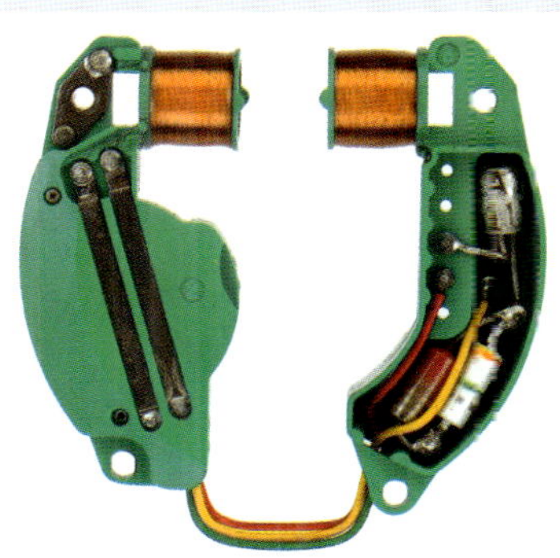

5.4. Electronic module of the Bulova Accutron caliber 214

iron shields. The magnets oscillated in and out of fixed coils at the rate of the tuning fork's resonant frequency.

The coil windings consisted of 80 meters of enamel-insulated, 15 micrometer copper wire. One of the coils consisted of 8100 turns and was meant to excite the tuning fork. The other coil consisted of 6000 turns for tuning fork excitation plus 2000 turns acting as a sensor for switching the transistor on and off.

The tuning fork's vibration amplitude was stabilized; it automatically returned to the normal value after a perturbation. Too high an amplitude resulted in the transistor being blocked for a longer time. Too small an amplitude allowed more current to flow through the coils. The sensor part of the winding allowed the correct synchrony of the transistor gate switching with the phase of the tuning fork prongs. The power efficiency of the entire system was between 80 and 90 percent.

A Tiny Ratchet Wheel

The tuning fork was meant to be both the oscillator and the motor of the new movement. Its back and forth motion thus had to be converted into rotation. For this purpose a thin, springy wire was attached at a right angle to one of the prongs; it acted as a driving pawl. A tiny prismatic jewel was glued to its end: it measured 0.18 x 0.18 x 0.06 mm. The pawl jewel engaged a 2.4 mm and just 0.04 mm thick ratchet wheel with 300 teeth with a height of just 0.025 mm. Each oscillation of the tuning fork resulted in the wheel being pushed forward by one tooth at a time (the above numbers refer to Hetzel's original Accutron caliber 214). In the Swiss prototypes, the pawls were made of the nickel-iron alloy "Elinvar," while "Nispan C" was used in the American-made Accutron. This was the alloy Bulova used for hairsprings.

The tiny prismatic jewels at the end of the pawls were essential for a proper functioning of the motor's motion. The ratchet wheel's motion was transferred to a conventional gear train driving the watch hands. A mercury oxide button cell energized the movement; the power consumption was 8 microwatt.

Critical Cutting

Cutting the beryllium bronze ratchet wheel was quite difficult in the 1950s, but Hetzel found an original solution to that problem too. He used a big, electromagnetically excited tuning fork to which he attached a diamond blade. The wheel revolved under the tool at the proper rate, so the teeth were cut perfectly in no time. This procedure was not suited for industrial production runs; in that case stacks of 100 wheels at a time were cut with a modified "Mikron" machine. Most watchmakers laughed when they saw the tiny, non-lubricated ratchet system: they knew the wheel would wear out rapidly. To everybody's astonishment this was not the case, even in the long term. Actually, lubrication was lethal, as it caused the system to stick.

The Accutron movement was self-starting, but it was characterized by a significant position error, as the resonance frequency of the tuning fork depended on its orientation with respect to the earth's gravitational field. The position error could be as high as ± 4.5 seconds per day. Another disadvantage was that strong shocks caused the pawl to jump over several teeth: this caused random errors. However, under normal circumstances, the daily rate was of the order of just 2 seconds. This corresponded to the performance of the best possible mechanical wrist-chronometers, which were much more expensive.

Hetzel completed eight prototypes with a diameter of 30 mm while he was in Bienne. Six of them were equipped with a 240 Hz tuning fork, the two others had a 360 Hz tuning fork. The latter showed the best performance. Those tuning forks consisted of the iron-chrome-nickel alloy "Elinvar" with a negligible thermoelastic coefficient.

All of the men's size tuning fork watches Bulova manufactured from 1960 to 1977 hummed at 360 Hz, which is somewhat lower than the standard tuning pitch of 440 Hz. For other Accutron models, either lower or higher frequencies were chosen, e.g. 480 Hz for the ladies' model "Mini-Accutron (caliber 2300) with a diameter of 19.5 mm. The Mosaba that was launched by Ebauches SA in 1969 featured a 300 Hz tuning fork. The "Omega Megasonic" that Hetzel developed after leaving the CEH hummed at 720 Hz.

Omega or Bulova USA?

In 1955, Hetzel developed an instrument for automatically adjusting the frequency of the tuning forks he used in his novel movement. But by then Bulova management lost patience: a whole series of prototypes had failed. It was discovered much later that this was due to faulty batteries. The CEO of Bulova Bienne, Lore Sandoz-Peter, had had enough. She ordered Hetzel to stop the nonsense, whereupon the project was transferred to the United States where the Bulova development laboratory in Jackson Heights (Long Island, New York) took over.

A new series of prototypes was assembled in 1957, but their performance was very disappointing. Arde Bulova seriously considered stopping the project, but he died quite unexpectedly. His nephew Harry Henshel took over and decided to continue the development of the tuning fork watch. A new test series of was built in early 1959; Bulova managers had to wear them in the utmost secrecy and to give them a really hard time.

While this was happening in Jackson Heights, Max

Hetzel was very unhappy in Bienne. His pet project had been stopped and transferred to the United States. The redeeming feature was that he had to fly to New York several times in order to solve problems that had surfaced as production was being prepared. Yet, he started looking for another position; he fairly informed Bulova Bienne about the situation and offered his services to Omega.

The New York Bulova management had also been informed and Harry Henshel definitely wanted to keep Hetzel. For this reason, Hetzel was offered a promotion to chief-physicist in March 1959 and a fully paid move to Jackson Heights with his family (he had two children); Hetzel accepted. Omega had already offered him a job, but it was less attractive than living in the United States for a couple of years. In June 1960, Hetzel started his new career in Jackson Heights, where he closely collaborated with Bulova vice president William O. Bennett.

Accutron Lore

Hetzel's contacts with Omega are the true core of an apocryphal story that is still believed and quoted even though it is totally false. Hetzel himself was quite clear about this. According to that story, Hetzel developed his tuning fork watch in total secrecy without anybody knowing about it. Then he visited various watch manufacturers and offered them his invention. Nobody was interested and, in fact, Omega literally threw him out in a very humiliating way. Only the American-owned Bulova realized the importance of the invention, with the result being that the Accutron ranks as an American innovation. All of this is extremely far from the truth.

As mentioned above, the report Hetzel had to write for Arde Bulova about the new electromechanical watches gave him the idea to use a transistor-excited tuning fork in a watch. This is why he started developing the new movement in 1952, with the formal authorization and encouragement of Bulova management. It was again Arde Bulova who wanted the tuning fork watch to be patented. That patent was granted in late 1953; according to Swiss law, it belonged to Hetzel's employer, i.e. Bulova Bienne. Omega would have loved to hire Hetzel in 1959. Charles L. Brandt, then human resources manager at Omega, sent Hetzel an offer in January 1959, one day after Hetzel had interviewed there. A copy of that letter is preserved in the Omega archives.

Omega management knew full well that their company would not be in a position to manufacture tuning fork watches: it was obvious that Bulova refused to sell patent licenses. Furthermore, Omega had quite different ideas, as negotiations with Battelle Geneva concerning a quartz watch were in full swing. The contract with Battelle would eventually lead to the "Megaquartz 2400," Omega's high frequency quartz wristwatch. The contract Omega set up for Hetzel explicitly stipulated that he would not work in the field of electronic watches. Otherwise, all his requests were granted, i.e. a yearly salary of 30,000 Swiss francs (which was rather high at the time), and that he would report directly to Omega management.

Jackson Heights, New York

A handwritten letter dated 4 March 1959 and signed by Max Hetzel has been preserved in the Omega archives. In it, Hetzel's apologized for not being able of accepting Omega's flattering offer. He had decided to stay with Bulova since he could simply not resist the perspective of spending a few years in the United States with his family. In Jackson Heights, Hetzel's first job was to repair the tuning fork watches of the American test series. Almost invariably, the problem was the completely useless oil on the beryllium bronze ratchet wheel. This oil became sticky and blocked the entire movement within a relatively short time. Further investigations showed that the whole ratchet mechanism was a problem, even without lubrication. Hetzel had only made provisions for a single, jeweled pawl that was attached to the tuning fork and turned the ratchet wheel. This wheel had to be prevented from turning backwards, which Hetzel did with a flat spring acting as a brake on the ratchet wheel.

Each and every watchmaker knows that a permanent brake is not feasible in a watch movement, as it eventually fatigues or wears out. But Hetzel was an electric engineer and had never heard of this. On the other hand, Egbert van Haaften, Hetzel's assistant in Jackson Heights, knew it all. He suggested a second, fixed pawl that would prevent the wheel from turning backwards. This was a real breakthrough: now the Accutron worked almost miraculously well and survived the most strenuous tests. During the first half of each cycle, the tuning fork pushed the ratchet wheel forwards by 1½ teeth. During the backswing, the ratchet wheel sprang back by one half tooth, whereupon it was blocked by the fixed pawl.

The result was that the ratchet advanced by one tooth per cycle of the tuning fork. The two pawls had to be adjusted to a phase difference of 1½ teeth. This adjustment was very simple: first the battery voltage was reduced to 0.95 volt, then an eccentric wheel was turned until the gear train started moving. To some extent, the second pawl made the motion of the gear train independent of the tuning fork vibration amplitude. Bill Mutter, who also closely worked with Hetzel in Jackson Heights, designed the gear train for the hour and minute-hands, as well as for the sweep seconds-hand.

Accuracy + Electronics

The brand name "Accutron" was derived from the words "Accuracy" and "Electronics." The complete movement consisted of just 27 parts, 12 of which were movable. In contrast, a conventional mechanical movement for an automatic caliber needs about 110 fixed parts and 26 movable parts. Even today, the dial-free "Spaceview" Accutrons, clearly showing the tuning fork, the coils, the transistor, and the resistor, are fascinating. The hour indexes were stamped on the watch glass. As a matter of fact, the "Spaceview" was meant to be a demonstration model that allowed the salesperson to explain the mechanism to potential customers. The result was that nearly everybody wanted to purchase the very "technical" version.

The "childhood diseases" of the tuning fork watch had all been cured by the end of 1959. Many of the thirty Bulova divisions in the United States that were involved with parts manufacturing and assembly now had to cooperate in order to get the Accutron production going. On 25 October 1960, Bulova president Omar N. Bradley, General Eisenhower's chief of staff during World War II, launched the Accutron tuning fork watch with the movement caliber 214 designed by Max Hetzel (later on, this caliber was called 2140).

The later variant 214 H (H for hack spring), was equipped with a lever to stop the oscillations of the tuning fork. It was the equivalent of caliber 2141 of 1962 with very minor differences. Caliber 214 HN (or 2142) had a 24-hour dial. Starting in 1968, Bulova exclusively used four digit caliber numbers. Watches equipped with caliber 214 did not have a crown: the hands were set with a folding bow in the case back. The battery case was there too, it could be opened with a coin in order to replace the battery. The calibers 2142 and 2143 were also derived from the original caliber 214. New caliber numbers were required for marking the transition from a pnp germanium transistor to an npn silicon transistor.

In 1965, Bulova launched caliber 218 with a height of just 4.9 mm. The diameter was 29.7 mm, the tuning fork frequency remained the same, i.e. 360 Hz. The ratchet wheel had 320 teeth. Numerous variants appeared: 218D, 2180, 2181, 2181F, 2181G, 2182, 2182F, 2182G, 2183, 2183F, 2183G, 2185, 2186, 218S, 218SC. The differences were minor and primarily concerned the date and day/date display functions. Dials with a 24-hour scale were also available.

In the variants Astronaut I and II, the hour-hand could be set independently of the other hands, which was quite useful when changing time zones. In the calibers 2180, 218D, 2181, 2182 the lever to stop the tuning fork was dispensed with. It was replaced by a spring that was controlled by the crown; its function

5.5. Bulova Accutron, first commercial caliber 214

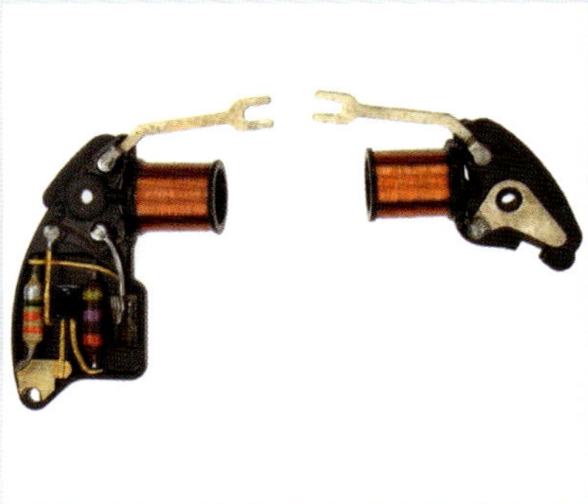

5.6. Electronic module and coils of the Bulova Accutron caliber 218

5.7. Bulova Accutron caliber 2186 bridge side

5.8. Bulova Accutron caliber 2186 with mechanical digital display

was to lift the pawl from the ratchet wheel. In caliber 218F, the crown actuated a switch that interrupted the current flow in the coils and the electronic circuit.

Caliber 218G with its mechanical digital display was quite interesting but impractical. Hours, minutes, and seconds were displayed by revolving discs. The rotation was continuous, which was quite disturbing in terms of the hour display, which should have been jumping. Shortly before the full hour it was difficult to say which numeral really applied.

The Accutron caliber type 219 that hit the marketplace in 1972, measured 28.75 mm x 4.4 mm. The frequency of the tuning fork had been maintained at 360 Hz, the ratchet wheel had 320 teeth. The tuning fork was now asymmetrical, only one arm being electronically excited which made production more efficient and allowed significant savings. Furthermore, the settings for the jewels were abandoned: jewels were now simply pressed into the main plate and bridges. Caliber 2190 did not make it into production, but the calibers 2191, 2192 and 2193 were mass produced.

The Mini-Accutron for ladies' watches was not developed in Jackson Heights but by Bulova Bienne under the direction of Hans Schaller. This caliber (230 or 2300) was launched in late 1971; it measured 19.4 mm (8¾ lines). With a date disc, the height was 5.2 mm; the tuning fork frequency was increased to 480 Hz. The tuning fork mechanism still acted as an oscillator and a motor, but the construction was quite different from that of Hetzel's old caliber 214.

The purpose of the flat coil on the main plate was not to generate a magnetic field: it was simply meant as a current conductor in a plane parallel to the tuning fork on the prongs of which small samarium-cobalt magnets were glued. The field of those magnets was oriented at a right angle to the plane of the coil: the resulting Lorentz force was parallel to the main plate and set the tuning fork in motion. Caliber 2300 was equipped with a silicon transistor from the very beginning. The variants were 2301, 2302, 2303, 2312 and 2313. The model with a diameter of 23.5 mm was designed for men's watches.

Bulova developed several other models of tuning fork calibers. Among those, the tonneau-shaped movement 2210 launched in 1973 only displayed hours and minutes. The dimensions were 19.4 mm x 17.4 mm x 4.40 mm; it thus was the second smallest Accutron caliber, only the ladies' watch caliber 2300 was slightly smaller. The oscillation frequency of the tuning fork was 440 Hz (this corresponds to the tuning pitch A). The semicircular arms of the tuning fork surrounded the battery. The ratchet wheel had

5.9. Bulova Accutron caliber 2191

5.10. Bulova Accutron ladies' caliber 2210

5.11. Bulova Accuquartz caliber 2242

5.12. Bulova Mini-Accutron caliber 2306

270 teeth, the first stage of the gear train was an Archimedean screw. This model was not very reliable, it was only produced in small numbers.

Caravelle

In the perspective of the 1960s and 1970s, the Accutron was rather expensive. Bulova needed to have other, less costly electronic watches in its catalog. This gap was filled by electrodynamic calibers that were purchased from outside vendors.

In 1970, Bulova received some 150,000 pieces of the transistorized electromechanical caliber LIP R50; Bulova gave it the designation Caravelle 7 OT. This 7¾-line caliber was mostly used in ladies' watches. Starting in 1967, Bulova equipped pocket watches and wristwatches with the 12½-line electrodynamic Citizen caliber X-8 and called it Caravelle 12 OTC; the Citizen caliber 0840 received the Bulova designation 12 OUC. Three years later, Bulova adopted the Citizen caliber 0830 and renamed it 12 OUCD. As a stopgap measure, the ESA-Dynotron caliber 9150 was used in Caravelle watches under the designation 12 UECD.

Grave Consequences

The Accutron was a sensation and made big headlines, particularly in Switzerland. For the first time in the history of horology, a watch had been designed that dispensed with the pendulum or torsion pendulum as a time base (Louis-Clément Breguet's clock of 1866 with a mechanically excited tuning fork had been forgotten).

It now became very obvious that new, fully electronic movements should urgently be developed. In January 1962, the "Centre Electronique Horloger" (CEH) was officially established. This is where the quartz prototypes Beta 1 and Beta 2 as well as the commercial caliber Beta 21 were developed within the next five years.

Bulova would have loved to equip the NASA-Astronauts of the Apollo Lunar expeditions with Accutrons. NASA already was a Bulova customer, since it had used five Accutron calibers in the Gemini capsules in 1967. Bulova president Omar Bradley tried everything, using the prestige of a retired five-star general. In vain: the Omega Speedmaster turned out to be much superior to the Accutrons in the extremely severe tests to which potential astronaut watches were subjected. Another point was that NASA wanted a purely mechanical wristwatch: there was a real danger that electronics might be affected by electromagnetic disturbances in space, particularly by sun flares. On the other hand, according to Wikipedia, the instrument panel clocks and timekeeping mechanisms in the Apollo mission spacecrafts were Accutrons.

Several Accutron watches made it into space between 1966 and 1970: astronauts of the "Explorer" and "Apollo" missions took them along as (perfectly legal) private watches. One of the instruments that were left on the moon by Apollo 17 astronauts (a seismometer) was equipped with an Accutron timer. Starting with 1974, Accutron movements with alarm and timer functions were ordered by NASA for the Spacelab program that followed the Apollo expeditions. In each of a total of 46 space missions, at least one Accutron was involved. An Accutron was used in the cockpit of many civilian and military aircraft, including the US president's Air Force 1.

Accuquartz

In 1971, Bulova USA engineers designed the "Accuquartz" caliber 224. This was a quartz wristwatch based on an Accutron 218 with a diameter of 29.7 mm and a height of 4.9 mm. The ratchet wheel had 320 teeth. The tuning fork was not used as a time base anymore, but was just a motor. It was excited by a 341.33 Hz signal that was derived from a quartz resonator (32,768 Hz). At the Basel Watch Fair of 1971, three Accuquartz calibers were shown. 2240 was a three-hand caliber, 2241 displayed the date, 2242 was a day/date caliber. Those 12¾-line calibers had a height of 5.3 mm (5.5 mm in the case of caliber 2242). The chip was made by Intersil, the yearly rate was indicated to be 3 minutes (i.e. less than one half second per day). Universal Genève used this caliber in their watches, the Company having a close relationship with Bulova; this ended with Bulova buying Universal.

There had already been a precedent for a quartz synchronized Accutron. Bulova Bienne designed a very pretty quartz clock on the base of an Accutron 214. It was powered by two batteries and used a hybrid circuit for dividing the quartz frequency all the way to 360 Hz. This small clock was exhibited at the 1964 Basel Watch Fair and was a sensation: in that year, it was the world's smallest quartz clock (see Chapter 6).

Hetzel's Swissonic

After the successful launch of the Accutron, Max Hetzel returned to Switzerland in 1963 where the Centre Electronique Horloger (CEH) in Neuchâtel had offered him a well-paid job. However, Hetzel played himself up as a prima donna, refusing to work with CEH director Roger Wellinger. After all, Hetzel had invented a completely new watch and had brought it into mass production. He was a real professional and he rated the CEH people as mere beginners and amateurs. In his eyes, developing a quartz wristwatch was an exercise in futility: it could simply not be done. For this reason, Hetzel was first given office space in the laboratory for conventional watch technology LSRH, but was officially "loaned out" to the CEH.

This is where Hetzel developed a new variant of his tuning fork movement. The mechanical resonator was ring-shaped and quite massive: it was placed at the very edge of the movement. The frequency was 480 Hz; due to the shape and mass of the resonator, the position error was smaller than in the case of the Accutron, but still not negligible: it was indicated to be ± 1.5 seconds per day. The ratchet wheel with its 200 teeth had a diameter of just 1.5 mm.

Hetzel named his new movement "Swissonic," it was decided that it should be mass produced. For this purpose, a special section of the CEH was created in Gals near Marin in 1966, Hetzel being its director until 1969. His first assignment was to install a pilot line that would produce 5000 Swissonic movements within three years. After a good start, CEH patent atternoys came to the conclusion that the Swissonic movement infringed upon Hetzel's own patents (that belonged to Bulova). The line was stopped after about 3000 pieces had been produced. They never hit the stores, but are avidly sought by collectors.

Without a patent license from Bulova, the commercial production of the Swissonic was impossible. The Gals pilot line thus was completely useless. Nevertheless, Omega bought it in 1969, moved it to Bienne and integrated it into their Elresor SA subsidiary (the name was later changed to SSIH Equipement SA). Hetzel had finally been hired by Omega; until 1976 he worked on ways to circumvent his own tuning fork watch patents. He succeeded with an entirely novel and quite robust tuning fork motor.

Hetzel's Megasonic

Hetzel's new motor consisted of an asymmetric tuning fork with a resonance frequency of 720 Hz. The actual micro-motor was housed in an oil-filled capsule attached to the long prong of the tuning fork. It measured just 6.5 mm x 2.5 mm, Hetzel named it "souris" (mouse). The capsule housed a 1.2 mm ratchet wheel with 180 teeth that rotated between two sapphire discs at the rate imposed by the vibrations of the tuning fork. Two ruby banking springs forced the wheel to rotate always in the same direction.

A four-pole platinum cobalt magnet on the ratchet wheel transferred its momentum magnetically to an adjacent disc with twelve alternating north and south poles. This disc was mechanically coupled to the gear train of the movement. Thanks to the high frequency

5.13. Swissonic prototype

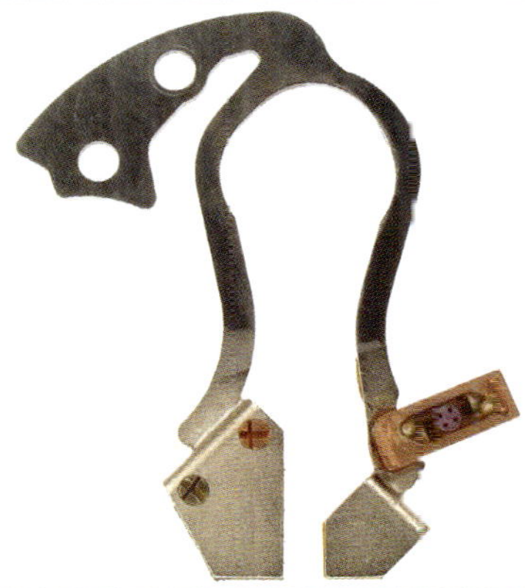

5.14. Omega Megasonic caliber 1220, tuning fork and motor

5.15. Omega Megasonic caliber 1220

5.16. Omega Electronic f = 300 Hz, caliber 1250

of the tuning fork (720 Hz instead of the usual 360 Hz in the Accutron) the rate was ± 1 second per day. The circuit comprised two complementary silicon transistors. Omega showed this extremely interesting construction to the media in 1973, mass production started in early 1974.

But in the end, only some 43,000 watches with the "Megasonic" movements (Omega caliber 1220 with date and caliber 1230 with day/date, 29.0 mm x 5.0 mm) were produced. Production was discontinued in 1975, as the magnetic coupling between the motor and the movement turned out not to be very reliable. Also, the rather thin tuning fork was subject to malfunctions. The Megasonic arrived in the marketplace at a time when the price of more accurate quartz movements dropped dramatically. For a few more years, Omega equipped some watch models with the Mosaba caliber 9162 from ESA, giving it the designation 1250.

CEH Project Alpha

In order to quantitatively simulate the behavior of mechanical oscillators (Project Alpha), CEH Director Roger Wellinger hired the civil engineer and mathematician Heinz Waldburger in the USA. Waldburger joined the CEH team in 1963. For a while, he worked in parallel (if not in competition) with Max Hetzel, literally the "Pope" of the tuning fork watch. In his function as group leader, Waldburger was allowed to consult with Hetzel who was quite friendly and open. He answered all the questions the younger colleague asked: tapping his long experience as an engineer was very useful. But he just answered questions and never volunteered any comments.

Waldburger soon figured out how to model mechanical resonators so they could be used in the real world of watches. In collaboration with an assistant at the Technical University in Zurich (ETH) he applied the method of finite elements in order to find the best possible shape for such a resonator. In this way, Waldburger developed the H-shaped resonator, in which both pairs of legs formed an elongated rectangle. A flattened figure 8 also turned out to be a good solution for a 300 Hz oscillator fitting into a watch movement.

Such systems were excited by electromagnets; the ratchet wheel and pawl mechanism transforming the tuning fork vibrations into a rotary motion was quite similar to that of the Accutron. But thanks to the symmetry of the oscillator, there was theoretically no position error with respect to gravity.

Because of the unusual geometry of the new oscillators, a high power mercury oxide battery had to be developed. It had to fit between the legs of the H. Leclanché in Yverdon took care of this development. However, the battery seals kept leaking, creating a big problem. It finally turned out that only round batteries can be sealed with near total reliability. By then the quartz movement project was close to its final breakthrough at the CEH. Project Alpha was cancelled in 1966; Waldburger left the CEH in that year.

Ebauches SA's Mosaba

Quite independently of the CEH and Bulova, engineers and technicians at the Ebauches SA laboratory in Neuchâtel developed a tuning fork movement with a significantly better performance than that of the Accutron. The project was started in 1961 under pressure from ETA, whose director Heinrich Stamm was convinced that the Accutron could be greatly improved upon. The team, led by André Beyner, succeeded in developing the "Mosaba" (an abbreviation of Montre sans balancier – Watch without a balance wheel). The dimensions were 29.0 mm x 4.8 mm (the height of the day/date model was 5.55 mm). Counterweights on the 300 Hz tuning fork made it nearly insensitive to position changes. The daily rate was ± 1 second and therefore was significantly better than that of the Accutron. The position error of the Mosaba was just ± 0.5 second per day.

Despite of those innovations (particularly the compensation of the position error), the Mosaba could not be produced and marketed without a license from Bulova. Thus, Ebauches SA contacted Bulova in the mid-1960s. The Ebauches team included Leo Dupasquier, François Dupasquier and André Beyner, while the Bulova-Team consisted of Harry Henshel, Sol Flick, Jules Sandoz, and Hans Schaller. The negotiations ended successfully in 1968.

The license cost Ebauches SA a cool million dollars, plus a royalty of several dollars for each Mosaba that was sold. According to some historical sources, Max Hetzel joined the CEH to develop the "Mosaba" that was subsequently produced by ETA in Grenchen and Ebauches Electroniques Marin (EEM). The truth is that Hetzel had absolutely nothing to do with the development of the Mosaba: this was a pure Ebauches SA project and it competed with the CEH's Project Alpha.

The Mosaba was shown to the media in 1967 and it was marketed under the "Swissonic-100" brand. The caliber numbers were ESA 9162 (with date) and ESA 9164 (with day/date). ETA produced the first batches, later on EEM in Marin took care of the electronic module while ETA made the mechanical parts. The frequency of the tuning fork was 300 Hz, the ratchet wheel had 300 teeth. Many watch brands bought the Mosaba caliber and cased it, e.g. Omega, Movado, Zenith, Longines and

IWC. A cheaper version without the Swiss made logo was developed for the Asian market, it was assembled in Hong Kong; the designations were 536.111 (with date) and 536.121 (with Day/Date).

At the Neuchâtel Observatory Concours of 1966, the Mosaba beat all the other tuning fork watches that also participated. But two years later, Hetzel's Swissonic with the massive, ring-shaped resonator he had developed at the CEH beat the Mosaba and was attributed the first prize.

The most remarkable Mosaba model certainly was caliber ESA 9210 with a day/date chronograph module supplied by Dubois Dépraz. A total of 21,000 pieces were produced for Omega, Longines, and Certina. Those tuning fork chronographs are quite rare and fetch high prices among collectors.

5.17. ESA caliber 9162 Mosaba

Hi-Sonic by Citizen

Towards the end of the 1960s, Citizen signed a joint venture agreement with Bulova. With 51% of the shares, Bulova was the majority shareholder of the joint venture the purpose of which was the production and marketing of tuning fork watches in Asia. This agreement was quite a natural development, as Citizen had started doing business with Bulova in 1960 and had supplied the Americans with large quantities of mechanical movements. It was only logical that Citizen acquired Bulova in 2008.

Citizen's first tuning fork watch was launched in 1972 under the brand Hi-Sonic 05 with the calibers 3700 (basic caliber with day/date), 3701, 3701B, 3710 date only), 3711, 3720 (three hands), 3721 and 3722. Those calibers were nearly identical with the Bulova caliber 218 (12½ lines x 5.5 mm). But there was an extra: a switch to disconnect the battery. The Bulova caliber 219 with a one-prong tuning fork was also manufactured by Citizen, but for unknown reasons it was a mirror image of the original.

5.18. Citizen Hi-Sonic caliber 3701B

5.19. Citizen Hi-Sonic caliber 3711A

The Bulova-Citizen joint venture for the production of the Accutron tuning fork turned out to be a strategic error. When the Hi-Sonic was launched, analog and digital quartz watches had already appeared in the marketplace. In the beginning, Citizen executives felt safe, as they had electronic watches. But this was not the electronics customers wanted. Citizen had to embark on a crash program in order to provide quartz watches to the customers.

Kamertoni, the Russian Copy

In the USSR, the First Moscow Watch Plant started developing an exact copy of the Accutron caliber 214;

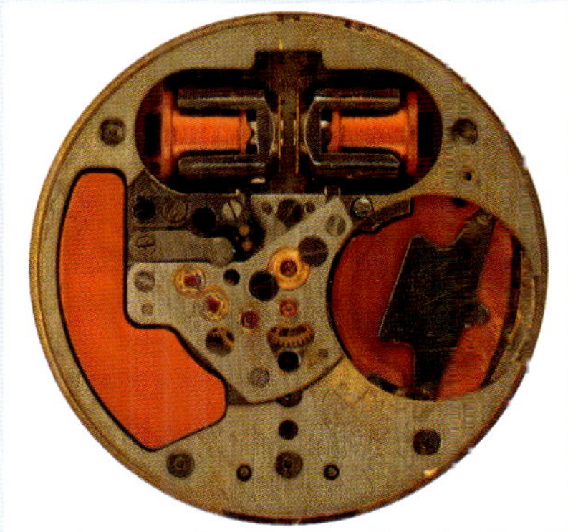

5.20. Kamertoni 2937, USSR imitation of Bulova caliber 214

production of caliber 2937 started in 1968. Watches with this movement were sold under the brands "Kamertoni" and "Slava Transistor." Nobody worried about obtaining a patent license from Bulova; nothing is known about production figures. One Kamertoni model was fitted with a date disc that ran nearly frictionless on six jewels. Quite obviously, Russian designers were not aware of the tremendous power reserve that characterized the tuning fork movement.

Cutting the ratchet wheel was quite a challenge. For this task, a near-vibrationless production unit had to be excavated in a Moscow suburb, well below the level of the subway. The Kamertoni is much sought by collectors, but original parts were disposed of or were used up. This is why there are only very few functioning units left. Accutron parts could be used, but a serious collector would never do such a thing.

Elton – Czech Prototypes

The engineers of the Czech watch plant PRIM in the Bohemian town Nové Mesto nad Metuji were fascinated by the Accutron. Soon after its launch, they carefully hand-built a series of 12 prototypes under the Elton brand. They were exact copies of the Bulova caliber 214. All the movement parts were made at the PRIM plant, but they probably are not interchangeable with Accutron parts. An industrial production was out of the question: The CSSR was then a communist country and hardly anybody would have been able to afford such an expensive timepiece.

Tianjin Watch Factory

For the Chinese, it was important to go through the full learning curve concerning electrical and electronic watches. This is why the Ministry for Light Industry ordered the Tianjin Watch Factory to design and build a tuning fork watch in the mid-seventies. This caliber was called SD2. It obviously was strongly "inspired" by Bulova's Accutron. It seems that there never was a mass production of this caliber, the Ministry was content with a small test series.

Universal Genève

In the 1960s, Universal Genève was a well-known and respected watch brand. It had been cooperating with Bulova for many years. Eventually Bulova bought out the company. Starting in 1967, Bulova supplied Universal with a limited number of caliber 218 tuning fork movements. Later on, Bulova supplied the quartz-synchronized caliber Accuquartz 2240 for men's watches. For the ladies' watches 47 and 48, Universal used caliber 2300 that had been developed by Bulova Bienne.

The Quartz Watch Wins It All

As soon as the quartz watch had proven its worth, the production of tuning fork watches was discontinued everywhere: this happened in 1977. In the relatively short product life of 17 years, some 4 to 5 million Accutrons were sold. In addition to that, Ebauches SA produced several hundred thousand Mosaba Swissonic movements that had been manufactured under a Bulova license from 1966 to 1976. They were cased by Baume & Mercier, Certina, Derby, Eterna, IWC, Longines, Omega, Movado, Tissot, and Zenith. Citizen Watch produced about a million "Hi-Sonic" watches on the base of a joint venture with Bulova covering the Accutron calibers 218 and 219.

Clocks with Mechanical Resonators

Starting with the mid 1960s, Copal and the Japan Electric Clock in Japan as well as Diehl and Junghans in Germany, produced table and wall clocks as well as alarm clocks in which the oscillator was an electronically excited spring or tuning fork. The frequency was either 100 Hz or 300 Hz. Unlike to the Accutron, the resonator was not used as a motor. The oscillations were transferred to the gear train by means of a so called Clifford escapement or a magnetic coupling.

This type of magnetic coupling was patented in the USA in 1967. Only one prong of the symmetrical tuning fork was electronically excited. A permanent magnet was attached to the end of the other prong that interacted with a wheel with magnetic teeth and turned it by one tooth for every cycle of the tuning fork's oscillation. Copal produced such movements: they are marked "Lic. Bulova." It is strange that Copal which had its own patent, had to pay a license fee to Bulova. The Omega caliber Megasonic with a magnetic coupling did not interfere with any Bulova patent.

6. *Frequency Standards and the First Quartz Clocks*

The fact that in the 1970s and 1980s the quartz watch almost fully replaced mechanical timekeepers has two simple reasons. Even the least expensive quartz watch is significantly more accurate than a certified mechanical chronometer. Furthermore, the latter is anywhere between a hundred and ten thousand times more costly. The quartz watch owes its accuracy to the remarkable physical properties of the quartz crystal. Mere silicon dioxide (SiO_2), it is very common and very low-priced, yet it may be used as an electronically excited, very stable resonator.

Indeed, quartz is piezoelectric: due to the way its atoms are packed, a mechanical force applied in certain directions causes electrical charges to appear at the ends of the crystal. The converse also applies: the shape of a quartz crystal is affected by a voltage that is applied to it. Those phenomena were carefully investigated around 1880 by Pierre Curie and his brother Jacques in Paris. Piezoelectric effects are not limited to crystals: even amorphous organic and biological substances may be piezoelectric.

If an alternating voltage with a given frequency is applied to a piezoelectric crystal, it will vibrate mechanically at the same frequency. The same crystal will generate an alternating voltage if it is subjected to mechanical vibrations. These effects are particularly strong at the sharply defined resonance frequency of each crystal. The resonance frequency is defined by the size of the crystal (or of the piece that is cut from it) as well as by the three-dimensional orientation of the cut.

The Electronic Frequency Standard of Horton and Marrison

By cutting and chemical etching, the resonance frequency of a piece of quartz may be adjusted in a very wide range, from a few kilohertz to many megahertz. Quartz also very effectively stabilizes an electrical signal with a frequency corresponding to its resonance frequency. The crystal can be excited by the approximate resonance frequency: it will emit a narrow-band, very stable signal, the frequency of which is exactly defined.

First Steps of a Revolution

The near-complete replacement of the traditional mechanical watch by the quartz watch in the early 1980s had dire consequences for the watch industry.

6.1. Horton and Marrsion's first frequency standard (Bell Laboratories, 1927)

6.2. Synchronous motor of the first quartz clock (Bell Laboratories, 1929)

6.3. Triply redundant quartz frequency standard of the Bell Telephone Company, 1930

It was forced to make a totally new start on a technologically very different base. Many companies did not survive what is called the quartz revolution.

The first step leading to that revolution was made in 1927 by two electrical engineers by the name of Joseph W. Horton and Warren A. Marrison. Both of them worked at the "Bell Lab," the research station of the Bell Telephone Company in New York City. On the base of work by Cady about stabilizing oscillator circuits, Horton and Marrison built a quartz stabilized frequency generator. Almost by accident, this frequency generator became the world's most accurate timekeeper in 1931.

Marrison was born in Canada; he obtained his B.S. degree at the Queen's University in Kingston and went to Harvard for an M.S. He was then hired by Western Electric who ran the Bell Lab. He stayed there until retirement and obtained more than 60 patents. He was an enthusiastic radio amateur, which became quite useful for his professional career. Marrison's colleague Horton on the other hand left Bell Labs in 1928 and joined a radio company before the quartz clock was completed.

Frequency Chaos

In the course of the 1920s, radio made a real breakthrough all over the world. In the United States and many other countries, the government licensed private companies to use certain parts of the electromagnetic spectrum for broadcasting. It was extremely important that each transmitter stuck strictly to its assigned frequency within a narrow bandwidth; otherwise it might interfere with neighboring transmitters.

Thus it was necessary to measure and control the frequency (or wavelength) of radio waves very accurately. For that purpose, highly reliable frequency generators were required. Horton and Marrison got the assignment to develop such a frequency generator, so Western Electric would be in a position to accurately measure the frequency of radio transmitters. Bell Telephone also had a direct interest in this work, as phone calls could be transmitted via lines or cable as well as via radio waves.

The device built by Horton and Marrison was the size of a cupboard and included a power supply of several kilowatts and a circuit comprising dozens of radio tubes with a quartz oscillator, various amplifiers, and dividing circuits.

A New Level of Precision

Electronic frequency generators had appeared in the early 1920s. They consisted of an electronic oscillator that excited a tuning fork to its resonance frequency of 100 Hz. Horton and Marrison took quite a different approach: their resonator was a mechanically and chemically very stable quartz crystal that could be excited with minor losses to the much higher frequency of 100,000 Hz.

With quartz they made just the right choice, which has by now been confirmed billions of times. The Bell engineers were aware of the temperature dependence of a quartz resonator. This is why they fitted it into a well insulated oven that was electrically heated to 40° C, ± 0.01°.

A simple 100,000 Hz oscillator with a platelet-shaped quartz resonator (in later versions of the frequency standard, a ring-shaped resonator was chosen) would have only had limited applications. Still higher frequencies were needed, which were available as harmonics (200,000 Hz, 400,000 Hz etc.) of the fundamental frequency. They just had to be filtered and amplified.

Furthermore, the fundamental frequency could also be divided stepwise by providing a resonant circuit with a lower frequency, let the two signals interfere and then extract the difference of the two frequencies. The latter could be divided again in the same way. Horton and Marrison called their apparatus "submultiple frequency generator." Unfortunately, only fragments were preserved for posterity. It was so accurate, that it soon replaced the old 100 Hz tuning fork oscillators. In a triply redundant system it became the frequency standard of the huge Bell telephone system.

The Crystal Clock

Marrison knew absolutely nothing about watchmaking, but in 1929 he remembered that time is the inverse of frequency (Horton had already left Bell Labs). By using the lowest frequency of his very stable generator (1000 Hz) to drive a synchronous motor, he would have a clock. All that had to be done was to add gears to drive hour, minutes- and seconds-hands.

This is how Marrison's "crystal clock," the world's first quartz clock, came into being. Marrison compared its rate with the best pendulum clocks built by Riefler and Shortt, on the base of the radio time signals of the US Naval Observatory. They were obtained by observing the transit of well-known stars. Marrison calculated that his clock had a rate of 0.3 second per year, which was an order of magnitude better than that of the best pendulum clocks. This was a true breakthrough and it had been set off by electronics.

In 1932, Marrison was granted a patent for the quartz clock (USP 1,788,533). He designed a second clock of this type with a 50 kHz quartz resonator, the frequency of which was divided to 5000 Hz. This signal drove a 100-pole synchronous motor that was connected to a clock movement.

Today, billions of people all over the world wear an amazingly accurate and low-priced quartz watch. We owe this to Horton and Marrison who literally invented the quartz timepiece in the late 1920s. This was a landmark in timekeeping. Within just a few years, quartz clocks were also designed in Germany (1932), Great Britain (1932), France (1936) and Japan (1937). They were constantly improved upon during World War II and the postwar period.

The Quartz Clocks of Scheibe and Adelsberger

In 1932, Udo Adelsberger and Adolf Scheibe, two physicists at the Physikalisch-Technische Reichsanstalt (PTR) in Berlin, designed the two first German quartz clocks (QI and QII). They started out with a 60 kHz quartz resonator that was lodged inside a double thermostat maintained at 36° C. Inductive feedback of low power transmitters allowed the three stage division of the 60 kHz quartz signal in the proportions 6:1, 10:1, and 3:1 all the way to 333 Hz. After amplification, this signal was fed a synchronous motor actuating a set of gears. The daily rate was ± 0.002 second. Time was not displayed with hands, but with a high speed recorder writing time marks on a paper strip. Some components of those first German quartz clocks have been preserved. They are exhibited at the town museum in Zeulenroda, which is located in Southeastern Thuringia. It is the birthplace of Adolf Scheibe.

As soon as QI and QII worked, Adelsberger and Scheibe started designing a pair of improved quartz clocks with the designations QIII and QIV. They had clearly recognized the weak point of the first design, i.e. the temperature dependence of the quartz resonator frequency. The temperature had to be held within the very narrow range of ± 0.002° C in a room where the temperature varied between +28° C and +12° C. For this reason, quartz bars with a 29 times higher mass and hundred times lower temperature coefficient were used. They were mounted in evacuated glass tubes.

After systematically comparing their four quartz clocks, Adelsberger and Scheibe concluded that the earth's rotation is subject to unpredictable fluctuations. At that time, the second was defined as 1/86,400th of the average duration of the solar day. The latter is not a constant, but varies by several milliseconds at irregular

6.4. Quartz clock of the Geodätisches Institut (Geodesic Institute) in Potsdam, built by F. Pavel and Werner Uhink, following plans for the Quartz Clock III by Udo Adelsberger and Adolf Scheibe at the PTR Berlin 1933-1934.

6.5. Quartz clock by Scheibe and Adelsberger at the Physikalisch-Technische Reichsanstalt Berlin (PTR), 1935

6.6. Quartz resonator by Adolf Scheibe (PTR Berlin, about 1930)

intervals in the course of the year. This was officially confirmed in 1948 by the Bureau International de l'Heure in Paris. Despite the war, the PTR physicists designed five more types of quartz clocks until 1945. They were installed on massive concrete blocks in the cellar of the Institute. The actual clocks were mounted in oblong wooden cases that sat on tennis balls for attenuating vibrations.

It is remarkable that commercial quartz clocks were available in Germany as early as 1938. They were developed by Lothar Rohde at the Physikalisch-Technisches Entwicklungslaboratorium (PTE) in Munich (the company was renamed Rohde & Schwarz after World War II). Rohde succeeded in miniaturizing the usually cupboard-size quartz clock to the size of an older hi-fi system. It was based on a quartz stabilized tuning fork. Such instruments were very costly, only big industrial and state-owned research laboratories could afford them. Yet, the Rohde quartz clock was still available in the 1950s.

Dye and Essen in Great Britain

In Great Britain, too, the motivation to design quartz clocks was given by the rapid development of radio: exactly defined frequencies had to be measured and generated. The astronomers who, until then, had been responsible for measuring time and the generation of time signals were less than enthusiastic about the quartz technology that actually endangered their jobs.

Royal Observatory versus NPL

In the early 20th century, the British Royal Observatory was responsible for the exact time in the UK. Like everywhere else, time was measured by observing the transit of stars at the meridian: this information was used to set precision pendulum clocks. Time signals were distributed via telegraph lines.

It is understandable that in the late 1920s the National Physical Laboratory (NPL) in Teddington near London and not the Greenwich Observatory was mandated by the Post Office to design a very accurate frequency generator. David William Dye was the project leader and Louis Essen, a fresh university graduate, became his assistant in 1929. Dye had quite some experience with electronically excited 1000 Hz tuning forks. The logical thing for him was to continue this line of work. Any integer multiple of the fundamental frequency could be filtered out and amplified.

Pulses at the rate of seconds were important for comparison with the astronomically generated pulses. They were obtained by a two stage frequency division to 1 Hz using synchronous motors. The tuning fork project ended in 1931. It turned out that tuning forks acting as clocks were about as accurate as the best precision pendulum clocks.

Dye was of course aware of Horton and Marrison's work in the USA and decided to design a similar quartz-controlled frequency standard. His resonator was a 10 cm diameter quartz ring that was cut in a direction perpendicular to the optical axis of the crystal; the resonance frequency was 20 kHz. This cut was characterized by very stable oscillations, but it was particularly sensitive to vibrations and temperature changes. Mounting this resonator also posed difficult problems.

Progress Oriented Royal Mail

Dye quite unexpectedly passed away in 1932 and Essen took over the quartz project, replacing the quartz ring by a less temperature sensitive cut. Careful measurements showed that the quartz clock was significantly better than pendulum clocks and even the astronomically derived time signals.

Within the next six years, Essen designed a grand total of six quartz clocks, the best of which had a rate of 0.33 second per year. By 1940, the Royal Mail started to use self-built quartz clocks. They followed the design Essen had developed and were built by the Post Office Research Station in Dollis Hill. By 1948, the total was 36 pieces. They were always used in groups of three and constantly compared to each other. This was more accurate than setting the clocks according to the Observatory's time signals.

Essen was aware of the fact that the quartz clock was not the last word of timekeeping. At the NPL, he built the UK's (and Europe's) first caesium stabilized quartz clock in 1955 and strongly advocated the introduction of an atomic time standard. Essen participated in formulating the new atomic definition of the second, as rates of one second per million years could be achieved with the caesium clock.

In 1967, all relevant authorities worldwide decided to redefine the second as the duration of 9,192,631,770 periods of the radiation corresponding to the transition between the two hyperfine levels of the ground state of the caesium-133 atom. Time had now become a purely physical dimension and was totally decoupled from astronomy. Caesium clocks and the less costly rubidium clocks have become absolutely indispensable for telecommunication and GPS navigation.

Jouaust and Decaux in France

The first French quartz clock was designed in 1936 by F. Jouaust and B. Decaux at the Etablissements Bélin, in cooperation with scientists at the Laboratoire National d'Electricité. It consisted of two massive electronic racks and can be seen today at the Musée du Temps in Besançon. The 100 kHz quartz resonator was bar-shaped. The device did display the exact time, but was really designed not as a clock, but as a frequency standard.

Quite generally, French scientists were quite skeptical about the performance of quartz devices, which was mostly due to the 1934 publications by the famous astronomer Nicolas Stoyko working at the Bureau International de l'Heure (BIH). Stoyko had mathematically analyzed and compared the rate of quartz clocks and precision pendulum clocks. His conclusion was that Scheibe and Adelsberger had made systematic errors in evaluating their quartz clocks. According to him, the rate of those clocks was only about 1/1000 of a second per day and not 1/10,000 as they had claimed.

According to Stoyko, the stability of quartz clocks was only superior to that of pendulum clocks in the short term. In the long term, both types of clocks were about equivalent. It turned out much later that Scheibe and Adelsberger had been right after all, their quartz clocks did indeed achieve a daily rate of 1/10,000 of a second.

World War II strongly impaired the further development of quartz clocks in France. But the postwar years were quite favorable: in 1948, the Laboratoire National de Recherche had four quartz clocks and the Paris Observatory had two of them. However, they were still set according to the astronomically synchronized pendulum clocks of the BIH!

Swiss Quartz Clocks for Industry and Academe

It seems that the negative attitude of the French, who thought that quartz clocks were not more accurate than precision pendulum clocks, also influenced scientists and engineers in Switzerland. This might explain the fact that first Swiss quartz clocks were not designed in one of the strongholds of the watch industry in western Switzerland, but at the Technical University in Zurich (ETH) – a full 15 years after Marrison had proven the concept in the USA.

6.7. Quartz clock by Dye and Essen at the National Physical Laboratory, 1936

6.8. Quartz clock of the National College of Horology in Clerkenhall, 1954

6.9. Quartz clock by Jouaust and Decaux, 1936. Exhibited at the Musée du Temps in Besançon (France)

AFIF and the University of Geneva

At the ETH, where the Department of Industrial Research, AFIF, was chaired by Prof. E. Baumann, graduate students F. Berger and H. Brandenberger started to design the very first Swiss quartz clock in 1944. It consisted of a stack of electronic racks measuring 220 cm x 54 cm x 40 cm and worked in 1948.

Three of those clocks were built. One of them stayed at the ETH (supposedly at the Electronics Institute chaired by Prof. Weber, but its traces were lost). The two other ones went to the Neuchâtel Observatory and the Horological Research Laboratory (Laboratoire Suisse de Recherches Horlogères, LSRH), respectively, also in Neuchâtel. Unfortunately nothing of those clocks was preserved: they were dismantled after being replaced by better and more compact designs. The rather expensive parts were re-used in other instruments.

The resonator was a quartz platelet measuring 27 mm x 11 mm x 2 mm oscillating at 100 kHz. Both sides were coated with silver and contacted by silver wires. This resonator was suspended in an evacuated glass envelope with a standard electronic tube socket. This unit was enclosed in a thick-walled aluminum container that was thermally insulated with glass fiber mats. Four wire heaters were mounted on the outside and formed a Wheatstone bridge.

Every second a weak current pulse was sent through the heaters. If needed, a relay was closed: this switched on the heater current for a short time, in order to keep the temperature of the quartz resonator at 50° C within 1/1000° C. The electronic oscillator for the quartz resonator and the amplifiers were also kept at a constant temperature; voltages were carefully stabilized. Aging processes in the electronic parts only had a negligible effect on the quartz frequency.

The resonator frequency of 100 kHz was divided down to 1 kHz (1000 Hz) by the electronic circuit and used for actuating a three-phase current synchronous motor that generated a very stable, 0.05 second pulse every second and two such pulses on the minute. Those pulses controlled an electric clock that displayed hour, minute, and second. The dividing circuit consisted of a 1000 Hz electronic oscillator the harmonics of which were amplified and synchronized with the signal of the quartz oscillator.

At the Geneva Observatory, P. Challande, A. Epitaux, and J. Freyman also designed a quartz clock, quite independently of the AFIF in Zurich; tests were started in 1947. The 100 kHz quartz resonator was stabilized within ± 0.01° C, the frequency being divided to 500 Hz by means of synchronized multivibrators. The 500 Hz signal was used for driving a synchronous motor that actuated a conventional clock movement. A second stage of division brought the frequency to 4 Hz that were used for reference. The average daily rate was ± 0.4 milliseconds.

Oscilloquartz

In 1949, F. Berger and H. Brandenberger, Prof. Baumann's postdocs at the AFIF, were hired by Ebauches SA in Neuchâtel in order to set up a new subsidiary called Oscilloquartz SA of which Berger was the CEO. Sydney de Coulon, one of the Ebauches directors and a true visionary had realized the enormous potential importance of electronics in horology at a rather early stage. The first products of the new company were quartz clocks that had been ordered by the Observatories in Paris, Brussels and Neuchâtel. It is clear that the know-how Berger and Brandenberger had created at the AFIF was put to good use for developing a new and greatly improved clock. The one bought by the Neuchâtel Observatory has been preserved and can be seen at the Musée International d'Horlogerie (MIH) in La Chaux-de-Fonds.

The second Oscilloquartz product was a quartz-stabilized reference frequency standard for the telephone company that ordered a whole series of them. The suppliers of telephone equipment also became important customers. Those devices weighed around twenty kilograms and were tailor made for the specific needs of each customer. The bar-shaped quartz resonators were made in house; they were cut from natural Brazilian quartz crystals. Lens-shaped high frequency quartz resonators were supplied by a vendor in France.

Oscilloquartz rapidly developed into the Swiss competence center for quartz resonators in the 1000 Hz to 100,000 Hz frequency range. Contact and support wires were soldered to the metalized surfaces of carefully ground and polished quartz platelets. They were mounted in evacuated glass tubes with a standard socket, which made them look like electronic tubes. This design was adopted by several other manufacturers of quartz resonators. In the 1960s, the Centre Electronique Horloger (CEH) in Neuchâtel and other pioneers of the quartz wristwatch purchased their miniaturized quartz resonators exclusively from Oscilloquartz. The company also supplied quartz resonators for precision clocks and marine chronometers, the best known example being the Ebauches-Oscilloquartz marine chronometer B-1390.

Oscilloquartz staff ignored the work of the Swiss physicist A. Carolus, a professor at the University of Freiburg (Germany). Carolus designed the first quartz tuning fork; in 1954 he showed it at the International Congress of Chronometry in Paris. He had gathered convincing data showing that it was far superior to the classical bar- or platelet-shaped quartz resonators in terms of shock resistance. Furthermore mounting it was extremely simple as the handle could be glued to any type of stable support.

In 1955, the Brush Company in Cleveland, Ohio developed a process for the industrial production of synthetic quartz crystals. At last there was competition for the natural quartz crystals that had to be imported from Brazil. All of a sudden, supply ceased to be a problem. However, the best possible cut and the shape of quartz resonators remained a subject of debate among specialists for quite some years.

Seiko engineers on the other hand were easy to convince and used tuning fork-shaped quartz resonators in all of their quartz wristwatches, including the very first prototypes that made quite a sensation at the 1967 chronometer contest of the Neuchâtel Observatory, the famous Concours d'Observatoire. Seiko stuck to this technical concept in the very long term, while Swiss manufacturers kept the expensive and delicate bar-shaped resonators for all too long. As already mentioned, a very constant temperature was essential for high precision quartz clocks and frequency standards, as the resonance frequency of a quartz crystal is strongly temperature dependent. Oscilloquartz purchased the thermostats used in their quartz clocks from outside vendors, but had to carefully check them, as a large percentage was not up to the specifications.

In the early 1960s, Oscilloquartz still produced quartz clocks and frequency standards with tube electronics, but the switch to transistors was achieved quite rapidly. The frequency generator type DB62 – a very successful product – was bought in large numbers by the national telephone company; it was fully transistorized and equipped solely with silicon transistors. DB62 synthesized quartz stabilized frequencies of 1 MHz, 100 kHz, 10 kHz, 100 Hz and 50 Hz.

In 1966, Oscilloquartz was confronted with a very special order. It came from the CEH in Neuchâtel, where several dozen engineers and physicists developed the first Swiss quartz wristwatch prototypes Beta-1 and Beta-2. Those watches needed a bar-shaped quartz resonator with a frequency of 8192 Hz small enough to fit into a wristwatch case. Oscilloquartz CEO Berger refused to take this order: he "knew," on the basis of his own theoretical work, that such resonators would never work due to overwhelming surface effects. But the CEH physicist Armin Frei managed to convince the excellent Oscilloquartz technician Jean Engdahl

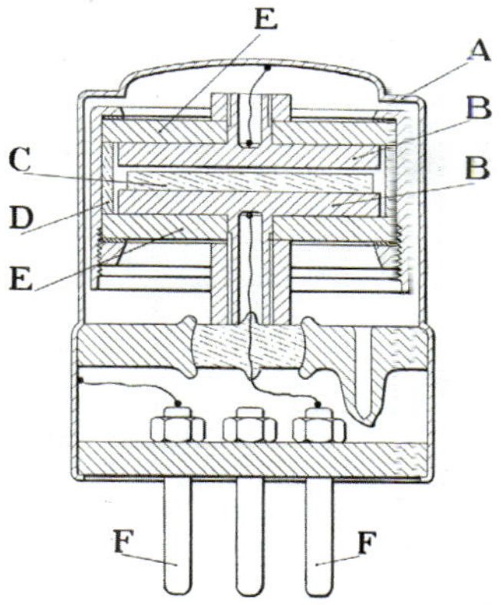

FIG. 7. — LABORATOIRE DE GRAMONT. — QUARTZ PIÉZO-ÉLECTRIQUE DANS SA MONTURE

A. *Ampoule métallique à vide.* — B. *Electrodes.* — C. *Lame de quartz.* — D. *Tube-entretoise de verre.* — E. *Rondelles isolantes en matière céramique.* — F. *Broches de contact.*

6.10. Disc-shaped quartz resonator (C) made by the Laboratoire de Gramont, 1937

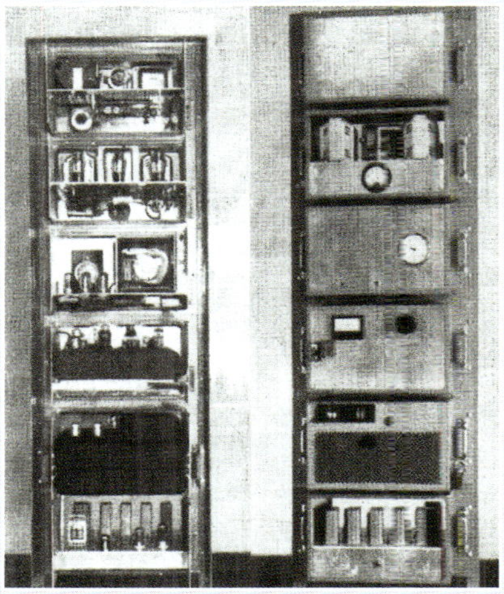

6.11. The first Swiss quartz clock, Department of Industrial Research (AFIF) at the ETH Zurich, 1948. Left back side, right front side

6.12. The first quartz clock built by Oscilloquartz in Neuchâtel in 1950. It is exhibited at the International Watch Museum in La Chaux-de-Fonds, Switzerland

to more or less illegally make a small series of those resonators with the equipment he had, i.e. precision diamond saws and polishing wheels.

Frei and his colleague Jean Hermann then had to singlehandedly solve all the problems posed by the metalizing, soldering, suspending and encapsulating the quartz bar. Miraculously enough, the resonator worked, and even with an excellent quality factor. Berger was surprised and agreed to produce 6000 pieces of the complete resonator for the industrial version Beta 21. Subsequently, Oscilloquartz also produced the cut and polished, 32 kHz tuning fork quartz resonator for the Rolex Oyster-Quartz. The Ebauches/Oscilloquartz marine chronometer B-1390 is described towards the end of this chapter.

Quartz Clocks for Sports Timing

Junghans

At the 1972 Olympic Games in Munich, Junghans and Longines were jointly given the sports timing job. Junghans got the most prestigious contest, i.e. the 100 meter dash. Junghans took this extremely seriously and made in-depth investigations to make sure the timekeeping was 100 percent honest – which was not appreciated by all athletes. It was found that at least 0.1 seconds elapse between the firing of the starter's pistol and the first kick at the instrumented starting block: a shorter reaction time is only achieved very rarely. The difference may be around 0.01 seconds.

It was then decided by international consensus that a reaction time of less than 0.1 second was equivalent to a false start. Junghans developed a special photo finish camera with Polaroid film for the 100 meter dash. A vertical slit moved in front of the film at about the speed of the sprinters. Time was thus displayed as a distance: the film clearly showed when each of the sprinters crossed the finishing line, despite the fact that the athletes stayed very close to each other. For purposes of redundancy, a light barrier was also installed.

The 400 meter dash was less critical, as the athletes were further apart at the finishing line that was filmed with a conventional movie camera displaying time marks. The quartz clocks used by Junghans at the Olympic Games were derived from the table clock caliber 610 Astro-Chron. The 12.8 kHz quartz resonator frequency was divided ten times by two with bistable multivibrators. The output frequency controlled a 12.5 Hz transistorized Atomat movement.

Longines

As far as sports timing was concerned, Longines was Omega's arch-rival for several decades. Longines was responsible for timing at fourteen Olympic Games. Starting with 1953, Longines used a home-built portable quartz timer with a 100 kHz resonator that it had developed jointly with Bernard Golay in Lausanne. Transistors were available at that time, but they were notoriously unreliable. It was safer to design the dividing circuit with tubes.

The Longines-Golay quartz clock was lodged in a sturdy wooden box. It received a very flattering certificate from the Neuchâtel Observatory, as a 24-hour rate of exactly zero was measured. This was a new record that proved that (within 24 hours) the Longines clock was at least as good as the Observatory instruments with a daily rate of ± 0.01 seconds. It was connected to the Longines timing instruments Chronocinégines, Chronotypogines, and Contifort. It was just a timer and did not display the exact time.

The heart of the instrument was a 100 kHz bar-shaped quartz resonator that was suspended in an evacuated glass tube. Due to the strong temperature dependence of the quartz frequency, the glass tube was mounted in a cylindrical heater the temperature of which was kept at 50° C ± 2° C even under the oftentimes very rough conditions of outdoors sports in any season and any weather. The heater was surrounded by three concentric layers of plastic foam alternating with silver plated glass tubes. The thermally insulated quartz resonator had a length of 15 cm and a diameter of 10 cm. Its price was around 4000 dollars while a 32 kHz quartz resonator for wristwatches now costs less than 10 cents.

An oscillator circuit kept the quartz resonator vibrating. The frequency division system was analog: after an amplification stage the frequency was first divided by 5, then by 4, twice by five and finally by 4. Each stage needed one of the functions of a double triode of the well-known type E88CC. At each stage the desired frequency was filtered out with a resonant circuit and again amplified.

The final frequency of the division chain was 50 Hz; it was amplified by a power pentode EL84 in order to obtain 220 Volt. This corresponded to the European line frequency and voltage, which was, of course, no coincidence. In an emergency, the timing instruments could be connected to the line, but, of course, the frequency was subject to uncontrollable variations and was a lot less stable than the quartz-controlled device. The instrument contained 8 electronic tubes including two rectifier tubes.

In a separate device, the quartz stabilized 50 Hz frequency controlled a synchronous motor actuating rotating drums on which were engraved numerals for minutes, seconds, tenths of a second, and hundredths of a second. Pressing a lever printed the numerals on a paper tape by means of an inked ribbon.

The first Longines quartz clock weighed about 10 kilograms; it was powered for 10 hours by 20 kilograms of 12-volt lead-acid batteries. They were charged by means of a separate power supply. A chopper and transformer supplied 220 volts AC; of course the clock could also be operated on the AC net. The whole system consisted of four massive wooden cases that together weighed nearly 78 kg, not counting the many cables.

The Longines tube quartz clock was in use for a full 15 years; in 1969/70 it was replaced by a much smaller, fully transistorized model. Very reliable, nearly temperature independent silicon transistors were available at that time; that meant a quantum jump of miniaturization. On this basis, Longines developed the highly miniaturized quartz clock and "Telesprint," a transistorized stop watch with light emitting diode (LED) display up to the hundredth of a second. The size was just 10 cm x 8 cm x 3.4 cm. The battery was inside the case.

Starting in 1969, Telesprint was produced in large series: it was quite inexpensive and greatly contributed to the downfall of mechanical stopwatches. In 1972, the individual transistors were replaced by integrated circuits. The next version was Tele-Longines 2002: now both the stopwatch function and the memory were fully digitized.

Omega

For the 1952 Olympic Games in Helsinki, Omega started developing a portable, battery-powered tube-type quartz clock in 1949: the Omega Timer. Time was recorded on film to the hundredth of a second.

The next Omega quartz stop watch was called Time Recorder (OTR-1). The rapidly rotating drums of the printer could be interrogated two to three times per second and were provided with a rattrapante mechanism. Whenever printing was required, the drums with the numerals stopped for a very short time but were brought to the correct position immediately afterwards, exactly compensating the time during which the drums had been stopped. With the OTR-1, the thousandth of a second could be obtained by interpolation.

Lead-acid batteries delivered the required power. The OTR-1 weighed 18 kg, the quartz resonator frequency was 10 kHz, the daily rate was 5/100 second. This electronic stopwatch ended up being standard issue in sports timing; it was normally operated with the Omega light barrier system. It stayed in use for more than ten years for timing automobile and motorcycle races, horse and bicycle races and in ski, skating, bob, canoe and rowing

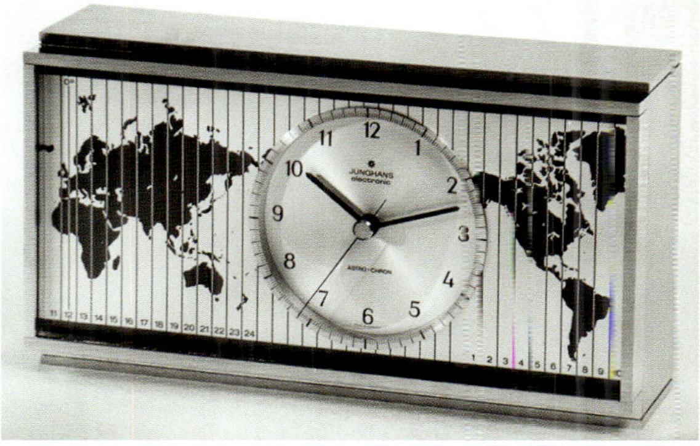

6.13. W 610 Astro-Chron by Junghans

6.14. Image taken by a movie camera coupled with a Junghans quartz clock for sports timing Olympic Games Munich, 1972

6.15. First Longines quartz clock for sports timing

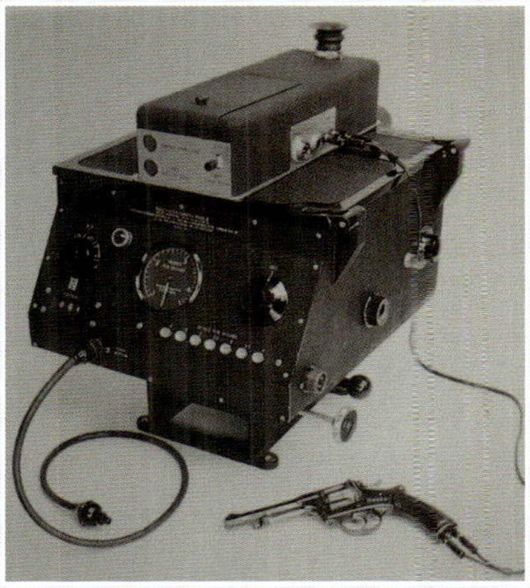

6.16. Omega's quartz-based timer for sports timing

contests. Furthermore, the OTR-1 was adopted by a large number of research laboratories and observatories all over the world.

OTR-2, the next Omega quartz clock was launched in 1964. It was fully transistorized and equipped with a 100 kHz quartz resonator; it routinely measured time to the thousandth of a second. OTR-2 obtained the homologation of the IAAF, but the official time for a race was still measured with a manual stopwatch. The quartz clock ran parallel to it and was used whenever a result was contested and the hundredth of a second was needed. The 1972 Olympic Games in Munich marked a breakthrough in that respect: now the quartz clock was officially adopted for sports timing.

The Omega Time Recorder OTR-3 was developed by Peter Tosin and Max Suter for measuring the thousandth of a second. It was launched in 1968, just in time for the Winter Olympics in Grenoble and the Olympic Games in Mexico-City. OTR-3 was used in combination with the Omega Time Subtractor that transformed the time difference between start and finish into the actual contest time. Other accessories were the Omegascope that indicated the speed of the athletes as well as the world record in each discipline, and the Omega Speed Recorder that calculated average speeds.

The OTR-6 quartz clock launched in 1974 weighed just 5 kg; it could be operated with lead-acid batteries or flashlight batteries. It could be operated with any of the current barrier systems. The quartz resonator was thermocompensated. OTR-6 displayed the hundredth of a second, the thousandth of a second could be obtained by interpolation. Up until 1985, well over 300 OTR-6 were produced. The electronics were microprocessor based and included a CMOS memory.

OTR-7 was the first "intelligent" minicomputer for sports timing. It was issued with plug-in modules, taking account of the specific needs of each and every sports discipline. OTR-7 was first used for the various alpine skiing races: athletics, bicycle racing and horse races followed.

The light and compact minicomputer OTR-8 was developed in 1978 by Jean-Paul Longchamp, Edgar Wyss and Max Suter. Several screens could be connected to it in order to display all the results of a given contest, with clock time and difference between start and finish.

Seiko

The Seiko "Crystal Chronometer" was originally developed as a marine chronometer and very accurate table clock. It was slightly modified for sports timing at the 1964 Olympic Games in Tokyo. In this configuration, the Crystal Chronometer could be triggered optically (light barrier) or electrically (gate, starting block). The size was 200 mm x 160 mm, it weighed 2.6 kg. The oscillator was a 10 cm long quartz bar vibrating in the flexural mode; the resonance frequency was 6369,388 Hz. It was cut from a natural quartz crystal and lodged in an evacuated glass tube.

The thermocompensated electronics consisted of 11 silicon transistors and 3 silicon diodes. Their functions were excitation of the quartz resonator, frequency division by 4, 4, 3 and finally 2, as well as supplying power to the 3 milliwatt synchronous motor. Battery status was indicated on the left side; just two D cells were needed, they had to be changed once a year. During this operation, the clock was not stopped, but continued working for 24 hours thanks to an extra, long life lithium battery. The daily rate was 0.2 second; operating temperature range was -10 and +50° C.

Small Clocks and Alarm Clocks

Starting with the 1960s, a great many different types of quartz wall clocks, small clocks, and alarm clocks appeared on the marketplace, some of them mass produced, without the ambition of being chronometers.

Bulova

Using an Accutron movement, Bulova Bienne staff under the technical direction of Hans Schaller developed a cute little quartz clock with a half-moon-shaped, transparent plastic case. The frequency of the quartz resonator was 11'520 Hz; five binary multivibrator division stages brought this frequency down to 360 Hz for the excitation of an Accutron tuning fork. The quartz resonator was supplied by Bulova USA: the company had an important quartz production unit in Woodside (New York), supplying industry and the military.

This clock was exhibited at the 1964 Basel Watch Fair and was a sensation, as it was the world's smallest quartz clock. Energy was supplied by two button cells in series, the power requirements were a mere 100 microwatt. The commercial version was rectangular and was called "Accuquartz." Bulova gave it away to high ranking visitors and particularly good customers, only small series were sold.

Citizen

In 1964, Citizen introduced its first line operated quartz tableclock with an analog display, the "Citizen Crystron;" it was equipped with caliber 58.000 with a 6.4 kHz quartz resonator. The daily rate was 0.5 seconds, power requirements were 3 milliwatt – quite high, but only bipolar integrated circuits were available at that time. A small, rechargeable battery made the clock immune against power outages.

The 1967 "Citizen Crystron QA-501" (another table-clock) used caliber 105.000; the life of the 1.5-volt battery was one year. QA-501 used the same quartz resonator as the "Citizen Crystron" of 1964, the daily rate was identical.

The "Citizen Crystron QK-101" wall clock was launched in 1970. It was equipped with caliber 48.000; the quartz resonator and the integrated circuit were the same as in model QA-501. The daily rate was 0.3 second, a slight improvement. The next table clock "Citizen Crystron QK-508" with caliber 78.000 also was launched in 1970, the specifications were the same as in model QK-101. The daily rate was improved to 0.2 second as long as the temperature stayed between 10° C and 30° C.

The "Citizen Crystron QK-151," launched in 1975, was first to incorporate a high frequency quartz resonator of 262,144 Hz. QK-151 used caliber 20.000 with a daily rate of 0.3 second. The CMOS type integrated circuit brought the current consumption down to 0.38 milli-amperes. The wall clock "Citizen Crystron QK-250" with caliber 40.000 launched in 1978 was another landmark. The rate was down to 0.15 second per day thanks to a 4 MHz quartz resonator.

Derby Swissonic

Production of Derby small clocks, alarm clocks, and pseudo-marine chronometers with gymbal mountings in a massive aluminum case was transferred to Marin in 1970 and continued until 1973 (see Chapter 1). The pseudo chronometers used a modified version of the problem prone 13-line quartz caliber 9180 by Ebauches Electroniques Marin; the mechanical part was supplied by A. Schild in Grenchen. In 1975 the Derby product line was discontinued. With prices ranging from sFr. 49 to sFr. 99 (even sFr. 499 for the chronometer models), profit margins were minimal. Furthermore, the Secticon clocks by Portescap were aggressively competing with Derby products.

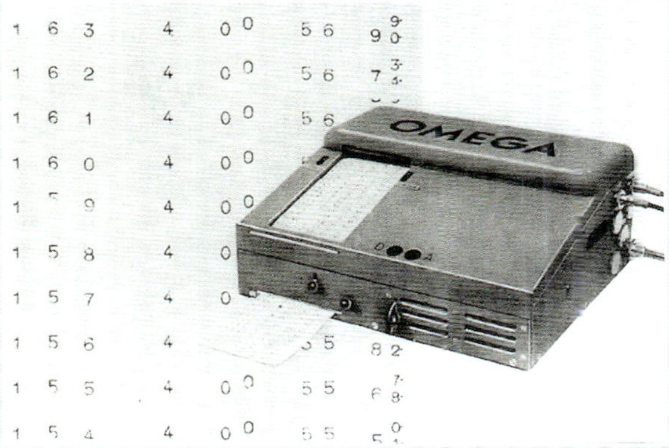

6.17. Omega Time Recorder with printer

6.18. Seiko's Crystal Chronometer

6.19. Quartz synchronized Bulova Accutron miniature clock

6.20. Quartz Crystron table clock by Citizen

Hanhart

Hanhart is a German company with Swiss roots. Founder Adolf Hanhart (1856-1932) ran a watch and jewelry store in Diessenhofen in the northeastern corner of Switzerland, just a few kilometers from Schaffhausen. As he repaired and sold a lot of pocket watches, he started dreaming of his own watch brand. But in the east of Switzerland, there was no horological infrastructure, except at IWC, which was quite self-sufficient. In 1902, for both cultural and geographical reasons, Hanhart moved to Schwenningen in the Black Forest region, a mere 50 kilometers from the Swiss border. At that time, Schwenningen was the center of the German clock and alarm clock industry.

Once established in Schwenningen, Hanhart manufactured both pocket watches and wristwatches, as well as stopwatches and chronographs. His son Willy (1902-1986) joined him in the business in 1920 and took over in 1932. Hanhart's main products were moderately priced and very robust stopwatches with pin pallet escapement for sports timing and industrial applications. Another hit were chronographs for soldiers and pilots.

As much of his production was bought by the Nazi Wehrmacht, Hanhart was blacklisted by the Allies towards the end of World War II. His plant was just about cleared in 1945, when LIP in Besançon bought Hanhart's mostly Swiss precision machines for a song. It took three years for a new start, the first products being the well-known and proven stopwatches, wristwatches, and chronographs. Within a decade, Hanhart rose to be the world leader for stopwatches.

In the 1960s, Hanhart diversified into transistorized alarm clocks with plastic cases and electromechanical movements. Permanent magnets were glued on the two-tiered balance wheel that was periodically repelled by a fixed coil. In 1972, a single-battery quartz synchronized version was developed in collaboration with Motorola in the USA: it was named caliber 520 and measured 60 mm x 74 mm. Unfortunately, it was not competitive and there were problems with the mechanical part of the movement.

Hanhart put a lot of effort into developing very cheap injection-molded plastic wheels. Such wheels were used in a new version of the old and nearly obsolete transistorized alarm clock movement of 1975. The new-old caliber 426 had the same dimensions as caliber 520 and needed only one B-type battery. This model was competitive with other brands' products, but now customers wanted quartz. Therefore, Hanhart developed the universal caliber 3505 that could be used both for quartz wall clocks and alarm clocks; Staiger supplied the motor. Production of this caliber started in 1978, i.e. relatively late. Furthermore, the dimensions (60 mm x 74 mm) were rather "generous," quite aside from the fact that the movement was rather noisy. Yet, some 4 million pieces were sold.

The next generation of quartz movements had to be smaller, quieter and cheaper. These three conditions were fulfilled by the automatically assembled caliber 3305 that was launched in 1982. This caliber was a big success as more than 40 million were sold in the course of a seven-year product life. The dimensions were 56 mm x 56 mm; a single AA battery powered the movement. The 3305 caliber was produced in Germany and several other countries under Hanhart-License, even in China.

The next models (calibers 300 and 400) were more compact and measured just 56 mm x 42 mm. But now the Chinese competition began to be felt, which was exacerbated by a massive devaluation of the dollar. In 1992 the price of a Chinese quartz clock movement reached 1 Deutschmark, which prompted Hanhart to discontinue this product. Henceforth, the company concentrated on electronic stopwatches and World War II pilot's watches with mechanical movements.

Hermle

For centuries, manufacturing grandfather clocks as well as wall clocks and table clocks was a specialty of the Black Forest; this industry employed some 25,000 people. In the "good old times," each and every house needed a time reference, usually a pendulum clock with striking mechanism. This has totally changed by now, the market for mechanical clocks and precision pendulum clocks has shrunk dramatically. Most households still need a time reference, but it usually is a radio-controlled wall clock or table clock made in the Far East and costing less than twenty dollars. The big mechanical clock has been demoted to a piece of luxury furniture creating an optical and acoustical atmosphere in high income homes. Its timekeeping function is quite secondary.

Coming from a family of watch and clock merchants, Franz Hermle started manufacturing wheels and pinions for the regional clock industry. In 1922, he set up shop in Gosheim on the western edge of the Schwäbische Alb, a range of wooded hills in southern Germany. Ten years later, Hermle diversified into complete clock movements. After World War II, new precision machinery could be bought with Marshall Plan money; production was resumed in 1946.

The transition to quartz was mastered without any major problems, as the company had gained valuable experience with electromechanical movements in the 1950s. However, by the early 1970s, Hermle trailed its competitors: aside from Junghans in neighboring Schramberg, Kienzle in Schwenningen as well as Kundo and Staiger in St. Georgen had been producing quartz movements and quartz clocks for quite some time.

In a first step, Hermle signed a cooperation agreement with Junghans: for a few years, Hermle would get quartz movements from Junghans and supply Junghans with mechanical movements. At that time, electronic movements were relatively simple, quite inexpensive and much more accurate than the best mechanical movements. The Junghans Quattro-Phon caliber 771 was particularly important for Hermle, as it was equipped with a microprocessor and generated four different tunes for the striking mechanism; Hermle called it caliber 1217.

Bartering with Junghans was just an expedience: it would be embarrassing if a customer found out that his Hermle clock was powered by a Junghans movement. In the early 1980s, Hermle decided to manufacture its own quartz movements. Quite understandably, the Hermle models bore a strong resemblance to the Junghans movements. The outside dimensions were identical, as the Hermle clocks had been adapted to the Junghans movements. Photolithographic 32,768 Hz quartz resonators were readily available; they had made their breakthrough in quartz wristwatch movements in the late 1970s. They are easy to recognize, as they are usually packed in a 6 mm x 2 mm metal tube.

In 1985, Hermle set up an automatic assembly line for quartz movements, which was quite a challenge. A whole set of new technologies had to be mastered, such as precision injection molding of plastics, automatic handling and soldering of printed wiring boards and bonding silicon chips. By 1986 everything seemed to work fine and the assembly line was successfully tested, but soon after everything was destroyed by a disastrous fire. To add insult to injury, another, even worse fire broke out three years later. This time, not only the assembly line for quartz movements, but also the production unit for manufacturing mechanical movements and the warehouses were totally destroyed. Everything had to be rebuilt.

Despite these setbacks, Hermle became the world leader for quartz clock movements. Hermle was cheaper than its competitors and produced 18,000 high quality movements per day. There were only two types: 2100, the basic movement measuring 60 mm x 60 mm, and 2200 with a non-functional pendulum. The variant 2214

6.21. Hanhart caliber 05

6.22. Hermle's basic movement caliber 2100

6.23. Hermle caliber 2200 with non-functional pendulum

6.24. Hermle caliber 2214 with pendulum drive and electronic chime was provided with a striking mechanism, caliber 2300 featured a torsion pendulum while the round caliber 2600 featured a sweep seconds-hand. The movement with striking mechanism was equipped with a memory chip for storing the algorithms generating the tunes (Bim-Bam, Westminster, Ave Maria, Whittington).

Hermle had been a latecomer in the field of quartz movements for clocks and could learn from the mistakes others had made: this explains the technical superiority of Hermle products. For example, Hermle used a very powerful stepping motor that actuated even long and heavy hands. This is why Hermle dominated the market for quartz clock movements from New Zealand all the way to Alaska, from Norway to South Africa. Selling quartz clocks meant dealing with new classes of customers. Department stores and furniture houses neither had watchmakers nor even the rudiments of horological culture. Non-functioning clocks were simply exchanged, Hermle had to take back the faulty pieces and repair them on their own premises.

The downfall of the Black Forest clock industry – which did not spare Hermle – was mostly due to the coming of China on the clock market. Japan never was a problem and even became a much appreciated customer. But there was no defense against the Chinese and Taiwanese producing at a tenth or even a twentieth of German prices.

The price war with China ended up destroying the market. Hermle sold its quartz movements for 1 Euro but could have purchased them for a sixth of that price in Asia. In 2007, the situation had become hopeless and Hermle decided to get out of the clock movement business. Only a very special quartz movement mostly made of metal was left: it is used for the "Astrolabe" clock showing the motion of earth and moon around the sun in three dimensions. All other Hermle clocks are equipped either with in-house mechanical movements, or with quartz movements supplied by external vendors.

Imhof

It is not clear whether the Arthur Imhof Company, manufacturer of small clocks and alarm clocks in La Chaux-de-Fonds, was founded in 1912 or 1924. But it was soon known for its high quality products. When quartz movements became available, Imhof bought the production rights for a clock movement that Portescap had developed and tested in small series, but never produced (see the section on Portescap). It used the stepping motor invented by Claude Oudet; Portescap brought it to industrial maturity and manufactured large numbers under the name Socrem.

Those motors made twelve steps per turn; the stator system consisted of magnetically soft, pure

6.25. Back side of the Imhof quartz miniature clock movement

6.26. Junghans Star-Quartz alarm clock

6.27. First radio-controlled solar table clock by Junghans (1986)

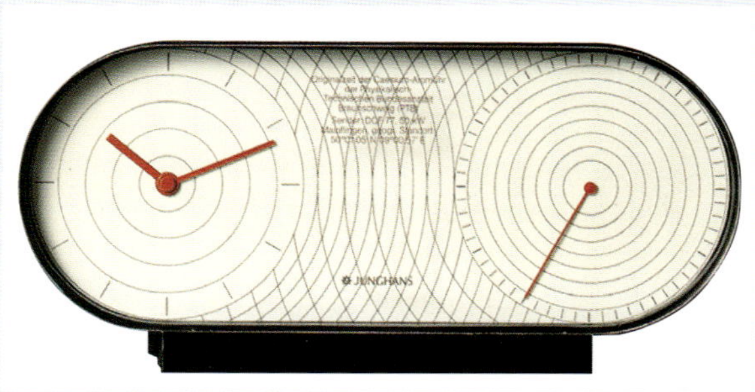

6.28. Junghans radio-controlled Star Quartz clock with caliber 743

iron. Imhof delivered their clocks to Portescap where the motor was installed. The complete movements (caliber 904 Q) were also checked on the spot and made ready for shipment. Those movements were a rather conventional construction with a 32 kHz quartz resonator and frequency division down to 1 Hz; the diameter was 49.5 mm, the height 13.6 mm. Aside from Imhof, this movement was also used in clocks by Ulysse Nardin and Montremo. It was extensively modified for the Omega marine chronometer designed for the French Navy. In 1988, Imhof was bought by Jean Roulet SA in Le Locle, which went bankrupt in 1999. After liquidation, Imhof disappeared from the commercial register in 2002.

Junghans

The "electrification" of Junghans clocks proceeded in two phases: balance wheel motors first, then quartz. In 1950s, Junghans developed several types of transistorized table clock movements with fixed coil and ferrite magnets on the two-tiered balance. Those designs were licensed from Hatot in Paris. They were sold under the brand names Atomat (caliber 794), Atovox, Atophon, and Atolux; the latter had a solar cell power supply.

The balance wheel oscillated with a frequency of 2.5 Hz; the daily rate of -10 to +20 seconds was not significantly better than the one of conventional mechanical clocks. However, the battery life was at least a year. As there was no main spring, the wheels and bearings were subject to much lower forces. Not only the case, but also the mainplate and the wheels could now be made of plastic by precision injection molding. This greatly simplified production and tremendously reduced costs.

Junghans management had no good reason to prioritizing the development of quartz clock movements as the electromechanical movements worked just fine. Yet, two prototypes dating back to the early 1960s have been preserved. The first used a flexural mode 8192 Hz quartz resonator: that frequency was divided to 2 Hz in 13 binary steps. This 2 Hz signal was used to stabilize a modified Atomat caliber 794, the balance of which now solely acted as a motor. Power was provided by a 9-volt rechargeable battery. The total volume was around 2000 cm^3; the daily rate was an astounding 0.1 second.

The second prototype used a 16,800 quartz resonator. Ten binary steps divided this frequency to 50 Hz AC for powering a classical synchronous motor. The volume of this movement was about 2800 cm^3

In 1963, a new development project was initiated. The objective was a commercial quartz movement for wall clocks, table clocks, and alarm clocks. The prototype worked in 1965, a year later production of caliber W610 started. The frequency of the quartz resonator was 12.8 kHz; ten binary dividing steps brought this down to 12.5 Hz, the frequency of a transistorized Atomat movement. A daily rate of 0.05 second was achieved. The dimensions of the cased movement were 105 mm x 155 mm x 57 mm.

Affordable integrated circuits did not exist yet, the electronics consisted of discrete components. This movement powered the luxury type, rectangular desk clock Astro-Chron. On both sides of the dial one could see the world's continents. Sales began in the fall of 1967, the price of 785 Deutschmark was extravagantly high. Only about 3000 pieces were sold, plus 5000 pieces of a less luxurious and thus less costly version.

The next model (W 750) was launched five years later in 1972, at a time when inexpensive 32 kHz quartz resonators and integrated circuits were available. A Lavet motor making six steps per turn further simplified the construction; the outside dimensions were 72 mm x 62 mm x 30 mm. The trend to miniaturization and reduced power requirements continued in 1975 with caliber W 751, despite the fact that now a 4.2 MHz high frequency quartz resonator was used, that required 22 binary division steps in CMOS-technology to reach 1 Hz. Wheels, hands, and case were the same as in the Atomat 736, but the movement was a full plastic construction. The outside dimensions were left unchanged.

In the following year (1976), the price of quartz movements reached that of transistorized, electro-mechanical movements with a balance wheel motor which therefore became obsolete. Production of caliber W 742 that required an ATO license was stopped. Its replacement was the quartz caliber 743 "Star Quartz" with a 4.2 MHz quartz resonator, an integrated circuit and a bipolar Lavet motor that made two steps per turn. This caliber was used in small clocks, alarm clocks, foldable alarm clocks and a radio alarm clock.

Junghans launched the next family of clock movements in 1980. The calibers W 756, W 757, and W 754 were only slightly smaller than W 751 and continued using a 4.2 MHz quartz resonator. The current consumption was much lower than that of the W 756 and W 757 (80 microampere versus 260 microampere for caliber W 743). Caliber W 754 had a non-functional pendulum and needed 130 microampere. Additional functions such a strike mechanisms and wake-up calls did not affect battery life.

The further development then concentrated on "fun" functions. Caliber 771 launched in 1980 was equipped with the electronic strike mechanism "Quattro-Phon" and a microprocessor. As suggested by the name, four different strike tunes were available: Bim-Bam, Westminster, Ave Maria, and Whittington. Hermle bore part of the development costs of this Junghans caliber 1217. The quartz alarm clock Junghans Minivox with caliber 759 was launched in 1981 and was quite original. The outside dimensions were 50 mm x 40 mm.

The development of Junghans quartz clock movements reached its apex in 1982 with caliber 738. Its height was just 15 mm; it was powered by a single AA battery and was mostly used for wall clocks. Mass produced photolithographic and very inexpensive 32 kHz quartz resonators (Statec technology) were now available and replaced the more expensive 4.2 MHz type.

The radio-controlled quartz caliber W 732 launched in 1985 was the last and ultimate milestone in the history of Junghans clock movements. Such clocks displayed "atomic" time and were always accurate within one second as long as they were within the range of DCF77 and could set themselves either once a day or on the hour. This transmitter now has three dedicated commercial caesium clocks that are continuously compared with the fountain-type caesium clocks built by the Physikalisch-Technische Bundesanstalt PTB in Braunschweig; their rate is 1 second in 50 million years. Caliber W 732 had a digital display, but an analog model was available in 1986.

Despite its very high price of about 1000 Deutschmark, the Junghans radio-controlled table clock was a success. This was the world's first industrially produced radio-controlled clock for private use. Outside dimensions were 265 mm x 130 mm x 85 mm; hours and minutes were displayed on the left, seconds were read off a separate dial on the right side of the case. The clock synchronized itself with DCF77 every hour. The movement was controlled by a 4.19 MHz quartz resonator. It was also used in a two-dial radio-controlled wall clock of 280 mm x 280 mm x 80 mm.

The Junghans radio-controlled table clock type RC launched in 1986/87 was characterized by a conventional three-hand display. Outside dimensions were 150 mm x 200 mm x 67 mm, but the upper part was shaped like a church window. Those clocks were often seen on the workbench of watchmakers, particularly in Switzerland where radio-controlled clocks never evolved beyond the prototype stage. The RC clock movement was equipped with a photolithographic 32 kHz tuning fork-shaped quartz resonator; the receiver section for DCF77 was switched on every hour on the hour. In the latter part of the 1980s, Junghans produced a wide variety of miniature radio-controlled alarm clocks with a digital liquid crystal display.

Kienzle

Jakob Kienzle (1859-1935) was a manufacturer of regulators and traditional Black Forest clocks. In 1883 he married the sole heiress of the old watchmaker family Schlenker. With his wife's money, Kienzle established the Schlenker and Kienzle Company, which allowed him to enormously extend the scope of his activities. By 1893 production had reached a grand total of 162,000 clocks and alarm clocks.

Grandfather clocks were still in the catalog, but became niche products when Kienzle started producing alarm clock movements the American way. This means that he produced large series of movement parts with precision machinery. The parts were sufficiently precise that they could be assembled without any adjustment or matching – which previously had been a time consuming chore. At the beginning of the 20th century, Kienzle had 1000 employees and produced a million "American" alarm clocks per year.

At the same time, Kienzle began producing pocket watches and, a few years later, wristwatches appeared in the Kienzle catalog. Another diversification was the car clock: famous car makers in Germany, France, and Great Britain became Kienzle customers. The extremely sturdy "heavy duty" wristwatch Kienzle launched in 1931 was extremely successful: no less than 25 million pieces were produced. At about the same time, Kienzle's car clock became an 8-day cockpit clock for airplanes. Many of them equipped World War II German fighter planes and bombers.

After the war, which had forced Kienzle to switch most of his production capacity to detonators for shells ad bombs, Kienzle developed the "Volks-automatic," an economical automatic wristwatch with pin pallet escapement. Kienzle grew to be the leading watch brand in Germany. The Swiss Bührle Group Oerlikon bought Kienzle in 1965, but sold it to the Kreidler Metalworks in 1966.

Kienzle survived the "quartz crisis" of the 1970s quite well, the production having been shifted to quartz movements. The new range of products included a digital wristwatch with LED display and quartz alarm clocks consisting mostly of modular, injection-molded plastic parts that could be assembled automatically. The Kienzle caliber 627 was a trendsetter and became the industry standard. In 1989, a brand new plant equipped with the most modern machinery was built in the Schwenningen Industrial Park. Suddenly, much less expensive Chinese alarm clocks appeared and gave Kienzle a lot of trouble. The company declared bankruptcy in 1997.

The Kienzle brand name, which was very well known in Germany, was sold to Highway Holding in Hong Kong. The equipment of the new plant was sold and moved to China lock, stock, and barrel. Soon, exact copies of the former Kienzle alarm clocks and brand new models appeared in the market. Low-priced, good quality wristwatches were added to the product mix. In 2002, the newly founded Kienzle AG in Hamburg took over the distribution of "made in Asia" Kienzle products in Germany.

Two businessmen in Hamburg – Stephan Kruse-Thamer and his partner Marco Hahn – both of them enthusiastic watch collectors, decided to bring the Kienzle brand back to Germany in 2007. Their aim was to revive the tradition of mechanical Kienzle watches. Unfortunately, the timing was badly chosen, as a severe economical downturn had just started. Business was bad and was not improved by integrating the Lacher watch company in Pforzheim, a respected brand since 1926 that had recently declared bankruptcy.

Kruse-Thamer and Hahn had intended to manufacture high quality mechanical watches in the upper price range under the brands Kienzle and Laco by Lacher. The new Kienzle Company had already hired 20 people and bet on the synergy between two well-known brands producing and distributing mostly "Made in Germany" products. The insolvency of Kienzle AG put an end to those ambitious plans. A new start of the brand with Swiss capital was made in 2012.

Kundo

The watchmaker Johann Obergfell (1871-1942) was in the employ of the watch manufacturer Matthias Bäuerle in the Black Forest region. In 1897 he had the unfortunate idea to set up the "Social-democratic Workers Union St. Georgen," for which he was immediately fired. In the longer term, this led to the founding of an important Black Forest company.

Because of his flirtation with socialism, Obergfell had not the slightest chance of ever finding a new job. Unemployment forced him into entrepreneurship and he began producing watch parts in his own house. He also invented and manufactured the straight gong for clocks; it was very succesful, even in the UK where the circular gong was universally used.

The Kundo Company was founded in 1918 when Obergfell went into a partnership with the watch engineer J.G. Kieninger. The first products were

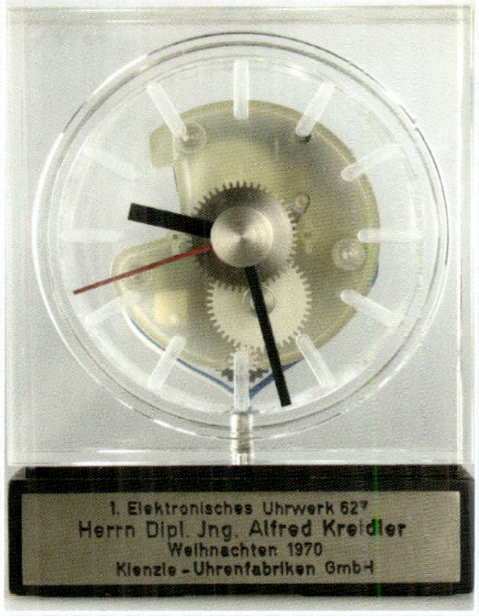

6.29. Electronic movement by Kienzle

6.30. Digital table clock Kundo 2000

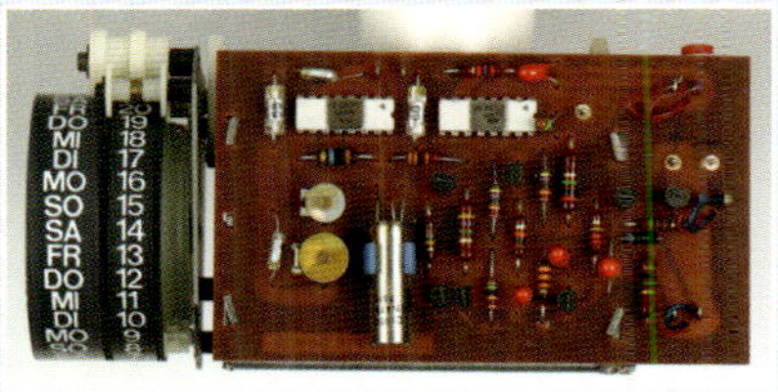

6.31. Electronic module and calendar system of the digital table clock Kundo 2000

6.32. Caliber 3000 quartz movement by Kundo

technical watch movements and household clocks; in 1923 clocks running for a full year with one winding and electrical clocks actuated by a synchronous motor were added to the product mix. Kundo grew very fast and developed into the biggest industrial company in St. Georgen.

Electronic technologies were adopted quite early: in 1955, Kundo launched the first electromechanical clock with transistor control – a worldwide first. Kundo was the largest purchaser of transistors in Germany, even though they cost 30 Deutschmark apiece.

The Kundo switch-clocks have even older roots; they were based on self-starting miniaturized synchronous motors: Kundo already produced such motors before the beginning of World War II. They led to a wide range of new products, from simple on-off switches all the way to mechanical switching programs that were used on a wide scale, primarily in washing machines and dishwashers.

The first Kundo quartz alarm clock with digital display and the possibility of setting the alarm time to the minute, was called Kundo 2000; it was launched in 1971. Kundo became the most important European manufacturer of this type of product. The movement was controlled by a 4.19 MHz quartz resonator and a CMOS chip. The 4 MHz resonators were much more stable and had a better shock resistance than the widespread 32 kHz flexural mode resonators. They also were very inexpensive, as they were used in enormous quantities in color television sets.

The Kundo program of digital clocks was expanded in 1972 with the Digilette model. Aside from the time in the 24-hour mode, it also displayed day and date. In the following year, the Tele-Quartz 3000 with analog display was launched. As the Munich silicon chip maker Eurosil had succeeded in changing over the integrated circuit from 3 volts to 1.5 volts, clock movements could now be powered by a single AA battery.

Kundo's first digital table clocks displayed time mechanically with moving metal tapes or flaps. The Transsegmenta model, launched in 1973, did away with mechanically moving parts as the display now consisted of light emitting diodes arranged in the usual 7 segment matrix. Hours, minutes and seconds were displayed continuously; a light tap on the case displayed date and month. Kundomatic 2000, a digital alarm clock suited for bedrooms and offices, was launched in 1974. The front side displayed hours, minutes, and seconds as well as the alarm time in hours and minutes.

In 1975, when it was learned that the Staiger watch plant (that was also located in St. Georgen) had switched its entire product line to quartz, Kundo followed the trend. Major competitors, such as Kienzle and Junghans, did the same later. Now the German clock industry was fully competitive with the very low-priced mechanical alarm clocks with which Czechoslovakia flooded the market. Furthermore, one had a superior, state of the art product.

Several clock manufacturers in the Black Forest area decided on the common development of a radio-controlled clock. But Kundo and Junghans got out of this venture and developed their own radio-controlled clocks that were launched almost simultaneously in 1986. The Kundo version was a table clock with an appealing, modern design. It displayed hours, minutes, seconds and date.

The UTS company (Uhrentechnik Schwarzwald) was founded in 1980 as a joint venture of Kundo and Staiger in view of the aggressive Japanese competition. Quartz clock movements were produced in Mariazell and in the neighboring village of Hardt, using fully automated assembly lines. The movements from both plants were identical, but were meant for clocks with different brands that competed in the marketplace. In the long run this was an awkward situation.

The long awaited answer came in 1992: the merger of both companies into Kundo-Staiger. Now clock movements were only produced in Hardt, while Kundo-Staiger diversified into new, very successful products such as heat meters, remote control of heating systems, and remote reading of gas and electricity meters.

Clock movements now exclusively used the 32 kHz technology with photolithographic quartz resonators from Statek. But, due to the fierce competition from China, production figures for those movements kept dropping after having reached an all-time high of 12 million pieces a year. In 2000, Kundo-Staiger was bankrupt, but would be recapitalized and reorganized into a new company. UTS, that had initially just supplied Kundo-Staiger, had been restructured to a subsidiary in 1995, and sold clock movements all over the world.

UTS was also restructured and is now responsible for the production and distribution of clock movements in Germany; its name has been changed to Uhrentechnik Schwarzwald Montageservice GmbH. After a management buyout, UTS now produces 2 to 3 million quartz movements for clocks at a new location in Dunningen near Rottweil where the automatic assembly line was moved. The required parts are mostly supplied by vendors in the immediate neighborhood.

The dimensions of those analog movements are fully standardized. UTS produces a round movement with a diameter of 53 mm and a height of 17 mm, as well as a square movement of 56 mm x 56 mm x 17 mm: the technology is identical. The square movement can be provided with a special back with a non-functional, electromagnetically actuated 1 Hz pendulum.

A market for German quartz clock movements still exists in Europe, despite of higher prices compared to those from China (80 to 90 cents versus 27 to 30 cents apiece in the Far East). The "Made in Germany" products have non-negligible advantages: very high quality standards and overnight delivery, even in small lots of at least 100 pieces, while whole container loads must be ordered from China.

Radio controlled movements are designed by Kundo XT, a division of OTG (Obergfell Technology Group) in St. Georgen, but produced in China. A particularly successful model is a receiver for time signals that is totally encapsulated in plastic and resists both water and vibrations. It is mounted under the bumper of cars and synchronizes the clock and the many microprocessors on board with DCF77 cesium time.

Mauthe

In the mid-19th century, a state of euphoria reigned in the Black Forest. Lots of money could be made with clocks, and even people without any knowledge or experience in this field started new clock plants. This was exactly what the grocer Friedrich Mauthe did in Schwenningen, where he opened a clock factory in 1844. Mauthe started with wall clocks that sold quite well, but soon diversified into alarm clocks, pendulum clocks, and spring-powered table clocks. Around 1900, the Mauthe Company had well over 1000 employees.

The production of wristwatches started only after World War II. The brand got a lot of attention and free publicity when it started manufacturing a special watch for the members of the 100,000 kilometer Volkswagen club. The company actually gave a watch to every Volkswagen owner who could prove that he had driven his Bug for 100,000 kilometers with the same engine.

Mauthe tragically missed the change to quartz in the early 1970s. A barter agreement with Junghans did not change anything: Mauthe would obtain Junghans quartz movements in exchange for mechanical movements that hardly anybody wanted and that Junghans had long discontinued.

Mauthe's last innovation was a vibrating alarm clock with a Junghans quartz movement. It was put under the pillow and discretely woke the owner without disturbing the bed partner. The product was called "Kiss-Kiss" but did not sell; Mauthe went into receivership in 1975. The local newspaper commented this with the headline "Double kiss with lethal consequences."

Portescap: Secticon and the Socrem Motor

In the 1960s, the Portescap Company in La Chaux-de-Fonds had yearly sales of well over 100 million Swiss francs, thanks to two very successful products: the Incabloc shock absorber for the balance wheel staff and the Vibrograf timing machine. The change to quartz movements seemed to be a minor problem, at least in the early days. Portescap was well acquainted with the quartz technology as it had been using it in the Vibrograf for quite some years. Furthermore, Portescap had started producing small electrical clocks with the brand name Secticon in 1959.

The Secticon movement was transistor controlled and consisted of plastic and metal modules that were simply plugged into each other for assembly. It was driven by a small DC motor with three windings and an ironless rotor. The speed had to be at least 400 rpm; any higher speed was acceptable as the motor's purpose was to wind a spring that gave the balance wheel a mechanical impulse at every vibration by means of a lever. This assured that the balance wheel was oscillating constantly. This process was synchronized with a set of contacts on the balance staff that made the transistor switch on and off. The daily rate was adjusted to ±1 or 2 seconds. A 1.5-volt flashlight battery powered the Secticon movement.

In the mid-1970s, Portescap started producing multipolar stepping motors for quartz wristwatch movements. This construction was based on magnetized domains on the rotor and star-shaped stators (Socrem); it was also considered for a quartz version of the Secticon clock. The rotor made twelve steps per turn and directly drove the seconds-hand; a wheel train took care of the reduction for the minute and hour-hands. The balance wheel was replaced by a 32 kHz quartz resonator; this frequency was electronically divided in 15 steps down to 1 Hz.

Even though Portescap perfectly mastered the quartz technology, the commercial version of the Secticon clock was launched with the electromechanical, transistor controlled movement. Only a small series of the quartz version was produced and used for extensive tests. The design was then sold to the Imhof company (see this section), which produced large numbers of the Portescap quartz movements for their table clocks and alarm clocks. The stepping motor used in these movements was a Socrem P15.1R. This movement was sold to many manufacturers of clocks, e.g. Ulysse Nardin and Montremo. Omega used it without the 32 kHz quartz resonator for the 4.19 MHz marine chronometer it developed for the French navy.

The Secticon clocks were a failure as long as their case was conventional and rather boring. Portescap then gave the architect Mangerotti in Milan the assignment to design original and entirely new Secticon cases, which he did. They are attractive even today, many owners still enjoy them but had the electromechanical movement replaced by quartz. However, the clock business was not really profitable for Portescap. Only about 10,000 pieces were sold per year; production was stopped in 1968.

The Secticon clocks had one positive effect, as they led to a new range of Portescap products: DC motors that were produced on the base of a licensing agreement with the Faulhaber company in Germany. The coils of the Faulhaber micromotors needed only a minimum of copper wire despite their very high torque. This was the technical advantage of Portescap micromotors that ended generating some 40 million Swiss francs in yearly sales. They were used in huge numbers in tape drives for big computers and in video cameras.

It was easy to magine that the Centre Electronique Horloger (CEH) in Neuchâtel had secretly started developing a quartz wristwatch in 1962. There were no information leaks, but miniaturizing quartz clocks was the obvious thing to do. Portescap had decided not to invest in this venture that was controlled by Ebauches SA with which Portescap did not have a particularly good relationship. Portescap managers were not really surprised when the first prototypes of quartz wristwatches (Beta 1 and Beta 2) were shown to the media in January 1968. The new electronic watches would need very small motors with minuscule power requirements, and Portescap could make an important contribution and find a new, promising diversification.

In May 1968, Portescap signed an agreement with the French research institute CETEHOR in Besançon with the aim of developing very small stepping motors for quartz wristwatches. This led Claude Oudet and his coworkers to develop a multipole motor that was strongly inspired by the motor Portescap had used in the Secticon clocks: it stood in direct competition with Lavet's bipolar motor. Ebauches de Savoie and the alarm clock manufacturer JAZ had already supported such a development in 1966. The technology is still being used by the manufacturer of micromotors in Soncebaux, to the north of Bienne. However, such motors are not ideal for watches: the bipolar Lavet motor has become standard in this respect. Billions of Lavet motors are produced each year, mostly in China.

Ebauches de Savoie did not have much staying power and soon stopped its participation in the multipole motor project. JAZ held on for a few more years but finally also gave up. Only Portescap was left and fully intended to mass produce the Oudet motor in several sizes, with diameters ranging from 2.5 mm to 8 mm. It is ironical that JAZ ended up buying a license from Portescap, as they needed motors for their small clocks. Portescap also worked with Enicar on the development of a quartz movement with a Socrem motor, but this project was abandoned. In addition, the cooperation with Timex was a success: Timex chose an open, six pole design and produced this motor by the millions under Portescap license.

The Socrem motor was quite ingenious. Depending on the size of the watch, configurations with four, six, eight, or twelve poles were used. A special device magnetized the required number of spots on the samarium-cobalt disc that served as rotor; they all had the same magnetic polarity. A six-pole motor had twelve magnetic spots on the periphery of the rotor and six more closer to the center. Whenever the coil received a current pulse, the outer magnetic spots were repelled by the field of the stator fingers and the rotor made one step and was then attracted to the closest inner spot and was blocked there until the next current pulse came.

Working with Ebauches SA was difficult. Portescap did get a mandate to develop a Socrem-type motor for a ladies' watch quartz movement from the Ebauches technical manager André Beyner. Ebauches engineers then used this design to develop a simpler, integrated construction for men's calibers: it bore some resemblance to the Timex design. But in the end, Ebauches SA opted for the Lavet motor. Portescap did not have enough customers for their watch stepping motors, so this product line was abandoned in the early 1980s. Instead, Portescap became a major producer of small motors for electronic products, particularly capstan-type motors for tape recorders and cassette recorders.

There also was a cooperation with SSIH-Omega; it was instigated by Hans Widmer and continued with his successor Hans Schaller. Omega was interested in the Portescap motor but ended up developing quite a similar, eight-pole design in its own laboratory. This inevitably led to a dispute: the problem was solved by a licensing agreement with Portescap.

The project of a multipole stepping motor was supported financially by a stimulation program of the Swiss federal government. An agreement was signed with ASUAG, SSIH, BBC, Faselec, and Portescap to support the development of electronic watch components to the tune of 15 million Swiss francs. This helped Portescap to set up the production of their eight-pole motor, mass production was started in 1975 by the Socrem Company, a Besançon subsidiary of Portescap. The diameter was 5 mm, the height 2.8 mm: this was the smallest stepping motor of the time. Early customers were the French company JAZ and the Piaget subsidiary Complications SA, while Timex bought a license and produced anywhere between 20 and 40 million Socrem-type motors in their own facilities. Overall, sales were disappointing. In the opinion of experts, the Socrem motor was unreliable, particularly in the small sizes, but the demise of the Socrem motor was due to its price. The less expensive and extremely rugged Lavet motor won the battle.

In the early 1970s, when digital watches with LED or LCD display appeared, many experts were convinced that it would eliminate the analog watch with the traditional hands. The next generation would be forced to buy digital watches as their children would not be capable of learning to read a watch with hands. Nevertheless, both Ebauches SA and Seiko bet on the quartz watch with an analog display. In the long run, they were proven right – except in the United States and many developing countries where the inexpensive "throwaway" digital watch is quite popular. It is a sad fact that, in those countries, children and young adults are often incapable of reading the hands on a dial: they are considered as unnatural.

Staiger

Gebrüder Staiger GmbH in St. Georgen (Black Forest) started out in 1898 as a family-owned business specializing in the production of insulating materials for the rapidly growing electrical illumination industry. In the 1920s, Staiger diversified into radio parts. After World War II, St. Georgen was part of the French occupation zone. As Staiger had been a big supplier of the German Wehrmacht, Staiger was stripped of 90 percent of its equipment; the authorities sold the confiscated machinery in France. This forced Staiger to make a totally new start.

With a few remaining machines and improvised gear, precision injection molding of plastics became the new pillar of the Staiger Company. Due to the importance of watch and clock making in the Black Forest, it was almost inevitable that Staiger started producing plastic cases for clocks and particularly for alarm clocks. This led to the production of complete alarm clocks in 1955. They were equipped with mechanical movements supplied by outside vendors in the region. In the 1960s, battery or AC powered electromechanical movements started displacing mechanical movements and rapidly acquired a major position in the Staiger product mix: electronics were obviously going to play an important role in watchmaking.

An important and innovative step for Staiger was the replacement of the relatively expensive balance wheel motor by a very simple construction that could be injection molded using Delrin, a kind of Nylon made by DuPont. The system consisted of a moving

6.33. Secticon clock movement by Portescap

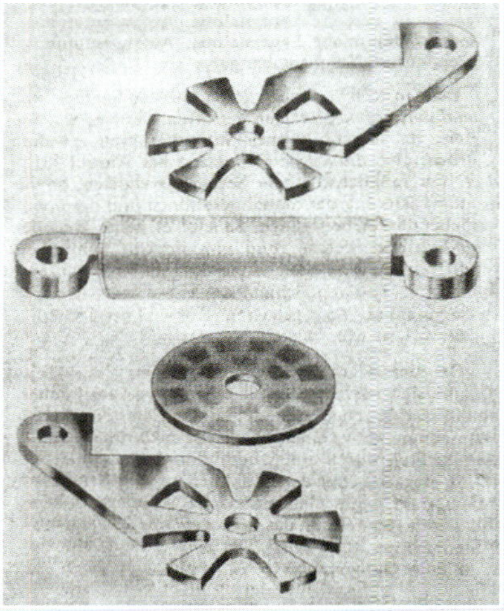

6.34. Main components of the six-pole Socrem motor by Portescap

6.35. Quartz movement, Chronetron CQ-2000. Staiger, St. Georgen, 1972

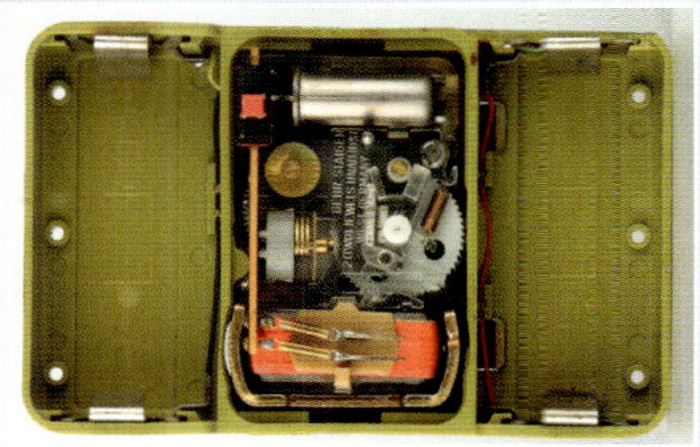

6.36. Quartz movement, Chronetron Quartz CQ-2002, Staiger, St. Georgen, 1973

coil and a coaxial, fixed permanent magnet. When the coil was energized, it started to turn, actuating a pawl that rotated the fourth wheel by one tooth. At the same time, the balance wheel was given a pulse that kept it oscillating. This motion periodically closed a set of contacts, which energized the coil.

In order to simplify the construction of the so-called Chrometron the balance wheel had an unusually low frequency of just 1 Hz. Later, this helped the development of a quartz movement, as now a 16 kHz flexural mode quartz resonator delivered 1 Hz pulses after 14 binary division steps. The chip was supplied by Intermetall. That company also developed the first prototype of the Staiger quartz caliber, but Staiger engineers were responsible for the commercial version. The quartz alarm clock movement was presented to the media in March 1971. It was called Chrometron-Quartz CQ-2000.

Just a year later, Staiger surprised everybody with a new model, the CQ-2001 with a 32 kHz quartz resonator and 15-fold binary division to 1 Hz. The rotating coil motor was unchanged. This design brought the price down by a hefty 50 percent. For the third generation of Staiger clock movements, the silicon chip producer Eurosil suggested the use of the less expensive 4.19 MHz quartz resonator from color TV sets. For the frequency division to 1 Hz, a 22-step CMOS Eurosil integrated circuit was available. The Chrometron caliber CQ-2002 was presented to the media in the summer of 1973. The IC needed 3 volts; the two batteries took up more space than the electronics and the moving coil motor. With this construction, the price of the Staiger clock movement decreased by another 50 percent.

The 4th generation Staiger quartz movement was launched in 1974 and needed only one battery. An Eurosil IC that only needed 1.2 volts now was available. The 5th generation movement followed in 1977; it had been completely redesigned, which made it simpler, smaller, and cheaper. Now the old moving coil motor turned out to be the most expensive part of the entire system. All five makers of electronic alarm clocks got together and obtained an important subsidy from the German Federal government. With this financial help, an extremely simple bipolar motor of the Lavet type was developed and brought to industrial maturity. It could be assembled automatically.

The 6th generation Staiger caliber of 1980 used the new motor and switched to a 32-kHz quartz resonator of the photolithographic Statek type. This also brought down the price of the IC, as only 15 division steps were required. This caliber was a true milestone: it was nearly perfect and was produced for nearly 20 years.

Radio control of the quartz clock meant another quantum jump in terms of timekeeping. Such a clock synchronized itself periodically with the time signal transmitter DCF77 in Mainfling near Frankfurt. The receiver and the ingenious opto-mechanical synchronization system were developed at the horological research institute in Stuttgart, working in close cooperation with engineers from all five companies. The first serially produced radio-controlled clocks of this type were available in 1990.

By now the Chinese competition had caught up technically and caused a constant erosion of clock prices: this included radio-controlled clocks. In order to generate volume, Staiger and the Kundo clock department merged in 1991. Alas, the Kundo-Staiger GmbH only survived for nine years: it went bankrupt in 2000.

UMF-Ruhla

The Watch and Machinery Factory UMF-Ruhla in Thuringia (see Chapter 9) launched its first quartz alarm clock in 1978 with the caliber number 64. The 4.19 MHz quartz resonator was purchased from the Japanese NDK, while Toshiba supplied the integrated circuit. The stepping motor of the anchor type was made in-house; it was the same construction that went into the quartz wristwatch movement caliber 28. The electromechanical buzzer was also made in-house.

The quartz alarm clock caliber 62 with a Lavet stepping motor and a ferrite rotor developed into an important product; it was launched in 1982. Now it was not necessary to rely on imports for the quartz resonator (32 kHz) and the integrated circuit: Glashütte had set up a quartz production line while Ruhla had become self-sufficient in CMOS ICs. The clock measured 56 mm x 56 mm x 20 mm; it was powered by an AA-type battery. Two variants existed: one with an ordinary beep, the other one with an intensity-modulated pulse tone and two volume levels. Later on, tunes stored on an IC became available. Caliber 62 was also used in an alarm clock for the deaf, in a stereo radio receiver, and in a car clock.

Caliber 63-04 was a LCD alarm clock with melody call. The Ruhla quartz alarm clock 18-39 also was quite successful. It was equipped with the movement module of the quartz wristwatch caliber GUB 18 by Glashütte. Hours and minutes were displayed with hands; the wake-up time was set and displayed on a LCD.

The AC powered LED table clock caliber 40 with a 4-digit display was launched in 1979. Caliber 49 followed; this round, radio-controlled table clock with a diameter of 152 mm was tuned to the West-German time signal transmitter DCF77. Aside from the time it also displayed day and date (on a LCD) as well as the signal strength of the transmitter.

VDO

In 1928, OTA Apparate GmbH and Deuta GmbH merged to form VDO. Up until the mid-1960s VDO was mainly known for their dashboard instruments. A clock was standard in most cars at that time; VDO supplied them to car makers but relied on outside vendors for electrically wound mechanical clocks. Their rate was ± 1.5 minutes per day. This was not what the customers expected after 1965, when quartz clocks were already quite common. VDO did not want to lose this important market and decided in 1966 to develop its own extremely sturdy and very accurate quartz car clock with a conventional analog display.

Philips/VALVO supplied the quartz resonator, the CMOS-type integrated circuit was ordered from SSS in the United States. Later on, VDO bought 25 percent of the SSS shares. Prototypes of the so-called Q4 clock were available in 1968; it was round and its height was 33 mm. In the following year, a test series of 1000 pieces was produced; large scale production of the Q4 clock started in 1970. By 1984, 30 million pieces had been sold.

In 1975, VDO replaced the rather noisy 50 Hz synchronous motor with a much quieter stepping motor. The trimmer capacitor for adjusting the 4.19 MHz quartz resonator was replaced by electronic adjustment. Many car makers eventually switched to clocks with a backlit digital LCD or a LED display, but even now millions of new cars are bought with a classical analog display. In view of this continuing demand, VDO redesigned its car clock in 2007: the new design is simpler and flatter than the previous models.

Having gained experience with the design of the Q4 car clock by the 1970s, VDO decided to diversify into battery operated quartz movements for clocks and alarm clocks. Starting in 1975, VDO supplied 100,000 movements per year to the Garant Company. The VDO Quarz Zeit GmbH was founded in 1977 and was responsible for the clock department which even produced a condensation proof bathroom clock.

VDO's attempt to create a strong European clock group by merging with the French Matra and several other Swiss, French, and German clock producers failed, as the restructuration of JAZ and Bayard would have cost a horrendous amount of money. Prices for quartz clocks kept dropping, mostly because of the very aggressive Chinese competition. For this reason, VDO gave up on clocks in 1983.

6.37. Quartz alarm clock, Ruhla Eurochron, caliber 62-02

6.38. Radio-controlled table clock, Eurochron caliber 49

6.39. Car clock movement, VDO, Frankfurt a. M., 1973

6.40. "Horloge lumineuse" by Patek Philippe

Chronometers and Marine Chronometers

Reasonably safe travel on the high seas was possible in the 18th century, when watchmakers in England succeeded in building true chronometers: they were indispensable for the longitude determination. The basic principle of stellar navigation had been known since the 16th century. Only two instruments were required: a sextant for measuring the height of sun and stars, and a very accurate watch. For a long time, only pendulum clocks were really accurate, and they were completely useless on board a ship that was constantly exposed to wind and waves. Things only changed after the invention of the balance wheel, which actually is a torsion pendulum. And ingenious watchmakers, such as John Harrison, Thomas Earnshaw, Thomas Mudge, and John Arnold in England, as well as Pierre Le Roy and Ferdinand Berthoud in France, were needed to design sturdy, truly seaworthy clocks with a daily rate of less than one second.

A ship's position is unequivocally given by latitude and longitude (the altitude is zero by definition). Determining one's latitude is simple: one must only measure the sun's height over the horizon at 12 noon local time when it reaches its culmination point. If the date is known, the latitude (north or south of the equator) can be easily calculated. The longitude is a different story: for its determination, the local time *and* the reference time at the zero meridian (Greenwich) must be accurately known.

The reference time must be brought along as the longitude is given by the difference between local time and the time at the zero meridian. This is why very accurate chronometers (so-called marine chronometers) are required for accurately keeping Greenwich Time during many weeks at sea. The same procedure is used for determining one's position with the height of the moon, bright planets, or stars. But a chronometer alone is useless: one also needs tables with the position of those celestial bodies at every hour of each day of the year at the reference meridian.

At the beginning of the 20th century, mechanical marine chronometers reached the limits imposed by physics. Their daily rate of 0.1 second allowed one's position to be determined with an error of less than 1400 meters after a month long sea voyage: this could be tolerated. But quartz chronometers appeared in the 1960s: all of a sudden, the best mechanical chronometer was demoted to a decorative item on the coffee table. Quartz clocks beat them easily by an order of magnitude. The radio-based navigation Loran system, time signals, inertial navigation, and GPS made the classical stellar navigation a thing of the past. Sextant and chronometer are only needed in the highly improbable case that all the multiply redundant electronic systems would fail at the same time. This is why quartz marine chronometers are still being made, but mostly for collectors and aficionados.

At the famous yearly Concours of the Neuchâtel Observatory, five classes of quartz timekeepers could participate (until 1967):

- Portable clocks with a volume of 8000 cm^3 or less
- Marine chronometers with a maximum volume of 1000 cm^3
- Ship's chronometers, B watches or torpedo boat watches (up to 200 cm^3)
- Pocket watches (up to 50 cm^3)
- Wristwatches (up to 3.7 cm^3)

Borg-Gibbs

The very first quartz marine chronometer was probably designed and built by the Borg-Gibbs Laboratories in Delavan (Wisconsin, USA). It was delivered to the U.S. Naval Laboratory on 12 June 1943, the development contract having been signed in September 1942. The clock consisted of two units: the first one resembled a classical marine chronometer with a small second at 6 o'clock and gimbal mounting in a 19.4 cm x 19.4 cm x 19.4 cm mahogany case with brass reinforced corners. This unit only contained a 60 Hz synchronous motor and a gear train for driving the hands. The electronic unit was much bigger and weighed about 12 kilograms: it was a ventilated metal box with transformers, filters and several radio tubes.

The 87,480 Hz quartz resonator was lodged in a well-insulated oven that was held at a constant temperature. The electronic circuit divided the quartz frequency by two and then six times by three in order to obtain 60 Hz. Normally, the clock was powered by 110 volts DC or AC, but 60 kilograms of lead-acid batteries were included, they made the clock and electronics autonomous for six hours. Two of those chronometers were built, but the daily rate was a disappointing 0.3 second. There was no advantage versus the classical mechanical marine chronometers that were much lighter and more compact. Hamilton built large numbers of them during the war years. They were excellent copies of the Ulysse Nardin chronometer type M.Gr.F.

Bulova

Electronic marine chronometers are normally based on a quartz clock. But Bulova launched a tuning fork-controlled chronometer in 1967. It consisted of three Accutron calibers 214 that were arranged in a circular pattern, 120 degrees from each other. This geometry compensated the position error of the three tuning forks of the system. The signals from the three tuning fork controlled oscillators were combined, so that the tuning forks were always excited with the average frequency. The entire system needed a shock-proof suspension in order to avoid errors due to missed teeth of the ratchet wheel.

The daily rate was significantly better than 1 second. Thus, the name "Accutron Marine Navigator" was misleading, as even a mechanical chronometer achieves 0.1 second per day. Yet, the product was quite successful, as it was relatively inexpensive and much more accurate than a mechanical watch. After the launch of the quartz synchronized Accutron caliber 2240 in 1973, the Accuquartz Marine Navigator followed. It needed a single Accuquartz movement, as the slightly modified tuning fork now only had the function of a motor. It was excited by the 341.33 Hz that were obtained from the 32,768 Hz quartz resonator after division by a factor of 96. The rate now was 0.1 second per day.

Glashütter Uhrenbetrieb – GUB

Glashütte had a lot of experience in the field of marine chronometers, as it had produced the so-called "Deutsche Einheits-Marinechronometer" in cooperation with Wempe in Hamburg during the war years. It was a truly excellent timepiece that was exactly copied by the Russians after the war. Glashütte had also produced the so-called B-watch (caliber 100), a ship's chronometer. As soon as the quartz technology had proven its worth in clocks, Glashütte (that now was part of the communist GDR) began producing electronic precision clocks for navigation purposes.

The 140 mm GUB quartz caliber 48 "Piezochron" was designed as a wall clock in ships' cabins; it was launched in 1975. Colored 3 minute sectors on the dial displayed the compulsory silent periods that are strictly reserved for reception of distress signals at sea. The functions were hours, minutes, and seconds. The motor made the seconds-hand jump every second, the 4.19 MHz quartz resonator guaranteed an extremely good daily rate of a few hundredths of a second. Production stopped in 1983.

There were two variants of the quartz marine chronometer caliber 71. Caliber 71-01 was equipped with a 32 kHz quartz resonator from Glashütte, while caliber 71-02 used a Japanese 4.19 MHz resonator. The dimensions of the case were 135 mm x 185 x mm x 150 mm; the diameter of the dial was 100 mm. The functions were hours, minutes, and seconds. Caliber 71-01 was driven by a half-second stepping motor, the integrated circuit was imported (type MTD 184P).

The motor of caliber 71-02 made one-second jumps; the IC (type U 124) was supplied by the Erfurt silicon foundry. The movement was energized by two type LR 20 batteries. The seconds-hand could be stopped for exact synchronization with the radio time signal. The GDR chronometer certificate or the certificate of the Hamburg test station DHI were optional. Caliber 71-01 was produced from 1975 to 1978; it was then replaced by caliber 71-02 that remained in production until 1990, i.e. after the fall of the Berlin wall. Both models guaranteed a rate of 0.01 second per day. The batteries (2 x 2 R20) were lodged in a special case that was strictly separated from the gear train and the electronics. When one set of two batteries was replaced, the movement continued working with the other set.

Golay

The Bernard Golay Company in Lausanne is mostly known for its cooperation with Longines for developing the Ultra-Quartz wristwatch that was launched in 1969. Actually, Golay had already worked with Longines for many years (see the section on Longines in this chapter). Golay developed a true quartz marine chronometer with the classical mahogany case in 1966. It won a prize at the Concours d'Observatoire in Neuchâtel in 1967.

Isatome

In 1939, Bertrand Voumard founded the Voumard Machines Company in Saint-Imier; it was soon renowned worldwide for its centerless grinding machines. In 1941, Voumard moved to La Chaux-de-Fonds; in 1956 a second plant was built in Hauterive, east of Neuchâtel. Voumard bought the Ebauches Sonceboz plant in 1964 and thus diversified into ebauches with pin pallet escapement of the Roskopf type; ebauches with lever escapement were also produced.

Voumard had actually been interested in watches since 1944, when he started to produce complete movements from Ebauches SA ebauches and cased them; he used the Germinal and Voumard brand names. Voumard bought participations in several companies making watch parts, e.g. Méroz Pierres, Prometal Boîtes, Angelus (small clocks), and Waltham Watch. He even developed his own shock absorber for the balance staff, but in view of the competition by Incabloc and KIF, this design remained at the prototype stage.

In 1963, Voumard had set up the Isatome division in Hauterive where his engineers Jean Shaad and Abraham Zaslawski developed an electrically wound, mechanical clock movement, the Isatron timing machine, and, finally, two quartz clocks that were called Isatome. It was a lead-acid battery-powered, portable but also AC compatible clock with a volume of 2.5 liters; it weighed about 5 kilograms.

Aside from the exact time, the clock delivered the quartz-controlled frequencies of 100 kHz, 10 kHz, 1 kHz, 500 Hz and 100 Hz. The latter frequency was amplified and supplied the synchronous motor driving the clock movement. Electrical minute and second pulses were available for driving slave clocks. Thanks to the excellent daily rate of 1 to 2 milliseconds, the Voumard clock could be used as a marine chronometer. From 1962 to 1967, it obtained the first prize four times and the second prize twice in its category (portable clocks with a volume of less than 6000 cm^3) at the Neuchâtel observatory.

One of the "secrets" of the Isatome clock was a small vacuum flask in which the heated, lens-shaped 1 MHz high frequency quartz resonator was kept at 70° C. Its aging was compensated by a very slowly rotating variable capacitor (it made one full turn every three years). The characteristic dial displayed hours, minutes and seconds; day, date, and month were displayed in windows. A perpetual calendar controlled all the functions. The frequency of the quartz resonator was divided by resonant circuits by a factor 10 at a time and re-amplified with transistors each time.

The Isatome clock was the first member of a family of industrial, quartz controlled time and frequency standards as well as master clocks. They all had basically had the same design, but were less compact than the clock. The quartz resonator was always thermally insulated the same way with a vacuum flask. The dial, too, was always the same. The master clock consisted of a standard 19 inch rack or several of them. Depending on the type, it delivered minute and second pulses for up to 1000 slave clocks in railroad stations, hospitals, and office buildings.

Even the Moscow and Beijing, railroad stations were equipped with Isatome master clocks; their daily rate was one millisecond. But eventually, manufacturing those quartz clocks was unprofitable and was discontinued in 1973. Isatome continued servicing those systems, some of which were still operating in the 1990s.

Leroy

The history of the nearly 250 years old Leroy company is extremely complicated; it was owned by three different dynasties of watchmakers with the same name, even though they were not even remotely related. In the early days, Leroy manufactured very luxurious and complicated pocket watches in Paris. An all time pinnacle was reached in 1902 when a pocket watch with 25 complications was designed for a very rich collector from Portugal. This watch can now be admired at the Musée du Temps in Besançon; however, it stopped operating long ago.

After 1890, Leroy shifted his interest more and more to precision pendulum clocks and marine chronometers. The advantages of electromechanical and electronic timekeepers were clearly understood at an early date. In 1953, a contact-controlled marine chronometer was designed in Besançon; it came out of the cooperation with Thomas Mercer Ltd. and Hatot. Three years later this chronometer was certified by the French navy. In order to eliminate the problems inherent with sparking, the "Chronostat," a transistorized version was developed in 1957. It was produced under an Hatot license and was certified as a marine chronometer in 1960. The next step (1963) was a chronometer with an electronically excited tuning fork. A year later, Leroy added a quartz resonator and dividing circuits to the marine chronometer with a balance wheel motor.

Longines

The cooperation of Longines with B. Golay S.A. in Lausanne began in the early 1950s with the development of a tube-operated quartz clock for sports timing (see the section on sports timing in this chapter). The next step was the development of the transistorized quartz caliber 800, the Chronostat IV. It won a prize as a deck watch at the Neuchâtel Observatory 1964, 1965, 1966 and 1967. Clocks with the caliber 800 won the first prize for unique designs in 1964 and 1965. Previous records were easily beaten at those contests: in 1965, caliber 800 got grade 0.15, in the following year it achieved 0.05 (0.00 would be perfect). At that time, mechanical chronometers obtained grades just below grade 2.0. In 1965 also, caliber 800 won the first prize for a series of four such chronometers; at constant temperature, the rate was just 0.005 second per day – more than enough for the classification as a marine chronometer.

6.41. Marine chronometer, Glashütter Uhrenbetrieb GUB

6.42. Marine chronometer by Golay Lausanne

6.43. Marine chronometer, LeRoy Chronostat

6.44. Marine chronometer, Vega 1525 by Omega with 4.19 MHz Quartz resonator

Golay designed the electronic part, Longines was responsible for the mechanical part. Wherever possible, parts of the 2.5 Hz/18,000 v/h contact controlled caliber 400 of 1983 were used (its dimensions were 28.00 mm x 5.70 mm), but of course neither the contacts nor the balance wheel were needed.

Caliber 800 was rectangular with the dimensions 56.0 mm x 64.0 mm, the height was 48.0 mm. A thermocompensated quartz resonator for 4° C, 20° C, and 36° C with a frequency of 12,000 Hz was used. The electronic circuit used 25 transistors on two printed circuit boards. The frequency was divided in five steps to 2 Hz, which were needed for driving the motor. Three mercury oxide cells supplied the power.

Longines used caliber 800 in pretty three-hand clocks measuring 70 mm x 67 mm x 80 mm. A special model with a watertight stainless steel case with 100 Hz and 2 Hz output won the second prize for marine chronometers at the Neuchâtel Observatory contest in 1967. There also was a variant in which caliber 800 was used as a master clock: it supplied 300 mA for operating a system of slave clocks.

Golay further miniaturized caliber 800; a three-hand clock using the new movement measured 50.0 mm x 50.0 mm x 10.0 mm. Its volume easily fulfilled the Observatory specification for a pocket chronometer (up to 50 cm3). In this category, the clock won second prize at the 1966 Neuchâtel contest. The winner was a pocket chronometer submitted by Seiko-Suwa.

Longines Ref. 8371 was a digital, AC operated thermocompensated quartz clock with a power consumption of 18 watt. It consisted of a 290 mm aluminum tube with a slit for the constantly illuminated nixie tube display. The functions were hours, minutes, seconds, and tenths of seconds (there was a model without the tenths that can make people nervous). The clock was asymmetrically mounted on a tripod. On the back side there was an outlet for a 2 Hz signal for operating slave clocks as well as four buttons for setting the time functions. The display could be accelerated thousandfold, fourfold, stopped, or advanced in steps of 1/10 second. Only a small number of those clocks was produced, collectors are willing to pay a high price for it.

Omega

In 1970, Omega developed the "Electroquartz" (caliber 1390), its first small quartz clock. The flexural mode resonator vibrated at 8192 Hz. It was supplied by Oscilloquartz and was identical with the one that was being used in the quartz caliber Beta 21. But in view of the enormous efforts that were needed for producing and launching the high frequency quartz wristwatch 2400, the small-clock project had a rather low priority.

The situation changed suddenly in 1978, when the French navy set up a contest for a quartz marine chronometer. Omega was one of 36 candidates – and won. The heart of caliber 1525 "Vega"was a 4.19 MHz movement that was combined with a 32 kHz Imhof clock movements with a Socrem stepping motor. The 32 kHz quartz resonator and the dividing circuit were removed. Instead, the signal of a ZT cut 4.19 MHz quartz resonator was divided to 1 Hz. This signal was fed into the motor of the Imhof movement the mechanical part of which was left unchanged.

The two 1.5-volt AA batteries had a life of three years. The instrument was extremely shock proof and sturdy: whatever the circumstances, the daily rate stayed at 0.01 second. The dial resembled that of the Patek Philippe marine chronometer; more particularly, Omega adopted the "sidewalk" concept around the periphery of the dial. The central seconds-hand hovered over this elevated section of the dial at a distance of just 0.1 mm: it was read practically without an error of parallax.

The clock movement was mounted on an aluminum plate that could be lifted off. Then one had access to the battery case with a small voltmeter for checking the battery voltage. The chronometer was synchronized with the radio time signal by means of a switch that either stopped the movement or doubled the rotation speed for a short time. The Omega part of the electronic circuit included a ZT cut 4.19 MHz flexural mode quartz resonator supplied by SSIH-Quartz, an Intersil-IC for 22fold division to 1 Hz and various parts that were soldered on a printed circuit board. There was an outlet for the 1 Hz signal that could be fed to slave clocks. In case of a short circuit at this output, an electronic fuse prevented the movement from stopping.

A grand total of 1300 pieces of Omega's marine chronometer was produced, half of them for the French Navy the other half for the civilian market. The navy series was provided with soft iron shielding inside the wooden box to protect the movement from magnetic fields and radar. The designer was Henri Schneider.

The chronometer had to endure 47 days of grueling tests at the Neuchâtel Observatory. They included abrupt changes of temperature, operation between -10° and +60° C, thermal shocks, intense magnetic and electrostatic fields as well as brutal mechanical vibrations. After that, the timekeeper obtained its certificate as a marine chronometer. But this was only the beginning: for a full year, a set of Omega chronometers were given more very hard tests on the aircraft carrier *Clémenceau*. But the extremely sturdy Omega marine chronometer came out of those tests with flying colors and was adopted as the official instrument of the French Navy. For many years it also was the present of the Swiss Federal government to foreign dignitaries such as kings, prime ministers etc., who officially visited the Country.

Oscilloquartz

In the section on Ulysse Nardin in this chapter we relate how it came to be that the legendary producer of (mechanical) marine chronometers in Le Locle started cooperating with the Oscilloquartz division of Ebauches SA in Neuchâtel. This resulted in two quartz marine chronometers being designed by electronics specialists of Oscilloquartz and Ebauches SA in 1961 on a mandate by Ulysse Nardin. The Neuchâtel Observatory tested them and gave them the first prize in marine chronometer section. In the following year, the Observatory also gave a prize to an Ulysse Nardin design in the class of portable clocks. Ulysse Nardin could not afford an extension of this contract, while Ebauches SA continued developing the new timepiece. One of them received an Observatory prize in 1963.

The smaller Ebauches quartz chronometer was certified as a marine chronometer in 1962 and won the first prize at the Observatory concours for a clock with a volume of 1000 cm^3 or less. Ebauches again submitted its marine chronometer to the Observatory in 1963, 1964, 1966, and 1967 for the Concours and obtained first prize 1964, 1966, and 1967. Those prizewinning calibers became a commercial product soon afterwards.

The Oscilloquartz-Ebauches marine chronometer B-1390 was launched in 1970. It displayed hours, minutes and seconds (the latter in steps of ½ second) on a 58 mm dial with a screw-on watch glass. The watertight case measured 135 mm x 112 mm x 90 mm and weighed 2.2 kilograms. It was ground out of a solid block, was extremely sturdy and had 11 mm thick aluminum walls; the bottom plate had a thickness of 6 mm, plus a soft iron magnetic shielding. Three alkali manganese batteries of the LR20 type gave the chronometer an autonomy of 400 days. The admissible temperature range was -10° to +50° C.

In the first model (caliber B-1390), the quartz crystal had a resonance frequency of 12 kHz and was thermocompensated for 4° C, 20° C, and 36° C by automatically connecting NTC resistors when those temperatures were reached. An analog, non-integrated circuit divided the 12 kHz frequency to 2 Hz (which explains the half-second jumps of the seconds-hand). There was an outlet on the front side for this frequency, and another one for 12 kHz quartz frequency. Two knobs were needed for synchronizing the chronometer with a radio time signal. The seconds-hand could be stopped with one of the knobs, while pressing the other knob accelerated it. The trimmer for adjusting the frequency of the quartz resonator was accessible through a hole in the dial at 12 o'clock, but the watch glass had to be removed to access it.

Model B-1390c saw the transition to digital electronics. The quartz resonator with a frequency of 8192 Hz was the same flexural mode crystal that had been used in the Beta 1 and Beta 2 quartz wristwatch prototypes and in the industrialized version Beta 21. However, the resonator was not suspended in an evacuated copper capsule, but in a glass tube that was inserted in a massive aluminum block. For the 13-step binary division to 1 Hz, two integrated circuits of the CMOS type from RCA were necessary. In both versions, the seconds-hand was driven by a stepping motor. The daily rate was less than 0.01 second.

The analogies with the Longines caliber 800 are, of course, not due to chance. At that time, there was only a small number of specialists for quartz chronometry and they all knew each other. Furthermore, they regularly met at the yearly meetings of the Swiss Chronometry Society. Therefore word quickly spread around about components that were both available and reliable, which circuits really worked well. Furthermore, there was no secrecy within the companies belonging to the ASUAG or SSIH holdings. It was perfectly acceptable to be "inspired" by the ideas of colleagues; after all, everybody sat in the same boat.

Yet, Oscilloquartz developed a pocket chronometer as in direct competition with the Longines caliber 800. This project only had a short life; it is not known whether prototypes were built. It seems that a galvanometer-type motor was planned; it acted on an anchor mechanism driving a toothed wheel.

Patek Philippe

In 1948, Patek Philippe started a research laboratory for the development of electromechanical and electronic watch movements. It was headed by

Georges Delessert, a step-brother of Patek Philippe president Henri Stern. Delessert started out with just two or three people.

First Projects

The team's first project was a quartz clock without any mechanically moving parts and without any contacts: tubes were to do the switching. The electronically controlled display consisted of small neon lamps that were arranged in three concentric rings (for hours, minutes, and seconds) around the 45 cm dial. This clock was finished in 1952; unfortunately its whereabouts are unknown; only fuzzy pictures exist. There are rumors that Philippe Stern bought it at auction for 20,000 Swiss francs, but Stern himself denies this. None of the Geneva museums has it in their inventory. The next project of the Division Electronique was a "nuclear" clock that was powered by the isotope strontium-90. The radioactive decay of this isotope causes electrons to be emitted: those electrons charged a capacitor that periodically fed an electric motor for winding the mainspring of the clock. This clock was exhibited at the 1955 "Atoms for Peace" congress in Geneva.

In 1956, Patek Philippe developed its first battery-operated quartz table clock. The case was a cylinder of Plexiglas with the dial on the top side; the volume was about one liter. In order to minimize the number of dividing steps, a rather low frequency, 10 kHz quartz resonator was used. Due to the strong temperature dependence of germanium transistors, only specially selected and artificially aged transistors were used.

Fortunately, the much superior and only minimally temperature dependent silicon transistors soon became available. They were used exclusively in the later versions of the Patek Philippe clocks. The electronic circuit consisted of two four stage division stages in analog technology that worked on slightly different frequencies; the outputs were combined in order to obtain 16 2/3 Hz by interference. After amplification, this signal was used to drive a 0.2 milliwatt miniature synchronous motor. The power source was a rechargeable nickel-cadmium battery.

The first prototype worked in 1956 and was exhibited at the Swiss Chronometry Congress in 1958. Later on, a small series of those clocks was produced; they bear the ST designation that is followed by a two figure number. One of them obtained the "US Award for Miniaturization 1956." This clock was certified as a marine chronometer at the Geneva and Neuchâtel Observatories; its daily rate was ± 0.01 second. It is not known how many of them were made.

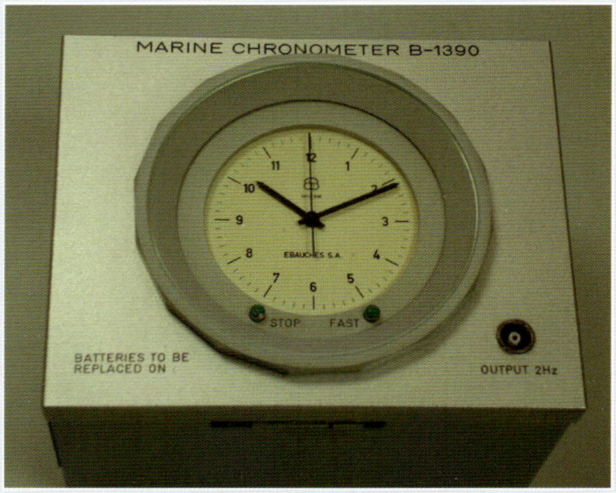

6.45. Marine chronometer, ESA Oscilloquartz B-1390

6.46. Quartz chronometer, Chronotome C by Patek Philippe

6.47. Naviquartz by Patek Philippe

6.48. Polaris, Chinese imitation of the Patek Philippe marine chronometer

Quartz Chronometers

This product range logically evolved into a true marine chronometer, the "Chronotome"; the prototype worked in 1958, commercial production followed two years later. The movement was neatly protected by a flat, spray-varnished aluminum case with a volume of 2400 cm^3; the battery had to be recharged after 120 hours. The power consumption was 30 milliwatts, very little for the time. The technical concept was practically the same as in the ST prototypes: the frequency of a 10 kHz quartz resonator was brought to 16 2/3 Hz by interference, for operating a synchronous motor.

There were two variants of the Chronotome: Type A with precision setting of the seconds-hand, Type B with an output for second-pulses to be used for driving a slave clock or acoustic control. Just 21 pieces of the "Chronotome" were made; they are characterized by a white regulator dial. Type A was certified as a marine chronometer by the Geneva Observatory.

The Chronotome C was launched in 1961 and looked quite similar to models A and B; the main difference was a black dial with separate sets of numerals for the minutes and the hours. The three hands were arranged coaxially. Production of a series of 30 pieces began in May 1962; the Chronotome C was officially launched in October of that year.

Technically, the Chronotome C was quite different from the A and B models, as the frequency given by the quartz resonator was divided from 10 kHz to 1 Hz. The latter frequency controlled an electromagnetic stepping motor. For this reason, the seconds-hand made jumps of a full second. In order to set the clock, one had to press a button that accelerated the movement by 1/100 of a second per second. There were outputs for a whole set of frequencies from 10 kHz to 1 Hz.

Several basic models were developed. Type CP was a marine chronometer powered by a set of six primary batteries with a life expectancy of a full year. Unfortunately, mercury oxide batteries were used in many cases; if they leaked (which happened all too often), the movement was bound to be badly corroded. The geodesy model CC was powered by rechargeable batteries and an AC power supply. Type CT could be operated on AC or DC. Type CTH had a thermostat for the quartz resonator. Type CCD was a battery operated plug-in unit for a standard rack; all controls were up front, including the outputs for the standard frequencies (1000 Hz, 100 Hz, 10 Hz, 1 Hz, and 0.1 Hz). In all cases, the daily rate was better than 0.03 second.

Less than one hundred Chronotome C were produced. The clock was very expensive – extreme quality has a price. Production of the Chronotome was discontinued in 1964, but it was still available on special order from Patek Philippe until 1970.

In March 1968, the Patek Philippe marine chronometer "Naviquartz" (reference numbers began with 1200) officially replaced the Chronotome. With its mahogany case and brass corners, it looked very classical. An early prototype had won first prize at the Neuchâtel Observatory contest in 1961. The Naviquartz Mark IIb (1974) and the Naviquartz Mark III (1978) were quite similar, but the later model had a smaller case.

Technical development led to higher frequency quartz resonators; the original 10 kHz frequency was increased to 16 kHz and finally to 32 kHz that still is the industry standard. The Naviquartz Mark III was even available with a 4.19 MHz quartz resonator. A mercury oxide battery powered the Naviquartz for a full year, but less than 30 pieces were produced.

Chronoquartz and Radio Control

In 1964 Patek Philippe launched the "Chronoquartz" marine chronometer. It was configured as a wall clock and was powered by the ship's power supply. Three different types were available: the very widespread standard mode E, the very rare series F with heated thermostat for the quartz resonator, and series G with stabilized power supply.

In November 1964, the Electronics Division moved from Patek Philippe Headquarters at the plush Rue du Rhône in Geneva to the Rue Eugène-Marziano in Carouge, just South of Geneva. Radio-controlled clocks then had the highest priority; the first one of them worked in September 1967. It was tuned to the 75 kHz time signal transmitter HEG in Prangins near Nyon on Lake Geneva. This transmitter marked each second with a click, each minute with a double click, each hour with a triple click and midnight with a quadruple click. The receiver just "knew" that a new minute or hour had started – which one it was had to be set manually if necessary. A much more sophisticated and unambiguous coding was introduced in 1985 for hours, minutes, seconds, day, date, and year, with automatic compensation for leap years and leap seconds.

Seiko

Seiko designed its first quartz clock at the Suwa laboratories in 1958; it had been ordered by the radio station CBC in Nagoya. Because of the tube electronics, the system was quite big: the height was 2.1 m, the width 1.3 m. This clock was followed in 1962 by the QC-6TM marine chronometer that still operated with tubes but was meant as a first step in miniaturization: it measured 43 cm x 45 cm x 20 cm and weighed 31 kg.

The transistor with its much smaller size and much lower power consumption was bound to win. The development of a tube-type marine chronometer was pushed forward in parallel with a transistorized, battery-operated version measuring just 160 mm x 160 mm x 39 mm. Three of the latter were sent to the Neuchâtel Observatory for the 1963 chronometer contest. They were rated 10th, 11th, and 12th in their class. This clock went into production in 1964; it was called "Crystal Chronometer" (calibers QC951 and QC952). As already mentioned in the section on sports timing, QC951 could also be used as a stopwatch. In the marine chronometer configuration, it won a prize at the Neuchâtel contest in 1963. In view of the 1964 contest, Seiko developed a set of significantly smaller quartz clocks (<200 cm^3) that rated as deck chronometers and won ranks 2 to 7 in Neuchâtel. That year, Longines won the first prize with the quartz chronometer it had developed in cooperation with Golay in Lausanne.

The frequency of the quartz resonator in the Crystal Chronometer was 12 kHz, which was divided to 2 Hz and fed into a micromotor with one second steps. A special model of the movement was launched commercially in 1965; it was called Crystal Timer SQCT 952 and was installed in the cockpit of the Shinkanzen bullet train. The daily rate of this version with a thermostat for the quartz was a mere 0.007 second. For the other commercial models, one could choose between daily rates of 0.05, 0.2, or 1 second, respectively: it was just a matter of price.

In 1965, Seiko sent 7 quartz deck watches to Neuchâtel and obtained ranks 3, 5, 8, 10, 14, and 15 –quite a performance. But the following year (1966) was disappointing, as the Seiko deck watches only obtained ranks 11, 14, 15, 17, and 18. Seiko Suwa still got a lot of attention in 1966, as it also had sent a set of quartz pocket chronometers that obtained ranks 1, 2, 3, 4, 5, 7, and 8 – a flawless performance. The volume of the movements was 18 cm^3; the case measured 48 mm x 46 mm x 14 mm. In 1967, Seiko Suwa got the first prize for pocket chronometers and also ranks 2, 3, 4, 5, 7, and 9 to 15. In 1962, Seiko designed the first quartz clock meant for individual households, type QC-OTV4. The integrated circuit was purchased from Texas Instruments.

Ulysse Nardin

For many decades, Ulysse Nardin in Le Locle had the flattering reputation of producing the world's absolute best marine chronometer. It was the famous type M.Gr.F. that had been designed in 1903.

6.49. Seiko Crystal Chronometer

6.50. Seiko Quartz Chronometer

6.51. Dial side of the Ulysse Nardin Electric

6.52. Movement of the Ulysse Nardin Electric

Merchant marines and navies of most countries used it until the early 1960s. This explains the anchor in Ulysse Nardin's logo.

After World War II it became known that in 1940, the US-Navy ordered a grand total of 10,500 marine chronometers from Hamilton in Pennsylvania. Between 1941 and 1945, 9300 pieces were actually delivered, while Ulysse Nardin produced 10,200 pieces in 100 years! The Navy stipulated that those chronometers had to be exact copies of the Ulysse Nardin M.Gr.F. Hamilton did a great job in industrializing this rather complicated movement. It even improved on the original, as it used a superior set of regulating organs: a cylindrical Elinvar hairspring with dual Phillips end curves and an anisotropic balance wheel. This balance wheel only weighed half as much as the split bimetallic compensation balance Ulysse Nardin used.

Elgin in Illinois also received an order for 3000 copies of the M.Gr.F. from the US Navy in 1941, but only about 250 pieces were produced. They did not meet the specifications of the Navy for the daily rate at different temperatures. In the end, nothing was delivered and the contract was cancelled.

In 1965, a retired Seiko engineer revealed that the marine chronometers used by the Japanese Navy during World War II also were copies of the Ulysse Nardin M.Gr.F. They were produced by Seiko in an unbelievably perfect quality: even the Ulysse Nardin logo was stamped on the mainplate. Seiko delivered 900 of those chronometers to the Japanese Admirality, but only a few of them survived the war. The logo and the Japanese characters misled even Swiss experts for years. The original consensus was that these were real, prewar M.Gr.Fs that the Japanese had provided with new markings. However, one of the few surviving Ulysse Nardin veterans who actually assembled M.Gr.F. chronometers is ready to be quoted with the statement that he can easily identify the Japanese copies. Be it as it may, there is some irony in the fact that during World War II, both the American and Japanese warships determined their position with copies of Ulysse Nardin's marine chronometers dating back to 1903.

With the coming of the much more accurate quartz marine chronometers, the mechanical version was obsolete by the early 1960s. The Neuchâtel Observatory stopped testing mechanical marine chronometers in 1975, as by now they were only bought by collectors. According to the Observatory's statistics, Ulysse Nardin had a de facto monopoly in the marine chronometer category for several decades. Of the grand total of 4504 certificates, 4324 of them were issued to Ulysse Nardin. Added to that are 1069 first prizes for large size chronometers and 747 first prizes for pocket chronometers and wrist chronometers.

Just eight years later, i.e. in 1983, Ulysse-Nardin was in the hands of the Ogival Company and operated under chapter 11. Only one employee was left, an excellent watchmaker by the name of Léo Berthoud. His friend Jean-Claude Jobin helped him out and was paid by the hour. The cash needed for survival was generated for quite some years by selling marine chronometers and deck watches to collectors. Those chronometers were assembled from spare parts until the supply was used up. When Ogival went bankrupt, there was no more future for Ulysse Nardin. A sad demise that was not unavoidable.

As early as the 1950s, Ulysse Nardin had developed a chronometer with electrical winding with a lever escapement instead of the usual detent escapement. A few years later, the company started its cooperation with Ebauches SA and its Oscilloquartz subsidiary in Neuchâtel: the intention was to develop a quartz marine chronometer in common. As already mentioned in the Oscilloquartz section, two prototypes were exhibited in Basel and won a prize at the Neuchâtel Observatory. Ulysse Nardin staff did not have any microelectronic know-how and the company had no available capital for starting a series production.

In addition, they would have to fight an aggressive competition. In Switzerland, Patek Philippe and Isatome produced quartz clocks with marine chronometer specifications. The communist Eastern block countries were supplied by GUB Glashütte, Seiko had their Crystal Chronometer. Ulysse Nardin gave up, but Ebauches SA/Oscilloquartz continued the development of an electronic marine chronometer and of a portable quartz chronometer (see the section on Oscilloquartz); this led to the marine chronometer B-1390.

In its desperate attempts to avoid bankruptcy with attractive consumer products, Ulysse Nardin launched two quartz tableclocks towards the end of the 1970s (they were called marine clocks). They looked like classical marine chronometers, but were mere imitations. The mahogany cases with their brass corners were just beautiful.

The bigger model measured 148 mm x 143 mm x 145 mm and was called a deck quartz clock (10.036 Q).

The watch part was suspended on gimbals and displayed hours, minutes and (sweep) seconds on a white dial with Roman numerals. This was the well-known Imhof movement with a diameter of 49.5 mm and a height of 13.6 mm. A classical construction with a 32 kHz flexural mode quartz resonator, powered by a mercury oxide battery. The temperature dependence of the resonator frequency was compensated by automatically connecting capacitors for 4° C, 12° C, and 36° C. The Socrem motor made 12 steps per turn.

The smaller model measured 37 mm x 109 mm x 71 mm; it was equipped with the same movement and also sported a beautiful mahogany case with brass corners. It was called "Torpilleur," caliber 10.040 Q. The yearly rate was ± 60 seconds as long as the temperature stayed between 15° C and 30° C. This means 0.16 second per day – far from today's COSC requirement of ± 0.07 second per day for a quartz chronometer.

6.53. Ulysse Nardin's Electronic Chronometer

6.54. Electrical drive of the mechanical marine chronometer by Wempe

6.55. Quartz-controlled marine chronometer by Wempe, Model 904

6.56. Wempe Marine Quartz sports chronometer

Wempe

In the 19th century, German ship-owners ordered their ships in Great Britain: they were delivered completely equipped, which included English marine chronometers. But as the German navy and merchant marine grew in importance, it became imperative to be self-supporting in terms of navigation instruments. A team of seven ship-owners got together in 1905 and founded the Chronometerwerke Hamburg GmbH. At that time, marine chronometers were a British specialty, the alternatives being Leroy in France and Ulysse Nardin in Switzerland.

None of the German watch producers had any experience with marine chronometers and local watchmakers had no experience with this kind of timekeeper: the necessary specialists first had to be trained. Much needed to be learned, but the challenge was mastered. Yet, the planned number of 240 chronometers per year was never reached. The new company stayed in the red; the situation even got worse after World War I and the world economic crises of the late 1920s. In 1938, the Chronometerwerke Hamburg were saved by being acquired by Herbert Wempe, whose father Gerhard Diedrich Wempe had started a watch and jewelry store in Elsfleth near Hamburg in 1878 – with a capital of 80 Deutschmark.

Being the owner of the Chronometerwerke, Wempe had attractive jobs for talented watchmakers. He could hire the best and they succeeded extremely well: the Wempe marine chronometers won many prizes. A very sturdy, patented new design was ready for production in 1942. As the war faring German navy had an enormous demand, production of the so-called Einheitschronometer was split between Wempe and A. Lange & Söhne in Glashütte. After World War II, the First Moscow Watch Factory started producing an exact copy of the Einheitschronometer for the Soviet Navy. Wempe also resumed production and sold its chronometers to the booming shipbuilding nations Japan and Korea.

In the 1960s, Wempe began integrating electrical and electronic features into their chronometers. As a first measure, a battery operated electrical motor was added to the Einheitschronometer movement: it wound the mainspring every 8 minutes. However, the massive power consumption and rapid wear of the 3-volt DC motor were big disadvantages of the new construction. Changing to mercury oxide batteries in 1963 did not solve the problem, their capacity was insufficient.

In the early 1970s, Wempe designed a quartz marine chronometer. The power consumption was moderate; the rate was 0.01 second per day: this was a tenfold improvement over the best mechanical chronometers. A 4.19 MHz thermocompensated quartz resonator with AT cut was used with 21 fold binary division in order to generate half-second pulses that actuated the motor. The active part of the electronics consisted of 60 transistors. The circuit was powered by 8 LR20 alkali-manganese batteries. The dial had a diameter of 95 mm; the chronometer weighed 5.0 kg including the mahogany case. But the generalized introduction of the GPS eliminated the need for timepieces for navigation purposes: even quartz chronometers were now obsolete.

Wempe still produces small series of mechanical marine chronometers for collectors and aficionados. A less costly possibility is the "Sports chronometer" that looks exactly like a real marine chronometer. The 185 mm x 185 mm x 185 mm mahogany case, the gimbal suspension and the 95 mm dial are just like the original. However, the sweep seconds-hand "betrays" the electronic imitation, as the original Wempe marine chronometer has a small second at 6 o'clock.

Wempe launched the Sports chronometer in the 1970s, but at that time, the movement was electromechanical, with a transistor controlled balance wheel motor. The daily rate of 2 seconds would have been totally inadequate for navigation, but this was not the purpose of this timepiece. In the 1980s, the electromechanical movement was replaced by a much more accurate quartz movement.

Today, the Wempe Chronometerwerke is among the world leaders for clock systems aboard ships. They are based on high grade quartz movements that are synchronized with time signal transmitters or the GPS satellite system.

Swiss Master Time Centers

Master clock/slave clock systems are so important that they are produced in many industrialized countries. France, Germany, Great Britain, Japan, Russia, Switzerland, the United States, etc. all have their own master clock "culture." This could be the subject of a book of its own. Here, we limit ourselves to the Swiss systems, the technology being basically the same everywhere.

Favag

For decades, the big, orange red Favag plant was the landmark of Monruz to the east of Neuchâtel. For a long time, Favag was a division of Hasler AG in Bern, an important manufacturer of telephone equipment. Favag's main product were telephone relays. In 1987, Hasler merged with Autophon and Zellweger Uster, two important competitors, to form the Ascom Holding. The Favag building was sold to the Italian watch and jewelry firm Bulgari.

The origin of Favag is a company founded in the mid-1800s by Matthias Hipp (1813-1893), the inventor of an original and amazingly accurate electromagnetic pendulum clock (see Chapter 1). Hipp was born in Kaufbeuren near Ulm in Southern Germany. Due to an accident in his childhood, he was slightly handicapped and was advised to learn a trade where he could sit. For this reason, Hipp was apprenticed to a watchmaker in his hometown. He then worked in Ulm, St. Gall, and St. Aubin (Switzerland) before returning to his native Kaufbeuren. In 1840, he set up his own shop in Reutlingen, where he invented his electromagnetic pendulum clock, a 1/1000th of a second chronograph, and a telegraph system that transmitted alphanumeric characters. Those inventions earned him quite a reputation. In 1852, he was nominated general manager of the Swiss federal telegraph works in Bern and of the rapidly expanding telegraph system in Switzerland.

In 1960, he founded his own company called "Hipps Telegrafenfabrik" in Neuchâtel where he had met his wife. Aside from electrical clocks, he produced signal and safety systems for the rail roads. He invented an electrical motor and a recording speedometer for locomotives. His close friend Adolph Hirsch, the first director of the Neuchâtel Observatory, gave him important assignments for electromechanical and electro-optical devices. This included a system for transmitting the Observatory's time signal via the telegraph wires to major centers of the watch industry in the Neuchâtel area and in neighboring cantons. After Hipp's death, the company was managed by his associates de Peyer and Favarger under the name Favarger & Co. After being bought out by Hasler in 1927, it was renamed Favag SA.

After World War II, Favag gained notoriety in the field of master time centers. It was the biggest Swiss producer of such systems many of which were exported. When Ascom was created, the Favag Clock Division was sold to Moser-Baer in Sumiswald, who took over some 25 Favag employees. Major Favag products – such as the quartz master clock E80 and the "Crystal Time" table clock – approximately corresponded to Moser-Baer products and were discontinued.

Moser-Baer/Mobatime

Wilhelm Moser (1905-1974) was the baker's son in the Bernese town of Sumiswald. He was expected to take over his father's bakery, but Moser Junior was so enormously fascinated by technology that he was allowed to be apprenticed to a local clock factory belonging to Jakob Gottlieb Bär.

This plant traced its origins back to a very skilled, purely autodidactic clockmaker by the name of Johann Leuenberger (1778-1865), his sons and grandsons. After finishing his apprenticeship Moser stayed on with Bär and married the owner's daughter. When Bär passed away quite unexpectedly in 1928, Moser ran the plant until Jakob Bär's son Gottlieb had completed his education and was old enough to take over.

In 1937, Moser went into business for himself in the attic of his father's bakery where he started developing and selling electrical master time centers under the Moser-Baer brand. He produced master clocks (with or without time signals), slave clocks, platform clocks, and even antique-looking Sumiswald clocks. Moser started dabbling in electronics on the basis of an assignment by the Swiss military who asked for an ultra-fast telegraph system with the "sensational" transmission rate of 1000 characters per minute.

During World War II and in the following years, Moser-Baer developed electromechanical master time centers with two master pendulum clocks that checked each other's rate. They generated pole changing minute pulses and second pulses that controlled large numbers of slave clocks.

Prototypes of quartz master clocks were developed in 1960 by Moser's son-in-law, the electrical engineer

6.57. Favag pendulum master clock

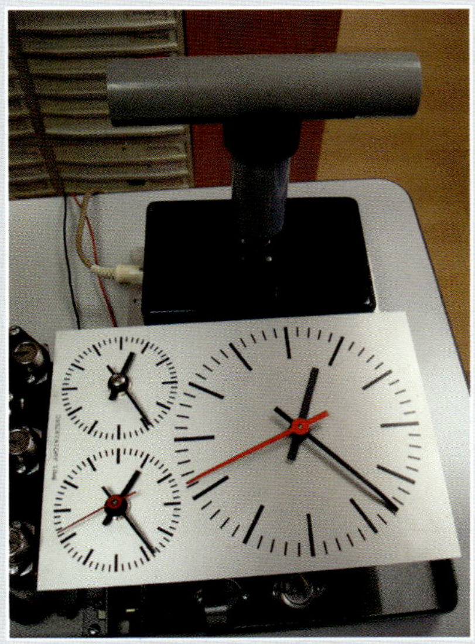

6.58. Mobatron master clock by Moser-Baer with receiver for the Prangins time signals (1969)

Albert Augstburger. They were first exhibited at the Basel Fair in order to test the potential market, and launched commercially in 1966. Moser-Baer's first generation of quartz master clocks had the designation H10 and H11. H10 only controlled slave clocks; H11 also actuated bells or horns. They used a bar-shaped flexural mode quartz resonator supplied by Oscilloquartz. Flip-flops with germanium transistors were used in the dividing circuits that delivered 50 Hz, the European AC frequency, to drive a 3 watt synchronous motor. A conventional gear train actuated the hands. Slave clocks were fed pole changing 24-volt DC minute pulses.

The H16 Moser-Baer quartz master clock was also launched in 1966. 32 kHz quartz resonators continued to be used; this frequency was divided all the way to 1 Hz to drive the master clock. The second pulses were amplified in order to drive up to 80 slave clocks. They were fed pole changing, 24-volt DC second pulses. Minute pulses were generated by the master clock that controlled up to 300 slave clocks. The daily rate was ± 0.1 second within the temperature range of 15 to 25° C.

The Moser-Baer Quartz master clocks H10 to H16 were quite successful: more than 500 of each type were produced a year. Most of them were exported to Sweden where the telephone company L.M. Ericsson took care of the worldwide distribution.

Starting in 1970, long-wave receivers for synchronizing Moser-Baer quartz master clocks with the Swiss time signal transmitter HBG became available. With the Neuchâtel Observatory and Patek Philippe, Moser-Baer was among the few producers of receivers for HBG signals.

The next generation of Moser-Baer master clocks was launched in 1972. They bore the designations HA10, HB10, HN10, and HN11. Their main difference was the power source. HA stood for built-in rechargeable batteries, HB for external batteries, HN for AC supply, power reserve, and electrical pulse storage. The frequency dividers now were integrated CMOS circuits. Hours and minutes were displayed with hands; the seconds were digital with LED display. Silicon transistors were used in the output amplifier.

Further milestones were: the introduction of 4.19 MHz quartz resonators (1975); an additional LCD time display (1981); the transition to microprocessor technology (1984); and the switch to the German time signal transmitter DCF77 (also 1984).

In 1989, Moser-Baer bought the Favag clock division from Ascom, which brought new customers and a wider range of products. In 1994, Moser-Baer

6.59. The Mobatron circuit architecture

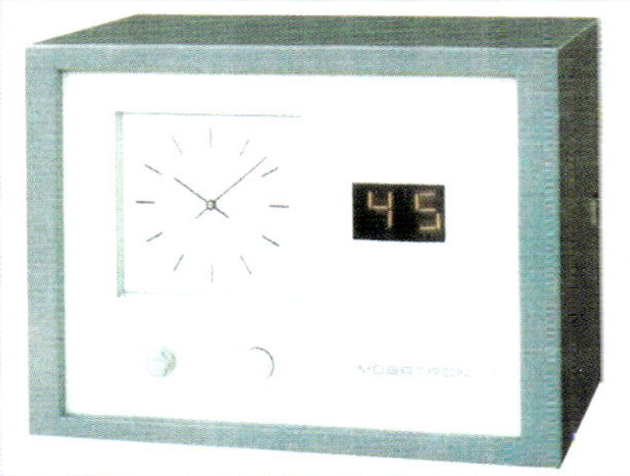

6.60. Mobatime IC HN10 by Moser-Baer for the control of slave clocks, attendance recorders, and calendar system (1972)

6.61. Small Patek Philippe master clock

6.62. Patek Philippe master clock MIC 40

became a minority shareholder of Elekon s.r.o. in Vyskov, Czechia, a maker of LED digital clocks and low-price slave clocks in plastic cases, which are made there for the entire Mobatime group.

With the purchase of Diem SA Time Centers, the successor of the Patek Philippe Division Electronique, Mobatime acquired still another competitor in 1996. Until mid-2000, the Diem systems were produced by Mobatime. Finally, Moser-Baer bought the Bürk Zeitsysteme Company in Schwenningen (Germany) in 1997 and renamed it Bürk Mobatime GmbH.

Patek Philippe

Patek Philippe's first master time center was derived from the Chronotome C quartz chronometer (see the section Chronometers and marine chronometers). A special socket for the Chronotome C amplified the second pulses so they could be used to control slave clocks and other systems. Chronotome C thus served as a master clock; the typical daily rate was ± 0.03 second.

MI Time Centers

The know-how Patek Philippe had acquired with marine chronometers and its accessories paved the way for the development of master clocks for time centers; they soon were an important product. In 1969, the master clock series MI was introduced (MI stood for Modules Intégrés, i.e. integrated modules). The individual building blocks or modules could be combined to form a great variety of systems with a master clock and an almost unlimited number of slave clocks. The dial was still the one that had been designed for the Chronotome.

The MI modules included an AC power supply with transformer, rectifier, and filter, an emergency power supply with rechargeable batteries, an amplifier with outputs for standard frequencies, a receiver for long wave radio time signals including a dipole ferrite antenna, and, of course, the clock unit with at least one master clock and control units for the slave clocks. Optional items were a heated thermostat for the quartz resonator and a digital display with nixie tubes. The MI concept was a very clever innovation, as the needs of each individual customer could be met very flexibly at fixed prices. Furthermore, changing the configuration was very easy: additional modules could simply be plugged in. Repairs consisted of replacing the damaged unit: this did not affect the clock that always kept working.

Until 1973, Patek Philippe used the programmable switch clock of the Inducta Company (Wittwer-Bär in Gwatt near Thun) as an accessory. Only the front plate was modified in order to make it a standard module that could be integrated into an MI system. Starting in 1973, the Inducta module was replaced by a similar system called ST4/1 that Patek Philippe had developed in-house. A gong and an amplifier for microphone announcements were available in modular form. Music could be also distributed through the time center. Patek Philippe never produced slave clocks for the MI systems, but bought them ready made from Inducta.

The MI system was technically perfect; it sold quite well despite its high price. Yet, Patek Philippe decided to introduce the more economical MS version that still was technologically at the forefront but did without luxurious features. Patek Philippe's MS version had a distinct edge on the relatively low-priced master time centers of its competitors.

Multiple Redundancy

The master time center Patek Philippe designed for the Federal Building in Bern (the seat of the Swiss Federal Government) was the biggest one of its kind. It was installed in 1971 and was equipped with a receiver for time signals that synchronized three independent quartz clocks. The minutes and seconds of each unit were constantly checked and compared with one another. If there was a difference, the presumably faulty unit was switched off and an alarm rang. More than 1000 slave clocks obtained signals from the master clock system. Each of them could be checked directly from the master time center and were stopped or reset if necessary. Other big master time centers were installed in railroad stations, airports, office buildings, and hospitals.

The MIC series with integrated CMOS-circuits was launched in 1979 while production of the MI master time centers continued. Time was displayed with hands; BCD outputs were provided for computers. In 1982, Patek Philippe launched the series MIC-P with a receiver for the German time signal transmitter DCF77. This was the first master clock with a microprocessor. Another new product was a relatively economical table clock with digital time display. It was called "Tele-quartz" and was very successful; production started in 1983.

DIEM and Mobatime

Henri Stern, senior president of Patek Philippe, was fascinated by electronic clocks and watches; the Division Electronique always received his special attention. This division did not make huge profits but it generated free publicity for the brand. In 1988 Henri Stern's son Philippe decided to put all his eggs in the basket of the venerable tradition of purely mechanical timepieces. An electronic division was not compatible with the new-old corporate image. He gave up the brand Patek Philippe Electronic; in 1990 he sold the division in equal parts to the employees and external investors.

Thus, Diem SA (the abbreviation for Division Electronique Marziano) was founded. It continued producing master time centers and other electronic products at the Rue Eugène-Marziano in Carouge, South of Geneva. Only the logo was changed. After a few years it became obvious that the rapid progress of microelectronics constantly required massive investments. In 1996, Diem was sold to Moser-Baer AG in Sumiswald.

Wittwer-Bär

The history of the Inducta master time centers is particularly interesting. Its roots go back to the inventor and industrialist Martin Fischer (1867-1947) from Zurich. It seems that he was apprenticed to a watchmaker, but as a young man he became known as a daredevil mountaineer, hunter, and photographer. In 1899 he was granted the Swiss patents 19701 and 19702 for "batteryless electrical clock system.".

For the production of those clock systems, Fischer founded the "Aktiengesellschaft Magneta, Fabrik electrischer Uhren" in 1901; the seat of the company was Zurich. As soon as everything worked fine, Fischer switched to automobile manufacturing and started producing one- and two-place Turicum roadsters in Niederuster. By 1912 he had already sold 1000 cars.

After his adventure with automobiles, Fischer switched back to his Magneta master time centers. The company grew rapidly and opened sales offices and factories in Berlin, Singen, Paris, London, Glasgow, New York, Vienna, and Lochau near Bregenz. Nine types of master clocks were developed that could control up to 500 slave clocks. Fischer also produced a water tight, spring drive regulator for use on board of ships. The diameter of his slave clock dials ranged from 20 cm to 200 cm. There was no movement, but simply two electromagnets with a pawl moving a toothed wheel whenever the magnets received a current pulse.

The Magneta time centers were particularly successful in the United States. They were installed in many railroad stations, public buildings, and hotels. For example, a Magneta slave clock was found in every room of the extremely luxurious Plaza Hotel in New

6.63. T3 master clock of the MI series

6.64. The biggest master clock ever built by Patek Philippe. It was designed for the Swiss Federal parliament building in Berne. It was equipped with a time signal receiver and three quartz clocks.

6.65. Giant Swatch built by Inducta in 1984 and suspended on the office building of the Commerzbank in Frankfort (Germany)

6.66. Flower clock by Inducta in Zürich

York. The famous opera singer Enrico Caruso stayed at the Plaza in 1907, just a month after it had been opened with great pomp. But the wonderfully regular clicks of the electromagnets marking the minutes drove Caruso crazy. He demolished his clock by beating on it with a shoe. This caused a short circuit that stopped all 245 slave clocks in the hotel. Whereupon the hotel's general manager apologized for the inconvenience to Caruso with a handwritten letter and a magnum bottle of French champagne.

The Magneta master clock was a precision pendulum clock that was powered by a massive weight of 20 kg to 200 kg, depending on the number of slave clocks. This weight had to be lifted to the starting position every 48 hours either manually or with an AC motor. The energy generated by the falling weight actuated an electrical generator (a so called inductor) that sent strong electrical pulses to the slave clocks. This means that the master clock generated its own electrical energy and needed neither batteries nor an AC connection, as specified in the original patents.

World War I brought great difficulties, as exporting Magneta clocks became nearly impossible. In 1920, the Landis & Gyr Company (known as L+G) that until then had mostly manufactured current meters, bought Fischer's company on the initiative of L+G president Karl Heinrich Gyr. Magneta was renamed Inducta and moved to Zug in central Switzerland. For many years, L+G produced master time centers and slave clocks on the base of the Magneta technology. In 1951, Inducta production was moved to the SAIA subsidiary in Murten. In 1961, the division was sold to Wittwer-Bär in Thun-Gwatt.

At that time, Wittwer-Bär's president was Hans Wittwer, the son-in-law of the tower clock manufacturer Adolf Bär who belonged to the Bär dynasty in Sumiswald. When it bought Inducta, Wittwer-Bär got all the plans, tooling, and a huge inventory of spare parts. Moving to Gwatt near Thun was no big problem, Wittwer-Bär continued working with the same suppliers and customers.

The master clocks "inherited" from Landis & Gyr continued to be precision pendulum clocks, but in the early 1970s it became obvious that one had to switch to quartz. After carefully evaluating the pros and cons, Wittwer-Bär managers came to the conclusion that the company should not develop its own quartz clocks but should buy them from an outside vendor. Patek Philippe's Division Electronique was the word's leader for that kind of product. It even cut and ground its own quartz resonators that were placed in an electrically heated thermostat and stabilized with extreme care. Henceforth, Wittwer-Bär bought their master clocks from Patek Philippe and supplied slave clocks that Patek Philippe never produced. The contract with Patek Philippe turned out to be a win-win situation and very good business for both parties.

The oil crisis of the 1970s in the wake of the Arab-Israeli war triggered a severe recession. Wittwer-Bär now needed an economical master time system for schools, hospitals, and factories with a tight budget. A new system was developed in-house, which also synchronized attendance recorders. There were two versions: HQM-1 generated minute pulses while HQS-1 generated second pulses. The new systems worked extremely well; thanks to an 8 MHz quartz resonator, the master clock was amazingly accurate.

The introduction of the summer time/winter regime posed a problem that was solved by synchronizing the systems with a time signal transmitter, i.e. either HBG in Prangins or DCF77 in Mainflingen near Frankfurt. Another option is the synchronization with the GPS-System. This is when Wittwer-Bär started to cooperate with Pierre Bonanomi's Precitel in Marin near Neuchâtel. Wittwer-Bär ended up buying Precitel in 2008.

Despite several recessions, Wittwer-Bär is still doing well – with a vastly different product range. Today, Wittwer-Bär employs some 40 people and operates in a modern building, replacing the old and decrepit farmhouse where the company started. Business is still good, but the product mix has totally changed. Nowadays, the main products are so-called publicity clocks, i.e. wall clocks that are enormously enlarged versions of the best known Swiss wristwatch models. There also are weatherproof versions for outdoor installation (they are called pocket watches) to signal watch stores.

Wiitwer-Bär still manufactures the classical master time centers, but only on special order. This product line has shrunk drastically: railroad companies have little money to spare and no new stations are built. Furthermore, the existing master time centers and slave clocks are so sturdy that they keep working for decades without any maintenance. Tower clocks too are in small demand, as no new churches are built anymore. For this reason, Wittwer-Bär decided to discontinue this type of product. World time clocks (with analog or digital display), as well as flower clocks, are still quite popular.

7. *The Eightfold Way to the Quartz Wristwatch*

CEH versus Seiko Suwa

The quartz wristwatch was invented at least eight times in Switzerland, Japan, and the United States. German and French engineers were not sleeping either, but their constructions came a little later. Switzerland definitely was the pioneer. Beta 1, the very first prototype, started ticking in the summer of 1967 at the Centre Electronique Horloger (CEH) in Neuchâtel. It was shown to the media in January 1968.

Beta 1 was very probably the world's absolutely first quartz wristwatch. Nobody remembers exactly when the first prototype started ticking at the CEH, but mid-July 1967 is a good guess. It is very likely that it was checked in the lab for several weeks. We know for sure that the CEH engineer and project manager Jean Hermann walked with it to the Neuchâtel Observatory and that the tests for the chronometer competition (the 1967 "Concours") started there on August 13th, 1967.

Seiko never revealed at what date its first quartz prototype worked. Documents confirm that five prototypes arrived by air mail at the Neuchâtel Observatory in early November 1967 and that tests began on November 11th. On the basis of the still extant Observatory books, the Swiss priority is assured. On the other hand, work that would lead to the Omega Megaquartz 2400 had already started in 1955, but a working prototype only became available in 1969.

The eight different technical concepts that are described in this chapter all led to working quartz wristwatch movements. One of them was superior to all the other ones and has hardly changed in the last four decades. Year after year, some three billion watch and timer movements are produced on the technological basis that was established in 1971. Three-quarters of the watches that are assembled with such movements (more than a billion of them per year) have an analog display, i.e. conventional hands. About 25 percent have a liquid crystal display (LCD). Light emitting diode displays (LED) are niche products.

7.1. Beta 1 by CEH, prepared for the Observatory contest of 1967, movement side

7.2. Beta 1, dial side

7.3. Beta 2, movement side

7.4. Beta 2, dial side

Beta 1 and Beta 2 by CEH

Gérard Bauer (1907-2000) who was president of the Fédération Horlogère (FH) for many years was a true visionary. He had been educated as a lawyer and did not understand electronics, but he felt intuitively that electronics were going to be of vital importance for the watch industry. In 1959, the FH scientific and technical director René Le Coultre was charged with working out a long range research project for a watch that would fully integrate the new avenues opened up by microelectronics. The new movement should go much beyond the already available or coming electromechanical and electronic horological devices, i.e. the balance wheel motor and the tuning fork motor.

The best candidate for managing the planned new research facility was an electrical engineer and physicist by the name of Roger Wellinger. He had done military service with Le Coultre and they had been colleagues for two years at the Technical University in Lausanne. Wellinger then went to Zurich, where he earned his master's and PhD degrees at the Swiss Federal Institute of Technology (ETH). He then emigrated to the United States, where he had obtained a postdoctoral position at the University of Illinois. He was put on the tenure track and subsequently was promoted to Research Assistant and Associate Professor. But after five years, Wellinger was tired of the prairie. In order to see mountains, he joined the General Electric Research Center in Schenectady (New York). This is where Le Coultre visited him in 1960 and convinced him that he should lead a team of physicists and engineers in Neuchâtel working on the development of electronic watches.

Wellinger went to Neuchâtel for a presentation and was formally offered the position as director of a new research institute that he would have to create. He asked for a five-year contract and a budget of 1.8 million Swiss francs per year; he got both. A few years later the budget would peak at 5 million francs per year. Wellinger returned to Switzerland with his family and reported for work on 1 March 1961. The new electronics lab only existed on paper and did not even have a name. Wellinger coined "Centre Electronique Horloger" (CEH); it was officially established on 30 January 1962 as a joint stock company.

The capital consisted of 480 shares of nominally 1000 Swiss francs. Ebauches SA took 190 shares, 15 went to the ASUAG holding, and 5 each to the Fabriques de Spiraux Réunies, the Assortiments Réunis, and the Fabriques de Balanciers Réunies. The Fédération Horlogère subscribed to 180 shares, while the Chambre Suisse de l'Horlogerie took 40 shares. The Communauté d'Horlogerie de Précision, IWC, Jaeger-LeCoultre, Mido, Omega, Portescap, Rolex, and Tissot each held 5 CEH shares. In 1973, the CEH was restructured and was left with just eight shareholders: ASUAG, Bulova, Complications SA (Piaget), Ebauches SA, FH, Rolex Geneva, Rolex Bienne, and SSIH.

Tough Beginnings in the Attic

Wellinger's first job was to make the CEH a reality, long before it was officially founded. The first priority was to recruit scientists, technicians, and administration personnel. Working out possible scientific and technical concepts of an electronic wristwatch was of equal importance. Wellinger's first office was an attic room in the building of the watch research institute LSRH. The "furniture" consisted of a very old table and a light bulb hanging from the ceiling; there was no phone and not even a chair!

Wellinger worked out three basic conditions that the new electronic watch had to fulfill. It had to be much more accurate than the traditional mechanical watches while having about the same volume. The energy consumption should be so low that an ordinary button cell would power it for at least a year. Furthermore, the majority of the parts should be commercially available.

It was deemed indispensable that the electronic circuitry should be concentrated on integrated circuits (IC) that needed to be developed. Simple integrated circuits were available at that time, but they cost about 800 US dollars apiece. Wellinger knew very well that very costly equipment was necessary for fabricating integrated circuits and that the yield would be abysmally low, at least in the beginning. But the IC technology had to be made watch-compatible regardless of the cost. Also one had to find an oscillator delivering a very stable frequency that would serve as a "pacemaker."

Wellinger taped big sheets of paper on the conference room walls and wrote down all potential parts of an electronic watch and the way they could be combined. Much later he learned that he had reinvented Fritz Zwicky's morphological analysis. This was a tool that allowed a theoretical exploration of all possible systems; most of them were dead ends and could be eliminated at an early stage.

Two projects were defined that were the most likely to succeed. Project Alpha was a greatly improved tuning fork movement with a very small position error. There were clear analogies with the "Accutron" that was launched by Bulova in 1960. This project was cancelled in 1966 because of possible infringement of Bolova patents. Wellinger resurrected Project Beta, which had stayed on the sidelines for a long time, but that he personally preferred: its objective was a quartz wristwatch.

Three to four times a year, Wellinger flew to the United States to recruit expat Swiss scientists who had specialized in microelectronics and solid state physics. He hired Max Forrer, Armin Frei, Raymond Guye, Kurt Hübner, Fritz Leuenberger, Rolf Lochinger, Henri

Oguey, and Heinz Waldburger for the CEH. The staff kept growing; at the end of the first year, the number of employees had risen to a full dozen, in 1967 their number exceeded 90.

Office space and, particularly, lab space were critical: the LSRH building being too small, several apartments were rented throughout the city. Money was not a problem, but nobody had thought of the fact that the scientists would ask for a raise once in a while. This led to "collisions" with CEH president Fritz Hummler: a few good people left the CEH, as they had been offered better paying jobs elsewhere.

A Joke in the Staircase

Electronics specialist Eric Vittoz started out with the design of a quartz clock that used a commercial quartz resonator made by the Oscilloquartz division of Ebauches SA. The electronic circuit consisted of individual silicon transistors – volume was not a problem at that stage. For the display, Vittoz used the dial and hands of a commercial electromagnetic clock and a microampere meter that moved at the pace of seconds.

This macro-prototype worked by the end of 1962 and was shown to the shareholders of CEH – they were impressed. Vittoz had succeeded in making the output of a quartz resonator compatible with the input of transistors – quite an accomplishment. It was also important to demonstrate that a few microamperes sufficed for powering the oscillator and frequency divider circuits.

The practical design of quartz wristwatch movements, for which the CEH would gain a worldwide reputation (Beta 1 and 2 were the prototypes, Beta 21 the production version), was "invented" in the CEH staircase in the fall of 1963 during an informal chat of Armin Frei with Rolf Lochinger: both were scientists at the CEH. In fact, there was no real invention, as the basic principle of a quartz frequency standard with divider circuits and a counter (watch hands or a digital display) had been known since the late 1920s. But the technology available in the 1960s was needed for a wristwatch-size quartz watch.

The most urgent priority was to miniaturize all the parts as much as possible. The total volume of the new movement was not allowed to exceed the limit imposed by the Observatory, i.e. 3.75 cm^3. The heart of the future quartz watch was the oscillator with a quartz resonator, the oscillator circuit and a fine tuning device for the quartz resonator. For all practical purposes, this unit was singlehandedly developed by Armin Frei.

Extensive calculations were needed for the quartz resonator, in order to find the best possible cut with the smallest dimensions. Finally, a quartz rod with a

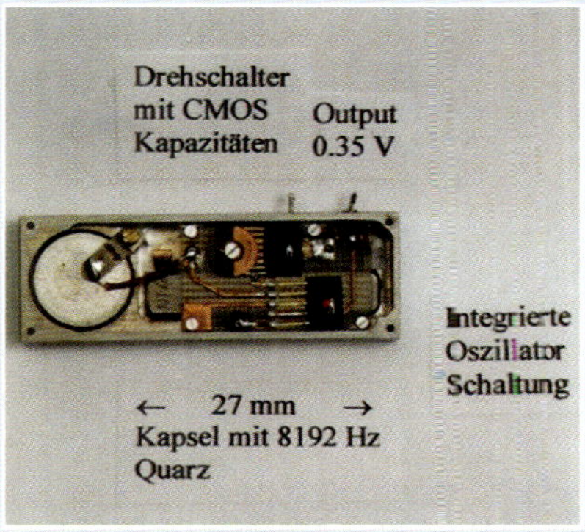

7.5. Oscillator for the Beta prototypes of the CEH

7.6. Integrated divider circuit for the Beta 2

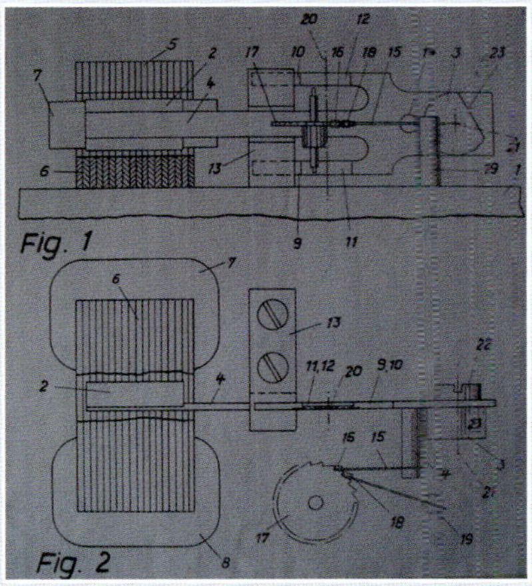

7.7. Schematic drawing of the Beta 2 vibrating motor

7.8. Prototype watch equipped with a Beta 21 quartz movement

rectangular section measuring 23.6 mm x 1.2 mm x 0.8 mm with a resonance frequency of 8192 Hz was chosen. It was encapsulated under vacuum.

At the Oscilloquartz division of Ebauches SA in Neuchâtel, Frei found a very smart technician by the name of Jean Engdahl who was ready to fabricate the required resonators by diamond sawing. Those rough bars were then carefully ground and polished at the CEH, in order to vibrate at just the right frequency. Finally, they had to be metalized by gold evaporation. Fine wires were then soldered on, so that the crystal could be suspended in its carefully degassed copper capsule. Encapsulating the resonator in an evacuated capsule made of degassed electrolytic copper turned out to be extremely difficult and full of nasty problems. For example, a solder formulation had to be found that could be fully degassed before use.

Bipolar IC-technology

The integrated divider circuit was particularly important. Its role was to reduce the quartz resonator's frequency to a sufficiently low level that it would drive a motor. In order to minimize the power consumption, the number of divider steps had to be as small as possible. At the same time, the motor had to operate at the highest possible frequency. The best solution was to use five binary steps in order to reduce the frequency of the quartz resonator from 8192 Hz to the motor frequency of 256 Hz.

The integrated circuit required for Project Beta not only needed to have a minimal power consumption, it also had to function at the 1.35 volts available from the mercury oxide batteries of the time. At such low voltages, there was no other choice than the bipolar configuration that consists of bipolar junction transistors. The latter are so named because their operation involves both electrons and holes. This type of IC normally worked with milliampere currents while in a watch movement circuit only microamperes could be tolerated, which corresponded to the leakage current of normal bipolar ICs.

Kurt Hübner who was responsible for the integrated circuits, managed to reduce their power requirements ten thousandfold by diffusing high valued resistors into the circuits. But one was still very far from the sensationally low power demand of CMOS circuits. Unfortunately their efficient operation required at least 10 to 20 volts. This was brought down to 1.2 volts eventually, but not before the early 1970s.

The development of a digital frequency divider based on flip-flop circuits was a true breakthrough: this device was far superior to any analog circuit. For the prototypes Beta 1 and Beta 2, circuits with two divider steps per chip were fabricated and inserted vertically into a special support on the printed circuit board that interconnected the chips and the other electronic components.

Unique Motors

Henri Oguey, another of the Swiss-American veterans that had been brought home by Wellinger, developed the micromotors that were needed for the analog display of the Beta quartz watches. This seemed to be a simple assignment, but the available power being extremely small, every possibility had to be considered.

The sarcastic comments made by the great tuning fork watch expert Max Hetzel were not particularly helpful. He was convinced that a quartz wristwatch would for ever remain a mirage. There simply was no motor with a power requirement in the microwatt range – and batteries could not deliver more power for a long time. Hetzel's opinion was confirmed by the Department for Industrial Research at the ETH in Zurich. Calculations had shown that a millimeter-size synchronous motor would have an absurdly low power efficiency. Obviously, it was impossible to drive such a motor with a button cell.

Despite of all this, Oguey developed a sturdy, shock-proof electromagnetic vibration motor for the Beta 2 system. It consisted of two coils that were magnetically connected with a soft iron yoke. A Ticonal magnet moved back and forth in the gap between the coils, in tune with the current pulses coming from the chain of frequency dividers. This magnet was glued to a nickel-iron lamella that ended with a counterweight and was suspended by two flexible arms. A pawl with a prismatic ruby was attached to the lamella and turned a ratchet wheel with 256 teeth one tooth at a time.

A 1 Hz stepping motor for the Beta 1 prototype was developed at the same time. It consisted of two stationary coils on an iron yoke. A toroidal flat coil moved back and forth in the gap. An anchor was attached to the coil and acted on a kind of escape wheel that transformed the back and forth motion of the anchor into a rotary motion. Beta 1 had the same overall dimensions as Beta 2, i.e. 28.0 mm x 25.0 mm x 5.3 mm.

The 1967 Chronometer Contest

In the Beta 2 configuration, the CEH quartz movement worked for a year and a half before the mercury oxide battery had to be replaced. In the Beta 1 configuration, a fourteen fold division of the 8192 Hz quartz resonator frequency was necessary. This reduced battery life to half a year – still plenty for the tests at the Neuchâtel Observatory that lasted six weeks.

In early 1967, it became certain that the quartz wristwatch would work. The miniaturized subsystems had been completed; all that remained was to integrate them into a complete movement. This was not really easy, but CEH management decided early in the project that each working prototype would immediately be brought to the Observatory for the tests in view of the 1967 "concours." The electronics engineer Jean Hermann was asked to organize the logistics of the Beta 1 and Beta 2 prototypes and to supervise the last phases of the movement assembly. This was done by an excellent young watchmaker by the name of François Niklès.

Hermann very carefully read the Observatory rules for the contest. While doing so he realized that the tests at 4° C and 36° C were a potential problem, as the frequency of a quartz resonator is strongly temperature-dependent. This could have easily been corrected with temperature-dependent resistors, as their capacity is also temperature-dependent. But this could not be done in practice, as the parts with the necessary impedance were not available.

Instead, Hermann chose to automatically connect a capacitor for the high test temperature and another one for the low one to the quartz resonator. This could be done automatically with temperature-dependent resistors that switched the capacitors on and off. Optimizing everything also meant choosing quartz resonators with a temperature curve that peaked at 28° C. Thanks to those "tricks," Beta prototypes received the first ten ranks in the 1967 Observatory contest. The prototypes that Seiko had submitted were beaten flat, simply because the Japanese had not thought of thermocompensation.

Aging Problems

The natural aging of the quartz resonator was another problem, as it causes a shift of the temperature dependence. Hermann very carefully measured this parameter: the quartz resonators and their capacitors were chosen with just the right characteristics in function of time, so that they would reach the optimum values during the tests.

In mid-July 1967, the first Beta 1 prototype worked to everyone's satisfaction: as mentioned above, the quartz resonator's frequency was divided to 1 Hz: those pulses energized an anchor-type stepping motor. In order to figure out the aging trends, the quartz was observed during 3 to 4 weeks. After final assembly, the movement was mounted into the standard aluminum frame requested for the tests.

On 10 August or thereabouts, Hermann walked to the Observatory with the Beta-1 – this took about 15 minutes. The books of the Observatory have been fully preserved. According to the 1967 volume, the 45 day tests of the first CEH quartz wristwatch prototype started on 13 August. When Hermann proudly arrived with the new movement, the Observatory Director Jacques Bonanomi asked him what kind of grade one should expect for the new chronometer. Hermann made a rapid mental calculation: at that time, the absolutely best mechanical chronometers achieved grades around 2. Quartz should be an order of magnitude better, which should result in a grade around 0.2. Bonanomi thought this was a joke and roared with laughter. He stopped laughing when well under 0.2 was achieved: the best CEH prototype of 1967 got grade 0.152 (zero would be perfect).

At the Neuchâtel Observatory, Bonanomi had created a new contest category for quartz wristwatches earlier in the year. The idea was not to irritate the traditional watchmakers: quartz movements were bound to be much better than mechanical chronometers, but nobody knew how much better. At the Geneva Observatory, this precaution was not taken: all chronometers of a given size range were in the same category. For the 1967 Geneva contest, the CEH sent in a Beta 1 and a Beta 2 with the result that they beat the absolutely best mechanical chronometers in a most embarrassing way.

According to Hermann himself, he was not clearly aware of the fact that he was carrying the world's absolute first quartz wristwatch movements in his briefcase and that they would lead to revolutionary changes in the watch industry. He was glad that each of the quartz movements had been working flawlessly for several weeks. He was concerned that the Japanese colleagues, particularly at Seiko, might come up with something similar. Everybody knew that they had obtained ranks 10, 11, and 12 in the category of electronic marine chronometers in 1963. In the following year they were classified second to seventh in the category of quartz deckwatches. And in 1966 they had sent pocket watch size quartz chronometers to Neuchâtel and obtained ranks 2 to 8.

In January 1968, when the results of the 1967 contest were made public, Hermann realized how right he had been to be worried. Seiko had participated with five wristwatch size quartz movements; four of them won a prize and obtained the places 11, 13, 14, and 15. Also in 1967, Seiko was first in the category of electronic pocket chronometers.

By November 1967, Hermann had walked another ten times to the Observatory, where he brought a total of six Beta 2 and five Beta 1 prototypes – the former with frequency division to 256 Hz. This frequency drove Henri Oguey's vibration motor with pawl and ratchet wheel. Six weeks after the Beta prototypes were brought to the Observatory, Hermann would return them to the CEH, with the certificate that had been filled out in longhand.

The Observatory requested that the name of the precision timer responsible for a given movement be mentioned on the certificate. Of course Hermann had done the "electronic timing" on all the 1967 prototypes. It was decided to distribute the privilege of appearing as a precision electronic timer among ten of the CEH engineers and technicians. Hermann himself is registered as the precision timer of the number 5 movement. The other certificates mention R. Challandes (twice), P.-A. Andrey, H. Oguey, N. Grelat, M. Jeannot, C.-A. Dubois, C. Frossard, J.-C. Droz and D. Paratte.

A Swiss Success Story

In December 1967 a meeting was called to inform the CEH shareholders of the sensational results that had been obtained with the CEH quartz prototypes at the Observatory. During the discussion that followed the presentation, one of the CEH scientists asked a very interesting question: what would the watch industry do if the price of a high grade electronic movement dropped to 3 Swiss francs apiece?

The audience was so surprised and horrified that nobody uttered a word. Only CEH president Fritz Hummler found that this would be great. Ten years later the price of 3 francs was indeed reached, with catastrophic consequences for the Swiss watch industry. But in 1967 nobody could imagine that the daily rate of the best mechanical chronometers could be easily beaten by an order of magnitude – at the price of the cheapest pin pallet watch.

In January 1968, the results of the Observatory competition of 1967 were communicated to the press. The Beta movements of both types were not only far better than mechanical calibers; they were also six times better than the tuning fork movements the Observatory had checked in 1967. The media all over the world echoed this astounding success of Swiss research.

Omega's Megaquartz 2400

In 1954, after the publication of Hetzel's patent covering the tuning fork watch, Omega executives realized that electronic watches might revolutionize the watch industry. Producing tuning fork watches was out of the question: Omega could not afford to look like an imitator of Bulova technology; it had to be on the market with its own, much superior development. One of the possibilities was to miniaturize the quartz clock to wristwatch size.

At that time, Omega did not have any microelectronic competence. For this reason, representatives of the SSIH holding (that included Omega, Tissot, Lanco, and other brands) negotiated a cooperation agreement with the Battelle Research Institute in Geneva in 1956. Interestingly enough, the threat and opportunity of microelectronics had been understood at an even earlier date by Ebauches SA executives. Battelle was contacted as early as 1955 by Ebauches. The idea was the development of an electronic watch movement, but nothing came of it.

Battelle was of course happy that the SSIH now took the place of Ebauches SA. In 1956, both sides signed an ultra-secret agreement with the objective of developing a quartz wristwatch. The costs were to be distributed between Omega (67%), Tissot (20%), Lemania (6.5%), Marc Favre (3.5%), and Rayville Blancpain (3%) and escalated rapidly. In 1957, SSIH was billed 88,000 Swiss francs, half a million in 1960, more than a million in 1965-67.

Battelle engineers Peter Döme and Jakob Lüscher took care of the project and were responsible for it from the very beginning. Starting from zero, they had to invent everything themselves. With one exception, nothing was available commercially. Blind alleys were inevitable and much time was lost. Döme and Lüscher were not permitted to talk about their work with colleagues and within the SSIH holding nobody was in a position to see whether real progress was being made.

Cooperation with the ETH in Zurich

In 1969, Professor Hans H. Günthard, who was responsible for physical methods at the Institute of Organic Chemistry of the Swiss Federal Institute of Technology (ETH) in Zurich, was hired as a consultant by SSIH/Omega. Günthard followed the project in a very meticulous way. He wrote detailed and very critical reports with suggestions for improvements – all of them were totally ignored by Battelle. The minutes of the meetings are still extant. They show that SSIH insisted on the rapid development of a commercial quartz caliber. Battelle, on the other hand, considered the mandate as an adventure and loved having interesting advanced research paid for by its customers.

A good example is the way the integrated frequency divider circuit was developed. This was a very critical component and very little progress seemed to be made in Geneva. Hans Widmer, Omega's technical director, contacted the brilliant Swiss physicist Jean Hoerni, who was working in the United States. Hoerni definitely was one of the world's leading semiconductor specialists. He had worked with Shockley at the famous Shockley Barn in California, the starting point of Silicon Valley. Hoerni did not particularly appreciate Shockley's management style; he quit with seven colleagues. Eventually he founded the Intersil Company in California that "spawned" the Munich subsidiary Eurosil in 1973 (SSIH was a shareholder). Intersil was among the pioneers of the low-voltage CMOS technology. Battelle did not appreciate Widmer's intervention at all, but he could not be intimidated. Thanks to him, the project finally started to move.

After 13 years of development, the very first prototype of the Megaquartz 2400 worked in 1969. In that same year, Omega took over the project (instead of the SSIH Holding) and became the direct contract partner of Battelle. Two years later, Omega cancelled the cooperation contract and took the development of the commercial model in its own hands. The product reached the market in 1973, exactly at the time when the price of Japanese quartz watches kept dropping. On the other hand, the Omega Megaquartz 2400, with its phenomenal monthly rate of less than one second (it was easily certified as an electronic marine chronometer), was extravagantly expensive (retail price was 3500 Swiss francs). It also was power-hungry, as it needed a new battery every six months. Less than 10,000 pieces were eventually sold. A very ingenious piece of technology the development of which started too early and that became available to consumers too late. A good part of the development costs – around 12 million Swiss francs – had to be written off.

High Frequency versus Low Frequency

After the SSIH/Battelle contract had been signed in 1956, the Battelle engineers started worrying about the quartz resonator. The bar-shaped, low frequency, flexural mode resonator did not look like a good solution as it was neither shockproof nor free of position errors. Instead, Döme and Lüscher chose a high frequency, shear mode resonator of the type that was already widely used in telecommunications and TV sets. It looked very much like an aspirin tablet. The project was thus stuck with a high frequency resonator, which probably was a mistake. The divider circuits ended up being very complicated and power-hungry. But at least the quartz resonator was commercially available, while everything else had to be invented ad hoc.

The available shear mode quartz resonator oscillated at the frequency of 2.359 Megahertz. Driving a motor meant dividing this frequency by a very high factor – and the technology of frequency division was still underdeveloped at that time. After many trials, a hybrid frequency division scheme was chosen; first two analog division steps by a factor of 6 each with resonant circuits, then 16 binary division steps all the way down to 1 Hz.

In the commercial models, the analog steps were eliminated by the CEH that had received a subcontract. The first part of the new divider chain consisted of dynamical circuits that divided the quartz resonator's

7.9. The Omega prototype caliber 1500

7.10. Prototype watch with the Omega caliber 1500 (1969)

7.11. Omega Megaquartz 2400 with the caliber 1510. It was certified as a marine chronometer

7.12. Seiko pocket chronometer. In 1967, it won the first prize of the Neuchâtel observatory contest.

frequency 12 consecutive times by three. This was followed by 16 binary division steps down to 1 Hz.

The Battelle people were real pioneers working on the frontiers of what was physically possible. They also had to explore many variants which was costly in terms of time and money. After many years of work, everything converged to the concept of the future "Megaquartz 2400" with the 2.4 MHz shear mode quartz resonator, frequency division with CMOS circuits and an anchor-type motor that drove the gear train. Not a coincidence: the CMOS circuit was supplied by Hoerni's Intersil; later on the chip was made by Eurosil.

The motor ran with positive 1 Hz current pulses that flowed though a coil surrounded by four platinum-cobalt permanent magnets. An anchor acting on the wheel train was mounted on the coil and moved sideways each time the coil was energized. A hairspring brought the coil back to the starting position after each current pulse. Optimizing this system to make it work under the most adverse conditions was a tremendous accomplishment. Ten times enlarged models made of plywood were very helpful; so was a stroboscope lamp.

A Grandiose Technical Dead End

At the International Chronometry Congress of 1969 in Paris, Omega exhibited a working prototype of the "Megaquartz 2400." It was a cone-shaped monster with two mercury oxide batteries, delivering the 2.7 volts that the CMOS IC needed. But it took another four years for developing the commercial version that worked with 1.35 volts.

A second version that was very close to the final, commercial design was shown to the media in 1970. It was called "Elephant," certainly a fitting name. Production started in 1972, but the Megaquartz 2400 was officially launched in 1973. Developing a date display mechanism had been particularly demanding. Conventional mechanisms required about five times more energy than the motor could supply. Peter Döme ended up developing a clever system of four pins that moved the date disc and practically needed zero power: it is still in use.

When the Omega Megaquartz 2400 with caliber 1510 at last reached the market, prices for conventional quartz watches with silver oxide battery, 32 kHz quartz resonator, CMOS divider and Lavet-type stepping motor were rapidly decreasing. Production of the standard Megaquartz model was limited to 1000 pieces, the same applied to the certified chronometer caliber 1515. In 1974, 1000 pieces of the marine chronometer (certified by the Besançon Observatory) were issued; later in the year a second series of 7000 pieces followed with caliber 1516.

The Megaquartz 2400 was a most remarkable technical achievement, but a commercial failure. Production was discontinued in 1975 after only about 10,000 pieces. The remaining stock was sold to Omega staff for 350 francs apiece, i.e. 10 percent of the retail price. This had the unexpected consequence that several barber shops in Bienne (Omega's hometown) now sold the Megaquartz for 400 francs, while they cost 3500 francs in the nearby watch store. This was widely reported by the local papers and generated unwanted publicity for Omega.

The Concours Caliber of Seiko Suwa

As was already mentioned in the first section of this chapter, eleven quartz wristwatch prototypes developed by the CEH won the first ten places at the Neuchâtel Observatory's chronometer contest of 1967. The Japanese competition had not been sleeping and had working prototypes of a different design within weeks of the CEH. Thus, the Observatory received five quartz wristwatch movements from the Seiko development laboratory in Suwa in early November 1967; tests started on 11 November. The movements were tested the regular way for six weeks and obtained ranks 11, 13, 14, 15, and 16 (the daily rate of the last one, W008, was 0.096 second which caused it to be disqualified). The technical version of the Seiko Suwa design was rapidly developed, so a (very small) commercial series could be launched in Tokyo on the day after Christmas 1969. Two generations of quartz watch movements later, Seiko had a design that could be produced very inexpensively and was a world wide winner.

Quartz – A Real Challenge

In the course of the 1950s, Japanese watch manufacturers became aware that the mechanical watch had just about reached the limits imposed by physics and production technology. For this reason, Seiko Suwa launched the project Gojukyu-A in 1959 and placed it under the direction of Tsuneya Nakamura, a mechanical engineer. Nakamura started out with a team of 7 or 8 engineers, but it eventually grew to about 30 people. Several professors of the Tokyo University who cooperated as consultants recommended the development of a radio-controlled wristwatch. However, Nakamura decided not to follow up on this idea, since the range of time signal transmitters is quite limited. Also, each of those transmitters operates with different codes.

All possible ways of designing electronic timepieces were explored. Nakamura liked Max Hetzel's tuning fork watch, but Bulova refused to sell licenses on its patents. The only truly promising alternative was the quartz watch. The microelectronic part was no problem: Japan had a solid infrastructure and produced high grade germanium and silicon transistors. Integrated circuits were in the course of development. There were no mental blocks regarding buying new technologies, particularly from promising startups in Silicon Valley.

Nakamura was well aware of the fact that he was embarking on a very risky adventure, but he was deeply convinced that the quartz wristwatch was the watch of the future. Nearly everybody wants a watch that displays the exact time and there was a good chance that quartz would beat the old mechanical watch by at least an order of magnitude. Yet he had to deal with rather violent opposition. High ranking Seiko managers could not understand why one should undermine the good profits that the conventional watch generated, with a totally new electronic watch the fabrication of which would require enormous investments.

Nakamura succeeded in convincing the skeptics. If Seiko had a quartz wristwatch that was superior to all other watches by an order of magnitude, this was bound to generate prestige for the brand and have positive consequences for the overall sales record. It was absolutely fundamental that only Seiko personnel and Seiko facilities were involved in the project. If everything was not kept totally secret until working prototypes were available, imitators certainly were bound to appear and damage Seiko's reputation of always being on the forefront of technology.

Seiko's Nakamura probably was not the only scientist who thought of a wristwatch size quartz timekeeper, so no time could be wasted. It was known within Seiko Suwa that Daini Seikosha (now Seiko Instruments, SII), that is also controlled by the Hattori family, was developing a quartz marine chronometer and a quartz wristwatch. It had also become apparent that some of the most important Swiss watch movement manufacturers were financing the Centre Electronique Horloger (CEH) in Neuchâtel – that quite obviously would be working on electronic watches: quartz could not possibly have escaped their attention.

"Crystal Chronometer" and Consequences

In view of the 1964 Olympic Games in Tokyo, Nakamura's team developed the "Crystal Chronometer" (calibers QC 951 and 952), a quartz table clock and stopwatch that could be triggered optically or electrically (see Chapter 6). Three chronometers of this type won a prize at the Neuchâtel Observatory's 1963 contest in the marine chronometer category. The original design had been slimmed down to a volume of slightly less

7.13. Rear of the Seiko pocket chronometer of 1967

7.14. Dial side of the Seiko wrist chronometer developed for the Neuchâtel observatory contest, 1967

7.15. Movement side of the Seiko wrist chronometers of 1967

7.16. The Seiko wrist chronometer of 1967 (cased)

than 1000 cm^3 in order to comply with the rules of the Observatory. Seiko was the only non-Swiss competitor and obtained the ranks 10, 11, and 12.

Masatoshi Tohyama, a graduate of the Chiba Technical University played a leading role in the development of Seiko's Crystal Chronometer and other quartz sports timing instruments. He had the signal honor of making a demonstration of the new electronic stopwatches in the presence of Crown prince Akihito, the present Emperor of Japan.

As mentioned above, Seiko kept sending ever smaller quartz chronometers to the Neuchâtel Observatory: deck watches in 1964 and 1965, pocket chronometers in 1966, and finally wrist chronometers in 1967, the culmination of Nakamura's quartz project. Those wrist chronometers directly competed with the Beta 1 and Beta 2 prototypes developed by the CEH. They were the result of extremely hard work put in by Nakamura's team of about 30 physicists, engineers, and technicians. Their lab was open day and night all year long. All members of the team lived within easy walking distance, so they could return to the lab after dinner. When they finally came home dead tired, the janitor was bound to come by and signal an electrical short or a leak of cooling water that had to be fixed immediately.

Last Ranks at the Observatory

The five quartz prototypes Seiko Suwa sent to the Neuchâtel Observatory in November 1967 did not resemble wristwatches; they were tiny rectangular clocks, i.e. miniaturized versions of the pocket chronometers of 1966. The movement measured 30.0 mm x 23.5 mm, the height was 5.3 mm, which resulted in a volume of 3.74 cubic centimeters – exactly within the Observatory's specifications. The frequency of the tuning fork-shaped quartz resonator was 8192 Hz. It had been mechanically cut and ground to approximately the right frequency. The fine tuning was done with a bank of ceramic capacitors on the right side of the movement. 29 capacitors were also needed for the divider circuit.

The electronically excited quartz resonator frequency (8192 Hz) was divided in 13 binary steps by a circuit that had been carefully soldered together on a ceramic platelet using 84 miniature resistors and 76 "naked" transistors; the whole was potted in black plastic. The output consisted of 1 Hz pulses that were amplified before being fed into the stepping motor. The latter consisted of two coils on a soft iron core and a platinum-cobalt magnetic rotor that moved back and forth at the 1 Hz pace. An anchor attached to the magnet drove a kind of escape wheel. The seconds-hand was mounted directly on its staff. A conventional gear train took care of the reduction for the minute and hour-hands.

As already mentioned, Seiko only obtained ranks 11, 13, 14, and 15 at the Neuchâtel Observatory. This was primarily due to the tests at 4° C and 32° C, as no precautions had been taken to compensate for the temperature dependence of the quartz resonator.

Priority Claims

Nobody knows at what exact date the world's first quartz wristwatch movements started ticking in Neuchâtel and Suwa. The priority claim must therefore be based on the extant documents, i.e. the books of the Neuchâtel Observatory with carefully dated and signed entries. On this basis, the technical priority definitely belongs to the CEH: the world's first quartz wristwatch was the above-mentioned Beta 1, Observatory tests having started on 13 August 1967, while they started on November 11 1967 for the five Seiko Suwa prototypes. On the other hand, the commercial priority belongs to Seiko, as 100 hand-assembled Seiko Astron quartz watches in a gold case were available in the best Tokyo stores on 26 December 1969. The price was 460,000 Yen, the cost of a small car.

Obtaining the last ranks at the Neuchâtel contest certainly was a bitter pill. But Nakamura's main interest was not to beat new records, but to develop a sturdy, economical quartz watch that would work flawlessly for years and that could be produced industrially in large numbers. Even before the prototypes worked, Nakamura assembled a small team of engineers that would concentrate on production related parameters of the future watch.

An early and intense study of those requirements gave Seiko an important lead while in Switzerland, hardly anybody could make up his mind whether quartz wristwatches had a future or not. Several precious years were lost with useless debates on this subject.

In order to give the quartz watch an autonomy of a year or more before the button cell had to be changed, it was of foremost importance to use low power circuits. Bipolar integrated circuits did not fulfill this condition. They were used (in a modified form) in the Beta prototypes simply because there was nothing else. Seiko did not accept any compromises in this respect; this is why a hybrid circuit with 76 individual transistors was chosen. While this was not a terribly elegant solution, the leakage current was extremely low.

Key Technologies

Three technologies were of decisive importance for the further development of Seiko quartz watches: the tuning fork-shaped photolithographic quartz resonator, the low voltage CMOS IC and the Lavet-type stepping motor. Those choices were made by three of Nakamuras best people: Kinji Fujita, Masahiro Kurita, and Kenichi Ushikoshi.

After graduating in electrical engineering, Fujita joined Seiko Suwa in 1960. He developed the prototypes of the "Crystal Chronometer." He was then made responsible for miniaturizing it to wristwatch size. Fujita's colleague Kurita joined the Suwa team in 1962: Seiko owes him the extremely compact parts layout in the Astron follow-up caliber 35SQ. Kurita realized that the Lavet-type stepping motor was far superior to any other design and insisted in having it drive the seconds-hand directly, so it jumped from second to second. This concept met fierce opposition, as the motion of the seconds-hand in mechanical watches gives the impression of being continuous. But Kurita prevailed: with full second steps the watch owner could show that he was wearing a modern quartz watch.

Ushikoshi started his career in Suwa in 1966. He belonged to the quartz prototype team and took part in the assembly of the Neuchâtel prototypes of 1967. He introduced the tuning fork quartz resonator that was first made mechanically and was later replaced by the photolithographic variety. Ushikoshi also developed some of the special machines that were to be used for the mass production of quartz movements.

In 1972, Seiko raised the frequency of the quartz resonator from 8192 Hz to 16,384 Hz, later to 32,768 Hz. The latter value became the industry standard as it is the best compromise in terms of size and power requirements.

In the 1960s, the best material for making permanent magnets was the very expensive platinum-cobalt: there was no alternative, the rotor of the stepping motor had to be made from it. But soon the much superior and much less costly samarium-cobalt became available, which allowed a further size reduction of the stepping motor and the design of elegant, slim quartz wristwatches. The Lavet motor with a samarium-cobalt rotor needed so little power that the battery life could be extended to two years.

CMOS Electronics and the Tuning Fork Quartz Resonator

The new low voltage, low power CMOS technology was the ideal way to integrate all the electronic functions of the quartz wristwatch (oscillator, divider, motor drive) on a single chip. Seiko secured this technology by injecting capital in a startup company named Intersil in Silicon Valley (it had been founded by the famous Swiss-American physicist Jean Hoerni). Seiko also bought an Intersil license, so the watch circuits could be produced in Japan. It is remarkable that Seiko immediately recognized the enormous advantages of the Intersil CMOS technology, while the Swiss were not interested, at least for the time being.

Another Californian startup that Seiko "discovered" was Statek. Jürgen Staudte, the company's founder, developed a photolithographic process for producing tuning fork quartz resonators. Seiko executives immediately understood the enormous advantages of this process compared to conventional cutting, grinding and polishing of quartz lamellae. Seiko bought a 15 percent share of Statek and a production license for 1.5 million dollars. The young company urgently needed the money to buy automated production equipment; mass production began in 1973 and eventually brought down the price of a quartz resonator by a factor of a thousand. The only disadvantage of the Statek quartz was a relatively low quality factor around 50,000. With bar-shaped, flexural mode resonators 220,000 was easily achieved, but 50,000 was an acceptable value.

Daini Seikosha's Caliber 3922A

The Hattori family not only controls Seiko Epson (to which Suwa belongs), but also Seiko Instruments (formerly Daini Seikosha), where Kouji Kubota assembled a team of about 20 people with the objective of developing their own quartz wristwatch – a project that had been in the wings for several years. The caliber was called 3922A, and it was clear at the very beginning that it should be equipped with a tuning fork quartz resonator and an integrated circuit of the CMOS-type; the prototype worked in 1970.

This specific caliber was not mass produced, but allowed to accumulate valuable experience that came to good use in Daini's later activities, as the company grew into one of Seiko's large producers of quartz movements. The idea of a quartz wristwatch goes back to the early 1960s; the feasibility had been carefully evaluated with professor Takami of the Todai University who acted as a consultant to Daini Seikosha.

Takami calculated, on a purely theoretical basis, what kind of performance one should expect from various sizes of quartz clocks and watches. This work guided the choice of technologies used in the sports timing equipment developed for the Olympic Games of 1964 in Tokyo.

A Quartz Wristwatch

As the Olympic projects were being completed, interest shifted to a quartz wristwatch. Both Daini Seikosha and Suwa Seikosha developed prototypes of quartz wristwatches. Each of those companies assembled a team of 20 to 30 people for its project. Daini's team was led by Kouji Kubota, while the Suwa team reported to Tsuneya Nakamura (see the preceding section). Those groups worked in full independence, but occasional meetings were called for exchanging ideas and avoiding too much duplication.

Practically nothing transpired about the work done in Switzerland. The existence of the Centre Electronique Horloger (CEH) in Neuchâtel was known, but a very strictly enforced policy of secrecy completely prevented leaks. In the eyes of the Japanese, it was only logical that the CEH people developed a quartz wristwatch, but this could not be confirmed before the official presentation of Beta 1 and Beta 2 in January 1968.

From the very beginning of the quartz wristwatch project, Kubota bet on integrated circuits of the CMOS type. They were far superior to the widely available bipolar IC that switched very rapidly but needed a lot of power. The CMOS-IC on the other hand switched more slowly, but had extremely small power requirements. CMOS-ICs were available in the early 1960s, but they worked with voltages around 30 volts. The problem was to find low voltage CMOS.

At that time, only a few American companies were at the forefront of CMOS technology. In April 1967 Kubota flew to California with his colleague Masatoshi Tohyama, where he met the Swiss physicist Jean Hoerni (1925-1997), a co-founder of Fairchild and later the founder of Intersil in Cupertino, California. Hoerni was invited to Japan and returned with an order for a 14-step binary CMOS divider circuit operating at 1.35 volts (the voltage of mercury oxide button-type cells). The power requirements were to be less than 7.5 Microwatt, the chip would measure 4 mm x 4 mm x 1.3 mm. Seiko would pay $190,000 for the development of this IC that had to be delivered within 15 months.

CMOS for Wristwatches

By 1967, CMOS-IC had made great progress and was available for operation at 6 volts. Hoerni and his colleague Luc Bauer managed to bring this down to a little over 1 volt by precision doping of silicon by ion implantation. Intersil delivered the chip ordered by Seiko in January 1969. It exactly fulfilled the conditions stipulated by Seiko: it divided 16,384 Hz to 1 Hz, in 14 binary steps. However, the yield was very poor: out of a production batch of several hundred IC, just a few were up to the specifications.

In May 1970, Daini Seikosha had CMOS circuits for wristwatch movements. For the time being, just the divider was CMOS; the oscillator and motor drive needed a separate, bipolar IC. Caliber 3602 used this combination, the movement still included a flexural mode quartz resonator. Caliber 3602 was rectangular; the dimensions were 25 mm x 28 mm x 5.45 mm.

Further development led to caliber 3922A with 13 jewels; it was launched in 1972. This was the first wristwatch movement with full CMOS electronics. The dimensions were 25.0 mm x 29.7 mm x 5.59 mm. It was powered by two silver oxide batteries in parallel. A light emitting diode at 1 o'clock on the dial started to flash when the batteries needed to be replaced. This movement worked fine, but due to its dimensions, the watches produced with it were far from elegant.

Kubota was not fully satisfied with what had been achieved so far. He had a semiconductor laboratory installed so CMOS chips could be produced in-house under Intarsil license. This is how the CMOS circuits for the calibers 36SQC and 39SQW were developed. Simple divider circuits were not enough: Kubota wanted extra functions such as alarm and chronograph. This led to the production of microprocessor chips with which any desired extra function could be generated with software alone.

Longines Ultra-Quartz

The 1950s and 1960s were critical years for the watch industry. New technologies were going to trigger a revolution. In the course of the next decades, the traditional mechanical watch would be almost fully replaced by electronic timepieces. At that time, a good part of Longines stock was owned by Ebauches SA, which essentially controlled Longines and also shared technologies with it. In 1960, Ebauches had launched the electromechanical caliber L4750, but Longines engineers were not impressed. Longines had the means and competent staff to develop this kind of caliber on their own. Yet, the company still hired a consultant, the Geneva physicist Raphaël Boolsky.

Calibers 399 and 400

The first result of this work was the contact-controlled electrodynamic Longines caliber 399. It was much too voluminous for a wristwatch and was not industrialized. But it led to an improved, miniaturized version, caliber 400 that was exhibited at the Basel watch fair of 1963. While nearly every contact controlled electrodynamic movement of that time used a balance wheel motor, the electromagnet of the Longines caliber 400 moved a pawl and a stepping wheel as well as a pin that gave the balance a push each time the coil was energized.

Cams on the balance wheel arbor closed the contact at each vibration. Those contacts were a problem: whether they were made of platinum, gold, or silver, sparking burned them sooner or later. It should be remembered that this problem particularly affected the Elgin calibers 722 and 725 (see Chapter 2) and certainly contributed to Elgin's demise. Watches with caliber 400 were extensively tested, but almost invariably failed after just a few months. The lesson was learned: there was no production of caliber 400, but at least it generated important information that was put to good use in the Longines quartz calibers.

The Accutron, the tuning fork watch that Bulova launched in 1960, did not overly impress Longines watchmakers and engineers. They were convinced that the future belonged to a fully electronic caliber using a quartz resonator as a high frequency oscillator. This was the way marine chronometers, high precision clocks and deck watches had evolved to ever smaller sizes. It was not terribly difficult to guess that the wristwatch would go through a similar evolution: the quartz wristwatch was bound to come.

With the quartz deck watch caliber 800, Longines played an active role in this development. Caliber 800 had been launched in 1964 as the result of a close cooperation with Bernard Golay SA in Lausanne. The dimensions of caliber 800 were 64 mm x 56 mm x 48 mm. The frequency of the quartz resonator was 12,000 Hz; a circuit with 25 transistors divided this down to 2 Hz for driving the motor. Caliber 800 was energized with three mercury oxide batteries; a gear train drove the conventional analog display. Caliber 800 obtained the grade 0.15 at the Neuchâtel Observatory in 1964 and thus established a new record, bettering the extant best grade that Seiko had received two years earlier.

7.17. Quartz wafer with Statek tuning fork quartz resonators

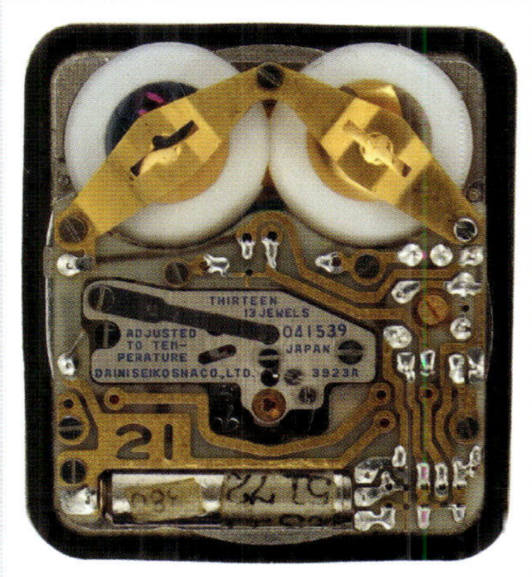

7.18. Daini Seikosha's first quartz movement caliber 3923A

7.19. Lateral view of the caliber 3923A

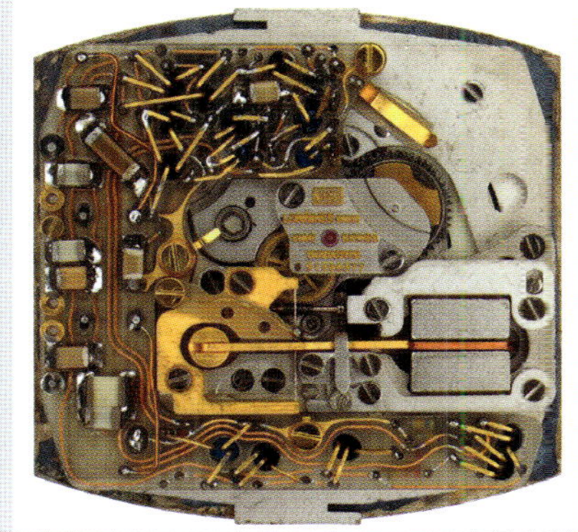

7.20. The quartz caliber Longines 6512 Ultra-Quartz

Devil's Island

Aurèle Maire, Longines' technical director, belonged to the CEH council at the time when the quartz calibers Beta 1 and Beta 2 as well as the commercial version Beta 21 were being developed. When it became clear that project Beta was going to be a success but that the product would be very expensive, Maire decided to develop a different, very simple quartz caliber that Longines alone would have.

In the Longines/Golay design, the quartz resonator was only used as a frequency standard in a control loop. This kind of circuit could easily be wired up with a relatively small number of individual components. A frequency divider was not needed in this scheme; consequently the then exotic and very expensive integrated circuit technology could be dispensed with. This would bring down the price considerably. The electronic module was developed and fabricated by Bernard Golay SA in Lausanne. This company produced its own quartz clocks at that time. Longines contributed the motor and the mechanical part of the movement that would be called Ultra-Quartz.

The cooperation of Longines with Golay had already begun in the 1950s. The first project was the lead-acid battery-operated quartz clock that still used electronic tubes; it was designed for sports timing (see Chapter 6). In 1964, i.e. ten years later, Golay and Longines had a marine chronometer with a volume of less than 200 cm^3; it was tested at the Neuchâtel Observatory. It was equipped with a 12 kHz flexural mode quartz resonator; this frequency was electronically divided to 2 Hz for driving a micromotor. In the following year (1965), Golay and Longines surprised the world with a quartz pocket chronometer; its volume was less than 20 cm^3. The daily rate of 0.01 second was simply amazing.

In 1966 and 1968, Golay-Longines won the first prize for electronic pocket watches at the Neuchâtel Observatory. But long before that, the managers of the cooperation were sure that they were able to develop a further miniaturized electronic movement for a quartz wristwatch. The major problem was the motor that had to transform 100 Hz to 200 Hz vibrations into a rotary motion to drive a gear train.

The development of such a motor was started in early 1963 at an ultra secret Longines Research Laboratory on the so-called "Ile du Diable" (Devil's Island). This sounds sinister, but it was simply an appendix to the old Longines factory on the shores of the Suze River the bed of which had been made into a channel – without any islands. Not even the Longines technical department was allowed to know what was going on and what kind of projects were pursued on Devil's Island.

At least thirty different vibratory motors with minimal power requirements were investigated. It was important that they did not fall under the extremely broad Bulova patent. Winding small coils with enameled copper wire with a diameter of 10 to 13 micrometers turned out to be a real challenge.

This nearly invisible wire was heated with hot air during the winding process; this made the enamel sticky. The only quartz resonators available at that time were of the bar-shaped, flexural mode type made by the Ebauches subsidiary Oscilloquartz. The frequency was 8192 Hz. They were suspended in a flat, evacuated metal case measuring 26 mm x 6 mm x 3 mm. Such a resonator cost 80 Swiss francs apiece.

Developing the Ultra-Quartz

It was not terribly difficult to design a vibratory motor with a good yield – 98 percent was finally achieved. The problem was to transform the vibrations into the rotary motion that was required for driving the gear-train. The solution ended up being the same one that Max Hetzel had chosen for the Bulova Accutron and that was used for the Beta 2 prototype and the Beta 21 commercial quartz watch designed at the CEH. The ratchet wheel had 170 very small teeth with a height of just 0.024 mm. It was pushed forward by a pawl attached to the motor. A fixed pawl prevented the ratchet wheel from turning backwards.

The construction of the motor forced the ratchet wheel to be mounted at a right angle to the plane of the main plate. The rotary motion thus had to be turned by 90 degrees. This was done by attaching a worm gear to the arbor of the ratchet wheel, this worm acted directly on the teeth of the third wheel. The moving part of the motor was an iron lamella to which a 13 mm long arm with a movable weight was attached so that the vibration frequency was tuned exactly to the required 170 2/3 Hz.

Non-negligible friction occurred at the interface between the worm gear and the third wheel. The oil normally used for the jewel-bearings was too thick: it had to be diluted with a non-volatile solvent in order to lower the viscosity. The U-shaped electronic module with a cybernetic control circuit designed by Golay consisted of a printed circuit on which were mounted 14 transistors, 19 resistors and 7 capacitors; the soldering was done manually.

Two separate oscillator circuits excited the 8192 Hz quartz resonator and the 170 2/3 Hz motor. The control circuit compared the two frequencies; each oscillation of the motor had to correspond to 48 oscillations of the quartz resonator. If the coincidence was not exact, a correction signal was generated that slightly adjusted the oscillator frequency of the motor circuit.

The complete "Ultra-Quartz" movement needed about 10 microamperes. A mercury oxide button-type battery with a capacity of 140 mAh gave the "Ultra-Quartz" an autonomy of at least 18 months. The movement was rectangular and measured 33 mm x 28.4 mm x 5.1 mm. The daily rate was 0.17 second or 5 seconds per month. At least one movement of the Ultra-Quartz test series has been preserved.

The production version of the Longines Ultra-Quartz was shown to the media in Geneva in August 1969; this was followed by a massive advertising campaign. The CEH initiated a confrontation, insisting that Longines had copied their Beta 1 and Beta 2. But Longines could clearly prove that its design had practically nothing in common with that of the CEH prototypes. Watches with the Ultra-Quartz movement were produced in small series; their total number is estimated at slightly over 2000 pieces. As is often the case with a totally new product, delivery to the stores was delayed due to production-related problems. After having so massively advertised quartz, Longines urgently needed a substitute as a stopgap. The result was that the first Longines quartz watches that were on sale in 1971 used a Beta 21 movement from the CEH, despite the previous, bitter dispute.

The Longines/Beta 21 watches were called "Quartz-Chron." They are easily recognized, as they have a crown, while with the "Ultra-Quartz" time and date were set from the back of the case with a knob actuated with a folding bow. Due to the "dollar crisis" and collapsing export markets, production of the "Ultra Quartz" was discontinued in 1973.

New Challenges for the Research Lab

The next project on Devil's Island was an electronic watch with a liquid crystal display (LCD), without any mechanically moving parts. For this watch, Longines worked with Texas Instruments that was supposed to deliver the integrated circuit. There was no dial, but a metalized surface with an 8 character display of the dynamic scattering type in the shape of a Swiss cross. It displayed hours, minutes, seconds, day, and date. In 1972, Longines won an "IR 100" innovation award for this original design.

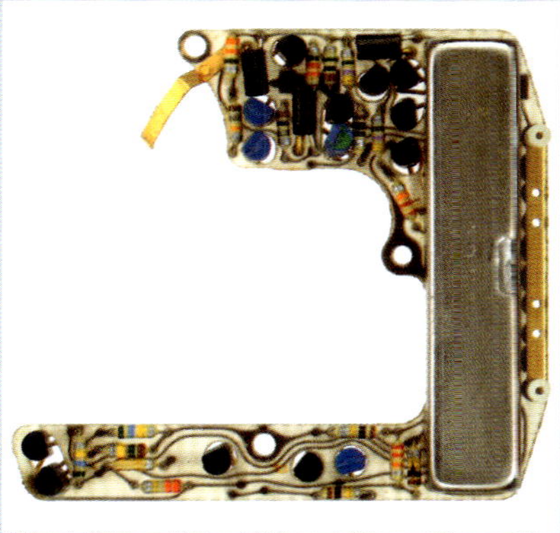

7.21. Electronic module of the Longines Ultra-Quartz. Front side, the quartz resonator is on the right

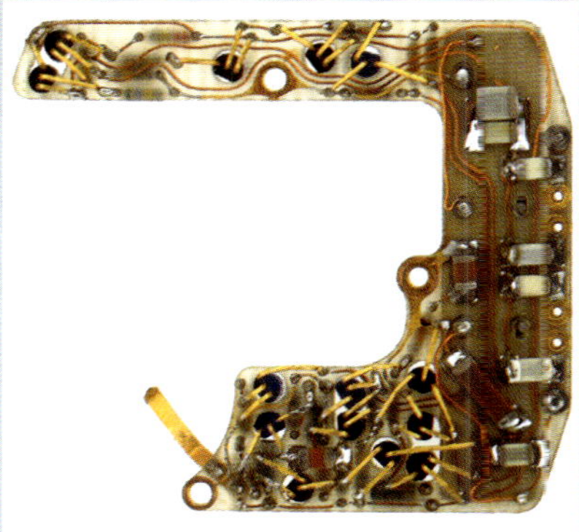

7.22. Back side of the Longines Ultra-Quartz electronic module

7.23. The Longines Ultra-Quartz

7.24. Case bottom of the Longines Ultra-Quartz

This American innovation prize rewarded the 100 best and most original new products of every year.

Unfortunately, there was a huge delay on the American side, as Texas Instruments could not (or would not?) deliver the integrated circuits. Production of this interesting watch was cancelled. At any rate, two batteries were needed for the obsolescent 3-volt technology. Another development that never made it beyond the prototype stage was called "Combo": a fully digital design with both a liquid crystal display (LCD) and a light emitting diode display (LED). Time was read off the LED after touching a button, while the usual chronograph functions (Stop, Lap and Split) were displayed on the LCD.

GP 350 by Girard-Perregaux

Charles-E. Virchaux, CEO of Girard-Perregaux (GP) in La Chaux-de-Fonds correctly guessed in the mid-1960s that the CEH in Neuchâtel was developing a quartz wristwatch. The CEH-membership being very costly, Virchaux decided to rather invest in the GP Company and to develop his own quartz wristwatch. He was convinced that this could be done at a fraction of the CEH cost. Once more, he was right: Georges Vuffray (b. 1940), his newly hired engineer developed the famous quartz caliber GP 350 in less than five years, all the way to actual production with never more than five engineers and technicians.

In September 1966, Georges Vuffray, a recent graduate in electrical engineering, reported for work at the Girard-Perregaux (GP) watch plant in La Chaux-de-Fonds: it was his first job. GP chief executive and major shareholder Charles-E. Virchaux told Vuffray that his assignment was the development of a quartz wristwatch and that he would have to work in absolute secrecy. Vuffray states today that he would never have accepted this assignment if he had known that it would be so horribly difficult.

As Vuffray had no preconceived ideas, he thought that this might be a lot of fun and started the job with youthful enthusiasm. GP already had a nicely instrumented electronics lab; it had been installed by Vuffray's colleague Jean-Pierre Jeanneret during the previous year. In the beginning, the two engineers worked alone, but as they made good progress, they hired several technicians who all contributed to the success of the project. Inevitably, Vuffray's team fluctuated during the next seven years; from 1965 to 1972 the following specialists worked with him: Pierre-André Girardbille, Jean-Marie Donzé, Jean Richardet, Denis Barberat, Daniel Huguenin, Pierre Schaffroth, Jean-Michel Jayet and Jean Metzener.

Vuffray decided that in order to get familiar with electronic horology, he would design a quartz clock with no size limit, just to have something that worked. Quartz resonators in evacuated and sealed glass tubes were available from the American ITT in the frequency range of 100 kHz. Exciting the resonator with an oscillator circuit was a minor problem. The frequency divider was more difficult. Integrated 10 x divider circuits were available, but five of them were necessary in order to get 1 Hz pulses for driving the motor. At 68 Swiss francs apiece, this added up to a great deal of money. For actuating the gear train and the watch hands, Vuffray chose a motor that was used in the big electromechanical clocks installed in industrial plants.

The "Elcron" Quartz Chronometer

As soon as the big clock functioned well, Vuffray and his team started the development of smaller quartz clocks with chronometer specifications. This resulted in a nice looking small clock with a brass case, which was called "Elcron I." The heart of the new device was a commercially available 15,625 Hz quartz oscillator that was used as a scanning device in TV sets. A six-fold division by five resulted in 1 Hz pulses. The frequency division was achieved with analog, monostable transistor circuits. The bipolar integrated circuits that were available at that time were not an option, as their power requirements were excessive. But thanks to transistors with very small leakage currents, the current consumption of the Elcron clock was limited to a few tens of microamperes.

Quartz resonators, transistors, and mercury oxide button batteries were readily available. The major problem in designing a quartz wristwatch was the motor driving the gear train and the hands. Obviously, Vuffray had to put a lot of effort in the design of such a motor with the smallest possible power consumption. Vuffray chose one of the many variants of the stepping motor with a magnetic rotor (made from ferrite first, samarium-cobalt later) and a fixed coil.

The rotor was laser-pierced by a specialized company; the axle had to be glued in the hole due to the brittleness of the magnetic material. The cylindrical coil was located in the air gap between rotor and stator. It was wound on a plastic core that was molded by the Tissot Watch Company and generated a transverse magnetic field. The stator was a soft iron ring with lateral notches creating an asymmetry that defined the rest position of the rotor.

This concept was maintained in all the early GP quartz calibers. Vuffray calls it "Laplace's law" in order to differentiate it from the Lavet motor. Laplace's law states that the force exerted on a current-carrying conductor equals the vectorial product of current, length of the conductor, and induction. Vuffray's version of the Laplace motor had a good yield and its power consumption was in the microwatt range. Nevertheless, the Lavet motor was adopted by practically all electronic watches, despite magnetic leaks. Its big advantage was that the magnetic field was generated by a long, slender coil that could be lodged just about anywhere within the movement. This allowed the design of very thin calibers.

In Vuffray's watch motor, each current pulse caused the rotor to turn by 180 degrees, always in the same direction if the polarity of the 1 Hz pulses alternated between positive and negative. This meant that alternating positive or negative current pulses had to be generated at the one second rate. The motor used in the small clock had a diameter of 6 mm and a height of 5 mm. It was not really suitable for a wristwatch, but it was very close. Vuffray got his first patent for this development.

The next GP clock was "Elcron II," a quartz controlled small clock with the same movement as "Elcron I." The front part of the case was made of brass, the rest was plastic. It was again powered by two 1.35-volt mercury oxide batteries. An important series of those neat little clocks was produced by GP in the early 1970s. Their daily rate was a fabulous ± 0.02 second.

Friends in America

Vuffray was not overly impressed when the CEH made a presentation of its quartz wristwatch prototypes Beta 1 and Beta 2 in January 1968. After all, he had made a late start with a very small team and never had the ambition of developing the world's first quartz wristwatch. It was useful to know about the competition and what they had been up to, but in the GP lab, work continued as usual. The next objective was to miniaturize the Elcron clock and bring down the volume of the movement to the value defined by the Neuchâtel observatory for a wristwatch, i.e. 3.75 cm^3.

As to the components, Vuffray had to use whatever industry had to offer. In the early days, "naked" 8192 Hz quartz resonators were supplied by the French CSF, while integrated circuits could be purchased from the French company S.E.P.E, a subsidiary of Thomson-Houston-Hotchkiss-Brandt (later, the two companies merged to form Thomson-CSF).

The important power consumption of the available integrated circuits was a major problem: such circuits were practically mandatory in a wristwatch. Very low power CMOS-circuits could be purchased in France, but they ran at relatively high voltages and not at 1.5 volts to 3 volts – a precondition for their use in a wristwatch.

A manipulated photograph of a non-functional "prototype" of a GP quartz wristwatch was shown at the 1970 Basel Watch Fair under the name Elcron. The caliber was exhibited too, but in the back where hardly anything could be seen. French components had been used throughout; the integrated circuit for the frequency divider was CMOS, but for 3 volts which meant that two silver oxide batteries had to be used in series. The oscillator was not integrated on the divider chip, it had been made up as a hybrid thick film circuit.

After Basel 1970, Vuffray clearly realized that he was getting nowhere with his French partners. They never could supply what he really needed. Furthermore, the French watch industry had intervened at a high ministerial level: in their opinion, French companies had no business cooperating with the Swiss competitors. After all, Thomson and CSF were big suppliers of the French armed forces, which meant that they were subsidized by the French taxpayer. Advanced French technology should be reserved for France.

In view of this situation, Vuffray started to contact the big American electronics companies. Later in 1970, he flew to the United States where he visited some twenty different companies. Motorola in Phoenix, Arizona, was the best qualified potential supplier of low voltage CMOS circuits, and Motorola executives were interested in collaborating with GP. This gave them a foot in the door in Switzerland and would allow them to build a strong position in the quartz watch circuit market. Within an amazingly short time, Motorola engineers integrated all the electronic functions of a quartz watch on a single chip, i.e. oscillator, 15-step digital divider, and motor drive.

Vuffray also visited Reeves-Hoffman in Carlisle, Pennsylvania, a specialist for quartz resonators. He was offered bar-shaped, flexural mode resonators in evacuated capsules. The best compromise between the size of the quartz and the number of divider steps, was a frequency of 32,768 Hz, that later would become the world standard. Two years later, GP adopted cut and ground tuning fork-shaped quartz resonators that could be purchased from the Japanese NDK. GP kept using them until Ebauches SA acquired the Statek license for photolithographic tuning fork quartz resonators and began having them produced by the ETA subsidiary Micro-Crystal in Grenchen.

It is amusing that Vuffray and Virchaux went to the United States together in 1971 and actually visited Statek in Orange, California, not far from Los Angeles. The two Swiss were just a bit early, as Statek founder Jürgen Staudte still worked in his garage and only had prototypes of his revolutionary quartz resonators that he produced just like integrated circuits. Mass production would only begin in 1973.

The GP 350 quartz caliber

The first working prototype of the GP quartz wristwatch movement with American components was completed in 1971. It consisted of a Motorola CMOS chip working at 1.5 volts that required only one battery. Add to that a 32 kHz quartz resonator from Reeves-Hoffman and the in-house made Laplace motor. This defined the technical concept of the Girard-Perregaux quartz caliber; it received the designation GP 350.

The mechanical part of GP's electronic movement (main plate, bridges, gear train) was supplied by Jaeger-LeCoultre, which belonged to the Saphir Group, together with Favre-Leuba. At that time, Girard-Perregaux did not produce any ebauches, but purchased them from outside vendors. Complete prototypes of the first GP quartz movement were tested by the Neuchâtel Observatory in 1971 and could be seen in detail at the 1972 Basel Watch Fair. The diameter was 30 mm, the height 7.2 mm; battery life was only one year but improved with later models.

The "GP Quartz" collection was also launched at the 1972 Basel Fair. Of course those watches were equipped with the GP 350 caliber. In the early models, the Reeves-Hoffman quartz resonator was mounted the normal way, as the battery was very thick which dictated a height of 7.2 mm. The next caliber, GP 352, was equipped with a thinner battery (3.6 mm); the resonator was then mounted sideways. The rotor in the motor was made of samarium-cobalt and Motorola supplied a new integrated circuit with significantly lower power consumption. Thanks to

7.25. Quartz caliber GP 352 by Girard-Perregaux

7.26. Quartz wristwatch with the GP 350 caliber by Girard-Perregaux

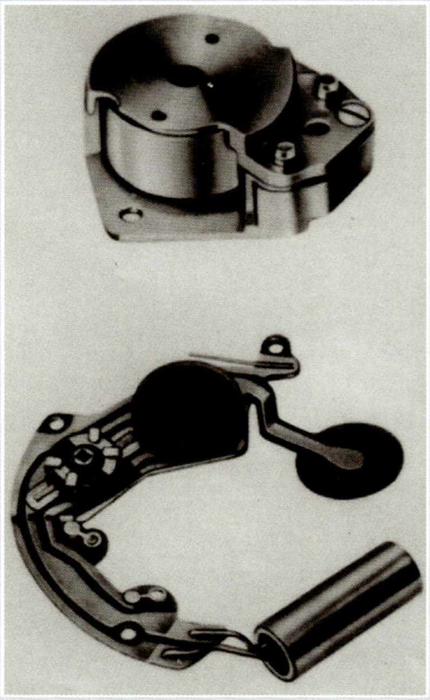

7.27. Motor and electronic module of the GP 641 caliber

these innovations, the battery life was extended to two years.

A modular construction was developed for the GP 350 and the models that followed. The watchmaker could remove each of those modules and replace it with just a screwdriver and tweezers. The GP 350 and its successors, i.e. the calibers 351, 352, 353, and 354 were quite successful; some 50,000 pieces were produced. They all used the same chip, the same quartz resonator and the same motor. The main difference was a different calendar mechanism and the positioning of the trimmer for fine tuning of the quartz frequency.

Jaeger-Le Coultre used the GP quartz movements in their own "Marine Quartz" collection. They had contributed the mechanical part on the base of Girard-Perregaux designs. Small series were also purchased by Favre-Leuba; their electronic watch was launched under the name "Quartz Raider." In 1972, complete watches equipped with the GP 350 quartz caliber and a stainless steel case were available in watch stores for just 480 Swiss francs, while IWC charged 45,000 francs for a platinum watch with a Beta 21 movement. This was the world's most expensive watch – excluding jewelry watches that have no upper price limit. Actually, the gold IWC quartz model was less expensive; it sold for just 8500 francs.

The Pulsar

The presentation of the first quartz wristwatches by the Swiss CEH and the Japanese Seiko in January 1968 meant a rude awakening for the Hamilton Watch Co. in Lancaster, Pennsylvania. At that time, Hamilton Watch Co. was the successor of Hamilton Watch, a company that had been founded in 1892 and had a rather convoluted history.

The extant historical documents clearly show that early on Hamilton acquired the flattering reputation of a manufacturer of relatively small series of sturdy pocket watches of truly outstanding quality: they were often compared to railroad watches. By the time World War I broke out, Hamilton was established as "America's Patek Philippe." General Pershing, the American supreme commander, and his staff used Hamilton watches for the very critical timing of artillery fire and the movement of troops at the front. Other well known Hamilton customers were the Antarctica explorer Admiral Richard Byrd, the stratosphere-balloonist Auguste Piccard, American airmail companies, the US Navy, and the US Air Corps. During World War I, most of Hamilton Watch's production potential was reserved for manufacturing military watches and marine chronometers.

History repeated itself in World War II when Hamilton manufactured no less than 9300 marine chronometers for the US Navy. They were perfect (or even slightly improved) copies of Ulysse Nardin's world famous marine chronometer M.Gr.F. In 1957, Hamilton launched the Hamilton Electric, the first commercially available electromechanical wristwatch (see Chapter 2). After the war, Hamilton Watch started on an aggressive diversification program that made it into a true conglomerate. Hamilton Watch Co. was transformed into the Holding HMW Industries with the divisions Hamilton Watch & Clock, Wallace Silversmiths and Metals; the latter was a weird collection of companies with somewhat related interests.

Futuristic Digital Clocks for Stanley Kubrick

Science-fiction author Arthur Clarke and movie producer Stanley Kubrick visited Hamilton Watch in 1966. This was a sensation and the earliest incentive to develop digital watches. Kubrick and Clarke were working on their movie *2001 – A Space Odyssey* and needed futuristic table clocks and wristwatches that looked very different from anything that existed at that time.

Hamilton Watch agreed to design and fabricate the so-called "Odyssey Clock" gratis, but the company would be mentioned in the movie. The clock that was developed in the Hamilton Watch laboratory had a matte-black, aerodynamically shaped case with an elliptical cross-section. Hours, minutes, and seconds were displayed digitally with luminous numerals generated by nixie tubes. The designers were the Hamilton employees Ken Derr and Edward Speakes.

The numerals were displayed at the rate given by a Hamilton Electric caliber. It generated 150 current pulses per minute, according to the balance wheel frequency of 2.5 Hz. This signal was digitized and used for the control of the numerals zero to nine of six nixie tubes. Volume and power consumption were no problem in this case.

Two Space Odyssey clocks were assembled by hand in the Hamilton lab and sent to England where the movie was produced. When the shooting was finished in 1968, the clocks were sent back to the United States. Unfortunately, one of them was totally demolished in transit. The surviving Space Odyssey clock now stands in the office of John Bergey, former technical director of Hamilton Watch and Time Computer. It is unfortunate that he had it transformed into a quartz clock with light emitting diode display. Apparently, the sockets for the nixie tubes are still in place. Ironically, absolutely all the scenes in the film where the Hamilton clock could be seen were eliminated by the editor. Developing the Space Odyssey clock was an exercise in futility.

The astronauts in the movie were wearing the above-mentioned futuristic, digital watches with a rubber bracelet and an aluminum case that was bent on one side. GMT time, date and month were displayed in windows. However, those "Odyssey Watches" were mere dummies without a movement. They appeared many times in the movie.

CMOS Saves the Day

The eleven Beta 1 and Beta 2 prototypes developed at the CEH (Centre Electronique Horloger in Neuchâtel) as well as the four prototypes of the future Seiko Astron from Suwa, that together had obtained ranks 1 to 15 at the 1967 Neuchâtel Observatory contest, were analog quartz wristwatches. It was obvious to Hamilton Watch managers that Switzerland and Japan had gained a headstart in this type of product and that it was nearly impossible to catch up with.

Hamilton Watch engineers carefully studied everything that was published about the Beta 1 and Beta 2 prototypes. They learned in particular that the CEH had used modified bipolar integrated circuits with the inherent disadvantage of an inordinately high power consumption per divider step. This is why Hamilton decided that the CEH design had no future. Yet, nobody seemed to know how one could improve on it.

A conference in Washington, D.C., changed everything. An RCA scientist gave a talk about a new type of integrated circuit with extremely low power consumption. RCA people called them "COSMOS," the abbreviation for "complementary MOS" technology; today they are called "CMOS."

Those really new circuits had a metallic gate that had been evaporated on an insulating layer of silicon dioxide. Current flow between source and drain could be controlled by a mere voltage, the control current being less than a millionth of a microampere. Combining two complementary MOS transistors (pnp and npn, respectively) practically eliminated leakage currents in the non-conducting condition. But at that time, CMOS needed an operating voltage of at least 15 volts – ten times the voltage of a standard watch battery.

Experiments with Analog Technology

The only circuit RCA had realized so far in COSMOS technology was a 15-step binary divider. A 15-fold division by two was exactly what Hamilton Watch needed in order to bring down the kilohertz frequency of a quartz resonator to the few Hertz needed by a miniature electric motor.

Hamilton engineers visited the RCA research laboratory in Princeton, New Jersey, and declared that with the RCA COSMOS circuit they were in a position to develop an electronic wristwatch that was far more accurate than the record holder of the time, Bulova's "Accutron." The RCA people thought this was fine, but a divider circuit working at 1.5 volts was unthinkable. By stretching everything, 3 volts might be possible. Hamilton agreed to pay $27,000 for this development.

As soon as the CMOS divider circuits arrived, Dick Walton and his team at the Hamilton development lab fabricated ten prototypes of a quartz wristwatch with a conventional analog display. A vacuum-encapsulated AT-cut quartz resonator was purchased from the Clock Crystals Company in Massachusetts. The frequency was 262,144 Hz; one could bring this down to 32 Hz with the 15 divider steps available on the RCA circuit. This was just right for a d'Arsonval motor that oscillated back and forth and actuated a ratchet wheel. The energy source was a two-cell, 3-volt zinc-carbon battery made in-house. Silver oxide batteries were not yet available, but they appeared in the trade later. A simple calculation showed that a watch of this type would cost $ 1200 retail. One thus had to find a less expensive solution: a purely electronic watch with a digital display was a possibility. Hamilton had already gained a lot of experience in this field with the digital clock for Stanley Kubrick. Experiments had also been done with liquid crystal displays for the Odyssey project, but the life and readability of the dynamic scattering displays available at that time were disappointing.

A Digital Display with Lamps

After the design of the Odyssey clocks, the idea of a digital wristwatch was not inconceivable for Hamilton Watch engineers. An in-house contest was launched in order to generate ideas for such a watch. In 1968, John Bergey suggested a mechanical movement actuating discs with the hour and minute numerals inside the watch. A miniature electric lamp would illuminate those discs; optical fibers would transfer the image to the side of the watch where it could be read off.

In the development lab, Dick Walton designed a quartz watch with a digital display consisting of a matrix of miniature lamps. The required decoders were already available in CMOS technology, but the power consumption of the lamps was absurdly high.

But did the display of an electronic watch have to be on all the time? After all, the display was only needed when one wanted to know the time. This is what led to the idea of a watch that could be read after touching a button. The concept of a lamp-matrix quartz watch was patented, but nothing came of it. A display with power saving light emitting diodes looked a lot more promising.

Cooperation with Electro/Data

Work on the "solid state watch" had just started in Lancaster, when the Hamilton Watch president was informed by a lawyer in Dallas, Texas, that Electro/Data, a small company on the outskirts of the city had developed a digital quartz clock with a light emitting diode display. Electro/Data was ready to miniaturize it to wristwatch size. This way, Hamilton would have a totally new type of watch without any mechanically moving parts and a bright, luminous display.

John Bergey and Dick Walton flew to Dallas and visited with Electro/Data in Garland, east of the city. The electrical engineer Willie Crabtree showed them his "Breadboard" of a digital clock with a dot matrix LED. He had developed it in cooperation with the company founder George Thiess. The prototype used commercially available integrated CMOS circuits supplied by RCA.

The light emitting diodes were behind a dark red sheet of Plexiglas that hid the rather chaotic-looking upper part of the movement: just the orange-red light of the LED came through. Frequency divider and decoder consisted of a grand total of 44 integrated circuits in 6-volt CMOS technology, but specially selected circuits worked at 4.5 volts. The whole was powered by three zinc-carbon cells.

Hamilton Management decided to give Electro/Data the mandate to develop a miniaturized version for a wristwatch and to produce a first series of 500 such movements. This cost Hamilton Watch some $330,000. Electro/Data not only had highly advanced microelectronic know-how, it was also well equipped for bonding silicon chips and for testing circuits and miniaturized systems. Hamilton would distribute the new watch and take over production later. The volume of the case that was to contain the new movement had to correspond to that of a Rolex "Submariner."

Presentation of the "Wrist Computer"

On March 13, 1970, the first wristwatch-size prototype of an LED quartz movement worked in the Electro/Data laboratory. In the following weeks, five more of those prototypes were assembled. The 44 small integrated circuits with 3474 transistor functions were interconnected with gold bonding wire. This was a nightmare: with so many connections the probability of failure was extremely high. Furthermore the hand-made batteries kept leaking.

The prototypes assembled by Electro/Data in April 1970 were rectangular with rounded edges and measured 33.5 mm x 35.6 mm (36.0 mm x 38.5 mm with the frame). The height was 8.70 mm (5.70 mm without the battery holder). Three of the prototypes with the Hamilton brand on the case were shipped to Lancaster. On May 5, 1970, Johnny Carson showed one of them in his extremely popular television show "Tonight." He put it aside with a slight thump – with the result that only two working prototypes remained.

They were shown to the media on May 6, 1970, during a press conference in the plush New York restaurant "Four Seasons." One of the prototypes failed during the presentation; the remaining functioning watch broke down soon afterwards. The Hamilton team returned to Lancaster with three broken prototypes, but they had started a revolution in the way time was displayed and triggered a huge turmoil throughout the United States.

Case Design

The digital electronic watch was made into a planetary sensation by the media, millions of Americans saw it on the most popular TV show. The Emperor of Abyssinia, the King of Jordan, the Shah of Iran, Roger Moore, Sammy Davis, Jr., and many other celebrities placed orders for one of those "Time Computers" with the "Pulsar" brand right away; they had to wait two years for delivery.

For further development and production of the new watch, the Pulsar Division was made into the Time Computer Inc. (TCI) subsidiary of HMW Industries by the end of 1972. Hamilton Watch had already been transformed into a subsidiary of the same holding in 1971.

The case of the New York prototypes had been designed by Edgar Zweibold at the Star Watchcase Company in Michigan. But the Swiss case designer Jean Wuischpard (1914-2006) was responsible for the first commercial version, the Pulsar P1. Wuischpard was born in Geneva and emigrated to the United States; Hamilton hired him and made him responsible for the case department.

One often hears that the case of the commercial Pulsars was designed by the famous sculptor Ernest Trova in Saint Louis. Trova did indeed get a mandate to design such watchcases, but Time Computer managers did not like Trova's sketches, which are extant. They do not bear the slightest resemblance to the case of the P1 Pulsar that was available in the stores in April 1972. Wuischpard had helped out in a hurry, but had created a true classic. He subsequently designed the cases of practically all American Pulsar models until 1977 when production was stopped.

The cases designed by Wuischpard were produced by Star Watchcase and Jacobi-Bender. The latter Company also supplied part of the metal watch bands; the balance was purchased from the Japanese Bear Company. The cases for the European Pulsar versions were made by Piquerez that also designed the case for the American P4 Pulsar.

The Pulsar

John Bergey found the clever name "Pulsar" purely by accident. One late evening, when he walked through the deserted offices, he saw a popular astronomy magazine on the desk of one of the employees. Bergey leafed through it and found an article on the rapidly rotating neutron stars that emit sharply focused radiation pulses in extremely regular intervals. Astronomers call them "pulsars" (the abbreviation of "pulsating stars"). This was just the right name for the new watch; it was registered as soon as possible.

The display's brightness was automatically adapted to the ambient light level in three steps (100%, 50%, and 12.5%). But only the first Pulsar model, with the Electro/Data module 25-IC that was in the stores on April 4, 1972, had this feature. It was the word's first serially made watch without mechanically moving parts. The later production models had a simpler, two-step brightness control.

Time on Demand

Because of the very high power demand of light emitting diodes, the Pulsar only displayed hours and minutes after pushing a button: then the LED were switched on for 1.25 seconds. The seconds appeared if the button was kept depressed. With a Pulsar, one theoretically was the master of one's own time. This slogan was fine as long as one did not carry heavy objects with both hands. The problem was solved with the Auto Command function that was introduced in 1975: now the time was also displayed after flicking the wrist.

When necessary, the Pulsar display was set with a small platinum-cobalt magnet that was integrated into the clasp. This magnet closed reed relays in the movement, one each for hours and minutes. After assembly, the case was filled with dry nitrogen and hermetically sealed. There were no bushings; even the pushers were purely magnetic – which meant that the case was perfectly tight. The numerous Pulsars that still are in perfect working order after 30 to 40 years are proof. However, no Pulsar ever made it to the Moon: the rumors to that effect are apocryphal.

Production in Lancaster

While Electro/Data was developing the very first Pulsar, Time Computer started its own model. It was rather obvious that in view of the large number of integrated circuits, the Electro/Data module was not going to be sufficiently reliable. But thanks to Electro/Data's small production, the Pulsar could be launched on April 4, 1972: the survival of Hamilton was at stake. Admittedly, Time Computer lost money in the early phases of the Pulsar P1. The watch was only available with an 18 carat solid gold case at a cost of $2100 apiece, $150 more than a gold Rolex. The margin of the retailer was 50 percent, i.e. $1050. With the remaining $1050 one had to pay Electro/Data's module that cost $ 700, plus the gold case, the bracelet, and the ruby glass. This definitely was not profitable.

Originally, Time Computer had planned to produce 400 gold Pulsars with the Electro/Data modules. But demand was so high that this number was increased to 450 pieces. For sales people and representatives, 37 Pulsars with gold plated case and bracelet were made. It did not make sense to give those people real gold watches. Just 4 modules were kept back as a source of spare parts.

The development of the Time Computer Pulsar module was greatly aided by the fact that in 1972, RCA integrated the entire divider and decoder on a single chip; the 3-volt CMOS technology was now available and 3 volts were needed anyway for the light emitting diodes. The Pulsar thus always needed two silver oxide button cells.

Toward the end of 1972, production of the Time Computer Pulsar module started in Lancaster. TCI then let it be known that they would take back all the Pulsars and replace the Electro/Data module by one made in Lancaster. The new modules had the caliber number 201; they were round, their diameter was 34.4 mm, the height was 7.75 mm. They could easily be fitted into the P1 Pulsar cases. Module production steadily increased and reached 10,000 pieces per month in 1975. Now the Pulsar became very profitable, as the production cost of caliber 201 dropped to $74 apiece, later on to only $10. The original Electro/Data modules (type 25-IC) were destroyed at the Lancaster plant; less than 10 pieces have survived.

The weak point of the Lancaster modules was the trimmer for adjusting the frequency of the quartz resonator. Only a small, cylinder-shaped trimmer was used at first, but its capacity range was insufficient. A bigger trimmer was adopted for approximate tuning, fine tuning then proceeded with the smaller trimmer. Later, the trimmer was replaced by a fixed capacitor.

Exotic Components

The first LED with a dot matrix for the Electro/Data prototypes were supplied by Hewlett Packard (HP). As they had been designed for instruments, HP developed a smaller LED, the matrix of which was left unchanged. Later on, Time Computer purchased 7 segment LED from Monsanto and Litronix. At first, they were covered with a perforated foil to simulate the old dot matrix. As this led to an appreciable loss of light, TCI switched to the unmodified, 7 segment LED.

The quartz was a 32,768 Hz flexural mode resonator in an evacuated capsule; it was supplied by Clark Crystals in Massachusetts. The electronic oscillator maintained the vibrations of the quartz resonator. The 32 kHz frequency was divided 14 times by 2, down to 2 Hz. The latter frequency plus 4, 8, 16 and 32 Hz were needed for the decoder that controlled the LED dots or segments to make up the numerals for hours, minutes and seconds.

The dark red filter through which the lit numerals were visible effectively hid the chaotic looking inside of the movement. It was made of Plexiglas for the prototypes; the P1 Pulsar model used synthetic red sapphire from Switzerland. With the P2 model, TCI introduced the gold ruby glass made by Corning.

The pushbutton used to switch on the display actuated a small, disc-shaped magnet that controlled a reed relay from Hermetic Industries. These actually were the exact same magnets that had been used for the balance wheel motor of the Hamilton Electric. There still was a big stock of those magnets and they were put to good use in the Pulsar. Hermetic also supplied the mercury switch activating the display with a flip of the wrist (Auto Command), e.g. on model P4.

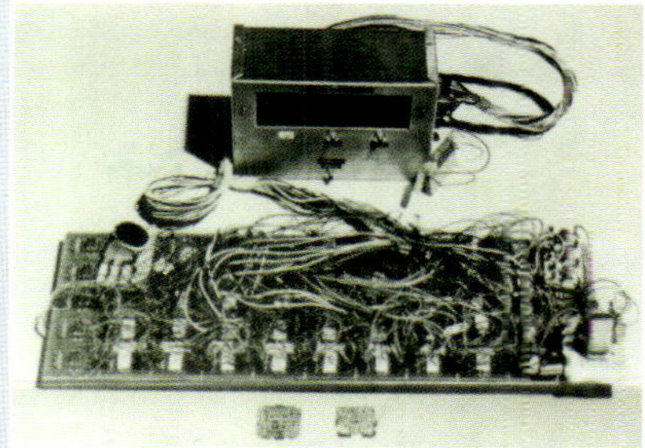

7.28. Pulsar breadboard by Electro/Data

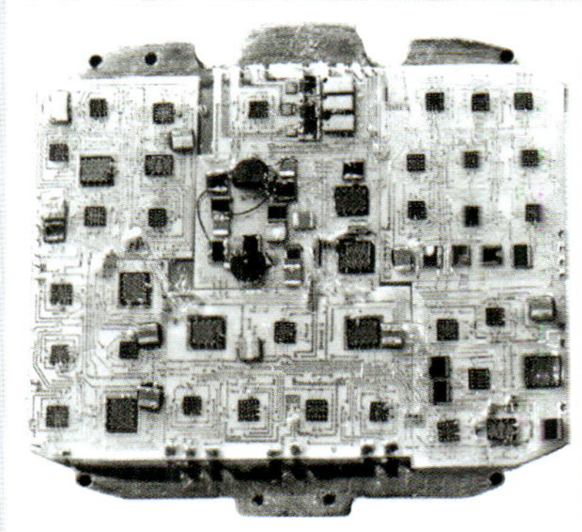

7.29. The 44 IC Pulsar circuit by Electro/Data

7.30. LED display of the 44 IC Pulsar with dot matrix

7.31. Pulsar Prototype by Electro/Data that was shown to the media in May 1970 in New York

A Cult Object for the Rich

The Time Computer module of the Pulsar production model P1 was supplied with energy by two silver oxide batteries in series. The current consumption was only 6 microampere, but jumped to 2.5 milliampere when the display was switched on. With the exception of the quartz resonator, the trimmer and the magnetically controlled reed relays for setting the display, all electronic components were potted in the main plate consisting of black plastic. Time Computer guaranteed that the Pulsar was at least ten times more accurate than any mechanical or electronic watch of the time, including Bulova's Accutron.

With a price of $2100 apiece, the Pulsar was extravagantly expensive; yet the first series was quickly sold out. Tiffany and Neiman-Marcus made most of the sales, thousands of pieces were ordered after the first series was gone. The Pulsar rapidly became a cult object; customers were literally fighting to get one. After the solid gold model with the Electro/Data module (later with the Time Computer module), TCI launched the gold filled Pulsar ($1275), and the steel Pulsar ($275).

AFIF/Neosonic

In the mid-1960s, seven Swiss watch manufacturers decided to join efforts in the electronic sector and to develop their own quartz wristwatch movement. The Department of Industrial Research (AFIF) of the Swiss Federal Institute of Technology (ETH) was given the mandate to do the actual development work that started in early 1965; a prototype was presented to the media in 1972.

Walter Born, CEO of the Roamer Watch Company was convinced that the electronic watch had a bright future. Seiko in Japan had sent ever smaller quartz marine chronometers and deck watches to the Neuchâtel Observatory where they participated in the famous chronometer contest. Sooner or later the last step to the quartz wristwatch would be made. Born called himself a "travelling salesman" looking for people with similar convictions who would develop an electronic wristwatch in common.

Born found several quite important partners for Roamer: Certina, Oris, Manufactures Suisses Réunies (MSR), Revue Thommen, and Rolex-Bienne. Born was named president of a consortium of the above watch brands. Once he had secured the financing, Born consulted with the AFIF in Zurich, which agreed to do the job in early 1965. As the future movement slowly took shape, the consortium founded a common subsidiary called Neosonic AG.

The project was quite a challenge. Aside from mercury oxide button-type batteries, there were no commercially available components for quartz watches: everything had to be invented or reinvented from scratch. The quartz resonator was a real headache, as it had to fit into the case of a wristwatch.

A small quartz resonator naturally has a high resonance frequency, which meant many steps of electronic division in order to obtain 1 Hz. Digital divider steps were state of the art at that time, but their power consumption was of the order of several milliwatts while the battery could only supply a few microwatts. The AFIF electronics specialists thus concentrated on the three critical components: quartz resonator, divider circuit and motor.

The Quartz Resonator

A conventional, flexural mode quartz resonator was chosen. In the early phases of the project, the resonance frequency was 16,384 Hz; that value was eventually doubled to 32,768 Hz. Quartz bars of the correct dimensions were cut, ground, and polished by a specialized company in Bienne. The electrical contact pads on the quartz bar were made of evaporated gold; four spiral-shaped elastic holding and contact wires were soldered to the quartz.

Each of those wires was shaped into a quarter turn circular segment. They were connected to two parallel, very stable connecting wires that were sealed into a short glass rod. Sealing the wires was done with a very strong, focussed light – lasers were not available at reasonable prices at that time.

The quartz resonator was inserted into a glass tube and evacuated; the tube and the rod with the contacts were also sealed with a powerful light in the vacuum jar. The overall dimensions of the glass encapsulated quartz resonator were 22 mm x 4 mm. Thanks to the elastic suspension, the resonator was shock proof up to 5000 g.

Divider Circuit and Motor

In the mid-1960s, only bipolar integrated circuits were available for low voltages. But their power consumption and leakage currents were far too high for an electronic watch. The main difficulty was to reduce the power consumption of a bipolar integrated circuit by a factor of thousand. The solution was a so-called lateral transistor with several emitters of different length. This system was used for powering the 15 required divider steps of the flip-flop-type. In this scheme, the length of the emitters and a resistor with 15 taps parallel to the set of emitters had the effect of halving the current flowing through each of the segments of the lateral transistor compared to the current of the neighboring segment.

The Neosonic stepping motor consisted of soft iron stator parts and a magnetic rotor that made 180 degree jumps for each current pulse in the coil – always in the same direction.

The entire movement had a current consumption of 11 to 13 microampere. The daily rate was less than 0.1 second. Developing the Neosonic prototype cost the consortium just 2.6 million Swiss francs, almost an order of magnitude less than the development costs of the Beta 21 and the Omega Megaquartz 2400, respectively.

7.32. The quartz caliber AFIF/Neosonic

Unhappy End

The AFIF/Neosonic watch was quite sturdy and reliable; a very successful media presentation was made by team leader André Greuter in 1972. The movement had a diameter of 28.0 mm and a height of 5.4 mm. The echo in the press was very positive: the prestigious Swiss daily "Neue Zürcher Zeitung" reported at length about the technology of the new movement. But shortly afterwards, the Neosonic consortium was dissolved: none of the partners were interested in actually producing the movement.

One of the reasons was that production models of the quartz movement Beta 21 developed by the CEH in Neuchâtel were shown at the 1972 Basel Watch Fair. The GP 350 quartz caliber by Girard-Perregaux, the Omega Megaquartz 2400 and the third generation of industrially produced Seiko quartz movements were not far behind.

Five prototypes of the Neosonic movement were assembled in the AFIF laboratory. Apparently, only one of them was saved for posterity. It is presently owned by a private collector in Texas. In 1978, Neosonic AG was deleted from the commercial register.

PART II

The Worldwide Development of Quartz Movements

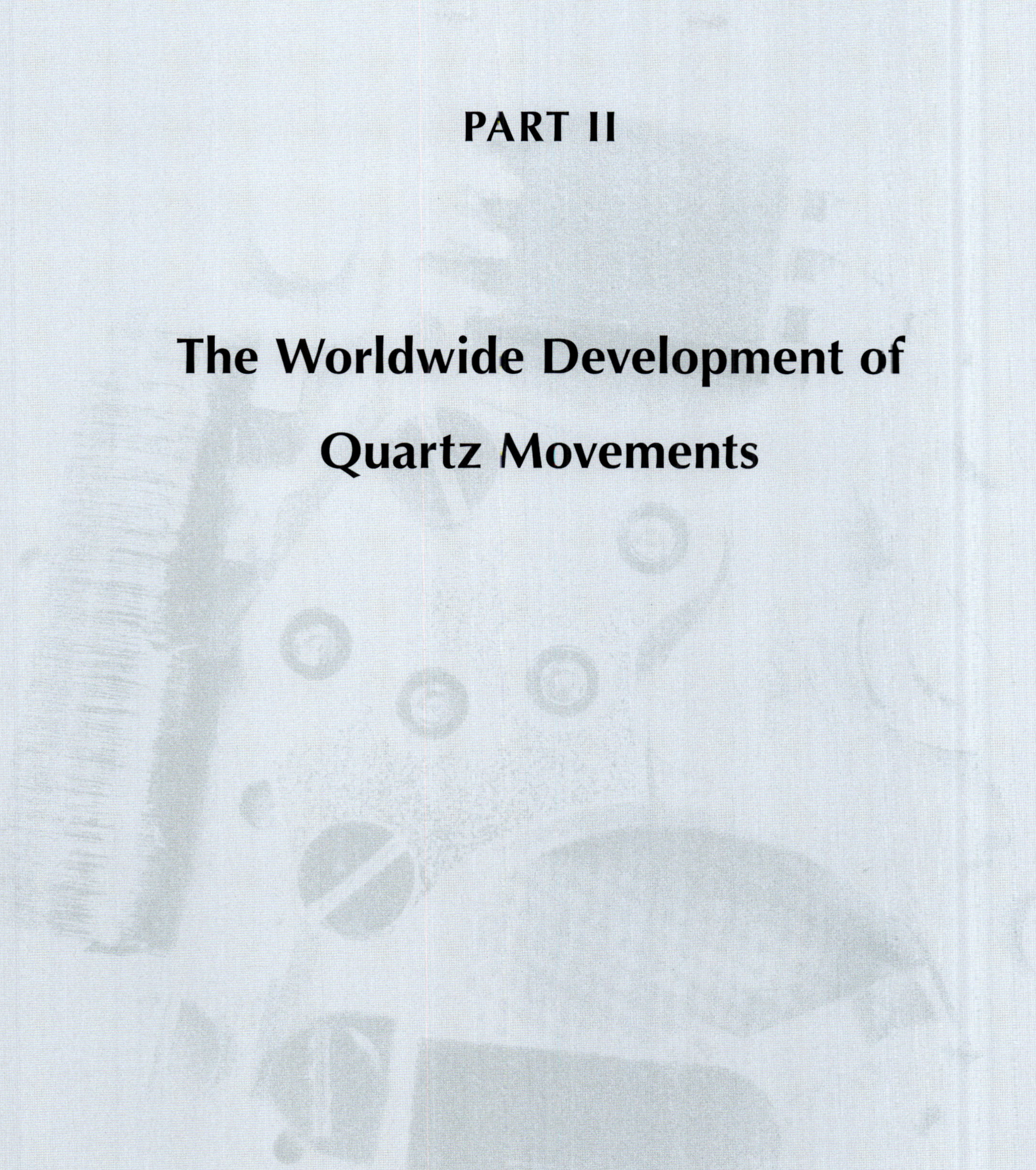

8. *Quartz Movements from Switzerland*

The quartz revolution deeply hurt the Swiss economy. It developed very quickly and triggered the worst crisis that the Swiss watch industry ever had to endure. A few figures may be useful to illustrate this: In 1973, Switzerland produced 71 million watches and watch movements, 1 million of which were quartz; this kept 77,000 people employed. In 1984, employment was reduced to 30,000; Switzerland exported 47 million watches and movements, 31 million of which were electronic, 9 million mechanical (with jewel pallet escapement) and 7 million Roskopf (with pin pallet escapement). Roskopf watches and movements came from old stocks, as the entire pin pallet watch industry had been wiped out by 1982. Digital watches with liquid crystal display (LCD) or light emitting diode display (LED) began as a purely American innovation and Switzerland did very little to compete with the US electronic industry. This probably was a blessing in disguise. The reasons will be explained in the course of this chapter.

Liquid Crystal Displays

In the 1970s, the digital watch with liquid crystal display (LCD) went through a boom phase all over the world. But the euphoria was quite short lived, particularly in Europe. Only two Swiss companies actually manufactured such watches; many others just cased commercially available modules. At first, they used dynamic scattering-type displays that were imported from the United States. These were soon replaced by greatly superior displays of the twisted nematic type (TN). Interestingly enough, the latter were invented in Switzerland and even produced there, albeit only for a short time.

The basic research that would lead to the TN LCD was done in the laboratories of the pharmaceutical giant Hoffmann La Roche (Roche) in Basel. In the 1960s, diversification was believed to be the magic wand that would bring new sources of big profits. This led Roche into a close cooperation with Brown-Boveri (BBC, today ABB) in the field of medical electronics. About twenty common projects were identified, one of them being displays using liquid crystals.

A 19th Century Discovery

Liquid crystals were discovered many decades before practical applications were found for them. It all began in 1888, when Friedrich Reinitzer, an Austrian botanist working at the Prague University investigated cholesteryl benzoate. As he was determining the melting point, he observed that at first, a cloudy liquid formed. It cleared up completely after heating to a significantly higher temperature. This is the typical behavior of liquid crystals that Reinitzer just had discovered.

Liquid crystals consist of rod- or leaf-shaped organic molecules forming perfectly normal crystals at low temperatures. If they are heated, they will eventually melt, forming a liquid phase. Yet the attracting forces between the molecules are so strong, that some measure of order is preserved, albeit much less than in the solid state. In many cases, the molecules remain parallel to each other and are stacked like matches in a box.

At temperatures above the melting point, liquid crystals exhibit the mobility of a liquid while still preserving the optical anisotropy of a crystal. The molecules can easily slide along each other, but rotational motion is strongly inhibited. The latter is possible above the second melting point (the so-called clearing point T_c), well above the first melting point (T_m). Now the molecules are free to rotate along their axes of symmetry and move in all three dimensions: the liquid is clear and optically isotropic. The operating temperature range of liquid crystals thus lies between the melting point and the clearing point.

Dynamic Scattering

In the mid-1960s, George H. Heilmeier who was then working at the RCA David Sarnoff Research Laboratory in Princeton, New Jersey, discovered that electrical fields strongly affect the transmission, scattering, and polarization of 10- to 20-micrometer thick liquid crystal films. This discovery led Heilmeier to develop a first generation of liquid crystal displays (LCD) based on the principles of dynamic scattering.

In those devices, a film of liquid crystals was sealed between two thin glass plates. Numerals and letters could be made up by electrically addressing a pattern of segments consisting of both transparent and conductive indium tin oxide (ITO) on the inner side of the upper glass plate. The lower glass plate was coated with a homogenous ITO film on the inside and with gold or chromium on the outside. As long as no voltage was applied between the structured and the unstructured ITO films, the whole system was transparent and only the reflecting metal was visible.

Due to their strong dipole moment, liquid crystal molecules have the tendency to agglomerate, forming oriented, birefringent domains that are isotropically distributed and coupled elastically. Applying a voltage causes the connected ITO segments to appear milky-white on the mirror-like background. This is due to the fact that the domains are aligned by the electric field while ions that were added to the preparation are attracted to the electrodes and cause turbulent hydrodynamic motions of the liquid crystals, which causes them to scatter reflected light.

The drawback of dynamic scattering displays is that the ions chemically react with the electrodes. If for example the cell contains sodium chloride, chlorine gas forms at the anode while sodium hydroxide forms at the cathode – which inevitably shortens the life of the display. Furthermore, the contrast of such displays is rather weak and highly directional.

Twisted Nematic Cells

After being hired by Roche to get the LCD project going, the physicists Martin Schadt and Wolfgang Helfrich were looking for ways to develop liquid crystal displays based on purely electro-optical effects. The idea was to reorient the molecules without any chemical reactions. This is how the twisted nematic cell (TN) was invented.

The starting point was an observation made in France as early as 1911: a thin film of liquid crystals has the properties of a polarizing filter. If it is placed in a microscope, rotating the instrument's polarizing filter causes periodic extinction and brightening of transmitted light. The explanation is that a stack of the rod-shaped liquid crystal molecules can form a helix acting as a waveguide for visible light. If such helices are arranged between two crossed polarizing filters that cause total extinction, the additional polarization by the liquid crystals eliminates the extinction and some light is transmitted.

To make practical use of this effect, it is necessary to immobilize the liquid crystal molecules that come in contact with the surface of the two glass plates forming the cell. This may be achieved by coating the two glass surfaces with plastic and rubbing the plastic with a velvet cloth. The rubbing direction on one of the glass plates must be at a right angle to the rubbing direction on the other plate. In that case the liquid crystal molecules will form a helix that is rotated by 90 degrees along the height of the cell. The waveguide effect of such a helix critically depends on the orientation of the twisted molecules on parallel planes. 1.5 volts at the electrodes causes a slight tilt of the molecules that eliminates the waveguide effect.

A twisted nematic cell (TN) consists of two crossed polarizing filters between which the helix-forming

8.1. SGT Quartz watch with liquid crystal display of the Dynamic Scattering type

8.2. LCD quartz watch by Texas Instruments with Twisted Nematic (TN) display

8.3. Martin Schadt, inventor of the TN display demonstrating liquid crystals

8.4. Early prototype of an alphanumeric LCD of the TN type

liquid crystals are arranged. If no voltage is applied to the electrodes, the entire surface of the display is bright. If a voltage is applied, the optical effect of the helices is eliminated, the energized segments of the display are dark. The inverse effect – light segments on a black background – is easy to achieve.

Selling Patent Licenses

Thanks to those new technologies, the liquid crystal display became a very attractive product, not only for digital watches but also for many types of instruments, replacing the traditional galvanometer-type analog display. A group of Roche chemists was given the assignment to synthesize a wide range of substances forming liquid crystals, particularly cyanoesters and cyanoalkyl Schiff bases. It was also discovered that the best TN displays were obtained with mixtures of anywhere between two and ten different kinds of liquid crystal molecules. All this progress did not make liquid crystals a profitable business, even if their price was around 15,000 Swiss francs per kilogram. Only a few milligrams were needed for an individual display, a few kilograms supplied the entire world for a year.

When Seiko Epson showed a great interest in acquiring the basic Roche liquid crystal patent, the Company decided to sell patent licenses for the production processes of liquid crystals. The Japanese companies Seiko, Sharp, and Casio were among the takers. They continued developing the technology and made tremendous advances.

Videlec – A Short Lived Venture

In 1974, a plant built by Brown Boveri in Othmarsingen near Lenzburg (canton Aargau) started producing liquid crystal displays using the preparations Roche had synthesized. Philips bought into the venture at the rate of 50 percent of the shares, which allowed a considerable production increase. The common BBC/Roche/Philips subsidiary was called Videlec. LCD displays of the TN type for digital watches and all kinds of instruments were produced there until 1984. In that year, Philips bought the BBC and Roche shares of Videlec, transferred it to Eindhoven in the Netherlands and closed it in 1986.

Neither BBC nor Roche or Philips foresaw the enormous commercial potential of liquid crystal displays. Their further development exclusively took place in Asia, where the now ubiquitous flat screens for laptop and tablet computers, TV sets, smart phones, and e-books were developed. Today, the sales generated by liquid crystal displays are of the order of 100 billion dollars and they keep growing.

The linear photopolymerization of plastic monomers was a very significant progress in the LCD production technology. As mentioned above, liquid crystal molecules tend to orient themselves on directionally deformed polymers. With the new method, the awkward manual or robotized rubbing of the polymer film could be dispensed with. Stannic acid esters are particularly well suited for this purpose, as they are easy to crosslink with light.

If such a monomer is irradiated with polarized ultraviolet light, only the double bonds parallel to the plane of polarization of the light are activated and lead to polymerization. The polymer is thus oriented in one direction. This orientation can be structured with masks. If the display is subdivided into many small domains with different orientations, the angle dependence of light reflection in LCDs can be eliminated.

Technology Transfer to Asia

For many years, Roche and Merck were the main suppliers of liquid crystal preparations. But in 1994, Roche decided to fully concentrate on its core business, i.e. pharmaceuticals and to sell off all the diversification ventures. However, the liquid crystal division was finally profitable; it was separated from Roche in 1996 by creating a spin-off by the name of Rolic near Basel. That company was sold to the Chip-Bonder Essec but was acquired by a private investor after Essec merged with the Oerlikon-Bührle Group in 1997. Rolic is still in business with LCD for special applications.

Merck bought all the Roche patents covering liquid crystals. Merck presently holds 60% of the worldwide production facilities for liquid crystals. Japanese and Chinese companies provide the remaining 40 percent. Today, simple liquid crystal displays for digital watches, instruments, and laptop computers are exclusively made in China – on the base of Japanese technology that was first transferred to Hong Kong, the Shenzhen Zone north of Hong Kong, and finally to Mainland China. In Switzerland, only the Swatch Group subsidiary EM-Microelectronic Marin still produces dials with integrated LCD displays as well as LCDs for instruments; they were developed under Roche license.

Swiss Quartz Calibers

ASSA – A. Schild SA

ASSA was one of the three founding fathers of Ebauches SA. Towards the end of the 1970s, it was given the mandate to concentrate on the development and the production of small quartz calibers for ladies' watches such as 5½ x 6¾ lines. Consequently, the development of the quartz caliber 977 was started. In 1979, ASSA merged with ETA that took over the movement design department and finished the development of caliber

977.001; production started in 1982. It had the exact same dimensions as the hand-wound mechanical caliber ASSA 277 for easy substitution, i.e. 5½ x 6¾ lines and a height of 3.6 mm. This movement was state of the art with a silver oxide battery, a 32 kHz tuning fork quartz resonator, CMOS circuit and Lavet stepping motor.

Audemars Piguet

Since it was founded in Le Brassus in 1875, Audemars Piguet rates as one of the classical watch manufacturer of the Vallée de Joux. Yet for many years, it was not formally a manufacturer, even though it had an excellent technical department where AP calibers were designed all the way to the prototype stage.

The production of ebauches, however, was left to Jaeger-LeCoultre (JLC), the neighbors in the Le Sentier, of which AP was a controlling stockholder at the rate of 40 percent of the share capital. This made a lot of sense, but *de iure* AP did not fulfill all the criteria of a manufacturer, even though the adjustment, the assembly, and the casing of the movements were done in Le Brassus.

In 2000, Vodaphone sold to Richemont the majority shareholding position in Jaeger-LeCoultre. Richemont then also bought all the JLC shares owned by Audemars Piguet. After a "grace period," AP could no longer depend on its former manufacturer of movement blanks and revert to the status of a true manufacturer. This was not too difficult, as the manufacturing facilities of the young watchmaker genius Giulio Papi in Le Locle were controlled by AP. Papi's former company was called Audemars Piguet Renaud & Papi (APRP); however, its production capacity was insufficient.

A new plant was built next to the old APRP and equipped with the most modern machinery; production started there in 2003 and almost fully covered the company's needs for high grade mechanical movements. But soon production was further expanded in Le Brassus, where another plant was inaugurated in late 2008; it is dedicated to the production of movement blanks. The legendary "Hotel des Landes," just next door, was bought by AP at the same time and converted into a cafeteria for the employees.

Most AP's watches were equipped with mechanical movements even after the breakthrough of quartz. However, the company had no qualms about using electronic movements when this was indicated. AP did not produce such movements, but purchased them from high prestige manufacturers such as Omega, Jaeger-LeCoultre and Girard-Perregaux.

Some quartz calibers were bought as kits and assembled in-house, others were procured ready for casing. The first quartz movement purchased by AP was the high frequency Omega caliber 1510 that

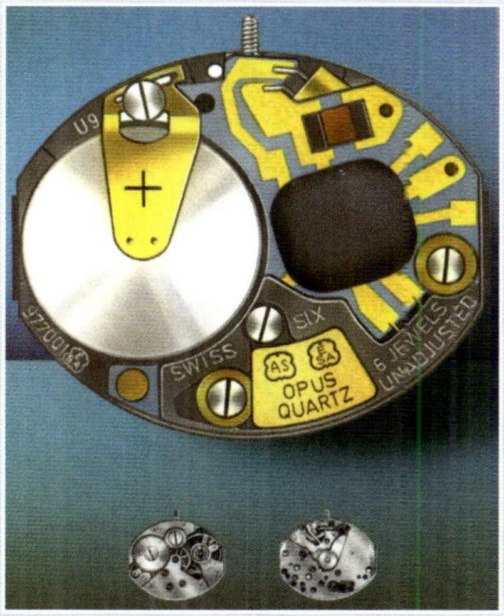

8.5. ASSA quartz movement, 5 ½ x 6 ¾ lines, caliber 977.001

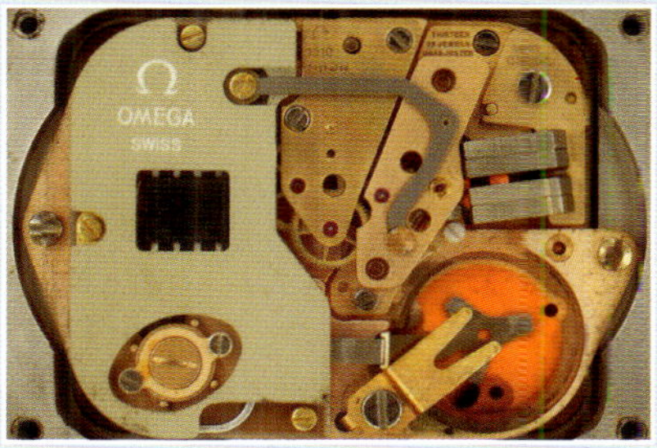

8.6. The first Audemars Piguet Quartz caliber was the Omega caliber 1510.

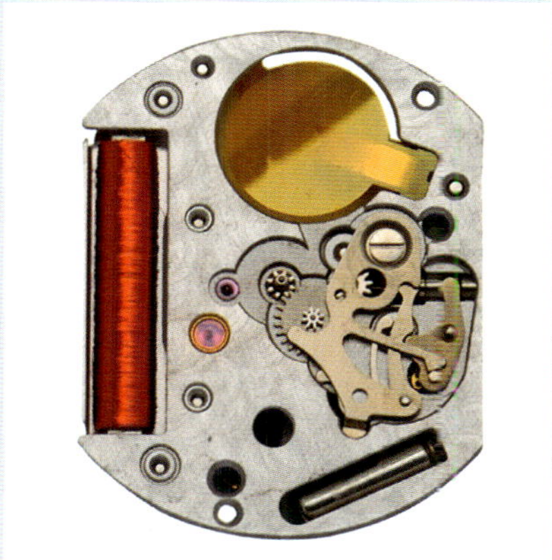

8.7. The Audemars Piguet caliber 2508 was the Jaeger-LeCoultre caliber 608/609.

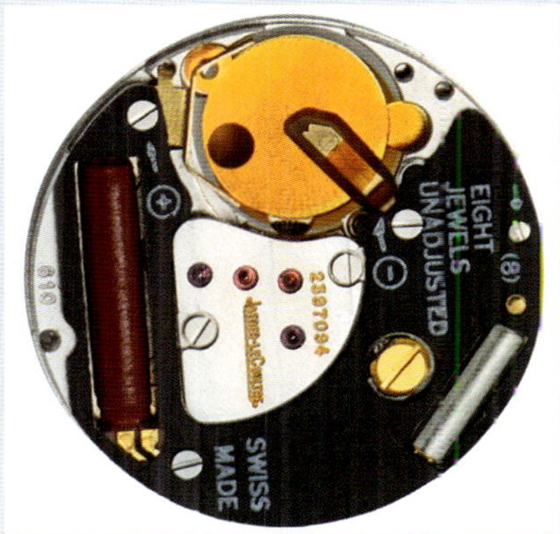

8.8. The Audemars Piguet caliber 2610 was the Jaeger-LeCoultre caliber 610.

had been developed for the "Megaquartz 2400"; the Besançon Observatory certified it as a marine chronometer. AP gave it the designation 2510 and used it from 1974 to 1975.

The next AP quartz movement (caliber 2511, 8 jewels) was also purchased from Omega where it was called 1330 "Ferrari." The dimensions were 10½ lines x 3.50 mm. It was equipped with a flexural mode quartz resonator and an 8-pole SSIH stepping motor that was quite similar to Portescap's Socrem motor. It made a step every 2.5 seconds, the lifetime of the mercury oxide battery was 16 months. The AP caliber 2511 was mostly used for the quartz version of the Royal Oak with a platinum case.

Caliber 2502 with 7 jewels was used by AP from 1979 to 1985. The dimensions of this shaped movement were 17.20 mm x 13.50 mm x 2.50 mm (including the battery). The 32 kHz quartz resonator was of the photolithographic Statek type. Thanks to digital frequency tuning, no trimmer was necessary in conjunction with the quartz resonator. Current consumption was just 1 microampere, the Lavet motor made a step every 10 seconds. The silver oxide battery had a life of one year. AP caliber 2502 actually was the JLC caliber 602.

The 11½-line pair of calibers 2506-2505 with 10 jewels and a date disc had a height of 2.5 mm (3.15 mm with the battery) and was equipped with a Lavet motor making one step every ten seconds. This reduced the power consumption considerably: battery life was 33 months. Caliber 2506 was JLC 606 and was used from 1982 to 1988. The closely related caliber 2505 was JLC 605 and was cased in AP watches from 1983 to 1987.

The extra flat shaped caliber 2508 (1983 to 2006) with 7 jewels and the dimensions 17.20 mm x 13.50 mm x 1.60 mm. Thanks to the ten second Lavet motor, current consumption could be reduced to 0.70 microampere; battery life was 33 months. AP 2508 was the JLC caliber 608.

The two-hand AP caliber 2601 with 3 jewels was purchased from JLC (caliber 601) from 1983 to 2005; AP still holds an important stock. This was a small shaped movement with the dimensions 11.70 mm x 9.70 mm x 1.80 mm. Thanks to the 1 minute Lavet motor and a new integrated circuit, current consumption had been reduced to 0.4 microampere which extended the battery life to three years. The hands were set with a pusher at 3 o'clock: there was no crown.

The round 17.20 mm x 2.50 mm AP two-hand caliber 2711 with 8 jewels and date disc was equipped with a 1 second Lavet motor. The current cosumption was 1.2 to 1.5 microamperes; depending on the height of the battery, its life was 24 or 30 months, respectively. This caliber was used by AP from 1984 to 1988. It was purchased from Girard-Perregaux and was its caliber 711.

AP produced quartz watches with the extra-flat caliber 2610 from 1988 to 2003; the diameter was 16.20 mm, the height 1.9 mm. This was JLC's caliber 610, a two-hand caliber with 8 jewels, a date disc and a 5 second stepping motor. The current consumption was 0.7 microampere which resulted in a battery life of 2 years. The AP quartz caliber 2612 (1989-2003) was also purchased from Jaeger-LeCoultre (caliber JLC 612). The dimensions were 20.80 mm x 2.10 mm; this was a two-hand caliber with date disc.

The extra flat quartz caliber AP 2710 only had a short product life (2003 to 2005); it replaced caliber 2610. This was a round, 16.20 mm x 1.90 mm two-hand movement with date disc, 7 jewels and a 5-second Lavet motor. Current consumption was a mere 0.40 microampere, the battery life was 38 months. It was developed by Girard-Perregaux as caliber 1010A specifically for Audemars Piguet.

Starting in 2004, caliber 2712 (20.80 mm x 1.90 mm) was used by AP. This larger size version of caliber 2710 also was a two-hand movement with date disc and a 5-second Lavet motor; the current consumption was 0.4 microampere. This was an exclusive Girard-Perregaux development for Audemars Piguet; the GP caliber number was 1310A.

The AP caliber 2713 was purchased from Cartier where it was marked with the Cartier logo and received the caliber number 688. However, Cartier never manufactured quartz movements: in this case it had purchased Jaeger-LeCoultre's caliber 619 and renamed it. One must assume that Cartier had too many of those movements in stock and sold part of them to Audemars Piguet.

Baumgartner Frères Grenchen (BFG)

The Roskopf-type pin pallet watch industry has the doubtful reputation of collapsing in record time: it was literally wiped off the Swiss industry scene without leaving a trace. In 1975, Switzerland exported 22 million Roskopf watches; by 1980 this figure had doubled to 44 million. Five years later, exports had dropped to a mere 1.5 million – and they were old stock, as all producers had closed years earlier. This industry branch was literally condemned to death in the early 1980s, when the price of quartz movements reached the level of Roskopf movements. As a quartz movement was the much better product with a superior quality, the pin pallet had no chance.

Production dropped abruptly to zero in 1983. The same happened in eastern France, where a sizable pin pallet watch industry was located.

The Roskopf industry had indeed seen the writing on the wall but failed to react adequately and fast enough. Heroic attempts to diversify into low-price quartz movements failed abysmally. Such products hit the marketplace much too late and the industry did not really believe in them. It is a sad fact that the ingenious but hopelessly obsolete Roskopf watch disappeared with a bang after 120 very successful years.

The end of Baumgartner Frères (BFG) in Grenchen traumatized an entire Swiss canton. The Company had been established in 1899 and was already very important before World War II. After the war, rather arbitrary legal restrictions were lifted and BFG grew to be the largest Swiss producer of Roskopf movements and watches. When the company celebrated its 75th year, it employed 1700 people, some 1000 in Grenchen.

In that year, quartz wristwatches were already produced in rapidly growing numbers, but those mostly Japanese watches were very expensive, they were considered as rather exotic niche products. Hardly anybody in Europe would have guessed how much their price could potentially drop. BFG executives – they were mostly recruited among family members – were convinced that their product had a bright future, as its price was going to be unbeatable even in the very long term. Indeed, BFG seemed to have trump cards, particularly with its caliber 866. It was launched in 1953, well over 300 million pieces were produced.

The irony is that BFG did not ignore microelectronics. Its engineers familiarized themselves with the technology in the early 1970s; by 1975 they started to design three quartz calibers the components of which were mostly made in-house – strictly in line with the unwritten rule that BFG had to be as autonomous as possible. The first small series of BFG quartz movements were sold in 1977, at a time when Ebauches Electroniques Marin (EEM) and ETA were already producing quartz calibers in rather large numbers.

The production of BFG quartz movements started in the late 1970s. At that time, the ferocious price war opposing American and Japanese electronics manufacturers that were producing ever increasing quantities of LCD quartz watches at ever shrinking prices was in full swing. The Japanese won: by 1983 American electronics firms were no longer making modules and watches. But it was a Pyrrhic victory: within the next few years, the production of LCD watches had to be integrally dislocated from Japan to China.

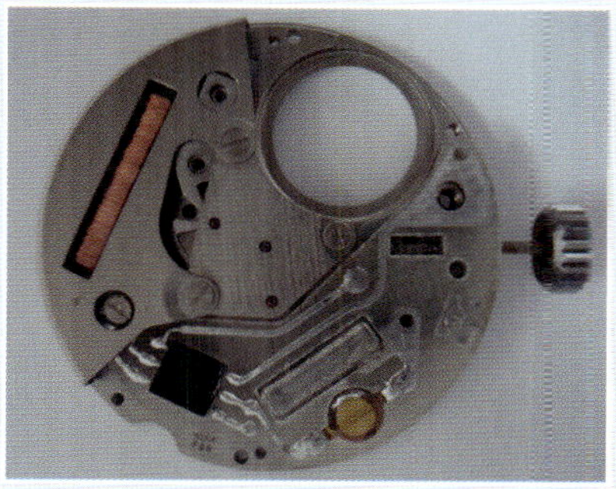

8.9. Quartz movement by Baumgartner Frères (BFG) caliber 888

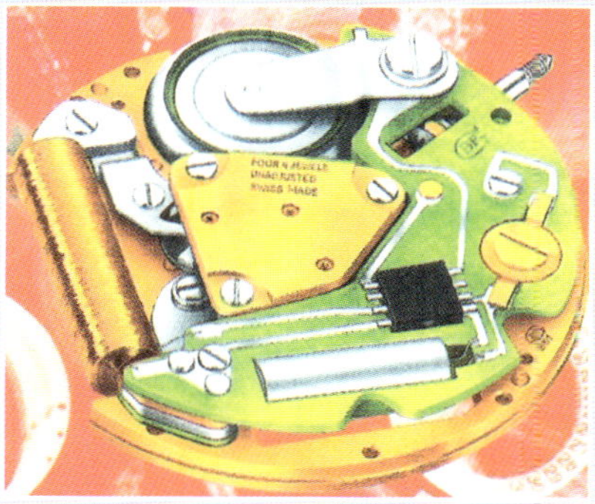

8.10. BFG quartz movement caliber 851

8.11. Buttes Watch Company (BWC) movement with an LCD by Optel

8.12. BWC watch with an LCD by Optel

Some American manufacturers still had big stocks of digital modules and watches. Almost inevitably they turned up in Hong-Kong where they were liquidated at ridiculous prices. The market was oversaturated and there was no market for BFG quartz movements: a profitable production was impossible. At the same time, the market for mechanical pin pallet movements also collapsed: this signaled the demise for BFG.

The three BFG quartz movements were well designed and very sturdy. But they were rather late: management had never given them the priority they deserved. BFG 888 with 4 jewels had a diameter of 13¾ lines. Its height of 5.55 mm (including day/date display and sweep seconds-hand) was very far from record-breaking, but fully in tune with the BFG concept of sturdiness and ease of fabrication. Everything was state of the art: a 32,768 Hz quartz resonator, a 15-step binary divider all the way to 1 Hz, a Portescap/Socrem stepping motor and a silver oxide battery. The coil and the motor were made in-house; all electronic components were supplied by outside vendors.

Caliber 851, BFG's second quartz movement for men's watches, had a diameter of 11½ lines and was provided with a sweep seconds-hand and day/date display; the height was 5.25 mm. It used practically the same components as caliber 888. The design was quite clever and optimized for sturdiness.

The BFG 6¾ x 8 lines movement for two-hand ladies' watches (caliber 915) was trimmed for low power consumption. The motor received a current pulse every 30 seconds, which advanced the minute-hand by half a minute. Thanks to this trick, battery life was two years despite a rather small battery.

Buler

The Economic Swiss Time group (EST) was part of the SSIH holding and included the brands Agon, Buler, Continental and Ferex. EST purchased various quartz modules for the group; the Swiss made logo was not relevant in the case of economical quartz watches. In early 1973, SSIH bought several hundred thousand 4-digit LED movements from Frontier Electronics, a US-Japanese joint venture in California. Aside from Buler those modules were also used by other SSIH brands such as Hamilton, Omega and Tissot. In the following year, EST bought 100,000 3½-digit LCD modules from Nippon Electric (NEC). The price of this type of module was close to $20.

Buler and Continental also used hundreds of thousands "Alpha" LCD modules made by NEC, with date and a second time zone as extra functions. Nippon Electric designed the first chip that controlled both an analog and an LCD display. This was the heart of the "single chip" ana-digi caliber "Gamma" that was imported by the EST subsidiary TRICE and used in watches with the Buler, Certina and Tissot brands.

BWC

In the late 1960s, every manager in the Swiss watch industry realized that a technical revolution was coming. The Bulova Accutron transistorized tuning fork watch was a relatively expensive electronic niche product, but rumors about quartz wristwatches were persistent and rather ominous. It was quite obvious that the Centre Electronique Horloger (CEH) in Neuchâtel developed electronic movements, but no details leaked out. Later on it was reported that several American electronics companies were developing quartz calibers with a digital liquid crystal display (LCD). The first were launched in 1971.

Edwin Volkart, the owner and CEO of the Buttes Watch Company (BWC) wanted to know what he could possibly do about this and visited Optel in New Jersey. He managed to obtain a set of LCD modules and was the first Swiss exhibitor to show this kind of product at the 1972 Basel Watch Fair. BWC had designed a special case for these Optel modules with a dynamic scattering-type liquid crystal display. This watch was called Ditronic, the German watch manufacturers Arctos and Provita were also involved in its design and distribution.

However, the early cooperation with Optel turned into a disaster. The 5000 modules that Volkart had ordered for the ad hoc community BWC-Cyma-Para-Provita were ridiculously expensive and delivery was delayed by more than a year. By the time the Ditronic reached the marketplace, the competition was more technically advanced and significantly cheaper. In the end, only 200 to 300 Ditronic watches were sold. They cost 1200 Deutschmark and most of them were returned to BWC because of various defects. They had to be repaired under warranty, which cost a lot of money. BWC was not the only company suffering from the disastrous performance of the Optel LCD caliber: Wyler, Delvina, Glycine, and Milus also were also affected.

When the 13¾ lines x 8.5 mm LCD digital module made by Arctos-Ph.Weber was available, BWC made a new start, but history repeated itself: the Arctos caliber LCQ 575 was expensive, much too thick, and prone to break down. Despite of the cooperation and the financial participation of the Ormo watch plant in Pforzheim, the Arctos module too was a source of deep losses for BWC.

The digital watch business made the turnaround to profits when Volkart found four independent partners with whom big orders could be placed in common. They were Pronto and Aubry Frères, both in Le Noirmont, Paul Junod in Bienne, and Claude Guilgot in Geneva.

In the 1970s, they founded SIPRA (Société Industrielle de Production et d'Achat), a purchasing cooperative that obtained significant quantity discounts. The members agreed to inform each other of new developments. This allowed BWC to produce a high grade digital quartz watch that was very successful, particularly in Germany, and opened the door for other BWC watches.

Thanks to SIPRA, BWC always had the newest products. The digital euphoria did not last: analog quartz suddenly was in great demand. That watch type could be easily supplied; assembly just meant casing complete movements. The quality of the incoming movements and modules was meticulously checked: due to SIPRA's purchasing power, the suppliers did not dare shipping anything but the best grades.

In the mid-1970s, the German market did not absorb any mechanical watches; only quartz sold. BWC managed to switch the production to quartz analog watches within three months. BWC acquired the reputation of being the specialist for this type of watch. Within the next few years, the BWC product mix consisted of 50 percent mechanical watches and 50 percent quartz watches; eventually the percentage of quartz timepieces rose to 90 percent. Now the movements were bought in Swiss quality either from EEM in Marin or ETA in Grenchen.

As a result, BWC was overstaffed: assembling quartz watches took considerably less time than assembling mechanical watches. Volkart reluctantly had to reduce the personnel by 25 to 30 people, a third of his entire staff.

CEH

After the spectacularly successful development of the Beta 1 and Beta 2 quartz prototypes and their triumph at the Neuchâtel Observatory (see Chapter 7), it may seem strange that not one Swiss watch manufacturer was interested in launching a quartz watch on his own, using the movements that resulted from a common financial engagement. All of a sudden, everybody wanted to show off with their own, hopefully even more advanced technology. In 1969, Longines surprised everybody with the Ultra-Quartz that had been developed in a great hurry in cooperation with Bernard Golay and worked without an integrated circuit. In 1970, Omega made a presentation of the final "Megaquartz 2400" prototype.

The other CEH shareholders lost valuable time with useless and endless discussions: they had been competitors for a long time and had developed a refined culture of quarreling, even though they were part of the same holding (ASUAG or SSIH). Nobody could imagine that within ten years the mechanical watch would be a niche product and that almost everybody would wear a quartz watch .

8.13. Beta 21 by CEH

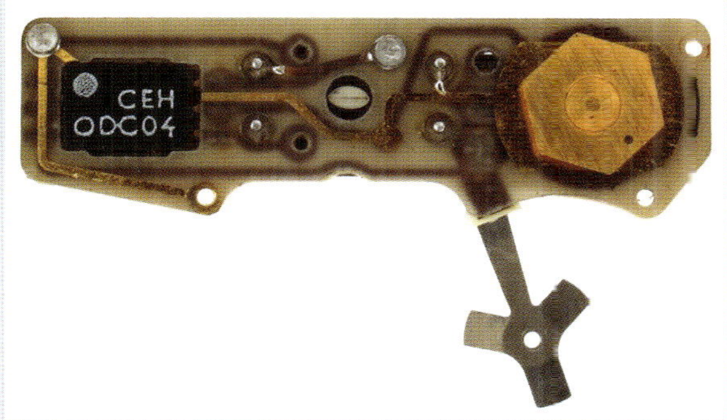

8.14. Electronic module of the Beta 21 cal ber

8.15. Back side of the Beta 21 electronic module with quartz resonator

8.16. Bulova 2467 Thermatron, developed by the CEH

After the 1967 chronometer contest at the Neuchâtel Observatory, the CEH decided to completely redo the Beta 2 prototype and develop it into something that could be fabricated in series. Nothing was holy: the whole movement had to be reinvented. An important step was the integration of the oscillator, the frequency divider, and the motor drive into a single bipolar chip with 110 transistors: using a 1.35-volt mercury oxide battery, the current consumption was 12 microamperes. The other components too were carefully checked for potential optimization.

This work was severely hindered by administrative harassment and endless rivalries between the various brands of the ASUAG group. Epic fights broke out for the mandate to fabricate even the most banal components for the industrial version of the Beta 2. Finally, a "Communauté d'Intérêt Bêta" was created. The community had the following members: Fédération Horlogère, Ebauches SA, Fabriques d'Assortiments Réunies, Fabriques de Balanciers Réunies, Fabriques de Spiraux Réunies, Bulova, Credos, Complications SA, Communauté d'Horlogerie de Précision BHP, Elgin, Enicar, IWC, Jaeger-LeCoultre, Longines, Omega, Patek Philippe, Rado, Rolex Bienne, Rolex Geneva, and Zenith.

This organization was going to manufacture 6000 pieces of the new quartz movement with the designation Beta 21. Despite the loss of efficiency, this very modest production was split up between three different locations: Ebauches Electroniques (first in temporary facilities in Saint-Blaise east of Neuchâtel, then in the newly built EEM plant in Marin); Omega Bienne; and a newly created facility in Yverdon that was run jointly by Jaeger-LeCoultre, Longines, Patek-Philippe, and Rolex. One thousand Beta 21 were actually available at the time the 1970 Watch Fair opened its doors in the spring. Sixteen Swiss watch brands exhibited wristwatches with the quartz caliber Beta 21.

The total production run of just 6000 movements was sold to the members of the Communauté d'Intérêt Bêta for the special price of 700 Swiss francs apiece. The biggest orders were issued by Jaeger-LeCoultre (150 pieces), Patek Philippe (750 to 1000 pieces), Omega (600 pieces) Rolex (600 pieces), and Longines (700 pieces). Longines ended up taking only 200 pieces (after all it had the Ultra-Quartz), but Jaeger-LeCoultre gladly took the extra 500 pieces. The quartz resonator of the Beta 21 was supplied by the Ebauches subsidiary Oscilloquartz, the vibratory motor was made by Omega, Ebauches SA supplied the gear train, and the bipolar integrated circuit came from the CEH semiconductor lab in Neuchâtel .

The major source of trouble was the vibratory motor of the Beta 21: it actually was a tuning fork resonating at 256 Hz. This frequency selective component was very sensitive to periodically reoccurring shocks (hedge cutting, shoe polishing, all kinds of sports) and made phase errors that significantly affected the daily rate. Just like the Accutron, the motor's tuning fork also was affected by a non-negligible position error. Quite a few people who bought a watch with a Beta 21 movement noticed that it was amazingly accurate as long as it was not worn. This performance deteriorated significantly when it was placed on the wrist.

Further Developments at the CEH

Another weak point of the Beta 21 and the early quartz movements from Ebauches Electroniques Marin was the trimmer capacitor that was needed for adjusting the frequency of the quartz resonator. It was exclusively supplied by a small American company and shipments were always delayed. Furthermore, the capacity of this trimmer was affected by temperature, humidity and aging processes that kept shifting the frequency and impaired the daily rate.

This problem was elegantly solved by Eric Vittoz and his team at the CEH. They developed the digital frequency alignment that eliminated the need for a trimmer. In this scheme, the frequency of the quartz resonator was purposefully set a little higher than 32,768 Hz at the fabrication stage. Extra cycles were suppressed after movement assembly, thus tuning the resonator exactly to the required frequency.

For this type of circuit, non-volatile storage elements of the EEPROM type were indispensable. The development of this component started quite early at the CEH and Eric Vittoz was also deeply involved. The basic principle of this type of memory is a transistor with an extra "floating gate" into which electric charges can be injected; about 40 volts are required. Those charges affect the current flowing through the transistor and are quite stable: they stay where they are for many years.

Beta 3 and Beta 4

The development of CMOS circuits that worked with the 1.35 volts supplied by mercury oxide batteries received a high priority at the CEH. Fritz Leuenberger's metal gate CMOS technology was put to good use for developing a 20-stage digital frequency divider that could be supplied for a full year by a small button-type battery. This truly ingenious circuit was patented by the CEH, but Motorola copied it – which was discovered by CEH staff. The dispute was settled out of court, Motorola paid an indemnity of $160,000.

In November 1970, CEH staff made a presentation of Beta 3, their new quartz movement. The major innovation was a rather involved integrated circuit of the CMOS type with about 500 transistors. The frequency of the quartz resonator (it was still a 32 kHz flexural mode device) was not critical: any value between 32,768 Hz and 32,772 Hz was acceptable. As mentioned above, a trimmer capacitor was needed previously for exactly adjusting the frequency to the required value. It was good news that this very unreliable component could be dispensed with thanks to digital frequency tuning.

The stepping motor of Beta 21 needed current pulses with alternating polarity at the rate of 0.5 Hz. This was achieved by 16-fold division of the quartz resonator frequency. The division actually continued to 1/32 Hz: the latter frequency was needed for the inhibition control circuit that suppressed pulses and set the frequency fed into the divider circuit exactly at 32,768 Hz.

The number of pulses that had to be suppressed was determined by the Vibrograf timing machine with its very accurate, thermocompensated quartz timebase. Seven switches (they actually were screws) could be combined to suppress a maximum of 127 pulses per second. This allowed correcting the daily rate by up to 10 seconds per day.

Many a watchmaker might be overtaxed by a row of screws that had the function of switches and that he had to combine in order to set the inhibition. But Beta 3 was not meant as a commercial movement. For this reason, anything but the electronic part was left virtual. Just designing, fabricating, and testing the electronics was quite a remarkable achievement. For the first time, a low voltage IC was used; this was a test bench for Fritz Leuenberger's metal gate CMOS technology. It was very useful for the development of the digital frequency adjustment circuit.

In a more advanced version, control of the inhibitor was left to a microprocessor. This led to the presentation of Beta 4 in 1972, still another prototype. This movement used a high frequency quartz resonator (524,288 Hz or 10^{19} Hz) with a tolerance of +1.5%. An external frequency generator could be connected inductively to the watch circuit and generated a set of three pulses that were separated by exactly one second. The first pulse switched on the learning circuit on the integrated circuit while the second pulse erased the previously set number of suppressed pulses from the quartz resonator. The number of pulses delivered by the quartz resonator was then measured during exactly one second. This number was put in the memory and was used to control the inhibition.

As long as no dependable non-volatile memory chips were available, a tiny, very long-lived lithium battery with a solid state electrolyte was used to hold the memorized information during a few hours, whenever the watch movement's battery needed replacing. Non-volatile EEPROM memories or flash memories started to be available towards the end of the 1970s. Now information in the memory was preserved without an external source of power. The microelectronics department of Piaget worked very closely with CEH staff at that time and used elements taken from the Beta 3 and Beta 4 calibers for the design of its calibers 7P (1976) and 8P (1981).

The CEH also worked feverishly at the development of radio-controlled watches (see Chapter 4), but the Swiss watch industry was not interested in this kind of timepiece. It was argued that Switzerland should only produce watches that worked anywhere in the world and not only in the limited radius of a time signal transmitter. The latter were to be found on each continent, but the frequencies and codes were quite different. The radio-controlled watch was left to Junghans and Ruhla in Germany, to Citizen and Casio in Japan, and finally to various Chinese manufacturers. Such watches are still a niche market, but their popularity is rapidly growing.

Gamma

Research at the CEH was strongly influenced by the concept of the metallic resonator that had been invented by Max Hetzel – and not just in the early days. This is documented by project gamma that was started in 1970, when the development of the commercial quartz caliber Beta 21 was progressing rapidly. The objective was a ladies' caliber with a diameter of 20 mm and a height of 4.8 mm. The oscillator was a metal bar that was excited in the torsional mode. Theoretically, torsion is not accompanied by any volume change: this would prevent unnecessary heat generation and allow for an energy-efficient operation.

The torsional oscillator was also expected to be characterized by a very good quality factor and zero position error. The torsional motion of the oscillator could easily be converted in a rotation with a double system of ratchets working in push-pull. However, it was soon noticed that it was not possible to excite the bar in a purely torsional mode: there were many other vibrational modes that used up energy. This effect was minimized by optimizing the torsion bar geometry, but not sufficiently to make this design a success.

No prototypes were fabricated at this stage, but Rolex and Jaeger-LeCoultre were still interested. In 1972, they gave the CEH a mandate to develop the torsional mode movement; Henry Oguey was put in charge of the project. But just a year later, progress of quartz technology was so spectacular that the project was terminated.

Delta and FAN

The idea of a quartz watch with a permanent, pseudo-analog luminous display was formulated in 1970 and patented in 1973. A prototype was designed at the CEH in 1972 (project Delta). In order to read the time, it was not necessary to push a button the Pulsar way. In the earliest version, time was displayed in the analog mode by the flashing of light emitting diodes that simulated the tips of watch hands. Two dots indicated the hour, one dot the minute. The seconds were not displayed, but the LED on the tip of the simulated hands flashed in one second intervals.

The display needed 60 x 5 light emitting diodes of the gallium arsenide phosphide/gallium arsenide type that emit orange red light with a wavelength of 635 nm. Those LEDs were arranged in concentric circles on the dial: the light jumped to the next position every minute ("minute-hand") and twelve minutes, respectively ("hour-hand"). In 1974, they yielded 0.12 lm/A. The diodes that were actually switched on received very short current pulses that generated light pulses at the rate of 25 per second, which the eye could not resolve. The LED just gave the impression of being switched on and off every second. In order to allow the emission of a maximum of light, a hemispherical lens made by electro-erosion was arranged on top of each LED in 1976.

The energy source of the movement was a nickel-cadmium battery that was charged with twelve wedge-shaped solar cells forming a circle in the center of the dial. Quite evidently, the power output was largest in full sunlight; the diodes could then be operated at full strength which meant a current of several milliamperes. On the other hand, the diodes could be operated at minimum strength in the dark: 10 microamperes sufficed. Ten years of work were needed until all the elements of the Delta design were perfected.

Starting in 1978, much more efficient LED based on gallium arsenide on sapphire and a coating of silicon carbide were available; they emitted blue green light with a wavelength of 495 nm. The new LED had been developed at the CEH with the help of Hewlett-Packard know-how and were made by epitaxial growth in the gas phase. A yield of 1.3 lm/A was achieved; the combination with solar cells allowed for pseudo-continuous operation.

The dial with a radial arrangement of 60 x 5 diodes was left unchanged, but now three diodes marked the hours, two marked the minutes, and a luminous dot jumping from second to second at the periphery indicated the seconds. A demonstration prototype was delivered to Piaget's subsidiary Complications SA in 1980, but otherwise the watch industry was not interested. Development work was stopped in 1981.

In parallel with the Delta caliber, CEH engineers started working on a somewhat similar design called FAN (for "Forme Analogique", caliber 7035) in 1975. This was a Rolex mandate (see the section on Rolex in this chapter). In the FAN, the display by dot-shaped LED was not continuous, solar cells were thus superfluous. Time and date were displayed Pulsar-like strictly on demand by pushing a button.

Thermatron

In 1982, Bulova launched the "Thermatron," a quartz watch that ran with body heat; the secret that this was a CEH development was well-kept. This was the world's first quartz wristwatch with a thermoelectric power source. A total of 704 bismuth telluride thermoelements, connected in series and arranged in the four corners of the movement, generated 175 millivolts. This was boosted electronically to 1.15 volts in order to charge a silver oxide / cadmium battery.

The temperature difference of 1.5° C between the skin and the upper part of the watch generated 10 to 20 microwatts – more than enough to power a quartz movement. This design was very ingenious but the Thermatron was quite thick and very expensive. There were problems with the battery too: it was supposed to keep the watch running for a year, but it was much less in practice.

Integrated Circuits

It is rather interesting that in the 1970s two so-called silicon foundries were established in Switzerland. Their objective was to supply integrated circuits to the watch industry and other branches of industry working with microelectronics. The origin of the two Swiss silicon foundries – Ebauches Electroniques Marin (EEM) and Faselec in Zürich – can be easily traced back to the first Swiss quartz movement.

Faselec (it was called Fasec in the early days), was founded jointly in 1966 by a subsidiary of the Fédération Horlogère FH, the Centre Electronique Horloger (CEH), Ebauches SA and quite a few other Swiss companies, such as Autophon, Brown Boveri, Contraves, Landis & Gyr, Philips Zurich, Philips Eindhoven, and Zellweger Uster. Philips held 55 percent of the Faselec shares. Yet, the watch industry set this in motion, as it rightfully asked itself who was going to supply the specialized miniature integrated circuits it needed in quartz watches.

The location of the Faselec production facility was the subject of endless discussions. Initially, a joint laboratory with a pilot line was supposed to be built in the Neuchâtel area (Pierre-à-Bot, Bevaix, or Marin). Philips already produced transistors in Zurich. This plant that was managed by Camille Margna was integrated into Faselec, Margna taking the reins at Faselec; he was its first director. Philips decided that this plant could not be moved to the west of Switzerland. It was much more convenient to move Kurt Hübner's semiconductor group at the CEH to Zurich. The CEH violently opposed this scheme, as it wanted to have full control of its semiconductor facility and refused to transfer staff to Zurich.

A compromise was finally found: Faselec was to be located at the transistor plant in Zürich. The essential CEH employees, Kurt Leuenberger and Eric Vittoz, stayed there but would closely cooperate with the Zurich group. More particularly, they developed a silicon gate CMOS technology that was brought to commercial maturity in Zurich. Peter Gross, the former CEH secretary, was promoted to the post of Faselec administrator. The electrical engineer and semiconductor specialist Heinz Rüegg, who had returned in 1967 from a seven-year stay in the United States, was the first scientist at Faselec and soon was promoted to head of the Research and Development department. He gave the team in Neuchâtel the mandate to develop a silicon gate CMOS technology.

CMOS versus Bipolar

It soon became obvious that each Faselec shareholder had different ideas about the circuits that should be developed and produced. As time went on, Faselec and the CEH drifted apart. More particularly, Faselec refused to produce the bipolar integrated ODC4 for Beta 21. The extremely small number of required circuits – some 6000 pieces – could only have been produced with an atrociously bad yield. The CEH ended up producing the ODC4 in Marin with the EEM facilities, but the silicon wafers were cut, bonded, potted and checked by Faselec.

The slow "divorce" of Faselec and EEM was favored by the Ebauches SA policy to be self sufficient both in mechanical and electronic watch components. For this reason, it built the giant development and production complex Ebauches Electroniques Marin (EEM) that started operating in 1970. Kurt Hübner was given the position of director. Production there started with a second generation of quartz movements that had precious little in common with the Beta 21.

The new movements were, of course, going to use CMOS circuits that worked with little over 1 volt and had an extremely small power consumption. Faselec actually was going to produce such circuits on the base of the know-how developed at the CEH, but was still very far from production. Faselec insisted in selling bipolar integrated circuits that were not really suited for quartz watches.

The Hughes License

Ebauches SA evidently was in a hurry. As Faselec could not deliver, it was decided to buy a CMOS production license from Hughes Electronics in California. The Swiss-American physicist Hans Dill, who was working with Hughes, was "bought along" in the package deal and most successfully organized the technology transfer to EEM in 1975. This was interpreted as a breach of contract by Faselec, since the members of the consortium had committed themselves not to produce their own integrated circuits. The "Hughes affair" generated a big controvery, but at he end of the day the storm blew itself out.

In the amazingly short time of 8 months, EEM succeeded in transferring the entire infrastructure needed for mass producing metal gate CMOS watch circuits. A yield of well over 90% was achieved almost from the beginning. Ironically, Faselec speeded up things in 1976 and soon was ready to produce CMOS circuits in the advanced silicon gate technology that had been developed by CEH and Philips-Eindhoven staff. The watch industry (and particularly ETA in Grenchen) liked this a lot, as it now had two sources for its integrated circuits and could negotiate good prices. The Swiss watch movement manufacturer Ronda turned into a very good Faselec customer, as did the German alarm clock industry in the Black Forest

The integrated circuits produced by both EEM and Faselec were designed for quartz movements with an analog display. Only Faselec produced the chips needed for digital watches, but the series were small, as this product never caught on in Switzerland. Digital watches also needed liquid crystal displays (LCD), for this reason, Brown Boveri started producing them near Lenzburg in 1974, as mentioned above. This BBC-Philips joint venture was called Videlec.

For a while, Ebauches SA, Brown Boveri and Faselec jointly produced the first fully electronic Swiss watch modules. Major customers were Nepro, Mondaine, and Tchibo. But the series were small and added up to just a few thousand pieces; EEM finally took over this production while Faselec stopped producing silicon chips in 1993. EEM (that was renamed to EM Microelectronic Marin), still produces low power-low voltage chips for horological and non-horological applications.

Delvina

Very little is known about Delvina, a Geneva-based watch manufacturer. After World War II, Delvina exported mechanical watches to Great Britain and the United States. When quartz movements became available in 1972, Delvina launched its model Quartz Instantime: it used the round LCD module from Optel (that was also purchased by BWC – see that section) with a 3½-digit dynamic scattering display with flashing dots at the rate of 1 Hz. Time was set by turning the crown at 3 o'clock.

Ebauches Bettlach

Ebauches Bettlach was a subsidiary of the Ebauches SA holding and was one of the most important Swiss manufacturers of pin pallet movements of the Roskopf type. Towards the end of the 1970s, it became clear that the much more precise quartz movement was becoming less and less expensive and would sooner or later replace the pin pallet movements. For this reason, Ebauches Bettlach developed an very low-price three-hand quartz movement with a stopping device for the seconds-hand, in cooperation with Ebauches SA's research laboratory in Neuchâtel). It was given the caliber designation 870.

Caliber 870.101 was the basic movement, caliber 870.111 was fitted with a date disc. The dimensions were 26.5 mm x 4.80 mm. The construction followed the precepts of the Roskopf technology that had contributed to the tremendous growth of the town of Grenchen. The caliber included a bottom plate, pillars, and bridges: it was not designed to be original and imaginative, but used the classical concept of the analog quartz movement. It was optimized for ease of assembly and a low price.

Peter Bachmann who headed the Bettlach engineers realized that the Lavet stepping motor could be drastically simplified and had his design patented. Major characteristics were a decentralized coil and a plastic-coated samarium cobalt rotor. Two pins directly engaged a wheel and turned it by one tooth every second. This motor was extremely sturdy but rather noisy: it bears some analogies with the Lavet motor developed for the Swatch. In 1981, a test series of caliber 870 was produced, but it was judged as being too thick and (paradoxically) too expensive. Ebauches Bettlach was closed in 1982. It reopened in 1987 as a plastics processing center for ETA, producing among other things the bracelets for the Swatch watch.

Ebauches Electroniques Marin (EEM)

As mentioned in Chapter 7, the Centre Electronique Horloger (CEH) in Neuchâtel was jointly created by the Swiss watch industry in 1962 with the aim of developing electronic watches. In 1967, its quartz movement prototypes Beta 1 and Beta 2 won ranks 1 to 10 and 12 at the chronometer contest of the Neuchâtel Observatory. The Japanese Seiko won the next places with very different prototypes.

Ebauches SA (the major subsidiary of the powerful ASUAG Holding) realized that urgent work was needed and started the construction of Ebauches Electroniques Marin, EEM (today EM Microelectronic-Marin), some 8 km east of Neuchâtel. This huge complex cost over 78 million Swiss francs and started operating in 1970. The idea was to be as self-sufficient in electronic watches as the group had been in mechanical timepieces. This concept had been worked out by André Beyner, the Ebauches SA technical director. Kurt Hübner transferred from the CEH as the first director of EEM.

EEM used the ESA brand (an abbreviation of Ebauches SA) and consisted of three departments. Major products were integrated circuits, liquid crystal displays, the electromechanical caliber "Dynotron," Derby table clocks with an alarm function, and printed circuit boards. Until 1985, the EEM design department developed and produced a large number of quartz calibers. The "Micro-Crystal" production facility for 32 kHz tuning fork-shaped quartz resonators of the Statek type started operating in 1979 as a subsidiary of ETA in Grenchen. Silver oxide watch batteries had been produced by Renata in Itigen near Basel since 1972; the company was part of the ASUAG group. All this meant that Ebauches SA was truly self-sufficient in the electronic watch sector.

The Beta 21 caliber developed by the CEH as the commercial version of Beta 2 certainly was a remarkable achievement, but it was never mass-produced – even though Ebauches SA paid a good part of the total costs, which are estimated to have been around 20 million Swiss francs. The Beta 21 ended up as ESA caliber 9173 in 1970. It was a rectangular movement with the dimensions 29.0 mm x 24.3 mm x 7.5 mm. But very soon the consensus of the experts was that it was too expensive, too big, too thick, too complicated and excessively power-hungry.

Early Analog Quartz Movements

In Marin, Ebauches SA started a crash program with the objective of developing a quartz movement that really was state of the art. The result of those efforts was the "Swissonic 1000" (caliber 9180, 13 lines x 5.75 mm) with 7 jewels and a trimmer. The variants were 9181 with date and 9183 with day/date. It was shown to the public at the 1971 Basel Watch Fair. Compared to the Beta 21, the quartz resonator frequency had been increased from 8 kHz to 32 kHz. At first, the 15-step binary division to 1 Hz was achieved with a bipolar integrated circuit from Faselec. But due to its excessive power consumption, it was soon replaced by an Eurosil CMOS Chip. ASSA in Grenchen fabricated the plate and the other metal parts, EEM was responsible for the electronic module and final assembly.

Several compromises had to be made for the 9180 caliber family. The quartz resonator was of the bar-shaped, flexural mode type made by the Ebauches SA subsidiary Oscilloquartz: it was costly and far from shockproof. The 8-pole stepping motor with 4 fixed coils and 8 cobalt-platinum magnets was purchased from LIP. As the rotor was attracted by the coil, the bearing was soon worn out if it was not well lubricated. The watch research institute LSRH in Neuchâtel had to develop the G5 high-pressure oil just for this application. The energy source was a 1.35-volt mercury oxide battery. Nobody was really happy with the 9180 calibers, but they were still produced by the hundreds of thousands. About 60 watch brands cased them – there simply was no alternative. Even a variant with an alarm function was developed for the second generation of Derby Swissonic clocks and alarm clocks.

In 1971, EEM launched the shaped caliber 9240 specifically for ladies' watches. Dimensions were 7¾ x 9½ lines x 6.0 mm. The variants were 9241 with date and 9243 with day/date; the stepping motor again was supplied by LIP.

Calibers with a Lavet Stepping Motor

The LIP motor was quite unsatisfactory; this led Ebauches SA to establish a team of very competent motor specialists who soon found an excellent replacement. It was the stepping motor that had been invented and patented in 1937 by the French engineer Marius Lavet, who worked with Léon Hatot. It is remarkable that the Seiko engineers came to the same conclusion and filed a number of new patents for this motor that came to be known as the Seiko motor.

The Lavet/Seiko motor had several important advantages: production was relatively simple and the coil could be placed anywhere within the movement. The magnetic field was led to the samarium-cobalt rotor by means of soft iron shims. The rotor turned by 180 degrees per current pulse. The magnets were supplied by BBC Brown Boveri (today ABB) and TDK

8.17. Ebauches Bettlach, caliber 870.111

8.18. ESA quartz caliber 9180

8.19. ESA quartz caliber 9240

8.20. ESA quartz caliber 9362

in Japan. The first EEM movement with a Lavet motor was launched in 1973; it was the 11-line caliber 9360 with a height of 5.5 mm. Variants were 9361 with date and 9362 with day/date.

Company rules forced EEM to purchase plates from FHF Fontainemelon, A. Schild (AS), or ETA (both in Grenchen). As they had a monopoly, those Ebauches SA subsidiaries charged exorbitant prices. This forced EEM to find a better solution and to develop plastic plates that were made in Le Landeron by precision injection molding. In order to give them the required stiffness, the plastic parts incorporated a "Kovar" grid, i.e. an alloy consisting of 54% iron, 29% nickel and 17% cobalt. The thermal expansion of this alloy was nearly identical to that of the plastic. Unfortunately, an unhappy choice was made: the polymer was brittle. It had to be replaced by a PPS-based thermoplastic resin.

Caliber 9220, a 6¾ x 8 lines movement that only displayed hour and minute, was launched in 1975. This was the first EEM caliber with a plastic plate. The 32 kHz quartz resonator was still bar-shaped. But the next caliber with the designation 9222 was equipped with a mechanically cut tuning fork-shaped quartz that was purchased from the Japanese NDK. The energy saving variant 9225 of 1975 with a 5 second stepping motor still used frequency tuning by trimmer capacitor. The height of the 9220 family calibers was 4.60 mm.

Caliber 923.031 was launched in 1978: this was a 6¾ x 8-lines crownless movement with a height of 2.95 mm. It was set with a pusher in steps of one minute or one hour. The stepping motor received a current pulse every 20 seconds. The unusual caliber number is explained by the fact that in 1977/78, the Ebauches group introduced six digit caliber numbers in all its subsidiaries, more particularly for the big movement producers A. Schild (AS), Ebauches Electroniques Marin (EEM), and Fabrique d'Horlogerie de Fontainemelon (FHF). The three last figures after the dot are a code. In the first digit, 0 stands for two hands, 1 for three hands, 3 for three hands and a bigger plate, everything else being unchanged. The meaning of the second numeral after the dot is as follows: 0 = no date display, 1 = date, 2 = day/date. The third numeral points to a variant: the slightest change had to be signaled to the servicing watchmakers.

All calibers of the 944 family had a diameter of 11½ lines; serial production started in 1979. Caliber 944.011 with a height of 3.1 mm displayed hours and minutes. 944.111 featured a seconds-hand and a date disc, 944.311 was set electronically, 944.112 was powered by a long-life lithium battery. The height of caliber 944.121 was 3.8 mm; it displayed time and day/date. 944.131 had the same dimensions but displayed the weekday written in full.

The 6¾ x 8 lines x 2.95 mm caliber 923.031 (1981) was equipped with a 20-second stepping motor. This was a crownless movement that was set electronically with a pusher in steps of one minute or one hour, respectively.

The two-hand Caliber 924.001 (1981) measured 5½ x 6¾ lines x 3.0 mm and featured the electronic correction, but this was performed with a crown, which many customers preferred to a pusher. The same remark applies to caliber 924.002; with 3.5 mm it was slightly thicker. Caliber 924.003 was set with a pusher; caliber 924.004 was equivalent to 924.002. Caliber 924.601 was the 6¾ x 8 lines x 3.0 mm model of the 924 family that was launched in 1981. The potentially universal caliber 925.111 with a diameter of 8¾ lines never saw mass production.

In 1982, ETA took over all still existing Ebauches SA subsidiaries and was made responsible for design and production of quartz movements for the entire group. Henceforth, EEM in Marin would only produce silicon chips of the low power/low voltage type as well as liquid crystal displays.

At the time of the transfer from Marin to Grenchen, the two 10½-line calibers 928.111 x 2.8 mm (with date disc) and 928.121 x 3.40 mm (with day/date) were still in the pipeline and were launched in 1983. However, two more 10½-line calibers in this family were discontinued. They were 928.411; the height of this three-hand caliber with date disc was 2.75 mm. Caliber 928.421 was thicker (3.25 mm) and indicated day and date.

Digital Quartz Movements

In cooperation with Texas Instruments (TI), EEM developed a fully electronic movement with a cross-shaped liquid crystal display of the dynamic scattering type (Swissonic 2000, Klepsydra, ESA 9260). A prototype was exhibited at the 1972 Basel Watch Fair; this model was going to be launched under the Longines brand. The project was cancelled, as TI could not (or would not?) supply the chip. In the following years, Marin shifted to digital quartz movements with twisted nematic LCD or with an LED display.

LED Movements

The Swiss watch industry did very little in the field of quartz movements with a light emitting diode display (LED). The ESA caliber 9290 launched in 1973 was of the "casquette" type, as the slanted LED was attached to the movement at 6 o'clock. A key had to be pushed for switching on the display of hours and minutes or of seconds and date. There was no mass fabrication, but caliber 9290 was the precursor of the caliber family 9291-9298. In 1976, Bulova, Mido, Jaz, Derby, and Eterna used caliber 9292 (30.00 mm x 11.00 mm) in their wristwatch with a near-vertical display. There was a version with stopwatch functions and 10 minute countdown with a resolution of 0.1 second. The entire circuit was integrated on a single chip by Micro Power Systems in California. The GP-LED movement was developed by Girard-Perregaux independently of the ESA 9289; similarities are purely accidental.

In 1975, ESA launched caliber 9370, a 6¾ x 8-lines LED movement with only 2 digits as well as the 11½-line caliber 9380 with 3½ digits, both with a vertical display. They were used by Borel, Certina, Dugena, Jaz, Jovial, Mido and Sohorec. The CMCS circuits in those calibers were among the first that were produced in Marin under the Hughes license. Two digits were enough for displaying the time in caliber 9370: when the key was depressed, hours and minutes appeared in sequence.

The LED caliber ESA/EEM 937.681 with a 3½-digit LED (6¾ x 8 lines, 5.0 mm) and two batteries was also launched in 1975 and was used in elegant ladies' watches. The equivalent movement for men's watches was ESA 9380 (11½ lines x 5.40 mm). Both time and day/date were displayed.

Dominating LCD Technology

As already mentioned, LED modules were just a niche product in Marin. On the other hand, ESA produced LCD modules by the millions – mostly with a 3½-digit twisted nematic display. Some of them also had an alarm and a chronograph function; they were quite successful in Hong Kong and the United States. The first caliber of this type was the ESA 9310, 12½ lines x 5.00 mm. It was launched in 1975 and displayed time or day/date. The next models were the calibers 9312 (11½ lines) and 9315 (12½ lines). Caliber 9346 displayed hour, minute, second and date with 8 digits. The calibers 9312, 9315 and 9346 could be stopped in order to synchronize them with a time signal.

8.21. ESA quartz caliber 536.121

8.22. ESA quartz Alarm caliber 926.311

8.23. ESA quartz caliber 944.111

8.24. ESA LED caliber 9290, front side

Caliber 934.611 (13 lines x 4.90 mm) was also launched in 1975. It featured a large size, square-shaped illuminated 7½-digit LCD for time and day/date. Calibers 934.711 and 934.712 were the same size, but included the chronograph function; 934.712 featured a 7½-digit with extra-big numerals. The same applied for the larger size caliber 934.831 (13 lines x 4.9 mm, 1975). Caliber 934.832 with a height of 4.90 mm, 3½ digits and big numerals could be switched from the hour/minute mode to day/date or seconds. Caliber 934.842 was the same size, but was illuminated with tritium gas light cells.

The LCD of caliber 934.912 (13 lines x 4.9 mm, 1975) was illuminated with a micro light bulb and displayed time, day/date and alarm time; the alarm was a musical tune. The 6-digit LCD of the rattrapante chronograph caliber 942.711 (12½ lines x 5.00 mm, 1976) displayed either time, month and date or stopped time with a resolution of 0.01 second. The ladies' watch calibers 945.731 (9 lines, x 5.2 mm, 1976) and 945.831/832 (9 lines x 3.50 mm) with an illuminated 3½-digit LCD also displayed the time or day/date and seconds, respectively.

In the latter part of the 1970s, Marin developed and produced several additional LCD quartz movements. Caliber 900.231 (1977) was an 11½ lines x 4.30 mm hybrid module with 4-digit LCD as well as a conventional analog display controlled by a 5 second stepping motor. Caliber 900.911 (also 1977) was the same size but had an alarm function. This was also the case with caliber 901.471, while caliber 901.071 indicated the time in a second time zone. Caliber 932.161 (7¾ lines x 4.6 mm) with the stopwatch function, displayed hour/minute or day/date. Caliber 932.051 featured a pseudo-analog display.

"Pseudo-analog" means that watch hands were simulated on a liquid crystal display. Early examples were the 13-line caliber 931.771 of 1980 with 6 digits and a pseudo-analog display on an extra dial, as well as the 7¾ lines x 4.6 mm caliber 932.051 "Golden Quartz" (4.6 mm) with gold colored background (1981). The displays were either made in Marin or in Singapore, but the esthetic aspects left much to be desired. Watches with a pseudo-analog display never made a breakthrough. The same applies for caliber 938.381 that was launched in 1981. It combined a 6-digit LCD with "pseudo hands."

In 1981, EEM Marin launched caliber 939.261: this was a complete 28 mm x 21 mm rectangular watch. It was advertised as the world's flattest digital watch. The module's height was 0.95 mm, 1.5 mm with

8.25. ESA LED ladies' watch quartz caliber 9370

8.26. ESA LED caliber 9380, front side

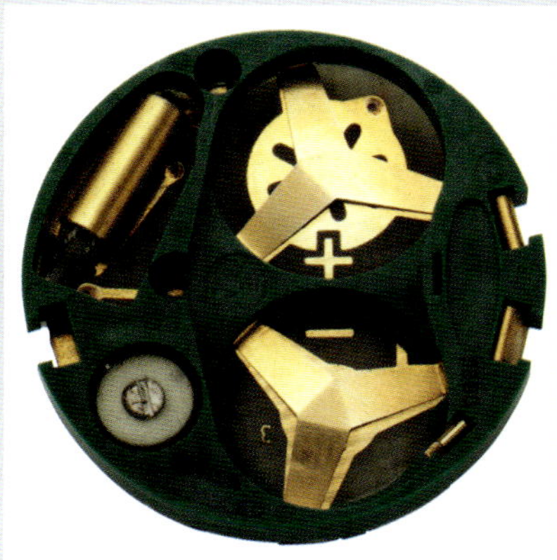

8.27. ESA LED caliber 9380, back side

8.28. ESA LCD caliber 9312, front side

the battery. Construction and shape were somewhat similar to the Delirium of 1979 (see that section). However, the module was not integrated into the back of the case. Activating the two pushers on the front displayed either hour and minute or second and date. Caliber 939.262 looked very similar but featured a stopwatch function for 24 hours: stopped time was read off the 4-digit LCD.

Electronic Complications

One of the last generations of LCD calibers from Marin included the 11-line caliber 946.621 (6 digits, date, chronograph) and the similar caliber 946.921 with the extra functions date, alarm and chronograph (both were launched in 1978). The big, 12½-line combo caliber 939.261/262 featured a 4-digit LCD, a sweep seconds-hand and the stopwatch function.

The 11 line, two-hand caliber 901.471 was still another ana-digi movement Marin developed in 1980, but only to the prototype stage. Date and alarm time were displayed digitally on the LCD. All setting operations were performed with the crown; the plate was made of Marinium, i.e. polyphenylene sulfide (PPS, Ryton) reinforced with a metal grid.

Marin launched its last LCD calibers in 1980-81. They were particularly sophisticated as they had to stand up to the low-priced competing products from Hong Kong. The 13-line calibers 931.571/572, 931.671 and 931.771 featured the alarm and chronograph functions. Caliber 931.671 with the "Memosail" function was a special model for sailing regattas. Four balls on the dial disappeared in sequence. This allowed the participants to adequately prepare themselves for passing the starting line at the right time. Marin had to develop a special chip just for this new function.

The LCD multifunction calibers with backlit LCD developed in 1981 included caliber 21.001 (11½ lines x 4.5 mm) with time, day/date in three languages, countdown, modulated alarm sound, second time zone, and chronograph. The 13 lines x 4.8 mm caliber 21.008 featured an LCD with 12 digits and 5 flags that displayed the same functions characterizing caliber 21.001 plus LAP stopwatch with 0.01 second resolution Caliber 22.010 had the same dimensions and functions as caliber 21.008. However, the LCD only featured 6 digits and 4 flags, plus a pseudo-analog display of hour and minute.

The LCD watch adventure ended with huge losses for EEM. The costs of developing and producing so many different calibers were enormous while export figures were disappointing. Not one of the important

8.29. ESA LCD caliber 9312, back side

8.30. ESA LCD caliber 9315, back side

8.31. ESA LCD caliber 934.611, front side

8.32. ESA LCD caliber 934.611, back side

Swiss watch brands was truly successful with LCD watches and disappointingly small series were exported. Only the Japanese Casio did quite well with this type of product – this has not changed to this day. Marin did sell huge numbers of LCD modules to Hong Kong, but mostly at liquidation prices.

Modutek and Microma

EEM-Marin exported large series of integrated circuits for LCD modules to Hong Kong. This helped to make full use of the production capacity. The reason for this demand was the fact that American producers of LCD watches and modules had waged a truly bloody price war against each other (see Chapter 15). By 1983, prices had dropped to such ridiculous levels that everybody gave up. The field of digital watches was left to Hong Kong where growth rates were spectacular.

The LCD module was ideally suited to the former British Crown Colony with an inexhaustible supply of low-priced labor. Its people were very diligent, hard working, meticulous, and willing to learn. The required know-how was simple; components – quartz resonator, integrated circuit, printed circuit, LCD, and battery – were easily available and cost very little. Soon the Hong Kong price of a standard LCD module dropped to less than one Swiss franc. Of course, assembling those modules cost very much less than in Switzerland.

The reaction of EEM-Marin was to establish the Modutek Hong Kong subsidiary in 1978. This was a joint venture with Stelux, later with Bloch H.K. Modutek employed 300 people and produced LCD modules that were sold worldwide under the MDK brand.

Intel had created the Microma brand for its LCD modules and solid state watches in 1972. This brand was acquired in 1978 by the private label manufacturer Endura (it was part of the ASUAG group, today it is a corporation of the Swatch Group); the price was $25,000. In Switzerland, Microma produced under the Microma Swiss brand, the modules were purchased from Ebauches Electroniques Marin. Modutek modules from Hong Kong were mostly exported to the United States. But Modutek accumulated huge losses and went bankrupt in 1982. The Microma subsidiary in the United States was also closed. EEM-Marin was relatively successful with motor modules for ana-digi watches; one of the major customers was Casio. Those exports were of great importance for EEM, but did not figure in the official statistics.

Even before the fusion of ASUAG and SSIH was decided in late 1983, Ebauches SA had been taken over by its former subsidiary ETA. Design and production

8.33. ESA LCD caliber 934.831, front side

8.34. ESA LCD caliber 934.831, back side

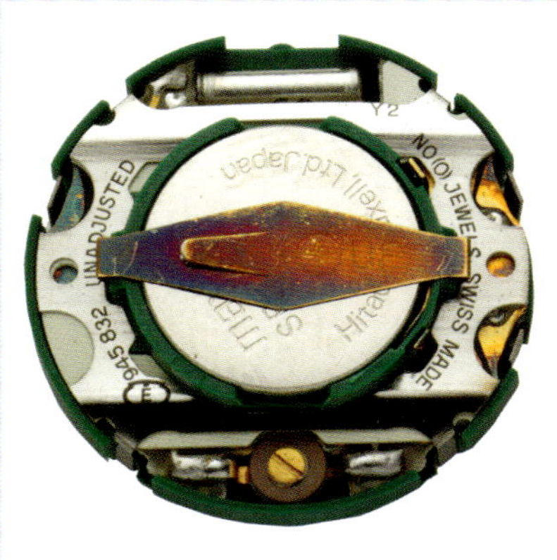

8.35. ESA LCD caliber 945.832, back side

8.36. ESA LCD caliber 942.711, front side

of quartz movements was then transferred from Marin to Grenchen; ETA was now fully responsible for this type of product. The technical design departments in Marin and Fontainemelon were now superfluous; half of the Marin team had to be terminated.

Ebosa

The Glocker family business Ebosa in Grenchen was one of the large manufacturers of pin pallet Roskopf movements (Baumgartner Frères Grenchen BFG, Ebauches Bettlach EB, Brac and Ronda were bigger). In the 1970s Ebosa developed a full range of simple and sturdy quartz calibers, but they could not prevent the demise of the Company. Ebauches Electroniques Marin and ETA were already well entrenched as producers of electronic wristwatch movements.

The Ebosa caliber family 700/701 "Bluebird" consisted of 12 round movements with 32 kHz tuning fork quartz resonator, CMOS integrated circuit, trimmer and silver oxide battery. Their height was an uniform 2.8 mm, they were launched in 1986. Bluebird calibers were available in three sizes. The smallest size was 8¾ lines with the models 700.101 (2 hands), 700.111 (3 hands, sweep seconds-hand), 701.101 (2 hands and date) and 701.111 (3 hands and date). Four calibers of 10½ lines had the exact same functions: 700.201 (2 hands), 700.211 (3 hands), 701.201 (2 hands and date), and finally 701.211 (3 hands and date). The 11½-line calibers were meant for big men's watches: 700.301 (2 hands), 700.311 (3 hands), 701.301 (2 hands and date) and 701.311 (3 hands and date). All Bluebird calibers were equipped with a one second stepping motor; they were mostly exported to Hong Kong.

Ebosa closed in 1991; the profile turning and cutting shops were purchased by ETA, a close neighbor. But two years later, ETA transformed the recently closed Venus plant in Moutier into a big watch-part plant. The Ebosa building was not needed any more and was just used for storage. The former Ebosa cafeteria was purchased by ETA, as it was located very close to its main plant.

Enicar Quartz

The Enicar watch manufacture was founded in La Chaux-de-Fonds in 1913 and established subsidiaries in Lengnau and Bienne. In the early 1970s, the Company used several quartz analog movements in its watches. Enicar was part of the Beta Community and acquired an unknown number of Beta 21 movements.

8.37. ESA LCD caliber 942.711, back side

8.38. ESA LCD ultra-flat caliber 939.262, front side

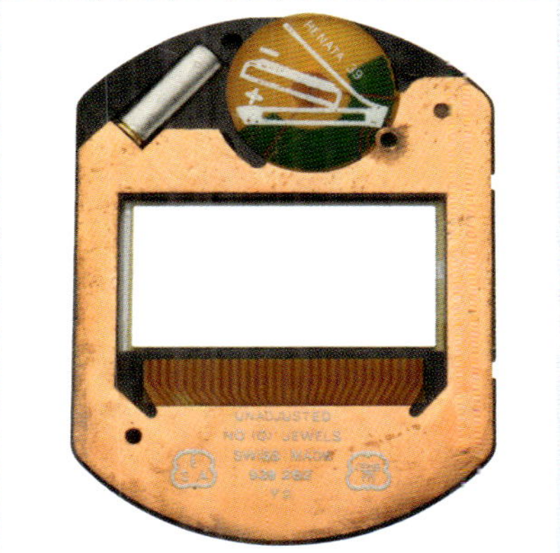
8.39. ESA LCD ultra-flat caliber 939.262 back side

8.40. Enicar Quartz caliber 320

In 1974, Enicar launched an 11½ lines x 6.57 mm quartz movement; it was part of the 320 Super Quartz family. This probably was Enicar's own design as it does not resemble any other quartz movement of the time and is found exclusively in Enicar watches. There was a version with 2 hands (320), 3 hands (321), date (323), and day/date (324). A 32 kHz SSIH quartz resonator with trimmer adjustment controlled the RCA CMOS divider IC and (indirectly) the 1 second stepping motor. The latter was connected to the battery via two flexible printed lanes, the printed circuit was contacted with screws; the central screw served to attach the stepping motor.

Enicar also sold watches with a 6-digit LCD caliber in 1976: it is caliber 64100 of the Japanese Orient (30.0 mm x 8.80 mm). In 1977, Orient launched the very similar 3½-digit caliber 63100 "Touchtron II" (30.9 mm x 5.65 mm). A sensor field had to be touched in order to switch on the display of time or date on the LED. The dimensions of caliber 63100 were similar to those of caliber 64100. For a while Enicar planned to develop another analog quartz movement in cooperation with Portescap, but nothing came of it.

ETA

In 1932, the Urs Schild watch manufacture joined the ASUAG Holding. According to Company rules, it was not allowed to produce both ebauches and watches. Theodor Schild, son of the owner, split the company in two parts: the Eterna watch plant and the ebauches plant ETA. The two companies were located in different wings of the same building and Schild supervised both of them from his corner office. ETA of course continued supplying very high grade ebauches to companies within the ASUAG group and to competitors.

The transition to quartz inevitably had profound implications for the Swiss watch industry and for the ASUAG group in particular. Sales of mechanical ebauches suffered a sharp downturn, the demand for hand-wound movements literally collapsed. FHF in Fontainemelon and ASSA in Grenchen were the biggest losers. The Roskopf pin pallet industry collapsed – this included the ASUAG subsidiary Ebauches Bettlach (see above).

Ten thousands of employees lost their job. An automatic mechanical movement consists of about 150 parts while a quartz movement needs less than 100. At the same time, the number of sold pieces suffered a sharp downturn. However, many new, technically demanding jobs were created, such as precision injection molding of plastics, winding of coils using very thin copper wire, and working with integrated circuits, printed circuits, and magnetic materials.

Furthermore, the market demanded smaller and slimmer quartz movements. ESA caliber 9162 had a height of 5.5 mm; in later calibers this was reduced to 3.7 mm, finally the industry standardized movement height to 2.5 mm, 2.0 mm and 1.5 mm. Those values were attained while maintaining or even improving the daily rate of the movements, something unheard of in the mechanical sector. So far, the rule had been that a reduction of size and/or thickness of a movement automatically meant a significant loss of performance.

In the critical 1970s, ETA was led by Fritz Scholl who had established a very strong position for his company within Ebauches SA and ASUAG. ETA's trump cards were the extremely reliable automatic movements 2671, 2824, and 2892 that were very profitable even during the worst part of the crisis. ETA ended up having to shoulder the huge deficits of other ASUAG companies such as Fontainemelon (FHF), Ebauches Bettlach (EB), Ebauches Electroniques Marin (EEM), A. Schild (ASSA), and even Ebauches SA (ESA).

Scholl therefore could do whatever he wanted, as the entire group depended on him. More particularly, ETA started producing its own quartz movements without the formal authorization of Ebauches SA. The idea was to offer the ETA customer sets of mechanical and electronic movements with the exact same dimensions, so that they were interchangeable without the slightest problem. This gave the watch producer much flexibility, as all the exterior parts of a given watch model fitted with both types of movement.

An (analog) quartz watch thus looked exactly like a mechanical one, except that the seconds-hand jumped from second to second in the electronic movement. The ticking was different too. An important point was that quartz movements were almost invariably supplied fully assembled: the watch producer just had to insert the battery and case the movement. Kits or ebauches of quartz watches were a rarity even in the early days of the new era.

There was a problem with precision injection molding. With this technology, movement parts were obtained that needed absolutely no machining. This meant that their cost was much lower than that of metal parts. Ebauches had decided that Marin was going to be responsible for plastic parts in the entire group. But Scholl was convinced that ETA had to have its own facility. He secretly bought an injection machine and installed it in the attic; the payment was hidden in the building maintenance budget. This had the long range consequence that the Swatch watch with its plastic case was designed and produced by ETA.

Technical Challenges

As already mentioned, the production of analog quartz movements demanded the mastering of entirely new technologies. In the Lavet motor for example, the air gap in the soft iron yoke bringing the coil's magnetic field to the stator was very critical. The iron shims were stamped and cut, but the air gap had to be milled in order to achieve the required precision. Such motors were so strong that the alignment of the gear train and the tolerances of the pivots in the jewel bearings were a lot less critical than previously.

New problems kept appearing: for example, the trimmer capacitor for fine tuning of the quartz resonator's frequency was sensitive to changes of temperature and humidity; the adjustment might even shift mechanically. Those trimmers were in short supply and had to be imported either from the United States or Japan. Fortunately, the digital frequency adjustment eliminated this part.

Mechanically cut tuning fork-shaped quartz resonators with a height of 0.3 mm were not produced in Switzerland and had to be imported from the Japanese NDK; they cost over 5 Swiss francs apiece. They were replaced by photolithographic quartz resonators in 1980, when Micro Crystal in Grenchen started production under a Statek license. The cost dropped to 0.10 Swiss francs apiece.

As related above, silicon gate integrated circuits of the CMOS type were originally supplied by Faselec in Zurich. Their power consumption was significantly less than that of the metal gate circuits produced by EEM in Marin. In the early 1980s, Marin switched to the silicon gate technology; ETA now had two sources for this critical part.

From the very beginning, ETA purchased silver oxide/ zinc batteries from Renata, a company that belonged to the group. It was noticed that the contact springs had to exert a minimum pressure otherwise the current flow might be interrupted. It happened occasionally that the highly corrosive electrolyte (potassium hydroxide) leaked out of the battery and ruined an entire movement. Thanks to a rigorously enforced quality control program, such defects became extremely rare.

For a time, there was a shortage of printed circuit boards; they consist of copper plated fiber reinforced plastic: the production capacity installed in Switzerland was totally insufficient. The Isola Company in Breitenbach finally enlarged its facilities. Those boards hat to be processed with extreme care: fragments of glass fibers getting into a movement could have dire consequences.

The Vacuumschmelze Hanau was of existential importance for ETA: Hanau supplied the two types of soft iron that were needed for the stator and the core of the coil. In former days, there had been a similar dependence for the alloy needed for making hairsprings. ETA had to learn giving the soft iron just the right heat treatment. Furnaces with very precise temperature control were needed for this process.

The First ETA Quartz Movements

As mentioned above, Ebauches Electroniques Marin (EEM) was set up as a subsidiary of Ebauches SA (ESA) and began operating in 1970. At that time, Marin was the only manufacturer of quartz movements in the ASUAG group. Later on, this task was spread all over the Ebauches SA subsidiaries: FHF-Fontainemelon was responsible for the 8¾, 10½, and 11½-line calibers as well as for the all important 6¾ x 8- lines segment. A. Schild was attributed the 5½ x 6¾-line movements as well as other sizes for ladies' watches. ETA had to be content with the big calibers for men's watches. Nevertheless, ETA decided to produce the quartz equivalents of its own mechanical calibers. This included both men's and ladies' calibers. According to the ASUAG rules, ETA was not allowed to produce small calibers but did it anyway.

ETA's first quartz movement for men's watches was the 11½ lines x 5.50 mm three-hand caliber 9360 (a date disc was added in 1975); the variant with day/ date display was 9362. The team leader of this project was the later ETA general manager Anton Bally. Those calibers were equipped with a one-second Lavet stepping motor and a bar-shaped quartz resonator supplied by Oscilloquartz. This was the quartz version of the automatic caliber ETA 2878. ETA acquired its quartz-know-how from a whole group including specialists working with Ebauches SA, EEM and ETA.

It seems strange had this ETA caliber received the designation ESA 9360, as Marin had very little to do with it. The point was that Ebauches general management had ordered that all quartz movements produced in the group were going to be ESA calibers, regardless of where they were designed and produced. After long deliberations, subsidiaries such as ETA were allowed to punch their own logo on the plate. After ETA took over all the former Ebauches subsidiaries, all calibers bore ETA designations. Low-priced or obsolete calibers were sold without the Swiss made declaration and without the ETA logo in Asia, mostly in Hong Kong.

As a next step, Bally designed caliber 940.111 (12½ lines x 3.7 mm with date). This was the quartz version of his mechanical masterpiece, the classical automatic movement 2892. However, the quartz version needed only 7 jewels, versus the 23 jewels in the mechanical movement. Furthermore, the very low-priced sintered aluminum oxide polycrystalline bearings made by Comadur SA could be used in quartz movements. The friction was higher, but the strong stepping motor took care of that.

ETA started to produce caliber 940.111 in 1976 with a special authorization of Ebauches management in series of 1000 pieces a day. Success was overwhelming, as this was the flattest commercially available quartz caliber of its time. It actually started the ETA Flatline family of movements. It was mostly used in high-priced watches and particularly in gold watches. Caliber 900.231 (11½ lines x 4.5 mm) was launched in 1977 and was often used in ana-digi combo watches with a liquid crystal display for the chronograph and alarm functions. It was distributed by EEM in Marin; production was stopped after ETA took over EEM.

More Caliber Twins

As a next step, ETA developed two quartz movements for ladies' watches – which in principle was illegal. The first one was caliber 950.001 (7¾ lines x 3.2 mm) with the exact same dimensions as ETA's hand-wound caliber 2512; the Lavet motor advanced in steps of 5 seconds. Caliber 950.001 became an important part of the Flatline collection; over a million pieces were produced. In 1978, ETA started the production of caliber 951.111, also with a diameter of 7¾ lines but a height of 4.8 mm; it was the electronic twin of the mechanical movement ETA 2671.

In 1978, ETA started producing caliber 977.001, the electronic twin of the ASSA hand-wound caliber of 5½ x 6¾ lines. ETA now routinely used Japanese 32 kHz tuning fork quartz resonators, CMOS circuits from Marin (later from Faselec in Zurich), printed circuit boards made of glass fiber reinforced plastic and Lavet stepping motors.

Also in 1978, ETA engineers designed caliber 952.111 with date disc and one-second stepping motor. It was the electronic twin of the very sturdy automatic caliber 2824 (11½ lines x 4.6 mm). The variant 952.121 was slightly higher (4.95 mm) because of the day/date discs and the stop second function. Caliber 954.111 was the quartz version of calibers 2892-A2 (11½ lines x 3.7 mm), i.e. the smaller version of the original 12½-line caliber 2892. In 1979, ETA launched the 8¾-line caliber 959.001; with a height of just 1.5 mm it was considered as ultra-flat. It was used mostly in elegant gold watches.

The design department of Fontainemelon FHF developed the quartz twin of the extremely successful caliber ST 69 (6¾ x 8 lines, x 3.5 mm). It was given the designation 961.101 and was first produced by ETA in Grenchen.

The caliber families 955 (10½ lines or 11½ lines) and 956 (7¾ lines or 8¾ lines) were launched in 1980. The height of the 10½-line calibers 955.411 (3 hands) and 955.431 (3 hands, Long Life) was 2.50 mm; this rose to 3.0 mm in caliber 955.421 (3 hands and date).

The 11½-line calibers 955.101 (3 hands), 955.111 (3 hands and date) and 955.031 Long Life all were 2.5 mm thick; this value rose to 3.0 mm with caliber 955.121 (3 hands, day/date).

The 956 family included the 7¾-line caliber 956.111 (3 hands, the height was 2.50 mm), 956.121 (3 hands and date, 3.00 mm) and 956.031 (Long Life, 2.50 mm). The 8¾-line calibers of this family were 956.411 (3 hands), the Long Life variant 956.431 (both of them 2.5 mm) and 956.421 (3 hands and date, 3.0 mm).

The caliber families 955 and 956 are still in production – after three decades. However, they are equipped with state of the art chips, most of the other components have also been improved. For many years they were the most important quartz calibers for the Swiss made watches of the Swatch Group.

In 1985, ETA developed a rectangular ana-digi LCD digital watch (caliber 988.331: 25.5 mm x 22.0 mm) with a height of 3.80 mm. An LCD display replaced the dial. The display was available in several colors and could even be made invisible. The biggest challenge was to pierce a hole into the LCD for the stem. This caliber was used by Breitling, Omega, and Rado.

Low-priced and High-Priced Quartz Movements

In 1981, EEM started distributing the calibers 926.301 and 926.311 (with day/date and alarm). Their dimensions were 11½ lines x 3.6 mm. The Lavet motor made a step every 2 seconds. ETA adopted them for its program, but it bore the ESA stamp. It was produced in very small series until 2008 for use in table clocks and travelling alarm clocks.

The 5½ x 6¾ lines x 2.95 mm caliber 927.101 was also designed by EEM and adopted by ETA for integration into its program in 1982. This was the two-hand version of caliber 927.001 with a 5 second stepping motor and a stop second device. Caliber 927.401 – also a 5½ x 6¾ lines movement, featured a calendar disc that increased the height by 0.5 mm.

The biggest calibers of this family were 947.111 (3 hands, date, 11½ lines x 3.10 mm) and 947.121 (3 hands, day/date, 11½ lines, 3.80 mm). The 13 lines x 3.80 mm caliber 947.131 with date and day written in full was even bigger. The common denominator of calibers 926 to 947 was a Marinium plate. As already mentioned, this material consisted of the plastic polyphenylene sulfide (PPS – Ryton); it was reinforced with a metal grid. The chips used in those calibers had been bonded by the chip-bonding method that was based on gold bumps on a Kapton film. It was meant to be more economical than the traditional bonding with gold wires.

The calibers 957.001 (2 hands) and 957.011 (2 hands and date) were ETA's first analog quartz movements with the alarm function. Mass production of those relatively costly calibers started in 1982, but they were not particularly user-friendly as handling was quite complicated: they did not sell well.

The height of the 9 line caliber 210.001 (1982) was only 0.98 mm; this rated as extremely flat. It was rather costly and was mostly used in gold watches. Caliber 279.001 (5½ x 6¾ lines x 1.80 mm, 1983) was designed specifically for ladies' watches. This was a very small and flat movement. Its predecessor was the ETA caliber 979.001 (5½ x 6¾ lines x 1.80 mm); it was very costly and was mostly used in gold watches.

The Big 500 Caliber Family

The 500 family included several low-cost calibers such as 536.121 (11½ lines x 5.5 mm, 1982) with day/date (this actually was a simplified variant of caliber 9362). Correspondingly, the 7½ lines x 4.3 mm caliber 551.111 with date disc was a simplified version of 951.111, while the 11½ lines x 3.70 mm caliber 554.111 with date disc had formerly been sold as caliber 954.112.

The 555 family group launched in 1982 included the round caliber 555.411 (10½ lines x 2.75 mm, with date), the variant 555.421 with day/date, and the very similar caliber 555.419; the moon phase indicator boosted its height to 3.35 mm. Day/date alone brought it to 3.25 mm; this changed the designation to 555.422. The 555 group also included 11½-line calibers: 555.111 x 2.95 mm (with date) and 555.121 x 2.95 mm with day/date.

Among the economy versions derived in 1982 from movements of the 955 and 956 families, one must mention the 7¾ lines x 2.75 mm calibers 556.111 (3 hands), 556.112 with date, and 556.121 x 3.45 mm with day/date; they were meant for ladies' watches. Caliber 556.119 (3.35 mm) featured a moon phase indicator. The round 8¾-line caliber 556.415 with date disc (2.75 mm) belonged to the same caliber class.

The 500 family also included shaped movements such as 561.001 (2 hands, 6¾ x 8 lines x 2.75 mm) with a 20-second stepping motor, and 561.101 (3 hands, same dimensions). Very large series of these calibers were sold in Asia. The two-hand caliber 581.001 and 581.011 (with a small seconds-hand) were developed by SEFEA, a French watch movement producer in Annemasse east of Geneva that Ebauches SA had acquired in 1967. SEFEA was specialized in 8¾-line calibers. The height of the calibers 581.001

8.41. ETA caliber 9362

8.42. ETA caliber 940.111

8.43. ETA caliber 926.311

8.44. ETA caliber 955.112

and 581.011 were 3.2 mm and 3.7 mm; starting with 1983, they were only sold with the ESA designation.

In 1981, ETA adopted caliber 588.001 (5½ x 6¾ lines x 3.6 mm); it had been designed by A. Schild (AS) and featured a 5-second Lavet motor. This caliber was sold without a punch declaring its origin.

Caliber 976.001 (5½ x 6¾ lines x 1.95 mm, 1983) featured a 20-second Lavet motor. The height of the more economical caliber 977.001 was 3.6 mm, while the height of the intermediate caliber 978.001 was 2.5 mm. Thanks to the 20-second stepping motor, the current consumption was only 0.5 microampere; this caliber was very successful.

ETA 979.001 was another 5½ x 6¾-line caliber, but its height was a mere 1.80 mm which made it very expensive; it was mostly used in gold watches. The more economical caliber 980.104 (5½ x 6¾ lines x 2.2 mm) featured a one-second Lavet motor. This latter caliber had a very simple end of life indicator. When the battery voltage dropped to 1.4 volts, the sweep seconds-hand made 4 second steps. Caliber 980.151 had the same specifications, but featured a small seconds-hand.

The 200 and 400 Caliber Families

In 1984, ETA launched caliber 211.101, an ultra-flat, shaped movement rounded on two sides with the dimensions 19.4 x 23.3 mm x 1.50 mm with date disc. This made it very expensive: it was mostly used in gold watches. The round 11½-line caliber 255.111 with a height of 1.95 mm also featured a date disc; the day/date variant was caliber 255.121 with a height of 2.45 mm. The round 8¾ lines x 1.95 mm caliber 256.111 was designed for ladies' watches. The day/date feature of caliber 256.121 added 0.50 mm to the height. Caliber 280.001 was a 6¾ x 3¾ lines x 2.05 mm universal movement with a 20-second Lavet motor. ETA "inherited" this caliber from Longines when its manufacturing facilities were closed in 1985. This movement was mostly sold in the United States for use in jewelry watches.

The 200 family included a few particularly interesting calibers, e.g. 201.001: This actually was the FHF miniature caliber Mini 101 (or 101.001, 4¼ lines x 2.25 mm) that ETA took over in 1985. It is still in production. Another very small caliber ETA took over from FHF was the Baguette 202.001 (3¾ x 10 lines x 3.5 mm). ETA developed it on the base of the FHF caliber 102.001. The Lavet motor made a step every 30 seconds in order to minimize the power consumption.

ETA designed the ana-digi quartz movement 958.331 for the Tissot Two-Timer: mass production started in 1986. This was a rectangular movement with the dimensions 24.0 mm x 20.5 mm x 2.0 mm. Time could be displayed both with hands and digitally on an LCD. A newly developed Lavet motor would turn in either direction. A wide choice of functions was available on the LCD: day/date, chronograph, timer, and alarm. The caliber was assembled automatically on a circular installation.

The 400 caliber family consisted of very inexpensive plastic movements that were sold in enormous quantities: some 100 million pieces were produced. In the early days, these were Swiss made calibers with the designations 400.001, 400.101 and 400.411; production started in 1985. Soon, production was transferred to ETA plants in Asia. In 1991, the 400 family was replaced by the very similar 800 family that is assembled automatically in Switzerland. These movements were designed fully in plastic, the sizes were 6¾ x 8 lines, 8¾ lines, 10½ lines, and 11½ lines; the technology was strongly influenced by the Swatch.

Caliber 450.111 (11½ lines x 3.10 mm with date disc) and 450.121 with day/date also were very inexpensive and were designed as the ultimate weapon for Swiss made watches.

Dizzying Heights

The 12½-line quartz caliber "Flatline" (Caliber 940.111) rated as a sensation, as its height was just 3.60 mm: at that time (1976) this was considered as exceptionally flat. A year later, ETA beat its own record with the 7¾ ladies' caliber 950.001: its height was 3.20 mm. This triggered the epic battle with Citizen and Seiko as to the world's flattest analog quartz watch. Existing and newly established records were literally pulverized in rapid sequence.

In June 1978, Citizen launched Exceed Gold 7900, a rectangular quartz watch with the dimensions 23.7 mm x 20.0 mm; the height was 4.1 mm – no great achievement. However, the movement's height was a diaphanous 0.98 mm, which admittedly rose to solid 2.25 mm with the battery. Five months later, in November 1978, Seiko followed with a watch the height of which was 2.5 mm; the movement (caliber 9320A) was 0.03 mm thinner than the Citizen caliber 7900. It was available commercially under the Seiko Credor brand with a round or a rectangular case; the price was a cool million yen.

The Citizen engineers responded by redesigning their caliber 7900 and equipping it with a much flatter battery. In October 1979, they had a 3.0 mm watch with a movement (caliber 7901) the height of which was 1.18 mm (including the battery).

So far, the race had been a strictly Japanese affair, but in 1978, ETA decided to break the "sound barrier" of 2.00 mm for the height of a complete watch. Gedalio Grinberg, the CEO von North American Watch had told his friend Ernst Thomke (who was then ETA general manager) in no uncertain terms that Swiss quartz movements were too thick and had too high a power consumption, which caused a short battery life. This was a challenge: Thomke immediately gave his engineers the task of developing the world's flattest watch.

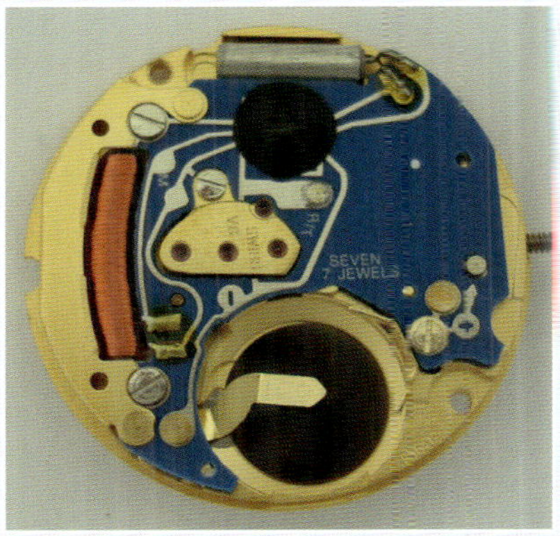

8.45. ETA caliber 956.412

The Movement Is a Watch

André Beyner, then technical manager of Ebauches SA in Neuchâtel and his chief designer Maurice Grimm liked Thomke's idea. They thought up a revolutionary watch (they called it "Delirium") that did not "contain" a movement, but "was" a movement integrated into a case. In other words the case was part of the movement. The inner side of the rectangular protective container made of 18 carat gold was machined to serve as the main plate. The movement was mounted from bottom to top all the way to the dial and hands. One just had to add a gold bezel with a wafer thin sapphire plate (0.24 mm) to close the watch. It was discovered much later that this design had been invented and patented in 1890 – for mechanical watches.

8.46. ETA caliber 210.001

In order to create space for the hands, the center of the dial was bent inward – a truly original idea. The thickest components had to be located at the periphery, i.e. the stepping motor (1.45 mm), the battery (0.9 mm), the coil (1.7 mm), and the quartz-resonator (1.4 mm). The integrated circuit was very flat anyway.

The samarium cobalt rotor had a ruby bearing on both sides of its arbor. These rubies were pressed into brass bushings, that themselves were pressed into the motor's stator. The five wheels of the gear-train were all flying, there were no bridges. Their arbors were recessed so they could rotate around stationary pins with a flat head similar to a nail's head; this prevented the wheels from moving out of the pins. The losses due to friction were rather high, but in this case, a short battery life was acceptable.

8.47. ETA caliber 256.111

Delirium I, II, III, and IV

The most challenging issue was to design a functioning watch based on the Delirium concept: ETA solved this problem. A team of engineers led by Anton Bally literally worked 24-hour days for weeks on end – and succeeded. The world's flattest quartz watch was officially named "Delirium Tremens" or "Delirium I" (caliber 999.001). It was shown to the media on

8.48. ETA caliber 988.333

14 January 1979 in Tokyo, Munich, Milano, Paris, and New York. The Delirium could easily be pulled through a sliding gauge set at 2.0 mm: its overall height really was just 1.98 mm. The cost was around $8000. The "launching carriers" were Concord, Eterna, and Longines; Barthelay, Chopard, Corum, DeLaneau, IWC, Loyal-Time, Movado, Sarcar, and Schlegel & Plana followed.

However, development of the Delirium continued. The height of the Delirium II was only 1.44 mm: it was also made into a skeleton watch. Omega quite independently developed its "Dinosaure" in 1980. Its height was 1.48 mm, which made it almost identical with that of the Delirium II. The Omega Dinosaure followed the "Mystérieuse" concept: its center was fully transparent, with "floating" hands, but it could be fitted with a gold dial.

In the early 1980s, the Delirium group developed the slightly smaller ladies' Delirium III with a height of 1.68 mm. Of course Seiko engineers were not sleeping during these hectic times. In February 1980 they had a 1.54 mm quartz watch with a 0.89 mm movement.

Delirium IV: The Winner

The Delirium IV with a height of 0.98 mm followed, still in 1980. In order to shed a full millimeter from the Delirium I, there were no hands and no dial. They were replaced by two wafer-thin sapphire discs on which hands were stamped; they were actuated at the periphery. Just two wheels were needed; the height of the battery – a special Renata construction, was 0.80 mm. It was inserted into the back without a cover.

The Delirium IV had to be worn very carefully so it would not warp. Some 50 to 60 pieces were made for Concord, Eterna, and Longines, the price was $60,000. ETA engineers worked on a Delirium with a total height of 0.5 mm, but this work was stopped as it became clear that Seiko and Citizen had given up the race. A few 0.5 mm "Deliriums" were actually made but they were a public relations gag, as they were miniaturized hourglasses. The Delirium prototypes are preserved for posterity at the ETA Museum in Grenchen that unfortunately it is not accessible to the public.

The cases of the various Delirium models were made of yellow gold, white gold, or platinum. Milling the case back was extremely difficult, particularly in platinum. Polycrystalline diamond tools that are ideal for machining platinum did not exist yet.

It is interesting to note that Seiko considered the ETA victory as an unacceptable disgrace and kept working on a still thinner watch. The result was revealed in 1989: the height of the Seiko caliber 9A85A was 0.85 mm and so beat the Delirium by 0.13 mm – without the battery. If this was measured also, the total height grew to 1.43 mm – a pyrrhic victory. Furthermore, the 9A85A was just a movement and not a complete watch, which was the case with the 0.98 mm Delirium IV.

Fabrique d'Horlogerie de Fontainemelon (FHF)

The Ebauches Holding (Ebauches SA or ESA in Neuchâtel), was created in 1926 by the merger of the three most important manufacturers of blank movements or ebauches. They were Fabrique d'Horlogerie de Fontainemelon (FHF), A. Schild (ASSA Grenchen,) and A. Michel AG (also in Grenchen). This was a rather loose conglomerate: its members still aggressively competed with each other. According to the statutes, ESA had important functions in the field of market watching and coordination. Yet, relationships between the three members of the holding were unsatisfactory and everybody was doing his own thing.

An Innovative Ebauches Manufacturer

The central R & D facilities of Ebauches SA in Neuchâtel were financed by the members, but it had very little influence on them. In 1970, Peter Döme was hired as a research coordinator in order to eliminate the numerous duplications within the group. In 1975, the ASULAB was founded, a research laboratory that would work both in the field of classical watch technologies (under André Beyner) and the new disciplines such as microelectronics and optoelectronics (under Hugo Wyss).

FHF Fontainemelon (whose CEO was Denis Robert) was known for many years for its universal, hand-wound 6¾ x 8-lines caliber ST 69; the oscillation frequency of the balance wheel was 3 Hz (21,600 vibrations per hour). The other important FHF caliber was the 11½-line caliber ST 96, also manually wound. FHF was a vertically integrated group within Ebauches SA: it consisted of a big plant in Fontainemelon, Ebauches Fleurier, Ebauches Peseux, a profile turning plant in Corgémont, a pinion plant in Fontaines and movement assembly in Isérables (Valais); employment was about 1200.

A most remarkable achievement of FHF was the automatic assembly of the mechanical 6¾ x 8 caliber ST 69. This line had been designed in cooperation with the French Yema and the Kelek Company in La Chaux-

8.49. ETA caliber 999 Delirium I (1.98 mm)

8.50. ETA Delirium IV (0.98 mm) in the sliding gauge

8.51. FHF caliber 935.112

8.52. FHF caliber 961.001

de-Fonds. By 1973, the line consisted of a system of 20 horizontal transfer machines that permitted production on a huge scale; they cost nearly a million Swiss francs apiece.

Later on, specialized machines were built that produced brass plates or bridges; they directly fed the assembly lines. The transistorized controls were developed by a highly specialized section within Ebauches SA. The machines were running in streams of oil that was continuously being filtered and recycled in the basement: the oil inventory was 180,000 liters. Nobody realized that this was a time bomb until the oil caught fire in September 1995: the plant was totally destroyed, but it was rebuilt in record time.

First Quartz Calibers

Beginning in 1975, sales of mechanical movements literally collapsed and the situation for FHF was becoming critical. ESA arranged for FHF to take over the production of analog quartz movements in the sizes 6¾ x 8, 8¾, 10½ and 11½ lines. Quite naturally, FHF adopted the established quartz technology that has not changed significantly up to now. That meant a silver oxide battery, a tuning fork quartz resonator, a CMOS integrated circuit and a Lavet stepping motor.

Daniel Paratte was the engineer who designed the first FHF quartz calibers. The bar-shaped flexural mode quartz resonator was supplied by the Ebauches SA subsidiary Oscilloquartz. As this type of resonator was sensitive to shocks, it was soon replaced by the Japanese NDK tuning fork-shaped resonator, whereupon the FHF technical director René Widmer was accused of high treason. But Widmer ignored this, as he was only interested in the quality of his products.

The electronic module with the oscillator, the binary frequency divider, and the amplifier could be purchased ready-made and potted in plastic from EEM in Marin. The Lavet stepping motors were developed by Eric Klein, a specialist who was transferred from Ebauches SA to Fontainemelon.

Starting in 1976, FHF Fontainemelon was in the market with round quartz calibers of the 935 family, i.e. the two-hand caliber 935.002 (8¾ lines x 3.90 mm), 935.102 with sweep seconds-hand (8¾ lines x 3.90 mm), 935.111/112 (similar to 935.102 but with date disc which brought the height to 4.50 mm), and 935.121/122 (8¾ lines x 5.00 mm) with day/date. In 1977/78, FHF also produced the 10½-line calibers 935.311/312 (three hands, 4.50 mm, date) and 935.322 (similar to 935.312 with day/date, 5.0 mm).

8.53. FHF caliber 963

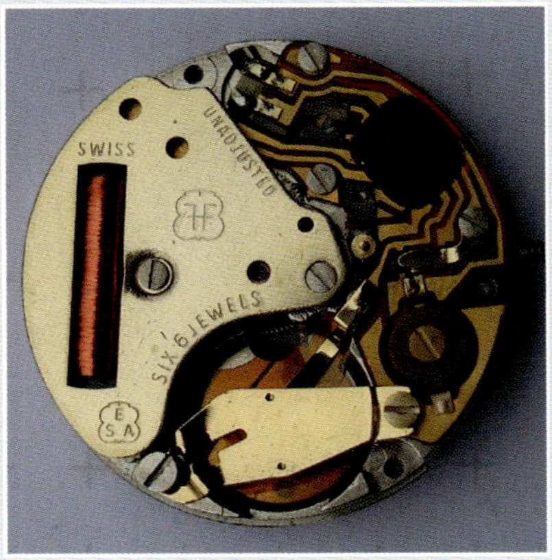

8.54. FHF caliber 965

8.55. FHF Baguette 102.001

8.56. FHF Mini 101.001

FHF's great hope were the 6¾ x 8 calibers 961.001 (two hands) and 961.101 (three hands, sweep seconds-hand, 6 pulses per minute), both with a height of 3.50 mm. They were launched in 1978, could be used in both ladies' and men's watches, and were meant to replace the mechanical ST 69. Caliber 961.001 was produced by ETA in Asia and sold without a logo, just with the caliber designation. Many Swiss watch producers thought the height of the 961 calibers was excessive. A slimmer alternative became available in 1982: the 579 calibers with a height of 2.75 mm. But EEM Marin and not FHF developed and produced them.

The 965, 960, and 963 Families

Caliber 965.001 (8¾ lines x 3.20 mm) with two hands and 6 pulses per minute was designed for ladies' watches; it was launched in 1978. The calibers 965.101/102 were equipped with three hands, variants were 965.111/112 (8¾ lines) with date or day/date, respectively. The heights were 3.80 mm and 4.30 mm.

The 10½-line caliber 965.311 with three hands and date disc had a height of 3.80 mm; the height of the day/date variant 965.321 was 4.30 mm. The two latter calibers were launched in 1979.

In 1979, FHF Fontainemelon launched the two-hand baguette quartz movement 102.001 (9.0 mm x 21 mm, two pulses per minute); the height was 3.50 mm. In the following year, the 11½-line caliber ESA-FHF 960.111 became available. The dimensions were identical with those of the mechanical caliber ST 96. This three-hand caliber with a 10 second stepping motor as well as the 961 calibers kept the automatic transfer machines busy. A considerable reduction of staff or a partial closure of FHF thus could be avoided.

Caliber 963.111 (11½ lines, three hands and date disc) and the variant 963.121 with day/date both had a height of 3.85 mm. The variant 963.123 had the same dimensions, but the power consumption was reduced by 30%.

Bonding on Kapton Film

An original technology for bonding the chips in the electronic modules of the 960 and 963 families was developed by FHF in cooperation with the Cicorel Company in Crissier (Vaud). Perforated 35 mm Kapton tapes were the starting point. The perforations were similar to the ones found on the classical photo film and allowed a correct and precise indexing of the tapes on which a pattern of hemispherical metal drops had been made. The chip was placed on the tape so that the "bumps" touched the contact pads of the chip. Soldering was done by ultrasound. When ETA took over FHF it was decided that the tape automated bonding was not economical. The classical bonding with gold wire was reintroduced.

The smallest FHF-Quartz was the Mini 101.001, a 4¼-line caliber with a height of 2.0 mm (1979). It was used in ladies' watches of the luxury class. The first series cost 1500 Swiss francs apiece. The 2.25 mm variant 201.001 was launched in 1982. ETA ended up adopting this caliber in 1983; it is still manufactured with the same caliber number.

When ASUAG and SSIH merged in 1983, all Ebauches SA subsidiaries that were not profitable and/or indispensable were closed. This was the fate of the ebauches manufacturing plants Bettlach, Peseux, Tramelan, Tavannes, Fleurier, and Venus in Moutier. What was left was integrated into ETA: FHF changed to "ETA Fabrique 20."

Girard-Perregaux

Girard-Perregaux (GP) was the result of the merger of the Geneva Company J.F. Bautte & Cie and Girard-Perregaux in La Chaux-de-Fonds in 1906. Together with his brother Numa, Constant Girard (1825-1903) had established Girard & Cie in 1852. After marrying Marie Perregaux, who was born in a family of highly distinguished watchmakers, Constant Girard changed the company name to Girard-Perregaux (GP).

Girard-Perregaux was an extreme perfectionist and always tried to realize technical function in a highly esthetic way. He also was very fond of pink gold (with 20.5% copper and 4.5% silver). He not only used it for cases, dials and crowns, but also for plates, bridges, springs, wheels, escapement parts and balance wheel screws.

The World Exhibition of 1889 in Paris granted Girard-Perregaux a gold medal for his masterpiece "Tourbillon sous Trois Ponts d'Or". On the back of the movement, three bar-shaped bridges were arranged in perfect symmetry. Their function was to bear the jewels for the barrel, the minute wheel and the tourbillon, respectively.

The Graef Dynasty

Louis-Constant Girard-Gallet, eldest son of the founder Constant Girard-Perregaux, succeeded his father at the helm of the company in 1903. Unfortunately, he was not up to mastering the challenges of 1918, i.e. a severe economic crisis and the collapse of the Russian market after the October Revolution. In 1928, Girard-Perregaux was sold to the watchmaker and jeweler Otto Graef (1862-1943). His sons Willy (1893-1984) and Paul (1894-1981), their sons-in-law Charles-Edouard Virchaux (1916-1994) and Jean-Edouard Friedrich (1912-1999) led the Company to new heights: watch production grew to 200,000 pieces a year.

In 1966, Virchaux hired Georges Vuffray who had just graduated from engineering school. His assignment was to develop a quartz wristwatch for GP. Vuffray succeeded with just one colleague and up to four technicians (see Chapter 7). The GP 350 quartz caliber (30.0 mm x 7.45 mm) featured a quartz resonator from Reeves-Hoffman, an integrated circuit from Motorola and a Laplace motor built in-house. Production started in 1971, the caliber was quite successful.

The next calibers GP 351, 352, and 354 were only slightly modified. However Caliber 353 was definitely thinner: its dimensions were 30.0 mm x 5.80 mm. The height reduction by 1.66 mm was obtained by mounting the quartz resonator sideways. In the early days, Jaeger-LeCoultre produced the mechanical parts; by 1978 GP had become self-sufficient in that respect.

GP-LED

Having successfully developed the analog quartz watch family GP 350, the electronics group was given the assignment to design a digital quartz watch of the luxury class. The LED display that had been pioneered by the "Pulsar" of Time Computer was chosen. It is esthetically much superior to the liquid crystal display (LCD), but time is only displayed on demand upon pressing a key.

The "GP LED" caliber 395 (25.9 mm x 25.6 x 7.30 mm) with a vertical four-digit LED was first shown to the media in 1976. It was powered by two batteries. The functions were hour, minute, second, month, and date. The case and the bracelet consisted of Makrolon plastic that was styled in an ultramodern way. Gold plated brass and stainless steel were also available: the plate was plastic.

This caliber was characterized by eleven transistors forming a neat row on one side of the movement. They were used for controlling the LED that needed several milliamperes: it was not possible to integrate them in one circuit. Eleven amplifiers were needed: four of them for the choice of a digit and seven more for switching on and off the 7 segments of the display. The display was sequential, one digit after the other, but at such a high rate that the eye could not follow the motion.

Pressing the pusher on the right switched on the display of hour and minute, pressing twice displayed day and date, pressing three times displayed the seconds for a minute. All the displays could be set with the pusher on the left. Calibers 395 and 397 were slightly modified variants.

The "casquette" caliber GP LED of the 390 family was a purely GP design. Similarities with the ESA caliber 9292 are purely accidental; GP never worked with Ebauches SA. The LED display was supplied by the American company Litronix. A similar caliber with an LCD display never made it beyond the prototype stage, there was no serial production.

The GP-LED was quite successful in the early days, but the market deteriorated later on, particularly in view of the standard digital watch whose price literally collapsed. Yet, parts are still available for servicing the GP-LED and all its variants.

Analog Calibers

The digital technology was just an episode; GP microelectronics were centered on analog watches. In 1976, GP launched the strongly miniaturized 5¾-line caliber GP 520 (13.40 mm x 2.95 mm). It was quite a challenge to develop such a small two-hand caliber for a ladies' watch. The jewelers insisted on getting it, but serial production was disappointing, there were many returns because of malfunctions.

However, caliber GP 510 (15.50 mm x 1.95 mm) launched in 1980 performed very well. As to the motor, Vuffray's Laplace motor was replaced by the simpler and less costly Lavet motor starting with the caliber family 500. Lavet motors were available from Seiko. Photolithographic quartz resonators (Statek type) were used instead of the flexural mode resonators from Reeves-Hoffman when they became available from Micro-Crystal in Grenchen in 1979.

Statek-type resonators are recognized easily, as they were packed in a 6 mm x 2 mm metal tube. The origin of the integrated circuits used in the second generation of GP quartz watches is not always known. This was the time of changing over from Motorola to an European manufacturer, which probably was Eurosil, but ITT is not excluded.

The only caliber of the 600 family was 641; production started in 1978. This was a 9 line caliber with the dimensions 20.50 mm x 3.70 mm. It was mostly used in men's watches. Unfortunately, grave problems occurred with some of the plastic parts that aged rapidly and literally disintegrated.

The 700 caliber family was launched towards the end of the 1970s and included the calibers 700, 705, 711, 721, and 731. At that time, GP was owned by Desco von Schulthess. Sales plummeted; total watch production was on the order of 3000 pieces. GP produced quartz watches of the finest quality, but most markets did not accept them. The watch brand Maurice Lacroix also belonged to the Desco group. For this reason, GP produced the 700 calibers in two styles, one for its own use and one for Maurice Lacroix. They were technically identical; the difference was simply the decoration of the movements.

The calibers of the 711/721/731 family were very similar, only the diameters differed. The dimensions were as follows: caliber 711: $7\frac{3}{4}$ lines x 2.50 mm; caliber 721: $8\frac{3}{4}$ lines x 2.50 mm; caliber 731: $10\frac{1}{2}$ lines x 2.50 mm). The functions were hour, minute, second, and date. In 1986, due to computerized bookkeeping, they were renamed to 7100, 7200 and 7300; the technical specifications were left unchanged. Calibers 700 ($9\frac{1}{2}$ lines x 4.20 mm) and 705 ($11\frac{1}{2}$ lines x 4.20 mm) had many parts in common.

The further development of the GP quartz technology led to the calibers 360, 380, and 382 in 1980. The diameter of caliber 360 was 17.2 mm ($7\frac{3}{4}$ lines); the height was 2.5 mm. It featured 8 jewels and displayed hour and minute. The hour-hand could be set in steps of one hour when changing time zones. The Lavet motor received a current pulse every 5 seconds from the CMOS circuit. A trimmer capacitor was used for the fine tuning of the quartz resonator's frequency. The three-hand caliber 380 (27.0 mm x 2.5 mm) also featured a date disc; the Lavet motor received current pulses every second. Caliber 382 was slightly smaller (25.60 mm x 2.5 mm) than caliber 380, but had the same functions; this was a 9 jewel caliber.

In the early 1980s, Girard-Perregaux produced the very sophisticated "Equation" watches with complications such as chronograph, perpetual calendar, signs of the zodiac, and moon phases. A quartz movement caliber 731 was used as a motor for the non-modular mechanical complications. There were models with three hands, three hands and moon phases, three hands and double moon phases (for the northern and southern hemispheres), signs of the zodiac, seasons, months, solstices, and equinoxes combined with a perpetual calendar. The 641 variant displayed the date by means of a pallet stone. Those calibers were extremely well done, even the battery cover was decorated with Geneva stripes. They could be personalized with the owner's initials.

The Equation Ladies' watches were driven by quartz movements of the 600 family, but had nothing in common with the old caliber 641. Three variants were available: with seconds-hand and

8.57. GP quartz caliber 350

8.58. GP caliber 352-544

8.59. GP LED caliber 395, front side

8.60. GP LED caliber 395, back side

date, with moon phases and the zodiac, and the "Equation Espace" with seasons, month, solstices, and equinoxes – all on a polychrome disc. There were no pushers, everything could be set with the crown that had three positions.

In 1998, Francis Besson acquired Girard-Perregaux; by 1992 the company was in the black. He then sold it to Luigi Macaluso (1948-2010), a businessman from Turin. Macaluso put the main accent on mechanical calibers. Yet, in 2004 a whole series of state of the art quartz movements was designed for GP's own needs in ladies' watches and the OEM market for high-price watches. Three new GP basic movements are presently available in a number of variants.

Girard-Perregaux's quartz caliber GP 350 celebrated its 40th birthday in 2010. For this occasion, the new 11½-line quartz caliber GP 13500 was unveiled at the Geneva luxury watch show SIHH. This movement is a very robust all metal construction with a conventional electronic part: 32 kHz quartz resonator, division to 1 Hz, Lavet motor for driving the three hands plus date disc and a silver oxide battery.

Golay µ-Quartz

Jean-Claude Bornay and André Guignard at the Bernard Golay Company in Lausanne, Jean-Pierre Golay at the Record Watch in Tramelan, and Roland Dubois at Les Assortiments Réunis in Le Locle presented their common development at the 1972 Basel Watch Fair and at the 47th congress of the Société Suisse de Chronométrie later in the year. They had designed the Golay µ-Quartz, a watch with a "soul."

An interesting word for what actually was a quartz synchronized electrodynamic watch: the balance wheel motor was driven by the 4 Hz signal that was obtained by dividing the 32 kHz quartz frequency. The balance wheel consisted of two discs with four permanent magnets.

Using the word "soul" certainly was a clever marketing gag. However, a mechanical or an electromechanical movement has about as much soul as a car's gearbox. Presumably the tick-tack of the escapement was supposed to mean soul, while the tick-tick-tick of the stepping motor in an analog quartz watch was defined as soulless.

Everybody is entitled to his own definition of soul, but a realistic one would be the sum of the activities of the central nervous system. There certainly are fascinating analogies between the switching processes among neurons and the ones in a silicon

8.61. GP caliber 705

8.62. GP caliber 738

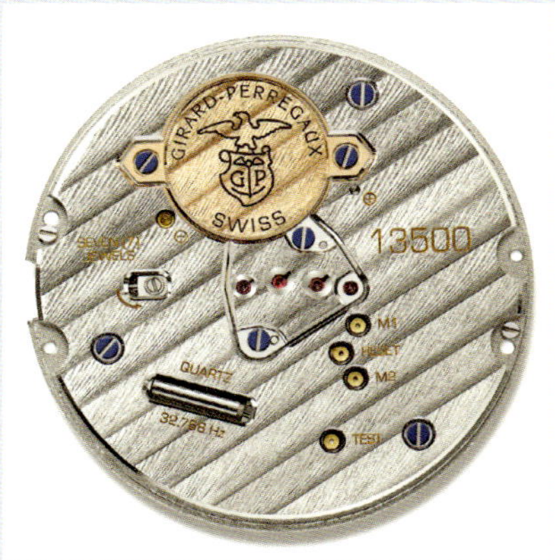

8.63. GP caliber 13500

8.64. Golay µ-Quartz caliber 7743

microprocessor. Everybody has learned that en ensemble of microprocessors can act in an irrational way, but this rarely happens with a watch, as the circuits are relatively simple.

The Golay μ-Quartz was launched in 1973 by Corum, Helvetia and Record: it looked attractive and interesting. The 13½-line movement with a height of 5.6 mm resulted in a big watch. The date was set with a lateral crown; an extra crown on the bottom of the case was needed for setting the hands. The watch had one major disadvantage: due to the combination of electromechanical parts and electronics, it was rather expensive. It was difficult to sell at a time when Far Eastern quartz watches (no soul, but many extra functions) were available at a much lower price. The Golay μ-Quartz displayed only hour, minute, second, and date; barely 3000 pieces were sold. Then production had to be stopped, as there was no demand. Golay went bankrupt soon afterwards.

Heuer

For decades, Heuer was mainly known for its chronographs. At the 1972 Basel Watch Fair, Heuer unveiled the world's first LED-digital pocket stopwatch. The "Microsplit 800" weighed just 100 grams, the functions were start-stop, split, and lap to the 1/100 of a second for 24 hours minus 1/100 second. It cost 1500 Swiss francs; just a year later, the price had dropped to 590 Swiss francs. In 1974, the LCD version Microsplit 420 was launched; it cost just 190 Swiss francs. The Microsplit 417 achieved a resolution of 1/1000th of a second. Heuer was a true pioneer, but the early involvement in the semiconductor technology meant that the company had to go through the entire learning curve, a costly undertaking. The products rated as sensational, but for a long time the profit margin was near zero.

Heuer CEO Jack Heuer was fully convinced that the future belonged to the electronic watch. In 1975, he launched the world's first LED-LCD quartz chronograph "Chronosplit." Short times (up to 9 hours) could be displayed with a resolution of 1/100 of a second on the LED. The LCD only displayed hour, minute, and second; at that time liquid crystals were too slow for the 1/100 second.

The "Chronosplit Manhattan GMT" was launched in 1977 and made big waves. The case was an elongated steel hexagon with a three-hand quartz movement and a date disc (ironically this was the Citizen caliber 8515; there was nothing equivalent

8.65. Heuer Microsplit 302

8.66. Heuer Chrono LCD Ladies' model caliber 01

8.67. Heuer Chronosplit LCD-LED caliber 100, front side

8.68. Heuer Chronosplit LCD-LED caliber 100, back side

available in Switzerland), plus a digital quartz movement with LCD. The latter displayed stopped time to the 1/100 of a second or optionally the date.

At the 1978 Basel Watch Fair, Heuer launched a true ana-digi wristwatch with stopwatch functions. The "Carrera Twin" Model 371 included a two-hand analog display and a 4-digit LCD displaying hour, minute, second, a second time zone, and stop time up to 24 hours minus 1/100 of a second with a resolution of 1/100 of a second and the functions start-stop, split, and lap. The resolution of model 413 was 1/1000th of a second.

Jaeger-LeCoultre

The quartz revolution significantly affected Jaeger-LeCoultre (JLC) in the Vallée de Joux. The return to stability was only achieved after the German VDO company had acquired JLC (together with IWC) in 1978. Considerable investments were necessary in order to introduce the quartz technology; JLC would not have been able to self finance this project.

JLC's first quartz caliber was the CEH caliber Beta 21, of which JLC received 200 pieces. This was followed by the three-hand caliber GP 350 by Girard-Perregaux launched at the 1971 Basel Watch Fair, two years before production started. JLC gave it the caliber designation 350. The calibers based on the GP 350 that JLC integrated into its collection also kept their GP designation. Jaeger-LeCoultre manufactured the mechanical part (plate, bridges, wheels) anyway, as Girard-Perregaux was not at that time equipped for this kind of production.

GP caliber 350 with a diameter of 30 mm and a massive height of 7.2 mm featured a bipolar integrated circuit from Motorola and a bar-shaped quartz resonator from Reeves-Hoffman (USA). The so-called Laplace-Motor was a construction of GP electronics engineer Georges Vuffray (see Chapter 7). This movement was powered by a mercury oxide battery. Jaeger-Le Coultre used caliber 350 as well as the later calibers 351, 352 (in 1972) and 353 (in 1973) for its watches of the "Master Quartz" collection.

JLC always had been a fully verticalized manufacturing facility. Wanting to be independent in every respect, JLC set up its own electronics department. This is where the 600 family of quartz calibers was designed, with the help of a free license to VDO patents. As already mentioned, VDO had taken control of the Saphir Group with JLC and Favre-Leuba in 1978. VDO itself was controlled by the Schindling

8.69. Heuer Chronosplit LCD caliber 102, front side

8.70. Heuer Manhattan GMT caliber 104, front side

8.71. Heuer Manhattan GMT caliber 104, back side

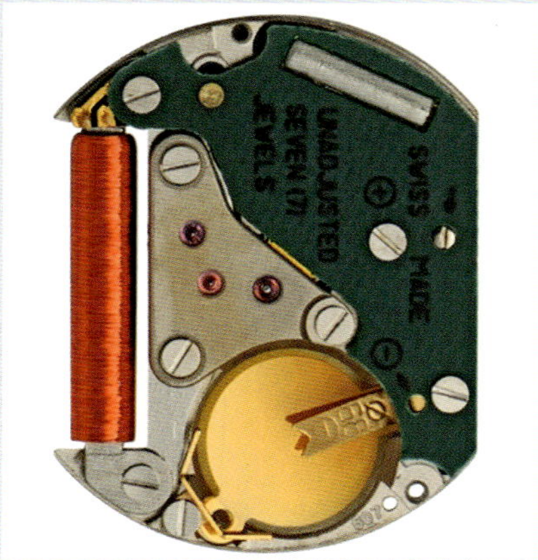

8.72. Jaeger-LeCoultre caliber 607

family, many members of which just loved watches. For this reason, they had planned to add a watch division to their company for quite some years.

In 1966, VDO started developing an analog car quartz watch; prototypes were available two years later. It was called Q4 and mass production started in 1970. Q4 was very successful and was modified for use in wall clocks and alarm clocks. The next objective logically was the wristwatch: the start was made in 1977 with a participation in Jaeger SA in Paris that controlled Jaeger-LeCoultre in the Vallée de Joux. The sequel was the acquisition of IWC in Schaffhausen in 1978.

In VDO's opinion, the very aggressive Japanese competition could only be met by a company with European dimensions. This led to the creation of the watch holding Matra/VDO that included the following members: Jaeger SA, Jaeger-LeCoultre, Favre-Leuba, VDO Quartz Zeit, Garant, IWC, Mauthe, Bayard, JAZ, Yema International, and Framelec. This was going to be the third largest watch group in Europe, after ASUAG and SSIH.

While the merger was being prepared, it became apparent that some of the companies of the future group were in serious financial trouble. Saving them would have required a huge amount of capital that VDO was not willing to provide. Matra and VDO formally separated in 1981 and went their own ways. On the VDO side, the cooperation with Jaeger-LeCoultre from 1980 to 1989 resulted in the design of a very important family of quartz movements with photolithographic quartz resonator and Lavet motor. This was the "high end" of quartz technology with the following calibers in chronological order:

8.73. Jaeger-LeCoultre caliber 608

8.74. Jaeger-LeCoultre caliber 611

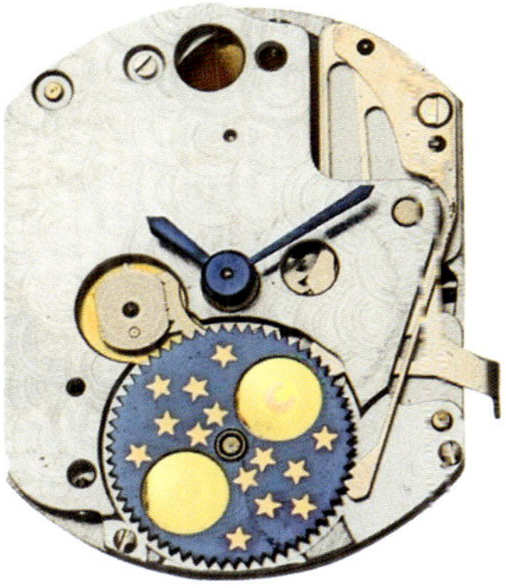

8.75. Jaeger-LeCoultre caliber 617

8.76. Jaeger-LeCoultre caliber 619

602: 17.2 mm x 13.5 mm x 1.98 mm, extra flat two-hand shaped movement (1980).

603: 17.2 mm x 13.5 mm x 2.50 mm, economical model of caliber 602 (1980). At least 100,000 pieces were produced with the Lavina brand and sold to numerous watch brands. 7 jewels, 10 second stepping motor, digital frequency tuning, tuning fork quartz of the Statek type.

605: 25.94 mm x 2.25 mm, round, two-hand movement with date disc and one-second Lavet motor (1981).

606: 25.94 mm x 2.25 mm, three-hand version of caliber 605 with sweep seconds-hand and date disc (1982).

601: 11.7 mm x 9.7 mm x 1.8 mm, extra flat two-hand, shaped movement with 3 jewels and one-minute stepping motor (1982). It is characterized by a curved coil on the side of the movement. Caliber 601 was designed for the Mini-Reverso; the hands were set with a pusher at 3 o'clock.

608: 17.2 mm x 13.5 mm x 1.6 mm, shaped movement for the Reverso Ladies' model with 7 jewels. Battery life was extended to 2.5 to 3 years (1983).

607: Slightly simplified version of caliber 608 with the same dimensions and a plastic bridge. Was also produced for Van Cleef & Arpels (1983).

609: Economy model of caliber 608 with the dimensions 17.2 mm x 13.6 mm x 1.6 mm. Caliber 609 was sold under the Lavina brand and was not formally part of the Jaeger-LeCoultre collection. 609 was launched in 1983 and discontinued in 1989.

628: 20.8 mm x 2.35 mm, round movement with moon phase indicator, small seconds-hand and 24-hour display for a second time zone (1984).

610: Round two-hand caliber 16.36 mm x 1.9 mm with date disc, 8 jewels, five-second stepping motor (1985).

611: Three-hand version of caliber 610. The sweep seconds-hand served as a battery life indicator, steps of 4 seconds meant that replacement was needed.

612: 20.7 mm x 1.9 mm, round two-hand caliber with date disc, derived from the smaller caliber 610 (1985).

613: Three-hand version of caliber 612 (1985).

617: Derived from caliber 607, 17.2 mm x 13.5 mm x 2.2 mm, two-hand movement with moon phase indicator (1986).

618: Derived from caliber 608, 17.2 mm x 13.5 mm x 2.2 mm, two-hand movement with moon phase indicator (1986).

619: Derived from caliber 617 with moon phase indicator and calendar by extra hand, 17.2 mm x 13.5 mm x 2.48 mm (1986).

620: 17.2 mm x 13.5 mm x 2.48 mm with moon phase indicator and calendar by extra hand (1986).

621: 15.82 mm x 2.35 mm with small seconds-hand at 6 o'clock (1988).

615: Two-hand movement, derived from caliber 605. 28.00 mm x 2.25 mm (1989).

616: Two-hand movement, derived from caliber 606. 25.94 mm x 2.25 mm (1989).

623: Derived from caliber 621, 15.82 mm x 2.35 mm, with date and moon phase indicator (1989).

The 630 family "Mecaquartz" featured mechanical complications that were driven by a quartz movement. This was not a modular construction; both technologies were fully integrated in order to obtain a minimum height. Calibers of this type are still in production.

630: One of the smallest chronographs 23.68 mm x 3.7 mm with two counters (30 minutes and 12 hours), date at 4.30, small seconds-hand at 6 o'clock, optional moon phase indicator. This caliber is still in production and used in the Master Compressor ladies' model. (1987).

631: Same dimensions as caliber 630, no moon phase indicator (1987).

632: Only prototypes were made, the objective was a Mecaquartz chronograph with alarm, but the caliber saw no serial production.

633: Presumably only one prototype was ever fabricated; this also was a chronograph with an alarm function (1991).

Polara Dress

In the mid-1970s, a JLC watch with an LED display existed. It was called Polara Dress LED; the electronic module with 5 functions was supplied by Hughes Aircraft. JLC does not have any information on this caliber. Presumably it was developed by the JLC U.S. representative on its own initiative. At that time, a watch company was only credible in the United States if it also had watches with digital movements.

Jean d'Eve

The Le Phare manufacturer was founded in La Chaux-de-Fonds in 1888 by Charles-Ami Barbezat. It is known mostly for its very innovative brand Jean d'Eve. Such watches were equipped with quartz movements supplied by outside vendors at an early date, as their accuracy was appreciated. However, it was considered irritating that a new battery was needed every 2 to 4 years. For this reason, Jean d'Eve decided to develop a quartz watch without a battery.

The result was the "Samara." This truly novel watch was unveiled in 1988 at the Basel Watch Fair to celebrate the 100th anniversary of the company. Its heart was a microgenerator with a diameter of 8.8 mm and a height of 2.7 mm. Production started several years later due to problems with the Samara movement.

The microgenerator consisted of a 14-pole magnetic rotor and a stationary coil. The rotor was driven by a flywheel that was similar to the rotor of an automatic movement. The motions of the wrist were thus used to generate electricity that was used to charge a long-life lithium ion battery.

It is well known that the yield of a microgenerator is very low at low rpm. In order to improve on that, the old principle of intermediate energy storage was applied. The rotor did not actuate the generator directly, but bent a spring. When a given tension was reached, the spring discharged very rapidly and turned the generator's rotor by 180 degrees – at a speed corresponding to 15,000 rpm. With a fully charged battery, the Samara could work for several months.

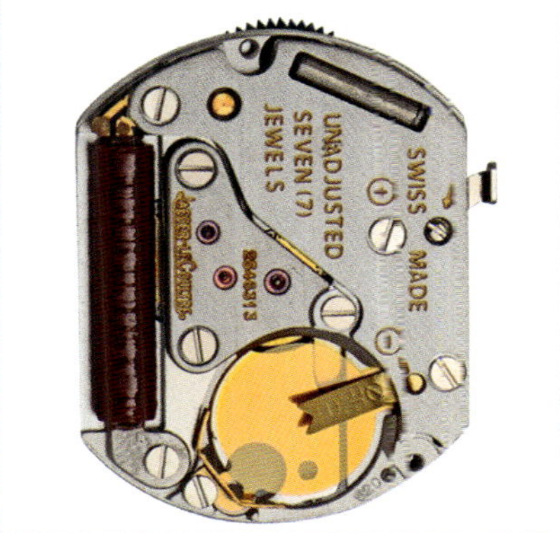

8.77. Jaeger-LeCoultre caliber 620

8.78. Jaeger-LeCoultre caliber 633, movement side

8.79. Jaeger-LeCoultre caliber 633, front side

8.80. Samara by Jean-d'Eve

Longines

After the launch of the "Ultra-Quartz" (caliber L.6512) in 1969 (see Chapter 7), nine years passed before Longines made a new start to design and produce its own quartz movements.

Analog Calibers

The mid- to late 1970s saw the establishment of the "classical" quartz technology that has not been significantly changed since. It is used by practically all manufacturers of quartz movements in Switzerland, India, China, and Japan. The major components are a 32 kHz tuning fork quartz resonator of the Statek type, a Lavet stepping motor, and a silver oxide battery. Caliber L.950 that Longines launched in 1978 does indeed follow the above scheme. The diameter was 25.6 mm (11½ lines), it featured a sweep seconds-hand and a date disc. The height was 2.95 mm; this made it the World's flattest quartz movement.

The shaped caliber L.960 (13.00 mm x 15.30 mm x 2.30 mm) was launched in 1980 and was produced in two extra variants with slight adjustments of the dimensions. They were meant for ladies' watches. Caliber L.961 (9.00 x 15.00 mm) "Baguette" was characterized by a height of 1.75 mm. So was variant L.962 with the dimensions 12.00 x 15.00 mm. The functions were the same as with L.950, the technology was identical.

Longines launched the L.970 caliber family, with a height of 1.95 mm, in 1981. These calibers were modular, which permitted the production of various models with different functions while always using the same basic caliber. Caliber L.970 (with sweep seconds-hand) and L.974 11½ lines (no seconds-hand) were round, while calibers L.976, L.977 and L.978 were 8¾ x 10½ lines shaped calibers. They either had a sweep seconds-hand or a small seconds-hand. Fast setting of date and time zone without stopping the movement were already routine at that time.

In the early 1980s the 1.55-volt silver oxide battery made its breakthrough, which marked the rapid demise of the mercury oxide battery. Main advantages of the "newcomer" were a higher voltage as well as unproblematic, non-toxic and recyclable components. In order to further extend the battery life that had reached two to three years, much hope was put in the lithium battery. Its self-discharge is very small while its energy density is much higher than that of the silver oxide battery. The disadvantage of the lithium battery is its volume: battery life easily reaches four to five years, but at the cost of the overall height

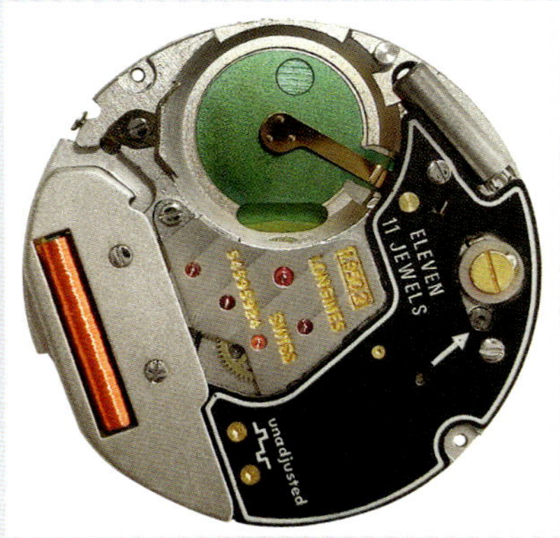

8.81. Longines caliber L950

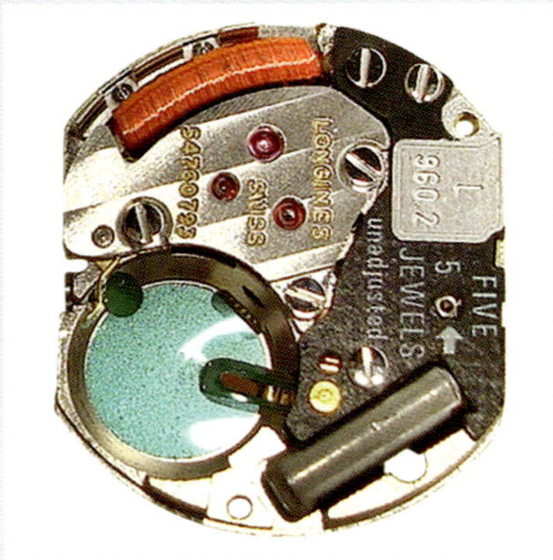

8.82. Longines caliber L960

8.83. Longines caliber L961

8.84. Longines caliber L970

of the watch it supplies. A lithium battery does not fit into a standard quartz movement but must be stored in the case back. Due to the higher voltage (3 volts) integrated circuits designed for this voltage must be used in the movement.

Longines chose just that way: In order to offer the consumer a watch with a very long battery life, it launched the 11½-line calibers L.270 and L.273 in 1982: they are equipped with 3-volt lithium batteries. Caliber L.270 featured a sweep seconds-hand while L.273 had no seconds-hand, but the battery life was a phenomenal 10 to 12 years.

8.85. Longines caliber L978

Digital Calibers

The quartz movements mentioned so far were all equipped with hands, i.e. an analog display. However, in the 1970s, Longines also proposed digital calibers to its customers, with either a light emitting display (LED) or (in most cases) a liquid crystal display (LCD).

Longines' first digital movement, L.775.1, was a 13½ lines x 7.95 mm caliber with a 3½-digit LCD; it was launched in 1975. It displayed hour and minute; dots flashing at 1 Hz indicated the sequence of seconds. The L.776 and L.777 LCD calibers (1975) were technically similar calibers; the display could be illuminated on demand and also showed the date. The modules were bought in the United States under the designations IDS 101, 102, and 103. IDS was the brand of the American module maker Integrated Display Systems in Montgomeryville, Pennsylvania.

8.86. Longines LCD caliber L776-1, front side

The display of the 13-line caliber L.778.1 (1976) with day-date and seconds could be illuminated by pressing a key. This actually was the ESA caliber 9315 from Marin. The Longines LED caliber L.782.1 (1976) was the 13 lines x 5.5 mm Sanyo caliber DWM 2005 that was produced in Japan under a Pulsar license.

8.87. Longines LCD caliber L776-1, back side

The 13-line caliber Gemini II LCD/LED (L.785.1) with a rounded-rectangular shape was launched in 1976. The functions were hour and minutes or seconds and date: they were displayed permanently on an LCD. The LED displayed hour, minutes, seconds, month, and date on demand; the pusher was located at 1:30. The electronics module was made by Hughes Aircraft. (Caliber MPD 219 with a high frequency quartz resonator, 786,432 kHz).

8.88. Longines LED caliber L782-1

The 13-line chronograph module L.780.1 (1977) with a 6-digit LCD and 12 functions, including rattrapante and stopwatch for an hour with a resolution of 1/100 second, was the EEM caliber ESA 942.711 from Marin. The ladies' watch caliber L.782 (1977) had the same functions as L.775. ESA 934.711 (1977) changed to L.787.1 in Saint-Imier; it was a 13-line chronograph module. The 7½-digit LCD displayed 16 functions, including chronograph rattrapante to 1/10 of a second for 20 hours. Another LCD caliber for ladies' watches was the 8¾ lines x 3.50 mm L.723.1; it was launched in 1977 and actually was ESA 945.831/832. Aside from hour and minute, the LCD displayed month, day/date and second.

VHP: a Quantum Leap

A standard quartz movement typically has a daily rate of a few tenths of a second; a maximum of 1 second may be tolerated. Since the early seventies no significant progress had been made in this respect. The daily rate mentioned above was mostly due to the temperature dependence of the quartz resonator's frequency. High temperatures increase the frequency: the watch gains time, while low temperatures lower the frequency: the watch loses time. Longines solved this problem in 1984 with its caliber L.276 VHP (VHP means Very High Precision). It was used in the Conquest-type Longines watches; their daily rate was a phenomenal 0.01 to 0.02 second.

This literally was a quantum jump and was achieved by thermocompensation of the quartz resonator. A separate, 262 kHz high frequency quartz resonator served as a thermometer. The standard 32 kHz quartz resonator was left unchanged, but units were chosen with a resonant frequency slightly above the nominal value of 32,768 Hz. Of course the high frequency resonator was also temperature dependent: in function of its frequency, one or more pulses of the 32 kHz resonator were suppressed, so that it always oscillated exactly at 32,768 Hz.

The VHP caliber was the swan song of the Longines movement design department. A few months after the presentation, the company lost its status as a manufacturer. Henceforth, Longines had to use ETA movements. ETA continues producing thermo-compensated movements to this day: they need a rather complicated (and costly) integrated circuit that was developed by ASULAB. They are the most accurate standard quartz movements and easily fulfill the COSC specifications of electronic chronometers, i.e. a daily rate ± 0.07 seconds. Despite of their very high precision, thermocompensated movements are used only marginally by the Swiss watch industry because of their high cost.

When Longines' movement design department was closed, ETA adopted the extra flat, rounded-rectangular Longines caliber 960 (5½ x 6¾ lines x 1.75 mm) and integrated it into its production program. It received the caliber number ETA 280.001.

8.89. Longines Day and Night caliber L785-1

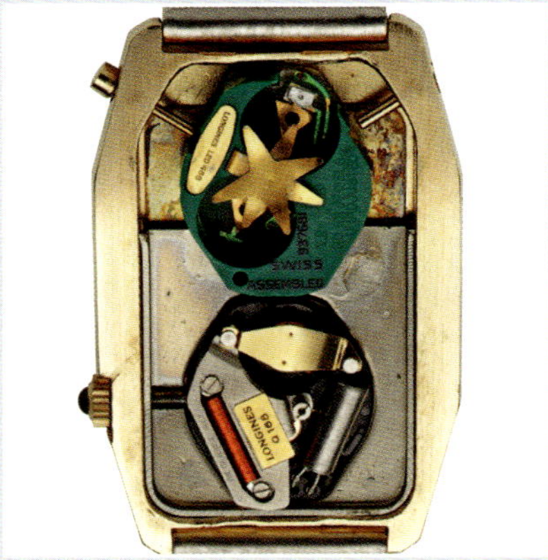

8.90. Longines LED ana-digi calibers Q16 and LED 468

8.91. Mondaine LCD Solar MW 310

8.92. Mondaine MW 11340

Collectors occasionally run into rather strange Longines quartz watches. These are models that are totally unknown by Longines archives and no information is available on them. As a rule, such watches are driven by calibers that Longines never purchased. This is due to the fact that during the quartz crisis of the 1970s Longines sold the rights to its name to Wittnauer for the United States and to Westinghouse for Asia and other countries.

Longines had already collaborated with Wittnauer in the 19th century; Wittnauer bought large quantities of movements from Longines and cased them in the United States, using exterior parts that were just right for the American taste. After 1972, Wittnauer was entitled to case movements of any kind; the Longines brand figured on the dial of such watches as well as the logo with the winged hourglass – just like original Longines watches from Saint-Imier.

Westinghouse had bought similar rights for Asia: its Longines quartz watches were mostly assembled in Korea and sold as low-priced items all over Asia. This seriously damaged the brand's reputation. Large numbers of "Longines watches" were on the market that had absolutely nothing in common with the brand's product philosophy.

This truly untenable situation ended after the ASUAG-SSIH merger in 1983, when Longines bought back its own brand – a costly transaction. In 1984, Longines could again appear with a unified marketing concept all over the world. The calibers that Longines either developed on its own or purchased from outside vendors are listed at the end of the book *Au Coeur d'une Vocation Horlogère* by Patrick Linder, Editions des Longines (2007).

Mondaine

The Mondaine Watch Company in Zurich accessed watch electronics on a rather tortuous path. In the 1960s, the company was offered the design of a kind of tuning fork watch, using a piezoelectric transducer that was brought to resonance with a transistor circuit. The oscillations were transformed into a rotary motion by a pawl/ratchet wheel system. This ended up being a blind alley, no commercial product came of it.

First Digital Calibers

The results of the 1967 chronometer contest at the Neuchâtel Observatory became known in early 1968. Quartz wristwatch prototypes had been submitted both by the Centre Electronique Horloger (CEH) in Neuchâte and Seiko in Suwa: they had obtained the highest 15 ranks. Mondaine had to react in some way, but it was clear that in terms of analog quartz watches the company was hopelessly behind. The only solution was the fully electronic digital movement, without any mechanically moving parts. With this kind of product, there was hope to access the electronic age very rapidly.

In 1969, Mondaine established a contact with Hughes Electronics in Scotland. Prototypes of a 3½-digit quartz movement with a point-matrix LED display and a convex ruby red cover glass had been developed there. This movement was integrated into a simple, round watch. Mondaine was very much interested and developed the design of a much nicer case with pushers. Unfortunately, not a single module was delivered to Zurich as Hughes cancelled the deal. Hughes in California had decided to get into the digital watch business and had the priority for this invention.

Mondaine thus had to develop digital watches on its own, which took time. Things improved when Brown Boveri (BBC) started producing LCD in 1973 on the base of Roche patents. Mondaine engineers combined such displays with a chip supplied by Solid State Scientific (SSS) in Pennsylvania (later on by Eurosil in Munich) and a 32 kHz bar-shaped quartz resonator from Oscilloquartz. The transition to photolithographic quartz resonators was made as soon as they were available. On this basis, Mondaine was in a position to start producing digital quartz watches.

The first digital watch by Mondaine was called Digi Quartz 100. Its 3½-digit LCD displayed hour, minute, second, and day/date in combination with a 4-year calendar. The complete watches of the Digi Quartz line were assembled at the Mondaine plant in Zurich. LCD are invisible at night: this disadvantage was circumvented by illuminating the display by means of a miniature lamp. It was supplied by Chicago Miniature Lamp and was switched on by pressing a key on the case. The current consumption was enormous – about 20 mA versus a few microamperes for the quartz module. Thus, a high drain battery had to be built into the watch just for the illumination.

The Epic Cooperative

In terms of LCD quartz watches, Mondaine was one of the earliest brands with this product. The product was very successful and was even exported to the United States where Digi Quartz watches were sold in department stores such as Sears and J.C.Penney. In order to be able to purchasing large quantities (with extra discounts) of the components, Mondaine set up the purchasing cooperative Epic with Nepro in La Chaux-de-Fonds and Diehl in Schramberg (Black Forest).

The Digi Quartz Series 100 was discontinued in 1975. An improved model – Series 200 – had been developed in the previous year. The functions were the same, but time could now be displayed in the 24-hour mode, an optional European calendar was also available. The module could be supplied with solar cells that charged a buffer battery.

Mondaine launched a second generation of "Digi" quartz watches in 1975. There were four families of models: Digi-Trio Series 300, Digi-Stop Series 350, Digi-Alarm Series 360, and Digi-Lady Series 250. Variants with a different set of functions were differentiated by a lower case letter after the caliber designation.

Common denominators were 12-hour mode display (with optional 24-hour mode in some models), display of hour, minute, second, month, and date according to the U.S. mode (month first, then date; some models also featured the European mode, date first, then month). All models had a four-year calendar that was reprogrammable until 2100 and an optional rechargeable battery for solar cell operation in the 300 Series. The latter series displayed the weekday alphanumerically.

Light for LCDs

The 200 and 250 modules featured a 4-digit LCD that could be lit by a miniature lamp. The 300, 350, and 360 modules also featured a lamp, but it illuminated a 6-digit display. Owners of 350/360 Series had the option of setting the daily rate of their watch to ± 0.05 second per day by suppressing oscillations of the quartz resonator. This meant doing manually what the thermocompensated movements do automatically. The 350/360 series movements would have fulfilled the very severe COSC requirements for electronic chronometers (± 0.07 second per day); but those rules did not exist yet. Eurosil supplied both the conventional divider chips of Series 200 and 300 and the microprocessor chips of the 350 and 360 series. The design of those microprocessors had been developed in the United States.

The Series 360 watches not only displayed time, date, and calendar, they also could be used as a stopwatch with a resolution of 1/100 of a second in the modes stop, split and lap. Series 360 also featured the memory functions Digi-Trimm, Alarm and Executive Reminder (Countdown). An option for the series 200 to 360 (but not for the caliber 250) was a set of solar cells. Series 300 alone featured a semi-automatic setting system for the display. When changing time zones, the hours could be set forwards and backwards in steps of one hour.

For Mondaine, the digital watches of the series 200 to 360 opened new markets in Germany, particularly the mail-order houses. The Swiss watch and jewelry stores followed, but only for a few months. The reason was that the largest Swiss discount supermarket chain Migros rapidly grew into the most important customer for Mondaine LCD watches. This resulted in an instant boycott of Mondaine products by the specialized stores. Nevertheless, Mondaine now had a foot in the door for digital watches in Switzerland. So far, the company had exported its entire production of pin pallet and jeweled pallet watches.

Very soon, the world was literally flooded with low-price digital watches from Hong Kong. The American producers who had dominated this market in the 1970s had waged brutal price wars against each other until 1983 when everybody gave up and left the product to the Chinese. The first symptom of this dramatic story was the collapse of the price of quartz resonators. Mondaine bought its first ones for 27 Swiss francs apiece, but shortly afterwards the price dropped to a fifth of that. The photolithographic tuning fork quartz resonators made by Statek in California triggered another dramatic drop. Finally, the simplest quartz resonator could be purchased wholesale for less than 0.10 francs.

In Europe, the digital watch had only a short product life; sales bottomed-out in the early 1980s when there just was no more demand for this kind of timepiece. Henceforth, Mondaine concentrated its efforts on quartz watches with an analog display: a first prototype had already been designed in 1975. This was a kind of electrification of the 11½-line ETA automatic caliber 2834. The battery replaced the barrel, the motor supplied by Audemars in the Tessin was installed at the place of the escapement and the balance wheel. Even a solar cell version was designed in 1977.

The Audemars motor proved to be unreliable; consequently, the electrified 2834 never was mass-produced. Mondaine had to purchase whatever was available in the marketplace, and this was the second Swiss quartz movement (the first one had been the Beta 21). The second caliber was designed and produced by Ebauches Electroniques Marin; its designation was 9180. The motor was made by LIP and generated very high friction forces in the bearings. EEM 944.111 (3 hands and date, 11½ lines, 1979) was a great improvement. Starting with 1980, ETA supplied extremely reliable analog quartz movements. Mondaine purchased important lots of the calibers 955.111 (3 hands and date, 10½ lines, 1980) and 956. 411 (3 hands, 8¾ lines, 1980).

M-Watch

The Mondaine M-Watch was created as a reaction to the Swatch. The Swiss supermarket chain Migros would have loved to sell the Swatch after it was launched in 1983. However, ETA general manager Ernst Thomke made delivery depend on two ironclad conditions: a fixed price, dictated by ETA that was not to be discounted, and marketing exclusively under the Swatch brand. Migros CEO Pierre Arnold decided this was unacceptable. So Mondaine was asked if it possibly could develop something similar, but at a lower price. Arnold had the vision of a "Volks quartz watch" that would sell for 38 Swiss francs.

After the Migros-Mondaine contract was signed, 28 days were left for designing a completely new watch and producing 500 pieces – they would be handed out to journalists after the launching press conference at the end of February 1983. The cooperation of the Mondaine subsidiary Constructa was vital for this project, as it had a lot of experience in injection molding of plastic cases for low-priced fashion watches.

The designers came up with a simple black ABS case with crown protection in a men's and a ladies' size. An inexpensive, standard quartz movement was installed from the front. This either was an 11½-line Ebauches Electroniques Marin caliber 944.111 with three hands and a date disc or caliber 444 made by Polos (a Mondaine subsidiary). The ladies' model used caliber FHF 965.101 (Fontainemelon, 8¾ lines, no date disc). The black dial was decorated with luminous white indexes; the luminous hands were also white.

The battery was accessible from the back of the case, the customer could easily replace it himself. This was a distinct analogy with the Swatch watch, but otherwise there were no common denominators with the Swatch that was wrapped in secrecy until everything was unveiled on the 1st of March 1983. The construction of the M-Watch was quite conventional: case, bracelet, and movement were independent entities. In the Swatch, case and movement were integrated and had been designed for automatic assembly. Once it was assembled, the Swatch could not be reopened; it was not repairable and was disposed of in case of failure. The M-Watch on the other hand was repairable, but repairs were limited to exchanging the movement that itself was not repairable.

As will be related later in this chapter, the Swatch acquired cult status and was a runaway success all over the world. Total production added up to several hundred million pieces – and counting. Everything was much more modest with the M-Watch. The design was registered internationally, the Migros supermarket chain had the exclusive rights in Switzerland, and it was available in the stores in May 1983. For the rest of the year, sales had been estimated at a total of 20,000 ladies' and men's models. This was a huge underestimation, actual sales were 180,000 pieces. Orders were growing rapidly and reached a maximum of 330,000 pieces a year. Inevitably, the standard model evolved into a wide variety of models, but basically an M-Watch still is an inexpensive analog quartz watch with a movement from ETA, Ronda, or ISA. As of today, a grand total of about 8 million pieces has been sold.

Nepro

In 1972, the Californian Microma (an Intel subsidiary) started to supply the first electronic modules for digital watches with a liquid crystal

8.93. Mondaine LCD module of the 200 family

8.94. Mondaine Solar Quartz digital watch

8.95. Microma LCD module DSM, first model, front side. Used exclusively by Nepro in Europe, 1974

8.96. Microma LCD module, first model, back side

display (caliber 360/I with two batteries). The Italian-Swiss watch producer Paolo Spadini, established in La Chaux-de-Fonds, bought the exclusive rights for this product for his Nepro brand in Europe. He had those modules mounted in simple stainless steel cases with a rectangular window for the liquid crystal display.

8.97. Nepro LCD watches with Day/Date display

Timelec

At the 1973 Basel Watch Fair, Spadini was the first Swiss watch producer exhibiting complete digital wristwatches with the improved Microma caliber 360/II with only one battery. In 1976, Spadini launched the ladies' model "Lady Quartz," also with a Microma module. Despite the exorbitant price of 1000 Swiss francs (at least in the early days), these watches sold like hot cakes. Nobody seemed to care at that stage that the liquid crystal display of the dynamic scattering type had little contrast, was hard to read at an angle and often gave up after only a few weeks due to leakage of the liquid crystals. Furthermore, battery life was limited to one year.

8.98. LCD module MSA 088, back side

Those problems were solved with twisted nematic LCD and more sophisticated CMOS circuits. Spadini and his friend Fritz Demuth were greatly encouraged by this turn of events and set up the Timelec Company in Uster (near Zurich). Timelec employment reached 100: in the latter part of 1974 the plant started producing complete Swiss made "solid state" quartz wristwatches that were equipped with home-made electronic modules ("N-Quartz").

8.99. LCD module MSA 088, front side

Those watches featured bar-shaped quartz resonators supplied by SSIH and twisted nematic LCD from Brown Boveri (BBC) in Lenzburg. The electronic chip was purchased from Eurosil in Munich, the European subsidiary of Intersil, the company that the Swiss-American genius Jean Hoerni had founded in California. Eurosil was managed by Luc Bauer, son of the former FH director Edgar Bauer who had had the early vision of microelectronics playing an important role in watches.

Timelec produced the round men's watch caliber 900.085 with the dimensions 30.3 mm x 4.5 mm; two batteries were needed, one of them just for lighting the LCD at night. The module for ladies' watches (caliber 960.077) measured 24.0 mm x 4.5 mm. There was no miniature lamp; only one battery was needed. The 3½-digit LCD displayed either hour and minute, date and month, or minute and second. A second ladies' watches module (caliber 900.071) was square with rounded corners; the dimensions were 18.5 mm x 18.5 mm x 5.2 mm.

8.100. MSA watches by Nepro

Three round men's watches calibers with a 6-digit LCD, a diameter of 30.3 mm and a height of 4.5 mm followed, all of them with two batteries, one of them for the light. Caliber 900.088 featured 6 functions: hour, minute, second, month, date, and day (with English abbreviations). Instead of the 12-hour display mode for the hours, the 24-hour mode was available as an option. Caliber 900.092 featured 5 functions: hour, minute, second, and date; the month was displayed in the setting mode. Caliber 900.094 featured the same six functions as caliber 900.088, but the day was indicated by a triangle that moved along the width of the LCD within a week. The abbreviation of the weekdays could be stamped on the dial in any language.

Caliber 900.046 rated as extra-flat, at least at that time: its height was 3.8 mm; the diameter was 27.0 mm. The 3½-digit LCD displayed hour, minute, second, month, and date. Caliber 900.045 was identical in every respect with 900.046, but the height was only 3.0 mm. The diameter of caliber 900.079 (also with a 3½-digit display) was 24.0 mm, the height was 3.6 mm. The date could be displayed either the European or the American way.

Timelec production initially was 1000 modules per month, but demand was enormous and the output soon reached 5000 pieces a month. Part of the production was exported, the best German customer was Dugena. In early 1975, Spadini took up the challenge of LED watches with solar cells. He bought 3000 Litronix modules and mounted then in a case that was angled in the 3-9 axis. The LED was up front, the solar cells were located further back. This watch was called Nepro Solid State and was exhibited at the 1975 Basel Watch Fair. The mail order house Quelle distributed the watch in Germany under the Meister Anker brand. The case was either bright or black stainless steel.

Spadini called his second LED watch XJ-S LED, it was launched in 1976. The case reminded one of the rear part of the Jaguar sports car XJ-S – Spadini was driving one of them at the time. The electronics module was supplied by Fairchild in the U.S. Pressing the key on the case displayed hour, minute, second, day/date, and month. Nepro also used Frontier modules in some of his LED watches.

MSA

The production capacity of Timelec soon reached its limit and it was geared on the efficient production of modules with 3½ digits. In 1976, Spadini established a second production facility 1976 in La Chaux-de-Fonds; he called it MSA (Modules Electroniques SA). This is where modules with a six-digit display were assembled and tested; kits were produced for certain customers. MSA grew to be one of the largest European producers of LCD modules and exported them worldwide. MSA employed 150 persons and produced up to 30,000 modules per month.

After a truly spectacular rise and a couple excellent years, the market for digital quartz watches literally collapsed in 1978. The American producers had engaged in a ruinous price war. Between 1973 and 1980, the price of a digital module went from $300 to $3 or less. Within a few years, all American producers closed; only Hong Kong survived thanks to its unlimited supply of cheap labor.

8.101. Nepro Solar LED

8.102. Nepro Solar LED, back side

8.103. Nouvelle Lemania quartz chronograph, caliber 181, movement side

8.104. Nouvelle Lemania quartz chronograph, caliber 181, dial side

Spadini just managed to avoid bankruptcy, but he had to terminate the entire workforce of Timelec and MSA and close both companies. Unfortunately, the digital watch "adventure" resulted in a cumulative loss of 6 million Swiss francs for both of Spadini's companies.

Nouvelle Lemania

The Nouvelle Lemania can be traced back to the watchmaking genius Alfred Lugrin (1858-1920). He came to the Vallée de Joux in his early teens as a herdsman, watching the cows summering in the valley. Watches fascinated him and he taught himself watchmaking as a pure autodidact, except that a local company taught him the fine art of filing. Very rapidly, he acquired the reputation of being the worldwide authority for complications, particularly for chronographs. He established the flourishing Lemania Company in L'Orient.

Two years after Lugrin's death, the company was merged into the SSIH Group (with Omega, Tissot, and Lanco). The Omega Speedmaster Chronograph, known as the "Moon Watch," was designed in the 1950s by Lemania engineers. This watch accompanied the astronauts of the Mercury, Gemini, and Apollo missions. In 1981, Lemania split from SSIH under the terms of a management buyout and was relaunched as Nouvelle Lemania. Eleven years later it was purchased by the Breguet Group, which itself was absorbed by the Swatch Group in 1999. The Manufacture Breguet built a huge new plant where mechanical movement parts for the top brands of the Swatch Group are produced with the newest generation of numerically controlled machines.

It is not generally known that during the Nouvelle Lemania time, when Claude Burkhalter was its CEO, the caliber designer Jean-Daniel Dubois was given the assignment to develop a top class quartz chronograph. This is how caliber 181 came into being. It featured three counters and an LCD on which the hundredths of a second were displayed digitally. The dimensions were 28.5 mm x 6.5 mm including the dial. It seems that this was the only quartz movement Nouvelle Lemania ever produced. Production started in 1985 but stayed quite small.

Omega

In the mid-1960s, three different types of electrical wristwatches were available: calibers with a balance wheel motor with either a fixed coil or a moving coil, and the tuning fork watch "Accutron" by Bulova. Omega also tried to develop an electromechanical watch, but apparently the performance of the prototype was not satisfactory.

Omega was a shareholder of the Centre Electronique Horloger (CEH) in Neuchâtel; in 1970, the company was supplied with a limited number of the CEH quartz caliber Beta 21 (see Chapter 7). They received the Omega caliber designation 1300 and were cased in Electroquartz watches.

8.105. Omega quartz caliber 1310 Fourmi

8.106. Omega quartz caliber 1320 Cigogne

8.107. Omega quartz caliber 1333 Leopard

8.108. Omega quartz caliber 1343 Elan

Many an expert thought that the height of the Beta 21 (7.50 mm) was excessive. For this reason, Omega developed two calibers known as Beta 22 with a height of 6.2 mm (caliber 1301) and 5.50 mm (caliber 1302) respectively (without the sweep seconds-hand and the date disc). They were significantly flatter than the Beta 21, but were left unchanged otherwise. Omega produced a total of 12,000 pieces of both calibers.

Error-Prone Futurology

In 1973, Omega celebrated its 125th year with a plush congress in Montreux on Lake Geneva, attended by 1300 persons from 70 countries, including all agents of the brand. Three major talks were given with the themes of mechanical watches (speaker was Pierre-Luc Gagnebin), the tuning fork watch (speaker was Max Hetzel), and the quartz watch (speaker was Hans Widmer, Omega technical director).

After that, groups were formed; they were asked to work out a consensus as to the future technical development of the wristwatch. The result was that the mechanical watch should be given the highest priority in terms of research, development, and production technology. Major investments should only be made in this sector. The tuning fork watch rated second: expensive timepieces of this kind had a niche market and testified the technical competence of Omega. The quartz watch had absolutely no chance; the great majority of the congress participants recommended that this silly article should completely disappear from the Omega production program.

This proves once more that the future cannot be predicted, not even by a consensus of well informed people. A few years later the tuning fork watch was forgotten, the mechanical watch fought for its survival and would shrink to 0.5 percent of worldwide watch production – thanks to the transformation into a jewelry-type luxury object with a marginal instrumental function. The much despised quartz watch had won a brilliant victory: in terms of precision and price, it was and remains absolutely unbeatable. This continues to be the case in the 21st century: a credible alternative is nowhere in sight. The rather costly thermocompensated and radio-controlled quartz watches are quite popular, but only occupy niche markets.

After the painful experience with the high frequency quartz watch, Megaquartz 2400 (see Chapter 7), Omega learned its lesson and switched rather rapidly to state-of-the-art quartz movements in tune with established industrial standards, i.e. 32 kHz technology. Omega remained autonomous in every respect; millions were invested to create the subsidiary SSIH Quartz, with up to 40 employees. This is where vacuum-encapsulated quartz resonators of the flexural mode type were produced. The switch to the tuning fork-type was made, but quite late in the game.

8.109. Omega quartz caliber 1352 Sauterelle

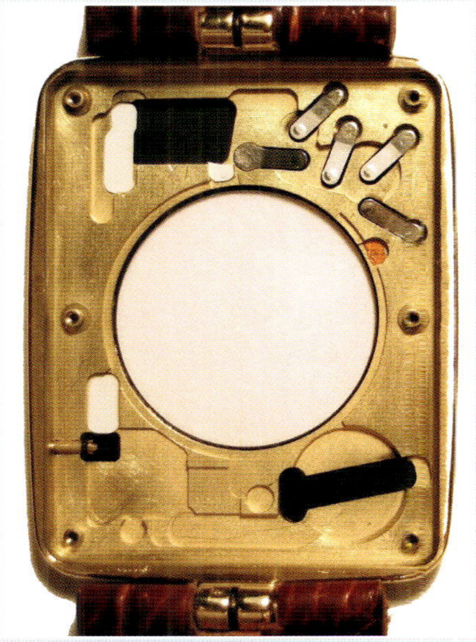

8.110. Omega quartz caliber 1357 Magique, back side

8.111. Omega quartz caliber 1357 Magique, front side

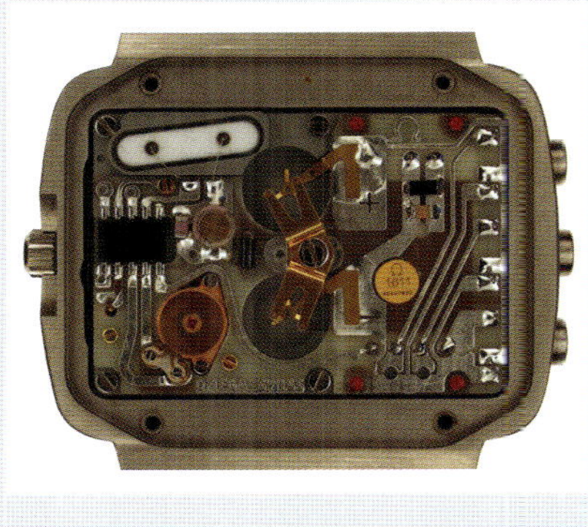

8.112. Omega quartz caliber 1611 Albatros

The bar-type resonator led to a dead end. In 1973, mass production of photolithographic tuning fork resonators of the Statek-type started. They used less space and were much less costly than any other type of quartz resonator. Yet, SSIH Management succeeded in selling its obsolete quartz technology to the Soviets, which allowed them to legally produce flexural mode bar-type resonators. For this reason, Russian quartz watches featured very high quality but ridiculously expensive quartz resonators until well into the eighties.

With Varta in Germany, Omega created a joint venture for the production of silver oxide watch batteries. Omega had been a shareholder of the "silicon foundry" Intersil and of its European subsidiary Eurosil. Thanks to this connection, the SSIH always was supplied with the newest types of CMOS integrated circuits. Complicated circuits were purchased in Japan, particularly for a digital watch with a touch sensitive sensor for accessing the watch functions (1978). Nippon Electric thus became an important supplier.

A New Start with 32 kHz Technology

In 1973, Omega surprised everybody with a quartz movement that had been designed in its own laboratory. This was the famous caliber 1310 "Fourmi" (Megaquartz 32 kHz) with a diameter of 13 lines (29 mm) and a height of 6.35 mm; well over 200,000 pieces were produced. The quartz resonator was of the 32 kHz flexural mode-type with a shock-proof suspension that was being produced photolithographically. It cost about 10 Swiss francs apiece: in the longer run, it was replaced by the Statek-type tuning fork resonator with a hundred times lower price. The movement was driven by a patented Omega stepping motor with a horizontal arbor and a worm gear.

The extremely sturdy, nonmagnetic caliber 1310 displayed time and day/date; its yearly rate was less than a minute. A pusher actuated setting steps of plus or minus 1 second, which allowed the precise synchronization with a time signal. When changing time zones, the hour-hand was set in steps of one hour without affecting the minute and seconds-hand. A variant that was launched in 1977 with only a date disc was caliber 1315.

The flattened-round calibers 1320/1325 "Cigogne" (8¾ lines x 6 mm) were designed for ladies' watches. This caliber marked the transition to an eight-pole stepping motor that Omega produced under a CEH license. Henri Oguey had designed and patented this construction, which was developed for the Beta project but was never used by the CEH. Only Omega bought a license.

The star-shaped rotor of the CEH motors originally was made of platinum cobalt, which at that time was the best, albeit very expensive, magnetic material. It was eventually replaced by the much less costly and magnetically superior samarium cobalt. The arms of the star were all magnetized axially with the same polarity. The rotor was located between two soft iron stators, while the coil was mounted coaxially below the stator-rotor system and formed a kind of pot. This construction did not allow very flat movements. When fashion switched to ultra-flat watches, SSIH too had to adopt the Lavet motor; this was first done with caliber 1360.

The Omega caliber 1330 "Ferrari" with a diameter of 23.3 mm (10½ lines) and a height of 3.5 mm still belonged to the generation of quartz movements with the CEH motor. It was launched in 1976 and allowed the design of relatively flat, elegant watches – at least according to the standards of the time. It was very successful; Rayville-Blancpain produced 215,000 pieces for Omega. The CMOS-type integrated circuit functioned with just 1.5 microwatts. The system for independently setting hour, minute and second with pushers was quite sophisticated. This caliber (with only slight modifications) was purchased by Audemars Piguet for use in the Royal Oak with a platinum case; AP gave it the designation 2511.

During the Olympic Games of 1976 in Montreal, Omega launched the "Chrono-Quartz" caliber 1611 "Albatros" with analog display of hour and minute. There were two LCD for displaying the second and the stopwatch functions. This was the world's first ana-digi caliber; the gear train was taken from caliber 1320. With the chronograph, time intervals were measured with a resolution of 1/100 of a second; the functions lap and split were also available.

In 1977 and 1978 the calibers of the 1340 family "Elan" were launched. Their diameter always was 25.6 mm (11½ lines). The height of caliber 1342 and the chronometer version 1343 with date was 4.2 mm. Caliber 1345 and its chronometer version caliber 1346 featured the day/date function, which raised the height to 5.0 mm. The energy source consisted of two mercury oxide batteries. Total production of caliber 1340 was around 15,000 pieces.

Interesting enough, the Elan caliber 1343 was used as a platform for Omega's last venture in the high frequency field. In 1978, Alphonse Zumsteg, at SSIH Quartz, produced a 4.19 MHz quartz resonator of the AT type, named SQA 4.19 for the Vega chronometer. The 4.19 MHz wristwatch movement just needed a new chip to adapt it to the 8-pole stepping motor. The rest of the Elan movement was left practically unchanged, but was given the caliber number 1522. At least two of the prototypes have survived to this day. The project, led by Michel Burdet, was cancelled in 1980 in view of the very high current consumption of the device and the insufficient shock resistance of the stepping motor. Preliminary studies of movements with 8.4, 12.6, and 16.8 MHz quartz resonators were also abandoned.

Caliber 1350 "Sauterelle" (5½ x 6¾ lines x 3.35 mm) and the variant 1353, with a trimmer for fine tuning of the quartz frequency, were launched in 1977. This was the first quartz movement for small ladies' watches and the world's smallest quartz movement. Its volume of 0.533 cm^3 was the equivalent of two Swiss five cent coins. The quartz resonator frequency was fine-tuned with pencil strokes, but this procedure did not work dependably. This is why the variant caliber 1533 with a trimmer was launched in 1979. It took the micro motor 12 minutes to make one turn. A grand total of 637,000 Sauterelle calibers were produced.

Caliber 1351 (1978) was round, but the technology corresponded to that of the Sauterelle 1350/1353; the diameter was 17.2 mm (7¾ lines). There was no crown, the hands were set electronically by means of pushers.

The baguette quartz caliber 1352 (1978) with the dimensions 9.0 mm x 21.0 mm and a height of 3.35 mm was absolutely unique. The technology corresponded to that of caliber 1350. In order to save energy, the minute-hand just made one step per minute. This resulted in a battery life of three years. Total production of caliber 1352 amounted to 17,000 pieces.

Caliber 1360 was developed in 1978 specifically for the ladies' Seamaster. The dimensions were 8¾ lines x 5.1 mm.

Caliber 1332 (1979) "Léopard" (25,6 x 28 mm x 2.9 mm) was also available in a chronometer version (1980, caliber 1333). Caliber 1336 with date was even flatter (2.45 mm); with the day/date feature, the height rose to 2.9 mm (caliber 1337). Production of the calibers 1336 and 1337 started in 1981.

Thanks to a height of just 1.9 mm, the 8¾-line caliber 1365 "Criquet" (1980) was extremely successful: nearly 600,000 pieces were produced. The massive height reduction compared to previous calibers was due to the introduction of the Lavet motor. In order to save space, the coil was curved. All further Omega quartz calibers also featured Lavet motors.

The dimensions of caliber 1387 "Grillon" (1981) were 13.0 x 15.15 mm x 2.3 mm. Fine tuning of the quartz resonator's frequency was achieved with a ceramic capacitor the capacity of which could be increased with pencil strokes. Total production of the "Grillon" caliber was around 500,000 pieces.

Omega's answer to the ultra-flat Delirium calibers (1.98 to 0.98 mm) designed by ETA was a quartz watch with a height of 1.48 mm (1.35 mm for a limited number of collector pieces); it was launched in 1980 and was called caliber 1355 "Dinosaure." Wafer-thin round sapphire plates with a wheeled rim and black markings replaced the hands. The next models (1356 and 1357 "Magique") were much thicker (27 mm x 40 mm x 2.6 mm). Omega's last analog quartz movement was caliber 1450 "Ecureuil". It was a 5½ x 6¾ lines caliber with a height of 2.1 mm. All following Omega calibers were provided by ETA.

8.113. Omega LED caliber 1602

8.114. Omega LCD caliber 1620 Faucon, front side

8.115. Omega LCD caliber 1620 Faucon, back side

8.116. Omega Memomaster caliber 1632, front side

SSIH Analog Quartz Calibers with a 32 kHz Resonator

Horizontal stepping motor with worm gear

1310	Fourmi 29 mm, h= 6.35 mm day/date, 1973
1315	Fourmi with date 1977

CEH motor

1320/25	Cigogne 18.2 mm x 20 mm, h= 6.0 mm 1975
1330	Ferrari 23.3 mm, h= 3.5 mm 1976 (Assembled by Rayville-Blancpain)
1332	Léopard 25.6 mm x 28.0 mm, h = 2.9 mm 1979
1333	Léopard chronometer 1980
1336	Léopard date flat, h = 2.45 mm 1981
1337	Léopard day/date, h = 2.70 mm 1981
1342	Elan 26.5 mm, h = 4.2 mm 1977 (Assembled by Tissot)
1343	Elan Chronometer date, h = 4.2 mm 1977
1345	Elan day/date, 25.6 mm, h = 5.2 mm
1346	Elan day/date chronometer
1350	Sauterelle 13.0 mm x 15.15 mm, h=3.35 mm 1977
1351	Sauterelle round 17.2 mm, h=3.35 mm 1978
1352	Sauterelle 9.0 mm x 21.0 mm, h = 3.35 mm 1978
1353	Sauterelle w. trimmer 13.0 mm x 15.15 mm, h = 3.35 mm 1979
1360	Faon, Ladies' Seamaster 19.4 mm, h = 5.1 mm 1978
1370	Tatou 25.6 mm, h = 4.8 mm (Tissot 3031-2037)

Lavet motor

1355	Dinosaure 27.0 mm x 35.0 mm, h = 1.48 mm or 1.35 mm 1980
1356	Magique 27.0 mm x 40.0 mm, h = 2.6 mm, 1981
1357	Magique slightly modified 1981
1365	Criquet 19.0 mm, h = 1.9 mm with pusher
1387	Grillon 13.0 mm x 15.15 mm, h = 2.3 mm, 1981
1450	Ecureuil 13.0 mm x 15.15 mm, h = 2.1 mm, 1983
1611	Albatros ana-digi chrono 26.0 mm x 37.4 mm, h = 8.55 mm 1976

8.117. Omega Memomaster caliber 1632, back side

8.118. Omega Sensor Quartz caliber 1640, front side

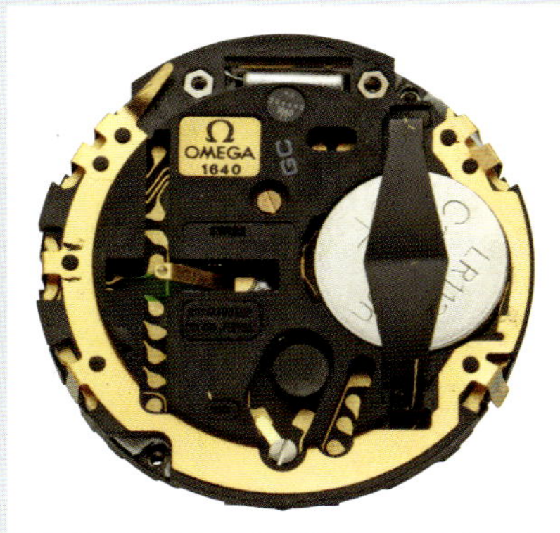

8.119. Omega Sensor Quartz caliber 1640, back side

8.120. Oris prototype, 8¾ lines, caliber 830

Digital Calibers

From 1972 to 1974, Omega obtained about 30,000 Pulsar modules with light emitting diodes (LED) from the HMW subsidiary Time Computer. They were given the caliber numbers 1600 and 1601. Omega also developed its own LED caliber, but it did not see production. Caliber 1602 and 1603 used components supplied by the American company Frontier.

Omega then concentrated on modules with a liquid crystal display (LCD). The first one was caliber 1615 (1976, "Cobra") with a diameter of 29.0 mm and a height of 4.45 mm. The 6-digit display was supplied by Optel, the Chip by Eurosil. Variant 1616 could be switched to a 24-hour display mode. Caliber 1620 "Faucon" followed in 1977, the dimensions also were 29.0 mm x 4.45 mm. The alarm function made it caliber 1632 "Chouette" (1978): this was a Frontier module.

Caliber 1635 "Martinet" (1981) was designed for ladies' watches; the dimensions were 17.0 x 24.0 mm with a height of 6.0 mm. It was designed specifically for the Swiss National Ladies' Ski Team; special functions were alarm and rattrapante chronograph. The same dimensions characterized caliber 1637 "Alouette" with just the alarm function; the module was supplied by Frontier.

Caliber 1640 "Sensor Quartz," Okapi (29.2 mm x 6.2 mm) was launched in 1980. Aside from hour, minute, and second, it displayed the date and offered the usual chronograph functions. Furthermore it featured an acoustic hour signal, two alarms, two agenda functions, and time in another time zone. The chip was supplied by Nippon Electric. The ultimate movement in the Omega LCD collection was exhibited at the 1981 Basel Watch Fair. Caliber 1655 "Equinoxe" (20.0 mm x 24.0 mm x 4.8 mm) featured a two-side analog and digital display, the functions of caliber 1640 plus a countdown.

Starting with 1981, Omega used more and more ETA quartz movements. After the SSIH-ASUAG merger in 1983, ETA was the sole source of movements. The Omega design office and the R&D department were closed in 1984. Some of the personnel moved to Grenchen where ETA had offered them new jobs. None of the Omega quartz calibers could be integrated into the ETA-collection.

Oris

Documents concerning electromechanical or electronic wristwatches start appearing in the Oris archives in 1960. They were a reaction to the launching of the Accutron tuning fork watch by Bulova. This activity was triggered by Walter Born, CEO, and Max Meier, technical director, of Roamer Watch. They were looking for partners joining forces with Roamer to introduce electronics in their watches. In this way, a consortium was formed. Members were Büren Watch, Certina, MSR, Revue Thommen, Roamer, Oris, and Rolex Bienne. Prof. E. Baumann of the department for industrial research (AFIF) of the ETH in Zurich agreed to do the development work. A five-year contract was signed in October 1960 (see Chapter 7.8, AFIF/Neosonic).

Actual work started on 1 January 1961; the first objective was the development of an electromechanical movement that would be followed by an electronic movement. Each member of the consortium agreed to pay AFIF the sum of 25,000 Swiss francs per year. In a first phase, possible configurations of a movement with a balance wheel motor were studied in great detail. This was strictly theoretical work, no prototypes were fabricated.

A meeting was called every year to inform the members of the progress that was made. As the first five-year period was nearing its end, an extension was voted until 1971. At this stage, the idea of an electromechanical movement had been abandoned, the electronic alternative looked much more promising. André Greuter, an electrical engineer at the AFIF took over the project. This led to the presentation of the AFIF/Neosonic quartz caliber in 1972.

In early 1966, Oris had become a shareholder of the Centre Electronique Horloger (CEH) and recommended that AFIF and CEH join forces in view of the common objective of an electronic wristwatch. This was not deemed acceptable by the other members of the consortium who wanted to continue their own project. They even agreed to increase the yearly dues to 50,000 francs. Oris did not agree and left the consortium.

In 1971, Oris joined the General Watch Corporation GWC, (the watch brand conglomerate of ASUAG) and started equipping some of its watch models with the transistorized electromechanical Ebauches caliber 9150 Dynotron. But quartz was quite obviously the future: GWC management gave Oris the mandate to develop a digital watch movement with liquid crystal display for the Swiss market.

The Oris physics laboratory developed several prototypes with components that were bought from EEM in Marin. It was found that Nippon Electric (NEC) supplied complete modules with a 32 kHz quartz resonator. In 1974, the quartz caliber LCD NEC-6001 (Oris 772) with a 9½ x 12 lines module and an illuminated 5½-digit liquid crystal display was fully developed and tested, the commercial version was available in 1975.

Four different display programs could be chosen at the touch of a key: hour, minute, second or hour, minute, date, with a double dot flashing between hour and minute indicating the rate of seconds. An energy saving mode could also be activated with the key and another display option was month and date. A silver oxide battery powered the module. Watches with the NEC module were assembled by Oris in Hölstein. They were exclusively sold under the Oris brand, retail price was around 70 to 100 Swiss francs at the end of the 1970s. The watch was quite successful, about 50,000 pieces were sold per year, but prices kept eroding dramatically. An alternative to the NEC caliber was the MDK caliber 121.001 MDK (Oris 768).

Many people still preferred the conventional analog display with hands; they were not forgotten. Starting with the mid-seventies, Oris had a collection of analog quartz watches with movements that were supplied by the ASUAG subsidiaries Fabrique d'Horlogerie de Fontainemelon (FHF), A. Schild SA (AS), ETA, and Ebauches Electroniques Marin (EEM and ESA). Some calibers were bought from France Ebauches (FE), ISA, Modutek (MDK), PUW, and Ronda (RL). The choice was very wide and added up to about 90 calibers; they all were given Oris designations. The display was mostly analog, but Oris also used calibers of the ana-digi- and LCD-type. A few Oris calibers (521, 522, 541, 765, 766 and 999) were purchased as complete watches with the Oris brand.

Starting with 1973, Oris had 25 different models of digital watches with a wide spectrum of functions in its catalogs. The modules were bought mostly from NEC in Japan (see above), but also in the U.S. and Korea: Oris just cased those modules. In many cases, the origin of the unmarked modules cannot be reconstructed. One model featured an LED display, but LCD was vastly in the majority.

In 1977, GWC gave Oris the mandate to develop a low-cost 8¾-line quartz analog caliber within two years. There was the possibility of extending this project to an 11½-line men's watch caliber. A smaller ladies' watch caliber in the 5½ x 6¾ lines class would also be considered. Manufacturing costs should not exceed 28 to 32 Swiss francs apiece.

Oris fabricated the plate and the wheels of the 8¾-line caliber 820. The electronic components were purchased on the OEM market. Ten prototypes were supposed to be ready by the end of 1977, but actually only two hand-assembled calibers were fabricated; one of them is exhibited in the Oris Museum in Hölstein. It was called Oris Quartz Analogue Economique.

The early prototypes were equipped with the SSIH six-pole stepping motor. A Lavet motor was purchased for the final version: its coil could be placed anywhere on the plate. A production of 1 million pieces per year had been planned to supply all the members of GWC. In 1978, Ebauches SA got wind of this project and stopped it immediately: at that time, Ebauches Electroniques Marin and ETA in Grenchen were in a position to supply any type of quartz movement.

Oris quit the GWC conglomerate in 1982; a management buyout assured the continuity of the brand. Efforts to develop quartz calibers were stopped. After 1991 there were no new Oris quartz watch models; in 1995, the sale of Oris watches with a quartz movement was stopped. Since that time, Oris exclusively uses mechanical ETA movements, some with special complications.

From 1979 to 1995, Oris produced a colorful plastic watch that had some similarities to the Swatch. It was called "Fiesta" and was made in two sizes and several variants for the case and the dial – a total of 22 different models. The movement was a separate entity; it was not integrated as was the case with the Swatch. The movements were supplied by Ronda (the models with moon phase indicator) or by ETA and PUW. Saros SA was a subsidiary of Oris that manufactured private label quartz watches. They could be marked with the Fiesta brand name if the customer asked for it.

Piaget

Piaget is mostly known for its extremely high grade mechanical calibers that are used in elegant luxury watches. It is less well known that Piaget is one of the early pioneers of watch microelectronics. Even though Piaget does not communicate it on a wide scale, the company created its own quartz culture that is very much alive. This implies among other things that each and every Piaget quartz watch will always be repaired, regardless of its age: mechanical or electronic.

Piaget's interest in advanced technologies is a tradition that is as old as the company itself. When Ebauches SA launched its tuning fork movement "Mosaba" in 1968 (under a Bulova license), customers had to be members of a sort of "Upscale Club." It goes without saying that Piaget was a member. Yet, Piaget decided not to produce watches with the Mosaba caliber, but left it to its subsidiary Baume & Mercier that made it into a big success.

The reason was that Piaget had something much more sophisticated in mind. Via still another subsidiary (Complications SA) it was a shareholder of the Centre Electronique Horloger (CEH) in Neuchâtel that had been created in 1962. In the summer of 1967, the world's first quartz wristwatch movements (Beta 1 and Beta 2) began ticking in the CEH laboratories.

The commercial version Beta 21 was presented in 1970 at the Basel Watch Fair – in Piaget cases (among others). Piaget used the big new caliber in two gold watches. There is no question that the Beta 21 was an admirable achievement, but the market clamored for smaller and flatter quartz movements. In order to be an informed partner in this new venture, Piaget established its own electronics division that was led by Gabriel Piaget, a direct descendent of the company founder Georges Piaget.

8.121. Piaget quartz caliber 7P3

The Quartz Caliber 7P

Working in close cooperation with the CEH at the Piaget design office in Bienne, Gabriel Piaget and his colleague John Corlet developed the quartz caliber family 7P. It was launched in 1976 after three years of very hard work. The product life was 12 years – an excellent record. The dimensions were 20.5 mm x 3.1 mm; the crown was hidden in back. The 7P family consisted of the calibers 7P1, 7P2, and 7P3. 7P1 was a prototype, 7P2 was sold in small series, while 7P3 (without a seconds-hand) was mass produced – on the Piaget scale. The hour-hand could be moved forwards and backwards without affecting the minute-hand – an important feature when changing time zones.

8.122. Piaget quartz caliber 8P

The 7P family used an unusual type of quartz resonator with a frequency of 524,288 Hz and digital frequency tuning. This quartz was supplied by Oscilloquartz in Neuchâtel; the yearly rate was just a few seconds. The frequency division required a large number of stages, but, thanks to an integrated circuit with 900 transistors designed by Walter Hammer, the power consumption was quite moderate. Battery life was more than a year.

8.123. Piaget quartz caliber 30P

The concept of a "low voltage/low power" metal gate CMOS-technology had been developed by Fritz Leuenberger at the CEH, based on a mandate received from Faselec in Zürich. This technology was subsequently adopted by Faselec. Eric Vittoz and Walter Hammer designed quite a few circuits with a high priority for watch circuits with oscillator, binary frequency divider, motor drive, and digital frequency tuning.

As a next step, Leuenberger, Mougahed Darwish, and Kurt Hübner developed an inherently superior silicon gate CMOS technology. The transistor gates consisted of electrically conducting, metal doped poly silicon. It was also adopted by Faselec and EEM Marin who industrialized the process.

8.124. Piaget quartz caliber 212P, chronograph

Caliber 7P was the first commercial 500 kHz-quartz caliber with silicon gate and inhibition. The latter term implied that the quartz resonator oscillated at a slightly higher frequency than the nominal value.

Pulses were suppressed in function of the difference to the nominal value. Thus, the trimmer capacitor, a permanent troublemaker could be eliminated. The chip was made by Faselec. A ZT cut was chosen for the quartz resonator, due to its minimal temperature dependence.

The 7P caliber used an 8-pole Portescap stepping motor. Piaget bought the motors as kits and assembled them in La Côte-aux-Fées. In order to save energy, the motor made a step every 12 seconds. Integrated circuits with nonvolatile memory were not yet available for the early 7P calibers. A small cylindrical lithium battery with a life of 10 years was thus developed by Leclanché in Yverdon specifically for holding the information for digital frequency tuning on the chip when the mercury oxide battery had to be replaced.

The quartz resonator thus never stopped. However, the emergency battery was eliminated for caliber 7P3, as nonvolatile memory chips had by now become available. Their power consumption was so small that battery life was extended to three years.

Type 7P3 quartz movements were used in ladies' watches with dials made of semi-precious stones such as onyx, lapis lazuli, jade, opal, malachite, coral, mother of pearl, turquoise, tiger eye, etc. No numerals or indexes were to be found on this type of dial. Time thus could be read only with an approximation of ± 1 to 2 minutes, a paradox for such a highly accurate watch.

The Polo collection of gold or platinum watches, with horizontal stripes on the dial, the case, and the integrated metal band, was launched in 1976 and turned out to be of enormous importance for the Piaget brand. Polo watches featured quartz movements of the 7P line that were amazingly rugged despite being very flat. Any sport was possible while wearing them.

The Quartz Caliber 8P

The tonneau-shaped, 16.0 mm x 14.5 mm x 1.95 mm caliber 8P2 with a 32 kHz photolithographic quartz resonator was launched in 1981. It stayed in production until 1999; the last version was 8P3 with a height of 2.05 mm.

Thanks to improved CMOS circuits and non-volatile memory for digital frequency tuning, the current consumption was reduced by 50 percent compared to 7P2, the average value being 0.25 μA. Adjustment of the daily rate was done by introducing screws into a row of five holes. In this manner, capacitors were connected to the quartz resonator in steps corresponding to 0.22, 0.44, 0.88, 1.76, and 3.52 seconds per day.

Piaget mostly used the relatively small, extra-flat calibers of the 8P line in jewelry watches (including ring watches). Pulling the crown to the middle position allowed setting the hour-hand in steps of one hour in either direction. If the crown was pulled to its end position, the movement was stopped and the minutes could be set in steps of one minute.

The motor was a special construction by Portescap. It consisted of two pairs of slightly offset flat coils and cross shaped stator. The samarium-cobalt rotor turned in either direction, depending on the polarity of the current pulses; each step corresponded to 90 degrees.

The CEH electronics engineer Eric Vittoz played a central role in developing digital frequency tuning. The reference frequency needed for initialization was stored in the "quartz timer," a device that was developed by Piaget and the CEH in common. Total production of the 8P family of calibers was around 160,000 pieces.

The 30P Family

The Piaget 30P caliber family was launched in 1986 and was discontinued two years later, after a total production of just 2000 to 3000 pieces. Yet, this was a very sophisticated, very flat movement (2.5 mm) with three Lavet motors, date disc, and perpetual calendar. The family consisted of three pairs of calibers 30P1 (with a sweep seconds-hand) and 31P1 (no seconds-hand), both with a diameter of 20.5 mm. The 32P1 (with a sweep seconds-hand) and 33P1 (no seconds-hand) were slightly bigger (23.5) mm, while the diameter of 34P1 (with a sweep seconds-hand) and 35P1 (no seconds-hand) was 25.5 mm. Caliber 300P was identical with 30P1 but was equipped with another chip. Caliber 33P was the prototype of 33P1.

All six commercial 30P calibers featured a photolithographic Statek-type tuning fork 32 kHz quartz resonator. The "quartz timer" was necessary for initialization of digital frequency tuning. One of the motors actuated the date disc, the second one the seconds-hand, the third one the hour- and minute-hands. The integrated circuit consisted of 25,000 transistors. Battery life was 2 years for the calibers with a sweep seconds-hand and five years for the calibers without a seconds-hand.

The standby pusher in the back of the case switched off the motors but measuring time continued with a much lower power consumption. After actuating the pusher once more, the hands and date disc went to their correct positions in terms of time and date.

Quartz Chronographs

For the "Mundial" of 1986 in Mexico City, Piaget developed caliber 34P with an unusual module. A special chronograph counter displayed the status of both 45 minute half-times and of the 15 minute pause. Those chronographs were a mandate from Cartier; they were reserved for the teams of the final match, only 40 pieces were produced.

Toward the end of the 1980s, the chronograph regained its popularity and there was a lot of competition. Success in this field depended on offering real innovations. Piaget obliged and launched caliber 202P in 1989. This quartz chronograph with three Lavet motors featured a perpetual calendar as well as the stop, lap, and split functions, but it was only a prototype.

The production model was caliber 212P with the functions hour, minute, second, counters for 30 minutes and 12 hours, as well as lap and split. The dimensions of this round caliber were 23.8 mm x 2.55 mm. The chronograph pushers were needed for setting the hands. Caliber 212P was the world's smallest chronograph with a perpetual calendar. Caliber 222P had the same dimensions and the same functions as caliber 212P, but also featured a date disc. It was launched in 1992 and is still being produced, albeit only in very small series.

Replacement Calibers

The extremely fast miniaturization of the transistors on integrated circuits led to a situation that Piaget would not have been able to predict; replacements for the calibers 7P2 and 7P3 as well as 8P2 and 8P3 could no longer be produced toward the end of the 1990s.

Piaget thus gave Eric Klein's design office in Neuchâtel the mandate to develop replacement calibers with top modern electronic components. They had, of course, to have the exactly same dimensions and (if possible) the exact same functions as the original calibers of 1976 and 1981, respectively. These are the calibers 757P to replace the 7P calibers, 857P to replace 8P2, and 858P to replace 8P3. The latter caliber is 1/10 of a millimeter thicker than 8P2; it therefore can also replace the mechanical movement 9P2.

Production of the calibers 757P and 857P/858P started in 1999 at the Richemont Valfleuri plant. They are not meant for new Piaget watches and are strictly reserved for the repair service. The new calibers are not quite identical with the originals, as customers prefer the conventional mechanical setting mechanism: many of them could not become accustomed to electronic setting. The new chips are supplied by EM-Electronic Marin, while the Statek-type quartz resonator is purchased from the ETA subsidiary Micro Crystal, the Lavet motor from ETA.

It is quite amazing that a venerable watch manufacture such as Piaget adopted microelectronics with the greatest ease while so many other companies were apprehensive of the new technology. Piaget had a long tradition of embracing innovations. For example, big risks were taken by heavily investing in the extra flat automatic caliber 12P and precious stone dials. But those adventures paid off handsomely. Boldly making the transition to microelectronics and establishing a Piaget quartz culture were done in exactly the same spirit.

Piaget was a true electronics pioneer: towards the end of the 1970s, two thirds of the Piaget watches had a quartz movement. This did not mean that the lovable tradition of micromechanics was abandoned; it actually attenuated the cultural shock to which Piaget watchmakers were exposed. Many of them had to learn entirely new trades, even though modules with the quartz resonator were supplied by the CEH. The modules had to be fitted with an integrated circuit and connected to the mechanical part of the movement.

Watchmakers who had become perfect with files, tweezers, and screwdrivers had to learn working with the microamperemeter and the oscilloscope. The chips were purchased "naked" from the CEH so that bonding and potting could be done in La Côte-aux-Fées and preserve a maximum number of jobs. Even the coils were wound in the Piaget plant.

Production of Piaget quartz calibers continued well into the 1990s in series of a few hundred pieces. But from then on new watches (only ladies' models) were fitted with ETA movements, which was much more economical.

A Piaget quartz watch that is sent to La Côte-aux-Fées will always be repaired, even when the series were small, as in the case of the Beta 21 and 30P calibers. Piaget has a large inventory of spare parts and old movements that can be cannibalized. Even integrated circuits from the early 1970s are still available. Customers who send in watches to be repaired are routinely offered an exchange for a mechanical movement, but very few are receptive.

The young watchmakers working in the Piaget service department are proud to continue the electronic tradition of the house and an important piece of Piaget product culture. They must be extremely polyvalent, as they have to repair both mechanical and electronic movements.

Today, close to 70 percent of Piaget watches again are equipped with mechanical movements. Yet Piaget is a member of the CSEM, the organization that was created by reorganizing the CEH. The two other industrial members (Rolex and the Swatch-Group) are much bigger. But Piaget management always wants to be informed on the newest trends in the field of micro and nanotechnologies.

Roamer Micro Quartz

In 1888, Fritz Meyer (1858-1926), who was established in Soleure, started fabricating cylinder escapements with half a dozen workers. This type of escapement was still being used in the 1930s in the movements of inexpensive watches. Meyer was quite successful and diversified into other movement parts. In 1904, Meyer took a partner; the company was registered as Meyer & Stüdeli (MST) and produced complete movements with cylinder, pin pallet, or lever escapement. By 1923, production reached a million movements and complete watches a year. MST was a true manufacturer and was near-totally independent of outside vendors, even for unusual parts.

MST's Roamer model was one of the early, dependably waterproof watches and was a runaway success. Consequently, the company had its name changed to Roamer Watch Co.; in the late 1960s employment was 1500. Roamer management was quite aware of the coming of the quartz watch and set up a cooperation agreement with the American General Time Corp. (it was part of the Westclox Holding) in the U.S. The objective was the development of a 30 mm quartz analog caliber. In 1972, the three-hand Roamer Micro Quartz was launched in Basel. The integrated circuit was purchased from RCA; the flexural mode quartz resonator was the well-known Oscilloquartz design. The Roamer Micro Quartz calibers bore the designations MST 900, 901, and 902.

Later, Roamer switched to quartz movements that were purchased from Bulova and the ASUAG group. Starting with 1976, Roamer used the 10½-line Fontainemelon caliber 935.312. Roamer always had innovative products and attractive prices, which allowed the company to remain independent. However, the Hong Kong Chung Nam Watch Group started buying Roamer stock in 1993. Roamer then switched to ISA movements, as ISA belonged to Chung Nam.

Rolex

André Heiniger, second Rolex president and successor of the founder Hans Wilsdorf, was a true visionary. His opinion was that the originally very costly quartz watch would soon be totally banal. This already had happened with transistor radios, TV sets, and pocket calculators.

Top quality mechanical movements would always remain expensive and exclusive due to the large amount of highly qualified labor that is required for manufacturing the parts and assembling them. The inescapable fact that a mechanical device can only tell time approximately could be easily hidden by writing "Superlative Chronometer, Officially Certified" (i.e. the COSC-Certificate) on the dial. This means a daily rate of +6/-4 seconds per day. In due time, every watch brand in the "upscale" sector copied Heiniger's concept. Wealthy people don't need an instrument that tells time: they want a beautiful and exclusive object on their wrist.

Analog Quartz Calibers

Rolex totally ignored microelectronics until the early 1970s; as a shareholder of CEH it obtained 320 Beta 21 calibers out of the 6000 that were actually produced. Furthermore, Rolex bought 650 pieces of the Beta 22 version produced by Omega; they are known as the Rolex caliber 5100. In addition, Rolex Bienne was part of the consortium that financed the Neosonic-AFIF adventure with the known, sad end. Rolex Geneva was not involved.

After this easy beginning, it was quite clear to Heiniger that Rolex had to be independent in the realm of microelectronics too. In 1971 he hired René Le Coultre (b. 1918) who was then technical director of the Fédération Horlogère (FH). Le Coultre headed the technical department with a staff of 49 including 10 engineers. His first activity was to set up a top modern electronics lab with 13 people. He was then in a position to design quartz equivalents of Rolex's mechanical calibers. This was not too difficult: at that time, Rolex only produced a ladies' watch movement with two hands, a version with three hands, and a men's watch caliber with sweep seconds-hand and date or day/date.

The quartz equivalent of the men's watch caliber came in two versions. The functions of caliber 5035 (29.75 mm x 6.35 mm) were hour, minute, sweep seconds-hand, and date. Caliber 5055 (29.75 mm x 7.11 mm) featured the day/date function which explains the difference in height. The quartz resonator's frequency was adjusted with a trimmer capacitor covering a range ± 2 to 3 seconds per day. The rate was less than 1 minute a year, i.e. 0.17 second per day. The Rolex quartz calibers were launched in 1977; product life was an amazing 26 years.

Both calibers featured a tuning fork-type, mechanically cut 32 kHz quartz resonator made by NDK in Japan. It was replaced at the earliest possible date by the photolithographic type supplied at first directly by Statek in Orange (California) and, later on, by Micro Crystal in Grenchen. ETA had signed a licensing agreement with Statek; delivery from Grenchen started in 1978, Rolex was one of the earliest customers.

The CMOS integrated circuit Rolex needed was supplied by Ebauches Electroniques Marin, while the anchor-type motor was purchased from FAR (Fabriques d'Assortiments Réunies). Silver oxide batteries were available from the Swiss Renata company as well as from American and German suppliers. Total production of the calibers 5035/5055 is exactly known: 105,097 – not terribly much considering that the product life was 26 years! Each and every one of them was certified by COSC as an electronic chronometer. Those movements were decorated with Geneva stripes exactly like a mechanical Rolex movement, they featured 11 jewels.

In the mid-1980s, the Rolex quartz movements were thoroughly redesigned and modernized. The result of this work were calibers 5235 (with date) and 5255 (with day/date). Caliber 5235 with a diameter of 28.10 mm and a height of 5.40 mm was equipped with a Faselec chip that included digital frequency tuning, a Lavet stepping motor from ETA and a 11.6 mm 3-volt lithium battery. The same applied for the day/date caliber 5255 (29.90 mm x 5.80 mm). These were among the best conventional quartz calibers that were ever designed – unfortunately, they never saw mass production.

Instead, the famous Oysterquartz watch models were equipped with the quartz calibers 5035 and 5055. They sold very well: they were less costly and at least ten times more accurate than the mechanical Oyster. However, the cases were somewhat different: Heiniger would not tolerate that a quartz watch looked similar to a classical mechanical Rolex. The production was only about 4000 pieces a year – not much marketing effort was made to promote the Oysterquartz – aside from a very original ad showing both Everest climbers Edmond Hillary and Reinhold Messner. Hillary had worn a mechanical Rolex on his ascent, Messner an Oysterquartz. The text just said:

8.125. Roamer Micro Quartz, caliber 900

8.126. Rolex Oysterquartz

8.127. Rolex Quartz caliber 5055

8.128. Rolex FAN

"In 1953 they needed Rolex watches and oxygen to climb Everest. In 1978 they did it without the oxygen." The retailers were not at all motivated to sell the Oysterquartz, as the cost was significantly less than the mechanical Oyster and therefore meant less profit. There are tales about potential customers who literally had to beg for an Oysterquartz.

The Rolex quartz caliber for ladies' watches (caliber 6035) with sweep seconds-hand and date disc had the exact same dimensions as the Ladies' Datejust 19.79 mm x 5.00 mm). The frequency of the 32 kHz Micro Crystal tuning fork quartz resonator was fine-tuned with a trimmer. The CMOS circuit was purchased from Faselec, the Lavet motor from Seiko. The energy source was a 7.90 mm silver oxide battery. 30 prototypes of caliber 6035 were built but there was no mass production.

However, the Cellini quartz models were equipped with caliber 6620 (no seconds-hand): this caliber was directly derived from caliber 6035. In July 1983, 20 prototypes of caliber 6620 were available for tests that took a long time; serial production started as late as 1987. The diameter was 8¾ lines (19.80 mm), the height 2.5 mm. The parts were standard Rolex issue, as mentioned above. In 1990 production of caliber 6621 started, this time the trimmer was replaced by an inhibition circuit. Production of this caliber continues to this day, total production so far was well over 100,000 pieces.

Rolex developed several technically advanced quartz movements that never got beyond the prototype stage. The most interesting of them certainly was a thermocompensated quartz caliber that was developed in 1985. Design studies were made with extremely stable high frequency (1.2 MHz and 2.4 MHz) quartz resonators with the ZT cut. The CEH produced those resonators and delivered 1000 pieces in 1984. In 1986, Rolex built 50 prototypes but there was no production, even though the yearly rate was just a few seconds.

Another very ingenious quartz caliber with a perpetual calendar had the same fate. It was set in a particularly easy way; it also featured a 2.4 MHz quartz resonator with ZT cut as well as a standard 32 kHz resonator. As the ZT quartz and its divider circuit needed a lot of power, it was only switched on every 10 minutes for 10 seconds in order to set the 32 kHz frequency.

An extraordinary rate and a battery life of 10 years were achieved with a 3-volt lithium battery that measured 22.0 mm x 2.5 mm. The 30 mm caliber featured three motors for the second, the minute and hour, and the day/date function, respectively. This design was patented, the patent became public domain in 2011. A test series of 400 pieces was assembled, but there was no production; none of those prototypes ever left the Rolex premises.

The Pseudo-Analog Caliber FAN

In the mid 1970s, Heiniger let it be known that Rolex was interested in new and original quartz watch designs. His technical director René Le Coultre immediately recalled the CEH design of caliber Delta, a pseudo-analog, solid state movement with light emitting diodes and a solar cell power supply, for which no watch brand so far had shown any interest (see the section CEH in this chapter). This was the starting point of the FAN project (FAN = Forme Analogique). The Rolex electronic team was enthusiastic, even though chances were small that such a watch would ever be produced in series. Working for Rolex presupposes strong nerves and the will to produce "l'Art pour l'art." Engineers who want to see their designs cut into metal immediately, should not work for Rolex.

In 1975, Le Coultre "borrowed" Raymond Vuilleumier from the CEH in order to find vendors in the United States who might become suppliers of parts for the FAN caliber. Veuilleumier had worked for six years with General Electric and had built up a comprehensive network of microelectronic firms all over in the USA. Furthermore, Vuilleumier had been in charge of the Delta caliber project. Le Coultre also decided to take along his electronics specialist Edmond Zaugg, who also was an enthusiastic radio amateur.

The trio travelled to the United States three times between 1975 and 1976. They were received everywhere with utmost courtesy: the name Rolex opened any door. The subcontracts for FAN could be placed with the best qualified firms. Thus, the FAN-dial module was supplied by Hewlett Packard in Palo Alto, and the multiple-layer connection module for the light emitting diodes was designed by Ceramic Systems in Sorrento Valley near San Diego. The CEH was responsible for the integrated circuit, Rolex would provide system integration and assembly.

The CEH Delta concept thus evolved into the Rolex FAN caliber 7035; the dimensions were 30.0 mm x 8.00 mm with a pseudo-analog LED display. The hours were displayed with 4 light emitting diodes in a row, the minutes with 7 such diodes, the seconds with a set of 60 diodes along the edge of the dial: they were lit one after the other within a minute in the clockwise direction. Thus, a radial matrix of 60 by 7 LED was needed. As a help to the user, a pusher was provided that lit two diodes at 12 o'clock and one diode at 3, 6, and 9 o'clock.

The date indication, consisting of the conventional 7-bar LED matrix, was arranged in the center of the dial. In order to energize it for 2 seconds, a key had to be pushed twice; a third pulse switched on the number of the month. A photodiode next to the date display controlled the intensity of the LED in function of the surrounding light level.

There was no need for solar cells, as the continuous operation of this movement was unthinkable: a key to switch on the display Pulsar-like was indispensable. Rolex built just five prototypes with the components that were supplied by the CEH and the American companies mentioned above; they worked in August 1978.

Developing the FAN had cost a cool million Swiss francs, but Rolex was not much poorer for that. When Heiniger decided that such a watch did not correspond to the Rolex product philosophy, the FAN project was cancelled and the development costs were written off. Publications on this subject were not authorized.

Ronda

The Ronda Company was founded in 1944 by William Mosset (1909-1985) in Lausen near Basel; it produced pin pallet and jewel pallet ebauches. When the quartz technology had proven itself as a viable alternative to micromechanics in the early 1970s, Ronda made the transition to microelectronics. The required components – silver oxide batteries, quartz resonators, CMOS-type integrated circuits, and Lavet stepping motors were available commercially. They were mounted on the modified plate of mechanical movements.

To that point, Ronda had produced most of the parts of its movements in its own plant, but the company began to depend strongly on outside vendors. Ronda bought a patent license for producing the Seiko version of the Lavet stepping motor, the samarium-cobalt rotor was purchased from Audemars in the Tessin. By 1978, chips were supplied by the Philips subsidiary Faselec, when the power-saving CMOS technology was introduced there.

8.129. Ronda quartz caliber 1173

8.130. Ronda quartz caliber 1377

8.131. Ronda Harley, caliber 375

8.132. Ronda Harley, caliber 7875

Early Quartz Calibers

Caliber 1373 was the first quartz caliber by Ronda; it was exhibited at the 1973 Basel Watch Fair, but it was just a Seiko module. A year later, Ronda developed caliber 1377 (13½ lines x 6.0 mm). The integrated circuit and the quartz resonator (first bar-type, then tuning fork-type) were supplied by Motorola. Production reached 10,000 pieces a month.

The first Ronda quartz movement that was fully developed in-house was caliber 671 (6¾ x 8 lines, motor pulses every 10 seconds). At the 1977 Basel Watch Fair, Ronda exhibited the 11½-line calibers 373, 375, and 377 (11½ lines x 4.2 mm, 1 meaning two hands, 3 three hands, 5 date, 7 day/date), and caliber 871 (8¾ lines x 3.7 mm). Price erosion accelerated dramatically: in early 1978, the 11½-line movement sold for 40 Swiss francs. At the end of the year the price had gone down to 26 francs, soon afterwards it hit 20 francs and even less.

The 1979 innovations were the 8¾-line movements 873 and 875. A 10½-line movement on the 873-875 base was now produced with the caliber numbers 1073 and 1075. In that year, a total of 450,000 quartz movements was produced, quartz watches were discontinued; the total production generated 12.1 million francs.

The End of the Mechanical Era

When Eric Mosset took over the company in 1989, the first priority was to secure Ronda's survival. Not an easy task in the face of the continuing price war that had actually been started by the Seiko-Citizen competition. At that time, the price for a quartz movement was 6 Swiss francs, but it fell by 30 percent in a very short time. This forced many manufacturers to close.

In the 1990s, only mass production could save the day. The 6¾ x 8 line, three-hand caliber 763 was of utmost importance in the high volume business. The classical technology with photolithographic tuning fork quartz, CMOS chip, and Lavet motor had reached full maturity. Ronda survived thanks to specialties such as small movements, movements with a small seconds-hand, and movements with a date disc.

The product range consisted of calibers in the sizes 5½ x 6¾ lines, 8¾ lines, 10½ lines, 11½ lines, and 12 lines. In the mid-1980s, Ronda introduced the Harlekin 1 technology, a sandwich construction with a very thin plate (0.6 mm) and inserts of different heights made of injection molded plastic. Those basic calibers have hardly changed in the past twenty years.

Quality and Diversity

The astonishing fact that Ronda – a pure quartz movement manufacturer – survives and thrives in the face of an extremely aggressive Chinese competition and even sells the bulk of its production in Hong Kong is due to several factors. The superior Ronda quality that stays stable even for very large series is an important asset and justifies higher prices. Ronda long ago discontinued its low-priced lines. Instead, it manufactures about 60 high quality calibers with a full-metal design.

For example, Ronda manufactures the Slimtech family with a height of 1.9 mm, including date disc and small seconds-hand. Two other product families, Normtech and Powertech, are characterized by a height of 2.5 mm and 3.0 mm, respectively. Ronda supplies movements to most of the remaining independent watch producers. Ronda's old Roskopf image is long forgotten: the brand stands for innovative products with an excellent price-performance ratio. One example is a big date chronograph that allows for unique designs.

The new 15-line Mastertech Line is ideally suited for the modern, really big watches with a nicely balanced display. Counters and other accessories are located at the expected place and are not concentrated in the center as happens if one tries to design big watches with small movements. Mechanical retrograde functions driven by a quartz movement are very popular nowadays. The wide scale of movements is due to the Ronda platform technology; the same basic movement may be combined with modules for many different functions.

SGT

One of the remarkable pieces that was exhibited at the 1972 Basel Watch Fair was a fully electronic watch with a liquid crystal display signed SGT (Société des Garde-Temps) in La Chaux-de-Fonds. SGT was created by the merger of a number of medium-size watch producers: Degoumois/Avia, Vuilleumier, Fleurier, and Solvil & Titus. Those companies acquired in common a third of the share capital of the American Waltham in Waltham, Massachusetts.

In 1970, more watch brands joined the conglomerate: Sandoz, Invicta, Elgin, Illinois, Helvetia, Isméca, and Gasser-Ravussin. By 1973, SGT had advanced to the third largest watch group in Switzerland (after ASUAG and SSIH). It went bankrupt in 1981, the brands owned by SGT were sold individually.

The SGT quartz watch has an interesting history, as its development was financed in 1972 by SGT and five German watch producers: Adolf Rapp (Adora) in Schwäbisch-Gmünd, and Eppo, Exquisit, Ormo, and Para, all of them in Pforzheim. The SGT subsidiary Waltham Watch and Optel Corp. in New Jersey were the technical actors behind the scene. The dynamic scattering type 3½-digit LCD was suboptimal in terms of contrast and reading angle, but nothing else was available. Hours and minutes were displayed in a rectangular, 25 mm x 14 mm window. In the center, two dots flashed 60 times a minute. In 1974 SGT launched the caliber ERC2, with a much superior LCD of the twisted nematic type. This caliber was cased by 21 Swiss and 4 American watch brands.

The SGT watch had no mechanically moving parts, aside from the crown that was used for setting the time and changing over to another time zone. The time base was a 32 kHz quartz resonator; the frequency divider and the decoder needed for controlling the display were integrated on a single chip. The power source was a silver oxide battery with a life of at least a year. The rate was better than ± 5 seconds per month.

Sicura

The important Roskopf manufacturer Sicura in Grenchen (that bought Breitling in 1979) understood the signs of the time and switched over to microelectronics in the course of the 1970s. In order not to lose time, quartz modules and LED were purchased in the United States and assembled in Grenchen. This is how the so-called "Superman" was created; it was launched in 1976. This watch featured two totally independent movements: a conventional automatic caliber that was connected to a micro generator, and an LED module. The required power was delivered by a nickel-cadmium battery that was recharged by the micro generator. Thus, the watch displayed the exact time, day and night.

Before this, watches of this type had drained their battery rather quickly. The Sicura model solved this problem with its micro generator actuated by motions of the wrist and with a rechargeable battery – that unfortunately had a propensity to leak.

In 1976-77, Sicura also produced the LED digital watch "Chronoflash" with the Frontier module 2006 that featured the chronograph functions. The inherent disadvantage of this type of watch was that two hands were required to display time. The LED lit only after touching a key, otherwise the power consumption would have been untenable. For this reason, Sicura

8.133. SGT square caliber with LCD

8.134. Sicura Solara LC with Economic Swiss Time (Hong Kong) module, front side

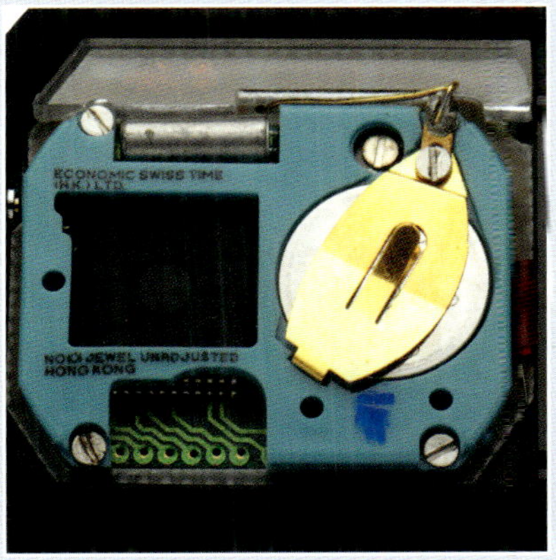

8.135. Sicura Solara LC, back side

8.136. Sicura Quartz Solar with FHF module

looked far and wide for liquid crystal displays that stayed on permanently thanks to their very low power consumption.

Unfortunately, the LCD available from Optel at that time were of the dynamic scattering type. Later, Videlec in Lenzburg produced much superior twisted nematic LCD that were really tight. But they were rather expensive for low-cost Sicura products. Yet, the solid state electronic watch was a status symbol; a credible manufacturer needed it in his collection.

Sicura was among the pioneers of solar cell-powered quartz watches. The silver oxide battery was replaced with a nickel-cadmium rechargeable battery: its life was a respectable five years. In 1976, Sicura launched the "Solar Star Quartz:" it rated as a sensation at the Basel Watch Fair. The light of a candle was enough to make it work – even when the battery was removed. In 1977, Sicura unveiled the analog model VIP 2000 "Sikato;" the solar cells were located on the right side of the case. It was equipped with the baguette quartz caliber Fontainemelon (FHF) 102.001. In 1978, Sicura launched the "Chrono Alarm" with chronograph and alarm functions; the module was imported from Japan. Sicura's product line also included analog solar quartz watches with Ronda or Seiko movements. Seiko mostly supplied movements with a date disc; later on, ETA had an equivalent product.

Swatch

Successful inventions have many fathers – inevitably, this was also the case with the Swatch. This low-cost, high-grade watch was launched in the United States in 1982 and in Switzerland in 1983. The plastic case watch with the unmistakable look was also the start of a revolution in production technology: automation. This was due to the fact that the Swatch was designed to be assembled dial side up and needed no bridges, the case bottom being used as the plate. This greatly facilitated the automated assembly. Design, development and production were concentrated at ETA in Grenchen.

Impossible Objective History

The history of the Swatch must be divided in three parts. When the concept was created and developed (1980-1983), very few people at ETA believed that such a "funny watch" would ever work, let alone that it would become a runaway success. After the launch and the first waves of spectacular sales (1983-1985), many former skeptics jumped on the bandwagon and pretended they had helped to create this miracle. Very clever marketing made the Swatch into a fashion article, into "the world's craziest watch." After 1985, the Swatch became a cult object all over the world, yearly production reached the tens of million pieces.

The triumph was perfect when the entire SMH group (that had been created by the merger of ASUAG and SSIH) changed its name to "Swatch Group" in 1989. The product grew into a giant that needed more and more employees and managers. Quite understandably, their perception of the Swatch was quite different from that of the inventors. And everybody has his own, subjective version of the honest-to-goodness history of the Swatch. One can at least try to illuminate an enormously complex object such as the Swatch from many different angles. Inevitably the question marks will never be eradicated.

The Delirium Vulgaris Project

With the 0.98 mm Delirium IV that was launched in 1980, the Swiss watch industry (or rather ETA) won the race for the world's flattest watch. This did not generate many new millions in terms of sales, but the Swiss had shown their competence at the uppermost level of quartz watches. ETA unquestionably was the leader, which actually was rather strange as by statute ETA was a producer of movement blanks and movements. It was not supposed to produce complete watches, as this would make it into a competitor of its own customers. Actually, the four versions of the Delirium were something very special, as the movement was the watch. ETA thus found a back door for producing watches, while pretending that it was an assignment it got from the "launching carriers" Concord, Eterna, and Longines.

The Delirium was very expensive, but as the number of pieces was small, it could not solve the industry's financial problems. After the Delirium, the focus shifted to the low-price sector, which had to be taken in storm. It was obvious that the industry needed a low-cost watch that could be produced by the millions and would generate big profits. This was the only way to preserve and modernize the watch industry's infrastructure. Once more, ETA was going to produce a complete watch: all the better if it again pretended to be a movement integrated into a case.

ETA general manager Ernst Thomke having defined the boundary conditions, the "Delirium vulgaris" project was launched in January 1979 (Delirium vulgare would be better Latin). The idea was to design a low-cost Delirium in which the gold case and the sapphire glass would be replaced by plastic. ETA engineers worked out a complete set of specifications. Basic concepts

8.137. Swatch Display

were formulated: undemountable, non-repairable, very simple, very low cost, waterproof to 30 meters, easily replaceable battery. There was no ambition to establish new records as to the height, about 8 mm would do. Production costs should be well below 10 Swiss francs apiece. The project went through several names: Delirium vulgaris changed to Delirium popularis (lousy Latin again!), then to S+Watch (for Swiss watch, 1981) and finally to Swatch (1982).

Ebauches SA and ETA

In 1979, the idea was not to invent the quartz wristwatch, as this had already been done in 1967. The plastic watch was not something new either; the "Flipper" by Fortis with a pin pallet movement and the "Fiesta" by Oris with a low-cost quartz movement had been rather successful products in the 1960s and 1970s, respectively. The objective was to produce an attractive, high grade, sturdy, and waterproof watch for 5 to 10 Swiss francs in Switzerland. This meant reducing production costs by two-thirds, which could not possibly be achieved with an evolutionary development of the classic quartz movement. This meant reinventing the quartz watch – a true revolution. A number of near-holy watchmaker dogmas, such as repairability, had to be thrown over board.

8. 138 - 8.146: Swatch models for men and ladies that were shown at the press conference on March 1st, 1983 in Zurich

8.138. Swatch of 1983

8.139. Swatch of 1983

Apparently, nobody within the ASUAG group was interested in the specifications of a revolutionary new watch: the list was buried in a file cabinet. Yet, Ebauches SA in Neuchâtel did accept the challenge of the Delirium vulgaris: in July 1979, a team headed by André Beyner and René Besson worked out the design of a rectangular plastic watch with the dimensions 34 mm x 30 mm x 4 mm and an independent movement that was not integrated with the case back. This design missed the specifications by far; besides, it was quite ugly. One hears rumors of other attempts to develop a Delirium vulgaris within the group, but they got nowhere. The project was stalled, but resuscitated in a totally unexpected way: the "culprits" were two ETA engineers named Elmar Mock (b. 1954) and Jacques Muller (b. 1947).

Two Improbable Actors

After obtaining his degree at the watchmaker school in Bienne, Elmar Mock was hired by ETA in Grenchen in 1976, even though the quartz crisis was worsening rapidly and more and more people were being laid off. Fritz Scholl who was then ETA's general manager saw that precision injection molding of plastic would play an important role in the next generation of watches. He secretly bought a small injection molding machine, declared it as laboratory supply in the ETA bookkeeping records, and hid it in the attic of the ETA main building.

This was where Mock taught himself plastic technology. It became quite obvious that he did not understand the scientific and technological aspects of what he was doing. He decided to enroll in a special one year curriculum at the technical school in Windisch, a town in the eastern part of Switzerland. ETA management approved the leave of absence under the condition that Mock would return to Grenchen within a year. Mock temporarily moved to Windisch and graduated as a plastics engineer in 1978.

When Mock returned to Grenchen, Fritz Scholl had retired and Ernst Thomke had taken his place. Thomke was a true visionary who immediately knew what kinds of product ETA needed. He also was a born leader and a very demanding but extremely fair boss. Mock was given three technicians who helped him develop precision molding of small watch parts that were indispensable for electrical insulation. He produced parts for the legendary "Delirium I," the world's first watch with a height of less than 2 millimeters, and for other ETA calibers.

Elmar Mock's friend and colleague Jacques Muller is a citizen of Bonfol in the Ajoie, close to the French border, but grew up in Tavannes further to the south. After high-school graduation he just couldn't make up his mind whether he wanted to be a surgeon or a pilot. The compromise was to enroll at the technical school in Saint-Imier and to study microtechnology on the fast track. Later, he still would have a chance to change his mind. He graduated at the age of 21 years in 1968.

Employment offers abounded for freshly graduated engineers, but they all came from the watch industry. Muller had decided early that watchmakers did not understand the physics of what they were doing and were just a bunch of amateurs. As he had no choice, he accepted a job with Ebauches Tavannes (that belonged to Ebauches SA). Surely he would find a job more to his liking within the next six months.

Those six months developed into ten years in the course of which Muller got a very thorough training in designing electromechanical and electronic movements. Ebauches Tavannes closed in 1978, but Muller was offered a job with ETA in Grenchen, where he designed several ultraflat calibers in record time. Mock and Muller met at a conference in Germany. As the talks were rather boring, the two Swiss engineers often retired to a bar where they elaborated dream projects about the utilization of plastic in watchmaking. Those dreams materialized in the form of a close cooperation in which Muller provided the horological background while Mock contributed his newly acquired know-how in plastics technology. Seven common patents grew out of this.

Precision Injection Molding

Mock could not guarantee the specifications of the plastic parts he had to produce for ETA with the low-cost injection machine Fritz Scholl had bought. He decided to launch a request for a high precision Netstal machine that cost half a million Swiss francs. The procedure for authorizing such an investment was very elaborate. Finally, Thomke was given the stack of forms for approval. Inevitably, he wanted to know exactly what this was all about. He asked Mock to meet him in his office at 1 p.m. on 27 March 1980. At 11 a.m. on this day, Mock was informed about the meeting by Thomke's secretary. Mock had exactly two hours to find really good reasons why ETA should purchase the Netstal machine.

This was an emergency: Mock literally ran to Muller's office. In common, they elaborated the concept of an ultra low-price plastic watch; the bottom of the case would serve as plate. The coil and the stator of the stepping motor would be fastened there. In this early design, an intermediate level made of glass fiber-reinforced plastic was needed for wheel train bearings and the integrated circuit. The plastic watch glass was going to be welded to the case to make it waterproof and give the case extra rigidity. This defined the basic concepts of the Swatch. The birth of the Swatch therefore is known very accurately: 27 March 1980 between 11 a.m. and 1 p.m.

Muller made a quick drawing of a section of the future watch on graph paper and colored it with pink and blue pencils he happened to have on his desk. Mock pocketed the paper a few minutes before 1 p.m. and found the big boss in a very bad mood. His opinion was that quite obviously Mock had gone insane in asking for half a million in order to purchase a completely useless machine. What could be done with it anyway? Mock pulled Muller's drawing out of his pocket and explained that the machine could be used to produce very low cost plastic watches.

Thomke looked at the drawing and immediately realized that this was a stroke of genius. Thanks to this technology, ETA would be able of producing and selling a complete watch with full impunity, as this was a movement integrated into a case. Yet, Mock had to hear that leading movement designers in the group such as André Beyner and René Besson had worked for years on a similar design. Thus, the perspectives were not encouraging.

Development of a Concept

After the meeting with Thomke, Mock's supervisor, Urs Giger, appeared in a very irritated mood. What business did Mock have to show the big boss an absolutely silly drawing that didn't even have an ETA number and had no relation with any ETA project? Apparently, Mock wanted to fool around with plastics – this was his chance. Thomke had given firm orders that Mock and his colleague Muller would immediately stop working on whatever they were involved in: henceforth, they would report directly to him in person. They were given six months to develop their plastic watch and deliver working prototypes. Furthermore, they would have an absolute priority in whatever supplies, parts and services they needed. As an aside, Mock learned that the purchase of the Netstal machine had been approved.

Muller and Mock rolled up their sleeves and started working. The 18-hour workday, six to seven days a week, was more the rule than the exception, but developing the intimately entwined electronics and micromechanics of an entirely new watch was glorious fun. Sure enough, Mock and Muller enjoyed the full "support" of numerous specialists at ETA and other ASUAG plants. However, their help mostly consisted in trying to kill the ideas of the inventors.

8.140. Swatch of 1983

8.141. Swatch of 1983

8.142. Swatch of 1983

8.143. Swatch of 1983

Nearly everybody was convinced that the Muller-Mock watch would never work and if it did, nobody would buy it. But Thomke's support for his two "madmen" never faltered.

One must remember that the entire Swiss watch industry (including ETA) was going through a devastating crisis. Every month more and more people lost their job; it was important to be on the safe side and avoid dealing with the Muller-Mock duo. Everybody knew that their insane project was doomed to fail. When this actually happened, chances were good that their entourage would also be pulled into the maelstrom and be terminated.

It is not really astonishing that the entire ETA establishment kept its distance as soon as Mock and Muller had received their assignment from Thomke. This was actually the best thing that could have happened to them. Nobody interfered with their work, nobody bothered them, bureaucrats and jugglers of obscure norms were far away. Mock and Muller were left alone and developed their watch in record time.

Variants of the Project

On 1 July 1980, a first evaluation of the project was made. Muller's presentation consisted of no less than 8 variants of the Delirium vulgaris that were thoroughly evaluated. The common denominators were a one-piece plastic case, two or three hands, and a date disc. Mock suggested a watch glass made of methyl methacrylate (Plexiglas), while ABS (an alloy of acrylonitrile, butadiene, and styrene) would be ideally suited for the case. The watch glass could be easily welded thermally or ultrasonically on ABS. The new watch would also feature an easily replaceable plastic bracelet.

As to the battery, either a silver oxide-type (1.5-volt) or a lithium-type (3-volt) was suitable: silver oxide was the winner. The motor had to be a variant of the design that had been invented and patented by Marius Lavet in 1936. This motor had been vastly improved and repatented by Seiko in Japan. Muller opted for a radially magnetized samarium-cobalt rotor with a plastic coating.

The motor's stator also had an important mechanical function: it served as high precision metallic holder for the gear train's bearings. The bearings would all be flying, there were no bridges. Just the rotor had to have bearings at both ends; for this purpose, it was inserted into a tiny plastic box with two open sides. Much effort was put in reducing costs in an extreme way, saving even a fraction of a cent rated as a triumph.

First Prototypes

Mock and Muller continued with their computing and drawing. In the fall of 1980 they assembled their first prototype, which was an unmitigated disaster. The hands did turn, but anticlockwise. In addition to that, the watch stopped at the slightest provocation. Thomke extended the deadline and ordered that the caliber should be entirely redesigned.

In January 1981, a lot of ETA employees bitterly complained that Mock and Muller's lab stank to high heaven. Unfortunately for the colleagues, Mock had to experiment with many different plastics and tried to melt, glue, and weld them. It was, of course, extremely important to find just the right type of ABS for the case. Nobody could decide what colors to choose, but at least the concept of a battery case that was accessible from the rear was adopted unanimously.

Swiss people do not throw away batteries, let alone complete watches. The battery had to be replaceable easily by the customer, who was expected not to pollute the environment. This might be done elegantly with a battery case cover with a bayonet catch. The only tool required for opening the cover was a coin. Yet, after the Swatch was launched, just a small market for replacement batteries came into being. Whenever a Swatch stopped working, it was put away in a drawer and a new one was bought that was fully in line with the fashion trends of the time.

In December 1981, five prototypes were working; the help of Mock's father, who was an excellent watchmaker and worked with Eterna, had been invaluable. Muller and Mock had developed a revolutionary new watch – from the idea to the first working prototypes for a mere 800,000 Swiss francs. Inevitably, many details were improved in the course of time, particularly to ease the automated assembly, but to this day, the basic design is still the same as in 1982. Getting the automated assembly to work was quite another thing; it cost at least 17 million.

Several months were needed to eliminate early problems and to adapt it to unusual production technologies. For example, finding the best injection point so that the mold for the case would fill rapidly and homogeneously turned out to be extremely difficult. As cases were urgently needed for demonstration models, they were laboriously milled with a numerically controlled machine.

The mechanism for setting the hands and changing the date was a very tough problem. The arbor ending with the crown had three positions and was very complicated. This mechanism had been inspired by a Cyma clock produced in the 1930s. Machining this arbor cost a full Swiss franc, making it the most expensive part in the entire movement. The tolerances were extremely critical and it was difficult to hold them in a huge, automatically machined series. This problem was solved in March 1982 by a colleague Alfonse Bron working in the ETA design department.

A single defective part meant that a pre-series of 10,000 pieces had to be scrapped. Those watches had been manually assembled for the U.S. launch in Dallas in the fall of 1982. One thing worked without a hitch: digital frequency tuning. Thanks to it, the prescribed daily rate of one second per day could easily be achieved. Even so, there was room for improvement, but the time allotted to adjustment on the assembly line could not be extended beyond 2.7 seconds.

Planning the Production

Thomke had hoped to launch a first commercial series in the spring of 1982, but that was wishful thinking. At least the name of the new watch was finalized: it would be Swatch. The potential of a Combo Swatch, a Swatch with a moon phase display and a Swatch pocket watch were evaluated, but such special models were left to the future.

For the time being, ETA would produce a simple model with or without a date disc. In December 1982, a series of 300 Swatch were manually assembled for observation in actual wear. A third of them were returned; the problems mostly concerned the case and the watchband. This system was immediately redesigned.

New problems kept surfacing. It is well known that inventing something is always the easy part. Sure enough, it requires enormous efforts and unusual creativity, the DNA of which is a rare gift of nature. Quite often, creativity thrives under chaotic frontier conditions. However, the difficult part is to convert inspiration into a quality item that may be produced in enormous quantities at low cost and with a minimum of waste. In this phase no geniuses are required, but experienced, meticulous engineers, exactly defining every parameter and leaving nothing to chance.

An interesting example of an unforeseeable problem was a non-negligible frequency shift of the quartz resonator after ultrasonic welding of the watch glass. It turned out that the ultrasound tore off tiny particles of the gold metallization on the quartz lamella, which inevitably affected the resonance frequency. Mock designed brutal shaking machines in order to artificially age the quartz resonators before

8.144. Swatch of 1983

8.145. Swatch of 1983

8.146. Swatch of 1983

8.147. Swatch Chrono, 1990

movement assembly. This removed potentially loose parts of the metallization before the frequency was adjusted to 32,768 Hz. The side effect was 20 to 30 percent waste; for a while, domestically made quartz resonators were in short supply, many of them had to be purchased in Japan.

Quality Control

Implementing rigorous quality control measures was expensive but drastically reduced waste. Muller and Mock had defined each and every production step while they were designing the Swatch, so there were no surprises. However, a simple error in a production tool caused the loss of 20,000 watches; the wheel driving the date disc was off tolerance.

The Swatch could neither be opened nor repaired: this meant that meticulous quality control measures had to be implemented for all watch parts before assembly. Paradoxically enough, the low-priced Swatch required much tighter tolerances than high-priced movements that were assembled manually. For the same reason, the production tools and systems had to be maintained in perfect condition on a tightly controlled schedule.

Each and every Swatch sold on the Swiss market that was returned under warranty was cut open and carefully analyzed in order to find out what went wrong. Thanks to statistical analysis of the defects, weak points in the assembly line could be pinpointed and eliminated. The result was that waste was considerably reduced while the number of faulty watches fell drastically. The correction measures that were introduced in the ongoing production process greatly contributed to making the Swatch an amazingly reliable timepiece.

Paradigm Change Due to Marketing

In April 1982, prototypes of the production version of the Swatch were available: they worked flawlessly. However, the production price was calculated to be 17.70 Swiss francs, including assembly (4.40 francs). Thomke was furious and gave firm orders that the total production price had to be lowered to between 8 and 10 francs. For the production planning office, this meant going back to square one. This time, the productivity gains that were expected from production automation were included in the calculations. After automation had actually been implemented, it turned out that it had a much greater impact than had been hoped for: assembly cost of a Swatch dropped to 0.30 franc per watch.

In December 1982, the Swatch at last could be shown to the media: the first press conference was held in Dallas, Texas. For this historical event, ETA supplied a series of 10,000 manually assembled Swatch. The event unfortunately backfired, as the colors were much too restrained or even ugly. Franz Sprecher wired a lengthy report to Grenchen, reminding ETA that the Swatch had not been designed as a cheap watch for the Third World, but as a fashion item for the industrialized countries. The collection had to be renewed completely every six months: bright, flashy colors and funny to stunning decorations were the order of the day. Max Imgrüth ("mad Max") had the merit of implementing this concept with great vigor. Jacques Irniger and his assistant Konstantin Theile, starting from scratch, built up the very lucrative Swatch market in Switzerland, France, Germany, and Italy.

A highly creative design team was found with Jean and Käthy Robert in Zurich: the two worked miracles. At long last flashy colors and insane patterns appeared on the dial, the case, and the watchband – from African art all the way to Einstein's famous formula $E = mc^2$. The Swatch rapidly acquired the status of "the world's craziest watch," one never had enough of them. In the early days, ETA used the designation 500+three code digits; later, these numbers were only used for men's models. For ladies' models, designations such as 505.111 were used.

Swiss Euphoria

In Switzerland, the Swatch was presented to the media on 1 March 1983 at a four-star Zurich hotel. Quite simple, mostly three-hand models (some with day/date) were shown. This fundamentally new watch triggered an incredible euphoria. On the next day, the high prestige Swiss daily *Neue Zurcher Zeitung* printed a full page article on the Swatch technology. The authors were Jacques Muller and Elmar Mock – not by chance.

Stores that sold the Swatch were emptied in no time. Everybody, but absolutely everybody, was wearing a Swatch, from the garbage picker all the way to the bank president. Soon something totally unexpected happened: the Swatch became a collectible item, quite a few collectors subscribed in advance to every new collection. ETA had to print Swatch books with pictures of all the models ever made.

ETA communicated very openly and with full transparency about the Swatch and its plastic construction. The inside of the case back doubled up as a plate, major parts being welded to it ultrasonically. Assembly consisted of packing the pre-assembled, completely wired electronic module with the quartz resonator, the integrated circuit, and the battery

8.148. Pop Swatch

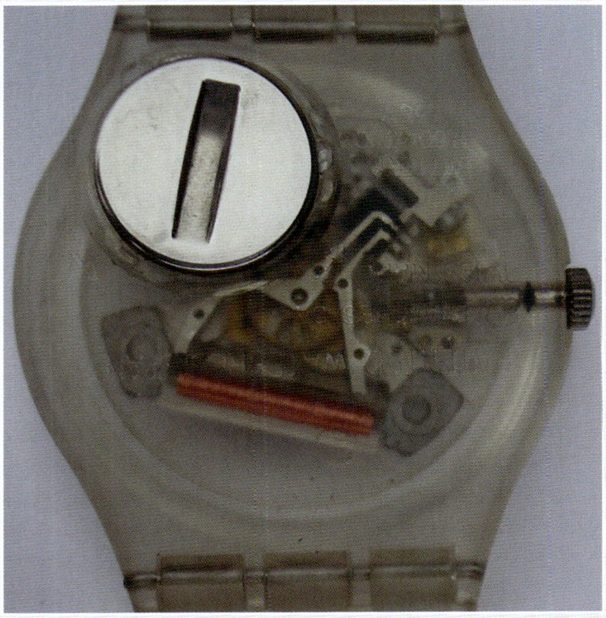

8.149. Jellyfish Swatch, issued to celebrate the first million Swatch watches

8.150. Swatch Scuba

8.151. Swatch Irony Aluminum

holder in the case bottom and welding it in place. The "printed circuit" consisted of a metal grid cut out appropriately; it also served as a support for the coil and the soft iron stator that itself held the pinions of the wheel train.

The samarium-cobalt rotor was supplied by Comadur in Le Locle (an ASUAG subsidiary) and was coated with polyoxymethylene (POM); it directly drove the sweep seconds-hand. The calendar mechanism and the setting mechanism also were pre-assembled. Final assembly of the movement was first done manually: automation was introduced gradually until finally full automation was achieved. For a long time, the last manual operation was soldering the fine copper wires of the coil to the supporting grid. The two parts of the watchband were connected to the case by a simple hinge mechanism; they could be replaced within seconds.

The integrated circuit of the early Swatch series was supplied by Faselec in Zurich in SOT8 packaging. Later, Ebauches Electroniques Marin took over the production of the digitally tuned integrated circuit, also in the SOT8 configuration, as it was soldered to the grid that both had electrical and mechanical functions; there was no printed circuit. Renata in Itigen near Basel (an ASUAG subsidiary) supplied the battery. ETA produced all mechanical parts of the Swatch, including the ABS case and the dial. Kyburz & Cie. in Marin made the methylmethacrylate watchglass; the hands were purchased from Universo in La Chaux-de-Fonds.

The Swatch Era

Marianne Egli was in charge of international public relations for the Swatch. Her team had the unique privilege of nursing an entirely new product that became a worldwide hit with explosive growth rates. In the course of three decades, the Swatch had to be reinvented many times so it would make ever-new headlines. The classical Swatch was followed by the Pop-Swatch, the Swatch-Chrono, the Scuba Swatch, the Stop-Swatch, the Platinum Swatch, the Swatch-Solar, the Swatch Automatic, the Swatch Musicall, the X-Large Swatch, the Swatch Irony with steel or aluminum case, the Swatch Access, the Swatch Skin, the Swatch Beat, the Square Swatch, and special models designed by artists as limited editions. In 2013, the Swatch celebrated its 30th anniversary: by then, a grand total of 400 million Swatches had been sold.

Development work on automating the Swatch assembly had, of course, started long before the Dallas and Zurich presentations. The fully robotized, high precision assembly line of small parts was an absolute first for industry worldwide. Everybody was in a great haste, as sales grew exponentially. The money invested in this project was well spent: at long last, the Swiss watch industry had a low cost mass product for the bottom of the price pyramid.

Automated assembly of the Swatch was a big quantum jump. The know-how that was acquired with this technology was soon transferred to conventional quartz movements, which meant tremendous cost reductions. Towards the end of the 1980s, equipping new Swatch derivatives with conventional quartz movements cost less than using the original Swatch technology.

Using the plastic case bottom as a plate was only an episode in the history of technology; at any rate, it had already been patented in 1890, as was discovered eventually. Fifty percent of the Swatches put on the marketplace today still correspond to the 1983 design. The Swatch technology was not really revolutionary: the actual revolution was the fully automated movement assembly. It was bound to come, but the Swatch triggered and greatly accelerated this development.

Many Fathers

The Swatch had many fathers: it was the result of a huge and very costly team effort. However, there is no question about the fact that Elmar Mock and Jacques Muller were the prime movers. And without Ernst Thomke's unwavering support, they would have gotten nowhere. Many ETA and Ebauches SA employees also played an important role in the development of the Swatch production technology, marketing and management. Those names are listed below, they were taken from a list published by the Swatch group: it may not be complete.

A. Armellino, A Bally, H. Baumann, C. Besançon,
A. Beyner, M. Bieri, P. Bischof, A. Bosisio, F. Bron,
R. Dinger, M. Egli. E. Favre, O. Franck, F. Füllemann,
D. Gaspoz, K. Gerzner, U. Giger, H. Gigi, F. Grindberg,
Y. Guérin, E. Gunzinger, B. Hammond, U. Hauerter,
W. Hügli, F. Kauffungen, P. Marchand, P.-A. Meister,
K. Renfer, P. Renggli, P. Salathé, J.C. Schaffner,
B. Schmid-Müller, H. Schneider, T. Schönenberger,
H. Sollberger, H. Sommer, D. Spahr, F. Sprecher,
M. Traub, P.Triebold, E. Weber, H. Widmer, H. Zaugg,
H. Zimmermann, G. Zoppoletto, G. Zubler.

The worldwide racket that was triggered by the Swatch had positive repercussions for the entire Swiss watch industry. The story about the man who wanted to buy a Swatch and left the store with a Patek

Philippe may be apocryphal. Nevertheless, the Swatch meant the end of a 10-year, devastating crisis that had cost some 60,000 jobs. It was also the beginning of a new era: in 1984 the Swiss watch was back on a growth path with a fully rejuvenated image. ETA never officially published production figures, but according to educated guesses, up to 23 million Swatch were sold in the record year.

As the Swatch is a collectible item, many early models have been preserved for posterity. Nobody pretended that the Swatch would last for ever: a few years was all that was needed for a fashion article, yet the sometimes rapid decay of the ABS plastic case and of the watchband took the manufacturer by surprise. It does indeed happen after 20 or 30 years that cracks appear in the cases and particularly in the watchbands, which eventually leads to the decay of the structure. These are rare cases, but museums are confronted with the provocative question whether the Swatch is really suitable as a museum piece. No risks would be taken if the Swatch were kept under an inert gas in the dark, but this is not the way museums operate.

Precious metals, stainless steel, copper, bronze and brass will last for thousands of years. Certain minerals such as zircons seem to last "forever" – some of them are nearly as old as our planet, i.e. well over 4 billion years. Plastics (i.e. polymers) are very different. They are based on the chemistry of carbon that is also called "organic" for historical reasons. They age, mostly under the influence of the oxygen in the air and ultraviolet radiation. In the long term they lose their strength and decay: a heap of dust is all that eventually remains. Everything organic has the same fate – this of course includes us humans.

Tissot

The first Tissot watches were manufactured in 1853. In that year Charles-Félicien Tissot-Daguette (1804-1873) and his son Charles-Emile (1830-1910) opened shop in Le Locle; remarkably enough this was the son's idea. He had been apprenticed to a watchmaker and had learned the watch business in the course of a long stay with his uncle in New York. He then accepted the hardships of extensive travelling and thus opened both the American and the rapidly growing Russian market. Tissot's very sturdy and reliable savonnette pocket watches were greatly appreciated on both sides of the Atlantic.

Charles-Emile Tissot belonged to the small minority of Swiss managers who understood, at an early date, that watches should not be manufactured in kitchens but in modern factories. It was indispensable to introduce the American technologies, i.e. semi-

8.152. Swatch Skin, transparent model

8.153. The inventors of the Swatch: Jacques Muller (left) and Elmar Mock

8.154. Tissot ana-digi caliber 2045

8.155. Tissot quartz caliber 2030 Lapin

automated mass production of identical, high precision parts. This vision materialized in a very modern plant that was built in 1907 on the hill overlooking Le Locle from the south. This still is Tissot's home – after many transformations and additions. In 1912, production shifted to wristwatches that subsequently were made waterproof and non-magnetic. During the World Economic Crisis, Tissot merged with Omega in 1930. This was the nucleus of the powerful SSIH group that subsequently incorporated the low-price watch manufacturer Lanco in Soleure and several important makers of pin pallet watches in the Basel area.

Unfortunately, there was no coherent brand policy within the group nor clear concepts as to the future of the group. The development of the mechanical plastic watch "Syntal" with the Astrolon caliber, most of the parts of which were made of polymers, ended up being written off as a costly adventure. Trying to compete with quartz watches that cost less and were much more accurate was an exercise in futility. At any rate the required, advanced plastics technology was not sufficiently developed at that time.

Tissot did not have the critical mass that would have allowed it to make the transition to quartz on its own. Thus, the company tried to survive through alliances and calibers purchased from outside vendors. Tissot actually ventured into electronics in 1973 when it began to case the tuning fork movement "Mosaba" made by Ebauches SA (Tissot calibers 2010 and 2020). In the same year, Tissot started a cooperation with Lanco and Hamilton with the objective of developing a digital quartz movement with a liquid crystal display of the dynamic scattering type. It was exhibited at the Basel Watch Fair but never went beyond the prototype stage.

Tissot was remarkably successful with an analog quartz movement that the R&D department of SSIH (it was Omega R&D until 1972) had developed. This was caliber 5500 with a date disc and electronic setting of minutes and seconds by means of a pusher in the crown. The bar-type quartz resonator was supplied by the SSIH plant in Bienne. Tissot renamed the caliber 2030 "Lapin" (rabbit, 28.0 mm x 6.4 mm); it just featured 3 jewels and was used in the low-price watches of the group. The mechanical part was taken from an old Tissot caliber. Production started in 1973; after the Tissot manufacture was closed, Lanco produced the caliber for a few more years.

Despite this success, 1976 was a devastating year for Tissot, the balance sheet was deep in the red. The crisis had reached alarming proportions and Tissot decided to give up designing and manufacturing mechanical movements in Le Locle. This activity was

8.156. Universal Genève, quartz caliber 1-74

8.157. Zenith caliber 47.1, front side

8.158. Zenith Caliber 47.1, back side

8.159. Circuit of the Zenith Caliber 47.1

concentrated with Omega for the entre group. Tissot now used Omega- and Lanco-made movements for mass produced watch models. For special designs in smaller series, movements or modules were bought wherever they were available, i.e. in Switzerland, France, the United States, and Japan.

Between 1977 and 1985, Tissot watches were fitted with well over 200 different calibers. Hoping for a big marketing success, just about every conceivable technology was tried: LED modules from Frontier and Nortec, LCD-calibers from Casio, SSIH, and EEM Marin, as well as analog quartz calibers from EEM Marin, ETA, FHF Fontainemelon, and Omega. The critical mass for a profitable product was never reached, while the repair service was confronted with an enormous number of calibers and even more different watch models: it was a nightmare. This devastating scattering of the inherent strength of the brand was stopped with the ASUAG-SSIH merger in early 1984. Now Tissot was forced to buy only from ETA. The only exceptions were a couple of ladies' quartz calibers by A. Schild that continued manufacturing such calibers even though it had already merged with ETA in 1979.

In 1979, Tissot gave SSIH R&D the mandate to develop caliber 2031 on the base of the "Lapin" design. This was the SSIH caliber 5520 "Tatou 1" (25.6 x 4.8 mm). Omega adopted it for its own watches in a slightly slimmer version with 4 jewels: this was caliber SSIH 5522 "Tatou 2", but Omega called it caliber 1370. After the ASUAG-SSIH merger, Tissot was made responsible for good quality economy watches with a remarkable price-performance ratio. Soon Tissot reemerged with highly original and very successful products such as Two-Timer, Rockwatch and Woodwatch.

For the Tissot Two-Timer that was launched in 1986, ETA developed the rectangular 24.0 mm x 20.5 mm x 2.0 mm ana-digi quartz movement 958.331. This was the result of the ETA-SSIH R&D cooperation that had started even before the big merger. A newly designed Lavet motor worked in both directions. Time was displayed with hands and on an LCD. All other functions such as day/date, chronograph, timer, and alarm were displayed on the LCD. A total of seven functions were accessible by just turning the crown. EEM Marin supplied the rather complex integrated circuit.

Universal Genève

Descombes & Perret (later Perret & Berthoud) was founded in 1894 in Le Locle and specialized in chronographs quite early. The company moved to Geneva in 1934 and changed its name to Universal Watch Co. Ltd. (or Universal-Genève). The Aero-Compax Chronograph was so successful that manufacturing had to be moved to a subsidiary in Les Ponts-de-Martel. Brand awareness was enormously boosted in 1954 with the Polerouter that had been designed by Gérald Genta. SAS pilots wore it on the North Pole route from Copenhagen to Alaska and California.

Bulova purchased Universal in 1967 and supplied it with caliber 218 tuning fork movements. Universal used them in its Unisonic Models. Quartz calibers appeared in the 1970s. In 1977, Universal launched a remarkably flat two-hand quartz watch; its movement caliber 1-74 (10½ lines x 3.45 mm) is not found in any other watch. Closer examination revealed that this actually was a Bulova caliber that Universal modified by the simple expedient of shifting the components. In that same year, Bulova sold Universal to a group of investors who sold it to the Stelux Group in Hong Kong in 1989.

Zenith

Zenith completely stopped manufacturing mechanical movements in 1976. In that same year, Zenith launched a very advanced ana-digi quartz movement (caliber 47.0 and 47.1) with two hands and digital LED display of second and date. The hands could be set in steps of one hour when changing time zones. The LED module was purchased from Armitron. Unfortunately, this movement was expensive; furthermore, operating it was done with four pushers, which was too complicated for the average consumer.

Zodiac

Ariste Calame in Le Locle founded Zodiac in 1882; he soon became known for the high quality movements he designed himself. In the 1950s, Ariste's grandson Pierre Calame (b. 1934) gradually replaced his own calibers by movements supplied by ETA, ASSA, and Fontainemelon. Zodiac's last mechanical innovation was the SST Astrograph with an automatic movement; it was launched in 1972. Instead of hands, the Astrograph was fitted with concentric discs with triangular markers for hour, minute, and second. Ebauches SA supplied Zodiac with electromechanical and electronic movements. The well-known quartz version of the Astrograph was equipped with the ESA quartz caliber 9362.

9. *Quartz Movements from Germany*

Arctos

In 1923, Philipp Weber and his partner, Jakob Aeschbach, opened a workshop in the German watch-city Pforzheim. They specialized in assembling blank movements as subcontractors. By 1941, Weber was the sole owner and registered his company as Philipp Weber Uhrenfabrik GmbH. In the 1940s Weber employed 900 people. The company acquired an excellent reputation for its extremely sturdy military watches, but the production facilities were destroyed by Allied bombing at the end of World War II. After reconstruction, the company's name was changed to Arctos Uhrenfabrik Philipp Weber KG in 1947. It produced good quality watches using ebauches from Switzerland and Germany. Philipp Weber's son Werner took over in 1962 and moved the company to a large new building for the production of watches, watchstones, and watchcases.

The tuning fork watch "Accutron" by Bulova was launched worldwide in December 1960. Arctos rapidly managed to secure a sample; its watchmakers took it apart and were greatly impressed. It soon became obvious that Arctos was not in a position to fabricate something similar. The high precision machinery that would be needed was simply not available: one had to find another avenue to watch electronics. Everywhere in the horological world, the tuning fork watch triggered the same questions: how could something comparable or superior be designed without infringing on Bulova patents and without having to machine extremely small ratchet wheels?

The Compulsory Quartz Watch

For several years, Arctos managers didn't quite know what to do. In 1967, a meeting was held with engineers from Valvo (a German electronics company). The consensus was that it should be possible to design a quartz wristwatch. It is remarkable that in Switzerland, Japan, and the United States watch and electronics engineers came to the same conclusion: the quartz watch was the only way to beat Bulova.

Quartz clocks serving as time and frequency standards had existed since the end of the 1920s, but they were cupboard size stacks of electronic racks. That meant that they would have to be miniaturized by a factor 10,000 in order to fit into the case of a wristwatch. Valvo agreed to develop the electronic part, i.e. the oscillator for maintaining the vibrations of a quartz resonator, the frequency dividing circuit, and the motor control. Arctos would take care of the gear train and all other mechanical parts.

Bar-shaped 16 kHz quartz resonators could be supplied by Siemens (later that frequency was doubled to 32 kHz). They were provided with a shockproof suspension and were encapsulated under vacuum. The motor was designed by Arctos watchmakers: this was an anchor-type device moving back and forth at the rate of 1 Hz and acting on a kind of escape wheel that engaged the gear train. Power was supplied by a mercury oxide battery. The prototype, with a diameter of 30.4 mm, was presented to the public and the media in February 1971, just three years after prototype wristwatch movements by the CEH in Neuchâtel and Seiko in Suwa had won the first 15 places at the Neuchâtel Observatory. One month after Arctos, Junghans too exhibited a quartz caliber at the Hannover Fair.

The development of the Swiss and Japanese prototypes had been very expensive. At the Centre Electronique Horloger (CEH) in Neuchâtel up to 90 engineers and scientists had been working for five years. Seiko managed with 20 to 30 people, but it took them seven years to reach their goal. Both projects cost several million Swiss francs. Interestingly, the Arctos quartz caliber 375 (13½ lines x 6.5 mm) that was designed by master watchmaker Alfred Leiter cost next to nothing. Valvo's cooperation was paid for with precision metal parts that the company needed. Only relatively small series of the "Arctos-Quartz" were produced. Assembly involved a lot of manual labor, and for this reason the watch was rather expensive.

Nobody was in a position to predict the incredibly fast development of microelectronics and the 100-fold price erosion of quartz movements within just a few years. It was logical for Arctos owner Philipp Weber to proceed with the quartz watch project – despite dire warnings. Serial production began in 1972. The Arctos Quartz cost 700 Deutschmark at retail and some 30,000 pieces were actually sold. The production model used a very low power CMOS circuit made by ITT, power was supplied by a silver oxide battery that had to be replaced once a year. The quality was flawless and some of those watches are still being worn.

Digital Quartz Movements

To celebrate its fiftieth anniversary, Arctos developed a digital quartz watch with a liquid crystal display in 1973 (caliber LCQ 575, 13¾ lines x 8.5 mm). The first prototype used an Optel LCD, it was exhibited in 1972 at the Electronika Fair in Munich; deliveries started in the spring of 1974. This was the first watch of this kind in Europe; in the United States and in Japan, digital LCD watches had been available since 1972.

For the first series of 10,000 Arctos digital watches, the LCD was supplied by Electrovac in Vienna. They cost 22 Deutschmark apiece and operated at 12 volts. Later, LCD were purchased from Siemens and Videlec in Lenzburg. The integrated circuits were bought from various manufacturers, mostly Eurosil in Eching near Munich.

LCQ 675 was launched in 1975. Next to hour and minute, it displayed the date and an AM/PM indication. A year later, Arctos started producing the LCQ 775 model. The dimensions were 30.8 mm x 8.5 mm and the LCD could be illuminated with a light emitting diode. The movement operated at 3 volts, thus two silver oxide batteries were needed. Caliber LCQ 790 was almost identical to caliber 775, but the 12-hour mode could be switched to the 24-hour mode; production started in 1977. The multifunction alarm-stop-chronograph LCQ 880 was developed simultaneously with LCQ 790. Next to hour, minute, and second, it displayed day/date and month. The alarm could be set a month ahead of time, the stopwatch had a resolution of 1/100 of a second.

The Arctos digital watch project was partly financed by BWC Buttes Watch Company (see Chapter 8) and Provita Franz Schnurr in Pforzheim. The earliest Arctos LCD cost 1200 Deutschmark while a regular automatic watch could be purchased for less than 100 Mark. Yet about 1000 pieces were sold. The later versions cost much less; tens of thousands of pieces were sold.

Unfortunately most of those watches were returned under warranty, which greatly affected the credibility of the technology. Furthermore, American, Japanese, and, finally, Chinese (Hong Kong) modules were a severe competition, as they were slimmer and less costly. Prices continued to fall; neither Arctos nor the Swiss and American manufacturers could win this murderous price war. After 1980 the LCD digital quartz watch was gradually left to the Chinese. In the end, everybody who had not gone bankrupt went back to analog quartz watches. However, watch production was now limited to casing, the movements being

9.1. Arctos caliber 375, mechanical module

9.2. Arctos caliber 375, electronic module

9.3. Arctos LCD DSM caliber LCQ 575, front side

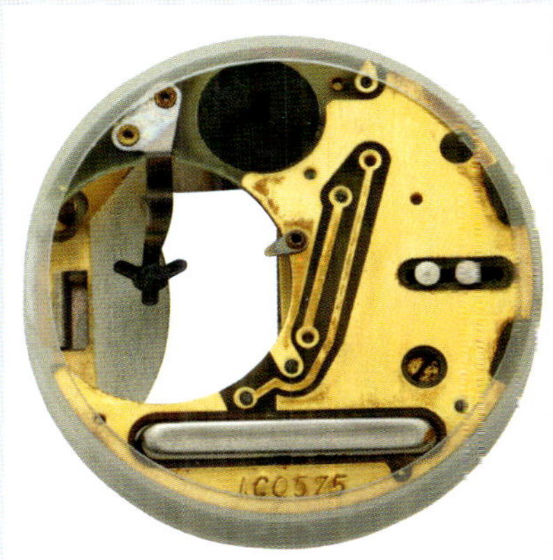

9.4. Arctos LCD DSM caliber LCQ 575, back side

purchased from PUW or, preferably, ETA. If the price was comparable, almost everybody bought the Swiss alternative.

Bifora

Schwäbisch-Gmünd, some 50 km east of Stuttgart was never bombed during World War II. For this reason, the Bifora Watch Company's production facilities were left unscathed. Several remarkable and very successful calibers were developed in the postwar period, e.g. the 13-line Bimag Caliber (1951) and the $12\frac{1}{2}$-line automatic caliber (1955). In the 1960s and the 1970s, Bifora was the most important German watch manufacturer: yearly production rose to 1.5 million pieces.

Bifora developed quartz movements at an early date; caliber B12 was presented in 1971 and production started two years later. This was a 13 lines x 5.9 mm three-hand caliber with a day/date window. In the early days, bar shaped 32 kHz resonators were purchased from Motorola, later the Japanese NDK tuning fork resonators won out. The CMOS integrated circuit was supplied by Motorola and Intersil. In 1975, Bifora started to equip ladies' watches with the three-hand caliber B20 (9 lines x 5.9 mm).

In the mid-seventies, quartz movements still were relatively costly. Bifora thus needed a low-cost, non-quartz alternative and simply continued to produce its electromechanical calibers B11 (13 lines x 5.9 mm) and B10 ($8\frac{1}{4}$ lines x 5.9 mm). The oscillator was a horseshoe-shaped, 5 Hz resonator operated in the torsional mode. Coils at the end of the horseshoe arms interacted with stationary magnets. The simpler versions B9 and B8 used the same resonator (see Chapter 3).

A large new plant was built for producing quartz movements, but this was Bifora's demise. Starting in 1977, Japan and Hong Kong flooded the world market with technically mature and ever less expensive quartz movements. Bifora went bankrupt; the machinery was sold to the Indian entrepreneur Hiranand Gajira and moved to India.

Cristalonic

The Cristalonic Company in Ottobrunn near Munich was founded by a former employee of the aerospace group Messerschmitt-Bölkow-Blohm (MBB) by the name of Zielinsky. He had invented a solar-powered quartz watch that he called "Solar Quartz."

9.5. Bifora quartz caliber B12

9.6. Bifora quartz caliber B14

9.7. Bifora quartz caliber B20

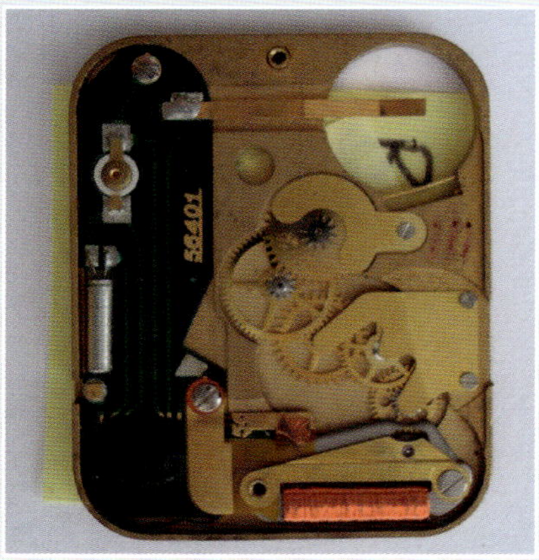

9.8. Bifora quartz caliber B17/18

Cristalonic was one of the pioneers of this type of watch. The solar cells on the front side charged a small nickel-cadmium battery; it was introduced at the 1975 Basel Watch Fair. Cristalonic produced just one easily recognizable LCD module and used it in all the watch models that were sold under the Cristalonic brand. This module was never sold to other brands.

It is not known who manufactured Cristalonic's TN field-effect LCD: it is not found in any other watch brand. On the other hand, the 32 kHz quartz resonator of the flexural mode type was certainly supplied by Siemens. The functions were either hour and minute, month and day, or hour and minute, month and day, or minute and second. A different model that also displayed the temperature was shown in 1976, but only prototypes were fabricated. The earliest model was a gold colored watch with a black plastic case and a plastic bracelet. In later models, case and watchband were made of carefully finished stainless steel.

Dugena

Dugena is an abbreviation of "Deutsche Uhrmacher-genossenschaft Alpina" in Eisenach: the company moved to Berlin in 1927. After World War II and the division of Berlin, Dugena found itself in the Soviet Zone – a rather awkward situation. In 1948, management succeeded in moving to Darmstadt. After the relocation to West Germany, the company grew rapidly. In the early 1970s, Dugena watches were sold in over 2000 German watch stores.

The quartz revolution in the early 1970s was first met by purchasing quartz modules and casing them. The Dugena Stratos Bertone model featured LED modules by National Semiconductor (NSC) or Hughes; the functions were hour, minute and second. Dugena also used the Hughes high frequency module (786,432 Hz).

Towards the end of the 1980s, Dugena was acquired by the Egana watch conglomerate in Hong Kong. A new collection was launched: it consisted of round watches with mechanical movements made in China with "small" complications. After Egana's liquidation, Heinz W. Pfeifer (the former Egana general manager in Europe) acquired Dugena through a merger of Dugena with Pfeifer's company Nova Tempora Watches and Jewelry in Dietzenbach. The well-known Dugena brand was relaunched with a collection consisting of both electronic and mechanical watches, all of them being made in Asia.

9.9. Cristalonic Solar LCD, front side

9.10. GUB caliber 31

9.11. GUB caliber 38

9.12. GUB caliber 39

Glashütte – GUB

Next to Ruhla in Thuringia, Glashütte in Saxony (Glashütter Uhrenbetrieb GUB) was the second production center of quartz movements in the German Democratic Republic (GDR, it was also called East Germany). The two companies closely worked together, but Glashütte had its own design department for quartz watches. Ruhla always manufactured the complete printed circuit as well as critical parts of the gear train. Complete movements were usually assembled in Ruhla using semi-automatic lines based on the Lecureux technology.

Glashütte specialized in the mechanical part of analog quartz movements, i.e. plate, bridges, wheels, and the Lavet motor. The tuning fork-type quartz resonators were first imported from NDK in Japan, but in 1982 Glashütte acquired an NDK license and produced those quartz resonators on its own premises. Integrated circuits of the CMOS type were supplied by Ruhla. Glashütte used silver oxide/zinc batteries in all its watch models.

Analog Quartz Calibers

Production of caliber 31 (6¾ x 8 lines x 4.8 mm) with 5 jewels started in 1978. It was mainly used in ladies' watches, there was no seconds-hand. The power consumption was less than 3 microamperes, the motor made a step every 5 seconds. The integrated circuit (type U 117) was supplied by Erfurt. The NDK quartz resonator measured 2 mm x 6 mm and was imported from Japan. Production was discontinued in 1981.

Caliber 38 with 5 jewels was introduced in 1981 and also measured 6¾ x 8 lines, but the height had been reduced to a mere 3.5 mm. New types of integrated circuits and motors were introduced with this caliber. Current consumption was less than 2 microamperes. This caliber was fully designed in Glashütte, but production was shared 50-50 with Ruhla as demand was enormous. Yearly production actually was over 1.2 million pieces. The dial of the watches that were sold in the GDR was marked "Glashütte," while the export models featured the logo and brand of each customer.

GUB caliber 38 was replaced in 1986 by caliber 39. This also was a 6¾ x 8 lines movement, but due to the demand for ever slimmer watches, the height was reduced to 2.8 mm. In other respects it was identical with the predecessor. Production continued until after the fall of the Berlin wall, i.e. 1990.

9.13. GUB caliber 31

9.14. Ladies' watch with GUB caliber 31

9.15. GUB caliber 30

9.16. Ladies' watch with GUB caliber 30

Caliber 30 was the last GUB analog three-hand caliber and measured 6¾ x 8 lines x 2.28 mm. Current consumption was reduced to just less than 1.3 microamperes. The current pulses for the motor were set to the absolute minimum of 3.9 microseconds. Whenever the motor failed to react, an extra 5.8 microsecond pulse was generated. The Statek-type quartz resonator was imported from Japan and measured 1.5 mm x 5 mm. Production started in 1990, but was limited to 5000 pieces that were reserved for export to the United States, but that order was cancelled. The watches with caliber 30 were then declared as official watches of the GDR communist party congress, but that congress never took place as the GDR collapsed and was subsequently annexed by West Germany.

A Single LCD Caliber

GUB Glashütte produced just one digital quartz movement with a liquid crystal display: it was the round 21.9 mm x 4.5 mm caliber 33 (1984). Current consumption was less than 2 microamperes; the integrated circuit type MJ 121 or CHL 5-4 was supplied by Integral Minsk. The 2 mm x 6 mm quartz resonator was made in-house.

Caliber 04 was produced from 1985 to 1988; this was the first ana-digi quartz movement made by GUB. It was called "Duo-Quartz;" the dimensions were 29.5 mm x 5.0 mm. Time was displayed in the 24-hour mode on a 4-digit LCD. The CMOS chip type U132+U117 from Erfurt controlled both the analog and the digital display that were arranged one on top of the other. The mechanical part of the analog display was taken from a slightly modified caliber 38. The LCD displayed hour, minute, second, day/date and month. The quartz resonator was the 2 mm x 6 mm NDK license construction, current consumption was around 4 microamperes.

Other GUB Calibers

Between 1988 and 1990, GUB produced the hybrid ana-digi chronograph caliber 18. Hour and minute were displayed with hands actuated by a modified caliber 39. The 6-digit LCD displayed the hour (on the 24-hour scale) as well as minute and second. Month, alphanumeric weekday, and day/date were displayed by pushing a key. The alarm ran for 30 seconds. In the stopwatch mode with the lap and split functions, minute and second were displayed for a maximum of 30 minutes with a resolution of 0.1 second. A tone signal accompanied start and stop; the

9.17. Men's watch with the GUB ana-digi caliber 18, movement side

9.18. Men's watch with GUB ana-digi caliber 18, dial side

9.19. Men's watch with GUB ana-digi caliber 04, movement side

9.20. Men's watch with GUB ana-digi caliber 04, dial side

acoustic hour signal could be switched off. Whenever longer times (up to 12 hours minus 1 second) were stopped, the chronograph displayed hour, minute and second. Current consumption was 2.2 microamperes.

In order to be complete, the quartz timer caliber 8-63-50 must be mentioned. It could be set to switch on and off any kind of device once a day. Caliber QS was more sophisticated; there were two types: QS01 featured a daily program, while QS 02 was equipped with an additional weekend switching function. Both of those time switches were produced from 1986 until 1990.

Junghans

Junghans was founded in 1861; its first quartz movement was designed in the mid-1960s. This was a clock for sports timing; this product's heyday were the 1972 Olympics in Munich, sports timing having been attributed jointly to Junghans and Longines (see Chapter 6).

Astro-Quartz

On the basis of the know-how available in-house, Junghans launched a quartz wristwatch project as early as 1961, although it was attributed a rather low priority. The situation changed suddenly in 1968, when the Neuchâtel Observatory released the results of the 1967 chronometer contest. Apparently, both the CEH in Switzerland and Seiko Suwa in Japan had developed functioning prototypes of quartz wristwatch movements. There was no time to lose, Junghans prototypes were assembled in record time. They were based on a 8192 Hz tuning fork-shaped quartz resonator and a 13-step dividing circuit by Intermetall. It was originally planned to produce this caliber in common with PUW, Para and Bifora, but this did not work out.

The next quartz project led to the Junghans "Astro-Quartz" that was unveiled at the Hannover Trade Fair in 1971. It was equipped with the quartz caliber W 666 (13½ lines x 5.80 mm). The time base was an 8192 Hz tuning fork-shaped quartz resonator. This was a rather unusual construction, but at least it avoided the inherent problems of the flexural mode bar-shaped resonator (poor shock resistance, critical suspension at the vibratory nodes). The first 25 pieces were made in-house; the resonators for the production series that started in 1972 were supplied by outside vendors.

The Junghans caliber W 666.00 (1972) featured a Siemens quartz resonator while Motorola supplied the

9.21. Junghans caliber W666

9.22. Junghans caliber 667.20

9.23. First Junghans radio-controlled watch Mega 1

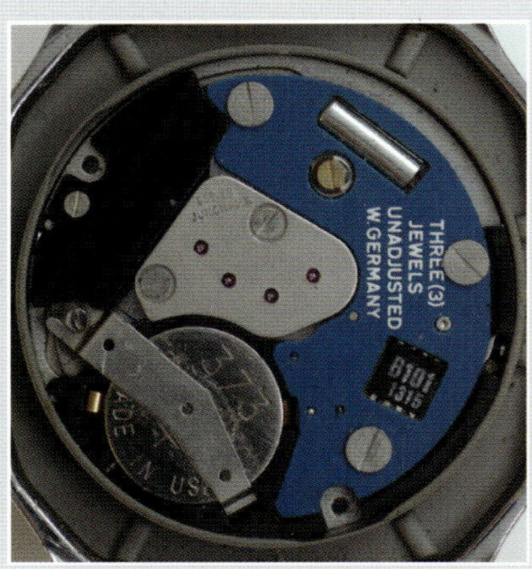

9.24. Junghans caliber 669.20

quartz for caliber W 666.02 (1974). In both cases, the frequency was quadrupled to the industry standard of 32,768 Hz. Later, tuning fork quartz resonators were purchased from the Japanese NDK (caliber W 666.25). Integrated circuits were purchased from Intersil in the United States. The anchor-type motor with a magnetic stator and a moving coil on the rotor acted on a ratchet wheel with 60 teeth that was directly connected to the seconds-hand. Current consumption was 15 microamperes.

The production costs of caliber W 666 being rather high, a different and much less costly construction was developed. Caliber W 667 (12 lines x 5.60 mm) was launched in 1972 and used a 32 kHz tuning fork quartz resonator. Depending on the model, the resonator's frequency was either divided to 0.5 Hz in 16 steps, or to 1 Hz in 15 steps. A Lavet stepping motor was used for the first time. The available variants were W 667.00 without a date disc, W 667.20/1 with date and W 667.30 with day/date.

Quartz as a Winner

Junghans satisfied its high-strung ambitions in the realm of chronometry with the high frequency quartz caliber W 667.26 (27.1 mm x 5.7 mm, 1978); production was limited to 2000 pieces. The quartz resonator just measured 6.5 mm x 2.2 mm x 0.4 mm and oscillated with a frequency of 4,194,304 Hz. This was almost twice the frequency of the Omega Megaquartz 2400.

This caliber was rather expensive, but its daily rate being 0.08 second, it fulfilled the specifications of a marine chronometer (it would not quite have met the present-day COSC specifications of ± 0.07 second per day for an electronic chronometer). This was a three-hand caliber with date disc; the motor was of the Lavet type. Thanks to a silicon gate CMOS circuit supplied by Eurosil, current consumption was limited to 12 microamperes: thus a battery life of at least one year could be guaranteed.

Total production of the calibers W 666 and W 667 was about 40,000 pieces. Those calibers signaled the end of an era for Junghans, as the manufacture of mechanical calibers was discontinued. Production of the ladies' watch quartz caliber W 632 (6¾ x 8 lines x 3.0 mm) started in 1977. As the consumers preferred ever slimmer watches, caliber W 669 was launched in 1981; the dimensions were 11½ lines x 2.75 mm.

In the mid-1970s, a respectable watch manufacturer had to have an LED watch in its catalog. Junghans obliged with the Time Commander caliber 613, the module being purchased from Fairchild in the U.S. Pressing the key once displayed hour and minute, pressing twice displayed date and weekday, and pressing three times displayed the seconds. In 1976, Junghans also supplied a pocket stopwatch with a 4 MHz quartz resonator and an LED display (caliber S/W 196 B). The resolution of this stopwatch was 0.01 second. Calibers 616 and 617 were Junghans digital movements with a liquid crystal display (LCD).

Cooperation with PUW

The most successful Junghans quartz movement was the result of a common project with PUW. Caliber W 632 measured 6¾ x 8 lines x 3.6 mm (4.2 mm with the date disc). At first it was equipped with a CMOS-type integrated circuit by Eurosil, but this was eventually replaced by a much less costly Korean imitation supplied by HanDok. Production started in 1977 and totaled 10 million pieces when the caliber was discontinued in 1988.

The round ana-digi quartz caliber W 633.86 (19.0 mm x 4.2 mm) with a date disc was also designed in cooperation with PUW. Time was displayed with hands; all other functions were displayed on an LCD. The smallest quartz caliber by Junghans was W 644. It was a 5½ x 6¾ lines caliber for small ladies' watches; the height was a mere 2.2 mm; it was launched in 1982.

Radio-Controlled Watches

As far as radio-controlled clocks and wristwatches were concerned, Junghans was the unquestioned pioneer and innovator. It is true that Léon Hatot in Paris had invented a clock that synchronized itself with the time signals transmitted from the Eiffel Tower at noon and at midnight in the 1930s. However, miniaturization was not feasible in the tube electronics era.

In Germany, radio-controlled clocks and watches were only developed after 1959, when the long wave time signal transmitter DCF77 (77.5 kHz) started operating in Mainflingen near Frankfurt. In the early days, the transmitter was operated by the Federal Post Office, but used an extremely accurate quartz clock that had been developed by the Physikalisch-Technische Bundesanstalt (PTB) in Braunschweig. Starting in 1967, this clock was synchronized with a PTB caesium clock. In 1969, the DCF77 transmitter was placed under the responsibility of PTB.

On June 5 1973, DCF77 started transmitting complete time information (a time telegram) every minute. This allowed the development of self-synchronized radio-controlled clocks and watches. In 1985, Junghans presented a table clock with a digital LCD display that synchronized itself with DCF77. In the following year, a model with an analog display was launched. The time telegrams included the full date and the year; it automatically compensated for the introduction of winter time and summer time, as well as leap years and leap seconds.

Mega 1 and Mega-Solar

Within a short time, Junghans miniaturized its radio-controlled table clock to radio-controlled alarm clocks and finally to the world's first radio-controlled wristwatch, "Mega 1." Prototypes started ticking in 1985; commercial production had to wait until 1989 when ceramic case watches with the digital caliber 026 were launched. Its dimensions were 28.0 mm x 26.0 mm x 6.1 mm. In 1986, Junghans also launched an ana-digi radio-controlled watch; the dimensions of the movement caliber 025 were 26.0 mm x 24.0 mm x 5.1 mm. In 1993, the combination with solar cells was called "Mega-Solar" (caliber 021); the dimensions were 34.0 mm x 6.5 mm. A watch of this type continues to operate as a normal quartz watch when it gets out of range of DCF77. The error accumulating during this time is automatically corrected as soon as the watch gets back in range.

The most important components of a modern radio-controlled watch are the ferrite antenna inside the case and an integrated circuit with the receiver for time signals. A microprocessor is needed for converting the DCF77 time telegrams into a reference signal that is compared with the step pulses actuating the Lavet motor. The adjustment is made by suppressing or adding extra pulses. The comparison is based on a reference point that may be set automatically or manually. This is only necessary when the watch is first switched on, or after replacing the battery.

Pallas

In the early 1970s, Pforzheim and Schwäbisch-Gmünd were the home of a number of medium- sized watch companies, mostly family owned. Individually, they lacked the strength to develop and produce their own quartz calibers. Adora, Otero/Eppo, Exquisit, Ormo, Para, Stowa, Laco, and Arctos, therefore, formed an alliance in 1971 in order to procure a common LCD digital caliber. This was a conventional construction with a 32 kHz quartz resonator, integrated divider

9.25. Junghans caliber 633.86

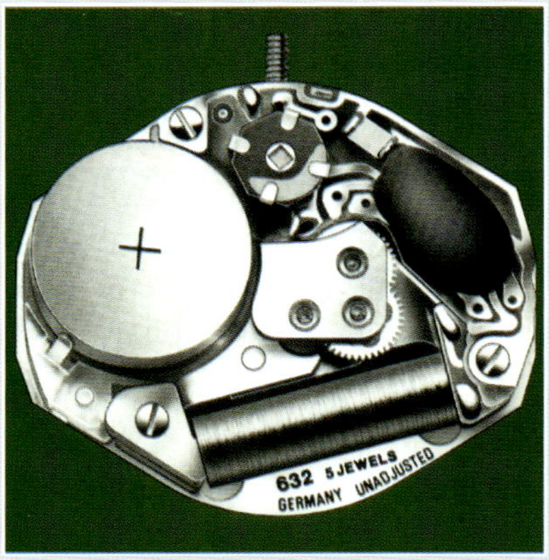

9.26. Junghans caliber 632

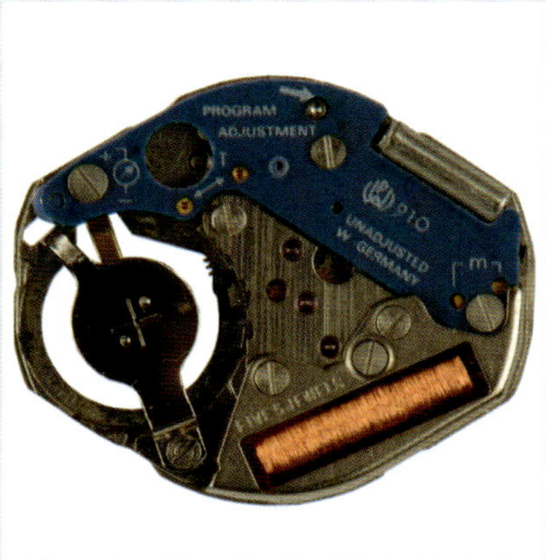

9.27. PUW caliber 910

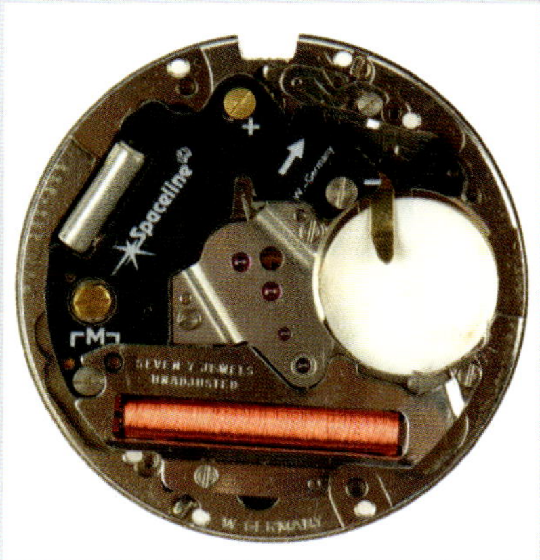

9.28. PUW caliber 255

and decoder and a 4-digit LCD from Optel. Optel actually supplied the complete module; production of watches with this caliber started in 1972 (see Chapter 8, section BWC).

PUW

The entire German watch industry was fascinated by the first electronic timepiece: the "Accutron" by Bulova. This tuning fork watch was launched in 1960 and particularly impressed PUW as the company had excellent relations with Bulova and supplied large numbers of mechanical movements to the Bulova assembly plant in the Virgin Islands. Quite a few watch industry executives started worrying about the long-term survival of the mechanical timepiece. Obviously, the electronic watch would be much more accurate.

Porta Models

The first three "Accutron" models were so profitable for Bulova that it offered to purchase PUW. However, only Bulova shares were offered in payment, which was not acceptable to PUW management. Yet, PUW was gradually forced to act. In 1966, it finally launched calibers 1000 and 1001 (12½ lines x 5.9 mm) with an electrodynamic balance wheel motor (i.e. the moving coil type). They were given the name "Porta Elechron;" the balance wheel frequency was 4 Hz (28,800 vibrations per hour). The spring-based contact system was good for at least 2 billion switching operations, which corresponded to 8 years of continuous operation.

The Porta-Elechron concept was extended to the ladies' watch caliber 3000 (6¾ x 8 lines x 4.85 mm). In this caliber too, the coil was mounted on the balance wheel while the three magnets were fixed. Serial production started in 1973, by 1976 some 30,000 movements of this kind had been produced. They worked quite well but generated little profit for PUW.

Calibers of the PUW 2000 and 2500 series were launched in 1971 and were called Porta-Electronic. Their diameter was 12½ lines, the height depended on the functions: it was 6.55 mm (2501/2509 with date disc) or 7.05 mm (2502/2508 with day/date). Those electrodynamic movements probably were the most advanced in the marketplace. They were first to replace the single transistors by an integrated circuit supplied by ITT in Freiburg. Its main function was controlling the balance wheel motor and regulating its amplitude.

The "Quartz Shock"

In 1968, both Switzerland and Japan had presented prototypes of quartz wristwatches. It was imperative for PUW not to fall behind in this respect; the main competitors in Pforzheim, Arctos and Timex-Laco, were known to work on such calibers. In order to save time, the 4 Hz electromechanical caliber PUW 2500 was synchronized with a quartz resonator. Caliber PUW Porta 5000 (5001 with date) was launched in 1972; the dimensions were 12½ lines x 5.90 mm. A bar-shaped 65,536 Hz Motorola quartz resonator was used, a CMOS-IC by Intermetall/Motorola divided this frequency down to 4 Hz.

In 1970, PUW president Klaus Wehner managed to buy an "Astron" Seiko quartz watch. Two watch generations later, Seiko secured the technical concept that would be adopted throughout the world: a 32 kHz photolithographic quartz resonator, binary division to 1 Hz with a CMOS circuit, a Lavet stepping motor, and a silver oxide-zinc battery.

The quartz crisis also stimulated the German Watch Industry Association to organize a conference in Munich in 1976. The idea was to set up an automated quartz caliber production facility for the entire industry; costs were estimated at 22 million Deutschmark. With this plant, everybody could profit from rock bottom prices due to mass production. An arm of the German government would pay a third of the costs if the watch industry and Siemens took care of the other two-thirds. Unfortunately, this project went nowhere, as the many watch producers were unwilling to cooperate.

Cooperation with Junghans

After the Munich conference, PUW and Junghans decided that they would cooperate in developing a 6¾ x 8 lines x 3.6 mm universal caliber. PUW called it caliber 632, it was launched in 1977 (see the Junghans section). The integrated circuit was first supplied by Eurosil, but replacing it by a Korean circuit reduced costs to the point that the caliber became competitive even in Hong Kong.

PUW and Junghans then opted for a second, even flatter 6¾ x 8 caliber: its height was going to be just 2.5 mm, including a sweep seconds-hand. For unknown reasons, Junghans decided to cancel the project, which forced PUW to continue by itself and launch the caliber with the designation 900 – it was a sensational success.

In the course of the 1990s, PUW began to lose its customers, who now produced their own quartz calibers since the components were so readily available. Others started buying on the cheap in Japan and in China. Timex for example, that had formerly bought up to 45 percent of the PUW movements, started its own production in Besançon and in Cebu on the Philippines, which made the company self-sufficient.

The bloody price war that Seiko and Citizen waged in the late 1970s further accelerated PUW's demise. By 1982/83 prices had declined so much that PUW began issuing red balance sheets. Enormous efforts were made to rationalize and automate production. Thus, disaster was barely averted. In 1983, PUW developed the 10½-lines x 2.5 mm caliber 241 that was meant as the movement for an imitation of the LIP/Roger Tallon watch design. Only a few prototypes resulted from that project. Yet, caliber 241 strongly inspired the 11½-line caliber 255. Several variants were produced, caliber 255.290 even had a moon phase indicator.

PUW Is Acquired by ETA

In order to open new markets, PUW CEO Klaus Wehner contacted René Guilgot, the owner of the CYMA Company in Le Locle. His idea was to develop a very flat quartz movement, as Raymond Weil, who always wanted to produce particularly elegant watches, was already clamoring for such a caliber, but he did not quite have the required volume. PUW had the personnel and the facilities (e.g. CAD) to develop such an ultra-flat movement within a very short time.

Wehner bought a very flat Seiko watch and had it taken apart. The quartz movement was characterized by a very unusual size: 5½ x 8 lines x 1.2 mm. This served as a source of inspiration for the PUW engineers, who soon came up with the PUW caliber 111 – it was almost a copy of the Seiko design. Raymond Weil and René Guilgot were very enthusiastic about it.

Yet, there was a problem: PUW was a German company, a watch with a German movement could not be declared Swiss made, which Raymond Weil absolutely wanted. In order to be in line with the Swiss made requirements, the stepping motor, the coil, and the complete printed circuit were made by a PUW subsidiary in Lausanne by the name of PISA (Porta Instruments SA). Consequently, caliber 111 ended up being more expensive than anticipated, but it was an honest Swiss made caliber, even though PUW supplied the mechanical part. CYMA assembled the movements in Le Locle. Raymond Weil bought large numbers of them, even though the price was 41 Deutschmark apiece – this was a unique product; nothing even remotely comparable was on the market.

9.29. PUW caliber 5000

9.30. PUW caliber 104

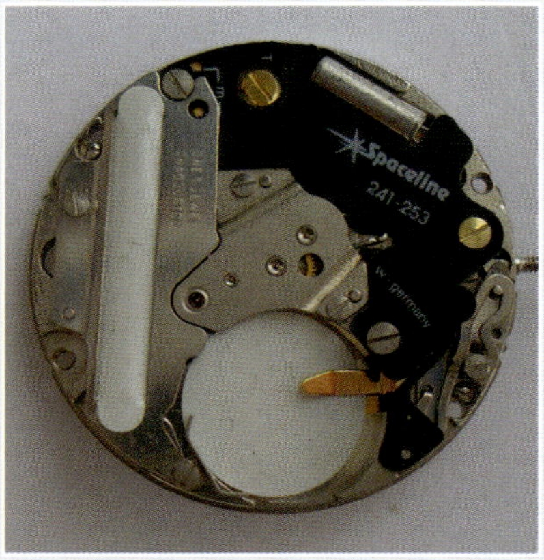

9.31. PUW caliber 241-253

9.32. PUW quartz alarm caliber 681

PUW then developed caliber 100 with a height of just 1.5 mm; it was available with a sweep seconds-hand or with a small seconds-hand and was very successful. Thanks to those innovations, PUW ended up the year 1988 with one of the highest profits the company had ever made. This inevitably had the consequence that taxes would go up.

The banks recommended that PUW acquire Rodi and Wienenberger (ROWI), a Pforzheim company that was close to bankruptcy and had accumulated an enormous debt load. ROWI produced watchcases, bracelets, and dials. It was mostly known for bracelets with the "Elastoflex" and "Fixoflex" brand. PUW bought the company for one symbolic Deutschmark; PUW profits could now be compensated by ROWI's losses.

But PUW did not manage to get ROWI back in the black. PUW itself was close to bankruptcy and was eventually sold to the SMH (Swatch Group today) in 1990. PUW was renamed to ETA Uhrwerke GmbH Pforzheim. For many years it was an important partner for the ETA group, as its specialty was high precision embossing tools. However, ETA Pforzheim was closed in 2005.

The funny thing was that ETA adopted one of the PUW quartz calibers for its own production program. Of all things, it was the "Seiko copy" caliber PUW 111; its designation was changed to ETA 282. The dimensions and the mechanical part stayed as they were, but the electronics were redesigned to make them "ETA-compatible".

UMF – Ruhla

As already mentioned in the Glashütte section, the entire East German watch industry was integrated into the conglomerate "Kombinat Mikroelektronik Erfurt" in 1978. Aside from Ruhla and Glashütte, this also included the watchcase plant in Weimar; the group employed 12,000 people and was centrally managed by Ruhla. Before that, Ruhla and Glashütte had been independent companies, even though they closely cooperated. Since 1952, Ruhla had been known as VEB Uhren and Maschinenfabrik Ruhla (UMF-Ruhla).

In 1969, UMF-Ruhla received the mandate to develop a quartz wristwatch movement. The engineers studied the question and came to the conclusion that the Lavet motor was ideally suited for this kind of caliber. The problem was that this motor was hardly known in East Germany. Aside from that, it seemed that this technology was blocked by recent Seiko and ETA patents. Prototypes and a test series of 100 pieces were fabricated, but they used an anchor-type motor.

This movement was further developed and was ready for production in 1975 as caliber 29 with the dimensions 13¼ lines x 5.5 mm. Assembling it involved a lot of manual labor, which made the caliber expensive. Besides, the test series of 100 pieces revealed that it was not as reliable as had been expected.

The Quartz Calibers 28 and 14

A new start was made: this time the very successful mechanical caliber 24 was redesigned into a quartz movement. It was named caliber 28 (14½ lines x 7.6 mm) and was equipped with a compact stepping motor. The 32 kHz quartz resonator was of the flexural mode at first, but the advantages of the tuning fork quartz were soon recognized, even though they had to be imported from NDK in Japan. This was not the only "Western" part: the CMOS dividing circuit had to be purchased from Motorola in the United States. Silver oxide batteries were first supplied by West German firms; in the mid-1980s a license agreement was concluded with Varta, so that the battery factory in Pima (Saxony) could produce watch batteries. The license allowed the production of the types SR 44 (Varta 357), SR 42 (Varta 344), SR 41 (Varta 384), and SR 48 (Varta 309). The extra flat Ruhla calibers 13 and 12 needed special batteries that were directly purchased from Varta.

The Cosmonauts Sigmund Jähn and Valeri Bykowski successfully tested caliber 28 in space in August 1978. However, the caliber just was too expensive and only small series were produced. The variant 28-39 with an electromechanical buzzer was used in a travel alarm clock (1977); the variant 28-32 was provided with a special case with a magnet and served as a dashboard clock (1978).

Development work went on: the next Ruhla quartz caliber was of the 3-hand type with a date disc. The designation was caliber 14 (11½ lines x 5.3 mm), it was available in the stores by 1979. Thanks to the 32 kHz NDK tuning fork quartz, an integrated circuit from VEB Mikroelektronik in Erfurt, and a Lavet stepping motor that had been developed in-house, caliber 14 was significantly better than the previous caliber 28. A variant was the wrist alarm clock caliber 14-39 with a central alarm clock hand that was set with the crown; a small alarm clock bore the same designation; both were available in 1980.

The Quartz Caliber 13

In 1984 Ruhla launched its 11½-line caliber 13 with 7 jewels and a Lavet stepping motor. The height was 3.5 mm including a date disc. The 32 kHz tuning fork quartz resonators now came from Glashütte where production under NDK license had started in 1982.

Caliber 13 was a winner: it was reliable and extremely sturdy. There were three types: 13-30 (two hands), 13-32 (two hands and sweep seconds-hand), and 13-33 (two hands, sweep seconds-hand, and date). About a million pieces were produced year after year until the Berlin wall fell; this production was highly automated. Erfurt supplied CMOS wafers with the dividing circuits; separation, bonding, and encapsulation were done by Ruhla. Among the interesting watches equipped with caliber 13, one should mention the army diver's watch 13-32 with a special steel case, screw-down crown, and a hardened watch glass. Production started in 1986; in the same year, Ruhla launched the 13-31 ship chronometer that used specially selected components; the monthly rate was ± 10 seconds.

Development of the quartz caliber 12 started in 1987. At that time, the public wanted small and flat watches. For this reason, the dimensions of caliber 12 were 10½ lines x 2.2 mm. Current consumption was just 1 microampere, the tuning fork quartz had a diameter of 2.0 mm. It was planned to equip one variant of caliber 12 with a lithium-iodine battery with a life of 8 years.

Enormous investments with many original innovations were necessary for setting up the production line. But this development was abruptly interrupted by the political events of 1989: production of caliber 12 had just had started but was stopped in early 1990. Only test series were produced; the total was around 50,000 pieces. An assorted ladies' watch caliber was attributed the designation 02 (8¾ lines by 2.2 mm), but it did not evolve beyond the prototype stage.

Digital Watches

Aside from the analog calibers mentioned above, Ruhla produced about half a million digital quartz calibers with a 6-digit liquid crystal display (LCD) per year. They were reserved exclusively for the GDR market. A prototype of caliber 27-01 worked in 1977; its diameter was 29.5 mm, the height was 4.0 mm (5.2 mm with the battery). It won a gold medal at the Leipzig Fair in 1977; production started in 1978. The functions were hour, minute, second, month, day and date.

9.33. UMF Ruhla caliber 29

9.34. UMF Ruhla caliber 28

9.35. UMF Ruhla caliber 13, front side

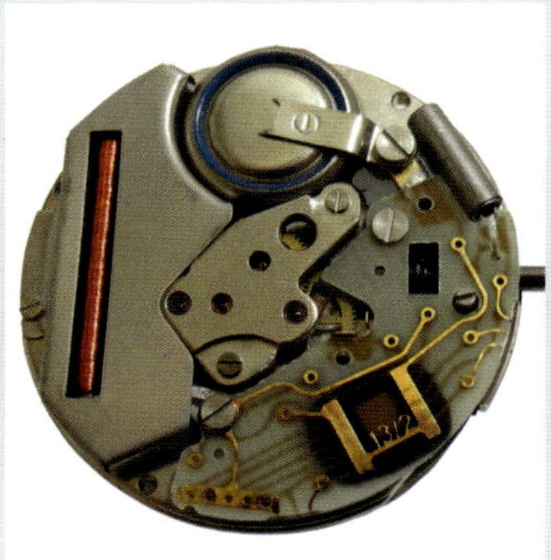

9.36. UMF Ruhla caliber 13, back side

Caliber 22 was ready for production in 1979; the dimensions were the same as those of prototype 27-01, but the stopwatch function had been added; time resolution was 0.1 second for model 27-01 and 0.01 second for model 27-02. Production of caliber 19 was started in 1981: this was the first LCD caliber for men's watches that had been fully developed in the GDR and was produced exclusively with parts that were made there. The previous LCD watches from Ruhla had used electronic modules from the Japanese companies Sharp or Orient. Next to time and day/date, the stopwatch function was available with a resolution of 0.1 second (caliber 19-04) or 0.01 second (caliber 19-06). Dimensions were 29.5 mm x 6.2 mm, including the battery.

Production of caliber 16 with 5 functions and GDR electronics started in 1982, in parallel with caliber 29 with 8 functions including stopwatch; it too was equipped with a locally made integrated circuit. The diameter of both calibers was 29.5 mm; the height, including the battery, was 6.2 mm. The variant for ladies' watches was caliber 33 (1984), also with a GDR made integrated circuit. Those calibers were available commercially in 1984. The diameter was 21.0 mm; the height was 4.0 mm (4.5 mm with the battery).

The LCD technology used by UMF-Ruhla was a Japanese design for which the GDR had acquired a license. The television electronics plant in Berlin took care of production. The "solid state" basic caliber 15 was exhibited at the Leipzig Spring Fair in 1983. The diameter of the module was 29.5 mm, the height was only 3.2 mm, including the battery. The display could be fitted with a tiny lamp for illumination of the LCD. Aside from hour, minute, and second, extra functions were month and date or month and day/date. Caliber 15-51 (1984) featured a total of 7 functions and solar cells, caliber 15-21 featured 8 functions including an alarm.

The GDR government had decided that the digital timepiece was the watch of the future, while experts from industry considered it as a relatively short-lived fashion article. They were right. The Gardé Company that succeeded UMF Ruhla after the political upheaval and the subsequent integration of the GDR into West Germany kept LCD just in a radio-controlled watch (caliber 14). Two LCD are integrated into the dial and digitally display day and date, the calendar week, the stop function and the alarm time.

9.37. UMF Ruhla military diving watch with caliber 13-32

9.38. UMF Ruhla caliber 12

9.39. UMF Ruhla LCD caliber 15-21 front side

9.40. UMF Ruhla LCD caliber 15-21 back side

10. Quartz Movements from France

Most of the major watch manufacturers in France were latecomers in terms of quartz wristwatches. This neglect cost them dearly: in the longer term, the once flourishing French watch industry disappeared for all practical purposes. Just the brand names and the design offices are left, otherwise French watches are made in China – or in Switzerland if they bear a high prestige brand name.

Cattin

In 1926, Emile Cattin (1904-1979) started manufacturing watches in Morteau; they sold under the Mortima brand. After the interruption caused by World War II, Cattin managed to restart production within a short time. In 1947, he produced 30,000 watches that were mostly sold on the domestic market, but the Spanish and American markets were important too. Mortima Watches used low-cost, extremely sturdy pin pallet movements of the Roskopf type.

In the course of the 1970s, it became abundantly clear that quartz was to be taken very seriously and might even replace the traditional mechanical watch. In 1979, Cattin gave the CETEHOR (see this section) the mandate to develop two analog quartz calibers with 32 kHz photolithographic quartz resonators; production started in 1980-1981. The three-hand caliber C 80 (11½ lines x 3.10 mm) generated 1 second pulses for the Lavet motor. The version C 80cal featured a calendar disc, which increased the height to 3.50 mm. The smaller caliber C 81 was designed for ladies' watches. This was a 6¾ x 8 lines movement with a height of 2.55 mm; the Lavet motor was fed six current pulses per minute.

The C 86 family followed in 1986, but Cattin most probably purchased those calibers from France Ebauches. They were available with two hands or three hands and a stopping device for seconds-hand synchronization. All of them featured one-second pulse Lavet motors. The dimensions of caliber C 86 800 were 8¾ lines x 2.50 mm; the variant C 86 810 with a date disc had the same dimensions.

Calibers C 86 1000 and C 86 1010 (with a date disc) were 10½ lines x 2.50 mm. The biggest of those calibers was C 86 1100 (C 86 1110 with date disc); its dimensions were 11½ lines x 2.50 mm. After the fatal traffic accident of the Mortima general manager Gruey in 1986, Cattin was slowly liquidated and was declared insolvent by 1990. A street in Morteau reminds us of the Cattin founder; his plant building was transformed by the city authorities and is now the home of many small businesses.

CETEHOR

The "Centre Technique de l'Industrie Horlogère" (CETEHOR), a state-owned institution in Besançon, was founded in 1945. Its objective was to help small and medium size French companies by giving them access to a top modern scientific and technological infrastructure. Research and development mandates, as well as literature and patent searches, were performed for the watch industry. The library and documentation center were available to everybody. The strength of CETEHOR was microelectronics, metallurgy, chemistry, and physics. The control service was of prime importance for checking the quality and the daily rate of movements and complete watches.

In the 1990s, the big name brands in French horology stopped producing in France, as this was getting too expensive. They just kept their design offices and ordered watches made according to those plans in China. For CETEHOR this had the consequence that most of its services were not needed any more. For this reason, CETEHOR merged with the CFHBJO (Comité Français pour l'Horlogerie, Bijouterie, Joaillerie et Orfèvrerie), forming a new organization named FRANCECLAT in 2008. The objectives of the CETEHOR have not basically changed, but the work is now centered on the needs of the jewelry industry that is supported with analytical services and metallurgical investigations.

Eurowatch

Towards the end of the 1980s it became clear, that the production of standard quartz movements might become impossible: the Chinese could beat any price and trying to compete with them was an exercise in futility. Yet, the Domi Company, which had been founded by Dominique Laval and specialized in low-price watches, made a last, heroic attempt. In Laval's opinion, even the lowest grade quartz movement from France Ebauches was still too costly. He decided that he could do a lot better on his own.

The objective was to design and produce in France a family of ultra low-priced quartz movements that would cover a very wide range of the French watch's requirements. Only three sizes were required: 6¾ x 8, 8¾ and 11½ lines with the options date or day/date for the two larger sizes. Instead of metal, precision injection molded plastic parts should be used whenever possible. The specifications made it obvious that a Swiss company should do the design. Conseilrey in La Chaux-de-Fonds was the ideal partner. It employed an extremely gifted caliber designer, Michiel Groothuis.

In order to make his dreams true, Domi created a subsidiary by the name of Eurowatch. However, it soon became obvious that the project was too ambitious for Domi's size. It needed a partner: the logical choice was Ambre in Morteau, which had supplied Domi with large quantities of low-cost watches for quite some years. Ambre had been founded in Morteau in 1965 as a trading company for jewelry and watches. In order to be able of satisfying the customers' needs, Ambre diversified into complete watches; the quartz movements were mostly purchased from France Ebauches. Its success was amazing: within a few years, Ambre was the number one watch producer in France.

Ambre Takes Over

In the late 1980s, Ambre was selling about 3 million watches a year. Its managers felt uneasy about their strong dependence on France Ebauches and were happy to help develop a second source. Ambre joined the Eurowatch project as a partner in 1990. However, Ambre was so much stronger than Domi that within just a few months it found itself in charge of the whole project and was taking care of the Eurowatch administration.

Robert Anguenot, former general manager of Montrélec was put in charge of Eurowatch. Production of the many high precision plastic movement parts was farmed out to Universo in Crêt-du-Locle, the Swiss industry's major producer of watch hands that had established a top modern plastics department as a diversification. The requested precision was at the very limit of industrial feasibility, but Universo made a success of it. Productivity was high; for example, six complete plates were ejected for each operation cycle. No jewels were necessary: holes in the plastic were either self-lubricating or needed just a tiny droplet of oil.

The gear train was made of metal in the early days and was supplied by Pignons Affolter in Malleray (canton of Bern). Later, Universo succeeded in making plastic wheels and pinions as well. The Lavet motor with a plastic-coated rotor was made by Micromag in Nyon. The electronics module with a quartz resonator and an integrated circuit on a printed circuit board were purchased at first from Let Sang Hong in the Shenzhen Zone north of Hong Kong; later Hana in Bangkok was found to be more reliable.

10.1. Cattin caliber C80, 10½ lines, front side

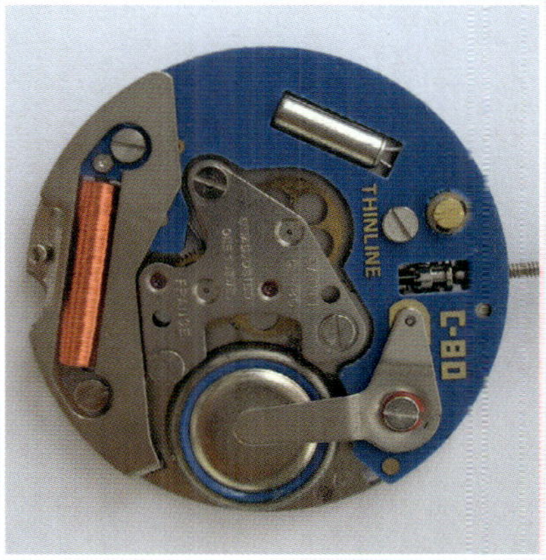

10.2. Cattin caliber C80 back side

10.3. Euromouvement 6¾ x 8 lines, back side

10.4. Euromouvement 6¾ x 8 lines, front side

Toward the end of 1990, two of the three planned Eurowatch calibers ($6^3/_4$ x 8 lines and $11^1/_2$ lines) were fully designed and tested; they were given the brand Euromouvement. Production began in 1991, the $6^3/_4$ x 8 caliber with the designation 630 was a great success; more than 5 million pieces were delivered to Ambre and Domi within the next five years. The $11^1/_2$-line caliber 1130 also fully proved itself, but just a few hundred thousand pieces were produced. On the other hand, the $8^3/_4$-line caliber suffered from "childhood diseases," including stability problems. Prototypes of the final version were available in 1996, but it never saw mass production. All three Eurowatch calibers were designed in such a way that most of the parts were the same in all of them and that plastic was used wherever possible. It is interesting to compare them with the ETA 800 family that was based on the same concepts and was commercially available in 1991.

Assembly China and Mauritius

It was impossible to assemble the Eurowatch calibers in France or in Switzerland, as costs were simply too high. Much lower prices were available in China; for this reason the Let San Hong assembly company was given the job. An office was set up in Ecole Valentin north of Besançon, where four to five employees controlled the parts, assembled the kits, sent them to China, and checked the assembled movements that arrived from there. The entire production was delivered to Ambre; Euromouvements were not sold to third parties.

However, it was soon clear that Let San Hong's quality was not up to specifications. In 1992, Eurowatch found a local partner on Mauritius Island and set up its own assembly plant there. This subsidiary by the name of Ambrex employed 200 people and assembled 250,000 movements as well as 50,000 watches a month. The dials were produced locally while the plastic or brass cases were imported from Hong Kong. The Eurowatch plastic movements cost about 2.50 Swiss francs apiece, while the price of complete watches was 5 to 6 francs.

Obviously, this was too much. At Conseilrey, Groothuis worked very hard at further simplifying the construction. For example, the screws were integrally replaced by ultrasonic welding. This meant that Euromouvements were not repairable; they were simply exchanged if there was a problem. Enormous efforts were made in order to bring down the production price to 2 Swiss francs apiece. At that time this was still far too much, as complete quartz movements were available in China for slightly over 1 franc.

The consequence was that the production of Eurowatch movements was stopped in 1996. The plant in Mauritius was remodeled; it is still in operation today, but a staff of 50 suffices for assembling watches of various brands belonging to the Ambre group. All components are imported from China, but the exterior parts are made according to Ambre designs.

France Ebauches

In 1967, the largest French producers of blank movements began to merge. Parrenin and Lorsa formed Lofrance. Ebauches Cupillard (Villers-le-Lac), Femga (Annemasse), Fernand Girardet (Morteau), as well as Ebauches Jeambrun and Tecnic Ebauches (both of them in Maîche), merged to form France Ebauches. The initiative for this bold move was taken by Yves Jeambrun. Femga was closed in the course of the merger.

At the very beginning, France Ebauches employed 485 people; ten years later that number had risen to nearly one thousand. In this period, sales grew by a factor of six; export sales were 19 times higher than in 1968. Research and development were concentrated in Besançon-Palente, while four centers were responsible for production: Maîche, Villers-le-Lac, Valdahon, and Hôpitaux-Neufs (where wheels and pinions were the specialty). Later, the entire production was concentrated in Valdahon and Maîche.

In 1975-1976, France Ebauches gave the CETEHOR in Besançon the mandate for a feasibility study of an analog quartz movement. On this basis, the first prototype of a quartz caliber was developed. It was called "La Convention" as it had been developed on the base of an agreement between France Ebauches and the CETEHOR. This movement featured a Lavet motor with an air gap in the stator and a mechanically cut 32 kHz tuning fork quartz resonator. This type of resonator had been invented almost simultaneously by Seiko, NDK, and Professor A. Karolus in Freiburg. i.Br. (Germany).

Electronic Modules from Seiko

Production of the 8000 family of France Ebauches analog quartz movements started in 1978-1979. These were round, state of the art movements that were derived from existing mechanical movements. The plate, bridges, and wheel train were maintained practically without any changes. The resulting electronic movement directly replaced the former mechanical device; casing was no problem since the dimensions were exactly maintained.

In order to save time and money, France Ebauches management decided to purchase the complete electronic module, with the quartz resonator and the integrated circuit, from Seiko. Mechanically cut tuning fork quartz resonators were used in the first series, but Seiko changed over to the much more economical photolithographic type (Statek technology) as soon as possible. On the base of a licensing agreement signed in early 1978, Seiko also supplied France Ebauches with the Lavet motor technology it had patented. Those motors were of the one-second type.

A few months later, France Ebauches launched the 8000-8100 caliber family that was based on the mechanical caliber 5600. The 11½-line calibers were 8020 (2 hands and date, the height was 3.25 mm), 8121 (3 hands and date, 4.50 mm), 8122 (3 hands and day/date, 4.95 mm), 8022 (2 hands and date, 4.95 mm, dial side this caliber looked exactly like 8122), and 8126 (3 hands, month, date, stop second, 4.95 mm). The dimensions of caliber 8128 were 13 lines x 4.95 mm (3 hands and day/date, the weekday was written in full).

In mid-1978, France Ebauches launched the two-hand 5½ x 6¾ caliber 6820 for ladies' watches that was based on the mechanical caliber 68. The low-price 9000 family with zero jewels consisted of just two calibers: 9930 (11½ lines x 3.20 mm, 3 hands), and 9931 (11½ lines x 3.20 mm, 3 hands and date).

Technology Transfers

Soon, France Ebauches was capable of manufacturing its own stepping motors including the coil, but a Seiko license was still necessary, even though the Lavet motor had been invented in France as early as 1936. Seiko had improved this motor in a decisive way and very cleverly patented it. France Ebauches purchased the samarium-cobalt rotors from Audemars in the Tessin. After expiration of the contract with Seiko, integrated circuits and quartz resonators were bought on the international OEM market in the Far East. The electronic module was assembled first in Hong Kong, then in China.

As mentioned above, the early France Ebauches quartz calibers were based on its most successful mechanical calibers. The customer was free to choose a watch with either a mechanical or an electronic movement, as the dimensions and the exterior parts were identical. The costs of assembling quartz calibers had a decisive influence on the price; therefore, part of this operation was transferred to low wage countries such as Mauritius and China. Swiss made movements, on the other hand, were assembled by the France Ebauches subsidiary FEBA in La Chaux-de-Fonds.

In 1982, France Ebauches signed a technology transfer agreement with the Indian watch company Titan that is part of the powerful Tata Group. This allowed Titan to produce some of the particularly attractive France Ebauches quartz calibers. These were the calibers 7020, 7021, and 7022 (8¾ lines), 7120, 7121, and 7122 (10½ lines), and 7220, 7221, and 7222 (11½ lines). Production started in 1987 in Bangalore.

10.5. Euromouvement, 11½ lines

10.6. France Ebauches caliber 6820, 5½ x 6¾ lines

10.7. France Ebauches FEBA Swiss made, 11½ Lines

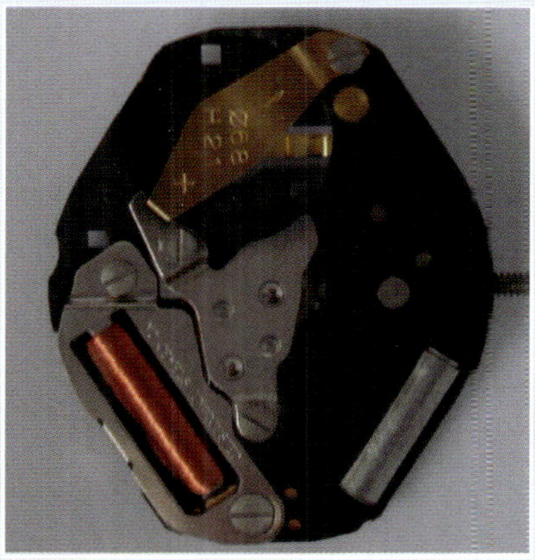

10.8. France Ebauches caliber 6920, 6¾ x 8 lines

The 5000 and 6000 Caliber Families

Production of the 5½ x 6¾ lines x 2.50 mm two-hand ladies' watch caliber 5020 began in 1982. The current consumption of 0.3 to 0.8 microampere was extremely low. In 1984, a whole set of 5½ x 6¾ lines calibers was launched. They were 5120 (two hands, 3 jewels) and 5130 (three hands, 5 jewels), the height of both of them was 2.60 mm. The two-hand caliber 5180 with 3 jewels featured an electronic setting of the hands, which increased the height to 2.65 mm. The two-hand caliber 5820 was particularly sturdy; it was characterized by a height of 3.26 mm and 1 jewel. The battery life was 45 months.

The 6000 family of 6¾ x 8 lines movements was also launched in 1982. The members were 6120 (2.60 mm, two hands, 20 seconds stepping motor), and 6130, same dimensions but with a one-second stepping motor. Several years later, in 1991, calibers 6220 (two hands) and 6230 (three hands) were announced. The innovation was an extra-thin plate made of a metal-plastic-metal sandwich; Citizen in Japan had developed something quite similar. A very low cost alternative for 6120-6130 was the caliber pair 6920-6930 with a plastic plate and zero jewels.

Caliber 6320 also was a 6¾ x 8 lines x 3.60 mm two-hand caliber, but with a 20-second stepping motor and without any jewels. It was interchangeable with the France Ebauches mechanical caliber 163. The construction was extremely simple as the price had to be very low. Caliber 6380 was quite similar to caliber 6320, but the hands could be set electrically. The dimensions of the two-hand caliber 6820 were 5½ x 6¾ lines x 3.60 mm; it was equipped with the same motor as the other calibers of the 6000 family and featured 1 jewel.

The product life of the 6000 family was relatively short, as it had a grave defect, the reasons of which never were quite clear. The Statek-type quartz resonators used in those calibers easily withstood strong shocks, while weak shocks detached them from their capsule. The current consumption was similar to that of the 5000 calibers, but with extra functions it could reach 3 microamperes.

The Successful 7000 Family

The 7000 caliber family was launched in 1980 and was very successful; production continued until 2007 and totaled about 100 million pieces. All calibers used the same motor and featured 4 jewels. The seconds-hand could be stopped for precise synchronization; current consumption was about 1 microampere. These were round movements with the usual diameters: 8¾ lines x 3.10 mm for the calibers 7020 (three hands)

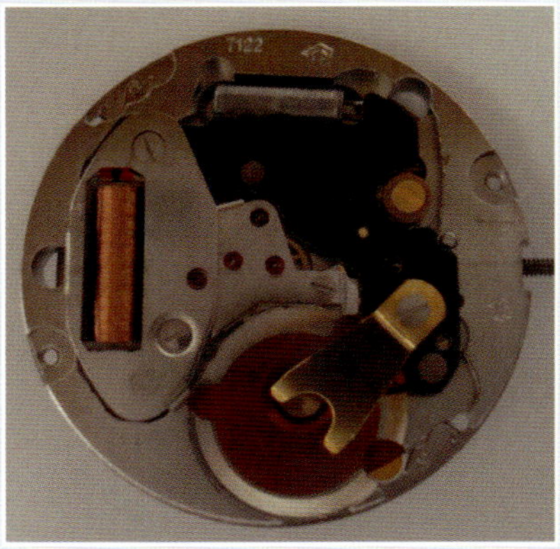

10.9. France Ebauches caliber 7122, 10½ lines

10.10. France Ebauches caliber 9930, 10½ lines

10.11 France Ebauches caliber 11030, 11½ lines

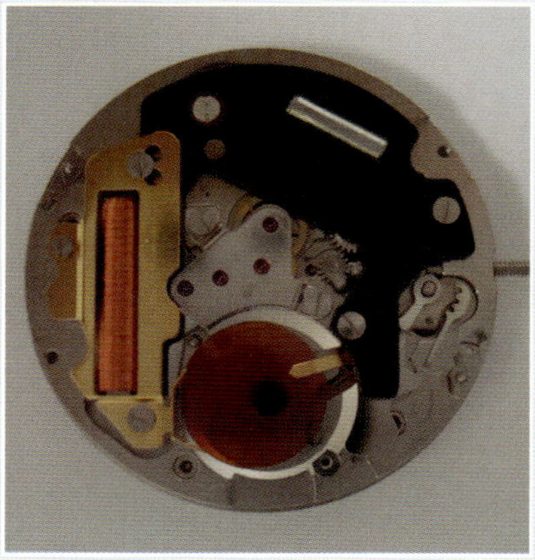

10.12. France Ebauches Caliber 11500, 11½ lines

and 7021 (three hands and date). The height of caliber 7022 (three hands and day/date) was 3.55 mm, that of caliber 7023 with an extra hand for the date was 3.30 mm. The dimensions of calibers 7027 (with a day/night indicator), 7037 (with 12-hour and 24-hour display), and caliber 7029 (world's smallest caliber with a moon phase indicator) were 8¾ lines x 3.60 mm.

The designations of the 10½-line calibers of the same family followed the same rules, as for caliber 7120 (three hands, 3.0 mm) and 7121 (three hands and date, 3.0 mm). The height of caliber 7122 (three hands and day/date) was 3.55 mm. That of calibers 7127 with date and day/night indicator, and 7129 with moon phase indicator was 3.60 mm.

Construction of all the 11½-line calibers of the 7000 family was very similar. Their height was mostly a standard 3.10 mm, which applied for caliber 7210 (two hands and small seconds-hand), 7220 (three hands), and 7221 (three hands and date). Extra functions increased the height to 3.55 mm for calibers 7222 (three hands and day/date) and 7224 with a disc 24-hour display. The day/night indicator plus date disc of caliber 7227, as well as the date and day/date discs – weekday written in full at 12 o'clock – of caliber 7228 also resulted in a height of 3.55 mm. Caliber 7229 featured a moon phase indicator and caliber 7237 had a 24-hour display with an extra hand; those extra functions increased the height to 3.60 mm.

Last Developments

Starting in 1986, France Ebauches used five-digit caliber designations. The first of these were the calibers 70200 (8¾ lines x 3.10 mm, 1 jewel) with a small seconds-hand, 70210 (8¾ lines x 3.10 mm, 1 jewel) with a small seconds-hand and date, 70220 (8¾ lines x 3.55 mm) with a small seconds-hand and day/date, and 70290 (8¾ lines x 3.60 mm) with a small seconds-hand, date and moon phase indicator.

Very similar calibers were developed in the 10½-line class: 71210 (3.10 mm, small seconds-hand and date, 1 jewel), 71220 (3.55 mm, small seconds-hand and day/date, 1 jewel), and 71290 (3.60 mm, small seconds-hand, date and moon phase indicator).

The 10000 caliber family of 10½ lines with 5 jewels was launched by France Ebauches in 1987; the height of the simpler calibers was a uniform 2.60 mm. This included calibers 10020 (two hands), 10021 (two hands and date), 10030 (three hands), and 10031 (three hands and date). The height rose to 3.10 mm for caliber 10032 (three hands and day/date, 3.1 mm). The 11000 series of 11½-line calibers with a height of 2.60 mm included the calibers 11021 (two hands and date), 11030 (three hands), and 11031 (three hands and date). Three hands and day/date (caliber 11032) raised the height to 3.10 mm.

The 11000 family was further extended by caliber 11522 (3.60 mm, three hands, indication of date and month by extra hands on counter dials), caliber 11536 (three hands, indication of date, weekday and month by extra hands on counter dials), 11539 (day and date by extra hands and moon phase indicator), and caliber 11549 with day, date and month display with extra hands plus moon phase indicator. The height of these calibers was 3.60 mm; they were characterized by a black piece of plastic protecting the quartz resonator and the printed circuit. Caliber 11599 added an extra function: an extra hand making a revolution every 12 hours and 25 minutes serving as an indicator of high and low tide, in addition to the moon phase indicator.

France Ebauches only made three 13-line calibers with 4 jewels: 7321 (three hands and date, 3.10 mm), 7322 (three hands and day/date, 3.60 mm), and 7328 (three hands, date, fully written out weekday, 3.60 mm).

The last phase of development lasted until 1994 and concerned the low-price 2000 family of calibers with one jewel. They were 2731 (7¾ lines), 2831 (8¾ lines), 2031 (10½ lines), and 2131 (11½ lines), all of them with three hands, date disc, and a uniform height of 2.65 mm. The plate was a sandwich of brass and plastic for precise placing of the bearings and easy connection of the battery. The current consumption was a mere 0.9 microampere.

Bankruptcy

Unfortunately, success of the rather clever France Ebauches quartz caliber designs was not overwhelming. Japan, Switzerland, and China had started their production years before and were well established on the international market. The French domestic market was too small and the price war with China was fearful. China produced low quality movements, but they were so inexpensive that up to 40 percent waste could be tolerated.

France Ebauches went through a phase of reduced production and personnel reduction but this was not enough to stave off disaster. Insolvency had to be declared in late March 1994. Five months later, the company was reorganized after a management buyout; its name became Société Nouvelle France Ebauches. A year later, China Resources in Hong Kong bought a controlling interest, changed the name to France Ebauches Microtechniques, but withdrew in 1999. The Chinese company Wellgain then acquired a majority of the shares.

Technotime

In 2000, the Swiss subsidiary of France Ebauches declared bankruptcy. The independent Swiss group Technotime with headquarters in La Chaux-de-Fonds took over in 2001 and also purchased whatever was left in France. Most of it was closed, but the production unit in Valdahon, where high quality quartz calibers with many different special functions were manufactured, continued operating. Those calibers had been designed by Daniel Huot-Marchand and one or two colleagues at the France Ebauches headquarters between 1976 and 2000. Starting in 2001, Technotime continued producing those calibers in Valdahon.

Despite the new Swiss management, the company was in great difficulties by 2009; that year, the share capital was written off to zero. The Hong Kong Chung Nam group that controlled ISA in Villers-le-Lac and Les Brenets bought the remaining Technotime assets, closed the Valdahon plant and sold the machines at auction, thus eliminating a competitor. However, Technotime resurfaced in 2011 in Les Brenets, close to the French border, as a supplier of mechanical movements with complications.

ISA

Today, ISA has plants both in France and Switzerland and is controlled by the Hong Kong Chung Nam group. Thus, the company is Swiss-French-Chinese, but its history was fascinating from the very beginning. It dates back to the Fabrique d'Ebauches de Sonceboz (F.E.S.) that Alfred Chopard (1907-1964) founded in 1957, Sonceboz being located between Bienne and Saint-Imier. There had been another company of the same name since 1849, but it was closed in the 1930s by Ebauches SA. The buildings were sold to Sonceboz SA, a successful manufacturer of small electric motors to this day.

Voumard the Investor

Soon after it was established, Ebauches Sonceboz experienced financial trouble. Chopard needed more capital and that was provided by his close friend Bertrand Voumard (1909-2004), the owner of Voumard Machines. In exchange, Voumard received 49 percent of the shares of F.E.S. Bertrand Voumard owned Voumard Machines (that was closed in 2010); the company was known all over the world for its centerless grinding machines. But the problems with F.E.S. persisted and got so bad that Chopard committed suicide in 1964. In 1965, Voumard purchased the rest of the shares of F.E.S.

At that time, F.E.S. produced Roskopf-type ebauches, i.e. caliber E.S. 54 with 17 jewels and E.S. 55 with 1 jewel. Those calibers were also available with a jewel pallet escapement (E.S. 95 with 17 jewels). All of them had the same dimensions; 12 lines x 4.20 mm. Prices were quite low and demand was good: up to half a million ebauches were produced monthly. In 1966, Voumard Montres in Hauterive, east of Neuchâtel, was established to assemble part of those ebauches to complete movements.

In 1969, Voumard gave the Bureau Herard in La Chaux-de-Fonds the mandate to develop a new mechanical caliber. This was the Voumard 2000, the crown of which was hidden in the case back. The construction was quite unusual; curing its early problems took an inordinately long time and was very costly. Whereupon Voumard reorganized his group: Voumard Montres and Ebauches Sonceboz were made into independent companies, but they were still owned by Bertrand Voumard. Quartz as a diversification was seriously considered, but Voumard Montres and F.E.S. survived exclusively with mechanical calibers up to the end of the 1970s.

Quartz Clocks

The group's wristwatch calibers have a long history. In 1963, Voumard Machines established the Isatome subsidiary that developed a small battery-powered clock in 1965. Its yearly rate was a sensational ± 0.5 second; until 1968 it won the first prize in its category (serial prize) every year at the Neuchâtel Observatory (see Chapter 6).

A quartz master clock with a superbly thermostatted quartz resonator was derived from the small clock. Jean Shaad and Abraham Zaslawsky, two top-rate engineers, were responsible for this development. They also developed an electromechanical wristwatch movement in 1971, but it did not evolve beyond the prototype stage.

Quartz Wristwatch Calibers at Last

In the second half of the 1970s is became obvious that the future belonged to the quartz wristwatch: F.E.S., too, had to have such calibers in its catalog. Pierre Vorpe in Sombeval/Sonceboz was given the mandate to develop a 13-line quartz prototype with date disc. His company was specialized in wheels and pinions; F.E.S. supplied the plate and bridges. The choice of the motor was suboptimal; Vorpe chose a Motorola-type that was far from slim and raised the height of his caliber to 7.3 millimeters.

Other important parts were a silver oxide battery and a Motorola 32 kHz quartz resonator; alternatives were available from Motorola, Oscilloquartz, SSIH, Toyocom, and NDK. The 16-step CMOS-divider chip could be purchased from Motorola, Intersil, or Toshiba; tests were made with the products of all of them. The new caliber was ready for evaluation in 1976, it was called "Quartz Buteco" or "01" (caliber 1382). Quite a bit of publicity was made about it, but there simply were no orders. Voumard Montres declared bankruptcy in 1977, F.E.S. bought the machinery as well as the brands ISA, Isatome, Isatron, and Isator. F.E.S. also took over the Isatome laboratory.

There was no time to lose: in 1977 the company urgently needed a quartz movement. The objective of the new development was a very simple, extremely sturdy construction that corresponded to the "Roskopf-culture" that prevailed in Sonceboz. Furthermore, it was deemed necessary to use the existing Fleury-transfer machines and semi-automatic carousel machines for assembly.

On the basis of those specifications, Shaad and Zaslawsky developed caliber 1198, an 11½ lines x 3.25 mm quartz movement with a sweep seconds-hand. With a date disc, the height increased to 3.70 mm, with day/date it reached 4.00 mm. Production started in 1979. In 1987, a moon phase indicator became available as an option. The wheel train was purchased from Aska in Arch, South of Grenchen; Voumard owned stock of this company and took control in 1992.

The electronics module with the quartz resonator, dividing circuit, coil, and battery was bought as a "black box" from Ebauches Electroniques Marin; the Lavet motor was made in Sonceboz. In the beginning, the trimmer was also made in-house, but the construction was somewhat unstable. A commercial trimmer was used instead; it was eventually replaced by a fixed capacitor and finally by digital frequency tuning.

Caliber 1198 was launched in 1979 and it was amazingly successful: so far, 200 million pieces have been made and ISA still has three versions of it in its catalog. The torque of the motor is so high, that it can easily actuate clock hands. Of course, a 30 year-old movement is hopelessly obsolete, but it is still being bought due to its legendary sturdiness. It is the preferred movement for nurses' clip watches that are nearly indestructible, even if they are dropped on a hard floor.

China Gains a Foothold

Due to the continued erosion of quartz movement prices, a lower price source had to be found. Hana in Bangkok turned out to be the ideal supplier – for a limited time. After Prestige Time had been established in Hong Kong, it was cheaper to buy the parts and assemble them there. Finally, purchasing and assembly were moved to continental China.

10.13. Quartz caliber Buteco Vorpe 1382

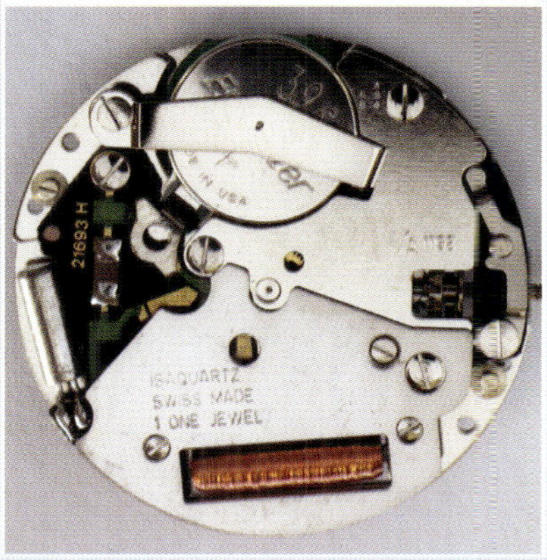

10.14. Isaswiss caliber 1198

10.15. Isaswiss caliber 125

10.16. Isaswiss caliber 389

The Lavet motor coil had a similar history. After ISA had purchased the Bulova plant in Villers-le-Lac in 1984, a coil-winding machine was found there and was put to good use. It was automated; a dozen copies were made by the in-plant machine shop. Isafrance could then make its own coils, but soon this procedure turned out to be too expensive. So coil winding was first transferred to Hong Kong and finally to mainland China.

Super-Flat Calibers

Trying to diversify in the realm of high quality super-flat calibers (125 and 127) became an expensive "hobby." The caliber designer Matthys in La Chaux-de-Fonds had developed a mechanical caliber with a height of 1.60 mm. All the pinions and the barrel were "flying,"i.e. they were provided with miniature ball bearings on one side only. Bertrand Voumard was very enthusiastic; in 1980 he set up a workshop in La Chaux-de-Fonds for developing corresponding quartz calibers.

A total of four such movements were developed with a height of 1.60 mm: caliber 160 (9 lines, round), 150 (6¾ x 8 lines), 120 (either 9 lines square or 10½ lines round). Unfortunately, those movements were not reliable: the project had to be written off in 1982.

Three years later, two excellent F.E.S. caliber designers solved the problems of the ultra-flat calibers and actually further reduced the height to 1.40 mm. Two calibers of this type were quite successful: Isatronic 125 (rectangular, 2 hands, 13.0 mm x 15.5 mm, 8 jewels) and 127 (round, 2 hands, 7¾ lines, 8 jewels). In 1995, another ultra-flat, round caliber was added to the collection: 129 (2 hands, 9 lines, 6 jewels).

F.E.S. Changes to ISA

Two smaller quartz movements were designed as alternatives to the 198 caliber of 11½ lines. These were caliber 369 (6¾ x 8 lines, 3 hands) and the 8¾-line pair 386 (two hands) and 389 (three hands). Without a date disc, the height was 2.75 mm; the date disc added 0.53 mm, while the day/date combination resulted in a height of 3.55 mm. Those calibers were very successful.

F.E.S. had been known for its Roskopf-type ebauches, and low-price jewel pallet movements. By the early 1980s, the inexpensive mechanical movements had been completely replaced by much more accurate and reliable quartz movements. The company thus needed a new corporate image. In 1986, ISA (derived from Isatome) was chosen as a new name and the logo with the hexagonal letter Q surrounding "ISA" was created.

10.17. ISA caliber K 63-301

10.18. ISA caliber U700 Radio

10.19. ISA caliber 815

10.20. ISA caliber 317

The China Connection

Starting with the 1970s, the best customer of F.E.S was Georges von Burg, a native of Bienne who had settled in Hong Kong with his Swiss Chinese International Holdings Ltd. This was why F.E.S. established a subsidiary for the production of movements in Hong Kong (they were Roskopf at first, then quartz). The name of this company was Prestige Time; it was a joint venture of Voumard and the Swiss Chinese.

Later, Prestige Time moved north to Shenzhen in the Special Economic Zone. Production in Asia rapidly gained in importance as the price of quartz movements kept eroding. Soon, a profitable production was only possible in Hong Kong and mainland China; the high wages in Europe marginalized the Old World and forced Switzerland into the expensive caliber classes.

More and more volume was needed in order to remain competitive. However, Swiss Chinese was opposed to the idea of merging with Voumard. Chung Nam, who was the best customer of Swiss Chinese, finally bought its shares of Prestige Time. Chung Nam is a billion-dollar company that is controlled by the Chong family in Hong Kong.

Isafrance and Isaswiss

In 1984, the ISA group purchased the Bulova plant in Villers-le-Lac, just a few kilometers from the Swiss border near Le Locle. This was where Bulova had produced tooling for the punching and embossing machines in the Bienne plant, until it gave up production of watch movements in Switzerland.

Isafrance made a new start in Villers-le-Lac by producing quartz movement parts. Most of them were assembled either in China or on Mauritius Island. An assembly plant for Swiss made movements was opened in 1987 in Les Brenets, just a short distance from Villers-le-Lac, where the parts were produced, but solidly on the Swiss side of the border. At first it was called Isatech, then Isaswiss; another assembly plant in Martigny started with an independent establishment that specialized in assembling the 1.4 mm calibers. This shop was eventually integrated into the ISA group. In 1997, Chung Nam bought the controlling majority of the ISA group.

In the early 1990s, ISA calibers split into several families of analog quartz movements in the size ranges from 5½ x 6¾ lines all the way to 13 lines, with or without special functions such as date, day/date, moon phase indicator, second time zone on a 24-hour scale, and alarm. Starting with 1995, even an ana-digi caliber (ISA K63-301) and a radio-controlled module (U700-K63) were available.

The high grade Isatronic line was developed from 1985 to 1990. Assembly was difficult, which is why this line was discontinued in 1996. The most interesting movement was the two-counter chronograph caliber 815 (11½ lines x 3.25 mm or 3.7 mm with date), launched in 1991. Other options were date with an extra hand and moon phase indicator. This movement was developed by the DETRA design office and was often sold with a compass made by ISA that was worn on the bracelet where the magnetic field of the "Birotor" Lavet motor did not interfere with the needle. The time base was a 32 kHz photolithographic quartz resonator of the Statek type.

Isaslender, Isaquartz and Isatron

The Isaslender line consisted of medium priced yet high grade calibers; all of them were gold plated, featured 5 jewels, and were Swiss made. Isaslender included shaped movements (6¾ x 8 lines) for tonneau and rectangular watches. Round calibers were available in the 7¾-, 8¾-, 10½-, and 11½-line sizes and all had a date disc. Their uniform height of 2.5 mm made them fully compatible with products of competing brands.

The Isaquartz line (that included the famous veteran caliber 1198) was primarily meant for private label watches and publicity watches; some of the total of 14 models featured special functions such as big date or a high torque stepping motor. The line included the following calibers: 258 (5½ x 6¾ lines x 2.0 mm); 268 (6¾ x 8 lines x 2.50 mm); 369 (6¾ x 8 lines x 2.75 mm); 389 (8¾ lines x 2.75 mm for the three-hand model, 3.20 mm with date, 3.55 mm with day/date or day/night indicator); 307 (10½ lines x 3.20 mm for the three-hand model with date, 3.60 with day/date or date plus a day/night indicator); and 317 (11½ lines x 3.20 mm with three hands with or without date disc, 3.60 mm with day/date or date plus day/night indicator). The chronograph caliber 815 that was mentioned above also was part of the Isaquartz line.

The Isatron line was launched in 1997; the height of the calibers with a diameter of 7¾, 8¾, or 10½ lines was a uniform 1.9 mm. The Isaspecialty line was also launched in 1997 and included models with solar cells, an electroluminescent dial, date by an extra hand, and special models for regatta sailors. The 11½-line three-counter chronograph calibers 8151 and 8152 with date window and weekday indication with an extra hand also belonged to the Isaspecialty line.

ISA calibers of the 1990s could be disassembled and repaired; they were available in a metal-plastic or a full metal version. Calibers in the metal-plastic version (calibers 307 to 317) featured a so-called metal heart (Coeur métallique) on which the most critical parts were mounted, thus assuring that the tolerances were fully met.

Kelton-Timex

Due to the fact that Kelton belonged to the American Timex group, the calibers developed in the Kelton Besançon laboratory are described in the Timex section of Chapter 15.

LIP

LIP undoubtedly was the quartz pioneer in France; its competitors saw the light much later. This is why the three-stage development of an electronic module by the CETEHOR for LIP is of considerable historical importance. This module was called "plaquette électronique;" the prototypes were called "Montre No. 1," "Montre No. 2," and "Montre No. 3." The second one was integrated into the LIP caliber R 032; production began in 1973. Caliber R 033 was launched in 1975-76, but changes were minor.

Development at the CETEHOR

All three electronic modules developed by the CETEHOR were based on a crescent-shaped printed circuit with a diameter of 31 mm or 32 mm, respectively, onto which the quartz resonator, the integrated circuit, and the trimmer were soldered. Those modules were meant to be combined with an existing mechanical movement in which the battery replaced the barrel while the stepping motor took the place of the escapement. LIP did not transform one of its own calibers, but one that was purchased from Durowe in Germany: it was the automatic caliber 7426. Its geometry could easily be modified to make it into a quartz caliber.

Development of Montre No.1 started in 1970: major components were a home-made tuning fork-shaped 16 kHz quartz resonator, an American trimmer (ERIE 518002), and a bipolar integrated circuit (ITT type SAJ 190) with 14 dividing steps to 1 Hz. The mechanical part included a Socrem motor type 131-B that was fed a 7.8 millisecond pulse from the integrated circuit every second. The energy source was a mercury oxide battery (1.35 volts); the current consumption was 8.5 microamperes. This electronic module was not made into a complete quartz movement, as the second CETEHOR prototype looked more promising.

Indeed, the CETEHOR module "Montre No. 2" basically corresponded to the electronics of the LIP quartz caliber R 032 "Exachron" with a diameter of 32 mm. The flexural mode 32 kHz quartz resonator was supplied by Oscilloquartz in Neuchâtel (Type 2VHA6). A prototype with an integrated circuit from Motorola was shown to the media in September of 1971. Three patents covering the design had already been filed by LIP in 1969.

Caliber R 032

In November 1971, a modified prototype with an European Chip was available: it was called R 032 "Isochron." Small series were actually used in commercial watch models produced by LIP in 1973. The diameter of caliber R 032 was 13 lines, the height was 6.3 mm, and it featured a sweep seconds-hand and a date disc. The 16-step integrated circuit of the commercial model was the Motorola CMOS type MC 2160 AF (later on it was MTD 160F). An alternative was the Sexocem type MTC 501.

Whatever its source, the integrated circuit delivered 32 millisecond pulses with alternating polarity at the rate of 1 Hz. However, the Socrem motor 131-B that was used at first required unipolar pulses. This was achieved by a pulse generator controlled by a capacitor discharge. The additional circuit consisted of three diodes, three transistors, and a mylar-insulated capacitor. Current consumption was 8.5 microamperes, the battery was of the 1.35-volt mercury oxide type. The tuning fork quartz resonator was made by LIP.

In March 1972, the third CETEHOR module "Montre No. 3" worked. The quartz resonator was the already mentioned, 32 kHz home-made LIP tuning fork-type. This time, a bipolar integrated circuit ITT type SAJ 220 H with 15 steps was used; it delivered unipolar 1 Hz pulses with a duration of 7.8 milliseconds for the Socrem motor type 131-B. The trimmer was supplied by Erie (type 518002). Current consumption was 12 to 15 microamperes.

In 1973 and 1974, production of the R 032 caliber was less than 1000 pieces. Caliber R 033 only slightly differed from caliber R 032: a date disc and a different trimmer capacitor were added. Production began to deserve this word: from August 1975 to January 1976, the output was 10,000 pieces.

LIP T24 – ESA 9240

LIP started developing a smaller quartz caliber ($7\frac{3}{4}$ x $9\frac{1}{2}$ lines x 6.0 mm) in the early 1970s. The designation was LIP T24; for Ebauches SA (ESA) it was caliber 9240. This was the first LIP caliber that involved the cooperation with Ebauches SA. This cooperation had actually started in 1967, when the Swiss heavily invested in LIP. T24 was further refined to the ESA caliber 9241, production of which started in 1971. There are no records that LIP itself produced this caliber; apparently it was supplied by Ebauches Fleurier. In the LIP 1975-1976 catalog, one also finds two 13-line digital quartz movements with LED, but they were purchased from National Semiconductor: R 01 with 3 functions (Novus 300.01) and R 02 (Novus WM 04) with 4 functions.

The LIP engineers developed their quartz calibers under dramatic circumstances: the former president and majority shareholder Fred Lip (1905-1996) resigned in 1971, two years later LIP was bankrupt. The heavily left st union on its side started one of the longest and bitterest strikes in the social history of France.

This strike must be subdivided into several phases; it ended in 1985 with a total victory of the union and is still celebrated in leftist circles. However, the result corresponded to the old bonmot "operation successful, patient dead." The "big victory" resulted in LIP being totally liquidated, with the loss of 1500 jobs. The LIP brand survived under more clement skies in the southwest of France (Gers). A service center was set up in Lectoure to the northwest of Toulouse. LIP watches with old and new designs made according to LIP specifications are distributed from there.

Lorsa

Lorsa (the abbreviation for "L'Horlogerie de Savoie"), in Annemasse close to the Swiss border east of Geneva, was a latecomer in the French watch industry, since it was only started in 1941. Actually, the brand had existed previously, but, as it was not used for a long time, it reverted to public domain and was re-activated by a new ebauches plant. One might think that this was not particularly remarkable, as there had already been a watch industry in the Haute Savoie next to the watch city Geneva in the early 18th century. It was started in 1720 by a Frenchman by the name of Claude Ballaloud who had learned watchmaking in Nuremberg.

A Fresh Start

In the valley of the Arve River between Geneva and the city of Cluses, watch movement parts and exterior parts were manufactured in large quantities; most of this production was exported to nearby Geneva. The Geneva watch manufacturers were happy to outsource their gold and nickel plating as well as the assembly of ebauches to low-wage France. However, the World Economic Crisis of the 1930s practically destroyed the Savoie watch industry that was totally dependent on nearby Switzerland. The Arve Valley kept producing precision parts, but they were mostly meant for the electronics, telecommunications, and automobile industry.

10.21. LIP quartz prototype caliber R32

10.22. LIP caliber R33

10.23. LIP designer watch by Roger Tallon with caliber R33

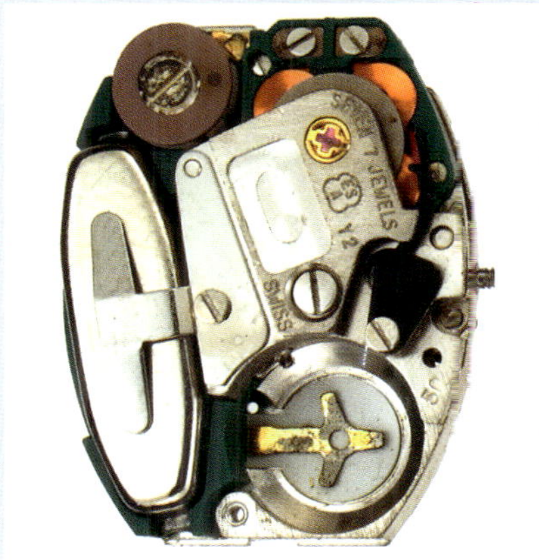

10.24. LIP caliber T24 – ESA caliber 9240

Before the outbreak of World War II, only one producer of blank movements and a few etablisseurs were left in the Annemasse region. After the north of France had been occupied by German forces in 1941, the Haute Savoie found itself in the so-called Zone Libre, but the Swiss border was hermetically closed and it was impossible to establish any contacts with the watch industry in the French Jura, which was located in the occupied zone. The sad remains of the Savoie watch industry had absolutely no future and were condemned to disappear.

Help came from a totally unexpected side. A young Paris-born businessman by the name of Robert Bouchet (1903-1972) moved to Annemasse just before World War II started, as he had been offered a job with a small watch manufacturer. However, his boss was called to arms within a few months and spent the rest of the war as a P.O.W. in Germany. Bouchet now was at the helm of a company that had practically no technically competent people: they had all been mobilized. Also, it was impossible to buy machinery and watch parts in Switzerland. Bouchet had a choice: either close the plant or do absolutely everything in-house.

It goes without saying that the young and ambitious Bouchet chose to go ahead. It was clear to him that if he wanted to manufacture watches, the production of ebauches had the highest priority. A few industrial watchmakers from the north who were stranded in the Zone Libre could be recruited. Used machinery was purchased from all over, production tooling was hand made, and an ebauches production unit was installed in a rented apartment. Bouchet closely worked with a mechanical engineer from Lorraine by the name of Clément Burth, who had no intention of returning to the occupied sector.

Lorsa Ebauches

Bouchet and Burth had no knowledge of watchmaking. Yet, they succeeded within a short time to reinvent horology without any preconceived ideas. Evidently, collaborating with various local producers of precision parts was very beneficial. This was the birth hour of the Lorsa Company; for the next four years it supplied ebauches to the few remaining watch manufacturers in the Zone Libre.

In the postwar period, there was an enormous and apparently insatiable demand for watches. Lorsa grew by leaps and bounds; soon it was Europe's largest ebauches manufacturer – not counting Switzerland. Yearly production reached 6 to 7 million ebauches. Inevitably, competitors emerged; soon the Annemasse region produced 50 percent of the total French output of ebauches. Robert Bouchet passed away in 1972; Clément Burth took control until the founder's son Philippe Bouchet (b. 1941) was old enough to manage the company that employed some 400 workers.

In 1977, Lorsa gave the CETEHOR the mandate to develop a 6¾ x 8-line quartz caliber with a Lavet motor. Only prototypes of this movement were made, as the price would have been 140 French francs apiece. This was far too much for the French market.

Again in cooperation with the CETEHOR, Lorsa developed an 11½-line analog quartz movement with sweep seconds-hand and date. Its designation was HS QS; production started in 1978, but did not go beyond 2000 to 3000 pieces. The QS-caliber was equipped with a multipolar motor developed by the CETEHOR. Most of those movements were sold to the Petit Company in Charquemont. There were few other customers, as France was just beginning to be flooded with low-price quartz movements from Japan, Hong Kong and China. HS QS was deemed too expensive; by 1980 Lorsa was insolvent and had to be liquidated.

Montrélec and Framelec

As already mentioned, the Centre Electronique Horloger (CEH) in Neuchâtel and Seiko in Suwa (Japan) both made presentations of their first quartz movement prototypes in January 1968. Actually, the daily rates of this totally new type of movement had already pulverized all previous world records at the Neuchâtel Observatory contest in 1967. Few people in France had predicted this, but several institutions and firms succeeded in developing similar systems within a relatively short time. However, the industrial production of quartz movements at acceptable costs was quite another thing. The government authorities soon realized that without a collective effort, the project was doomed to failure.

A Broad Partnership

Six watch industry partners for a common project were found: Cattin, Cupillard-Rième, Dodane, Jaz, LIP, and Yema International. The electronic giant Thomson CSF, the CETEHOR, and the Portescap-subsidiary Socrem joined the club. The institutional members were the "Groupement des Fabricants d'Ebauches de Montres" and the "Société de Développement Horloger (SDH)." In 1969, they collectively founded the "Montrélec" company in Morteau. Its objectives were the development and production of quartz movements.

Pierre Vovelle was at the helm of the new company, which was heavily subsidized by the French State. Its objective was the coordination (and sometimes coercion) of a number of vendors in order to produce large numbers of identical quartz calibers at a low cost and in a limited range of sizes. The capacity of Montrélec was intended to cover the needs of the entire French watch industry. Chances for success were good, as most of the important French watch manufacturers were members of Montrélec. The big exception was France Ebauches.

Three prototypes of the analog caliber Q400 (11½ lines) were shown at the 1973 meeting of the "Société Chronométrique de France." The functions of Q400 were hour, minute, second, and day/date. This caliber had been developed by Patrick Augereau and Jean-Claude Châtelain at the CETEHOR. Cattin and Parrenin agreed to produce the gears that had been specifically developed for quartz movements. The CETEHOR also developed plastic wheels and pinions, but only Eurowatch would eventually use them – in Swiss made quality!

A Stillbirth

The project moved very slowly; by mid-1974, the first experimental series of 100 Q400 movements had not yet been completed at the CETEHOR in Besançon (at that time, the general manager was Paul George, a former LIP executive). The government authorities were exerting massive pressure, but to no avail. Problems kept appearing, as several unhappy choices had been made, including a 32 kHz flexural mode quartz resonator that was made by Frésard and Cheval Frères in Ecole Valentin under SSIH license; fine tuning was done with a trimmer capacitor. The Socrem motor too was a problem, as it was expensive and lacked sturdiness. In the end, only 80 pieces were made; twenty of them were sold as souvenirs to the members of the Montrélec community.

This was precisely the time when the digital watch was rapidly gaining in importance. All of a sudden, the interest for analog calibers waned. Production of the Q400 caliber did start, but only a few thousand pieces were actually sold. Many of them were returned under warranty; very often the problems were due to the Socrem motor that was not reliable in the small sizes needed for watches.

Montrélec was rapidly reoriented towards digital movements with liquid crystal display. In 1977, an assembly unit was started in Morteau. According to plans, its capacity was supposed to reach 120,000 units a year, but actually only a few tens of thousands were produced. Those modules were too costly and their quality was marginal. The best customers were Jaz, Herma, and Lov, the latter two belonging to the Anguenot group.

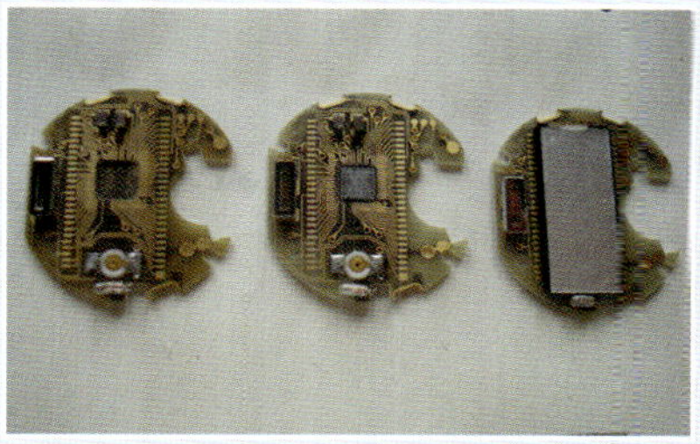

10.25. Assembly steps on the display side of the Montrélec LCD caliber MTC 936

10.26. Back side of the Montrélec LCD caliber MTC 936

10.27. Assembly steps for the Montrélec LCD caliber MTC 75

10.28. Back side of the Montrélec LCD caliber MTC 75

Four different LCD digital calibers were developed; three of them were mass produced. These were the 8¾-lines x 4.7 mm module MTC75 with an illuminated 3½-digit display; the functions were hour, minute, and dots flashing at the 1 Hz rate. Pressing a key activated the display of seconds as well as month and date. Another possibility was the display of hours and minutes or month and date.

The 11½-line, 6-digit module MTC936 with a height of 4.1 mm featured a CMOS integrated circuit from Thompson-CSF, with the functions oscillator, divider, and decoder. The yearly rate usually was less than a minute. On the LCD, hour, minute, and second were displayed permanently, but the second alternated with the date. Another mode was hour, minute, and date, the latter alternating with the weekday. MTC75 and MTC936 were both developed by Montrélec, while the 3½-digit men's watch caliber MTC61 (13 lines x 4.8 mm) was developed by CETEHOR on the base of a mandate from Montrélec. Plans also called for a 6-digit chronograph, but nothing materialized. MTC61 was quite successful; several ten thousands of pieces were sold.

The silver oxide battery lasted for a year if the night light was not switched on for more than 10 seconds per day. Without the lamp, the module's current consumption was 4.5 microamperes. The functions were hour, minute, second, and month plus date. The calendar was programmed for 4 years. Two versions were made with different integrated circuits, but the functions were exactly the same. In order to use the available space, the integrated circuit had to be mounted on the outside face of the electronic module in one of the versions.

Framelec Takes Over

Several members of the Montrélec consortium refused to buy the new calibers, as they were expensive and not reliable. Instead, those watch producers bought their movements in Switzerland or Japan, where proven, high quality calibers were available. Most importantly, all customers clamored for Lavet motors, while the analog Montrélec caliber was equipped with the expensive and unreliable Socrem motor, the small sizes of which were particularly bad.

Starting in 1978, Montrélec had a direct competitor by the name of Framelec (abbreviation for France montre électronique). It was a newly formed joint venture of Herma-Lov (Anguenot group) and Charpier-Rième (Cupillard-Rième group); JAZ was in charge. Framelec's offer consisted in the Montrélec-LCD digital calibers MTC75 (8¾ lines) and MTC936 (11½ lines). On the analog side, Seiko movements were available: Framelec bought them in kit form and had them assembled in Villers-le-Lac.

Framelec then gave Ebauches Electroniques Marin (a subsidiary of Ebauches SA) the mandate to develop two analog calibers with a diameter of 6¾ x 8 lines (FR 206) and 11½ lines in two versions: FR 305-1 (with date) and FR 305-2 (with day/date). Total production of each of those calibers was several ten thousands of pieces. In 1981, Framelec took over Montrélec, but operated it just for a few months, as Framelec itself was taken over by Matra Horlogerie, whereupon Montrélec was immediately closed. Framelec survived until 1984, then it too was closed by Matra.

Parrenin

The immediate postwar period was a very prosperous time for the French watch industry. Watches were hard to get, and, like most of its competitors, Ebauches Parrenin in Villers-le-Lac literally boomed. However, just a few years later, the market was saturated and Parrenin's slow decline started. In the 1960s, bankruptcy was barely avoided several times. The problems were an incompetent management as well as a design department that was incapable of developing first class calibers.

As opposed to Switzerland, France lacked truly ingenious caliber designers. Management all too often refused to take risks and to commercially push through interesting innovations. The French market was fairly big and the still existing colonial empire was even bigger. Thanks to this enormous market, a good living could be made without any technical innovations.

Great Achievements and Decline

Things did not improve, even though Parrenin's mechanical department kept developing new systems that allowed an unheard of precision in the manufacture of movement parts. Thus, it became possible to reproducibly fit the ruby bearings in the plate and the bridges within ± 5 micrometers of the specifications.

The CETEHOR in Besançon checked the French ebauches and compared them with Swiss ebauches made by ETA and Fontainemelon. Those tests showed that Parrenin was at the top in many ways. Parrenin also was the pioneer of dry lubrication with molybdenum sulfide: only the escape wheel was oiled conventionally. However, it did not do much good to manufacture mediocre calibers with extreme precision, to make them service-friendly and to give them a long life: they still were mediocre calibers.

The Parrenin caliber HP60 won first prize several times at the Besançon Observatory's chronometer contest. However, this caliber had a basic defect that nobody noticed as long as only the daily rate in seconds was considered. The consumer found out all too soon that his chronometer indeed had a rate of just a few seconds per day, but that the minutes were wrong.

Production automation was an important issue for Parrenin. FHF (Fabrique d'Horlogerie de Fontainemelon) had been the real pioneer in this respect. The linear transfer machines used for the production of the Fontainemelon calibers 69 and 96 brought a quantum leap in terms of productivity. Parrenin needed something similar, but did not make it, despite heroic efforts. Enormous investments were made, but to no avail: Parrenin's production costs were still too high.

A New Management

After all those failures, Parrenin's technical director resigned in 1964. He was succeeded by René Binétruy who had been in charge of the mechanical department and soon advanced to general manager. It was quite a challenge, as the company stood on the verge of bankruptcy. The banks refused to extend the credit line and the company just managed to pay the salaries at the end of the month.

Things slowly improved thanks to the efforts of the new manager. Enormous efforts were made to improve productivity and reduce costs; production was concentrated on just a few simple and low-cost Roskopf pin pallet calibers. Precision injection molding of plastics was introduced very successfully: soon, plastic parts were made by the hundred thousands. Old calibers were modernized and two new, very decent calibers were developed.

Operating alone was not feasible in the longer term. Cupillard, Jeambrun, Tecnic Ebauches, and Femga did the right thing and merged to France Ebauches. Parrenin tried to merge with Lorsa (L'Horlogerie de Savoie), but only a lose alliance was accepted by the partner. The group was called Lofrance, but the two companies remained independent. Lorsa had bought a controlling majority of the Parrenin shares but was dead-set against a true merger. Not even the technical cooperation worked. Each partner concentrated on certain calibers. Parrenin specializing in pin pallet movements while Lorsa preferred jewel pallets. Under those circumstances, productivity gains were small.

Cooperation with Oudet and Stcherbatcheff

Parrenin management realized that the electrification of the wristwatch was inescapable and began to worry about the small motors that could be used in a watch. A cooperation was thus started with two professors at the Besançon technical school: Claude Oudet and Georges Stcherbatcheff.

Stcherbatcheff was a native Russian who had managed to escape to France during the October Revolution. This was where he invented and patented a clever linear motor. It was based on a magnet inside

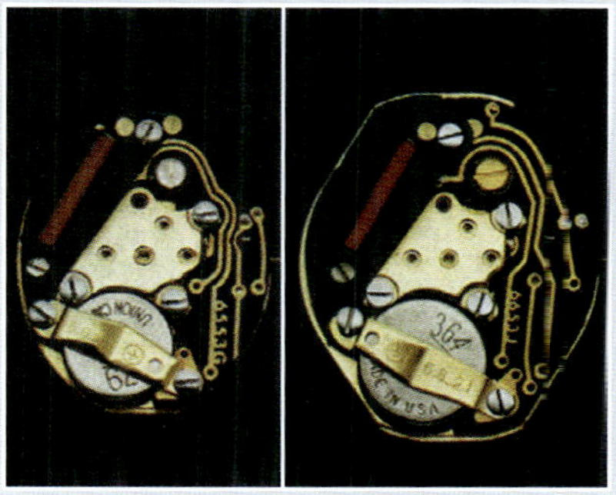

10.29. Parrenin caliber 6553 (5½ x 6¾ lines, left) and 6653 (6¾ x 8 lines, right)

10.30. Parrenin caliber 6871 (8¾ lines)

10.31. Parrenin caliber 6884 with moon phases (10½ lines, dial side)

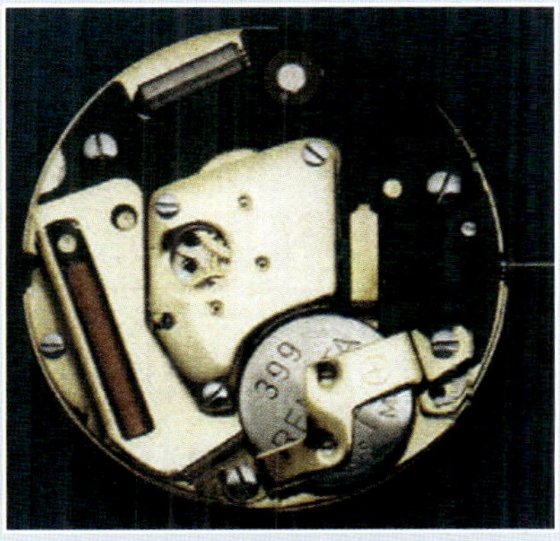

10.32. Parrenin caliber 6640 B (11½ lines)

a coil. A current pulse generated a repulsive force and set the magnet in motion. This system was very energy-efficient – just what was needed in a wristwatch.

However, transforming the linear motor into a rotating motor was difficult. Of all things, a ball track system generating a lot of friction was chosen. Yet, the clock manufacturer Imhof in La Chaux-de-Fonds was interested; after checking out a prototype, Imhof bought the patent rights and adapted the system to an electromechanical clock movement of its own production.

Stcherbatcheff's life ended tragically. He was an enthusiastic mountaineer, and one day he did not return from a hike in the Montblanc area and never reappeared. All kinds of rumors circulated: the Soviet secret service KGB was accused of having abducted him, as his brains were needed back home. Reality was much more pragmatic. Twenty years later, Stcherbatcheff's badly mutilated corpse was found with his ID card and other papers at the foot of a glacier. Obviously, he had fallen into a crevice and did not manage to get out, so he miserably froze to death.

Using the Stcherbatcheff concept to design an energy-efficient stepping motor was a real challenge. Lorsa decided to industrialize the motor: it gave up after having wasted a lot of money. The clock manufacturer JAZ took over and also failed. Finally, Portescap in La Chaux-de-Fonds, working with Claude Oudet as a consultant, developed the multipolar Socrem stepping motor on the base of the Stcherbatcheff concept.

The Socrem motor was quite ingenious, but it was rather complicated, which made it expensive. It worked fine, but it was not reliable in the small sizes needed for wristwatches. In the end, the Lavet motor prevailed. The Japanese were first to understand its inherent superiority and reinvented it. The Seiko patents could not be circumvented, so for many years, the production of Lavet motors was not possible without a Seiko license.

Quartz Calibers

Parrenin managers were not overly worried when the CEH and Seiko Suwa unveiled their wristwatch-size quartz movements in January 1968. These were exotic things that certainly had a sensational daily rate, but were bound to be very expensive and would at best occupy a niche market. At any rate, Parrenin did not have the means to develop something similar. Hardly anybody realized the enormous price reduction potential of this brand new technology.

Less than ten years later, quartz calibers were rapidly invading the core territory of mechanical movements that were more expensive and less accurate. Parrenin signed an agreement with the CETEHOR covering the design of a whole family of quartz movements. The very first one was based on Parrenin's pin pallet Roskopf caliber 64-1640: the mechanical part was combined in 1980 with an electronic module designed by the CETEHOR to form caliber 6649. With the experience gained on this base, production of the 65/66 family could proceed: these were the calibers 6553 (three hands, 5½ x 6¾ lines x 2.85 mm), and 6653 (three hands, 6¾ x 8 lines x 2.85 mm). The variants with two hands were identical in terms of dimensions; their designations were 6552 and 6652, respectively. Caliber 6660 was slightly flatter (6¾ x 8 lines x 2.75 mm, 8 pulses per minute).

The caliber family 6870 consisted of 8¾ lines, three-hand movements with a uniform height of 3.50 mm. The variant 6871 was equipped with a date disc, caliber 6872 displayed day and date. The 10½-line caliber 6881 featured a date disc, its twin 6882 displayed day and date.

The three-hand, 11½-line calibers with a seconds-hand stopping device were given the designations 6640 A and 6640 B; the height was 3.35 mm, the B variant was for watches with hands turning counterclockwise. Caliber 6641 A featured a date disc that increased the height to 3.90 mm. Other variants were: 6643/3 with two hands, small seconds-hand, and date, 6643/4 with three hands, small seconds-hand, and date; 6624 with three hands and a 24-hour display; 6625, which was 6624 with a date disc; and 6626 that featured three hands and a day/night indicator with the sun or moon appearing at 12 o'clock.

The 7000 caliber family never evolved beyond the prototype stage. After Lorsa declared bankruptcy in 1980, Parrenin was acquired by the Hong Kong Remex group, which sold it to ETA in 1992. It was eventually integrated into the Fabrique de Fournitures de Bonnétage (FFB) that ETA had bought in 1991. None of the Parrenin calibers was continued by ETA, which transformed the plant into a production center for small metal parts.

SEFEA

For many years, SEFEA (Société d'Exploitation de Fabriques d'Ebauches Annemasse) was the most important ebauches producer in the Haute Savoie, but eventually Lorsa took its place. In 1967, SEFEA was acquired by the Swiss Ebauches SA and henceforth specialized in 8¾-line calibers with sweep seconds-hand. This actually was the Fontainemelon caliber FHF 36, that SEFEA imported as blank movements, assembled into complete movements, and sold mostly to Italy where they were cased by jewelers who had their own, local brand.

At that time, exporting Swiss ebauches was not allowed, except to France on the basis of special agreements. Italy, thus, managed to obtain Swiss movements via the SEFEA loophole. It is probably not a pure accident that after the takeover by Ebauches SA, SEFEA's general manager was Jean-Didier Bauer, a nephew of Gérald Bauer, president of the Fédération Horlogère (FH) in Bienne.

SEFEA produced five analog quartz movements, all of them with a seconds-hand stopping device: 961.001 and 961.101 (6¾ x 8 lines x 3.50 mm and 4.10 mm, respectively); 965.101 and 965.111 (8¾ lines x 3.20 mm and 3.80 mm, respectively); and 965.311 (10½ lines x 3.80 mm). These were ETA movements that were produced in Annemasse. There was a design office on the premises; ETA gave it the mandate to develop the 8¾-line calibers 581.001 and 581.011 with a height of 3.2 mm and 3.7 mm. Today, SEFEA is a supplier of Swatch Group companies and produces movement parts for ETA and assembles electronic systems for other Swatch Group companies.

Yema

Yema in Besançon did not miss the forebodings of the quartz revolution. In late 1971, its catalog featured LED watches with 5 functions: the modules were supplied by the American Fairchild. LCD modules were first purchased from the Japanese NEC; later the Korean HanDok was a major supplier. Starting in 1973, Yema analog movements were assembled from Bifora kits. However, Bifora was badly hurt by the aggressive Japanese competition and went bankrupt in 1977.

For a while, Yema considered producing its own analog quartz movements as a partner of France Ebauches. A joint venture by the name of Quartzelec was established (it was also known as UFIDEL). But this project was quite short-lived, as Yema succeeded in signing an agreement with Ebauches Electroniques Marin (EEM).

EEM was a true pioneer and established Switzerland's first "silicon foundry" (see Chapter 8). Yema would buy EEM quartz movements in kit form and assemble them in its own premises in Besançon. This was quite exceptional, as normally EEM only sold complete movements. Yema always purchased its quartz calibers and never designed its own. Eventually, Seiko took over Yema, but preserved its identity; it is only logical that Yema exclusively used Seiko movements after 1982.

11. Quartz Movements from Great Britain

Sinclair

Almost all Sinclair products were black; small wonder that Sinclair's LED digital watch featured a pitch black dark rectangular plastic case (28 mm x 45 mm) and was called "The Black Watch." The display was located on the upper part of the case; two big pushers integrated into the case were used for setting hour and minute. A key at 3 o'clock switched on the LED.

The Sinclair watch was launched in 1974 and was available completely assembled or as a kit that the customer assembled himself. The plate's dimensions were 24 mm x 24 mm. The module was fully conventional and consisted of two silver oxide batteries, a 32 kHz quartz resonator of the flexural mode type (Motorola MTQ 32) with a trimmer, a CMOS integrated circuit with 2000 transistors (Intersil ICM 7202), and a 4 digit-LED (hp 5082).

The Sinclair Radio Watch, that was launched in late 1985, consisted of stainless steel and plastic, but it was not black. Its three-part construction was quite unique, flexibility was achieved with hinges. Thus, the Sinclair Radio Watch automatically adapted itself to any wrist. The lower part contained the watch module and an illuminated LCD. The central part accommodated a miniature loudspeaker while the upper part housed the FM receiver; the frequency was displayed by an LCD.

The antenna was integrated into the bracelet; the battery was hidden in the clasp. Each of the three parts of the case was provided with a crown on the right side, it was used in combination with the three pushers under the LCD. In this manner the time, the volume, and the FM frequency could be set. The best reception was achieved when the left arm was held straight up, which could be rather tiring.

Sinclair had planned to mass produce the Radio Watch, but, soon after production had started, the company found itself in financial difficulties. It was purchased by Amstrad, which stopped the project. The result was that the Radio Watch is among the rarest Sinclair products and is avidly sought by collectors.

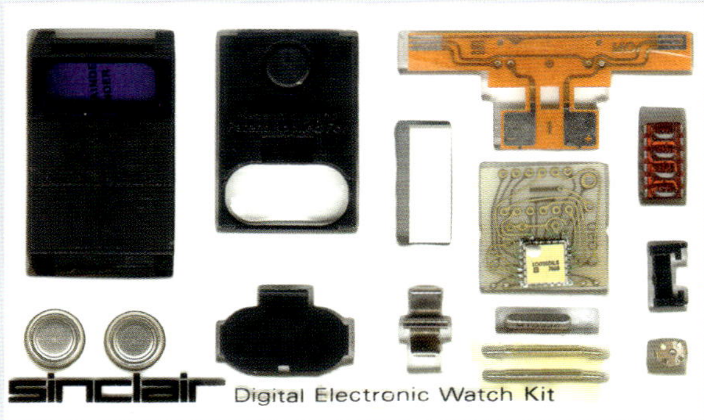

11.1. Kit for the Sinclair Black Watch

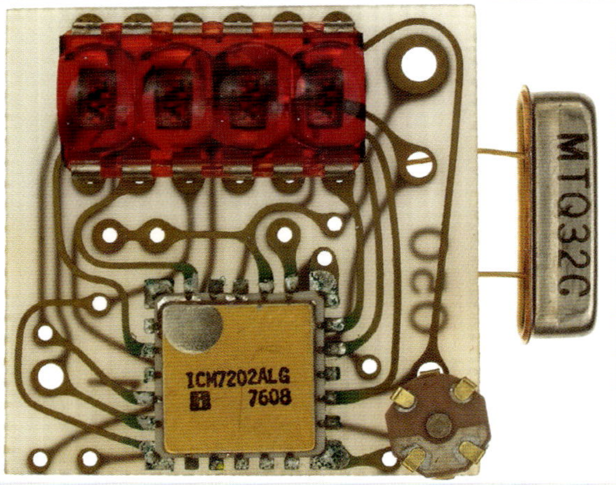

11.2. Electronic module of the Black Watch

11.3. Sinclair Radio Watch

12. Quartz Movements from Japan

The Japanese watch industry is characterized by two totally different corporate cultures. Casio, Citizen, and Seiko, which have been established in the watch business for many decades, have an acute sense of history. They own complete collections of the electromechanical and electronic calibers that they developed – including the full documentation – and are happy to inform media people and visitors.

Things are very different with the electronics companies that participated in the short-lived boom of digital watches (LED and LCD types). Hitachi, NEC, Ricoh, Sanyo, and Sharp – formerly very big producers of digital watches – all gave up on this product in the early 1980s, exactly like their American counterparts. Price erosion had been so severe that the field was left to low wage Asian countries.

In the Japanese electronics companies, it seems that absolutely everybody ignored the fact that at one time they produced electronic watch modules. When pressed hard, the public relations people will tell you that no documents relative to watches have been preserved: there are no pictures, no catalogs, no samples – absolutely nothing concerning their former horological activities.

The explanation is a reflection of Japanese mentality: abandoning a product may be interpreted as a failure, and failure means a loss of face. This is the worst possible thing that can happen to a Japanese: for this reason, anything related to it is mentally repressed. It is a sad fact that the electronics companies' extremely interesting venture into the field of watches can only be documented in a fragmentary way.

Casio

The Casio Company was founded in 1942 by Tadao Kashio (1917-1993) in Tokyo. Soon his three brothers had to help him manage the rapidly growing business. In 1957, Casio almost became a household word, as it had developed a very successful relay-type calculator. The production of watch movements and watches started in 1974 and was meant as a diversification within the field of microelectronics. Today, Casio is known all over the world for computers and calculators, electronic musical instruments, digital cameras, electronic dictionaries, and mobile phones. Casio produces about 32 million watches per year; they generate 14 percent of the company's sales.

Watches as a Diversification

Casio decided to enter the watch business in the early 1970s. The trouble was that not one Japanese or European watch company had the slightest intention of supplying movements or exterior parts to a potential competitor. Casio thus had to develop everything on its own. The beginning was made with LCD digital watches, as they were quite simple and required no watchmaking know-how. In 1981, quartz analog watches were added to the product-mix. By that time, Casio had grown into a major force in the watch business and had to be taken very seriously.

Ricoh agreed to cooperate and supplied Casio with analog calibers. In 1985-86, Casio developed its own digi-ana calibers; 1990 marked the start of analog calibers that had been fully developed in-house. Watch manufacturers normally speak of ana-digi (or combo) watches, when an analog movement is combined with a digital one. Casio people however insist on the word digi-ana, as in their perspective, the digital part is more important.

From the very beginning and to this day, Casio watch models always looked distinctly sporty and youthful. Metal cases were used until 1978; after that, plastic cases were introduced. Casio designed them in its own offices, but the actual production was farmed out to specialists in precision injection molding.

The G-Shock line was launched in 1983: it was and is a runaway success, as those watches have achieved cult status. They are equipped with an LCD quartz movement in a container that is guaranteed to be waterproof to a depth of 100 meters. This container is inserted in a massive armor made of rubberlike plastic. The watch survives a free fall of 10 meters onto concrete.

Casiotron

Casio's first watch model was called "Casiotron" and was launched in 1974. It was equipped with the QW02 module (QW means Quartz Watch); the dimensions were 31.0 mm x 7.2 mm. The plate consisted of glass fiber reinforced epoxy laminate. The functions were AM/PM time as well as month and day/date. A pusher stopped the watch for exact synchronization with a time signal. The CMOS integrated circuit with oscillator, frequency divider and decoder was purchased from OKI. The LCD was imported from the Swiss Videlec (see Chapter 8).

The quartz resonator was supplied by Toyocom and was encapsulated under vacuum in a 13.5 mm x 3.5 mm case. This still was a flexural mode resonator, but the frequency of 32,768 Hz had established itself as the industry standard. Both the chip and the LCD needed 3 volts; for this reason two Union Carbide silver oxide batteries in series were provided. From the very beginning of its watch effort, Casio adopted the concept of designing both the movement and the exterior parts, but of purchasing the majority of the parts and assembling them on its own premises.

It may seem odd that Casio started its watch business with caliber QW02, while caliber QW01 was launched as late as 1975. This was a precedent for the non-chronological sequence of Casio caliber numbers. This was due to the fact that quite often the design of several movements was started at the same time, each project being given a sequential number. Technical as well as commercial factors often meant that a caliber with a low number was used long after calibers with a higher number had already been launched. Furthermore, not all designs that made it to the prototype stage received the go-ahead for actual production. This explains the gaps in the sequence of QW numbers.

Caliber QW01 (31 mm x 6.5 mm) was launched in 1975: it featured a 1.5-volt integrated circuit and a 3-volt LCD, both by Hitachi. As there was only one battery, a voltage doubler was necessary for the LCD. QW01 used the same flexural mode quartz oscillator as QW02. These were Casio's first watch calibers, production continued for a full year.

More and More Functions

Microprocessor watches by the name of "Casiotron" were first marketed by Casio in 1975. They used caliber QW03 with a diameter of 31 mm. The chip was very probably purchased from Nippon Electric. The functions were the same as in QW02.

Caliber QW04 (31 mm x 7.5 mm) with a Hitachi integrated circuit was also launched in 1975. The mechanically cut tuning fork quartz resonator supplied by NDK (Nihon Dempa Kogyo Co. Ltd. Tokyo) was used. This type of resonator was the industry standard until the end of the 1970s, at least in Japanese quartz watches. It was then replaced by the much less costly photolithographic Statek-type tuning fork resonator.

The smaller ladies' model caliber QW06 Casiotron Lady (21.5 mm x 6.2 mm) was launched in 1976. Aside from the time, it also displayed the month and day/date by actuating a pusher. After QW06 there was a big gap in the caliber numbering; actually, several new calibers were brought to the prototype stage, but

12.1. Casio LCD caliber QW02, front side

12.2. Casio LCD caliber QW02, back side

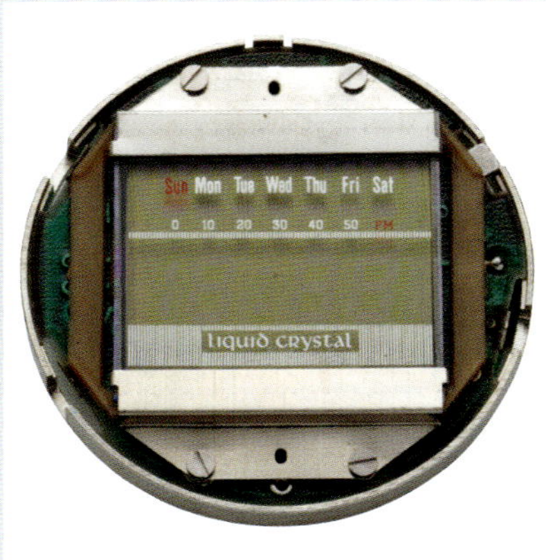

12.3. Casio LCD caliber QW04, front side

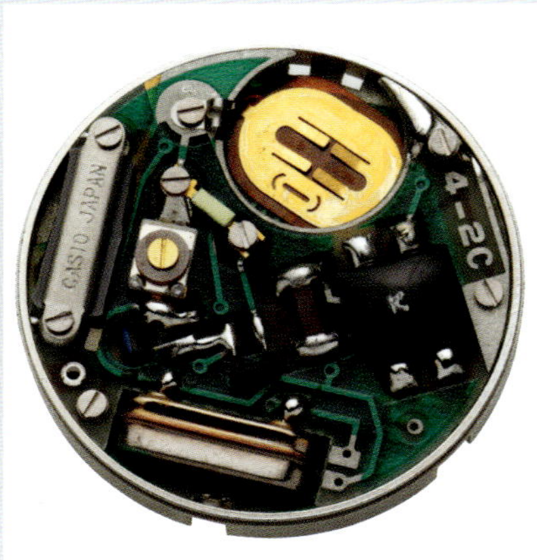

12.4. Casio LCD QW04, back side

were not produced in series. Thus, caliber QW17 (31 mm x 7.5 mm) in the watch model Casiotron X-1 was launched in 1976. It was equipped with a stopwatch function to 1/10 second, a timer, and world time with six time zones that could be set individually. Aside from the time, the movement displayed month, date, and day.

QW17 was followed by the calibers 18, 19, and 20, but they were similar to QW16 and QW17. Those calibers had the same functions and were only different in terms of details such as the height.

In 1977, caliber QW31 followed with an integrated circuit from OKI and a Hitachi quartz resonator. The diameter was 27.5 mm, the height 4.2 mm. In the same year, Casio launched the 28.5 mm x 5.2 mm caliber QW21 with a Hitachi chip and the same functions as QW17. Still in 1977, QW37/38 was launched, a true chronograph with short time measurement to 1/100 of a second. The diameter was 30.5 mm, the height 8.5 mm. Many of the older Casio calibers were characterized by two baseplates made of glass fiber-epoxy laminate; screws made the electrical connections. For this reason, such Casio calibers were not record flat, but were extremely robust, which was deemed to be more important.

Crystal Sound

Caliber QW25, brought to market in 1978, was called "Crystal Sound"; it featured an alarm function with a piezoelectric tone generator. The quartz resonators of the photolithographic type were now available and allowed further miniaturization: the length was just 6 mm, the diameter 2 mm.

Between QW25 and QW65, several calibers were produced. They all displayed the time, month, and day/date, and only differed in their size. In chronological order these were the calibers QW50, 52, 53, 22, 51, 49, 45, 46, 47, 60, 26, 86, and 35. Only some of them went into mass production, but those that did were used for various types of watches for many years.

3-Volt Electronics

In 1980 Casio launched caliber QW65 (C-80), its first movement with 3-volt electronics and a lithium battery by Panasonic. For a diameter of 29.5 mm, a height of 6.0 mm was required. From then on, this concept was used in men's watches that had not to be particularly flat; this allowed a battery life of 5 to 7 years. QW65 was a chronograph with a time resolution of 1/100 of a second. The smaller ladies' watches by Casio continued to be equipped with a 1.5-volt battery and suitable electronics; battery lifetime was 2 years.

12.5. Multifunction wrist watch Casiotron Alarm-Chronograph, Casio, 1975

12.6. Casio LCD caliber CW06, back side

12.7. Casio LCD caliber CW06, front side

12.8. Casio Multifunction LCD caliber S14, back side

Caliber QW133 of 1980 was the first calculator watch by Casio. On the 8-digit LCD, the results of calculations of all four basic modes were displayed, aside from time, month, and day/date. Further development led to QW166, that included countdown, daily alarm with two tunes to choose from, bells, an hourly alarm, and a weekly alarm. In the calculator mode, chain and mixed calculations could be performed.

Also in 1980, caliber QW82 (QW81 was identical with it in terms of dimensions) became available; it was an alarm watch with 12 different tunes. The lithium battery had a life of at least two years. The diameter was 29.5 mm, the height 7.0 mm.

In the 1980s data storage, altimeter, barometer, thermometer, UV meter, compass, depth meter, pulse, blood pressure, distance, speed, and used calories became new features. Towards the end of the 1980s the batteryless caliber 520 was launched; it was equipped with solar cells. For recharging after an operating time of 7 days in the dark, just 10 minutes in full sunlight were necessary.

Radio Control

The often multi-colored LCD was used by Casio not only for numbers and letters, but also for graphics, e.g. for barometric pressure or heart rate in function of time. Also available were functionless animations. Today, Casio supplies a specialized digital watch for just about any kind of sport.

Since 2000, Casio has become very active in the field of radio-controlled wristwatches, "Wave Ceptor," without neglecting special functions, all the way to calculator, radio, data storage, digital music, TV remote control, or garage door opener. Such special functions need special sensors and microprocessors; they have nothing to do with the horological functions of the watch.

This also applies to the Casio radio-watch TM 100, the Casio camera watch, and the Casio watch with GPS satellite receiver, Navi PRT-1GP1. The latter synchronizes itself with the cesium clock signals of the GPS system and is even more accurate than a classical radio-controlled watch. However, nobody notices the difference as the seconds-hand makes jumps of one second.

Citizen

An early Japanese competitor of Seiko, Citizen Watch Co. launched a battery operated watch with a balance wheel motor in 1966; it was called "X-8" (caliber 0800). It was already equipped with a transistor acting as a switch for the balance wheel motor and worked according to the electrodynamic concept of fixed coil and moving magnets. For good reasons, Citizen skipped the concept of the contact controlled balance wheel motor.

Fixation on the Mechanical Tuning Fork

In 1970, Citizen signed a joint venture contract with Bulova concerning the production of the tuning fork watch "Accutron" that had already been launched in the U.S. in 1960; this proved to be a mixed blessing. At the time, producing the Accutron in Japan was considered as a tremendous privilege. So far, Bulova had only licensed the Swiss Ebauches SA to manufacture their own, improved version of the tuning fork watch, but Citizen had excellent relations with Bulova and had supplied them with mechanical movements for many years.

Until 1975, Citizen produced about a million tuning fork watches under the brand name "Hi-Sonic." In the beginning, the movement was equipped with a symmetrical tuning fork. Later, for cost reasons, the transition was made to an asymmetric system with only one coil.

The very accurate but expensive analog quartz watches by Seiko as well as the American digital quartz watches that were mass produced since 1972 did not particularly impress Citizen management. The same applied to the Swiss Quartz caliber "Beta 21;" it was considered as something exotic and very expensive. Just like Bulova, Citizen thought they were safe with their version of watch electronics. Why bother with crazy things?

In-House Quartz Technology

Soon the price of quartz watches began to drop much faster than anybody had anticipated: electromechanical and tuning fork watches were being rapidly displaced. Citizen had to establish a crash program in order to make up the lost time and produce its own quartz calibers. Thus, the company was not among the pioneers of the quartz wrist watch.

In the beginning, Citizen engineers tried to proceed on their own way. Starting with a bar-type, 16,384 Hz quartz resonator, this frequency was divided ten times by two, which allowed the synchronization of the 16 Hz balance wheel motor of a modified electrodynamic caliber of the X-8 family. Excitation of the quartz resonator and control of the motor coils was achieved with an integrated circuit of the CMOS type.

This is how the quartz caliber 8800 came into being; it was used in the three-hand watch "Crystron" with day/date display that was launched in August

1973. The movement had a diameter of 11½ lines (25.6 mm) and a height of 5.55 mm. Compared to the tuning fork watch, the daily rate was improved by nearly a factor 7, i.e. from ±2 seconds per day to ±0.3 seconds per day. Selected models had a daily rate of just ±0.16 second per day. The battery lifetime was guaranteed to be at least one year (see also Chapter 3).

Crystron Mega

The quartz synchronized electrodynamic movement was just a stopgap. Citizen absolutely needed watches with "real" quartz movements. The problem at that time was the low frequency quartz resonator that could not yet be obtained commercially. This was solved the same way Omega had done with their Megaquartz 2400: a lens-shaped, commercially available high frequency quartz resonator was used.

Citizen chose a 4 MHz quartz resonator used in telecommunications with a frequency of 4,194,304 Hz that was brought down to 32,768 Hz by sevenfold division by 2 with a CMOS circuit, i.e. dynamic-type flipflops on a SOS-Chip (silicon on sapphire). The following 15-stage division to 1 Hz was achieved with a conventional CMOS-Chip; a 6 pole stepping motor transformed the pulses into a rotary motion. The accuracy of the 11¾ lines x 5.17 mm caliber 8650A was extreme, the rate being just ± 3 seconds per year. The functions were time with a sweep seconds-hand and day/date with rapid correction. Due to the many division stages, battery life was limited to 12 months.

The Citizen Quartz Crystron Mega, with the caliber 8650A, was brought to market in December 1975 in an eight-sided gold case. Just 3000 pieces were produced, the price was the rather extravagant sum of $15,000. In June of the following year, caliber 8650B, with a considerably longer battery life, appeared in the stores. A more economical model with a steel case was launched in 1979; it was equipped with caliber 7370D, the height of which was 4.96 mm.

CMOS-IC and Tuning Fork Quartz

Parallel to launching the Crystron Mega, Citizen introduced the quartz technology that has been the world standard ever since. The analog display, 32 kHz quartz watches were still sold under the name "Crystron," but without the suffix "Mega." They were equipped with the calibers of the 8600 family (11¾ lines x 5.17 mm). The resonator was a 32 kHz quartz tuning fork that had been obtained by cutting and polishing a quartz platelet. Further major components were a CMOS circuit for frequency division to 1 Hz, a Lavet-type stepping motor and a silver oxide battery.

12.9. Citizen Quartz caliber 8630A

12.10. Citizen Crystron caliber 8700A

12.11. Citizen Crystron caliber 7300C

12.12. Citizen Crystron 4 Mega caliber 7370D

Caliber 8620 (11¾ lines x 5.17 mm) with sweep seconds-hand and day/date was extremely successful. It was launched in 1975 and was produced in large series. Caliber 8629A (the follow-up models were 8630A, 8640A, and 8650A) launched in 1976 was equipped with four square solar cells on the dial, thanks to which the life of the silver oxide battery might be extended to 5 years. This was the world's first analog solar wristwatch. The Citizen-Quartz watches of the Series 8600 were already equipped with photolithographic tuning fork quartz resonators of the Statek type.

The ana-digi Citizen watches of the 8900 series had a very original design. Under the 6-digit LCD for time, day/date, alarm, and chronograph to 1/100 of a second of caliber 8910 (1979, 30.0 mm x 4.97 mm), an accessory dial was located on the left with a small two-hand watch and a further LCD for the alarm time.

The ana-digi watch launched in 1980 with caliber 8920 also had timer functions. The height of this round 30 mm caliber was 4.9 mm. With a height of just 2.28 mm, the rounded rectangular ana-digi caliber 8930 with the dimensions 26.2 x 23.2 mm was much flatter, but the extra functions were just day/date, alarm, and hour signal.

The "Time Track" model launched in 1980 with caliber 8940 measuring 26.0 x 27.4 x 4.55 mm was a conventional ana-digi watch with a big two-hand analog display (with 20-second motor steps) and a narrow LCD underneath. Stopped time to 1/100 of a second, alarm time, and timer time, as well as day/date were displayed digitally. The corresponding ana-digi caliber 8960 for ladies' watches (also with a 20-second stepping motor) measured 18.0 mm x 15.0 mm, the height was 2.50 mm. The follow-up caliber 8980 measured 26.0 x 26.0 mm x 4.05 mm. The additional functions were day/date, second time zone, alarm, chime, stopwatch, and thermometer. A one-second stepping motor was used.

Lithium Batteries

The Citizen calibers of the 1400 family that were launched in 1980 were equipped with 3-volt lithium batteries and the appropriate chips. The most important calibers were 1400A with day/date and 1410 with date, while 1429A and 1430A were three-hand watches.

The analog calibers of the 2000 family were produced mostly for ladies' watches. For example, the 6¾ x 8 lines x 2.8 mm caliber 2030A used digital frequency adjustment instead of a trimmer. The same applied for the bigger, 22.6 mm x 22.0 mm x 3.8 mm caliber 2100A. Particularly small watches were using the 13.2 mm x 9 mm x 1.87 mm caliber 2200 with a 20-second stepping motor; it also featured the digital frequency adjustment. The universally usable rectangular, 19.0 mm x 16.0 mm x 2.5 mm caliber 2400 was equipped with a 20-second stepping motor.

The rectangular caliber 2730 E, F launched in 1982 without special functions was equipped with a one-second stepping motor; the dimensions were 18.0 mm x 18.0 mm x 1.9 mm. Caliber 2850 of 1982 was similarly equipped; the dimensions were 21.5 mm x 21.7 x 2.58 mm. The general purpose calibers 2931 (14.5 mm x 13.5 mm x 2.15 mm, 1983) and 2951 (15.15 mm x 13.0 mm x 2.3 mm, 1984) were characterized by an extremely low current consumption of 0.68 microampere.

As was already mentioned, Citizen was using the photolithographically made tuning fork quartz resonators of the Statek type by the end of the 1970s. It was much smaller than the standard tuning fork resonators and contributed to the height reduction of a 23.7 mm quartz movement to 0.98 mm that Citizen achieved in 1978. This was the world's flattest watch movement and was built into a remarkably elegant watch with a height of just 2.1 mm, the "Exceed Gold" with an 18 Kt gold case.

The Smallest Ladies' Watch Movement

The launch of the world's smallest ladies' watch also happened in 1980. It was part of the Citizen collection "Exceed Gold;" the diameter was just 12 mm. The quartz movement that was used (caliber 1500) measured 7.0 mm x 9.0 mm x 2.0 mm and therefore had a volume of 126 mm^3. The Mini 101 by Fontainemelon (a subsidiary of the Ebauches Group) was round with a diameter of 9.5 mm and a height of 2.25 mm; it thus had a slightly higher volume of 159.5 mm^3. However, Citizen beat its own record in 2002, when it launched caliber G720; the dimensions were 5.0 mm x 6.0 mm x 2.5 mm; this corresponds to a volume of 75 mm^3.

The purely analog alarm chronograph with caliber 3530A (13¾ lines x 3.6 mm) was launched in 1984. This was a classical three-counter chronograph with additional stop, alarm, and timer modes. The movement included three Lavet stepping motors.

The integrated circuits needed for Citizen quartz watches were always bought from outside vendors. As watch calibers with special functions needed a microprocessor, Citizen set up a microelectronics design office, but had the integrated circuits made by a silicon foundry in Japan.

Digital Watches

In 1975, Citizen launched a LED-watch with the designation CQ; the module was from Sanyo. The functions were hour, minute, and second. With a pusher at 9 o'clock, month and day could be displayed too. The pusher for switching on the LED was placed at 2 o'clock.

Citizen was among the early producers of solid state watches with a liquid crystal display. In 1974, it launched the "Crystron LC," a quartz watch with a 5½-digit field effect LCD. It displayed time including second, month, date, and AM/PM. The 9010 caliber and the other calibers of the 9000 movement family had no moving parts; they were square with cut-off edges; the dimensions were 26.4 mm x 26.0 mm; the height was 7.0 mm.

The battery could be replaced without opening the case back; it had a life of 2 years. The most important parts were a brass plate with a height of 1.7 mm with openings for the battery, the NDK quartz resonator with a trimmer, and the CMOS-type integrated circuit from Mitsubishi Denki. A printed circuit connected the components. This construction was very high grade, but also rather complicated. Watches with caliber 9010 sold for extravagant 1790 Deutschmark in Germany.

The ladies' model 9050A was equipped with a 3½-digit display for hour and minute, the dimensions were 18.2 mm x 17.2 mm x 5.22 mm. Model 9060 was equipped with solar cells in 1977 and changed to caliber 9069. The rounded rectangular caliber 9060A measured 27.0 mm x 22.6 mm x 7.43 mm. The illuminated 5½-digit LCD also displayed month and day/date.

In 1977, Citizen launched the Quartz Calculator LC, the world's first calculator watch with dual digital display (caliber 9090; in 1978 it changed to caliber 9130). The tiny pushers around the bezel had to be actuated with a special pen; calculations were possible up to 8 digits. In 1978 a similar model was introduced with the stopwatch function (caliber 9190). The calculator handled the four basic operations, square root, and percent; there also was a temporary data storage. Caliber 9140, launched in 1978, looked like a calculator with mini-pushers under the LCD, the calculator functions were the same as with caliber 9130.

12.13. Citizen caliber 1500. From 1980 until 2002 this was the world's smallest quartz caliber. The diameter of the 100 Yen coin on the left is 22.4 mm.

12.14. Citizen quartz caliber 6110A (1987)

12.15. Citizen multizone radio controlled wristwatch caliber 7400A (1993)

12.16. Citizen LCD module caliber 9010A

Multi Alarm

Watches of the Quartz Crystron Multi-Alarm type were equipped with movements of the 9100 family. Caliber 9100A (1978, 12 lines x 6.06 mm) with illuminated LCD had the functions time, chronograph, alarm, timer, countdown timer, month, and day/date. Two alarm times could be set to the minute. The modes of five functions were displayed on the LCD.

In the following year, Citizen launched the model Multi Alarm II with caliber 9101A that included the lap function (intermediate time). For the alarm signal, two different modes were available; in addition to the timer, a countdown function was available for 12 hours minus one second. At the hour, a signal beeped, which was not very much appreciated in movie theatres and concert halls. Caliber 9180 (11½ lines x 5.0 mm), with illumination, displayed time and day/date as well as the year. Short times could be stopped to 1/10 of a second.

The 9200 series calibers were used for watches of the type "Quartz Multi-Alarm Four Front;" the series was launched in 1979. Caliber 9200A (13 lines x 5.15 mm) was equipped with four pushers in order to change the mode of the alarm and the timer. Major functions were hour, minute, second, month, day, and date. The 13-line caliber 9240A launched in 1979 was called "Slim." Its height was 3.95 mm; the 5½-digit LCD could be illuminated. Additional functions were chronograph, month, day/date, two alarm times, hour signal, and countdown timer.

The follow-up caliber 9250A (8¼ lines x 4.73 mm) was put on the market in 1980 and was equipped with the following functions: chronograph, month, day/date, and second time zone. In 1981, another 12½ lines x 5.15 mm caliber of the same family with the designation 9570A was introduced. The illuminated LCD was divided in two parts that displayed day/date and hour, or minute and second, respectively. There were 7 small windows indicating the mode of melody-chime, alarm, stopwatch, timer, second time zone, alphanumeric character memo, and AM/PM indicator.

Citizen watches of the Digital Alarm Chronograph Motor Sports type were launched in 1984. The caliber code D046 means that it is a Miyota-construction. The case was made of black plastic; a 3½-digit LCD was located in the upper part of the dial, and below it was a small, 2-digit LCD. The rest of the available surface was taken up by three circular counters, the graphic display of which was part of an LCD. Thus, stop, timer and alarm modes were displayed in a pseudo-analog way.

Hamazawa

Very little is known about the Hamazawa Ltd. Company. It seems that it was founded in 1954 and produced large numbers of low-price mechanical movements. The best known Hamazawa product was the automatic caliber 5026A. It was redesigned by the Japan CBM Corp. and produced in a version that included as many plastic parts as possible, e.g. bridges, certain wheels, and the date disc. This caliber (named 2604) was mostly used in the Q&Q brand watches that were assembled in Korea. All these measures reduced costs, yet the price of quartz watches kept eroding and the competition was getting ever fiercer. Caliber 2604 had the same fate as the (nearly) all-plastic mechanical watch Astrolon developed by Tissot: it was a total failure.

The technical cooperation of Hamazawa with Epson started in the 1970s, although only indirectly in the watch sector. Finally, in 1983, Hamazawa merged with two companies, one of which was an important supplier of watchcases for Seiko Epson. Three years later, Epson absorbed the entire group, whereupon Hamazawa disappeared as a corporate entity and watch brand name.

It is still remarkable that, in the late 1970s, Hamazawa succeeded in making the transition from mechanical watches to quartz, albeit not in an enduring way. Several Hamazawa LCD digital quartz movements were launched that seemed to have been "inspired" by Seiko calibers. One of them was a 28.2 mm x 4.8 mm caliber with a silver oxide battery. The quartz resonator was of the photolithographic Statek type. The functions were set with three pushers and the following combinations were possible: hour, minute, second, and day/date; hour, minute, and second; or date, month, and year. The LCD was of the 5½-digit type; a moving marker indicated the day, another marker showed whether it was AM or PM. Those modules were exported in large numbers to China.

In the early 1980s, Hamazawa produced analog quartz movements for Seiko Suwa; this was a mandate of Seiko general management that needed Pulsar brand quartz movements. The brand had been acquired by Seiko in the course of Time Computer's liquidation.

Hitachi

Kabushikigaisha Hitachi Seisakusho (abbreviated to Hitachi) was founded in 1910 in Tokyo. The company started with electric motors and soon developed into a globally active company in the field of electrical technology and electronics. It thus rose to be the Japanese equivalent of General Electric, Siemens, and ABB. Today, Hitachi has around 400,000 employees generating sales of some $100 billion.

Like so many other electronics companies, Hitachi tried to diversify into digital watches in the early 1970s: no watchmaking know-how was needed for developing and producing them. LED displays were left out, but millions of a 3½-digit, 13-line LCD modules made of black plastic were produced from 1975 to 1985. The LCD were supplied by Hitachi's own facilities.

Hitachi even produced complete watches. The most remarkable model was launched in 1974 under the "Accurist" brand. This was a chrono-alarm with a 6-digit LCD; the resolution was 1/100 second. Hitachi also used the "Beltime" brand for a 6-digit LCD watch. Aside from the time, it also displayed month and date. Hitachi closed its watch department in 1985, as unbeatably priced LCD watches from Hong Kong and Mainland China appeared on the market.

NEC Corporation (Nippon Electric)

The Nippon Electric Company officially changed its name to NEC Corporation in 1983, but old habits tend to persist: to many people, the company still is known as "Nippon Electric." Whatever its name, the company has a long and rich history. It started in 1898 with the production of telephones and switchboards. The next diversification was electronic tubes around 1925, which were followed by high power radio transmitters and facsimile equipment. After World War II, Nippon Electric started producing transistors. Around 1960, a silicon foundry was established as the starting point for the production of integrated circuits. Today, NEC is one of the worldwide leaders in this field. Its global employment is around 140,000; yearly sales are of the order of 42 billion dollars.

In the 1970s, NEC started producing liquid crystal displays and CMOS chips for digital watches. From there, it was only a small step to complete LCD watch modules. Casio was one of the early Japanese customers. In Switzerland, Oris and Buler purchased NEC modules, while Omega bought specialized chips. In the wake of the catastrophic collapse of LCD digital watch prices and the shift of production centers from the United States and Japan to Hong Kong and Mainland China, NEC decided to get out of the watch business in the early 1980s.

12.17. Citizen LCD watch with the caliber 9010A

12.18. Citizen Crystron ladies' model with the module LC 9030A

12.19. Citizen Calculator caliber 9190A, front side

12.20. Hitachi LCD Alarm Chrono HT54002L

NEC does not seem to have kept the specifications or samples of the modules it produced by the millions. One can only try to reconstruct those modules on the basis of indications of former customers. There definitely was a basic 13-line module with a 3½-digit LCD, as well as a module called "Alpha" with date and a second time zone as extra functions. Nippon Electric also developed a chip that controlled both an analog movement and an LCD display. This was the heart of the "Single Chip" ana-digi caliber "Gamma;" large numbers of these were used in low-price Swiss watches.

Orient Watch

The Orient Watch Co. Ltd. started in 1901: in that year, Shogoro Yoshida opened a watch store in Tokyo. He was succeeded by his son, who was also called Shogoro and started to assemble small lots of excellent quality watches. After World War II, the successful company was bought by an engineer named Yoshikara Suganuma, who integrated it into his Tama Equipment Co. The watch subsidiary was named Orient Watch Co. Ltd. in 1950. As he needed capital, Suganuma sold 20 percent of the Orient shares to Seiko Watch. Orient grew very fast and established itself as the third largest watch company in Japan. However, it was relegated to rank 4 by Casio in the 1970s.

The OEM Market

By the end of the 1980s, Orient's yearly production had grown to some 11 million mechanical and electronic movements; it eventually peaked at 20 million pieces. They were, of course, used in the Orient brand watches, but large numbers were sold on the OEM market in Hong Kong. As prices for quartz movements kept eroding, this business turned into a liability and Orient abandoned it. This was a good opportunity for reorganizing the company that was moved from Tokyo north to Yuzawa City in Akita prefecture.

Over the years, Seiko had bought shares of Orient Watch; in 2008, the company was established as a 100 percent Seiko subsidiary. The Orient brand was maintained, and now Orient watches are produced in Japan, China and Brazil. In 2003, the Japanese production was concentrated in Ugo, also in Akita prefecture, at the Orient Technical Center (OTC), where premium quality watches are assembled, mostly under the "Royal Orient" subbrand. Regular Orient watches are appreciated all over the world for their particularly good price-performance ratio. These are automatic or electronic watches with a conventional analog display and a sporty or classically elegant look.

The big Orient plant in Yuzawa, where watch movements were produced initially, has been converted to the production of printers and integrated circuits. Lower cost Orient watches are assembled in Shenzhen, north of Hong Kong. The Brazilian plant exclusively produces watches for the domestic market. The Japanese Orient Watch – a Seiko subsidiary – has absolutely nothing to do with the Korean Orient Co. Ltd. that produces watches under various brand names, particularly Chagal.

Orient developed many quartz calibers in its own design department: others were purchased, mostly from Seiko. This is why Orient digital watches were available as early as 1972: the movements were made by Seiko. Orient's own digital calibers were launched in 1975. As Seiko bought more and more Orient shares, development and production of digital calibers were discontinued in 1983. On the other hand, Orient's own analog calibers (B0 and C7 series) continued to be produced until 1999. Since the beginning of the 21st century, Orient has purchased its watch movements mostly from Seiko Epson; special models are supplied by Seiko Instruments and Miyota.

Digital Calibers

Only a few collectors know that, in the 1970s, Orient had a full range of LCD and LED watches. Due to the enormous demand for this kind of timepiece, development proceeded as a crash program in cooperation with Sharp, which was very experienced in the field of microelectronics. The first solid state calibers made by Orient were launched in 1975. Caliber 62000 (28.5 mm x 9.81 mm) featured a simple but unusually big 3½-digit LCD; the quartz resonator was a mechanically cut tuning fork.

All the functions of the 62000 caliber family were integrated into a CMOS chip supplied by Sharp. The functions were hour, minute, and AM/PM. AM was indicated by 1 Hz flashing of the lower dot between the numerals, and PM by both dots flashing. Many variants were produced in the course of time; they differed by the two last numerals of the caliber designation. Their dimensions and functions were always identical with those of the basic model.

Still in 1975, Orient watches with a 6-digit LCD became available, with the second also displayed permanently. Caliber 64100 (30.0 mm x 8.80 mm) was particularly successful; among many others, it was used by Enicar in Switzerland and Dugena in Germany. Then Orient launched the 66000 caliber family with a diameter of 29.0 mm and a 4-digit LCD; the functions were the same as those of the 62000 family.

The "Touchtron" models of the 68000 family (30.9 mm x 9.27 mm) were launched in 1976. They were Orient's first LED movements; hour and minute or the date were displayed upon touching the bezel. In the following year, Orient launched the similar but much flatter 3½-digit caliber 63100 "Touchtron II" (30.9 mm x 5.65 mm). A sensor field had to be touched in order to switch on the display of time or date. The dimensions of caliber 63100 were similar to those of caliber 64100, but the 5½-digit display of the latter featured the seconds. Enicar was a good customer for this movement.

All later digital calibers by Orient used LCD displays, e.g. the 6-digit caliber 65100 with a diameter of 28.8 mm and a height of 6.15 mm. Hour, minute, and second, or hour, minute, and date, or month and weekday could be displayed.

Chronographs and Calculators

Caliber 61100 "Chronograph 20" with a stopwatch function for 9 minutes and 59 seconds was introduced by Orient in 1977. Aside from hour, minute, and second, it displayed the date or the month and day/date; the dimensions were 30.9 mm x 5.80 mm. Caliber 69100 was also launched in 1977; it was nearly square (31.7 mm x 32.0 mm), the height was 6.90 mm. It displayed hour, minute, and second, but it could also be used as a calculator with the four basic operations.

Production of the 28.8 mm x 6.90 mm caliber 65800 started in 1978. The functions were the same as those of caliber 61100 but without the stopwatch. Power was supplied by solar cells and a buffer battery. In 1979, Orient launched the 29.0 mm x 5.75 mm caliber 60100. It was used in the "Digital Ace" watches. The very wide LCD displayed hour, minute, second, and AM/PM, or month and day/date.

The "Chrono Deluxe" caliber 67110 (29.5 mm x 6.15 mm, 1979) had the same functions as the Chronograph 20. The same applied for caliber 67810 (29.5 mm x 7.20 mm) that was equipped with solar cells; it was called "Sunshine Chrono Deluxe."

12.21. Orient Quartz Crystal, caliber 0291C

12.22. Orient LCD caliber 61100, front side

12.23. Orient LCD Caliber 61100, back side

12.24. Orient LED Touchtron caliber 63100, front side

Starting in 1979, Orient used microprocessors in its multifunction watches. For example, caliber 70100 "Music Alarm Digital III" (29.0 mm x 6.06 mm) not only displayed hour, minute, second, and date, but was also equipped with a timer, the stopwatch function, and a music-alarm. The rectangular caliber 71100 (17.6 x 17.0 mm x 5.80 mm) of the "Digital Lady" ladies' watch model displayed hour, minute, and date, or month and day/date, or second and weekday.

Alarm

The height of the Orient caliber 72110 "Chrono Slim" was just 3.80 mm. The functions were hour, minute, second, and date, or month and Day/Date plus the stopwatch function. The shaped caliber 73000 (30.5 mm x 11.53 mm x 5.60 mm) was meant for use in pendant watches; the 4-digit LCD just displayed hour and minute and a 1 Hz beat by flashing dots. The last movement that Orient launched in 1979 was caliber 74100 (29.5 mm x 4.75 mm). The functions were hour, minute, second, and weekday, or month and day/date, plus stopwatch and alarm.

Caliber 76100 (23.8 mm x 6.08 mm) was available in 1980. This was the first small LCD ladies' watch caliber by Orient, it was called "Alarm Chronograph Uni." The special functions were stopwatch, timer and alarm with a new sound. The "Alarm Dual Chronograph" (29.0 mm x 4.98 mm) caliber 78100 was also launched in 1980. The functions were the same as in caliber 76100, but two times could be stopped independently of each other. The setting mode was considerably simplified in comparison with previous models. This system was used in all further models until 1983.

In 1981, Orient launched no less than nine new movements. The first of them was caliber 75100 (28.0 mm x 5.72 mm) "Memory Alarm" displaying hour, minute, second, day/date, alarm, timer, stopwatch and memo. Orient's first ana-digi caliber 77100 (29.0 mm x 6.75 mm) "Double Face" displayed time with simulated hands on an LCD as well as digitally. The digital display was also needed for the alarm, the stopwatch and the timer.

Caliber 79100 "Music Alarm" (28.0 mm x 5.35 mm) had the same functions as caliber 75100, but four different tunes could be set for the alarm. Caliber 80000 for pendant watches was remarkably small: the dimensions were 9.7 mm x 16.5 mm x 5.13 mm. The options were hour-minute, month-date, or just the second.

Super Slim

The "Super Slim" caliber 81000 also was rectangular, but considerably bigger than caliber 80000, the dimensions being 24.0 mm x 27.0 mm. The height (without the battery) was just 3.2 mm; yet aside from time and day/date, the special functions alarm, timer, and stopwatch were available. The "Chronograph 820" with the caliber number 82000 was even more sophisticated. The dimensions of this movement were 27.0 mm x 5.10 mm; it featured all the functions of caliber 81000, plus perpetual calendar, the lap function, and the possibility of timing two different events simultaneously.

Caliber 83000 "Digital Lady" (19.5 mm x 5.45 mm) was designed for sporty ladies' watches. The functions were time, day/date, alarm, stopwatch, timer, and dual timer. Caliber 86000 with a diameter of 28 mm displayed hour, minute, second, and date, or month and day/date. The extra functions were stopwatch and alarm. The ninth and last digital caliber Orient launched in 1981 was number 87000 (27.0 mm x 27.0 mm x 4.80 mm). This was a so-called "game watch." It was available in three different styles with one game each: Spacewars (caliber 87200), Combat (caliber 87201) and Baseball (caliber 87202).

Starting in 1982, Orient LCD calibers were designated with a letter plus a 4-digit number. Thus, caliber L0300 with the same dimensions as the 87000 calibers was available in two styles: L0300 with tennis and L0301 with boomerang. Caliber M1000 was mainly used for pendant watches. The dimensions were 19.0 mm x 9.9 mm and the functions were the same as those of caliber 80000 of 1980. Caliber N2400 "Multi Sports" was also launched in 1982. This was a shaped caliber with the dimensions 25.0 mm x 27.0 mm x 5.2 mm; the special functions were alarm, timer, stopwatch and dual time.

Caliber P3000 was launched in 1983; the diameter was 16.5 mm and the height 5.15 mm, which made it ideal for ladies' watches. The functions again were the same as those of caliber 80000. Production of the square caliber Q4000 with the dimensions 27.0 mm x 27.0 mm x 4.90 mm started in late 1983; the special functions were timer, stopwatch, and dual time. Orient stopped the development and production of digital calibers in 1983.

Analog Calibers

Orient started producing analog quartz calibers in 1977. Caliber designations followed a simple five-digit code system. The first two numerals or a letter plus a numeral identified the caliber series, while the third numeral stood for the type of movement: 0 for two hands; 5 for three hands; 6 for two hands and date; 7 for three hands and date; 9 for three hands and day/date. The last two digits identified minor modifications that were introduced in order to simplify production. They had to be communicated to customer service, even though the dimensions and the functions stayed exactly the same.

The first analog quartz movements by Orient belonged to the 51000 family (18.0 mm x 3.80 mm). Several types were available from the beginning: caliber 51000 with two hands; caliber 51500 with three hands; caliber 51700 with three hands and date; and caliber 51900 with three hands and day/date. In 1978, Orient launched the 52000 caliber family with the calibers 52720 and 52920; the diameter was 25.6 mm, the height 4.40 mm.

The diameter of calibers 54710 and 54910 also was 25.6 mm, but the height was a mere 3.50 mm. They were used in the 1979 watch models "Orientron Just." There was a simple procedure for synchronizing the seconds-hand with a time signal. The hour-hand could be set without affecting the minute-hand. In the following year (1980), Orient launched the calibers 56710 and 54910 (25.6 mm x 3.50 mm).

Until 1980, Orient only produced round quartz analog movements. In later years, the plates often were ground flat on the sides, sometimes also on top and bottom in order to gain space in the case. Such calibers are characterized by two or even three dimensions and the height.

Shaped Movements

The shaped calibers 53530 and 53030, with the dimensions 15.3 mm x 18.2 mm x 3.0 mm, were launched in 1980. Caliber 53530R (1981) with the same dimensions as caliber 53530 was a gag, as the hands moved counterclockwise. The hands of the calibers 58510 and 58710 (2.50 mm, 1981) moved clockwise, but the moon phase indicator ran counterclockwise. Caliber 58410 (1982) was a modified caliber 58310; its dimensions also were 22.4 mm x 20.0 mm x 2.50 mm. The hour-hand made a full revolution within 24 hours, instead of the usual 12 hours.

In 1982, Orient started a new system of designations consisting of a letter and four digits, starting with the calibers A0520 and A0020. These were shaped calibers for ladies' watches with the dimensions 13.3 mm x 15.0 mm x 2.65 mm. The B series with calibers B1520 and B1020 was also launched in 1982; the

12.25. Orient LED Touchtron caliber 6310C, back side

12.26. Orient LCD caliber 64100, front side

12.27. Orient LCD caliber 64100, back side

12.28. Orient LCD caliber 66100, front side

dimensions were 15.3 mm x 17.6 mm x 2.50 mm. The laterally flattened series C with the dimensions 22.0 mm x 20.0 mm deserved their status as "ultra-flat," as their height was a mere 1.98 mm. Production started in 1983, the family consisted of the three-hand caliber C2520 (C2720 with date) and the two-hand variant C2020, that was also available with a date disc (C2620).

Caliber D3520, launched in 1984 (23.7 mm x 21.0 mm x 2.85 mm), was an elliptical caliber with three hands; it was available with date (D3720) and day/date (D3920). Orient also launched the three-hand caliber E4510 for ladies' watches in 1984 (13.0 mm x 15.15 mm x 2.85 mm), the variant E4010 featured only two hands. The shaped calibers of the A5 series (1985) were somewhat bigger, their dimensions were 15.3 mm x 18.2 mm x 2.85 mm. Caliber A5510 was the three-hand version, A5010 had only two hands.

The slightly flattened shaped calibers launched by Orient in 1986 were B6500 (three hands), B6700 (three hands and date), and B6000 (two hands) with the dimensions 22.0 mm x 20.0 mm x 1.90 mm. They were designed for very flat, elegant watches. The same applied for Series C7 (1987) with the same dimensions, but a height of only 1.75 mm. Next to the three-hand basic version C7510 (22.5 mm x 18.5 mm x 20.0 mm) the variants C7010, C7710, and C7610, as well as the R-Version with hands rotating counterclockwise, were available. The shaped caliber D8510 (13.0 mm x 15.15 mm x 1.85 mm) for ladies' watches featured three hands; D8010 was the two-hand version.

1987 was the last year Orient launched new quartz analog movements. These were the three-hand caliber B0500 (15.3 x 18.2 mm) and caliber C7E00 "Multi Eye" with three hands in the basic version. Several variants were produced, one of them with a moon phase indicator. The dimensions were 23.3 mm x 20.0 mm x 2.85 mm. As mentioned in the introduction, production of some of the C7 calibers continued until 1999, but no new calibers were developed by Orient.

Ricoh

Today, Ricoh is globally known as a Japanese producer of copying machines, fax equipment, printers, and digital cameras. Few people in the company remember that Ricoh and its predecessor were established in 1938 as a watchmaking firm. As the war was approaching, the major products of the young Takano Precision Industries were clocks for

12.29. Orient LCD caliber 66100, back side

12.30. Orient LED Touchtron caliber 68000, front side

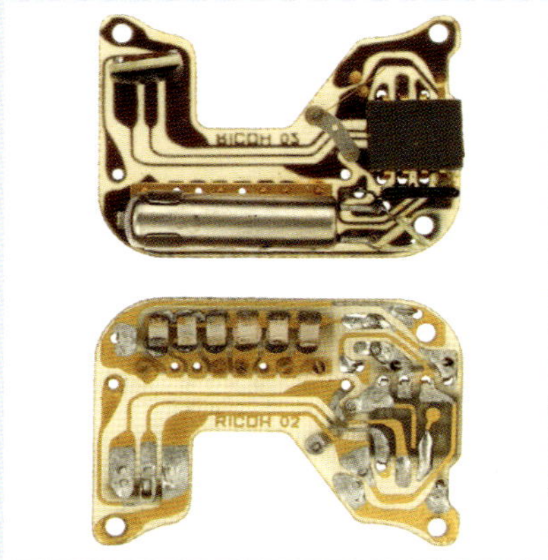

12.31. Ricoh electronic module, caliber 550

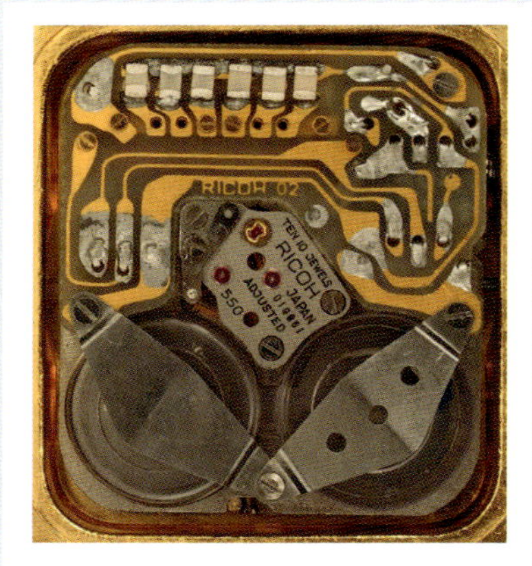

12.32. Ricoh quartz caliber 550

ships, trucks, and tanks as well as detonators for shells and bombs.

Shortly before the end of the war, the Takano production facilities were totally destroyed by American bombs. Yet, before the end of 1945, Takano had resumed the production of clocks, record players, and spinning machines. Military supplies followed – this time for the American occupation forces.

Micromechanics and Riquartz

Takano began manufacturing mechanical wristwatches in 1957. Within just a couple of years, the Ricoh brand gathered so much strength that the company was given that name. The transition to electronic watches started with a cooperation agreement with Hamilton. This led to the production of the electrodynamic Hamilton-Ricoh 555 and of the identical Vantage Electric 130E in 1962 (see Chapter 2).

In the 1970s, Ricoh evolved into a big producer of 32 kHz analog quartz movements. The first caliber of this kind was the Riquartz 550; production started in 1972. This was a very sturdy, rectangular three-hand movement; the dimensions were 22.0 mm x 24.5 mm (11½ x 12½ lines) with a window at 6 o'clock for the day/date display.

The CMOS integrated circuit was supplied either by AMI (American Microsystems Inc.) or Hughes Aircraft, and the 32 kHz tuning fork-type quartz resonator was made by NDK. For fine frequency tuning, six ceramic capacitors could be connected to the quartz with a bank of screws. The six-pole stepping motor featured a platinum-cobalt rotor. Two serially connected silver oxide batteries supplied the necessary power.

Small-Size Calibers

The Riquartz 570 caliber with 10 jewels and day/date display was introduced in 1973; the dimensions were 27.0 mm x 5.50 mm. European customers were Nivada in Switzerland and Dugena in Germany. Starting with 1974, several calibers for ladies' quartz watches followed, including the Riquartz 560 (15.15 mm x 13.0 mm x 2.0 mm) with a 20-second Lavet stepping motor, which was the flattest analog quartz movement of its time. The tonneau-shaped two-hand caliber 530 with the dimensions 16.0 x 13.5 mm x 4.0 mm featured a one-minute stepping motor.

The Riquartz 510 was still another ladies' watch caliber; it was launched in 1976. This was a round two-hand caliber with a diameter of 22.0 mm and a height of 4.3 mm. The hands were actuated by a six-pole, 60-second stepping motor with an axially magnetized rotor that directly moved the minutes wheel. The tuning fork-type quartz resonator was supplied by NDK, the integrated circuit by Hitachi. The plate consisted of a brass-steel sandwich; the steel part with two jewels was on the dial side and assured an excellent magnetic shielding. Stopping the seconds-hand and setting the other hands were done electronically by means of two pushers.

The men's watch caliber 580 (25.6 x 4.5 mm, 1975) was a three-hand caliber with day/date display at 3 o'clock and electronic stopping of the seconds-hand. Caliber 590A (27.0 mm x 5.5 mm) was extremely sturdy; in Europe it was cased by the German Dugena.

The round two-hand caliber 632 was characterized by a diameter of 26.0 mm and a height of 3.1 mm. Caliber 640/641 (24.0 x 4.7 mm) without or with a date disc, respectively, featured a sweep seconds-hand. The round caliber 680 (26.0 mm x 3.6 mm) with day/date and electronic setting followed. The last Riquartz caliber (the number was 701) featured three hands and a date disc; it was launched in the early 1980s7. The dimensions of this tonneau-shaped caliber were 24.0 mm x 21.0 mm x 2.6 mm.

Digital Chronographs

Until 1975, Ricoh only produced analog quartz watches. In that year, Ricoh launched two digital chronographs with a 6-digit field-effect (or TN) LCD. The movements were purchased from outside vendors. Caliber 840 also displayed time in a second time zone. The stopwatch mode with a resolution of 1/100 of a second included the split and lap functions. The design of the Riquartz Alarm Chrono was similar, but it also featured an alarm function for 12 hours minus 1 second. Caliber Riquartz LCD 811 was launched in 1978 and displayed time, AM/PM, month and weekday.

In the 1980s, Ricoh so successfully diversified into copiers, scanners, and beamers, that the old watch business looked less and less attractive. Also, Ricoh could neither find a partner in Switzerland nor big customers for its analog and digital movements. The company ended up producing only low-cost quartz movements for export to Hong Kong, but this was not profitable either. Consequently, Ricoh completely abandoned the watch business in the 1990s.

Sanyo

In former days, Sanyo was a big player in the field of entertainment electronics, household appliances, watches, and mobile phones. However, financial difficulties forced the company to sell most of its subsidiaries and to fully concentrate on lithium batteries and solar cells. In 2008, Sanyo merged with Panasonic whose product mix was ideally complemented by the remaining Sanyo products.

A Sanyo company museum exists in Osaka and a few Sanyo watches are exhibited there, but no information has been preserved about them and the watches may not be removed from their showcase. The information still available about Sanyo watches thus is very fragmentary.

From 1976 to 1990 Sanyo successfully participated in the digital watch boom, with orange-red or green light emitting diodes (LED) or liquid crystal displays (LCD). In the early days, even Pulsar components were used in LED calibers that Sanyo assembled in Japan for the OEM market. Two different pushers switched on either time or date, weekday, and month. This module was quite successful and was used by many watch brands, such as Citizen, Dugena, Klaiber, Longines, Hamilton, and Sears. (What is meant here is Hamilton Watch in Lancaster, PA that was controlled by the SSIH group in Switzerland since 1972 and still produces Swiss watches under the Hamilton brand. Hamilton had already stopped producing watch movements in 1969, it just cased movements that were imported from its Swiss subsidiary Buren in Büren an der Aare.)

Another LED watch by Sanyo was launched in 1976. It featured a pusher at 12 o'clock that switched on the display of hour and minute; continued pressure on the pusher displayed the seconds. Pressing the other pusher at 6 o'clock displayed date and month. Time setting was done by means of mechanically actuated contacts and/or magnetically actuated reed relays. Sanyo also produced a smaller LED module for ladies' watches and used it in its own, good looking watches, which were provided with an integrated metal bracelet.

12.33. Ricoh quartz caliber 510

12.34. Ricoh quartz caliber 570

12.35. Ricoh quartz caliber 580

12.36. Ricoh quartz caliber 590

The Sanyo LED modules were famous for their excellent quality; they were nearly identical with the Pulsar modules of the same size. The reason was that Sanyo had signed a special know-how license with Time Computer/Pulsar and obtained both the technical drawings and the specifications. There was a lively and fruitful exchange of information; Sanyo people came to Lancaster, Pennsylvania, and Time Computer people went to Osaka in order to advise on the licensed production. For example, Sanyo was licensed to fabricate the magnetic pushers patented by Time Computer for switching on the display. All other licensees were obliged to use a different and not necessarily elegant switch.

Between 1976 and 1980, Sanyo also produced smaller quartz movements with a square liquid crystal display for ladies' watches. Those calibers had been developed for the OEM market, but they were also used in Sanyo men's watches with a square, gold plated case. The "Chrono Lithium" with a 3-volt lithium battery was also sold under the Sanyo brand. It was launched in 1980 and featured an 8-digit LCD displaying time, day/date, and month. The stopwatch function had a range of 100 hours with a resolution of 1/100 of a second.

In the late 1980s, Sanyo was a big player in the field of LCD digital watches with solar cells. The buffer was a small nickel-cadmium battery (1.25-volt) with a lifetime of 7 years. The "Amorton PMS 300" model was launched in 1988, it featured the stopwatch and alarm functions. The next model available in 1989 was called "Sanyo Solar DS 502". It was equipped with a 6-digit LCD; the solar cells covered more than half of the upper surface of the case. Extra functions were stopwatch, alarm and day/date.

12.37. Sanyo LCD Lithium

12.38. Sanyo LED ladies' watch caliber, back side

12.39. Sanyo LED ladies' watch caliber, front side

12.40. Sanyo Big LED (29.4 mm), back side

Seiko

When talking about Seiko-Quartz calibers, one must differentiate between Suwa Seikosha (which is part of Seiko Epson), in Kami Suwa in central Japan, and Daini Seikosha in Tokyo (today Seiko Instruments Inc., SII). Both companies are controlled by the Hattori family. Their products were always developed quite independently from each other and reflect different corporate cultures. In this chapter, we will abbreviate to "Suwa" and "Daini" respectively.

Commercial Priority

The world's first commercial quartz watch was launched on 26 December 1969 in Tokyo under the name "Astron." It was equipped with the Suwa caliber 35SQ with a diameter of 30.0 mm and a height of 5.3 mm. The tuning fork-shaped 8192 Hz quartz resonator and the hybrid circuit with 76 MOS-transistors, 29 capacitors, and 84 resistors on a ceramic platelet were technically similar to the ones used in the five prototypes that were sent to Neuchâtel in November 1967 for the Observatory's chronometer contest. They obtained the ranks 11, 13, 14, and 15 – one was disqualified (see Chapter 7). However, the contest prototypes were typical of the thinking of engineers and physicists, while the commercial caliber 35SQ bore the signature of experienced and highly perfectionist watchmakers.

The hand made electronic circuit of caliber 35SQ was, of course, not an industrial product; actually just 100 such watches were ever produced. With an 18 carat gold case, their price was 450,000 yen in 1969, which was the price of a small car. Two years later, steel-cased models became available for 175,000 yen (model without date, caliber 35SQ), and 185,000 yen (model with date, caliber 35SQC).

Each of these watches was individually assembled from A to Z by a specially qualified watchmaker. Commercial Astron watches were allowed a daily rate of one second, but ±0.2 seconds were achieved routinely. A three-pole Lavet motor with the coil on the side of the movement was fitted with soft iron shims conducting the magnetic field to the platinum-cobalt magnetic rotor.

Analog Calibers

In January 1971, Seiko Suwa watches fitted with caliber 3500/3502 (the latter with a date disc) became available. They were equipped with a 16,384 Hz quartz resonator of the tuning fork type. The hybrid

12.41. Sanyo Medium LED (25.4 mm), front side

12.42. Sanyo Medium LED (25.4 mm), back side

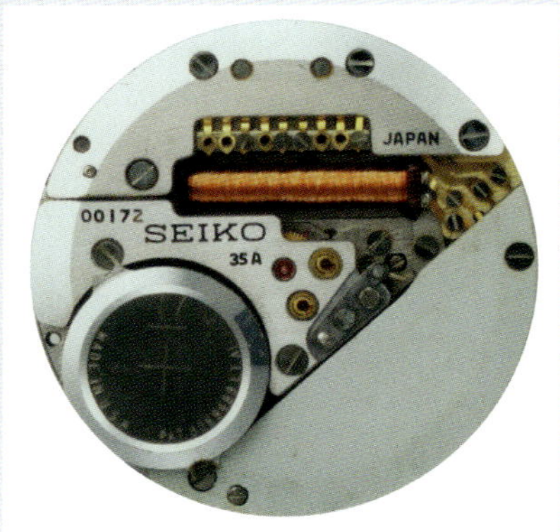

12.43. Seiko 35A Astron, world's first commercially produced quartz wristwatch, launched on December 26, 1969

12.44. Seiko Quartz watch with caliber 35A

circuitry of caliber 35SQ was replaced in caliber 3502 by a 14-stage binary CMOS-IC by Intersil. The quartz frequency was adjusted with exchangeable ceramic capacitors. Until 1971, 1800 pieces of caliber 3502 were produced; there was a recall due to possible technical defects. The price per piece was $1250.

The rectangular calibers of Series 36 (25.0 mm x 28.0 mm x 5.45 mm) were developed by Daini Seikosha (independently from Suwa). The calibers 36SQ and 36SQC (the latter with a date disc) were produced from 1970 to 1972. They were equipped with a 16 kHz quartz resonator of the bar type, a CMOS-IC from Intersil and an anchor-type motor. Power was supplied by two parallel silver oxide batteries with a life of 3 years. Cal 3602 was manufactured at the rate of 300 pieces per month; it was sold on the Japanese and American market.

The round Suwa caliber series 38 of 1971 was still equipped with a 16 kHz tuning fork quartz; it stayed in production until 1973. Caliber 3823 was characterized by a diameter of 25.6 mm, a height of 5.3 mm and 7 jewels. The monthly rate was ± 5 seconds. The CMOS divider circuit was already manufactured in-house, under Intersil license. The oscillator and motor drive circuit were still bipolar. Only one silver oxide battery was used, its life was one year.

Several variants were produced. Caliber 3823A launched in 1973 was equipped with 9 different ceramic capacitors with which the quartz resonator frequency could be adjusted by means of metal screws acting as connectors. This was called V.F.A. (Very Fine Adjustment). Caliber 3802A featured a date disc; 3803A displayed day/date. Both were launched in 1972. The latter caliber was equipped with 13 capacitors that could be added individually, each of them resulted in a slight frequency shift of the quartz resonator corresponding to a rate shift of 0.4 second per day. After adjustments, Seiko guaranteed a monthly rate of ± 5 seconds. The operating temperature range was -10°C to +60° C.

Suwa and Daini

The caliber designation posed problems with most big manufacturers. Seiko was no exception, particularly since there were two Seikos. Certain number series were reserved for Suwa, others for Daini. However, Seiko produced quartz movements under other brand names such as Morioka, Shiojiri, S.EPSON, and SII: this did not alleviate the confusion. The brand names were stamped into the plates but quite often they are difficult to read.

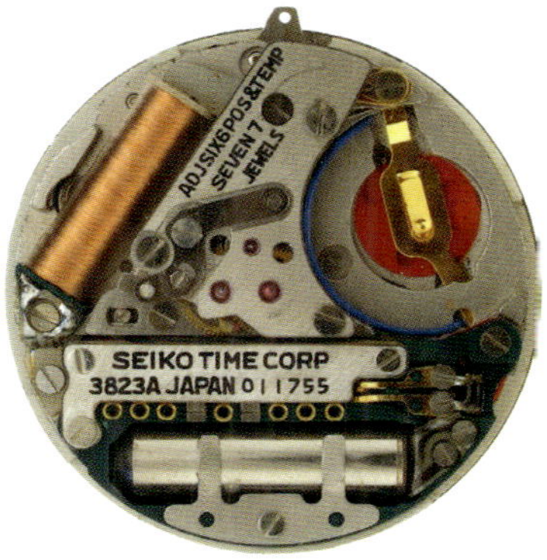

12.45. Seiko Quartz caliber 3823A (1971)

12.46. Seiko quartz caliber 3819

12.47. Seiko quartz caliber 4303

12.48. Seiko quartz caliber 4633A

Caliber 03SQ developed by Suwa featured a 16 kHz quartz resonator and was the first analog quartz ladies' watch caliber. The dimensions were 17.5 mm x 21.0 mm, the height being 5.2 mm.

Daini Seikosha was responsible for the Series 39 calibers that was produced from 1972 to 1974. Those rectangular calibers with 13 jewels and date disc were powered by two silver oxide batteries in parallel. The size of Caliber 3922A was 25.0 mm x 29.7 mm, the height was 5.59 mm. Caliber 3923A (with the day/date function) was fitted with a light emitting diode at 1 o'clock that flashed as long as the battery voltage was sufficient. This was called positive battery control.

The $10\frac{3}{4}$ lines x 5.6 mm three-hand caliber 0822A (0823A with a date disc) was launched in 1974 and was meant for use in both men's and ladies' watches. The 32 kHz tuning fork quartz resonator was now made by Seiko and was fine-tuned by an exchangeable fixed ceramic capacitor. This caliber was equipped with a CMOS-type integrated circuit and a bipolar stepping motor with a samarium-cobalt rotor. Basically this technical design was already optimized and would displace all others within a short time. The only change that economics were going to impose was the photolithographic quartz resonator of the Statek type: Seiko adopted it at the earliest possible time by buying into the Statek company, supplying it with capital and setting up a production facility in Japan under license.

The "Dress Watch" of 1974-1976 was a flat and elegant men's quartz watch equipped with Series 4100 calibers from Suwa. By leaving out the sweep seconds-hand, the height was reduced to 3.8 mm; power requirements and price were reduced, too. With a sweep seconds-hand, the height increased to 4.0 mm; the size of the rectangular calibers 4100 and 4110 was 17.2 mm x 19.4 mm.

Ladies' Watch Calibers

The diameter of the round calibers for ladies' watches was 19.4 mm (calibers 4120/4130), it was 22.0 mm for caliber 4122A; the date disc increased the height to 4.2 mm (without the battery). The power source was a 3.6 mm thick silver oxide battery with a diameter of 7.9 mm. The bipolar stepping motor with a disc-shaped samarium-cobalt rotor made a step every 5 seconds. The 32 kHz tuning fork quartz resonator with fine-tuning by trimmer was made in-house.

Series 41 opened the way for small and flat quartz calibers for ladies' watches. Series 43 calibers were launched in 1975 and were kept in production until

12.49. Seiko Quartz 3003 with caliber 4823A

12.50. Seiko Twin Quartz, caliber 9923A

12.51. Seiko Alarm Quartz, caliber 7223A

12.52. Seiko caliber 7M22 AGS

1977. The size was 16.0 mm x 18.3 mm; the height was 3.50 mm to 4.24 mm, depending on the type. The height of caliber 4300 was 3.50 mm. The dimensions of the round caliber 4312A were 25.6 mm x 4.11 mm; it indicated the date just as caliber 4325 did, while the day/date function characterized caliber 4316A. Series 43 was continued in 1978 with caliber 4336. The dimensions of this rectangular three-hand caliber with day/date display were 25.6 mm x 23.7 mm x 4.24 mm. The round day/date variant 4336A was characterized by a diameter of 25.6 mm and a height of 4.24 mm.

The three-hand caliber 4843 (Suwa) for the Seiko Grand Quartz of 1975 (25.0 mm x 26.0 mm x 4.8 mm) displayed day/date in a window at 3 o'clock. The monthly rate was ± 5 seconds, the need for battery replacement was signaled by two-second jumps of the sweep seconds-hand.

The rectangular two-hand caliber 4720 with rounded edges and 7 jewels by Suwa (13.6 mm x 16.0 mm x 3.8 mm) was equipped with a 2.5-seconds stepping motor; the battery life was 2 years.

Precision Injection of Plastics

In 1978, Daini Seikosha introduced an interesting innovation, the Twin Quartz caliber 9256 with a diameter of 26.0 mm and a height of 3.4 mm. The second quartz resonator served as a temperature sensor for thermocompensation: the rate of such calibers was ± 0.02 seconds per day.

Precision plastic injection molding was extensively used in this caliber; such parts needed absolutely no extra machining. The end of battery life was indicated by two-second jumps of the seconds-hand. This caliber was very successful and remained in production until 1981 in a wide range of variants that included the caliber families 9200, 9600, and 9700. It was used by Suwa for its 9400 and 9900 caliber families with Twin Quartz.

Also in 1978, Daini Seikosha introduced its flattest analog caliber. This tonneau-shaped caliber was called "Eagle;" its dimensions were 22.0 mm x 20.0 mm. Without the battery, its height was 1.9 mm; with the battery the height rose to 2.2 mm. The electronic frequency adjustment was done in two steps. The coarse adjustment caused a rate change of up to 1.04 seconds per day; the fine level adjusted the daily rate within 0.2 seconds.

Daini caliber 7223, launched in 1978, featured an alarm function that was set by turning the flange with an extra crown at 2 o'clock. The diameter was 29.0 mm; the height (with battery) was 4.3 mm. Fine adjustment of the quartz frequency was achieved with a trimmer; the battery life was 3 years.

In 1978, Seiko Suwa launched the two-hand 3 jewels analog caliber 9320 with the dimensions 20.0 mm x 22.0 mm with a height of 0.90 mm (without the battery) and 0.98 mm (with the battery). The minute-hand made 20-second jumps. The magnetic field for the Lavet-motor was generated by two very thin coils mounted next to each other.

Ultra-Flat and Combo Calibers

The dimensions of Suwa-caliber 9983 "Superior" of 1978 with 5 jewels were 26.0 mm x 3.7 mm. Aside from the time, it also displayed day/date. This thermocompensated analog caliber guaranteed a rate of 5 seconds per year. The temperature sensor was a second 32 kHz tuning fork quartz.

Another ultra-flat caliber measuring 18 mm x 3 mm with a height of 0.98 mm was introduced by Daini Seikosha in 1979. It equipped a ladies' watch with a thickness of just 1.78 mm. With an 18 K gold case and diamonds the price was 1.5 million yen.

H127 of 1979 was the first ana-digi caliber by Suwa-Seiko with a 6-digit LCD and a two-hand analog display. The size was 30.3 mm x 27.2 mm, the height 5.3 mm. The displays were synchronized; the monthly rate was ± 10 seconds. Extra functions were month and day/date, chronograph, and stopwatch for 12 hours, the first 20 minutes being measured at 1/100 of a second. The digital display flashed at the end of the 3-year battery life.

In 1980, Daini Seikosha launched the ana-digi caliber H249 with two hands and an LCD at 6 o'clock. The size was 27.2 mm x 22.7 mm, the height 3.6 mm, including the buzzer. The LCD displayed day/date, stopwatch to 1/10 of a second, alarm time, and second time zone. The hour signal could be switched on and off.

Suwa also launched an ana-digi caliber in 1980 with 7 jewels and the caliber number H357. The front side was dominated by a three-hand analog display, above which a slim 6-digit LCD was arranged, for time, alarm time, and stopped time up to 12 hours with a resolution of 1/100 of a second.

A Daini ladies' caliber of the same kind with the size 21.0 mm x 19.2 mm and a height of 3.3 mm was launched in 1981 under the designation E029. The analog part was a three-hand movement, the LCD displayed day/date and the alarm time.

Chronographs

Caliber 7A28 of 1983 was the first quartz analog chronograph by Suwa. This was a round, three-hand caliber with 15 jewels with the dimensions 29.0 mm x 3.5 mm. The accessory dials displayed minutes, seconds, and units of 5 hundredths of a second. Further functions were countdown and split. The follow-up calibers 7A38 and 7A48 of 1984 also displayed day/date and moon phases; for this reason, the height was 4.4 mm.

In 1983, Suwa launched the tiny ladies' caliber 1E70 with 5 jewels. The diameter was 8.5 mm, the height 2.4 mm. The minute-hand made jumps of 20 seconds, battery life was 2 years, the monthly rate ± 15 seconds.

Under the Epson brand (and not Seiko!), a black Wrist Computer with the caliber RC20 was launched in 1984. It allowed bi-directional communication with not more than 1000 signs per message.

In 1986, Daini Seikosha launched the first quartz caliber with a hand-wound microgenerator: the caliber number was 8T23. Turning the crown generated electrical energy that was stored in a supercap. The maximum autonomy was 3 days, the dimensions were 26.0 mm x 25.0 mm x 3.4 mm.

Caliber 7741 by Daini Seikosha was launched in 1985, the dimensions were 24.4 mm x 20.0 mm x 1.6 mm, the yearly rate was only ± 20 seconds. The bar-shaped quartz resonator vibrated in flexion and torsion; its frequency was 196,608 Hz.

Multifunction Calibers

Model 0634 of 1975 was the world's first digital multifunction caliber. Aside from displaying hour, minute, second, date, and AM/PM, it also featured chronograph functions including lap to 1/10 of a second for 60 minutes. The mode was chosen by turning the crown. The dimensions of the round caliber were 28.1 mm x 8.8 mm without the battery. The plate consisted of a 0.9 mm layer of fiberglass reinforced plastic that was completed by a 0.5 mm layer of the same material wherever necessary.

The LCD-Series 06 continued with the calibers 0644 (1975) and 0664 (1977) with the functions hour, minute, second, AM/PM, day/date, month, and year. The diameter of the module was 27.0 mm, the height 8.5 mm. Setting was done, as in caliber 0644, by means of two lateral pushers: the pushers were arranged under the LCD in caliber 0664. Caliber 0674, with the same dimensions, was also launched in 1977.

12.53. Seiko LCD watch with caliber 0614A

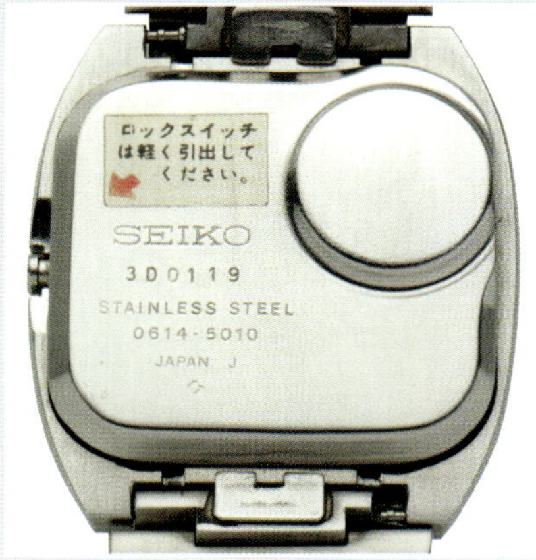

12.54. Seiko LCD watch with caliber 0614A, back side

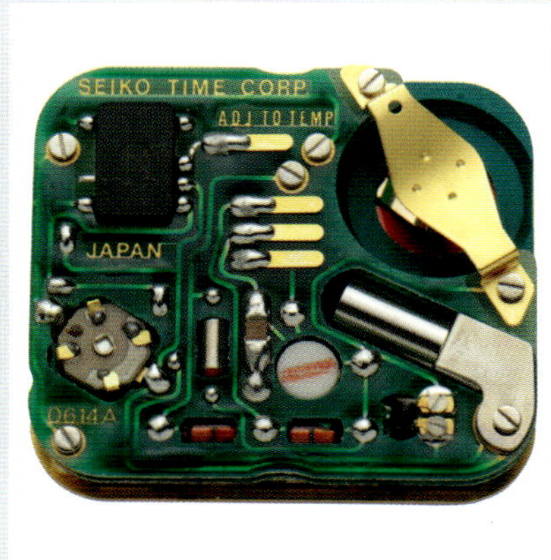

12.55. Seiko LCD module, caliber 0614A

12.56. Seiko LC caliber 0624, front side

The functions of time and a perpetual calendar until 2009 were displayed on the illuminated 6-digit LCD. A watch with this caliber was worn by James Bond in the movie *The Spy Who Loves Me*. However, it was equipped with a fax printer writing on a paper tape, which of course never existed in a real watch.

Starting with 1973, Daini Seikosha also produced a series of digital watches with LCD. According to the culture of the house, the earlier ones were still equipped with 32 kHz quartz resonators of the flexural mode bar type. This included the round caliber 0114 with a diameter of 28.6 mm and a height of 8.30 mm. Hour and minute were displayed on a 3½-digit LCD, a ribbon extending from left to right all the way to 60 displayed the seconds. Abbreviations of the weekdays were placed on top of the display: the framed weekday was the actual one. The date was displayed in a separate window. Caliber 0124 of 1974 had the same functions as caliber 0114.

Caliber M154 with a diameter of 27.0 mm and a height of 6.1 mm was also launched in 1974. The functions were time, day/date (weekday on a ribbon), and year. Caliber 0662A, with the same dimensions also, was launched in 1976. It featured two pushers under the LCD. The functions were hour, minute, second, AM/PM, and day/date.

The two-hand multifunction caliber 8M26 was launched by Daini Seikosha in 1988. It was equipped with two motors; the functions were a stopwatch with a resolution of 1/5 second, an alarm for 12 hours, and a countdown system for 1 hour. The mode was displayed in a window. The dimensions were 26.0 mm x 3.0 mm.

Caliber 7G21 was illuminated with light emitting diodes emitting white light with an excellent yield, thus dispensing with superluminova numerals and hands. Caliber 7M22 (27.6 mm x 4.2 mm) was launched by Suwa in 1988. It was the first watch of the kinetic family "Golden Rotor" with 5 jewels, two hands and day/date discs.

The basic construction was similar to an automatic-caliber. The microgenerator entrained by the rotor charged a supercap that stored energy for 72 hours. The diameter was 27.6 mm, the height 4.2 mm. With the hand-wound predecessor 8T23 of 1986, the crown had to be turned for several minutes to fully charge the supercap. Full charge was indicated by an LED at 6 o'clock.

Caliber 9A85 developed in 1989 by Suwa was the world's flattest two-hand watch movement with the

12.57. Seiko LC caliber 0624, back side

12.58. Seiko Chrono caliber 0634, front side

12.59. Seiko Chrono caliber 0634, back side

12.60. Seiko LC Alarm, caliber A159, front side

dimensions 18.0 mm x 20.0 mm. The height without the battery was 0.85 mm, with the battery it reached 1.43 mm. The monthly rate was ± 15 seconds.

Digital LCD Calibers

In May 1970, the American Time Computer Company (a Hamilton subsidiary) surprised the world with the "Pulsar," the first quartz wristwatch without any mechanically moving parts. The digital light emitting diode display (LED) was switched on by pushing a button. In the following year, Seiko succeeded in developing a similar movement that was equipped with a permanently switched on 6-digit liquid crystal display (LCD) of the Twisted Nematic type (TN). Four pushers were arranged under the LCD for setting hours, minutes, seconds, and switching on the light.

On the basis of this sturdy and long-lived prototype, Seiko-Suwa developed the commercial caliber 0614 with an illuminated, 6-digit LCD displaying hours, minutes, and seconds (1973). The size was 29.2 mm x 26.0 mm (13 x 11½ lines), the height was 8.3 mm. Two integrated circuits were needed, one for the oscillator and the frequency division from 32 kHz to 1 Hz at 1.5 volts, the other for controlling the display that needed 6 volts. This voltage was obtained by two doublings of the battery voltage. By 1974, all functions were integrated into a single integrated circuit.

Caliber 0624 was launched in 1974. This was a completely new design with the dimensions 30.0 mm x 28.0 mm x 8.5 mm. The number of pushers on the front side had been reduced to two, but there was a lateral pusher at 3 o'clock. The crown was needed for setting the time. The follow-up calibers in the same series also maintained the two buttons on the front side. The display was illuminated by a tiny bulb that was switched on by pressing an extra button. This system was maintained for a very long time. Seiko finally switched to backlighting their LCD with an electroluminescent foil in 1995.

Daini Seikosha also developed LCD watches. Caliber 0514, their first movement of this kind with a 3½-digit display, was brought to market in 1973. Three pushers arranged under the display allowed setting hours, minutes, and seconds, and switching on the light. Starting with the 05 family, Daini only used tuning fork-shaped 32 kHz quartz resonators.

12.61. Seiko LC Alarm, caliber A159, back side

12.62. Seiko LC caliber 0124, front side

12.63. Seiko LC caliber 0124, back side

12.64. Seiko World Time LCD with caliber A239

Time Zones

The first Seiko digital caliber with two time zones was launched in 1975, under the designation 0139. The size was 28.5 mm x 29.2 mm, the height was 6.41 mm. Hour and minute were displayed in the lower part of the LCD; above it, the weekday was displayed by framing the abbreviation; seconds were displayed on top. Pressing the crown displayed time in another time zone. The date appeared when the crown was turned. The dimensions of caliber 0439A of 1976 were 27.0 mm x 5.5 mm. The 6-digit LCD displayed hour, minute, and second, or AM/PM, month, and date.

The A-Series of Suwa was started in 1976 with caliber A156. The diameter was 30.1 mm; time and day/date were displayed. The multifunction caliber A159 was launched in 1977; its diameter was 28.9 mm, the height was 7.0 mm. Aside from time, day/date and month, this caliber featured a 12-hour stopwatch function and an alarm.

Also in 1977, Daini Seikosha launched the ladies' watch caliber F023 with a $3\frac{1}{2}$-digit LCD and two pushers for setting hour and minute and for displaying month and day/date. Caliber F051 with a 6-digit LCD followed in 1977; aside from the time it displayed AM/PM, month, and day/date; the names of the weekdays could be set in 5 languages.

Seiko Suwa launched a digital alarm-chrono with an illuminated display in 1977. The dimensions of caliber A158 were 28.3 mm x 6.50 mm; it displayed the time as well as month, day/date, and AM/PM. The stopwatch measured time for up to 12 hours with a resolution of 1/10 second. Aside from that, an alarm was available. The rather similar sister model A159 of 1978 also featured an hourly signal.

Suwa's World Timer caliber M158 was launched in 1977: the diameter was 27.0 mm, the height 6.1 mm. It displayed time in 29 time zones as well as day/date. The pusher on the upper right set the time zones counterclockwise, the one on the bottom right worked clockwise. The monthly rate was ± 10 seconds.

The Electrochrome Display

Caliber F623A of Daini Seikosha (1980) was unique, as it featured a dark blue electrochromic digital display (ECD) of hour and minute (with AM/PM indication) on a white background. Despite the excellent contrast, this type of display was not satisfactory, the reaction time being too slow. However, the ladies' watch caliber L221, also launched in 1979, was very successful. Its dimensions were 14.0 mm x 17.0 mm by 3.9 mm. The 6-digit LCD displayed time and day/date; the calendar needed to be reset only once a year in February.

The dimensions of World Timer A239A by Daini (1979) were 28.0 mm x 6.0 mm (without the battery). This caliber displayed time in the AM/PM-mode as well as day/date until 2009 in 28 time zones (including the half-hour time zones). The movement also featured an alarm. The monthly rate was less than ± 15 seconds. Caliber M718A by Suwa was another World Timer of the same kind, day/date being programmed until 2099.

The alarm-chronograph caliber A156 was also launched in 1979 by Suwa. It was equipped with silicon solar cells that were arranged above the 6-digit LCD. They charged a silver oxide battery. Time, stopwatch, and alarm time were displayed. The caliber's diameter was 36.1 mm, the height being 6.5 mm; the monthly rate was ± 10 seconds.

Calculator Watches

C359 was the designation for an alarm calculator watch with hour signal that was launched by Suwa in 1979. It was based on the calculator module C153. The result of the four basic arithmetic operations was displayed on the 8-digit LCD. The calculator also featured percentage calculations. Hour, minute, second, month, day/date, and alarm time were displayed. Four pushers were necessary for setting all the functions; the dimensions were 27.1 x 26.2 mm, the height was 5.9 mm. The dimensions of the follow-up caliber C439, that was launched in 1980, were 30.0 mm x 21.9 mm x 5.3 mm. Eight digits were available for calculations. Additional functions were alarm (5 different times), hour signal, stopwatch to 1/10 of a second, day/date, month, and year. The 20 pushers needed for all the functions were arranged under the LCD and were actuated with a stylus.

Daini Seikosha started using solar cells in 1980 in the caliber A258. The diameter was 23.0 mm, the height was 4.9 mm without the battery. A258 was a chronograph with daily alarm, weekly alarm, interval timer, and stopwatch to 1/100 of a second. Month and day/date were also displayed.

The digital caliber G757 by Suwa, launched in 1980 was quite innovative. The dimensions were 27.0 mm x 28.5 mm x 5.9 mm. A slim 6-digit LCD was at the bottom, while a big, pseudo-analog seconds scale was on the upper left. The seconds appeared as small indexes at the periphery until the minute was over and the accessory dial cleared. Extra functions were timer, alarm, second time zone, and stopwatch.

Seiko continued the production of LCD-watches well into the 1980s. A typical creation of that time was caliber D031 that was launched by Daini in 1980. It featured the alternating display of time, weekday, and month, as well as month and date on an illuminated LCD. The dimensions were 25.6 mm x 19.0 mm, the height was 3.9 mm. Caliber D229 (1981, 28.3 mm x 5.1 mm without the battery) also had the functions stopwatch (10 hours with a resolution of 1/100 of a second), alarm, timer, display of battery life, and illumination.

In 1982, Suwa launched a television watch, the screen being a relatively big active matrix LCD on top of the T001 caliber with the dimensions 37.9 mm x 34.3 mm x 5.9 mm. The receiver was located in a much bigger, flat box where earphones could be plugged in. A slim LCD was located on top of the screen for displaying hour, minute, second, and date. The battery life was 2 years. The *Guinness Book of World Records* designated caliber T001 as the world's smallest TV set.

Exotic Functions

The "Talking Watch" caliber Y823 (1982, 30.7 mm x 7.0 mm) displayed hour, minute, and second on an LCD; acoustical time indication in clear speech was an option. Other functions were timer, stopwatch, and countdown. The rate was better than ± 15 seconds per month.

The multi-timer caliber A827/A828 of 1983 (28.0 mm x 4.3 mm without the battery) displayed the time and the complete calendar; it also featured alarm, stopwatch, hour signal, and multi-timer functions. Time intervals of 3, 5, 10, 15, 20, 40, 45, and 60 minutes could be set. There also was a timer for up to 2 hours and 59 minutes in units of one minute. The LCD was illuminated. For the sister caliber A828, the accent was on the tachymeter functions, with preset distances of 0.2, 0.4, 1.6, 10, 50, and 100 km, plus any distance up to 699.9 km in steps of 0.1 km.

World Timer caliber Y951 was of special interest, as it was launched under the Pulsar brand that had been acquired from Hamilton. A big LCD replaced the dial and time was displayed in 26 time zones indicated by the city code, including local time that was displayed with three hands.

A radio receiver caliber (W601) for VHF (88-108 MHz) operating with earphones was launched by Daini Seikosha in 1983. The size was 32.0 mm x 30.0 mm, the height was 8.9 mm. Time was displayed with an accuracy of ± 20 seconds per month, the battery life was 2 years.

12.65. Seiko ECD caliber F623, front side

12.66. Seiko ECD caliber F623, back side

12.67. Seiko Memo LCD with caliber M516

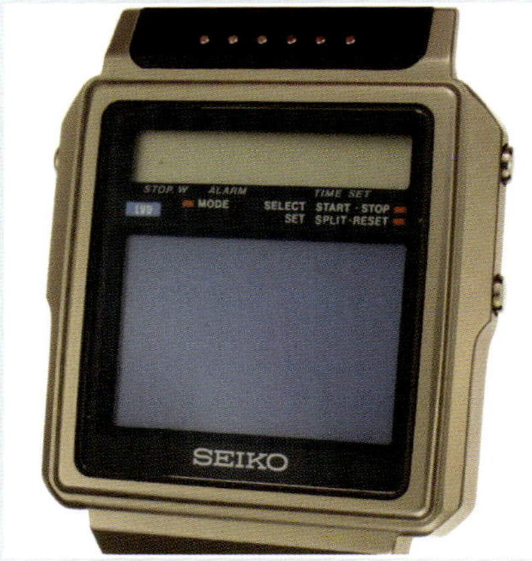

12.68. Seiko TV

Another special caliber by Daini Seikosha was the Memo-Watch UW01A of 1984. The functions were local time, universal time, complete calendar, calculator, stopwatch to 1/100 of a second, alarm, hour signal, and storage of two memos of not more than 1000 characters. Power was provided by a 3-volt lithium battery. The diameter of this caliber was 30.9 mm, the height was 7.0 mm. In order to record memos and operate the calculator, the watch had to be laid on a special external keyboard; the connection was wireless. The display on the LCD featured numbers moving from left to right.

In 1984, Suwa launched its memo-watch S501; the dimensions of this caliber were 30.0 mm x 30.0 mm x 7.5 mm. Additional functions were day/date, daily alarm, weekly alarm, hour signal, and world time. The memo-function stored up to 80 alphanumeric characters.

Sharp

Sharp, a globally active electronics company was founded in 1912 in Osaka. The name was derived from the first product of the young company: an "Ever Ready Sharp Pencil," i.e. an automatic pencil. Radio receivers followed and, after World War II, TV sets and microwave ovens were added to the product mix.

In 1964, Sharp launched a transistorized table calculator. Pocket calculators with a liquid crystal display (LCD) and IBM compatible laptop computers followed. Today, Sharp is among the world leaders for solar cells and LCD. The company employs 60,000 people, its yearly sales are well over $30 billion.

In the 1970s, LCD watch modules were a logical diversification of the LCD calculators. Unfortunately, Sharp did not keep any information concerning those products that were discontinued in 1985.

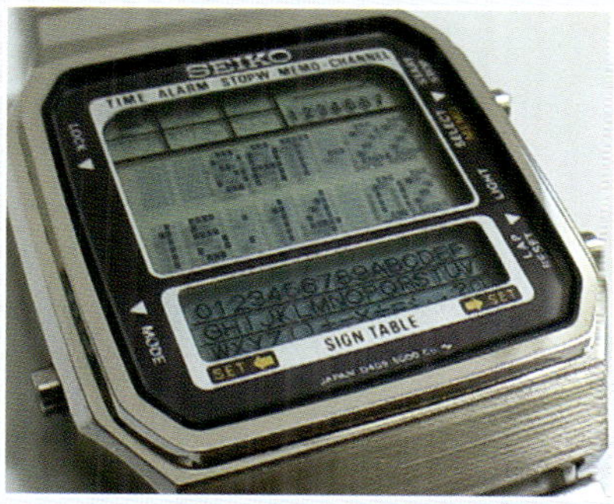

12.69. Seiko Memo LCD watch with caliber D409

13. Quartz Movements from Eastern Europe and the USSR

Quartz Movements from Bulgaria

Bulgaria never had a watch industry, yet the Bulectronic Company produced several digital calibers. The first one was an LED caliber, launched in a watch in 1975. There is a controversy as to whether this was a truly Bulgarian product or a license production of the USSR LED caliber developed in 1973 at the Pulsar laboratory in Moscow and launched as caliber 3051 in 1976; production remained semi-official, but was quite high. The Buletronic plant was located in Botevgrad Prava, the birthplace of the Bulgarian Communist leader Tador Zoskov. In 1979, the Soviet Soyuz mission 33 brought the Bulgarian astronaut George Ivanov into space. For this occasion, Buletronic launched a digital LCD watch specially designed for the Interkosmos program. Those watches were perceived as the ultimate luxury, only prominent party members could afford them.

The switch to LCD modules was made rapidly, but the modules were purchased from Sanyo and cased in Bulgaria. This was a prestige item, primarily made for propaganda purposes to "prove" how advanced Socialist technology was. At the XIth congress of the Communist Party in Bulgaria, all delegates received such a watch with the Pravel brand. It is not known how big the actual production was. At any rate, history repeated itself in 1981 when a Buletronic watch, with the LCD caliber Model 20, was produced specifically for the XIIth Congress of the Bulgarian Communist Party and was handed out as a souvenir to the attending members.

Quartz Movements from Croatia

Around 1980, when Croatia still was part of the Yugoslav Republic, the electronics company Selk in Kutina developed a 3½-digit quartz watch with an illuminated LCD. The brand name was SELK QUARTZ. The functions were hour and minute, or month and date, or just the seconds. Those functions were switched on with the two upper pushers on the right side of the case, while the illumination was controlled with the lower pusher.

The dimensions of the round "Model 20" module were 29.3 mm x 5.1 mm. It consisted of two plastic discs of the same size that were joined with screws. The upper disc consisted of glass fiber reinforced epoxy resin; the LCD was fastened on the outside, while the two batteries, the printed circuit, the quartz resonator, and the integrated circuit were found on the inside. The lower plate just acted as protection. It is not known where those unmarked components were purchased.

Quartz Movements from Poland

There never was a watch industry in Poland; after World War II and until the Berlin Wall fell, Poland imported its watches almost exclusively from the Soviet Union. Yet, two Polish companies developed their own LED watches: this was the only kind of quartz watch that could not be purchased in the USSR.

Thus, the Compu Chron Company launched an LED watch in 1976. It was based on a high frequency module by Hughes Aircraft with a 786,432 Hz quartz resonator. The functions were hour, minute, second, month, and date.

The Unitra Varel Calculator Company in Warsaw launched its LED watch in 1975. This was meant as a purely commercial diversification; Unitra Varel used a Sanyo LED caliber that was produced under Pulsar license in Japan. Unitra was a conglomerate of companies founded in 1956. It produced most of the Polish electronic hardware until 1986.

Quartz Movements from Romania

Just as in Bulgaria, Croatia, and Poland, there was neither a watchmaking tradition nor a watch industry in Romania. Yet, there was plenty of microelectronics know-how and this was important for the development and production of digital watches. It is small wonder that Romania was capable of taking the decisive steps towards an LCD quartz watch at the end of the 1970s. Only one Rumanian manufacturer is known: the Intreprindera Mecanica Fina in Bucharest that launched the Optimef – IMF products.

A round, 3½-digit LCD watch is known to collectors. The dimensions of this module manufactured in 1981 were 21.0 mm x 7.5 mm. It consisted of two superimposed plastic discs, the lower of which carried the two batteries, while the electronic components and the LCD were mounted on the upper plastic disc. The Japanese NDK supplied the tuning fork-type quartz resonator while the integrated circuit was purchased from Intersil, the LCD from Beckman.

Thanks to an unusually large battery with a diameter of 11.5 mm and a modest current consumption of 1.3 microamperes, the watch was theoretically supposed to run for five years before the battery had to be replaced. The functions were hour and minute, and pushers were provided for switching on the display of month and date or the second alone.

Quartz Movements from Czechoslovakia

In western Europe, no one noticed that a watch industry was established in Czechoslovakia after World War II. The reasons were mostly nationalistic: the country refused to depend on watches imported from the unloved (to say the least) Soviet Union. Starting from zero, the production of watch movements and complete watches with the PRIM brand was started in the 1950s in the small Bohemian city of Nové Mesto nad Metuji. Evidently, these were mechanical movements.

The first was launched in 1957 and was called PRIM 50. Its design had been strongly "inspired" by the hand-wound caliber LIP R25. The next caliber, PRIM 66, borrowed from the design of the ETA caliber 1080. In 1978, the ladies' caliber PRIM 80 and the automatic movement PRIM 96 were launched. PRIM engineers were greatly impressed by the Bulova Accutron caliber 214. It was carefully copied, but just a dozen prototypes with the ELTON brand were hand made in 1966 (see Chapter 5).

Analog quartz calibers with the PRIM brand were launched in 1982, i.e. at a time when the technology had reached its full maturity and was informally standardized. The main components were a 32 kHz tuning fork-type quartz resonator, a CMOS chip, a Lavet motor, and a silver oxide battery. The electronic chip type U118F was purchased in the German Democratic Republic, and the quartz resonator was first supplied by OMIG in Poland, then by Seiko Epson in Japan. The samarium-cobalt rotor of the stepping motor and the trimmer were also imported from Japan.

The first PRIM quartz caliber was given the designation PRIM 100; the dimensions were 21.0 mm x 4.8 mm; it was meant to be used in ladies' watches with a sweep seconds-hand. Caliber 110 (that only reached the prototype stage) with the same dimensions featured a date disc that increased the height to 5.5 mm. The pinions were provided with ruby bearings.

The PRIM calibers 200 (1982, no date), 210 (1984, with date), and 220 (with day/date) – all with a sweep seconds-hand – were characterized by a diameter of 25.6 mm and a height of 4.8 mm, 5.5 mm, or 5.8 mm, respectively. The calibers 200 and 210 were available in the stores in 1982 and 1984, respectively. They were quite successful, despite their relatively high price; production continued well into 1991. However, caliber 220 never went beyond the prototype stage.

A second family of Czech analog quartz calibers was developed from 1987 to 1988. It included the PRIM quartz caliber 150 (21.0 mm x 3.3 mm – no date) and caliber 160 (21.0 m x 3.8 mm – with date). Two 23.3 mm movements

13.1. Buletronic LCD model 20, front side

13.2. Buletronic LCD model 20, back side

13.3. PRIM caliber 100

13.4. PRIM caliber 200

followed: caliber 230 with a height of 3.3 mm, no date, and caliber 240 with date (the height was 3.8 mm), both with a sweep seconds-hand.

The 25.6 mm movements designed specifically for men's watches were calibers 250, 260, and 270. Their height was 3.3 mm for caliber 250 (no date), 3.8 mm for caliber 260 (with date) and 4.1 mm for caliber 270 (with day/date), all with a sweep seconds-hand.

To be able to offer a truly complete collection, PRIM also needed digital quartz watches in its catalog. However, the demand was rather marginal, which excluded a production in Czechoslovakia. For this reason, PRIM purchased LCD watches from Elektronika, a subsidiary of the electronics company Integral Minsk in Belarus; the PRIM brand was stamped on the dial right there. A grand total of 72,580 pieces was imported between 1983 and 1984 when digital watches were fashionable.

The PRIM LCD digital watches were available in three styles. The simple men's watch featured a 4-digit LCD that displayed the hour in the 24 hour mode, the minute, and day/date (Elektronika module 5-204/30351, 29.0 mm x 4.8 mm). There also was a smaller ladies' watch model with the same functions (Elektronika module 5-203A/22356, 22.0 mm x 7.0 mm). The most sophisticated module (29.0 mm x 5.0 mm) equipped a men's watch with an illuminated 6-digit LCD for the functions of hour, minute, second, and weekday on a bar following the lower edge of the LCD. This was the Elektronika caliber 5-206/30353 (see the section "Quartz Movements from the USSR").

The design of the second generation of PRIM analog quartz movements (150 to 270) was finalized, the plans were completed, and prototypes were fabricated. However, there was no production as the political system totally changed in 1989. The borders were suddenly opened and goods could now be bought from anywhere in the world. PRIM found itself in an awkward situation. The few wealthy people immediately bought Swiss watches, and for all the others, Chinese watches became available, at prices well below those of the PRIM timekeepers.

Yet, watch production continued at a reduced level at the PRIM plant. From 1967 to the 1990s, total production was about 15 million watches, 98 percent of which were mechanical. This means that some 300,000 quartz watches were produced. PRIM maintained two service centers in Czechoslovakia where all its watches were repaired, whether under warranty or later; there never was a shortage of spare parts.

Quartz Movements from the USSR

Before the October Revolution of 1917, Russia only produced tower clocks and beautiful mantle clocks, but practically no pocket watches and wristwatches, at least not on an industrial scale. There were a few highly talented Russian watchmakers who fabricated extremely luxurious, individually crafted pocket watches for the Imperial Court and the upper tier of aristocracy. We also know of at least one genius level Russian watchmaker. His name was M.S. Bronnikov; he literally carved the movement and the case of his watches using birchwood and ivory. Just the escapement, the mainspring and the hairspring were made of steel.

Self-Support by Order

Before the October Revolution of 1917, watches for every day use were imported, mainly from Switzerland; Russia actually was the fourth largest customer of the Swiss watch industry. Well known brands such as Buhré, Moser, Gabus, Patek Philippe, and Tissot maintained distribution centers in St. Petersburg, Moscow, and Kiev. As Russia levied rather high import duties, the system of assembling movements on the basis of so-called chablons and casing them locally grew to be very important. After the Revolution, the Russian watch market collapsed. Imports from Switzerland were limited to small quantities of marine chronometers and deck watches that were indispensable for the war effort: the young Soviet Union had practically no foreign currency.

This situation was intolerable. In 1927, the Communist Party issued a decree according to which the Soviet Union was ordered to become self-sufficient in terms of watches. This looked great on paper, but in the real world, things were quite different. The warehouses with Swiss watch parts having been confiscated, it was not too difficult to assemble new watches using them. Once the supply was gone, however, one had to resort to "Reverse engineering." This simply meant that Western products were faithfully copied 1:1, and patents and intellectual property were irrelevant. This strategy worked out reasonably well with alarm clocks: by 1928, the famous caliber 10 by Junghans was faithfully copied. The quality was terrible, but at least those clocks worked.

Pocket watches and wristwatches were quite another matter. A massive technology transfer was indispensable, and it became necessary to resort to aid from the arch-enemy of socialism: the United States. Two American watch plants that were on the verge of declaring bankruptcy – Ansonia in Connecticut and Duebner-Hampden in Ohio – were acquired for a song. Machinery, tools, spare parts, plans and some of the workers (on a temporary contract) were moved to Moscow. The First Moscow Watch Factory was set up in an old tobacco plant.

The next challenge was to train Russia's first 4000 industrial watchmakers. About forty German watchmakers with a communist party membership book were recruited from the Black Forest region, in Ruhla and in Glashütte. A few extra came from Switzerland. Thanks to them, Soviet watch production started exactly as planned in 1930; a year later, the Second Moscow Watch Factory was established.

Tough Beginning

In the beginning, only pocket watches were produced: the parts were bigger and the precision requirements were less extreme. Yet, in the early days, the proportion of rejects was a horrendous 85 percent. Things slowly improved, particularly after Switzerland officially authorized the export of precision lathes and milling machines in 1935. By then, the American machines obtained from Ansonia and Duebner-Hampden were totally worn out.

There was an enormous demand for watches among post-revolutionary Russians. The Soviet watch industry grew very fast and soon the two Moscow factories reached the limits of their capacity. New plants were needed and, in 1933 the Pensensky watch factory was established in Pensa, some 500 kilometers southeast of Moscow. It manufactured watches with the brands Sarja, Pobieda, and Tegrov. In 1940, the Uglitch watch factory followed, located on the Volga, 200 kilometers northeast of Moscow. Still another watch factory was established in Petrodworez near Leningrad (St. Petersburg today) specifically for the Raketa brand. When World War II began, the USSR watch production had reached half-million pieces per year.

Production rapidly shifted to sturdy clocks for army trucks and tanks, as well as deck watches and marine chronometers for the navy. Of course, detonators for artillery shells had the highest priority, and they were made by the millions. As the German armies were getting dangerously close to Moscow, the strategically important First Moscow Watch Factory was moved to Slatoust, south of the Urals. Later, this facility was used for the production of table clocks, sports timers, and alarm clocks with the Agat brand. Some sections were moved to Chelyabinsk for the Molnija brand, and to Tchistopol, where the Vostok military watches were manufactured.

Many Production Centers

The Minsk Watch Plant was established in 1946 in what was then the Belarus Soviet Republic. It had originally been planned as an assembly facility, using parts that were supplied by other plants but due to the great demand, Minsk was upgraded as a production center for the Lutch watch brand in 1955. By 1950, the Soviet watch industry had reached pre-war production figures and kept growing rapidly, fully concentrating on the civilian market. The main accent had shifted to wristwatches: the backlog demand was enormous. For this type of manufacture, the precision machinery that had been confiscated in Glashütte (Saxony) and brought to Moscow proved to be extremely useful.

At that time, the most important caliber was a copy of the LIP R26. The Soviet version was called K-26 and it proved to be incredibly sturdy. Many watches with this movement still work fine after 50 or 60 years, much to the delight of their owners or their heirs. The First Moscow Watch Factory used the K-26 in its Pobieda brand watches until 1953. The brand was then transferred to the Second Moscow Watch Factory, to Petrodvorets, and to ZIM in Samara. In Petrodvorets, the caliber was fitted with imitation Incabloc shock absorbers.

In the course of time, more watch production centers were built. Starting in 1959, they all had to report to the central research institute for the watch industry NII-Tchasprom in Moscow (NII was the abbreviation for Nautschno Issledovatelskii Institut) where about 1500 people worked. NII-Tchasprom's role was to design mechanical calibers and analog electronic calibers for the entire Soviet watch industry. Those calibers were then assigned to one or several watch factories that took care of production. The various production centers were solely responsible for the exterior parts. Each brand had its own product profile while using only a very limited number of movements – which of course greatly simplified both production and after sales service.

NII-Tchasprom also certified the Soviet marine chronometers and manufactured special electronic timekeepers for use in space. Digital quartz calibers were only designed and produced by Elektronika, which was founded in 1974 as a section of the Integral Minsk electronics company.

The new watch production facilities that were built in the postwar period were as follows: Orlov near Moscow (Jantar brand), Serdobsk (Majak brand), Rostov-on-Don (Witjas alarm clocks), Riga (LCD alarm clocks, in cooperation with the Alpha electronics plant), and Selenograd/Novosibirsk, which specialized in electronic games with an alarm clock function and LCD travel alarm clocks. See below for a list of the Soviet watch and alarm clock producers with the brands they used. Some of those brands were used by several production centers. It often happened that a given model was produced at several places when demand was particularly strong.

The Soviet watch industry was grouped into several production associations, but some plants reported directly to certain ministries. For example, the production association "Tchasprom" was responsible for NII-Tchasprom and the brands Slava, Lutch, Sarja, Molnija, Vostok, Raketa, Agat, and Majak. The Poljot brand was subordinated to the Ministry for Electrotechnical Industry and Equipment, the Aviation Industry Ministry controlled Witjas, while Elektronika and Alpha had to report to the Electronics Industry Ministry.

One of the weak points of the Soviet watches was dials. They were so ugly and poorly made that watches with such

Name	Brands
1st Moscow Watch Factory	Poljot, Pobieda, Kirovskie, Signal, Sportivnie, Rodina, Moskva, Sputnik, Stolichnie, Strela, Sturmanskie, Mayak
2nd Moscow Watch Factory	Slava, Pobieda, Kama, Molnia, Cornavin, Salyut
Tchistopol	Raduga, Vostok, Kremlovskie, Sputnik, Amphibia, Yantar, Saturn, Mir, Precision, Kama, Uran, Tchaika, Albatros, Generalskie, Admiralskie, Ofizirskie, Buran, Svet, Kolos, Neptun, Troyka, Pobieda, Drushba, Rubin, Cornavin
Uglitch	Tchaika, Volga
Pensa	Zarja, Zvesda, Vesna, Aurora, Kometa, ZIF, Mechta
Maslenikova	Pobieda, ZIM
Tcheljabinsk	Molniya, Ural, Lutch
Wladimir	Vesna
Rostow	Vitjas
Minsk	Lutch, Sekonda, Zaria
Serdobsk	Majak
Zlatust	Agat
Integral Minsk	Elektronika
Pinsk (Branch of Integral)	Kamertoni
Samara	Pobieda
Petrodvorets	Raketa, Pobieda, Leningrad, Svet, Rossija, Neva, Majak, Start, Baltika
Riga	Alpha: LCD alarm clocks
Selenograd/Novosibirsk	LCD travel alarm clocks

dials could not be sold outside the Communist Block, even at dumping prices. The USSR had to resort to using meager reserves of Western currency to import good-looking dials, mostly from Pforzheim in Western Germany. Those dials were of course solely used for watches that were exported to western Europe.

When the USSR collapsed in 1990, there were eighteen production centers for watches, movements and alarm clocks. In addition, more than a dozen plants produced movement parts, exterior parts, and special tools for watchmakers. Production capacity was around 80 million movements and watches per year, but that was never fully used; in the record year, about 75 million pieces were produced.

After the demise of communism and the opening of markets, hardly anybody in the former Soviet Union went on buying Soviet watches. By 2009 the Uglitch, Second Moscow Watch Factory, Chelyabinsk, Petrodvorets, Riga, Serdobsk, and Selenograd/Novosibirsk plants were closed. A few of the old production centers changed to importing quartz movements and exterior parts, and just assembling watches of their brands; this is what Vitebsk did with the Wimpel brand. In Orel, the Yantar brand specialized in clocks with mechanical or electronic movements, from the alarm clock all the way to grandfather clocks with chimes.

Inspiration from the West

When the time came to switch from mechanical to electromechanical and electronic watch movements, the USSR once more resorted to copying know-how that had been created in western Europe and North America. The Epperlein prototype calibers of 1952-1956 were faithfully copied by the Second Moscow Watch Plant. However, only prototypes were made; there was no production – for perfectly good reasons. A few watches were assembled with those movements and tested in the field. Strangely enough, those movements were cased in copies of the Hamilton Electric. The Hamilton 500 actually looked much more promising than the Epperlein (see Chapter 2). A copy of that movement reached the production stage in 1958, the cases continued to be copied from Richard Arbib's original design.

When transistor-controlled electrodynamic movements appeared in the West, the USSR was content with copying just the Junghans caliber 600 of 1967 with the dimensions 30.8 mm x 5.4 mm (see Chapter 3). The quality of this copy was flawless: it was used in watches with the Poljot, Slava, Raketa, Lutch, Sarja, Tegrov, and Tchaika brands and considerable quantities were exported. In the 1970s, this balance wheel motor caliber was provided with a quartz control (caliber 3055, see below). The design was similar to that of the Junghans movement, but the construction was unmistakably Russian.

As far as quartz was concerned, scientists and engineers at the NII-Tchasprom (the well-known professor D. D. Malkin was among them) reached the conclusion that miniaturizing the quartz clock to wristwatch dimensions should be practically feasible. They were quite independent of their colleagues at Seiko Suwa, the Centre Electronique Horloger (CEH) in Neuchâtel, and the Battelle Research Institute in Geneva. A big research program was started, and, by 1977, Poljot in Moscow started the production of quartz wristwatches – exactly ten years after the first Beta 1 and Beta 2 calibers had started to tick in Neuchâtel.

The Poljot electronic watches were assembled manually, whereas automation was an important issue in Japan in the earliest days. Automatic assembly lines were very expensive and cost several hundred million dollars; they had to be amortized via the price of the watches. The USSR relied on manual work, so investments in equipment were minimal. The result was that Soviet quartz movements were exported for an average price of $3 apiece while the Japanese product cost $4.

For this reason, the Soviet quartz movements were serious competitors of the Japanese calibers, particularly in the Asian markets. Eventually, though, the Japanese paid for their assembly lines: production costs dropped, which allowed drastic price reductions, all the way to $1.50 apiece. Now the Russians had a problem: heroic efforts were made in terms of efficiency and partial automation, which brought the price of a quartz movement down to $1.30 apiece.

The Japanese answer was to transfer the production of ordinary quartz movements to China, where low wages allowed further price reductions of 45 percent. In the early 1990s, this led to the "capitulation" of Poljot and the other Russian watch plants that had survived the collapse of communism. By 1995, the production of analog quartz movements had completely stopped in the territory of the former Soviet Union: they were now imported from China (low grades) and Japan (high grades).

Soviet Digital Quartz Movements

The Integral Electronics Company in Minsk (Belarus Republic) was established in 1961. For some years it covered half the world's needs of silicon chips. Integral doubled its production in each of the last three five-year plans of the Soviet era. Today, Integral mostly produces customer specific integrated circuits and "translates" circuit designs of its customers into silicon (this is called foundry services).

Integral chips are designed both for industrial electronics and entertainment electronics, but the latter are actually made by the Kamertoni subsidiary in Pinsk in the South of Belarus. Kamertoni also produces agricultural machinery and printed circuits. Another subsidiary is

13.5. Elektronika LED watch, 1976

13.6. Elektronika LED module

13.7. Elektronika LCD module, front side (1.5-volt technology)

13.8. Elektronika LCD module, back side

Electromodul in Molodechno near the Polish border. Its main products are rectifiers, diodes, hybrid integrated circuits, and voltage stabilizers. Elektronika in Minsk was established in 1978 as the Integral subsidiary that is fully responsible for digital watches.

Integral purchases 200 mm semiconductor quality silicon crystals on the world market, separates them into wafers, and generates the requested circuit by photolithography using up to 28 masks. The wafers are then sliced into chips that are bonded and potted. Integral supplied the Chinese industry with up to 27 million chips per month until the Chinese became self-sufficient for simple, mass produced integrated circuits.

The Elektronika subsidiary is located in the Integral manufacturing complex in Minsk and was created specifically for the production of digital watches. Due to the aggressive competition from China, this activity has been considerably reduced, but Elektronika has found a whole new range of successful products, e.g. solid state memories for laptop computers, smart cards, electric meters, medical instruments, POS terminals, printers for cash registers, liquid crystal displays (LCD), and metrology instruments. The integrated circuits are supplied by the neighbor Integral that uses 0.35 micron technology for the production of silicon gate CMOS circuits.

The Pulsar Research Institute

The very first USSR digital watches were assembled at the Pulsar Research Institute of the Ministry for Electronics Industries. According to Russian sources, the Pulsar name was chosen long before the presentation of the Pulsar LED watch by Hamilton. Actually, Pulsar developed the first Soviet transistors and integrated circuits. The Soviet era of quartz watches began with the prototypes from the Pulsar institute. Accordingly, only digital quartz watches were produced in the early days: analog quartz watches started to be available to Russian consumers in 1977.

The first LED digital wristwatch prototype started working in May 1973 at the Pulsar Laboratory. The transistor department of Integral immediately was given the assignment to prepare mass production of this type of watch, but the LED needed to be replaced by a liquid crystal display (LCD).

This was done, but Pulsar engineers continued experimenting with LED and even managed to convince the colleagues at Elektronika to cooperate. Starting in 1976, series of a few ten thousand LED watches were produced, supposedly for wear tests and marketing tests. Later, even bigger series of the 30 mm caliber 3051 were produced. Hour and minute were displayed in the 24-hour mode on the 4-digit display. Alternatively, day/date could be called up, but the weekday was purely numeric (1 for Monday, 2 for Tuesday, etc.).

13.9. Elektronika LCD module, front side (3-volt technology)

13.10. Elektronika LCD module, back side

13.11. SU Quartz, caliber 3050

13.12. SU Quartz, caliber 1936

In 1977, an LED watch in the 26 mm ladies' size was developed (caliber 2651). The functions were the same as in the men's model 3051. There is a controversy about the size of that production that seems to have been "unofficial;" the experts' educated guess is several hundred thousand pieces. It is certain that those products were quite successful. Today, they are often offered for sale on the Internet, even in lots of 100 pieces at very reasonable prices – not what one would pay for something really rare.

Integrated Circuits and Quartz Resonators

The Transistor Company's major challenge was to develop chips that would excite the quartz resonator, divide its frequency from 32,768 Hz to 1 Hz and decode that signal in order to correctly address the 7 bars that made up the numerals of the LED or LCD displays. Originally, four integrated circuits were needed for those tasks. Later, everything was integrated on a single chip.

The bar-type flexural mode quartz resonators PB-72 were supplied by the Phonon Company in Moscow. Quartz resonators of the same type were also purchased from the Swiss SSIH/Omega. Eventually, Phonon decided to set up a top modern assembly line on the base of Omega know-how. However, the change to the superior mechanically cut tuning fork-type resonator had to be made. Part of this fabrication was eventually farmed out to China. The exact same resonators were also made in Uglitch, so that watch manufacturers had a second source. Eventually, Uglitch took over all research and development activities in the field of quartz resonators.

In Saratow, an old electronic tube factory was converted to a production facility for liquid crystal displays. The State Institute for Capacitors and Resistors Girecond in Leningrad developed the trimmer for fine tuning of the quartz resonator's frequency. The Battery Research Institute in Novosibirsk developed the button cells for LCD watches. They were first based on mercury oxide, but in 1976, the shift was made to silver oxide. Finally, the piezoelectric elements for alarm signals were produced in Stalingrad (Volgograd today). Kamertoni, a specialized company in Pinsk was responsible for LCD watch case production. Some of the best Soviet scientists acted as consultants.

Coordinating those various institutes, companies and scientists to achieve a common objective was a rather novel venture. The Ministry for Electronic Consumer Goods led this coordination and did an outstanding job. By 1974, all problems had been solved, prototypes of an LCD digital caliber worked fine, the caliber was ready for production.

Elektronika: An Integral Subsidiary

No time was lost setting up the production in Minsk; the first series came off the assembly line in 1975. Transistor produced 1.6 million modules until the end of 1978. However, in October 1978, the Ministry issued a decree ordering Integral to create a company that would be fully dedicated to the production of LCD digital watches.

On January 1st 1979, the Integral subsidiary Elektronika was started. In its first year, the new company produced 1.5 million watches. In the following two years, production kept growing; in 1991, shortly after the disintegration of the USSR, it reached an all-time high of 6.65 million. In that year, Elektronika created its own Fotech subsidiary for the in-house production of liquid crystal displays that so far had been supplied by the Saratow plant. Within 30 years, Elektronika produced a grand total of 74 million digital watches.

After the political upheaval, things deteriorated rapidly, going from bad to worse. By 1995, production had dropped to less than a million LCD watches. In 2008 production shrank to 20,000 pieces, but has recovered considerably since. Continuing with the production of digital LCD watches in Minsk at the rate of a few hundred thousand pieces a year only makes sense because modules for customer service are still needed in large numbers. The entire digital watch business is now concentrated in Minsk, as the watch activities of the Kamertoni subsidiary in Pinsk were stopped. Liquid crystal displays, which formerly were a mere diversification, have become a major product. Today, Elektronika produces more than 10 million LCD per year; they have replaced the old electromechanical galvanometer-type instruments in practically all applications.

The cooperation of Elektronika with the East German watch producer UMF Ruhla is interesting from a historical point of view. UMF produced machinery for the precision injection molding of plastic watchcases. Elektronika needed those machines, as they had to be replaced periodically due to extreme wear and tear. UMF Ruhla was "paid" in integrated circuits from Elektronika.

UMF Ruhla and Elektronika cooperated in developing the electronic stopwatch type 56 with an LCD digital display for sports timing. The resolution was 1/100 of a second. Elektronika contributed the circuit and UMF Ruhla was responsible for the plastic case. Both Minsk and Ruhla produced this stopwatch and the Soviet production was 40,000 pieces per month. Another Minsk-Ruhla product was a pulse meter. The sensor was worn on an elastic belt around the chest, the data were transmitted by microwave to the display unit on the wrist.

Leading Personalities

The names of the leading scientists and engineers who were instrumental in developing digital LCD watches at Integral Minsk and Elektronika are hardly known in the West. Yet, they did a tremendous job under difficult circumstances. This is why they are mentioned here. Alexander Shokin headed the Ministry for Electronic Industries in the 1970s and was an enthusiastic supporter of the LCD watch project. Vladislav Kolesnikov, also with the rank of minister, was Shokin's colleague and deputy. Pjotr Gaydenko headed Integral from 1975 to 1993; Integral Chief Engineer Valery Tsaryov also made important contributions.

In the mid-1980s the transition was made from mechanically cut quartz resonators (first the bar-type from Phonon in Moscow, then the tuning fork-type from Pskov), to the photolithographic-type in the standard tubular 2 mm x 6 mm container. A still smaller type (1.5 mm x 5 mm) was ready for mass production, but this project was cancelled due to the collapse of the Soviet Union.

Functions of the Elektronika Quartz Calibers

Between 1975 and 1990, the USSR Ministry for Consumer Goods developed three generations of LCD digital quartz movements in cooperation with Integral Minsk and the aforementioned component suppliers. This resulted in 29 commercial calibers with various diameters from 32 mm to 18 mm. They are designated by means of a six-digit code, where 5 means Integral/ Elektronika, while the first two numerals after the hyphen indicate the diameter in millimeters. The last three numerals (sometimes followed by a letter) are the actual caliber designation.

The basic movement of the first generation, produced from 1975 to 1982, was characterized by a diameter of 32.0 mm and a height of 8.4 mm that was gradually reduced to 7.0 mm. The second generation was produced between 1980 and 1986; there were two types with the dimensions 30.0 mm x 6.4 mm, and 28.0 mm x 4.3 mm. The third generation was launched in 1985 ; the basic movement was characterized by the dimensions 28.0 mm x 3.6 mm. The 29 commercial digital calibers the USSR watch industry produced between 1975 and today were all derived from those three generations of prototypes. They were used in solid state wristwatches, alarm clocks, and stopwatches and are listed below.

5-30530: 6-digit LCD; hour (in the 24-hour mode) and minute, second in smaller digits, or month and date, display of the weekday with seven black squares on the lower part of the LCD; the symbol (first letter) for the current day was framed in black on white (or rather gray).

5-30530K: 6-digit LCD; pocket watch caliber; all functions as in 5-30350.

5-18351: 4-digit LCD; no light; hour in the 24-hour mode, minute. Other functions available with pushers: minute and second, or numerical weekday and date, or hour in the 12 hour mode plus AM/PM indication and minute, or month and date, or just seconds.

5-22356: Basic functions as in 5-18351, but illuminated 4-digit LCD; battery accessible on the case back via a threaded cover.

5-30351: Functions as in 5-22356, illuminated 4-digit LCD; 12-hour or 24-hour mode; no battery cover.

5-18351B: 4-digit LCD; illuminated display; hour in 24-hour mode and minute. Replaced 5-18351 in ladies' watches. Functions accessible by pusher: second, month, date and light. Battery accessible on the case back via a threaded cover; adjustable rate.

5-18351.1: Functions as in 5-18351B; some new components in the module.

5-18391: 4-digit LCD; no illumination; module for tube-shaped key ring pendant. Hour display in 24-hour mode, minute. Functions accessible by pusher: second, month and date. Adjustable rate.

5-18394: Often used in children's watches with plastic case. 4-digit LCD; no illumination; hour display in 24-hour mode, minute. Functions accessible by pusher: second, month and date. Adjustable rate.

5-30351A: 4-digit LCD; illuminated display; hour in 24-hour mode or 12-hour mode, minute, second, month, date.

5-29351B: Functions as in 5-30351A; hour only available in the 24-hour mode, plus minute. Second, month and date available by pusher. Battery accessible on the case back via a threaded cover, adjustable rate.

5-29391: Functions as in 5-30351A; module for wide, rectangular watches.

5-30353: 6-digit LCD; last numerals on the right smaller. Illuminated display. Functions: hour in 24-hour mode or 12-hour mode, minute, and second, or hour, minute, and date, or month, day, and year. Display of the weekday by seven squares in a row in the lower part of the LCD.

5-30353A: 6-digit LCD; functions as in 5-30353. Battery accessible on the case back via a threaded cover.

5-29360: 6-digit LCD; functions as in 5-30353: hour (12-hour or 24-hour mode), minute, second, weekday, month, date, and year.

5-30354: 6-digit illuminated LCD; 4 big digits on the lower level, 2 smaller digits on the upper right. Display of the weekday by framing of one of the seven abbreviations on the upper level of the LCD. Functions: hour, minute, second, or weekday, month, date, top of the seconds-hand, 12-hour or 24-hour mode. Stopwatch to the 1/10 of a second; countdown.

5-30355: 6-digit illuminated LCD; 24-hour mode, minute, second, weekday, month, date, countdown, alarm, signal for set alarm, hour signal; battery accessible on the case back via a threaded cover.

5-30357: 8-digit LCD on two levels. On the lower level, the two last digits were smaller (e.g. for the second). Date and 7 letters for the weekdays, current day underlined on the upper level. 24-hour display, minute, second, weekday, month, date, stopwatch to 1/100 of a second, countdown, alarm, alarm setting signal, hour signal; battery accessible on the case back via a threaded cover.

B6-03: Simple design for incorporation in apparatus and instruments. The functions were hour, minute, second, month, and date.

5-29359: 6-digit illuminated LCD. Hour, minute, second, weekday, month, and date.

5-30364: 6-digit illuminated LCD; two numerals on the right smaller for the seconds. 24-hour mode only, minute, and second. Alarm, alarm setting signal, hour signal, battery accessible on the case back via a threaded cover.

5-29367: 6-digit illuminated LCD; hour in 24-hour mode, minute, second, weekday, month, date. Extra functions: stopwatch, music alarm with 2 tunes, alarm setting signal, hour signal.

5-29358: 6-digit LCD with two levels; abbreviations of the weekdays on the lower level, calendar of the current month. Functions: hour, minute, second, weekday, date, month, and year. Adjustable rate.

5-29361: 10-digit LCD on two levels. Hour in 12-hour or 24-hour mode, minute, and second on the lower level, abbreviated weekday and date, or month and date, music alarm, hour signal; adjustable rate.

5-29366: 8-digit LCD on two levels. Hour, minute, and second on the lower level, date on the upper right, underlined first letter of the weekday, countdown. Battery life: 3 years.

5-29367: 6-digit LCD on one level. Hour, minute, second, music alarm, stopwatch.

2-11 Pocket size, battery powered digital watch in a rectangular plastic case. Display of hour, minute, and second; alarm on a 6-digit LCD.

2-14 Digital alarm clock; same functions as 2-11.

56 Digital sports timer developed in cooperation with UMF-Ruhla. The time resolution was 1/100 of a second.

In 2003, Integral started marketing an interesting special product called Polimaster PM1203M. It was an analog watch (with the Ronda caliber 763) that featured a Geiger-Muller counter. The gamma dose was displayed on an LCD; the range was 0.01 to 9999.99 microsievert per hour in the energy range of 0.06 to 1.5 MeV. The prototype was fully digital, in the commercial version an analog time display was preferred. The size of the case was 50 mm x 45 mm x 20 mm, the complete watch weighed 100 grams. Production was limited to 15,000 pieces; most of them were exported to the United States.

Analog Quartz Calibers

As mentioned in Chapter 7, the quartz wristwatch was invented at least eight times. The contributions that Soviet scientists made are usually ignored, as they were held secret and were not communicated in international journals. Also, it is a fact that Soviet horological products more often than not were "inspired" by Western technology. Quite understandably, this was not communicated either.

As we have seen, the first Soviet quartz watches were of the digital type with an LED or LCD display; they were available in the stores by 1975. The analog quartz technology was something quite new. In order to familiarize themselves with it, Soviet engineers first developed the quartz-synchronized electromechanical, 30 mm caliber 3055 with a balance wheel motor and a height of 7.5 mm. The two-level balance wheel featured two banana-shaped pieces of soft iron; the coils were fixed. Caliber 3055 featured day and date discs. Lutch in Minsk produced it from 1975 to 1981. Current consumption was a horrendous 18 microamperes. The daily rate was guaranteed to be ± 2 seconds per 24 hours; this was the accepted rate for all Soviet quartz watches.

The Uglitch watch plant in Tchistopol supplied 32 kHz quartz resonators. In the early days, these were of the bar-shaped, flexural mode type. They were replaced by the mechanically cut tuning fork type, and finally by the photolithographic Statek type in the standard 2 x 6 mm container. A production line for integrated circuits was installed at the Lutch plant; the coil and the other parts of the Lavet motor were also made in Minsk. The samarium-cobalt rotor was supplied by Kusa in Moscow, a company that had specialized in sapphire watch stones. This company still exists, but it was renamed LLC Instruments in 2000.

The first Soviet quartz wristwatch with an analog display bore the designation 3050, but only an experimental series of 1000 pieces was produced in 1977. This was a development ordered by the Ministry for Electrotechnical Industries that approved the design. Production started in 1977 under the Slava brand at the Second Moscow Watch Factory. Caliber 3055 was quite a conventional three-hand construction with 7 jewels and jumping day/date display; the dimensions were 30.5 mm x 7.5 mm. The stepping motor made 8 steps per minute, the quartz resonator was the 32,768 bar-shaped type supplied by Uglitch. Battery life was limited to 12 months.

Obsolete Technologies

It might be useful to remember that Statek in Orange, California, started producing photolithographic tuning fork quartz resonators in 1973. The price was a small fraction of the previous designs made from conventionally processed quartz. However, the Soviet system did not reward risk-taking: technologies that worked were still in use long after they had become hopelessly obsolete and uneconomical. Thus Statek-type resonators were not available in the USSR before the early 1980s.

Mass production of the 30 mm caliber 3056/3056A (30.95 mm x 4.2 mm) was started in 1979 by Uglitch. The stepping motor made eight 45 degree steps per second. At that time, the one-second Lavet motor had been adopted by most Western brands. Uglitch had to comply; there were no more changes in that respect later on. Caliber 3056/3056A was also manufactured by the First Moscow

13.13. SU Quartz, caliber 3056

13.14. SU Quartz, caliber 3056A

13.15. Men's watch with caliber Lutch 2350, dial side

13.16. Men's watch with caliber Lutch 2350, movement side

Watch Factory (under the Poljot brand), while the Second Moscow Watch factory and Petrodvorez used the Slava brand. Caliber 3056A did not feature a date disc, while 3056 displayed both day and date, which halved battery life from 24 months to 12 months.

By the early 1980s, the Soviet watch industry was finally equipped for the production of smaller quartz calibers. Thus, Lutch launched the 19.9 mm x 16.0 mm x 3.8 mm caliber 1956 with a single integrated circuit, a Lavet motor, and 8 jewels in 1980. Tchaika and Sarja followed in 1981 with the same caliber. With those calibers, the final transition to Statek-type 6 mm x 2 mm quartz resonators was made. Current consumption was 3.5 microamperes, which explains the short battery life of just 12 months.

Flat Calibers

In 1985, Lutch, Raketa and Tchaika simultaneously launched the tonneau-shaped men's watch caliber 2356. The dimensions were 24.0 mm x 20.0 mm x 2.1 mm, it was thus remarkably flat. Current consumption still was 2 microamperes, which allowed a battery life of 18 months. With this caliber, the modular construction typical for Soviet watch movements – electronics, motor, and wheel train – had proven its worth and was maintained to the end. The integrated circuit, the quartz resonator, and the printed circuit were potted in plastic and formed a "black box." In later models, the Statek-type resonator was not potted but was left visible in an opening.

A 23 mm three-hand caliber with the designation 2350 and the dimensions 23.9 mm x 20.0 mm x 3.14 mm was launched by Lutch, Slava, and Tchaika in 1984. It looked just like caliber 2356, but featured non-jumping day/date discs. Current consumption had been brought down to 1.7 microamperes, which increased battery life to 24 months. The daily rate was guaranteed to be ± 1 second. Production continued until 2003.

Still another 23 mm movement was caliber 2360. The dimensions were the same as those of caliber 2356, but the height was 3.14 mm. In addition to the functions of caliber 2356, caliber 2360 had an interesting feature: the speed of the day and date discs was accelerated towards midnight. The variant 2361 featured an hour alarm and 7 jewels. Production ran from 1985 all the way to 2003.

A family of 24 mm calibers launched in 1986 by Poljot and other brands was characterized by the dimensions 24.0 mm x 20.3 mm and 7 jewels. The heights were 2.5 mm (no date, caliber 2456), 3.5 mm with date disc (caliber 2460), and 3.5 mm with day/date display (caliber 2450). The quartz resonator's frequency was fine-tuned with a trimmer capacitor. Battery life was 18 months for all three calibers. An important innovation characterized the gear train, as an involute tooth system developed by NII-Tchasprom was used for the first time in the 24 mm caliber family. The frequency of the quartz resonator was adjusted with a fixed capacitor. The daily rate was guaranteed to be ± 1 second.

The 24 mm caliber family was completed towards the end of the 1980s by caliber 2466. The dimensions were 24.0 mm x 20.3 mm x 3.5 mm. Aside from a date disc, this caliber featured an alarm function that was set with a separate hour-hand, and an hour signal that could easily be switched off. The owner of such a watch could himself adjust the daily rate to ± 0.7 second. Battery life was 24 months. This caliber was produced by the Minsk Watch Plant (Lutch), the First Moscow Watch Factory (Poljot), the Second Moscow Watch Factory (Slava), and the Uglitch Watch Plant (Tchaika).

Calibers for Ladies' Watches

The 6¾ x 8 lines three-hand caliber 1656 was quite flat with its height of 2.3 mm. The current consumption was still over 1 microampere; the battery life was limited to 18 months. Caliber 1656 was launched in 1986 and was used in men's and in ladies' watches. The variant with a date disc was designated as caliber 1660; its height was 3.2 mm.

The 5½ x 6¾ lines x 2.3 mm two-hand caliber 1358 was developed for really small ladies' watches and ultra-flat men's watches. The current consumption was 1.7 microamperes; the tiny battery lasted just 13 months. Caliber 1358 was launched in 1986 by the First Moscow Watch Factory, but it was also produced under the Lutch, Poljot, and Tchaika brands.

Another 5½ x 6¾ lines three-hand ladies' watch caliber 1356 with a height of 2.9 mm and 5 jewels (no date disc) was launched by Lutch in 1988; it was produced by Uglitch until 2003. Current consumption was 1.2 microamperes, battery life was 18 months.

Magnetic Shielding

Remarkably enough, the Lutch calibers, particularly, were provided with an excellent magnetic shielding. Even the flat coil of caliber 2356 was shielded with a U-shaped soft iron plate. This sturdy and low-price caliber with zero jewels was kept in production until 2003.

The 6¾ x 8 lines caliber 1656A, with a height of 2.5 mm, was also launched in 1988 by Lutch; production lasted until 2002, Uglitch had it in its product mix as well. Current consumption was 1.2 microamperes; the battery life was 18 months. In order to achieve a good magnetic shielding, the back of the movement was entirely covered by a soft iron foil, with openings for the battery and the motor module. This system had been introduced with caliber 1356 and kept in all later models. Variant 1656H was produced by Poljot in Moscow and Zarja in Pensa; the specifications were the same as those of caliber 1656A, but different components were used.

Still another 6¾ x 8 lines caliber with the designation 1685 was produced by Lutch from 1989 to 2002. The dimensions were the same as those of caliber 1656A, but it featured a moon phase indicator at 12 o'clock and a semi-circular window for the date at 6 o'clock. Not only the current date, but also the previous day and the next day were thus visible. The height was 3.16 mm; the current consumption being 2 microamperes, battery life was a short 14 months.

Combo Calibers

A watch designed by Lutch with the combo caliber 2964-1H was of particular interest. The elongated octagonal case was made of black plastic, and the bezel was gold plated. The analog part was a three-hand caliber with a 4-digit LCD at 6 o'clock for displaying day/date, alarm time, and timer. The movement module was rectangular with dimensions of 30.8 mm x 23.3 mm x 5.2 mm; production started in 1990 and ended in 1996. Due to the hefty current consumption of 3.65 microamperes, battery life was limited to 14 months.

Another rectangular combo caliber bore the designation 2968, which was mostly produced by Poljot. The dimensions were 29.8 mm x 23.0 x 2.85 mm. Hour, minute, and second (with a sweep seconds-hand) were conventionally displayed by hands. The 6-digit LCD displayed the hour in the 12-hour or 24-hour mode, plus minute, two alarm times, month, weekday, date, and stopped time up to 1 hour with a resolution of 1/100 second.

After the political change of 1990, it took the privatized Lutch 9 years until it launched new quartz movements. The first one was the two-hand caliber 2358 with the same dimensions as caliber 2356. It was so sturdy that it could be used in watches for the blind. Lutch listed it in its catalog until 2003. Caliber 23858 was also introduced in 1999 with dimensions of 23.0 mm x 19.6 x mm x 3.2 mm. The dial of watches with caliber 23858 made it resemble a chronograph, but the counters are actually day and date displays combined with a 24-hour dial and a small seconds-hand. Caliber Lutch 2370 nearly had the same layout as caliber 23858 and the same dimensions, but it featured a day-night indicator. Product life was short and lasted from 2001 to 2003.

Dwindling Production

Two decades after the disintegration of the USSR, very little is left of the once powerful Soviet watch industry. Nobody seems to know the extent of the present Russian watch production. It certainly dropped in a spectacular way from the all-time record of 75 million pieces down to maybe a few hundred thousand pieces. One of the reasons for this sad downfall is certainly the total lack of support by state authorities.

Even at the highest levels of government, a domestic production of watches is deemed unnecessary. After all, the world's best watches can be imported and paid for with natural gas and oil of which the country has gigantic reserves. Trade is much more profitable than production anyway. The consequence is that almost all of the former Soviet watch plants have closed. The few survivors are the brands Poljot, Zarja, and Vostok in Russia and Lutch in Belarus.

Poljot and Zarja were purchased by an investor named Makharov; they are part of his holding Mak-Time that exclusively produces mechanical movements. During the Soviet era, Makharov had a leading position with Poljot. After the political change he founded his own company and earned a considerable amount of money with good looking gold watches. Thus, he had the means for acquiring the Poljot and Zarja brands; the latter's specialty were ladies' watches with small size movements.

The Tchistopol Watch Plant with the Vostok brand is much reduced in size but has managed to remain a true manufacturing facility that produces mechanical movements and watches equipped with them. A few more watch producers exist that design cases and equip them with purchased movements. The most important one is Zolotoja Vremija, which purchases its movements from the remaining Russian suppliers.

Watches with Imported Movements

A few small companies that produce watches with either Swiss or Japanese quartz movements also survived. The most important one is Vitebsk with the Wimpel brand. It uses Miyota movements, the dimensions of which exactly correspond to those of former Soviet calibers, e.g. 1356, 1656, 1658, and 2350. In some cases, plastic adapter rings have to be fitted.

In Minsk (Belarus), the Minsk Watch Plant with the Lutch brand also survived. Next to Poljot, Lutch was the Soviet pioneer for quartz movements but fully concentrated on analog calibers. Manufacturing still goes on in the old Minsk factory buildings; both the company museum and the archives were fully preserved. During the Soviet era, Lutch closely cooperated with Poljot. For some quartz calibers, component production was shared; they were fully exchangeable regardless of the place and time they were produced.

The political changes of 1990 forced Lutch to shrink dramatically. Formerly, about 10,000 people worked at Lutch, but 90 percent of these jobs were eliminated. After the borders opened, the newly formed countries that previously were part of the USSR were literally swamped with cheap Chinese watches. Thus, Lutch lost most of its former markets.

Assembly in Minsk

Thanks to a core of outstanding people, Lutch maintained its reputation as a strong, high quality brand with a tradition of 50 years operating in the field of medium priced watches. During the Soviet era, the Minsk Watch Plant produced some 4 million pieces per year; today the output is around 360,000 pieces. Two thirds are quartz; the rest is mechanical.

Lutch quartz watches are assembled in Minsk, but the movements are Miyota from Japan, or of Chinese origin for low-cost watches. Mechanical movements are assembled semi-automatically and are manufactured in-house. Plans to add a modern, 11-line automatic caliber to the existing line of products were put on ice after Franck Muller of Switzerland became the majority stockholder (see below). It is only logical that Lutch will have to adopt the Franck Muller technology.

Aside from Lutch watches, the Minsk Watch Plant also manufactures company watches and private label watches. Diversification in car dashboard watches, precision metal, and plastic parts, smart cards, and control units for electrical vehicles such as trolleybuses and fork lifts are also important for the company's survival. Printed circuits are developing into a major diversification.

In 2010, the Swiss luxury watch manufacturer Franck Muller acquired 80 percent of the Minsk Watch Plant's shares. The transaction was reported to have cost about $12 million, which paid for the debts Lutch had accumulated in the previous years. Franck Muller is negotiating with the government for the remaining 20 percent also. It was originally planned to sell the old plant near the Minsk downtown area and build a new factory in the nearby countryside. At the time of writing, nothing had been done yet in that respect due to the lack of capital.

Production continues in the old facilities of which a large part are not in use. Lutch has received numerically controlled machinery from Franck Muller in Switzerland, which allows the manufacture of complicated stainless steel watchcases. But the company still suffers from an acute shortage of modern equipment and highly qualified personnel.

14. Quartz Movements from South Korea

Very little is known about the early phases of Korea's horological history (in this chapter, Korea of course means South Korea). It seems that the first Korean clock was built in 1934: it was a clepsydra (Chagyoungnu) that was built by Chang Yong-sil. Mechanical clocks were crafted after the end of the Korean War in 1953. Until then, 100 percent of watches and clocks had been imported, almost exclusively from Japan that had annexed all of Korea in 1910.

Imported Movements

The Korean YoungMyun Industry (now Orient Co., not to be confused with the Japanese Orient Watch Co. Ltd., a Seiko subsidiary), started in Seoul in 1959. In the early days, YoungMyun cased Japanese movements. The gold plated cases and the leather watchbands were made locally; this was much less expensive than paying import duty on complete watches.

In terms of wall clocks and table clocks, the Korean pioneer was the Dae-Han Electric Light Company that was established in 1961. It, too, relied on imported movements, but by 1967, Dae-Han produced up to 95 percent of clock parts in Korea. Thus the very first truly Korean timepieces were Dae-Han clocks. Wristwatch movements were still imported throughout the 1970s and cased locally. Those movements were strictly mechanical until 1974, when quartz movements appeared; their proportion in the product-mix grew quickly. By then, all exterior parts were made in Korea: they represented 50 percent of a watch's total value.

The Korean state authorities actively supported the precision industry, which led to spectacular growth figures of 30 percent per year. As everybody wanted to have his piece of cake, the watch industry fragmented and up to 60 companies were fighting each other mercilessly.

Fast Growth

By 1972, Korea was already producing 1.28 million watches, but only five of the manufacturers could boast a production of more than 35,000 pieces. With half a million watches per year, Orient was ahead of everybody else. Ten years later, Korean watch production had risen to 8 million pieces per year, plus 2.2 million wall and table clocks. The domestic

14.1. LED module Kuatron 20.3 mm, front side

14.2. LED module Kuatron 20.3 mm, back side

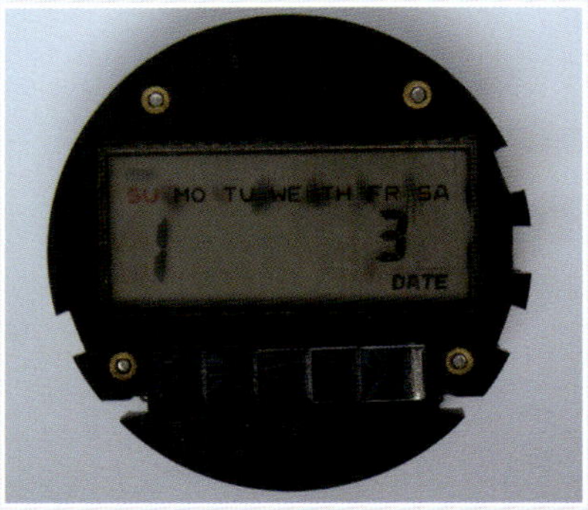

14.3. Gold Star solar LCD chronograph module, front side

14.4. Gold Star Solar LCD chronograph module, back side

market was fully saturated and could, at most, absorb 2.5 to 3 million watches per year. Thus, most of the production had to be exported, the major destination being Third World countries. The quality was average, which meant low-prices and low profit margins.

Orient was the uncontested number one watch producer in Korea; the next ranks were held by Samsung Precision, Citizen (with two subsidiaries in Korea), Anam Industry, and HanDok. The factory price of watches produced in 1982 was $5 to $7. These were primarily fashion watches with an analog display and an imported movement. All the exterior parts and the button-type batteries were made in Korea.

It was quite clear that Korea was not in a position to drastically improve the quality of its watches on its own. This is why the major Korean producers signed cooperation and licensing agreements with leading Japanese and European companies in the early 1980s: Orient with Rado and Omega; Samsung with Seiko and Longines; and HanDok with Bulova and PUW. It is interesting that Samsung manufactured various Longines Conquest models (even with gold cases), including the VHP line with thermocompensated quartz movements. Now the leading Korean watch manufacturers were suddenly present in the high-price sector with special brands: Orient with Galaxy; Samsung with Kappa Prime; Anam with Caritas; and HanDok with Odin. In the mid-price range, Orient became well known with the Chagal brand, Samsung with Kappa, Anam with Alpex, and HanDok with Dolphin.

Korean Digital Watches

In 1983, American electronics companies completely stopped producing digital modules and digital watches (most of them with a liquid crystal display, LCD). Hong Kong (including the Special Economic Zone and Mainland China) "inherited" the technology and suddenly held the dominant position in mass produced, low-cost digital watches. At one time, up to 1400 Hong Kong watch producers cased the enormous stocks of American digital modules that had found their way to the Far East. Some tried their luck with more expensive analog watches, including mechanical ones, as the Swiss and French pin pallet Roskopf watch industry had been eliminated by Japanese quartz watches just at that time.

In the late 1970s, Hong Kong-made digital modules with LCD (rarely with LED) were exported to Korea, where the fabrication of such modules had also started. Those modules were used for the production of complete digital watches with simple plastic or

14.5. Olympos LED module 15K, front side

14.6. Olympos LED module 15K, back side

14.7. Gold Star caliber 909, front side

14.8. Gold Star caliber 909, back side

metal cases. However, the Korean plastics industry was still behind Japan and even Hong Kong in terms of quality and design. The situation was quite different with clocks: most of the parts were produced locally in quite a good quality; Korean clocks had acquired a very good reputation, but their design was still a bit old-fashioned.

By the end of the 1970s, several Korean companies were mass producing LCD modules. No horological know-how was required and the components (battery, quartz resonator, integrated circuit, liquid crystal display) were readily available. Printed circuits were made in Korea. Samsung had started producing the required ceramic platelets, which gave the company a solid base for the production of zirconium oxide, aluminum oxide, and hard metal parts. Korean liquid crystal displays were also available. Thus, the production of digital wristwatches grew rapidly and reached 30 million pieces in 1988. The economic crisis in Southeast Asia and lower tariffs had a strong impact on the Korean watch industry: LCD watch production dropped to 20 million pieces in the early 1990s. About 75 percent were absorbed by the domestic market, so the Korean export position was rather weak.

In the long term, three globally active Korean watch manufacturers survived: Orient, Samsung, and Romanson. They have completely shifted their production to conventional, analog quartz watches. The movements are imported from China, Japan, and Switzerland (where Ronda is the main supplier). The exterior parts, for which Korean industry has a long experience, have achieved perfect parity with the foreign designs. Other companies such as HanDok and Olympos Electronic are interesting for historical reasons.

HanDok

The HanDok Industry Corporation in Inchon on the Korean west coast, about 40 km from Seoul, was established in 1966. Its first product was wigs for export to industrialized countries. Five years later, the first HanDok plant was built in Inchon. In 1973 the company diversified into trawler fishing and paper manufacture. HanDok began watchmaking in 1975, and by 1979 the company produced liquid crystal displays and digital watch modules. Another diversification was the construction and operation of vacation villages on Song Island off Inchon.

HanDok assembled complete digital and analog watches; the movements for the latter were imported from PUW in Germany. In exchange, HanDok supplied PUW with low-price integrated circuits that

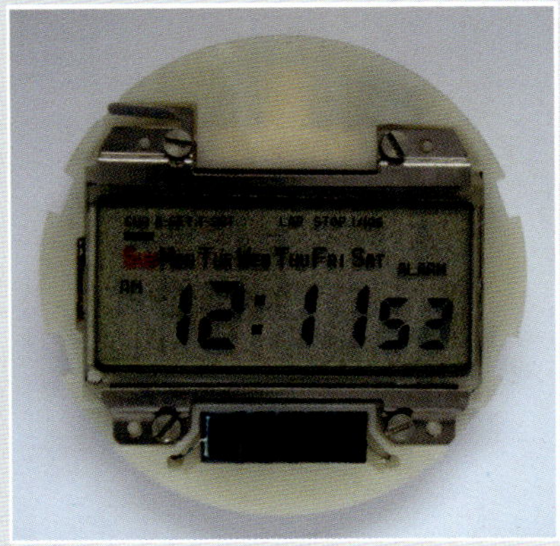

14.9. LCD Solar Alarm Chron module G704A, front side

14.10. LCD Solar Alarm Chron module G704A, back side

14.11. Olympos LCD module, front side

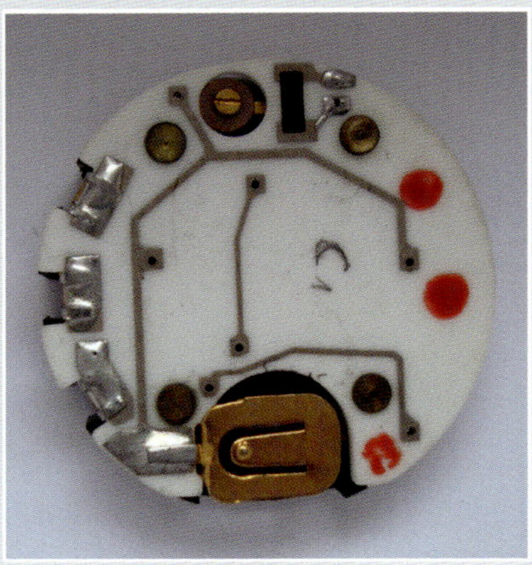

14.12. Olympos LCD module, back side

were purchased from a silicon foundry in Korea. The importance of watch manufacture rapidly grew: in the 1980s, watches, computers, and peripheral equipment made a 60 percent contribution to total company sales.

At that time, HanDok's employment was about 4000. Watch technology was the reason for the company's spectacular growth in the field of microtechnologies and precision machinery. By 1987, HanDok was the most important exporter of watch products: their value was around $60 million, while sales of liquid crystal displays totaled another $12 million.

As mentioned above, HanDok imported movement kits for its analog watches from Germany. However, the company had to pay extravagantly high import duties of 25 percent. In addition to that, the German currency kept increasing in value, which made HanDok products more expensive.

In order to maintain its export volume, the company's R&D division instituted a very strict quality control and modernization program in the HanDok plants and in those of its suppliers. Furthermore, HanDok started to produce its own watch cases, which made its products competitive in Europe and in North America. The brands used were HanDok, Odin, Fashion Time, and Dolphin.

In 1984, HanDok built two new plants in Banwol (Gyeoggi Province). However, the Chinese competition developed into a major problem in the 1990s. HanDok's LCD and watch case divisions were sold in 1992 and 1994, respectively. The rest of the company merged with Woolrie Car Sales that changed its name to Daewoo in 1997 and was itself absorbed by the Ssang Yong Motor Company.

Olympos Electronic

In 1977, Olympos Electronic in Seoul started with the production of a 3½-digit, illuminated, round LCD quartz module with the dimensions 25.4 mm x 6.9 mm and the functions hour, minute, second, month and date. The functions were chosen with three pushers on the right side of the case. One could even choose an alternate display of time and month plus date for 2 seconds each. Battery life was 38 months without the use of the LCD illumination.

The Japanese NDK supplied the tuning fork-shaped 32 kHz quartz resonator; the origin of the trimmer, the integrated circuit and the LCD are not known. The module consisted of a plastic disc for positioning the battery and the LCD, a 1 mm ceramic disc with the printed circuit and the electronics, and a metal cover with a window for the LCD.

Orient

As already mentioned, the Orient Co. Ltd. was the most important Korean watch producer in the 1970s. It started in 1959 as YoungMyon and assembled watches using imported movements and cases of its own fabrication, which were also sold to other Korean watch producers. In 1969, the company's name was changed to Orient. Its specialty was watchcases. In 1981, cases were manufactured that were waterproof to a depth of 100 meters. Later, Orient started producing cases made of sintered steel and stainless steel that could be coated by PVD or CVD. Movements were always purchased from outside vendors. The Galaxy brand was introduced in 1984, but production was eventually transferred to China.

Romanson

Romanson is the best known Korean watch brand even though it was established as late as 1988. It produces about 1.6 million watches per year, 70 percent of which are exported to more than 60 countries. The great majority of the company's watches incorporate Ronda quartz movements of the "Swiss parts" variety. The more expensive models are equipped with ETA or Japanese movements. The exterior parts are made mostly in Korea, but they are the work of a Scandinavian designer working for Romanson. Sub-brands are Elevé, Tulip, Trofish, and Phil.

Samsung, SWC

Samsung is the most important industrial group in Korea and one of the world's largest electronics producers. Company employment is close to 250,000. The Samsung Watch Co. Ltd (SWC Co. today) was established in 1983 as a subsidiary of Samsung. Thanks to cooperation agreements with Longines and Seiko, only high quality quartz watches were produced from the very beginning. Starting in 1986, they were exported under the Samsung brand. In 1997, Samsung bought the Geneva watch brand Haas & Co. that developed into the most important brand of the group; other brands are Samsung, SWC, and D.Factory. Today, Samsung produces more than a million watches per year.

15. Quartz Movements from the United States

Digital Calibers

The phenomenon of the American digital watch is quite unique in the history of watchmaking; furthermore, it lasted only a dozen years, i.e. from 1970 to 1983. The Pulsar, the very first digital wristwatch displaying time on demand with light emitting diodes (LED) started it all. Prototypes were shown to the media in 1970, serial production started in 1972. This enticed more and more American electronics companies to develop solid-state watches with a permanent liquid crystal display (LCD).

No horological know-how was needed for this type of product, which had no mechanically moving parts aside from pushers for setting the time. Those companies had no experience in watchmaking but were rather ambitious in declaring that they were going to destroy the traditional Swiss and Japanese watch industries. This did not work out too well: by 1983 the last American manufacturer of digital watches was out of business.

The Pulsar as a Trigger

The Pulsar by Time Computer (a subsidiary of HMW Industries, HM referring to Hamilton Watch – see Chapter 7) started it all. Inevitably, the required components were supplied to Time Computer by various electronics companies. The idea that they themselves might make money by producing digital watch modules or even complete watches quickly caught on. Optel and the Intel subsidiary Microma were among the first and the big and not quite so notable names in U.S. electronics followed suit – in all 20 to 30 companies. Another 50 bought modules and used them for the production of their own digital watches.

Soon after they had launched this totally new product, the electronics companies realized that they had not paid much attention to several crucial factors. For one thing, it was far from enough to produce a little black box that told the exact time. The exterior parts had to be sturdy enough to withstand hard wear and tear and still look good. A case, a watch glass and a watchband were indispensable. Instead of a dial, a cover plate with a window for the LCD gave the watch a kind of face.

The requirements for the exterior parts were particularly severe when the watch had be waterproof. There were not enough producers in the United States with the required capacity. In Europe on the other hand the case makers heartily welcomed this new inflow of orders. The French case makers in particular profited from the demand from across the Atlantic and enjoyed a short boom. German and Swiss case makers also profited from the new situation.

Marketing and sales are very important in the watch business, but these terms were almost unknown to the electronics industry. Many new electronic watch manufacturers also had to learn about warranty and after sales service. Digital quartz modules were practically non-repairable, so the entire module had to be replaced if a problem appeared. In the early days, this was rather costly, as the price per piece was quite high.

A Short Boom

The American manufacturers of digital watches that could be identified with certainty are listed below. One often hears that there were about 40 of them, but this number is probably exaggerated – 20 or, at most, 30 true manufacturers should be closer to the truth. Many more companies were content with just purchasing modules and casing them. However, the list given below is certainly incomplete.

Most of those companies are no longer in existence or they may be involved with different products. Hardly anybody remembers the short-lived quartz watch "adventure." Documents and prototypes were discarded long ago or may have found their way into private collections. What is certain is that those companies waged a merciless price war. Lower prices forced very high growth rates upon manufacturers; this led to massive overproduction and to a spectacular collapse of prices.

Between 1973 and 1980, the price of a complete digital module dropped from well over $300 to $3 or even less. Two to three years later, all of the U.S. manufacturers had stopped their domestic production and left this article to Hong Kong, China (including Taiwan), and Korea. Thanks to low wages, the Far East acquired a monopoly position for digital modules and watches. Since that time, the United States has imported digital quartz watches from Asia: by 1983, the number had risen to 53 million pieces. In that same year, the United States also imported some 8 million quartz analog watches, most of them from Japan and Switzerland. But China soon caught up at the low-price end of the market; today it dominates the field even in the mid-price range.

In the industrialized countries, LCD digital movements are used in low-price fashion watches and in (mostly Japanese) multifunction watches and radio-controlled watches. The use of LCD watches in sports (where Casio and Timex are the major players) is also quite important. The lowest quality modules are found in the truly cheap watches with plastic or alloy case that are sold in the streets of Third World cities for a few dollars apiece. China has a monopoly for this product too. LED watches have almost disappeared; very special models are occasionally launched as original niche products.

American digital watches, as well as their European and Asian counterparts, basically all had the same circuit design and very similar architectures, regardless of whether their display was of the liquid crystal type (LCD) or of the light emitting diode type (LED). The electronically excited quartz resonator generated a very stable 32,768 Hz signal that was divided by two 15 times down to 1 Hz by means of an integrated circuit.

Those second pulses were fed into the decoder (usually on the same integrated circuit). The decoder energized the required luminous spots or bars on the display; in other words it switched on the luminous elements that formed the required numerals.

Manufacturers of LED and LCD Modules

The surviving literature concerning the LED and LCD modules of the 1970s and 1980s consists of handbooks for watchmakers, explaining to them how to replace batteries and set the display. Defective modules were simply replaced, they were not meant to be disassembled and repaired. The differentiation between true manufacturers, who assembled their modules using components of their own or supplied by vendors, and the companies that just cased purchased modules is very difficult. The number of manufacturers that produced critical parts such as displays and integrated circuits and used them to assemble modules or even complete watches was quite small.

The following sections include the most important American electronics companies that diversified into digital quartz modules or digital watches, respectively, in the early 1970s, but gave up the watch business between 1977 and 1983. The list summarizes the companies that purchased LED and/or LCD modules and assembled watches on this basis. In many cases, only the name of the company and the type of product they specialized in have been preserved.

American Microsystems AMI

AMI manufactured both LED and LCD modules with 5 functions (hour, minute, second, and day/date). As far as AMI is concerned, the transition from LCD of the dynamic scattering type to the much superior twisted nematic (TN) type (it is also called field-effect) in the early 1970s is of special interest. Everybody soon followed suit; the modules looked exactly alike, but the TN display offered much better contrast and readability. A controversy exists as to the company that produced the first TN displays: it was either AMI or Optel. We only know that Gruen used the AMI module type 606 as early as 1973 in its Teletime watch.

AMI's headquarters were located in Santa Clara, California, but the modules were assembled in Korea and in the Philippines. It is likely that AMI was ahead of Optel in terms of field effect LCD, but only by a few months. AMI also produced watches with an LED display and a 3-volt lithium battery, but the module was purchased from National Semiconductor. Such watches are very rare.

Bowmar

The Bowmar Company in Acton, Massachusetts, started out with LED displays; in 1971 it decided to diversify into pocket calculators with an 8-digit LED display. The "Bowmar Brain" was the first pocket calculator available in the United States. As prices for calculators eroded rapidly, Bowmar was hard pressed for revenue, even though it was producing its own integrated circuits in a plant it built in Arizona in 1974.

A further diversification into digital watches could not save the company, but as a last resort this was tried. In 1974, Bowmar launched a 3½-digit LED Module with the dimensions 14 lines x 8.0 mm. It was characterized by two silver oxide batteries in series and a bar-shaped, 32 kHz quartz resonator that was presumably supplied by Reeves-Hoffman. A single CMOS integrated circuit handled the oscillator, divider, and decoder functions. Pressing the two pushers on the right side of the case switched on the display of hour and minute or of date and second. The pusher on the left side in combination with the pushers on the right was used to set the time. Tiffany was one of the early customers: nevertheless Bowmar went into receivership in 1976 and was liquidated.

Commodore

Commodore was established in 1954 in Toronto (Canada); its first product was a typewriter. Later, it diversified into mechanical and electronic calculators. In the 1970s, the company moved to King of Prussia near Philadelphia, Pennsylvania, and developed into one of the major suppliers of electronic components for quartz watches. In 1977, Commodore became known as the pioneer of low-cost personal computers and established itself as the world leader. The PC division was located in Palo Alto, California. The company declared bankruptcy in 1994, which was primarily due to management errors.

Commodore also manufactured low-cost digital watches of the LED and LCD types, but this was not a profitable line of business. In 1976, the company subcontracted 26.8 mm x 6.2 mm LED modules in South Korea; the components were bought locally or in Japan. The 32 kHz quartz resonator was made by NDK; when the pusher was pressed, the 4-digit LED with two central dots first displayed hour and minute, then the running seconds. Pushing twice displayed month and date.

The module's construction was quite conventional: a disc-shaped ceramic printed circuit with the LED was glued to a plastic disc with openings for the two batteries, the quartz resonator and the trimmer. The retail price of complete watches with a plastic case and the LED module was less than $15.

In 1978, Commodore contributed to the decline of the digital watch by passing the 10 dollar mark with a collection of 15 such watches that were sold in blister packs in department stores, food markets, drugstores, and electronics shops. The price range was $7.95 to $19.95. An 8-digit LCD displayed hour, minute, second, and day/date. There was no repair service: defective watches were simply replaced.

Fairchild Semiconductor

Fairchild Semiconductor was established in 1957 in South Portland, Maine, but soon moved to Mountain View and Palo Alto in California. The founders, Gordon Moore and Robert Noyce, had left the Shockley Semiconductor Laboratory, where William B. Shockley, one of the co-inventors of the transistor, had developed a novel diode for telephone switching that was a commercial failure.

In 1959, during his time with Fairchild, Noyce invented the integrated circuit. Almost simultaneously, but completely independently, Jack Kilby at Texas Instruments made the same invention that would revolutionize our everyday life. After ten years in court, the patent was attributed to Noyce.

Almost inevitably, Fairchild played an active role in the digital watch boom, both with LED and LCD calibers. Patent licenses had to be purchased from

15.1. Gruen caliber 606, AMI module

15.2. Fairchild LED medium size (21.8 mm), front side

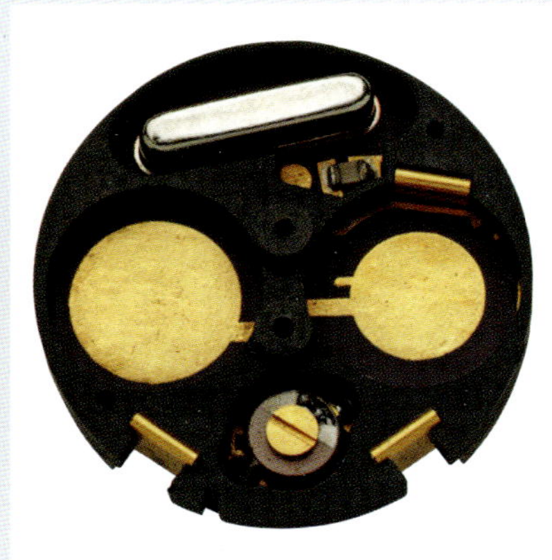

15.3. Fairchild LED medium size (21.8 mm), back side

15.4. Fairchild LED large size (29.2 mm), back side

Time Computer; the modules were assembled in Hong Kong. The LED men's watch caliber (29.3 mm x 6.1 mm) was cased by many famous brands such as Croton Time, Elgin, Junghans, Ness Time, Nepro, Solitron, Tiffany, and Yema.

In 1975, complete watches with the Fairchild brand became available; Exetron was a sub-brand. Two million pieces were sold in the first year alone. The 3½- or 4-digit LCD displayed time, month, and date. The module used a flexural mode quartz resonator from Motorola and a CMOS integrated circuit from the Exetron subsidiary. The folded, epoxy resin printed circuit was quite unusual. The half with the integrated circuit featured wiring on both sides; the other side with the display was normal. The two batteries and the quartz resonator were lodged in recesses in a plastic disc. The configuration of the Fairchild LCD caliber 0203A was quite similar to that of the LED caliber.

Exetron, the Fairchild subsidiary in Santa Clara, California, produced LED watches with the usual five functions; they were assembled in Singapore using Fairchild modules. Starting in 1976, the bar-type quartz resonator was replaced by a mechanically cut tuning fork-type supplied by the Japanese NDK. The LED was purchased from Litronix.

Practically all manufacturers of LED modules used a very similar design. It almost invariably consisted of a ceramic disc with the display on one side, while the integrated circuit and the wiring were on the other side. A much thicker plastic disc was attached to the rear: the two batteries, the quartz resonator and the trimmer were lodged in recesses.

In late 1976, Exetron launched 28.8 mm x 4.6 mm, 3½-digit LCD modules with two 1 Hz flashing dots; they were assembled in Singapore. It is not known which company supplied the illuminated LCD, but the quartz resonator certainly was by Motorola. A single CMOS-IC took care of all electronic functions. Normally, hour and minute were displayed; the upper pusher on the right side of the case switched on month and date or the seconds. The lower pusher switched on the LCD light. Time was set with a third, recessed pusher

In 1978, the Fairchild LCD module was provided with 6 serially connected solar cells with a surface of 80 mm^2; they were mounted under the watch glass and increased the module's height to 7.7 mm. In bright sunshine, they generated up to 2.5 microamperes. As silver oxide batteries can only be partly recharged, their life was just extended by the solar cells.

Frontier

The LED modules by Frontier in Newport Beach, California, were quite similar to the ones made by Hughes Aircraft and other manufacturers, as they all had to buy patent licenses from Time Computer. The Frontier 3½-digit caliber 2002 launched in 1975 (13 lines x 8.2 mm) featured a horizontally mounted bar-type quartz resonator at 12 o'clock, the two batteries in the middle, and the trimmer at 6 o'clock. Actuating the pusher once displayed hour and minute for 1½ seconds and then the continuous seconds. Pushing twice displayed date and month.

Another, recessed pusher was used for setting the display. The height of the ceramic disc with the wiring, on which the display and the integrated circuit were soldered, was 0.6 mm. As usual, the batteries, the quartz resonator, and the trimmer were lodged in a thick plastic disc in the rear. The power consumption of the module was only 2 microamperes, but it rose to 15 to 50 milliamperes when the display was switched on.

Frontier is interesting to collectors of LED watches, as it produced a 29.4 mm x 8.4 mm module that featured a stopwatch function with a resolution of 1/100 of a second (caliber 2006). The Swiss companies Beltime, Breitling (in the model Navitimer LED), Ibico, and Sicura used this model in attractive chronographs with a screw-on bezel. The module was powered by two button-type batteries in series; the 32 kHz bar-type quartz resonator was supplied by Reeves-Hoffman. Frontier produced its own CMOS integrated circuits. Three of them were required for caliber 2006; they were potted in black plastic including the connectors.

The pusher on the right side of the case activated the display for the five usual functions plus the weekday. The chronograph was controlled by two pushers on the left side of the case. If an interval of one minute or more was stopped, hours and minutes were displayed first, then the seconds and hundredths of a second. The LAP function was also available

The 29.4 mm x 7.5 mm LED module 2007 by Frontier not only offered the six classical functions, but also an alarm. The sound was generated by a piezoelectric disc. The alarm sounded in two stages; the first signal lasted one second; three seconds later the signal was repeated for ten seconds while the LED dots in the middle of the display flashed. The alarm was stopped by pressing the pusher on the upper right twice. Beltime in Switzerland used this module in its alarm watches. Lanco produced beautiful watches with either the stopwatch or the alarm module by Frontier; the design was similar.

Hewlett-Packard

In 1976, Hewlett-Packard (HP) launched its calculator module HP-01: this was the company's sole incursion into the digital watch business. Production facilities were located in Central America and in Asia. Both LED and LCD variants of the HP calculator module for men's watches were developed, but only the LED type was actually produced.

The HP-01 Calculator Watch was most remarkable. The functions were time, calculator, alarm (to the second), stopwatch (to the 1/100 of a second), timer, and calendar for 200 years. The watch featured 28 tiny pushers that were actuated with a special pin that was supplied with the watch. Another, smaller pin could be snapped into the watch band.

The module featured 6 integrated circuits and 3 batteries, two of them for the LED display, one for the electronic circuit. The quartz resonator's frequency was 38,400 Hz: it is found in no other watch and its origin is unknown. The watch was non-magnetic, shock proof, and waterproof to 10 meters. The yearly rate was less than 30 seconds. The case and the bracelet were made of stainless steel; a gold filled variant was also available.

The HP-01 was launched in 1977 and was available in the stores until 1990. Collectors fully agree on the fact that the HP-01 was by far the best calculator watch that was ever made: it was vastly superior to all the others. The only (slight) disadvantage was the center of gravity of the watch that was rather high. This had the effect that the watch tended to turn around the wrist.

From the commercial point of view, the HP-01 was a failure. It was launched in the year when Time Computer announced insolvency and the fad for LED watches had evaporated. In all, 12,000 pieces were produced, but only a small percentage of them was actually sold. HP had no sales network or structure that would have allowed a dialogue with the retailers. HP ended up selling the watches to its staff for a symbolic price and gave them away to employees that celebrated an anniversary.

Hughes Aircraft

Hughes Aircraft (or simply Hughes) in Newport Beach, California, produced LED, LCD, and LED/LCD watch modules from 1973 to 1976. Complete watches with the Hughes brand did exist, but they were meant as special rewards for employees. Production in California and Mexico reached 100,000 modules per month in 1975. Hughes, too, had to acquire patent licenses from Time Computer.

The first Hughes LED caliber-type 201 (29.3 mm x 7.72 mm) featured two silver oxide batteries, a Hughes

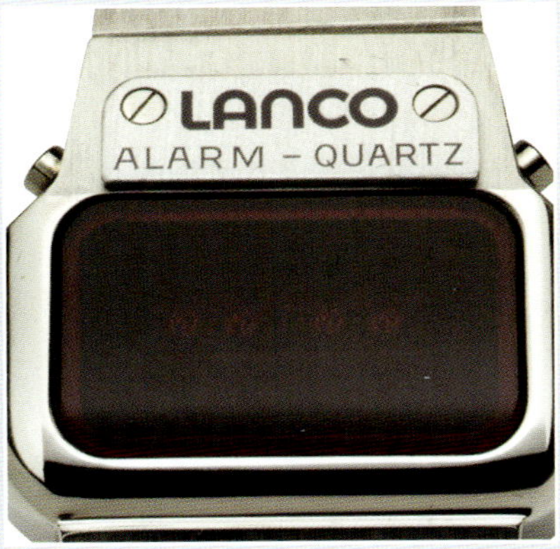

15.5. Frontier LED Alarm caliber 2007 (in a Lanco watch)

15.6. Frontier LED module caliber 2002

15.7. Frontier LED module caliber 2006 (in a Sicura Chronoflash)

15.8. HP Calculator 01

high frequency quartz resonator (786,432 Hz) and two CMOS integrated circuits also made by Hughes. One was needed for frequency division, decoding, and multiplexing, the other for the oscillator and the first divider stages.

The LED display of time, day/date, and month was controlled by bipolar integrated circuits that were available at very low prices. In 1974, Hughes made the transition to CMOS technology, which would have been available even in 1973. However, the chip would have been bigger and significantly more expensive. The calendar function was compensated for leap years.

The module consisted of a sandwich of discs. The upper one was made of aluminum oxide ceramic: the printed circuit interconnected the quartz resonator, the integrated circuits and the display (4 digits and AM/PM indication). A thin brass plate with a rectangular window for the display covered the module. The ceramic disc was glued to a plastic disc with recesses for the two batteries, the quartz resonator and the trimmer. With the LCD module 202, Hughes changed to the conventional 32 kHz technology and CMOS circuits. The following calibers were made by Hughes until 1976:

202 (LCD, men's)	29.5 mm x 9.3 mm, 4 functions.
203 (LED, ladies')	elongated with a height of 7.2 mm, 2 functions.
205/215 (LED, men's)	29.5 mm x 7.72 mm, 5 functions.
206/217 (LED, men's)	25.4 mm x 6.35 mm, 5 functions.
209/221 (LED, men's)	29.5 mm x 7.72 mm, 6 functions, 24 h mode and leap year indicator as options.
214 (LCD, men's, 1974)	29.5 mm x 9.3 mm, 5 functions.
219 (LED/LCD men's)	29.5 mm x 9.3 mm, 4 functions.
222/233 (LED, men's)	29.5 mm x 7.72 mm, 5 functions.
224 (LED, ladies')	elongated, with a height of 7.2 mm, 5 functions.

15.9. Hughes Aircraft LED caliber 201, front side

15.10. Hughes Aircraft LED caliber 201, back side

15.11. Hughes Aircraft LED for ladies' watches caliber 203

15.12. Hughes Aircraft LED calculator

226/236 (LED, ladies')	20.7 mm x 5.7 mm and 20.7 mm x 7.5 mm respectively, 5 functions.
228/238 (LED, men's)	29.5 mm x 7.72 mm, 6 functions, 24 h hour mode and leap year indicator as options.
232/243 (LED, men's)	25.4 mm x 6.35 mm, 5 functions, one programmable message.
233/245 (LED, men's)	25.4 mm x 6.35 mm, 6 functions, 24 h mode and leap year indicator as options.
235/262 (LED, men's	1973), 29.5 mm x 7.72 mm, 5 functions, one programmable message (5 words of 5 letters).
242/248 (LED, men's)	25.4 mm x 6.35 mm, 5 functions, 1 programmable message.
246/249 (LED, men's)	29.5 mm x 7.72 mm, 5 functions, 1 programmable message.
252 (LED, ladies' 1974)	20.3 mm x 5.7 mm or 20.3 mm x 7.5 mm, 6 functions, 24 h mode and leap year indicator as options.

The most important customers for Hughes LED modules were Benrus, Bulova, CompuChron (a Japanese-American company), Concord, Digitime, Helbros, Jaeger-LeCoultre, Longines, Ricoh, and Wittnauer.

In 1973 too, Hughes launched a square LED-ladies' caliber with the dimensions 12 mm x 12 mm. The two batteries were located in extensions of the watch band between the horns. In 1974, Hughes launched a round ladies' watch caliber with a diameter of 20.2 mm; both batteries were now integrated into the movement.

Starting in 1974, Hughes also produced TN-LCD modules that came from its own fabrication. At the same time, Hughes changed from the 32 kHz bar-type quartz resonators to the tuning fork type and also adopted standardized printed circuits; both were produced in-house.

15.13. Hughes Aircraft LCD, front side

15.14. Hughes Aircraft LCD, back side

15.15. Litronix LED large size, front side

15.16. Litronix LED large size, back side

Hughes was ahead of everybody else in the field with integrated circuits that operated at low voltage thanks to ion implantation. Hughes guaranteed a monthly rate of ± 15 seconds, but the practice showed that the actual performance was much better.

Hughes launched an LCD chronograph module (caliber 1348, 29.3 mm x 8.8 mm) with a 786 kHz quartz resonator in 1976. In Switzerland, it was used by Beltime. The illuminated liquid crystal display featured two levels with the same size numerals. In the normal mode the watch displayed hour, minute, and second or date, pressing the pusher on the right replaced hour and minute by month and date. Pressing the pusher on the left changed the watch into a chronograph with a resolution of 1/100 of a second for 59 minutes and 59 seconds.

Hughes invested heavily into the production of digital watch modules, as they were extremely popular and the profit margins were high. Up to 500 persons worked in the Hughes Newport Beach watch plant; in 1975, production reached 100,000 modules per month. Certain types of modules were assembled in Mexico. However, competition (particularly from Texas Instruments and Motorola) soon drove prices down. Swiss, Japanese, and Hong Kong manufacturers also made big inroads; within a few years, the price of a digital watch had dropped to $10.

The weakness of American digital watches was the case: they lacked the know-how to waterproof it. Also, the many brands wanted to be in line with fashion trends, which meant frequent changes of case models. In 1977, Hughes tried its luck with a calculator watch. It featured an 8-digit LED, the keys were much bigger than those of competing products, but this could not save the watch division. Hughes closed it in 1978, before the losses went out of control – which was a wise decision.

Integrated Display Systems IDS

IDS was located in Montgomeryville, Pennsylvania. It was controlled by Bulova and produced digital modules for Bulova and other watch manufacturers, e.g. Longines in Switzerland.

Litronix

Litronix in Cupertino, California, was founded by an engineer who had developed the first LED displays when he was with Monsanto. Litronix soon became an important supplier of light emitting diodes and displays. The Pulsar P1-modules made by Time Computer were equipped with Litronix LED.

Later in the 1970s, Litronix developed into one of the most important manufacturers of LED modules (there were at least three different ones, with either 2 or 5 functions). Litronix LED displays were sold worldwide; the Swiss Nepro was one of the early customers. Litronix even produced complete LED watches with the Litronix brand. Unfortunately, there were quality problems, but so many of those watches were sold that functioning ones can still be found.

Litronix LED modules were all quite similar and were state of the art. For example, caliber 058-2032 (29.1 mm x 7.0 mm) consisted of the usual ceramic-plastic sandwich. The display was attached to outside of the ceramic disc, and on the inside one found the printed circuit and the integrated circuits. The batteries, the quartz resonator, and the trimmer were lodged in recesses of the plastic disc. Litronix was one of the earliest customers of the Statek corporation that produced photolithographic tuning fork quartz resonators. Some of the Litronix modules were assembled in Singapore.

Micro Display Systems MDS

Micro Display Systems (MDS) was established in Dallas, Texas, in 1973 with the objective of producing LCD digital watches. They were marketed under the Sensor and Savant brands. From the very beginning, MDS relied on quality and design, not the lowest possible price. This was why MDS watches cost $50 to $170 and they were only sold in qualified jewelry and watch stores. MDS gave a five-year warranty on all its products.

MDS used 4- and 6-digit LCD that were backlit with tritium gas cells (Betalight). The company advanced to the third largest manufacturer of LCD watches in the United States. Most of the modules had 5 functions, a pusher switched from time to day and date. Special models included the stopwatch function. Quartz resonators were purchased from outside vendors, particularly from Seiko.

MDS watches ran on just one silver oxide battery. The dimensions of the modules were 26.0 mm x 8.7 mm or 30.5 mm x 5.9 mm for the chronograph models. An integrated circuit of the CMOS type designed by MDS handled all electronic functions. The plate was plastic, the LCD was contacted by means of zebra stripes.

Microma

Gordon Moore and his lifelong friend and colleague Robert Noyce founded the Fairchild Semiconductor Corp. in 1957. This is where Noyce invented the integrated circuit in 1959. After a fight with management, Moore and Noyce left Fairchild in 1968 and established their own company; they called it Intel Corp. Their first product was the DRAM memory chip. In 1973 they made a real breakthrough with the 4004 microprocessor that could perform 60,000 floating point operations per second, which was unheard of at that time. Intel developed into a giant; there hardly is a laptop or PC that does not bear the sticker "Intel inside." This means an absolutely invisible but indispensable Pentium processor – of the eighth or ninth generation.

There were failures too, particularly LCD quartz watches that were launched in 1972 by the Intel subsidiary Microma in Santa Clara, California. Microma also sold LCD modules that were cased in Switzerland by Paolo Spadini's Nepro Company in La Chaux-de-Fonds.

The dimensions of the first Microma module were 13½ lines x 8.6 mm. Major components were two silver oxide batteries in series, a 32 Hz quartz resonator of the bar type with fine tuning by trimmer, a CMOS integrated circuit, and a 3½-digit liquid crystal display that was potted in plastic with the last divider stages and the decoder. The 1 Hz rate was given by two flashing dots between the hour and minute numerals. The LCD required 15 volts. Not everything could be handled with just one integrated circuit, so a few discrete components such as a transistor, a diode, an inductivity, two capacitors, and four resistors were necessary.

The next model was launched in 1974, its dimensions were 13½ lines x 8.55 mm. Only one battery was needed, the current consumption was a hefty 28 microamperes. As in the previous model, hour and minute were displayed permanently: pressing the pusher on the right side of the case displayed the seconds.

Microma also launched a ladies' watch module of 9¾ lines x 7.9 mm with a 3½-digit LCD in 1974. Only one battery was needed, the current consumption had been reduced to 9 microamperes. The quartz resonator was of the usual bar type with a trimmer for fine tuning. Only hour and minute were displayed, flashing dots between the numerals provided the 1 Hz rhythm. The display could be set with a magnet that was supplied with the watch and that actuated reed relays above and below the display.

The Synchroquartz line was introduced in 1976 and was very successful; retail prices ranged from $29.95 to $49.95. The 30.3 mm x 6.8 mm module had been improved considerably and featured a big, illuminated 6-digit LCD. This allowed a constant display of the date and timing of up to 30 minutes with a resolution of a tenth of a second. The 32 kHz quartz resonator of the bar type and the LCD were now produced in-house, Intel supplied the integrated circuit.

15.17. Microma LCD DSM, first model, front side

15.18. Microma LCD DSM, first model, back side

15.19. Microma LCD ladies' watch module, front side

15.20. Microma LCD ladies' watch module, back side

One of the two batteries supplied energy to the module, the second to the LCD. The miniature lamp was switched on by the pusher on the left side of the case. The two pushers on the right were needed for activating the timer functions (including flyback, intermediate times, and deductions for breaks). A recessed pusher was used for setting the display. The module was made of white plastic with recesses for the two batteries, the quartz resonator and the trimmer. The ceramic disc with the electronics and the display was glued to it.

Microma was one of the pioneers of the LCD digital module. But inevitably, aggressive competition appeared, as other electronics companies also started to produce LCD modules and complete digital watches. Intel executives realized at an early time that there was a problem. Production was stopped in 1978; the Microma brand was sold to the Swiss Endura company for $25,000 (see the Nepro section in Chapter 8). For a while, Endura sold watches with the "Microma Swiss" brand; the modules were supplied by ESA. The brand has not been used since the 1980s.

Motorola

The Galvin Manufacturing Corp. was established in Illinois in 1928 and hit a home run two years later: the Motorola car radio. It was sinfully expensive, but everybody just had to have a car radio: business was great. Very soon, the company took the name of its most successful product. Motorola developed into a highly diversified electronics company specializing in mobile radio equipment for police, the military, and aerospace. The manufacture of integrated circuits was started quite early with a strong accent on memory chips and microprocessors.

In the early 1970s, Motorola decided to diversity into LCD watch modules and complete watches. The company built a large production facility in Phoenix, Arizona, but closed it in November 1974 before it had produced a single module. Motorola management realized that they had no chance in the watch sector in the face of Far Eastern competition.

Henceforth, Motorola produced its modules in Taiwan and in the Philippines. Huge quantities of the ladies' watch caliber 218 were made in 1976 for Timex. This was a rectangular module with an illuminated LCD; the functions were hour, minute, second, AM/PM, month, and date. Another very successful module by Motorola was the men's watch 29.5 mm caliber; the functions were similar to caliber 218.

National Semiconductor

National Semiconductor (or just National) was founded in Connecticut in 1959 but moved to Santa Clara, California in 1967. Next to Intel and Hughes, National developed into one of the most important chip producers in Silicon Valley with major plants in Sunnyvale. National bought important competitors such as Fairchild Semiconductor and Cyrix, but sold them in the late 1990s. Henceforth, the company specialized in analog semiconductor components; employment is about 7600.

National Semiconductor's involvement in the digital watch business started in 1973 with the round LED module WM-01 (29.1 mm x 7.4 mm). The 32 kHz bar-type quartz resonator with fine tuning by trimmer was supplied by Reeves Hoffman. In the early days, LCD of the dynamic scattering type were used. The functions were hour, minute, and second. Caliber WM-01, with a diameter of 25.3 mm and a height of 6.48 mm and the same functions, was significantly smaller. Caliber WM-03 was the same size as WM-02, but displayed hours in the 24-hour mode. The same applied for caliber WM-05 while caliber WM-06 also displayed time in another time zone.

The modules and the National watches with the Novus brand were assembled in Malaysia. The excellent quality cases were designed by Pierre Essig and were supplied by the Swiss company Miserez & Piquerez. The lower price Exelar brand watches featured the same modules, but the cases were much simpler. Caliber WM-08 was launched in 1974, the functions were the same as in WM-01. Caliber WM-09, with time, day/date, and month functions, was launched in 1975.

In the same year, National Semiconductor issued the round caliber LCM01/731, which marked the switch to liquid crystal displays of the twisted nematic type. This module was used in watches with the National Semiconductor brand. The functions were time, AM/PM, date, month, and a light for the display. In Singapore, National assembled the LCD caliber LCM04 with a tuning fork quartz resonator and chronograph functions in 1977. Stopped time was displayed with a resolution of 1/100 of a second; the split and lap functions were also available.

In 1977, National launched a very sophisticated chronograph; it was followed in 1978 by a speaking watch and a calculator watch with a covered keyboard and an 8-digit LCD (CW-01). This product, too, was sold under the National Semiconductor brand. The dimensions of the rectangular module were 37.7 mm x 29.2 mm x 8.7 mm; the components' support was glass fiber reinforced plastic. Energy was supplied by two button-type batteries in series, time base was a 32 kHz tuning fork quartz by NDK.

The microprocessor was much bigger than the integrated circuit for the watch functions, which were hour, minute, second, month, and date. The calculator not only handled the four basic operations, but cumulated no less than 25 functions such as power, square roots, trigonometric functions, logarithms, all of them with floating point. The keyboard consisted of 20 tiny keys that were operated with the tip of a ball point pen.

The chronograph with the module LCM04 (30.5 mm x 6.1 mm) was assembled in Singapore. The illuminated display was divided in two levels. The upper one featured two digits and alphanumeric segments for the weekday, and the lower one had 4 digits for hour and minute; between them figured the indications AM, PM, or lap Normally, the LCD displayed hour, minute, date, and weekday; the alternative to the date was the second. The stopwatch measured the 1/100 of a second; the lap function was also available.

The LCM07A LCD module was assembled in Singapore and was energized by a very long life 2.8-volt lithium battery supplied by Matsushita; it was launched in 1978. Batteries of this type were nearly the same diameter as the module (29.0 mm), and they were lodged in the case bottom. Despite great efforts to reduce the height of such modules, the lithium battery added millimeters. This resulted in a height of 6.1 mm – almost the value of the early modules, but, of course, battery life was greatly extended. The very low self-discharge rate of the lithium battery was another advantage.

The module featured the five usual functions, but could be set in such a way that hour, minute, month, and date were constantly displayed in sequence. A single CMOS-LSI chip made by National took care of all the electronic functions. National manufactured its own liquid crystal displays and continued to use the classical, bar-shaped quartz resonators even though the much smaller and less costly photolithographic resonators of the Statek type had become available in 1973.

By the end of the 1970s, the price war had reduced profit margins for digital quartz watches to near zero. This led National to diversify into analog quartz watches with the NSC brand that were sold for $40 to $90 apiece and yielded reasonable profits. The movements were supplied by Citizen in Japan.

Ness Time

The Ness Time Company (and its subsidiary Solid State Time) in Mountain View, California, was founded in 1971 by Gordon L. Ness. It was among the earliest

15.21. NSC LED ladies' watch module, caliber WM16, front side

15.22. NSC LED men's watch module, caliber WM01

15.23. NSC LCD caliber LCM01A, back side

15.24. Ness LCD, back side

producers of quartz modules with a liquid crystal display of the dynamic scattering type. The switch to twisted nematic LCD was made as soon as possible; nevertheless, modules made by Ness were quite different from those of other manufacturers.

Business was very good in the first years; by 1974, Ness was still considered as a company with a spectacular growth rate in the field of LCD digital watches. Ness even manufactured a solar quartz watch; the nickel-cadmium batteries were potted in red plastic and could not be replaced, but they were supposed to outlast the watch anyway. Nevertheless, the company was insolvent in the spring of 1976. Several factors contributed to the spectacular downfall of Ness Time. It had the ambition of manufacturing its own ceramic substrates, but the development costs and required investments were too much for a company of that size that was undercapitalized. Furthermore, the profit margins kept eroding.

As a relatively small company, Ness Time had a rather low priority among the suppliers of electronic parts and watch cases. Consequently, Ness Time never delivered its watches on time and lost one customer after another. Eventually, Gordon Ness could not even pay his telephone bill, so his line was cut – that was the end.

Nortec

LED watches by Nortec in Santa Clara, California, featured three different modules:

Standard WM-1	29.03 mm x 8.13 mm, 5 functions;
Unisex WM-2	25.3 mm x 6.48 mm;
Petite WM-3	20.32 mm x 5.33 mm.

Nortec manufactured those three modules in its own facilities, but Tissot was the only foreign brand that purchased them.

Optel

The Optel Corporation in Princeton, New Jersey, is known as one of the pioneers of LED and LCD displays. Optel developed both types in parallel. First prototypes were available in 1970, but were not shown to the media – which Hamilton Watch did in connection with the Pulsar. Optel gained momentum when it succeeded in hiring Nunzio Luce and his colleagues Louis Zanoni, George Graham, and Joel Goldmacher; all three of them had previously worked at the David Sarnoff RCA Research Institute in Princeton. The team developed and built one of the world's first watches with a liquid crystal display, without any mechanically moving parts (apart from pushers for setting). In cooperation with the

15.25. Nortec caliber 4005, front side (used by Tissot)

15.26. Nortec caliber 4005, back side

15.27. Optel LCD Quartzmatic DSM, back side

15.28. Optel LED caliber 70450

Solid State Scientific Corporation, a clock movement was miniaturized to the size of a wristwatch. In 1971, Omega bought a participation in Optel in order to secure its supply of liquid crystal displays and digital modules. Luce rose to the presidency of Optel Corp.; later he established his own company, SEI International, that developed and manufactured LCD digital calibers and quartz analog calibers.

In the early days, Optel manufactured liquid crystal displays of the dynamic scattering type; they were replaced by the much superior twisted nematic type in 1972. Optel sold those displays in the OEM market, but also manufactured complete digital watch modules including a light. One of the most important of them was the 3½-digit caliber 726 that Optel launched in 1972.

In 1973, Optel issued its caliber R 705 with a similar LCD, but the display could be changed from hour and minute to second and AM/PM with a pusher; the dimensions were 13½ lines x 7.0 mm. The quartz resonator was of the bar type; 1.5 -volt CMOS integrated circuits were used for the oscillator/divider and the decoder, respectively. Optel produced complete watches for the private label market and also sold them under its own brand.

Parallel to the LCD modules and watches, Optel sold LED calibers that were assembled in Malaysia using American components. Depending on the pushers, hour, minute, second, day/date, and month could be displayed. In 1973, Optel launched the LED caliber H1 with the functions hour, minute, second, day/date, and month.

Pulsar

The Pulsar P1 story is found in Chapter 7. Here we relate the further development of this extraordinary watch until 1977 when production was stopped. The fate of the Time Computer Company should also be of interest.

After the tremendous initial success of the Pulsar, the Swiss watch group SSIH (Omega-Tissot-Lanco) started buying into Hamilton Watch via its Geneva subsidiary Aetos. The participation was 17 percent at the beginning, but rose to 51 percent in May 1974. In 1971, Hamilton Watch changed to the status of a subsidiary of HMW Industries, which itself was the legal successor of the old Hamilton Watch Co. that had been founded in 1892. The HMW subsidiary Time Computer Inc. (TCI) and not Hamilton Watch was responsible for the Pulsar project starting with 1972, but SSIH now had a direct link to HMW Industries. This was why TCI supplied Omega with tens of thousands of Pulsar modules; they were used in watches with the Time Computer brand.

15.29. Optel module 726

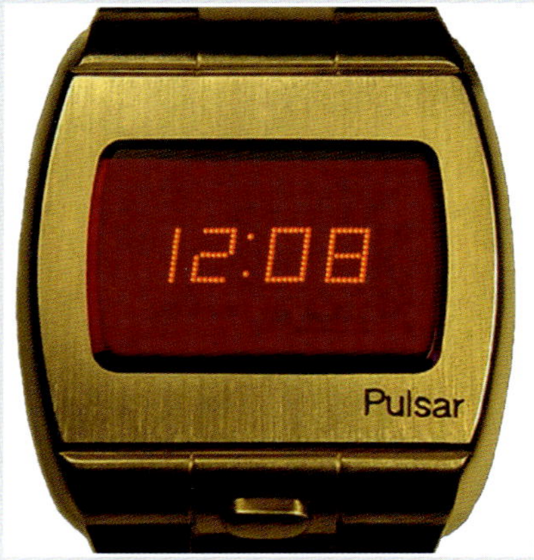

15.30. Pulsar P1 with gold case

15.31. Pulsar module by Time Computer with one IC

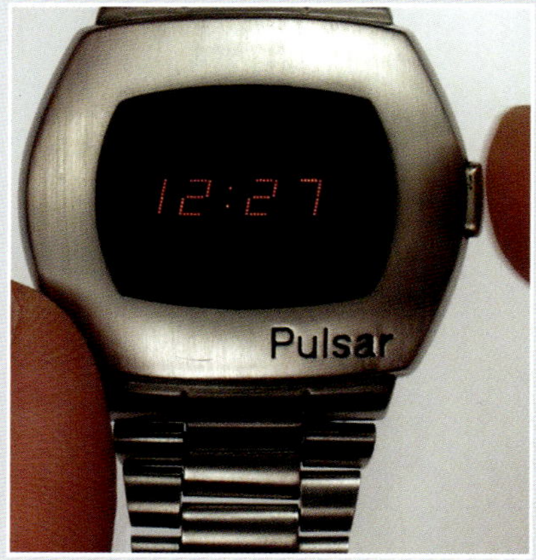

15.32. Pulsar P2

The Pulsar patent license was sold to the most of the important Japanese watch manufacturers, i.e. Citizen, Seiko, Toshiba, and Sanyo. The LEDs were first bought from Monsanto and later from Litronix. Reeves-Hoffman supplied large numbers of quartz resonators. The chips were purchased from RCA. Later, as the demand literally exploded, Hewlett-Packard and other silicon foundries were used as second sources. The ruby glass was supplied exclusively by Corning Glass.

Pulsar Men's Watch Calibers

The designation of the Pulsar modules made by Time Computer is not fully logical; they are described here in chronological order.

44-IC: As related in Chapter 7, the Pulsar story began with six prototypes that featured 44 integrated circuits each. Three of the prototypes were shown to the media during the press conference of 1970 in New York. The functions were hour, minute, and second. The LED was supplied by Hewlett-Packard. The dimensions were 35.6 mm x 26.3 mm; the height is not known, as none of the prototypes seems to have survived.

25-IC: A total of 491 pieces of the Electro/Data caliber with 25 integrated circuits were produced, 450 of which were provided with gold cases and were available in selected stores in April 1972. The LED was supplied by Monsanto. The functions were hour, minute, and second. The dimensions of the rounded-rectangular module were 33.5 mm x 35.6 mm or 36.0 mm x 38.5 mm (including the frame); the height was 7.75 mm.

201A (1972): 34.4 mm x 7.75 mm. All Pulsars with caliber 25-IC that had been supplied by Electro/Data were recalled by Time Computer and replaced by a much more reliable module that had been developed in-house; this was caliber 201A at first, then caliber 201B. A single COSMOS chip by RCA replaced the 25 small integrated circuits that Electro/Data had needed. This module was also used in the Pulsar model P2. The quartz frequency was adjusted with a trimmer capacitor. The watch was set with a magnet that acted on reed relays in the module. The functions were hour, minute, and second.

201B (1972): 34.4 mm x 7.75 mm. This variant of caliber 201A was called "Dual Trimmer" because it featured one of the usual flat trimmers and a second cylindrical trimmer that allowed a very precise fine tuning of the quartz resonator's frequency. Setting was done magnetically, the functions were the same as in module 201A.

15.33. Pulsar P2 Date

15.34. Pulsar case back with setting magnet

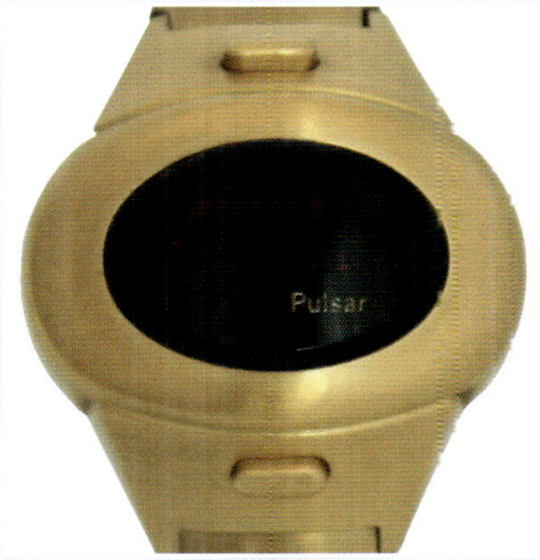

15.35. Pulsar Ladies' Auto Set

15.36. Pulsar Dress model with milanaise bracelet

301 (1973): 34.4 mm x 7.75 mm. The functions were hour, minute, second, and date. Setting was done magnetically. This module was used in the Pulsar P3.

102 (1974): 25.4 mm x 6.2 mm. This round caliber was called "Men's Dress." Time was set with a magnetic pin (Magnet Set). The LED was supplied by Litronix. The functions were hour, minute, second, and date.

104 (1975): 25.4 mm x 6.2 mm. Men's Auto Set. Setting was now done with pushers. The functions were hour, minute, second, and date.

401 (1975): 30.2 mm x 6.8 mm. The functions were hour, minute, second, and date. Setting was done magnetically. Module 401 was the first with the Auto Command function. This meant that time was not only displayed by pressing a key, but also by a flick of the wrist. The date was only available by pusher. This module was used in the Pulsar P4.

601 (1975): 25.4 mm x 20.2 mm x 6.45 mm. This rectangular module was developed for the Pulsar P5 with angular display. Only two prototypes were made, but there was no production. The functions were hour, minute, second, and date.

901D (1975): 33.5 mm x 34.8 mm x 6.63 mm. This was the first Pulsar Calculator watch with a dot matrix display. The functions were hour, minute, second, day/date, and the four basic arithmetic operations.

402 (1976): 30.2 mm x 6.80 mm. The functions were hour, minute, second, date, and auto-command. The Auto-Set concept allowed setting by just touching the key. This module was used in the Pulsar P4.

403 (1976): 30.2 mm x 6.80 mm. The functions were hour, minute, second, day/date, and Auto Command. This module was also used in the Pulsar P4.

406 (1976): 30.2 mm x 6.80 mm. The functions were hour in the 24-hour mode, minute, second, date, and Auto Command. This module too was used in the Pulsar P4.

901L (1976): 33.5 mm x 34.8 mm x 6.63 mm. Pulsar Calculator Watch with 6-digit segment display. The functions were: hour; minute; second; day/date; the four basic arithmetic operations with the possibility of making chain calculations, percentage and margin calculations; and data storage. The C pusher (clear) was on the lower right of keyboard.

106 (1977): Men's Dress. Dimensions and functions were the same as in caliber 104, but the usual orange-red LED were replaced by green ones.

501 (1977): 25.4 mm x 5.98 mm. This module was called Touch-Command Dress. Time was displayed by simply touching and not by pressing the key. The functions were hour, minute, second, and date.

502 (1977): 25.4 mm x 5.98 mm. This module, too, was called Touch-Command Dress. The functions were hour, minute, second, date, and Auto Command.

902 (1977): 33.5 mm x 34.8 mm x 6.63 mm. Pulsar Calculator watch with segment display. The functions were hour, minute, second, day/date, Auto Command, and calculator functions as in 901L. The C-key (clear) was on the upper right of the keyboard.

903 (1977): 33.5 mm x 34.8 mm x 6.63 mm. Pulsar calculator watch with segment display. The functions were hour, minute, second, day/date, calculator functions as in caliber 901L, Auto Command, and Double Flick. The latter function meant that a flick of the wrist switched on the time, while a second flick displayed the date. The C-key (clear) was on the upper right of the keyboard. This is the rarest Pulsar module; most collectors do not even know that it exists. Only three pieces are known. They were used in gold cases made exclusively for Tiffany.

910 (1977): 34.95 mm x 7.92 mm. The Pulse/Time module featured two light barriers for measuring the oscillations of the skin: they correspond with the pulse. The functions were hour, minute, second, date, and pulse.

LCD-Pulsar (1977): 26.7 mm x 5.52 mm. This module was purchased by the Japanese Ricoh. The illuminated, 4-digit LCD also displayed AM/PM. The functions were hour, minute, and seconds by flashing dots. The case was designed by Jean Wuischpard and was waterproof to 30 meters. Time was set with the pusher at 1 o'clock; the pusher at 8 o'clock switched on the miniature light bulb for illuminating the LCD. There was a gold-filled version (5140) and a stainless steel version (3140), both with a braided metal watch band. The Pulsar LCD Dress was the last model produced by Time Computer. After 1977 there was no more market for luxury class digital watches.

Pulsar Ladies' Watch Calibers

101 (1974): 25.4 mm x 6.20 mm. Ladies' Magnet Set with magnetic setting. The functions were hour, minute, second, and date; Hughes supplied the LED.

103 (1975): Ladies' Auto Set. Time was set with the pushers. The functions were hour, minute, second, and date.

801 (1977): 20.6 mm x 5.68 mm. Ladies'/Men's Touch Command.

Pulsar Watch Models

P1: The very first Pulsar. 450 pieces with an 18 carat gold case were sold. They were originally equipped with the 25-IC module by Electro/Data. Toward the end of 1972 the recall started, the objective being to replace the Electro/Data module by the Time Computer module 201A or 201B. Those two modules had been developed in-house by Time Computer and were also produced in its own facilities. "Hamilton" was still engraved on the back of many of the gold cases. They had been ordered that way and it would have been too expensive to replace them. All later models featured "Pulsar" on the front. Model P1 featured only one pusher and displayed hour, minute, and second.

P2: First production model by Time Computer in Lancaster. A key switched on the display of hour, minute, and second. The steel model with the name "Astronaut" was worn by James Bond in the movie "Live and Let Die."

P3: Two keys, four functions, i.e. hour, minute, second, and date. Two case variants exist: the old case with an extra pusher and the new case with two keys (P3 Date Command and Date 2, respectively). The case could be either gold, gold filled, or steel. Biggest production of all Pulsar models.

P4: Auto Command Series. The display was switched on with the key and also with a flick of the wrist. The functions were hour, minute, second, date, and Auto Command. The Big Time model was only made with a steel case. Classic, another model, was available in gold, gold filled, and steel (the gold variety is very rare).

Several different modules were used for the Pulsar P4, including module 401 with a magnet to set the display and module 402 Auto-Set. The display was set by just tapping on the keys. Module 403 displayed day/date in addition to the Auto Command and Auto Set. This module only was available in the Big Time and Classic cases. Module 406 in the Executive style displayed time in the 24-hour mode.

P5: "Angled Display" (September 1974). The LED was at an angle, as is always the case with driving watches. Only two prototypes were made, there was no production.

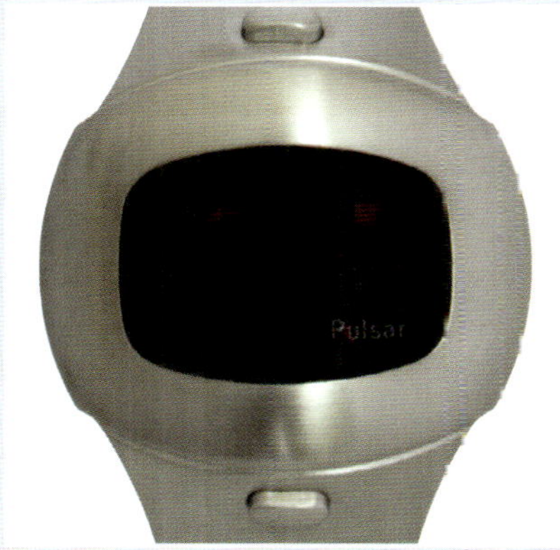

15.37. Pulsar P4

15.38. Pulsar Calculator

15.39. Pulsar Pulse/Time Computer

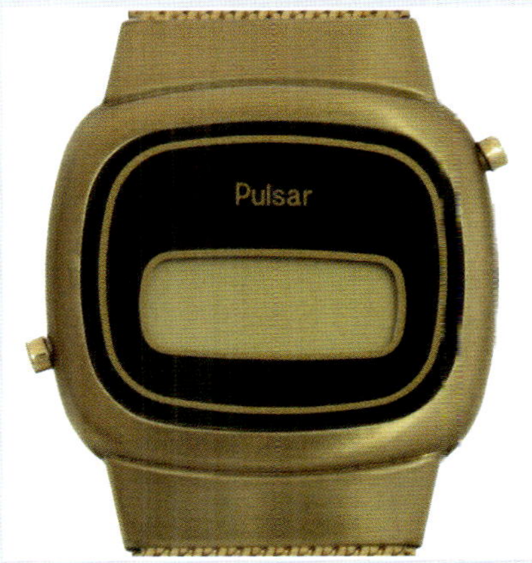

15.40. Pulsar LCD

The Pulsar Calculator and the Pulse/Time model were not given P-numbers. The production of the calculator model (only available in massive gold) was originally limited to 100 pieces, the price was $3950, but due to the big demand, four more series of 100 pieces each were manufactured. They were exclusively sold in the United States. U.S. president Gerald Ford wanted such a watch for Christmas, but his wife thought that it was too expensive. Each piece of this "Limited Edition" was numbered (001 to 500) and delivered with a certificate of authenticity signed by Time Computer CEO, John M. Bergey.

In addition, thousands of Pulsar Calculator Watches with 18K gold or steel cases were manufactured and sold in the United States and abroad. They were not part of the Limited Edition, were unnumbered, and were delivered in a leather "Flip Top Box" with a special pin for actuating the keys of the calculator.

The Pulse/Time Pulsar did not measure the pulse very accurately, but it was an interesting instrument. A total of 100 pieces were made with a solid gold case, somewhere between 10,000 and 15,000 pieces were sold with a steel case.

The P3L ladies' watch was also made in a version with a wider watch band for Asian men with a small diameter wrist, it was called P3L Men's.

Starting in 1976, the competition by American, Taiwanese and Hong Kong companies was felt more and more acutely. Their LED watches cost only a fraction of the price of a Pulsar. Pulsar production, which had grown to a maximum of 150,000 pieces in 1975, shrank dramatically. Only 10,000 pieces were sold in 1977, whereupon Time Computer stopped the production. Between 1972 and 1977, a grand total of 493,500 Pulsars were made.

In Europe, the Pulsar was sold by Pulsar SA, a distribution company in Neuchâtel. The cases were slightly different from the American originals, particularly the ladies' models. The gold and gold filled cases were made by Piquerez in La Chaux-de-Fonds. The European Calculator Watch too was recognizably different from American model.

In Zurich, the Pulsar was sold by Beyer, a famous luxury watch store on the elegant shopping mile Bahnhofstrasse. The American press took notice with spiteful remarks. Quite obviously, the gnomes of Zurich had been passed on the right: superior American technology was bringing the watch industry back to the United States. Many American newspapers reproduced a cartoon showing the Statue of Liberty with a Pulsar on its wrist.

The Fall of the Pulsar

The Pulsar lost its status as a cult object within just a few years. American electronics companies such as Fairchild, Hughes Aircraft, Microma (an Intel subsidiary), National Semiconductor, and Texas Instruments started to mass produce much less expensive digital watches and digital modules with LED, soon also with permanently active LCD. In 1976, 30 million were sold, and a year later that number rose to 42 million.

Between 20 and 30 American manufacturers assembled watches with their own digital LED and LCD modules; some of them were assembled in Central America or Asia. One often reads that there were about 40 manufacturers of digital watches, but this number is almost certainly inflated. On the other hand, at least 50 American companies made watches with their own brands using modules they purchased from the true manufacturers.

The price war between American producers of digital watches was only a foretaste of what would happen in the early 1980s. Prices literally collapsed as Japan, Korea, Taiwan, and Hong Kong started producing digital watches (mostly of the LCD type) in ever increasing numbers. This soon led to single digit retail prices and ridiculous production prices that were well below $1 in 1986.

Long Survival of Time Computer

In July 1977, the Rhapsody jewelry company in Philadelphia acquired the brands "Pulsar," "Time Computer," and "Big Time," as well as the inventory of Pulsar watches from Time Computer. For about a year, Rhapsody supplied Pulsar watches to the big network of retailers in the United States – as long as the supply lasted. About 50,000 Pulsars of all types were sold through these channels. This included many watches that Pulsar SA in Neuchâtel had returned. In this manner, Pulsars with slightly different European cases made their way to the United States.

In 1978, the Californian watch trader David Rosenaur bought the remaining inventory of Pulsars from Rhapsody. In addition, Seiko in Japan bought the "Pulsar" and "Time Computer" brands. At that time, the Pulsar brand was only used for Seiko's LCD calculator watches with an alarm function. Later on, "Pulsar" became the Seiko brand name for medium priced analog quartz watches: this is still the case today. However, the "Time Computer" brand was never used by Seiko. Eventually the American Pulsar collector Dennis Klein re-registered it for LED watches of his own design; only prototypes have been made so far.

The Pulsar Service Center

A company by the name of Time Computer continued operating after 1977. It collected patent license fees from many domestic and foreign companies and used the money to pay for a repair and warranty center in Lancaster. The service center took over the entire stock of Pulsar modules and spare parts. This included the inventories of all the other service centers as well as part of the Time Computer machinery.

The Pulsar Service Center was headed by Eddie Radcliff who had been in charge of the former Time Computer Customer Services. Radcliff and an assistant formed the personnel of the Time Computer Company until 1982. In that year, Radcliff purchased the Service Center and continued operating it as his own company. He also was the U.S. representative of the Seiko Pulsar watches, even though the collection exclusively consisted of analog quartz watches and bore no analogy to the original Pulsars.

Radcliff had "inherited" thousands of steel Pulsar cases which he knew he would never use. He sold them for scrap, but not before he had removed the magnets in the pushers. They were small platinum-cobalt discs that he sold to a platinum recycling company. Hamilton Watch had used such magnets in the balance wheel motor of the "Hamilton Electric." At the time of the Pulsar launching, Hamilton still had a big inventory of such magnets that were put to good use in the Pulsar pushers.

For a time, Radcliff employed 13 former Time Computer employees. They assembled new Pulsar modules on the basis of the available stock of spare parts. Defective Pulsars that were sent to the Service Center were opened in order to replace the complete module. After the spare part stock had been practically exhausted, only one lady was left on the payroll; she helped Radcliff do the repair work.

Eddie Radcliff retired in 1986, but his son Greg took over the business and stayed in Lancaster. He has almost run out of parts and purposefully makes himself unavailable, as he only services Pulsars from the Lancaster vicinity. The Time Computer Company disappeared in the early 1980s. By 1984, HMW had sold all its former divisions, only the name HMW Industries was left; it was finally erased from the registry of commerce.

Hamilton Watch Today

Like Time Computer, Hamilton Watch originally was a subsidiary of HMW Industries. By 1969, it had stopped producing watch movements: it was more economical to import them from Switzerland. The assembly of watches continued in Lancaster in the 1970s, but only partly. SSIH in Bienne, Switzerland, had been the majority shareholder of Hamilton since 1974, controlled it, and finally integrated it into the group. After the ASUAG-SSIH merger in late 1983, all Hamilton Watch buildings were sold. The manufacturing complex in Lancaster, with the two clock towers, was transformed into apartment buildings. Hamilton watches are still available, but they are fully "Swiss made." The headquarters of Hamilton, a company of the Swatch Group, is in Lengnau between Soleure and Bienne.

When Hamilton Watch was sold to SSIH, this did not include Time Computer (TCI), another subsidiary of HMW Industries that never fell into Swiss hands. Today, Hamilton – a Swatch Group brand – considers itself the rightful heir of Time Computer. The 40th anniversary of the Pulsar was celebrated with the Pulsomatic, an LCD watch with a microgenerator. The case is strongly reminiscent of the classical Pulsar designs by Jean Wuischpard.

The Pulsar was a watch without any mechanical moving parts. The Pulsomatic thus "betrays" the original design, as it features the rotor of an automatic watch that winds a mainspring and actuates a microgenerator that charges the battery. This is the power source of a conventional LCD watch movement. Of course, nobody prevents the Swiss Hamilton from doing that, as Time Computer and HMW Industries have long disappeared as business entities. Aside from that, few people know the difference between HMW holding and its subsidiary Hamilton Watch.

Ragen Semiconductors

Roger Riehl, the president of Riehl Electronics Corp. in Troy, Ohio, launched the Synchronar 2100 in 1973. This was one of the earliest LED watches with solar cells. The integrated circuit of the CMOS type was supplied by Ragen Semiconductors in Whippany, New Jersey, that eventually merged with Riehle. Ness Time in Mountain View, California, took care of the Synchronar distribution until Ness went bankrupt in 1976.

The solar cells were arranged on the roof-shaped top of the watch. The display was switched on by a sliding magnet on the right. Pushing backwards displayed hour and minute, pulling forwards displayed date and weekday. A similar sliding magnet on the left activated the display of seconds. The same sliding magnets were used to set the watch. The dot matrix LED was located between the horns in a recess, so that time could be read even in bright sunlight. The dimensions of the rectangular module were 30.6 mm x 30.1 mm x 9.0 mm.

Two button-type nickel-cadmium batteries were charged by the solar cells whenever the light intensity was sufficient. If time was read off the Synchronar twenty times a day, the batteries were fully charged by just ten minutes of sunlight. Several display options were available for the 4-digit LED; the customer could decide himself what priority hour and minute (in the 24-hour mode), day/date, or second should have.

Riehl's vision was a highly accurate multifunction watch with a very good contrast under all circumstances: it would never require any service work. It was designed to be insensitive to extreme temperature variations, and was shockproof and waterproof even at great depths. In other words, Riehl's ambition was the ultimate watch that would last a lifetime. This justified the very high price of $500. The calendar was programmed to the year 2100. However, Riehl knew very well that his watch would not last ten years. The weak points were the rechargeable nickel-cadmium batteries and the solar cells that do not last for ever either.

The first version of the Synchronar was launched in 1972; the models Mark II, Mark III, and Mark IV were available in 1975, 1977, and 1981, respectively. Mark IV was characterized by the following functions: hour, minute, second, second time zone, eternal calendar until 2100, programmed change from summertime to wintertime and vice-versa, alarm, stopwatch, and flashing timer. The owner could set the rate of his watch within ± 8 seconds per year.

The case of the original Synchronar consisted of the plastic Lexan. It was perfectly sealed; there were no openings at all. Riehl assumed that the nickel-cadmium buffer batteries would never need to be replaced, which of course was wrong. In order the replace the batteries, the case had to be cut open. Then one would find out that the module was completely filled with a silicone sealer.

Switching on the LED, changing the functions, and setting the display were done with sliding magnets actuating switches inside the case. Not everybody liked the feel of plastic, which is why a thin sheet of stainless steel was glued to the back of the case in later models. All Synchronar watches featured a thin metallic cap that covered the case (except for the openings, particularly for the solar cells). This cap was either stainless steel,

15.41. Ragen Synchronar LED Mark II

15.42. Texas Instruments LED caliber 100, front side

15.43. Texas Instruments LED caliber 100, back side

15.44. Texas Instruments module 8251

titanium, or brass with a black PVD-coating; it could also be gold plated. The design of the case was so original that Riehl was granted a patent for it. The Mark IV of 2002 still looked exactly like the 1972 original Synchronar. Unfortunately, the quality of the lugs and watch bands was not very good, which could lead to the loss of the watch. The magnetic actuators, too, were not sufficiently sturdy.

The Synchronar 2100 soon established itself as a classic and was manufactured for several decades, which is unique in the history of American digital watches. When he passed away in 2005, Roger Riehl was still working on the Mark V that was supposed to feature many new functions. It is unfortunate that many collectors and Synchronar fans lost their watch when they tried to have it serviced by Riehl. In old age, Riehl accepted watches for repair and pocketed the fee that had to be paid in advance, but never returned the watches. Instead, he dismantled them and used the parts for the development of his Mark V. Riehl's son Barry is known to have sold many of those parts on eBay. The second son Howard helped his father in the workshop, but disappeared in 2005. The total Synchronar production is estimated to have been around 50,000 pieces.

Siliconix

Siliconix Inc. in Santa Clara, California, was an important semiconductor producer in the mid-1970s. The company diversified into digital watches when it acquired Harper Time & Electronics in Mountain View, California. Siliconix continued manufacturing the digital stopwatches Harper had developed, but discontinued the wristwatch LCD modules.

Suncrux

In the mid-1970s, Suncrux in Cupertino, California was an important supplier of liquid crystal displays and 3½-digit LCD modules. The latter were smaller and flatter (e.g. 24.0 mm x 4.9 mm) than what was more or less standard in the industry. Suncrux also used 32 kHz Japanese NDK tuning quartz resonators and CMOS integrated chips that incorporated all the electronic functions of the module, i.e. oscillator, divider, and decoder.

In the early days, the display was conventionally set with pushers. Certain models made in the mid-1970s featured a rotating crown with four stable positions. With caliber DM 401, Suncrux returned to the universally used pushers in late 1976, as consumers were used to them. Suncrux modules were very accurate; some of them would be certified as electronic chronometers by the Swiss COSC, even though the quartz resonators were not thermocompensated.

The construction of caliber DM 401 was characteristic for Suncrux and some other brands. The 0.25 mm thick steel bottom plate had openings for the battery, the trimmer, and a test contact for checking the electronics. Four pillars connected it to an epoxy resin disc with the electronic components, the negative battery connector, and the contact springs for the pushers. Another plastic disc that was fastened to the pillars with screws carried the liquid crystal display. This construction was very stable and sturdy, which was characteristic for all Suncrux modules.

Caliber DM 600 was launched in late 1978 and featured the usual five functions; the dimensions were just 20.2 mm x 4.5 mm: this caliber was even more compact than previous models. It seems strange that Suncrux – a truly progressive company – used mechanically cut tuning fork 32 kHz quartz resonators with fine tuning by trimmer capacitor at a time when the photolithographic Statek resonators were already being mass produced and electronic frequency tuning was well known. Illumination of the 3½-digit LCD with a minibulb also looked archaic, but presumably Suncrux had substantial stocks of those items and needed to get rid of them. On the other hand, Suncrux used the newest generation of integrated circuits so that the current consumption of its modules was only 2.4 microamperes – a remarkably low value at that time.

In 1975/76, Suncrux launched a watch with a pseudo-analog display and sold it under the La Croix brand. This means that the hands of an analog watch were simulated on a liquid crystal display. Texas Instruments patented this concept in 1979, but Suncrux had come up with a very similar design much earlier. Suncrux also developed a similar ladies' watch. Only small series were produced, as watches of this kind had a rather low acceptance, mostly for esthetic reasons. It is simply not possible to simulate real, good looking hands on an LCD. An interesting aside is that Timex had a strong patent for "analog LCD displays" that were marketed as "Solid State Analog." The major stockholders of Suncrux were the Japanese companies Toshiba, Dainippon Toryo, Dainippon Printing; they controlled the company. The American electronics company VBC was a minority stockholder.

Texas Instruments

The Texas Instruments Company (TI for short) was established in Dallas, Texas, in 1941. It achieved international fame by launching the world's first transistor radio in 1954. However, the invention of the integrated circuit by Jack S. Kilby four years later was much more significant. The consequence was that Texas Instruments developed into one of the most important producers of silicon chips. As far as consumer products were concerned, TI became known among scientists, engineers, and students as the supplier of calculators. They are still available in many different styles and with a mind-boggling diversity of functions. In 1973, Texas Instruments did what most of its competitors did: it launched a five-function digital watch module with LED displays and eventually switched to the LCD technology.

Texas Instruments started its "watch adventure" with the 3½-digit LED module in the TI watch model 100 that was sold under the Wells brand. There were just three functions: hour, minute, and second. The dimensions were 14 lines x 7.3 mm. The display was set with a two-position switch on the back of the case: H stood for hours, M for minutes. The actual setting was done with a pusher. In 1974, Texas Instruments launched the models 101, 102, 103, and 104 that also displayed month and date. The dimensions were 27.5 mm x 6.6 mm. Two button cells were integrated into the module.

In the early models, Texas Instruments still used bar-shaped, flexural mode quartz resonators, but they were eventually replaced by tuning fork-shaped resonators supplied by the Japanese NDK. The watch model with a steel case was first to break the $100 barrier – on the way down of course. In 1975, TI launched three even more economical models (500, 600, and 700), all of them with the same LED module with a diameter of 20.2 mm. Next to hour, minute, and second they displayed date and month.

For the 100-type models, the intensity of the light emitting diodes was controlled in relation to the intensity of ambient light by means of a light sensitive resistor. Two batteries were provided, one each for the electronic circuit and the LED. TI produced a total of 14 models under the Texas Instruments brand, all of them with five functions and perpetual calendar. The rate was guaranteed to be less than a minute a year.

In mid-1976, prices eroded rapidly: the retail price of LED watches with a plastic case dropped to $19.95. Modules for such watches with a diameter of 26.1 mm and a height of 8.3 mm were produced in enormous quantities and were extremely simple. They consisted of a printed circuit with the CMOS type integrated circuit and an NDK tuning fork quartz resonator that were potted in plastic. Recesses lodged the two batteries, which were connected in series, and the trimmer. When the pusher was depressed, the 3½-digit LED displayed hour and minute, then the second. Pressing twice displayed month and date. Another, recessed pusher was needed for setting the display.

Several factors contributed to Texas Instruments' success with digital watches. Most importantly, TI produced its own integrated circuits and concentrated all electronic functions on a single CMOS chip in I^2L-technology. Texas Instruments also rigorously enforced a very tight quality control program; this resulted in a very low percentage of warranty claims. For the brands of other companies, 30 to 50 percent was not unusual.

In 1976, TI launched a new LED watch collection with 23 models; retail prices ranged from $19.95 to $29.95. Seven of them featured metal cases (TI-401/402); the other 16 models sported brightly colored polysulfone cases (TI 501/502). There were numerous variants in the 400 model line, but all were the same size: 28.0 mm x 6.4 mm. The architecture also was very similar. For example, the base plate of caliber 451-2 was plastic with a recess for the battery and the quartz resonator. The 3½-digit LED with a dot matrix and the integrated circuit were soldered to the printed circuit on an epoxy resin base.

Pushing the key once displayed hour and minutes; pushing twice displayed month and date. A continuous pressure started the seconds. A recessed pusher was used for setting the display. Motorola supplied the bar-type quartz resonator.

TI engineers were well aware of the disadvantages of LED watches and were searching for alternatives. A brand new watch project was started on the basis of the twisted nematic liquid crystal technology in collaboration with the Japanese Hitachi. With an experienced partner on its side, Texas Instruments hoped to be able to launch an interesting product in a very short time. And so it happened: in early 1978 TI launched the "LCD Quartz" collection with a 3½-digit display of hour and minute or month and date. In the countries where it was permissible, the LCD display was backlit with tritium gas light cells.

In addition to its conventional LCD modules and still in 1978, TI launched an alarm chronograph with the functions time, AM/PM, day/date, month, alarm, and stopwatch to 1/10 of a second. Another LCD chronograph with a resolution of 1/100 of a second was sold under the Texas Instruments brand, but actually the complete watch was supplied by the Korean HanDok company. It was owned by Duk Cho, a very interesting player in both digital and analog watches until the mid-1980s (see Chapter 14).

Still in 1978, TI purchased the module of an alarm watch that seemed to have been inspired by the Seiko caliber 159 of 1977, but was made by HanDok in Korea. The module's dimensions were 29.3 mm x 7.9 mm and it used a tuning fork quartz resonator of 3 mm x 8 mm that probably was supplied by the Japanese NDK. The integrated circuit was supplied by Toshiba; it also generated the alarm signal. The transducer was a 9.4 mm x 2.0 mm cylindrical capsule that was mounted above the LCD. The latter was of the 6-digit type (for displaying hour, minute, and second); a black bar underlined the current weekday. The display was illuminated by two tritium gas light cells or a miniature light bulb.

Normally, the LCD displayed the time. Pressing the upper key on the right displayed date and month. The stopwatch function (up to 30 minutes with a resolution of 1/10 of a second) was activated with the lower key on the right; the LAP function was also available. Alarm time (hour and minute) was selected with the left hand side key; it was armed with the keys on the right. The 55 dB alarm signal lasted 10 seconds, but could be stopped immediately with the lower key on the right.

Texas Instruments evolved into one of the largest producers of liquid crystal displays of the TN type. The leakage problem was solved by fusing the edges of the two glass plates with the help of a low melting glass powder. This was a fundamental breakthrough and launched LCDs to their leading position in digital watches. The basic problem with liquid crystal displays of the TN type was that the polarizing filters absorbed 50 to 75 percent of the incoming light. The consequence was that liquid crystal displays always gave the impression of being "dark," which many people do not like. This problem was solved quite recently with the "electronic paper" displays, which are reflective rather than refractive devices.

Yet, TI was very successful with its TN LCD watches, at least until the end of the 1970s. This prompted the company to manufacture a wide range of LCDs in many different sizes for non-watch applications. In addition to the conventional digital LCD, TI developed a pseudo-analog display. Radially arranged LCD sectors simulated the hour and minute-hands of a conventional analog watch. The watch with this kind of display was called "Starburst," and a presentation was made at the Basel Watch Fair in 1979. The dimensions of this round module 30.5 mm x 7.6 mm. Energy was supplied by a 3-volt lithium-manganese battery in the case back.

The 32 kHz tuning fork-shaped quartz resonator was supplied by NDK. Normally, hour and minute were displayed; the other functions – minute and second, date and weekday, and stopwatch for 24 hours with a resolution of 1/10 of a second – were activated with three pushers. The pseudo-analog display needed 120 connections and was esthetically unpleasing. Furthermore, watches with this type of display cost nearly $300 apiece – an outrageously high price for a digital watch.

Still another type of radial LCD display was not significantly better: the hour scale consisted of wedges, an additional one was opened every ten minutes, while the minutes were displayed by a simulated hand that reached the periphery of the dial. Day and date were displayed digitally by pressing a key. For some years, Texas Instruments produced up to 400,000 liquid crystal displays per month; 8 to 10 different types were available. Most of them were simple models with 3½ or 6 digits.

15.45. Texas Instruments LCD, back side

15.46. Timex LED SSQ, front side

15.47. Timex LED SSQ, back side

15.48. Timex LCD SSQ, back side

During the short boom phase of digital watches, Texas Instruments grew into the third or fourth largest American producer of watch modules. The company was determined to be in business with a wide variety of watches with many different functions, from the calculator watch all the way to the jewelry watch. This probably was overly ambitious, as liquid crystal displays simply are not esthetically appealing. The introduction of phase filters for color effects and of integrated circuits generating constantly moving geometric patterns did not save the TI watch division either.

It finally became obvious to Texas Instruments that there was no market for high-priced digital watches – exactly what Time Computer had already experienced with the Pulsar. In the long term, digital displays found their niche in the sector of multifunction watches and low-priced watches. Texas Instruments was one of the last American electronic companies to abandon the watch business, and it did it with a bang in 1983: all 2800 employees of the watch division were terminated.

Timex

Timex is mostly known for its quartz analog movements. Yet, in the 1970s, the company was significantly involved in the digital watch business (LED and LCD type). The Irvington (New York) Timex laboratory was responsible for the development, and components were mostly bought from outside sources while the complete watches were assembled in Taiwan.

As far as LED watches were concerned, Timex launched three men's and one ladies' caliber between 1973 and 1976. Two serially connected button-type batteries were the energy source. Upon pressing a key, the men's watch calibers displayed hour and minute, or date and weekday, or seconds. Occasionally, Timex used LED calibers that were purchased from Hughes. Success was not overwhelming: customers thought that a Timex watch for $125 was just too expensive. This appreciation did not change significantly when the price was lowered to $85. About 50 percent of the Timex LED watches were returned under warranty, which did not improve sales either.

In 1974, Timex made a new start with a collection of LCD calibers. Very large quantities were assembled in Taiwan first, in the Philippines later (Manila, then Cebu). Bar-shaped quartz resonators from Korea were used in the beginning, but were replaced by the photolithographic Statek-type as soon as it became available. The CMOS chip also was purchased in Korea. The 3½-digit illuminated LCD displayed hour, minute, and AM/PM. It was made in a former RCA production unit in New Jersey, which Timex had acquired. Timex digital watches with an LCD display were marked "SSQ" (Solid State Quartz).

15.49. Uranus LED ladies' watch caliber

15.50. Uranus LED Message

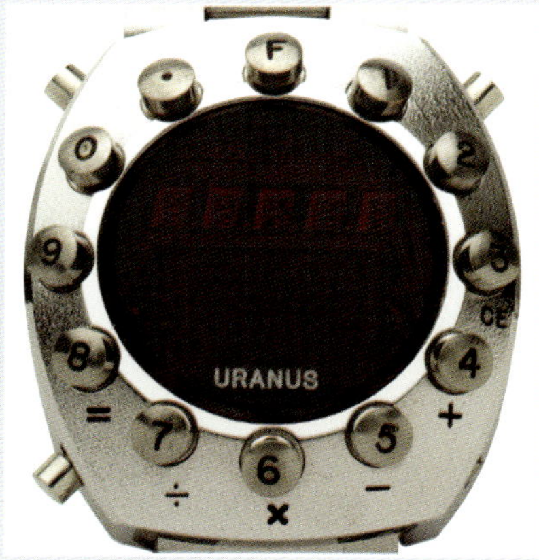

15.51. Uranus Solar Calculator LED

15.52. Uranus Solar LED prototype

By 1983, all the electronics companies that had participated in the digital watch boom of the 1970s had given up. The low-cost digital watch survived quite well, but it had become an almost exclusively Chinese and Korean product. Timex managed to survive in this business with highly specialized niche products. More particularly, it launched the "Triathlon" in 1984. This was the first mass produced digital watch with a programmable microprocessor. It was followed in 1985 by the Timex "Ironman" that featured all the functions sportsmen and non-sportsmen wanted, i.e. hour, minute, second, day/date, month, and stopwatch with up to 8 intermediate times. For a while, the Ironman was the most successful watch in the United States. Both the Triathlon and the Ironman were initially poduced by HanDok in Korea and then moved to the Timex Cebu plant in the Philippines.

Uranus

The Uranus Company in Port Chester, New York, is of particular interest. To collectors of LED watches it is one of the great favorites. Unfortunately, precious little information is available on Uranus. It was one of those small companies that hoped to gain a foothold in the field of digital watches. What is remarkable is that Uranus developed an LED watch practically simultaneously with the Pulsar of Time Computer. However, Uranus was late in filing patents, and, by the time it did, the Time Computer design was already patented. It would have been impossible to launch such a watch without infringing on the Hamilton or Time Computer patents.

Only prototypes were made of the Uranus LED modules. They were presumably developed in 1971-72 and obviously were ahead of their time. Two kinds of prototypes are known, one of them with solar cells. This module featured no less than 53 small integrated circuits. They were arranged on both sides of a pertinax plate and interconnected with hundreds of bonding wires. Needless to say that this system was very fragile; only a few of the surviving prototypes are still in working order; defective units are unrepairable.

Uranus also designed a calculator watch; two models are known with a round or octagonal case, respectively. Solar cells extended the lifetime of the silver oxide battery; the alternative was a nickel-cadmium battery that was charged with a simple power supply. Both models were available with either solar cells or the rechargeable battery.

Manufacturers of Digital Quartz Watches

The following list (certainly incomplete) covers American companies that assembled watches of their own brand, using modules or parts purchased from outside vendors. Most of them did not produce any parts themselves.

Advance (Redford, MI)

Advance was primarily a digital clock company and used LED modules with 5 or 6 functions, mostly from Hughes Aircraft in its wristwatches.

Alcor (Flushing, NY)

Alcor watches used LED modules with 3, 5, or 6 functions. Those modules were purchased from Hughes Aircraft.

Armitron (Long Island City)

The Armitron watch brand was launched in 1956 by the E. Gluck Corp. in Long Island City, New York. Today, Armitron is among the ten most important watch brands in the United States. The most recent catalog lists only analog watches with quartz or automatic movements. However, in the 1970s Armitron had a big choice of digital LED watches: the modules were purchased from HanDok and Litronix. Around 1977, the market for watches of this kind collapsed, but Armitron rapidly switched to LCD digital watches. The required modules were purchased on the OEM market.

Arnex Quartz (Carlstadt, NJ)

Arnex used LED modules with 2, 4, 5, and 6 functions that were purchased from Hughes Aircraft.

Avatar (Cupertino, CA)

Avatar used LED modules with 2, 4, or 5 functions that were made by Hughes and Litronix.

Benrus (Ridgefield, CT)

Benrus (the abbreviation for Benjamin Lazarus, the company's founder) assembled watches using LED modules with 4, 5, or 6 functions. Starting with 1973, those modules were purchased from Texas Instruments. Later, Hughes was the preferred supplier. The Benrus Pop-Up Watch with a vertical display was quite unique. Normally, the watch was integrated into the watch band and was invisible. Pressing a key made the display pop up; the LED was switched on at the same time.

Bulova (Jackson Heights, NY)

The LED watch model Bulova Computron with caliber 2403.10 was launched in the United States in 1973. Pressing the key for a short time displayed hour and minute. If the pressure were maintained, the display switched to seconds. The Accuquartz Digital LED 28 was presented at the Basel Watch Fair in 1975; it used a Hughes caliber. The dimensions were 29.34 mm x 7.0 mm, the quartz resonator was of the bar type with a frequency of 786,432 Hz. The functions were hour, minute, second, month and date.

Caliber Bulova 244 launched in 1975 featured a 3½-digit LED. The module was supplied by the Korean Handok. The ladies' watch caliber 2535.10 was also presented at the 1975 Basel Watch Fair. It displayed hour, minute, second, date, and month.

Still in 1975, Bulova launched a LCD watch with an illuminated display, but only a small series was produced (caliber 2260). The numerals for hours and minutes were separated by an arrow flashing at the 1 Hz rate. The case was similar to the one Bulova had used for its first LED watch and was quite complicated. In 1976, Bulova added another LCD model with the designation 099A; it was mass produced at the Bulova Bienne plant using a 6-digit module. The functions were hour, minute, second, date, and month. A second Swiss-made model, that was significantly less costly, featured a 3½-digit LCD .

The Bulova Computron 229 started a whole series of calibers with a vertically mounted LED in desk-shaped watches. The module was nearly square, the edge length was 26.6 mm, the height 7.80 mm. Bulova caliber X was launched in 1976. It featured a vertically mounted 3½-digit LED. Caliber Y (1978) was similar, but it featured a 6-digit LED. Both calibers displayed hour, minute, second, day/date, and month.

Chronex (Garland, TX)

George Thiess established two new companies after Electro/Data fired him 1971 – even though he had been one of the founders. Thiess had hired Willie Crabtree, his most important colleague, who at that time was working with Texas Instruments. Unfortunately, Thiess had no controlling majority in his own company, but depended on external investors. As often happens, the majority shareholders got rid of the initiator with the result that Electro/Data was insolvent a year later.

Immediately after leaving Electro/Data, Thiess established Electro Research Inc. It manufactured LCD digital quartz table clocks, wall clocks, and bracket clocks. The display was illuminated.

In 1973, Thiess founded still another company called Chronex. It manufactured LED watches with gold, silver,

15.53. Bulova Accuquartz LCD caliber 2260, back side

15.54. Bulova Driver LED, caliber 2544-10

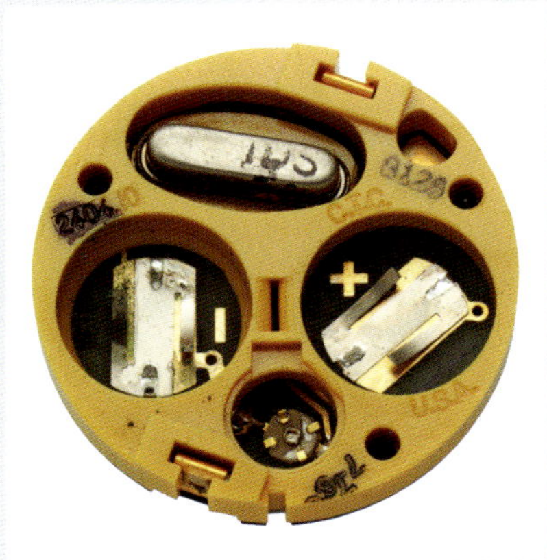

15.55. Bulova LED caliber 2404-10, back side

15.56. Gillette LED chronograph, back side

and steel cases. A pendant LED watch for ladies, that was worn on a gold chain, featured a mercury switch. If the watch was tilted by 90 degrees in the horizontal position the display was switched on. The Chronex company still exists as a subsidiary of the department store chain Sears-Roebuck; it services watches that were bought at Sears.

Collins

It seems that Collins bought one of the divisions of Electro/Data, when it became insolvent. The modules used by Collins were supplied by Hughes. Collectors know about a 6-digit stopwatch, a 6-digit alarm watch, and 3½-digit LCD watches.

Concord

Concord is known as one of the "launching carriers" of the (analog) Delirium (it was developed by ETA) with a height of just 1.98 mm for the complete watch. Concord was also active in digital watches and used four different LED modules with 5 or 6 functions. The modules were supplied by Hughes.

Cronex (Santa Clara, CA)

Cronex used a five function LED module that very probably was supplied by Hughes.

Croton Time, (Croton-on-Hudson, NY)

Four different LED modules with 5 or 6 functions were purchased from Fairchild.

Datatime (Palo Alto and Sunnyvale, CA)

In the 1970s, Datatime mandated the production of both LED and LCD modules in Hong Kong and had them cased there with its brand name. The dimensions of the modules were 29.4 mm x 6.1 mm. The 32 kHz bar-type quartz resonator was supplied by Motorola, with power provided by two serially connected silver oxide batteries. The CMOS chip with 39 connections was marked Datatime and took care of all the electronic functions. The 4-digit LED or LCD displayed hour and minute constantly; pushers were provided for switching to day/date or minute and seconds.

The Datatime "Monte Carlo" model, with a game function, was most unusual. Pushing the key on the left side of the case displayed GAMES instead of hour and minute. Each time the key on the right side was pushed, a randomly generated combination of three numerals from 1 to 6 appeared. Any combination of the three same numerals such as 111, 222, 333, etc., counted as a Jackpot. The DICE configuration used only two numerals. Finally, the ROULETTE configuration generated numbers from 0 to 36, as in "real" roulette. The time functions continued normally and could be switched back on at any time.

Duratime

Duratime watches were available with four different LED modules with 5 or 6 functions and with an LED/LCD module with 4 functions. They were presumably purchased from Fairchild Semiconductor.

Elgin (Elgin, IL)

The old Elgin Company had declared bankruptcy in 1965. In the 1970s, the brand was owned by Elgin National Industries, an engineering company. Watches were either purchased from private label vendors in Switzerland or (mostly) from Waltham in Massachusetts, which was itself partly owned by the Swiss Group SGT (see Chapter 8). Elgin is known to have used five different LED modules with 4, 5, or 6 functions, as well as one LED/LCD module with 4 functions. All of those modules were supplied by Fairchild Semiconductor.

Gillette (Boston, MA)

In the 1970s, the Gillette Company, which had been known mostly for its razors and razor blades, was fully determined to develop into a highly diversified industry group. In order to be at the forefront of technology, Gillette established two R&D laboratories. One of them was located in Los Angeles, California, and was headed by William Christopher. The second laboratory was in Neuchâtel, Switzerland; its director was the former CEH general manager Roger Wellinger.

In order to avoid problems with the Swiss watch industry, Wellinger stipulated in his contract that he would have nothing to do with electronic watches. The consequence was that Gillette LED watches were developed in Los Angeles. The first model was called "Gillette Reflex" and was launched in 1976. The watch was very plain looking; the module (29.4 mm x 8.6 mm) was purchased from Integrated Display Systems (IDS) in Pennsylvania and the watch was assembled in Mexico.

The module was equipped with the usual two button cells and a photolithographic 32 kHz Statek-type quartz resonator; the CMOS integrated circuit (probably) was supplied by Solid State Scientific (SSS). Pushing the key on the upper right displayed hour and minute, then second and date. A position switch could be activated so that the display was switched on whenever the watch was tilted in the forward direction.

The architecture of the module was quite conventional; the batteries were lodged in recesses at the rear, a 0.7 mm ceramic disc was glued to the front with the printed circuit, the integrated circuit, and the display. There was no trimmer as fine tuning of the quartz frequency was done electronically.

Another Gillette LED watch was launched in 1977. The case was quite original: a narrow strip of ruby glass using the whole width of the watch case concealed the LED; the display was activated by pressing a key. The module was again purchased from IDS and the functions were the same as those of the previous model, including the position switch. The electronics were completely potted in plastic. Time was set by accelerating or slowing down the rate of the seconds.

The construction of the position switch was quite interesting. The sensor was a metal ball that contacted two springs. If the wrist was moved within a certain time to an exactly defined position, the integrated circuit responded by switching on the display. This avoided many unwanted and battery draining activations of the display.

Helbros (New York, NY)

Helbros used LED modules with 4, 5, or 6 functions, one of them with an alarm, as well as an LED/LCD module with 4 functions. There was also a single LCD module with 2 functions. All of those modules were supplied by Hughes.

Innovative Time (Irvine, CA)

The brand used four LED modules with 5 or 6 functions, presumably from Hughes.

Jupiter Time Corp. (New York, NY)

The watches of this brand were equipped with four different LED modules with 5 or 6 functions, plus an LED/LCD module with 4 functions: all of those modules were probably supplied by Hughes.

Majesty

The majesty watches used four different LED modules with 5 or 6 functions. Hughes supplied those modules

Marcel Watch Co. (New York, NY)

The Marcel brand used seven different LED modules with 3, 4, 5, or 6 functions and one LED/LCD module with 4 functions. The origin of those modules is obscure, but Hughes is a good guess.

Mercury Time (Old Bethpage, NY)

The brand used five different LED modules with 5 or 6 functions. Their origin is not known.

Microsonic Digital (Philadelphia, PA)

Microsonic Digital used 8 different LED Modules with 2, 4, 5, or 6 functions including a module with the stopwatch function.

Quantum

Quantum was one of the brands used by the Gluck Company (see the section on Armitron).

Saturn (New York, NY)

Saturn used just one LED module of unknown origin with 4 functions.

Savant (Dallas, TX)

Savant was a watch brand of the module manufacturer Micro Display Systems MDS. It used MDC's LCD watch modules with 4, 5, or 6 functions and two chronograph modules with 4 or 6 functions.

Sensor (Dallas, TX)

Sensor used seven different LCD modules with 4, 5, or 6 functions including two chronograph modules with 4 or 6 functions.

Speidel (Providence, RI)

This division of TEXTRON (Providence, RI) was well known for its expansion watch bracelets and watches with a 5 function LED module and an LCD module, also with 5 functions.

Stanford Scientific

Stanford Scientific manufactured a pulse watch with two LCD in the 1970s. In the normal mode it displayed hour and minute on one of the LCD; pressing a key displayed the pulse frequency on the second LCD. The case design was copied from Heuer's Ford watch; Stanford Scientific never worried about licensing.

Timeband (Palo Alto, CA)

Timeband was one of the watch brands of Fairchild Semiconductor. It used a Fairchild LED module with 5 functions as well as an LCD module, also with 5 functions.

Unitron

Unitron is known for watches with two different LED modules with 5 or 6 functions, one of them with an alarm.

Waltham

Waltham Watch in Waltham near Boston made tremendous contributions to the American World War II effort, but missed the return to civilian watch production. Waltham was insolvent by 1957, which was the year the contact controlled electromechanical Hamilton Electric was launched commercially. The brand survived and collectors know of Waltham watches with a 5 function LED module. Waltham also used two LCD modules with 2 or 5 functions, respectively.

One must be aware of the fact that the Waltham brand was sold separately in Europe and the United States. In Europe, Waltham was one of the brands of the SGT watch group that declared bankruptcy in 1981. The Waltham brand was then acquired by the Heiwado Trading Company that mandated the production of Swiss-made luxury class Waltham watches. They were mostly exported to Japan where the brand has an excellent reputation. Many Japanese remember that their National Railroads (JNR) exclusively used Waltham watches until the end of World War II.

For North America on the other hand, the Waltham brand was purchased by the M.Z. Berger company that sells low-cost Asian Waltham watches to chain stores and department stores. Almost inevitably, Waltham LCD watches appeared on the American market in the course of the 1970s. More particularly, type LCD/3001 was launched in 1975, with an illuminated 4-digit LCD. It displayed hour and minute alternatively with month and date. Pushing a key displayed minute and second. The origin of this module is unknown.

Water Watch (El Paso, TX)

A rather odd thing appeared towards the end of the LCD euphoria: the Water Watch. It was invented in 1977 by Ventu Research in El Paso, Texas, and produced by Ching Cheong Electronics Ltd. in Aberdeen (Hong Kong): Patric Chau was the general manager. Power was not provided by the usual button-type battery, but by three zinc/manganese oxide cells connected in parallel. They were visible in the upper part of the polystyrene case. These were bore-holes into which zinc and manganese dioxide powders were pressed. It is assumed that a small quantity of ammonium chloride was added; in the presence of water, three classical Leclanché cells were formed.

The Water Watch was bought in the dry condition there was no buffer battery. In order to start it, a few drops of water had to be added into lateral openings that could not be closed. It seems that the Water Watch could also be started with beer, wine, coke, orange juice and even urine – any water-based liquid would do. Inevitably, this led to a lot of fun in bars and at parties. Once it had been started, the Water Watch ran for a long time, as the digital quartz module with a 3½-digit LCD needed only a few microwatts. If the cells dried out, adding water restarted the watch.

The case was made of transparent polystyrene with a black back; it was slightly angled in the 6-12 direction. Paradoxically, the Water Watch was not waterproof: it failed in many cases because water got into the module. It did not look particularly attractive, but, after all, who would argue: the factory price was only $4.50 apiece. It was an amusing toy and, several million Water Watches were sold well into the 1930s.

Westclox (Peru, IL)

The United Clock Company was founded in 1884 in Peru, Illinois, not far from Chicago. It had to declare bankruptcy twice before it developed into the world's largest producer of clocks. The corporate name became Western Clock Co. and the brand it adopted in 1909 was Westclox. Its products were wall clocks, table clocks, and alarm clocks of the Big Ben and Baby Ben Type. After the merger with the clock manufacturer Seth Thomas, the holding General Time was created in 1930. Wristwatches were added to the product mix in 1936, but they were of minor importance. In 1948, the 200 millionth timepiece was sold; in 1961 an additional plant was built in Mexico.

In the 1970s, Westclox switched to quartz movements for its clocks. It also started manufacturing digital wristwatches that were assembled in Canada. For example, Westclox launched the Quartzmatic LED ladies' watch in 1975; the functions were hour, minute and day/date. The case and the integrated watch band were made of steel. The pendant model featured a gold case and was worn on a gold chain.

In the same year, Westclox launched a digital men's watch with a 3½-digit LCD of the dynamic scattering type that was supplied by Optel. The functions were time and day/date, and it was also called Quartzmatic. The origin of the module is unknown; it is noticeably different from the widely used modules by Microma and Optel. The company moved to Georgia in 1979, but the competition from the Far East was getting fiercer all the time. In 2001, General Time closed all of its production facilities; six years later it even sold the Westclox brand and completely disappeared.

Windert (Los Angeles, CA)

Windert manufactured LED watches with no less than 13 different modules with 2, 3, 5, or 6 functions, including one with an alarm and one with a stopwatch. Windert also used an LED/LCD module with 4 functions supplied by Hughes.

Wittnauer (New Rochelle, NY)

In the Lady Polara model, Wittnauer used two LED modules with two digits and just 2 functions; they were made by Hughes in Mexico. The men's model Polara Quartz Digital was equipped with a module with 5 functions supplied by Pulsar Time Computer

Analog Quartz Calibers

Benrus

The Benrus Watch Co. Inc. (short for Benjamin Lazarus) was established in New York in 1924 as an import and trading company for Swiss watches. A subsidiary in La Chaux-de-Fonds purchased ebauches from ETA, Fontainemelon, A. Schild, Unitas, Venus, Peseux, Felsa, Recta, Imhof, Looping, Valjoux, and Aurore. The parts were nickel plated, after which movements were assembled and cased on the spot. By 1955, Benrus New York imported about a million watches per year with its own brand for distribution in the United States. Benrus thus was a classical "etablisseur" that produced Swiss watches for export to the United States.

In 1973, Benrus made a start in the quartz analog technology with caliber H 010. It was based on the mechanical, $11\frac{1}{2}$-line ebauche ETA 2620 with day/date, but was transformed into a quartz analog movement with a height of 7.0 mm using Motorola components, i.e. a 32 kHz bar-shaped quartz resonator, a trimmer, a CMOS divider circuit, and a stepping motor.

Those components were mounted on a small, black plastic plate with a printed circuit and formed the electronic module of the caliber. This Benrus quartz movement was comparable to caliber 2823 by Seiko or with the Riquartz 550 by Ricoh. It was used in the "Technipower" watch models. An LED digital watch was also part of the Benrus product mix; it was based on the Texas Instruments module 100.

Bulova

Being the sole owner of the patents for the "Accutron" tuning fork watch, Bulova did not expect that this revolutionary movement would be replaced by quartz movements within just a few years (see Chapter 5). However, Bulova did take notice of this development and was a shareholder of the Centre Electronique Horloger (CEH) in Neuchâtel. The first Bulova quartz model was presented at the 1970 Basel Watch Fair, equipped with the CEH caliber Beta 21.

In mid-1974, Bulova Bienne started developing an analog $11\frac{1}{2}$-line three-hand quartz movement with sweep seconds-hand; it was given the designation 2420 and was the first member of a whole family of calibers. Many months previously, Bulova Bienne had started to alert headquarters in New York. It was obvious in Switzerland that the quartz technology had made decisive breakthroughs and that it was going to prevail much sooner than one had thought so far. The tuning fork technology was obsolete and quartz calibers were urgently needed. As there was no response, Bulova

15.57. Water Watch

15.58. Benrus caliber H0101, mechanical module

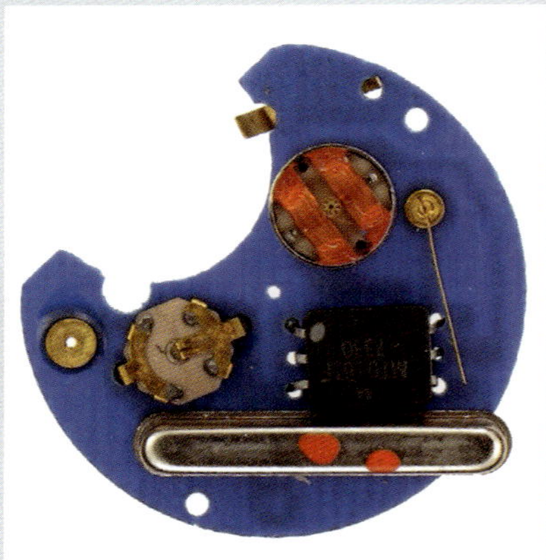

15.59. Benrus H0101, electronic module

15.60. Bulova Quartz caliber 2426-10

Bienne developed caliber 2420 at its own risk: it was launched in late 1976. The daily rate was better than 0.2 second.

Caliber 2420 was provided with the Accuset system, which allowed accelerating or slowing the seconds-hand in order to achieve a perfect synchronization with a time signal. Four variants of caliber 2420 were developed: 2421.10 with a stopping feature of the seconds-hand (the height was 4.55 mm); 2422.10 with a fast correction of the date disc; 2423.10 with day/date; and 2426.10 with the weekday indication in two languages. The height of those modules was 5.50 mm. The hour-hand could be set without influencing the minute and second.

The frequency of the bar-shaped quartz resonator was 32 kHz; it was made by the Bulova quartz division in New York. The U.S. company Synertec (a Fairchild spinoff) or Intersil supplied the integrated circuit. The 5-second bipolar stepping motor with a platinum-cobalt rotor was manufactured by Bulova Bienne under a CEH license. This was the motor that Henri Oguey had patented at the CEH. There were similarities with the Socrem-Portescap motor, but the variant Bulova chose featured a pot-shaped coil. The caliber had a current consumption of 6.6 microampere; consequently, battery life was about two years.

Almost at the same time, Bulova engineers in Jackson Heights, New York, developed a 6¾ x 8 lines two-hand quartz movement with a height of 4.5 mm that was meant to be used in ladies' watches. The designation was 2470 and serial production started in 1977. In 1979, Bulova Bienne launched the low-cost, conventional two-hand 5½ x 6¾ lines x 3.5 mm 323 kHz caliber 2500/2510. The Lavet motor coil at the edge of the plate was curved in order to save space. With just 1 microwatt, the power consumption was very low.

The Bulova Thermatron was developed at the CEH but was licensed exclusively by Bulova and launched commercially in 1982 as the "Bulova-Thermatron" with a thermoelectric power supply (see Chapter 8).

Timex

In early 1968, the sensational results of the Neuchâtel Observatory's chronometer contest were published. For the first time, quartz wristwatch calibers had participated. Eleven of them had been developed by the CEH, five others had been sent in by Seiko Suwa. The very first quartz wristwatches were launched by Seiko in 1969 and by the CEH in 1970: they made technological history

The Laco Heritage

At that time, all Timex had in terms of electrical watches were electrodynamic movements based on the Laco technology. Nevertheless, the electronic caliber Laco 872 (Timex caliber M86, 13½ lines x 5.63 mm, 3 Hz = 21,600 vibrations per hour) had been launched as early as 1966. It corresponded to the electrodynamic calibers Laco 870/871 (Timex M84), but the mechanical contact had been replaced by a transistorized circuit consisting of two silicon transistors (PNP and NPN), four resistors, and one capacitor. The current was interrupted by pulling the crown. Caliber 882/M87 was the date version of 872/M86.

The electrodynamic drive of the 3 Hz balance wheel motor featured the same coil on the balance wheel and the same system of permanent magnets as the older contact-controlled caliber. The transistor circuit was triggered by voltage pulses that were induced in the coil whenever the balance wheel was swinging over the stationary magnets.

Quartz Synchronization of M40

In the late 1960s, everybody was talking about quartz wristwatches. In order to stay credible, Timex just had to have quartz – and quickly! The fastest expedient was to quartz-synchronize the electrodynamic caliber Timex M40 (9¾ x 12¾ lines x 6.20 mm) made by Timex USA in Middlebury, Connecticut. For this purpose, the signal of a 49,152 Hz quartz resonator was electronically divided to 3 Hz. This was the frequency the balance wheel motor needed; the previous oscillator function of the balance wheel was now unnecessary. It was just a motor that was controlled by the quartz resonator. Strangely enough, the old contact control was chosen; with the previous transistor control, the price might have become excessive.

The first prototype of the Timex quartz synchronized caliber (it was eventually given the designation M62), consisted of the M40 movement and the quartz resonator on one side, the electronics on the other side. The latter was supplied by either Hughes, Microma, or RCA. This system was patented in the United States and in Germany in 1972. In the same year, Timex launched its first quartz watch with caliber M62. It was called "Time Machine" and featured a round, gold plated case; the price was $125.

The Calibers M63, M64 or M66 (no date, date, or day/date, respectively) were used in the "Time Machine." Their diameter was 13½ lines, the height 7.00 mm. As mentioned above, their base was the contact-controlled caliber series M40/41/42 that had been quartz-synchronized. The daily rate was 0.5 second, fine tuning was achieved with a resistive trimmer. This product line was maintained until 1980.

In the early days, Timex used bar-shaped flexural mode quartz resonators; after 1973 it switched to the photolithographic, tuning fork-shaped Statek type. Fine tuning was achieved with a capacitive trimmer. Eventually, the notoriously unreliable trimmer was eliminated. If the rate did not conform to the specifications, the entire electronic block was exchanged; this was very easy, as it was connected to the actual movement with just two screws. The product life of the quartz synchronized electrodynamic movement was short. The Seiko concept with a 32 kHz Statek-type quartz resonator, a CMOS divider circuit, and a Lavet stepping motor was so good that, by 1975, it had practically eliminated all others worldwide.

In parallel with the developments at the Timex headquarters, Timex Pforzheim (Germany) developed caliber M85 (13½ lines x 5.63 mm), that was quite similar to the American caliber M40. Quartz synchronization changed the designation to caliber M86/M87 (M87 featured a calendar disc). This line of development was not continued; production stopped in 1974.

Quartz Calibers with Rocking Motor

In the mid-1970s, Timex finally had "real" quartz movements. they had been developed in 1974 jointly by the Timex Laboratories in Besançon and Pforzheim. The 10-line calibers M43 and M44 were designed for two-hand ladies' watches without date or with a date disc; the height (including the battery) was 7.34 mm. The corresponding 13-line calibers M55, M56 and M58 for two-hand men's watches (no date, date or day/date, respectively – the height was 6.98 mm) featured a 32 kHz Statek tuning fork quartz resonator and a 15-step divider circuit to 1 Hz. The battery life was 19 months for the ladies' models and 32 months for the men's models. These calibers were in production from 1977 to 1980.

The weak point of these calibers was the rocking motor that had been developed in Besançon. This was a one-minute stepping motor that Timex patented. It consisted of an oscillating, triangular piece of plastic with a flat coil. A bar-shaped samarium-cobalt magnet was glued directly underneath it. Each time the coil received a current pulse, it was deflected by the magnet; a spring returned it back to the initial position.

Each rocking motion caused a driving pawl to push a ratchet wheel, moving it forward by one tooth. A second, stationary pawl prevented the wheel from rotating backwards. The construction was quite sturdy, but needed a lot of space: rocking motor calibers were quite thick. Other disadvantages were the relatively high power consumption and a rather short life. The batteries were supplied by Timex Wahington. Caliber M82 still used mercury oxide batteries, but for later calibers the switch was made to silver oxide cells.

15.61. Timex caliber M62

15.62. Timex caliber M63

15.63. Timex caliber M64

15.64. Timex caliber M160

Calibers with the Portescap Motor

Timex management knew that it needed the technically much superior Lavet motor. The problem was that Seiko had developed a particularly efficient version and held patents on it. ETA had made similar progress and the two companies agreed on a cost-free cross licensing agreement. The result was that Seiko and ETA had a de facto monopoly on the Lavet motor: it could not be produced without infringing on Seiko's patents.

In order to avoid the rather stiff licensing fees, Timex had to find something superior to the rocking motor. At any rate, the Besançon Timex plant would not have been in a position to manufacture the Lavet motor by the millions, as it was a delicate, high precision mechanism. Thus, the multipole Socrem stepping motor was chosen as the alternative. Socrem belonged to the Portescap Group in La Chaux-de-Fonds. The motor had been developed by Erich Jucker, a close friend and colleague of the Portescap president and majority shareholder Philippe Braunschweig. Socrem was managed by Claude Oudet, a former professor at the Besançon School of engineering.

The six-pole Portescap motor was characterized by a rather high torque of up to 1.5 µNm. The permissible tolerances were generous, which perfectly fit the machinery and tooling available at the Besançon plant. The Socrem motor consisted of two stamped, star-shaped pieces of soft iron that were connected magnetically with the coil. The disc-shaped rotor was made of plastic-bonded samarium-cobalt powder. Socrem motors were also available with 4, 8, or 12 poles.

Centering the three parts of the Socrem motor as well as the width of the air gap between stator and rotor were not critical. The rotor was coated with plastic and magnetized with six poles oriented at a right angle to the plane of the disc. Twelve additional poles of the same kind were located at the periphery of the rotor (see figure 6.34). A current pulse in the coil magnetized the poles of the stator; their interaction with the poles on the rotor caused the latter to turn by 60 degrees.

The Socrem motor normally turned only in one direction and made six steps per revolution. However special applications (such as in the Timex caliber M160 described below) required the motor to rotate in both directions. This could be done with a specially shaped current pulse that preceded the main pulse. The latter had the inverted polarity of the one needed for forward rotation. All parts of the Socrem motor were produced in Besançon and were assembled first in Taiwan, then in Cebu.

The two-hand caliber for ladies' watches M160/145 (M160 electroset with a pusher and no crown; M145 electroset with a crown) was produced from 1977 to 1978. It featured a bidirectional Portescap/Timex motor with the dimensions 6¾ x 8 lines x 4.1 mm. It was part of the Timex product mix from 1979 to 1987. The minute-hand made three steps per minute. For fine tuning, the trimmer capacitor was eliminated; it was replaced by a fixed capacitor that was individually chosen for each movement. The height of 4.1 mm including the battery certainly was a handicap; on the other hand the current consumption of 1.4 microamperes was reasonable; battery life was 40 months.

The three-hand caliber family M130 for men's watches featured a unidirectional Portescap motor. Production started in 1979-1980. Due to the current consumption of 4.6 microamperes, battery life was limited to 25 months. The dimensions of the calibers M135/137/139 (no date, date, or day/date, respectively) were 9½ x 12¼ lines x 3.50 mm. Caliber M139 indicated the moon phases, albeit not very accurately as the moon disc was controlled by the hour wheel. Caliber M140 was nearly identical with caliber M135.

The bigger calibers M132 (no date) and 133 (with date disc) were basically derived from M135. Their dimensions were 11¾ x 12¼ lines x 4.35 mm. There was no mass production as Timex Marketing objected to the height of those calibers and the short battery life.

For this reason Portescap and Timex jointly developed a significantly flatter, bidirectional Socrem motor in 1980/1981. It also made six steps per minute and equipped the flat, two-hand caliber M271 (11½ x 13½ lines x 3.97 mm with date). Thanks to the current consumption of 1.4 microamperes, battery life was extended to 56 months. The smaller two-hand caliber M272 (9 x 10¾ lines x 3.97 mm) for ladies' watches (no date) was equipped with a 40-month battery.

Production of those rectangular calibers continued from 1982 until 1988 but stayed modest: the height of 3.97 mm required costly movement parts. Assembly was rather demanding and could not be done in Cebu, though wages were much higher in Besançon. Setting was done with the recessed electroset pusher in the case back. The plate was very thin and was stabilized by contacting the case back.

The 18-pole Socrem motor, with 12 poles on the periphery of the rotor and 6 poles on the inside, had been specifically developed for Timex, but production was fully taken over by Timex; the numbers were huge and are estimated at 20 to 40 million. Portescap just supplied the heads for magnetizing the rotors.

Calibers with a Lavet Motor

In a third phase, Timex finally made the switch to the Lavet motor in the mid-1980s. Initially, the Seiko design was adopted, but by then licensing fees were much lower than originally. In this construction, the magnetic circuit had no air gap; the stator consisted of just one part. This simplified production and assembly.

However, the fact that one had to use two different kinds of soft iron complicated the issue. The core of the coil and the armature consisted of an iron type that reached saturation as late as possible so that they could tolerate a strong magnetic field before reaching saturation. The stator with a hole for the rotor had to reach saturation at an early stage, so the magnetic field lines left the stator and interacted with the samarium-cobalt rotor.

On both sides of the stator, at an angle of 45 degrees to its long axis, a half-moon-shaped recess was cut into the side of the borehole for the rotor. In the rest condition, the magnetic north-south direction of the rotor's magnetic field was oriented along the line connecting the two recesses. As soon as the coil received a current pulse, a magnetic field was set up in the stator, repelling the poles of the rotor that turned by 135 degrees; after the current pulse had died down, the rotor continued turning by 45 degrees and then stopped. Its magnetic axis was again oriented along the line connecting the recesses. The next current pulse had the opposite polarity, so that the rotor always turned in the same direction.

Venus

Timex was not quite happy with its own quartz calibers, at least as long as the energy-hungry Socrem motor was used. Timex-CEO Lem Tarshish and ETA general manager Ernst Thomke agreed to cooperate in developing an economical, general purpose quartz movement that would be ideally suited for Timex watches.

The 6¾ x 8 lines ETA-Timex caliber was called "Venus." Timex produced the parts in Besançon and assembled the movements in Cebu. This three-hand caliber with the designation M276 was extremely successful; starting in 1984 it was used in ladies' and men's watches for a full fifteen years. ETA produced it on its own, the basic concept was used in the "Ecoline" family of calibers.

Timex was very enthusiastic about the Venus caliber and suggested the development of more calibers of the same type; they were meant to replace the Timex calibers with the Socrem motor. What was needed first was a 10½- or 11½- line version of the Venus caliber, but for reasons of cost, ETA refused to get involved in a second joint development program with Timex. ETA was convinced that its existing 11½-line caliber "Jupiter" was just the right thing for Timex. It would have been easy for ETA to supply Timex with quartz movements from the existing assortment, if Timex, on its part, would supply parts from Besançon. This was not acceptable to Timex, as ETA calibers simply were too expensive. Furthermore, it was very important for Timex to produce its own, thoroughly modern quartz movements.

Home Made Lavet Motors

In order to circumvent the decisive Seiko patent and thus save the licensing fees, Timex started to produce Lavet motors with only the late saturation types oft iron in the magnetic circuit (including the complete stator). The overall efficiency of the motor suffered accordingly, but this could be tolerated. This was compensated by using lithium batteries with the full diameter of the movement's plate. The battery life was 7 to 8 years.

Watches with a lithium battery are relatively thick, but this was not a problem for Timex. The 3-volt output of lithium batteries was an advantage for the electroluminescent illumination of the dial (Indiglo), as it required about 80 volts. Starting from 3 volts instead of 1.5 volts saved one doubling step.

Timex finally achieved self-sufficiency in terms of quartz movements with the three-hand calibers M601, M602, and M603 (no date, with date, or day/date, respectively). Furthermore, very rigorous quality control standards were enforced. The dimensions of all three calibers were nearly identical (9¾ x 10½ lines). The height of M601 and M602 was the same (3.10 mm), while the height of M603 was 3.60 mm. The design had been worked out in Pforzheim; Besançon took care of the production. This line of calibers was maintained from 1987 to 1995.

In the second half of the 1980s, a Seiko license was obtained at a very low cost. The first Timex quartz caliber with a Seiko-type Lavet motor (now with two kinds of soft iron) was the ladies' watch caliber M650 (5½ x 6¾ lines x 3.00 mm); this was a two-hand caliber without a date disc. Production was started in 1986 and continued until 1993. The variant M651 actually was a completely new construction with a bigger battery; its life was 69 months instead of the former 51 months. Only prototypes of M651 were made; there was no production.

The dimensions of the ladies' watch three-hand caliber M653 were 5½ x 6¾ lines x 3.10 mm. It featured a Timex Lavet motor developed in Pforzheim. It was maintained in production from 1988 until 1996. Caliber M658 was nearly identical with M653, but this caliber was equipped with a relatively expensive Seiko integrated circuit; the series was quite small. It was very interesting from a technological point of view. The current pulses for the motor were as weak as possible. Whenever the motor did not react, a second, stronger pulse was generated that moved the rotor with certainty.

Caliber M659 was nearly identical with M654 (7¾ lines x 3.3 mm) that was produced from 1988 to 1996. Thanks to a more economical integrated circuit, battery life was extended from 23 to 39 months. The Timex caliber families 9XX and 3XX again used Lavet motors with only one kind of soft iron that did not need a Seiko license.

Combo Calibers

As far as combo movements (with both analog and digital display) are concerned, Timex had been content with purchasing calibers that were supplied as complete watches by the Korean HanDok. They combined the analog caliber PUW 900 and a digital module from HanDok. Starting in 1988, Timex also became self-sufficient for this kind of product. This is when the three-hand Combo caliber M629, M630 and M633 were launched; the LCD was supplied by Nippon Electric.

Caliber M629 consisted of a 6-digit display at 12 o'clock and featured the three-hand quartz analog movement M653: it replaced the old caliber M626 Handok-PUW900. In caliber M630, the 6-digit display was placed at 6 o'clock. In this case also, the analog part of M653 was used; it replaced M628 Handok-PUW900.

The combo caliber M633 with a 12-digit LCD at 6 o'clock "Aviator" also used caliber 653. This was the first combo movement with a lithium battery that extended battery life to 67 months. Caliber M633 replaced M708 Handok-PUW900. M634 was still another combo caliber; it was a variant of M633 with a multisegment display. All the combo calibers listed above were produced between 1990 and 1997.

Acknowledgments

We thank the many persons who kindly supplied us with information, picture material and old calibers. Without their cooperation this book could neither have been written nor illustrated.

AFIF: André Greuter; ***Audemars Piguet:*** Christophe Guhl, Martin Wehrli; ***Bulova:*** Hans Schaller; ***BWC:*** Theo Keller, Edwin Volkart; ***Casio:*** Kai-Christian Helms, Norutoshi Minami, Hideyuki Shoji, Atsushi Yamaoka; ***CEH:*** Walter Hammer, Jean Hermann, Henri Oguey, Luc Omlin, Eric Vittoz, Raymond Vuilleumier, Roger Wellinger; ***CENTREDOC*** Bégonia Tora; ***CETEHOR:*** Patrick Augereau, Jean-Marc Feuvrier; ***Citizen:*** Hisachi Kawahara, Tadashi Hanaoka, Haruhiko Higuchi, Tadashi Hasegawa, Urs Meier, Yoshio Miura; ***Derby Swissonic:*** André Hug; ***Deutsches Uhrenmuseum Furtwangen:*** Johannes Graf, Eduard C. Saluz; ***Ebosa:*** Paul Glocker; ***EEM:*** André Beyner, Jean-Claude Fatton, Kurt Hübner, François Jeanrenaud; ***ETA:*** Anton Bally; ***Eurowatch:*** Robert Anguenot; ***Fédération Horlogère:*** Vincent Grossenbacher, Philippe Wenger; ***FHF:*** René Widmer; ***France Ebauches:*** Patrick Augereau, Michel Huot-Marchand; ***Girard-Perregaux:*** Willy Schweizer, Georges Vuffray; ***Glashütte-GUB:*** Artur Kamp; ***Inducta/Wittwer-Bär:*** Eduard Spring; ***Integral Minsk:*** Dmitry V. Kasatov, Vladimir K. Remenchik, Ludmila V. Kajura; ***ISA:*** René Binétruy, Philippe Binétruy, Joseph Flores, Olivier Michalet, Claude Receveur; ***Isatome:*** Hugues Voumard, Vincent Zaslawsky; ***Jaeger-LeCoultre:*** Christine Giotto, Sébastien Vivas; ***Junghans:*** Sandra Müller, Gernot Stähle, Matthias Stotz; ***Korea:*** P. Christine Rangkyoung Im; ***Laco and Laco-Timex:*** Wolfgang Burkhardt, Joseph Egger, Willi Namyslo, Gerhard Stotz; ***LCD:*** Martin Schadt; ***Longines:*** Frédéric Donzé, Frank Vaucher; ***Lutch Minsk:*** Andrey Robertovitch, Stempen, Oleg Ryabkov (WTC Minsk), Taisiya V. Bogdashich; ***Mondaine:*** Ronnie Bernheim; ***Montrélec/Framelec:*** Robert Anguenot, Patrick Augereau; ***Moser-Baer/Mobatime:*** Albert Augstburger, Felix Wädensweiler; Musée International; d'Horlogerie; ***La Chaux-de-Fonds:*** Ludwig Oechslin, Jean-Michel Piguet; Musée de la Montre; ***Villers-le-Lac (F):*** Yves Droz; ***NAWCC*** Nancy Dyer; ***Nepro:*** Paolo Spadini; ***Omega:*** Hans Schaller, Henry Schneider; ***Orient Watch:*** Inoue Hiroyuki; ***Oris:*** Rolf Portmann; ***Oscilloquartz SA:*** Isabelle Romerio, Bernard Schlueter; ***Patek Philippe:*** Robert Anguenot, Hans van der Ijssel, Michael Schuldes; ***Piaget:*** Yves Bornand, John Carlet, Gérard Junod, Yves Piaget, Valérie Steck; ***Porta (PUW):*** Klaus Wehner; ***Portescap:*** Philippe Braunschweig, Erich Jucker; ***Prim Slava:*** Libor Hovorka, Jaroslav Rehor; ***Pulsar:*** John Bergey, Dennis Klein; ***Rolex:*** René Besson, René Le Coultre; ***Ronda:*** Michiel Groothuis, Jakob Leuenberger, Erich Mosset; ***Russia:*** Libor Hovorka, Artur Kamp, Vladimir Rogovets, Alexander Shorokhov; ***Seiko:*** Kaoru Iida, Tetsu Imada, Naoto Ueda, Robert Wilson; ***Sicura:*** Nathalie Frésard; ***Swatch:*** Anton Bally, Elmar Mock, Jacques Müller, Ernst Thomke; ***Technisches Museum Pforzheim***: Simone Flumm, Günter Gõtz; ***Timex:*** Wolfgang Burkhardt, Lou Galie, Joseph Egger, Willi Namyslo, Michel Plançon, Gerhard Stotz; ***Ulysse Nardin:*** Susanne Hurni, Rolf Schnyder, Léo Berthoud, Jean-Claude Jobin; ***UMF Ruhla:*** Artur Kamp, Klaus Mleinek; ***Water Watch:*** Daniel Dreifuss; ***Wempe:*** Carsten Petersen, Nadja Weisweiler; ***Zenith:*** Marc Roethlisberger.

In the final phases of preparing the manuscript for publication, a number of persons bent over backwards to help us achieve our goal, supplying us with last minute pictures and digging out super rare quartz calibers. In this connection, it is a pleasure to mention Patrick Augereau, Anton Bally, Hans Bantli, René Binétruy, Yves Bornand, Joseph Flores, Johannes Graf, Michiel Groothuis, Vincent Grossenbacher, Libor Hovorka, Tetsu Imada, C. Ishizaki, Artur Kamp, Dennis Klein, Gregory Maugain, Klaus Mleinek, Jean-Michel Piguet, Carsten Petersen, Rolf Portmann, Eduard C. Saluz, Martin Schadt, Willy Schweizer, Philippe Wenger, René Widmer, and Alphonse Zumsteg.

Particular thanks are due Christian Pfeiffer-Belli (Ebner Publishing Group) who enthusiastically supported the project, supplied important contacts, and worked out the technical concept of the book.

Heartfelt thanks are due Renate Ramm who took over very boring editing work, while Edda Wenzig found hundreds of mistakes and typing errors that none of us had noticed. In Japan, everything worked like a charm thanks to the linguistic and geographic assistance of my wonderful partner of 35 years Yoshiko Yasukawa. My friend and colleague Fred Stuber and his wife Audrey in North Haven, Connecticut, carefully reviewed the entire English manuscript with utmost care and greatly improved both the English and the typing.

Picture credits

The following initials are used:

AF	Armin Frei
AK	Artur Kamp
DK	Dennis Klein
DUF	Deutsches Uhrenmuseum Furtwangen
GR	Günther Ramm
HO	Henri Oguey
HS	Henri Schneider
HW	Hugo Wyss
JF	Joseph Flores
KH	Kurt Hübner
LH	Libor Hovorka
LFT	Lucien F. Trueb
MIH	Musée International d'Horlogerie La Chaux-de-Fonds
MS	Michael Schuldes
PS	Paolo Spadini
PW	Peter Wenzig
PW/R	Peter Wenzig/Copyright Reinhard Rohner
RB	René Binétruy
RLC	René Lecoultre
RW	Roger Wellinger
WU	Werner Uhink

The companies that kindly supplied us with graphic material from their archives or public relations department are listed with their corporate name.

Part I: FROM ELECTRICAL CLOCKS TO QUARTZ WRISTWATCHES IN 128 YEARS

Ch. 1: Earliest Electrical Clocks and Watches
1.1 to 1.12 DUF
1.13 LFT
1.14 and 1.15 DUF

Ch. 2: Fixed Coil Electromechanical Wristwatches
2.1 to 2.9 GR

Ch. 3: Moving Coil Electromechanical Wristwatches
3.1 to 3.16 GR
3.17 to 3.20 AK

Ch. 4: Transition to Electronics
4.1 GR
4.2 PW/R
4.3 GR
4.4 GR
4.5 PW/R
4.6 and 4.7 GR
4.8 GR
4.9 PW/R
4.10 to 4.19 GR
4.20 PW/R
4.21 to 4.25 GR

Ch. 5: The Tuning Fork Wristwatch
5.1 to 5.4 GR
5.5 PW/R
5.6 to 5.18 GR
5.19 PW/R
5.20 GR

Ch. 6: Stationary and Mobile Quartz Clocks
6.1 to 6.3 Internet
6.4 DUF/WU
6.5 and 6.6 DUF
6.7 to 6.10 DUF/Eduard C Saluz
6.11 Bull.Soc.Ch.Chrono
6.12 MIH
6.13 and 6.14 Junghans
6.15 LFT
6.16 and 6.17 Omega
6.18 and 6.19 LFT
6.20 to 6.24 DUF
6.25 LFT
6.26 to 6.28 Junghans
6.29 to 6.33 DUF
6.34 LFT
6.35 and 6.36 DUF
6.37 and 6.38 AK
6.39 and 6.40 DUF
6.41 to 6.43 LFT
6.44 HS
6.45 to 6.53 LFT
6.54 to 6.56 Wempe
6.57 DUF
6.58 and 6.59 LFT
6.60 Mobatime/Moser-Baer
6.61 to 6.64 MS
6.65 Swatch. Germany
6.66 Inducta Wittwer Bär

Ch. 7: The Eightfold Way to the Quartz Wristwatch
7.1 to 7.4 MIH
7.5 AF
7.6 KH
7.7 HO
7.8 RW
7.9 and 7.10 GR
7.11 Omega
7.12 to 7.16 Seiko
7.17 HW
7.18 to 7.24 GR
7.25 GR
7.26 PW/R
7.27 Girard-Perregaux
7.28 LFT
7.29 to 7.31 DK

Part II: QUARTZ MOVEMENTS FROM ALL OVER THE WORLD

Ch. 8: Quartz Movements from Switzerland
8.1 GR
8.2 to 8.7 LFT
8.7 and 8.8 Jaeger-LeCoultre
8.9 and 8.10 LFT
8.11 and 8.12.PW/R
8.13 to 8.16 GR
8.17 PW
8.18 to 8.20 GR
8.21 to 8.39 PW/R
8.40 and 8.41 GR
8.42 LFT
8.43 GR
8.44 to 8.48 LFT
8.49 GR
8.50 to 8.56 LFT
8.57 PW/R
8.58 LFT
8.59 and 8.60 PW/R
8.61 to 8.63 Girard-Perregaux
8.64 GR
8.65 to 8.70 PW/R
8.71 to 8.79 Jaeger-LeCoultre
8.80 to 8.85 LFT
8.86 to 8.92 PW/R
8.93 PW
8.94 Mondaine Watch
8.95 and 8.96 PW/R
8.97 PS
8.98 and 8.99 LFT
8.100 PS
8.101 and 8.102 PW/R
8.103 and 8.104 LFT
8.105 GR
8.106 PW/R
8.107 GR
8.108 PW/R
8.109 to 8.112 GR
8.113 to 8.119 PW/R
8.120 to 8.124 LFT
8.125 GR
8.126 to 8.128 RLC
8.129 PW/R
8.130 GR
8.131 and 8.132 LFT
8.133 PW/R
8.134 to 8.137 GR
8.138 to 8.147 Swatch
8.148 to 8.154 LFT
8.155 to 8.157 PW/R
8.158 to 8.160 GR

Ch. 9: Quartz Movements from Germany
9.1 and 9.2 GR
9.3 and 9.4 PW/R
9.5 GR
9.6 PW
9.7 PW/R
9.8 T1
9.9 PW/R
9.10 to 9.20 AK
9.21 GR
9.22 to 9.28 Junghans
9.29 GR
9.30 and 9.31 LFT

9.32 PW/R
9.33 to 9.40 AK

Ch. 10: Quartz Movements from France
10.1 to 10.10 LFT
10.11 PW/R
10.12 LFT
10.13 to 10.20 JF
10.21 PW/R
10.22 GR
10.23 PW/R
10.24 GR
10.25 to 10.28 LFT
10.29 to 10.32 RB

Ch. 11: Quartz Movements from Great Britain
11.1 to 11.3 GR

Ch. 12: Quartz Movements from Japan
12.1 to 12.4 PW/R
12.5 DUF
12.6 to 12.8 PW/R
12.9 GR
12.10 and 12.11 PW
12.12 PW/R
12.13 Citizen
12.14 PW
12.15 DUF
12.16 and 12.17 PW/R
12.18 PW
12.19 PW/R
12.20 PW
12.21 to 12.30 PW/R
12.31 and 12.32 GR
12.33 to 12.42 PW/R
12.43 and 12.44 GR
12.45 to 12.51 PW/R
12.52 GR
12.53 to 12.67 PW/R
12.68 GR

Ch. 13: Quartz Movements from Eastern Europe and the USSR
13.1 and 13.2 PW/R
13.3 and 13.4 LH
13.5 and 13.6 DUF
13.7 to 13.10 LFT
13.11 to 13.14 PW/R
13.15 and 13.16 DUF

Ch. 14: Quartz Movements from South Korea
14.1 to 14.12 LFT

Ch. 15: Quartz Movements from the United States
15.1 GR
15.2 to 15.7 PW/R
15.8 GR
15.9 to 15.28 PW/R
15.29 GR
15.30 and 15.31 DK
15.32 GR
15.33 to 15.40 DK
15.41 to 15.55 PW/R
15.56 PW
15.57 LFT
15.58 and 15.59 GR
15.60 PW/R
15.61 to 15.64 GR

Index